T0234588

Lecture Notes in Computer Science 9486

More information about this series at http://www.springer.com/series/7407

Zaixin Lu · Donghyun Kim · Weili Wu
Wei Li · Ding-Zhu Du (Eds.)

Combinatorial Optimization and Applications

9th International Conference, COCOA 2015
Houston, TX, USA, December 18–20, 2015
Proceedings

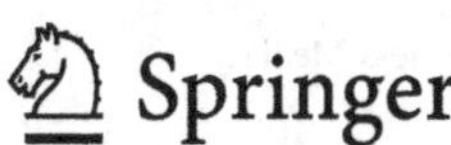

Editors
Zaixin Lu
Marywood University
Scranton, PA
USA

Wei Li
Texas Southern University
Houston, TX
USA

Donghyun Kim
North Carolina Central University
Durham, NC
USA

Ding-Zhu Du
University of Texas at Dallas
Richardson, TX
USA

Weili Wu
University of Texas at Dallas
Richardson, TX
USA

ISSN 0302-9743 ISSN 1611-3349 (electronic)
Lecture Notes in Computer Science
ISBN 978-3-319-26625-1 ISBN 978-3-319-26626-8 (eBook)
DOI 10.1007/978-3-319-26626-8

Library of Congress Control Number: 2015954595

LNCS Sublibrary: SL1 – Theoretical Computer Science and General Issues

Springer Cham Heidelberg New York Dordrecht London

Printed on acid-free paper

Springer International Publishing AG Switzerland is part of Springer Science+Business Media
(www.springer.com)

Preface

The 9th Annual International Conference on Combinatorial Optimization and Applications (COCOA 2015) was held during December 18–20, 2015, in Houston, Texas, USA. COCOA 2015 provided a forum for researchers working in the area of theoretical computer science and combinatorics.

The technical program of the conference included 59 contributed papers selected by the Program Committee from 125 full submissions received in response to the call for papers. All the papers were peer reviewed by Program Committee members or external reviewers.

The topics cover most aspects of theoretical computer science and combinatorics related to computing, including classic combinatorial optimization, geometric optimization, network optimization, optimization in graphs, applied optimization, complexity and game, and miscellaneous. Some of the papers will be selected for publication in special issues of *Algorithmica, Theoretical Computer Science, Journal of Combinatorial Optimization,* and *Computational Social Networks*. It is expected that the journal version of the papers will appear in a more complete form.

We thank all the people who made this meeting possible: the authors for submitting papers, the Program Committee members, and external reviewers for volunteering their time to review conference papers. We would also like to extend special thanks to the publication, publicity, and local organization chairs for their work in making COCOA 2015 a successful event.

September 2015

Zaixin Lu
Donghyun Kim
Weili Wu
Wei Li
Ding-Zhu Du

Organization

Program Chairs

Zaixin Lu	Marywood University, USA
Donghyun Kim	North Carolina Central University, USA
Ding-Zhu Du	University of Texas at Dallas, USA

Program Committee

Wolfgang Bein	University of Nevada, Las Vegas, USA
Gruia Calinescu	Illinois Institute of Technology, USA
Zhi-Zhong Chen	Tokyo Denki University, Japan
Xujin Chen	Chinese Academy of Sciences, China
Yongxi Cheng	Xi'an Jiaotong University, China
Ovidiu Daescu	University of Texas at Dallas, USA
Bhaskar DasGupta	University of Illinois at Chicago, USA
Vladimir Deineko	Warwick University, UK
Thang Dinh	Virginia Commonwealth University, USA
Ding-Zhu Du	University of Texas at Dallas, USA
Zhenhua Duan	Xidian University, China
Neng Fan	University of Arizona, USA
Bin Fu	University of Texas-Pan American, USA
Xiaofeng Gao	Shanghai Jiao Tong University, USA
Majed Haddad	University of Avignon, France
Juraj Hromkovic	ETH Zurich, Switzerland
Sun-Yuan Hsieh	National Cheng Kung University, Taiwan
Tsan-sheng Hsu	Academia Sinica, Taiwan
Hejiao Huang	Harbin Institute of Technology, China
Kazuo Iwama	Kyoto University, Japan
Donghyun Kim	North Carolina Central University, USA
Jie Li	University of Tsukuba, Japan
Minming Li	City University of Hong Kong, SAR China
Zaixin Lu	Marywood University, USA
Mitsunori Ogihara	University of Miami, USA
Xian Qiu	Zhejiang University, China
Suneeta Ramaswami	Rutgers University, USA
Rahul Vaze	The Tata Institute of Fundamental Research, India
Lusheng Wang	City University of Hong Kong, SAR China
Boting Yang	University of Regina, Canada
Hsu-Chun Yen	National Taiwan University, Taiwan

Contents

Classic Combinatorial Optimization

Geometric Optimization

Complexity and Game

Optimization in Graphs

Miscellaneous

Classic Combinatorial Optimization

Improved Algorithms for the Evacuation Route Planning Problem

Gopinath Mishra[1], Subhra Mazumdar[2], and Arindam Pal[2](✉)

[1] Advanced Computing and Microelectronics Unit, Indian Statistical Institute, Kolkata, India
gopianjan117@gmail.com

[2] Innovation Labs, TCS Research, Tata Consultancy Services, Kolkata, India
{subhra.mazumdar,arindam.pal1}@tcs.com

Abstract. Emergency evacuation is the process of movement of people away from the threat or actual occurrence of hazards such as natural disasters, terrorist attacks, fires and bombs. In this paper, we focus on evacuation from a building, but the ideas can be applied to city and region evacuation. We define the problem and show how it can be modeled using graphs. The resulting optimization problem can be formulated as an integer linear program. Though this can be solved exactly, this approach does not scale well for graphs with thousands of nodes and several hundred thousands of edges. This is impractical for large graphs.

We study a special case of this problem, where there is only a single source and a single sink. For this case, we give an improved algorithm *Single Source Single Sink Evacuation Route Planner (SSEP)*, whose evacuation time is always at most that of a famous algorithm *Capacity Constrained Route Planner (CCRP)*, and whose running time is strictly less than that of CCRP. We prove this mathematically and give supporting results by extensive experiments. We also study randomized behavior model of people and give some interesting results.

1 Introduction

Emergency evacuation is the process of movement of people away from the threat or actual occurrence of hazards such as natural disasters, terrorist attacks, fires and bombs. In this paper, we focus on evacuation from a building, though the ideas can be applied to city and region evacuation. We are motivated by the evacuation drill that regularly happens in our company Tata Consultancy Services. We are developing a system SMARTEVACTRAK [1] for people counting and coarse-level localization for evacuation of large buildings. Safe evacuation of thousands of employees in a timely manner, so that no one is left behind, is a major challenge for the building administrators. Time is the main parameter in our model. The travel time between different areas of the building is part of the input and the evacuation time is the output. In the following discussion, we use {*graph, network*}, {*node, vertex*}, {*edge, arc*}, and {*path, route*} interchangeably.

We have a building along with its floor plan. Employees are present in some portions (rooms) of the building. There are some *exits* on the floor. Every

Z. Lu et al. (Eds.): COCOA 2015, LNCS 9486, pp. 3–19, 2015.
DOI: 10.1007/978-3-319-26626-8_1

corridor has a *capacity*, which is the number of employees that can pass through the corridor per unit time. Every corridor also has a *travel time*, which is the time required to move from the start of the corridor to the end. The goal is to suggest a feasible route for each employee so that he can be guided to an exit. It must be ensured that at any time the number of employees passing through a corridor does not exceed it's capacity.

A complex building does not provide its occupants with all the information required to find the optimal route. In an emergency, people tend to panic and do not always follow the paths suggested by the algorithm. They are not given enough time to establish a cognitive map of the building. To address this issue, we need to model the behavior of people in emergency situations. We have proposed a simple randomized behavior model and analyzed it. The expected evacuation time comes out to be quite good. None of the previous works considered any behavior model of people.

2 Related Work

In this section, we give a summary of different algorithms for the evacuation route planning problem. Skutella [12] has a good survey on the network flows over time problem. The monograph by Hamacher and Tjandra [4] surveys the state of the art on the mathematical modeling of evacuation problems. Both these papers give a good introduction and comprehensive treatment to this topic.

The LP based polynomial time algorithm for evacuation problem by Hoppe and Tardos [5] uses the ellipsoid method and runs in $O(n^6T^6)$ time, where n is the number of nodes in the graph and T is the evacuation *egress time* for the given network. It uses time-expanded graphs for the network, where there are $T+1$ copies of each node. The expression for time complexity shows that it is not scalable even for mid-sized networks. Another disadvantage is that it requires the evacuation egress time (T) *apriori*, which is not easy to estimate. As the time complexity is a function of T, it is not a fully polynomial time algorithm.

One of the earliest algorithms by Lu et al. [8] is Capacity Constrained Route Planner (CCRP). CCRP uses Dijkstra's generalized shortest path algorithm to find shortest paths from any source to any sink, provided that there is enough capacity available on all nodes and edges of the path. An important feature of CCRP is that instead of a single value which does not vary with time, edge capacities and node capacities are modeled as time series (function of time). Here, we need to update edge and node capacities for each time period. The running time of CCRP is $O(p(m+n\log n))$, ($O(pn\log n)$ for sparse graphs, where $m = O(n)$) and space complexity is $O((m+n)T)$ ($O(nT)$ for sparse graphs). Here m and n denotes the number of edges and the number of vertices of the graph respectively, p denotes the number of evacuees, and T denotes the evacuation egress time. As space complexity is always at most the time complexity, the running time of CCRP is implicitly dependent on T. For sparse graphs, $nT \leq pn\log n$, *i.e.*, $T \leq p\log n$. So, for sparse graphs the evacuation egress time is at most $O(p\log n)$. The space complexity of $O(nT)$ and unnecessary expansion of source nodes in each iteration are two main disadvantages of CCRP.

To overcome the unnecessary expansion in each iteration, Yin et al. [14] introduced the CCRP++ algorithm. The main advantage of CCRP++ is that it runs faster than CCRP. But the quality of solution is not good, because availability along a path may change between the times when paths are reserved and when they are actually used.

Min and Neupane [11] introduced the concept of *combined evacuation time* (CET) and *quickest paths*, which considers both transit time and capacity on each path and provides a fair balance between them. Let there be k edge-disjoint paths $\{P_1, P_2, \ldots, P_k\}$ from source node s to sink node t. Then, the combined evacuation time is given by,

$$CET(\{P_1, P_2, \ldots, P_k\}) = \left\lceil \frac{p + \sum_{i=1}^{k} C_i T_i}{\sum_{i=1}^{k} C_i} \right\rceil - 1 \quad (1)$$

where C_i and T_i denotes the capacity and transit time of path P_i respectively, and p denotes the number of evacuees. Time required to evacuate p people via a path P having transit time T and capacity C is $T + \lceil \frac{p}{C} \rceil - 1$. So, P_i is said to be the *quickest path* if and only if $T_i + \left\lceil \frac{p}{C_i} \right\rceil - 1 \leq T_j + \left\lceil \frac{p}{C_j} \right\rceil - 1$, for all $j \in \{1, \ldots, k\} \setminus \{i\}$.

The formula for combined evacuation time not only gives an exact expression for the evacuation time, but it also gives the number of people that will be evacuated on each path. The intuition behind the concept of CET is that paths having lesser arrival time will evacuate more groups. This algorithm is known as QPER (Quickest Path Evacuation Routing). The algorithm finds all edge-disjoint paths between a single source and a single sink and orders them according to the quickest evacuation time (calculated using CET) and adds them one by one. The algorithm is fairly simple. It does not use time-expanded graphs and there is no need to store availability information at each time stamp, as only edge-disjoint paths are considered. But their algorithm is limited to single source and single sink evacuation problems. Besides these, the addition of paths is not consistent, i.e., a path added at some point of time may be removed by the algorithm at a latter point of time, in case removal makes the solution better.

The solutions produced by CCRP++ and QPER do not follow semantics of CCRP, i.e., the solution quality is not better than that of CCRP. Recently Gupta and Sarda [3] have given an algorithm called CCRP*, where evacuation plan is same as that of CCRP and it runs faster in practice. Instead of running Dijkstra's algorithm from scratch in each iteration, they resume it from the previous iteration.

Kim et al. [6] studied the contraflow network configuration problem to minimize the evacuation time. In the *contraflow* problem, the goal is to find a reconfigured network identifying the ideal direction for each edge to minimize the evacuation time, by reallocating the available capacity. They proved that this problem is NP-complete. They designed a greedy heuristic to produce high-quality solutions with significant performance. They also developed a bottleneck relief heuristic to deal with large numbers of evacuees. They evaluated the proposed approaches both analytically and experimentally using real-world data sets. Min and Lee [10] build on this idea to design a maximum throughput flow-based contraflow evacuation routing algorithm.

Min [9] proposed the idea of *synchronized flow* based evacuation route planning. Synchronized flows replace the use of time-expanded graphs and provides higher scalability in terms of the evacuation time or the number of people evacuated. The computation time only depends on the number of source nodes and the size of the graph.

Dressler et al. [2] uses a network flow based approach to solve this problem. They use two algorithms: one is based on *minimum cost transshipment* and the other is based on *earliest arrival transshipment.* They evaluate these two approaches using a cellular automaton model to simulate the behavior of the evacuees. The minimum cost approach does not consider the distances between evacuees and exits. It may fail if there are exits very far away. Problems also arise if a lot of exits share the same bottleneck edges. The earliest arrival approach uses an optimal flow over time and thus does not suffer from these problems. But the exit assignment computed by the earliest arrival approach may not be optimal.

There are some previous works which considered the behavior of people in an emergency. Løvs [7] proposed different models of finding escape routes in an emergency. Song et al. [13] collect big and heterogeneous data to capture and analyze human emergency mobility following different disasters in Japan. They develop a general model of human emergency mobility using a Hidden Markov Model (HMM) for generating or simulating large amount of human emergency movements following disasters.

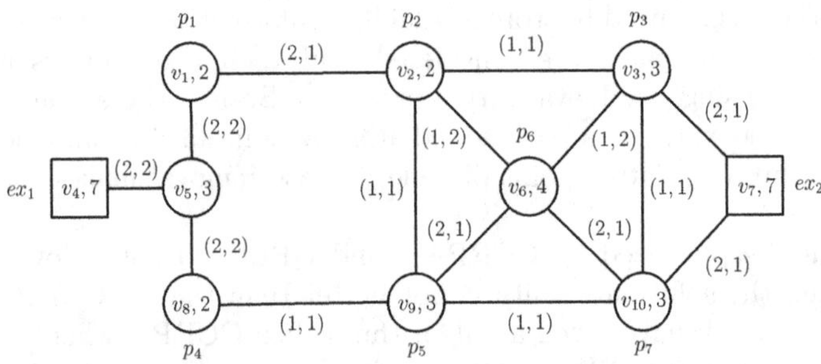

Fig. 1. A building graph, where vertices represented as squares denote exits. The vertex name and capacity are written inside a vertex. The edge capacity and travel time are written beside an edge. Persons residing on a vertex are specified beside that vertex.

3 Problem Definition and Model

The building floor plan can be represented as a graph $G = (V, E)$, where V and E are the set of vertices and edges respectively. The number of vertices and edges are n and m respectively. Nodes represent rooms, lobbies and intersection points and arcs represent corridors, hallways and staircases. Some nodes in the building having significant number of people are modeled as *source* nodes. The exits of a building are represented as *sink* nodes. Each node has a *capacity*, which is the maximum number of people that can stay at that location at any given time and an *occupancy*, which is the number of people currently occupying the location. Here, p is the total number of people who needs to be evacuated.

Each edge has a *capacity*, which is the maximum number of people that can traverse the edge per unit time and a *travel time*, which is the time needed to travel from one node to another along that edge.

Figure 1 shows a building graph that consists of 10 vertices and 15 edges. For each vertex v, it's name and the capacity are specified by a pair of the form $(v, c(v))$. A vertex representing an exit is drawn as a square, while the others are drawn as circles. For each edge e, the capacity and the travel time are specified on the edge by the pair $(c(e), d(e))$. The goal is to find the exit and the path (route) for each employee, subject to the constraint that the number of source-sink paths passing through an edge does not exceed the capacity of the edge at any unit time interval. The objective function we want to minimize is the total time of evacuation, that is the time at which the last employee is evacuated. Let's define this as the *evacuation time*. In the *quickest flow problem*, we are given a flow value f. We want to *minimize* the time T in which a feasible flow of value at least f can be sent from sources to sinks.

4 The Single Source Single Sink Problem

In this section, we focus on the single source single sink evacuation (SSEP) problem. In real life, single source single sink evacuation problem has many applications. For example, if all the people are in an auditorium, and there is only one exit in the building, we want to evacuate people as soon as possible, when there is an emergency. Throughout the rest of this paper, s denotes the source and t denotes the sink. Before proceeding further let's have some definitions.

Definition 1. Transit time of a path *is the sum of the transit times of all the edges in P from s to t, and is denoted as $T(P)$.*

Definition 2. Destination arrival time of a path *is the time required by a person to move from s to t using path P subject to prior reservations, and is denoted as $DA(P)$. In other words, we can say that $DA(P)$ is the sum of $T(P)$ and any intermediate delay. Note that $DA(P) \geq T(P)$.*

Definition 3. Capacity of a path *is the minimum of the capacities of all nodes and edges present in the path P, and is denoted by $C(P)$.*

Definition 4. *A node (edge) on a path P is called* saturated *if the capacity of the node (edge) equals the capacity of P.*

Definition 5. *Two paths P_1 and P_2 are said to be* distinct *if $V_1 \neq V_2$ or $E_1 \neq E_2$, where V_1, V_2 are the set of vertices and E_1, E_2 are the set of edges on the paths P_1 and P_2 respectively.*

4.1 Limitation of QPER Algorithm for SSEP

Using the concept of combined evacuation time, Min et al. [11] gave an algorithm $QPER$ for the single source single sink evacuation problem. Their algorithm works well when we have already discovered k edge-disjoint paths. In $QPER$, paths from s to t are added one by one in ascending order of quickest paths, and new CET is calculated after each path addition. But after addition of a path, the new CET may be less than the transit time of a previously added path. In that case, we have to delete those paths which have higher transit time than the current CET. This in turn increases the running time, since the addition of paths is not consistent.

We overcome the above limitations of the algorithm by adding paths in increasing order of transit time in each iteration till the transit time of the currently discovered path exceeds the CET of the previously added set of paths. Note that, we need not discover all possible paths from source to sink, since unlike $QPER$, if a path is added in any iteration, it will remain till the end. The CET after each iteration will be monotonically non-increasing.

4.2 Modified Algorithm for SSEP When We Are Given k Edge-Disjoint Paths

Let $P_1, P_2, \ldots, P_k$ be k edge-disjoint paths from s to t in ascending order of their transit time, *i.e.*, $T_1 \leq T_2 \leq \ldots \leq T_k$. We define, $S_i = \{P_1, \ldots, P_i\}$. We add paths to our set of routes ($\mathcal{R}$) in the following fashion.

1. $\mathcal{R} = \{P_1\}$.
2. $CET = CET(S_1)$.
3. Start with $i = 1$ Execute step 4 and 5 till $i \leq k$ and $T_{i+1} \leq CET$.
4. Add path P_{i+1} to R.
5. $CET = CET(S_{i+1})$ and $i \leftarrow i + 1$.
6. Return $\mathcal{R}$.

Lemma 1. *If $S_j = \{P_1, P_2, \ldots, P_j\}$, $j \leq k$ is returned as $\mathcal{R}$ by the above algorithm then*

1. $T_{l+1} \leq CET(S_l), 1 \leq l < j$
2. $CET(S_1) \geq CET(S_2) \geq \ldots \geq CET(S_j)$
3. $CET(S_j) \leq CET(S_l), j < l \leq k$.

Proof. Directly follows from the algorithm.

Lemma 2. *If $S_j = \{P_1, P_2, \ldots, P_j\}$, $j \leq k$ is returned as $\mathcal{R}$ by above algorithm then $T_1 \leq T_2 \leq \ldots \leq T_j \leq CET(S_j) \leq CET(S_{j-1}) \leq \ldots \leq CET(S_1)$*

Proof. Here $T_1 \leq T_2 \leq \ldots \leq T_j$ and by Lemma 1 $CET(S_j) \leq CET(S_{j-1}) \leq \ldots \leq CET(S_1)$. So, the only thing remains to prove is $T_j \leq CET(S_j)$. Let by contrary assume that $T_j > CET(S_j)$. By putting formula for $CET(S_{j-1})$ from Eq. (1) and then solving we get $T_j > CET(S_{j-1})$. By Lemma 1, $T_j \leq CET(S_{j-1})$. This is a contradiction.

Lemma 3. *If $S_j = \{P_1, P_2, \ldots, P_j\}$, $j \leq k$ is returned as $\mathcal{R}$ by above algorithm then $CET(S_j) \leq CET(S_j \setminus \{P_i\})$, $2 \leq i \leq j$.*

Proof. We will prove this statement by contradiction. Let $CET(S_j) > CET(S_j \setminus \{P_i\})$, which implies $T_i > CET(S_j)$ by putting formula for CET from equation-1. It is not possible by Lemma 2. Hence the claim holds.

Remark 1. The addition of paths by the above algorithm is consistent, i.e. if a path is added then it will remain till the end of the algorithm execution.

4.3 An Important Observation

In Fig. 2, ordered pair (C, T) denotes capacity and transit time of an edge. There are two paths P_1 and P_2 between s and t.

$$P_1 : s - B - C - E - G - t, C(P_1) = 4, T(P_1) = 19.$$
$$P_2 : s - A - C - E - F - t, C(P_2) = 6, T(P_2) = 23.$$

P_1 and P_2 are not edge-disjoint, but common edge CE has capacity of 10 i.e. $C(P_1) + C(P_2) = C(CE)$. So, flow can be sent through P_1 and P_2 in parallel and we may think like we have two copies of edge CE one having capacity 4, dedicated for P_1 and other one having capacity 6, dedicated for P_2. We name such set of paths as "virtually edge disjoint". Now it is easy to observe that to apply the formula of combined evacuation time on a set of paths, defined in Eq. (1), the necessary condition is they should be virtually edge disjoint rather than edge disjoint.

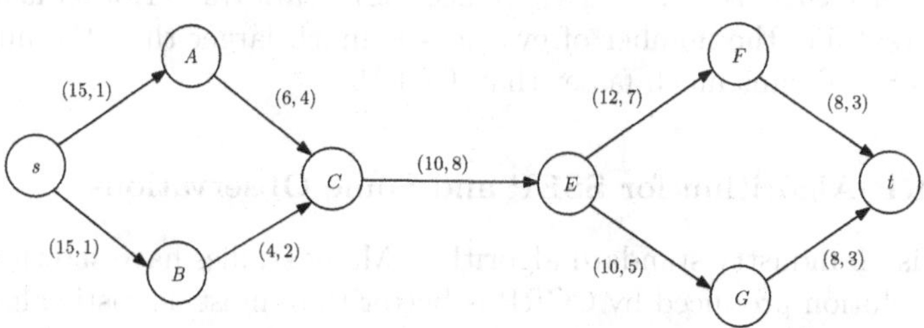

Fig. 2. An example to show that parallel flows can be sent on non edge-disjoint paths.

4.4 Our Algorithm for SSEP

The main idea of the algorithm is to find set of virtually edge disjoint paths one by one and calculate CET as in Sect. 4.2 after each path addition till it satisfies a required condition.

We discover paths one by one in the order of their transit time as follows. We find path P_1 along with its capacity C_1 having minimum transit time and decrease capacities of each node and path of P_1 by its capacity C_1 permanently and delete saturated nodes and edges. Let's say we have already added paths $\{P_1, P_2, \ldots, P_i\}$, $i \geq 1$, and updated the capacities of nodes and edges along with deletion of required saturated nodes and edges. Note that $P_1, P_2, \ldots, P_i$ are virtually edge disjoint. Hence formula of CET can be applied. In next iteration we discover a path P_{i+1} in residual graph iff t is reachable from s and $i < p$(see line number-4 in Algorithm 1). We add the discovered path P_{i+1} iff $T_{i+1} \leq CET(S_i)$(see line number-6 in Algorithm 1). As we delete saturated nodes and edges in each iteration when a path is added we discover paths in maximum of $m+n$ iterations i.e. at max $m+n$ paths and we are not going to discover more than p paths as each path can evacuate atleast one people. So, our algorithm restricts finding exponential number of possible paths from s to t . More clearly we discover at most $min(m+n, p)$ paths.

Here one may think of we are adding paths only based on transit time without considering capacity. Note that selection of a path for addition is based on transit time, addition of selected path is done if its transit time less than or equal to previously calculated CET, which is function of both capacities and transit times of previously added paths. So, our addition of paths to the solution is based on both transit time and capacities of paths implicitly.

4.5 Running Time Analysis of SSEP

From the above discussion it is clear that at most $min(m+n, p)$ paths will be discovered and equivalently our algorithm runs for at most $min(m+n, p)$ iterations. As each path discovery can be done in $O((m+n\log n)$ time, using well known Dijkstra algorithm for shortest path, our entire algorithm requires $O(\min(m+n, p)(m+n\log n)$ time. Assuming $m = O(n)$, this becomes $O(\min(n, p) \cdot n \log n)$, which is always at most $O(pn \log n)$. Recall that the time-complexity of CCRP is $O(pn \log n)$. Hence, SSEP always performs faster than CCRP. In real life, the number of evacuees is much larger than the number of vertices, so SSEP runs much faster than CCRP.

4.6 CCRP Algorithm for SSEP and Some Observations

CCRP [8] is an industry standard algorithm. Many studies have shown that the quality of solution produced by CCRP is better than most heuristic algorithms. We present the CCRP algorithm in simplified form, when there is a single source and a single sink.

Algorithm 1. Single Source Single Sink Evacuation Route Planner (SSEP)

Input: A graph $G(V, E)$ representing the network with designated source $s \in V$ and sink $t \in V$. Every node $v \in V$ has an occupancy and maximum capacity. Every edge $e \in E$ has a maximum capacity and transit time. Initially, all persons are in s.
Output: Evacuation route plan for each person.

begin
  Initialize $\mathcal{R} = \emptyset$ and $CET = \infty$.
  Initialize $i \leftarrow 0$.
  while *(t is reachable from s) and number of discovered paths* $\leq p - 1$ **do**
    Find the shortest path P_{i+1} from s to t in $G(V, E)$and let T_{i+1}, C_{i+1} be its transit time and capacity respectively.
    if $T_{i+1} \leq CET$ **then**
      $\mathcal{R} = \mathcal{R} \cup \{P_{i+1}\}$.
      $CET = CET(S_{i+1})$.
      Reduce capacity of each node and each edge of P_{i+1} by C_{i+1}.
      $V = V \setminus \{v : v \text{ is a saturated node of } P_{i+1}\}$.
      $E = E \setminus \{e : e \text{ is a saturated edge of } P_{i+1}\}$.
    end
    else
      break.
    end
  end
  Let $\mathcal{R} = \{P_1, P_2, \ldots, P_k\}$.
  Send x_i persons via $P_i, 1 \leq i \leq k$, where $T_i + \lceil \frac{x_i}{C_i} \rceil - 1 = CET$.
end

1. s is added to the priority queue. The nodes in priority queue are ordered based on the distance calculated from s during algorithm execution.
2. While there are evacuees in s, find the path P having minimum destination arrival time from s to t taking the capacity of the various nodes and edges into consideration.
3. Find capacity of P and reserve capacity along the path for a group of size equal to the minimum capacity.
4. If there are evacuees left at s, go to step 2.

Definition 6 (Group Size of a Path). *In each iteration of CCRP one path (say P_i) from s to t is discovered along with maximum number of people that can be evacuated through that path. This is defined as the* group size *of P_i for this iteration.*

For the below sections we denote T_i, C_i as transit time and group size of path P_i respectively.

Observation 1. *Let's consider execution of single source(s) single sink(t) evacuation network by CCRP algorithm. Let $P_1, P_2, \ldots, P_k$ be distinct paths(not necessarily edge-disjoint) from s to t discovered by CCRP such that $T_1 \leq T_2 \leq \ldots \leq T_k$. Here $A_i(T)$ is any permutation of $P_1(T), P_2(T), \ldots, P_i(T)$ and $P_j(T)$ is the path P_j with destination arrival time T.*

Phase 1: $A_1(T_1), A_1(T_1 + 1), \ldots, A_1(T_2 - 1)$

...

Phase i: $A_i(T_i), A_i(T_i+1), \ldots, A_i(T_{i+1}-1), i < k$

...

Phase k: $A_k(T_k), A_k(T_k+1), \ldots, A_k(T_k+\epsilon-2), A_k(T_k+\epsilon-1)$.

Here ϵ *is the maximum number of times any path is discovered in phase* k. *Note that* $\epsilon \geq 1$ *as* P_k *is discovered at least once.*

Number of times any path discovered in phase-k is either ϵ *or* $\epsilon - 1$. *It is because of the following argument. By definition of* ϵ *there exists a path (say* P_m*) discovered* ϵ *number of times. Let* P_l *is a path discovered less than* $\epsilon - 1$ *number of times. In this case CCRP algorithm would have returned* P_l *instead of* P_m*, because using path* P_l *some people can reach destination before or at time* $T_k+\epsilon-2$ *and* P_m *has earliest destination arrival time of* $T_k+\epsilon-1$.

Consider the point when all k paths have been returned $\epsilon - 1$ *times in phase* k. *Now we may not have enough evacuees such that CCRP will return each path once. We can add some virtual evacuees such that we will use all the paths exactly* ϵ *times in phase-k and for simplicity we can say* ϵ *is the number of times path* P_k *is returned by CCRP.*

Here it is easy to note that evacuation egress time $T_{Evac}^{CCRP} = T_k + \epsilon - 1$ *and it is independent of permutation of paths in any* $A_i(T)$. *So, fix a permutation i.e.* $A_i(T) = P_1(T), P_2(T), \ldots, P_i(T)$. *Fixing up this permutation doesn't affect the solution, but it will make the analysis easier.*

Observation 2. *Let* $P_1, P_2, \ldots, P_k$ *be distinct paths(not necessarily edge-disjoint) from s to t discovered by CCRP such that* $T_1 \leq T_2 \leq \ldots \leq T_k$. *Here* P_i *is the shortest path discovered after deletion of saturated nodes/edges of* $P_1, P_2, \ldots, P_{i-1}$.

Remark 2. Algorithm 1 finds a path even after we have deleted saturated nodes and edges of all previously discovered path, if it satisfies the conditions given on line numbers 4 and 6.

Observation 3. *Let's consider the sequence of paths as in Observation 1 with the fixed permutation of each* $A_i(T)$ *as explained. A path* P_i *may be returned in many iterations of CCRP. Group size returned in all iterations are equal possibly except last time when* P_i *is discovered(in phase* k*) in case we don't have enough evacuees left at s. This type of situation might happen only once as we are dealing with single source single destination network and it can happen in phase k after or while discovery of* P_k *for the first time. In such cases we can add some virtual evacuees to s so that group size of a path remains same in all iterations. It will not affect evacuation egress time but it will make the analysis easier.*

Remark 3. We can represent each path discovered by CCRP as an ordered pair of path and its group size. Algorithm 1 returns a path with maximum number of people who can travel by that path at any time. As each path is discovered only once, we can represent each path along with the capacity as an ordered pair.

4.7 Analysis of Algorithm 1

Lemma 4. *Let* $(P_1, C_1), (P_2, C_2), \ldots, (P_k, C_k)$ *be distinct paths (not necessarily edge-disjoint) from* s *to* t *in order of their transit time discovered by CCRP.*

1. *Number of iterations that will return path* P_i *is* $T_k - T_i + \epsilon$, $1 \leq i \leq k$, *where* ϵ *denotes number of iterations that returns path* P_k.
2. *Number of iterations that will return path* P_i *before phase* j *is* $T_j - T_i$, *where* $i \leq j \leq k$.
3. *The same paths will be returned by Algorithm 1, and* $T_1 \leq T_2 \leq \ldots \leq T_k$.

Proof. Parts (1) and (2) directly follows from Observation 1. For part (3), by induction we can prove that algorithm 1 finds each path $P_j, 1 \leq j \leq k$ with available capacity C_j.

Base Case: $j = 1$ i.e. (P_1, C_1) is added by Algorithm 1. This is obvious.

Inductive Step: Suppose paths $(P_1, C_1), \ldots, (P_j, C_j), 1 \leq j < k$ have been added by Algorithm 1. We have to prove that Algorithm 1 will also add (P_{j+1}, C_{j+1}).

Part 1: From Observation 2, P_{j+1} is the shortest path from s to t in residual graph i.e. if we delete saturated node(s) and/or edge(s) of the paths $P_1, P_2, \ldots, P_j$. Algorithm 1 also adds paths one by one after deleting saturated node(s) and/or edges(s) of previously discovered paths. So, structure of the graph remains same after addition of these j paths both in CCRP and Algorithm 1. So, P_{j+1} is also the best path w.r.t. transit time in residual graph according to Algorithm 1. As P_{j+1} is the best path in residual network either no paths will be added or P_{j+1} will be added to set of routes in Algorithm 1.

Let by contrary assume that Algorithm 1 doesn't add path P_{j+1} i.e. Algorithm 1 does not add any path. Clearly it may happen due to one of the two reasons i.e. either t is not reachable from s or number of paths discovered $= p$(line number-4 in Algorithm 1) or $T_{j+1} > CET(S_j)$(line number-6 in Algorithm 1).

Case 1(a): (t is not reachable from s)
As CCRP is able to find path P_{j+1}, t is reachable from s. Contradiction!

Case 1(b): (Number of paths discovered $= p$)
It is clear from CCRP Algorithm given in Sect. 4.6 that it does not discover more than p paths as in each path at least one people will be evacuated. As CCRP finds path P_{j+1}, number of paths discovered before discovery of P_{j+1} by Algorithm 1 can't be more than $p - 1$.

Case 2: ($T_{j+1} > CET(S_j)$)
Just come back to the point when CCRP adds path (P_{j+1}, C_{j+1}) for the first time. It can happen only in phase $j+1$. From Lemma 4 P_i is returned in $T_{j+1} - T_i, 1 \leq i \leq j < k$, iterations before phase $j+1$. As P_{j+1} discovered in phase $j+1$ for the first time total number of people evacuated through P_i before discovery of

P_{j+1} is at least $T_{j+1} - T_i$. As group size of path P_i is C_i, total number of people evacuated before discovery of P_{j+1} is at least $\sum_{i=1}^{j} C_i(T_{j+1} - T_i)$. As CCRP adds the path P_{j+1} we can say that still there are people to be evacuated. Also from Observation 3 virtual evacuees are added while or after addition of path P_k. So, total number of people evacuated before discovery of P_{j+1} is strictly less than p. Mathematically $\sum_{i=1}^{j} C_i(T_{j+1} - T_i) < p$, which implies $T_{j+1} \leq CET(S_j)$. Contradiction!

Part 2: Now one thing remains to prove is available capacity of the path P_{j+1} returned by Algorithm 1 is also C_{j+1}. If P_{j+1} doesn't share any node or edge with previously discovered path we are done. So, assume that there is some node or edge x which is common to both P_{j+1} and some P_i, $1 \leq i \leq j$. Here we argue considering x as a node and argument for x as an edge is same. Let t_n^k denotes time required to travel from s(source) to node n via path P_k with out intermediate delay. Observe that $t_x^{j+1} \geq t_x^i$. From observation 1 P_{j+1} is discovered in phase $j+1$ for the first time by CCRP algorithm. In phase $j+1$ consider $A_{j+1}(T_{j+1})$. P_i has been discovered once before discovery of P_{j+1} with its destination arrival time T_{j+1} i.e. it has made a reservation of C_i at x for the time instance t_x^{j+1} at node x. Now arrival time of evacuees via P_{j+1} to x is also t_x^{j+1}. At t_x^{j+1} we can not use that capacity of C_i for evacuees routing via P_{j+1}. In other words as if node x has dedicated capacity of C_i at time t_x^{j+1} for evacuees routing via P_i and that can't be used by evacuees routing via P_{j+1}. Here we have not assumed anything on i and x. For each such i and x, P_{j+1} can't use the capacity of C_i at time t_x^{j+1} at node x. It is equivalent to permanently decrementing the capacity of such x's by corresponding C_i, because from observation 1 whenever P_{j+1} is discovered prior to that a reservation of C_i must have been done at common node x(of P_i and P_{j+1}) by path P_i. Now come back to Algorithm 1. By induction each path P_i, $i \leq j$ is returned with capacity C_i. We find path P_{j+1} by decrementing the capacity of each path by C_i permanently. So, just before addition of P_{j+1} structure of the graph remains same w.r.t. capacity both in CCRP and Algorithm 1. From this discussion we can say that capacity of path P_{j+1} returned by Algorithm 1 is C_{j+1}.

Theorem 1. *The evacuation time of the solution given by Algorithm 1 is at most as that of the CCRP Algorithm for single source and single sink.*

Proof. Let $(P_1, C_1), (P_2, C_2), \ldots, (P_k, C_k)$ be distinct paths (not necessarily edge-disjoint) from s to t in order of their transit time (neglecting delays) discovered by CCRP. By Lemma 4, Algorithm 1 also returns the same set of paths. From Observation 1, we can say that evacuation time of CCRP is $T_{Evac}^{CCRP} = T_k + \epsilon - 1$. Evacuation time of Algorithm 1 is $CET(S_k)$. Also from Lemma 4, number of people that are evacuated through P_i is $C_i(T_k - T_i + \epsilon)$. As All people have been evacuated we can write $\sum_{i=1}^{k} C_i(T_k - T_i + \epsilon) \geq p$, which implies $T_{Evac}^{CCRP} \geq CET(S_k)$.

Theorem 2. *Upper bound on the evacuation time given by CCRP (hence by Algorithm 1) for single source single sink network is* $\lfloor \frac{p}{k} \rfloor + (n-1)\tau - 1$, *where* p *is the number of evacuees,* n *is the number of nodes in the graph,* τ *is the* *maximum transit time of any edge and k is the number of paths used by CCRP (and Algorithm 1).*

Proof. From Lemma 4, number of iterations executed by CCRP is $\sum_{i=1}^{k}(T_k - T_i + \epsilon) \leq p$, as in each iteration at least one person will be evacuated. Hence, $T_{Evac}^{CCRP} \leq \lfloor \frac{p}{k} \rfloor + (n-1)\tau - 1$.

5 Randomized Behavior Model of People

The idea of combined evacuation time [11] can be extended by considering probabilistic behavior of people. Suppose in an evacuation, people do not follow the paths suggested by Algorithm 1 (or CCRP). Let's say with probability $\alpha > 0$ a person follows suggested path and with probability $1 - \alpha$ he follows the shortest path (to the nearest exit). In this situation, we have to redistribute people via various paths. If we suggest x_i persons via $P_i, i \neq 1$, then the number of persons who will follow P_i and P_1 is αx_i and $(1-\alpha)x_i$ respectively (in expectation). The total number of people following P_1 and P_i are $x_1 + \sum_{i=2}^{k}(1-\alpha)x_i$ and $\alpha x_i, i \neq 1$ respectively. Expected time at which the last person will arrive at destination via P_1 is $T_1 + \frac{x_1 + \sum_{i=2}^{k}(1-\alpha)x_i}{C_1} - 1$. Expected time at which last person will arrive at destination via P_i is $T_i + \frac{\alpha x_i}{C_i} - 1, i \neq 1$
Let the expected evacuation time in this scenario be $E[T]$. Now we can write,

$$E[T] = \max\left(T_1 + \frac{(1-\alpha)n}{C_1} - 1, \max_{2 \leq i \leq k}\left(T_i + \frac{\alpha x_i}{C_i} - 1\right)\right).$$

$E[T]$ will be minimum when it satisfies the following equation,

$$\begin{aligned} E[T] &= T_1 + \frac{x_1 + \sum_{i=2}^{k}(1-\alpha)x_i}{C_1} - 1 \\ &= T_i + \frac{\alpha x_i}{C_i} - 1, 2 \leq i \leq k. \end{aligned} \tag{2}$$

where $\sum_{i=1}^{k} x_i = n$ and $x_i \geq 0, \forall i$. Solving the above equations we get,

$$E[T] = \frac{n + \sum_{i=1}^{k} C_i T_i}{\sum_{i=1}^{k} C_i} - 1 = CET(\{P_1, P_2, \ldots, P_k\}) \tag{3}$$

Expected evacuation time given by Eq. (3) doesn't depend on α. This is true and solution is feasible as long as $x_1 \geq 0$. But it is not always the case, specifically when $(1-\alpha)\sum_{i=2}^{k} x_i > C_1(T - T_1 + 1)$. So, implicitly evacuation time is dependent on α. In the following sections we give the algorithm that considers the randomized behavior of people along with analysis for expected evacuation time.

5.1 Lower Bound for Expected Evacuation Time

On expectation $x_1 + (1-\alpha)\sum_{i=2}^{k} x_i = \alpha x_1 + (1-\alpha)n$ number of people will be evacuated via path P_1. This is minimum when $x_1 = 0$ as $x_1 \geq 0$. So, lower bound for expected evacuation time is $T_1 + \frac{(1-\alpha)n}{C_1} - 1$.

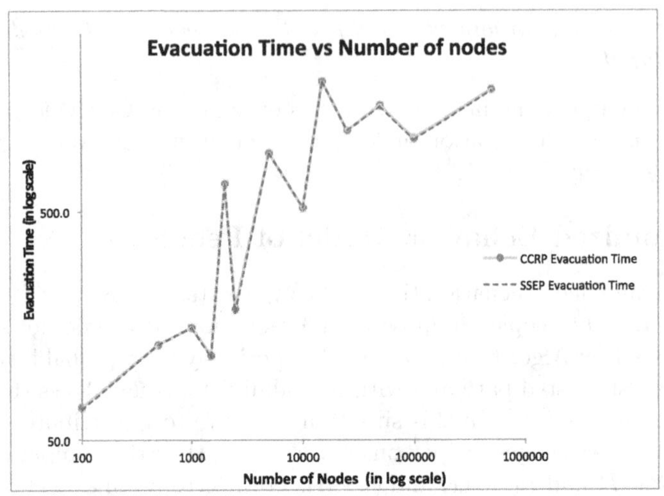

Fig. 3. Evacuation time vs number of nodes for SSEP and CCRP.

5.2 Algorithm for Randomized Behavior of People

Algorithm 2

1. Run Algorithm 1. Find CET and $x_1, x_2, \ldots, x_k$ using Eq. (2).
2. If $x_1 \geq 0$ then quit; else go to step 3. In this case, the expected evacuation time = CET.
3. Assign x'_1 to 0 and $x'_i = \frac{nx_i}{\sum_{j=2}^{k} x_j}, \forall i \neq 1$. In this case, the expected evacuation time $= T_1 + \frac{(1-\alpha)n}{C_1} - 1$.

Lemma 5. $x'_i < x_i$, $\forall i \neq 1$, *and* $\sum_{i=2}^{k} x'_i = n$.

Proof. Directly follows from the algorithm.

Lemma 6. *Above algorithm has a expected evacuation time of* $CET(\{P_1, P_2, \ldots, P_k\})$ *when it quits from step-2.*

Proof. In this case $x_1 \geq 0$. From the equation-4 also we can observe that $x_i \geq 0, \forall i \neq 1$. Hence the solution is feasible. So, we can safely say that the expected evacuation time is CET.

Lemma 7. *Above algorithm has a expected evacuation time of* $T_1 + \frac{(1-\alpha)n}{C_1} - 1$ *when it quits from step-3.*

Proof. In this case $x_1 < 0$ and by Lemma 5 $x'_i < x_i, i \neq 1$. For $i \neq 1$ x'_i number of people are suggested path P_i. Hence $T_i + \frac{px'_i}{C_i} - 1 < T_i + \frac{px_i}{C_i} - 1 < CET$, $i \neq 1$ and $T_1 + \frac{(1-\alpha)n}{C_1} - 1 > CET$.

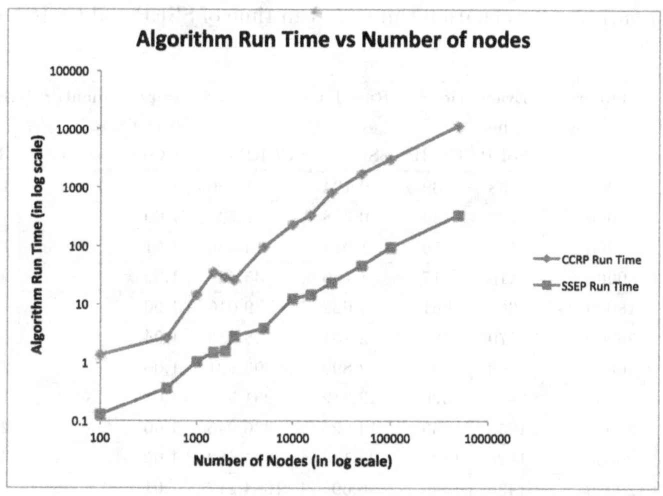

Fig. 4. Run time vs number of nodes for SSEP and CCRP.

Theorem 3. *In a single source single sink evacuation problem, if people follow the path suggested by Algorithm* 2 *with probability* α, *then the expected evacuation time is* $\max(CET, T_1 + \frac{(1-\alpha)n}{C_1} - 1)$ *and algorithm runs in* $O(\min(n, p) \cdot n \log n)$ *time.*

6 Experimental Results

6.1 Details of the Experiments

We executed the SSEP and CCRP algorithms on a Dell Precision T7600 server having an Intel Xeon E5-2687W CPU running at 3.1 GHz with 8 cores (16 logical processors) and 128 GB RAM. The operating system is Microsoft Windows 7 Professional 64-bit edition. We used the C/C++ network analysis libraries *igraph* and *LEMON* to implement the algorithms. We used *netgen* to generate synthetic graphs. The number of vertices in the graph varies from 100 to 500,000. The number of people varies from 3,000 to 120,000. The results are shown in Table 1. The graphs are plotted on a log-log scale.

6.2 Results

We show the variation of evacuation time and run time with number of nodes for SSEP and CCRP algorithms in Figs. 3 and 4 respectively. From Fig. 3, we can see that the evacuation time of SSEP is at most that of CCRP. It is evident from Fig. 4 that the running time of SSEP is much lower than that of CCRP. Hence, for all these instances SSEP clearly outperforms CCRP with respect to both evacuation time and run time. The absolute and relative amount by which SSEP performs better than CCRP is shown in Table 1.

Table 1. Comparison of evacuation time and run time of SSEP and CCRP algorithms

Number of Nodes (n)	Number of Evacuees (p)	Evacuation Time		Run Time		Improvement in SSEP over CCRP ($\frac{CCRP}{SSEP}$)	
		SSEP	CCRP	SSEP	CCRP	Evacuation Time	Run Time
100	3000	68	69	0.124	1.326	1.01	10.69
500	5000	130	130	0.358	2.73	1.00	7.63
1000	7000	155	156	1.014	14.586	1.01	14.38
1500	9000	115	117	1.466	35.443	1.02	24.18
2000	15000	661	661	1.622	29.016	1.00	17.89
2500	25000	179	186	2.761	25.739	1.04	9.32
5000	40000	903	903	3.899	93.521	1.00	23.99
10000	65000	517	520	12.012	231.535	1.01	19.28
15000	95000	1848	1853	14.025	336.946	1.00	24.02
25000	100000	1126	1128	23.134	815.682	1.00	35.26
50000	120000	1436	1446	46.69	1684.217	1.01	36.07
100000	110000	1032	1044	93.4952	3016.3005	1.01	32.26
500000	100000	1698	1720	344.341	11363.253	1.01	33.00

7 Conclusion and Future Work

In this paper, we have studied the evacuation route planning problem and given an improved algorithm for the single source single sink case. We theoretically showed that the SSEP algorithm performs better than the CCRP algorithm, both in terms of evacuation time and run time. This is also demonstrated by extensive experiments. We also analyzed a simple probabilistic behavior model of people. Here are some open problems which we would like to work in future.

- Design a system for real time monitoring of evacuation in a building using our indoor localization app [1].
- Extend this algorithm to the multiple source multiple sink case, and compare it's performance with CCRP and other algorithms.
- Develop a more sophisticated probabilistic behavior model of people for the case when they don't follow the routes suggested by the algorithm.
- Give good lower and upper bounds for the problem.

References

1. Ahmed, N., Ghose, A., Agrawal, A.K., Bhaumik, C., Chandel, V., Kumar, A.: SmartEvacTrak: a people counting and coarse-level localization solution for efficient evacuation of large buildings. In: IEEE International Conference on Pervasive Computing and Communication Workshops (PerCom Workshops), pages 372–377. IEEE (2015)
2. Dressler, D., Groß, M., Kappmeier, J.-P., Kelter, T., Kulbatzki, J., Plümpe, D., Schlechter, G., Schmidt, M., Skutella, M., Temme, S.: On the use of network flow techniques for assigning evacuees to exits. Procedia Eng. **3**, 205–215 (2010)

3. Gupta, A., Sarda, N.L.: Efficient evacuation planning for large cities. In: Decker, H., Lhotská, L., Link, S., Spies, M., Wagner, R.R. (eds.) DEXA 2014, Part I. LNCS, vol. 8644, pp. 211–225. Springer, Heidelberg (2014)
4. Hamacher, H.W., Tjandra, S.A: Mathematical Modelling of Evacuation Problems: A State of Art. Fraunhofer-Institut für Techno-und Wirtschaftsmathematik, Fraunhofer (ITWM) (2001)
5. Hoppe, B., Tardos, É.: Polynomial time algorithms for some evacuation problems. In: Proceedings of the fifth annual ACM-SIAM symposium on Discrete algorithms, pp. 433–441. Society for Industrial and Applied Mathematics (1994)
6. Kim, S., Shekhar, S., Min, M.: Contraflow transportation network reconfiguration for evacuation route planning. IEEE Trans. Knowl. Data Eng. **20**(8), 1115–1129 (2008)
7. Løvs, G.G.: Models of wayfinding in emergency evacuations. Eur. J. Oper. Res. **105**(3), 371–389 (1998)
8. Lu, Q., George, B., Shekhar, S.: Capacity constrained routing algorithms for evacuation planning: a summary of results. In: Medeiros, C.B., Egenhofer, M., Bertino, E. (eds.) SSTD 2005. LNCS, vol. 3633, pp. 291–307. Springer, Heidelberg (2005)
9. Min, M.: Synchronized flow-based evacuation route planning. In: Wang, X., Zheng, R., Jing, T., Xing, K. (eds.) WASA 2012. LNCS, vol. 7405, pp. 411–422. Springer, Heidelberg (2012)
10. Min, M., Lee, J.: Maximum throughput flow-based contraflow evacuation routing algorithm. In: IEEE International Conference on Pervasive Computing and Communications Workshops (PERCOM Workshops), pages 511–516. IEEE (2013)
11. Min, M., Neupane, B.C.: An evacuation planner algorithm in flat time graphs. In: Proceedings of the 5th International Conference on Ubiquitous Information Management and Communication, p. 99. ACM (2011)
12. Skutella, M.: An introduction to network flows over time. In: Cook, W., Lovász, L., Vygen, J. (eds.) Research Trends in Combinatorial Optimization, pp. 451–482. Springer, Heidelberg (2009)
13. Song, X., Zhang, Q., Sekimoto, Y., Shibasaki, R., Yuan, N.J., Xie, X.: A simulator of human emergency mobility following disasters: knowledge transfer from big disaster data. In: AAAI Conference on Artificial Intelligence (2015)
14. Yin, D.: A scalable heuristic for evacuation planning in large road network. In: Proceedings of the Second International Workshop on Computational Transportation Science, pp. 19–24. ACM (2009)

Improved MAXSAT Algorithms for Instances of Degree 3

Chao Xu[1], Jianer Chen[1,2], and Jianxin Wang[1](✉)

[1] School of Information Science and Engineering, Central South University, Changsha, People's Republic of China
jxwang@mail.csu.edu.cn

[2] Department of Computer Science and Engineering, Texas A&M University, College Station, USA

Abstract. The degree of a variable x_i in a MAXSAT instance is the number of times x_i and $\bar{x}_i$ appearing in the given formula. The degree of a MAXSAT instance is equal to the largest variable degree in the instance. In this paper, we study techniques for solving the MAXSAT problem on instances of degree 3 (briefly, $(n,3)$-MAXSAT), which is NP-hard. Two new non-trivial reduction rules are introduced based on the resolution principle. As applications, we present two algorithms for the $(n,3)$-MAXSAT problem: a parameterized algorithm of time $O^*(1.194^k)$, and an exact algorithm of time $O^*(1.237^n)$, improving the previous best upper bounds $O^*(1.2721^k)$ and $O^*(1.2600^n)$, respectively.

1 Introduction

The MAXIMUM SATISFIABILITY problem (MAXSAT) plays a key role in the study of computational optimization [9]. The *Strong Exponential Time Hypothesis* [10] conjectures that MAXSAT cannot be solved in time $O^*(2^{cn})$ for any constant $c < 1$, where n is the number of variables in the input instance, which is a CNF formula. Algorithms for MAXSAT and various restricted versions of MAXSAT have been studied extensively (see, for example, [6] and its references). Define the *degree* of a variable x_i in a MAXSAT instance to be the number of times x_i and $\bar{x}_i$ appearing in the formula. The (s,t)-MAXSAT problem is a well-known restricted version of the MAXSAT problem in which each clause in an instance contains at most s literals and each variable has degree bounded by t [12,14]. It is shown that $(n,2)$-MaxSAT problem can be solved in polynomial time [7].

This paper focuses on algorithms for the $(n,3)$-MAXSAT problem. Since the $(2,3)$-MAXSAT problem is NP-hard [12], the $(n,3)$-MAXSAT problem is NP-hard for $n \geq 2$. Exact and parameterized algorithms (e.g., [2,3,12,13]) have been extensively studied for the problem. Two main parameters used for evaluating the performance of the algorithms have been used: the number n of variables and the number k of satisfied clauses.

This work is supported by the National Natural Science Foundation of China, under grants 61173051, 61232001, 61472449, and 61420106009.

Z. Lu et al. (Eds.): COCOA 2015, LNCS 9486, pp. 20–30, 2015.
DOI: 10.1007/978-3-319-26626-8_2

The main results of the current paper are two new non-trivial reduction rules (R-Rules 7–8), which are based on the resolution principle. As applications, we propose two algorithms for the $(n, 3)$-MAXSAT problem in terms of the two main parameters.

We first give formal definitions of the problems we are focused on. *The* $(n, 3)$-MAXSAT *problem* asks for an assignment satisfying the maximum number of clauses in a given formula in which each variable has degree bounded by 3. The *(parameterized)* $(n, 3)$-MAXSAT *problem* consists of instances of the form (F, k), where F is a formula of the $(n, 3)$-MAXSAT and k is an integer, asking whether there is an assignment to the variables that satisfies at least k clauses in F.

The table in Fig. 1 lists the current literature on algorithms for the $(n, 3)$-MAXSAT problem. For comparison, we also include our result in the current paper in the table.[1]

Bound(n)	Bound(k)	Reference	Year
$O^*(1.732^n)$		Raman *et al* [14]	1998
$O^*(1.3248^n)$		Bansal, Raman [1]	1999
	$O^*(1.3247^k)$	Chen, Kanj [5]	2002
$O^*(1.27203^n)$		Kulikov [15]	2005
	$O^*(1.2721^k)$	Bliznets, Golovnev [3]	2012
$O^*(1.2600^n)$		Bliznets [4]	2013
$O^*(1.237^n)$	$O^*(1.194^k)$	this paper	2015

Fig. 1. Progress in $(n, 3)$-MAXSAT algorithms

Most algorithms for MAXSAT (as well as for $(n, 3)$-MAXSAT) are based on the branch-and-bound technique [8]. The *Strong Exponential Time Hypothesis* [10] conjectures, to some extent, a popular opinion that branch-and-bound is perhaps unavoidable to solve the MAXSAT problem and its variations. Therefore, how to branch more efficiently in algorithms solving $(n, 3)$-MAXSAT becomes crucial.

A contribution of the current paper is to show that the resolution principle [7] can be applied to solve the $(n, 3)$-MAXSAT problem, while keeping all variables of degree 3. It has been well-known that the resolution principle is a very powerful tool to solve the satisfiability problem [7]. In particular, variable resolutions in a CNF formula preserve the satisfiability of the formula. Unfortunately, variable resolutions cannot be used directly to solve the $(n, 3)$-MAXSAT problem in general case, since not all clauses are presumed to be satisfied by an optimal assignment to an instance of the $(n, 3)$-MAXSAT problem.

We begin with some preliminary definitions.

[1] Following the current convention in the research in exact and parameterized algorithms, we will use the notation $O^*(f)$ to denote the bound $f \cdot m^{O(1)}$, where f is an arbitrary function and m is the instance size.

A (Boolean) *variable* x can be assigned value either 1 (TRUE) or 0 (FALSE). A variable x has two corresponding literals: the *positive literal* x and the *negative literal* $\bar{x}$, which will be called the *literals* of x. A *clause* C is a disjunction of a set of literals, also regarded as a set of the literals. So, $C_1 = zC_2$ indicates that the clause C_1 is in consist of the literal z plus all literals in the clause C_2, and use C_1C_2 to denote the clause that consists of all literals that are in either C_1 or C_2, or both. Without loss of generality, we assume that a literal can appear in a clause at most once. A clause C is *satisfied* by an assignment if under the assignment, at least one literal in C gets a value 1. A (CNF Boolean) *formula* F is a conjunction of clauses $C_1, \ldots, C_m$, regarded as a collection of the clauses. The formula F is *satisfied* by an assignment to the variables in the formula if all clauses in F are satisfied by the assignment.

The *size* of a clause C is the number of literals in C. A clause is an *h-clause* if its size is h, and an *h^+-clause* if its size is at least h. A clause is *unit* if its size is 1 and is *non-unit* if its size is larger than 1. The *size* of a CNF formula F is equal to the sum of the sizes of the clauses in F.

A literal z is an *(i,j)-literal* in a formula F if z and $\bar{z}$ appear i times and j times in F, respectively. Thus, a variable x is of degree h if x is an (i,j)-literal such that $i + j = h$. An $(i,1)$-literal z is an *$(i,1)$-singleton* if $\bar{z}$ occurs in a unit clause $(\bar{z})$. When i is not critical, we also call an $(i,1)$-singleton simply a *singleton*. A variable of degree h (resp. at least h) is also called an *h-variable* (resp. *h^+-variable*).

A *resolvent* on a variable x in a formula F is a clause of a new form CD such that xC and $\bar{x}D$ are clauses in F. The *resolution* on the variable x in F, written as $DP_x(F)$, is a formula that is obtained by first removing all clauses that contain either x or $\bar{x}$ from F and then adding all resolvents on x into F.

2 Reduction Rules

A *reduction rule* converts, in polynomial time, an instance (F,k) of $(n,3)$-MaxSAT into another instance (F',k') with $k \geq k'$ such that (F,k) is a Yes-instance if and only if (F',k') is a Yes-instance. Note that a reduction rule can be acknowledged as a special case of branching steps.

We present 9 reduction rules, R-Rules 1–9. The reduction rules are supposed to be applied *in order*, i.e., R-Rule j cannot be applied until none of R-Rules i with $i < j$ is applicable. In the following, F is always supposed to be a conjunction of clauses.

The first three reduction rules are from [4].

R-Rule 1 ([4]). $(F \wedge (x\bar{x}C), k) \rightarrow (F, k-1)$, and $(F \wedge (x) \wedge (\bar{x}), k) \rightarrow (F, k-1)$.

R-Rule 2 ([4]). If there is an (i,j)-literal z in the CNF formula F, with at least j unit clauses (z), then $(F,k) \rightarrow (F_{z=1}, k-i)$, where $F_{z=1}$ is the formula F with an assignment $z = 1$ on the literal z.

Assume that R-Rule 2 is not applicable to F, and then each literal in F has its negation also in F. Thus, all variables are 2^+-variables. Under this condition, we can process 2-variables based on the resolution principle [7], whose correctness can be easily verified.

R-Rule 3 ([4]). For any 2-variable x, $(F\wedge(xC_1)\wedge(\bar{x}C_2),k) \rightarrow (F\wedge(C_1C_2),k-1)$.

Note that each variable appears at most 3 times in F. In case none of R-Rules 1–3 is applicable, every variable is a 3-variable. Moreover, for each $(2,1)$-literal z, there is no unit clause (z). Now we describe two reduction rules based on variations of the resolution principle, which are from [2].

R-Rule 4 ([2]). For a $(2,1)$-literal x in a 2-clause (xy) and the clause with $\bar{x}$ contains at least two literals, $(F_1 = F \wedge (xy) \wedge (xC_1) \wedge (\bar{x}D), k) \rightarrow (F_2 = F \wedge (yD) \wedge (\bar{y}C_1D), k-1)$.

After R-Rule 4, the degree of all variables in yD becomes 4. In order to keep all variables being 3-variables, a branching for variable y must be applied. Naturally, the branching vector is (3+1,1+1), whose root is 1.2721. However, since we only need consider $|D| \geq 1$, we can do better.

R-Rule 5 ([2]). If two variables x and y of degree 3 appear together in 3 clauses, then all these 3 clauses can be satisfied by assigning x and y properly.

Next rule is just a part from Corollary 1 in [2].

R-Rule 6 ([2]). If there are two clauses xyC_1 and $\bar{x}\bar{y}C_2$ in the formula F such that each variable appears in three times, then we can safely do resolution on x, such that F is replaced with $DP_x(F)$.

After R-Rule 6, R-Rule 1 must be followed, keeping that each variable appears three times in the obtained formula. The next 2 rules are based on resolution, but its transformation is non-trivial, i.e. there is no direct relation to other MAXSAT algorithms. To be convenient, let $maxsat(F)$ be the maximum number of clauses satisfied in F. Note that since R-Rule 5 is not applicable, any two 3-variables appear in at least 4 clauses.

R-Rule 7. For a CNF formula $F_1 = F\wedge(xyC_1)\wedge(x\bar{y}C_2)\wedge(\bar{x}D_1)\wedge(\bar{y}D_2)$, where x is a $(2,1)$-literal in F_1, $(F_1 = F\wedge(xyC_1)\wedge(x\bar{y}C_2)\wedge(\bar{x}D_1)\wedge(\bar{y}D_2),k) \rightarrow (F_2 = F \wedge (x'C_2) \wedge (x'D_2) \wedge (\bar{x}'C_1D_1), k-1)$.

Lemma 1. *R-Rule 7 converts the instance (F_1,k) of $(n,3)$-MAXSAT into an instance $(F_2,k-1)$ such that $(F_1 = F \wedge (xyC_1) \wedge (x\bar{y}C_2) \wedge (\bar{x}D_1) \wedge (\bar{y}D_2),k)$ is a Yes-instance if and only if $(F_2 = F \wedge (x'C_2) \wedge (x'D_2) \wedge (\bar{x}'C_1D_1), k-1)$ is a Yes-instance.*

Proof. For an optimal assignment satisfying $C_1 = 1$ or $D_1 = 1$ or $C_2 = D_2 = 1$, by reassigning properly to variables x, y, x', $maxsat(F_1) = maxsat(F) + 4$ and $maxsat(F_2) = maxsat(F) + 3 = maxsat(F_1) - 1$. For an optimal assignment satisfying $C_1 = D_1 = 0$ and $C_2 = 0$ (resp. $D_2 = 0$), $maxsat(F_1) = maxsat(F)+3$ and $maxsat(F_2) = maxsat(F) + 2 = maxsat(F_1) - 1$. Thus, for both cases $maxsat(F_2) = maxsat(F_1) - 1$. □

Now, we introduce the following new reduction rule.

R-Rule 8. For any two $(2, 1)$-literals xy both in two clauses (xyC_1) and (xyC_2), $(F_1 = F \wedge (xyC_1) \wedge (xyC_2) \wedge (\bar{x}D_1) \wedge (\bar{y}D_2), k) \rightarrow (F_2 = F \wedge (x'C_1) \wedge (x'C_2) \wedge (\bar{x}'D_1D_2), k-1)$.

Lemma 2. *R-Rule* 8 *converts the instance* (F_1, k) *of* $(n, 3)$-MAXSAT *into an instance* $(F_2, k-1)$ *such that* $(F_1 = F \wedge (xyC_1) \wedge (xyC_2) \wedge (\bar{x}D_1) \wedge (\bar{y}D_2), k)$ *is a Yes-instance if and only if* $(F_2 = F \wedge (x'C_1) \wedge (x'C_2) \wedge (\bar{x}'D_1D_2), k-1)$ *is a Yes-instance.*

Proof. Similar to the proof in Lemma 2, for an optimal assignment satisfying $D_1 = 1$ or $D_2 = 1$ or $C_1 = C_2 = 1$, by reassigning properly to variables x, y, x', $maxsat(F_1) = maxsat(F) + 4$ and $maxsat(F_2) = maxsat(F) + 3 = maxsat(F_1) - 1$. For an optimal assignment satisfying $D_1 = D_2 = 0$ and $C_1 = 0$ (resp. $C_2 = 0$), at least one clause is not satisfied, and $maxsat(F_1) = maxsat(F) + 3$ and $maxsat(F_2) = maxsat(F) + 2 = maxsat(F_1) - 1$. For both cases $maxsat(F_2) = maxsat(F_1) - 1$. □

After R-Rules 1–3, 5–8, for any 2-literal x, the two clauses with x, each of which has at least two literals, and each pair of distinct clauses share at most one common variables. According to this property, similar to R-Rule 5, we can prove that for each quadruple of 3-variables must appear in at least 6 clauses.

Lemma 3. *If any quadruple variables of degree* 3 *appear together in at most* 5 *clauses, then all these clauses can be satisfied by assigning properly.*

Proof. Suppose variables x_1, x_2, x_3 and x_4 of degree 3 are in 5 clauses. Note that $3 \times 4 = 12$ literals of x_1, x_2, x_3, x_4 must be in these clauses. We first choose a 2-literal, i.e. the literal occurring in two clauses, denoted by x_1. Let $x_1 = 1$, and two clauses x_1C_1, x_1C_2 are satisfied. Since R-Rules 5–8 are not applicable, for any literal x_i, no two clauses containing literal x_i share another common variables. Thus, there are at most 3 literals (of variables from x_2, x_3 and x_4) satisfied by $x_1 = 1$ in C_1C_2. In all, at most 6 literals (of variables from x_1, x_2, x_3 and x_4) are reduced by at this time. Then there are two cases to consider. One is that there is still a 2-literal in the remaining of 3 variables, denoted by x_2. Let $x_2 = 1$, and at most 2 extra literals (of variables x_3 and x_4) are satisfied. In all, there are $6 + 2 + 3 = 11$ literals (of variables from x_1, x_2, x_3 and x_4) satisfied. Thus, there are at least one literal not satisfied and denoted by x_3. Let $x_3 = 1$, and all clauses are satisfied. The other case is that each literal (of variables from x_2, x_3 and x_4) appearing in the remaining of three clauses is a 1-literal. Since no two clauses share two common variables, there is at most one unit clause in the remaining of the 3 clauses, and can be simultaneously satisfied by assigning x_2, x_3 and x_4 properly. □

Next, we introduce another reduction rule, which is implemented in practical MAXSAT solver MaxSatz [1].

R-Rule 9 ([1]). $(F_1 = F \wedge (xy) \wedge (\bar{x}) \wedge (\bar{y}), k) \rightarrow (F_2 = F \wedge (\bar{x}\bar{y}), k-1)$.

3 An $O^*(1.194^k)$-time Parameterized Algorithm

Branching on an instance (F, k) of $(n, 3)$-MAXSAT leads to a collection $\{(F_1, k-d_1), \ldots, (F_r, k-d_r)\}$ of instances of $(n, 3)$-MAXSAT, such that (F, k) is a Yes-instance if and only if at least one of $(F_1, k-d_1)$, ..., $(F_r, k-d_r)$ is a Yes-instance. Such a branching step is called a $(d_1, \ldots, d_r)$-*branching*, and the vector $t = (d_1, \ldots, d_r)$ is called the *branching vector* for the branching. Each branching vector corresponds to a polynomial (see, e.g. [5]), and has a unique positive root that is larger than or equal to 1. Let the root $\rho(t)$ of a branching be the root of the branching vector t.

Let t_1 and t_2 be two branching vectors. We say that the t_1-branching is *inferior* to the t_2-branching if $\rho(t_1) > \rho(t_2)$. Based on the branch-and-bound technique, if every branching step in the algorithm has its root bounded by a constant $c \geq 1$, then the running time of the algorithm is bounded by $O^*(c^k)$.

We called formula F *reduced* if none of R-Rules 1–3, 5–9 and Lemma 3 is applicable.

Next we first give two lemmas for branching.

Lemma 4. *F_0 is a reduced formula where each variable is a 3-variable. Let x be a 3-variable such that $F_0 = F \wedge (xy) \wedge (xC) \wedge (\bar{x}D)$, with $|D| \geq 1$. First apply R-Rule 4, and then: (1) if y is a $(2,1)$-literal in F_0, branch on y; (2) if y is a $(1,2)$-literal in F_0, branch with $y = 1$ and $y = D = 0$. As a result, we have a $(5,3)$-branching.*

Proof. Since R-Rules 1–3, 5–8 are not applicable, $|C| \geq 1$, and $y \cup C \cup D$ contains no same literals and no conflict literals (e.g. z and $\bar{z}$). By applying R-Rule 4, $(xy) \wedge (xC) \wedge (\bar{x}D)$ is replaced with $(yD) \wedge (\bar{y}CD)$. Consequently, only variables in yD become 4-variables. Since D is contained both in clauses (yD) and $(\bar{y}CD)$, both branchings with $y = 1$ and $y = 0$ can reduce D. So, the obtained formula is also a $(n, 3)$-MAXSAT formula. Next, we show it is not inferior to $(5, 3)$-branching.

Case 1: if y is a $(2,1)$-literal in F_0, then since R-Rules 5–8, except clause (xy), there is no other clause both containing variables xy. Hence, denote the clauses with variable y as $(xy), (yE_1), (\bar{y}E_2)$, where $|E_1| \geq 1$, and let $E_1 = y'E'$. Moreover, y is a $(2,2)$-literal after R-Rule 4 (i.e. $(yD), (\bar{y}CD), (yy'E_1'), (\bar{y}E_2)$, and k is reduced by 1). Thus, we simply branch on variable y. When $y = 1$, two clauses containing literal y are satisfied. At least one other variable y', either $y' \in D$ or $y' \notin D$, becomes a 2-variable. Therefore, R-Rules 1–3 are applicable and the number of satisfied clauses is 3. When $y = 0$, clauses $(\bar{y}E_2)$ and $(\bar{y}CD)$ are satisfied. Thus, at least one literal $z \in C$, noted that $|C| \geq 1$ and $z, \bar{z} \notin D$, becomes a 2-variable and R-Rules 1–3 become applicable again. Similarly, 3 clauses are satisfied. As a result, besides the satisfied clause by R-Rule 4 (where k is reduced by 1), we give a (4,4)-branching.

Case 2: if y is a $(1,2)$-literal in F_0, then after R-Rule 4, y becomes a $(1,3)$-literal. It is safe to branch with $y = 1$ and $y = D = 0$, when y is a $(1,3)$-literal and there is a clause (yD), because for each optimal assignment σ with $y = 0$ and $D = 1$, there is another assignment σ' by only replacing $y = 0$ with $y = 1$.

Note that $\bar{y}$ is a non-singleton, with clauses $(yD), (\bar{y}CD), (\bar{y}E_1)$ and $(\bar{y}E_2)$, because $|D| \geq 1$, and let $D = z_1 \cdots z_h$. Thus, we branch with: B1)$y = 0$, satisfying 3 clauses. Similarly, R-Rule 2 becomes applicable, since each literal in C appears at least 1 (by R-Rules 5–8) and at most 2 times in the obtained formula after R-Rule 4 and branch with $y = 0$. Therefore, at least 4 clauses are satisfied; B2)$y = 1, z_1 = \cdots = z_h = 0$, and at least a clause with y and a clause with $\bar{z}_1$ are satisfied. This is a $(5, 3)$-branching (besides R-Rule 4). □

Next, we show the branching is still better if R-Rule 4 is not applicable on a chosen variable.

Lemma 5. *F_0 is a reduced formula where each variable is a 3-variable. Let x be a non-singleton such that: $F_0 = F \wedge (xy_1y_2C_1) \wedge (xy_3y_4C_2) \wedge (\bar{x}y_5D)$. Branch on variable x, we have $(7, 2)$-branching and the obtained formula is a $(n, 3)$-*MaxSAT *formula.*

Proof. If $|D| \geq 1$, then let $D = y_6$. Since R-Rules 5–8 are not applicable, no two clauses share two common variables, so that x, y_1-y_6 are 7 different variables. Also, according to Lemma 3, there are at least 6 clauses containing at least one of 3-variables y_1, y_2, y_3 and y_4. When $x = 1$, all the 3-variables y_1, y_2, y_3 and y_4 become 2-variables. Hence, Rule 1–3 can be applied and the number of satisfied clauses is reduced by at least $2 + 4 = 6$. When $x = 0$, at least y_5 and y_6 become 2-variables, where R-Rules 1–3 are applicable. The number of satisfied clauses is reduced by at least $1 + 2 = 3$. This is a $(6, 3)$-branching.

Consider when $|D| \geq 0$ and y_5 is a $(2, 1)$-literal. When $x = 1$, y_5 is contained in unit clause (y_5). By R-Rule 2, $y_5 = 1$. Similarly, all the 3-variables $y_1y_2y_3y_4$ become 2-variables. The number of satisfied clauses is reduced by at least $2 + 1 + 4 = 7$. When $x = 0$, at least y_5 becomes a 2-variable, and the number of satisfied clauses is reduced by at least $1 + 1 = 2$. This is a $(7, 2)$-branching.

Consider when $|D| \geq 0$ and y_5 is a $(1, 2)$-literal. Similarly when $x = 1$, the number of satisfied clauses is reduced by at least $2 + 4 = 6$. When $x = 0$, no literal y_5 exists, so that $y_5 = 0$ according to R-Rule 2, and two other clauses containing y_5 are satisfied. This is a $(6, 3)$-branching.

In summary, since $\rho(7, 2) = 1.191$ and $\rho(6, 3) = 1.174$, branching on variable x, so that we have a $(7, 2)$-branching. □

Before showing the algorithm, we must first introduce a lemma.

Lemma 6. *(Bliznets [2]). If each variable of F appears once negatively and twice positively and all negative literals occur in unit clauses, then there is a polynomial time algorithm to F that returns an optimal assignment satisfying the maximum number of clauses in F.*

Summarizing all the discussions, we present our algorithm for the $(n, 3)$-MaxSAT problem in Fig. 2.

Algorithm (n,3)-MaxSAT-Solver(F, k)
INPUT: an instance (F, k) of $(n, 3)$-MAXSAT, where
each variable appears in three clauses
OUTPUT: an assignment to F that satisfies at least k clauses,
or report no such an assignment exists
1. apply R-Rules 1-3 and 5-9, in order, repeatedly until (F, k) is irreducible;
2. **if** all variables are singletons
then using Lemma 6 to solve it in polynomial time; return;
3. choose a non-singleton x: $(xC_1), (xC_2), (\bar{x}D)$ and $|D| \geq 1$;
4. **if** only one literal in C_1 or C_2 **then** using Lemma 4 for branching;
5. **else** using Lemma 5 for branching;

Fig. 2. The parameterized algorithm for $(n, 3)$-MAXSAT in time $O^*(1.194^k)$

Theorem 1. *The algorithm* **(n,3)-MaxSAT-Solver** *solves the* $(n, 3)$-MAXSAT *problem in time* $O^*(1.194^k)$.

Proof. The algorithm **(n,3)-MaxSAT-Solver** can be described as a search tree $\mathcal{T}$, where each node of T is an instance of the $(n, 3)$-MAXSAT problem. Each leaf of $\mathcal{T}$ corresponds to a reduced formula containing only singletons and their negations, where step 2 of the algorithm concludes with a decision. By Lemma 6, a leaf in the search tree $\mathcal{T}$ can be solved in polynomial time.

Each internal node of $\mathcal{T}$ is associated with an instance (F, k) and corresponds to an application of one of the branching rules in Lemmas 4 and 5, and its children correspond to the branches of the branching rule. By Lemmas 4 and 5, the root of each of the branching rules is bounded by 1.194, which is the root of the $(5, 3)$-branching. Now a simple induction shows that the search tree $\mathcal{T}$, i.e., the algorithm **Max-SAT-Solver** solves the $(n, 3)$-MAXSAT problem in time $O^*(1.194^k)$. □

4 An $O^*(1.237^n)$-time Algorithm for $(n, 3)$-MAXSAT

Note that R-Rules 2–3 and 5–9 reduce the number n of variables by at least 1, and R-Rule 1 reduces the size of the formula by at least 1. So, they can be applied in polynomial time, and we can also use them as the transformation rules for our exact algorithm in terms of n.

Next, we present our exact algorithm for $(n, 3)$-MAXSAT problem in Fig. 3. Note that a $(2, 1)$-literal z is a singleton, if $\bar{z}$ occurs in the unit clause $(\bar{z})$. A variable z is called a singleton, if literal z or $\bar{z}$ is a singleton.

Theorem 2. *The algorithm* **n3MaxSAT** *solves the* $(n, 3)$-MAXSAT *problem in time* $O^*(1.237^n)$.

Proof. The process of the algorithm **n3MaxSAT** can be depicted by a search tree $\mathcal{T}$ in which each node corresponds to an instance of the $(n, 3)$-MAXSAT

Algorithm n3MaxSAT(F)
INPUT: an instance (F, k) of $(n, 3)$-MAXSAT, where
each variable appears in three clauses
OUTPUT: an assignment to F that satisfies the maximum number of clauses
1. apply R-Rules 1-3 and 5-9 repeatedly until F is irreducible;
2. **if** all variables are singletons or there is no variable in F
then return corresponding result according to Lemma 6;
3. choose a non-singleton x: $(xC_1), (xC_2), (\bar{x}D)$ and $|D| \geq 1$;
4. **if** $|D| = 2$ (i.e. $D = yy'$) and at least one of yy' is not a singleton
then max(n3MaxSAT($F[x]$), n3MaxSAT($F[\bar{x}, \bar{y}, \bar{y}']$));
5. **else** max(n3MaxSAT($F[x]$), n3MaxSAT($F[\bar{x}]$));

Fig. 3. The exact algorithm for $(n, 3)$-MAXSAT in time $O^*(1.237^n)$

problem. Each leaf of the search tree $\mathcal{T}$ corresponds to a simplified instance for which step 2 of the algorithm concludes with a decision. Therefore, by Lemma 6, a leaf in the search tree $\mathcal{T}$ associated with instance (F, k) of $(n, 3)$-MAXSAT can be solved in polynomial time.

Each internal node of the search tree $\mathcal{T}$ either branch on x or branch with $x = 1$ and $x = y = y' = 0$. Remind that it is safe to branch $x = 1$ and $x = y = y' = 0$ in Step 4, because $\bar{x}$ is a 1-literal, for each optimal assignment σ with $x = 0$ and $y = 1$ (or $y' = 1$), there is another assignment σ' by only replacing $x = 0$ with $x = 1$.

Now we analyze the branching vector of each cases. Since R-Rule 2 and R-Rules 5–8 are not applicable, $|C_1| \geq 1, |C_2| \geq 1$ and there is no common variable among C_1, C_2 and D.

Case1. consider $y \in D$ is a 1-literal. (1) When $x = 1$, at least two variables from C_1C_2 become 2-variables and can be reduced by R-Rules 2–3. In all, the number of variables is reduced by at least 3. (2) When $x = 0$, by R-Rule 2, $y = 0$. Since R-Rules 2, 5–8 are not applicable to F, at least two other variables, except for x and y, are satisfied by $y = 0$, so that they become 2-variables. Thus, R-Rules 2–3 become applicable, and the number of variables is reduced by at least 4. This is a $(4, 3)$-branching.

Case2. consider $|D| \geq 3$, and all literals in D are 2-literals. (1) When branching $x = 1$, similar to Case 1, the number of variables is reduced by at least 3. (2) When $x = 0$, at least three variables become 2-variables, so that R-Rules 2–3 become applicable. The number of variables is reduced by at least 4. This is a $(4, 3)$-branching.

Case3. consider $|D| = 2$, and both literals yy' in D are singletons. (1) When branching $x = 1$, at least two variables, except xyy', become 2-variables and can be reduced by R-Rules 2–3. Moreover, clause $(\bar{x}yy')$ becomes (yy'), plus unit clauses $(\bar{y}), (\bar{y}')$. Thus, by R-Rule 9, replace $(yy'), (\bar{y}), (\bar{y}')$ with $(\bar{y}\bar{y}')$ and both

yy' are reduced by R-Rule 2. As a result, the number of variables is reduced by at least 5. (2)When $x = 0$, at least two variables become 2-variables, so that R-Rules 2–3 become applicable. The number of variables is reduced by at least 3. This is a $(5, 3)$-branching.

Case4. consider $|D| = 2$, and a literal y in D is not a singleton, so that Step 4 is applicable. (1) When branching $x = 1$, similar to Case 1, the number of variables is reduced by at least 3. (2) When $x = D = 0$, because y is a non-singleton, at least another variable becomes a 2-variable, so that R-Rules 2–3 become applicable. The number of variables is reduced by at least 4. This is a $(4, 3)$-branching.

Case5. consider $|D| = 1$, denoted by $D = y$, and y is a 2-literal.

Case5.1. consider $|C_1C_2| \geq 3$. (1) When branching $x = 1$, at least other three variables become 2-variables and can be reduced by R-Rules 2–3. Moreover, by R-Rule 2, $y = 1$. Thus, the number of variables is reduced by at least 5. (2) When branching $x = 0$, variable y is reduced by R-Rules 2–3. This is a $(5, 2)$-branching.

Case5.2. consider $|C_1| = |C_2| = 1$.

Case5.2.1. consider at least one literal of C_1C_2 is a (2,1)-literal. (1) When branching $x = 1$, similar to Case 1, at least 2 other variables except for xy are reduced. Moreover, since $|D| = 1$, (2,1)-literal y appears in a unit clause (y), so that set $y = 1$ by R-Rule 2. In all, the number of variables is reduced by at least 4. (2) When branching $x = 0$, variable y becomes a 2-variable and can be reduced by R-Rules 2–3. Also, at least one literal of C_1C_2 is a 2-literal, so that R-Rule 2 becomes applicable. The number of variables is reduced by at least 3. This is a $(4, 3)$-branching.

Case5.2.2. consider both the literals, denoted by z_1z_2, in C_1C_2 are $(1, 2)$-literals. (1) When branching $x = 1$, by R-Rule 2, $y = 1$, and $C_1 = C_2 = 0$; (2) When branching $x = 0$, then variable y can be reduced by R-Rules 2–3. There are two cases. If there is at least another variable reduced by $x = y = 1$ and $C_1 = C_2 = 0$, then the number of variables is reduced by $4 + 1 = 5$, where we have a $(5, 2)$-branching. If there is no other variable reduced by $x = y = 1$ and $C_1 = C_2 = 0$, then, when branching $x = 0$, variable y can be reduced by R-Rules 2–3 and one of variables z_1z_2 can be reduced by R-Rules 5–8, where we have a $(4, 3)$-branching. In all, this is a $(5, 2)$-branching.

In summary, the root of each branching rule is bounded by 1.237, which is the root of the $(5, 2)$-branching. Now a simple induction shows that the search tree T, i.e., the algorithm **n3MaxSAT** solves the $(n, 3)$-MAXSAT in time $O^*(1.237^n)$. □

5 Conclusion

In this paper we present two new reduction rules, so that each pair of distinct clauses has at most one variable in common in the reduced formula, which is also called a linear CNF formula [11]. Consequently, we improve parameterized algorithm from $O^*(1.2721^k)$ [2] to $O^*(1.194^k)$-time, and exact algorithm from $O^*(1.2600^n)$ [3] to $O^*(1.237^n)$-time for the $(n, 3)$-MAXSAT problem.

References

1. Bonet, M.L., Levy, J., Manya, F.: Resolution for max-sat. Artif. Intell. **171**(8), 606–618 (2007)
2. Bliznets, I., Golovnev, A.: A new algorithm for parameterized MAX-SAT. In: Thilikos, D.M., Woeginger, G.J. (eds.) IPEC 2012. LNCS, vol. 7535, pp. 37–48. Springer, Heidelberg (2012)
3. Bliznets, I.: A new upper bound for (n, 3)-MAX-SAT. J. Math. Sci. **188**(1), 1–6 (2013)
4. Chen, J., Kanj, I.: Improved exact algorithms for Max-SAT. Discrete Appl. Math. **142**, 17–27 (2004)
5. Chen, J., Kanj, I., Jia, W.: Vertex cover: further observations and further improvements. J. Algorithms **41**, 280–301 (2001)
6. Chen, J., Xu, C., Wang, J.: Dealing with 4-variables by resolution: an improved MaxSAT algorithm. In: Dehne, F., Sack, J.-R., Stege, U. (eds.) WADS 2015. LNCS, vol. 9214, pp. 178–188. Springer, Heidelberg (2015)
7. Davis, M., Putnam, H.: A computing procedure for quantification theory. J. ACM **7**, 201–215 (1960)
8. Gu, J., Purdom, P., Wah, W.: Algorithms for the satisfiability (SAT) problem: a survey. In: Satisfiability Problem: Theory and Applications. DIMACS Series in Discrete Mathematics and Theoretical Computer Science, pp. 19–152. AMS (1997)
9. Hochbaum, D. (ed.): Approximation Algorithms for NP-Hard Problems. PWS Publishing Company, Boston (1997)
10. Impagliazzo, R., Paturi, R.: On the complexity of k-SAT. J. Comput. Syst. Sci. **62**, 367–375 (2001)
11. Porschen, S., Speckenmeyer, E., Zhao, X.: Linear CNF formulas and satisfiability. Discrete Appl. Math. **157**(5), 1046–1068 (2009)
12. Raman, V., Ravikumar, B., Rao, S.S.: A simplified NP-complete MAXSAT problem. Inf. Process. Lett. **65**(1), 1–6 (1998)
13. Kulikov, A.S.: Automated generation of simplification rules for SAT and MAXSAT. In: Bacchus, F., Walsh, T. (eds.) SAT 2005. LNCS, vol. 3569, pp. 430–436. Springer, Heidelberg (2005)
14. Kratochvil, J., Savicky, P., Tuza, Z.: One more occurrence of variables makes satisfiability jump from trivial to NP-complete. SIAM J. Comput. **22**(1), 203–210 (1993)

Directed Pathwidth and Palletizers

Frank Gurski[1(✉)], Jochen Rethmann[2], and Egon Wanke[1]

[1] Institute of Computer Science, Heinrich Heine University Düsseldorf, 40225 Düsseldorf, Germany
frank.gurski@hhu.de

[2] Faculty of Electrical Engineering and Computer Science, Niederrhein University of Applied Sciences, 47805 Krefeld, Germany

Abstract. In delivery industry, bins have to be stacked-up from conveyor belts onto pallets. Given k sequences of labeled bins and a positive integer p. The goal is to stack-up the bins by iteratively removing the first bin of one of the k sequences and put it onto a pallet located at one of p stack-up places. Each of these pallets has to contain bins of only one label, bins of different labels have to be placed on different pallets. After all bins of one label have been removed from the given sequences, the corresponding place becomes available for a pallet of bins of another label. In this paper we introduce a graph model for this problem, the so called sequence graph, which allows us to show that there is a processing of some list of sequences with at most p stack-up places if and only if the sequence graph of this list has directed pathwidth at most $p - 1$.

Keywords: Computational complexity · Combinatorial optimization · Directed pathwidth · Stack-up systems · Palletizing systems

1 Introduction

We consider the combinatorial problem of stacking up bins from a set of conveyor belts onto pallets. This problem originally appears in *stack-up systems* that play an important role in delivery industry and warehouses. A detailed description of the practical background of this work is given in [3,15].

The bins that have to be stacked-up onto pallets reach the stack-up system on a main conveyor belt. At the end of the line they enter the palletizing system. Here the bins are picked-up by stacker cranes or robotic arms and moved onto pallets, which are located at *stack-up places*. Often vacuum grippers are used to pick-up the bins. This picking process can be performed in different ways depending on the architecture of the palletizing system (single-line or multi-line palletizers). Full pallets are carried away by automated guided vehicles, or by another conveyor system, while new empty pallets are placed at free stack-up places.

The developers and producers of robotic palletizers distinguish between single-line and multi-line palletizing systems. Each of these systems has its advantages and disadvantages.

Z. Lu et al. (Eds.): COCOA 2015, LNCS 9486, pp. 31–44, 2015.
DOI: 10.1007/978-3-319-26626-8_3

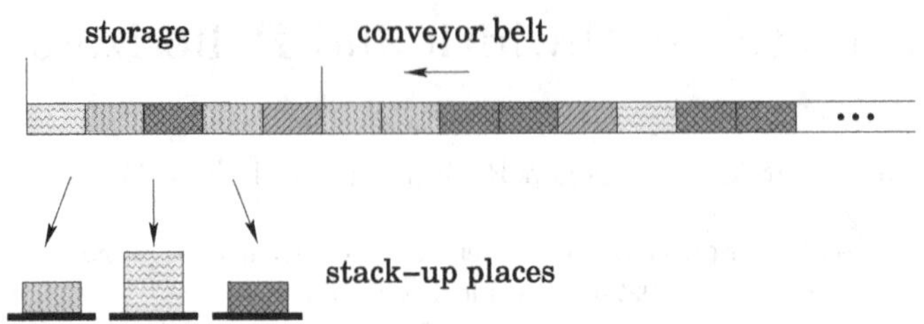

Fig. 1. A single-line stack-up system using a random access storage of size 5. The patterns represent the pallet labels. Bins with different patterns have to be placed on different pallets, bins with the same pattern have to be placed on the same pallet.

In *single-line palletizing systems* there is only one conveyor belt from which the bins are picked-up. Several robotic arms or stacker cranes are placed around the end of the conveyor. We model such systems by a random access storage which is automatically replenished with bins from the main conveyor, see Fig. 1. The area from which the bins can be picked-up is called the storage area. It is determined by the operation range of stacker cranes or robotic arms.

In *multi-line palletizing systems* there are several buffer conveyors from which the bins are picked-up. The robotic arms or stacker cranes are placed at the end of these conveyors. Here, the bins from the main conveyor of the order-picking system first have to be distributed to the multiple infeed lines to enable parallel processing. Such a distribution can be done by some cyclic storage conveyor, see Fig. 2. From the cyclic storage conveyor the bins are pushed out to the buffer conveyors. A stack-up system using a cyclic storage conveyor is, for example, located at Bertelsmann Distribution GmbH in Gütersloh, Germany. On certain days, several thousands of bins are stacked-up using a cyclic storage conveyor with a capacity of approximately 60 bins and 24 stack-up places, while up to 32 bins are destined for a pallet. This palletizing system has originally initiated our research.

If we ignore the task to distribute the bins from the main conveyor to the k buffer conveyors, i.e., if the filled buffer conveyors are already given, and if each arm can only pick-up the first bin of one of the buffer conveyors, then the system is called a FIFO palletizing system. Such systems can be modeled by several simple queues, see Fig. 3.

From a theoretical point of view, an instance of the FIFO Stack-Up problem consists of k sequences $q_1, \ldots, q_k$ of bins and a number of available stack-up places p. Each bin of q is destined for exactly one pallet. The stack-up problem is to decide whether one can remove iteratively the bins from the sequences such that in each step only one of the first bins of $q_1, \ldots, q_k$ is removed and after each step at most p pallets are open. A pallet t is called *open*, if at least one bin for pallet t has already been removed from the sequences, and if at least one bin for pallet t is still contained in the remaining sequences. If a bin b is removed from a sequence then all bins located behind b are moved-up one position to the front (cf. Sect. 2 for the formal definition).

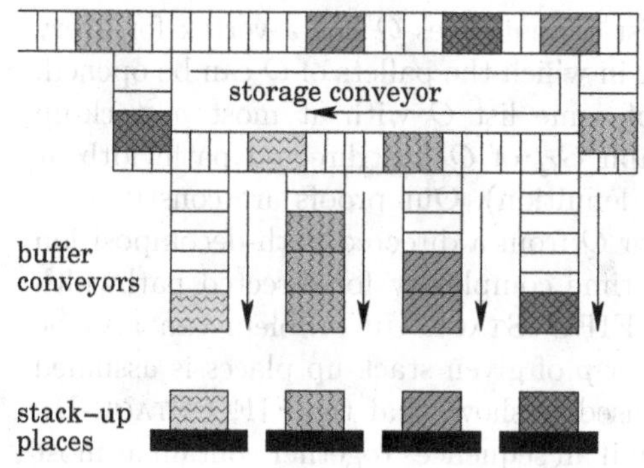

Fig. 2. A multi-line stack-up system with a pre-placed cyclic storage conveyor.

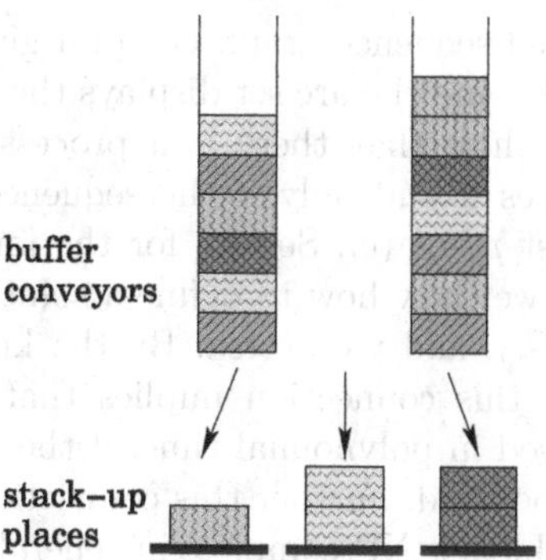

Fig. 3. The FIFO stack-up system analyzed in this paper.

Every processing should absolutely avoid blocking situations. A system is *blocked*, if all stack-up places are occupied by pallets, and non of the bins that may be used in the next step are destined for an open pallet. To unblock the system, bins have to be picked-up manually and moved to pallets by human workers. Such a blocking situation is sketched in Fig. 3. The system is blocked (cf. above for the definition), because the pallet for the bins which are striped from bottom left to top right cannot be opened and the pallets for the other three types of bins located on the stack-up places cannot be closed.

The single-line stack-up problem can be defined in the same way. An instance for the single-line stack-up problem consists of one sequence q of bins, a storage capacity s, and a number of available stack-up places p. In each step one of the first s bins of q can be removed. Everything else is defined analogously.

Many facts are known about single-line stack-up systems [15–17]. In [15] it is shown that the single-line stack-up decision problem is NP-complete, but can be solved efficiently if the storage capacity s or the number of available stack-up places p is fixed. The problem remains NP-complete as shown in [16], even if the sequence contains at most 9 bins per pallet. In [16], a polynomial-time *off-line* approximation algorithm for minimizing the storage capacity s is introduced. This algorithm yields a solution that is optimal up to a factor bounded by $\log(p)$. In [17] the performances of simple *on-line* stack-up algorithms are compared with optimal off-line solutions by a competitive analysis [2,5].

The FIFO Stack-Up problem seems to be not investigated by other authors up to now, although stack-up systems play an important role in delivery industry and warehouses. In [6] we have shown a dynamic programming solution and in [7] we have given several parameterized algorithms for the FIFO Stack-Up problem. A breadth first search solution combined with some cutting technique for the problem was presented in [8].

In this paper, we introduce a digraph model which leads to a close relation between the FIFO Stack-Up problem and the directed pathwidth. The so-called sequence graph G_Q of a given list of sequences Q has a vertex for every pallet and the arc set displays the order in which the pallets of Q can be opened. We show that there is a processing of some list Q with at most p stack-up places if and only if the sequence graph G_Q of Q has directed pathwidth at most $p-1$ (cf. Sect. 3 for the formal definition). Our proofs are constructive, i.e. we show how to define a solution for Q from a directed path-decomposition of G_Q, and vice versa. By the known time complexity for directed pathwidth [18] this connection implies that the FIFO Stack-Up problem can also be solved in polynomial time, if the number p of given stack-up places is assumed to be fixed. Further this connection is used to show, that the FIFO Stack-Up problem is NP-complete in general, even if all sequences together contain at most 6 bins destined for the same pallet. Due to the approximation result of directed pathwidth in [11] the optimization version of the FIFO Stack-Up problem can be approximated up to a factor of $\mathcal{O}(\log^{1.5} m)$, where m represents the number of pallets in all sequences. Our result extends the previously known areas of applications for directed pathwidth in graph databases and boolean networks, which have been shown in [4].

2 Preliminaries

Unless otherwise stated, k and p are some positive integers throughout the paper. We consider *sequences* $q_1 = (b_1, \dots, b_{n_1}), \dots, q_k = (b_{n_{k-1}+1}, \dots, b_{n_k})$ of pairwise distinct *bins*. These sequences represent the buffer queues (handled by the buffer conveyors) in real stack-up systems. Each bin b is labeled with a *pallet symbol* $plt(b)$. We say bin b is destined for pallet $plt(b)$. We use in our examples characters for pallet symbols. The set of all pallets of the bins in some sequence q_i is denoted by

$$plts(q_i) = \{plt(b) \mid b \in q_i\}.$$

For a list of sequences $Q = (q_1, \dots, q_k)$ we denote

$$plts(Q) = plts(q_1) \cup \dots \cup plts(q_k).$$

For some sequence $q = (b_1, \dots, b_n)$, we say bin b_i is *on the left of* bin b_j in sequence q if $i < j$. A sequence $q' = (b_j, b_{j+1}, \dots, b_n)$, $j \geq 1$, is called a *subsequence* of sequence $q = (b_1, \dots, b_n)$. We define $q - q' = (b_1, \dots, b_{j-1})$.

Let $Q = (q_1, \dots, q_k)$ and $Q' = (q'_1, \dots, q'_k)$ be two lists of sequences of bins such that each sequence q'_j, $1 \leq j \leq k$, is a subsequence of sequence q_j. Each such pair (Q, Q') is called a *configuration*. In every configuration (Q, Q') the first entry Q is the initial list of sequences of bins and the second entry Q' is the list of sequences that remain to be processed. A pallet t is called *open* in configuration (Q, Q'), if a bin of pallet t is contained in some $q'_i \in Q'$ and if another bin of pallet t is contained in some $q_j - q'_j$ for $q_j \in Q$, $q'_j \in Q'$. The *set of open pallets* in configuration (Q, Q') is denoted by $open(Q, Q')$. A pallet $t \in plts(Q)$ is called *closed* in configuration (Q, Q'), if $t \notin plts(Q')$, i.e. no sequence of Q' contains a bin for pallet t. Initially all pallets are *unprocessed*. After the first bin of a pallet t has been removed from one of the sequences, pallet t is either open or closed.

In view of the practical background, we only consider lists of sequences that together contain at least two bins for each pallet. Throughout the paper Q always denotes a list of some sequences of bins.

The FIFO Stack-Up Problem. Consider a configuration (Q, Q'). The removal of the first bin from one subsequence $q' \in Q'$ is called a *transformation step*. A sequence of transformation steps that transforms the list Q of k sequences into k empty subsequences is called a *processing* of Q.

Name: FIFO Stack-Up
Instance: A list $Q = (q_1, \ldots, q_k)$ of k sequences and a positive integer p.
Question: Is there a processing of Q, such that in each configuration (Q, Q') during the processing at most p pallets are open?

We use the following variables: k denotes the number of sequences, and p stands for the number of stack-up places. Furthermore, m represents the number of pallets in $plts(Q)$, and n denotes the number of bins in all sequences, i.e. $n = n_k$. Finally, $N = \max\{|q_1|, \ldots, |q_k|\}$ is the maximum sequence length. In view of the practical background, it holds $p < m$, $k < m$, $m < n$, and $N < n$.

It is often convenient to use pallet identifications instead of bin identifications to represent a sequence q. For n not necessarily distinct pallets $t_1, \ldots, t_n$ let $[t_1, \ldots, t_n]$ denote some sequence of n pairwise distinct bins $(b_1, \ldots, b_n)$, such that $plt(b_i) = t_i$ for $i = 1, \ldots, n$. We use this notion for lists of sequences as well. For the sequences $q_1 = [t_1, \ldots, t_{n_1}], \ldots, q_k = [t_{n_{k-1}+1}, \ldots, t_{n_k}]$ of pallets we define $q_1 = (b_1, \ldots, b_{n_1}), \ldots, q_k = (b_{n_{k-1}+1}, \ldots, b_{n_k})$ to be sequences of bins such that $plt(b_i) = t_i$ for $i = 1, \ldots, n_k$, and all bins are pairwise distinct.

For some list of subsequences Q' we define $front(Q')$ to be the set of pallets of the first bins of the queues of Q'.

Example 1. Consider list $Q = (q_1, q_2)$ of sequences $q_1 = (b_1, \ldots, b_4) = [a, a, b, b]$ and $q_2 = (b_5, \ldots, b_{12}) = [c, d, e, c, a, d, b, e]$. Table 1 shows a processing of Q with 3 stack-up places. The underlined bin is always the bin that will be removed in the next transformation step. We denote $Q_i = (q_1^i, q_2^i)$, thus each row represents a configuration (Q, Q_i).

Consider a processing of a list Q of sequences. Let $B = (b_{\pi(1)}, \ldots, b_{\pi(n)})$ be the order in which the bins are removed during the processing of Q, and let $T = (t_1, \ldots, t_m)$ be the order in which the pallets are opened during the processing of Q. Then B is called a *bin solution* of Q, and T is called a *pallet solution* of Q. The transformation in Table 1 defines the bin solution

$$B = (b_5, b_6, b_7, b_8, b_1, b_2, b_9, b_{10}, b_{11}, b_3, b_4, b_{12}),$$

and the pallet solution $T = (c, d, e, a, b)$.

During a processing of a list Q of sequences there are often configurations (Q, Q') for which it is easy to find a bin b that can be removed from Q' such that a further processing with p stack-up places is still possible. This is the case, if

Table 1. A processing of $Q = (q_1, q_2)$ from Example 1 with 3 stack-up places. There is no processing of Q that needs less than 3 stack-up places.

i	q_1^i	q_2^i	$front(Q_i)$	remove	$open(Q, Q_i)$
0	$[a, a, b, b]$	$[\underline{c}, d, e, c, a, d, b, e]$	$\{a, c\}$	b_5	$\emptyset$
1	$[a, a, b, b]$	$[\underline{d}, e, c, a, d, b, e]$	$\{a, d\}$	b_6	$\{c\}$
2	$[a, a, b, b]$	$[\underline{e}, c, a, d, b, e]$	$\{a, e\}$	b_7	$\{c, d\}$
3	$[a, a, b, b]$	$[\underline{c}, a, d, b, e]$	$\{a, c\}$	b_8	$\{c, d, e\}$
4	$[\underline{a}, a, b, b]$	$[a, d, b, e]$	$\{a\}$	b_1	$\{d, e\}$
5	$[\underline{a}, b, b]$	$[a, d, b, e]$	$\{a\}$	b_2	$\{a, d, e\}$
6	$[b, b]$	$[\underline{a}, d, b, e]$	$\{a, b\}$	b_9	$\{a, d, e\}$
7	$[b, b]$	$[\underline{d}, b, e]$	$\{b, d\}$	b_{10}	$\{d, e\}$
8	$[b, b]$	$[\underline{b}, e]$	$\{b\}$	b_{11}	$\{e\}$
9	$[\underline{b}, b]$	$[e]$	$\{b, e\}$	b_3	$\{b, e\}$
10	$[\underline{b}]$	$[e]$	$\{b, e\}$	b_4	$\{b, e\}$
11	$[]$	$[\underline{e}]$	$\{e\}$	b_{12}	$\{e\}$
12	$[]$	$[]$	$\emptyset$	-	$\emptyset$

bin b is destined for an already open pallet, see configuration (Q, Q_3), (Q, Q_5), (Q, Q_6), (Q, Q_7), (Q, Q_9), (Q, Q_{10}), or (Q, Q_{11}) in Table 1. In the following we show:

- If one of the first bins of the sequences is destined for an already open pallet then this bin can be removed without increasing the number of stack-up places necessary to further process the sequences.
- If there is more than one bin at choice for already open pallets then the order, in which those bins are removed is arbitrary.

To show the rules, consider a processing of some list Q of sequences with p stack-up places. Let

$$(b_{\pi(1)}, \ldots, b_{\pi(i-1)}, b_{\pi(i)}, \ldots, b_{\pi(\ell-1)}, b_{\pi(\ell)}, b_{\pi(\ell+1)}, \ldots, b_{\pi(n)})$$

be the order in which the bins are removed from the sequences during the processing, and let (Q, Q_j), $1 \leq j \leq n$ denote the configuration such that bin $b_{\pi(j)}$ is removed in the next transformation step. Suppose bin $b_{\pi(i)}$ will be removed in some transformation step although bin $b_{\pi(\ell)}$, $\ell > i$, for some already open pallet $plt(b_{\pi(\ell)}) \in open(Q, Q_i)$ could be removed next. We define a modified processing

$$(b_{\pi(1)}, \ldots, b_{\pi(i-1)}, b_{\pi(\ell)}, b_{\pi(i)}, \ldots, b_{\pi(\ell-1)}, b_{\pi(\ell+1)}, \ldots, b_{\pi(n)})$$

by first removing bin $b_{\pi(\ell)}$, and afterwards the bins $b_{\pi(i)}, \ldots, b_{\pi(\ell-1)}$ in the given order. Obviously, in each configuration during the modified processing there are at most p pallets open. To remove first some bin of an already open pallet is a kind of priority rule.

A configuration (Q, Q') is called a *decision configuration*, if the first bin of each sequence $q' \in Q'$ is destined for a non-open pallet, see configurations (Q, Q_0), (Q, Q_1), (Q, Q_2), (Q, Q_4), and (Q, Q_8) in Table 1, i.e. $\mathit{front}(Q') \cap \mathit{open}(Q, Q') = \emptyset$. We can restrict FIFO stack-up algorithms to deal with such decision configurations, in all other configurations the algorithms automatically remove a bin for some already open pallet.

If we have a pallet solution computed by some FIFO stack-up algorithm, we can convert the pallet solution into a sequence of transformation steps, i.e. a processing of Q. This is done by algorithm TRANSFORM shown in Fig. 4. Given a list of sequences $Q = (q_1, \ldots, q_k)$ and a pallet solution $T = (t_1, \ldots, t_m)$ algorithm TRANSFORM gives us in time $\mathcal{O}(n \cdot k) \subseteq \mathcal{O}(n^2)$ a bin solution of Q, i.e. a processing of Q.

Algorithm TRANSFORM

$q'_1 := q_1, \ldots, q'_k := q_k$
$j := 1$
$T' := \{t_1\}$
repeat the following steps until $q'_1 = \emptyset, \ldots, q'_k = \emptyset$:

1. if there is a sequence q'_i such that the first bin b of q'_i is destined for a pallet in T', i.e. $\mathit{plt}(b) \in T'$, then remove bin b from sequence q'_i, and output b
   Comment: Bins for already open pallets are removed automatically.
2. otherwise set $j := j + 1$ and $T' := T' \cup \{t_j\}$
   Comment: If the first bin of each subsequence q'_i is destined for a non-open pallet, the next pallet of the pallet solution has to be opened.

Fig. 4. Algorithm for transforming a pallet solution into a bin solution.

Obviously, there is no other processing of Q that also defines pallet solution T but takes less stack-up places.

3 Main Result

Next we show a correlation between the used number of stack-up places for a processing of an instance Q and the directed pathwidth of a digraph G_Q defined by Q. The notion of *directed pathwidth* (*directed treewidth*) was introduced by Johnson, Robertson, Seymour, and Thomas in [9].

This correlation implies that (1) the decision version of the FIFO STACK-UP problem is NP-complete, (2) a pallet solution can be computed in polynomial time if there are only a fixed number of stack-up places, and (3) the optimization version of the FIFO STACK-UP problem can be approximated up to a factor of $\mathcal{O}(\log^{1.5} m)$.

A *directed path-decomposition* of some digraph $G = (V, E)$ is a sequence $(X_1, \ldots, X_r)$ of subsets of V, called *bags*, that satisfy the following three properties.

(dpw-1) $X_1 \cup \ldots \cup X_r = V$

(dpw-2) for each arc $(u, v) \in E$ there are indices i, j with $i \leq j$ such that $u \in X_i$ and $v \in X_j$

(dpw-3) if $u \in X_i$ and $u \in X_j$ for some vertex u and two indices i, j with $i \leq j$, then $u \in X_\ell$ for all indices ℓ with $i \leq \ell \leq j$

The *width* of a directed path-decomposition $(X_1, \ldots, X_r)$ is $\max_{1 \leq i \leq r} |X_i| - 1$. The *directed pathwidth* of G, d-pw(G) for short, is the smallest integer w such that there is a directed path-decomposition for G of width w. For symmetric digraphs, the directed pathwidth is equivalent to the undirected pathwidth of the corresponding undirected graph [12]. For each fixed integer w, it is decidable in polynomial time whether a given digraph has directed pathwidth at most w, see Tamaki [18].

The *sequence graph* $G_Q = (V, E)$ for an instance $Q = (q_1, \ldots, q_k)$ of the FIFO Stack-Up problem is defined by vertex set $V = plts(Q)$ and the following set of arcs. There is an arc $(u, v) \in E$ if and only if there is a sequence $q_i = (b_{n_{i-1}+1}, \ldots, b_{n_i})$ with two bins b_{j_1}, b_{j_2} such that (1) $j_1 < j_2$, (2) $plt(b_{j_1}) = u$ (3) $plt(b_{j_2}) = v$, and (4) $u \neq v$.

If $G_Q = (V, E)$ has an arc $(u, v) \in E$ then $u \neq v$ and for every processing of Q, pallet u is opened before pallet v is closed. Digraph $G_Q = (V, E)$ can be computed in time $\mathcal{O}(n + k \cdot |E|) \subseteq \mathcal{O}(n + k \cdot m^2)$.

Example 2. Figure 5 shows the sequence graph G_Q for $Q = (q_1, q_2, q_3)$ with sequences $q_1 = [a, a, d, e, d]$, $q_2 = [b, b, d]$, and $q_3 = [c, c, d, e, d]$.

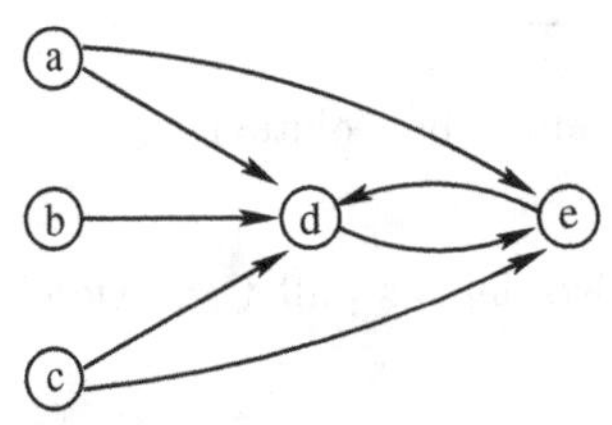

Fig. 5. Sequence graph G_Q of Example 2.

Before we give our main Theorems we want to emphasize that not every directed path-decomposition of a sequence graph G_Q immediately leads to a pallet solution. In Example 2 the sequence $(\{e, a\}, \{e, b\}, \{e, c\}, \{e, d\})$ is a directed path-decomposition of optimal width 1 for the sequence graph G_Q. But opening the pallets one after another leads (e, a, b, c, d), which is no pallet solution since pallet e cannot be opened at first and must be put on hold. Within the proof of Theorem 2 we show how to transform a directed path-decomposition of G_Q into a pallet solution for Q. Example 3 and Table 2 illustrate this process.

Theorem 1. *A processing* $(Q, Q_0), (Q, Q_1), \ldots, (Q, Q_n)$ *of* Q *with* $Q_0 = Q$, $Q_n = (\emptyset, \ldots, \emptyset)$, *and* p *stack-up places defines a directed path-decomposition* $\mathcal{X} = (open(Q, Q_0), \ldots, open(Q, Q_n))$ *for* G_Q *of width* $p - 1$.

Proof. We show that $\mathcal{X} = (open(Q, Q_0), \ldots, open(Q, Q_n))$ satisfies all properties of a directed path-decomposition.

(dpw-1) $open(Q, Q_0) \cup \cdots \cup open(Q, Q_n) = plts(Q)$, because every pallet is opened at least once.

(dpw-2) If $(u, v) \in E$ then there are indices i, j with $i \leq j$ such that $u \in open(Q, Q_i)$ and $v \in open(Q, Q_j)$, because v can not be closed before u is opened.

(dpw-3) If $u \in open(Q, Q_i)$ and $u \in open(Q, Q_j)$ for some pallet u and two indices i, j with $i \leq j$, then $u \in open(Q, Q_\ell)$ for all indices ℓ with $i \leq \ell \leq j$, because every pallet is opened at most once.

Since Q is processed with p stack-up places, we have $|open(Q, Q_i)| \leq p$ for $0 \leq i \leq n$, and therefore $\mathcal{X}$ has width at most $p - 1$. □

Theorem 2. *If there is a path-decomposition $\mathcal{X} = (X_1, \ldots, X_r)$ for G_Q of width $p - 1$ then there is a processing of Q with p stack-up places.*

Proof. For a pallet t let $\alpha(\mathcal{X}, t)$ be the smallest i such that $t \in X_i$ and $\beta(\mathcal{X}, t)$ be the largest i such that $t \in X_i$, see Example 3. Then $t \in X_i$ if and only if $\alpha(\mathcal{X}, t) \leq i \leq \beta(\mathcal{X}, t)$. If (t_1, t_2) is an arc of G_Q, then $\alpha(\mathcal{X}, t_1) \leq \beta(\mathcal{X}, t_2)$. This follows by (dpw-2) of the definition of a directed path-decomposition.

Instance Q can be processed as follows. If it is necessary to open in a configuration (Q, Q') a new pallet, then we open a pallet t of $front(Q')$ for which $\alpha(\mathcal{X}, t)$ is minimal.

We next show that for every configuration (Q, Q') of the processing above there is a bag X_i in the directed path-decomposition $\mathcal{X} = (X_1, \ldots, X_r)$, such that $open(Q, Q') \subseteq X_i$. This implies that the processing uses at most p stack-up places.

First, let (Q, Q') be a decision configuration, let $t \in front(Q')$ such that $\alpha(\mathcal{X}, t)$ is minimal, and let $t' \in open(Q, Q')$ be an already open pallet. We show that the intervals $[\alpha(\mathcal{X}, t), \beta(\mathcal{X}, t)]$ and $[\alpha(\mathcal{X}, t'), \beta(\mathcal{X}, t')]$ overlap, because $\alpha(\mathcal{X}, t) \leq \beta(\mathcal{X}, t')$ and $\alpha(\mathcal{X}, t') \leq \beta(\mathcal{X}, t)$.

(1) To show $\alpha(\mathcal{X}, t) \leq \beta(\mathcal{X}, t')$ we observe the following:
Since $t' \notin front(Q')$, there has to be a pallet $t'' \in front(Q')$ such that (t'', t') is an arc of G_Q. This implies that $\alpha(\mathcal{X}, t'') \leq \beta(\mathcal{X}, t')$. Since t is a pallet of $front(Q')$ for which $\alpha(\mathcal{X}, t)$ is minimal, we have $\alpha(\mathcal{X}, t) \leq \alpha(\mathcal{X}, t'')$ and thus $\alpha(\mathcal{X}, t) \leq \beta(\mathcal{X}, t')$.

(2) To show $\alpha(\mathcal{X}, t') \leq \beta(\mathcal{X}, t)$ we observe the following:
Let (Q, Q'') be the configuration in that t' has been opened.
   (a) $t \in front(Q'')$. Since t' is a pallet of $front(Q'')$ for which $\alpha(\mathcal{X}, t')$ is minimal, we have $\alpha(\mathcal{X}, t') \leq \alpha(\mathcal{X}, t)$ and thus $\alpha(\mathcal{X}, t') \leq \beta(\mathcal{X}, t)$.
   (b) $t \notin front(Q'')$. Then there is a pallet $t'' \in front(Q'')$ such that (t'', t) is an arc of G_Q. This implies that $\alpha(\mathcal{X}, t'') \leq \beta(\mathcal{X}, t)$. Since t' is a pallet of $front(Q'')$ for which $\alpha(\mathcal{X}, t')$ is minimal, we have $\alpha(\mathcal{X}, t') \leq \alpha(\mathcal{X}, t'') \leq \beta(\mathcal{X}, t)$.

Finally, let $(Q, \hat{Q})$ be an arbitrary configuration during the processing of Q. By the discussion above, we can conclude that for every pair $t_i, t_j \in open(Q, \hat{Q})$ the intervals $[\alpha(\mathcal{X}, t_i), \beta(\mathcal{X}, t_i)]$ and $[\alpha(\mathcal{X}, t_j), \beta(\mathcal{X}, t_j)]$ overlap, because t_i is opened before t_j or vice versa. Since all intervals $[\alpha(\mathcal{X}, t_i), \beta(\mathcal{X}, t_i)]$, $t_i \in open(Q, \hat{Q})$ mutually overlap, the cut of all these intervals is not empty, and so there is a bag X_j in the directed path-decomposition $\mathcal{X}$ such that $open(Q, \hat{Q}) \subseteq X_j$. □

Example 3. We consider the digraph G_Q for $Q = (q_1, q_2, q_3)$ with sequences $q_1 = [a, a, d, e, d]$, $q_2 = [b, b, d]$, and $q_3 = [c, c, d, e, d]$ from Example 2. Sequence $\mathcal{X} = (\{a, e\}, \{b, e\}, \{c, e\}, \{d, e\})$ is a directed path-decomposition of width 1, which implies the following values for α and β used in the proof of Theorem 2.

pallet t	a	b	c	d	e
$\alpha(\mathcal{X}, t)$	1	2	3	4	1
$\beta(\mathcal{X}, t)$	1	2	3	4	4

Table 2 shows a processing of Q with 2 stack-up places and pallet solution $S = (a, b, c, d, e)$. The underlined bin is always the bin that will be removed in the next transformation step. We denote $Q_i = (q_1^i, q_2^i, q_3^i)$, thus each row represents a configuration (Q, Q_i).

Table 2. A processing of Q with respect to a given directed path-decomposition for G_Q of Example 3.

i	q_1^i	q_2^i	q_3^i	$front(Q_i)$	$open(Q, Q_i)$
0	$[\underline{a}, a, d, e, d]$	$[b, b, d]$	$[c, c, d, e, d]$	$\{a, b, c\}$	$\emptyset$
1	$[\underline{a}, d, e, d]$	$[b, b, d]$	$[c, c, d, e, d]$	$\{a, b, c\}$	$\{a\}$
2	$[d, e, d]$	$[\underline{b}, b, d]$	$[c, c, d, e, d]$	$\{b, c, d\}$	$\emptyset$
3	$[d, e, d]$	$[\underline{b}, d]$	$[c, c, d, e, d]$	$\{b, c, d\}$	$\{b\}$
4	$[d, e, d]$	$[d]$	$[\underline{c}, c, d, e, d]$	$\{c, d\}$	$\emptyset$
5	$[d, e, d]$	$[d]$	$[\underline{c}, d, e, d]$	$\{c, d\}$	$\{c\}$
6	$[\underline{d}, e, d]$	$[d]$	$[d, e, d]$	$\{d\}$	$\emptyset$
7	$[e, d]$	$[\underline{d}]$	$[d, e, d]$	$\{d, e\}$	$\{d\}$
8	$[e, d]$	$[]$	$[\underline{d}, e, d]$	$\{d, e\}$	$\{d\}$
9	$[\underline{e}, d]$	$[]$	$[e, d]$	$\{e\}$	$\{d\}$
10	$[\underline{d}]$	$[]$	$[e, d]$	$\{d, e\}$	$\{d, e\}$
11	$[]$	$[]$	$[\underline{e}, d]$	$\{e\}$	$\{d, e\}$
12	$[]$	$[]$	$[\underline{d}]$	$\{d\}$	$\{d\}$
13	$[]$	$[]$	$[]$	$\emptyset$	$\emptyset$

4 Applications

4.1 Hardness Result

Next we will show the hardness of the FIFO Stack-Up problem. In contrast to Sect. 3 we will transform an instance of a graph problem into an instance of the FIFO Stack-Up problem.

Let $G = (V, E)$ be a digraph. We will assume that $G = (V, E)$ does not contain any vertex with only outgoing arcs and not contain any vertex with only incoming arcs. This is only for technical reasons and the removal of such vertices will not change the directed pathwidth of G, because a vertex u with only outgoing arcs can be placed in a singleton $X_i = \{u\}$ at the beginning of the directed path-decomposition and a vertex u with only incoming arcs can be placed in a singleton $X_i = \{u\}$ at the end of the directed path-decomposition, without to change its width.

Let $G = (V, E)$ be some digraph and $E = \{e_1, \ldots, e_\ell\}$ its arc set. The *queue system* $Q_G = (q_1, \ldots, q_\ell)$ for G is defined as follows.

(1) There are 2ℓ bins $b_1, \ldots, b_{2\ell}$.
(2) Queue $q_i = (b_{2i-1}, b_{2i})$ for $1 \leq i \leq \ell$.
(3) The pallet symbol of bin b_{2i-1} is the first vertex of arc e_i and the pallet symbol of b_{2i} is the second vertex of arc e_i for $1 \leq i \leq \ell$. Thus $plts(Q_G) = V$.

The definition of queue system Q_G and sequence graph G_Q, defined in Sect. 3, now imply the following proposition.

Proposition 1. *For every digraph G it holds $G = G_{Q_G}$.*

Lemma 1. *There is a directed path-decomposition for G of width $p-1$ if and only if there is a processing of Q_G with at most p stack-up places.*

Proof. By Proposition 1 we know that $G = G_{Q_G}$. If there is a directed path-decomposition for $G = G_{Q_G}$ of width $p-1$ then by Theorem 2 there is a processing of Q_G with at most p stack-up places. If there is a processing of Q_G with at most p stack-up places then by Theorem 1 there is a directed path-decomposition for G of width $p-1$. □

Theorem 3. *The* FIFO STACK-UP *problem is NP-complete.*

Proof. The given problem is obviously in NP. Determining whether the pathwidth of some given (undirected) graph is at most some given value w is NP-complete [10] and for symmetric digraphs a special case of the problem on directed graphs (cf. Introduction of [12]). Thus the NP-hardness follows from Lemma 1, because Q_G can be constructed from G in linear time. □

According to [1] it is shown in [13] that determining whether the pathwidth of some given (undirected) graph G is at most some given value w remains NP-complete even for planar graphs with maximum vertex degree 3. Thus the problem to decide, whether the directed pathwidth of some given symmetric digraph G is at most some given value w remains NP-complete even for planar digraphs with maximum vertex in-degree 3 and maximum vertex out-degree 3. Therefore by our transformation of graph G into Q_G we get sequences that contain together at most 6 bins per pallet. Hence the FIFO STACK-UP problem is NP-complete even if the number of bins per pallet is bounded. Thus we have proved the following statement.

Corollary 1. *The* FIFO STACK-UP *problem is NP-complete, even if the sequences of Q contain together at most 6 bins per pallet.*

4.2 Bounded FIFO Stack-Up Systems

In this section we show that the FIFO STACK-UP problem can be solved in polynomial time, if the number p of stack-up places or the number k of sequences is assumed to be fixed.

Fixed Number of Stack-Up Places. In [18] it is shown that the problem of determining the bounded directed pathwidth of a digraph is solvable in polynomial time. By Theorems 1 and 2 the FIFO STACK-UP problem with fixed number p of stack-up places is also solvable in polynomial time.

Theorem 4. *The* FIFO STACK-UP *problem can be solved in polynomial time, if the number p of stack-up places is fixed.*

Fixed Number of Sequences. Next we assume that the number k of sequences is fixed. In [8] we have shown that the FIFO STACK-UP problem can be solved by dynamic programming in time $\mathcal{O}(k \cdot (N+1)^k)$.

Theorem 5 ([8]). *The* FIFO STACK-UP *problem can be solved in polynomial time, if the number k of sequences in Q is fixed.*

Next we improve this result. Therefore we need two additional definitions. The position of the first bin in some sequence q_i destined for some pallet t is denoted by $\mathit{first}(q_i, t)$, similarly the position of the last bin for pallet t in sequence q_i is denoted by $\mathit{last}(q_i, t)$.

Theorem 6. *The* FIFO STACK-UP *problem is non-deterministically decidable using logarithmic work-space, if the number k of sequences in Q is fixed.*

Proof. We need $k+1$ variables, namely $pos_1, \ldots, pos_k$ and *open*. Each variable pos_i is used to store the position of the bin which has been removed last from sequence q_i. Variable *open* is used to store the number of open pallets. These variables take $(k+1) \cdot \lceil \log(n) \rceil$ bits. The simulation starts with $pos_1 := 0, \ldots, pos_k := 0$ and $open := 0$.

(i) Choose non-deterministically any index i and increment variable pos_i. Let b be the bin on position pos_i in sequence q_i, and let $t := plt(b)$ be the pallet symbol of bin b.
Comment: The next bin b from some sequence q_i will be removed.
(ii) If $\mathit{first}(q_j, t) > pos_j$ or $t \notin plts(q_j)$ for each $j \neq i$, $1 \leq j \leq k$, and $\mathit{first}(q_i, t) = pos_i$, then increment variable *open*.
Comment: If the removed bin b was the first bin of pallet t that ever has been removed from any sequence, then pallet t has just been opened.
(iii) If $\mathit{last}(q_j, t) \leq pos_j$ or $t \notin plts(q_j)$ for each j, $1 \leq j \leq k$, then decrement variable *open*.
Comment: If bin b was the last one of pallet t, then pallet t has just been closed.

If *open* is set to a value greater than p in step (ii) of the algorithm then the execution is stopped in a non-accepting state. To execute steps (ii) and (iii) we need a fixed number of additional variables. Thus, all steps can be executed non-deterministically using logarithmic work-space. □

Theorem 6 implies that the FIFO Stack-Up problem with a fixed number of given sequences can be solved in polynomial time since NL is a subset of P. The class NL is the set of problems decidable non-deterministically on logarithmic work-space. Even more, it can be solved in parallel in polylogarithmic time with polynomial amount of total work, since NL is a subset of NC_2. The class NC_2 is the set of problems decidable in time $\mathcal{O}(\log^2(n))$ on a parallel computer with a polynomial number of processors, see [14].

4.3 Approximation

In [11] it is shown that the directed pathwidth of a digraph $G = (V, E)$ can be approximated up to a factor of $\mathcal{O}(\log^{1.5}|V|)$. By Theorems 1 and 2 the optimization version of the FIFO Stack-Up problem can be approximated up to a factor of $\mathcal{O}(\log^{1.5} m)$.

5 Conclusion

In this paper, we have shown that the minimum number of stack-up places needed to solve the FIFO Stack-Up problem for some instance Q is equivalent to the directed pathwidth of the sequence graph G_Q of Q.

In our future work, we want to find online algorithms for instances where we only know the first c bins of every sequence instead of the complete sequences. Especially, we are interested in the answer to the following question: Is there a d-competitive online algorithm? Such an algorithm must compute a processing of some Q with at most $p \cdot d$ stack-up places, if Q can be processed with at most p stack-up places.

In real life the bins arrive at the stack-up system on the main conveyor of a pick-to-belt orderpicking system. That means, the distribution of bins to the sequences has to be computed. Up to now we consider the distribution as given. We intend to consider how to compute an optimal distribution of the bins from the main conveyor onto the sequences such that a minimum number of stack-up places is necessary to stack-up all bins from the sequences.

References

1. Bodlaender, H.L.: A tourist guide through treewidth. Acta Cybern. **11**, 1–23 (1993)
2. Borodin, A.: On-line Computation and Competitive Analysis. Cambridge University Press, Cambridge (1998)
3. de Koster, R.: Performance approximation of pick-to-belt orderpicking systems. Eur. J. Oper. Res. **92**, 558–573 (1994)

4. Dehmer, M., Emmert-Streib, F. (eds.): Quantitative Graph Theory: Mathematical Foundations and Applications. CRC Press Inc., New York (2014)
5. Fiat, A., Woeginger, G.J. (eds.): Online Algorithms: The State of the Art. LNCS, vol. 1442. Springer, Heidelberg (1998)
6. Gurski, F., Rethmann, J., Wanke, E.: Moving bins from conveyor belts onto pallets using FIFO queues. In: Huisman, D., Louwerse, I. (eds.) Operations Research Proceedings 2013, pp. 185–191. Springer, Heidelberg (2014)
7. Gurski, F., Rethmann, J., Wanke, E.: Algorithms for controlling palletizers. In: Proceedings of the International Conference on Operations Research (OR 2014), Selected Papers. Springer-Verlag (2015, to appear)
8. Gurski, F., Rethmann, J., Wanke, E.: A practical approach for the FIFO stack-up problem. In: An Le Thi, H., Dinh, T.P., Nguyen, N.T. (eds.) Modelling, Computation and Optimization in Information Systems and Management Sciences. AISC, vol. 360. Springer, Heidelberg (2015)
9. Johnson, T., Robertson, N., Seymour, P.D., Thomas, R.: Directed tree-width. J. Comb. Theory, Ser. B **82**, 138–155 (2001)
10. Kashiwabara, T., Fujisawa, T.: NP-completeness of the problem of finding a minimum-clique-number interval graph containing a given graph as a subgraph. In: Proceedings of the International Symposium on Circuits and Systems, pp. 657–660 (1979)
11. Kintali, S., Kothari, N., Kumar, A.: Approximation algorithms for digraph width parameters. Theor. Comput. Sci. **562**, 365–376 (2015)
12. Kitsunai, K., Kobayashi, Y., Komuro, K., Tamaki, H., Tano, T.: Computing directed pathwidth in $O(1.89^n)$ time. In: Thilikos, D.M., Woeginger, G.J. (eds.) IPEC 2012. LNCS, vol. 7535, pp. 182–193. Springer, Heidelberg (2012)
13. Monien, B., Sudborough, I.H.: Min cut is NP-complete for edge weighted trees. Theor. Comput. Sci. **58**, 209–229 (1988)
14. Papadimitriou, C.H.: Computational Complexity. Addison-Wesley Publishing Company, New York (1994)
15. Rethmann, J., Wanke, E.: Storage controlled pile-up systems, theoretical foundations. Eur. J. Oper. Res. **103**(3), 515–530 (1997)
16. Rethmann, J., Wanke, E.: On approximation algorithms for the stack-up problem. Math. Methods Oper. Res. **51**, 203–233 (2000)
17. Rethmann, J., Wanke, E.: Stack-up algorithms for palletizing at delivery industry. Eur. J. Oper. Res. **128**(1), 74–97 (2001)
18. Tamaki, H.: A polynomial time algorithm for bounded directed pathwidth. In: Kolman, P., Kratochvíl, J. (eds.) WG 2011. LNCS, vol. 6986, pp. 331–342. Springer, Heidelberg (2011)

Black and White Bin Packing Revisited

Jing Chen[1], Xin Han[2], Wolfgang Bein[3(✉)], and Hing-Fung Ting[4]

[1] Graduate School of Informatics, Kyoto University, Kyoto, Japan
chen@algo.cce.i.kyoto-u.ac.jp

[2] Software School, Dalian University of Technology, Dalian, China
hanxin@dlut.edu.cn

[3] Department of Computer Science, University of Nevada, Las Vegas, USA
wolfgang.bein@unlv.edu

[4] Department of Computer Science, The University of Hong Kong, Hong Kong, China
hfting@cs.hku.hk

Abstract. The *black and white bin packing* problem is a variant of the classical bin packing problem, where in addition to a size, each item also has a color (black or white), and in each bin the colors of items must alternate. The problem has been studied extensively, but the best competitive online algorithm has competitiveness of 3. The competitiveness of 3 can be forced even when the sizes of items are 'halved', i.e. the sizes are restricted to be in $(0, 1/2]$. We give the first 'better than 3' competitive algorithm for the problem for the case that item sizes are in the range $(0, 1/2]$; our algorithm has competitiveness $\frac{8}{3}$.

1 Introduction

We consider the *Black and White Bin Packing Problem* (*B&W*, for short) recently introduced by Bálogh et al. [1,2]. The input is a set of items with sizes in $(0, 1]$, furthermore each item is categorized as "black" or "white". The object is to pack the items into the minimum number of bins under the additional stipulation that no two items of the same color can be packed into a bin consecutively. In this paper we are interested in the online version of the problem, i.e. items arrive one by one according to a list L, and no information is given in advance. Thus the next item can be packed only into a bin where it fits and the last item already packed into that bin has the opposite color; if there is no such bin, the item must be packed into a new bin. This problem is a variant of the classical bin packing (Refer to e.g. [5,10–13]) which is a well known NP-hard problem [9].

As pointed out in Bálogh et al. [1] no algorithm can be competitive against an offline algorithm which can reorder items. Thus *B&W* is instead analyzed against the *restricted offline* algorithm, where items as before are given by a list L and packing has to be according to L, but in contrast to the online situation the order of the items, and the sizes and colors are known in advance. The absolute competitive ratio of algorithm A is then defined as

$$C_A = \sup_L \{A(L)/OPT_R(L)\},$$

Z. Lu et al. (Eds.): COCOA 2015, LNCS 9486, pp. 45–59, 2015.
DOI: 10.1007/978-3-319-26626-8_4

where $A(L)$ and OPT_R denote the number of bins required by online algorithm A and the restricted offline algorithm for list L.

Though in this paper we analyze competitiveness in terms of the absolute competitive ratio, we mention that online bin packing algorithms are sometimes analyzed in term of the asymptotic competitive ration defined by:

$$C_A^\infty = \lim_{n\to\infty} \sup_L \{A(L)/OPT_R(L) \mid OPT_R = n\}.$$

We review a number results relevant to our contribution; for a more complete exposition of the history of $B\&W$ we refer the reader to Bálogh et al. [1]. They proved that first fit (FF) is 3-competitive in the asymptotic sense and it was also shown that this is tight (for FF see [4,6,15]). Bálogh et al. [1] introduced algorithm Pseudo, which is 3 competitive in the absolute sense.

Furthermore Bálogh et al. [1,3] give a lower bound of 1.7213. This bound was improved to 2 by Dósa et al. [7], and it was shown that there is no online algorithm for $B\&W$ with asymptotic competitive ratio smaller than 2. However a large gap remains between the tight competitiveness of Pseudo and the lower bound of 2. Dósa et al. [7], have introduced the colorful bin packing problem, where items have a size from $(0, 1]$, and a color from color set C. This problem generalizes the black and white bin packing problem (where $|C| = 2$). They showed the method applied for $|C| = 2$ does not work for $|C| \geq 3$ and constructed an algorithm for $|C| \geq 3$ with absolute competitive ratio of 4. Veselý et al. [3], gave an absolutely 3.5-competitive algorithm for the colorful bin packing problem, and a lower bound of 2.5.

As mentioned Pseudo [1] is 3-competitive in the absolute sense; in fact it is the first such algorithm. For the parametric case, if items sizes are at most $1/d$ (for $d \geq 1$, d is an integer), the performance ratio of pseudo is $1 + \frac{d}{d-1}$. The 3 competitive ratio of Pseudo algorithm is tight, even items have size at most 1/2, as shown in [1]. Vesely [8,14], proved that the competitive ratio of the first fit algorithm for the $B\&W$ problem is at most 3. Furthermore Bálogh et al. [1], proved that in any parametric case, if items sizes are at most $1/d$ (for $d \geq 1$, d is an integer), the performance ratio of FF is at least 3.

Our Contribution. It has been conjectured that there be a "better than 3" competitive algorithm for $B\&W$. In this paper we settle this conjecture in the affirmative for the case when sizes are in $(0, 1/2]$.

2 Algorithm "Balance Between Stacks"

We recall the algorithm Pseudo given in [1]. Pseudo depends on a lower bound LB_1: For an input list L of n items, let $c_i = 1$ if the ith item is black and $c_i = -1$ if it is white, then LB_1 is:

$$LB_1 = \max_{1 \leq i < j \leq n} |\sum_{k=i}^{j} c_k| \tag{1}$$

Remark 1. ([1]) For any problem instance, LB_1 is a lower bound on the optimum, both in the online and restricted offline cases.

Algorithm Pseudo is as follows

Step 1: Render all items as "pseudo items" with size 0 but retained color, and pack these pseudo items using algorithm Any Fit, which will uses exactly LB_1 bins (also called stacks).

Step 2: Consider the original size of the pseudo items, and divide the contents of each such stack into subsequent bins of unit sizes, as soon as a bin would exceed 1, open a new bin.

The competitiveness of Pseudo is tight even when sizes are restricted to $(0, 1/2]$ (see [1]). The worst case for algorithm Pseudo occurs in the following situation: one Stack has many bins, but all the remaining Stacks only have one bin with little contents in the bin.

In our algorithm we use data structures, namely Stacks, Buffers and a Set C to avoid such unbalances. Our aim is to construct an algorithm for *B&W* with competitive ratio better than 3. The algorithm is called BAL.

2.1 Description of Algorithm BAL

We now give a description of our algorithm, while referring to Fig. 1 below. As shown in the Figure, we make use of data structures: Stacks and Buffers. Stacks contain bins which in turn contain the items. We define the *color of a Stack* as the color of the last item packed into this Stack. When an item x arrives, first find a Stack with the opposite color of x. Then pack x into the open bin in this Stack. Each Stack also has a *Buffer* which can contain at most two bins. The invariant for a Stack is that it carries at most three bins. When a bin is opened it is affiliated with the Stack where it was opened. Each bin is only affiliated with one Stack, but its affiliation may change during packing.

Given a Stack, let k be the number of bins in this Stack for $k \leq 3$; we call this an S_k Stack. The expression $Buff(S)$ is the Buffer of Stack S. An L_0 Stack is a special S_1 Stack, where one bin (dashed lines in Fig. 1) is not opened here but moved from another Stack. L_1 Stack is S_1 Stack throughout the algorithm and the one bin is opened in this Stack.

Furthermore we define set C to be a set of pairs of bins. There are two types of pairs, *couple* and *fat.* A *couple pair* is a pair of bins with different top colors, total contents of these two bins larger than 1, and each individual bin total contents at least $\frac{1}{5}$. A *fat pair* is a pair of bins which has a total contents of at least $\frac{6}{5}$. A *fat* pair may have the same top color. In the algorithm, whenever a new Stack is about to be opened, the set of C is checked first to create better balance. Unless C is empty, a pair of bins will be used for the opening of a new Stack.

In our algorithm, an S_1 Stack where $Buff(S_1) = 0$, and the one bin inside the S_1 Stack has contents less than $1/5$, is called a *low* S_1 Stack. S_1 Stacks that have $Buff(S_1) = 0$ are called unsaturated S_1 Stacks. An S_2 Stack where the second bin is not size collision with the first bin, and the contents of the second bin is less than 3/10 (Stacks marked by $*$ in our algorithm) is called an isolated Stack. Both S_1 Stacks with $Buff(S_1) = 0$ and marked S_2 Stacks are called unsaturated Stacks.

At any moment of the algorithm: if there is at most one unsaturated Stack, we record this state by a variable $State = 0$; if there are at least two unsaturated Stacks, and all the unsaturated Stacks have the same top color, we record this state by $State = 1$; if there are more than one unsaturated Stack, and at least two unsaturated Stacks have different top colors, we record this state by $State = 2$. In our algorithm, if there are at least two unsaturated Stacks, we try to keep $State = 2$ if possible. With these data structures, our algorithm runs as follows:

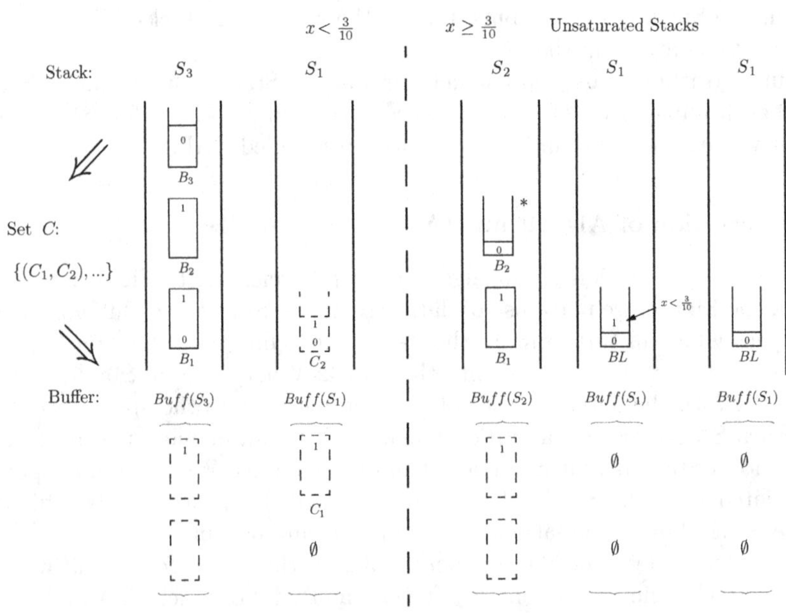

Fig. 1. Data structures of algorithm BAL

For an input item x, we first choose an existing Stack for x.

Step I: Choose a Stack for x.

We will choose Stack with opposite color for x. If $x \geq 3/10$ we will choose an S_1 Stack with $Buff(S_1) = 0$, or an isolated Stack (marked by $*$) first. If $x < 3/10$ we will first to insure that at least two unsaturated Stacks ($Buff(S_1) = 0$ Stacks and marked Stacks) have different top colors. If at least two unsaturated Stacks have different top colors, or at most one unsaturated Stack exists, we will choose Stack with the largest number of affiliated bins first. The details are as follows:

State = 0. In this case we choose Stack as follows:

If $x \geq 3/10$

If an unsaturated Stack with opposite color of x exists, choose it (S_1 Stack prior to marked Stack).

Else choose a Stack S with opposite color of x and the largest number of affiliated bins.

If $x < 3/10$, from all the Stacks that are not unsaturated, choose a Stack S with opposite color of x and the largest number of affiliated bins.

If an S_1 Stack with $Buff(S_1) = 1$ is chosen, and the bin in S_1 and bin in $Buff(S_1)$ have different colors then proceed as follows: If there is a marked S_2 Stack with opposite color of x, choose the S_2 for x, else choose S_1.

If the only Stack with opposite color of x is an unsaturated Stack, choose it.

State = 1. If all the unsaturated Stacks have different color with x, choose one as S (choose S_1 Stack with $Buff(S_1) = 0$ prior to marked Stack), update the value of *State*. Otherwise all unsaturated Stacks have same color with x. Choose a Stack S with opposite color of x and the largest number of affiliated bins.

State = 2. In this case at least two unsaturated Stacks have different colors. Therefore there must be one unsaturated Stack with opposite color of x.

If $x \geq 3/10$, choose one unsaturated Stack with opposite of x. Update the value of *State*.

If $x < 3/10$, from all the existing Stacks, choose an Stack S with opposite color of x and the largest number of affiliated bins (In this case S_3 or S_2 Stack with opposite color of x will be chosen prior to S_1 Stack). Update the value of *State*.

Step II: Pack x into Stack.

(1) If no existing Stack is available for x.

A new Stack will be opened.

If set C is empty: open a new Stack.

If S_1 Stack with $|Buff(S_1)| = 2$ exists, and the contents of the bin in S_1 has contents at least $1/5$. Then move the two bins in $Buff(S_1)$ to the Buffer of the new Stack. If x can be packed into one bin move this bin to new Stack as open bin. Otherwise open a new bin in the new Stack for x. (As Stack S_1 of Fig. 1.)

Else directly open a new bin in the new Stack for x.

If set C is not empty:

(a) If x can be packed into one bin of a pair in set C, pack x into it. Assume this bin is C_2, the other bin of the pair is C_1. After packing x, if C_2 has same top color with C_1, and C_1 has less contents, then move C_1 to the new Stack as open bin. Otherwise move C_2 to the new Stack as open bin. In both case, move the other bin to the Buffer of the new Stack.

(b) In this case x can not be packed into any bin of a pair in set C.

If $x < 1/5$, move a pair to the Buffer of the new Stack, and open a new bin in new Stack for x (Refer to Fig. 1). For such Stack with $Buff(S_1) = 2$, we can prove that the three bins have a total contents of at least $1 + 1/5$ (Refer to the analysis of Eqs. (10) and (12)).

If $x \geq 1/5$, then directly open a new Stack and open a new bin for x.

(2) Assume Stack S is chosen for packing x.

S is an S_1 Stack. Assume the bin in the Stack is B_1.

(*i*) $|Buff(S)| \neq 2$. If possible, pack x into the open bin B_1 in the Stack. If x can not be packed into B_1, just open a new bin B_2 to pack x. If $Buff(S) = 1$ move B_1 to $Buff(S)$.

(*ii*) $|Buff(S)| = 2$.

If the contents of bin B_1 is at least $2/5$, and a bin in the $Buff(S)$ has the same color of B_1 but contents smaller than B_1. In this case, assume the bin in $Buff(S)$ to be B_f bin, exchange B_1 with B_f, use B_f as open bin.

Pack x into the open bin in the Stack as possible. If x can not be packed into the open bin, open a new bin B_2 to pack x. If the pair in $Buff(S)$ is a *fat* pair and S_1 Stack with $Buff(S_1) = 0$ exists, move the pair to the Buffer of a S_1 Stack (*low* S_1 is preferable). Otherwise move the *fat* pair to set C.

S is an S_2 Stack. Assume the bins in the Stack are B_1 and B_2.

If x can be packed into B_2 bin.

Pack x into the open bin B_2 in the Stack. If the contents of the second bin is larger than $3/10$, remove the mark $*$ of S if exists.

If x can not be packed into B_2.

Remove the mark $*$ of the Stack if the mark exists.

Then check if x can be packed in to B_1, if so pack x into B_1 bin, and exchange B_1 and B_2, use B_1 as open bin. Otherwise open a new bin for x.

S is an S_3 Stack. As in Fig. 1 assume the three bins are B_1, B_2 and B_3.

Case: $|Buff(S)| = 0$. Pack x into B_3 bin if possible. If x can not be packed into B_3, then open a new bin for x, and move B_1 and B_2 to $Buff(S)$, S becomes an S_2 Stack.

Case: $|Buff(S)| = 1$. (Impossible for S_3 Stack.)

Case: $|Buff(S)| = 2$

(2.1) If there are at least two unsaturated Stacks, pack x into B_3 if possible. If x can not be packed into B_3, open a new bin for x. In this case B_2 and B_3 must be a *fat* pair (Lemma 1). Move B_2 and B_3 to the Buffer of a *low* S_1 Stack. If no *low* S_1 Stack exists, move the *fat* pair to set C.

(2.2) There is at most one unsaturated Stack.

If there is an S_1 Stack with $Buff(S_1) = 0$ then

(2.2.1) We change the S_3 Stack and the S_1 Stack with empty Buffer into a marked S_2 Stack and an S_1 Stack with $Buff(S_1) = 1$. (As the S_1 in Fig. 2.) Assume BL represents the bin in the S_1 Stack.

Let $min\{B_1, B_2\}$ be the bin of minimum contents between B_1 and B_2. Move $min\{B_1, B_2\}$ bin to the Buffer of S_1 Stack.

If B_3 and BL have the same top color, exchange B_3 and BL bin. Change the affiliation of BL to S, while B_3 and B_2 still affiliate to the S Stack. If BL has contents less than 3/10, mark Stack S by mark $*$. Pack x into the new open bin BL in S. S Stack becomes an S_2 Stack.

If B_3 and BL have different top colors, pack x into B_3 bin first then exchange B_3 and BL, do the same operations as above.

Else check bins B_2 and B_3 (Fig. 3).

(2.2.2) B_2 and B_3 have different top colors, and the contents of B_3 is at least $\frac{1}{5}$. By definition B_2 and B_3 is a *couple* pair. Then in this case we will check B_1 and B_2 bin first:

(a) B_1 and B_2 have different top colors. Pack x into B_3; if x can not be packed into B_3, then open a new bin for x, and move B_1 and B_2 to set C as a *couple* pair; S becomes a S_2 Stack.

(b) B_1 and B_2 have the same color. Let $min\{B_1, B_2\}$ be the bin of minimum contents between B_1 and B_2. Move $min\{B_1, B_2\}$ and B_3 as a *couple* pair (Lemma 3) to set C. Then open a new bin in S for x, and mark the Stack S by mark $*$ since the new bin is not size collision with B_1. Now S becomes a S_2 Stack.

(2.2.3) If B_2 and B_3 is not a *couple* pair.
The algorithm will pack x into bin B_3, if possible. If x can not be packed into bin B_3, then open a new bin for x, and move B_1 and B_2 to set C (B_1 and B_2 must be a pair by Lemma 4), S becomes an S_2 Stack.

We have the following lemmas.

Lemma 1. *Assume B_i is the open bin of an unsaturated Stack. Bin B_i has contents at least* 7/10, *if at least two unsaturated Stacks exist when it is closed.*

Proof. Let B_i be the open bin of a Stack S, where S is not an unsaturated Stack. Assume item x is chosen to be packed into Stack S, and x is too large for B_i. We must prove that it is not possible for $x \geq 3/10$. Assume that x is larger than 3/10, then in this case the algorithm chooses Stack S, only in the case that all the unsaturated Stacks have the same color with x. Let y be the top item of B_i bin, and z be the top item of any unsaturated Stack. If z comes after y then, z chosen the unsaturated Stack only when $z \geq 3/10$, contradicting the definition of an unsaturated Stack. Therefore z comes before y. Then by our algorithm y should choose an unsaturated Stack. At least two unsaturated Stacks have different top colors. Again a contradiction. Therefore if at least two unsaturated Stacks exist, it is impossible for the present item x to be larger than 3/10, when B_i is about to close. □

Lemma 2. *Assume S_2 is a marked Stack with two bins, B_1, B_2, and B_2 is an isolated bin (B_2 is not size collision with B_1). Let C_1 and C_2 be the last pair moved out from S_2 Stack, then C_1 and C_2 can only be in an L_0 or L_1 Stack or in Set C.*

The proof of the lemma will be in the journal version.

Lemma 3. *Bins $min\{B_1, B_2\}$ and B_3 in step (b) of* (2.2.2) *form a couple pair.*

Proof. In step (*b*) of (2.2.2), we know that B_2 and B_3 form a *couple* pair, and B_1 and B_2 have the same top color, we only need to prove that B_1 and B_3 form a *couple* pair. Since B_1 and B_2 have the same top color, therefore when the first item of the B_3 bin comes, the Stack is a S_2 Stack. According to our algorithm (step (2)) the algorithm will attempt to put the item into B_1 bin also, and only the case that the item is size collision with both bins then a new bin will be opened. Therefore we have B_1 and B_3 have total size large than 1. B_1 has contents at least 1/2 and B_3 has contents at least 1/5, they are a *couple* pair. □

Lemma 4. *In step (2.2.3), B_1 and B_2 is a couple pair or a fat pair.*

Proof. If the first item of B_3 is smaller than 1/5, then B_2 has contents at least $\frac{4}{5}$. Since bin B_1 has contents at least 1/2, B_1 and B_2 form a *fat* pair.

Now for the case that the first item x of bin B_3 is larger than 1/5. When the next item of x in B_3 appears, B_2 and B_3 is a *couple* pair at the time. If B_1 and B_2 have same top color then by (*b*) of (2.2.2), $min\{B_1, B_2\}$ and B_3 will be moved out as a *couple* pair. Therefore if $x \geq 1/5$, for the next items of x in the same Stack, algorithm can go to step (2.2.3) only under the condition that B_1 and B_2 have different top colors. B_1 and B_2 form a *couple* pair. Therefore in step (2.2.3), if B_1 and B_2 is not a *fat* pair, B_1 and B_2 must be a *couple* pair. □

Finally, we observe that an S_3 Stack cannot transition into an S_4 Stack, and an S_3 Stack can only transition into an S_2 Stack after generating a pair. And each pair in set C is generated when no *low* S_1 Stack exists.

3 Competitive Analysis of Algorithm BAL

We now analyze the performance of our algorithm. Let M be the total number of bins used by the algorithm at the end. Then we divide M into disjoint subsets; and analyze the total size of the bins in each subset. For example a pair of bins in set C has total size at least 1, while a *fat* pair by definition have total size at least $1 + 1/5$. If a *couple* pair is used to open a Stack and the Stack opened a new bin, then these three bins have a total size of at least $1 + 1/5$, since the new opened bin must has size collision with one bin of the pair, and the contents of the left bin is at least 1/5. Using these properties we can calculate the contents of the disjoint subsets; if we can assure that there are at least total size of $c \cdot M$, $c > 0$ then the competitive ratio of our algorithm is at most $1/c$.

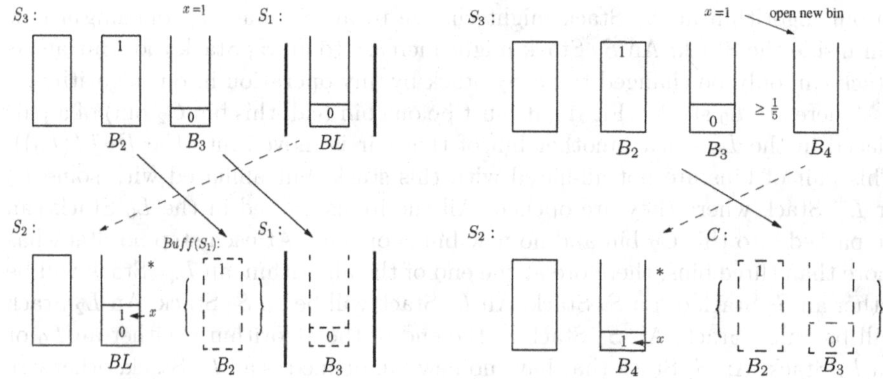

Fig. 2. $Buff(S_1) = 1$.

Fig. 3. *Couple pair.*

In the following sections we first we define a number of useful terms for the analysis of the algorithm, and introduce a case analysis. Then we develop a methodology of calculating contents for various kinds configurations. Finally we give the proofs of our performance ratio.

3.1 Terminology and Case Analysis

As defined before an S_k Stack has k bins ($k \leq 3$), an S_4 Stack has at least 4 bins but does not exist in our algorithm. We define $\bar{S}_k$ as the number of S_k Stacks at the end of the algorithm.

Let k be the number of bins affiliated to a Stack at the end of the algorithm, if $k \leq 3$ we call the stack an L_k Stack, else if $k \geq 4$ we call it an L_{4^+} Stack (L_{4^+} Stacks exist in the algorithm. Because when a Stack generated pairs, the pairs are moved out of the Stack but are still affiliated with this Stack). An L_1 Stack is a Stack that has only one affiliated bin at the end of the algorithm, it must be an S_1 Stack. But an S_1 Stack at the end of the algorithm is not necessarily an L_1 Stack. There may be one bin in the Stack and another bin in the Buffer, the two bins are bins of a pair that moved in from other Stack. This Stack is an L_0 Stack.

Lemma 5. *An S_1 Stack at the end of the algorithm can not be an L_{4^+} Stack.*

Proof. Assume an S_1 Stack at the end of the algorithm is an L_{4^+} Stack. Since there is only one bin left in the Stack at the end of the algorithm, the other bins affiliated with this Stack are moved out of the Stack. If the last bin moved out is only one bin, then this S_1 Stack is once an S_2 Stack. But in our algorithm no bin will be moved out from an S_2 Stack. Therefore it is impossible to move only one bin from the Stack at a time. Suppose the last bins moved are two bins, then this S_1 Stack is once an S_3 Stack. Where by our algorithm each operation on an S_3 Stack is insured to make the Stack to be an S_2 Stack when moving a pair of bins outside the Stack. Again a contradiction. □

In our algorithm an S_1 Stack might change to an S_2 Stack by opening a new bin inside the Stack. An S_2 Stack might increase to an S_3 Stack too. But an S_3 Stack can only be changed to an S_2 Stack by any operation in our algorithm.

There are L_0 stacks (Fig. 1), it must be one bin (call this bin C_2 bin) of a pair placed in the L_0 stack (another bin of the pair is moved into the $Buff(L_0)$). This pair of bins are not affiliated with this stack, but affiliated with some L_3 or L_4^+ Stack where they are opened. All the items packed in the L_0 Stack can be packed into this C_2 bin and no new bin is opened. At each step no Stack has more than three bins, therefore at the end of the algorithm, an L_{4+} Stack will be either an S_2 Stack or an S_3 Stack. An L_3 Stack will be an S_3 Stack. An L_2 Stack will be an S_2 Stack. An S_1 Stack at the end of the algorithm is either an L_0 or an L_1 Stack. An S_1 Stack that have no new bin opened is an L_0 Stack, otherwise it is an L_1 Stack. An L_1 Stack can only be an S_1 Stack at any moment. If the Stack is start with a C_2 bin, it will be moved to the Buffer of the Stack when a new bin is opened. As mentioned $\bar{S}_k$ is the number of S_k Stacks at the end of the algorithm, then we have

$$\bar{S}_1 = L_0 + L_1.$$

As defined $\bar{S}_2$ is the number of S_2 Stacks at the end of the algorithm, and $\bar{S}_3$ is the number of S_3 Stacks. Then we have:

$$LB_1 = \bar{S}_1 + \bar{S}_2 + \bar{S}_3 = L_0 + L_1 + \bar{S}_2 + \bar{S}_3, \tag{2}$$

As defined M is the total number of bins used by the algorithm, then we have

$$M = L_1 + 2\bar{S}_2 + 3\bar{S}_3 + T,$$

where T is the total number of bins in set C, bins in $Buff(S)$ and bins in Stack L_0 (C_2 bin as in Fig. 1). T can also be calculated as the total bins in dashed lines in Fig. 1 plus the bins in set C. Therefore we have $T = 2x$, where x is the number of pairs generated by the algorithm. Each pair can used to fill the Buffer of a Stack.

$$T = M - L_1 - 2\bar{S}_2 - 3\bar{S}_3. \tag{3}$$

Each Buffer of a Stack has a capacity of at most 2. Therefore at least the number of $x = min\{T/2, LB_1\}$ Stacks can be filled by T, these Stacks include L_k Stacks, $k \geq 2$. Therefore we can only assure that the number of S_1 Stack with $|Buff(S_1)| \neq 0$ is at most $x - \bar{S}_2 - \bar{S}_3$.

The first case is that T is large enough that can insure all S_1 Stacks ($\bar{S}_1 = L_0 + L_1$) will have $Buff(S_1) \neq 0$. That is

$$Case\ ①: \quad x \geq \bar{S}_1 + \bar{S}_2 + \bar{S}_3.$$

Consider the more general case: only part of S_1 can be insured to have $Buff(S_1) \neq 0$.

$$Case\ ②: \quad \bar{S}_2 + \bar{S}_3 < x < \bar{S}_1 + \bar{S}_2 + \bar{S}_3.$$

The third case is that

$$Case\ ③:\quad x \le \bar{S}_2 + \bar{S}_3.$$

Thus in the third case no S_1 stack can be insured have no empty Buffer, but in this case we can find the total number of bins is bounded.

3.2 A Function Calculating the Total Size of Bins

For a set $S = \{B_1, B_2, ...\}$, function $w(S)$ calculates the lower bound of the total size of the bins in S. Let V be the set of all M bins at the end of our algorithm, We divide V into disjoint subsets of $V_1, V_2, ...$, then the function $w(V_i)$ has the property:

$$OPT \ge \text{ total size of items in V } \ge \sum_i w(V_i). \tag{4}$$

The values of $w(x)$ for different set of bins are defined as follows:

If $V_i = \{B_1, B_2\}$, where B_1 and B_2 are two consecutive opened bins in the same Stack, or both bins have contents at least 1/2, then we have

$$w(\{B_1, B_2\}) = 1. \tag{5}$$

If $V_i = \{B_1, B_2\}$, where B_2 is not size collision with B_1 (B_2 is an isolated bin). Refer to step (2.2.1) or step (a) of (2.2.2) in our algorithm, that a pair of bins are moved out, and the next bin is not size collision with B_1 bin in the S_2 Stack. We have the following lemmas.

Lemma 6. *For an $S_2 = \{B_1, B_2\}$ Stack with an isolated bin B_2(B_2 is not size collision with B_1). Let C_1 and C_2 be the pair moved out, if C_1 and C_2 is moved out by step (b) of (2.2.2), B_1 and C_1 have same top color.*

Proof. As in our algorithm, if the isolated bin is generated in step (b) of (2.2.2), that B_1 and B_2 must have same color. □

Lemma 7. *For an $S_2 = \{B_1, B_2\}$ Stack with an isolated bin B_2, and S_2 is not marked (B_2 has contents larger than 3/10), let B_{f1}, B_{f2} be the two bins in the $Buff(S_2)$, we have:*

$$w(\{B_{f1}, B_{f2}\} \cup \{B_1, B_2\}) = 3/2 + 3/10 = 9/5. \tag{6}$$

Lemma 8. *For an $S_2 = \{B_1, B_2\}$ Stack with an isolated bin B_2, let C_1 and C_2 be the last pair moved out. If S_2 is still marked, and C_1 and C_2 are in set C or in an L_0 Stack at the end of the algorithm, then we have:*

$$w(\{B_{f1}, B_{f2}\} \cup \{B_1, C_1, C_2, \overset{*}{B_2}\}) = 2.5. \tag{7}$$

Proof. C_1 and C_2 have total size 1, B_1 is closed bin have size at least 1/2. B_{f1}, B_{f2} are a pair of bins and have total size 1. □

Lemma 9. *For an $S_2 = \{B_1, B_2\}$ Stack with an isolated bin B_2, let C_1 and C_2 be the last pair moved out. If S_2 is still marked, and C_1 and C_2 are in an L_1 Stack at the end of the algorithm. Let BL be the bin in L_1 Stack, then we have:*

$$w(\{B_{f1}, B_{f2}\} \cup \{B_1, C_1, C_2, \overset{*}{B_2}\} \cup \{BL\}) = 3. \tag{8}$$

Proof. BL bin must be opened later than C_1 and C_2 in Stack L_1. The size of (8) is calculated as follows: if BL is size collision with C_2 (have total size 1), then the total size of B_1 and C_1 is 1, therefore the total is 3. If C_1 and C_2 are moved into L_1 Stack by operation (2.2.1) then BL must be size collision with C_2.

If BL has size collision with C_1, then B_1 and C_1 must have the same color (otherwise B_1 and B_2 would have moved out as *couple* pair), and size of C_1 is no more than B_1. (We move the smaller bin with C_2 as *couple* in step (b) of (2.2.2)). Therefore BL and B_1 have total size at least 1, and also C_1 and C_2 have total size 1, therefore we have the above size equation. □

Lemma 10. *If $V_i = \{B_1, B_2, B_3\}$, where B_2 and B_3 are two consecutively opened bins in the same Stack, then we have*

$$w(\{B_1, B_2, B_3\}) = 1.5. \tag{9}$$

Proof. where B_3 is opened to pack an item which is size collision with B_2. Therefore the contents in the two bins is at least 1, B_1 is a closed bin and has contents at least 1/2.

Lemma 11. *If $V_i = \{C_1, C_2, BL\}$, where C_1, C_2 form a couple pair and BL is the bin of a L_1 Stack. The total size of the three bins is at least:*

$$w(couple \cup \{BL\}) = 1 + \frac{1}{5}. \tag{10}$$

Proof. BL is opened by size collision with one bin of the *couple* bins, by definition the top colors of the *couple* are different and each bin of a *couple* pair has contents at least 1/5. □

Regarding *fat* pairs we have following lemmas about the total size inside the bins:

Lemma 12. *If $V_i = fat = \{B_1, B_2\}$, where B_1, B_2 is a fat pair, we have*

$$w(fat) \geq \frac{6}{5}. \tag{11}$$

Proof. If a *fat* pair is generated by operation (2.2.3) from an S_3 stack, then the first item of the third bin B_3 of S_3 has a size $< \frac{1}{5}$. Refer to Fig. 3. Therefore the contents of B_2 has contents larger than $\frac{4}{5}$, while B_1 have contents larger than $\frac{1}{2}$ as a closed bin. If the *fat* pair is generated in operation (ii) of (2) in a S_2 Stack, then by Lemma 2, they have total size at least 6/5. □

We can assign one *low* L_1 bin with a *fat* pair. And we have

$$w(fat \cup \{BL\}) \geq \frac{6}{5}. \tag{12}$$

where BL is the bin of an L_1 Stack.

3.3 Case by Case Analysis of the Competitive Ratio of BAL

Lemma 13. *Case* ① : *if* $x \geq \bar{S}_1 + \bar{S}_2 + \bar{S}_3$*, then the competitive ratio of algorithm BAL is at most* $\frac{5}{2}$.

Proof. The number of pairs is larger than the number of Stacks. If set C is empty, then there is a number of LB_1 pairs in LB_1 Stacks. If the set C is not empty, then there is no *low* L_1 Stack. But there are potentially L_1 Stacks that have $Buff(L_1) = 0$, and contents at least 1/5. By operation of (2.2.1), the pairs in set C can not be generated later than any L_1 Stack that has $Buff(L_1) = 0$. These L_1 Stacks opened the Stack when the pairs in C are existing. Then it can only be the case of operation (*b*) of (1), the first item of the Stack is larger than 1/5 and can not be packed into any bin of the pairs. We can move a pair in set C to the $Buff(L_1)$, then the total size of the three bin is at least 6/5 and the bin in L_1 Stack is size collision with one bin in the $Buff(L_1)$. This kind of L_1 Stack just has the same property as the L_1 Stack with $Buff(L_1) = 2$ that are generated by our algorithm. Therefore at the end of the algorithm, if set C is not empty, and there are L_1 Stacks with $Buff(L_1) = 0$, we can simply move the pairs from set C to the Buffers of the L_1 Stacks. This new L_1 with $Buff(L_1) = 2$ still has the property for (8) and for the proof of Lemma 2.

Since the number of pairs is larger than the number of Stacks, if C is not empty we can move pairs to the empty Buffers of L_1 Stacks. Therefore we can assume that all L_1 Stacks have $Buff(L_1) \neq 0$ in Case ①.

We consider the contents of S_2 Stacks first. The size of an $\bar{S}_2$ Stack (B_1 and B_2 bin) is either by (5) that has total contents of at least 1, or B_2 is not size collision with B_1 (B_2 is an isolated bin). But B_1 B_2 can combine size by (7) or (8). We define y_2 to be the number of $\bar{S}_2$ Stacks that have the following properties: S_2 Stack has an isolated bin, the last moved out pair is either in set C or used by an L_0 Stack. The total size of these y_2 Stacks can be calculated by (7). We define y_2' to be the number of $\bar{S}_2$ Stacks that have the following properties: S_2 Stack has an isolated bin, the last moved out pair is used by an L_1 Stack. Then the total size of y_2 Stacks can be calculated by (8). We define y_2'' to be the number of $\bar{S}_2$ Stacks, that have an isolated bin but not marked at the end of the algorithm (isolated bin have contents 3/10), then these Stacks are calculating size by (6). All of these types of Stacks have $Buff(S_2) = 2$.

Let $x_1 = L_1 - y_2'$, then by Eqs. (10) and (12). These bins will have total contents at least:

$$x_1 \cdot w(\{C_1, C_2,\} \cup \{BL\}) \geq \frac{6x_1}{5}, \tag{13}$$

where BL is the bin of an L_1 Stack, and C_1, C_2 are a pair of bins from T. If the bins C_1 and C_2 are moved into the L_1 Stack by operation (2.2.1), the three bins will have total size of at least 1.5.

M bins are divided into subsets as follows:

$$\begin{aligned} M &= L_1 + 2\bar{S}_2 + 3\bar{S}_3 + T, \\ &= x_1 + y_2' + 2x_1 + 3\bar{S}_3 + T - 2x_1 + 2\bar{S}_2. \end{aligned}$$

By Eqs. (5) and (9), we have

$$w(T-2x_1)=\frac{T-2x_1}{2}.$$

Let $\bar{S}_2'=\bar{S}_2-y_2-y_2'-y_2''$, by (6), (7) and (8), we have

$$w(2\bar{S}_2+2y_2+2y_2'+y_2'+2y_2'')=\bar{S}_2'+2.5y_2+3y_2'+9y_2''/5.$$

Let $T'=T-2x_1-2y_2-2y_2'-2y_2''$, then we have:

$$\begin{aligned}w(M)&=w(x_1+2x_1)+w(3\bar{S}_3)+w(T')+w(2\bar{S}_2)\\&\geq\frac{6x_1}{5}+\frac{3\bar{S}_3}{2}+\frac{T'}{2}+\bar{S}_2'+2.5y_2+3y_2'+9y_2''/5.\end{aligned}$$

We have $OPT\geq w(M)$, therefore the competitive ratio is:

$$\begin{aligned}\frac{M}{OPT}&\leq\frac{x_1+2x_1'+3\bar{S}_3+T'+2\bar{S}_2}{6x_1/5+(3\bar{S}_3)/2+(T')/2+\bar{S}_2}\\&\leq\frac{3x_1+3\bar{S}_3+T'+2\bar{S}_2'+6y_2+7y_2'+4y_2''}{6x_1/5+3\bar{S}_3/2+(T')/2+\bar{S}_2'+2.5y_2+3y_2'+9y_2''/5}\\&\leq\frac{5}{2}.\end{aligned}$$

□

Lemma 14. *Case ② : If* $\bar{S}_2+\bar{S}_3<x<\bar{S}_1+\bar{S}_2+\bar{S}_3$*, then the competitive ratio of algorithm BAL is at most* $\frac{8}{3}$.

Lemma 15. *Case ③ : if* $x\leq\bar{S}_2+\bar{S}_3$*, then the competitive ratio of algorithm BAL is at most* $\frac{8}{3}$.

Proofs of the Lemmas 14 and 15 will be in the journal version. Combine the results of Lemmas 13, 14 and 15, we have the following theorem:

Theorem 1. *For the instance that all items have sizes in (0, 1/2], the competitive ratio of algorithm BAL is at most* $\frac{8}{3}$.

4 Concluding Remarks

We conjecture and have obtained partial results to show that our scheme can be used for the problem where items sizes are in $(0,\alpha]$, with $\alpha<1$, to obtain a competitive ratio smaller than $3-\omega(\frac{1}{\alpha})$. We note that the analysis, though similar to the analysis of BAL presented in the previous chapter, is more involved when item sizes are in $(0,\alpha]$. We also conjecture that there is a better than 3 competitive algorithm for *B&W*, and it is conceivable that further refinement of the ideas for BAL will yield such a desirable result.

Acknowledgment. Author Wolfgang Bein conducted this research while on sabbatical at Kyoto University, Japan. A sabbatical from the University of Nevada, Las Vegas and support from National Science Foundation grant IIA 1427584 is acknowledged.

References

1. Bálogh, J., Békési, J., Dósa, G., Epstein, L., Kellerer, H., Tuza, Z.: Online results for black and white bin packing. Theory Comput. Syst. **56**, 137–155 (2015)
2. Balogh, J., Békési, J., Dosa, G., Kellerer, H., Tuza, Z.: Black and white bin packing. In: Erlebach, T., Persiano, G. (eds.) WAOA 2012. LNCS, vol. 7846, pp. 131–144. Springer, Heidelberg (2013)
3. Bálogh, J., Békési, J., Galambos, G.: New lower bounds for certain classes of bin packing algorithms. Theor. Comput. Sci. **440–441**, 1–13 (2012)
4. Boyar, J., Dósa, G., Epstein, L.: On the absolute approximation ratio for First Fit and related Results. Discrete Appl. Math. **160**(13–14), 1914–1923 (2012)
5. Coffman, E.G., Garey, M.R., Johnson, D.S.: Approximation algorithms for bin packing: a survey. In: Hochbaum, D. (ed.) Approximation Algorithms. PWS Publishing Company (1997)
6. Dósa, G., Sgall, J.: First fit bin packing: a tight analysis. In: Proceedings of STACS 2013, LIPICS, vol. 20, pp. 538–549 (2013)
7. Dósa, G., Epstein, L.: Colorful bin packing. In: Ravi, R., Gørtz, I.L. (eds.) SWAT 2014. LNCS, vol. 8503, pp. 170–181. Springer, Heidelberg (2014)
8. Böhm, M., Sgall, J., Veselý, P.: Online colored bin packing. In: Bampis, E., Svensson, O. (eds.) WAOA 2014. LNCS, vol. 8952, pp. 35–46. Springer, Heidelberg (2015)
9. Garey, M.R., Johnson, D.S.: "Strong" NP-completeness results: motivation, examples, and implications. J. ACM **25**(3), 499–508 (1978)
10. Karmarkar, N., Karp, R.M.: An efficient approximation scheme for the one-dimensional bin-packing problem. In: Proceedings of the 23rd Annual Symposium on Foundations of Computer Science (FOCS 1982), pp. 312–320 (1982)
11. Lee, C.C., Lee, D.T.: A simple on-line bin packing algorithm. J. ACM **32**, 562–572 (1985)
12. Seiden, S.: On the online bin packing problem. J. ACM **49**(5), 640–671 (2002)
13. Ullman, J.D.: The performance of a memory allocation algorithm. Technical report 100, Princeton University, Princeton 695 (1971)
14. Veselý, P.: Competitiveness of fit algorithms for black and white bin packing. In: Middle-European Conference on Applied Theoretical Computer Science (2013)
15. Xia, B.Z., Tan, Z.Y.: Tighter bound of the First Fit algorithm for the bin-packing problem. Discrete Appl. Math. **158**(15), 1668–1675 (2010)

Local Search Algorithms for k-Median and k-Facility Location Problems with Linear Penalties

Yishui Wang[1], Dachuan Xu[1](✉), Donglei Du[2], and Chenchen Wu[3]

[1] Department of Information and Operations Research, College of Applied Sciences, Beijing University of Technology, 100 Pingleyuan, Chaoyang District, Beijing 100124, People's Republic of China
wangys@emails.bjut.edu.cn, xudc@bjut.edu.cn

[2] Faculty of Business Administration, University of New Brunswick, Fredericton, NB E3B 5A3, Canada
ddu@unb.ca

[3] College of Science, Tianjin University of Technology, Tianjin 300384, People's Republic of China
wu_chenchen_tjut@163.com

Abstract. We present two local search algorithms for the k-median and k-facility location problems with linear penalties (k-MLP and k-FLPLP), two extensions of the classical k-median and k-facility location problems respectively. We show that the approximation ratios of these two algorithms are $3+2/p+\epsilon$ for the k-MLP, and $2+1/p+\sqrt{3+2/p+1/p^2}+\epsilon$ for the k-FLPLP, respectively, where $p \in \mathbb{Z}_+$ is a parameter of the algorithms and $\epsilon > 0$ is a positive number. In particular, the $(3+2/p+\epsilon)$-approximation improves the best known 4-approximation for the k-MLP for any $p > 2$.

Keywords: Local search · Approximation algorithm · k-median · k-facility location · Penalty

1 Introduction

Facility location problem is one of the most important problems in the area of combinatorial optimization, and it has numerous applications in computer science, industrial engineering, and operations management etc. The uncapacitated facility location problem (UFLP) is a classical location problem, in which we are given a set of facilities $\mathcal{F}$ with $|\mathcal{F}| = n$, a set of clients $\mathcal{D}$ with $|\mathcal{D}| = m$, connection costs c_{ij} for all $i \in \mathcal{F}$ and $j \in \mathcal{D}$, and facility costs f_i for all $i \in \mathcal{F}$. The objective is to open some facilities $S \subseteq \mathcal{F}$ and connect each client to an opened facility, such that the total connection and facility cost is minimized. In the metric case, the connection costs c are in a given metric space $(\mathcal{F} \bigcup \mathcal{D}, c)$, satisfying nonnegativity, symmetry, and triangle inequalities. From now on we only consider metric connection costs.

Z. Lu et al. (Eds.): COCOA 2015, LNCS 9486, pp. 60–71, 2015.
DOI: 10.1007/978-3-319-26626-8_5

There are two important variants of the classical UFLP. The first is the k-median problem. In contrast to the UFLP, the k-median problem incurs no facility costs ($f_i = 0$ for all $i \in \mathcal{F}$), and opens no more than k facilities. The second is the k-facility location problem (k-FLP), which is similar to the k-median problem except that opening facilities may incur non-zero costs.

Both the k-median problem and the k-FLP are NP-hard. Therefore there have been many studies focusing on the design of approximation algorithms for these two problems. For the k-median problem, Charikar et al. [6] apply the LP-rounding technique to give a $6\frac{2}{3}$-approximation algorithm, the first constant approximation for this problem. Subsequently, several approximation algorithms are presented based on LP-rounding [5,9], primal-dual [10,11], and local search [2]. The currently best known approximation ratio of $2.611 + \epsilon$ is due to Byrka et al. [5] based on LP-rounding and primal-dual techniques. Jain et al. [10] prove that no algorithm can achieve approximation ratio better than $1 + 2/e \approx 1.735$ unless P = NP for the k-median problem.

For the k-FLP, the first approximation algorithm with ratio 6 is given by Jain and Vazirani [11], based on the primal-dual scheme. Jain et al. [10] further combine the greedy process and the factor-revealing LP technique to improve the approximation ratio to 4. Zhang [14] offers the currently best known approximation ratio $2 + \sqrt{3} + \epsilon$ based on local search technique. Since the k-FLP is an extension of k-median problem, $1+2/e \approx 1.735$ is also a lower bound for k-FLP.

This work incorporates penalty cost to consider the k-median problem with linear penalties (k-MPLP) and the k-FLP with linear penalties (k-FLPLP). Penalty cost $p_j \in \mathcal{D}$ is incurred whenever client j is denied service and this cost is linear, namely for any subset $T \subseteq \mathcal{D}$, $p(T) = \sum_{j \in T} p_j$. The objective is to minimize the total connection and penalty cost in the former problem and the total connection, facility and penalty cost in the latter. These two problems are evidently extensions of the k-median and the k-FLP where $p_j = +\infty$ for all $j \in \mathcal{D}$, and hence also NP-hard. The only extant result for these two problems is a primal-dual based approximation algorithm with ratio 4 for the metric k-MPLP due to Charikar et al. [7].

Combinatorial optimization problems with penalty cost have been widely investigated in the literature, including the facility location problem with penalties [7,8,12], the scheduling problem with rejection [4,13], and the price-collecting Steiner tree problem [1,3], among others.

In this paper, we offer a $(3+2/p+\epsilon)$-approximation algorithm for the metric k-MPLP and a $(2+1/p+\sqrt{3+2/p+1/p^2}+\epsilon)$-approximation algorithm for the metric k-FLPLP, where $p \in \mathbb{Z}_+$ is a parameter of the algorithms and $\epsilon > 0$ is a positive number, utilizing the local search techniques from Arya et al. [2] and Zhang [14], respectively.

The rest of this paper is organized as follows. In Sect. 2, we introduce the general local search algorithm. In Sects. 3 and 4, we present the local search algorithms for the k-MPLP and k-FLPLP respectively. In Sect. 5, we improve the algorithms in Sects. 3 and 4 to polynomial-time. All the proofs are deferred to the journal version of this paper.

2 General Local Search Algorithm

The main idea of local search algorithm is to move from solution to its neighbouring solution iteratively with improved cost. Formally, for any feasible solution X, define its *neighborhood* $N(X)$ and its cost $cost(X)$. Then a local search algorithm can be described as the following pseudo-code.

Local search algorithm $A1$

1. Give an initial feasible solution X_0.
2. $X \leftarrow X_0$.
3. While $\exists X' \subseteq N(X)$ such that $cost(X') < cost(X)$
$X \leftarrow X'$.
Endwhile
4. Return X.

For any problem instance I, the final solution produced by a local search algorithm is called a *local* optimal solution along with its *local* optimal value. Denote $local(I, X)$ and $global(I)$ as the local optimal value produced by a local search algorithm from the initial solution X and the global optimal value of the instance I respectively. The *local gap* of the local search algorithm is defined as

$$\sup_{I,X} \frac{\text{local}(I, X)}{\text{global}(I)}.$$

In Sects. 3 and 4, we will present two local search algorithms for the metric k-MPLP and k-FLPLP with local gaps at most $3 + 2/p$ and $2 + 1/p + \sqrt{3 + 2/p + 1/p^2}$ respectively.

3 Local Search for k-MPLP

For the location problems considered in this work, any subset of opened facilities $X \subseteq \mathcal{F}$ represents a solution because the assignment of clients to facilities afterwards can be easily achieved with the minimum connection cost once the opened facilities are fixed.

For the the k-median problem, Arya et al. [2] define the neighborhood of any feasible solution X as

$$N(X) := \{(X \backslash A) \cup B : A \subseteq X, B \subseteq \mathcal{F}, \text{ and } 1 \leq |\mathrm{A}| = |\mathrm{B}| \leq \mathrm{p}\} \quad (1)$$

where $p \leq k$ is a given positive integer. The local search operation $(X \backslash A) \cup B$ therein is called the *swap* of A and B, denoted as *swap*(A, B). Arya et al. [2] consider two types of swap, namely, single-swap ($p = 1$) and multi-swap ($p > 1$), with local gaps of 5 and $3 + 2/p$ respectively.

Note that there must exist an optimal solution O such that $|O| = k$ in the k-median problem as we can add facilities in absence of any opening cost. Thus, it is reasonable to start from a feasible solution with $|X| = k$ and apply the operation $swap(A, B)$ with $|A| = |B|$ in every step of the local search procedure.

We now apply the local search algorithm $A1$ to k-MPLP using the same neigborhood $N(X)$ defined in (1) with the penalized total cost:

$$cost(X) := cost_s(X) + cost_p(X),$$

where $cost_s(X)$ and $cost_p(X)$ are the connection and penalty costs respectively.

3.1 Analysis

The main idea to establish the local gap α between the local optimal solution X and the global optimal solution O, namely $cost(X) \leq \alpha cost(O)$, is to focus on some specific swaps between X and O. Each of these swaps satisfies an inequality due to the local optimality of X, and the sum of these inequalities leads to the desired local gap.

We follow similar analysis in Arya et al. [2] with the further complication of penalty cost. For our purpose, we view the penalty cost of each client j as the connection cost between j and a dummy facility d with $c_{dj} := p_j$. However these new connection costs may violate the triangle inequality, a crucial property in the analysis of [2] for the k-median problem. To overcome this hurdle, we divide $\mathcal{D}$ into four subsets according to whether a client is penalized in X or O. Only those clients that are not penalized in both X and O require the triangle inequality property to carry through the analysis.

From now on, we use d and d^* to represent the dummy facility in the solution X and O respectively. We need the following notations.

- $\sigma(j)$ and $\sigma^*(j)$: the facilities (including the dummy facility) serving the client j in the local optimal solution X and the global optimal solution O respectively.
- $\mathcal{D}_\sigma(A) := \{j \in \mathcal{D} : \sigma(j) \in A\}$ for $A \subseteq X \cup \{d\}$, and $\mathcal{D}_{\sigma^*}(B) := \{j \in \mathcal{D} : \sigma^*(j) \in B\}$ for $B \subseteq O \cup \{d^*\}$.
- $\mathcal{D}_B^A := \mathcal{D}_\sigma(A) \cap \mathcal{D}_{\sigma^*}(B)$.
- $\mathcal{D}_s := \mathcal{D}_\sigma(X)$, $\mathcal{D}_p := \mathcal{D}_\sigma(\{d\})$, $\mathcal{D}_s^* := \mathcal{D}_{\sigma^*}(O)$, and $\mathcal{D}_p^* := \mathcal{D}_{\sigma^*}(\{d^*\})$.
- X_j and O_j: the connection cost of the client j in X and O respectively for any unpenalized client j.
- $C(A) := \{y \in O : |\mathcal{D}_\sigma(A) \cap \mathcal{D}_{\sigma^*}(y)| > \frac{1}{2}|\mathcal{D}_{\sigma^*}(y) \backslash \mathcal{D}_p|\}$. We say that A captures facility y if $y \in C(A)$, and A captures B if $B \subseteq C(A)$.

For convenience, we abbreviate the notation of a single-element set $\{i\}$ to i whenever there is no confusion.

We call a facility $i \in X$ *good* If i does not capture any facility, i.e. $C(i) = \emptyset$; otherwise we call it *bad*. Let $X' = \{b_1, b_2, \cdots, b_{r-1}\}$ be the set of all bad facilities. We partition $X = \dot{\cup}_{t=1}^{r} A_t$ and $O = \dot{\cup}_{t=1}^{r} B_t$ based on a similar procedure in Arya et al. [2]:

Procedure of partitioning X and O for k-MPLP

For $t = 1$ to $r - 1$
  Let $A_t = \{b_t\}$.
  While $|A_t| < |C(A_t)|$
    Add a facility $i \in X \backslash (A_1 \cup \cdots \cup A_t \cup X')$ to A_t.
  End while
  Let $B_t = C(A_t)$.
End for
Let $A_r = X \backslash (A_1 \cup \cdots \cup A_{r-1})$, $B_r = O \backslash (B_1 \cup \cdots \cup B_{r-1})$.

It is easy to see that each of $A_1, \cdots, A_{r-1}$ contains exactly one bad facility; A_r contains only good facilities; and $|A_t| = |B_t|$ for $t = 1, \cdots, r$.

Now we consider some specific swaps between A_t and B_t for $t = 1, \cdots, r$. There are two cases depending on the number of facilities in A_t and B_t:

1. If $|A_t| = |B_t| \leq p$, we consider *swap*(A_t, B_t).
   1.1 $\mathcal{D}_\sigma(A_t) \cap \mathcal{D}_p^* \neq \emptyset$. In this case the clients in $\mathcal{D}_\sigma(A_i) \cap \mathcal{D}_p^*$ are penalized after the swap operation.
   1.2 $\mathcal{D}_\sigma(A_t) \cap \mathcal{D}_p^* = \emptyset$. In this case no client in $\mathcal{D}_\sigma(A_t)$ is penalized after the swap operation.
2. If $|A_t| = |B_t| = q > p$, we consider the single-swaps *swap*(i, o) for each good facility $i \in A_t$ and each facility $o \in B_t$. Note that there are $q-1$ good facilities in A_t and q facilities in B_t, implying that there are $q(q-1)$ swaps in total. Similar to the case 1, some clients in $\mathcal{D}_\sigma(i)$ may be penalized after the swap operation.

A local operation *swap*(A, B) may reassign a client in $\mathcal{D}_\sigma(A) \cup \mathcal{D}_{\sigma^*}(B)$ to another facility. We therefore need to bound the new connection cost by using the bijection $\pi : \mathcal{D}_{\sigma^*}(o) \backslash \mathcal{D}_p \mapsto \mathcal{D}_{\sigma^*}(o) \backslash \mathcal{D}_p$ for each $o \in O$. With a given partition of $X = \cup_{t=1}^r A_t$, the bijection π is constructed via the following process.

We partition $\mathcal{D}_{\sigma^*}(o) \backslash \mathcal{D}_p = \cup_{t=1}^r \mathcal{D}_{A_t}^o$. Renumber the clients in $\mathcal{D}_{\sigma^*}(o) \backslash \mathcal{D}_p$ as $\{j_1, \cdots, j_l\}$, such that clients in $\mathcal{D}_i^o$ for all $i \in X$ are numbered consecutively, and clients in $\mathcal{D}_{A_t}^o$ for all $t \in \{1, \cdots, r\}$ are numbered consecutively. If o is not captured by any facility, let $\pi(j_s) = j_{s'}$ for every $j_s \in \{j_1, \cdots, j_l\}$, where $s' = 1 + (s + \lfloor l/2 \rfloor - 1)$ modulo l. Otherwise, o must be captured by only one facility by definition; we denote this facility as i_1, and assume that $\mathcal{D}_{i_1}^o = \{j_1, \cdots, j_{l_1}\}$ without loss of generality. Since $l_1 > \frac{1}{2} l$, we can construct $|l - l_1|$ mutual mappings between j_s and j_{s+l_1} for $s = 1, \cdots, l - l_1$. For the remaining $|2l_1 - l|$ clients $\{j_{l-l_1+1}, \cdots, j_{l_1}\}$, let $\pi(j_s) = j_{s'}$ where $s' = 1 + (s + \lfloor (2l_1 - l)/2 \rfloor - 1)$ modulo $(2l_1 - l)$. It is easy to prove that bijection π has the following properties.

1. $\sigma^*(\pi(j)) = \sigma^*(j)$.
2. If $\sigma(\pi(j)) = \sigma(j)$, then $\sigma(j)$ captures $\sigma^*(j)$.
3. If $\sigma(j) \in A_t$, $\sigma(\pi(j)) \in A_t$ for some $t \in \{1, 2, \cdots, r\}$, then A_t captures $\sigma^*(j)$.
4. π is a bijection on $\mathcal{D}_s \cap \mathcal{D}_s^*$, and also on $\{j \in \mathcal{D}_s \cap \mathcal{D}_s^* : \pi(j) \notin \mathcal{D}_s(\sigma(j))\}$.

From the above properties we have the following two lemmas.

Lemma 1. *Given a facility $o \in O$, let A_t $(t \in \{1, 2, \cdots, r\})$ be a set such that $\mathcal{D}^o_{A_t} \neq \emptyset$ and does not capture o. The removal of A_t from X results in a new connection (or penalty) cost of each client $j \in \mathcal{D}^o_{A_t}$ being bounded by $O_j + O_{\pi(j)} + X_{\pi(j)}$.*

Lemma 2. *Given a facility $o \in O$, let facility i $(i \in X)$ be a set such that $\mathcal{D}^o_i \neq \emptyset$ and does not capture o. The removal of A_t from X results in a new connection (or penalty) cost of each client $j \in \mathcal{D}^o_i$ being bounded by $O_j + O_{\pi(j)} + X_{\pi(j)}$.*

To analyze the new cost after *swap*(A, B), we only need to focus on the clients in $\mathcal{D}_\sigma(A) \cup \mathcal{D}_{\sigma^*}(B)$. We consider four cases.

Case 1. $\sigma^*(j) \in B$ and $\sigma(j) \neq d$. Let $i^*(j)$ be the closest facility to j in the new solution. Then we have $c_{i^*(j),j} \leq O_j$ from $\sigma^*(j) \in B$. So we can use O_j to bound the new connection cost. In this case, the upper bound of the increased cost is $O_j - X_j$.

Case 2. $\sigma^*(j) \in B$ and $\sigma(j) = d$. In this case, $\sigma^*(j)$ must be the closest facility to j because $c_{ij} \geq p_j$ for all $i \in X$ and $p_j \geq O_j$. So the increased cost is $O_j - p_j$.

Case 3. $\sigma^*(j) \notin B \cup \{d^*\}$ and $\sigma(j) \in A$. We do not know which facility is closest to j in the new solution, and we can not use O_j or p_j to bound the cost because $\sigma^*(j)$ is not in the new solution. However, we know that $\mathcal{D}^{\sigma^*(j)}_A \neq \emptyset$ and $C(A) = B$ if $|A| > 1$, or $A = \{i\}$ such that i does not capture any facility. We use Lemmas 1 and 2 for these two cases respectively to obtain an upper bound of the increased cost: $O_j + O_{\pi(j)} + X_{\pi(j)} - X_j$.

Case 4. $\sigma^*(j) = d^*$ and $\sigma(j) \in A$. It is clear that j is penalized or connected to a facility i with $c_{ij} \leq p_j$. So $p_j - X_j$ is an upper bound of the increased cost in this case.

Combining the above four cases, we have the following theorem.

Theorem 1. *For the k-MPLP, the local gap of algorithm A1 with $N(X)$ defined in (1) is at most $3 + 2/p$.*

4 Local Search for k-FLPLP

For the k-FLP, adding or dropping facilities is not costless. For this reason, Zhang et al. [14] provide a local search algorithm for the k-FLP with a local gap of $2 + \sqrt{3}$, by introducing the new operations *swap, add*, and *drop*:

- *swap*(A, B): for a solution X, swap the subset $A \subseteq X$ and $B \subseteq \mathcal{F} \backslash X$, i.e. $X \leftarrow (X \backslash A) \cup B$.
- *add*(i): add the facility $i \in \mathcal{F} \backslash X$ to X, i.e. $X \leftarrow X \cup \{i\}$.
- *drop*(i): drop the facility $i \in X$, i.e. $X \leftarrow X \backslash \{i\}$.

Consequently, the neighborhood $N(X)$ with $|X| \leq k$ is defined as

$$N(X) := \{(X \backslash A) \cup B : A \subseteq X, B \subseteq \mathcal{F}, \text{ and } 1 \leq |\mathrm{A}| = |\mathrm{B}| \leq \mathrm{p}\} \cup \{X \cup \{i\} : i \in \mathcal{F} \backslash X, |X \cup \{i\}| \leq k\} \cup \{X \backslash \{i\} : i \in X\}. \tag{2}$$

In our work, we apply the local search algorithm $A1$ to the k-FLPLP, using $N(X)$ defined above and $cost(X) := cost_f(X) + cost_s(X) + cost_p(X)$ where $cost_f(X)$, $cost_s(X)$ and $cost_p(X)$ are the facility, connection and penalty costs respectively.

4.1 Analysis

Similar to the analysis for the k-MPLP in Sect. 3, we need to construct some local operations to establish the relationship between the local optimal solution and the global optimal solution. Our analysis is similar to that in Zhang [14] with the complication of penalty cost.

We first cite the following lemma from [14], modified to include penalty cost.

Lemma 3. *For a given $i \in X$, let $o \in O$ be the closest facility captured by i and o' another facility captured by i. For each client $j \in \mathcal{D}_i^{o'}$ such that $\pi(j) = \mathcal{D}_\sigma(i)$, the new connection (or penalty) cost of j is no more than $2X_j + O_j$ after $swap(i, o)$.*

To bound the facility cost, let $X' = \{b_1, \cdots, b_{|X'|}\}$ be the set of all bad facilities and partition X and O using the following procedure similar to that in [14]. The bijection $\pi : \mathcal{D} \mapsto \mathcal{D}$ with this partition can be constructed using the same method in Sect. 3 earlier. Let $X = \dot{\cup}_{t=1}^{r} A_t$ be this partition. Then Lemmas 1 and 2 also hold for this partition.

Procedure of partitioning X and O for the analysis of facility cost for k-FLPLP

For $t = 1$ to $|X'|$
  Let $A_t = \{b_t\}$ and $B_t = C(A_t)$.
  While $|A_t| < |B_t|$
    Add a facility $i \in X \backslash (A_1 \cup \cdots \cup A_t \cup X')$ (if it exists) to A_t.
  End while
End for
Let $B_{|X'|+1} = O \backslash (B_1 \cup \cdots \cup B_{|X'|})$.
If $|X \backslash (A_1 \cup \cdots \cup A_{|X'|})| \geq |B_{|X'|+1}|$
  $A_{|X'|+1} =$ the set of arbitrary $|B_{|X'|+1}|$ facilities in $X \backslash (A_1 \cup \cdots \cup A_{|X'|})$;
  $A_{|X'|+2} = X \backslash (A_1 \cup \cdots \cup A_{|X'|+1}) =: R$.
Else if
  $A_{|X'|+1} = X \backslash (A_1 \cup \cdots \cup A_{|X'|})$.
End if

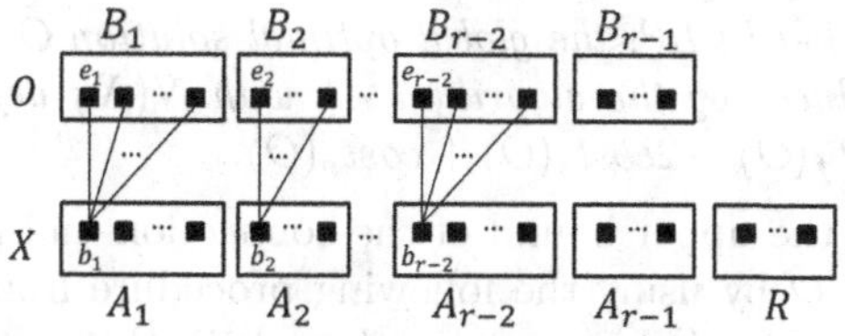

Fig. 1. Partition for the analysis of facility cost, case 1: $l_O \leq l_X$.

Let $l_O = |O|$ and $l_X = |X|$. The partition involves two cases: $l_O \leq l_X$ and $l_O > l_X$.

For $l_O \leq l_X$ (Fig. 1), let $X := A_1 \dot\cup \cdots \dot\cup A_{r-1} \dot\cup R$ and $O := B_1 \dot\cup \cdots \dot\cup B_{r-1}$. According to the partition procedure, each A_t ($t \in \{1, \cdots, r-2\}$) contains one bad facility denoted by b_t, while none of each $B_t = C(b_t)$ and A_{r-1} contains bad facility. Moreover, $|A_t| = |B_t|$ for $t \in \{1, \cdots, r-1\}$, and R is the remaining set such that $|R| = |X| - |O|$. Without loss of generality, assume that A_{r-1}, B_{r-1}, and R are non-empty. Denote e_t as the closest facility to b_t in B_t, and then consider the following local operations.

1. Consider *drop*(i) for facility $i \in R$.
2. Consider *swap*(b_t, e_t) for $(A_t, B_t), t \in \{1, \cdots, r-2\}$ and consider *swap*(i, o) for the remaining $|A \backslash \{b_t\}|$ pairs in which every facility is swapped only once.
3. Consider $|A_{r-1}|$ single-swaps *swap*(i, o) for $i \in A_{r-1}$ and $o \in B_{r-1}$ where every facility is only swapped once.

For $l_O > l_X$, let $X := A_1 \dot\cup \cdots \dot\cup A_r$ and $O := B_1 \dot\cup \cdots \dot\cup B_r$. Each A_t ($t \in \{1, \cdots, r-1\}$) contains one bad facility; each B_t is the corresponding captured set; A_r contains no bad facility; and B_r is the uncaptured facility set. Let P be the union of the sets formed by any $|B_t| - |A_t|$ facilities in $B_t \backslash \{e_1, \cdots, e_{r-1}\}$ for $t \in \{1, \cdots, r\}$. Note that there are two subcases: $A_r = \emptyset$ and $A_r \neq \emptyset$ (Fig. 2(a) and (b) respectively). For the facilities not in P, we apply swap operations similar to that of $l_O \leq l_X$, while for each facility $o \in P$, we apply the *add*(o) operation.

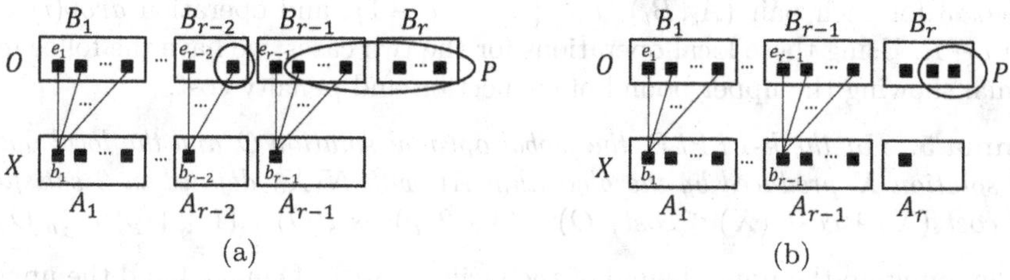

Fig. 2. Partition for the analysis of facility cost, case 2: $l_O > l_X$.

Combining the local operations specified earlier for these two cases (namely $l_O \leq l_X$ and $l_O > l_X$), the following lemma bounds the facility cost of X.

Lemma 4. *For the k-FLPLP, the global optimal solution O and the local optimal solution X produced by the algorithm A1 with N(X) defined in 2 satisfies that* $cost_f(X) \leq cost_f(O) + 2cost_s(O) + cost_p(O)$.

Next we analyze the upper bound of the connection and penalty cost of X. We partition X and O by using the following procedure from [14] for the case of $l_O \leq l_X$ (Fig. 3). Note that except for the set R, this partition is the same as that for k-MPLP in Sect. 3. So similar properties apply and Lemmas 1 and 2 also hold for this partition.

Procedure of partitioning X and O for the analysis of connection and penalty cost for k-FLPLP

For $t = 1$ to $|X'|$
  Let $A_t = \{i_t\}$.
  While $|A_t| < |C(A_t)|$
    Add a facility $i \in X\backslash(A_1 \cup \cdots \cup A_t \cup X')$ to A_t.
  End while
  Let $B_t = C(A_t)$.
End for
Let $B_{|X'|+1} = O\backslash(B_1 \cup \cdots \cup B_{|X'|})$;
  $A_{|X'|+1} =$ the set of arbitrary $|B_{|X'|+1}|$ facilities in $X\backslash(A_1 \cup \cdots \cup A_{|X'|})$.
Let $A_{|X'|+2} = X\backslash(A_1 \cup \cdots \cup A_{|X'|+1}) =: R$.

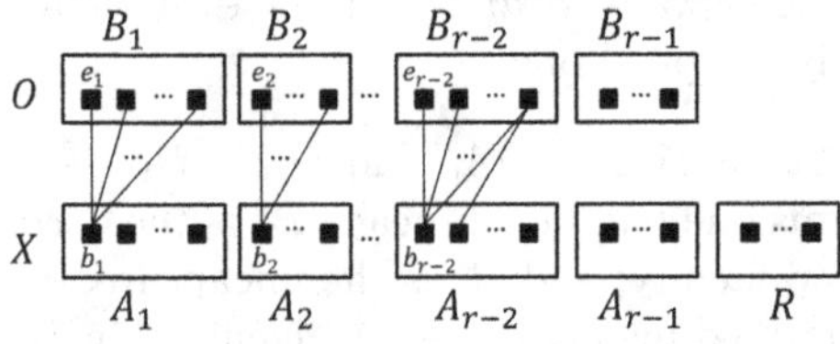

Fig. 3. Partition for the analysis of connection and penalty cost, case 1: $l_O \leq l_X$.

For the case of $l_X > l_O$, we apply operation *add*(o) for each $o \in O$. For the case of $l_X \leq l_O$, we consider the swap operations similar to that for k-MPLP in Sect. 3 for each pair (A_t, B_t), $t \in \{1, \cdots, r-1\}$, and operation *drop*(i) for each $i \in R$. Using these local operations for the two cases, we have the following lemma, showing the upper bound of connection and penalty cost.

Lemma 5. *For the k-FLPLP, the global optimal solution O and the local optimal solution X produced by the algorithm A1 with N(X) defined in 2 satisfies that* $cost_s(X) + cost_p(X) \leq cost_f(O) + (3 + 2/p)cost_s(O) + (1 + 1/p)cost_p(O)$.

Summing up the upper bound of the facility cost in Lemma 4 and the upper bound of the connection and penalty cost in Lemma 5, we get

$$cost(X) \leq 2cost_f(O) + \left(5 + \frac{2}{p}\right) cost_s(O) + \left(2 + \frac{1}{p}\right) cost_p(O).$$

Finally, we get the following theorem.

Theorem 2. *For the metric k-FLPLP, the local gap of the algorithm A1 with N(X) defined in 2 is at most* $5 + 2/p$.

4.2 Improve the Local Gap Using Scaling Technique

In Sect. 4.1, the local gap of the algorithm $A1$ is given by $\max(2, 5+2/p, 2+1/p)$, where the three numbers therein are the factors of facility, connection, and penalty costs of O respectively for bounding the cost of X. These three factors are different, implying that there are room to improve the local gap by scaling the input of the instance. We construct a new instance I' by scaling the facility costs in the original instance I, namely, in the instance I', $f_i' = \delta f_i, c_{ij}' = c_{ij}, p_j' = p_j$ for $i \in \mathcal{F}, j \in \mathcal{D}$, where $\delta > 0$ is a constant to be determined later.

Define $cost_f(I, X)$, $cost_s(I, X)$, $cost_p(I, X)$, and $cost(I, X)$ as the facility, connection, penalty, and total costs of the solution X for instance I respectively. From the analysis in Sect. 4.1, we know that after replacing the optimal solution O with any feasible solution S, Lemmas 4 and 5 still hold; that is, for any instance I with the local optimal solution X and any feasible solution S, we have

$$cost_f(I, X) \leq cost_f(I, S) + 2cost_s(I, S) + cost_p(I, S);$$

$$cost_s(I, X) + cost_p(I, X) \leq cost_f(I, S) + \left(3 + \frac{2}{p}\right) cost_s(I, S) + \left(1 + \frac{1}{p}\right) cost_p(S).$$

Now we have the following theorem.

Theorem 3. *For the k-FLPLP, the local gap of Algorithm A1 with N(X) defined in 2 to the scaled instance is at most* $2 + \frac{1}{p} + \sqrt{3 + \frac{2}{p} + \frac{1}{p^2}}$.

5 Polynomial-Time Algorithm for k-MPLP and k-FLPLP

The local search approach $A1$ is not a polynomial-time algorithm because it may not reduce sufficient cost at each local search step, resulting in exponential number of iterations. To solve the problem in polynomial-time, we modify the local search condition $cost(X') < cost(X)$ to $cost(X') < (1 - \epsilon/Q)cost(X)$ where $\epsilon > 0$ is a constant and Q is the number of local operations for the analysis of local gap. We call this modified algorithm $A2$. The following lemma is from [2].

Lemma 6 (Arya et al. [2]). *The number of local search steps in algorithm A2 is at most* $\log(cost(X_0)/cost(O))/\log\frac{1}{1-\epsilon/Q} \leq \log(cost(X_0)/cost(O)) \cdot \frac{1}{\epsilon \log e} Q$. *If the local gap of algorithm A1 is* α, *then the local gap of algorithm A2 is at most* $\frac{\alpha}{1-\epsilon} = \alpha + \epsilon'$, *where* $\epsilon' = \frac{\alpha\epsilon}{1-\epsilon} \sim \epsilon$.

It is easy to see that the number Q is at most $n^2 + n$ for both k-MLP and k-FLPLP. Moreover, at each local search step, $A2$ searches for at most $|N(X)| = O(n^p)$ solutions, and the time for calculating the cost is $O(mn)$. So by Lemma 6 we have the following theorem.

Theorem 4. *The time complexity of algorithm A2 is* $O(\frac{1}{\epsilon}\log(cost(X_0)/cost(O))\cdot m\cdot n^3)$ *for both k-MLP and k-FLPLP. The approximation ratio of algorithm A2 is* $3+2/p+\epsilon'$ *for k-MLP, and* $2+1/p+\sqrt{3+2/p+1/p^2}+\epsilon'$ *for k-FLPLP. In particular, when the parameter p is large enough, the approximation ratios for the two problems are* $3+\epsilon'$ *and* $2+\sqrt{3}+\epsilon'$ *respectively.*

Acknowledgements. The research of the first author is supported by Collaborative Innovation Center on Beijing Society-Buliding and Social Governance. The second author's research is supported by NSFC (Nos. 11371001 and 11531014). The third author's research is supported by the Natural Sciences and Engineering Research Council of Canada (NSERC) grant 283106. The fourth author's research is supported by NSFC (No. 11501412).

References

1. Archer, A., Bateni, M., Hajiaghayi, M., Karloff, H.: Improved approximation algorithms for prize-collecting Steiner tree and TSP. SIAM J. Comput. **40**, 309–332 (2011)
2. Arya, V., Garg, N., Khandekar, R., Meyerson, A., Munagala, K., Pandit, V.: Local search heuristics for k-median and facility location problems. SIAM J. Comput. **33**, 544–562 (2004)
3. Bienstock, D., Goemans, M.X., Simchi-Levi, D., Williamson, D.: A note on the prize collecting traveling salesman problem. Math. Program. **59**, 413–420 (1993)
4. Bartal, Y., Leonardi, S., Marchetti-Spaccamela, A., Sgall, J., Stougie, L.: Multiprocessor scheduling with rejection. SIAM J. Discrete Math. **13**, 64–78 (2000)
5. Byrka, J., Pensyl, T., Rybicki, B., Srinivasan, A., Trinh, K.: An improved approximation for k-median, and positive correlation in budgeted optimization. In: Proceedings of the 26th Annual ACM-SIAM Symposium on Discrete Algorithms, pp. 737–756 (2015)
6. Charikar, M., Guha, S., Tardos, É., Shmoys D.B.: A constant-factor approximation algorithm for the k-median problem. In: Proceedings of the 31st Annual ACM Symposium on Theory of Computing, pp. 1–10 (1999)
7. Charikar, M., Khuller, S., Mount, D.M., Narasimhan, G.: Algorithms for facility location problems with outliers. In: Proceedings of the 20th Annual ACM-SIAM Symposium on Discrete Algorithms, pp. 642–651 (2001)
8. Gupta, N., Gupta, S.: Approximation algorithms for capacitated facility location problem with penalties (2014). ArXiv: 1408.4944
9. Czumaj, A., Mehlhorn, K., Pitts, A., Wattenhofer, R.: A dependent LP-rounding approach for the k-median problem. In: Charikar, M., Li, S. (eds.) ICALP 2012. LNCS, vol. 7391, pp. 194–205. Springer, Heidelberg (2012)
10. Jain, K., Mahdian, M., Markakis, E., Saberi, A., Vazirani, V.V.: Greedy facility location algorithms analyzed using dual fitting with factor-revealing LP. J. ACM. **50**, 795–824 (2003)
11. Jain, K., Vazirani, V.V.: Approximation algorithms for metric facility location and k-median problems using the primal-dual schema and Lagrangian relaxation. J. ACM. **48**, 274–296 (2001)
12. Li, Y., Du, D., Xiu, N., Xu, D.: Improved approximation algorithms for the facility location problems with linear/submodular penalties. Algorithmica **73**, 460–482 (2015)

13. Shabtay, D., Gaspar, N., Kaspi, M.: A survey on offline scheduling with rejection. J. Sched. **16**, 3–28 (2013)
14. Zhang, P.: A new approximation algorithm for the k-facility location problem. Theor. Comput. Sci. **384**, 126–135 (2007)

A $(5.83 + \epsilon)$-Approximation Algorithm for Universal Facility Location Problem with Linear Penalties

Yicheng Xu[1], Dachuan Xu[1](✉), Donglei Du[2], and Chenchen Wu[3]

[1] Department of Information and Operations Research, College of Applied Sciences, Beijing University of Technology, 100 Pingleyuan, Chaoyang District, Beijing 100124, People's Republic of China
ycxu@emails.bjut.edu.cn, xudc@bjut.edu.cn

[2] Faculty of Business Administration, University of New Brunswick, Fredericton, NB E3B 5A3, Canada
ddu@unb.ca

[3] College of Science, Tianjin University of Technology, Tianjin 300384, People's Republic of China
wu_chenchen_tjut@163.com

Abstract. In the universal facility location problem, we are given a set of clients and facilities. Our goal is to find an assignment such that the total connection and facility cost is minimized. The connection cost is proportional to the distance between each client and its assigned facility, whereas the facility cost is a nondecreasing function with respect to the total number of clients assigned to the facility. The universal facility location problem generalizes several classical facility location problems, including the uncapacitated facility location problem and the capacitated facility location problem (both hard and soft capacities). This work considers the universal facility location problem with linear penalties, where each client can be rejected for service with a penalty. The objective is to minimize the total connection, facility and penalty cost. We present a $(5.83 + \epsilon)$-approximation local search algorithm for this problem.

Keywords: Local search · Approximation algorithm · Universal facility location · Penalty

1 Introduction

Facility location is one of the most classical and active research topics of combinatorial optimization. In the universal facility location problem, we are given a set of clients $\mathcal{D}$ and a set of facilities $\mathcal{F}$. Each client j served by facility i pays a connection cost c_{ij}, which is assumed to be metric (i.e., nonnegative, symmetric, and satisfying triangle inequalities). Let $f_i(u_i)$ denote the facility cost which is a nondecreasing left-continuous function with respect to its allocated capacity u_i and $f_i(0) = 0$ where u_i depends on the total amount of clients served by facility

Z. Lu et al. (Eds.): COCOA 2015, LNCS 9486, pp. 72–81, 2015.
DOI: 10.1007/978-3-319-26626-8_6

i. The objective is to assign every client to a facility such that the total facility and connection cost is minimized subject to the facility capacity constraint.

This problem generalizes several classical facility location problems such as the uncapacitated facility location problem and capacitated facility location problem (both hard and soft capacities). In the uncapacitated problems we pay a fixed cost for opening each facility, which can serve any number of clients. In the hard-capacitated problems each facility has an upper bound on the amount of demand it can serve. In the soft-capacitated problems each facility has a capacity, but we are allowed to open multiple copies of each facility.

As special cases of universal facility location problems, uncapacitated and soft-capacitated facility location problems have been investigated intensively in the literature based on linear programming relaxation technique. However, this technique has been ineffective in dealing with hard-capacitated or universal cases, mainly because no linear programming formulation of bounded integrality gap was known for hard-capacitated case, until An et al. [2] propose a new multi-commodity-flow relaxation for the capacitated facility location problems, resulting in the first constant approximation ratio of 288 via semi-rounding technique.

On the other hand, local search technique has been effective in handling these problems. Under the assumption of uniform capacities, Korupolu et al. [7] and Chudak et al. [5] give an approximation algorithm with constant approximation guarantee based on local search. The currently best approximation ratio for the hard capacitated facility location problem with uniform capacities is 3, achieved by Aggarwal et al. [1].

For nonuniform hard capacities, Pal et al. [10] give a local search algorithm that achieves a constant approximation ratio of $(9 + \epsilon)$. They propose such operations as *add*, *close* and *open* in their algorithm. Zhang et al. [15] extend their *close/open* operation to a more general *multi-exchange* operation, achieving $(5.83 + \epsilon)$ approximation ratio. It is the first local search algorithm in which the analysis is proved to be tight.

For the universal facility location problem, Mahadian and Pal [9] give the first constant approximation ratio of $(7.88 + \epsilon)$ based on local search algorithm. They employ *add* and *pivot* operations in their work. Vygen [12] improves this result to $(6.702 + \epsilon)$-approximation by extending the *pivot* operation. The currently best approximation ratio is $(5.83 + \epsilon)$ by Angel et al. [3].

Facility location problems with penalties is another extension of the basic facility location problems and have been studied for many years (e.g. [4,13,14]). Both linear and submodular penalties problems have been investigated by Li et al. [8] with approximation ratios 2 and 1.5148, respectively. Gupta and Gupta's work [6] on capacitated facility location problem with linear penalties achieves an approximation ratio of $(5.83 + \epsilon)$ for uniform case and $(8.532 + \epsilon)$ for nonuniform case.

In this paper, we consider the universal facility location problem with linear penalties which generalizes the universal facility location problem and the (capacitated) facility location problem with linear penalties. We design a local search based $(5.83 + \epsilon)$-approximation algorithm which maintains the same approximation ratio for the universal facility location problem. Comparing with the universal facility location problem, we explore the penalty structure in the so-called flow decomposition and exchange graph (cf. Sect. 3).

The rest of this paper is organized as follows. In Sect. 2, we present the local search algorithm for the universal facility location problem with linear penalties. In Sect. 3, we analyze the algorithm and obtain the $6 + \epsilon$ approximation ratio. We further improve the ratio to $5.83 + \epsilon$ in Sect. 4. All the proofs are deferred to the journal version of this paper.

2 Local Search Algorithm

The universal facility location problem with linear penalties can be formulated as the following program.

$$\begin{aligned} \min \ & \sum_{i\in\mathcal{F}} f_i(u_i) + \sum_{i\in\mathcal{F},j\in\mathcal{D}} c_{ij}x_{ij} + \sum_{j\in\mathcal{D}} p_j z_j \\ \text{s.t.} \ & \sum_{i\in\mathcal{F}} x_{ij} + z_j = 1, \quad \forall j \in \mathcal{D}, \\ & \sum_{j\in\mathcal{D}} x_{ij} \le u_i, \quad \forall i \in \mathcal{F}, \\ & x_{ij}, z_j \in \{0,1\}, \quad \forall i \in \mathcal{F}, j \in \mathcal{D}. \end{aligned} \tag{1}$$

Here $f_i(u_i)$ denotes the facility cost function with respect to its allocation u_i and c_{ij} denotes the connection cost from client j to facility i. The set of facility is $\mathcal{F}$ and the set of clients is $\mathcal{D}$. p_j is the penalty cost of client j. The decision variables are x, u, z, where x represents the assignment, u represents the allocation and z_j denotes whether client j is served. Under the assumption that $f_i(\cdot)$ is nondecreasing, the optimal solution $S^* := (x^*, u^*, z^*)$ must satisfy $u_i^* = \sum_{j\in\mathcal{D}} x_{ij}^*$ for each facility i. Therefore, it suffices to denote any solution of the above program by (x, z).

2.1 Operations

The set of operations that will be used in our algorithm are defined as follows.

- ADD(s,δ): Given any solution $S := (x, z)$, add capacity u_s of facility s by δ, and assign clients optimally by solving a linear program. Note that when given allocation u, the program (1) above is a transportation problem with integer parameters, and hence equivalent to its linear relaxation. This operation is widely used in both universal and hard-capacitated facility location problems to bound the service cost. In our work, this "ADD" operation will be used to bound the total service and penalty cost (Lemma 4).
- OPEN(s,δ): Given any solution $S := (x, z)$, increase the capacity u_s by δ from some unserved clients or some other facilities i_1,i_2, ... to s via the shortest path. In other words, capacities of i_1,i_2, ... will decrease.

- CLOSE(t,δ): Given any solution $S := (x, z)$, decrease the capacity u_t from t to some other facilities i_1,i_2, ... via the shortest path or penalize. In other words, capacities of i_1,i_2, ... will increase.
- OPEN-CLOSE(s,t,δ_s,δ_t): Given any solution $S := (x, z)$, increase the capacity u_s by δ_s from some unserved clients or some other facilities i_1,i_2, ... (may include t) to s via the shortest path. Decrease the capacity u_t from t to some other facilities i_1,i_2, ... (may include s) via the shortest path or penalize.

2.2 Polynomial-Time Proof

In this subsection, we will show how to compute the minimum cost of the operations in polynomial time. Note the operation OPEN-CLOSE(s,t,δ_s,δ_t) generalizes the OPEN(s,δ) and CLOSE(t,δ) operation. Thus it suffices to prove the OPEN-CLOSE(s,t,δ_s,δ_t) and ADD(s,δ) operations run in polynomial time.

For ADD(s,δ), let S be the current solution and Y the solution after the ADD(s,δ) operation. We define our estimated cost as

$$c^S(s,\delta) := c_s(Y) + c_p(Y) - c_s(S) - c_p(S) + f_s\left(\sum_{j\in\mathcal{D}} x_{sj} + \delta\right) - f_s\left(\sum_{j\in\mathcal{D}} x_{sj}\right),$$

where $c_s(Y)$ denotes the service cost and $c_p(Y)$ denotes the penalty cost of Y. One can obtain minimum $c_s(Y)+c_p(Y)$ by solving a transportation problem, i.e.,

$$\begin{aligned}
\min\ c_s(Y) + c_p(Y) := & \sum_{i\in\mathcal{F},j\in\mathcal{D}} c_{ij}y_{ij} + \sum_{j\in\mathcal{D}} p_j z_j \\
\text{s.t.}\ & \sum_{i\in\mathcal{F}} y_{ij} + z_j = 1, \qquad \forall j \in \mathcal{D}, \\
& \sum_{j\in\mathcal{D}} y_{ij} \le \sum_{j\in\mathcal{D}} x_{ij}, \qquad \forall i \in \mathcal{F}\setminus\{s\}, \\
& \sum_{j\in\mathcal{D}} y_{sj} \le \sum_{j\in\mathcal{D}} x_{sj} + \delta, \\
& 0 \le y_{ij}, z_j \le 1, \qquad \forall i \in \mathcal{F}, j \in \mathcal{D}.
\end{aligned}$$

Lemma 1. *Given $\epsilon > 0$ and $t \in \mathcal{F}$, let x be a feasible solution to a given instance of the problem. We can find a $\delta \in \mathcal{R}^+$ with $c^S(t,\delta) \le -\epsilon c(x)$ or decide that no $\delta \in \mathcal{R}^+$ exists for which $c^S(t,\delta) \le -2\epsilon c(x)$ in polynomial time.*

To guarantee the polynomial running time of OPEN-CLOSE operation, we have the following lemma.

Lemma 2. *Given the current solution $S := (x, z)$, we can, in polynomial time, find s, t, δ_s, and δ_t such that the cost of* OPEN-CLOSE(s,t,δ_s,δ_t) *is minimized and the minimum cost can be computed in polynomial time in terms of n and m, where $s,t \in \mathcal{F}$, $0 \le \delta_s, \delta_t \le n$, $|\mathcal{F}| := m$, and $|\mathcal{D}| := n$.*

2.3 Algorithm

Let $\epsilon > 0$ be any fixed small constant. Starting with any feasible solution, apply ADD, OPEN, CLOSE and OPEN-CLOSE operations iteratively until the cost of iterated solution can no longer decrease. Then the following lemma implies that we can in polynomial time find an approximate local optimal solution. Note the approximate local optimal solution is only $(1 + \epsilon)$ worse than local optimal one.

Lemma 3. *If we can find* ADD, OPEN, CLOSE *and* OPEN-CLOSE *operation with estimated cost at most* $-c(S)/p(n, \epsilon)$, *where* $p(n, \epsilon)$ *is a suitably chosen polynomial in* n *and* $1/\epsilon$, *then the algorithm terminates after at most* $O\left(p(n, \epsilon) \log \frac{c(S)}{c(S^*)}\right)$ *operations, where* S *denotes the initial solution and* S^* *denotes the optimal solution.*

3 Analysis

To bound the connection and penalty cost of the approximate local optimal solution, we utilize the ADD operation.

Lemma 4. *For all* $\epsilon > 0$, *let* S *and* S^* *be a feasible solution and the optimal solution respectively to a given instance. Let* $c^S(t, \delta) \geq -\frac{\epsilon}{n} c(S)$ *for all* $t \in \mathcal{F}$ *and* $\delta \in \mathcal{R}_+$, *where* n *denotes the number of facilities. Then* $c_s(S) + c_p(S) \leq c_s(S^*) + c_p(S^*) + c_f(S^*) + \epsilon c(S)$.

For bounding facility cost, we recall the concepts of flow decomposition and exchange graph. For a network graph, flows can be represented either on arc or on path and cycle. Flow decomposition technique is to transform a flow from arc representation to path and cycle representation. This can be done in polynomial time respect to the number of nodes and edges (e.g., [11]).

Let S and S^* be the approximate local optimal solution and optimal solution, respectively. We view them as flows in a bipartite graph with vertices corresponding to facilities and clients. For S, the arc direction is from the facilities to the clients, and for S^* the arc direction is opposite. Each edge (i, j) carries $x(i, j) - x^*(i, j)$ units of flow. Negative flow indicates the reverse direction. Applying flow decomposition on this bipartite graph we obtain paths and cycles. A path can only start at a facility and end at a facility, since every client has the same outdegree and indegree. If we are only concerned about the origin and destination, we obtain the so-called exchange graph with only facility vertices. This technique also provides a feasible way to reassign demands from an origin to a destination along the path.

We introduce a dummy facility N for S to denote the penalty facility with zero facility cost and p_j service cost for client j. The corresponding dummy facility for S^* is N^*. We want to replace all capacities in S with capacities in S^* via the exchange graph. We call a path starting at $i \in \mathcal{F}$ and ending at N^* a penalty path. Consider one such path P starting at $s \in \mathcal{F}$. Let s' be the facility just before N^* on this path and j be a client of s'. Define $Pen_j(s, s')$ as the set of all such paths. Let $w(Pen_j(s, s'))$ be the total flow along these paths.

To distinguish paths in different sets, let $N^*_{s',j}$ be the destination of a penalty path corresponding to s' and j. In other words, we introduce at most $|\mathcal{F}||\mathcal{D}|$ dummy copies of facilities instead of one.

The following transportation problem describes the transferring from the approximate local optimal solution $S := (x, z)$ to the optimal solution $S^* := (x^*, z^*)$.

$$\begin{aligned}
\min \quad & \sum_{s\in U^+, t\in U^-} c_{st} y(s,t) + \sum_{s\in U^+, s'\in\mathcal{F}, j\in\mathcal{D}} (c_{ss'} + p_j) y(s, N^*_{s',j}) \\
\text{s.t.} \quad & \sum_{t\in U^-} y(s,t) + \sum_{s'\in\mathcal{F}, j\in\mathcal{D}} y(s, N^*_{s',j}) = u_s - u^*_s, && \forall s\in U^+, \\
& \sum_{s\in U^+} y(s,t) = u^*_t - u_t, && \forall t\in U^-, \qquad (2)\\
& y(s,t) \ge 0, && \forall s\in U^+, t\in U^-, \\
& 0 \le y(s, N^*_{s',j}) \le w(Pen_j(s,s')), && \forall j\in\mathcal{D}, s\in U^+, s'\in\mathcal{F},
\end{aligned}$$

where $U^+ := \{i\in\mathcal{F} : u_i > u^*_i\}$, $U^- := \{i\in\mathcal{F} : u_i < u^*_i\}$.

Lemma 5. *Let opt_{tp} be the optimal value of the transportation problem. We have $opt_{tp} \le c_s(S) + c_p(S) + c_s(S^*) + c_p(S^*)$.*

Now we know that a local optimal with respect to ADD operation always reduces the service and penalty cost. We argue that whenever S has a large facility cost, there will exist an OPEN-CLOSE operation that can improve the cost of S.

Consider the optimal solution y to the transportation problem. Without loss of generality we assume the set of edges with nonzero flow of y forms a forest. Since we can make the cost of a cycle (if exists) zero by taking augmenting techniques afterwards, we are able to remove all cycles one by one.

So we are only concerned about replacing the capacity u of every facility with u^*. We say the OPEN-CLOSE operations that decreases the capacity of some facilities from u_i to u^*_i is *closed*, and that increases some facilities from u_i to at most u^*_i is *opened*. Note that only facilities in U^- can be opened and only facilities in U^+ can be closed. We root each tree at an arbitrary facility in U^-. Now we search for OPEN-CLOSE operations that close facilities in U^+ exactly once and open facilities in U^- as few times as possible.

For a vertex $t\in U^-$, let T_t be the subtree of depth at most 2 rooted at t in the forest. Define $K(t)$ as the set of children of t. Note that the odd level (i.e. U^-) in the tree may contain a penalty facility $N^*_{s',j}$ for some $s'\in U^+$ and $j\in\mathcal{D}$. So let $K(s)$ denote the children of s excluding the penalty facilities. We will analyze every such T_t.

For $t\in U^-$ that is not the penalty facility, i.e., $\forall s'\in U^+, j\in\mathcal{D}, t\neq N^*_{s',j}$, we consider the following cases.

1. For each $s \in NDom(t)$, define $Rem(s) := (y(s,t) - y(s,w(s)))^+$. Order the facilities in nondecreasing sequence of $Rem(s)$, say $1, 2, \ldots, l$. For $i = 1, 2, \ldots, l-1$, Close s_i and open $K(s_i) \cup S(s_{i+1})$; that is, apply operation CLOSE(s_i,$y(s_i, \cdot)$). Note that this operation will increase the capacities of facilities in $K(s_i) \cup S(s_{i+1})$.
2. For the rest of facilities in $K(t)$, i.e. s_l and $Dom(t)$:
   when t is strong. Open $\{t\} \cup K(s_l)$ and close $Dom(t) \cup \{s_l\}$; that is, apply operation OPEN-CLOSE(t,s_l,$y(s_l,t) + \sum_{s \in Dom(t)} y(s,\cdot)$,$y(s_l,\cdot)$).
   when t is weak but there exists $h \in Dom(t)$ such that $y(h,t) \geq \frac{1}{2}y(\cdot,t)$.] Close h and open $K(h) \cup \{t\}$; then close $\{s_l\} \cup Dom(t) \backslash \{h\}$ and open $K(s_l) \cup \{t\}$. That is, apply operations CLOSE(h,$y(h,\cdot)$) and OPEN-CLOSE(t, s_l, $y(s_l,t) + \sum_{s \in Dom(t) \backslash \{h\}} y(s,\cdot)$, $y(s_l,\cdot)$).
   when t is weak and no $h \in Dom(t)$ exists such that $y(h,t) \geq \frac{1}{2}y(\cdot,t)$. If $Dom(t) \neq \varnothing$, we will prove that there exists a facility $\overline{s} \in Dom(t)$ such that the rest of $Dom(t)$ can be partitioned into two parts D_1 and D_2, satisfying $y(\overline{s},t) + \sum_{s \in D_1} y(s,\cdot) \leq y(\cdot,t)$ and $y(s_l,t) + \sum_{s \in D_2} y(s,\cdot) \leq y(\cdot,t)$. In other words, operations OPEN-CLOSE(t,$\overline{s}$,$y(\overline{s},t) + \sum_{s \in D_1} y(s,\cdot)$,$y(\overline{s},\cdot)$) and OPEN-CLOSE ($t$,$s_l$,$y(s_l,t) + \sum_{s \in D_2} y(s,\cdot)$,$y(s_l,\cdot)$) are feasible.

These operations are illustrated in Figs. 1, 2, 3 and 4. If there exist $s' \in U^+$ and $j \in \mathcal{D}$ such that $t = N^*_{s',j}$, then consider operation OPEN(t,$y(\cdot,t)$).

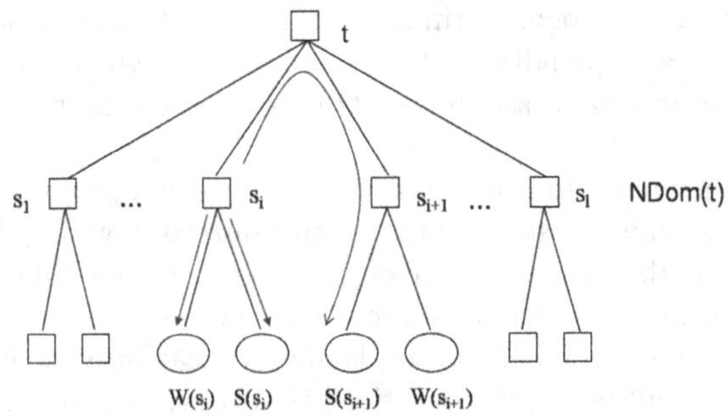

Fig. 1. CLOSE(s_i,$y(s_i,\cdot)$)

Lemma 6. *The operations used in the transferring are all feasible, and we close each facility in U^+ exactly once, open each facility in U^- at most 3 times and the transportation cost of the transfer is bounded by twice the optimal flow y of the transportation problem.*

Therefore the following lemma holds.

Lemma 7. $c_f(S) \leq 4c_s(S^*) + 4c_p(S^*) + 5c_f(S^*)$.

Combining Lemma 7 with the upper bound for $c_s(S)$, we have

Theorem 1. $c(S) \leq 5c_s(S^*) + 5c_p(S^*) + 6c_f(S^*)$.

The last theorem yields an approximation ratio of $(6 + \epsilon)$.

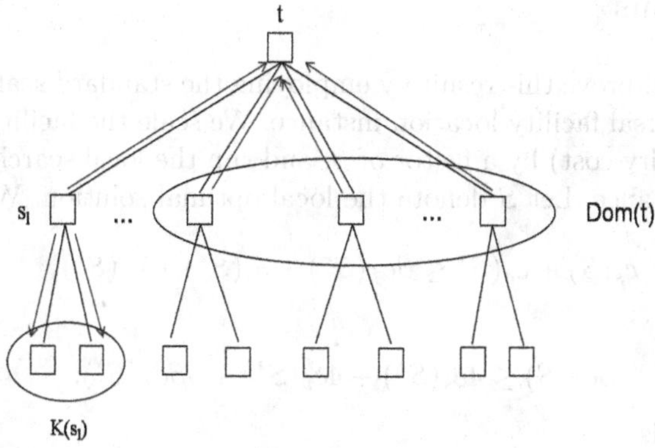

Fig. 2. OPEN-CLOSE(t,s_l,$y(s_l,t)+\sum_{s\in Dom(t)} y(s,\cdot)$,$y(s_l,\cdot)$)

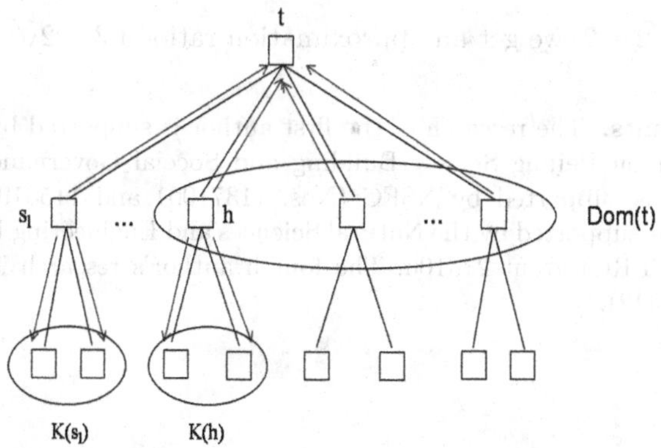

Fig. 3. CLOSE(h,$y(h,\cdot)$) and OPEN-CLOSE(t,s_l,$y(s_l,t)+\sum_{s\in Dom(t)\setminus\{h\}} y(s,\cdot)$,$y(s_l,\cdot)$)

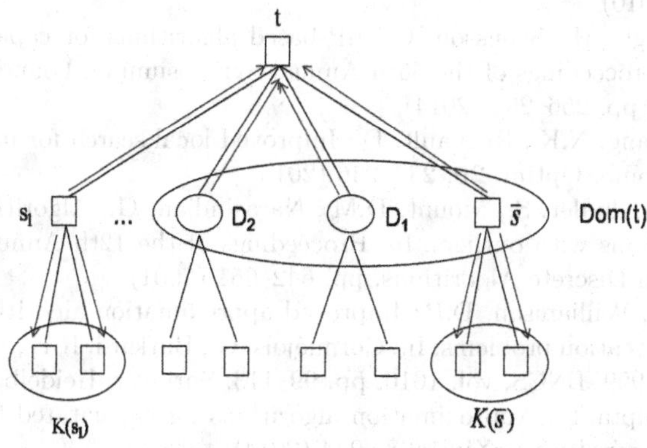

Fig. 4. OPEN-CLOSE(t,$\overline{s}$,$y(\overline{s},t)+\sum_{s\in D_1} y(s,\cdot)$,$y(\overline{s},\cdot)$) and OPEN-CLOSE ($t$,$s_l$,$y(s_l,t)$ $+\sum_{s\in D_2} y(s,\cdot)$,$y(s_l,\cdot)$)

4 Discussions

We can slightly improve this result by employing the standard scaling technique. Consider a universal facility location instance. We scale the facility cost (or unit service and penalty cost) by a factor of β, and run the local search algorithm on the modified instance. Let S denote the local optimal solution. We have

$$c_s(S) + c_p(S) \le \beta c_f(S^*) + c_s(S^*) + c_p(S^*),$$

and

$$\beta c_f(S) \le 4c_s(S^*) + 4c_p(S^*) + 5\beta c_f(S^*).$$

Summation yields

$$c(S) \le \left(\frac{4}{\beta} + 1\right) c_s(S^*) + \left(\frac{4}{\beta} + 1\right) c_p(S^*) + (5 + \beta) c_f(S^*).$$

Selecting $\beta = 2\sqrt{2} - 2$, we get an approximation ratio of $3 + 2\sqrt{2} \approx 5.83$.

Acknowledgements. The research of the first author is supported by Collaborative Innovation Center on Beijing Society-Building and Soccial Governance. The second author's research is supported by NSFC (Nos. 11371001 and 11531014). The third author's research is supported by the Natural Sciences and Engineering Research Council of Canada (NSERC) grant 283106. The fourth author's research is supported by NSFC (No. 11501412).

References

1. Aggarwal, A., Anand, L., Bansal, M., Garg, N., Gupta, N., Gupta, S., Jain, S.: A 3-approximation for facility location with uniform capacities. In: Eisenbrand, F., Shepherd, F.B. (eds.) IPCO 2010. LNCS, vol. 6080, pp. 149–162. Springer, Heidelberg (2010)
2. An, H.C., Singh, M., Svensson, O.: LP-based algorithms for capacitated facility location. In: Proceedings of the 55th Annual Symposium on Foundations of Computer Science, pp. 256–265 (2014)
3. Angel, E., Thang, N.K., Regnault, D.: Improved local search for universal facility location. J. Comb. Optim. **29**, 237–246 (2015)
4. Charikar, M., Khuller, S., Mount, D.M., Narasimhan, G.: Algorithms for facility location problems with outliers. In: Proceedings of the 12th Annual ACM-SIAM Symposium on Discrete Algorithms, pp. 642–651 (2001)
5. Chudak, F.A., Williamson, D.P.: Improved approximation algorithms for capacitated facility location problems. In: Cornuéjols, G., Burkard, R.E., Woeginger, G.J. (eds.) IPCO 1999. LNCS, vol. 1610, pp. 99–113. Springer, Heidelberg (1999)
6. Gupta, N., Gupta, S.: Approximation algorithms for capacitated facility location problem with penalties. arXiv:1408.4944 (2014)
7. Korupolu, M.R., Plaxton, C.G., Rajaraman, R.: Analysis of a local search heuristic for facility location problems. J. Algorithms **37**, 146–188 (2000)

8. Li, Y., Du, D., Xiu, N., Xu, D.: Improved approximation algorithms for the facility location problems with linear/submodular penalties. Algorithmica **73**, 460–482 (2015)
9. Mahdian, M., Pál, M.: Universal facility location. In: Di Battista, G., Zwick, U. (eds.) ESA 2003. lncs, vol. 2832, pp. 409–421. Springer, Heidelberg (2003)
10. Pal, M., Tardos, E., Wexler, T.: Facility location with nonuniform hard capacities. In: Proceedings of the 42nd Annual Symposium on Foundations of Computer Science, pp. 329–338 (2001)
11. Ahuja, R.K., Magnanti, T.L., Orlin, J.B.: Network Flows: Theory, Algorithms, and Applications. Prentice Hall, Englewood Cliffs (1993)
12. Vygen, J.: From stars to comets: improved local search for universal facility location. Oper. Res. Lett. **35**, 427–433 (2007)
13. Xu, G., Xu, J.: An LP rounding algorithm for approximating uncapacitated facility location problem with penalties. Inf. Process. Lett. **94**, 119–123 (2005)
14. Xu, G., Xu, J.: An improved approximation algorithm for uncapacitated facility location problem with penalties. J. Comb. Optim. **17**, 424–436 (2009)
15. Zhang, J., Chen, B., Ye, Y.: A multiexchange local search algorithm for the capacitated facility location problem. Math. Oper. Res. **30**, 389–403 (2005)

Variants of Multi-resource Scheduling Problems with Equal Processing Times

Hamed Fahimi and Claude-Guy Quimper(✉)

Université Laval, Quebec City, Canada
hamed.fahimi.1@ulaval.ca, claude-guy.quimper@ift.ulaval.ca

Abstract. We tackle the problem of non-preemptive scheduling of a set of tasks of duration p over m machines with given release and deadline times. We present a polynomial time algorithm as a generalization to this problem, when the number of machines fluctuates over time. Further, we consider different objective functions for this problem. We show that if an arbitrary function cost $c_i(t)$ is associated to task i for each time t, minimizing $\sum_{i=1}^{n} c_i(s_i)$ is NP-Hard. Further, we specialize this objective function to the case that it is merely contingent on the time and show that although this case is pseudo-polynomial in time, one can derive polynomial algorithms for the problem, provided the cost function is monotonic or periodic. Finally, as an observation, we mention how polynomial time algorithms can be adapted with the objective of minimizing maximum lateness.

1 Introduction

We explore several variants of the problem of scheduling, without preemption, tasks with equal processing times on multiple machines while respecting release times and deadlines. More formally, we consider n tasks and m identical machines. Task i has a release time r_i and deadline $\bar{d}_i$. All tasks have a processing time p. Without loss of generality, all parameters r_i, $\bar{d}_i$ and p are positive integers. Moreover, we consider the time point $u_i = \bar{d}_i - p + 1$, by starting at which a task i oversteps its deadline. We denote $r_{\min} = \min_i r_i$ the earliest release time and $u_{\max} = \max_i u_i$ the latest value u_i. A solution to the problem is an assignment of the starting times s_i which satisfies the following constraints

$$r_i \leq s_i < u_i \qquad \forall i \in \{1, \dots, n\} \tag{1}$$

$$|\{i : t \leq s_i < t + p\}| \leq m \qquad \forall t \in [r_{\min}, u_{\max}) \tag{2}$$

The completion time of a task C_i is equal to $s_i + p$. From 1, we obtain $C_i \leq \bar{d}_i$.

Following the notations of [9], this problem is denoted $Pm \mid r_j; p_j = p; \bar{d}_j \mid \gamma$ where γ is an objective function. The problem is sometimes reformulated by dividing all time points by p, resulting in tasks with unit processing times [13, 14]. However, this formulation does not make the problem easier to solve, as release times and deadlines lose their integrality. Without this integrality, greedy

Z. Lu et al. (Eds.): COCOA 2015, LNCS 9486, pp. 82–97, 2015.
DOI: 10.1007/978-3-319-26626-8_7

algorithms, commonly used to solve the problem when $p = 1$, become incorrect. Indeed, when the greedy scheduling algorithms choose to start a task i, they assume that no other tasks arrive until i is completed. This assumption does not hold if release times can take any rational value.

We explore several variations of this scheduling problem. Firstly, we solve the problem when the number of machines fluctuates over time. This models situations where there are fewer operating machines during night shifts or when fewer employees can execute tasks during vacation time or holidays. Then, we consider the problem with different objective functions. For an arbitrary function $c_i(t)$ associated to task i that maps a time point to a cost, we prove that minimizing $\sum_{i=1}^{n} c_i(s_i)$ is NP-Hard. This function is actually very general and can encode multiple well known objective functions. We study the case where all tasks share the same function $c(t)$. This models the situation where the cost of using the resource fluctuates with time. This is the case, for instance, with the price of electricity. Executing any task during peak hours is more expensive than executing the same task during a period when the demand is low. We show that minimizing $\sum_{i=1}^{n} c(t)$ can be done in pseudo-polynomial time and propose improvements when $c(t)$ is monotonic or periodic. The periodicity of the cost function is a realistic assumption as high and low demand periods for electricity have a predictable periodic behavior. Finally, we point out how the problem is solved in polynomial time with the objective of minimizing maximum lateness.

The paper is divided as follows. Section 2 presents a brief survey on existing algorithms related to the scheduling problem, the basic terminology and notations used in this paper, and the objective functions of interest. Section 3 solves the case where the number of machines fluctuates over time and shows how to adapt an existing algorithm for this case, while preserving polynomiality. Section 4 shows that minimizing $\sum_{i=1}^{n} c_i(s_i)$ is NP-Hard. Sections 5 and 6 consider a unique cost function $c(t)$ that is either monotonic or periodic and present polynomial time algorithms for these cases. Finally, as an additional remark, we show how to adapt a polynomial time algorithms for minimizing maximum lateness.

2 Literature Review, Framework and Notations

2.1 Related Work

Simons [13] presented an algorithm with time complexity $O(n^3 \log\log(n))$ that solves the scheduling problem. It is reported [9] that it minimizes both the sum of the completion times $\sum_j C_j$, and the latest completion time $C_{\max}$ (also called the makespan). Simons and Warmth [14] further improved the algorithm complexity to $O(mn^2)$. Dürr and Hurand [4] reduced the problem to a shortest path in a digraph and designed an algorithm in $O(n^4)$. This led López-Ortiz and Quimper [11] to introduce the idea of the scheduling graph. By computing the shortest path in this graph, one obtains a schedule that minimizes both $\sum_j C_j$ and $C_{\max}$. Their algorithm runs in $O(n^2 \min(1, p/m))$.

There exist more efficient algorithms for special cases. For instance, when there is only one machine ($m = 1$) and unit processing times ($p = 1$), the problem is equivalent to finding a matching in a convex bipartite graph. Lipski and Preparata [10] present an algorithm running in $O(n\alpha(n))$ where α is the inverse of Ackermann's function. Gabow and Tarjan [5] reduce this complexity to $O(n)$ by using a restricted version of the union-find data structure.

Baptiste proves that in general, if the objective function can be expressed as the sum of n functions f_i of the completion time C_i of each task i, where f_i's are non-decreasing and for any pair of jobs (i, j) the function $f_i - f_j$ is monotonous, the problem can be solved in polynomial time. Note that the assumption holds for several objectives, such as the weight sum of completion times $\sum w_i C_i$. A variant of the problem exists when the tasks are allowed to miss their deadlines at the cost of a penalty. Let $L_j = C_j - d_j$ be the *lateness* of a task j. The problem of minimizing the maximum lateness $L_{\max} = \max_j L_j$ (denoted $P \mid r_i, p_i = p \mid L_{max}$) is polynomial [15] and the special case for one machine and unit processing times (denoted $1 \mid r_j, p_j = p \mid L_{max}$) is solvable in $O(n \log n)$ [8].

Möhring et al. [12] study the case where no release times or deadlines are provided and the processing times are not all equal. For the case that their problem is not resource-constrained, they consider the objective of minimizing costs per task and per time, as it is considered in this paper. They establish a connection between a minimum-cut in an appropriately defined directed graph and propose a mathematical programming approach to compute both lower bounds and feasible solutions. The minimum-cut problem is the dual of the maximum flow problem that will be used in this paper.

Bansal and Pruhs [3] consider preemptive tasks, a single machine, no deadlines, and distinct processing times. Tasks incur a cost depending on their completion time. They introduce an approximation for the general case and an improved approximation algorithm for the case that all release times are identical.

2.2 Objective Functions

Numerous objective functions can be optimized in a scheduling problem. We consider *minimizing costs per task and per time*, in which case executing a task i at time t costs $c(i, t)$ and we aim to minimize the sum of costs, i.e. $\sum_{i,t} c(i, s_i)$. Such an objective function depends on the release time of the task. In the industry, that can be used to model a cost that increases as the execution of a task is delayed. For instance, $c(i, t) = t - r_i$. Then, we consider *minimizing task costs per time*, in which case executing any task at time t costs $c(t)$ and we want to minimize $\sum_i c(s_i)$. An alternative common objective function is to minimize the sum of the completion times. In the context where the tasks have equal processing times, a solution that minimizes the sum of the completion times necessarily minimizes the sum of the starting time. We consider these two objectives equivalent. Finally, we consider minimizing the maximum lateness $L_{\max} = \max_i L_i$.

2.3 Network Flows

Consider a digraph $\overrightarrow{N} = (V, E)$ where each arc $(i, j) \in E$ has a flow capacity u_{ij} and a flow cost c_{ij}. There is one node $s \in V$ called the source and one node $t \in V$ called the sink. A *flow* is a vector that maps each edge $(i, j) \in E$ to a value x_{ij} such that the following constraints are satisfied.

$$0 \leq x_{ij} \leq u_{ij} \tag{3}$$

$$\sum_{j \in V} x_{ji} - \sum_{j \in V} x_{ij} = 0 \qquad \forall i \in V \setminus \{s, t\} \tag{4}$$

The min-cost flow satisfies the constraints while minimizing $\sum_{(i,j)\in E} c_{ij}x_{ij}$.

A matrix with entries in $\{-1, 0, 1\}$ which has precisely one 1 and one -1 per column is called a *network matrix*. If A is a network matrix, the following linear program identifies a flow.

$$\text{Maximize } c^T x, \text{ subject to } \begin{cases} Ax = b \\ x \geq 0 \end{cases} \tag{5}$$

There is one node for each row of the matrix in addition to a source node s and a sink node t. Each column in the matrix corresponds to an edge $(i, j) \in E$ where i is the node whose row is set to 1 and j is the node whose column is set to -1. If $b_i > 0$ we add the edge (i, t) of capacity b_i and if $b_i < 0$ we add the edge (s, i) of capacity $-b_i$ [16].

The residual network with respect to a given flow x is formed with the same nodes V as the original network. However, for each edge (i, j) such that $x_{ij} < u_{ij}$, there is an edge (i, j) in the residual network of cost c_{ij} and residual capacity $u_{ij} - x_{ij}$. For each edge (i, j) such that $x_{ij} > 0$, there is an edge (j, i) in the residual network of cost $-c_{ij}$ and residual capacity x_{ij}.

To our knowledge, the *successive shortest path algorithm* is the state of the art, for this particular structure of the network, to solve the min-cost flow problem. This algorithm successively augments the flow values y_{ij} of the edges along the shortest path connecting the source to the sink in the residual graph. Let $N = \max_{(i,j)\in E} |c_{ij}|$ be the greatest absolute cost and $U = \max_{i \in V} b_i$ be the largest value in the vector b. To compute the shortest path, one can use Goldberg's algorithm [6] with a time complexity of $O(|E|\sqrt{|V|}\log N)$. Since at most $|V|U$ shortest path computations are required, this leads to a time complexity of $O(|V|^{1.5}|E|\log(N)U)$.

2.4 Scheduling Graph

López-Ortiz and Quimper [11] introduced the *scheduling graph* which holds important properties. For instance, it allows to decide whether an instance is feasible, i.e. whether there exists at least one solution. The graph is based on the assumption that it is sufficient to determine how many tasks start at a given time. If one knows that there are h_t tasks starting at time t, it is possible to determine which tasks start at time t by computing a matching in a convex bipartite graph (see [11]).

The scheduling problem can be written as a satisfaction problem where the constraints are uniquely posted on the variables h_t. As a first constraint, we force the number of tasks starting at time t to be non-negative.

$$\forall\, r_{\min} \leq t \leq u_{\max} - 1 \qquad h_t \geq 0 \tag{6}$$

At most m tasks (n tasks) can start within any window of size p (size $u_{\max} - r_{\min}$).

$$\forall\, r_{\min} \leq t \leq u_{\max} - p \qquad \sum_{j=t}^{t+p-1} h_j \leq m, \qquad \sum_{j=r_{\min}}^{u_{\max}-1} h_j \leq n \tag{7}$$

Given two arbitrary (possibly identical) tasks i and j, the set $K_{ij} = \{k : r_i \leq r_k \wedge u_k \leq u_j\}$ denotes the jobs that must start in the interval $[r_i, u_j)$. Hence,

$$\forall\, i, j \in \{1, \ldots, n\} \qquad \sum_{t=r_i}^{u_j-1} h_t \geq |K_{ij}| \tag{8}$$

Some objective functions, such as minimizing the sum of the starting times, can also be written with the variables h_t.

$$\min \sum_{t=r_{\min}}^{u_{\max}-1} t \cdot h_t \tag{9}$$

To simplify the inequalities (6)–(8), we proceed to a change of variables. Let $x_t = \sum_{i=r_{\min}}^{t-1} h_i$, for $r_{\min} \leq t \leq u_{\max}$, be the number of tasks starting to execute before time t. Therefore, the problem can be rewritten as follows.

$$\forall\, r_{\min} \leq t \leq u_{\max} - p \qquad x_{t+p} - x_t \leq m, \qquad x_{u_{\max}} - x_{r_{\min}} \leq n \tag{10}$$

$$\forall\, r_{\min} \leq t \leq u_{\max} - 1 \qquad x_t - x_{t+1} \leq 0 \tag{11}$$

$$\forall\, r_i + 1 \leq u_j \qquad x_{r_i} - x_{u_j} \leq -|K_{ij}| \tag{12}$$

These inequalities form a system of difference constraints which can be solved by computing shortest paths in what is call the *scheduling graph* [11]. In this graph, there is a node for each time point t, $r_{\min} \leq t \leq u_{\max}$ and an edge of weight k_{pq}, connecting the node q to the node p for each inequality of the form $x_p - x_q \leq k_{pq}$.

The *scheduling graph* has for vertices the nodes $V = \{r_{\min}, \ldots, u_{\max}\}$ and for edges $E = E_f \cup E_b \cup E_n$ where $E_f = \{(t, t+p) : r_{\min} \leq t \leq u_{\max} - p\} \cup \{(r_{\min}, u_{\max})\}$ is the set of *forward edges* (from inequalities (10)), $E_b = \{(u_j, r_i) : r_i < u_j\}$ is the set of *backward edges* (from inequality (12)), and

$E_n = \{(t+1,t) : r_{\min} \leq t < u_{\max}\}$ is the set of *null edges* (from inequality (11)). The following weight function maps every edge $(a,b) \in E$ to a weight:

$$w(a,b) = \begin{cases} m & \text{if } a + p = b \\ n & \text{if } a = r_{\min} \wedge b = u_{\max} \\ -|\{k : b \leq r_k \wedge u_k \leq a\}| & \text{if } a > b \end{cases} \quad (13)$$

Theorem 1 shows how to compute a feasible schedule.

Theorem 1 (López-Ortiz and Quimper [11]). Let $\delta(a,b)$ be the shortest distance between node a and node b in the scheduling graph. The assignment $x_t = n + \delta(u_{max}, t)$ is a solution to the inequalities (10)–(12) that minimizes the sum of the completion times.

The scheduling problem has a solution if and only if the scheduling graph has no negative cycles. An adaptation [11] of the Bellman-Ford algorithm finds a schedule with time complexity $O(\min(1, \frac{p}{m})n^2)$, which is sub-quadratic when $p < m$ and quadratic otherwise.

In the next sections, we adapt the scheduling graph to solve variations of the problem.

3 Variety of Machines

Consider the problem where the number of machines fluctuates over time. Let $T = [(t_0, m_0), \ldots, (t_{|T|-1}, m_{|T|-1})]$ be a sequence where t_i's are the time points at which the fluctuations occur and they are sorted in chronological order and m_i machines are available within the time interval $[t_i, t_{i+1})$. This time interval is the union of a (possibly empty) interval and an open-interval: $[t_i, t_{i+1}) = [t_i, t_{i+1} - p] \cup (t_{i+1} - p, t_{i+1})$. A task starting in $[t_i, t_{i+1} - p]$ is guaranteed to have access to m_i machines throughout its execution, whereas a task starting in $(t_{i+1} - p, t_{i+1})$ encounters the fluctuation of the number of machines before completion. Therefore, no more than $\min(m_i, m_{i+1})$ tasks can start in the interval $(t_{i+1} - p, t_{i+1})$. In general, a task can encounter multiple fluctuations of the number of machines throughout its execution. Let $\alpha(t) = \max\{t_j \in T \mid t_j \leq t\}$ be the last time the number of machines fluctuates before time t. At most $M(t)$ tasks can start at time t.

$$M(t) = \min\{m_i \mid t_i \in [\alpha(t), t + p)\}. \quad (14)$$

From 14, we conclude that no more than $\max_{t' \in [t, t+p)} M(t')$ tasks can start in the interval $[t, t+p)$. Accordingly, one can rewrite the first inequality of the constraints (10)

$$x_{t+p} - x_t \leq \max_{t \leq t' < t+p} M(t') \quad (15)$$

and update the weight function of the scheduling graph.

$$w(a,b) = \begin{cases} \max_{a \leq t' < a+p} M(t') & \text{if } a + p = b \\ n & \text{if } a = r_{\min} \wedge b = u_{\max} \\ -|\{k : b \leq r_k \wedge u_k \leq a\}| & \text{if } a \geq b \end{cases} \quad (16)$$

It remains to show how the algorithm presented in [11] can be adapted to take into account the fluctuating number of machines. This algorithm maintains a vector $d^{-1}[0..n]$ such that $d^{-1}[i]$ is the latest time point reachable at distance $-i$ from the node $u_{\max}$. In other words, all nodes whose label is a time point in the semi-open interval $(d^{-1}[i+1], d^{-1}[i]]$ are reachable at distance $-i$ from node $u_{\max}$. Let a be a node in $(d^{-1}[i+1], d^{-1}[i]]$ and consider the edge $(a, a+p)$ of weight $w(a, a+p)$. Upon processing this edge, the algorithm updates the vector by setting $d^{-1}[i - w(a,b)] \leftarrow \max(d^{-1}[i - w(a,b)], b)$, i.e. the rightmost node accessible at distance $-i + w(a,b)$ is either the one already found, or the node $a+p$ that is reachable through the path to a of distance $-i$ followed by the edge $(a, a+p)$ of distance $w(a, a+p)$.

To efficiently perform this update, the algorithm evaluates $w(a, a+p)$ in two steps. The first step transforms T in a sequence $T' = [(t'_0, m'_0), (t'_1, m'_1), \ldots]$ such that $M(t) = m'_i$ for every $t \in [t'_i, t'_{i+1})$. The second step transforms the sequence T' into a sequence $T'' = [(t''_0, m''_0), (t''_1, m''_1), \ldots]$ such that $w(t, t+p) = m''_i$ for all $t \in [t''_i, t''_{i+1})$. Interestingly, both steps execute the same algorithm.

To build the sequence T', one needs to iterate over the sequence T and find out, for every time window $[t, t+p)$, the minimum number of available machines inside that time window. If a sequence of consecutive windows such as $[t, t+p), [t+1, t+p+1), [t+2, t+p+2), \ldots$ have the same minimum number of available machines, then only the result of the first window is reported. This is a variation of the *minimum on a sliding window problem* [7] where an instance is given by an array of numbers $A[1..n]$ and a window length p. The output is a vector $B[1..n-p+1]$ such that $B_i = \min\{A_i, A_{i+1}, \ldots, A_{i+p-1}\}$. The algorithm that solves the minimum on a sliding window problem can be slightly adapted. Rather than taking as input the vector A that contains, in our case, many repetitions of values, it can simply take as input a list of pairs like the vector T and T' which indicate the value in the vector and until which index this value is repeated. The same compression technique applies for the output vector. This adaptation can be done while preserving the linear running time complexity of the algorithm.

Once computed, the sequence T' can be used as input to the *maximum on a sliding window problem* to produce the final sequence T''. Finally, the Algorithm 1 simultaneously iterates over the sequence T'' and the vector d^{-1} to relax the edges in $O(|T| + n)$ time. Since relaxing forward edges occurs at most $O(\min(1, \frac{p}{m})n)$ times [11], the overall complexity to schedule the tasks is $O(\min(1, \frac{p}{m})(|T| + n)n)$.

Algorithm 1. RelaxForwardEdges($[(t''_1, m''_1), \ldots, (t''_{|T''|}, m''_{|T''|})], d^{-1}[0..n], p$)

```
t ← r_min, i ← n, j ← 0
while i > 0 ∨ j < |T''| do
    if i − m''_j > 0 then  d^{-1}[i − m''_j] ← max(d^{-1}[i − m''_j], t + p)
    if j = |T''| ∨ i ≥ 0 ∧ d^{-1}[i − 1] < m''_{j+1} then
        i ← i + 1
        t ← d^{-1}[i]
    else
        j ← j + 1
        t ← t''_j
```

4 General Objective Function

We prove that minimizing costs per task and per time, i.e. $\sum_{i,t} c_i(s_i)$ for arbitrary functions $c_i(t)$ is NP-Hard. We proceed with a reduction from the INTERDISTANCE constraint [2]. The predicate INTERDISTANCE($[X_1, \ldots, X_n], p$) is true if and only if $|X_i - X_j| \geq p$ holds whenever $i \neq j$. Let $S_1, \ldots, S_n$ be n sets of integers. Deciding whether there exists an assignment for the variables $X_1, \ldots, X_n$ such that $X_i \in S_i$ and INTERDISTANCE($[X_1, \ldots, X_n], p$) hold is NP-Complete [2]. We create one task per variable X_i with release time $r_i = \min(S_i)$, latest starting time $u_i = \max(S_i)$, processing time p, and a cost function $c_i(t)$ equal to 0 if $t \in S_i$ and 1 otherwise. There exists a schedule with objective value $\sum_{i,t} c_i(s_i) = 0$ iff there exists an assignment with $X_i \in S_i$ that satisfies the predicate INTERDISTANCE, hence minimizing $\sum_{i,t} c_i(s_i)$ is NP-Hard.

The NP-hardness of this problem motivates the idea of studying specializations of this objective function in order to seek polynomial time algorithms.

5 Monotonic Objective Function

Let $c(t) : \mathbb{Z} \to \mathbb{Z}$ be an increasing function, i.e. $c(t) + 1 \leq c(t + 1)$ for any t. We prove that a schedule that minimizes $\sum_i s_i$ also minimizes $\sum_i c(s_i)$. Theorem 1 shows how to obtain a solution that minimizes $\sum_i s_i$. Lemma 1 shows that this solution also minimizes other objective functions. Recall that h_t is the number of tasks starting at time t.

Lemma 1. *The schedule obtained with Theorem 1 minimizes* $\sum_{a=t}^{u_{\max}-1} h_a$ *for any time* t.

Proof. Let $(a_1, a_2), (a_2, a_3), \ldots, (a_{k-1}, a_k)$, with $a_1 = u_{\max}$ and $a_k = t$, be the edges on the shortest path from $u_{\max}$ to t in the scheduling graph. By substituting the inequalities (10)–(12), we obtain $\delta(u_{\max}, t) = \sum_{i=1}^{k-1} w(a_i, a_{i+1}) \geq \sum_{i=1}^{k-1}(x_{a_{i+1}} - x_{a_i}) = x_t - x_{u_{\max}}$. This shows that the difference $x_t - x_{u_{\max}}$ is at most $\delta(u_{\max}, t)$ for any schedule. It turns out that by setting $x_t = n + \delta(u_{\max}, t)$, the difference $x_t - x_{u_{\max}} = \delta(u_{\max}, t) - \delta(u_{\max}, u_{\max}) = \delta(u_{\max}, t)$ reaches its maximum and therefore, $x_{u_{\max}} - x_t = \sum_{a=t}^{u_{\max}-1} h_a$ is maximized. □

Theorem 2. *The schedule of Theorem 1 minimizes $\sum_{i=1}^{n} c(s_i)$ for any increasing function $c(t)$.*

Proof. Consider the following functions that differ by their parameter a.

$$c_a(t) = \begin{cases} c(t) & \text{if } t < a \\ c(a) + t - a & \text{otherwise} \end{cases} \tag{17}$$

The function $c_a(t)$ is identical to $c(t)$ up to point a and then increases with a slope of one. As a base case of an induction, the schedule described in Theorem 1 minimizes $\sum_{i=1}^{n} s_i$ and therefore minimizes $\sum_{i=1}^{n} c_{r_{\min}}(s_i)$. Suppose that the algorithm minimizes $\sum_{i=1}^{n} c_a(s_i)$, we prove that it also minimizes $\sum_{i=1}^{n} c_{a+1}(s_i)$. Consider the function

$$\Delta_a(t) = \begin{cases} 0 & \text{if } t \leq a \\ c(a+1) - c(a) - 1 & \text{otherwise} \end{cases} \tag{18}$$

and note that $c_a(t) + \Delta_a(t) = c_{a+1}(t)$. For all t, since $c(t+1) - c(t) \geq 1$, we have $\Delta_a(t) \geq 0$.

If $c(a+1) - c(a) = 1$ then $\Delta_a(t) = 0$ for all t and therefore $c_a(t) = c_{a+1}(t)$. Since the algorithm returns a solution that minimizes $\sum_{i=1}^{n} c_a(s_i)$, it also minimizes $\sum_{i=1}^{n} c_{a+1}(s_i)$.

If $c(a+1) - c(a) > 1$, a schedule minimizes the function $\sum_{i=1}^{n} \Delta_a(s_i)$ if and only if it minimizes the number of tasks starting after time a. From Lemma 1, the schedule described in Theorem 1 achieves this. Consequently, the algorithm minimizes $\sum_{i=1}^{n} c_a(s_i)$, it minimizes $\sum_{i=1}^{n} \Delta_a(s_i)$, and therefore, it minimizes $\sum_{i=1}^{n} c_a(s_i) + \sum_{i=1}^{n} \Delta_a(s_i) = \sum_{i=1}^{n} c_{a+1}(s_i)$.

By induction, the algorithm minimizes $\sum_{i=1}^{n} c_\infty(s_i) = \sum_{i=1}^{n} c(s_i)$. □

If the cost $c(t)$ function is decreasing, i.e. $c(t) - 1 \geq c(t+1)$, it is possible to minimize $\sum_{i=1}^{n} c(t)$ by solving a transformed instance. For each task i in the original problem, one creates a task i with release time $r'_i = -u_i$ and latest starting time $u'_i = -r_i$. The objective function is set to $c'(t) = -c(t)$ which is an increasing function. From a solution s'_i that minimizes $\sum_{i=1}^{n} c'(s'_i)$, one retrieves the original solution by letting $s_i = -s'_i$.

6 Periodic Objective Function

6.1 Scheduling Problem as a Network Flow

Theorem 1 shows that computing the shortest paths in the scheduling graph can minimize the sum of the completion times. We show that computing, in pseudo-polynomial time, a flow in the scheduling graph can minimize $\sum_{i=1}^{n} c(s_i)$ for an arbitrary function $c(t)$.

The objective function (9) can be modified to take into account the function c. We therefore minimize $\sum_{t=r_{\min}}^{u_{\max}-1} c(t)h_t$. After proceeding to the change of variables $x_t = \sum_{i=r_{\min}}^{t-1} h_i$, we obtain $\sum_{t=r_{\min}}^{u_{\max}-1} c(t)(x_{t+1} - x_t)$ which is equivalent to

$$\text{maximize } c(r_{\min})x_{r_{\min}} - \sum_{t=r_{\min}+1}^{u_{\max}-1} (c(t) - c(t-1))\, x_t - c(u_{\max} - 1)x_{u_{\max}}$$

We use this new objective function with the original constraints of the problem given by Eqs. (10)–(12). This results in a linear program of the form $\max\{\boldsymbol{c}^T x \mid A\boldsymbol{x} \leq \boldsymbol{b}, \boldsymbol{x} \lessgtr 0\}$ which has for dual $\min\{\boldsymbol{b}^T y \mid A^T \boldsymbol{y} = \boldsymbol{c}, y \geq 0\}$. Every row of matrix A has exactly one occurrence of value 1, one occurrence of the value -1, and all other values are null. Consequently, A^T is a *network matrix* and the dual problem $\min\{\boldsymbol{b}^T y \mid A^T \boldsymbol{y} = \boldsymbol{c}, y \geq 0\}$ is a min-cost flow problem.

Following Sect. 2.3, we reconstruct the graph associated to this network flow which yields the scheduling graph augmented with a source node and a sink node. An edge of capacity $c(r_{\min})$ connects the node $r_{\min}$ to the sink. An edge of capacity $c(u_{\max} - 1)$ connects the source node to the node $u_{\max}$. For the nodes t such that $r_{\min} < t < u_{\max}$, an edge of capacity $c(t-1) - c(t)$ connects the source node to node t whenever $c(t-1) > c(t)$ and an edge of capacity $c(t) - c(t-1)$ connects the node t to the sink node whenever $c(t-1) < c(t)$. All other edges in the graph (forward, backward, and null edges) have an infinite capacity. Figure 1 illustrates an example of such a graph.

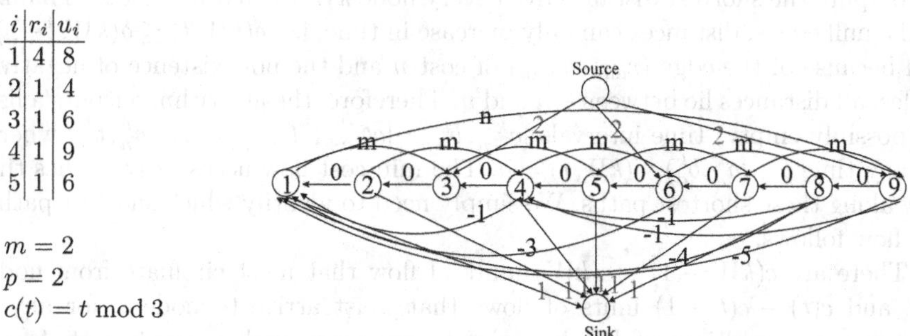

Fig. 1. A network flow with 5 tasks. The cost on the forward, backward, and null edges are written in black. These edges have unlimited capacities. The capacities of the nodes from the source and to the sinks are written in blue. These edges have a null cost.

The computation of a min-cost flow gives rise to a solution for the dual problem. To convert the solution of the dual to a solution for the primal (i.e. an assignment of the variables x_t), one needs to apply a well known principle in network flow theory [1]. Let $\delta(a, b)$ be the shortest distance from node a to node b in the *residual graph*. The assignment $x_t = \delta(u_{\max}, t)$ is an optimal solution of the primal. The variable x_t is often called *node potential* in network theory.

Consider a network flow of $|V|$ nodes, $|E|$ edges, a maximal capacity of U, and a maximum absolute cost of N. The successive shortest path algorithm computes a min-cost flow with $O(|V|U)$ computations of a shortest path that each executes in $O(|E|\sqrt{|V|}\log N)$ time using Goldberg's algorithm [6]. Let $\Delta c = \max_t |c(t) - c(t-1)|$ be the maximum cost function fluctuation and $H = u_{\max} - r_{\min}$ be the horizon. In the scheduling graph, we have $|V| \in O(H)$, $|E| \in O(H + n^2)$, $N \in O(n)$, and $U = \Delta c$. Therefore, the overall running time complexity to find a schedule is $O((H - p + n^2)(H)^{3/2}\Delta c \log n)$.

6.2 Periodic Objective Function Formulated as a Network Flow

In many occasions, one encounters the problem of minimizing $\sum_{i=1}^{n} c(s_i)$ where $c(s_i)$ is a periodic function, i.e. a function where $c(t) = c(t + W)$ for a period W. Moreover, within a period, the function is increasing. An example of such a function is the function $c(t) = t \bmod 7$. If all time points correspond to a day, the objective function ensures that all tasks are executed at their earliest time in a week. In other words, it is better to wait for Monday to start a task rather than executing this task over the weekend. In such a situation, it is possible to obtain a more efficient time complexity than the algorithm presented in the previous section.

Without loss of generality, we assume that the periods start on times kW for $k \in \mathbb{N}$ which implies that the function $c(t)$ is only decreasing between $c(kW-1)$ and $c(kW)$ for some $k \in \mathbb{N}$. In the network flow from Sect. 6.1, only the time nodes kW have an incoming edge from the source. We use the algorithm from [11] to compute the shortest distance from every node kW to all other nodes. Thanks to the null edges, distances can only increase in time, i.e. $\delta(kW, t) \leq \delta(kW, t+1)$, and because of the edge $(r_{\min}, u_{\max})$ of cost n and the nonexistence of negative cycles, all distances lie between $-n$ and n. Therefore, the algorithm outputs a list of (possibly empty) time intervals $[a^k_{-n}, b^k_{-n}), [a^k_{-n+1}, b^k_{-n+1}), \ldots, [a^k_n, b^k_n)$ where for any time $t \in [a^k_d, b^k_d)$, $\delta(kW, t) = d$. The min-cost flow necessarily pushes the flow along these shortest paths. We simply need to identify which shortest paths the flow follows.

There are $c(kW-1) - c(kW)$ units of flow that must circulate from node kW and $c(t) - c(t-1)$ units of flows that must arrive to node t, for any t that is not a multiple of W. In order to create a smaller graph with fewer nodes, we aggregate time intervals where time points share common properties. We consider the sorted set S of time points a^k_i and b^k_i. Let t_1 and t_2 be two consecutive time points in this set. All time points in the interval $[t_1, t_2)$ are at equal distance from the node kW, for any $k \in \mathbb{N}$. The amount of units of flow that must reach the sink from the nodes in $[t_1, t_2)$ is given by

$$\sum_{j=t_1}^{t_2-1} \max(c(j)-c(j-1), 0) = c(t_2-1)-c(t_1-1)+\left(\left\lfloor \frac{t_2-1}{W} \right\rfloor - \left\lceil \frac{t_1}{W} \right\rceil + 1\right)(c(W-1) - c(0)) \quad (19)$$

Consequently, we create a graph, called the *compressed graph*, with one source and one sink node. There is one node for each time point kW for $\frac{r_{\min}}{W} \leq k \leq \frac{u_{\max}}{W}$.

There is an edge between the source node and a node kW with capacity $c(kW - 1) - c(kW)$. For any two consecutive time points t_1, t_2 in S there is a time interval node $[t_1, t_2)$. An edge whose capacity is given by Eq. (19) connects the interval node $[t_1, t_2)$ to the sink. Finally, a node kW is connected to an interval node $[t_1, t_2)$ with an edge of infinite capacity and a cost of $\delta(kW, t_1)$. Figure 2 shows the compressed version of the graph on Fig. 1.

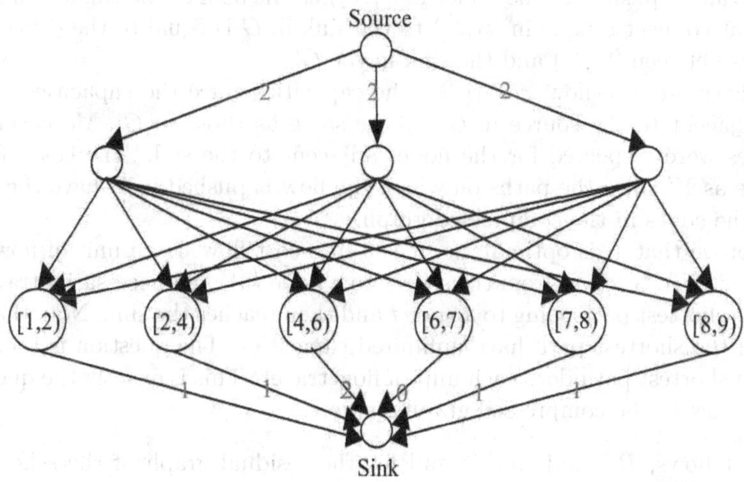

Fig. 2. The compressed version of the graph on Fig. 1.

Computing a min-cost flow in this network simulates the flow in the scheduling graph. Indeed, a flow going through an edge $(kW, [t_1, t_2))$ in the compressed graph is equivalent, in the scheduling graph, to a flow leaving the source node, going to the node kW, going along the shortest path from node kW to a time node $t \in [t_1, t_2)$, and reaching the sink.

Theorem 3. *To every min-cost flow in the compressed graph corresponds a min-cost flow in the scheduling graph.*

Proof. Let G be the scheduling graph and G' be the compressed graph. Let Y' denote a min-cost flow in G'. We show how to obtain a min-cost flow Y in G whose cost is the same as the cost of Y'.

Consider an edge $e_j = (kW, [t_1, t_2))$ in G' which conveys a positive amount of flow, say f. In the scheduling graph G, it is possible to push f units of flow along the shortest paths from kW to the nodes within the interval $[t_1, t_2)$. It suffices to see how one can retrieve Y from Y', presuming it is initially null. This is done by considering all incoming flows to $[t_1, t_2)$ and manage to spread them over the edges of G. We start with the node t_1 and consider the shortest path P from kW to t_1 in G. The amount of flow that can be incremented is the minimum between f and the amount of flow that t_1 can receive. Then, we increment the amount of flow on the extended path in G, which connects the source to P and connects P to the sink.

If the capacity of t_1 is reached, we decrement f by the amount of flow which was consumed and we move to the next node in the interval. Now, there remains f units of flow for the nodes within the interval $[t_1 + 1, t_2)$. By repeating the same instruction for the rest of the nodes in $[t_1, t_2)$ and for every edge in G' that carries a positive amount of flow, we obtain the flow Y. It is guaranteed that all the flow can be pushed to the nodes in $[t_1, t_2)$ as the sum of the capacities of the edges that connect a node in $[t_1, t_2)$ to the sink in G is equal to the capacity of the edges between $[t_1, t_2)$ and the sink in the G'.

Furthermore, the flow Y satisfies the capacities since the capacities on the edges adjacent to the source in G are the same as those in G'. Moreover, the capacities were respected for the nodes adjacent to the sink. The cost of Y is the same as Y' since the paths on which the flow is pushed in Y have the same cost as the edges in the compressed graph.

We prove that Y is optimal, i.e. it is a min-cost flow. Each unit of flow in a min-cost flow in G leaves from the source to a node kW and necessarily traverses along the shortest path going to a node t and then reaches the sink. Note that the edges on the shortest path have unlimited capacities. The question is therefore on which shortest path does each unit of flow travel? This is exactly the question that the flow in the compressed graph answers. □

In what follows, RG and RG', stand for the residual graph of the scheduling graph G and the residual compressed graph G'.

Lemma 2. *Let t be a node in the residual scheduling graph RG and $[t_i, t_{i+1})$, such that $t_i \leq t < t_{i+1}$, be a node in the residual compressed scheduling graph. The distance between node kW and t in RG is equal to the distance between kW and $[t_i, t_{i+1})$ in RG'.*

Proof. We show that for any path P' in the residual compressed graph RG', there is a path P in the residual graph RG that has the same cost. From Lemma 3, we know that for a flow in the compressed graph G', there is an equivalent flow in the original graph G. Consider a path P' from a node kW to an interval node $[t_i, t_{i+1})$. By construction of the compressed graph, for each edge of this path corresponds a path of equal cost in the residual graph RG'. Consequently, there is a path in G' that goes from node kW to any node $t \in [t_i, t_{i+1})$ with the same cost as the path going from kW to $[t_i, t_{i+1})$ in G.

Consider a path P in the residual graph RG going from a node k_1W to a node t. Suppose that this path contains exactly one edge in RG that is not in G. We denote this edge (a, b) and the path P can be decomposed as follows: $k_1W \rightsquigarrow a \rightarrow b \rightsquigarrow t$. The edge (a, b) appears in the residual graph RG because there is a positive amount of flow circulating on a shortest path $S : k_2W \rightsquigarrow b \rightarrow a \rightsquigarrow u$ to which the reversed edge (b, a) belongs. Let Q be the following path in the residual graph RG: $k_1W \rightsquigarrow u \rightsquigarrow a \rightarrow b \rightsquigarrow k_2W \rightsquigarrow t$. Let l be the function that evaluates the cost of a path. We prove that Q has a cost that is no more than P and that it has an equivalent in the residual compressed graph RG'.

$$\begin{aligned} l(Q) &= l(k_1W \rightsquigarrow u \rightsquigarrow a \rightarrow b \rightsquigarrow k_2W \rightsquigarrow t) \\ &\le l(k_1W \rightsquigarrow a \rightsquigarrow u \rightsquigarrow a \rightarrow b \rightsquigarrow k_2W \rightsquigarrow t) \end{aligned}$$

In the residual graph, the paths $a \rightsquigarrow u$ and $u \rightsquigarrow a$ have opposite costs, hence $l(a \rightsquigarrow u \rightsquigarrow a) = 0$.

$$\begin{aligned} &= l(k_1W \rightsquigarrow a \rightarrow b \rightsquigarrow k_2W \rightsquigarrow t) \\ &\le l(k_1W \rightsquigarrow a \rightarrow b \rightsquigarrow k_2W \rightsquigarrow b \rightsquigarrow t) \end{aligned}$$

In the residual graph, the paths $b \rightsquigarrow k_2W$ and $k_2W \rightsquigarrow b$ have opposite costs, hence $l(b \rightsquigarrow k_2W \rightsquigarrow b)$.

$$= l(k_1W \rightsquigarrow a \rightarrow b \rightsquigarrow t) = l(P)$$

The path Q has an equivalent in the residual compressed graph RG'. Indeed, the sub-paths $k_1W \rightsquigarrow u$ and $k_2W \rightsquigarrow t$ are edges in RG' whose cost is given by the shortest paths in G. The path $u \rightsquigarrow a \rightarrow b \rightsquigarrow k_2W$ is the reverse of path S. Since S is an edge in G' and there is a flow circulating on S, the reverse of S also appears in RG'. Consequently, the path P can be transformed into path Q that has an equivalent in the compressed residual graph. If P contains more than one edge that belongs to RG but not G, then the transformation can be applied multiple times.

Since a path in RG has an equivalent path whose cost is not greater in RG' and vice-versa, we conclude that a node kW is at equal distance from all the other nodes in either graph. □

Notice that the above lemma implies that after computing the min-cost flow in the compressed graph, one sets the value for x_t to the shortest distance between an arbitrary but fixed node kW to the interval node that contains t.

Let $H = u_{\max} - r_{\min}$ be the horizon, we need $\Theta(\frac{H}{W})$ calls to the algorithm in [11] to build the compressed graph in $O(\frac{H}{W}n^2 \min(1, \frac{p}{m}))$ time. As in Sect. 6.1, the successive shortest paths technique, with Goldberg's algorithm [6], computes the maximum flow. The compressed graph has $|V| \in O(\frac{H}{W} + n^2)$ nodes, $|E| \in O(\frac{H}{W}n^2)$ edges, a maximum absolute cost of $N \in O(n)$, and a maximum capacity of $U = \Delta c = C(W-1) - c(0)$. Computing the values for x_t requires an additional execution of Goldberg's algorithm on the compressed graph. The final running time complexity is $O\left(\left(\left(\frac{H}{W}\right)^{2.5} + n^5\right) \Delta c \log(n)\right)$ which is faster than the algorithm presented in the previous sections when the number of periods is small, i.e. when $\frac{H}{W}$ is bounded. In practice, there are fewer periods than tasks: $\frac{H}{W} < n$.

7 Additional Remark

Consider the case where tasks have due dates d_i and deadlines $\bar{d}_i$. One wants to minimize the maximum lateness $L_{\max} = \max_i \max(C_i - d_i, 0)$ while ensuring that tasks complete before their deadlines. To test whether there exists a schedule with maximum lateness L, one changes the deadline of all task i for $\min(\bar{d}_i, d_i + L)$. If there exists a valid schedule with this modification, then there exists a schedule with maximum lateness at most L in the original problem. Since the maximum lateness is bounded by $0 \leq L \leq \lceil \frac{np}{m} \rceil$, a well known technique consists of using the binary search that calls at most $\log(\lceil \frac{np}{m} \rceil)$ times the algorithm in [11] and achieves a running time complexity of $O(\log(\frac{np}{m}) n^2 \min(1, \frac{p}{m}))$.

8 Conclusion

We studied variants of the problem of non-preemptive scheduling of tasks with equal processing times on multiple machines. We presented polynomial time algorithms for different objective functions. We generalized the problem to the case that the number of machines fluctuate through time.

References

1. Ahuja, R.K., Magnanti, T.L., Orlin, J.B.: Network Flows: Theory, Algorithms, and Applications. Prentice Hall, Upper Saddle River (1993)
2. Artiouchine, K., Baptiste, P.: Inter-distance constraint: an extension of the all-different constraint for scheduling equal length jobs. In: van Beek, P. (ed.) CP 2005. LNCS, vol. 3709, pp. 62–76. Springer, Heidelberg (2005)
3. Bansal, N., Pruhs, K.: The geometry of scheduling. SIAM J. Comput. **43**(5), 1684–1698 (2014)
4. Dürr, C., Hurand, M.: Finding total unimodularity in optimization problems solved by linear programs. Algorithmica (2009). doi:10.1007/s00453-009-9310-7
5. Gabow, H.N., Tarjan, R.E.: A linear-time algorithm for a special case of disjoint set union. In: Proceedings of the 15th annual ACM symposium on Theory of Computing, pp. 246–251 (1983)
6. Goldberg, A.V.: Scaling algorithms for the shortest paths problem. SIAM J. Comput. **24**(3), 494–504 (1995)
7. Harter, R.: The minimum on a sliding window algorithm. Usenet article (2001). http://richardhartersworld.com/cri/2001/slidingmin.html
8. Horn, W.A.: Some simple scheduling algorithms. Naval Res. Logistics Q. **21**(1), 177–185 (1974)
9. Leung, J.Y.-T.: Handbook of Scheduling: Algorithms, Models, and Performance Analysis. CRC Press, Boca Raton (2004)
10. Lipski Jr, W., Preparata, F.P.: Efficient algorithms for finding maximum matchings in convex bipartite graphs and related problems. Acta Informatica **15**(4), 329–346 (1981)
11. López-Ortiz, A., Quimper, C.-G.: A fast algorithm for multi-machine scheduling problems with jobs of equal processing times. In: Proceedings of the 28th International Symposium on Theoretical Aspects of Computer Science (STACS 2011), pp. 380–391 (2011)
12. Möhring, R.H., Schulz, A.S., Stork, F., Uetz, M.: Solving project scheduling problems by minimum cut computations. Manage. Sci. **49**(3), 330–350 (2003)

13. Simons, B.: Multiprocessor scheduling of unit-time jobs with arbitrary release times and deadlines. SIAM J. Comput. **12**(2), 294–299 (1983)
14. Simons, B.B., Warmuth, M.K.: A fast algorithm for multiprocessor scheduling of unit-length jobs. SIAM J. Comput. **18**(4), 690–710 (1989)
15. Simons, B.: A fast algorithm for single processor scheduling. In: 2013 IEEE 54th Annual Symposium on Foundations of Computer Science, pp. 246–252. IEEE (1978)
16. Wolsey, L.A., Nemhauser, G.L.: Integer and Combinatorial Optimization. Wiley, New York (2014)

Geometric Optimization

The Discrete and Mixed Minimax 2-Center Problem

Yi Xu[1(✉)], Jigen Peng[1,2], Yinfeng Xu[3,4], and Binhai Zhu[5]

[1] School of Mathematics and Statistics, Xi'an Jiaotong University, Xi'an 710049, People's Republic of China
xy.clark@stu.xjtu.edu.cn

[2] Beijing Center for Mathematics and Information Interdisciplinary Sciences, Beijing 100048, People's Republic of China

[3] School of Management, Xi'an Jiaotong University, Xi'an 710049, People's Republic of China

[4] The State Key Lab for Manufacturing Systems Engineering, Xi'an 710049, People's Republic of China

[5] Department of Computer Science, Montana State University, Bozeman, MT 59717, USA

Abstract. Let P be a set of n points in the plane, the discrete minimax 2-center problem ($DMM2CP$) is that of finding two disks centered at $\{p_1, p_2\} \in P$ that minimize the maximum of two terms, namely, the Euclidean distance between two centers and the distance of any other point to the closer center. The mixed minimax 2-center problem ($MMM2CP$) is when one of the two centers is not in P. We present algorithms for solving the $DMM2CP$ and $MMM2CP$. The time complexity of solving $DMM2CP$ and $MMM2CP$ are $O(n^2 \log n)$ and $O(n^2 \log^2 n)$ respectively.

Keywords: 2-center problem · Farthest point Voronoi diagram · Computational geometry

1 Introduction

Facility location problems can be seen as the model for applications in a diverse set of fields including public policy, locating fire stations in a city, locating base stations in wireless networks, clustering of documents and so on. It has been extensively studied over the past years. The facility location problem is to choose the location of facilities to minimize the cost of satisfying the demand for certain commodity. Sometimes the location problem is associated with the costs for locating the facilities, as well as the transportation costs for distributing the commodities. This paper considers two new facility location problems: the discrete and mixed minimax 2-center problem.

The 2-center problem for a planar point set P, i.e., finding two congruent closed disks whose union covers the point set and the radius is as small

Z. Lu et al. (Eds.): COCOA 2015, LNCS 9486, pp. 101–109, 2015.
DOI: 10.1007/978-3-319-26626-8_8

as possible, has also been extensively studied. It is a special case of the general k-center problem, where we need to find k congruent closed disks whose union covers P and the radius is as small as possible. When k is part of the input, the problem is known to be NP-complete [12]. For the 2-center problem, Jaromczyk and Kowaluk first gave a deterministic algorithm with running time $O(n^2 \log n)$ [7]. Eppstein gave an improvement with a randomized algorithm running in $O(n \log^2 n)$ expected time [8]. In a major breakthrough, Sharir showed that the planar 2-center problem can actually be solved in near-linear time. The time bound of their algorithm is $O(n \log^9 n)$ time [3] and finally the algorithm was further improved by Chan in $O(n \log^2 n \log^2 \log n)$ time [4]. Yet another version of the 2-center problem is under the so-called streaming model, i.e., the points in P appear in sequence and we need to maintain the two centers right after a point arrives. A factor-5.611 approximation was designed in [14].

The discrete 2-center problem ($D2CP$) is defined as follows: covering P by the union of two congruent closed disks whose radius is as small as possible, and whose centers are two points in P. It has also been considered. The first near-quadratic algorithm was proposed in [9] and finally improved to $O(n^{\frac{4}{3}} \log^5 n)$ time in [2].

While much has been done on this classical problem, little has been done in some practical variations. Some interesting results were provided by Ho et al. [1] and Gudmundsson et al. [6]. Ho et al. studied the geometric minimum-diameter spanning tree ($MDST$) of P that is a tree which spans P and minimizes the Euclidean distance of the longest path. They showed that there always exists an $MDST$ which is either a monopolar or a dipolar. The more difficult dipolar case can be computed in $O(n^3)$ time in [1]. The cubic time algorithm has been improved to $O(n^{\frac{17}{6}})$ time by Chan [10]. Gudmundsson et al. [6] studied the minimum sum dipolar spanning tree ($MSST$), which mediates between the minimum-diameter dipolar spanning tree and the discrete 2-center problem in the following sense: find two centers p_1 and p_2 in P that minimize the sum of their distance plus the maximum distance of any other point to the closer center. They showed that the $MSST$ can be solved in $O(n^2 \log n)$ time.

Note that in the dipolar case, the $MDST$ is to find two centers $\{p_1, p_2\} \in P$ such that $r_1 + r_2 + d(p_1, p_2)$ is minimized, where $d(p_1, p_2)$ is the Euclidean distance between p_1 and p_2, r_1 and r_2 are the radii of two disks centered at p_1 and p_2 whose union covers P. Using the notation above, the discrete 2-center problem consists of finding two centers $\{p_1, p_2\} \in P$ such that $max\{r_1, r_2\}$ is minimized; the $MSST$ consists of finding two centers $\{p_1, p_2\} \in P$ such that $d(p_1, p_2) + max\{r_1, r_2\}$ is minimized.

In this paper we study two generalizations of the 2-center problem: the discrete minimax 2-center problem ($DMM2CP$) and mixed minimax 2-center problem ($MMM2CP$). The minimax 2-center problem is that of finding two centers $\{p_1, p_2\}$ with radius r_1, r_2 respectively such that $max\{r_1, r_2, d(p_1, p_2)\}$ is minimized. If the two centers are not in P, we call the problem the standard minimax 2-center problem ($SMM2CP$); if one of the two centers is in P, we call the problem the mixed minimax 2-center problem ($MMM2CP$) and if both the two centers are in P, we call the problem the discrete minimax 2-center problem ($DMM2CP$). $DMM2CP$ is similar to the $MSST$ because both of them are interested in when the two centers do not only serve their customers, but also frequently exchange goods or personnel between themselves. We show that the way solving the $MSST$ can also be applied to the $DMM2CP$. Furthermore, we consider the mixed minimax 2-center problem that one of two centers is not in P. This is of importance, e.g., considering the construction cost of the facility location, the minimax 2-center location problem with cost constraint is that, suppose building a new facility costs M_1, rebuilding (changing one facility into a new one) a facility costs M_2 and the budget is M, under the budget constraint, how to choose two centers is necessary (Note that M_1 is greater than M_2). Here we consider the minimax two center location problem as following:

$$\begin{cases} if\ M \geq 2M_1, & \text{correspoding to the } SMM2CP \\ if\ M_1 + M_2 \leq M < 2M_1, & \text{correspoding to the } MMM2CP \\ if\ M < M_1 + M_2, & \text{correspoding to the } DMM2CP \end{cases}$$

In Sect. 2, we discuss the $DMM2CP$ which can be solved by using the data structure in [6]. In Sect. 3, we show some properties and consider the minimum enclosing disk with constraint that the center must be lying on a given circle or arc, which can be solved in $O(n \log n)$ time. Finally the $MMM2CP$ can be solved in $O(n^2 \log^2 n)$ time using $O(n)$ space. In Sect. 4, we give a conclusion.

2 The Discrete Minimax 2-Center Problem

For a planar point set P, the discrete minimax 2-center problem ($DMM2CP$), as a variation of the discrete 2-center problem, is defined as follows: finding two disks centered at $\{p_1, p_2\} \in P$ that minimize the maximum of two terms, namely, the Euclidean distance between two centers and the distance of any other point to the closer center. It is simple to give an $O(n^3)$ time algorithm. Just go through all $O(n^2)$ pairs $\{p, q\}$ of input points and compute $\forall m_i \in P \setminus \{p, q\}, min\{max\{d(p, q), min\{d(m_i, p), d(m_i, q)\}\}\}$. Note that $d(m_i, p) < d(m_i, q)$ means m_i is in the closed halfplane h_{pq} that contains p and is delimited by the perpendicular bisector b_{pq} of p and q. Let P_p (resp. P_q) be the point set with all the points in it are closer to p (resp. q) (Points on b_{pq} can be assigned to either p or q). Clearly $min\{max\{d(p', p), d(q', q), d(p, q)\}\}$ where $p' \in P_p$ and $q' \in P_q$ is the solution of $DMM2CP$ for P. It is clear that we only need to compute the farthest point f_p (resp. f_q) to p (resp. q) in P_p (resp. P_q).

From [6], Gudmundsson et al. considered the minimum sum dipolar spanning tree, which can be seen as finding two centers p, q in P that minimize the sum of $d(p, q)$ plus the maximum distance of any other point to the closer center. They presented a solution by using a new data structure, built upon a balanced red-black search tree that solves f_p for each ordered pair $\{p, q\}$ in a batch. For fixed p as one center, computing f_p for every $q \in P \setminus \{p\}$ takes $O(n \log n)$ time by using their data structure. It can be implied that the data structure can also be applied to the $DMM2CP$: for each fixed center p, compute the f_p for each q

in $P \setminus \{p\}$. After that, for each ordered pair $\{p, q\}$, compute $max\{f_p, f_q, d(p, q)\}$. Then $DMM2CP$ can also be solved in $O(n^2 \log n)$ time.

Algorithm 1. The Algorithm of Discrete Minimax 2-Center Problem

Phase I: Compute all f_p
for each $p \in P$ **do**: Compute all the farthest points f_p
end for $\{p\}$
Phase II: Search for $DMM2CP$
for each $\{p, q\} \in P$ **do**: $d_{pq} \leftarrow max\{d(p, f_p), d(p, q), d(q, f_q)\}$
end for $\{p, q\}$
Return d_{pq} minimum with p and q.

Theorem 1. *Let P be a planar point set of n points, the $DMM2CP$ can be found in $O(n^2 \log n)$ time using quadratic space.*

3 The Mixed Minimax 2-Center Problem

In this section, we consider the minimax 2-center problem in the mixed case where only one center is in P. First we give some notations: for a fixed point p and fixed radius r, let $D(p, r)$ denote the disk with center p and radius r, $C(p, r)$ denote the circle bounding $D(p, r)$. Let set $P_{p,r}$ consist of the points in $D(p, r)$ and $P^T_{p,r}$ be the complementary set. The points located on $C(p, r)$ are also in $P_{p,r}$. The center and the radius of the minimum enclosing disk of $P^T_{p,r}$ are o^s_p and r^s_p respectively. We call the 2-center problem with one center in p the mixed 2-center problem ($M2CP$). For fixed p and r, let the other disk of the solution of $MMM2CP$ be $D(o^m_p, r^m_p)$. For each fixed center p, we first go through the points $p_i \in P \setminus \{p\}$ $(i = 1, 2, ..., n-1)$ in order of non-decreasing distance from p. For a trivial $O(n^2 \log n)$ time implement of this procedure, for each p and $d(p, p_i)$ as the fixed center and radius, we compute the optimal solution for $MMM2CP$. We show some geometry properties of the optimal solution of $MMM2CP$ for fixed center and radius.

3.1 The Structure of the Optimal Solution

First we show two properties between $MMM2CP$ and $M2CP$.

Lemma 1. *For a fixed center p and radius r, $max\{r, r^m_p, d(p, o^m_p)\} \geq max\{r, r^s_p\}$.*

Lemma 2. *For a fixed center p and radius r, if p and o^s_p are in one of $D(p, r)$ or $D(o^s_p, r^s_p)$, then $MMM2CP$ and $M2CP$ have the same optimal solution.*

Proof. The optimal radius of $M2CP$ is $max\{r, r^s_p\}$, and $max\{r, r^s_p\} \geq d(p, o^s_p)$. As r^s_p is the radius of the minimum enclosing disk of set $P^T_{p,r}$, from Lemma 1, we have $r^s_p \leq r^m_p$. Then we choose o^s_p and r^s_p as the center and radius of the other disk of $MMM2CP$ respectively. The $MMM2CP$ and $M2CP$ have the same optimal solution. □

From Lemma 2, we have

Lemma 3. *For a fixed center p and radius r, the optimal solution of MMM2CP is either when p and o_p^m are in one of $D(p,r)$ or $D(o_p^m, r_p^m)$ which is same to M2CP, or when one center lies on the boundary of another disk.*

Proof. From Lemma 2, if p and o_p^s are in one of $D(p,r)$ or $D(o_p^s, r_p^s)$, the optimal solution equals to the solution of $M2CP$. We choose o_p^s and r_p^s as the center and radius of o_p^m and r_p^m.

Otherwise, we show that if p and o_p^s are not in one of $D(p,r)$ or $D(o_p^s, r_p^s)$, then p or o_p^m must lie on the boundary of the other disk. We can enlarge the radius r_p^s into $r_p^s + \delta$ where δ is a positive number so that $P_{p,r}^T$ is still located in $D(o_p^s, r_p^s + \delta)$. We move o_p^s towards p until $C(o_p^s, r_p^s + \delta)$ hits at least one point $q_1 \in P_{p,r}^T$ which satisfies $d(o_p^s, p) = r_p^s + \delta$ or $d(o_p^s, p) = r$. We denote the new center o_p^s as p_1^*. During the move, the distance between p and p_1^* is decreasing. There exists a δ such that $max\{r_p^s + \delta, d(p, p_1^*), r\}$ is minimized.

At this time, either p_1^* lies on $C(p,r)$ which means $d(p, p_1^*) = r$, or p lies on $C(p_1^*, r_p^s + \delta)$ which means $d(p, p_1^*) = r_p^s + \delta$. Because for δ, $max\{r_p^s + \delta, d(p, p_1^*), r\}$ is minimized, $\forall \gamma > \delta$, $max\{r, d(p, p_1^*), r_p^s + \gamma\} > max\{r, d(p, p_1^*), r_p^s + \delta\}$. Finally we choose p_1^* as the center o_p^m and $r_p^s + \delta$ as the radius r_p^m. Figure 1 shows these cases. The blue bold circle is the boundary of the fixed disk $D(p,r)$; the green thin circle is the boundary of the other disk of the solution of $M2CP$; the red dash circle is the boundary of the other disk of the solution of $MMM2CP$. □

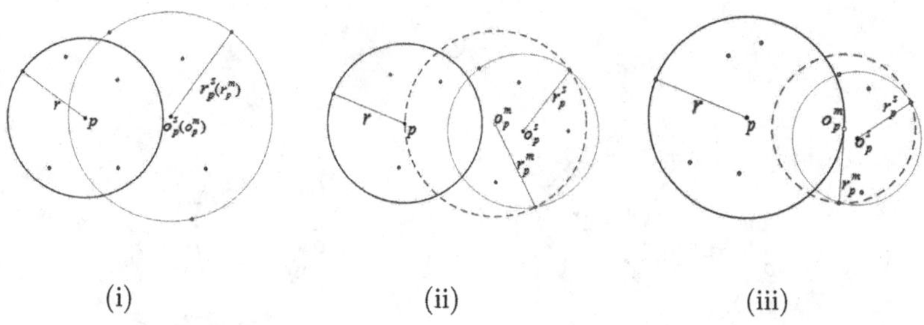

(i) (ii) (iii)

Fig. 1. (i) p and o_p^m are in one disk (ii) p is on $C(o_p^m, r_p^m)$ (iii) o_p^m is on $C(p,r)$

3.2 Solving $MMM2CP$ for a Fixed p and r

We first show that how to find the other center and radius of $MMM2CP$ for a fixed center p and radius r. From Lemma 2, we know that if p and o_p^s are both in $D(p,r)$ or $D(o_p^s, r_p^s)$, then $M2CP$ and $MMM2CP$ have the same solution.

We turn to solve the case where one center lies on the boundary of the other disk. We have the following theorem which is similar to the Lemma 1 in [13].

Theorem 2. *For a planar point set P and a fixed circle or arc whose center is o and radius is r, the minimum enclosing disk $D(o^*, r^*)$ covering set P with center on the given circle or arc, satisfies that either when at least two points are on $C(o^*, r^*)$, or when one point p is on $C(o^*, r^*), o, o^*$ and p must be collinear and p is the farthest point of o in P.*

Proof. If the minimum enclosing disk $D(o^*, r^*)$ covering set P with center on the given circle or arc is defined by at least two points a, b on $C(o^*, r^*)$, then a, b are the farthest points to o^*. Otherwise there is only one point p on $C(o^*, r^*)$.

Suppose o, o^* and p are not on a line. As p is the only one point on $C(o^*, r^*)$, we move the center o^* along $C(o, r)$ in order that $\angle oo^*p < \angle oo'^*p$, where o'^* is also on $C(o, r)$ so that P is still located in $D(o'^*, r^*)$. With center o'^* and radius r^*, all the distance between the points in P and o'^* is smaller than r^*, which implies that we can reduce the radius of the disk to maintain that the disk can still cover P. With center o'^*, the radius of the minimum enclosing disk covering P is smaller than r^*. This contradicts to that $D(o^*, r^*)$ is the minimum enclosing disk covering P with center on $C(o, r)$. Figure 2(i) shows that if o, o^* and p are not on a line, then there exists a disk that covers P with center on $C(o, r)$ but the radius is smaller than r^*. The disk with the dash circle can also cover P. Figure 2(ii) shows that o, o^* and p are on a line, then $D(o^*, r^*)$ is the minimum enclosing disk with center on $C(o, r)$ and the disk with dash circle can't cover P.

Suppose p is not the farthest point to o in set P, let f_o be the farthest point to o, then $d(o, f_o) > d(o, p)$. As $d(o^*, p) > d(o^*, f_o)$, by the triangle inequality, $d(o, p) = d(o^*, p) + d(o, o^*) > d(o^*, f_o) + d(o, o^*) = d(o, f_o)$. This contradicts to the assumption. □

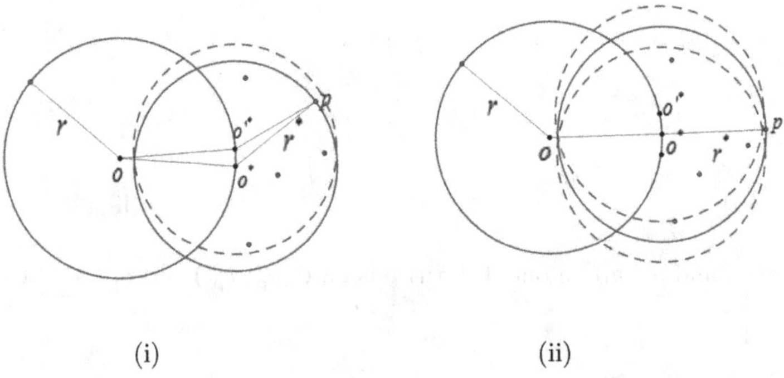

(i) (ii)

Fig. 2. (i) o, o^*, p is not on a line (ii) o, o^*, p is on a line

From Lemma 3, for a fixed center p and radius r, $D(o_p^m, r_p^m)$ is the minimum enclosing disk covering set $P_{p,r}^T$ with the constraint that either p lies on $C(o_p^m, r_p^m)$ or o_p^m lies on $C(p, r)$. If p lies on $C(o_p^m, r_p^m)$, we know that $D(o_p^m, r_p^m)$ is the minimum disk covering $p \cup P_{p,r}^T$. Otherwise o_p^m lies on $C(p, r)$.

For the first case, we can compute the farthest point Voronoi diagram of $p \cup P^T_{p,r}$ and compute the minimum enclosing disk with p on its boundary. This can be done in $O(n \log n)$ time [11]; for the second case, from Theorem 2, if there are more than one point in $P^T_{p,r}$ on $C(o^m_p, r^m_p)$, we can compute the intersection points of $C(p,r)$ and farthest point Voronoi diagram of $P^T_{p,r}$. We can then find the minimum distance between the intersection points and their farthest points. Otherwise there is only one point which is the farthest point of p in $P^T_{p,r}$ on $C(o^m_p, r^m_p)$. We compute the farthest point f_p to p, then find the intersection point w of $C(p,r)$ and l_{pf_p} which is the line crossing p and f_p. Let R_{f_p} denote the region in which all the points have the same farthest point f_p. If w lies in the region R_{f_p}, then we find the optimal center. Otherwise there must be at least two points on $C(o^m_p, r^m_p)$. As there are at most $O(n)$ lines of the farthest point Voronoi diagram of $P^T_{p,r}$ and $O(n)$ intersection points, then this can be done in $O(n \log n)$ time. These two cases can be solved in $O(n \log n)$.

Lemma 4. *For a fixed p and r, the MMM2CP can be solved in* $O(n \log n)$ *time.*

Algorithm 2. The Algorithm of Mixed Minimax 2-Center Problem for a Fixed p and r

---

For a fixed p and r, by using linear programming in [5], compute the center o^s_p and radius r^s_p of the minimum enclosing circle of $P^T_{p,r} := P \setminus \{p, p_1, ..., p_i\}$ where $d(p, p_i) \leq r$.
if $d(o^s_p, p) \leq max\{r^s_p, r\}$, **then**
    The $MMM2CP$ and $M2CP$ have the same solution. Return $r^m_p = max\{r^s_p, r\}$.
end if
(1) Compute the minimum enclosing disk $P^T_{p,r} \cup \{p\}$ with p on its boundary. Let the radius be r_1.
(2) Compute the farthest point f_p to p in set $P^T_{p,r}$, and the intersection point w of $C(p,r)$ and l_{pf_p}. If w is in the region of R_{f_p}, then w is o^m_p. Let $r_2 = d(p, f_p) - r$. Otherwise go to next step.
(3) Compute the intersection point of the farthest point Voronoi diagram of $P^T_{p,r}$ and $C(p,r)$. Compute the distance between the intersection points and their farthest points, choose the minimum distance, denoted as r_3.
(4) Return $r^m_p = min\{r_1, r_2, r_3\}$.

3.3 The Monotone Property

In the preprocessing, we have $d(p,p_1) \leq d(p,p_2) \leq ... \leq d(p,p_{n-1})$. Suppose that $D(p,r_i)$ is the fixed disk where $r_i = d(p,p_i)$, let $D(o^m_i, r^m_i)$ denote the other disk of the $MMM2CP$ whose center is o^m_i and radius is r^m_i. We discuss the monotone property under two cases.
Case 1. $r_i \geq r^m_i$. In this case, we have:

Lemma 5. *If* $r_i \geq r^m_i$, *then* $\forall k > i, r_i < max\{r_k, r^m_k, d(p, o^m_k)\}$.

Proof. If $r_i \geq r_i^m$, from Lemma 3, the optimal solution of $MMM2CP$ is either when the two centers are in $D(p, r_i)$ or when o_i^m is on $C(p, r_i)$. $\forall k > i, r_k \leq max\{r_k, d(p, o_k^m), r_k^m\}$. Thus, $r_i < max\{r_k, r_k^m, d(p, o_k^m)\}$. □

Case 2. $r_i < r_i^m$. We handle this case with the following two lemmas.

Lemma 6. $\forall k < i, r_k < r_k^m$.

Proof. Suppose that there exists a point p_k with $k < i$, such that $r_k \geq r_k^m$. Then the union of the two disks $D(p, r_k) \cup D(o_k^m, r_k^m)$ can cover the set P. So, $D(p, r_i) \cup D(o_k^m, r_k^m)$ can also cover the set P. Then, $r_i \geq r_k^m$. The optimal solution of $MMM2CP$ for a fixed point p and radius r_i is less than or equal to r_i. It contradicts to the optimal solution r_i^m. □

Lemma 7. *If* $r_i < r_i^m$, *then* $\forall k < i, r_i^m \leq r_k^m$.

Proof. Suppose that there exists a point p_k with $k < i$ such that $r_k^m < r_i^m$. Then we can enlarge the radius r_k into r_i, the union of $D(p, r_i)$ and $D(o_k^m, r_k^m)$ can cover the set P. In this case, the solution of optimal radius of $MMM2CP$ with a fixed center p and radius r_i is $max\{r_i, r_k^m, d(p, o_k^m)\}$. From Lemma 6, we have $r_k < r_k^m$, so $d(p, o_k^m) \leq r_k^m$. With a fixed center p and radius r_i, $D(p, r_i) \cup D(o_k^m, r_k^m)$ covers the set P and the solution of $MMM2CP$ is $max\{r_i, r_k^m\}$. As $r_k^m < r_i^m, r_i < r_i^m$, again, it contradicts to the optimal solution $r_i < r_i^m$. □

From Lemma 4, for a fixed p and radius r, the $MMM2CP$ can be found in $O(n \log n)$ time. From Lemmas 5 and 7, by using binary search, in $O(\log n)$ time, the $MMM2CP$ can be solved in $O(n \log^2 n)$ time for a fixed p. Then going through all the points in P, the total complexity for solving the $MMM2CP$ is $O(n^2 \log^2 n)$.

Theorem 3. *Let P be a planar point set of n points, the $MMM2CP$ can be solved in $O(n^2 \log^2 n)$ time using $O(n)$ space.*

4 Conclusion

We consider two facility location problems called the discrete minimax 2-center problem ($DMM2CP$) and the mixed minimax 2-center problem ($MMM2CP$). We show that the $DMM2CP$ can be solved in $O(n^2 \log n)$ time using quadratic space and the $MMM2CP$ can be solved in $O(n^2 \log^2 n)$ time using $O(n)$ space. So far, we have not been able to find any algorithm for the $SMM2CP$. This is an interesting problem for further research.

Acknowledgment. This work is supported by NSF of China under Grants 61221063 and Program for Changjiang Scholars and Innovative Research Team in University under Grant IRT1173.

References

1. Ho, J.M., Lee, D.T., Chang, C.H., Wong, C.K.: Minimum diameter spanning trees and related problems. SIAM J. Comput. **20**(5), 987–997 (1991)
2. Agarwal, P.K., Sharir, M., Welzl, E.: The discrete 2-center problem. Discrete Comput. Geom. **20**, 287-305 (2000)
3. Sharir, M.: A near-linear algorithm for the planar 2-center problem. Discrete Comput. Geom. **18**, 125–134 (1997)
4. Chan, T.M.: More planar two-center algorithms. Comput. Geom. Theory Appl. **13**, 189–198 (1999)
5. Megiddo, N.: Linear-time algorithms for the linear programming in R^3 and related problems. SIAM J. Comput. **12**, 759–776 (1983)
6. Gudmundsson, J., Haverkort, H., Park, S.M., Shin, C.S., Wolff, A.: Facility location and the geometric minimum-diameter spanning tree. Comput. Geom. **27**, 87–106 (2004)
7. Jaromczyk, J., Kowaluk, M.: An efficient algorithm for the Euclidean two-center problem. In: Proceedings 10th ACM Symposium on Computational Geometry, pp. 303–311 (1994)
8. Eppstein, D.: Faster construction of planar two-centers. In: Proceedings of the 8th ACMCSIAM Symposium Discrete Algorithms, pp. 131–138 (1997)
9. Hershberger, J., Suri, S.: Finding tailored partitions. J. Algorithms **12**, 431–463 (1991)
10. Chan, T.M.: Semi-online maintenance of geometric optima and measures. SIAM J. Comput. **32**(3), 700–716 (2003)
11. Shamos, M., Michael, I., Hoey, D.: Closest-point problems. In: 16th Annual Symposium on IEEE Foundations of Computer Science, pp. 151–162 (1975)
12. Megiddo, N., Supowit, K.: On the complexity of some common geometric location problems. SIAM J. Comput. **13**, 1182–1196 (1984)
13. Roy, S., Bardhan, D., Das, S.: Base station placement on boundary of a convex polygon. J. Parallel Distrib. Comput. **68**, 265–273 (2008)
14. Poon, C.K., Zhu, B.: Streaming with minimum space: an algorithm for covering by two congruent balls. Theor. Comput. Sci. **507**, 72–82 (2013)

Approximation Algorithms for Generalized MST and TSP in Grid Clusters

Binay Bhattacharya[1], Ante Ćustić[1], Akbar Rafiey[1](✉), Arash Rafiey[1,2], and Vladyslav Sokol[1]

[1] Simon Fraser University, Burnaby, Canada
{binay,acustic,arafiey,arashr,vsokol}@sfu.ca
[2] Indiana State University, Terre Haute, IN, USA
arash.rafiey@indstate.edu

Abstract. We consider a special case of the generalized minimum spanning tree problem (GMST) and the generalized travelling salesman problem (GTSP) where we are given a set of points inside the integer grid (in Euclidean plane) where each grid cell is 1×1. In the MST version of the problem, the goal is to find a minimum tree that contains exactly one point from each non-empty grid cell (cluster). Similarly, in the TSP version of the problem, the goal is to find a minimum weight cycle containing one point from each non-empty grid cell. We give a $(1+4\sqrt{2}+\epsilon)$ and $(1.5+8\sqrt{2}+\epsilon)$-approximation algorithm for these two problems in the described setting, respectively.

Our motivation is based on the problem posed in [6] for a constant approximation algorithm. The authors designed a PTAS for the more special case of the GMST where non-empty cells are connected end dense enough. However, their algorithm heavily relies on this connectivity restriction and is unpractical. Our results develop the topic further.

Keywords: Generalized minimum spanning tree · Generalized travelling salesman · Grid clusters · Approximation algorithm

1 Introduction

The *generalized minimum spanning tree problem* (GMST) is a generalization of the well known *minimum spanning tree problem* (MST). An instance of the GMST is given by an undirected graph $G = (V, E)$ where the vertex set is partitioned into k *clusters* V_i, $i = 1, \ldots, k$, and a weight $w(e) \in \mathbb{R}^+$ is assigned to every edge $e \in E$. The goal is to find a tree with minimum weight containing one vertex from each cluster.

The GMST occurs in telecommunications network planning, where a network of node clusters need to be connected via a tree architecture using exactly one node per cluster [9]. More precisely, local subnetworks must be interconnected by a global network containing a gateway from each subnetwork. For this

Supported by NSERC Canada.

Z. Lu et al. (Eds.): COCOA 2015, LNCS 9486, pp. 110–125, 2015.
DOI: 10.1007/978-3-319-26626-8_9

inter-networking, a point has to be chosen in each local network as a hub and the hub point must be connected via transmission links such as optical fiber, see [14]. Furthermore, the GMST has some applications in design of backbones in large communication networks, energy distribution, and agricultural irrigation [10].

The GMST was first introduced by Myung, Lee and Tcha in 1995 [14]. Although MST is polynomially solvable [7], it was shown in [14] that the GMST is strongly NP-hard and there is no constant factor approximation algorithm, unless P = NP. However, several heuristic algorithms have been suggested for the GMST, see [9,10,16,17]. Furthermore, Pop, Still and Kern [18] used an LP-relaxation to develop a 2ρ–approximation algorithm for the GMST where the size of every cluster is bounded by ρ.

In [6], Feremans, Grigoriev and Sitters consider the *geometric generalized minimum spanning tree problem in grid clusters*, GGMST for short. In this special case of the GMST, a complete graph $G = (V, E)$ is given where the set of vertices V correspond to a set of points in the planar integer grid. Every non-empty 1×1 cell of the grid forms a cluster. The weight of the edge between two vertices is given by their Euclidean distance. Figure 1 depicts one instance of the GGMST.

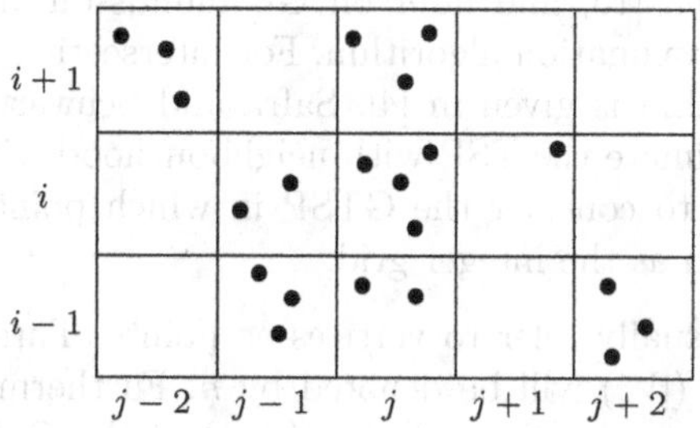

Fig. 1. An GGMST instance with $n = 21$ points and $N + 1 = 8$ non-empty cells, which are connected and fit into a 3×5 sub-grid

We say that two grid cells are connected if they share a side or a corner. Furthermore, we say that a set of grid cells is connected if they form one connected component. The authors in [6] show that the GGMST is strongly NP-hard, even if we restrict to instances in which non-empty grid cells are connected and each grid cell contains at most two points. Furthermore, they designed a dynamic programming algorithm that solves in $\mathcal{O}(l\rho^{6k}2^{34k^2}k^2)$ time the GGMST for which the set of non-empty grid cells is connected and fits into $k \times l$ sub-grid. (Note that the algorithm is polynomial if k is bounded.) Moreover, the authors used this algorithm to develop a polynomial time approximation scheme (PTAS) for the GGMST for which non-empty cells are connected and the number of non-empty cells is superlinear in k and l. The GGMST instances are often used to test heuristics for the GMST which, in light of the results in [6], is not adequate. The objective of this paper is to develop this topic further and to design a simple approximation algorithms for the GGMST and of its variants without restricting only to connected and dense instances.

Analogously as the GMST and the GGMST, the *generalized travelling salesman problem* (GTSP) and the *geometric generalized travelling salesman problem in grid clusters* (GGTSP) can be defined. The GTSP was introduced by Henry-Labordere [11] and is also known in the literature as *set TSP*, *group TSP* or *One-of-a-Set TSP*. This problem has many applications, including airplane routing, computer file sequencing, and postal delivery, see [2,12,13]. Elbassioni, Fishkin, Mustafa and Sitters [5] considered the GTSP in which non-empty clusters (i.e. regions) are disjoint α-fat objects with possibly varying size. In this setting they obtained a $(9.1\alpha + 1)$-approximation algorithm. They also give the first $\mathcal{O}(1)$-approximation algorithm for the problem with intersecting clusters (regions). Note that in the GGTSP, fatness of each cluster is 4 (each cluster is a square).

As a special case of the GTSP we can look at each geometric region as an infinite set of points. This problem, called the *TSP with neighbourhood*, was introduced by Arkin and Hassin [1]. In the same paper they present constant factor approximation algorithm for two cases in which the regions are translates of disjoint convex polygons, and for disjoint unit disks. For the general problem Mata and Mitchell [15] and later on Gudmundsson and Levcopoulos [8], gave an $\mathcal{O}(\log n)$-approximation algorithm. For intersecting unit disks an $\mathcal{O}(1)$-approximation algorithm is given in [4]. Safra and Schwartz [19] show that it is NP-hard to approximate the TSP with neighbourhood within $(2 - \epsilon)$. In this context, it is natural to consider the GTSP in which points are sitting inside geometric objects such as the integer grid.

Notation. We will usually refer to vertices as points. Throughout this paper, the number of points ($|V|$) will be denoted by n. Furthermore, N denotes the number of edges in every feasible solution (tree) of the GGMST, i.e. N is the number of non-empty cells minus 1. The edge between two points u and v will be denoted by $e_{u,v}$. We naturally extend the notation for the weight to sets of edges and graphs, i.g. the weight of a tree T is denoted by $w(T) = \sum_{e \in T} w(e)$, where $e \in T$ means that e is an edge of T. We assume that every point is in just one cell, i.e. points on the cell borders are assigned to only one neighbouring cell by any rule. An optimal solution of the GGMST will be denoted by T_{opt} throughout this paper.

Our results and organization of the paper. The main result of this paper is a $(1 + 4\sqrt{2} + \epsilon)$-approximation algorithm for the GGMST. We do not assume any restrictions on connectivity, density or cardinality of non-empty cells. The algorithm is presented and analyzed in Sect. 2. A lower bound for the weight of an optimal solution in terms of N is used to prove the approximation quality of the algorithm. Section 3 is devoted to proving this lower bound. Lastly, in Sect. 4 we use our GGMST algorithm to develop an approximation algorithms for the GGTSP.

2 The GGMST Approximation Algorithm

In this section we present a $(1+4\sqrt{2}+\epsilon)$-approximation algorithm (Algorithm 3) for the GGMST. Main part of the algorithm is Algorithm 1 which we describe next.

Algorithm 1. $\left(1+4\sqrt{2}+\frac{2\sqrt{2}}{w(T_{opt})}\right)$-approximation alg. for the GGMST

$T \leftarrow$ solution of the MST problem on non-empty cells (where the distance
  between a pair of cells is the length of the shortest edge between them);
$G \leftarrow$ the graph consisting of the set of edges (and points) that correspond to the
  edges in T;
for all cells C that contain more than one point from G **do**
  $C_G \leftarrow$ the set of points from G that are in C;
  $p \leftarrow$ point from C that is a median for C_G;
  Replace C_G by p, i.e. reconnect to p all edges of G that enter C;
end
return G;

Algorithm 1 is divided into two parts; in the first part we solve an MST instance defined as follows: non-empty cells play the role of vertices, and the weight of the edge between two cells C_1, C_2 is the smallest weight edge e_{p_1,p_2} where $p_1 \in C_1$ and $p_2 \in C_2$. Let T be an optimal tree of such MST instance, and let graph G be the set of edges (with its endpoints) of the original GGMST instance that correspond to the edges of T. Note that G has N edges and spans all non-empty cells but it can have multiple points in some cells. In the second part of the Algorithm 1 (i.e. the **for** loop), we modify G to obtain the GGMST feasibility, by iteratively replacing multiple cell points by a single point p. We choose point p to be the one that has the minimum sum of distances to other points of G that are in the corresponding cell.

Next we present an upper bound for solutions obtained by Algorithm 1 in terms of the number of edges N.

Theorem 1. *Algorithm 1 produces a feasible solution T_A of the GGMST such that $w(T_A) \leq w(T_{opt}) + \sqrt{2}N - \sqrt{2}$, where N is the number of edges of T_A.*

Proof. Denote by G_0 the non-feasible graph obtained in the first part of the algorithm, i.e. the first version of graph G. Then the weight of the solution T_A obtained by the algorithm is equal to $w(G_0) + ext$, where ext is the amount by which we increase (extend) the weight of G_0 in the second part of the algorithm. Note that $w(G_0) \leq w(T_{opt})$, as G_0 is an optimal solution of the problem for which T_{opt} is a feasible solution (find a minimum weight set of edges that spans all non-empty cells, with all GGMST edges being allowed). In the rest of the proof we will bound the value of ext.

In every run of the **for** loop we replace the set of points C_G with p. In doing so, every edge $e_{q,c}$, $c \in C_G$ from G, is replaced by $e_{q,p}$. From the triangle inequality we get that $w(e_{q,p}) - w(e_{q,c}) \leq w(e_{c,p})$. Hence, the increase (extension) of the weight of G in every run of the **for** loop is less or equal than $\sum_{c \in C_G} w(e_{c,p})$. Instead of bounding such absolute values, we will bound its average per edge adjacent to the corresponding cell. More precisely, we will calculate an average extension per *half-edge* assigned to the corresponding cell. Namely, every edge will be extended at most two times, once on each endpoint, so we can look at each extension as an extension of a half-edge. Furthermore, note that edges that contain leafs will be extended only on one side. We will use this fact to assign half-edges that contain leafs to other cells to lower their average half-edge extension. To every cell C, we will assign $|C_G| - 2$ leaf half-edges. Intuitively, we can do this because every node v of a tree *generates* $\deg(v) - 2$ leafs. Formally, it follows from the following well know equality:

$$|V_1| = 2 + \sum_{i \geq 2} |V_i|(i-2), \tag{1}$$

where $V_i = \{v \in V\colon \deg(v) = i\}$, and V is the set of vertices of a graph.

Then for a cell C the average extension per assigned half-edges is bounded above by

$$\frac{\sum_{c \in C_G} w(e_{c,p})}{|C_G| + (|C_G| - 2)}. \tag{2}$$

Note that the maximum distance between two cell points is $\sqrt{2}$. Since points from C_G are candidates for p, it follows that $\sum_{c \in C_G} w(e_{c,p}) \leq \sqrt{2}(|C_G| - 1)$. Hence, (2) is bounded above by

$$\frac{\sqrt{2}(|C_G| - 1)}{2|C_G| - 2} = \frac{\sqrt{2}}{2}.$$

Hence, in average, every half-edge (except 2 leaf half-edges, see (1)) is extended by at most $\sqrt{2}/2$. Note that this average bound is a constant, i.e. does not depend on C. Now *ext* can be bounded by

$$ext \leq \frac{\sqrt{2}}{2}(2N - 2) = \sqrt{2}N - \sqrt{2}. \tag{3}$$

Finally, we can bound the solution T_A of the algorithm by

$$w(T_A) \leq w(G_0) + ext \leq w(T_{opt}) + \sqrt{2}N - \sqrt{2}.$$

□

The following theorem gives a lower bound for the optimal solution in terms of the number of edges N. Section 3 is dedicated to proving the theorem.

Theorem 2. *If T_{opt} is an optimal solution of the GGMST on $N+1$ non-empty cells, then $N \leq 4w(T_{opt}) + 3$.*

Now from Theorems 1 and 2 the following approximation bound for Algorithm 1 follows.

Corollary 1. *Algorithm 1 produces a feasible solution* T_A *of the GGMST such that* $w(T_A) \leq (1+4\sqrt{2})w(T_{opt}) + 2\sqrt{2}$.

Note that, due to the constant $2\sqrt{2}$, Corollary 1 does not gives us a constant approximation ratio for Algorithm 1. Namely, the approximation ratio that we get is equal to $1 + 4\sqrt{2} + \frac{2\sqrt{2}}{w(T_{opt})}$. Next we focus on improving Algorithm 1 so that $\frac{2\sqrt{2}}{w(T_{opt})}$ is replaced by arbitrary small $\epsilon > 0$. Note that the optimal solution weight does not necessarily increase with the increase of the number of points n, namely all points can be in the same cells. Hence we cannot use the standard approach. However, the following two facts will do the trick. First, note that the weight of the GGMST optimal solution increases as the number of non-empty cells increases. Second, given a spanning tree structure of non-empty cells T, we can in polynomial time find the minimum weight GGMST feasible solution T' with the same tree structure as T (i.e. there is an edge in T' between two cells if and only if these two cells are adjacent in T). Next we design one such dynamic programming algorithm (see Algorithm 2).

Given an GGMST instance, let T be a spanning tree of the complete graph where the set of vertices correspond to the set of non-empty cells. Denote by X_i the set of points inside cell C_i. We observe T as a rooted tree with C_r as its root. If C_i is a leaf of T then the weight $W(z)$ of each point z in set X_i is set to zero. If C_i is not a leaf then T has some children $C_{i_1}, \ldots, C_{i_k}$ and the weight for points inside sets $X_{i_1}, \ldots, X_{i_k}$ has already been computed. Then for each point p in cell C_i (set X_i) we compute:

$$W(p) = \sum_{j=1}^{k} \min_{q \in X_{i_j}} \{W(q) + w(e_{p,q})\}$$

Algorithm 2 computes $W(p)$ for all $p \in C_r$. Note that it is easy to adapt Algorithm 2 to store selected points at each step.

Now we have all ingredients to design a $(1 + 4\sqrt{2} + \epsilon)$-approximation algorithm, see Algorithm 3. Note that $1 + 4\sqrt{2}$ is approximately equal to 6.66.

Theorem 3. *For any* $\epsilon > 0$, *Algorithm 3 is a* $(1 + 4\sqrt{2} + \epsilon)$*-approximation algorithm for the GGMST.*

Proof. If $N \leq 15$ or $N \leq 10\sqrt{2}/\epsilon$, then we can enumerate all spanning trees on $N + 1$ non-empty cells, and apply Algorithm 2 on each of them. That will give us an optimal solution in polynomial time.

Assume $N > 15$ and $N > 10\sqrt{2}/\epsilon$. By Corollary 1 it follows that Algorithm 1 will produce a solution T_A such that

$$w(T_A) \leq \left(1 + 4\sqrt{2}\right) w(T_{opt}) + 2\sqrt{2}. \tag{4}$$

Algorithm 2. Optimal GGMST solution for a given spanning tree of cells

Data: A spanning tree T of non-empty cells
Result: An optimal weight of the GGMST tree with the same structure as T
Choose an arbitrary cell C_r as the root of T;
for each leaf C_i of T **do**
  for each $p \in X_i$ **do**
    $W(p) = 0$;
  end
end
$CurrentLevel$ = height of T;
while $CurrentLevel \geq$ root level **do**
  for each node C_i of $CurrentLevel$ **do**
    Let $C_{i_1}, \ldots, C_{i_k}$ be children of C_i in T;
    for each $p \in X_i$ **do**
      $W(p) = \sum_{j=1}^{k} \min_{q \in X_{i_j}} \{W(q) + w(e_{p,q})\}$;
    end
  end
  $CurrentLevel = CurrentLevel - 1$;
end
return $\min_{p \in X_r} W(p)$;

Algorithm 3. $(1 + 4\sqrt{2} + \epsilon)$-approximation algorithm for the GGMST

if $N \leq 15$ **or** $N \leq 10\sqrt{2}/\epsilon$ **then**
  Output minimum weight solution obtained by Algorithm 2 on all spanning trees of non-empty cells;
else
  Run Algorithm 1;
end

From Theorem 2 and $N > 15$ it follows that $1 \leq 5w(T_{opt})/N$. Applying that on the rightmost element of inequality (4) we get

$$w(T_A) \leq \left(1 + 4\sqrt{2}\right) w(T_{opt}) + \frac{10\sqrt{2}}{N} w(T_{opt}),$$
$$\leq \left(1 + 4\sqrt{2} + \frac{10\sqrt{2}}{N}\right) w(T_{opt}).$$

Now from $N > 10\sqrt{2}/\epsilon$ it follows that

$$w(T_A) \leq \left(1 + 4\sqrt{2} + \epsilon\right) w(T_{opt}),$$

which proves the theorem. □

3 The Lower Bound Proof

This section is entirely devoted to proving Theorem 2 which gives us a lower bound on the weight of an optimal solution. The lower bound is expressed in terms of the number of edges N.

Throughout this section we identify 1×1 grid cell with its coordinates (i, j), where $i, j \in \mathbb{Z}$ is the row and the column of the cell inside the infinite integer grid. For example, in Fig. 1, cell $(i, j+1)$ contains one point which is near its upper right corner.

We start by proving lower bounds for trees of small size.

Lemma 1. *The weight of any subtree of T_{opt} with four edges is at least* 1.

Proof. Consider a subtree T' of T_{opt} with four edges. Let H denote the set of the five cells that contain vertices of T'. Note that there will be two cells in H with coordinates (i, j) and (i', j') such that $|i - i'| \geq 2$ or $|j - j'| \geq 2$. Hence, Euclidean distance between a vertex from the cell (i, j) and a vertex from the cell (i', j') is a least 1. This implies $w(T') \geq 1$. See Fig. 2 for an example. □

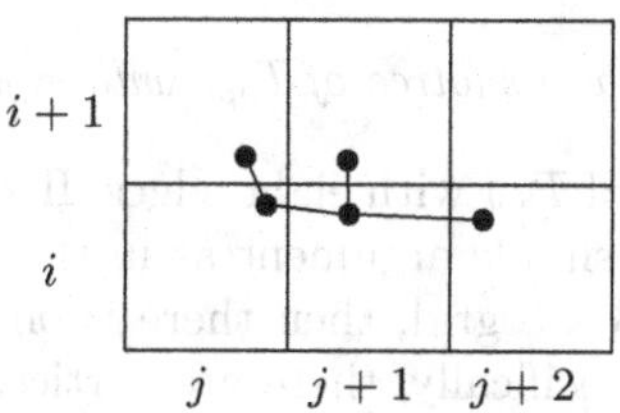

Fig. 2. An example of a tree T' with four edges

Lemma 2. *The weight of any subtree of T_{opt} with seven edges is at least $\frac{1}{3}(2\sqrt{6}+\sqrt{6-3\sqrt{3}})$ (which is greater than 1.93).*

Proof. Let T' be a subtree of T_{opt} with seven edges. If T' does not fit in any 3×3 sub-grid of the original grid, then there are two vertices u, v of T' which are from cells with coordinates (i, j) and (i', j') such that $|i - i'| \geq 3$ or $|j - j'| \geq 3$. In that case $w(e_{u,v}) \geq 2$ and therefore $w(T') \geq 2$.

Next we consider the case when T' fits into 3×3 grid. Since T' has eight vertices, at least three of them are in the corner cells of a 3×3 grid. Without loss of generality we assume that these three vertices are vertex v in cell (i, j), vertex u in cell $(i+2, j)$ and vertex y in cell $(j+2, i)$. Let P be a shortest path in T' from v to u and let Q be the shortest path in T' from v to y. Note that $w(e_{v,u}) \geq 1$ and $w(e_{v,y}) \geq 1$. If P and Q do not have a common vertex apart from v, then $w(T') \geq 2$. Thus we are left with the case when P and Q have a common vertex other than v, which we denote by x.

First we assume that P and Q do not go through the point in cell $(i+1, j+1)$. In this case, up to symmetry, one of the configurations depicted in Fig. 3(a,b)

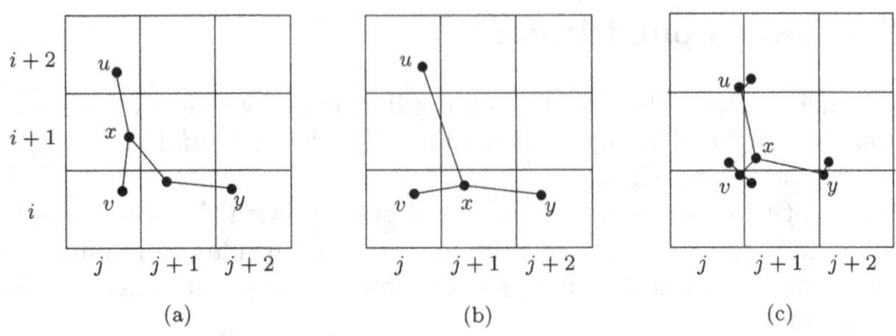

Fig. 3. Layouts of P and Q

occurs. However, it is clear that $w(e_{v,x}) + w(e_{x,y}) + w(e_{x,u}) \geq 2$ and hence $w(T') \geq 2$.

Lastly, we observe the case when vertex x is in cell $(i+1, j+1)$. Then $w(P \cup Q)$ is at least $w(e_{x,v}) + w(e_{x,u}) + w(e_{x,y})$, which is minimized when x is the Fermat point for the three corners of cell $(i+1, j+1)$ and T' has the structure depicted in Fig. 3(c). Therefore it can be computed that $w(T') \geq \frac{1}{3}(2\sqrt{6} + \sqrt{6 - 3\sqrt{3}}) > 1.93$. □

Lemma 3. *The weight of any subtree of* T_{opt} *with eight edges is at least* 2.

Proof. Let T' be a subtree of T_{opt} with eight edges. If T' does not fit in any 3×3 sub-grid then by the same simple argument as in the proof of Lemma 2 we get $w(T') \geq 2$. If T' fits in a 3×3 grid, then there is one vertex of T' in any cell of such 3×3 grid. More specifically, there are vertices in cells (i, j), $(i+2, j)$, $(i, j+2)$ and $(i+2, j+2)$ from which easily follows that $w(T') > 2$. □

Lemma 4. *The weight of any subtree of* T_{opt} *with nine edges is at least* $1 + \sqrt{3}$.

Proof. Let T' be a subtree of T_{opt} with nine edges. If T' does not fit in any 4×4 sub-grid of the original grid, then there are two vertices u, v of T' which are in cells with coordinates (i, j) and (i', j') such that $|i - i'| \geq 4$ or $|j - j'| \geq 4$. In that case $w(e_{u,v}) \geq 3$ and therefore $w(T') \geq 3 > 1 + \sqrt{3}$.

Next we consider the case when the smallest rectangular sub-grid that contains T' is of the size 4×4, and let (i, j) be the bottom left corner cell of such 4×4 grid. In that case there are four (not necessarily distinct) vertices u, v, x, y of T' that for some $i \leq i', i'' \leq i+3$ and $j \leq j', j'' \leq j+3$ lie in cells $(i', j), (i, j'), (i'', j+3), (i+3, j'')$, respectively. Let P be the shortest path in T' from u to x and let Q be the shortest path in T' from v to y. Let us observe the union of paths P and Q. This union is a set of k edges we denote by e_ℓ, $\ell = 1, \ldots, k$. Let us denote by x_ℓ and y_ℓ the lengths of projections of e_ℓ on x-axis and y-axis, respectively. Then

$$w(P \cup Q) = \sum_{\ell=1}^{k} \sqrt{x_\ell^2 + y_\ell^2}. \tag{5}$$

Since distance between projections of u and x on x-axis is at least 2 and distance between projections of v and y on y-axis is at least 2, it follows that $\sum_{\ell=1}^{k} x_\ell \geq 2$ and $\sum_{\ell=1}^{k} y_\ell \geq 2$. Hence, (5) is minimized when $k = 1$ and $x_1 = y_1 = 2$ with minimal value being $2\sqrt{2}$. Therefore we get $w(T') \geq 2\sqrt{2} > 1 + \sqrt{3}$.

Lastly, we consider the case when T' fits into a rectangular sub-grid R of dimensions smaller than 4×4. Without loss of generality we can assume that R is of the size 4×3, and let (i, j) be the bottom left corner cell of R. Note that there are at least two vertices of T' that are in corner cells of R. Without loss of generality we assume that vertex v is in cell (i, j). Next we distinguish remaining cases with respect to the position of the second corner point which we denote by u.

Case 1. Vertex u is in cell $(i, j + 2)$. As there are ten vertices in T', one of them must be in cell $(i + 3, j')$ for some $j \leq j' \leq j + 3$. Denote such vertex by y. By calculating the Fermat point x it can be seen that weight of the Steiner tree containing u, v and y is at least $2 + \sqrt{3}/2$ which is greater than $1 + \sqrt{3}$, see Fig. 4(a).

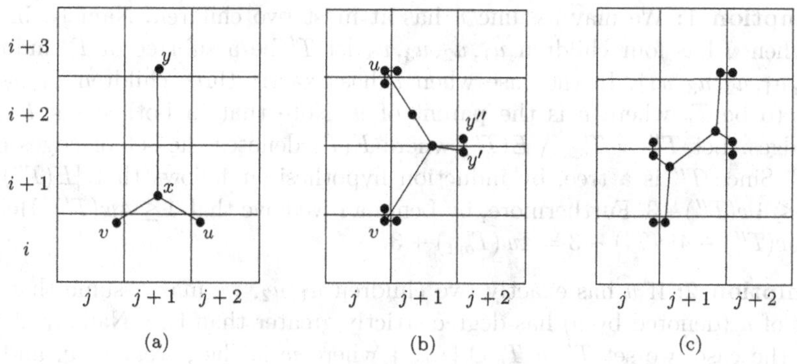

Fig. 4. T' configurations cases

Case 2. Vertex u is in cell $(i + 3, j)$. We can assume that there are no vertices of T' in cells $(i, j + 2)$ or $(i + 3, j + 2)$ as then **Case 1** applies. Then there must be vertices y', y'' in T' in cells $(i + 1, j + 2)$ and $(i + 2, j + 2)$. Hence, $w(T')$ must be at least as the weight of the Steiner tree that contains right upper corner of cell (i, j), right bottom corner of cell $(i + 3, j)$ and left bottom corner of cell $(i + 2, j + 2)$. By calculating the Fermat point, one can see that such Steiner tree has weight $1 + \sqrt{3}$, hence $w(T') \geq 1 + \sqrt{3}$. In Fig. 4(b) subtree T' has the configuration that mimics such Steiner tree.

Case 3. Vertex u is in cell $(i + 3, j + 2)$. We can assume that there are no vertices of T' in cells $(i, j + 2)$ or $(i + 3, j)$ as then **Case 1** or **Case 2** apply. In this case minimal weight T' mimics the Steiner tree that contains right upper corner of cell (i, j), left bottom corner of cell $(i + 3, j + 2)$, right bottom corner of cell $(i + 2, j)$ and left upper corner of the cell $(i + 1, j + 2)$, see Fig. 4(c). It is easy to calculate that the weight of such Steiner tree is $\sqrt{5+2\sqrt{3}}$ which is greater than $1+\sqrt{3}$. □

Now we are ready to prove Theorem 2.

Proof (of Theorem 2). We will proof the theorem by induction on N. Recall that N is the number of edges in T_{opt}.

By Lemmas 1, 3 and 4, theorem holds for $N \leq 13$. Next we assume that theorem holds for all trees with number of edges strictly less than N.

We will perform the induction step as follows: through exhaustive case study we will show that there always exist a subtree T' of T_{opt} for which $w(T')$ is greater or equal to number of edges of T' divided by 4, and if we remove from T_{opt} the edges of T', it remains connected. In that case, by induction hypothesis the bound for T_{opt} holds.

We observe T_{opt} as a rooted tree, and given a vertex v of T_{opt}, we denote by T_v the maximal subtree of T_{opt} rooted at v.

Let u be a non-leaf vertex of T_{opt} with maximum number of edges in its path to the root.

Assumption 1: We may assume u has at most two children. Namely, in the case when u has four children u_1, u_2, u_3, u_4 let T' be a subtree of T_u induced by $\{u, u_1, u_2, u_3, u_4\}$. In the case when u has exactly three children u_1, u_2, u_3 set T' to be T_v where v is the parent of u. Note that in both cases T' has four edges. Let $T'' = T_{opt} \setminus E(T')$ where $E(T)$ denotes the set of edges of a tree T. Since T'' is a tree, by induction hypothesis it follows that $|E(T'')| = N-4 \leq 4w(T'')+3$. Furthermore, by Lemma 1 we have that $4 \leq 4w(T')$. Hence, $N \leq 4w(T'') + 4w(T') + 3 = 4w(T_{opt}) + 3$.

Assumption 2: If u has exactly two children u_1, u_2, we may assume that the parent of u (denoted by v) has degree strictly greater than two. Namely, if this is not the case, we set $T' = T_v \cup \{e_{v,w}\}$ where w is the parent of v, and we set $T'' = T_{opt} \setminus E(T_w)$. Since T' has four edges and T'' is a tree, by induction hypothesis for T'' and Lemma 1 we obtain the bound.

Case 1: Vertex u has exactly two children u_1, u_2. Then by **Assumption 2** v has at least two children. By the choice of u, the number of edges in any path from v to a leaf in T_v is at most 2. Let w' be another child of v. By **Assumption 1** w' has at most two children. Also note that we can assume that w' has at least one child. Otherwise the subtree T' induced by $\{w', v, u, u_1, u_2\}$ has four edges, hence by removing the edges of T' from T we can apply the induction hypothesis and obtain the bound.

Case 1.1: Vertex v has another child w''. In this case using the same arguments as above it can be shown that w'' must have exactly one or two children. Note that subtree T' induced by v, u, u_1, u_2 together with $T_{w'}, T_{w''}$ has at least seven edges and at most nine edges. Therefore, Lemmas 2, 3 or 4 can be applied for each of the cases. Furthermore, for the remaining subtree $T_{opt} \setminus E(T')$ the induction hypothesis can be applied to obtain the bound.

Case 1.2: Vertex v has only two children w', u. Let w be the parent of v. We can assume that w' has exactly one child, otherwise the subtree T' induced by the vertices of T_v and vertex w has exactly seven edges, hence we could use Lemma 2. If the degree of w is two, then let T' be the subtree induced by T_w together with the edge $e_{w,y}$, where y is the parent of w. T' has seven edges and therefore, the result follows. Now, we may assume that w has another child v'. Let $T_1 = T_v$ and observe that T_1 has 5 edges. Let $T_2 = T_{v'}$. By the same argument used for T_v, we conclude that T_2 has at most five edges. Let $T' = T_1 \cup T_2 \cup \{e_{w,v'}, e_{w,v}\}$. If T_2 has zero, one or two edges, then T' has at least seven and at most nine edges, and hence the bound follows. If T_2 has four edges then by induction hypothesis on $T_{opt} \setminus E(T_2)$ and by applying the Lemma 1 on T_2, we obtain the bound. It remains to consider the cases when T_2 has three or five edges. If T_2 has three edges, then we add edge $e_{w,v'}$ to T_2 and now the new tree has four edges, hence we can apply the same arguments as before. We are left only with the case when T_2 has five edges. In this case $w(T_2) \geq 1$, according to Lemma 1, and also $T_3 = T_1 \cup \{e_{w,v'}, e_{v,w}\}$ has seven edges. By Lemma 2, either $w(T_3)$ is at least 2, or it has the structure depicted in Fig. 3(c), and it is clear that every edge incident to the tree in Fig. 3(c) is grater than, say 0.5. Hence, in either case $w(T') \geq 3$. Since T' has twelve edges the bound is obtained by induction hypothesis on $T_{opt} \setminus E(T')$.

Case 2: Vertex u has exactly one child u_1.

Case 2.1 Vertex v has another child w'. In this case $T_{w'}$ has depth at most 1. If w' has more than one child, then from **Case 1** (w' instead of u) we are done. If w' has one child (denoted by w_1), then the subtree induced by $\{u_1, u, v, w', w_1\}$ has four edges and we are done.

We continue by assuming that w' has no child. If v has another child $w'' \notin \{u, w'\}$, then as we argued for w', we can assume that w'' has no child. However, in this case subtree induced by $\{u_1, u, v, w', w''\}$ has four edges and we are done. Therefore we can assume that v has exactly two children w' and u. Let w be the parent of v. Then the subtree induced by u_1, u, v, w', w has four edges and we are done.

Case 2.2: Vertex v has only child u. Let w be the parent of v. W can assume that v has a sibling node v', as otherwise we can remove the four edge subtree induced by $\{u_1, u, v, w, z\}$, where z is the parent of w. Furthermore, we can assume that v' has a child u', as otherwise we can remove the four edge subtree induced by $\{u_1, u, v, w, v'\}$.

Case 2.2.1: Vertex u' has no child but has a sibling u''. We can assume that no child of v' has a child, as we can observe such case as an instance of **Case 2.2.3**. Furthermore, we can assume that u' and u'' are only children of v'. Otherwise, in the case when v' has more than three children, there would exist a subtree of $T_{v'}$ with four edges that we could remove. Furthermore, in the case when v' has exactly three children, we can remove $T' = T_{v'} \cup \{e_{w,v'}\}$.

Hence we are left with the case when u'' is the only sibling of u'. In the case v and v' are only children of w, we can remove seven edge subtree $T' =$

$T_w \cup \{e_{w,z}\}$, where z denotes the parent of w. Lastly, we consider the case when there exist third child of w denoted by v''. From the assumptions and solved cases above, we can assume that $T_{v''}$ has at most two edges, hence subtree $T' = T_v \cup T_{v'} \cup T_{v''} \cup \{e_{w,v}, e_{w,v'}, e_{w,v''}\}$ has seven, eight or nine edges, therefore we can remove it.

Case 2.2.2: Vertex u' has no child nor sibling. In the case there exists a third child of w, from the assumptions and solved cases above if would follow that we can assume that it has only one child which has no child. In that case thee would exist a subtree of T_w with four edges that we can remove. Hence, we can assume that w has no other children besides v and v'. Then T_w is a path with five edges. If $w(T_w)$ is grater than $5/4$, we can remove it and we are done. Otherwise it must be similar to the structure depicted in Fig. 5, i.e. with a path of approximate size 1 alongside a border of a cell, and with remaining vertices grouped at the endpoints of such path. Note that in that case, edge $e_{w,z}$ must be big enough so that $w(T_w \cup \{e_{w,z}\})$ is greater than $6/4$. Hence we can remove $T_w \cup \{e_{w,z}\}$ and by induction hypothesis obtain the bound.

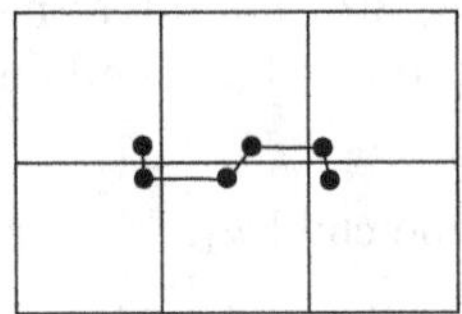

Fig. 5. A short path with five edges

Case 2.2.3: Vertex u' has a child u'_1. Note that from the assumption on maximality of depth of u, u'_1 has no children. As we solved **Case 2.1**, we can assume that u'_1 has no siblings. Furthermore, we can assume that there is no sibling of u' that has a child, as in that case there would exist subtree of $T_{v'}$ with four edges that we could remove. Now in the case that u' has more than one sibling, again, there would exist subtree of $T_{v'}$ with four edges that we could remove. In the case that u' has exactly one sibling, subtree $T' = T_{v'} \cup \{e_{w,v'}\}$ can be removed. We are left with the case when both T_v and $T_{v'}$ are paths with two edges. In the case there is a third child of w, denoted by v'', from the solved cases above if follows that we can assume that $T_{v''}$ is also a path with two edges. In that case there is a subtree of T_w with nine edges that can be removed. In the case there is no third child of w, the seven edges subtree $T' = T_w \cup \{e_{w,z}\}$ (with z being the parent of w), can be removed and the bound obtained. We considered all the cases, therefore proving the theorem. □

4 Approximation of the GGTSP

Our approximation algorithms for the GGMST can be used to obtain approximation algorithms for the geometric generalized travelling salesman problem on grid clusters (GGTSP) using standard methods.

Algorithm 4. $(2+8\sqrt{2}+2\epsilon)$-approximation algorithm for the GGTSP

Data: Instance I of the GGTSP
Result: Generalized travelling salesman tour
1 $T_A \leftarrow$ output of Algorithm 3 on I;
2 $G_E \leftarrow$ Eulerian graph obtained by doubling all edges in T_A;
3 $\mathcal{ET} \leftarrow$ an Euler tour of G_E;
4 $\mathcal{C} \leftarrow$ a GGTSP tour obtained by going along $\mathcal{ET}$ and skipping repeated vertices;
5 **return** $\mathcal{C}$;

We start with the approach of shortcutting a double MST, presented in Algorithm 4 and analyzed next.

By removing one edge from a GGTSP tour, one obtains a GGMST tree, hence $w(T_A)$ is less than $(1+4\sqrt{2}+\epsilon)OPT$, where OPT is the weight of an optimal solution of the GGTSP. Therefore, $w(G_E)$ is less than $2(1+4\sqrt{2}+\epsilon)OPT$. Due to triangle inequality, shorcutting the Euler tour in line 4 of the algorithm does not increase the weight. Hence, Algorithm 4 is a $(2+8\sqrt{2}+2\epsilon)$-approximation algorithm for the GGTSP. Note that $2+8\sqrt{2}$ is approximately equal to 13.31.

Next we use the approach from the famous Christofides $\frac{3}{2}$-approximation algorithm for the metric TSP, see [3]. This approach will give us 0.5 decrease of the approximation ratio. We give a sketch of the algorithm and the analysis, and leave details to the reader.

We start by running Algorithm 1 on the GGTSP instance. Let T_G be the resulting tree. Note that $w(T_G)$ is less or equal than $(1+4\sqrt{2})OPT+2\sqrt{2}$, where OPT is the weight of an optimal solution of the GGTSP. Let S be a set of non-empty cells that contain a vertex of T_G with an odd degree. Note that $|S|$ is even. Let M be a minimum perfect matching among cells in S, where the distance between two cells $C_1, C_2 \in S$ is the smallest distance between two points p_1, p_2 among all $p_1 \in C_1$, $p_2 \in C_2$. It is not hard to show that $w(M) \leq \frac{1}{2}OPT$. Let M_G be the set of edges e_{t_1,t_2} for which t_1, t_2 are vertices of T_G and there exist an edge $e_{p_1,p_2} \in M$ such that p_1 and t_1 are in the same cell and p_2 and t_2 are in the same cell. Note that $w(M_G) \leq \frac{1}{2}OPT + N\sqrt{2}$, and hence by Theorem 2 we get that $w(M_G) \leq \frac{1}{2}OPT + 4\sqrt{2}OPT + 3\sqrt{2}$. By merging M_G and T_G we obtain an Eulerian graph, and by shortcutting one of its Euler tours we obtain a GGTSP tour with weight at most $(\frac{3}{2}+8\sqrt{2})OPT + 5\sqrt{2}$. By similar approach as in Algorithm 3 and Theorem 3, we can get rid of $5\sqrt{2}$ error, and obtain a $(\frac{3}{2}+8\sqrt{2}+\epsilon)$-approximation algorithm for every $\epsilon > 0$.

5 Conclusions

We presented a simple $(1+4\sqrt{2}+\epsilon)$-approximation algorithm for the geometric generalized minimum spanning tree problem on grid clusters (GGMST) and $(1.5+8\sqrt{2}+\epsilon)$-approximation algorithm for the geometric generalized travelling salesman problem on grid clusters (GGTSP).

To obtain guarantied approximation ratios for our algorithms, we used the following lower bound on the optimal solution: Every tree with N edges that contains at most one point from any 1×1 grid cell is of size at least $\frac{N-3}{4}$. Obtaining a tight lower bound in terms of the number of edges would decrees guaranteed approximation ratios of our (and other similar) algorithms. Moreover, it would be an interesting result on its own.

Acknowledgment. We would like to thank Geoffrey Exoo for many usefull discussions.

References

1. Arkin, E.M., Hassin, R.: Approximation algorithms for the geometric covering salesman problem. Discrete Appl. Math. **55**(3), 197–218 (1994)
2. Bovet, J.: The selective traveling salesman problem. Papers Presented at the EURO VI Conference, Vienna (1983)
3. Christofides, N.: Worst-case analysis of a new heuristic for the travelling salesman problem. In: Traub, J.F. (ed.) Symposium on New Directions and Recent Results in Algorithms and Complexity, p. 441. Academic Press, Orlando, Florida (1976)
4. Dumitrescu, A., Mitchell, J.S.B.: Approximation algorithms for TSP with neighborhoods in the plane. In: Symposium on Discrete Algorithms, pp. 38–46 (2001)
5. Elbassioni, K.M., Fishkin, A.V., Mustafa, N.H., Sitters, R.A.: Approximation algorithms for Euclidean group TSP. In: Caires, L., Italiano, G.F., Monteiro, L., Palamidessi, C., Yung, M. (eds.) ICALP 2005. LNCS, vol. 3580, pp. 1115–1126. Springer, Heidelberg (2005)
6. Feremans, C., Grigoriev, A., Sitters, R.: The geometric generalized minimum spanning tree problem with grid clustering. 4OR **4**, 319–329 (2006)
7. Garey, M.R., Johnson, D.S.: Computers and Intractability: A Guide to the Theory of NP-Completeness. W. H Freeman, New York (1979)
8. Gudmundsson, J., Levcopoulos, C.: Hardness result for TSP with neighborhoods. Technical report LU-CS-TR:2000–216, Department of Computer Science, Lund Unversity, Sweden (2000)
9. Golden, B.L., Raghavan, S., Stanojevic, D.: Heuristic search for the generalized minimum spanning tree problem. INFORMS J. Comput. **17**(3), 290–304 (2005)
10. Jiang, H., Chen, Y.: An efficient algorithm for generalized minimum spanning tree problem. In: Proceedings of Genetic and Evolutionary Computation Conference, pp. 217–224 (2010)
11. Henry-Labordere, A.L.: The record balancing problem: a dynamic programming solution of a generalized traveling salesman problem. RIBO **B–2**, 736–743 (1969)
12. Laporte, G.: The traveling salesman problem: an overview of exact and approximate algorithms. Eur. J. Oper. Res. **59**, 231–247 (1992)
13. Laporte, G., Asef-Vaziri, A., Srikandarajah, C.: Some applications of the generalized traveling salesman problem. J. Oper. Res. Soc. **47**, 1461–1467 (1996)
14. Myung, Y.-S., Lee, C.-H., Tcha, D.-W.: On the generalized minimum spanning tree problem. Networks **26**(4), 231–241 (1995)

15. Mata, C.S., Mitchell, J.S.B.: Approximation algorithms for geometric tour and network design problems. In: Proceedings of the 11th Annual Symposium on Computational Geometry, pp. 360–369. ACM (1995)
16. Oncan, T., Corseau, J.F., Laporte, G.: A tabu search heuristic for the generalized minimum spanning tree problem. Eur. J. Oper. Res. **191**, 306–319 (2008)
17. Pop, P.C.: The generalized minimum spanning tree problem. Ph.D. Thesis, University of Twente (2002)
18. Pop, P.C., Still, G., Kern, W.: An approximation algorithm for the generalized minimum spanning tree problem with bounded cluster size. In: Proceedings of the First ACiD Workshop. Texts Algorithms vol. 4, pp. 115–121 (2005)
19. Safra, S., Schwartz, O.: On the complexity of approximating TSP with neighborhoods and related problems. In: Di Battista, G., Zwick, U. (eds.) ESA 2003. LNCS, vol. 2832, pp. 446–458. Springer, Heidelberg (2003)

Covering, Hitting, Piercing and Packing Rectangles Intersecting an Inclined Line

Apurva Mudgal and Supantha Pandit(✉)

Department of Computer Science and Engineering,
Indian Institute of Technology Ropar, Nangal Road,
Rupnagar 140001, Punjab, India
{apurva,supanthap}@iitrpr.ac.in

Abstract. We consider special cases of *set cover*, *hitting set*, *piercing set*, and *independent set* problems for axis-parallel squares and axis-parallel rectangles in the plane, where the objects are intersecting an *inclined line*, or equivalently a *diagonal line*. We prove that for axis-parallel unit squares the hitting set and set cover problems are NP-complete, whereas the piercing set and independent set problems are in P. For axis-parallel rectangles, we prove that the piercing set problem is NP-complete, which solves an open question from Correa et al. [Discrete & Computational Geometry (2015) [3]]. Further, we give a $n^{O(\lceil \log c \rceil + 1)}$ time exact algorithm for the independent set problem with axis-parallel squares, where n is the number of squares and side lengths of the squares vary from 1 to c. We also prove that when the given objects are unit-height rectangles, both the hitting set and set cover problems are NP-complete. For the same set of objects, we prove that the independent set problem can be solved in polynomial time.

Keywords: Covering · Hitting · Piercing · Packing · Inclined line · Diagonal line · NP-complete · Exact algorithm · Unit-height · Rectangles · Squares

1 Introduction

The four problems - set cover, hitting set, piercing set, and independent set - are NP-hard even for simple geometric objects like disks, squares, rectangles and many more. For the reason that it is computationally hard to get efficient algorithms, researchers have explored special cases of these problems. One of the special cases is when all the given geometric objects intersect a given diagonal line. In this paper, we consider this special case of the above four problems where the geometric objects are axis-parallel unit squares, squares, unit-height rectangles, and rectangles in the plane.

Partially supported by grant No. SB/FTP/ETA-434/2012 under DST-SERB Fast Track Scheme for Young Scientist.

Z. Lu et al. (Eds.): COCOA 2015, LNCS 9486, pp. 126–137, 2015.
DOI: 10.1007/978-3-319-26626-8_10

We are given a set P of points, a set $\mathcal{O}$ of objects, and a diagonal D such that all the objects in $\mathcal{O}$ are intersecting the diagonal D. The *Set Cover Problem (SCP)* is to find a subset $\mathcal{O}' \subseteq \mathcal{O}$ of objects with minimum cardinality that covers all the points in P. In the *Hitting Set Problem (HSP)*, the goal is to find a subset $P' \subseteq P$ of points with minimum cardinality that hits all the objects in $\mathcal{O}$. When the point set is not given as a part of the input for *HSP*, the resulting problem is known as the *Piercing Set Problem (PSP)*. In the *Independent Set Problem (ISP)*, a set $\mathcal{O}$ of objects is given, the goal is to find a subset $\mathcal{O}' \subseteq \mathcal{O}$ of objects with maximum cardinality such that any pair of objects in $\mathcal{O}'$ do not intersect.

Assume that $\mathcal{O}$ is a set of axis-parallel rectangles and the diagonal D makes an angle $180° - \theta$, $0 < \theta < 90°$ with the x-axis. We can apply a *rotation* such that the diagonal D is parallel to the x-axis and all the rectangles are tilted at the same angle θ with respect to the x-axis.

Previous Work: Chepoi and Felsner [2] first consider *PSP* and *MIS* where given geometric objects are axis-parallel rectangles intersecting an axis-monotone curve. They give a factor 6 approximation algorithm for both these problems. Recently, Correa et al. [3] consider *PSP* and *ISP* on axis-parallel rectangles. They show that *ISP* for axis-parallel rectangles is NP-complete even when each of the rectangles is touching the diagonal line at a corner. Further, they give factor 2 and factor 4 approximation algorithms for *ISP* and *PSP* respectively. They optimally solve *ISP* in quadratic time by improving a cubic time algorithm of Lubiw [10] when the axis-parallel rectangles are on one side of the diagonal and touch the diagonal at a single point.

By the result of Chan and Grant [1], it follows that *HSP* and *SCP* with axis-parallel rectangles touching a diagonal are APX-hard. Erlebach and van Leeuwen [5] show that *SCP* with axis-parallel unit squares (and hence *HSP*, since for unit squares *SCP* and *HSP* are dual to each other) can be solved in polynomial time when the unit squares intersect a fixed number of horizontal lines.

Fraser et al. [6] prove that the minimum hitting set and minimum set cover problems on unit disks are NP-complete, when the set of points and the set of disk centers lie inside a strip of any non-zero height. Recently, Das et al. [4] give a cubic time algorithm for the maximum independent set problem on disks with diameter 1 such that the disk centers lie inside a unit height strip.

Our Contributions: We summarize our contributions in Table 1.

2 Set Cover and Hitting Set Problems

2.1 Unit Squares Intersecting a Diagonal Line

In this section, we prove that *SCP* and *HSP* with axis-parallel unit squares are NP-complete, when the diagonal D makes an angle 135° with x-axis. We give a reduction from a NP-complete problem: *vertex cover in planar graphs with maximum degree 3 (PVC(3))* [7]. In this reduction we assume that D is horizontal and squares are tilted.

Table 1. Our contributions are shown in bold colored text. (*n is the number of squares with side lengths in $[1, c]$.)

Geometric Objects	*SCP*	*HSP*	*PSP*	*ISP*
Axis-Parallel Unit Squares	NP-complete **Theorem 1**	NP-complete **Theorem 1**	P **Theorem 5**	P **Corollary 2**
Axis-Parallel Squares	NP-complete **Theorem 1**	NP-complete **Theorem 1**	Open question	$n^{O(\lceil \log c \rceil + 1)*}$ time exact algorithm **Theorem 6**
Axis-Parallel Unit-Height Rectangles	NP-complete **Theorem 2**	NP-complete **Theorem 2**	Open question	P **Theorem 7**
Axis-Parallel Rectangles	APX-hard Chan et al. [1]	APX-hard Chan et al. [1]	NP-complete **Theorem 4**	NP-complete Correa et al. [3]

The reduction is similar to the reduction of Fraser et al. [6] for the hitting set problem on unit disks and point inside a strip. The construction can be done in two *phases*. Phase I is identical to Fraser et al. [6]. In this phase, we are given an instance G (Fig. 1(a)) of *PVC(3)* with different x-coordinates of vertices. We reduce G into another instance G'' (Fig. 1(b)) of *PVC(3)* through an intermediate instance G' of *PVC(3)* (Fig. 1(b) without pink colored vertices) by adding *dummy vertices* to the edges of G. The graph G' satisfies the following *properties*: (i) G' is embedded inside a horizontal strip, (ii) there is no degree three vertex in G' whose all incident edges connect to it either from left or from

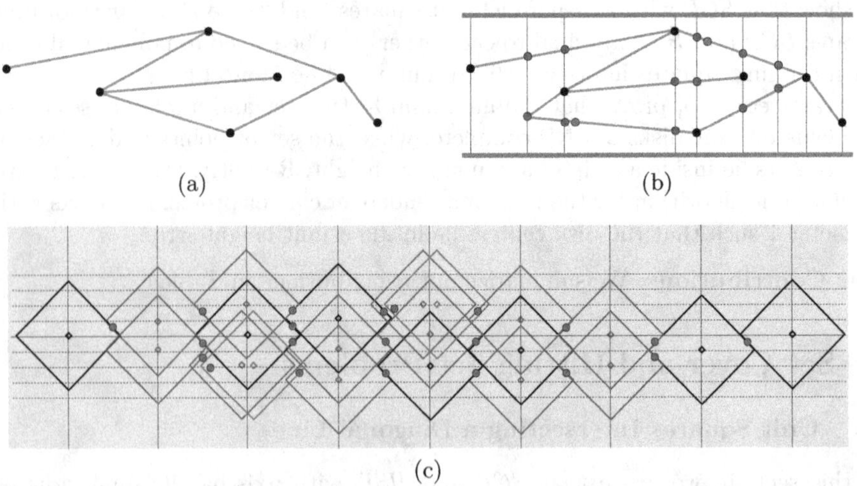

Fig. 1. (a) An instance of vertex cover problem on planar graph of degree at most 3. (b) After adding dummy vertices in Phase I. (c) The *SCP* instance created with unit squares and points in Phase II from the planar graph in (b). (Color figure online)

right side, (iii) vertical line passing through any vertex of G' does not intersect an edge of G' in the middle, and (iv) the difference between the number of vertices on two vertical lines passing through vertices of G' with consecutive x-coordinates is at most 1. Finally, G'' is constructed from G' by adding one extra dummy vertex to each edge of G containing an odd number of dummy vertices. Note that edges of G'' may have a single dummy vertex (pink colored vertex in (Fig. 1(b)) which does not lie on any vertical line.

In Phase II, we incorporate our ideas. Here we first reduce graph G' to $\mathscr{H}$ (Fig. 1(c)), an instance of *SCP* with unit squares, as follows. For each vertex in G' we take a unit square and for each edge we take a point in $\mathscr{H}$. The different types of arrangements of the squares and points in $\mathscr{H}$ for the vertices in two consecutive vertical lines and edges between these two lines in G' are shown in Fig. 2. The centers of the squares for the vertices of G' on a vertical line are on a vertical line in $\mathscr{H}$ and two consecutive centers are $h = \frac{\sqrt{2}}{\kappa n}$ distance apart, where n is the number of vertices in G'' and κ is a suitable constant. This ensures that all squares will intersect D. Two consecutive vertical lines containing the centers of the squares are $\sqrt{2} - h$ distance apart. The reduction from G'' to $\mathscr{H}$ is as follows. Let (a, b) be an edge of G'. Suppose a dummy vertex d is added inside (a, b) in G''. We take one unit square s (see pink colored squares in (Fig. 1(c)) for d. Next we remove the point for edge (a, b) and add two points, one for edge (a, d) and other for edge (d, b), such that these two points are covered by s.

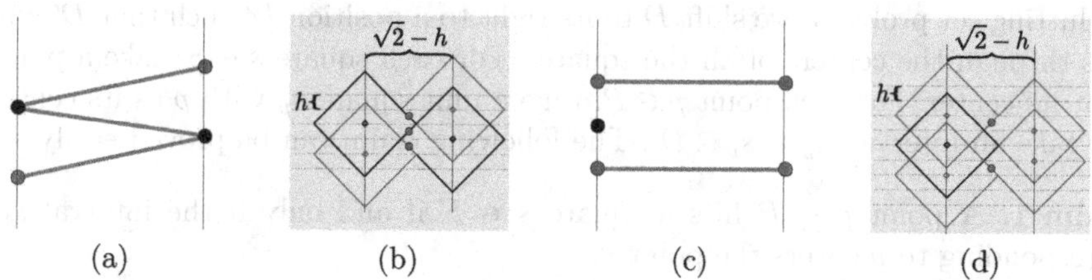

Fig. 2. (a) One configuration in G', (b) gadget of (a) in $\mathscr{H}$, (c) another configuration in G', and (d) gadget of (c) in $\mathscr{H}$. The other types of gadgets can be made by either modifying or extending the gadgets (b) and (d). (Color figure online)

The proof of correctness is straightforward. Finding a vertex cover in G'' is equivalent to finding a set cover in $\mathscr{H}$. Further, since hitting set and set cover for unit squares are self-dual, we have the following theorem.

Theorem 1. *The minimum set cover and the minimum hitting set problems with unit squares intersecting a diagonal line are* NP*-complete.*

2.2 Unit-Height Rectangles Intersecting a Diagonal Line

We prove that *SCP* and *HSP* for axis-parallel unit-height rectangles are NP-complete. The reduction is similar to the reduction described in Sect. 2.1. Applying transformations $x' = x$ and $y' = \delta y$, where δ is chosen appropriately, to

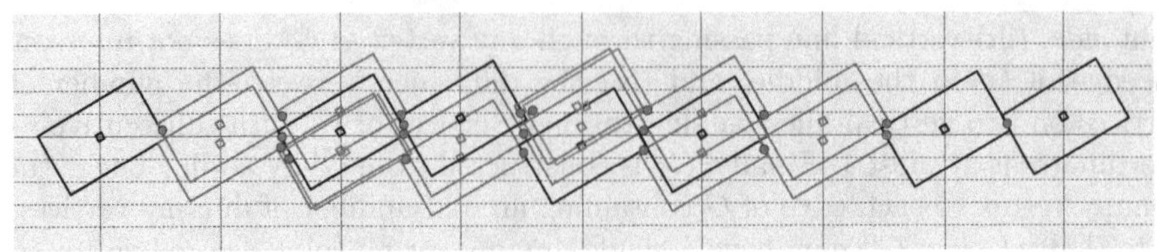

Fig. 3. An instance of *SCP* with unit-height rectangles generated for the graph in (Fig. 1(b)). (Color figure online)

the reduction in Sect. 2.1 gives NP-completeness for diagonal making any angle between 90° and 180° with the x-axis. We depict the NP-completeness construction in Fig. 3 for the *PVC(3)* instance G in Fig. 1(a) where the diagonal makes an angle 150° with the x-axis.

Theorem 2. *The minimum set cover and the minimum hitting set problems with unit-height rectangles intersecting a diagonal line are* NP*-complete.*

2.3 Unit Squares Touching a Diagonal Line

Let P be a set of points and S be a set of axis-parallel unit squares such that squares in S intersect a diagonal at a single point from the right side. We consider the hitting set problem. we shift D to its right to a position D' such that D' will pass through the centers of all the squares. For each square $s \in S$, take a point p_s at its center. For each point $p \in P$, take a unit square s_p with p as its center and take an interval $i_{s_p} = s_p \cap D'$. The following claim can be proved easily.

Claim 1. A point $p \in P$ hits a square $s \in S$ if and only if the interval i_{s_p} corresponding to p covers the point p_s.

Hence, *HSP* with unit squares is equivalent to the problem of covering points on a real line by intervals. Similarly, we can reduce *SCP* to the problem of hitting intervals by point on a real line. Since both the problems can be solved in polynomial time using standard algorithms, we have the following theorem.

Theorem 3. *The minimum set cover and the minimum hitting set problems with unit squares touching a diagonal line can be solved in polynomial time.*

3 Piercing Set Problem

3.1 Rectangles Intersecting a Diagonal Line

In this section, we prove that *PSP* with axis-parallel rectangles is NP-complete. We give a reduction from the NP-complete problem *Rectilinear-Planar-3-SAT* [8] which is another form of *Planar-3-SAT* [9] problem. The reduction involves ideas from Correa et al. [3]. We slightly modify the *Rectilinear-Planar-3-SAT* problem as follows. Let ϕ be a 3-SAT formula with every clause containing exactly 3 literals. The variables are placed on a diagonal line. The clauses shaped "┗┻" or "┳┓" connect to the variables either from the left or from the right side of the diagonal such that the shapes do not intersect with each other (see Fig. 4(a)). The goal is to find a satisfying assignment for the formula ϕ. We take α to be the maximum number of clauses which connect to a variable of ϕ either from left or from right. Take $t = 2\alpha + 1$.

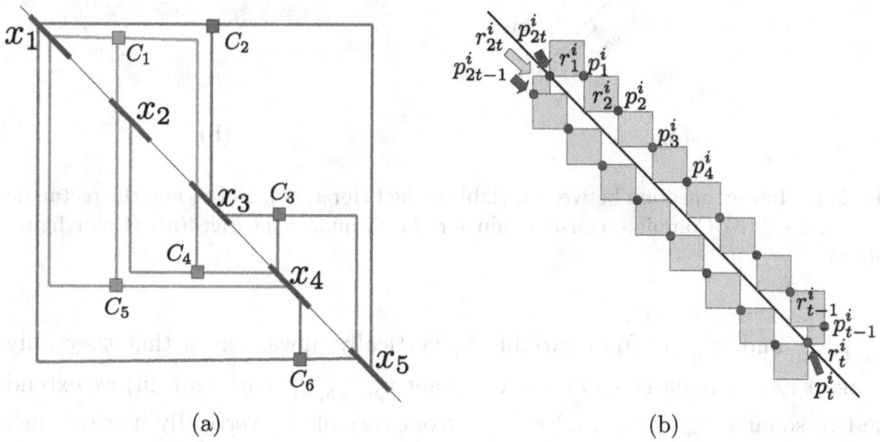

Fig. 4. (a) Representation of a modified *Rectilinear-Planar-3-SAT* formula $\phi = C_1 \wedge C_2 \wedge C_3 \wedge C_4 \wedge C_5 \wedge C_6$, where $C_1 = (x_1 \vee \overline{x}_2 \vee \overline{x}_3)$, $C_2 = (\overline{x}_1 \vee x_3 \vee \overline{x}_5)$, $C_3 = (\overline{x}_3 \vee x_4 \vee \overline{x}_5)$, $C_4 = (\overline{x}_2 \vee \overline{x}_3 \vee x_4)$, $C_5 = (x_1 \vee x_2 \vee \overline{x}_4)$, and $C_6 = (x_1 \vee \overline{x}_4 \vee x_5)$. (b) Structure of gadget for a variable x_i. (Color figure online)

For each variable x_i, we take $2t$ squares with t squares each on the left and right side of the diagonal D (see Fig. 4(b)). Two consecutive squares intersect at a point except for pairs r^i_{t-1}, r^i_t and r^i_{2t-1}, r^i_{2t} which intersect on a portion of their boundaries. We choose a single point from each intersection and call these points as *representative points*. Let $P^i_V = \{p^i_1, p^i_2, \ldots, p^i_{2t}\}$ be the $2k$ representative points for variable x_i. Now to pierce the variable squares we can select points from the set P^i_V. Since the representative points form a cycle of length $2t$, there are two optimal piercing sets of points $\{p^i_1, p^i_3, \ldots, p^i_{2t-1}\}$ and $\{p^i_2, p^i_4, \ldots, p^i_{2t}\}$ for each variable gadget.

Now consider a clause C which connects to the variables x_i, x_j, and x_k from the right side of D. Define three binary variables b_i, b_j, and b_k as follows: $b_l = 0$, if x_l occurs as a negative literal in C, otherwise $b_l = 1$, for $l \in \{i, j, k\}$. We take a rectangle r_c for clause C. Let clause C be the n_1-, n_2-, and n_3-th clause for x_i, x_j, and x_k respectively, when the clauses from the right side of D connecting to these variables are ordered from left to right. We place the rectangle r_c such that it satisfies the following *conditions* (see Fig. 5(a)): (i) it covers only the representative point $p^i_{2n_1+b_i-1}$ from variable x_i, (ii) we extend the two squares

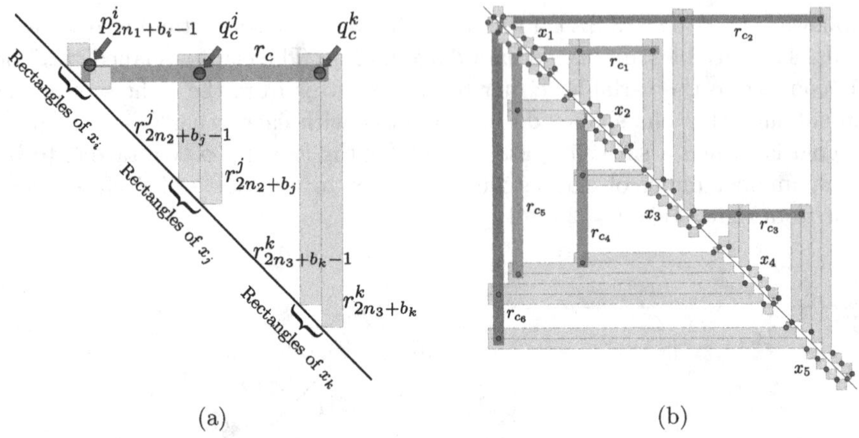

Fig. 5. (a) Interconnection between variable gadgets for x_i, x_j, and x_k and the rectangle for clause C. (b) Complete construction for the formula π in Fig. 4(a). (Color figure online)

$r^j_{2n_2+b_j-1}$ and $r^j_{2n_2+b_j}$ from variable x_j vertically upward such that they only intersect r_c at a point q^j_c and move the point $p^j_{2n_2+b_j-1}$ to q^j_c, and (iii) we extend the two squares $r^k_{2n_3+b_k-1}$ and $r^k_{2n_3+b_k}$ from variable x_k vertically upward such that they only intersect r_c at a point q^k_c and move the point $p^k_{2n_3+b_k-1}$ to q^k_c. Note that after extension some of the variable squares become rectangles. From now on we call all the variable squares as rectangles. Similarly, we make the construction for the clauses that connect to the variables from left.

In Fig. 5(b) we demonstrate the above construction for the formula shown in Fig. 4(a). Therefore, from formula ϕ with n variables and m clauses we construct an instance $\mathscr{R}$ of piercing set with $2tn+m$ rectangles intersecting a diagonal line. We can observe that any point p in the plane can be replaced by a representative point which hits at least all the rectangles hit by p. Therefore, any optimal piercing set can pick points from the representative points. Now we prove the correctness of the construction.

Theorem 4. *The minimum piercing set problem with rectangles intersecting a diagonal line is* NP*-complete.*

Proof. We shall prove that ϕ is satisfiable iff $\mathscr{R}$ has a piercing set of at most tn points. First, a satisfying assignment of ϕ is considered. From the gadget of x_i, select odd-numbered representative points if x_i is false, otherwise select even-numbered representative points. Clearly, these tn selected points hit all the variable as well as clause rectangles. Next, the rectangles for one variable are disjoint from those for any other variable. Then, any tn size solution must select t alternate representative points from each variable. Hence, we can set binary values to each variable based on which of the two optimal solutions of t representative points is selected for that variable. Finally, a clause rectangle is hit iff the alternate representative points selected for some variable make the binary value of a literal present in the clause 1. ■

3.2 Unit Squares Intersecting a Diagonal Line

In this section, we give an exact algorithm for *PSP*, where the objects are axis-parallel unit squares. Let S be a set of n unit squares intersecting a diagonal line D. Assume D is horizontal by a rotation. Let H be a horizontal strip of height $2\sqrt{2}$ such that the diagonal D divides it into two equal parts. We further partition strip H into rectangular *regions* of length $\sqrt{2}$ and height $2\sqrt{2}$. Now we remove all regions which are not intersected by one of the given squares. Let $R_1, R_2, \ldots, R_r$ be the remaining regions sorted according to x-coordinate. Note that r is at most $2n$, since a unit square can intersect at most two regions. We now take a dummy region R_0 before R_1 and a dummy region R_{r+1} after R_r. Let S_i be the subset of squares that intersect region R_i. Also let S_i^{in} be the set of squares in S_i whose centers are inside R_i and let $S_i^{out} = S_i \setminus S_i^{in}$ be the set of squares in S_i whose centers are outside R_i.

Lemma 1. *The maximum number of points required to pierce the squares in S_i^{in} for any region R_i is at most* 8.

Proof. Note that the dimension of region R_i is $\sqrt{2} \times 2\sqrt{2}$. Take 8 squares $D_1, D_2, \ldots, D_8$ of length $\frac{1}{\sqrt{2}}$, four in a column and two in a row, such that together they completely cover R_i. Since the diagonal lengths of the squares are exactly 1, the center c_i of square D_i is at distance less than or equal to $\frac{1}{2}$ from any other point of D_i. Thus, c_i pierces all squares in S_i^{in} whose centers lie in D_i. Hence, the 8 points $c_1, c_2, \ldots, c_8$ form a piercing set of size 8 for S_i^{in}. ■

Lemma 2. *Any optimal piercing set contains at most* 24 *points from region R_i.*

Proof. Observe that $S_i^{out} \subseteq S_{i-1}^{in} \cup S_{i+1}^{in}$. Therefore, by Lemma 1 the squares in S_i can be hit by 24 points, 8 points each from regions R_{i-1}, R_i, and R_{i+1}. Since points inside R_i can only hit squares in S_i, if an optimal solution OPT contains more than 24 points from region R_i, we can replace them by 24 points in regions R_{i-1}, R_i, R_{i+1} without leaving any square to be hit. This contradicts the assumption that OPT was an optimal piercing set. ■

Let the squares in S_i divide R_i into m_i *subregions*. Since any two squares can intersect in at most 2 points, $m_i = O(|S_i|^2) = O(n^2)$. Let P_i be a set of m_i points, with one point picked from each of the m_i subregions. We can assume that optimal piercing set is a subset of $\bigcup_{i=0}^{r+1} P_i$.

For $0 \leq i \leq r$, let $T(i, U_1, U_2)$ where $U_1 \subseteq P_i$ and $U_2 \subseteq P_{i+1}$ denote the cost of an optimal piercing set H for the squares which lie completely inside $\bigcup_{j=i}^{r+1} R_j$ such that $H \bigcap P_i = U_1$ and $H \bigcap P_{i+1} = U_2$. Note that by Lemma 2, we can assume that both U_1 and U_2 have at most 24 points. $T(i, U_1, U_2)$ satisfies the following recurrence:

1. If $U_1 \bigcup U_2$ does not pierce all squares which lie completely inside $R_i \bigcup R_{i+1}$, then $T(i, U_1, U_2) = \infty$.
2. Otherwise,

$$T(i, U_1, U_2) = \min_{U_3 \subseteq P_{i+2},\ |U_3| \leq 24} T(i+1, U_2, U_3) + |U_3|$$

The optimal piercing set is now given by $\min_{U_1 \subseteq P_0, U_2 \subseteq P_1} T(0, U_1, U_2)$ and can be obtained from the above recurrences by dynamic programming.

We now analyze the time required to find the optimal piercing set. As each P_i has $O(n^2)$ points, the number of 24 element subsets of P_i is $O(n^{48})$. Therefore, the number of subproblems $T(i, U_1, U_2)$ defined above are at most $O(n^{97})$. Each subproblem depends on $O(n^{48})$ smaller subproblems. By allowing an extra bookkeeping cost the total time taken to compute the optimal piercing set is $n^{O(1)}$. This leads to the following theorem.

Theorem 5. *The minimum piercing set problem with unit squares intersecting a diagonal line can be solved in polynomial time.*

4 Independent Set Problem

4.1 Squares Intersecting a Diagonal Line

In this section, an exact algorithm for *ISP* with axis-parallel squares is proposed. We are given a set S of n axis-parallel squares with integer side lengths in $[1, c]$. The squares are intersecting a diagonal D. We make the diagonal parallel to x-axis by performing a rotation on the given input. The idea of the algorithm is similar to the algorithm of Das et al. [4] for the maximum independent set problem on unit disks such that disks do not overlap and hit a horizontal line. In their algorithm, they show that at most 4 pairwise independent unit disks can intersect a vertical line. However, in this case we partition the squares in S into $k = \lceil \log c \rceil + 1$ groups $g_1, g_2, \ldots, g_k$, where the i-th group g_i contains squares with side lengths in the semi-open interval $[2^{i-1}, 2^i)$ and show that at most $16k$ pairwise independent squares can intersect a given vertical line.

Lemma 3. *There are at most* 16 *pairwise independent squares whose side lengths are in* $[1, 2)$ *such that they intersect both* D *and a vertical line* L.

Proof. Let R be a square of length $2\sqrt{2}$ with D and L as middle horizontal and vertical lines. Observe that the centers of the squares that intersect both L and D should be inside R. Now we partition the region R into 16 congruent squares $D_1, D_2, \ldots, D_{16}$ arranged in a grid such that there are exactly 4 squares in each row and each column respectively. The center c_j of D_j is at most $\frac{1}{2}$ unit far from any other point inside D_j and hence at most one square with center inside D_j is in the independent set. Thus, any independent set has size at most 16. ■

The following corollary directly follows from the above lemma.

Corollary 1. *There are at most* $16k$ *pairwise independent squares, a maximum of 16 squares from each group* g_i*, which intersect* D *and a vertical line.*

Now consider vertical lines through every corner point of the squares. Let $L_1, L_2, \ldots L_r$ be these vertical lines sorted from left to right. Add two extra vertical lines, L_0 to the left of L_1 and L_{r+1} to the right of L_r, such that no square in S can intersect them. Let Sq_i be the set of squares from S that intersect the vertical line L_i for $i = 1, 2, \ldots, r$. For $0 \leq i \leq r+1$, the subproblem $T(i, Sq'_i)$ denotes the size of the optimal independent set S^* of squares such that: (i) squares in S^* intersect the closed region bounded by L_0 and L_i, (ii) $Sq'_i \subseteq Sq_i$ is the set of squares in S^* that intersect L_i. Note that from Corollary 1 we have $|Sq'_i| \leq 16k$. Now the following recurrence is satisfied by the above subproblem:

1. If any two squares in Sq'_i intersect, then $T(i, Sq'_i) = -\infty$.
2. Otherwise, let Sq'_{i-1} be a subset of squares from Sq_{i-1} such that the squares in Sq'_i which intersect L_{i-1} are in Sq'_{i-1} and squares in Sq'_i do not intersect with the squares in Sq'_{i-1}. Then,

$$T(i, Sq'_i) = \max_{Sq'_{i-1} \subseteq Sq_{i-1}, |Sq'_{i-1}| \leq 16k} \{T(i-1, Sq'_{i-1})\} + |Sq'_i \setminus Sq'_{i-1}|$$

The optimal independent set can be calculated by solving $T(r+1, \emptyset)$.

Running Time: There are $4n$ corner points and hence r is $O(n)$. By Lemma 3, at most $16k$ squares from S can intersect a vertical line L_i. Observe that there are at most $O(n^{16k+1})$ subproblems. Further, at most $O(n^{16k})$ subproblems are considered in the right hand side of the above recurrence. Therefore, to compute an optimal independent set $n^{O(k)}$ i.e., $n^{O(\lceil \log c \rceil + 1)}$ total time is required. Hence, we have the following theorem.

Theorem 6. *The maximum independent set problem with squares intersecting a diagonal line can be solved in* $n^{O(\lceil \log c \rceil + 1)}$ *time, where* n *is the number of squares and the side lengths of the squares are in* $[1, c]$.

For unit squares we can take c as 1 and hence we have the following result.

Corollary 2. *The maximum independent set problem with unit squares intersecting a diagonal line can be solved in polynomial time.*

4.2 Unit-Height Rectangles Intersecting a Diagonal Line

In this section, we prove that *ISP*, where the given objects are axis-parallel unit-height rectangles can be solved in polynomial time using a dynamic programming algorithm similar to that described in Sect. 4.1. However, a large number of unit-height rectangles from any optimal solution may now intersect a vertical line. To address this problem we do the following.

Let M be a real number. We now perform the following transformation to every point (x, y) in the plane: $y' = y$ and $x' = Mx$. We choose M such that after the above transformation the width of each unit-height rectangle becomes at least 1. Note that after this transformation the angle made by D with x-axis will increase and D becomes almost parallel to x-axis for large M.

Lemma 4. *Let D be a diagonal line and D_1, D_2 be two parallel lines on either side of D at unit vertical distance from D. Let r_1, r_2 be two axis-parallel unit-height rectangles that intersect D. Then r_1, r_2 intersect each other iff they have an intersection point in the area bounded by D_1 and D_2.*

Proof. Let q be a point to the right of D_1 such that r_1 and r_2 contain q. Further, assume that r_1 and r_2 do not have any intersection point inside the region bounded by D_1 and D_2. Clearly the vertical distance from D to q is greater than 1. Now take a horizontal line L_q through q and let it intersect D_1 at a point, say q' (see Fig. 6(a)). The vertical distance between D and q' is exactly 1. So no unit-height rectangle with left boundary to the right of the vertical line through q' and intersecting D can cover q. Since r_1 and r_2 intersect at q, the left boundaries of both r_1 and r_2 should start before or at q'. Hence both have q' also as an intersection point. This gives a contradiction. ■

Now by Lemma 4, for computing optimal independent set we need to consider only the *portions* of the unit-height rectangles inside the strip formed by D_1 and D_2. Next, we apply a rotation such that D becomes parallel to x-axis.

Lemma 5. *Let L be a vertical line segment of length 2 such that D partitions L in two equal parts. Then at most 18 pairwise independent unit-height rectangles, each having width at least 1, can intersect both D and L.*

Proof. Let R be a rectangle of size $\sqrt{5} \times (2+\sqrt{5})$ with D as the middle horizontal line. Place the vertical line L such that D splits L into two equal segments and the distances from left and right boundaries of R to L are equal. Let r be a unit-height rectangle which intersects L and D. Let p be a point on $L \cap r$. Take a line L' through p with the same slope as the left boundary of r. Further, take two rectangles r_1 and r_2 each of size $\frac{1}{2} \times 1$ such that the left boundary of r_1 and the right boundary of r_2 coincide with the unit segment $L' \cap r$ (see Fig. 6(b)). Since the width of r is at least 1, at least one of r_1 or r_2 is fully contained inside r. Let r fully contain r_1. Since r_1 covers p, r_1 is fully contained inside R. The area of R is $5+2\sqrt{5}$ and that of r_1 is $\frac{1}{2}$. Since each unit-height rectangle intersecting both D and L has at least $\frac{1}{2}$ unit of area inside R, at most $\lfloor 10+4\sqrt{5} \rfloor$ i.e., 18 independent unit-height rectangles can intersect L and D. ■

We apply the same dynamic programming algorithm described in Sect. 4.1 with L_i's as the vertical line segments of length 2 bisected by D and passing through every corner of the portions of unit-height rectangles inside the region bounded by D_1 and D_2. As each portion has at most 6 corner points, the number of L_i's is $O(n)$. By Lemma 5, each L_i intersects at most 18 portions in an independent set and hence by an analysis similar to that in Sect. 4.1, we can say that the total time taken to compute the optimal independent set is $n^{O(1)}$.

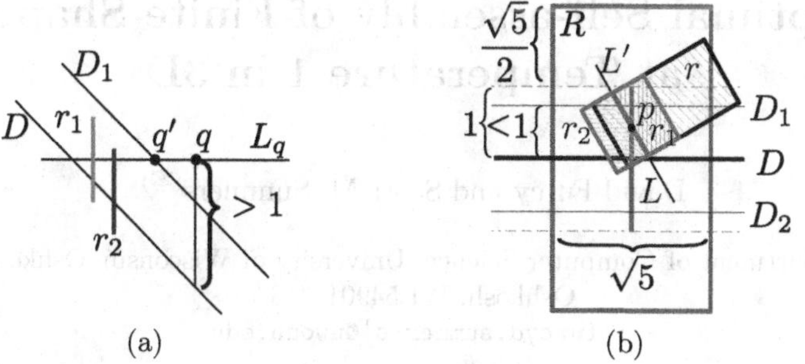

Fig. 6. (a) Proof of Lemma 4. (b) Proof of Lemma 5. (Color figure online)

Theorem 7. *The maximum independent set problem with unit-height rectangles intersecting a diagonal line can be solved in polynomial time.*

Note that M can be very large. Therefore, we will not calculate the actual value of M. Instead, we will take M as a variable and execute the above algorithm symbolically on a computer.

Acknowledgements. The second author would like to thank Subhas C. Nandy and Aniket Basu Roy for helpful discussions. We express our gratitude to the anonymous reviewers for reviewing our manuscript.

References

1. Chan, T.M., Grant, E.: Exact algorithms and APX-hardness results for geometric packing and covering problems. Comput. Geom. Part A **47**(2), 112–124 (2014)
2. Chepoi, V., Felsner, S.: Approximating hitting sets of axis-parallel rectangles intersecting a monotone curve. Comput. Geom. **46**(9), 1036–1041 (2013)
3. Correa, J., Feuilloley, L., Pérez-Lantero, P., Soto, J.A.: Independent and hitting sets of rectangles intersecting a diagonal line: algorithms and complexity. Discrete Comput. Geom. **53**(2), 344–365 (2015)
4. Das, G.K., De, M., Kolay, S., Nandy, S.C., Sur-Kolay, S.: Approximation algorithms for maximum independent set of a unit disk graph. Inf. Process. Lett. **115**(3), 439–446 (2015)
5. Erlebach, T., Jan Van Leeuwen, E.: PTAS for weighted set cover on unit squares. In: Serna, M., Shaltiel, R., Jansen, K., Rolim, J. (eds.) APPROX 2010. LNCS, vol. 6302. Springer, Heidelberg (2010)
6. Fraser, R., Lopez-Ortiz, A.: The within-strip discrete unit disk cover problem. In: CCCG 2012, pp. 53–58, Charlottetown, Canada, August 8–10, 2012 (2012)
7. Garey, M.R., Johnson, D.S.: The rectilinear steiner tree problem is NP-complete. SIAM J. Appl. Math. **32**(4), 826–834 (1977)
8. Knuth, D.E., Raghunathan, A.: The problem of compatible representatives. SIAM J. Discrete Math. **5**(3), 422–427 (1992)
9. Lichtenstein, D.: Planar formulae and their uses. SIAM J. Comput. **11**(2), 329–343 (1982)
10. Lubiw, A.: A weighted min-max relation for intervals. J. Combinatorial Theor. Ser. B **53**(2), 151–172 (1991)

Optimal Self-assembly of Finite Shapes at Temperature 1 in 3D

David Furcy and Scott M. Summers(✉)

Department of Computer Science, University of Wisconsin–Oshkosh, Oshkosh, WI 54901, USA
{furcyd,summerss}@uwosh.edu

Abstract. Working in a three-dimensional variant of Winfree's abstract Tile Assembly Model, we show that, for an arbitrary finite, connected shape $X \subset \mathbb{Z}^2$, there is a tile set that uniquely self-assembles into a 3D representation of X at temperature 1 with optimal program-size complexity (the program-size complexity, also known as tile complexity, of a shape is the minimum number of tile types required to uniquely self-assemble it). Moreover, our construction is "just barely" 3D in the sense that it only places tiles in the $z = 0$ and $z = 1$ planes. Our result is essentially a just-barely 3D temperature 1 simulation of a similar 2D temperature 2 result by Soloveichik and Winfree (SICOMP 2007).

1 Introduction

Self-assembly is intuitively defined as the process through which simple, unorganized components spontaneously combine, according to local interaction rules, to form some kind of organized final structure. While examples of self-assembly in nature are abundant, Seeman [21] was the first to demonstrate the feasibility of self-assembling man-made DNA tile molecules. Since then, self-assembly researchers have used principles from DNA tile self-assembly to self-assemble a wide variety of nanoscale structures, such as regular arrays [25], fractal structures [8,19], smiling faces [18], DNA tweezers [26], logic circuits [15], neural networks [16], and molecular robots [11]. What is more, over roughly the past decade, researchers have dramatically reducing the tile placement error rate (the percentage of incorrect tile placements) for DNA tile self-assembly from 10 % to 0.05 % [2,7,8,19].

In 1998, Winfree [24] introduced the abstract Tile Assembly Model (aTAM) as an over-simplified, combinatorial, error-free model of experimental DNA tile self-assembly. The aTAM is a constructive version of mathematical Wang tiling [23] in that the former bestows upon the latter a mechanism for sequential "growth" of a tile assembly starting from an initial seed. Very briefly, in the aTAM, the fundamental components are un-rotatable, translatable square "tile

S.M. Summers—This author's research was supported in part by UWO Faculty Development Research grant FDR881.

Z. Lu et al. (Eds.): COCOA 2015, LNCS 9486, pp. 138–151, 2015.
DOI: 10.1007/978-3-319-26626-8_11

types" whose sides are labeled with (alpha-numeric) glue "colors" and (integer) "strengths". Two tiles that are placed next to each other *bind* if both the glue colors and the strengths on their abutting sides match and the sum of their matching strengths sum to at least a certain (integer) "temperature". Self-assembly starts from a "seed" tile type, typically assumed to be placed at the origin, and proceeds nondeterministically and asynchronously as tiles bind to the seed-containing assembly one at a time. In this paper, we work in a three-dimensional variant of the aTAM in which tile types are unit cubes and tiles are placed in a *non-cooperative* manner.

Tile self-assembly in which tiles may bind to an existing assembly in a non-cooperative fashion is often referred to as "temperature 1 self-assembly" or simply "non-cooperative self-assembly". In this type of self-assembly, a tile may non-cooperatively bind to an assembly via (at least) one of its sides, unlike in *cooperative* self-assembly, in which some tiles may be required to bind on two or more sides. It is worth noting that cooperative self-assembly leads to highly non-trivial theoretical behavior, e.g., Turing universality [24] and the efficient self-assembly of $N \times N$ squares [1,17] and other algorithmically specified shapes [22].

Despite its theoretical algorithmic capabilities, when cooperative self-assembly is implemented using DNA tiles in the laboratory [2,13,19,20,25], tiles may (and do) erroneously bind in a non-cooperative fashion, which usually results in the production of undesired final structures. In order to completely avoid the erroneous effects of tiles unexpectedly binding in a non-cooperative fashion, the experimenter should only build nanoscale structures using constructions that are guaranteed to work correctly in non-cooperative self-assembly. Thus, characterizing the theoretical power of non-cooperative self-assembly has significant practical implications.

Although no characterization of the power of non-cooperative self-assembly exists at the time of this writing, Doty et al. conjecture [6] that 2D non-cooperative self-assembly is weaker than 2D cooperative self-assembly because a certain technical condition, known as "pumpability", is true for any 2D non-cooperative tile set. If the pumpability conjecture is true, then non-cooperative 2D self-assembly can only produce simple, highly-regular shapes and patterns, which are too simple and regular to be the result of complex computation.

In addition to the pumpability conjecture, there are a number of results that study the suspected weakness of non-cooperative self-assembly. For example, Rothemund and Winfree [17] proved that, if the final assembly must be fully connected, then the minimum number of unique tile types required to self-assemble an $N \times N$ square (i.e., its *tile complexity*) is exactly $2N - 1$. Manuch et al. [12] showed that the previous tile complexity is also true when the final assembly cannot contain even any glue mismatches. Moreover, at the time of this writing, the only way in which non-cooperative self-assembly has been shown to be unconditionally weaker than cooperative self-assembly is in the sense of *intrinsic universality* [4,5]. First, Doty et al. [4] proved the existence of a universal cooperative tile set that can be programmed to simulate the behavior of any tile set (i.e., the aTAM is intrinsically universal for itself). Then, Meunier et al. [14] showed, via a combinatorial argument, that there is no universal non-cooperative tile set that can be programmed to simulate the behavior of an arbitrary (cooperative) tile set. Thus, in the sense of intrinsic universality, non-cooperative self-assembly is strictly weaker than cooperative self-assembly.

While non-cooperative self-assembly is suspected of being strictly weaker than cooperative self-assembly, in general, it is interesting to note that 3D non-cooperative self-assembly (where the tile types are unit cubes) and 2D cooperative self-assembly share similar capabilities. For instance, Cook et al. [3] proved that it is possible to deterministically simulate an arbitrary Turing machine using non-cooperative self-assembly, even if tiles are only allowed to be placed in the $z = 0$ and $z = 1$ planes (Winfree [24] proved this for the 2D aTAM). Cook et al. [3] also proved that it is possible to deterministically self-assemble an $N \times N$ 3D "square" shape $S_N \subseteq \{0, \ldots, N-1\} \times \{0, \ldots, N-1\} \times \{0, 1\}$ using non-cooperative self-assembly with $O(\log N)$ tile complexity (Rothemund and Winfree [17] proved this for the 2D aTAM). Furcy et al. [9] reduced the tile complexity of deterministically assembling an $N \times N$ square in 3D non-cooperative self-assembly to $O\left(\frac{\log N}{\log \log N}\right)$ (Adleman et al. [1] proved this for the 2D aTAM), which is optimal for all algorithmically random values of N. Given that it is possible to optimally self-assemble an $N \times N$ square in 3D using non-cooperative self-assembly, the following is a natural question: Is it possible to self-assemble an arbitrary finite shape in 3D using non-cooperative self-assembly with optimal tile complexity?

Note that the previous question was answered affirmatively by Soloveichik and Winfree [22] for the 2D aTAM, assuming the shape of the final assembly can be a scaled-up version of the input shape (i.e., each point in the input shape is replaced by a $c \times c$ block of points, where c is the *scaling factor*). Specifically, Soloveichik and Winfree gave a construction that takes as input an algorithmic description of an arbitrary finite, connected shape $X \subset \mathbb{Z}^2$ and outputs a cooperative (temperature 2) tile set T_X that deterministically self-assembles into a scaled-up version of X and $|T_X| = O\left(\frac{|M|}{\log |M|}\right)$, where $|M|$ is the size of (i.e., number of bits needed to describe) the Turing machine M, which outputs the list of points in X. In the main result of this paper, using a combination of 3D, temperature 1 self-assembly techniques from Furcy et al. [9] and Cook et al. [3], we show how the optimal construction of Soloveichik and Winfree can be simulated in 3D using non-cooperative self-assembly with optimal tile complexity. Thus, our main result represents a Turing-universal way of guiding the self-assembly of a scaled-up, just-barely 3D version of an arbitrary finite shape X at temperature 1 with optimal tile complexity.

2 Definitions

In this section, we give a brief sketch of a 3-dimensional version of the aTAM along with some definitions of scaled finite shapes and the complexities thereof.

2.1 3D Abstract Tile Assembly Model

Let Σ be an alphabet. A 3-dimensional *tile type* is a tuple $t \in (\Sigma^* \times \mathbb{N})^6$, e.g., a unit cube with six sides listed in some standardized order, each side having a *glue* $g \in \Sigma^* \times \mathbb{N}$ consisting of a finite string *label* and a non-negative integer *strength*. In this paper, all glues have strength 1. There is a finite set T of 3-dimensional tile types but an infinite number of copies of each tile type, with each copy being referred to as a *tile*.

A 3-dimensional *assembly* is a positioning of tiles on the integer lattice $\mathbb{Z}^3$ and is described formally as a partial function $\alpha : \mathbb{Z}^3 \dashrightarrow T$. Two adjacent tiles in an assembly *bind* if the glue labels on their abutting sides are equal and have positive strength. Each assembly induces a *binding graph*, i.e., a "grid graph" (sometimes called the *adjacency graph*) whose vertices are (positions of) tiles and whose edges connect any two vertices whose corresponding tiles bind. If τ is an integer, we say that an assembly is *τ-stable* if every cut of its binding graph has strength at least τ, where the strength of a cut is the sum of all of the individual glue strengths in the cut.

A 3-dimensional *tile assembly system* (TAS) is a triple $\mathcal{T} = (T, \sigma, \tau)$, where T is a finite set of tile types, $\sigma : \mathbb{Z}^3 \dashrightarrow T$ is a finite, τ-stable *seed assembly*, and τ is the *temperature*. In this paper, we assume that $|\text{dom } \sigma| = 1$ and $\tau = 1$. An assembly α is *producible* if either $\alpha = \sigma$ or if β is a producible assembly and α can be obtained from β by the stable binding of a single tile. In this case we write $\beta \rightarrow_1^{\mathcal{T}} \alpha$ (to mean α is producible from β by the binding of one tile), and we write $\beta \rightarrow^{\mathcal{T}} \alpha$ if $\beta \rightarrow_1^{\mathcal{T}^*} \alpha$ (to mean α is producible from β by the binding of zero or more tiles). When $\mathcal{T}$ is clear from context, we may write $\rightarrow_1$ and $\rightarrow$ instead. We let $\mathcal{A}[\mathcal{T}]$ denote the set of producible assemblies of $\mathcal{T}$. An assembly is *terminal* if no tile can be τ-stably bound to it. We let $\mathcal{A}_\Box [\mathcal{T}] \subseteq \mathcal{A}[\mathcal{T}]$ denote the set of producible, terminal assemblies of $\mathcal{T}$.

A TAS $\mathcal{T}$ is *directed* if $|\mathcal{A}_\Box [\mathcal{T}]| = 1$. Hence, although a directed system may be nondeterministic in terms of the order of tile placements, it is deterministic in the sense that exactly one terminal assembly is producible. For a set $X \subseteq \mathbb{Z}^3$, we say that X is uniquely produced if there is a directed TAS $\mathcal{T}$, with $\mathcal{A}_\Box [\mathcal{T}] = \{\alpha\}$, and dom $\alpha = X$.

2.2 Complexities of (Scaled) Finite Shapes

The following definitions are based on the definitions found in [22]. We include these definitions for the sake of completeness.

A *coordinated shape* is a finite set $X \subset \mathbb{Z}^2$ such that X is connected, i.e., the grid graph induced by X is connected. For some $c \in \mathbb{Z}^+$, we say that a *c-scaling* of X, denoted as X^c, is the set $X^c = \{(a, b) \mid (\lfloor a/c \rfloor, \lfloor b/c \rfloor) \in X\}$. Intuitively, X^c is the coordinated shape obtained by taking X and replacing each point in X with a $c \times c$ block of points. Here, the constant c is known as the *scale factor* (or *resolution loss*). Note that a c-scaling of an actual shape is itself a coordinated shape.

Let X_1 and X_2 be two coordinated shapes. We say that X_1 and X_2 are *scale-equivalent* if $X_1^a = X_2^b$, for some $a, b \in \mathbb{Z}^+$. We say that X_1 and X_2 are *translation-equivalent* if they are equal up to translation. We write $X_1^a \cong X_2^b$ if X_1^a is translation-equivalent to X_2^b, for some $a, b \in \mathbb{Z}^+$. Note that the three previously defined relations are all equivalence relations (see the appendix of [22]). We will use the notation $\widetilde{X}$ to denote the equivalence class containing X under the equivalence relation $\cong$. We say that $\widetilde{X}$ is the *shape* of X. While $\widetilde{X}$ is technically a set of coordinate shapes, we will abuse the notation $\left|\widetilde{X}\right|$ and say that it represents the size of coordinate shape $X \in \widetilde{X}$, i.e., $\left|X^1\right|$.

We will now define the tile complexity of a 3D shape. However, we will first briefly define Kolmogorov complexity of a binary string x, relative to a universal Turing machine U. We say that the *Kolmogorov complexity* of x relative to U is $K_U(x) = \min\{|p| \mid U(p) = x\}$, where, for any Turing machine M, $|M|$ denotes the number of bits used to describe M, with respect to some fixed encoding scheme. In other words, $K_U(x)$ is the smallest program that outputs x (see [10] for a comprehensive discussion of Kolmogorov complexity).

Relative to a fixed universal Turing machine U, we say that the *Kolmogorov complexity of a shape* $\widetilde{X}$ is the size of the smallest program that outputs some $X \in \widetilde{X}$ as a list of locations, i.e., $K_U\left(\widetilde{X}\right) = \min\left\{|p| \,\middle|\, U(p) = \langle X\rangle \text{ for some } X \in \widetilde{X}\right\}$. Beyond this point, we will assume U is a fixed universal Turing machine and therefore will be omitted from our notation.

The *3D tile complexity* at temperature τ (often referred to as *program-size complexity*) of a shape $\widetilde{X}$ at temperature τ is

$$K^{\tau}_{3DSA}\left(\widetilde{X}\right) = \min\left\{ n \,\middle|\, \begin{array}{l} \mathcal{T} = (T,\sigma,\tau), |T| = n \text{ and there exists} \\ X \in \widetilde{X} \text{ such that } \mathcal{T} \text{ uniquely produces } \alpha \\ \text{such that } X \times \{0\} \subseteq \text{dom } \alpha \subseteq X \times \{0,1\} \end{array} \right\}.$$

3 Main Theorem

The main theorem of this paper describes the relationship between the quantities $K\left(\widetilde{X}\right)$ and $K^1_{3DSA}\left(\widetilde{X}\right)$. This relationship is formally stated in Theorem 1. Note that the main result of [22] describes the relationship between $K\left(\widetilde{X}\right)$ and $K^2_{SA}\left(\widetilde{X}\right)$, where $K^2_{SA}\left(\widetilde{X}\right)$ (see [22]) is the tile complexity of the 2D shape $\widetilde{X}$ at temperature 2. In this section, assume that $\widetilde{X}$ is an arbitrary finite shape.

Theorem 1. *The following hold:* $K\left(\widetilde{X}\right) = O\left(K^1_{3DSA}\left(\widetilde{X}\right) \log K^1_{3DSA}\left(\widetilde{X}\right)\right)$ *and* $K^1_{3DSA}\left(\widetilde{X}\right) \log K^1_{3DSA}\left(\widetilde{X}\right) = O\left(K\left(\widetilde{X}\right)\right)$.

We will prove Theorem 1 in Lemmas 1 and 2.

Lemma 1. $K\left(\widetilde{X}\right) = O\left(K^1_{3DSA}\left(\widetilde{X}\right)\log K^{\tau}_{3DSA}\left(\widetilde{X}\right)\right)$.

Proof. Soloveichik and Winfree [22] showed that $K\left(\widetilde{X}\right) = O\left(K^{\tau}_{SA}\left(\widetilde{X}\right)\log K^{\tau}_{SA}\left(\widetilde{X}\right)\right)$, which still holds when K^{τ}_{SA} is replaced with K^{τ}_{3DSA}, for any $\tau \in \mathbb{Z}^+$.

The main contribution of this paper is the following lemma, the proof of which mimics the proof of $K^2_{SA}\left(\widetilde{X}\right)\log K^2_{SA}\left(\widetilde{X}\right) = O\left(K\left(\widetilde{X}\right)\right)$ from [22]. We give the details of the proof here for the sake of completeness.

Lemma 2. $K^1_{3DSA}\left(\widetilde{X}\right)\log K^1_{3DSA}\left(\widetilde{X}\right) = O\left(K\left(\widetilde{X}\right)\right)$.

Proof. Let s be a program (Turing machine) that outputs a list of points contained in some coordinated shape $X \in \widetilde{X}$. We develop a temperature 1 3D construction that takes s as input and outputs a TAS $\mathcal{T}_{\widetilde{X}} = \left(T_{\widetilde{X}}, \sigma, 1\right)$ such that $\mathcal{T}_{\widetilde{X}}$ uniquely produces an assembly whose domain is some shape $X \in \widetilde{X}$ and $\left|T_{\widetilde{X}}\right| = O\left(\frac{|s|}{\log|s|}\right)$. This construction is discussed in Sect. 4.

Now suppose that s is the smallest Turing machine that outputs the list of points in some coordinated shape $X \in \widetilde{X}$. In other words, s is such that $K\left(\widetilde{X}\right) = |s|$. Let $\mathcal{T}^*_{\widetilde{X}} = \left(T^*_{\widetilde{X}}, \sigma, 1\right)$ be the TAS produced by our construction, when given s as input. Observe that, for any TAS $\mathcal{T} = (T, \sigma, 1)$ in which the shape $\widetilde{X}$ uniquely self-assembles, we have $K^1_{3DSA}\left(\widetilde{X}\right) \leq |T|$. Then, for some constant $c \in \mathbb{Z}^+$, the following is true:

$$
\begin{aligned}
K^1_{3DSA}\left(\widetilde{X}\right)\log K^1_{3DSA}\left(\widetilde{X}\right) \leq & \left|T^*_{\widetilde{X}}\right| \log\left|T^*_{\widetilde{X}}\right| \leq c\frac{|s|}{\log|s|}\log\frac{|s|}{\log|s|} \\
= & \; c\frac{|s|}{\log|s|}\left(\log|s| - \log\log|s|\right) \leq c\,|s|.
\end{aligned}
$$

Thus, $K^1_{3DSA}\left(\widetilde{X}\right)\log K^1_{3DSA}\left(\widetilde{X}\right) = O\left(K\left(\widetilde{X}\right)\right)$.

4 Main Construction

In this section, we give an overview of our main construction.

4.1 Setup

Our construction represents a Turing-universal way of guiding the self-assembly of a scaled-up, just-barely 3D version of an arbitrary input shape X at temperature 1 with optimal tile complexity. Therefore, we let U be a fixed universal Turing machine over a binary alphabet, with a one-way infinite tape (to the right), such that, upon termination, U contains the output – and only the output – of the Turing machine being simulated on its tape, perhaps padded to the left and right with 0 bits (the tape alphabet symbol 0) and the tape head is reading the last bit of its output. We also assume that, to the left of the leftmost tape cell, there is a special left marker symbol #, which can be read by U but can neither be overwritten nor written elsewhere on the tape. In general, if M is a Turing machine and w is an input string such that $M(w) = y$, then, upon termination, $U(\langle M, w\rangle)$ leaves exactly a string of the form $\#0^*y0^*$ on its tape with its tape head reading the last bit of y (# is eventually converted to a 0 bit). Our construction is programmed by specifying the program to be executed by U.

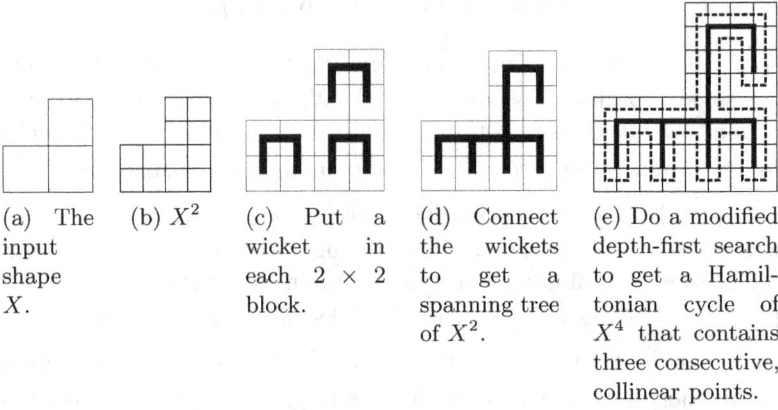

(a) The input shape X. (b) X^2 (c) Put a wicket in each 2×2 block. (d) Connect the wickets to get a spanning tree of X^2. (e) Do a modified depth-first search to get a Hamiltonian cycle of X^4 that contains three consecutive, collinear points.

Fig. 1. An overview of the algorithm for plotting out a Hamiltonian cycle of an input shape (scaled up by a factor of 4), such that the cycle has three consecutive, collinear points. Although a Hamiltonian cycle is hard to compute, in general, it is clear that, for an arbitrary finite shape X, a Hamiltonian cycle, as described above, of X^4, can be computed in polynomial time.

The input to U is a program p, appropriately encoded as $\langle p\rangle$, using some fixed encoding scheme. The output of p is used to guide the self-assembly of our construction. We assume p is actually the concatenation of two programs s and p_{hc}, where s, the input to our construction, is a program that outputs the list of points in X and p_{hc} is a fixed program (independent of X) that uses the output of s as its input and builds a special Hamiltonian cycle H of X^4, along which there are three consecutive, collinear points. The *seed block*, which is described in the next subsection, is defined as the middle point of an arbitrarily chosen triplet of consecutive, collinear points of H (by the way we construct H, there is always at least one such triplet of points). We further assume that p outputs H – and only H – as a sequence of pairs of bits, such that, 00, 10, 01 and 11 correspond to "no-move", "left", "right" and "straight", respectively and possibly padded to the left with at least two no-moves and to the right with an even number of no-moves (see Fig. 1 for an overview of how H is constructed). Thus, p satisfies $|p| = |s| + |p_{hc}|$.

4.2 Seed Block

The (upper portion of what will eventually become the) seed block of our construction grows from a single seed tile and carries out the following three logical phases: decoding, simulation and output. These three phases are depicted in Fig. 2 with vertical, zig-zag and diagonal patterns, respectively.

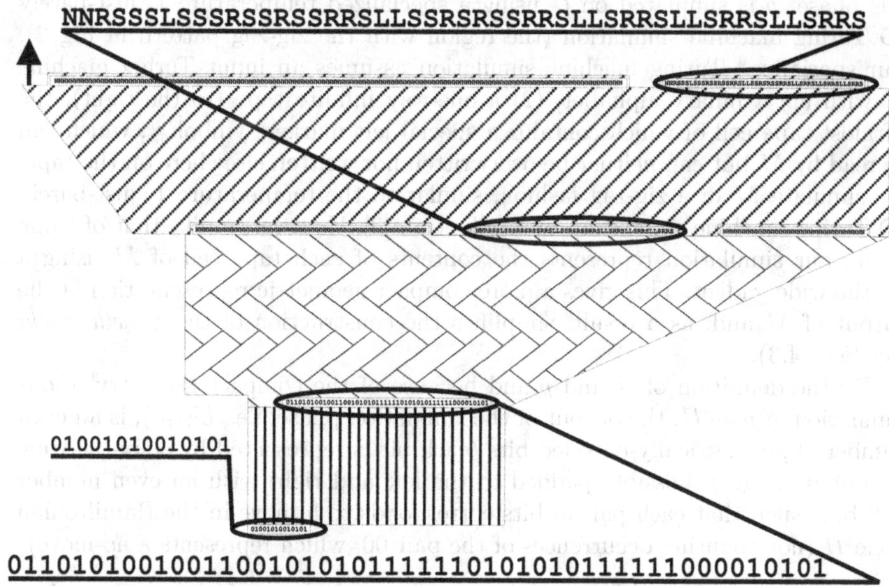

Fig. 2. Self-assembly of the upper portion of the seed block consists of three logical phases. In the first phase (the region filled with vertical lines), the bits of p are decoded using the 3D, temperature 1 optimal encoding scheme of Furcy et al. [9] (the encoded bits of p are depicted as the shorter binary string). The decoded bits of p (the longer binary string) are input to a fixed universal Turing machine U. Then, in the second phase, the simulation of p on U is carried out (in the region with the zig-zag pattern). We require that $U(\langle p \rangle)$ evaluates precisely to the sequence of moves in the Hamiltonian cycle of X^4, padded to the left and right with an even number of 0 bits (the boxes that are not encircled in this figure). In other words, we require that $U(\langle p \rangle)$ evaluates to a string of the form $(00)^*(00|10|01|11)^*(00)^*$, with the tape head of U reading the second bit in the last move of the Hamiltonian cycle. Finally, in the third phase (in the region with the diagonal line pattern), the moves in the Hamiltonian cycle are shifted to the right. Self-assembly of the first growth block begins from the upward-pointing arrow. Note that the moves in the Hamiltonian cycle are listed in the grey boxes and we use the characters 'N', 'L', 'R' and 'S' to represent "no-move", "left", "right" and "straight", respectively.

Decoding. The first phase is the decoding phase. In the decoding phase, the bits of $\langle p \rangle$ are decoded from a $O(\log |\langle p \rangle|)$-bits-per-tile representation to a 1-bit-per-tile representation (actually, we end up with a 1-bit-per-gadget representation, which is sufficient to maintain the optimality of our construction). To accomplish this, we use the 3D, temperature 1 optimal encoding scheme of Furcy et al. [9]. When the decoding phase completes, the decoded bits of $\langle p \rangle$ are advertised in a one-bit-per-gadget representation along the top of the optimal encoding region (the rectangle with the vertical lines in Fig. 2).

Simulation. Once the bits of $\langle p \rangle$ are decoded, the simulation phase begins. In this phase, p is simulated on U using a specialized temperature 1, just-barely 3D Turing machine simulation (the region with the zig-zag pattern in Fig. 2). Our specialized Turing machine simulation assumes an input Turing machine M with (1) a binary alphabet, (2) a one-way infinite tape (to the right), the leftmost tape cell of which contains a special left marker symbol #, which can be read by M but can neither be overwritten nor written elsewhere on the tape. We simulate M in a zig-zag fashion, similar to the temperature 1, just-barely 3D Turing machine simulation by Cook et al. [3]. However, unlike that of Cook et al., our simulation represents the contents of each tape cell of M using a six-tile-wide gadget. This gives a more compact geometric representation of the output of M and, as a result, simplifies the construction of the *growth blocks* (see Sect. 4.3).

By the definition of U and p and because of the compact geometry of our simulation of p on U, the output of the simulation phase, i.e., $U(\langle p \rangle)$, is an even number of geometrically-encoded bits (each bit is represented by a six-tile-wide bit-bump gadget), possibly padded to the left and right with an even number of 0 bits, such that each pair of bits corresponds to a move in the Hamiltonian cycle H (not counting occurrences of the pair 00, which represents a no-move).

Output. In order to satisfy certain geometric constraints, which are required by the growth blocks, after the simulation phase of p on U is complete, the (final) output phase begins. In the output phase, we use a special, constant-size tile set to shift the geometrically-encoded bits of H to the right, so that the bits of H are in a right-justified position along the top of the seed block (the region with diagonal lines in Fig. 2). For each right-shift, we add a pair of 0 bits to the left, which ensures that the upper portion of the seed block will be wider than it is taller. After the output phase of the seed block, self-assembly of the first growth block, which is always to the north in our construction, as guaranteed by p, begins from the left side of the top of the upper portion of the seed block (see the upward-pointing arrow in Fig. 2).

Scale factor. Let W_{decode} and H_{decode} be the maximum horizontal and vertical extent, respectively, of the seed block after the decoding phase (the rectangle with vertical lines pattern in Fig. 2) completes. From [9], we know that $W_{decode} > H_{decode}$. Next, in the simulation phase, the tape grows to the right by two tape cells for each transition. Each transition is comprised of two rows of gadgets, which are wider than they are tall (six tiles versus four tiles). Also, we may assume that, during the simulation phase, p is programmed to initially scan the input from left-to-right and then from right-to-left before beginning. Let W_{sim} and H_{sim} be the maximum horizontal and vertical extent, respectively, of the seed block after the simulation phase (the zig-zag pattern in Fig. 2) completes. Then we have $W_{sim} > H_{sim}$. Finally, in the output phase, as the output bits of the simulation phase are shifted to the right, two tape cells are added to the left for each shift. Each shift is comprised of two rows of gadgets, which, like the simulation gadgets, are wider than they are tall (six versus four). Therefore, let W_{sb} and H_{sb} be the maximum horizontal and vertical extent, respectively, of the seed block after the simulation phase (the diagonal pattern in Fig. 2) completes. Then we have $W_{sb} > H_{sb}$. From this we may conclude that the seed block, once completely filled in by the last growth block (see Fig. 5a) will be a square. The scale factor of our construction is W_{sb}.

4.3 Growth Blocks

Each growth block has a single input side, which reads the remaining moves in the Hamiltonian cycle and a single output side, which advertises the same remaining path but with its first move erased. This first move determines the position of the output side in relation to the input side. In this section, we assume that the input side of the growth block is its south side (the construction simply needs to be rotated for the three other possible positions of the input side). So, if the first (erased) move in the remaining path is a right turn, then the output side of the growth block is its east side. We describe the construction for this case here (see Fig. 4). The overview figure for the growth blocks uses gadgets whose structure is explained in Fig. 3.

The growth block starts assembling in its southwest corner and progresses in a zig-zag pattern. The first row of gadgets, moving from left to right, starts by copying all of the leading no-moves, of which there are exactly two in Fig. 4 but generally many more. Once the first actual move is found, a set of gadgets specific to its type is activated. In the case of a right turn, all of the moves are shifted by one position to the right (and the first move is replaced by a no-move). The last move in the remaining path is advertised at the bottom of the output (or east) side of the block. Then the construction switches direction and moves from right to left, simply copying the shifted path, which completes the first iteration. In each subsequent iteration, the left-to-right pass shifts the whole

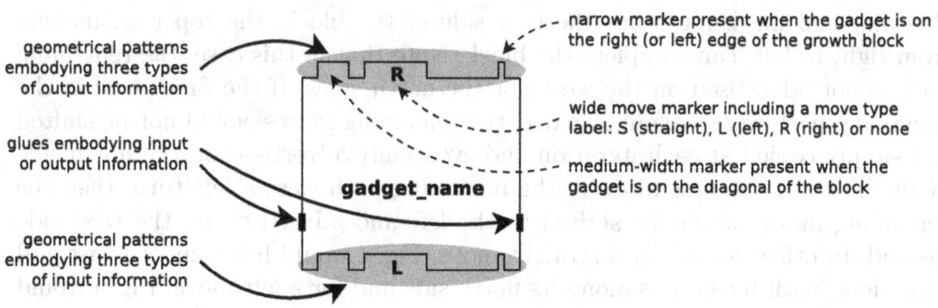

Fig. 3. Key to the representation of our gadgets

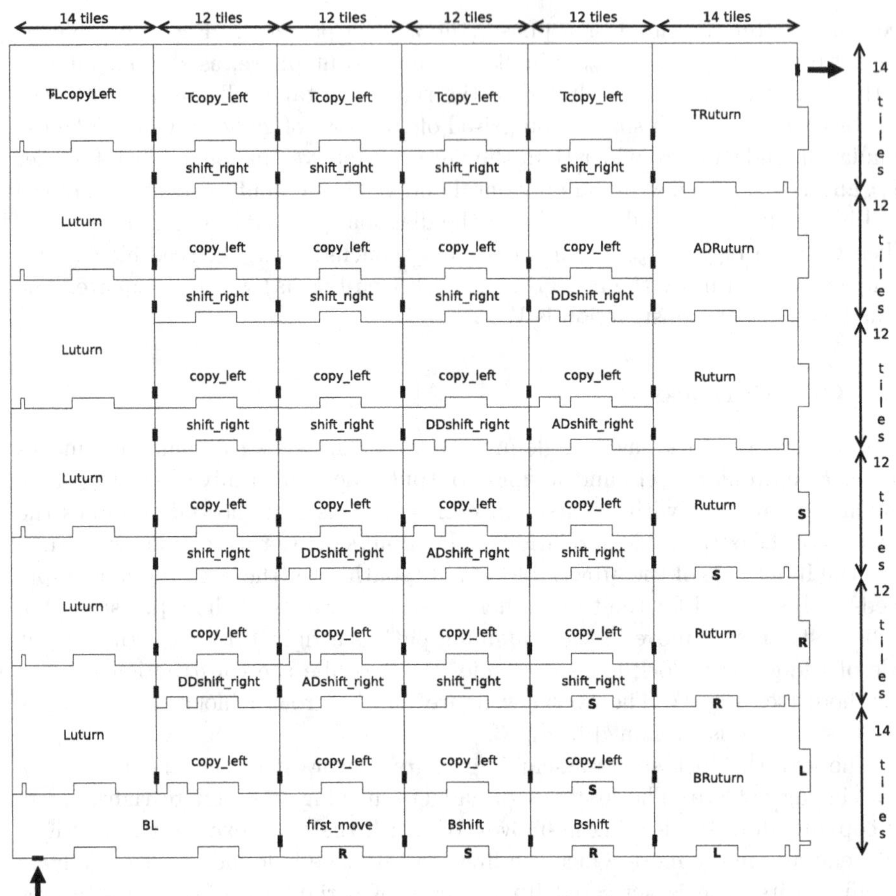

Fig. 4. Overall construction of a growth block whose input path starts with a right turn

path to the right by one position and advertises one more move on the output (or east) side. In addition to the right shift, each zig-zag iteration moves a diagonal marker by one position to the right, starting from the southwest corner. Once this diagonal marker reaches the east side of the block, the top row, moving from right to left, can complete the block. Note that in this case, the remaining path is not advertised on the west nor the north sides. If the first move in the remaining path were a straight move, the remaining moves would not be shifted but simply copied at each iteration and eventually advertised on the north side of the block. If the first move in the remaining path were a left turn, then the remaining moves would be shifted to the left and advertised on the west side instead. In other words, for a straight move, Fig. 4 would have smooth east and west sides, with bit-bumps along its north side and for a left move, Fig. 4 would have smooth east and north sides, with bit-bumps along its west side.

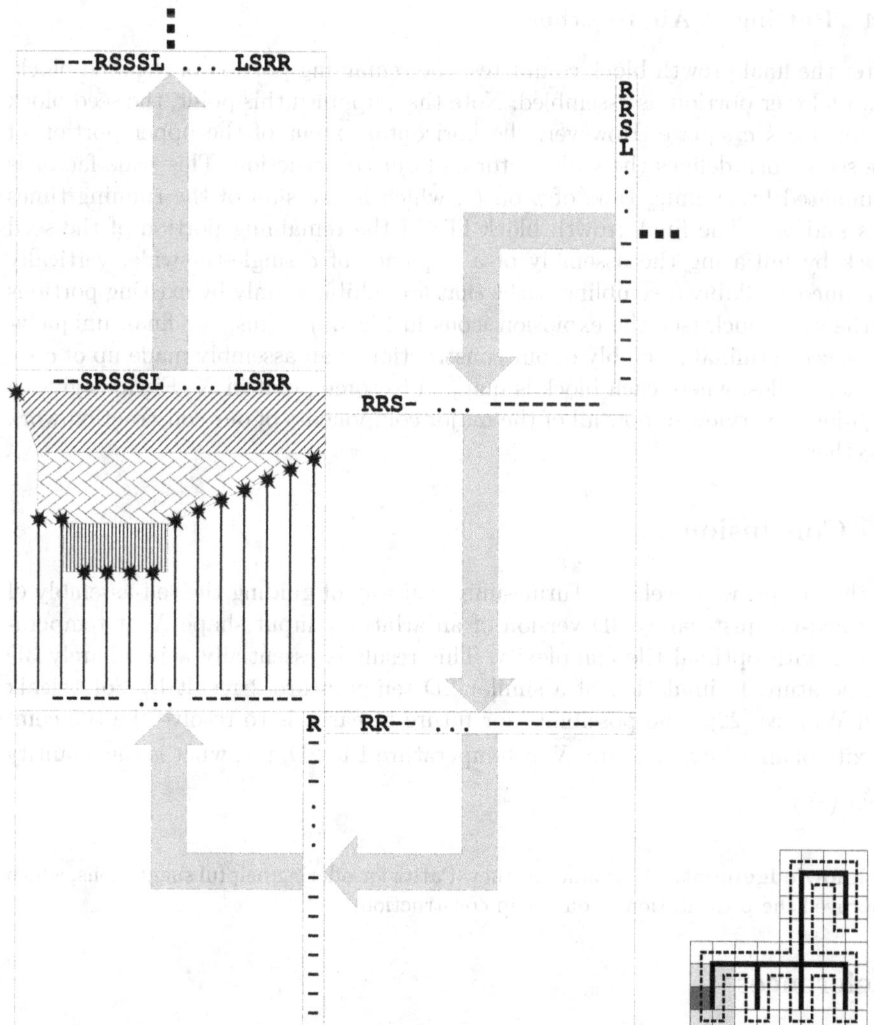

(a) In the upper portion (pattern-filed regions) of the seed block, the points in X are decoded and a special Hamiltonian cycle H of X^4. As moves of H are carried out within each growth block, the bits of H are propagated to the next growth block but the move that was just executed is erased (depicted as dashes). The last growth block assembles the remaining portion of the seed block with a sequence of single-tile-wide paths that assemble northward until running into the existing portion of the seed block.

(b) The grey squares indicate the portion being assembled in Figure 5a. The darker square is the seed block. Self-assembly always proceeds north from the seed block.

Fig. 5. Putting it all together. In this figure, we depict the moves in the Hamiltonian cycle of X^4 as '-', 'L', 'R' and 'S' for "no-move", "left", "right" and "straight", respectively. In our construction, these moves are represented using the pairs of bits 00, 10, 01 and 11, for "no-move", "left", "right" and "straight", respectively.

4.4 Putting It All Together

After the final growth block completes, the remaining portion of the seed block, i.e., its lower portion, is assembled. Note that, up until this point, the seed block is not a $c \times c$ square. However, the horizontal extent of the upper portion of the seed block defines the scale factor c of our construction. This scale factor is dominated by running time of p on U, which is the sum of the running times of s and p_{hc}. The final growth block fills in the remaining portion of the seed block by initiating the assembly of a sequence of c single-tile-wide, vertically and uncontrollably assembling paths that are inhibited only by existing portions of the seed block (see the explosion icons in Fig. 5a). Thus, the final, uniquely-produced terminal assembly of our construction is an assembly made up of $c \times c$ blocks of tiles, where each block is mapped to some point in X. Figure 5 gives a high-level overview of how all of the major components of our construction work together.

5 Conclusion

In this paper, we develop a Turing-universal way of guiding the self-assembly of a scaled-up, just-barely 3D version of an arbitrary input shape X at temperature 1 with optimal tile complexity. This result is essentially a just-barely 3D temperature 1 simulation of a similar 2D temperature 2 result by Soloveichik and Winfree [22]. One possibility for future research is to resolve the tile complexity of an arbitrary shape $\widetilde{X}$ at temperature 1 in 2D, i.e., what is the quantity $K^1_{SA}\left(\widetilde{X}\right)$?

Acknowledgement. We thank Matthew Patitz for offering helpful suggestions, which improved the presentation of our main construction.

References

1. Adleman, L.M., Cheng, Q., Goel, A., Huang, M.-D.A.: Running time and program size for self-assembled squares, STOC, pp. 740–748 (2001)
2. Barish, R.D., Schulman, R., Rothemund, P.W., Winfree, E.: Aninformation-bearing seed for nucleating algorithmic self-assembly. Proc. Natl. Acad. Sci. **106**(15), 6054–6059 (2009)
3. Cook, M., Fu, Y., Schweller, R.: Temperature 1 self-assembly: deterministic assembly in 3D and probabilistic assembly in 2D. In: Proceedings of the 22nd Annual ACM-SIAM Symposium on Discrete Algorithms (2011)
4. Doty, D., Lutz, J.H., Patitz, M.J., Schweller, R.T., Summers, S.M., Woods, D.: The tile assembly model is intrinsically universal. In: Proceedings of the 53rd Annual IEEE Symposium on Foundations of Computer Science, FOCS 2012, pp. 302–310 (2012)
5. Doty, D., Lutz, J.H., Patitz, M.J., Summers, S.M., Woods, D.: Intrinsic universality in self-assembly. In: Proceedings of the 27th International Symposium on Theoretical Aspects of Computer Science, pp. 275–286 (2009)

6. Doty, D., Patitz, M.J., Summers, S.M.: Limitations of self-assembly at temperature 1. Theor. Comput. Sci. **412**, 145–158 (2011)
7. Evans, C.: Crystals that count! physical principles and experimental investigations of DNA tile self-assembly, Ph.D. thesis, California Institute of Technology (2014)
8. Fujibayashi, K., Hariadi, R., Park, S.H., Winfree, E., Murata, S.: Toward reliable algorithmic self-assembly of DNA tiles: a fixed-width cellular automaton pattern. Nano Lett. **8**(7), 1791–1797 (2007)
9. Furcy, D., Micka, S., Summers, S.M.: Optimal program-size complexity for self-assembly at temperature 1 in 3D. In: Phillips, A., Yin, P. (eds.) DNA 2015. LNCS, vol. 9211, pp. 71–86. Springer, Heidelberg (2015)
10. Li, M., Vitányi, P.: An Introduction to Kolmogorov Complexity and Its Applications, 3rd edn. Springer, New York (2008)
11. Lund, K., Manzo, A.T., Dabby, N., Micholotti, N., Johnson-Buck, A., Nangreave, J., Taylor, S., Pei, R., Stojanovic, M.N., Walter, N.G., Winfree, E., Yan, H.: Molecular robots guided by prescriptive landscapes. Nature **465**, 206–210 (2010)
12. Manuch, J., Stacho, L., Stoll, C.: Two lower bounds for self-assemblies at temperature 1. J. Comput. Biol. **17**(6), 841–852 (2010)
13. Mao, C., LaBean, T.H., Relf, J.H., Seeman, N.C.: Logical computation using algorithmic self-assembly of DNA triple-crossover molecules. Nature **407**(6803), 493–496 (2000)
14. Meunier, P.-E., Patitz, M.J., Summers, S.M., Theyssier, G., Winslow, A., Woods, D.: Intrinsic universality in tile self-assembly requires cooperation. In: Proceedings of the 25th Annual ACM-SIAM Symposium on Discrete Algorithms (SODA), pp. 752–771 (2014)
15. Qian, L., Winfree, E.: Scaling up digital circuit computation with DNA strand displacement cascades. Science **332**(6034), 1196 (2011)
16. Qian, L., Winfree, E., Bruck, J.: Neural network computation with DNA strand displacement cascades. Nature **475**(7356), 368–372 (2011)
17. Rothemund, P.W.K., Winfree, E.: The program-size complexity of self-assembled squares (extended abstract). In: STOC 2000: Proceedings of the Thirty-Second Annual ACM Symposium on Theory of Computing, pp. 459–468 (2000)
18. Rothemund, P.W.K.: Folding DNA to create nanoscale shapes and patterns. Nature **440**(7082), 297–302 (2006)
19. Rothemund, P.W.K., Papadakis, N., Winfree, E.: Algorithmic self-assembly of DNA Sierpinski triangles. PLoS Biol. **2**(12), 2041–2053 (2004)
20. Schulman, R., Winfree, E.: Synthesis of crystals with a programmable kinetic barrier to nucleation. Proc. Natl. Acad. Sci. **104**(39), 15236–15241 (2007)
21. Seeman, N.C.: Nucleic-acid junctions and lattices. J. Theor. Biol. **99**, 237–247 (1982)
22. Soloveichik, D., Winfree, E.: Complexity of self-assembled shapes. SIAM J. Comput. **36**(6), 1544–1569 (2007)
23. Wang, H.: Proving theorems by pattern recognition - II. Bell Syst. Tech. J. **XL**(1), 1–41 (1961)
24. Winfree, E.: Algorithmic self-assembly of DNA, Ph.D. thesis, California Institute of Technology, June 1998
25. Winfree, E., Liu, F., Wenzler, L.A., Seeman, N.C.: Design and self-assembly of two-dimensional DNA crystals. Nature **394**(6693), 539–44 (1998)
26. Yurke, B., Turberfield, A.J., Mills, A.P., Simmel, F.C., Neumann, J.L.: A DNA-fuelled molecular machine made of DNA. Nature **406**(6796), 605–608 (2000)

Line Segment Covering of Cells in Arrangements

Matias Korman[1], Sheung-Hung Poon[2], and Marcel Roeloffzen[3,4(✉)]

[1] Tohoku University, Sendai, Japan
mati@dais.is.tohoku.ac.jp
[2] School of Computing and Informatics, Institut Teknologi Brunei, Brunei, Brunei Darussalam
sheung.hung.poon@gmail.com
[3] National Institute of Informatics (NII), Tokyo, Japan
marcel@nii.ac.jp
[4] JST, ERATO, Kawarabayashi Large Graph Project, Tokyo, Japan

Abstract. Given a collection L of line segments, we consider its arrangement and study the problem of covering all cells with line segments of L. That is, we want to find a minimum-size set L' of line segments such that every cell in the arrangement has a line from L' defining its boundary. We show that the problem is NP-hard, even when all segments are axis-aligned. In fact, the problem is still NP-hard when we only need to cover rectangular cells of the arrangement. For the latter problem we also show that it is fixed parameter tractable with respect to the size of the optimal solution. Finally we provide a linear time algorithm for the case where cells of the arrangement are created by recursively subdividing a rectangle using horizontal and vertical cutting segments.

1 Introduction

Set cover [3] is one of the most fundamental problems of computer science. This problem is usually formulated in terms of hypergraphs: the input of the problem is a hypergraph $\mathcal{H} = (X, \mathcal{F})$ where $\mathcal{F} \subseteq 2^X$ is a collection of subsets of X, and we aim for a subset $\mathcal{F}' \subseteq \mathcal{F}$ of smallest cardinality that *covers* X (i.e., $\cup_{F \in \mathcal{F}'} F = X$). This problem is known to be NP-hard and even hard to approximate [3,6,8].

Given its importance, it is not surprising that this problem has been studied extensively. In most cases, the set $\mathcal{F}$ is given implicitly (this is specially true when considering geometric variants of the problem). For example, in the well-known k-center problem [5] we want to cover a set S of n points with unit disks. In the hypergraph definition, this is equivalent to $X = S$ and $\mathcal{F}$ is the collection of subsets of S that can be covered with a single unit disk.

Sometimes the relationship between X and $\mathcal{F}$ is much more involved. For example, in the *discrete center problem*, we only consider the disks whose center is a point of S. Akin to the discrete variant of the k-center problem, in this paper

M. Korman—Partially supported by the ELC project (MEXT KAKENHI No. 24106008).

Z. Lu et al. (Eds.): COCOA 2015, LNCS 9486, pp. 152–162, 2015.
DOI: 10.1007/978-3-319-26626-8_12

we study a geometric setting where the elements X and sets $\mathcal{F}$ are defined by the same geometric primitives. Specifically, we study the problem of covering the cells of an arrangement of line segments L with segments of L. Given a set L of line segments in the plane a *cell* in the arrangement of L is defined as a maximally connected region that is not intersected by any segments of L. Essentially the cells are the 'empty'—not intersected by segments of L—regions in the arrangement defined by L. Now let C denote the set of all cells in the arrangement of L. We say that a cell $c \in C$ is *covered* by a line segment $\ell \in L$ if and only if ℓ is part of the boundary of c. Similarly c is covered by a set L' of line segments if and only if there is a segment $\ell \in L'$ that covers c. The goal is then to find a minimum-size set $L' \subset L$ that covers all cells of C. We call this the *line-segment covering* problem.

The problem can also be viewed as a guarding problem. In the traditional art gallery problem, the goal is to place guards so that the guards together see the whole gallery (often a simple polygon). Many variants of this have been studied. Bose *et al.* [2] study guarding and coloring problems between lines. They provide results for several types of guards and objects to guard, such as guarding the cells of the arrangement with the lines, or guarding the lines by selecting cells. Their results however do not extend to line segments as they use properties of the lines that do not hold for line segments. To the best of our knowledge covering cells in an arrangement of line segments with the segments has not been studied before.

We study three different variants of this problem. First, in Sect. 2 we show that the line-segment covering problem is NP-hard, even when all segments of L are axis-aligned. In Sect. 3 we consider a slightly different variant, where we are required to cover only rectangular cells, those defined by four line segments. For this variant we show that the NP-hardness reduction still works. However, we show that this variant is fixed parameter tractable with respect to the size of the optimal solution. In Sect. 4, inspired by subdivisions induced by KD-trees, we study a variant where the line segments define a type of rectangular subdivision. That is, an axis aligned rectangle that is recursively subdivided with horizontal or vertical line segments, similar to the subdivision defined by a KD-tree [1]. For this case we show that an optimal cover can be computed in linear time, assuming that the partitioning is given as a tree-structure defined by the splitting lines.

2 NP-hardness for Rectilinear Line Segments

In this section we show that the line-segment covering problem is NP-hard, even if the input consists of only horizontal and vertical line segments. We reduce the problem from PLANAR 3SAT [7]. An input instance for the 3SAT problem is a set $\{x_1, x_2, \ldots, x_n\}$ of n variables and a Boolean expression in conjunctive normal form. That is, the expression is a conjunction of clauses $\Phi = c_1 \wedge \ldots \wedge c_m$ such that each clause c_i is a disjunction of three literals (a variable or negation of a variable). The problem is then to decide if there is a truth assignment for the variables so that Φ is true. In PLANAR 3SAT we impose further restrictions by looking at the representation of Φ as a bipartite graph with variables and clauses

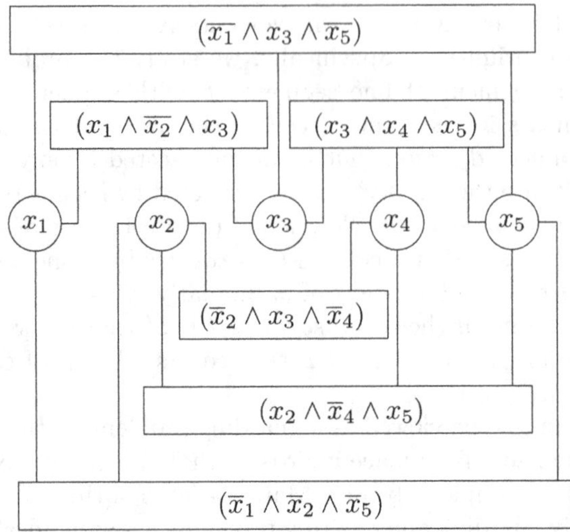

Fig. 1. PLANAR 3SAT problem instance along with a planar embedding.

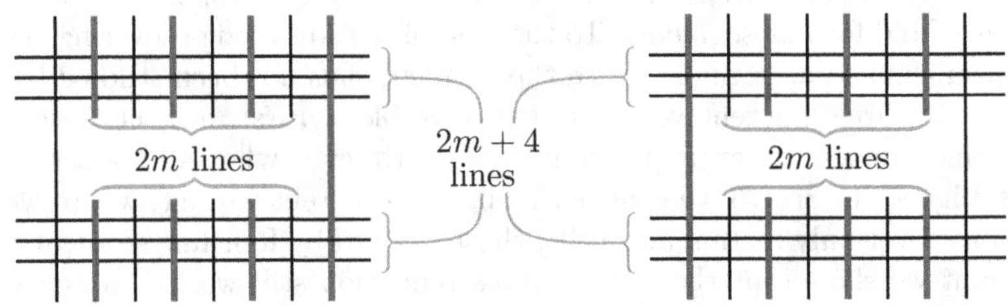

Fig. 2. Gadget for a variable with a true (left) and false (right) assignment. Red (thick) edges show the two possible covers with $m + 1$ segments (Color figure online).

as vertices. A variable-node v is connected to a clause-node c if and only if v occurs in c. In PLANAR 3SAT we assume that this graph is planar. Specifically we assume that a planar embedding is given that places all variable-nodes on a horizontal line and all clause-nodes above or below this line (see Fig. 1). We also assume that no variable appears more than once in any clause (that is, the above described bipartite graph is a proper graph and not a multigraph).

It is well-known that the PLANAR 3SAT problem is NP-hard [7]. Also note that it is easy to see that the line-segment covering problem is in NP. Indeed, given a possible covering, we can construct the arrangement of line segments and verify in polynomial time that indeed all cells are covered. In the remainder of this section we provide a polynomial time reduction and prove its correctness.

2.1 Reduction

For each variable in Φ we create a gadget consisting of $4m+8$ horizontal segments and $4m + 2$ vertical segments. The leftmost and rightmost vertical line stab all

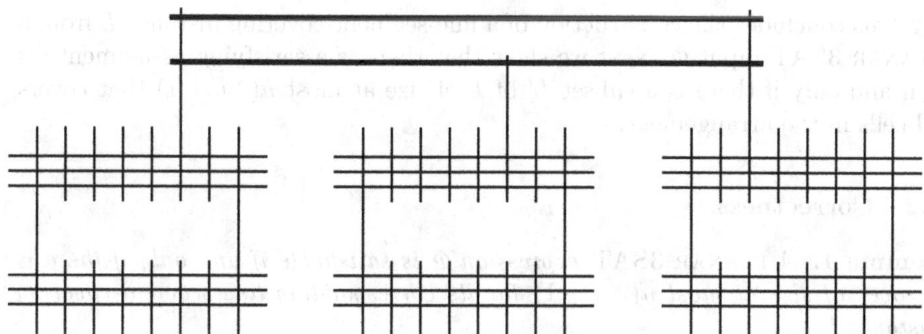

Fig. 3. A clause gadget used in showing NP-hardness in Sect. 2, parts are variable gadgets that connect to the edges of a clause gadget (marked with thicker line segments).

horizontal lines of the gadget, whereas $2m$ of the other lines stab the top $2m+4$ horizontal lines and $2m$ stab the lower $2m+4$ horizontal lines as illustrated in Fig. 2. As we describe later, some of the vertical line segments may be further extended (above or below) to connect to the clause gadgets, but the horizontal segments will not cross any segments of other gadgets.

Intuitively speaking, we will show that any covering of the variable gadgets must choose one every other vertical segment, including either the right or leftmost segment. The choice of using either the rightmost or leftmost vertical segment is equivalent to assigning the variable to be true or false. The clause gadget will create additional cells that will be covered *for free* (without selecting additional line segments) provided that at least one variable satisfies the clause.

Let $s_1^{(i)}, \ldots, s_{2m}^{(i)}$ be the vertical segments created in the gadget for variable x_i (numbered from left to right). We would like to sort the clauses $c_1, \ldots, c_j$ in which x_i occurs in the order in which they appear on the embedding of the PLANAR 3SAT instance. However, this is not well-defined (since it is not always clear when a segment goes before another), so we proceed as follows: let $c_1, \ldots, c_{j'}$ be the clauses that contain variable x_i and are embedded above the line containing all variable nodes, sorted in clockwise order of their connections to x_i. Similarly, let $c_{j'+1}, \ldots, c_j$ be the clauses that are embedded below the line (this time in counter-clockwise order). We define the ordering of the clauses around x_i as the concatenation of both orderings. Since we have $2m$ vertical line segments for each clause and m clauses we ensure that any vertical segment of a variable gadget is extended only towards a single clause, and any clause is associated to exactly three segments.

We now detail the gadget associated to clause $c = \ell_i \wedge \ell_j \wedge \ell_k$ (for $i < j < k$), where ℓ_i is a literal of variable x_i (similarly, ℓ_j and ℓ_k are literals of variables x_j and x_k, respectively). First, we extend the three segments associated to clause c (above or below depending on where c is placed in the embedding). We extend the segments associated to variables x_i and x_k slightly further than the segment of x_j. We complete our transformation by adding two horizontal segments that create a rectangle with the three extended segments, see Fig. 3.

This concludes the construction of a line-segment-covering instance L from a PLANAR 3SAT input Φ. Next we show that there is a satisfying assignment for Φ if and only if there is a subset L' of L of size at most $n(2m+1)$ that covers all cells in the arrangement.

2.2 Correctness

Lemma 1. *A* PLANAR 3SAT *expression* Φ *is satisfiable if and only if there is a cover of size at most* $n(2m+1)$ *for its corresponding line-segment-covering instance* L.

Proof. First we prove that given a satisfying assignment for Φ we can cover L with $n(2m+1)$ segments. If a variable is true, then we select the rightmost vertical segment, and for each set of $2m$ segments that only intersect the top or bottom we select the odd ones counting from the leftmost segment starting at one, see also Fig. 2. If the variable is false we select the leftmost longer segment and the even ones from the sets of shorter segments.

Next we show that all cells are covered. We consider three types of cells: cells in the interior of the grids created of the variable gadgets are called *variable cells*, the single rectangular cell associated to a clause gadget is called *clause cell*; any other cell (included the unbounded one) that is created with our construction is simply called an *other cell*.

Since we have selected one every other segment, clearly all variable cells are covered. The fact that the variable assignment satisfies all clauses implies that at least one of the three vertical segments defining a clause cell has been selected, so the clause cell is covered. Hence, all clause cells are also covered. Finally, for the remaining cells it suffices to see that each such cell always has two consecutive vertical segments of a variable gadget in its boundary. Indeed, Such cells are only created when connecting clauses and variables and in particular, their left and right boundaries are created by those extensions. Thus, when walking along the boundary of any such cell, we will find the next vertical segment of the variable gadget (or the predecessor in case the segment was the last one). One of the segments must have been selected, so also these cells are covered.

The reverse statement is similar. Assume that we have a cover L' for L of size $n(2m+1)$. First observe that each variable gadget needs at least $2m+1$ selected line segments to cover its interior cells. To achieve this we must select either the left or rightmost segment, after which there is a unique cover for the remaining cells that uses only $2m$ segments. Covering the cells within the variable gadgets with fewer than $2m+1$ segments is not possible, so any cover consist of exactly $2m+1$ segments per variable gadget. Furthermore, none of these segments can be reused between different variable gadgets. Thus, we conclude that each clause gadget must be covered by the lines selected from the variable gadgets. We create a variable assignment for each variable as before, depending if the leftmost or rightmost segments has been selected.

Since L' covers all cells, it must also cover the clause cells, which implies that at least one of the three vertical segments has been selected. Equivalently, this implies that in each clause the choice of assignment of the variables makes at least one of its literals true and the formula Φ is satisfiable.

Since the reduction is easily computed in polynomial time we conclude the following result.

Theorem 1. *Given a set L of axis-aligned line segments, it is NP-hard to find a minimum-size set $L' \subseteq L$ so that for each cell of the arrangement at least one of its defining segments is in L'.*

3 Covering Only Rectangular Cells

From the above reduction we can see that the main difficulty of the problem lies in covering the rectangular cells. Thus, in this section we turn our attention to a variant of the problem in which segments are axis-aligned and we are not required to cover all cells, but only those that are rectangles. That is, cells whose boundary is formed by exactly four line segments. First we briefly argue that this variant is also NP-hard by adapting the NP-hardness proof in the previous section. Then we show that the problem is fixed parameter tractable (FPT) with respect to k, the number of segments in the optimal solution.

3.1 NP-hardness

The hardness almost follows from the construction of Sect. 2. Indeed, clause cells are the only critical part of the reduction that need to be modified. Instead, we create the clause gadget with 6 segments as shown in Fig. 4. This modified gadget contains three rectangular cells. Note that incoming segments from variables can cover at most two of these cells, but at least one segment must be added so as to cover the intermediate rectangular cell. This additional edge can cover two cells, either the left and middle cells, or the middle and right cells. Thus, it follows that we can find a covering of all rectangular cells of this modified instance with $n(2m+1)+m$ segments if and only if the associated PLANAR 3SAT instance is satisfiable, and thus this variation is also NP-hard.

Theorem 2. *Given a set L of axis-aligned line segments, it is NP-hard to find a minimum-size set $L' \subseteq L$ so that for each rectangular cell of the arrangement at least one of its defining segments is in L'.*

3.2 FPT on the Size of the Optimal Solution

Next we show that the problem is fixed parameter tractable (FPT) with respect to k, the size of the optimal solution. Our aim is to compute a kernel of small size (or conclude that there is no solution of size at most k).

Since we want to cover only rectangular cells we can represent each cell by an *associated subset* (or subset for short) $C = \{\ell_1, \ell_2, \ell_3, \ell_4\}$ with the four bounding line segments as its elements. This reduces the line segment covering problem to

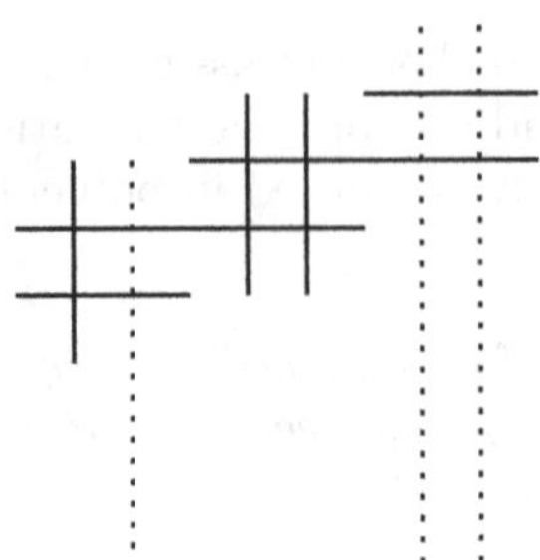

Fig. 4. Modified clause gadget. The three dashed vertical segments connect to the corresponding variable gadgets. This new gadget creates three rectangular cells that can be guarded with one additional segment if and only if the variable assignment satisfies the clause.

a hitting set problem for a collection $\mathcal{C}$ of subsets of size four. Our approach is to reduce the number of subsets to consider; first to a set $\mathcal{C}_1$ where for any two line segments there are at most $2k$ subsets that contain both these line segments; then to a set $\mathcal{C}_2$ where for any single line segment there are at most $2k^2$ subsets containing it. First we prove the following lemma.

Lemma 2. *Let ℓ, ℓ' and ℓ'' be any three line segments in L. There are at most two subsets in $\mathcal{C}$ containing all three line segments, ℓ, ℓ' and ℓ''.*

Proof. Since the arrangement is rectilinear, two lines, say ℓ and ℓ' are parallel and the other is orthogonal to these. This means that any cell having all three line segments ℓ, ℓ' and ℓ'' on its boundary must span the strip between ℓ and ℓ'. However at most two such cells can also be adjacent to the third line segments ℓ''.

We start reducing our problem instance by looking at pairs of line segments. Specifically we count for every pair of line segments how many subsets contain both. Then for any pair ℓ, ℓ' shared in more than $2k$ subsets we add the subset $\{\ell, \ell'\}$ and remove all subsets containing both ℓ and ℓ'. Let $\mathcal{C}_1$ denote this reduced subset.

Lemma 3. *A set $L' \subset L$ of line segments, with $|L'| \leq k$ is a minimum-size cover of $\mathcal{C}_1$ if and only if it is a minimum-size cover of $\mathcal{C}$.*

Proof. Clearly the claim holds if $\mathcal{C} = \mathcal{C}_1$. Thus, from now on we assume that $\mathcal{C}^+ = \mathcal{C}_1 \backslash \mathcal{C}$ and $\mathcal{C}^- = \mathcal{C} \backslash \mathcal{C}_1$ are two nonempty sets that contain all elements that were added and removed from $\mathcal{C}$, respectively.

Now assume that L' with $|L'| \leq k$ is a minimum size cover for $\mathcal{C}_1$. Observe that all subsets of $\mathcal{C}^-$ must also be covered, since for every subset $C \in \mathcal{C}^-$ there is a subset $\{\ell, \ell'\}$ such that both ℓ and ℓ' occur in C (and $\{\ell, \ell'\} \in \mathcal{C}^+$). It follows that L' is a also a cover for $\mathcal{C}$. To show that L' is of minimum size assume for a contradiction that a smaller set L'' is also a cover for $\mathcal{C}$. Clearly, L'' covers $\mathcal{C} \cap \mathcal{C}_1$. Now take any set $\{\ell, \ell'\}$ in $\mathcal{C}^+$, if neither ℓ nor ℓ' is part of L'', then we claim that $|L''| > k$. Indeed, we introduced $\{\ell, \ell'\}$ into $\mathcal{C}_1$ only when more than $2k$ subsets in

$\mathcal{C}$ contain both ℓ and ℓ'. If neither ℓ nor ℓ' are part of L'', then Lemma 2 implies that every other line can cover at most two of these subsets. In particular, the cardinality of L'' will be larger than k, contradicting with the fact that L'' is smaller than L'. A similar argumentation shows that any minimum-size cover of $\mathcal{C}$ is also a minimum-size cover of $\mathcal{C}_1$, which concludes the proof.

Now we further reduce our problem instance to a set $\mathcal{C}_2$ as follows. We count for each line how many subsets of $\mathcal{C}_1$ contain it. Then for each line ℓ that has more than $2k^2$ subsets containing it we replace all these subsets by subset $\{\ell\}$.

Lemma 4. *A set L', with $|L'| \leq k$ is a minimum-size cover for $\mathcal{C}_1$ if and only if it is a minimum-size cover of $\mathcal{C}_2$.*

Proof. As before, it suffices to consider the case in which $\mathcal{C}_1 \neq \mathcal{C}_2$. Let $\mathcal{C}_1^+ = \mathcal{C}_2 \backslash \mathcal{C}_1$ and $\mathcal{C}_1^- = \mathcal{C}_1 \backslash \mathcal{C}_2$ denote the sets that were added and removed from $\mathcal{C}_1$ to create $\mathcal{C}_2$. Since all sets in $\mathcal{C}_1^-$ contain a line of one of the singleton sets of $\mathcal{C}_1^+$, a set L', with $|L'| \leq k$ that is a minimum size cover of $\mathcal{C}_2$ is also a cover of $\mathcal{C}_1$. To show that L' is also a minimum-size cover for $\mathcal{C}_1$, assume for a contradiction that there is a smaller cover L'' for $\mathcal{C}_1$. The lines of L'' also cover $\mathcal{C}_2 \backslash \mathcal{C}_1^+$. Therefore, if L'' is not a cover of $\mathcal{C}_2$, then there must be a subset $\{\ell\} \in \mathcal{C}_1^+$ that is not covered by L''. Recall that $\{\ell\}$ was added to $\mathcal{C}_2$ because there were more than $2k^2$ sets in $\mathcal{C}_1$ that contain ℓ. By construction of $\mathcal{C}_1$, no line $\ell' \neq \ell$ can cover more than $2k$ of these sets (otherwise, the pair $\{\ell, \ell'\}$ would have been added to $\mathcal{C}_1$). Thus, we conclude that L'' has more than k segments, a contradiction.

In a similar way we can prove that a subset L', with $|L'| \leq k$, that is a minimum-size cover for $\mathcal{C}_1$ is also a minimum size cover for $\mathcal{C}_2$, thus, concluding the proof.

Lemma 5. *If $|\mathcal{C}_2| > 2k^3$, then there is no cover of size at most k for $\mathcal{C}$.*

Proof. Proof of this claim follows from Lemmas 3 and 4 and the fact that each segment can only cover at most $2k^2$ sets of $\mathcal{C}_2$.

Now we look at the computational aspect of generating $\mathcal{C}_2$. Both reduction steps from $\mathcal{C}$ to $\mathcal{C}_1$ and from $\mathcal{C}_1$ to $\mathcal{C}_2$ require counting subsets. Here we use the fact that the subsets are of size at most 4 to show that this can be done in linear time using hash tables. Note that linear time here is in the size of $\mathcal{C}$, the size of the arrangement, which may be quadratic with respect to the number of line segments.

Lemma 6. *The set $\mathcal{C}_2$ can be constructed in time $O(n \log n + C)$, where n is the number of segments in L and C is the number of cells in the arrangement induced by L.*

Proof. To reduce $\mathcal{C}$ to $\mathcal{C}_1$ we count for every pair of line segments ℓ, ℓ' the number of subsets of $\mathcal{C}$ that contain both. We do this by making a single pass over all subsets of $\mathcal{C}$ and maintaining for each pair ℓ, ℓ' how many subsets contain them thus far. To avoid having to initialize counts for all pairs ℓ, ℓ', which may be more than linear in the size of $\mathcal{C}$ we use a hash table and create a new count whenever we encounter a new pair of line segments. Each subset contains at most 4 segments and at most 6 pairs of segments, so we can process it in $O(1)$ time. After counting we can go through all pairs of line segments with non-zero count and for each pair ℓ, ℓ' that is contained in more than $2k$ subsets we remove the sets (which is easily done by storing which sets contain the pair) and add a new subset $\{\ell, \ell'\}$. To compute $\mathcal{C}_2$ from $\mathcal{C}_1$ we can use a similar construction.

Theorem 3. *For the problem of finding a minimum-size cover for all rectangular cells in an arrangement of n axis-parallel line segments L we can either find a kernel of size $O(k^3)$ or conclude that no solution of at most size k is possible. Moreover, the algorithm runs in $O(n \log n + C)$ time, where C is the number of cells in the arrangement induced by L.*

Since we now have a kernel of size $O(k^3)$ it follows that the problem is fixed parameter tractable [4].

Corollary 1. *Given a set L of line segments the problem of finding a minimum size set $L' \subseteq L$ that covers the rectangular cells of the arrangement of L is fixed parameter tractable with respect to the size k of the optimal solution.*

4 Rectangular Subdivisions

Although the problem of finding a minimum set of covering segments is NP-hard, there are special cases where the problem can be solved in polynomial time. One such case is that the input line segments form a special type of rectangular subdivision. The rectangular subdivisions we consider are those defined by a KD-tree [1]. That is, a recursive subdivision of a rectangle using horizontal or vertical splitting segments (see Fig. 5). Note that the segments that have only an endpoint on the boundary of a cell are not considered part of the boundary of that cell. From now on we refer to such a subdivision simply as a *rectangular subdivision.*

A rectangular subdivision provides a clear tree-structure which we can use to compute an optimal covering in a bottom up fashion. Each node in the tree is associated to some rectangle. For a leaf-node this rectangle is a cell of the arrangement, whereas for an interior node its associated rectangle r is formed by the union of the rectangles associated to its children. We also associate an interior node of the tree with a horizontal or vertical segment that splits its associated rectangle into two rectangles associated to its children.

Although the above FPT approach works we show there is a much faster exact algorithm for rectangular subdivision. Without loss of generality we assume that the subdivision is given as a binary tree (the KD-tree structure)[1]. We show that an optimal covering can then be computed in linear time. Note that in this definition the outer face is covered if and only if at least one of the edges of the bounding rectangle is in the cover, and any other cell is covered if one of the segments defining its boundary is chosen as a covering segment.

[1] If the structure is not given as a binary tree, but a more general subdivision structure such as a double-connected edge list, we can construct the tree in linear time.

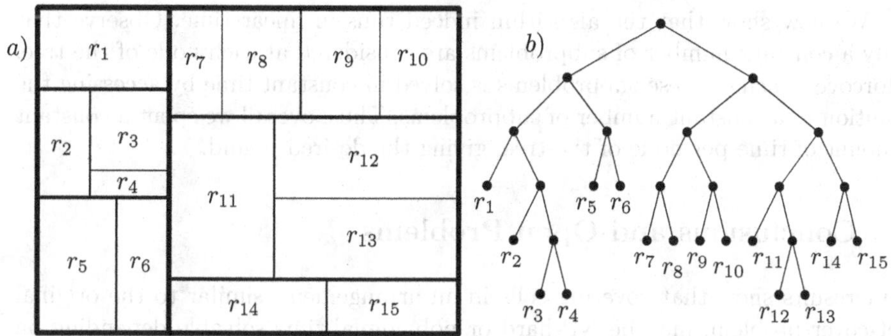

Fig. 5. (a) A rectilinear binary space partition within a rectangle. (b) A possible tree representing the partition.

Theorem 4. *Given the tree structure of a rectangular subdivision, where each node stores the rectangle it represents and its splitting segment, we can compute in linear time a segment-cover of the cells with a minimum size.*

Proof. Starting at the leaves, we compute an optimal covering in a bottom up fashion. For each node v of the tree we compute the solution to sixteen different subproblems and store these solutions in the corresponding nodes. Let R be the rectangle associated to a given node, and let $\{s_1, \ldots, s_4\}$ be the four segments that define its boundary. For any subset $S' \subseteq \{s_1, \ldots, s_4\}$, we consider the subproblem of finding the smallest covering of all the cells within R that contains the segments of S'. For each such subproblem its cardinality as well as how it is constructed from solutions of its children (this second part is needed to reconstruct the optimal solution).

Clearly, if v is a leaf the optimal cover is simply S' (unless $S' = \emptyset$ in which case there is no solution). For an interior node we proceed as follows: let v be an interior node of the tree and R the rectangle stored at v. Without loss of generality assume that R is split by a vertical line ℓ. The children of v are v_{left} and v_{right} with corresponding rectangles R_{left} and R_{right}. We must compute a minimum-size cover for each possible choice of the top, left, bottom and right edges of R. Note that a fixed choice of boundary edges for r already forces a choice of three edges for R_{left} and R_{right}. Only their shared edge ℓ is not fixed by the choice of boundary edges of r. However, we can simply try both options and see which results in the overall better cover of R. That is, we first assume ℓ is not part of the cover and retrieve our already computed solutions for R_{left} and R_{right} for the current selection of boundary edges. Then we do the same when we do pick ℓ and choose the solution with the smallest number of edges. This results in an optimal solution since the only edges shared by R_{left} and R_{right} are ℓ and the top and bottom edge of R. We consider all possible selections of these edges. For each selection the two subproblems of finding an optimal cover for the R_{left} and R_{right} are independent, so we can reuse our previously computed solutions.

We now show that the algorithm indeed runs in linear time. Observe that only a constant number of subproblems are considered at each node of the tree. Moreover, each of these subproblems is solved in constant time by accessing the solution to a constant number of subproblems. Thus, overall we spent a constant amount of time per node of the tree, giving the desired bound.

5 Conclusions and Open Problems

Our results show that covering cells in an arrangement, similar to the original set-cover problem, may be NP-hard or polynomial-time solvable depending on various restrictions. It may be interesting to investigate further variants of the problem to see which restriction makes the problem polynomial-time solvable—this may be due to the lack of intersections between line segments or due to the tree-structure of the subdivision. It would also be interesting to see if good approximations are possible for the more general case or even the case with line segments of arbitrary orientations.

References

1. de Berg, M., Cheong, O., van Kreveld, M., Overmars, M.: Computational Geometry: Algorithms and Applications, 3rd edn. Springer, Heidelberg (2008)
2. Bose, P., Cardinal, J., Collette, S., Hurtado, F., Korman, M., Langerman, S., Taslakian, P.: Coloring and guarding line arrangements. Discrete Math. Theoret. Comput. Sci. **15**(3), 139–154 (2013)
3. Cormen, T.H., Leiserson, C.E., Rivest, R.L., Stein, C.: Introduction to Algorithms, 2nd edn. McGraw-Hill Science/Engineering/Math, New York (2001)
4. Downey, R.G., Fellows, M.R.: Fundamentals of Parameterized Complexity, 1st edn. Springer, London (2013)
5. Feder, T., Greene, D.: Optimal algorithms for approximate clustering. In: Proceedings of the Twentieth Annual ACM Symposium on Theory of Computing (STOC 1988), pp. 434–444 (1988)
6. Feige, U.: A threshold of ln n for approximating set cover. J. ACM **45**(4), 634–652 (1998)
7. Lichtenstein, D.: Planar formulae and their uses. SIAM J. Comput. **11**, 329–343 (1982)
8. Raz, R., Safra, S.: A sub-constant error-probability low-degree test, and a sub-constant error-probability PCP characterization of NP. In: Proceedings of the Twenty-ninth Annual ACM Symposium on Theory of Computing (STOC 1997), pp. 475–484 (1997)

An Improved On-line Strategy for Exploring Unknown Polygons

Xuehou Tan[1,2(✉)] and Qi Wei[1]

[1] Dalian Maritime University, Linghai Road 1, Dalian, China
[2] Tokai University, 4-1-1 Kitakaname, Hiratsuka 259-1292, Japan
tan@wing.ncc.u-tokai.ac.jp

Abstract. We present a new, on-line strategy for a mobile robot to explore an unknown simple polygon P, so as to output a so-called *watchman route* such that every interior point of P is visible from at least one point along the route. The length of the robot's route is guaranteed to be at most 6.7 times that of the shortest watchman route that could be computed off-line. This significantly improves upon the previously known 26.5-competitive strategy. A novelty of our strategy is an on-line implementation of a previously known off-line algorithm that approximates the optimum watchman route to a factor of $\sqrt{2}$. The other is in the way the polygon exploration problem is decomposed into two different types of the subproblems and a new method for analyzing its cost performance.

1 Introduction

Visibility-based problems of guarding or searching have received much attention in the communities of computational geometry, robotics and on-line algorithms. Finding stationary positions of guarding a polygonal region P of n vertices is the well-known *art gallery problem*. The *watchman route problem* asks for a shortest route along which a mobile robot can see the whole polygon P [1,5–7].

When a point s on the boundary of P is given, the shortest watchman route through s can be computed in $O(n^4)$ time [6,7]. It was later improved to $O(n^3 \log n)$ [3]. A linear-time approximation algorithm for the watchman route problem has also been proposed [5], which reports a watchman route guaranteed to be at most $\sqrt{2}$ times longer than the shortest watchman route through s.

In the *polygon exploration problem*, a starting point s on the boundary of the (unknown) polygon P is given. A robot with a vision system that continuously provides the visibility of its current position walks to see (or explore) the whole region of P, starting from s. When each point of P has been seen at least once, the robot returns to s. We are interested in a *competitive* exploration strategy that guarantees that the route of the robot will never exceed in length a constant times the length of the shortest watchman route through s. For the problem of exploring unknown simple polygons, Deng et al. were the first to claim that a competitive strategy does exist, but the constant is estimated to be in the

This work was supported by JSPS KAKENHI Grant Number 15K00023.

Z. Lu et al. (Eds.): COCOA 2015, LNCS 9486, pp. 163–177, 2015.
DOI: 10.1007/978-3-319-26626-8_13

thousands [2]. A factor of 133 was later given by Hoffmann et al., and further improved to $18\sqrt{2}+1 \leq 26.5$ [4]. Since the known lower bound is smaller than 1.207 [2], it is conjectured that the competitive factor is far below 10 [4].

In this paper, we present a new, on-line strategy for a mobile robot to explore an unknown simple polygon. An important observation is that the known off-line $\sqrt{2}$-approximation algorithm can be used to build blocks of the on-line strategy. Next, we reduce the polygon exploration problem to the subproblems of exploring two different types of reflex vertices. Although the reduction is done in a way similar to that of [4], we show that the exploration of the same type of reflex vertices can be evaluated together. More specifically, a subproblem is solved by resembling the $\sqrt{2}$-approximation algorithm for that watchman route subproblem. For this purpose, we present a paradigm for implementing on-line the off-line approximating techniques. Furthermore, we present a new method to evaluate the total cost of the solutions to the subproblems of exploring the same type of reflex vertices. With these new ideas, we are able to prove that an unknown polygon can be explored by a route of length at most $4\sqrt{2}+1 < 6.7$ times that of the shortest watchman route through s. This gives a significant improvement upon the previously known 26.5-competitive strategy [4].

2 Preliminaries

A polygon is simple if it has neither holes nor self-intersections. Let P be a simple polygon with a point s on its boundary. A point $p \in P$ *sees* the other point $q \in P$ if P contains the line segment $\overline{pq}$ that has p and q as its two endpoints.

A vertex of P is *reflex* if its internal angle is strictly larger than π; otherwise, it is *convex*. The polygon P can be partitioned into two parts by a "cut" C that starts at a reflex vertex v and extends an edge incident to v until it first hits the polygon boundary. The part of P containing s and including C itself is called the *essential part* of C. We denote by $P(C)$ the essential part of the cut C. The cut C is said to be a *visibility cut* if it produces a convex angle at v in $P(C)$. Also, we call v the *defining vertex* of C.

We say a visibility cut C_j *dominates* the other cut C_i if $P(C_j)$ contains $P(C_i)$. If C_j dominates C_i, any route that visits C_j will automatically visit C_i. We also say a point p *dominates* cut C if p is not contained in $P(C)$ (i.e., p lies in $P - P(C)$). A visibility cut is called an *essential cut* if it is not dominated by any other cuts. The watchman route problem is then reduced to that of finding the shortest route intersecting or visiting all essential cuts [1,5].

For two arbitrary points a and b inside the polygon P, we denote by $\pi(a,b)$ the shortest path between a and b, which does not cross the boundary of P. The *shortest path tree* of s, denoted by $SPT(s)$, consists of all shortest paths from s to the vertices of P. The vertices touching a shortest path from the right are called the *right reflex vertices*, or shortly, *right vertices*. The *left reflex vertices* or *left vertices* can be defined accordingly [4].

A region D inside P is said to be a *relatively convex polygon* if the shortest path between any two points of D is contained in D [4]. Also, a polygonal chain

H inside P is *relatively convex* if the region bounded by H and the shortest path between two endpoints of H is a relative convex polygon in P. The *relative convex hull* of a set of points inside P is defined as the boundary of the smallest, relative convex polygon containing the set of the given points.

For a route R, we denote by $|R|$ the length of the route R. In this paper, we denote by W_{opt} the shortest watchman route, and W_{app} the watchman route that is computed by the off-line approximation algorithm [5].

Let S denote an on-line exploration strategy. A vertex is said to be *discovered* if it has ever been visible once from the robot, when the robot follows S to explore the polygon P. A left or right vertex is *unexplored* as long as its cut has not been reached, and *fully explored* thereafter. Denote by W_{rob} the robot's route produced by the strategy S. The competitive factor of the strategy S is then defined as the upper bound of $\frac{|W_{rob}|}{|W_{opt}|}$.

2.1 An Overview of the $\sqrt{2}$-approximation Algorithm

Let a and b denote two points in the same side of a line L. The shortest path visiting a, L and b in this order, denoted by $S(a, L, b)$, follows the reflection principle. That is, the incoming angle of $S(a, L, b)$ with L is equal to the outgoing angle of $S(a, L, b)$ with L. Denote by $L(a)$ the point of L closest to a, and b' the point obtained by reflecting b across L. See Fig. 1(a). The path consisting of the line segments $\overline{a\, L(a)}$ and $\overline{L(a)\, b}$, denoted by $S'(a, L, b)$, then gives a $\sqrt{2}$-approximation of the path $S(a, L, b)$ [5]. Generally, for a line segment l, denote by $l(a)$ the point of l closest to a. The path consisting of $\overline{a\, l(a)}$ and $\overline{l(a)\, b}$ is also a $\sqrt{2}$-approximation of the shortest path from a to b that visits l (Fig. 1(b)).

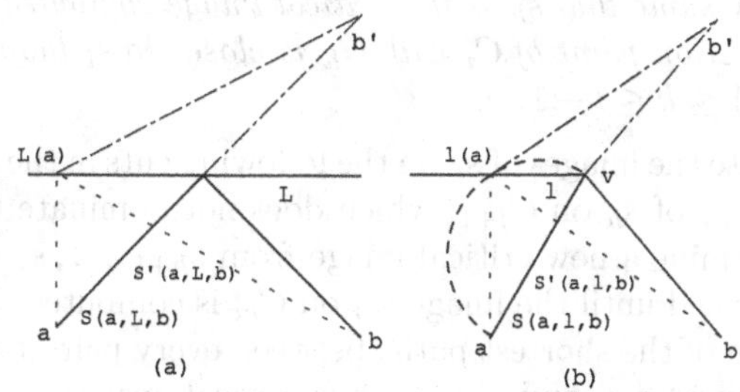

Fig. 1. The reflection principle and its approximation.

Let m be the number of essential cuts, and $C_1, C_2, \cdots, C_m$ the sequence of essential cuts indexed in clockwise order of their *left* endpoints, as viewed from s. Let $s = s_0 = s_{m+1}$. Also, let the edge containing s be the cuts C_0 and C_{m+1}, whose essential parts $P(C_0)$ and $P(C_{m+1})$ are defined as the polygon P itself. Given a point p in the polygon $P(C)$, we define the *image of* p *on the cut* C as the point of C that is *closest* to p inside $P(C)$ (in the geodesic distance).

Beginning with the point s, the approximation algorithm repeatedly computes the point of the cut C to visit next, which is closest to the endpoint e

of the currently found path and contained in the essential part of the cut on which e is [5]. Specifically, we first compute the images of s_0 on the cuts in the polygon $P(C_0)$. Let s_1 denote the image of s_0 on C_1, s_2 the image of s_0 on C_2, and so on. The computation of s_0's images is terminated when the image s_{i+1} does not dominate all the cuts $C_1, C_2, \ldots, C_i$ before it. See Fig. 2(a). Then, we choose a *critical* image s_h $(1 \leq h \leq i)$ from $s_1, s_2, \ldots, s_i$ as the first image (of the smallest index) such that the image of s_h on C_{i+1}, which is computed (and thus contained) in $P(C_h)$, dominates $C_{h+1}, \ldots, C_i$. For the polygon shown in Fig. 2(a), we have $i = 4$ and $h = 2$ (i.e., s_2 is critical in Fig. 2(a)).

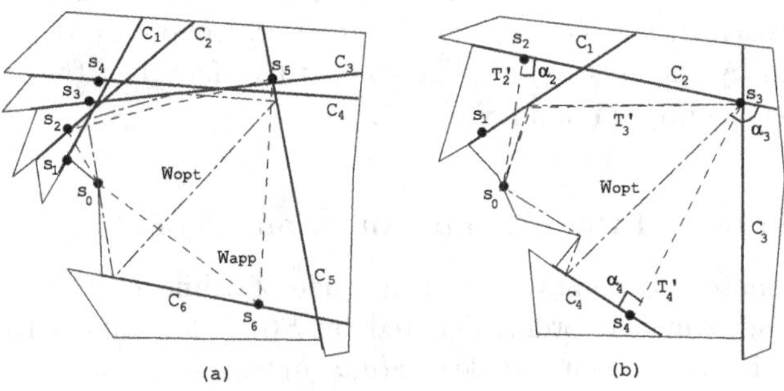

Fig. 2. Critical images and the routes W_{opt}, W_{app} and T_i'.

From the definition of the critical image $s_h (1 \leq h \leq i)$, C_h intersects all the cut(s) before it. Hence, the following observations can be made.

Observation 1. *If C_i does not intersect C_{i+1}, then the image s_i is critical.*

Observation 2. *Assume that s_h is the critical image computed above, and $h < i$. Then, the intersection point of C_i with C_h is closer to s_i than that of C_i with any other C_k, $h+1 \leq k \leq i-1$.*

Next, we compute the images of s_h on the following cuts in the polygon $P(C_h)$. When the image s_{j+1} of s_h on C_{j+1}, which does not dominate $C_{h+1}, \ldots, C_j$, is found, we can determine a new critical image from $s_{h+1}, \ldots, s_j$. This procedure is repeatedly performed until the image s_m on C_m is computed. The route W_{app} is the concatenation of the shortest paths between every pair of adjacent critical images. (We also consider s_0 and s_{m+1} as two critical images.)

The following results on W_{app} then hold.

Lemma 1 *(See [5]). Suppose that the image s_i on C_i is critical. If the route W_{opt} reflects on C_i, then s_i is to the left of the reflection point of W_{opt} on C_i.*

Lemma 2 *(See [5]). For any instance of the watchman route problem with a given starting point s, $|W_{app}| \leq \sqrt{2}|W_{opt}|$ holds.*

Lemma 3. *Let T_i be the portion of W_{opt} from s to the point of W_{opt} on C_i, which visits all the cuts $C_1, C_2, \ldots, C_i$ $(1 \leq i \leq m)$. If the (computed) image s_i on C_i is critical, then $|\pi(s, s_i)| \leq |T_i|$.*

Proof. Omitted in this extended abstract. □

2.2 An Overview of the 26.5-Competitive Strategy

In an unknown polygon, exploring a reflex vertex v requires a little care. Since the equation of the cut of v is not known, the point of the cut closest to the current position of the robot, say, a, cannot simply be computed. This difficulty can be overcome by using the circle spanned by v and by a. The intersection point of the circular arc with the cut is then the required point (see Fig. 1(b)).

Let D denote a convex region in the plane. Suppose that a photographer follows a path to take a picture of D that shows as large a portion of D as possible, but no white space or other objects. The photographer uses a fixed angle lens, say, of 90°. While the right angle is touching two vertices, u and v, of D, its apex follows the circular arc spanned by u and v. All points enclosed by the photographer's path, and no other, can see two points of D at the right angle; we call this point set the *angle hull* of D, and denote it by AH(D) [4].

Consider now the setting where D is contained in a simple polygon whose edges are considered as obstacles. In this case, the region D is defined as a relatively convex polygon in P. The photographer does not want any edges of P to appear in pictures; thus, the photographer's path may touch a vertex of P or overlap with a portion of the polygon edge. Again, call the set of all points enclosed by the photographer's path the *angle hull* of D, and denote it by AH(D).

Lemma 4 *(See [4]). Suppose that P is a simple polygon, and D is a relatively convex polygon inside P. The length of the perimeter of the angle hull $AH(D)$, with respect to P, is less than 2 times the length of D's boundary. Moreover, the bound of 2 holds if D is a relative convex chain that is contained in P.*

The strategy of Hoffmann et al. [4] recursively reduces the polygon exploration problem to two different types of subproblems: one for exploring a group of right vertices and the other for exploring a group of left vertices. Only such right (resp. left) vertices are gathered into a group that the shortest paths from the local starting point to them (called the *stage point* in [4]) make only right (resp. left) turns. The competitive factor for a subproblem of exploring right (resp. left) vertices is then proved to be $6\sqrt{2}$, by employing the structure of angle hulls. Next, groups of reflex vertices that are on sufficiently different recursive levels are classified into *three* categories [4]. Any two local routes for the subproblems of the same category are mutually invisible, except for their starting points. Hence, the total length of the robot's routes for all subproblems of a single category is at most $6\sqrt{2}|W_{opt}|$. Since the connection among all subproblems in three categories makes up to an additional path of length at most $|W_{opt}|$, putting all results together gives a competitive factor of $18\sqrt{2}+1 \leq 26.5$ [4].

In the rest of this paper, a polygon P is termed a *right polygon* (*left polygon*) if the shortest paths from s to all reflex vertices of P turn only right (left).

3 Exploring a Right Polygon

Assume that the robot can measure the distance to a point in its view. Denote by CP the current position of the robot. The following observation can be made [4].

Observation 3. *For a point p that had ever been seen, the robot knows the shortest path between p and CP, even when p is currently invisible from the robot.*

The robot in our strategy always selects the smallest discovered vertex to explore. Denote by *RightTarget* the list of the right vertices in clockwise order, which have already been discovered but not yet explored. Denote by r the head of *RightTarget*, which is the target vertex that the robot is going to explore. So, the value of r is updated as soon as a smaller right vertex becomes visible.

The on-line computation of the critical images is crucial to our strategy. To simplify the discussion, we first consider a simple situation in which no smaller vertex is found in the procedure of exploring r, and the vertex r is always visible from CP. Then, we complete the exploration strategy and give the performance analysis of our strategy.

3.1 How to Reach the Cut of a Right Vertex at the Wanted Point

Assume that CI is the critical image that has just been found. (The initial value of CI is s.) Let C be the visibility cut of a vertex v, which is currently reached by the robot (say, by a walk on the semicircle spanned by CI and by v). Denote by TI the image of CI on the cut C. So, $CI \neq TI$.

Starting from the point $CP(= TI)$, the robot walks to explore the target vertex r. Assume that no smaller right vertex is found in the procedure of exploring r, and the vertex r is visible from CP. In exploring r, the variable C as well as TI is dynamically changed, as soon as a previously visited cut is reached again. For ease of presentation, we may also use C' to represent such a previously visited cut and TI' the image of CI on C'. Except for the cut of r, all other cuts mentioned in this section had been visited by the robot's path from CI to TI (on the cut C). Assume also that TI (resp. TI') dominates the cuts having been explored along the path from CI to TI (resp. TI').

In our algorithm, we make use of the following two angle hulls. Denote by $AH(TI, r)$ the angle hull of the shortest path between TI and r. Although the cut of r has not been reached, the robot can walk on the *known portion* of $AH(TI, r)$. (Note that the unknown portion of $AH(TI, r)$ is only the semicircle spanned by TI and by r.) In particular, we denote by $Ah(TI, r)$ the portion of $AH(TI, r)$, which is contained in $Q(C)$. So, $Ah(TI, r)$ is completely *known* to the robot, and a part of the cut C appears on the boundary of $Ah(TI, r)$. Analogously, we denote by $AH(CI, r)$ the angle hull of the shortest path between CI and r, and $aH(CI, r)$ the portion of $AH(CI, r)$ that lies in $Q - Q(C)$.

To explore the vertex r, starting from TI $(= CP)$ on C, the robot initially walks on the boundary of $Ah(TI, r)$. Since r is larger than the defining vertex v of C, the left (right) point of $aH(CI, r)$ on C, is to the right (left) of TI (Fig. 3). The robot then walks to explore r by the following rules.

Rule 1. When the robot reaches the left point of $aH(CI, r)$ on the current cut C (which is a part of $Ah(TI, r)$), it changes to move on $aH(CI, r)$. Afterwards, if the robot ever reaches the right point of $aH(CI, r)$ along C, then it changes to move on the present hull $Ah(TI, r)$.

Rule 2. While the robot walks on $aH(CI, r)$, it may encounter a previously reached cut. In this case, the variable C is changed to that cut, and TI and $Ah(TI, r)$ are updated accordingly. Afterwards, the robot moves along the boundary of the new hull $Ah(TI, r)$.

Rule 3. While the robot walks on the cut C (a part of $Ah(TI, r)$), it may encounter an intersection point i of C with a previously reached cut C'. If i is on the hull $Ah(TI', r)$ (it is known to the robot), then the variables C and TI are set to C' and TI' respectively, and the robot walks along the boundary of the new hull $Ah(TI, r)$. Otherwise, nothing happens.

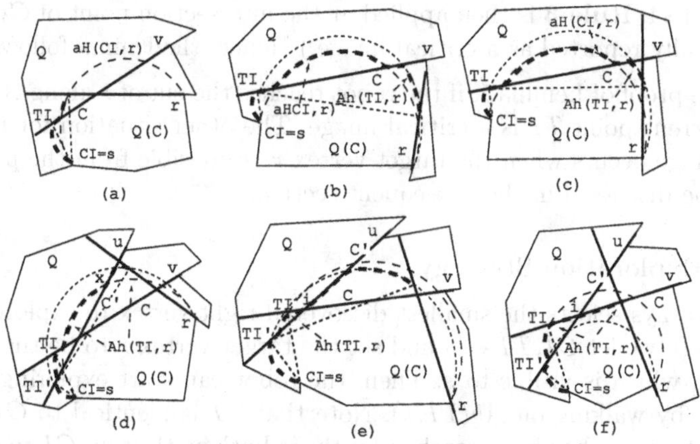

Fig. 3. Basic motions for exploring the target vertex r.

Lemma 5. *Assume that $TI(TI')$ dominates the cuts having been explored along the path from CI to $TI(TI')$. By Rules 1 $\sim$ 3, the robot can reach the cut of r at the same point as that is found by the approximation algorithm [5].*

Proof. Starting from the point TI on C, the robot first walks on $Ah(TI, r)$. If the robot reaches the left point of $aH(CI, r)$, it changes to move on $aH(CI, r)$ (**Rule 1**). If the cut of r is ever reached along $aH(CI, r)$, the (stopping) position of the robot gives the image of CI on the cut of r and we are done (Fig. 3(a)). If the robot moves back to the cut C (from $aH(CI, r)$), it then walks on $Ah(TI, r)$ again. In this case, the exploration of the vertex r may finish when the robot moves to the image of TI on the cut of r (Fig. 3(b)–(c)).

While walking on $aH(CI, r)$, the robot may encounter a previously visited cut C'. In this case, we let $C \leftarrow C'$ and $TI \leftarrow TI'$, and then, the robot walks on the new hull $Ah(TI, r)$ (**Rule 2**). This makes it possible to move to the image of TI' on the cut of r, as that image may dominate C (see Fig. 3(d)–(e)). While walking on $Ah(TI, r)$, the robot may reach an intersection point i of C with a previously visited cut C' (Observation 2). If i is on the hull $Ah(TI', r)$, then let $C \leftarrow C'$ and $TI \leftarrow TI'$, and the hull $Ah(TI, r)$ is thus updated (**Rule 3**).

In this case, the robot moves on the new hull $Ah(TI, r)$ and probably finishes the exploration of r at the image of TI' on the cut of r. See Fig. 3(e). Otherwise, the robot continues to walk on $Ah(TI, r)$, because it *knows* that the image of TI' on the cut of r doesn't dominate C and thus needn't be computed (Fig. 3(f)).

From **Rules 1 $\sim$ 3**, the cut of r is reached along either $aH(CI, r)$ or $Ah(TI, r)$. In the former case, the stopping position CP of the robot, which dominates all the cuts visited by the robot's path from CI to CP, is the image of CI on the cut of r (Fig. 3(a)). In the latter case, the image of the *current* point TI on the cut of r does not dominate all the cuts visited by the path from CI to CP, but it dominates the cuts between its cut and the cut of r (see Figs. 3(b)–(f)). From **Rules 2 $\sim$ 3** as well as Observation 2, we are sure that the image TI is a critical one. (One can see it in Fig. 2(a) by exploring the cut C_5, starting from $s4$. **Rule 3** is then applied at the intersection point of C_4 and C_2, and $s2$ is finally reported as a critical image.) Hence, the lemma follows. □

From the proof of Lemma 5, if the robot reaches the cut of r along $Ah(TI, r)$, then the current point TI is a critical image. The other situation for reporting a critical image occurs when the target vertex r is invisible from the point CP, which will be discussed in the subsequent section.

3.2 The Exploration Strategy

The robot always selects the smallest discovered right vertex to explore. At the very first step, we let $CI, TI \leftarrow s$ and set the target vertex r to be the smallest right vertex, which is visible to s. Then, the robot can start exploring towards the cut of r by walking on $Ah(TI, r)$. Note that TI is identical to CI only in the case that no cut has been reached by the robot's path from CI to CP.

In approaching the target vertex r, all newly discovered right vertices are added to *RightTarget*. The robot may also go across the cuts of some right vertices, which are larger than r. All of these vertices are removed from *RightTarget*. After r is fully explored, we update the variable C to the cut of r and the image TI accordingly, and delete the vertex r from *RightTarget*.

Let us now describe how to explore the target vertex r, if the value of r is allowed to be changed. Since it has been discussed in the procedure *ExploreRightVertex* of [4], we first give a brief review of *ExploreRightVertex*. It starts at a point, called the *base* point, and always selects the smallest of all discovered right vertices to explore. The target vertex r is explored by repeatedly performing the following motions: (i) the robot follows the clockwise oriented circle spanned by r and by the last vertex before CP on the shortest path from the base point to CP, and (ii) when the view to the current target vertex r gets blocked, the robot walks straight towards the blocking vertex (or follow the polygon boundary) until the motion (i) becomes possible again. It is also shown in [4] that if the robot reaches the cut of the target vertex at point c, the robot's path is part of the boundary of the angle hull of the shortest path from the base point to c, except for the segments leading to the blocking vertices.

For our strategy given in Sect. 3.1, each point CI can be considered as a base point. **Rules 1 $\sim$ 3** ask the robot to move on the boundary of $aH(CI, r)$ or

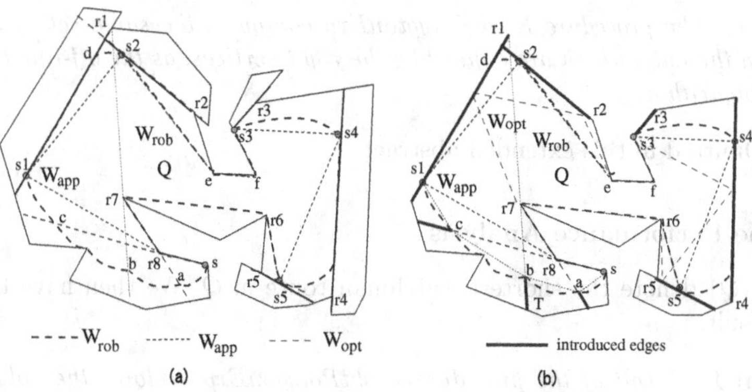

Fig. 4. Exploring a right-like polygon

$Ah(TI, r)$. By allowing the value of r in $aH(CI, r)$ and $Ah(TI, r)$ to dynamically change, motion (i) is already implied by **Rules 1 $\sim$ 3**. In the case that the robot loses sight of the new target vertex r, except for letting the robot walk straight towards the blocking vertex (motion (ii)), we report the current point $TI(= CP)$ as a critical image becasue the cut on which CP is cannot intersect its following the cuts (Observation 2). The following results can then be concluded.

Lemma 6. *Let r denote the smallest of the right vertices discovered by the mobile robot. Rules 1 $\sim$ 3 together with the two motions described above can be used to explore the target vertex r.*

By exploring the target vertex r repeatedly till the list *RightTarget* becomes empty, the polygon Q can then be explored. Let us now give our procedure *RightPolygonExp* for exploring Q. The point s and the list *RightTarget* of the right vertices, which are visible from s, are input of *RightPolygonExp*.

Procedure. *RightPolygonExp* (**in** *RightTarget*, **in** s)

1. Set $CI, TI, C \leftarrow s$. Assume that s is a special cut, with $Q(s) = Q$.
2. **While** *RightTarget* is not empty **do**
   (a) Set the target vertex r to the head of *RightTarget*. As soon as a smaller right vertex is discovered, the value of r changes to it. Moreover, a right vertex is removed from *RightTarget* when it is fully explored, or added to *RightTarget* when it becomes visible to the robot for the first time.
   (b) The robot walks on the boundaries of the current hull $Ah(TI, r)$ or $aH(CI, r)$ (by **Rules 1 $\sim$ 3**) to explore the target vertex r, and performs the following operations as soon as possible.
      i. If no right vertex is visible from CP, the robot walks clockwise on the shortest path with the turning points at polygon vertices until the target vertex r becomes visible again. In this case, the current point TI is reported as a critical image, and we let $CI \leftarrow TI$.
      ii. If the cut of r is reached along the boundary of $Ah(TI, r)$, the point TI is reported as a critical image and we let $CI \leftarrow TI$.
3. The robot walks on the shortest path back to the starting point s.

Lemma 7. *The procedure RightPolygtonExp computes the same set of critical images on the cuts, which are defined by the right vertices, as the off-line approximation algorithm.*

Proof. Omitted in this extended abstract. □

3.3 The Performance Analysis

Let $W_{opt}(Q)$ denote the shortest watchman route in Q. We then have the following result.

Theorem 1. *A call of the procedure RightPolygonExp explores the polygon Q by outputting a watchman route of length at most $2\sqrt{2}|W_{opt}(Q)|$.*

Proof. It follows from Lemma 7 that a call of *RightPolygonExp* (*RightTarget*, s) simulates on-line the off-line approximation algorithm [5]. Denote by W_{app} the route consisting of the shortest paths in Q', which connect every pair of consecutive critical images. Let $W_{rob}(Q)$ denote the robot's route produced by our strategy for exploring Q. To apply Lemma 4, we then construct an artificial polygon from Q. For each essential cut C, we cut off the portion $Q - Q(C)$ from Q, in the order of essential cuts. See Fig. 4(b). For every point at which the robot's route changes between two *different circular arcs*, we insert into Q two isometric edges between it and a boundary point such that the added edges are outside of $W_{rob}(Q)$ and on the line through two vertices of Q, which are on the shortest paths from s to some right vertices. The resulting polygon, denoted by Q', is simple. Both $W_{rob}(Q)$ and W_{app} are the relatively convex polygons in Q'.

We prove below $|W_{rob}(Q)| \leq 2|W_{app}|$. To this end, we claim that each part of $W_{rob}(Q)$ between two consecutive critical images, say, p and q, is no longer than twice the length of the shortest path between p and q in Q'. For any two consecutive points at which the artificial edges are added, we connect them using the shortest path. This gives a path between p and q inside Q', and we denote it by T. It is easy to see that T is a relatively convex chain in Q', and the portion of $W_{rob}(Q)$ between p and q is contained in the angle hull of the path T inside Q'. Thus, the portion of $W_{rob}(Q)$ between p and q cannot exceed in length $2|T|$. Since the path T is longer than the shortest path between p and q, our claim is proved. Hence, we can conclude that $|W_{rob}| \leq 2|W_{app}|$. Since $|W_{app}| \leq \sqrt{2}|W_{opt}(Q)|$ holds, the theorem follows. □

A polygon Q is said to be *right-like* if (i) some right vertices in the *initial* list *RightTargeT* may not be visible from s but to them the robot knows in advance the shortest paths, (ii) if a reflex vertex x happens to be encountered by the robot's route, then the right vertices that are visible from x are also added to the list *RightTarget*, and (iii) a call of *RightPolygonExp(RightTarget, s)* explores the polygon Q. The left-like polygons are defined analogously. For instance, the polygon shown in Fig. 4 is right-like, as vertex $r3$ is visible from vertex f that happens to be encountered by the robot's route, and thus, $r3$ is also explored. Also, we have the following result.

Theorem 2. *A call of RightPolygonExp(RightTarget, s) can explore the right-like polygon Q by outputting a watchman route of length at most* $2\sqrt{2}|W_{opt}(Q)|$.

Obviously, there is a symmetric procedure *LeftPloygonExp* that is identical to *RightPolygonExp*, except that left/right, clockwise/counterclockwise and small/large are exchanged.

4 Exploring a Simple Polygon

Our strategy for exploring a simple polygon P is mainly a recursive procedure that reduces the polygon exploration problem to the subproblems of exploring the groups of right vertices and the groups of left vertices. It differs from the strategy of Hoffmann et al. [4] in the following two sides. First, the explorations for the left vertices and for the right vertices are exchanged as soon as possible. This helps to save an additional path of length at most $|W_{opt}|$, which is used to connect the local starting points in the strategy of Hoffmann et al. [4]. Second, we give a new method to analyze the total cost of the solutions for exploring the right (left) vertices. This makes it possible to classify the subproblems into *two* categories, instead of *three* ones required in [4]. Excluding some special paths, the solutions for the subproblems of the same category together have a competitive factor $2\sqrt{2}$. Moreover, all the special paths together cannot exceed in length $|W_{opt}|$. Hence, the competitive factor of our strategy is $4\sqrt{2}+1$.

4.1 The 6.7-Competitive Strategy

Denote by *RightExplorationRec*(*LeftExplorationRec*) the recursive procedure for exploring a group of right (left) vertices. Suppose that the robot makes the very first clockwise tour from s to explore the discovered right vertices, which satisfy the definition of a right-like polygon. For those right vertices whose parents in the *known portion* of $SPT(s)$ are left vertices, they may be discovered by this tour but are not explored, because they do not belong to the right-like polygon starting from s. On the other hand, some left vertices may happen to be fully explored in this clockwise tour. Among those fully explored left vertices, we mark the ones that have the right vertices as their sons in $SPT(s)$.

The procedure *RightExplorationRec* can be given by modifying *RightPolygonExp* as follows. First, the essential cuts and critical images are now defined with respect to the set of explored right vertices. Second, we maintain during the clockwise tour a list *LeftList* of the discovered left vertices, which either have not been fully explored or have a right-vertex son in the known portion of $SPT(s)$. The list *LiftList* is kept as a *local* data structure in *RightExplorationRec.*

A new idea of our strategy here is to change to explore the discovered left vertices, as soon as it is possible. We say the *switching* condition is satisfied in performing *RightExplorationRec* if CP is on the cut of a *right vertex*, say, v, and at least one vertex of *LeftList* is the son of v in the known portion of $SPT(s)$. (The condition for switching to explore the right vertices is defined analogously.)

Whenever the switching condition is satisfied, the robot moves to v (if needed) and then makes a call of *LeftExplorationRec* with the initial list *LeftTarget*, whose elements are v's descendants (in $SPT(s)$) contained in *LeftList*.

Since *LeftExplorationRec* is a recursive procedure, all elements of *LeftTarget*, and everything behind, get explored by a call of *LeftExplorationRec*. After *LeftExplorationExp* terminates, the robot continues its clockwise tour, from the *break* point, so as to explore the remaining elements of *RightTarget*.

By symmetry, we will give only the detail of *RightExplorationRec*. Let s_r denote the point at which *RightExplorationRec* is invoked, and *RightTarget* the initial list of right vertices in clockwise order, to which the robot knows the shortest paths. The initial list *RightTarget* may contain some right vertices, which had been fully explored but have a left-vertex son in $SPT(s)$. So, if the target vertex was already explored, the robot walks along the shortest path to it. Also, we consider s_r as the largest element of *RightTarget* and an already explored vertex.

Procedure. *RightExplorationRec* (**in** *RightTarget*, **in** s_r)

1. Set $CI, TI, C \leftarrow s_r$. Assume that s_r is a special cut, with $P(s_r) = P$.
2. **While** *RightTarget* is not empty **do**
   (a) If the target vertex r was already explored, walk along the shortest path to it (at which the switching condition is always satisfied).
   (b) If r has not yet been explored, perform Step 2 of *RightPolygonExp* with the following modifications: (i) the essential cuts and critical images are defined with respect to the set of explored right vertices, and (ii) save in *LeftList* in counterclockwise order the discovered left vertices, which have not been fully explored or have a right-vertex son in $SPT(s)$.
   (c) Whenever the switching condition is satisfied in Step 2(a) or 2(b), do the followings:
      i. Walk along the current cut to its defining vertex (if needed).
      ii. Select (and delete) from *LeftList* the elements, which are the descendants of CP in $SPT(s)$, and put them in *LeftTarget*.
      iii. Call *LeftExplorationRec*(*LeftTarget*, s_l) by setting $s_l \leftarrow CP$.
      iv. Restart Step 2(a) or 2(b) from the breaking point.

By considering the point s as an already explored left/right vertex, the procedure for exploring the polygon P can be given as follows.

Procedure. *PolygonExploration* (**in** P, **in** s)

1. Set *RightTarget* (*LeftTarget*) to the list of the right (left) vertices, which are visible from s and ordered in clockwise (counterclockwise) order.
2. **If** *RightTarget* is not empty Call *RightExplorationRec*(*RightTarget*, s) **else** Call *LeftExplorationRec*(*LeftTarget*, s).

The robot's walk caused by Step 2 of a procedure *RightExplorationRec* or *LeftExplorationRec*, without considering further calls within it, clearly forms a closed curve. In Fig. 5(a), the route $R1$ is the very first tour for exploring right

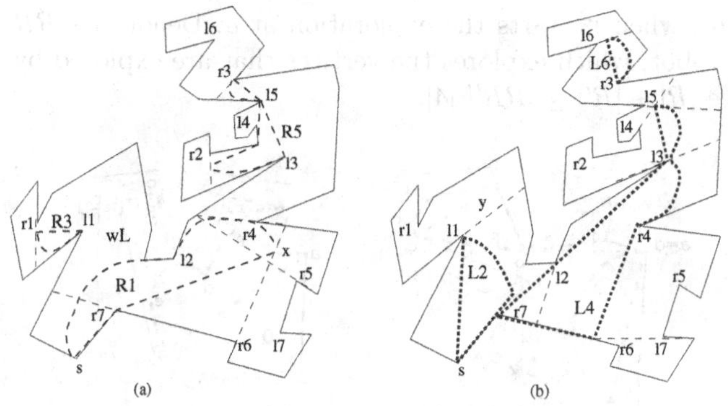

Fig. 5. Exploring an unknown polygon.

vertices. The switching condition is satisfied at the point s, as $l1$ belongs to *LeftList* and is the son of the right vertex s. So, the robot moves to explore $l1$ along the route $L2$, see Fig. 5(b). When the robot reaches the point $l1$, the switching condition is satisfied again. The robot walks to explore $r1$ along $R3$, and then moves back to $l1$ (along $R3$) and finally to s (along $L2$). Next, the robot moves to explore $r7$, $r4$, $r5$ and $r6$ along $R1$. When the robot reaches $r7$, it changes to explore the left vertices along $L4$, and so on.

Lemma 8. *A call of the procedure PolygonExploration outputs a watchman route.*

Proof. Omitted in this extended abstract. □

Lemma 9. *All local starting points, at which RightExplorationRec and LeftExplorationRec are called, have to be visited at least once by W_{opt}.*

Proof. Omitted in this extended abstract (see also [4]). □

By now, we can give the main result of this paper.

Theorem 3. *For a polygon P and a starting point s on its boundary, a call of PolygonExploration(P, s) explores P, which outputs a watchman route of length at most 6.7 times the length of the shortest watchman route through s.*

Proof. Denote by $WR_{rob}(WL_{rob})$ the union of the routes R of the robot, which are outputted in Step 2 of *RightExplorationRec* (*RightExplorationRec*). Also, denote by $WR_{opt}(WL_{opt})$ the optimal watchman route that visits the union of the cuts of right (left) vertices and starting points, which are visited by all routes in $WR_{rob}(WL_{rob})$. Then, $|WR_{opt}| \leq |W_{opt}|$ and $|WL_{opt}| \leq |W_{opt}|$ (Lemma 9).

Denote by R and R' two routes of WR_{rob}. Assume that the starting point a of R is immediately before the starting point b of R' on the boundary of P. We first give a method to evaluate the total length of R and R'.

Case 1. R and R' are mutually invisible, except for their local starting points. Suppose that a "clever" robot knows the shortest paths to the right vertices explored by R', when it starts the exploration at a. Denote by RR' the route of this clever robot, which explores the vertices that are explored by R and R'. Then, we have $|R| + |R'| \leq |RR'|$ [4].

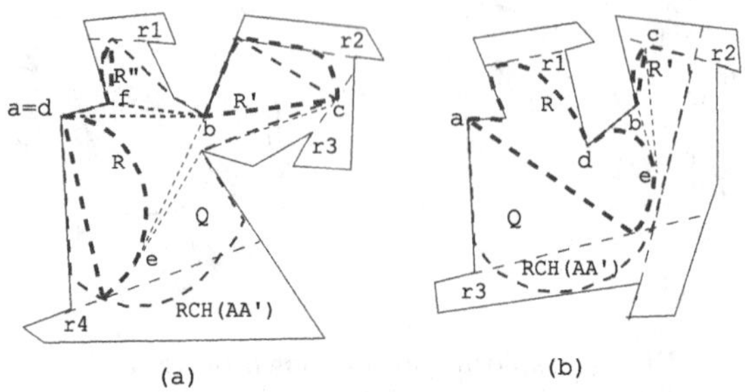

Fig. 6. Illustrating the proof of Theorem 3.

Case 2. R and R′ are not mutually invisible. From the definition of the right-like polygons, R cannot touch the starting point b of R' (see Fig. 6). Let C be the last cut explored by R', and c the point of C, at which R' reaches C. Since R and R' are not mutually invisible, we denote by d and e the two points on R such that d (e) is visible from b and closest to (furthest from) the point a along R. See Fig. 6. So, if the point d (e) is not a vertex of P, then the line through b and d (e) is tangent to R. This implies that the extension of the last segment of $\pi(c, b)$ intersects R. The angle formed by $\pi(b, e)$ and $\pi(b, c)$ at point b is then larger than $\pi/2$ and smaller than π. Since $\pi(b, d)$ is just the line segment $\overline{bd}$, we have $|\pi(b, d)| < |\pi(c, d)|$. Since the portion of R from d to e is relatively convex, it cannot exceed in length the route that consists of $\pi(d, b)$ and $\pi(b, e)$.

As in Case 1, denote by RR' the clever robot's route, which explores the vertices that are explored by R and R', starting from a. Denote by AA' the route that is computed by the off-line approximation algorithm, with respect to the cuts visited by RR'. See Fig. 6. The route consisting of the portion of R from a to d, the path $\pi(d, b)$, the portion of R' from b to c, the path $\pi(c, e)$ and the portion of R from e back to a is then relatively convex. Clearly, the length of this route cannot exceed $|RCH(AA')|$. See Fig. 6. By noticing the fact that $\pi(d, b)$ does not belong to R nor R', we are sure that $|R| + |R'| - |\pi(b, c)| \leq |RCH(AA')|$.

The above analysis can repeatedly be made to any two consecutive routes of WR_{rob}. So, except for the paths $\pi(b, c)$ used in Case 2, the total length of the routes in WR_{rob} cannot exceed the clever robot's route that visits the union of the cuts of right vertices and starting points, which are visited by all routes of WR_{rob}. Since the length of this clever robot's route cannot exceed $2|WR_{opt}|$, the total length of the routes outputted by all calls of *RightPolygonRec* is at most $2\sqrt{2}|W_{opt}|$. Analogously, except for the special paths $\pi(b, c)$, the total length of the routes outputted by all calls of *LeftPolygonRec* is at most $2\sqrt{2}|W_{opt}|$.

It remains to evaluate the total length of all the paths $\pi(b, c)$ used in Case 2. First, the starting points b as well as the cuts C are all different, no matter whether the paths $\pi(b, c)$ are introduced for the routes of WR_{rob} or WL_{rob}.

Second, W_{opt} passes through these points b (Lemma 9). Finally, c is a critical image, with respect to the cuts visited by R'. The total length of these paths $\pi(b, c)$ is then less than $|W_{opt}|$ (Lemma 3). Hence, we obtain the competitive factor $4\sqrt{2} + 1 < 6.7$. □

References

1. Chin, W.P., Ntafos, S.: Optimum watchman routes. IPL **28**, 39–44 (1988)
2. Deng, X., Kameda, T., Papadimitriou, C.: How to learn an unknown environment. In: Proceedings of the 32nd Annual Symposium on Foundations of Computer Science, pp. 298–303 (1991)
3. Dror, M., Efrat, A., Lubiw, A., Mitchell, J.S.B.: Touring a sequence of simple polygons. In: Proceedings of the 35th Annual ACM Symposium Theory of Computing, pp. 473–482 (2003)
4. Hoffmann, F., Icking, C., Klein, R., Kriegel, K.: The polygon exploration problem. SIAM J. Comput. **31**(2), 577–600 (2001)
5. Tan, X.: Approximation algorithms for the watchman and zookeeper's problems. Discrete Appl. Math. **136**, 363–376 (2004)
6. Tan, X., Hirata, T., Inagaki, Y.: An incremental algorithm for constructing shortest watchman routes. Int. J. Comput. Geom. Appl. **3**, 351–365 (1993)
7. Tan, X., Hirata, T., Inagaki, Y.: Corrigendum to an incremental algorithm for constructing shortest watchman routes. IJCGA **9**, 319–323 (1999)

Polynomial Time Approximation Scheme for Single-Depot Euclidean Capacitated Vehicle Routing Problem

Michael Khachay[1,2](✉) and Helen Zaytseva[2]

[1] Krasovsky Institute of Mathematics and Mechanics, Ekaterinburg, Russia
mkhachay@imm.uran.ru
[2] Ural Federal University, Ekaterinburg, Russia
zaytsevy_vse@mail.ru

Abstract. We consider the classic setting of Capacitated Vehicle Routing Problem (CVRP): single product, single depot, demands of all customers are identical. It is known that this problem remains strongly NP-hard even being formulated in Euclidean spaces of fixed dimension. Although the problem is intractable, it can be approximated well in such a special case. For instance, in the Euclidean plane, the problem (and it's several modifications) have polynomial time approximation schemes (PTAS). We propose polynomial time approximation scheme for the case of $\mathbb{R}^3$.

Keywords: Capacitated vehicle routing problem · Polynomial time approximation scheme · Iterated tour partition

1 Introduction

The Capacitated Vehicle Routing Problem is the well known special case of Vehicle Routing Problem, which belongs to the class of combinatorial optimization models widely adopted in operations research. The problem under consideration is closely related to Traveling Salesman Problem (TSP) [2,5,9,19] and, especially to its modification, the Multiple Traveling Salesmen Problem [12,13,16,17], which is of finding the minimum cost set of tours for a team of collaborating visitors. It is convenient, to give the substantial statement of the problem in terms of operations research. We are given by a set X of n points (customers), a distinguished point O outside X, called the depot as well as a distance function. The objective is to find a set of tours, each including the depot and at most q points in X, which covers all points in X and achieves the minimum total length.

For the first time, VRP was introduced by Dantzig and Ramser in [6]. They considered routing of a fleet of gasoline delivery trucks between a bulk terminal and a number of service stations supplied by the terminal. The distance between any two locations was given and a demand for a given product was specified for the service stations.

Z. Lu et al. (Eds.): COCOA 2015, LNCS 9486, pp. 178–190, 2015.
DOI: 10.1007/978-3-319-26626-8_14

In its simplest setting, the VRP can be defined as the combinatorial optimization problem of designing the least cost collection of delivery routes from the dedicated point (*depot*) to a set of geographically dispersed locations (*customers* or *clients*) subject to a set of constraints. In recent decades, this problem has been studied very extensively in papers on combinatorial optimization and operations research (see, e.g. [4,10,20]).

The CVRP problem contains the TSP problem as a special case arising when depot is among the customers and $q = n$. TSP problem is known to be NP hard, even in Euclidean space of fixed dimension [18]. Almost all known modifications of the Vehicle Routing Problem are NP-hard [14,15] as well[1], even being formulated in fixed-dimensional Euclidean space.

For this reason, the research on CVRP has focused on heuristic algorithms and approximation algorithms. The general metric case of CVRP for $q \geq 3$ has been shown to be APX-complete [3], that is there exists $\varepsilon > 0$ such that no $1+\varepsilon$ approximation algorithm exists unless $P = NP$.

The existence of a PTAS for the 2d-Euclidean version remains an active area of research. One of the first studies of two-dimensional Euclidean CVRP has been due to Haimovich and Rinnooy Kan [11], who presented several heuristics for the Euclidean CVRP, including a PTAS for the two-dimensional Euclidean CVRP with $q \leq c \log\log n$, for some constant c. Asano et al. [3] substantially improved this result by designing a PTAS for $q = O(\log n / \log\log n)$. They also observed that Arora's [2] PTAS for the two-dimensional Euclidean TSP implies a PTAS for the corresponding CVRP where $q = \Omega(n)$. Recently Das and Mathieu [7,8] proposed a quasi-polynomial time approximation scheme (QPTAS) for the two-dimensional[2] Euclidean CVRP for every q. Their algorithm combines the approach developed by Arora [2] for Euclidean TSP with some new ideas to deal with CVRP and gives a $(1+\varepsilon)$-approximation for the two-dimensional Euclidean CVRP in time $n^{(\log n)^{O(1/\varepsilon)}}$ (for any value of q). In the work [1] a new PTAS for all values of $q \leq 2^{\log^{\delta} n}$, where $\delta = \delta(\varepsilon)$ was presented. To the best of our knowledge, there is no PTAS for CVRP in d-dimensional Euclidean space for any fixed $d > 2$.

New polynomial time approximation scheme for the case of $d = 3$ extending the approach developed by Haimovich and Rinnooy Kan for the Euclidean plane is the main contribution of this paper.

The rest of the paper is organized as follows. In Sect. 2, we recall the general statement of CVRP along with its metric and Euclidean special cases. In Sect. 3, we give a short overview of the Iterated Tour Partition (ITP) heuristic for the metric CVRP, which is introduced in [11] and extensively used in our subsequent constructions. Although, all of these results are well known, we provide them with proofs to emphasize the most general case of CVRP, for which they remain valid. Further, in Sect. 4, we propose a new upper bound for an optimal value for the corresponding TSP-instance in three-dimensional Euclidean space. This result

[1] Although, for $q = 1$ or $q = 2$, CVRP can be solved to optimality in polynomial time.

[2] Using Arora's technique, this result can be extended onto d-dimensional Euclidean space for any fixed d.

helps us to propose a PTAS for the Euclidean CVRP in $\mathbb{R}^3$, which is presented in Sect. 5. Finally, in Sect. 6 we summarize the results obtained and discuss some open questions.

2 Problem Statement

We start with some necessary definitions and notation.

1. $X = \{x_1, \ldots, x_n\}$ is a set of customers, x_0 is a dedicated point (depot). $G^0 = (X \cup \{x_0\}, E, w)$ is a complete weighted undirected graph for $w : E \to \mathbb{R}_+$. Along with the graph G^0, we consider its subgraph $G = G(X)$ induced by the set X.
2. For each customer, denote transition cost from the depot x_0 to x_i by $r_i = w(x_0, x_i)$. Also, we denote the maximum cost by $r_{max} = max_i\{r_i\}$ and the average one by $\bar{r} = (\sum_{i=1}^{n} r_i)/n$.
3. W.l.o.g., we can assume that there is only one vehicle visiting all the customers in some number of tours. Each tour starts and finishes in the depot and visits at most q customers due to the capacity constraint.
4. Denote by $X_j \subseteq X$ the set of customers visited in the j-th tour, $|X_j| \leq q$, and $X_j^0 = X_j \cup \{x_0\}$. By condition, each customer has to be visited once, therefore $X_{j_1} \cap X_{j_2} = \varnothing$ for any $j_1 \neq j_2$.
5. For any tour $x_0, x_{i_1}, x_{i_2}, \ldots, x_{i_k}, x_0$, the cost of this tour is equal to the sum $r_{i_1} + w(x_{i_1}, x_{i_2}) + \ldots + r_{i_k}$.

The problem is, for a given graph $G^0 = (X \cup \{x_0\}, E, w)$, to find a cheapest set of tours visiting all the customers.

Further, we consider two important special cases of the problem: metric and Euclidean.

Metric CVRP. In this case, the weight function w meets the triangle inequality. For any vertices x_{i_1}, x_{i_2} and x_{i_3}, $w(x_{i_1}, x_{i_2}) \leq w(x_{i_1}, x_{i_3}) + w(x_{i_2}, x_{i_3})$.

Euclidean CVRP. In this case, the depot and all the customers locations are points in d-dimensional Euclidean space $X \cup \{x_0\} \subset \mathbb{R}^d$ and

$$w(x_i, x_j) = \|x_i - x_j\|_2.$$

For these cases, for any points x_{i_1} and x_{i_2}, the weight (cost) $w(x_{i_1}, x_{i_2})$ is *a distance* between them, and the cost of any tour can be naturally referred as *a length* of this tour.

3 Iterated Tour Partition Heuristic

To construct our polynomial time approximation scheme, we use Iterated Tour Partition (ITP) heuristic proposed in [11] for the Euclidean plane. We extend this result for the more general cases of CVRP.

The ITP heuristic relates the initial CVRP problem with TSP problem for the graph G. Consider an arbitrary Hamiltonian cycle H in the graph G. Starting from x_1, break this tour into $l = \lceil n/q \rceil$ disjoint segments such that each of them contains at most q customers. Then, connect the endpoints of any segment with the depot to provide a feasible solution for the initial CVRP. Performing the same procedure iteratively for any starting point x_i, we construct n feasible solutions $V_1, \ldots, V_n$ of the initial instance of CVRP; output the best (cheapest) among them (see Fig. 1).

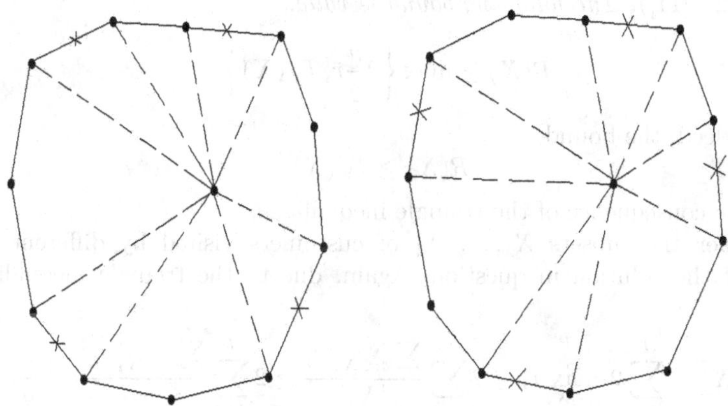

Fig. 1. Example of the ITP for $q = 3$

Let $T(X)$ be a cost of H, and $R_H(X)$ be a cost of the resulting set of vehicle routes.

Lemma 1 ([11]). *The following equation*

$$R_H(X) \leq 2 \left\lceil \frac{n}{q} \right\rceil \bar{r} + \left(1 - \frac{\lceil n/q \rceil}{n}\right) T(X) \tag{1}$$

is valid.

Proof. Indeed, each edge $\{x_{i_1}, x_{i_2}\}$ of the tour H is included $n - l$ times to the solutions $V_1, \ldots, V_n$ and l times is replaced with 'radial' edges $\{x_0, x_{i_1}\}$ and $\{x_0, x_{i_2}\}$ of costs r_{i_1} and r_{i_2}, respectively. Therefore, the cumulative cost C of all solutions $V_1, \ldots, V_n$ is defined by the following equation

$$C = 2l \sum_{i=1}^{n} r_i + (n - l)T(X).$$

Finally, since $R_H(X)$ is less or equal to the average cost of the solutions $V_1, \ldots, V_n$, we have

$$R_H(X) \leq \frac{C}{n} = 2l\bar{r} + (1 - l/n)T(X)$$

providing (1) by substitution $l = \lceil n/q \rceil$. Lemma 1 is proved.

Remark 1. The claim of Lemma 1 is valid for the most general statement of CVRP.

In the metric case, the lower bound of the cost of any solution of CVRP also can be obtained in terms of the corresponding TSP route. Indeed, as above, suppose that $R(X)$ denotes a length of an arbitrary solution V of metric CVRP. Taking its tours in any order, we can connect the last and the first customers visited by successive routes directly (excluding the depot) and construct a Hamiltonian cycle of length $T_V(X)$ in the graph G.

Lemma 2 ([11]). *The following bound is valid.*

$$R(X) \geq \max\left\{2\frac{n}{q}\bar{r}, T_V(X)\right\}. \tag{2}$$

Proof. Indeed, the bound

$$R(X) \geq T_V(X)$$

is a simple consequence of the triangle inequality.

Consider the subsets $X_1, \ldots, X_l$ of customers visited by different routes (tours) of the solution in question. Again, due to the triangle inequality, we have

$$R(X) \geq \sum_{j=1}^{l} 2 \max_{x_i \in X_j} r_i \geq 2\sum_{j=1}^{l} \frac{\sum_{x_i \in X_j} r_i}{|X_j|} \geq 2\sum_{j=1}^{l} \frac{\sum_{x_i \in X_j} r_i}{q} = 2\frac{n}{q}\bar{r}.$$

Lemma 2 is proved.

For a given instance of the metric CVRP, denote by $R^*(X)$ the optimum value (of this instance) and by $T^*(X)$ the length of an optimal Hamiltonian cycle in the graph G.

Theorem 1 ([11])

$$\max\left\{2\frac{n}{q}\bar{r}, T^*(X)\right\} \leq R^*(X) \leq 2\left\lceil\frac{n}{q}\right\rceil\bar{r} + \left(1 - \frac{1}{q}\right)T^*(X).$$

Proof. The upper bound can be obtained as a straight-forward consequence of Lemma 1. Indeed, let H^* be an arbitrary cheapest Hamiltonian cycle (of length $T^*(X)$) in the graph G. Then, using the claim of Lemma 1 and the evident inequality $\lceil n/q \rceil \geq n/q$, we have the following equation

$$2\left\lceil\frac{n}{q}\right\rceil\bar{r} + \left(1 - \frac{1}{q}\right)T^*(X) \geq 2\left\lceil\frac{n}{q}\right\rceil\bar{r} + \left(1 - \frac{\lceil n/q \rceil}{n}\right)T^*(X) \geq R_{H^*}(X),$$

which implies the required bound, since $R_{H^*}(X) \geq R^*(X)$.

On the other hand, let V^* be an optimal solution (of cost $R^*(X)$) of the given instance of the CVRP. Then, by Lemma 2, we have

$$R^*(X) \geq \max\left\{2\frac{n}{q}\bar{r}, T_{V^*}(X)\right\} \geq \max\left\{2\frac{n}{q}\bar{r}, T^*(X)\right\},$$

since $T^*(X)$ is an optimal value of the corresponding TSP. Theorem 1 is proved.

Further, we recall some basic definitions describing the performance of approximation algorithms for combinatorial optimization (minimization in the case under consideration) problems. Let OPT be an optimal value of a problem and APP be its approximate value obtained by some algorithm $\mathcal{A}$. Then *relative error* of the algorithm and its *approximation ratio* are defined by the equations

$$\varepsilon_{\mathcal{A}} = \frac{APP - OPT}{OPT} \quad \text{and} \quad \rho_{\mathcal{A}} = \frac{APP}{OPT},$$

respectively.

Theorem 1 gives us an ability to represent a performance of ITP-based approximation algorithms for metric CVRP by means of relative errors of heuristics used in approximation of the inner TSP problem. Indeed, suppose, we obtain a Hamiltonian cycle H, whose cost is at most $(1+\varepsilon)T^*(X)$. Then, by Lemma 1, for the cost $R_H(X)$ of ITP-based approximate solution of the CVRP, we obtain

$$R_H(X) \leq 2\left\lceil \frac{n}{q} \right\rceil \bar{r} + (1 - 1/q)(1+\varepsilon)T^*(X).$$

Lets analyze the approximation ratio of the resulting heuristic. If ρ_T - the approximation ratio of the algorithm for solving TSP, then

$$\frac{R_H(X)}{R^*(X)} \leq \frac{2\lceil \frac{n}{q} \rceil \bar{r} + (1 - l/n)\rho_T T^*(X)}{\max\left\{2\frac{n}{q}\bar{r}, T^*(X)\right\}}$$

$$\leq \frac{q}{n} + 1 + \left(1 - \frac{\lceil n/q \rceil}{n}\right)\rho_T \leq \frac{q}{n} + 1 + \left(1 - \frac{1}{q}\right)\rho_T, \quad (3)$$

since $\frac{n}{q} \leq \lceil \frac{n}{q} \rceil \leq \frac{n+q}{q}$.

If $q = o(n)$ then the right-hand side of equation (3) tends to $1 + \rho_H$ as $n \to \infty$. Therefore, in this case, any ρ-approximation algorithm for the metric TSP induces asymptotically $(1+\rho)$-approximation algorithm for the metric CVRP. For instance, the well-knonw 3/2-approximation Christofides algorithm accompanied by the ITP heuristic provides 5/2-approximation for CVRP.

Since the running time of the ITP is at most $O(n^2)$, the overall complexity of any ITP-based approximation algorithm is defined by the running time of the underline approximation algorithm for TSP. In particular, for Christofides algorithm, we get $O(n^3)$.

For the problem on the Euclidean plane there are polynomial-time approximation scheme, proposed by Arora [2]. Using this PTAS, one obtain, for any $c > 1$, a polynomial time approximation algorithm for 2d-Euclidean CVRP with asymptotic approximation ratio of $2 + \frac{1}{c}$.

4 Approximation of TSP in $\mathbb{R}^3$

In the paper [11], some upper bounds for an optimum of the TSP induced by the considered CVRP based on geometry of the plane were given. In our work, we

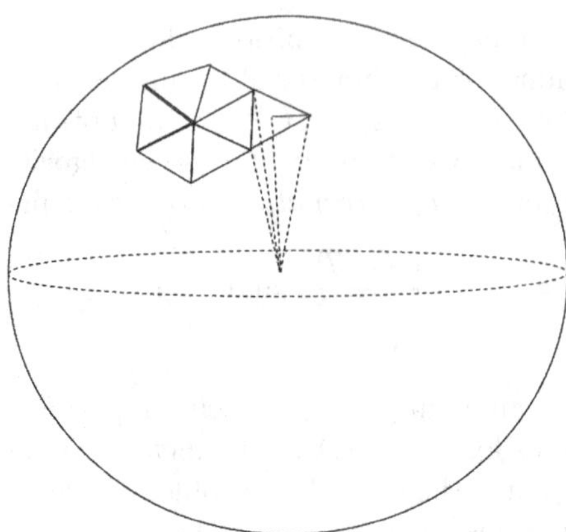

Fig. 2. Construction of TSP-tour

extend the technique proposed in [11] to the case of $\mathbb{R}^3$ and obtain the following result.

Theorem 2. *For some constant $C > 0$, the following bound*

$$T^*(X) \leq C\sqrt[3]{r_{max}(n\bar{r})^2}$$

is valid.

Proof. By condition, for any $\delta > 0$, all the customers lie in the sphere with radius $r = r_{max}(1+\delta)$ and center in depot. Than we construct the triangulation of sphere with $4h$ near equal equilateral triangles. Further we construct the cycle as follows: starting from the depot we go to the center of triangle, than go to the arbitrary chosen vertex of the triangle, walk round the triangle and return to the depot through the center of triangle (Fig. 2). On the next stage we have to add all the customers to our cycle with the double connection to the closest radius (Fig. 3).

Now we have to analyze the length of resulting path. The area of each of $4h$ triangle is equal to

$$A_\triangle \leq \frac{A_{sphere}}{4h} = \frac{4\pi r^2}{4h} = \frac{\pi r^2}{h}$$

On the other hand $A_\triangle = \frac{\sqrt{3}a^2}{4}$, where a is the side of triangle. So, $a \leq \frac{2\sqrt{\pi}r}{\sqrt{h\sqrt{3}}} = c_1\frac{r}{\sqrt{h}} = c_1\frac{r_{max}(1+\delta)}{\sqrt{h}}$, for some constant $c_1 > 0$. Then, the perimeter of any such a triangle is at most $c_2\frac{r_{max}}{\sqrt{h}}$ (again, for some independent constant c_2). Length of the path from the center of triangle to its vertex is also proportional to $\frac{r_{max}}{\sqrt{h}}$. Hence the length of the tour through each such a pyramid is at most

$$c_3\frac{r_{max}}{\sqrt{h}} + 2r_{max}.$$

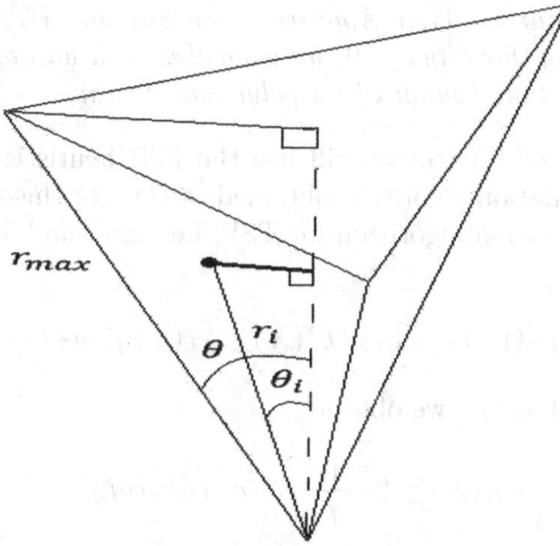

Fig. 3. Distance between the customer and the constructed tour

Distance (see Fig. 3) between each customer and the path constructed is at most

$$r_i \sin\theta_i \le r_i \sin\theta = r_i \frac{a}{\sqrt{3}(1+\delta)r_{max}}.$$

Therefore, this distance can be bounded from above by $c_4 \frac{r_i}{\sqrt{h}}$, for some constant $c_4 > 0$.

Total length of our tour is comprised of tour through all pyramids and links to all customers. Therefore, for optimal value $T^*(X)$ of TSP, we obtain

$$T^*(X) \le 4h(c_3 \frac{r_{max}}{\sqrt{h}} + 2r_{max}) + \sum_{i=1}^{n} 2\frac{r_i}{\sqrt{h}} c_4 = c_5 r_{max}\sqrt{h} + 8r_{max}h + \frac{n\bar{r}c_4}{\sqrt{h}}.$$

Hence,

$$T^*(X) \le c_6 r_{max} h + \frac{n\bar{r}c_4}{\sqrt{h}}, \tag{4}$$

since $\sqrt{h} \le h$ for any $h \ge 1$.

By taking $h = \sqrt[3]{(\frac{c_4 n\bar{r}}{2c_6 r_{max}})^2}$ in order to minimize the right-hand side of (4) we obtain the desired result. Theorem 2 is proved.

5 Polynomial Time Approximation Scheme for Capacitated Vehicle Routing Problem in 3-Dimensional Space

Using the technical result provided by Theorem 2, we construct a polynomial time approximation scheme for Capacitated Vehicle Routing Problem in 3-dimensional Euclidean space.

Definition 1. *Polynomial Time Approximation Scheme (PTAS) is a family of algorithms containing, for any $\varepsilon > 0$, an algorithm with an approximation ratio $1+\varepsilon$ and a running time bounded by a polynomial*[3] *in n.*

For construction of PTAS, we will use the ITP heuristic described above (denote the approximation algorithm obtained as H). For this algorithm and for any $(1+\tau)$-approximation algorithm for TSP, Lemma 2 and Theorem 2 implies that

$$R_H(X) \leq 2\bar{r}(n+q)/q+(1-1/q)(1+\tau)T^*(X) \leq 2\bar{r}(n+q)/q+(1+\tau)C\sqrt[3]{r_{max}(n\bar{r})^2}.$$

Denoting $c^H = (1+\tau)C$ we obtain

$$R_H(X) \leq 2\frac{n+q}{q}\bar{r} + c^H r_{max}^{\frac{1}{3}}(n\bar{r})^{\frac{2}{3}} \tag{5}$$

Under Lemma 2,

$$R^*(X) \geq 2\sum r_i/q \tag{6}$$

Consider the following heuristic $A(H)$:

(i). Enumerate customers by decreasing their distance from the depot

$$r_1 \geq r_2 \geq \ldots \geq r_n.$$

(ii). Take the set $X(k) = \{x_1, \ldots, x_{k-1}\}$ of *outside* customers and find the optimal solution (for CVRP defined for this subset and the same depot x_0).

(iii). Apply H to the set of *inside* customers $X \setminus X(k)$.

Theorem 3. *The proposed heuristic $A(H)$ is a polynomial time approximation scheme for 3d-Euclidean CVRP.*

Proof. We need to show that, for any $\varepsilon > 0$, the relative error $e^{A(H)}(X)$ of $A(H)$ satisfies the inequality $e^{A(H)}(X) \leq \varepsilon$.

To prove this equation, we obtain an upper bound for $e^{A(H)}(X)$ by the technique proposed in [11]. Indeed, consider any optimal solution of the given instance of CVRP and the sphere with radius r_k centered at the depot. For any i-th tour of this solution, by connecting endpoints of any outside sectors to each other and with the origin we obtain l_i separate outside subroutes and a single inside subroute (see Fig. 4), such that $\sum_i l_i \leq k-1$.

Therefore,

$$R^*(X(k)) + R^*(X \setminus X(k)) \leq R^*(X) + 4(k-1)r_k, \tag{7}$$

since the length of any chord is at most $2r_k$.

[3] The degree of such a polynomial along with its coefficients can depend on $1/\varepsilon$.

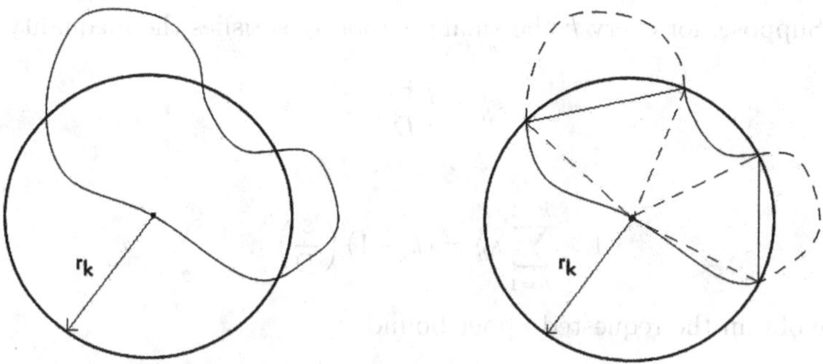

Fig. 4. Transformation of an optimal solution of CVRP

By construction, for any $A(H)$-based approximation algorithm,

$$R_{A(H)}(X) = R^*(X(K)) + R_H(X \setminus X(k)).$$

Since $R^*(X \setminus X(k)) \geq 2\sum_{i=k}^{n} \frac{r_i}{q}$, we get

$$R_{A(H)}(X) \leq R^*(X(k)) + R^*(X \setminus X(k)) + \left(R_H(X \setminus X(k)) - 2\frac{\sum_{i=k}^{n} r_i}{q}\right).$$

Combining this with (6) and (7) we obtain

$$\begin{aligned} e^{A(H)}(X) &= \frac{R_{A(H)}(X) - R^*(X)}{R^*(X)} \\ &\leq 2q(k-1)\frac{r_k}{\sum_{i=1}^{n} r_i} + \frac{q}{2}\frac{R_H(X \setminus X(k)) - 2(\sum_{i=k}^{n} r_i)/q}{\sum_{i=1}^{n} r_i} \qquad (8) \end{aligned}$$

Further, since

$$R_H(X \setminus X(k)) \leq \frac{2}{q}\sum_{i=k}^{n} r_i + 2r_k + c^H r_k^{\frac{1}{3}}(n\bar{r})^{\frac{2}{3}},$$

due to (5), we obtain

$$\begin{aligned} e^{A(H)}(X) &\leq 2q(k-1/2)\frac{r_k}{\sum_{i=k}^{n} r_i} + \frac{q}{2}c^H\left(\frac{r_k}{\sum_{i=k}^{n} r_i}\right)^{\frac{1}{3}} \\ &\leq 2qk\frac{r_k}{\sum_{i=k}^{n} r_i} + \frac{q}{2}c^H\left(\frac{r_k}{\sum_{i=k}^{n} r_i}\right)^{\frac{1}{3}}. \qquad (9) \end{aligned}$$

If we choose k so that (9) is less than ε we will have approximation scheme. All we have to do is to obtain the upper bound on k independent of n. Put $s_h = \sqrt[3]{r_h/\sum r_i}, A = 2q, 2B = c^H q/2$ and investigate the lower bound of inequality solutions.

$$Ahs_h^3 + 2Bs_h - \varepsilon \geq 0 \qquad (h = 1, \ldots, k-1). \qquad (10)$$

Indeed, take any constant $D > 0$, such that $1 - 2B/D > 0$.

Case 1. Suppose, for every h the smallest root s_h satisfies the inequality

$$s_h^3 \geq \left(\frac{\varepsilon}{D}\right)^3.$$

Therefore,

$$1 \geq \sum_{h=1}^{k-1} s_h^3 = (k-1)\left(\frac{\varepsilon}{D}\right)^3.$$

Hence we obtain the requested upper bound

$$k \leq \left(\frac{D}{\varepsilon}\right)^3 + 1. \tag{11}$$

Case 2. If there is h_0, such that $s_{h_0}^3 < \left(\frac{\varepsilon}{D}\right)^3$. The bound of inequality solutions is decreasing while h is rising, hence for every $h > h_0$ the same inequality $s_h^3 < \left(\frac{\varepsilon}{D}\right)^3$ is valid as well. Since $B > 0$, for every solution of (10), the rougher inequality takes place:

$$Ahs_h^3 + 2B\frac{\varepsilon}{D} - \varepsilon \geq 0$$

that is why

$$Ahs_h^3 \geq \varepsilon\left(1 - \frac{2B}{D}\right),$$

and

$$s_h^3 \geq \varepsilon\left(1 - \frac{2B}{D}\right)\left(\frac{1}{Ah}\right).$$

Hence,

$$1 \geq \sum_{h=1}^{k-1} s_h^3 \geq \frac{\varepsilon}{A}\left(1 - \frac{2B}{D}\right)\sum_{h=1}^{k-1}\frac{1}{h},$$

and, since

$$\sum_{h=1}^{k-1}\frac{1}{h} > \int_1^{k-1}\frac{1}{z}\,dz = \ln(k-1),$$

we obtain

$$1 \geq \frac{\varepsilon}{A}\left(1 - \frac{2B}{D}\right)\ln(k-1)$$

and

$$k \leq e^{\frac{A}{\varepsilon(1-2B/D)}} + 1. \tag{12}$$

So, we obtained the upper bound on k, that means that running time of finding optimal set of routes for outside customers doesn't depend on n. Since, other steps of heuristic we can do in polynomial time, we prove that $A(H)$ is polynomial time approximation scheme. Its running time depends on algorithm for solving TSP. Theorem 3 is proved.

Remark 2. All the arguments give above, carried out with the implicit assumption that right-hand sides of Equations (11) and (12) are at most n. Taking into account that, for any fixed $\tau > 0$, all of A, B and D are $\Theta(q)$, we conclude that the proposed algorithm based on $(1 + \tau)$-approximation for TSP is a PTAS for CVRP for $q = o(\ln n)$.

6 Conclusion

Extending the approach to construction approximation algorithms for CVRP on the basis of well-known Iterated Tour Partition (ITP) heuristic, we propose new polynomial-time approximation scheme for 3d-Euclidean case of the problem. The proposed approach seem to be expendable to the case of an arbitrary dimension $d \geq 3$, which will be done in forthcoming paper. Also, the future work can be concerned with extending the result obtained onto the case of Euclidean multi-depot CVRP of an arbitrary fixed dimension.

Acknowledgements. This research was supported by Russian Foundation for Basic Research, grants no. 13-01-00210 and 13-07-00181, Center of Excellence in Quantum and Video Information Technologies at Ural Federal University, and the Complex Program of Ural Branch of RAS, grant no. 15-7-1-23.

References

1. Adamaszek, C., Czumaj, A., Lingas, A.: PTAS for k-tour cover problem on the plane for moderately large values of k. Manuscript 1 (2009)
2. Arora, S.: Polynomial time approximation schemes for Euclidean traveling salesman and other geometric problems. J. ACM **45**(5), 753–782 (1998)
3. Asano, T., Katoh, N., Tamaki, H., Tokuyama, T.: Covering points in the plane by k-tours: a polynomial time approximation scheme for fixed k. IBM Tokyo Research (1996)
4. Caric, T., Gold, H.: Vehicle Routing Problem. InTech (2008)
5. Christofides, N.: Worst-case analysis of a new heuristic for the traveling salesman problem. In: Symposium on New Directions and Recent Results in Algorithms and Complexity, p. 441 (1975)
6. Dantzig, G., Ramser, J.: The truck dispatching problem. Manage. Sci. **6**, 80–91 (1959)
7. Das, A., Mathieu, C.: A quasi-polynomial time approximation scheme for Euclidean capacitated vehicle routing. In: Proceedings of the Twenty-first Annual ACM-SIAM Symposium on Discrete Algorithms, SODA 2010, pp. 390–403. Society for Industrial and Applied Mathematics, Philadelphia (2010). http://dl.acm.org/citation.cfm?id=1873601.1873634
8. Das, A., Mathieu, C.: A quasipolynomial time approximation scheme for Euclidean capacitated vehicle routing. Algorithmica **73**, 115–142 (2014)
9. Deineko, V.G., Klinz, B., Tiskin, A., Woeginger, G.J.: Four-point conditions for the TSP: the complete complexity classification. Discrete Optim. **14**, 147–159 (2014)

10. Golden, B., Raghavan, S., Wasil, E. (eds.): The Vehicle Routing Problem: Latest Advances and New Challenges. perations Research/Computer Science Interfaces, 1st edn. Springer, US (2008)
11. Haimovich, M., Rinnooy Kan, A.H.G.: Bounds and heuristics for capacitated routing problems. Math. Oper. Res. **10**(4), 527–542 (1985)
12. Khachai, M., Neznakhina, E.: Approximability of the problem about a minimum-weight cycle cover of a graph. Doklady Math. **91**(2), 240–245 (2015). http://dx.doi.org/10.1134/S1064562415020313
13. Khachai, M., Neznakhina, E.: A polynomial-time approximation scheme for the Euclidean problem on a cycle cover of a graph. Proc. Steklov Inst. Math. **289**(1), 111–125 (2015). http://dx.doi.org/10.1134/S0081543815050107
14. Kumar, S., Panneerselvam, R.: A survey on the vehicle routing problem and its variants. Intell. Inf. Manage. **4**, 66–74 (2012)
15. Lenstra, J., Rinnooy Kan, A.: Complexity of vehicle routing and scheduling problems. Networks **11**, 221–227 (1981)
16. Manthey, B.: On approximating restricted cycle covers. SIAM J. Comput. **38**, 181–206 (2008)
17. Manthey, B.: Minimum-weight cycle covers and their approximability. Discrete Appl. Math. **157**, 1470–1480 (2009)
18. Papadimitriou, C.: Euclidean TSP is NP-complete. Theoret. Comput. Sci. **4**(3), 237–244 (1997)
19. Sahni, S., Gonzales, T.: P-complete approximation problems. J. ACM **23**, 555–565 (1976)
20. Toth, P., Vigo, D.: The Vehicle Routing Problem. Monographs on Discrete Mathematics and Applications, SIAM (2001)

Network Optimization

A Fast and Effective Heuristic for Discovering Small Target Sets in Social Networks

Gennaro Cordasco[2], Luisa Gargano[1], Marco Mecchia[1], Adele A. Rescigno[1], and Ugo Vaccaro[1](✉)

[1] Department of Informatics, University of Salerno, Fisciano, Italy
uvaccaro@unisa.it

[2] Department of Psychology, Second University of Naples, Caserta, Italy

Abstract. Given a network represented by a graph $G = (V, E)$, we consider a dynamical process of influence diffusion in G that evolves as follows: Initially only the nodes of a given $S \subseteq V$ are influenced; subsequently, at each round, the set of influenced nodes is augmented by all the nodes in the network that have a sufficiently large number of already influenced neighbors. The question is to determine a small subset of nodes S (*a target set*) that can influence the whole network. This is a widely studied problem that abstracts many phenomena in the social, economic, biological, and physical sciences. It is known [6] that the above optimization problem is hard to approximate within a factor of $2^{\log^{1-\epsilon}|V|}$, for any $\epsilon > 0$. In this paper, we present a fast and surprisingly simple algorithm that exhibits the following features: (1) when applied to trees, cycles, or complete graphs, it always produces an optimal solution (i.e., a minimum size target set); (2) when applied to arbitrary networks, it always produces a solution of cardinality matching the upper bound given in [1], and proved therein by means of the probabilistic method; (3) when applied to real-life networks, it always produces solutions that substantially outperform the ones obtained by previously published algorithms (for which no proof of optimality or performance guarantee is known in any class of graphs).

1 Introduction

Social networks have been extensively investigated by student of the social science for decades (see, e.g., [32]). Modern large scale online social networks, like Facebook and LinkedIn, have made available huge amount of data, thus leading to many applications of online social networks, and also to the articulation and exploration of many interesting research questions. A large part of such studies regards the analysis of social influence diffusion in networks of people. Social influence is the process by which individuals adjust their opinions, revise their beliefs, or change their behaviors as a result of interactions with other people [11]. It has not escaped the attention of advertisers[1] that the process of social influence can be exploited in *viral marketing* [26]. Viral marketing refers

[1] and politicians too [4,24,29,31].

Z. Lu et al. (Eds.): COCOA 2015, LNCS 9486, pp. 193–208, 2015.
DOI: 10.1007/978-3-319-26626-8_15

to the spread of information about products and behaviors, and their adoption by people. According to Lately [23], "*the traditional broadcast model of advertising-one-way, one-to-many, read-only is increasingly being superseded by a vision of marketing that wants, and expects, consumers to spread the word themselves*". For what interests us, the intent of maximizing the spread of viral information across a network naturally suggests many interesting optimization problems. Some of them were first articulated in the seminal papers [21,22]. The recent monograph [7] contains an excellent description of the area. In the next section, we will explain and motivate our model of information diffusion, state the problem we are investigating, describe our results, and discuss how they relate to the existing literature.

2 The Model, the Context, and Our Results

Let $G = (V, E)$ be a graph modeling the network. We denote by $\Gamma_G(v)$ and by $d_G(v) = |\Gamma_G(v)|$, respectively, the neighborhood and the degree of the vertex v in G. Let $t : V \rightarrow \mathbb{N}_0 = \{0, 1, \ldots\}$ be a function assigning thresholds to the vertices of G. For each node $v \in V$, the value $t(v)$ quantifies how hard it is to influence node v, in the sense that easy-to-influence elements of the network have "low" $t(\cdot)$ values, and hard-to-influence elements have "high" $t(\cdot)$ values [20].

Definition 1. *Let $G = (V, E)$ be a graph with threshold function $t : V \rightarrow \mathbb{N}_0$ and $S \subseteq V$. An* activation process *in G starting at S is a sequence of vertex subsets*[2] $\mathsf{Active}_G[S, 0] \subseteq \mathsf{Active}_G[S, 1] \subseteq \ldots \subseteq \mathsf{Active}_G[S, \ell] \subseteq \ldots \subseteq V$ *of vertex subsets, with* $\mathsf{Active}_G[S, 0] = S$ *and*

$$\mathsf{Active}_G[S, \ell] = \mathsf{Active}_G[S, \ell - 1] \cup \Big\{u : \big|\Gamma_G(u) \cap \mathsf{Active}_G[S, \ell - 1]\big| \geq t(u)\Big\}, \; for\, \ell \geq 1.$$

A ***target set*** *for G is set $S \subseteq V$ such that* $\mathsf{Active}_G[S, \lambda] = V$ *for some* $\lambda \geq 0$

In words, at each round ℓ the set of active nodes is augmented by the set of nodes u that have a number of *already* activated neighbors greater or equal to u's threshold $t(u)$. The vertex v is said to be *activated* at round $\ell > 0$ if $v \in \mathsf{Active}[S, \ell] \setminus \mathsf{Active}[S, \ell - 1]$. The problem we study in this paper is defined as follows:

TARGET SET SELECTION (TSS).
Instance: A network $G = (V, E)$, thresholds $t : V \rightarrow \mathbb{N}_0$.
Problem: Find a target set $S \subseteq V$ of *minimum* size for G.

2.1 Related Work

The Target Set Selection Problem has roots in the general study of the *spread of influence* in Social Networks (see [7,18] and references quoted therein). For

[2] In the rest of the paper we will omit the subscript G whenever the graph G is clear from the context.

instance, in the area of viral marketing [17], companies wanting to promote products or behaviors might initially try to target and convince a few individuals who, by word-of-mouth, can trigger a cascade of influence in the network leading to an adoption of the products by a much larger number of individuals.

The first authors to study problems of spread of influence in networks from an algorithmic point of view were Kempe *et al.* [21,22]. However, they were mostly interested in networks with randomly chosen thresholds. Chen [6] studied the following minimization problem: Given a graph G and fixed arbitrary thresholds $t(v)$, $\forall v \in V$, find a target set of minimum size that eventually activates all (or a fixed fraction of) nodes of G. He proved a strong inapproximability result that makes unlikely the existence of an algorithm with approximation factor better than $O(2^{\log^{1-\epsilon}|V|})$. Chen's result stimulated a series of papers [1–3,5, 8–10,12–14,19,27,28,34] that isolated interesting cases in which the problem (and variants thereof) become tractable. A notable absence from the literature on the topic (with the exception of [16,30]) are heuristics for the Target Set Selection Problem that work for *general graphs*. This is probably due to the previously quoted strong inapproximability result of Chen [6], that seems to suggest that the problem is hopeless. Providing such an algorithm for general graphs, evaluating its performances and experimentally validating it on real-life networks, is the main objective of this paper.

2.2 Our Results

We present a fast and simple algorithm that exhibits the following features: **(1)** It always produces an optimal solution (i.e., a minimum size subset of nodes that influence the whole network) in case G is either a tree, a cycle, or a complete graph. These results were previously obtained in [6,27] by means of *different ad-hoc* algorithms. **(2)** For general networks, it always produces a solution S of cardinality $|S| \leq \sum_{v \in V} \min\left(1, \frac{t(v)}{d(v)+1}\right)$, matching the upper bound given in [1], and proved therein by means of the probabilistic method. **(3)** In real-life networks it produces solutions that outperform the ones obtained using the algorithms presented in the papers [16,30], for which, however, no proof of optimality or performance guarantee is known in any class of graphs. The data sets we use, to experimentally validate our algorithm, include those considered in [16,30].

It is worthwhile to remark that our algorithm, when executed on a graph G for which the thresholds $t(v)$ have been set equal to the nodes degree $d(v)$, for each $v \in V$, it outputs a *vertex cover* of G, (since in that particular case a target set of G is, indeed, a vertex cover of G). Therefore, our algorithm appears to be a new algorithm, to the best of our knowledge, to compute the vertex cover of graphs (notice that our algorithm differs from the classical algorithm that computes a vertex cover by iteratively deleting a vertex of maximum degree in the graph). We plan to investigate elsewhere the theoretical performances of our algorithm (i.e., its approximation factor); computational experiments suggest that it performs surprisingly well in practice.

3 The TSS Algorithm

In this section we present our algorithm for the TSS problem The algorithm, given in Fig. 1, works by iteratively deleting vertices from the input graph G. At each iteration, the vertex to be deleted is chosen as to maximize a certain function. During the deletion process, some vertex v in the surviving graph may remain with less neighbors than its threshold; in such a case v is added to the target set and deleted from the graph while its neighbors' thresholds are decreased by 1 (since they receive v's influence). It can also happen that the surviving graph contains a vertex v whose threshold has been decreased down to 0 (which means that the deleted nodes are able to activate v); in such a case v

---

Algorithm TSS(G)
Input: A graph $G = (V, E)$ with thresholds $t(v)$ for $v \in V$.

$S = \emptyset$
$U = V$
for each $v \in V$ **do**
    $\delta(v) = d(v)$
    $k(v) = t(v)$
    $N(v) = \Gamma(v)$
while $U \neq \emptyset$ **do**
    [Select one vertex and eliminate it from the graph as specified in the following cases]
    if there exists $v \in U$ s.t. $k(v) = 0$ **then**
        [Case 1: The vertex v is activated by the influence of its neighbors in $V - U$ only; it can then influence its neighbors in U]
        for each $u \in N(v)$ **do** $k(u) = \max(k(u) - 1, 0)$
    else
        if there exists $v \in U$ s.t. $\delta(v) < k(v)$ **then**
            [Case 2: The vertex v is added to S, since no sufficient neighbors remain in U to activate it; v can then influence its neighbors in U]
            $S = S \cup \{v\}$
            for each $u \in N(v)$ **do** $k(u) = k(u) - 1$
        else
            [Case 3: The vertex v will be influenced by some of its neighbors in U]
            $v = \texttt{argmax}_{u \in U} \left\{ \frac{k(u)}{\delta(u)(\delta(u)+1)} \right\}$
    [Remove the selected vertex v from the graph]
    for each $u \in N(v)$ **do**
        $\delta(u) = \delta(u) - 1$
        $N(u) = N(u) - \{v\}$
    $U = U - \{v\}$

Fig. 1. Pseudocode of the TSS algorithm.

is deleted from the graph and its neighbors' thresholds are decreased by 1 (since once v activates, they will receive v's influence).

In the rest of the paper, we use the following notation. We denote by n the number of nodes in G, that is, $n = |V|$. Moreover we denote:

- By v_i the vertex that is selected during the $n-i+1$-th iteration of the while loop in TSS(G), for $i = n, \ldots, 1$;
- by $G(i)$ the graph induced by $V_i = \{v_i, \ldots, v_1\}$
- by $\delta_i(v)$ the value of $\delta(v)$ as updated at the beginning of the $(n-i+1)-th$ iteration of the while loop in TSS(G).
- by $N_i(v)$ the set $N(v)$ as updated at the beginning of the $(n-i+1)-th$ iteration of the while loop in TSS(G), and
- by $k_i(v)$ the value of $k(v)$ as updated at the beginning of the $(n-i+1)-th$ iteration of the while loop in TSS(G).

For the initial value $i = n$, the above values are those of the input graph G, that is: $G(n) = G$, $\delta_n(v) = d(v)$, $N_n(v) = \Gamma(v)$, $k_n(v) = t(v)$, for each vertex v of G.

We start with the following two technical Lemmata.

Lemma 1. *Consider a graph G. For any $i = n, \ldots, 1$ and $u \in V_i$, it holds that*

$$\Gamma_{G(i)}(u) = N_i(u) \quad \text{and} \quad d_{G(i)}(u) = \delta_i(u). \tag{1}$$

Proof. For $i = n$ we have $d_{G(n)}(u) = d_G(u) = \delta_n(u)$ and $\Gamma_{G(n)}(u) = \Gamma_G(u) = N_n(u)$ for any $u \in V_n = V$.
Suppose now that the equalities hold for some $i \leq n$. The graph $G(i-1)$ corresponds to the subgraph of $G(i)$ induced by $V_{i-1} = V_i - \{v_i\}$. Hence

$$\Gamma_{G(i-1)}(u) = \Gamma_{G(i)}(u) - \{v_i\}, \quad d_{G(i-1)}(u) = \begin{cases} d_{G(i)}(u) - 1 & \text{if } u \in \Gamma_{G(i)}(v_i), \\ d_{G(i)}(u) & \text{otherwise.} \end{cases}$$

We deduce that the desired equalities hold for $i-1$ by noticing that the algorithm uses the same rules to get

$$N_{i-1}(u) = N_i(u) - \{v_i\}, \quad \delta_{i-1}(u) = \begin{cases} \delta_i(u) - 1 & \text{if } u \in N_i(v_i) = \Gamma_{G(i)}(v_i), \\ \delta_i(u) & \text{otherwise.} \end{cases}$$ □

Lemma 2. *For any $i > 1$, if $S^{(i-1)}$ is a target set for $G(i-1)$ with thresholds $k_{i-1}(u)$, for $u \in V_{i-1}$, then*

$$S^{(i)} = \begin{cases} S^{(i-1)} \cup \{v_i\} & \text{if } k_i(v_i) > \delta_i(v_i) \\ S^{(i-1)} & \text{otherwise} \end{cases} \tag{2}$$

is a target set for $G(i)$ with thresholds $k_i(u)$, for $u \in V_i$.

Proof. Let us first notice that, according to the algorithm TSS, for each $u \in V_{i-1}$ we have

$$k_{i-1}(u) = \begin{cases} \max(k_i(u)-1, 0) & \text{if } u \in N_i(v_i) \text{ and } (k_i(v_i) = 0 \text{ or } k_i(v_i) > \delta_i(v_i)) \\ k_i(u) & \text{otherwise.} \end{cases} \tag{3}$$

(1) If $k_i(v_i) = 0$, then $v_i \in \mathsf{Active}_{G(i)}[S^{(i)}, 1]$ whatever $S^{(i)} \subseteq V_i - \{v_i\}$. Hence, by the (3) any target set $S^{(i-1)}$ for $G(i-1)$ is also a target set for $G(i)$.

(2) If $k_i(v_i) > \delta_i(v_i)$ then $S^{(i)} = S^{(i-1)} \cup \{v_i\}$ and $k_{i-1}(u) = k_i(u) - 1$ for each $u \in N_i(v_i)$. It follows that for any $\ell \geq 0$,

$$\mathsf{Active}_{G(i)}[S^{(i-1)} \cup \{v_i\}, \ell] - \{v_i\} = \mathsf{Active}_{G(i-1)}[S^{(i-1)}, \ell].$$

Hence, $\mathsf{Active}_{G(i)}[S^{(i)}, \ell] = \mathsf{Active}_{G(i-1)}[S^{(i-1)}, \ell] \cup \{v_i\}$.

(3) Let now $1 \leq k_i(v_i) \leq \delta_i(v_i)$. We have that $k_{i-1}(u) = k_i(u)$ for each $u \in V_{i-1}$. If $S^{(i-1)}$ is a target set for $G(i-1)$, by definition there exists an integer λ such that $\mathsf{Active}_{G(i-1)}[S^{(i-1)}, \lambda] = V_{i-1}$. We then have $V_{i-1} \subseteq \mathsf{Active}_{G(i)}[S^{(i-1)}, \lambda]$ which implies $\mathsf{Active}_{G(i)}[S^{(i-1)}, \lambda + 1] = V_i$. □

We can now prove the main result of this section.

Theorem 1. *For any graph G and threshold function t, the algorithm TSS(G) outputs a target set for G.*

Proof. Let S be the output of the algorithm $TSS(G)$. We show that for each $i = 1, \ldots, n$ the set $S \cap \{v_i, \ldots, v_1\}$ is a target set for the graph $G(i)$, assuming that each vertex u in $G(i)$ has threshold $k_i(u)$. The proof is by induction on the number i of nodes of $G(i)$.
If $i = 1$ then the unique vertex v_1 in $G(1)$ either has threshold $k_1(v_1) = 0$ and $S \cap \{v_1\} = \emptyset$ or the vertex has positive threshold $k_1(v_1) > \delta_1(v_1) = 0$ and $S \cap \{v_1\} = \{v_1\}$.

Consider now $i > 1$ and suppose the algorithm be correct on $G(i-1)$, that is, $S \cap \{v_{i-1}, \ldots, v_1\}$ is a target set for $G(i-1)$ with threshold function k_{i-1}. We notice that in each among Cases 1, 2 and 3, the algorithm updates the thresholds and the target set according to Lemma 2. Hence, the algorithm is correct on $G(i)$ with threshold function k_i. The theorem follows since $G(n) = G$. □

It is possible to see that the TSS algorithm can be implemented so to run in $O(|E| \log |V|)$ time. Indeed we need to process the nodes $v \in V$ according to the metric $t(v)/(d(v)(d(v)+1))$, and the updates that follow each processed node $v \in V$ involve at most $d(v)$ neighbors v.

4 Estimating the Size of the Solution

In this section we prove an upper bound on the size of the target set obtained by the algorithm TSS(G) for any input graph G. Our bound, given in Theorem 2, matches the bound given in [1]. However, the result in [1] is based on the probabilistic method and an effective algorithm results only by applying suitable derandomization steps.

Theorem 2. *For any G, the algorithm TSS(G) outputs a target set S of size*

$$|S| \leq \sum_{v \in V} \min\left(1, \frac{t(v)}{d(v)+1}\right). \tag{4}$$

Proof. Let $W(G(i)) = \sum_{j=1}^{i} \min\left(1, \frac{k_i(v_j)}{\delta_i(v_j)+1}\right)$. We prove by induction on i that

$$|S \cap \{v_i, \ldots, v_1\}| \leq W(G(i)). \tag{5}$$

The bound (4) on S follows recalling that $G(n) = G$. If $i = 1$ we have $|S \cap \{v_1\}| = \min\left(1, \frac{k_1(v_1)}{\delta_1(v_1)+1}\right) = W(G(1))$. Assume now (5) holds for $i - 1 \geq 1$, and consider $G(i)$ and the node v_i. We have

$$|S \cap \{v_i, \ldots, v_1\}| = |S \cap \{v_i\}| + |S \cap \{v_{i-1}, \ldots, v_1\}| \leq |S \cap \{v_i\}| + W(G(i-1)).$$

We show that $W(G(i)) \geq W(G(i-1)) + |S \cap \{v_i\}|$. Recalling that $N_i(v_i)$ denotes the neighborhood of v_i in $G(i)$, we have

$$\begin{aligned}
&W(G(i)) - W(G(i-1)) \\
&= \sum_{j=1}^{i} \min\left(1, \frac{k_i(v_j)}{\delta_i(v_j)+1}\right) - \sum_{j=1}^{i-1} \min\left(1, \frac{k_{i-1}(v_j)}{\delta_{i-1}(v_j)+1}\right) \\
&= \min\left(1, \frac{k_i(v_i)}{\delta_i(v_i)+1}\right) + \sum_{v \in N_i(v_i)} \left[\min\left(1, \frac{k_i(v)}{\delta_i(v)+1}\right) - \min\left(1, \frac{k_{i-1}(v)}{\delta_{i-1}(v)+1}\right)\right]
\end{aligned}$$

Therefore, we get

$$\begin{aligned}
&W(G(i)) - W(G(i-1)) \\
&\qquad = \min\left(1, \frac{k_i(v_i)}{\delta_i(v_i)+1}\right) + \sum_{\substack{v \in N_i(v_i) \\ k_i(v) \leq \delta_i(v)}} \left[\frac{k_i(v)}{\delta_i(v)+1} - \frac{k_{i-1}(v)}{\delta_{i-1}(v)+1}\right]
\end{aligned} \tag{6}$$

We distinguish three cases according to the cases in the algorithm TSS(G).

- Suppose that Case 1 of the Algorithm TSS holds; i.e. $k_i(v_i) = 0$. By (6),

$$\begin{aligned}
W(G(i)) - W(G(i-1)) &= \sum_{\substack{v \in N_i(v_i) \\ k_i(v) \leq \delta_i(v)}} \left[\frac{k_i(v)}{\delta_i(v)+1} - \frac{k_i(v)-1}{\delta_i(v)}\right] \\
&\geq 0 = |S \cap \{v_i\}|.
\end{aligned}$$

- Suppose that Case 2 of the algorithm holds; i.e. $k_i(v_i) \geq \delta_i(v_i)+1$. By (6),

$$\begin{aligned}
W(G(i)) - W(G(i-1)) &= 1 + \sum_{\substack{v \in N_i(v_i) \\ k_i(v) \leq \delta_i(v)}} \left[\frac{k_i(v)}{\delta_i(v)+1} - \frac{k_i(v)-1}{\delta_i(v)}\right] \\
&\geq 1 = |S \cap \{v_i\}|.
\end{aligned}$$

- Suppose that Case 3 holds; i.e. $k_i(v_i) \leq \delta_i(v_i)$. In such a case we know that

$$\frac{k_i(v)}{\delta_i(v)(\delta_i(v)+1)} \leq \frac{k_i(v_i)}{\delta_i(v_i)(\delta_i(v_i)+1)}$$

for each $v \in \{v_i, \ldots, v_1\}$ and $S \cap \{v_i\} = \emptyset$. By this and (6) we get

$$
\begin{aligned}
W(G(i)) - W(G(i-1)) &= \frac{k_i(v_i)}{\delta_i(v_i)+1} + \sum_{\substack{v \in N_i(v_i)\\ k_i(v) \le \delta_i(v)}} \left[\frac{k_i(v)}{\delta_i(v)+1} - \frac{k_i(v)}{\delta_i(v)}\right] \\
&= \frac{k_i(v_i)}{\delta_i(v_i)+1} - \sum_{\substack{v \in N_i(v_i)\\ k_i(v) \le \delta_i(v)}} \frac{k_i(v)}{\delta_i(v)(\delta_i(v)+1)} \\
&\ge \frac{k_i(v_i)}{\delta_i(v_i)+1} - \frac{k_i(v_i)}{\delta_i(v_i)+1} = 0 = |S \cap \{v_i\}|. \qquad \square
\end{aligned}
$$

5 Proofs of Optimality

In this section, we prove that our algorithm TSS provides a unified setting for several results, obtained in the literature by means of different *ad hoc* algorithms. Trees, cycles and cliques are among the few cases known to admit optimal polynomial time algorithms for the TSS problem [6,27]. In the following, we prove that our algorithm TSS provides the *first* unifying setting for all these cases.

Theorem 3. *The algorithm TSS(T) returns an optimal solution for any tree T.*

Proof. Let $T = (V, E)$ and $n = |V|$. We recall that for $i = 1, \ldots, n$: v_i denotes the node selected during the $n-i+1$-th iteration of the while loop in TSS, $T(i)$ is the forest induced by the set $V_i = \{v_i, \ldots, v_1\}$, and $\delta_i(v)$ and $k_i(v)$ are the degree and threshold of v, for $v \in V_i$. Let S be the target set produced by the algorithm TSS(T). We prove by induction on i that

$$|S \cap \{v_i, \ldots, v_1\}| = |S_i^*|, \tag{7}$$

where S_i^* represents an optimal target set for the forest $T(i)$ with threshold function k_i. For $i = 1$, it is immediate that for the only node v_1 in $F(1)$ one has

$$S \cap \{v_1\} = S_1^* = \begin{cases} \emptyset & \text{if } k_1(v_1) = 0 \\ \{v_1\} & \text{otherwise.} \end{cases}$$

Suppose now (7) true for $i-1$ and consider $T(i)$ and the selected node v_i.

1. Assume $k_i(v_i) = 0$. We get $|S \cap \{v_i, \ldots, v_1\}| = |S \cap \{v_{i-1}, \ldots, v_1\}| = |S_{i-1}^*| \le |S_i^*|$ and the equality (7) holds for i.
2. Assume $k_i(v_i) \ge \delta_i(v_i) + 1$. Clearly, any solution for $T(i)$ must include node v_i, otherwise it cannot be activated. This implies that

$$|S_i^*| = 1 + |S_{i-1}^*| = 1 + |S \cap \{v_{i-1}, \ldots, v_1\}| = |S \cap \{v_i, \ldots, v_1\}|$$

and (7) holds for i.

3. Suppose now that $v_i = \mathbf{argmax}_{i \geq j \geq 1} \{k_i(v_j)/(\delta_i(v_j)(\delta_i(v_j)+1))\}$. In this case each leaf v_j in $T(i)$ has

$$\frac{k_i(v_\ell)}{\delta_i(v_\ell)(\delta_i(v_\ell)+1)} = \frac{1}{2}$$

while each internal node v_ℓ has

$$\frac{k_i(v_\ell)}{\delta_i(v_\ell)(\delta_i(v_\ell)+1)} \leq \frac{1}{\delta_i(v_\ell)+1} \leq \frac{1}{3}.$$

Hence the selected node is a leaf in $T(i)$ and has $k_i(v_i) = \delta_i(v_i) = 1$. Hence $|S \cap \{v_i, \ldots, v_1\}| = |S \cap \{v_{i-1}, \ldots, v_1\}| = |S^*_{i-1}| \leq |S^*_i|$. □

Theorem 4. *The algorithm TSS(C) outputs an optimal solution if C is a cycle.*

Proof. If the first selected node v_n has threshold 0 then clearly $v_n \notin S^*$ for any optimal solution S^*.
If the threshold of v_n is larger than its degree then clearly $v_n \in S^*$ for any optimal solution S^*. In both cases $v_n \in \mathsf{Active}[S^*, 1]$ and its neighbors can use v_n's influence; that is, the algorithm correctly sets $k_{n-1} = \max(k_n - 1, 0)$ for these two nodes.

If threshold of each node $v \in V$ is $1 \leq t(v) \leq d(v)$, we get that during the first iteration of the algorithm TSS(C), the selected node v_n satisfies Case 3 and has $t(v_n) = 2$ if at least one of the nodes in C has threshold 2, otherwise $t(v_n) = 1$. Moreover, it is not difficult to see that there exists an optimal solution S^* for C such that $S^* \cap \{v_n\} = \emptyset$.

In each case, the result follows by Theorem 3, since the remaining graph is a path on nodes $v_{n-1}, \ldots, v_1$. □

Theorem 5. *Let* $K = (V, E)$ *be a clique with* $V = \{u_1, \ldots, u_n\}$ *and* $t(u_1) \leq \ldots \leq t(u_{n-m}) < n \leq t(u_{n-m+1}) \leq \ldots \leq t(u_n)$. *The algorithm TSS(K) outputs an optimal target set of size*

$$m + \max_{1 \leq j \leq n-m} \max(t(u_j) - m - j + 1, 0). \tag{8}$$

Proof. It is well known that there exists an optimal target set S^* consisting of the $|S^*|$ nodes of higher threshold [27]. Being S^* a target set, each node u_j must activate, that is, $u_j \in \mathsf{Active}[S, i]$ for some $i \geq 0$. Assume $V = \{u_1, \ldots, u_n\}$ and $t(u_1) \leq \ldots \leq t(u_{n-m}) < n \leq t(u_{n-m+1}) \leq \ldots \leq t(u_n)$. Since the thresholds are non decreasing with the node index, it follows that:

- for each of the m nodes s.t. $t(u_j) \geq n$, it must hold $u_j \in S^*$, hence $|S^*| \geq m$;
- for each $j \leq n - |S^*|$, the node u_j activates if it gets, in addition to the influence of its m neighbors with threshold larger than $n-1$, the influence of $t(u_j) - m$ other neighbors, hence we have that $t(u_j) - m \leq j - 1 + (|S^*| - m)$ must hold;
- if $n - |S^*| + 1 \leq j \leq n - m$, then

$$t(u_j) - m - j + 1 \leq (n-1) - m - (n - |S^*| + 1) + 1 = |S^*| - m + 1.$$

Summarizing, we get,

$$|S^*| \geq m + \max_{1 \leq j \leq n-m} \max\left(t(u_j) - m - j + 1, 0\right).$$

We show now that the algorithm TSS outputs a target set S whose size is upper bounded by the value in (8). In general, the output S does not consist of the nodes having the highest thresholds.
Consider the residual graph $K(i) = (V_i, E_i)$, for some $1 \leq i \leq n$. It is easy to see that for any $u_j, u_s \in V_i$ it holds

(1) $\delta_i(u_j) = i$;
(2) if $j < s$ then $k_i(u_j) \leq k_i(u_s)$;
(3) if $t(u_j) \geq n$ then $k_i(u_j) \geq i$,
(4) if $t(u_j) < n$ then $k_i(u_j) \leq i$.

W.l.o.g. we assume that at any iteration of algorithm TSS if the node to be selected is not unique then the tie is broken as follows (cfr. point (2) above):

(i) If Case 1 holds then the selected node is the one with the lowest index,
(ii) otherwise the selected node is the one with the largest index.

Clearly, this implies that $K(i)$ contains i nodes with consecutive indices among $u_1, \ldots, u_n$, that is,

$$V_i = \{u_{\ell_i}, u_{\ell_i+1}, \ldots, u_{r_i}\} \tag{9}$$

for some $\ell_i \geq 1$ and $r_i = \ell_i + i - 1$.

Let $h = n - m$. We shall prove by induction on i that, for each $i = n, \ldots, 1$, at the beginning of the $n - i + 1$-th iteration of the while loop in TSS(K), it holds

$$|S \cap V_i| \leq \begin{cases} (r_i - h) + \max_{\ell_i \leq j \leq h} \max(k_i(u_j) - (r_i - h) - j + \ell_i,\ 0) & \text{if } r_i > h, \\ \max_{\ell_i \leq j \leq r_i} \max(k_i(u_j) - j + \ell_i,\ 0) & \text{if } r_i \leq h. \end{cases} \tag{10}$$

The upper bound (8) follows when $i = n$; indeed $K(n) = K$ and $|S| = |S \cap V(n)|$.
For $i = 1$, $K(1)$ is induced by only one node, let say u, and

$$|S \cap \{u\}| = \begin{cases} 1 & \text{if } k_1(u) \geq 1, \\ 0 & \text{if } k_1(u) = 0. \end{cases}$$

proving that the bound holds in this case.
Suppose now (10) true for some $i - 1 \geq 1$ and consider the $n - i + 1$-th iteration of the algorithm TSS. Let v be the node selected by algorithm TSS at the $n - i + 1$-th iteration. We distinguish three cases according to the cases of the algorithm TSS(G).

Case 1: $k_i(v) = 0$. By (i) and (9), one has $v = u_{\ell_i}, \ell_{i-1} = \ell_i + 1$ and $r_{i-1} = r_i$. Moreover, $k_i(u_j) = k_{i-1}(u_j) + 1$ for each $u_j \in V_{i-1}$. Hence,

$$
\begin{aligned}
|S \cap V_i| &= |S \cap V_{i-1}| \\
&\leq \begin{cases} (r_i - h) + \max_{\ell_i+1 \leq j \leq h} \max(k_{i-1}(u_j) - (r_i - h) - j + \ell_i + 1,\, 0) & \text{if } r_i > h, \\ \max_{\ell+1 \leq j \leq r} \max(k_{i-1}(u_j) - j + \ell + 1,\, 0) & \text{if } r_i \leq h, \end{cases} \\
&= \begin{cases} (r_i - h) + \max_{\ell_i \leq j \leq h} \max(k_i(u_j) - (r_i - h) - j + \ell_i, 0) & \text{if } r_i > h, \\ \max_{\ell \leq j \leq r} \max(k_i(u_j) - j + \ell, 0) & \text{if } r \leq h. \end{cases}
\end{aligned}
$$

Case 2: $k_i(v) > \delta_i(v)$. By (ii) and (9) we have $v = u_{r_i}$, $\ell_i = \ell_{i-1}$, $r_{i-1} = r_i - 1$. Moreover, $k_i(u_j) = k_{i-1}(u_j) + 1$ for each $u_j \in V_{i-1}$. Recalling relations (3) and (4), we have

$$
\begin{aligned}
|S \cap V_i| &= 1 + |S \cap V_{i-1}| \\
&\leq 1 + \begin{cases} (r_{i-1} - h) + \max_{\ell_{i-1} \leq j \leq h} \max(k_{i-1}(u_j) - (r_{i-1} - h) - j + \ell_{i-1},\, 0) \\ \qquad \text{if } r_{i-1} > h, \\ \max_{\ell_{i-1} \leq j \leq r_{i-1}} \max(k_{i-1}(u_j) - j + \ell_{i-1},\, 0) \\ \qquad \text{if } r_{i-1} \leq h, \end{cases} \\
&= \begin{cases} (r_i - h) + \max_{\ell_i \leq j \leq h} \max(k_{i-1}(u_j) + 1 - (r_i - h) - j + \ell_i,\, 0) \\ \qquad \text{if } r_i - 1 > h, \\ \max_{\ell \leq j \leq r_i - 1} \max(k_{i-1}(u_j) + 1 - j + \ell_i,\, 1) \\ \qquad \text{if } r_i - 1 \leq h. \end{cases} \\
&= \begin{cases} (r_i - h) + \max\{0, \max_{\ell_i \leq j \leq h} k_i(u_j) - (r_i - h) - j + \ell_i\} & \text{if } r_i > h, \\ \max\{0, \max_{\ell_i \leq j \leq r_i} k_i(u_j) - j + \ell_i\} & \text{if } r_i \leq h. \end{cases}
\end{aligned}
$$

Case 3: $0 < k_i(v) \leq \delta_i(v)$. By (ii) and (9) we have $v = u_{r_i}$, $\ell_i = \ell_{i-1}$, $r_{i-1} = r_i - 1$. Moreover, $k_i(u_j) = k_{i-1}(u_j)$ for each $u_j \in V_{i-1}$. Recalling that by (3) and (4) we have $t(u_r) < n$, which implies $r_i \leq h$, we have

$$
\begin{aligned}
|S \cap V_i| = |S \cap V_{i-1}| &\leq \max_{\ell_{i-1} \leq j \leq r_{i-1}} \max(k_{i-1}(u_j) - j + \ell_{i-1},\, 0) \\
&\leq \max_{\ell_i \leq j \leq r_i - 1} \max(k_i(u_j) - j + \ell_i,\, 0) \\
&\leq \max_{\ell_i \leq j \leq r_i} \max(k_i(u_j) - j + \ell_i,\, 0).
\end{aligned}
$$
□

6 Computational Experiments

We have extensively tested our algorithm TSS(G) both on random graphs and on real-world data sets, and we found that our algorithm performs surprisingly well in practice. This seems to suggest that the otherwise important inapproximability result of Chen [6] refers to rare or artificial cases.

6.1 Random Graphs

The first set of tests was done in order to compare the results of our algorithm to the exact solutions, found by formulating the problem as an 0–1 Integer Linear Programming (ILP) problem. Although the ILP approach provides the optimal solution, it fails to return the solution in a reasonable time (i.e., days) already for moderate size networks. We applied both our algorithm and the ILP algorithm to random graphs with up to 50 nodes. Our algorithm produced target sets of size close to the optimal; for several instances it found an optimal solution (Fig. 2).

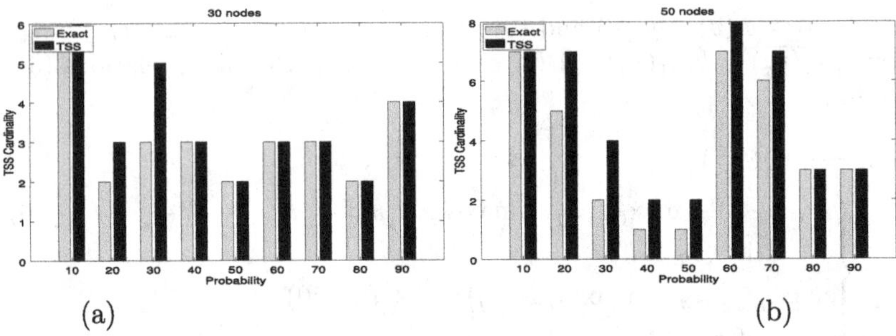

Fig. 2. Experiments for random graphs $G(n, p)$ on n nodes (any possible edge occurs independently with probability $0 < p < 1$). (a) $n = 30$, (b) $n = 50$ with $p \in \{10/100, 20/100, \ldots, 90/100\}$. For each node the threshold was fixed to a random value between 1 and the node degree.

6.2 Large Real-Life Networks

We performed experiments on several real social networks of various sizes from the Stanford Large Network Data set Collection (SNAP) [25] and the Social Computing Data Repository at Arizona State University [33]. The data sets we considered include both networks for which small target sets exist and networks needing larger target sets (due to the existence of communities, i.e., tightly connected disjoint groups of nodes that appear to delay the diffusion process). We compare the performance of our algorithm TSS toward that of the best, to our knowledge, computationally feasible algorithms in the literature. Namely, we compare to Algorithm *TIP_DECOMP* recently presented in [30], in which nodes minimizing the difference between degree and threshold are pruned from the graph until a "core" set is produced. We also compare our algorithm to the *VirAds* algorithm presented in [16]. Finally, we compare to an (enhanced) *Greedy* strategy, in which nodes of maximum degree are iteratively inserted in the target set and pruned from the graph. Nodes that remains with zero threshold are simply eliminated from the graph, until no node remains All test results consistently show that the TSS algorithm we introduce in this paper

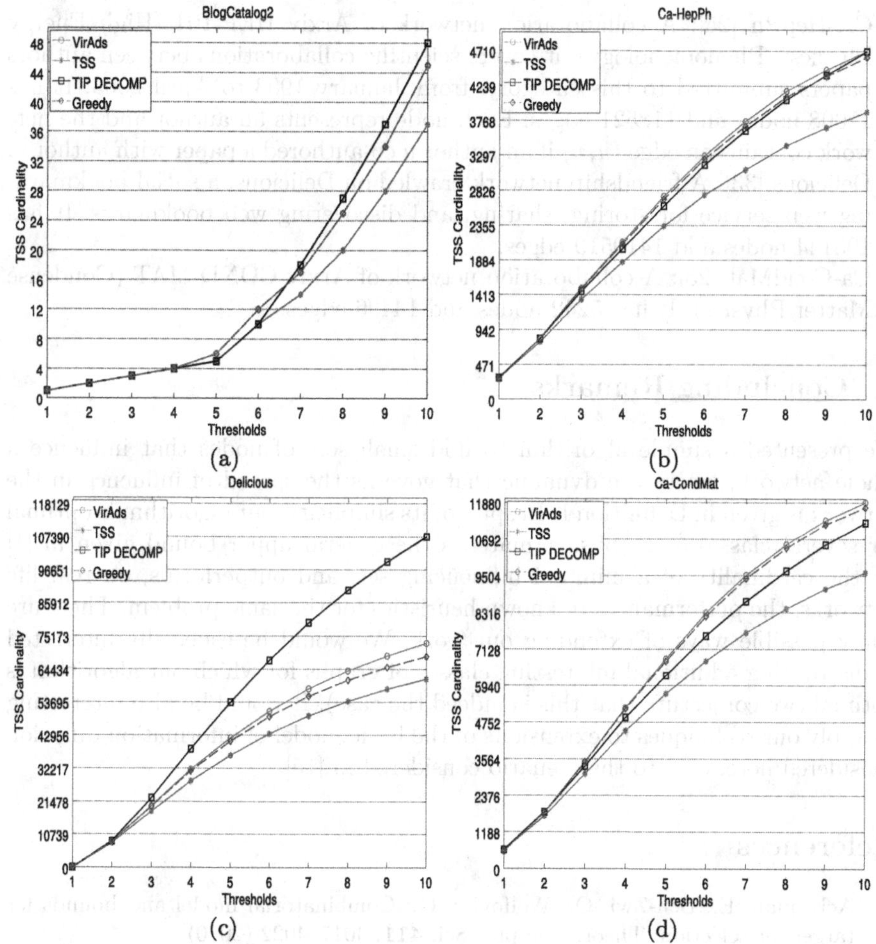

Fig. 3. Target set size in function of node thresholds. Following the scenario considered in [30], the threshold are kept constant among all nodes and set to an integer in the interval [1,10]; hence, comparisons are reported for this scenario. The datasets are (a) BlogCatalog2, (b) CA-HepPh, (c) Delicious, (d) Ca-CondMat.

presents the best performances on all the considered data sets, while none among TIP_DECOMP, VirAds, and Greedy is always better than the other two (Fig. 3).

The following data set were used in our testing. Additional experimental data will be given in the journal version of the paper.

- BlogCatalog2 [33]: a friendship network crawled from BlogCatalog. a social blog directory website which manages the bloggers and their blogs. It has 97,884 nodes and 2,043,701 edges. Each node represents a blogger and the network contains an edge (u, v) if blogger u is friend of blogger v.

- CA-HepPh [25]: A collaboration network of Arxiv HEP-PH (High Energy Physics - Phenomenology), it covers scientific collaborations between authors papers submitted to this category, from January 1993 to April 2003. It has 12008 nodes and 118521 edges. Each node represents an author and the network contains an edge (u, v) if an author u co-authored a paper with author v.
- Delicious [33]: A friendship network crawled on Delicious, a social bookmarking web service for storing, sharing, and discovering web bookmarks. It has 103144 nodes and 1419519 edges.
- Ca-CondMat [25]: A collaboration network of Arxiv COND-MAT (Condense Matter Physics). It has 5242 nodes and 14496 edges.

7 Concluding Remarks

We presented a simple algorithm to find small sets of nodes that influence a whole network, where the dynamic that governs the spread of influence in the network is given in Definition 1. In spite of its simplicity, our algorithm is optimal for several classes of graphs, it matches the general upper bound given in [1] on the cardinality of a minimal influencing set, and outperforms, on real life networks, the performances of known heuristics for the same problem. There are many possible ways of extending our work. We would be especially interested in discovering additional interesting classes of graphs for which our algorithm is optimal (we conjecture that this is indeed the case). It would be also interesting to apply our techniques to extensions of the basic model of information diffusion considered here, e.g., to the scenario considered in [15].

References

1. Ackerman, E., Ben-Zwi, O., Wolfovitz, G.: Combinatorial model and bounds for target set selection. Theor. Comput. Sci. **411**, 4017–4022 (2010)
2. Bazgan, C., Chopin, M., Nichterlein, A., Sikora, F.: Parameterized approximability of maximizing the spread of influence in networks. In: Du, D.-Z., Zhang, G. (eds.) COCOON 2013. LNCS, vol. 7936, pp. 543–554. Springer, Heidelberg (2013)
3. Ben-Zwi, O., Hermelin, D., Lokshtanov, D., Newman, I.: Treewidth governs the complexity of target set selection. Discrete Optim. **8**, 87–96 (2011)
4. Bond, R.M., et al.: A 61-million-person experiment in social influence and political mobilization. Nature **489**, 295–298 (2012)
5. Centeno, C.C., et al.: Irreversible conversion of graphs. Theor. Comput. Sci. **412**(29), 3693–3700 (2011)
6. Chen, N.: On the approximability of influence in social networks. SIAM J. Discrete Math. **23**, 1400–1415 (2009)
7. Chen, W., Lakshmanan, L.V.S., Castillo, C.: Information and Influence Propagation in Social Networks. Morgan & Claypool, San Francisco (2013)
8. Chopin, M., Nichterlein, A., Niedermeier, R., Weller, M.: Constant thresholds can make target set selection tractable. In: Even, G., Rawitz, D. (eds.) MedAlg 2012. LNCS, vol. 7659, pp. 120–133. Springer, Heidelberg (2012)
9. Chiang, C.-Y., Huang, L.-H., Yeh, H.-G.: Target set selection problem for honeycomb networks. SIAM J. Discrete Math. **27**(1), 310–328 (2013)

10. Chiang, C.-Y., Huang, L.-H., Li, B.-J., Wu, J., Yeh, H.-G.: Some results on the target set selection problem. J. Comb. Opt. **25**(4), 702–715 (2013)
11. Christakis, N.A., Fowler, J.H.: Connected: The surprising Power of our Social Networks and how they Shape our Lives. Little, Brown (2011)
12. Cicalese, F., Cordasco, G., Gargano, L., Milanic, M., Vaccaro, U.: Latency-bounded target set selection in social networks. Theoret. Comput. Sci. **535**, 1–15 (2014)
13. Cicalese, F., Cordasco, G., Gargano, L., Milanič, M., Peters, J.G., Vaccaro, U.: Spread of influence in weighted networks under time and budget constraints. Theor. Comput. Sci. **586**, 40–58 (2015)
14. Coja-Oghlan, A., Feige, U., Krivelevich, M., Reichman, D.: Contagious sets in expanders. In: Proceedings of SODA 2015, pp. 1953–1987 (2015)
15. Demaine, E.D., et al.: How to influence people with partial incentives. In: Proceedings of WWW 2014, pp. 937–948 (2014)
16. Dinh, T.N., Zhang, H., Nguyen, D.T., Thai, M.T.: Cost-effective viral marketing for time-critical campaigns in large-scale social networks. IEEE/ACM ToN **22**(6), 2001–2011 (2014)
17. Domingos, P., Richardson, M.: Mining the network value of customers. In: ACM International Conference on Knowledge Discovery and Data Mining, pp. 57–66 (2001)
18. Easley, D., Kleinberg, J.: Networks, Crowds, and Markets: Reasoning About a Highly Connected World. Cambridge University Press, Cambridge (2010)
19. Gargano, L., Hell, P., Peters, J., Vaccaro, U.: Influence diffusion in social networks under time window constraints. Theor. Comput. Sci. **584**, 53–66 (2015)
20. Granovetter, M.: Threshold models of collective behavior. Am. J. Sociol. **83**, 1420–1443 (1978)
21. Kempe, D., Kleinberg, J.M., Tardos, E.: Maximizing the spread of influence through a social network. In: Proceedings of the Ninth ACM SIGKDD, pp. 137–146 (2003)
22. Kempe, D., Kleinberg, J.M., Tardos, É.: Influential nodes in a diffusion model for social networks. In: Caires, L., Italiano, G.F., Monteiro, L., Palamidessi, C., Yung, M. (eds.) ICALP 2005. LNCS, vol. 3580, pp. 1127–1138. Springer, Heidelberg (2005)
23. Lately, D.: An army of eyeballs: the rise of the advertisee. The Baffler, 12 Septmeber 2014
24. Leppaniemi, M., et al.: Targeting young voters in apolitical campaign: empirical insights into an interactive digitalmarketing campaign in the 2007 Finnish general election. J. Nonprofit Public Sect. Mark. **22**, 14–37 (2010)
25. Leskovec, J., Sosič, R.: SNAP: a general purpose network analysis and graph mining library in C++ (2014). http://snap.stanford.edu/snap
26. Leskovic, H., Adamic, L.A., Huberman, B.A.: The dynamic of viral marketing. J. ACM Trans. Web (TWEB) **1**(1), 1–39 (2007). Article No 5
27. Nichterlein, A., Niedermeier, R., Uhlmann, J., Weller, M.: On tractable cases of target set selection. Soc. Netw. Anal. Min. **3**, 233–256 (2012)
28. Reddy, T.V.T., Rangan, C.P.: Variants of spreading messages. J. Graph Algorithms Appl. **15**(5), 683–699 (2011)
29. Rival, J.-B., Walach, J.: The use of viral marketing in politics: a case study of the 2007 French presidential election. Master Thesis, Jönköping University
30. Shakarian, P., Eyre, S., Paulo, D.: A scalable heuristic for viral marketing under the tipping model. Soc. Netw. Anal. Min. **3**, 1225–1248 (2013)
31. Tumulty, K.: Obama's viral marketing campaign. TIME Mag. (2007). http://content.time.com/time/magazine/article/0,9171,1640402,00.html

32. Wasserman, S., Faust, K.: Social Network Analysis: Methods and Applications. Cambridge University Press, Cambridge (1994)
33. Zafarani, R., Liu, H.: Social Computing Data Repository at ASU (2009). http://socialcomputing.asu.edu
34. Zaker, M.: On dynamic monopolies of graphs with general thresholds. Discrete Math. **312**(6), 1136–1143 (2012)
35. Zhang, H., Mishra, S., Thai, M.T.: Recent advances in information diffusion and influence maximization of complex social networks. In: Wu, J., Wang, Y. (eds.) Opportunistic Mobile Social Networks. CRC Press, Taylor & Francis Group (2014, to appear)

An Efficient Shortest-Path Routing Algorithm in the Data Centre Network DPillar

Alejandro Erickson[1(✉)], Abbas Eslami Kiasari[2], Javier Navaridas[2], and Iain A. Stewart[1]

[1] School of Engineering and Computing Sciences, Durham University, Science Labs, South Road, Durham DH1 3LE, UK
{alejandro.erickson,i.a.stewart}@durham.ac.uk

[2] School of Computer Science, University of Manchester, Oxford Road, Manchester M13 9PL, UK
{abbas.kiasari,javier.navaridas}@manchester.ac.uk

Abstract. DPillar has recently been proposed as a server-centric data centre network and is combinatorially related to the well-known wrapped butterfly network. We explain the relationship between DPillar and the wrapped butterfly network before proving a symmetry property of DPillar. We use this symmetry property to establish a single-path routing algorithm for DPillar that computes a shortest path and has time complexity $O(k\log(n))$, where k parameterizes the dimension of DPillar and n the number of ports in its switches. Moreover, our algorithm is trivial to implement, being essentially a conditional clause of numeric tests, and improves significantly upon a routing algorithm earlier employed for DPillar. A secondary and important effect of our work is that it emphasises that data centre networks are amenable to a closer combinatorial scrutiny that can significantly improve their computational efficiency and performance.

Keywords: Data centre networks · Routing algorithms · Shortest paths

1 Introduction

A *data centre network* (*DCN*) is the topology by which the servers, switches and other components of a data centre are interconnected and the choice of DCN strongly influences the data centre's practical performance (see, e.g., [13]). DCNs have traditionally been tree-like and *switch-centric*; that is, so that the servers are located at the 'leaves' of a tree-like structure that is composed entirely of switches and where the interconnection intelligence resides within the switches. Typical examples of such switch-centric DCNs are ElasticTree [9], Fat-Tree [4], VL2 [5], HyperX [3], Portland [14] and Flattened Butterfly [1]. However, it is generally acknowledged that tree-like, switch-centric DCNs have deficiencies when it comes to, for example, scalability with the core switches (at the 'roots' of the tree-like structure) quickly becoming bottlenecks.

Z. Lu et al. (Eds.): COCOA 2015, LNCS 9486, pp. 209–220, 2015.
DOI: 10.1007/978-3-319-26626-8_16

Alternative architectures have recently emerged and *server-centric* DCNs have been proposed whereby the interconnection intelligence resides within the servers as opposed to the switches. Now, switches only operate as dumb crossbars (and so the need for high-end switches is diminished as are the infrastructure costs). This paradigm shift means that more scalable topologies can be designed and the fact that routing resides within servers means that more effective routing algorithms can be adopted. However, packet latency can increase in server-centric DCNs, with the need to handle routing providing a computational overhead on the server. Nevertheless, server-centric data centres are now becoming commercially available. Typical examples of server-centric DCNs are DCell [7], BCube [8], FiConn [10], CamCube [2], MCube [15], DPillar [12], HCN and BCN [6] and SWKautz, SWCube and SWdBruijn [11]. An additional positive aspect of some server-centric DCNs is that not only can commodity switches be used to build the data centres but commodity servers can too: the DCNs FiConn, MCube, DPillar, HCN, BCN, SWKautz, SWCube and SWdBruijn are all such that any server only needs two NIC ports (the norm in commodity servers) in order to incorporate it into the DCN.

It is with the DCN DPillar that we are concerned here. In [12], basic properties of DPillar are demonstrated and single-path and multi-path routing algorithms are developed (along with a forwarding methodology for the latter). Our focus here is on single-path routing. The algorithm in [12] is appealing in its simplicity but for most source-destination pairs it does not produce a path of shortest length. We remedy this situation and develop a single-path routing algorithm that always outputs a shortest path and does so in linear time complexity. What is more, although the proof of correctness of our algorithm is non-trivial, the actual algorithm itself is a very simple sequence of numeric tests and consequently yields no implementation difficulties.

A pervasive theme within our work is that the design and performance of modern data centres can benefit significantly from additional combinatorial and mathematical analysis. Often, when new DCNs are proposed they are done so within a broader context so that the topology is considered as part of a wider and more practically-driven network environment. As such, the analysis is often empirical with a key aim being to demonstrate the practical viability of the DCN taking into account issues relating to, for example, infrastructure costs, traffic patterns, fault tolerance, network protocols and so on. Such presentations are often impressive in holistic terms but unavoidably basic in terms of combinatorial sophistication: the driver of practical viability means that there is a lessened inclination to optimize the various intrinsic components. It is once practical viability has been established that a closer combinatorial scrutiny can lead to improved performance (as we demonstrate here).

2 The DCN DPillar

We abstract the DCN DPillar as an undirected graph whose nodes represent the servers of the DCN DPillar and whose edges represent pairs of servers that are

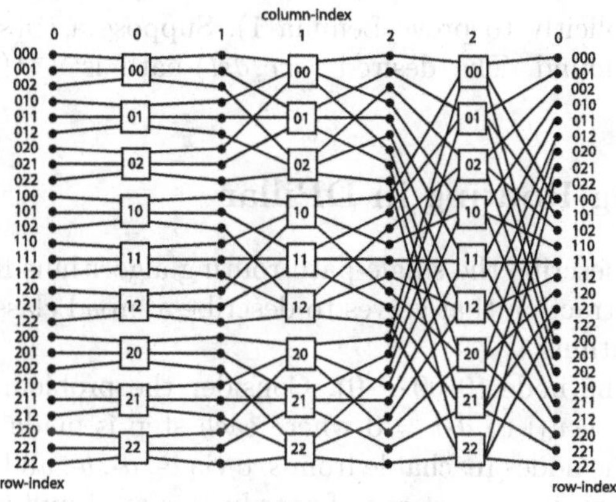

Fig. 1. Visualizing $\text{DPillar}_{6,3}$.

connected to the same switch. Representing a sever-switch-server connection of the DCN DPillar by one edge of the graph serves to reflect the negligible latency overheads encountered in a crossbar switch as compared to the overheads of routing through a server. We call this graph $\text{DPillar}_{n,k}$ when the DCN has *dimension* k and each switch has n-ports, where $2|n$. A node in *column* c with *row-index* $v_{k-1}v_{k-2}\cdots v_0$ is labelled $(c, v_{k-1}v_{k-2}\cdots v_0)$ with $0 \leq c < k$ and $0 \leq v_i < \frac{n}{2}$ for $0 \leq i < k$. $\text{DPillar}_{n,k}$ has 4 types of edges, called *clockwise edges (c-edges)*, *anti-clockwise edges (a-edges)*, *basic static edges (b-edges)*, and *decremented static edges (d-edges)*, which are of the following form:

$$\begin{aligned}
&(c)\text{: } ((c, v_{k-1}\ldots v_{c+1}v_cv_{c-1}\ldots v_0), (c+1, v_{k-1}\ldots v_{c+1} * v_{c-1}\ldots v_0)) \\
&(a)\text{: } ((c, v_{k-1}\ldots v_cv_{c-1}v_{c-2}\ldots v_0), (c-1, v_{k-1}\ldots v_c * v_{c-2}\ldots v_0)) \\
&(b)\text{: } ((c, v_{k-1}\ldots v_{c+1}v_cv_{c-1}\ldots v_0), (c, v_{k-1}\ldots v_{c+1} * v_{c-1}\ldots v_0)) \\
&(d)\text{: } ((c, v_{k-1}\ldots v_cv_{c-1}v_{c-2}\ldots v_0), (c, v_{k-1}\ldots v_c * v_{c-2}\ldots v_0)).
\end{aligned}$$

The c and b edges characterise the servers connected to the same column-c switch as the server represented by $(c, v_{k-1}v_{k-2}\cdots v_0)$, whilst the a and d edges characterise the servers connected to the same column-$(c-1)$ switch as the server represented by $(c, v_{k-1}v_{k-2}\cdots v_0)$. The DCN $\text{DPillar}_{6,3}$ can be visualized as in Fig. 1, where switches are shown and the left-most and right-most columns are actually the same columns, represented twice for clarity, with the graph $\text{DPillar}_{6,3}$ obtained by replacing each switch by a clique of edges on 6 nodes.

We shall rely on symmetry within $\text{DPillar}_{n,k}$; the term is used but not defined in [12], and the literature on DCN design tends to use 'symmetry' somewhat loosely. We omit the (straightforward) proof of the following result due to space constraints. Our notion of symmetry is node-symmetry (i.e., vertex-transitivity).

Lemma 1. *The graph $\text{DPillar}_{n,k}$ is node-symmetric.*

As a result of Lemma 1, the problem of routing from a node *src* to a node *dst*, is equivalent to the problem of routing from node $\phi(src)$ to the node $\phi(dst) = (0, 00\cdots 0)$, where ϕ is an automorphism of $\text{DPillar}_{n,k}$ (such automorphisms may be constructed explicitly to prove Lemma 1). Suppose a $(\phi(src), \phi(dst))$-path $p_0, p_1, \ldots, p_{\ell-1}$ is found. The desired (src, dst)-path is $\phi^{-1}(p_0), \phi^{-1}(p_1), \ldots,$ $\phi^{-1}(p_{\ell-1})$.

3 Abstracting Routing in DPillar

In this section we describe the single-path routing algorithms from [12] in order to motivate an abstraction that serves to describe a broad class of useful single-path routing algorithms.

Let $\mathbf{0}$ denote the node $(0, 00\cdots 0)$. Consider the problem of routing from $src = (c, v_{k-1}v_{k-2}\cdots v_0)$ to $dst = \mathbf{0}$ where each step is made using an edge of type c, a, b, or d. The nodes reachable from src via c-, a-, b-, and d-edges, given in Sect. 2, differ from src in at most one of coordinates $c-1$ and c of the row-index (and no others), and lie in one of columns $c-1, c$, or $c+1$. Any non-zero symbols v_i of the row-index of src must be 'fixed' in one of these steps in order to reach dst, which has row index $00\cdots 0$; v_i can only be fixed by visiting a column of the graph where coordinate i can be changed by one of the edges of type c, a, b, or d. Recall that edges of type c and b outgoing from src, in column c, enable us to change symbol v_c to whichever element of $\{0, 1, \ldots, n/2 - 1\}$ is desired; as such, we say that c- and b-edges *cover* the column they originate in. Edges of type a and d cover column $c-1$; as such, we say that a- and d-edges cover the anti-clockwise neighbouring column. In addition to fixing bits by covering the appropriate columns, the route may need to travel from column to column, via c- and a- edges, possibly without changing the row-index.

One of the routing algorithms detailed in [12] is to travel from src only along c-edges whilst changing a symbol v_i to 0, if necessary, at each step until dst is reached. After at most k steps, every column i with a non-zero coordinate v_i has been covered and fixed and the node $(j, 00\cdots 0)$ is reached, for some j. We then continue to travel along c-edges to increase j until $\mathbf{0}$ is reached and the route is complete. The other single-path routing algorithm of [12] is the anticlockwise analogue, where only a-edges are used. It is very easy to see (by looking at some typical source-destination examples) that this routing algorithm is by no means optimal and that more often than not much shorter paths exist (an upper bound of $2k-1$ on the lengths of paths produced was stated in [12]). For example, if one chooses to route with only c-edges (in a clockwise fashion) in DPillar$_{n,k}$ and the source is $\mathbf{0}$ and the destination is $(1, 10\ldots 0)$ then the routing algorithm in [12] yields a path of length $k+1$ because the $(k-1)$st column must be visited in order to change the $(k-1)$st coordinate; the a-edge algorithm in [12] yields a path of length $k-1$. Neither of these algorithms are optimal (when $k > 3$), however, since the shortest path is of length 2 and can only be achieved by following a d-edge and then a c-edge to yield the path $\mathbf{0}$, $(0, 10\ldots 0)$, $(1, 10\ldots 0)$.

3.1 Another Abstraction: The Marked Cycle

Observe in the above discussion that the need to cover column c and fix the cth coordinate of the row-index arises if, and only if, $v_c \neq 0$, but it does not matter here what other value v_c may take on. Consequently, for arbitrary nodes *src* and *dst*, we instead consider a walk on a cycle on k-nodes, $G_{n,k}(src, dst)$, with a node corresponding to each column of $\text{DPillar}_{n,k}$ in which we *mark* the ith node whenever the ith column must be covered in $\text{DPillar}_{n,k}$; that is, where the coordinates of the row-indices of *src* and *dst* differ. Let *src′* and *dst′* be the columns of *src* and *dst*. We abstract a path from *src* to *dst* in $\text{DPillar}_{n,k}$ by a sequence of *moves* in $G_{n,k}(src, dst)$ starting with node *src′* and ending at node *dst′*, where each move is analogous to a c-, a-, b-, or d-edge. From node c: (i) a *c-move covers* node c and moves to node $c+1$, (ii) an *a-move covers* and moves to node $c-1$, (iii) a *b-move covers* and stays at node c; and, (iv) a *d-move covers* node $c-1$ and stays at node c. We call $G_{n,k}(src, dst)$ a *marked cycle*.

It should be clear as to how moves in the marked cycle $G_{n,k}(src, dst)$ correspond to edges of type c, a, b, and d in $\text{DPillar}_{n,k}$ (and so to server-switch-server link-pairs in the DCN $\text{DPillar}_{n,k}$) with the coverage of a node in $G_{n,k}(src, dst)$ and a node of $\text{DPillar}_{n,k}$ being in direct correspondence. A *path* in $G_{n,k}(src, dst)$ is a sequence of moves leading from *src′* to *dst′* and corresponds to a path of the same length in $\text{DPillar}_{n,k}$ from node *src* to node *dst* (and *vice versa*). Consequently, in order to find a shortest (src, dst)-path in the DCN $\text{DPillar}_{n,k}$, it suffices to find a shortest (src', dst')-path in the marked cycle $G_{n,k}(src, dst)$ so that every marked node is covered by a move.

4 Routing in a Marked Cycle

We make some initial observations about shortest paths in a marked cycle, and then prove a structural result on shortest paths. Let *src* and *dst* be nodes of $\text{DPillar}_{n,k}$, in columns *src′* and *dst′*, respectively. Henceforth, ρ is a shortest path from *src′* to *dst′* in $G_{n,k}(src, dst)$; therefore, ρ is a sequence of moves. We denote a sequence of moves by strings of the (corresponding) letters c, a, b, and d so that, for example, *ccbaaa* represents two c-moves, followed by a b-move, followed by three a-moves. In addition i repeated symbols, say, $cc\cdots c$ can be written c^i so that $ccbaaa = c^2ba^3$.

We can often rule out certain consecutive pairs of moves in ρ. For example, if we have a subsequence *bd* then this has the same effect as *db*, and so we may suppose that a subsequence *db* within ρ is forbidden. We can achieve much more by arguing on the optimality (as regards length) of ρ; for example, suppose the subsequence of moves *ca* occurs in ρ. We can replace *ca* by a b-move so as to obtain a shorter path with identical coverage to ρ. This contradicts the optimality of ρ, so we may assume that *ca* does not occur in ρ. Similarly, *ac* can be replaced by a d-move, so we may assume *ac* does not occur in ρ. Continuing in this manner, we obtain two tables of move-pair replacements, given below, whose (i, j)th entries represent the following: the move in the i-th position of the first column followed by the move in the j-th position of the first row must be replaced by the move given in the (i, j)th entry of the table.

	b c
a	a d
b	b c

	a d
c	b c
d	a d

We describe the structure of paths resulting from similar arguments.

Lemma 2. *If ρ has length at least 3 then it must be of one of two forms:*

$$d^{\epsilon}c^{i_1}ba^{j_1}dc^{i_2}\dots c^{i_m}b^{\delta} \quad or \quad d^{\epsilon}c^{i_1}ba^{j_1}dc^{i_2}\dots a^{j_m}d^{\delta}, \tag{1}$$

for some $m \geq 1$, where $i_1, i_2, \dots, i_m, j_1, j_2, \dots, j_m > 1$ and where $\epsilon, \delta \in \{0, 1\}$;

$$b^{\epsilon}a^{i_1}dc^{j_1}ba^{i_2}\dots a^{i_m}d^{\delta} \quad or \quad b^{\epsilon}a^{i_1}dc^{j_1}ba^{i_2}\dots c^{j_m}b^{\delta}, \tag{2}$$

for some $m \geq 1$, where $i_1, i_2, \dots, i_m, j_1, j_2, \dots, j_m > 1$ and where $\epsilon, \delta \in \{0, 1\}$.

Proof. Omitted due to space constraints (it is a simple case analysis).

4.1 A Shortest Path has at Most Two Turns

If we have a c-move followed by a b-move followed by an a-move in ρ then we say that an *anti-clockwise turn*, or simply an *a-turn*, occurs at the b-move; similarly, if we have an a-move followed by a d-move followed by a c-move then we say that a *clockwise turn*, or simply a *c-turn*, occurs at the d-move. Note that if we have an a-turn in ρ then the node at which this turn occurs, i.e., the node that is covered by the d-move, must be marked in $G_{n,k}(src, dst)$ as otherwise we could delete the corresponding d-move from ρ and still have a sequence from src' to dst' covering all the marked nodes, which would yield a contradiction. Similarly, if we have a c-turn then the node at which this c-turn occurs, i.e., the node that is covered by the b-move, must be marked. We will use these observations later; but now we prove that any shortest path ρ must contain at most 2 turns.

Suppose that ρ is a shortest path and has at least 3 turns.

Case (a): Suppose that ρ is of Form (1) and has a prefix ρ' of the form $c^iba^jdc^lba$, where $i, j, l \geq 1$. By this we mean that ρ begins with i c-moves followed by a b-move followed by j a-moves followed by a d-move followed by l c-moves followed by a b-move followed by an a-move.

If $j < i$ then we can replace the prefix c^iba^jdc in ρ' with c^iba^{j-1} and still obtain the same coverage; this contradicts that ρ is a shortest path (note that we have actually only assumed so far that ρ has 2 turns). If $j = i$ then we can replace the prefix c^iba^idc in ρ' with dc^iba^{i-1} so as to obtain a contradiction (we have still actually only assumed that ρ has 2 turns). Hence, we must have that $j > i$. Suppose that $j \geq l > j - i$. We can replace the prefix $c^iba^jdc^l$ in ρ' with $a^{j-i}dc^jba^{j-l}$ so as to obtain a contradiction (we have still actually only assumed that ρ has 2 turns). Hence, $j > i$ and either $l \leq j - i$ or $l > j$.

Suppose that $l > j$. We can replace the prefix $c^iba^jdc^l$ in ρ' with $a^{j-i}dc^l$ so as to obtain a contradiction (we have still actually only assumed that ρ has 2 turns). Hence, we must have that $j > i$ and $l \leq j - i$. However, if we replace ρ'

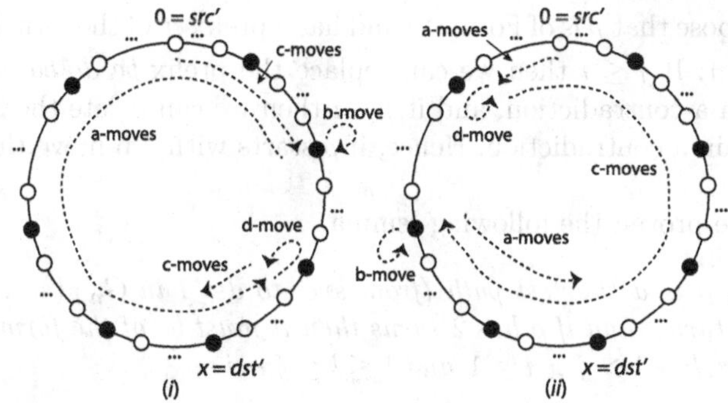

Fig. 2. Visualizing paths with 2 turns.

with $c^iba^jdc^{l-1}$ then we obtain a contradiction (here we do use the fact that ρ has at least 3 turns). So, ρ has at most 2 turns and if it has 2 turns then ρ is of the form $c^iba^jdc^l$ where $j > i$ and $l \leq j - i$.

We can say more if ρ has 2 turns. Suppose that $j \geq k - 1$. The b-move can be deleted from ρ' and we obtain a contradiction. Hence, if ρ has 2 turns then ρ is of the form $c^iba^jdc^l$ where $k - 1 > j > i \geq 1$ and $1 \leq l \leq j - i$. We can visualize ρ as in Fig. 2(i). The marked cycle $G_{n,k}(src, dst)$ is shown as a cycle where a black node denotes a node of B; that is, a node that needs to be covered by some path in $G_{n,k}(src, dst)$ (with $0 = src' \neq dst' = x$ in this illustration). The path ρ is depicted as a dotted line partitioned into composite moves.

Case (b): Suppose that ρ is of Form (1) and has a prefix ρ' of the form $dc^iba^jdc^lba$, where $i, j, l \geq 1$. If $j \leq i$ then we can replace the prefix dc^iba^jdc in ρ' with dc^iba^jc so as to obtain a contradiction, and if $j > i$ then we can delete the first d-move from ρ to obtain a contradiction. Hence, if ρ starts with a d-move then it has at most 1 turn.

Case (c): Suppose that ρ is of Form (2) and has a prefix ρ' of the form $a^idc^jba^ldc$, where $i, j, l \geq 1$. If $j < i$ then we can replace the prefix a^idc^jba in ρ' with a^idc^{j-1} so as to obtain a contradiction. If $i = j$ then we can replace the prefix a^idc^iba in ρ' with ba^idc^{i-1} so as to obtain a contradiction. Hence, $j > i$.

Suppose that $j \geq l > j - i$. We can replace the prefix $a^idc^jba^l$ in ρ with $c^{j-i}ba^jdc^{j-l}$ so as to obtain a contradiction. Suppose that $l > j$. We can delete the first occurrence of a d-move in ρ so as to obtain a contradiction. Hence, $l \leq j - i$. Note that if ρ has 2 turns then ρ is of the form $a^idc^jba^l$ where $j > i$ and $l \leq j - i$. Alternatively, suppose that ρ has at least 3 turns. We can replace the prefix $a^idc^jba^ldc$ in ρ with $a^idc^jbc^{l-1}$ so as to obtain a contradiction. Hence, ρ has at most 2 turns.

We can say more if ρ has 2 turns. Suppose that $j \geq k - 1$. The d-move can be deleted from ρ' and we obtain a contradiction. Hence, if ρ has 2 turns then ρ is of the form $a^idc^jbd^l$ where $k - 1 > j > i \geq 1$ and $1 \leq l \leq j - i$. We can visualize ρ as in Fig. 2(ii).

Case (*d*): Suppose that ρ is of Form (2) and has a prefix ρ' of the form $ba^idc^jba^ldc$, where $i, j, l \geq 1$. If $j \leq i$ then we can replace the prefix ba^idc^jba with ba^idc^ja so as to obtain a contradiction, and if $j > i$ then we can delete the first b-move from ρ to obtain a contradiction. Hence, if ρ starts with a b-move then it has at most 1 turn.

So, we have proven the following lemma.

Lemma 3. *If ρ is a shortest path (from src' to dst') in $G_{n,k}(src, dst)$ then ρ has at most 2 turns, and if ρ has 2 turns then it must be of the form $c^iba^jdc^l$ or $a^idc^jba^l$, where $k-1 > j > i \geq 1$ and $1 \leq l \leq j-i$.*

With reference to Fig. 2, the numerical constraints in Lemma 3 mean that there is no interaction or overlap involving the 2 turns in ρ.

5 An Optimal Routing Algorithm for DPillar

We now develop an optimal single-path routing algorithm for DPillar. We do this by finding a small set Π of paths (from src' to dst') in $G_{n,k}(src, dst)$ so that at least one of these paths is a shortest path. By Lemma 1, we may assume that $src = \mathbf{0}$ and $dst = (x, v_{k-1}v_{k-2}\ldots v_0)$, and by Lemma 2, we may assume that any shortest path has at most 2 turns.

5.1 Building Our Set of Paths When $x \neq 0$

We first suppose that $0 \neq x$. Let $B = \{i : 0 \leq i \leq k-1, v_i \neq 0\}$ (that is, the bit-positions that need to be 'fixed'). Suppose that $B \setminus \{0, x\} = \{i_l : 1 \leq l \leq r\} \cup \{j_l : 1 \leq l \leq s\}$ so that we have $0 < j_s < j_{s-1} < \ldots < j_1 < C < i_1 < i_2 < \ldots < i_r < k$ (we might have that either r or s is 0 when the corresponding set is empty). If $r \geq 2$ then define $\delta_l = i_{l+1} - i_l$, for $l = 1, 2, \ldots, r-1$, with $\delta = \max\{\delta_l : l = 1, 2, \ldots, r-1\}$; and if $s \geq 2$ then define $\epsilon_l = j_l - j_{l+1}$, for $l = 1, 2, \ldots, s-1$, with $\epsilon = \max\{\epsilon_l : l = 1, 2, \ldots, s-1\}$. Also: define $\Delta_0 = 1$ (resp. 0), if $0 \in B$ (resp. $0 \notin B$); and $\Delta_x = 1$ (resp. 0), if $x \in B$ (resp. $x \notin B$). We can visualize the resulting marked cycle $G_{n,k}(0, x)$ as in Fig. 3(*i*). Note that in this particular illustration $0 \notin B$ and $x \in B$; so, $\Delta_0 = 0$ and $\Delta_x = 1$. Of course, what we are looking for is a sequence of (a-, b-, c- and d-)moves that will take us from 0 to x in $G_{n,k}(0, x)$ so that all nodes of B have been covered.

In what follows, we examine different scenarios involving the number of marked nodes, r, and also the number of marked nodes, s. Each scenario for r contributes certain paths to Π as does each scenario for s. Note that perhaps the most obvious paths to consider as potential members of Π are the paths c^{k+x} and a^{2k-x} which have lengths $k+x$ and $2k-x$, respectively. So, we begin by setting $\Pi = \{c^{k+x}, a^{2k-x}\}$.

From Lemma 3, any shortest path ρ from 0 to x having 2 turns requires that $r \geq 2$ or $s \geq 2$ and that both nodes at which these turns occur are different from 0 and x and lie on the anti-clockwise path from 0 to x or on the clockwise path

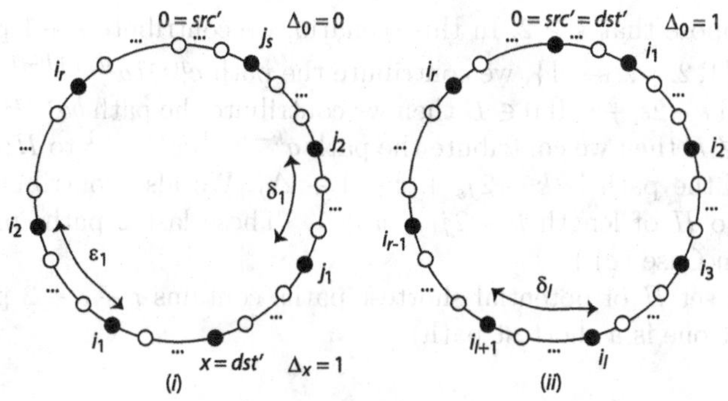

Fig. 3. Visualizing our notation.

from 0 to x, accordingly. Also, the node at which any turn occurs on a shortest path ρ is necessarily a marked node (irrespective of the number of turns in ρ).

Case (a): Suppose that $r = 0$. In this scenario, we contribute either the path $c^x b$ to Π, if $x \in B$, or the path c^x to Π, if $x \notin B$; either way, the length of the path contributed is $x + \Delta_x$.

Case (b): Suppose that $s = 0$. In this scenario, we contribute either the path ba^{k-x} to Π, if $0 \in B$, or the path a^{k-x} to Π, if $0 \notin B$; either way, the length of the path contributed is $k - x + \Delta_0$.

Case (c): Suppose that $r = 1$. In this scenario, we contribute 2 paths to Π. If $x \in B$ then we contribute the path $a^{k-i_1-1}dc^{k-i_1-1+x}b$ to Π, or if $x \notin B$ then we contribute the path $a^{k-i_1-1}dc^{k-i_1-1+x}$ to Π; either way, the length of the resulting path is $2k - 2i_1 + x - 1 + \Delta_x$. We also contribute the path $c^{i_1}ba^{i_1-x}$ to Π of length $2i_1 - x + 1$. There is potentially another path when $i_1 = x + 1$ and $x \in B$, namely $a^{k-x-1}dc^{k-1}$, but the length of this path is $2k - x - 1$ which is greater than $2k - x - 3 + \Delta_x$ which is $2k - 2i_1 + x - 1 + \Delta_x$ evaluated with $i_1 = x + 1$.

Case (d): Suppose that $s = 1$. In this scenario, we contribute 2 paths to Π. If $0 \in B$ then we contribute the path $ba^{k-j_1-1}dc^{x-j_1-1}$ to Π, or if $0 \notin B$ then we contribute the path $a^{k-j_1-1}dc^{x-j_1-1}$ to Π; either way, the length of the resulting path is $k - 2j_1 + x - 1 + \Delta_0$. We also contribute the path $c^{j_1}ba^{k+j_1-x}$ to Π of length $k + 2j_1 - x + 1$. There is potentially another path when $j_1 = 1$ and $0 \in B$, namely $a^{k-1}dc^{x-1}$, but the length of this path is $k + x - 1$ which is greater than $k + x - 3 + \Delta_0$ which is $k - 2j_1 + x - 1 + \Delta_0$ evaluated with $j_1 = 1$.

Case (e): Suppose that $r \geq 2$. In this scenario, we contribute $r+1$ paths to Π. For each $l \in \{1, 2, \ldots, r-1\}$, we contribute the path $a^{k-i_{l+1}-1}dc^{k-i_{l+1}-1+i_l}ba^{i_l-x}$ to Π of length $2k - 2\delta_l - x$. If $x \in B$ then we contribute the path $a^{k-i_1-1}dc^{k-i_1-1+x}b$ to Π, or if $x \notin B$ then we contribute the path $a^{k-i_1-1}dc^{k-i_1-1+x}$ to Π; either way, the length of the path is $2k - 2i_1 + x - 1 + \Delta_x$. We also contribute the path $c^{i_r}ba^{i_r-x}$ to Π of length $2i_r - x + 1$. (These last 2 paths mirror those constructed in Case (c).)

Case (f): Suppose that $s \geq 2$. In this scenario, we contribute $s+1$ paths to Π. For each $l \in \{1, 2, \ldots, s-1\}$, we contribute the path $c^{j_{l+1}} b a^{j_{l+1}+k-j_l-1} d a^{x-j_l-1}$ to Π of length $k - 2\epsilon_l + x$. If $0 \in B$ then we contribute the path $b a^{k-j_s-1} d c^{x-j_s-1}$ to Π, or if $0 \notin B$ then we contribute the path $a^{k-j_s-1} d c^{x-j_s-1}$ to Π; either way, the length of the path is $k - 2j_s + x - 1 + \Delta_0$. We also contribute the path $c^{j_1} b a^{j_1+k-x}$ to Π of length $k + 2j_1 - x + 1$. (These last 2 paths mirror those constructed in Case (c).)

Thus, our set Π of potential shortest paths contains $r+s+2$ paths (from which at least one is a shortest path).

5.2 Building Our Set of Paths When $x = 0$

Now we suppose that $x = 0$. We proceed as we did above and build a set Π of potential shortest paths. Let $B = \{i : 0 \leq i \leq k-1, v_i \neq 0\}$. Suppose that $B \setminus \{0\} = \{i_l : 1 \leq l \leq r\}$ so that we have $0 < i_1 < i_2 < \ldots < i_r < k$ (we might have that r is 0 when the corresponding set is empty). If $r \geq 2$ then define $\delta_l = i_{l+1} - i_l$, for $l = 1, 2, \ldots, r-1$, with $\delta = \max\{\delta_l : l = 1, 2, \ldots, r-1\}$. We define $\Delta_0 = 1$, if $0 \in B$, and $\Delta_0 = 0$, if $0 \notin B$. We can visualize the resulting marked cycle $G_{n,k}(0,0)$ as in Fig. 3(ii). Again, the most obvious path to consider is c^k (or a^k) which has length k. We begin by setting $\Pi = \{c^k\}$.

Case(a): Suppose that $r = 0$. In this scenario, we contribute the path b of length 1 (note that in this case the node 0 is necessarily marked as we originally assumed that we started with distinct source and destination servers in the DCN DPillar$_{n,k}$).

Case(b): Suppose that $r = 1$. If $i_1 = k-1$ then we contribute the path bd, if $0 \in B$, and the path d, if $0 \notin B$; either way, the path has length $1 + \Delta_0$. If $1 = i_1 \neq k-1$ then we contribute the path cba of length 3. If $1 \neq i_1 \neq k-1$ then we contribute 2 paths. The first of these paths is the path $b a^{k-i_1-1} d c^{k-i_1-1}$, if $0 \in B$, and the path $a^{k-i_1-1} d c^{k-i_1-1}$, if $0 \notin B$; either way, this path has length $2k - 2i_1 - 1 + \Delta_0$. The second of these paths is the path $c^{i_1} b a^{i_1}$ of length $2i_1 + 1$.

Case(c): Suppose that $r \geq 2$. In this scenario, we contribute $r+1$ paths to Π. For each $l \in \{1, 2, \ldots, r-1\}$, we contribute the path $a^{k-i_{l+1}-1} d c^{k-i_{l+1}-1+i_l} b a^{i_l}$ to Π of length $2k - 2\delta_l$. If $0 \in B$ then we contribute the path $b a^{k-i_1-1} d c^{k-i_1-1}$ to Π, or if $0 \notin B$ then we contribute the path $a^{k-i_1-1} d c^{k-i_1-1}$ to Π; either way, this path has length $2k - 2i_1 - 1 + \Delta_0$. We also contribute the path $c^{i_r} b a^{i_r}$ to Π of length $2i_r + 1$. (These last 2 paths mirror those constructed in Case (b).)

Thus, our set Π of potential shortest paths contains at most $r+1$ paths (from which at least one is a shortest path).

5.3 Our Algorithm

We now use our set Π of potential shortest paths so as to find a shortest path or the length of a shortest path. Our algorithm for $G_{n,k}(0,x)$ is as follows.

```
calculate B
if 0 ≠ x then
   L = min{k + x, 2k − x}
   calculate r, s, δ, ε, Δ_0 and Δ_x
   if r = 0 then L = min{L, x + Δ_x}
   if s = 0 then L = min{L, k − x + Δ_0}
   if r = 1 then L = min{L, 2k − 2i_1 + x − 1 + Δ_x, 2i_1 − x + 1}
   if s = 1 then L = min{L, k − 2j_1 + x − 1 + Δ_0, k + 2j_1 − x + 1}
   if r ≥ 2 then
      calculate δ      % we need only consider the maximal δ_l
      L = min{L, 2k − 2δ − x, 2k − 2i_1 + x − 1 + Δ_x, 2i_r − x + 1}
   if s ≥ 2 then
      calculate ε      % we need only consider the maximal ε_l
      L = min{L, k − 2ε + x, k − 2j_s + x − 1 + Δ_0, k + 2j_1 − x + 1}
else
   calculate r and δ
   if r = 0 then L = 1
   if r = 1 then
      if i_1 = k − 1 then L = 1 + Δ_0
      if 1 = i_1 ≠ k − 1 then L = 3
      if 1 ≠ i_1 ≠ k − 1 then L = min{2k − 2i_1 − 1 + Δ_0, 2i_1 + 1}
   if r ≥ 2 then L = min{k, 2k − 2δ, 2k − 2i_1 − 1 + Δ_0, 2i_r + 1}
output L
```

If we wish to output a shortest path then all we do is apply the above algorithm but remember which shortest path corresponds to the final value of L and output this shortest path (note that there may be more than one shortest path; exactly which path one obtains depends upon how one implements checking the paths of Π). The time complexity of both algorithms is clearly $O(k \log(n))$ (that is, linear in the length of the input).

It should be clear (using Lemma 2) that the different considerations for r and s exhaust all possibilities and that consequently the set of paths Π considered by the above algorithm is such as to contain a shortest path. Hence, our algorithm outputs the length of a shortest path from some source node to some destination node in DPillar$_{n,k}$. The validity of our algorithm was verified with a breadth-first search and we also found that it yields a 20–30 % improvement in the average length of a path over the algorithms given in [12], for various small parameters n and k; for example, DPillar$_{16,5}$, which has 163840 server-nodes, has an average shortest path length of 4.77, but employing the clockwise algorithm from [12] yields an average path length of 6.86.

6 Conclusions

In this paper we have developed an optimal and efficient single-path routing algorithm for the DCN DPillar and have shown that DPillar is node-symmetric. We feel that there are other areas where efficiency gains might be made; in particular, we intend to focus on multi-path routing in forthcoming research.

Acknowledgement. This work has been funded by the Engineering and Physical Sciences Research Council (EPSRC) through grants EP/K015680/1 and EP/K015699/1.

References

1. Abts, D., Marty, M.R., Wells, P.M., Klausler, P., Liu, H.: Energy proportional datacenter networks. In: Proceedings of the 37th Annual International Symposium on Computer Architecture, pp. 338–347 (2010)
2. Abu-Libdeh, H., Costa, P., Rowstron, A., OShea, G., Donnelly, A.: Symbiotic routing in future data centers. SIGCOMM Comput. Comm. Rev. **40**(4), 51–62 (2010)
3. Ahn, J.H., Binkert, N., Davis, A., McLaren, M., Schreiber, R.S.: HyperX: topology, routing, and packaging of efficient large-scale networks. In: Proceedings of the Conference on High Performance Computer Networking, Storage and Analysis, Article 41 (2009)
4. Al-Fares, M., Loukissas, A., Vahdat, A.: A scalable, commodity data center network architecture. SIGCOMM Comput. Comm. Rev. **38**(4), 63–74 (2008)
5. Greenberg, A., Hamilton, J.R., Jain, N., Kandula, S., Kim, C., Lahiri, P., Maltz, D.A., Patel, P., Sengupta, S.: VL2: a scalable and flexible data center network. SIGCOMM Comput. Comm. Rev. **39**(4), 51–62 (2009)
6. Guo, D., Chen, T., Li, D., Li, M., Liu, Y., Chen, G.: Expandible and cost-effective network structures for data centers using dual-port servers. IEEE Trans. Comput. **62**(7), 1303–1317 (2013)
7. Guo, C., Wu, H., Tan, K., Shi, L., Zhang, Y., Lu, S.: DCell: a scalable and fault-tolerant network structure for data centers. SIGCOMM Comput. Comm. Rev. **38**(4), 75–86 (2008)
8. Guo, C., Lu, G., Li, D., Wu, H., Zhang, X., Shi, Y., Tian, C., Zhang, Y., Lu, S.: BCube: a high performance, server-centric network architecture for modular data centers. SIGCOMM Comput. Comm. Rev. **39**(4), 63–74 (2009)
9. Heller, B., Seetharaman, S., Mahadevan, P., Yiakoumis, Y., Sharma, P., Banerjee, S., McKeown, N.: ElasticTree: saving energy in data center networks. In: Proceedings of the 7th USENIX Conference on Networked Systems Design and Implementation, pp. 249–264 (2006)
10. Li, C., Guo, D., Wu, H., Tan, K., Zhang, K., Lu, S.: FiConn: using backup port for server interconnection in data centers. In: Proceedings of INFOCOM, pp. 2276–2285 (2009)
11. Li, D., Wu, J.: On data center network architectures for interconnecting dual-port servers. IEEE Trans. Comput. **64**(11), 3210–3222 (2015)
12. Liao, Y., Yin, J., Yin, D., Gao, L.: DPillar: dual-port server interconnection network for large scale data centers. Comput. Netw. **56**(8), 2132–2147 (2012)
13. Liu, Y., Muppala, J.K., Veeraraghavan, M., Lin, D., Katz, J.: Data Centre Networks: Topologies, Architectures and Fault-Tolerance Characteristics. Springer, New York (2013)
14. Mysore, R.N., Pamboris, A., Farrington, N., Huang, N., Miri, P., Radhakrishnan, S., Subramanya, V., Vahdat, A.: Portland: a scalable fault-tolerant layer 2 data center network fabric. SIGCOMM Comput. Comm. Rev. **39**(4), 39–50 (2009)
15. Wang, C., Wang, C., Yuan, Y., Wei, Y.: MCube: a high performance and fault-tolerant network architecture for data centers. In: Proceedings of the International Conference on Computer Design and Applications, vol. 5, pp. V5-423-V5–427 (2010)

A Sensor Deployment Strategy in Bus-Based Hybrid Ad-Hoc Networks

Hongwei Du[1], Rongrong Zhu[1(✉)], Xiaohua Jia[2], and Chuang Liu[1]

[1] Department of Computer Science and Technology, Harbin Institute of Technology Shenzhen Graduate School, Shenzhen Key Laboratory of Internet Information Collaboration, Shenzhen, China
hwdu@hitsz.edu.cn, {hitzrr13,chuangliuhit}@gmail.com

[2] Department of Computer Science, City University of Hong Kong, Hong Kong, China
csjia@cityu.edu.hk

Abstract. Sensor deployment is one of the challenging issues in Wireless Sensor Networks (WSNs). Traditionally, sensors are deployed for sensing and transmitting data to Base Station (BS) for further data processing in cities. However, instead of sensor independently allocated, we apply the existing transportation infrastructure - Bus-based Ad hoc Network (BANETs) to assist the sensing and transmission of the WSNs. A novel hybrid scheme is derived to this hybrid network. Then we utilize the practical simulator ONE [1] to analyse our scheme and simulation indicates that the scheme can significantly reduce the number of sensors to be deployed, and the information collected from the city could reach the BS within a fixed time.

Keywords: Sensor deployment · Hybrid network · Bus-based Ad hoc Networks

1 Introduction

WSNs are composed of a large number of tiny, low-cost sensor nodes, which can communicate over short distances [2–4]. The sensor nodes of WSNs are spatially distributed for monitoring environmental conditions or events of concern, and cooperatively delivery the data to the BS in a multi-hop way [5]. BANETs consist of buses equipped with wireless communication devices, GPS, digital maps. Buses exchange information with other buses as well as with access points within their communication radius. In this paper, we utilize the hybrid network which consists of WSNs and BANETs to monitor the entire city, as shown in Fig. 1, urban information such as pollution, temperature, pressure or traffic conditions, can be captured by statically deployed sensor nodes and sensors carried by the bus. These data can then be forwarded to buses passing by and subsequently delivered to the BS periodically. The BS can be connected to a designated Web Server that makes it possible for authorized users to remotely access and configure the hybrid networks anytime, anywhere via traditional computers and mobile

Z. Lu et al. (Eds.): COCOA 2015, LNCS 9486, pp. 221–235, 2015.
DOI: 10.1007/978-3-319-26626-8_17

devices such as iPhone, iPad and so on. The problem considering in the hybrid network is to minimize the number of sensors for the sensing interest area. And indeed, it has become a challenging issue.

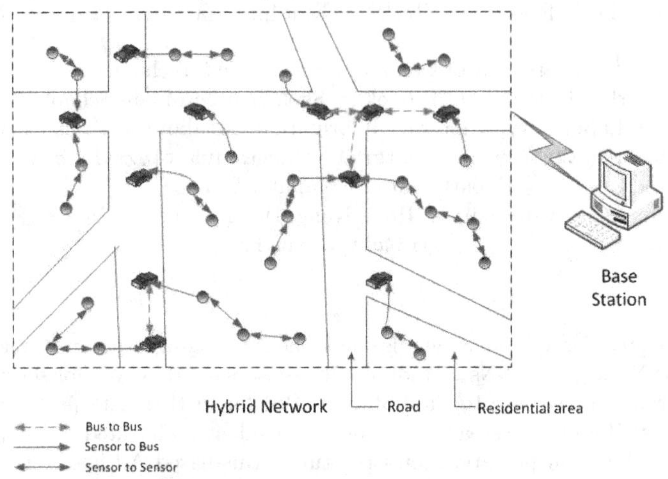

Fig. 1. Illustration of a brief hybrid network structure

Most literatures mainly consider using only WSNs to monitor the entire city. In WSNs, the sensor nodes cooperatively monitor the city information and delivery the data to the BS in a multi-hop way. However, a large amount of sensors needed to be deployed in the city to ensure coverage and connectivity in such environment. And most deployment methods assume that the sensing field is an open space and the ratio of the sensing radius to communication radius is fixed. In this paper we utilize the existing transportation infrastructure BANETs to help to monitor and deliver the data to the BS. Thus, compared with the method already exist, our scheme in this paper can save a lot of sensors needed to be deployed in the area of the city. In BANETs, the buses move regularly, and they have a fixed route for traveling. Our simulations show that the data collected from the entire city can be delivered to the BS within a certain amount of time.

In this paper, we study the problem of the efficient sensor deployment under a hybrid network structure. Then we analyze the average delivery delay to ensure the data collected by the BS is useful and valuable. Sensor deployment is a critical issue since it reflects the cost and detection capability of a WSN. The asymptotic optimality of strip-based deployment pattern to achieve one and two connectivity and full coverage was proved in [9]. Our method is based on the strip-based deployment pattern and it can adapt to the area with different shapes. The contributions of our research are summarized as follows:

1. A hybrid network structure is proposed that the WSNs and BANETs can monitor the city and route the application-specific information cooperatively to the BS within an effective time period.
2. An effective algorithm is proposed to deploy sensors in the sub-regions partitioned by the roads in the city. The shape of the sub-regions can be irregular and the relationship between the communication radius and sensing radius can be arbitrary. The simulations show that our method considerably reduces the total number of sensors in comparison with three conventional sensor deployment methods.
3. The ONE [1] is utilized to analyze the average delivery delay during which the BS collected the data from the all the sensors deployed in the city. And simulation indicates that the data collected can reach to the BS within a fixed time, which means that the data received is effective and valuable.

The remaining part of the paper is organized as follows: Sect. 2 presents the related work. In Sect. 3, we introduce the network model. Section 4 gives the definition of the problem and the assumptions. In Sect. 5, we present the schemes for sensor deployment. Section 6 presents our simulation results. Finally, Sect. 7 concludes the paper.

2 Related Work

Sensor deployment has been well studied over the past decades. According to the roles that the deployment nodes play, sensor deployment can be classified into two categories: deployment of ordinary nodes and deployment of relay nodes.

In the deployment of ordinary nodes, sensors can be placed exactly on carefully engineered positions, or thrown in bulk on random positions. Most work on deterministic deployment seeks to determine the optimal deployment pattern. According to the applications and goals of the WSNs, different optimality criteria are used. A common objective is to minimise the number of required sensors needed, subject to the coverage and connectivity constraints. It is well known that the triangle pattern is asymptotically optimal in terms of the number of discs needed to achieve full coverage [6]. This result has also been reproved in [7] using a different approach. This result naturally provides six connectivity only when $R_c \geq \sqrt{3}R_s$. However, values of R_c/R_s in practice can be an arbitrary positive number. Iyengar et al. [8] proved that the strip-based deployment pattern is asymptotically near optimal when $R_c = R_s$. The strip-based pattern, nodes are arranged along many parallel horizontal strips, and nodes in different horizontal strips can communicate via some nodes located as a vertical ribs. Bai et al. [9] extended the work by proving that this strip-based pattern is the asymptotic optimal pattern for an infinite network. The asymptotic optimality of strip-based deployment pattern to achieve one and two connectivity and full coverage was proved in [9]. Bai et al. [10] explored the asymptotically optimal pattern to achieve four connectivity and full coverage. Bai et al. [11] published new results, which proposed optimal sensor placement patterns to achieve coverage and k-connectivity ($k \leq 6$). Ishizuka and Aida [12] proposed three types of random placement models. They suggested that the placement model which assumes that sensors are uniformly distributed in terms of the network radius and angular direction from the sensor field centre is the best candidate for random node placement. Others (see [13] for an example) assume that all nodes follow identical Gaussian distribution.

The problem of relay node placement can be categorized into single-tiered and two-tiered, according to the data forwarding scheme adopted by the WSN. In single-tiered relay node placement, both relay nodes and ordinary sensor nodes participate in packet forwarding. In contrast, in a WSN with two-tiered relay node placement, only relay nodes can forward packets. In the following discussion, the transmission ranges for relays and ordinary sensors are denoted R and r, respectively. Cheng et al. [14] developed algorithms to place the minimum number of relay nodes and maintain the connectivity of a single-tiered WSN, under the assumption that $R = r$. The problem was modelled by the Steiner Minimum Tree with Minimum Number of Steiner Points and Bounded Edge Length (SMT-MSP) problem, which arose in the study of amplifier deployment in optical networks, and was proved to be NP-hard [15]. Based on a minimum spanning tree, Lin and Xue [15] developed an algorithm to solve the SMT-MST problem. They proved it to have an approximation ratio of 5, which Chen et al. [16] tightened to 4. Based on Lin and Xue's algorithm, Cheng et al. [14] proposed a different 3-approximation algorithm and a randomized 2.5-approximation algorithm. In order to provide fault-tolerance, Kashyap et al. [17] studied how to place minimum number of relays such that the resulted WSN is 2-connected, when relay nodes and ordinary sensor nodes have identical transmission range, i.e., $R = r$. Zhang et al. [18] improved the results of Kashyap et al. [17] by developing algorithms to compute the optimal node placement for networks to achieve 2-connectivity, under the more general condition that $R \geq r$. These algorithms aimed to minimize the number of relay nodes while providing fault-tolerance. Lloyd and Xue [19] developed algorithms to find optimal placement of relay nodes for the more general relationship $R \geq r$, under single-tiered and two-tiered infrastructures.

However, the context of the above deployment methods mainly assume that the sensing field is an open space and there is a special relationship between the communication range and the sensing range of sensors. In our hybrid network framework, the city area are divided by roads into many irregular subregions, if we use the conventional methods, it may result in coverage hole along the boundary. And in practical, the value of R_c/R_s can be an arbitrary positive number. So how to find an effective deployment method in our hybrid networks is a challenging research.

3 Network Model

In this section, we present the network model of the street map and the WSNs.

As shown in Fig. 2, the entire city of the street map is abstracted as a directed graph *G(V,E)*, *V* denote the set of intersections, for any two intersections *a* and

b, $(a,b) \in G$ if and only if there is a road segment connecting a and b and buses can travel from a towards b on that segment. The width of the road is d, and the city area is divided into many subregions.

As shown is Fig. 3, consider n sensors having the same sensing radius r_s and communication radius r_c are deployed in the entire city. We assume that the sensing range of a sensor is a disk with radius r_s, centered at a node. Let the undirected communication graph $G' = (V', E'(r_c))$ donate the entire sensor network topology, where the sensors act as vertices and an edge exists between any two sensors if the Euclidean distance between them is less than r_c. A path of sensors between a and b in the communication graph is called a communication path between the sensors a and b.

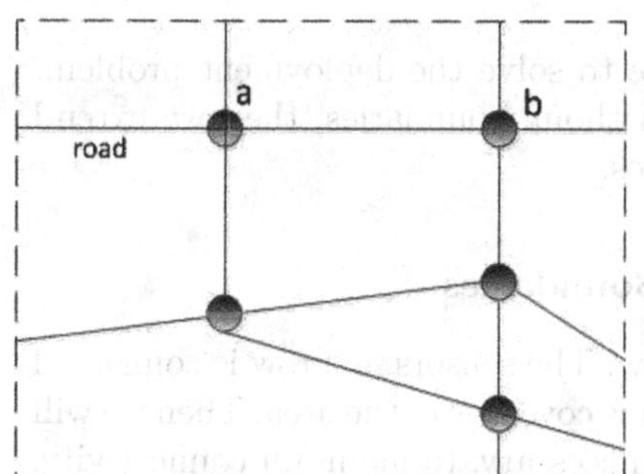

Fig. 2. Bus can travel from crossroads a to crossroads b, the dotted line denotes the region boundary

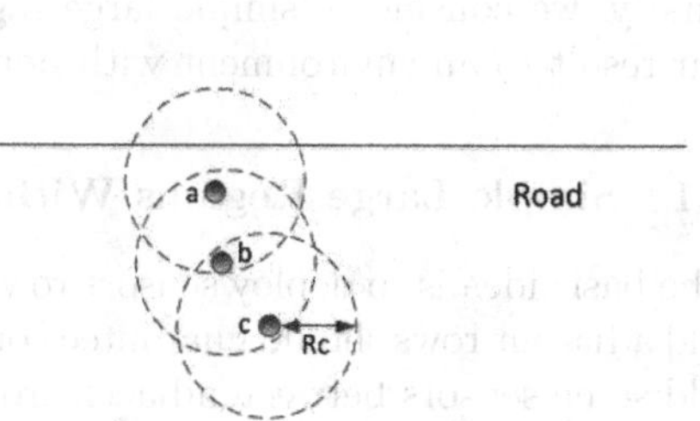

Fig. 3. Sensor a, b, c are connected, each of them can find a communication path to road

4 Problem Definition and Assumptions

In this section, we give the definition of our problem and the assumptions.

Assumptions. We assume that the bus are equipped with a sensor and run on the road at a speed of v, the sensing radius of the sensor is R_s, and communication radius is R_c. The city has N buses, each bus starts from the bus station and has its own bus routes and fixed departure time. The bus can accept data from the sensors deployed on the roadside. When two buses encounter, they can communicate with each other, here the encounter denotes that the Euclidean distance between the two bus is less than R_c. The BS is set in the bus station, and the bus can delivery the collected data to the BS only if the bus return to the BS.

Problem Definition. The problem of the effective sensor deployment under the hybrid network environment: Given the street map of a city, the problem is to deploy the minimum number of sensor nodes, such that: (1) the entire area Ω of the city can be totally covered by n+N sensors, where N is the number of sensors carried by the bus mentioned above, and (2) each of the deployed n sensors can find a communication path between itself and the bus on the road such that the data collected by the sensor can be forwarded to the bus.

After we solve the sensor deployment problem, we need to calculate the time period during which the BS received data from all the N+n sensors at least once. Since different applications have different demand for data timeliness, we should analyze the average delivery delay to ensure the data collected by the BS is useful and valuable.

5 Sensor Deployment Scheme

In this part, we propose an effective scheme to solve the deployment problem. Firstly, we consider a simple large region without boundaries, then we extend our result to an environment with boundaries.

5.1 Simple Large Regions Without Boundaries

The basic idea is to deploy sensors row by row. The sensors in a row is connected and adjacent rows should guarantee continuous coverage of the area. Then we will add some sensors between adjacent rows, if necessary, to maintain connectivity. Based on the relationship between r_s and r_c, we separate the discuss into two cases.

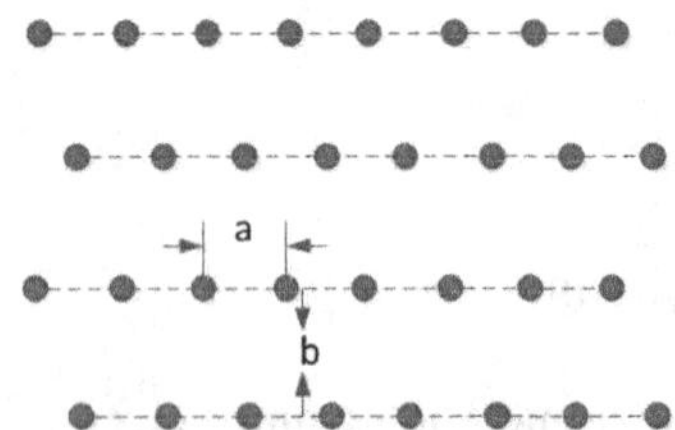

Fig. 4. $r_c \leq \sqrt{3}r_s$. Here, a = r_c, and b = $r_s + \sqrt{r_s^2 - \frac{r_c^2}{4}}$. The dots denote the sensor locations that form the horizontal strip, the dash line means the sensors it connected are connected.

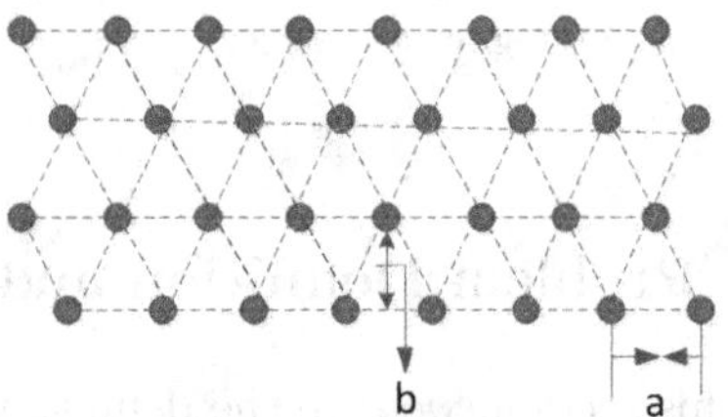

Fig. 5. $r_c > \sqrt{3}r_s$. Here, a = $\sqrt{3}r_s$, and b = $\frac{3r_s}{2}$. The red dots denote the sensor locations that form the triangular pattern, the dash line means the sensors it connected are connected (Color figure online).

Case 1: $r_c \leq \sqrt{3}r_s$. As shown in Fig. 4, sensors on each row are separated by a distance of r_c, so the connectivity of sensors in each row is already guaranteed. Since $r_c < \sqrt{3}r_s$, each row of sensors can cover a belt-like area with a width of $2 \times \sqrt{r_s^2 - \frac{r_c^2}{4}}$. Adjacent rows will be separated by a distance of $r_s + \sqrt{r_s^2 - \frac{r_c^2}{4}}$ and shifted by a distance of $\frac{r_c}{2}$ in an alternate way. With such an arrangement, the coverage of the whole area is guaranteed. Note that in the case of $r_c < \sqrt{3}r_s$, the distance between two adjacent rows is larger than r_c, so we need to place additional sensors between two adjacent rows, each separated by a distance no larger than r_c, to connect them. The optimality of this strip-based deployment pattern has been proved in [9].

Case 2: $r_c > \sqrt{3}r_s$. In this case, the triangular lattice [6] is the optimal deployment pattern to achieve both coverage and connectivity. As shown in Fig. 5, adjacent sensors are regularly separated by a distance of $\sqrt{3}r_s$, and both coverage and connectivity properties are satisfied.

5.2 Large Regions with Boundaries

In this paper, we consider the sensor deployment under the hybrid network structure, the city region are partitioned by the roads into may irregular convex and concave polygon region. In this section, we propose a method to deploy sensors in a region of arbitrary shapes.

Table 1. Coordinates of the six neighbors of a sensor in location (x,y)

Neighbor	$r_c \leq \sqrt{3}r_s$	$r_c > \sqrt{3}r_s$
N_1	$(x + r_c, y)$	$(x + \sqrt{3}r_s, y)$
N_2	$(x + \frac{r_c}{2}, y - \sqrt{r_s^2 - \frac{r_c^2}{4}} - r_s)$	$(x + \frac{\sqrt{3}r_s}{2}, y - \frac{3r_s}{2})$
N_3	$(x - \frac{r_c}{2}, y - \sqrt{r_s^2 - \frac{r_c^2}{4}} - r_s)$	$(x - \frac{\sqrt{3}r_s}{2}, y - \frac{3r_s}{2})$
N_4	$(x - r_c, y)$	$(x - \sqrt{3}r_s, y)$
N_5	$(x - \frac{r_c}{2}, y + \sqrt{r_s^2 - \frac{r_c^2}{4}} + r_s)$	$(x - \frac{\sqrt{3}r_s}{2}, y + \frac{3r_s}{2})$
N_6	$(x + \frac{r_c}{2}, y + \sqrt{r_s^2 - \frac{r_c^2}{4}} + r_s)$	$(x + \frac{\sqrt{3}r_s}{2}, y + \frac{3r_s}{2})$

Now, we modify the above solution for deploying sensors in a region with boundaries. Observe that in our solution, sensors are deployed in regular patterns. Table 1 summarize the coordinates of a sensor's six neighbours. The modified strip-based sensor deployment with boundaries (MSSDB) are described as follows:

In MSSDB, the input is the sensing radius r_s, the communication radius r_c, the boundary information of the irregular regions partitioned by the roads in the city, the output is the number of sensors we need to deploy. We first place a sensor in any location of the region, then the six locations that can potentially be deployed with sensors are determined according to Table 1. These locations are inserted into a queue Q. For each location *(x,y)* in Q, if *(x,y)* is outside the region, we delete the *(x,y)* from Q, otherwise we place a sensor node at *(x,y)*, then calculate the six locations of the six neighbours of *(x,y)* and insert them into Q if they have not been deployed with sensors. While Q is empty, the process is terminated.

Algorithm 1. MSSDB

```
Initialize: Q=NULL, N=NULL;
Place a sensor in any location of the region;
Calculate the coordinates of the six neighbours of the first deployed sensor node;
Insert the six locations into Q.
while Q ≠ NULL do
  for each (x,y)∈ Q do
    if (x,y) is inside the region then
      Place a sensor node at (x,y);
      Delete (x,y) from Q;
      Calculate the coordinates N {n1, n2, ..., n6) of the six neighbours of (x,y)};
      for each n ∈ N do
        if location n has not been deployed with sensors then
          Insert n intoQ;
        else
          Discard n;
        end if
      end for
    else
      Delete (x,y) from Q;
    end if
  end for
end while
```

Here is the time complexity analysis of the algorithm MSSDB, MSSDB is a simple greedy algorithm. A stack is used in the algorithm and the time complexity is related to the number of locations in the region. The number of locations in the region is related to the size of the target area Ω and the sensing radius r_s and communication radius r_c, so the complexity is $O(\frac{\Omega}{Min(r_c,r_s)^2})$. The time complexity of MSSDB is linear.

The above approach may leave two problems unsolved. First, as mentioned before, when $r_c < \sqrt{3}r_s$, we need to add extra sensors between adjacent rows to maintain connectivity. Second, some areas near the boundaries uncovered, as shown in Fig. 6. For the first problem, we don't need to add extra sensors between adjacent rows to maintain connectivity, because in our scheme, if each sensor can find a communication path between itself and the bus, then the data can be delivered to the BS by the bus. For the second problem, we separate the discuss into two cases.

Case 1: If the boundary of the region is the road, then the coverage hole can be healed by the bus, for the sensing range of the bus is large enough.

Case 2: If the boundary of the region is not the road, we need to place extra nodes to coverage the coverage hole. From the **observation** below, we can know that the largest diameter of the coverage hole is r_s, so we expand the boundary by the length of r_s, then we use the algorithm of MSSDB to deploy sensors

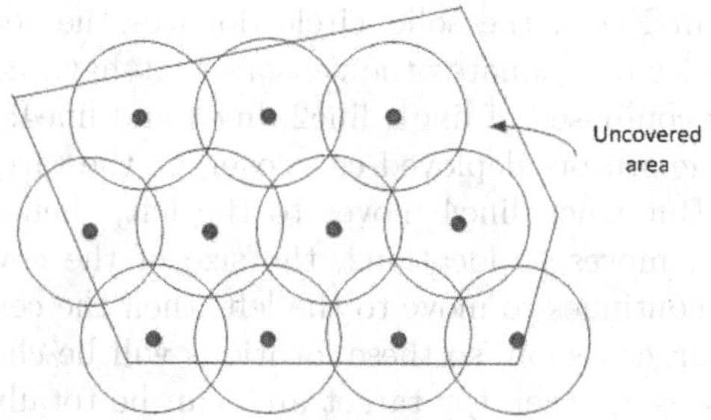

Fig. 6. Uncovered area around the boundary

within the expanded target area. Then the original target area can be totally covered by the sensors deployed in the expanded target area. This can be proved in the following way. The largest diameter of the coverage hole in the expanded target area is r_s, so the original target area must be covered totally.

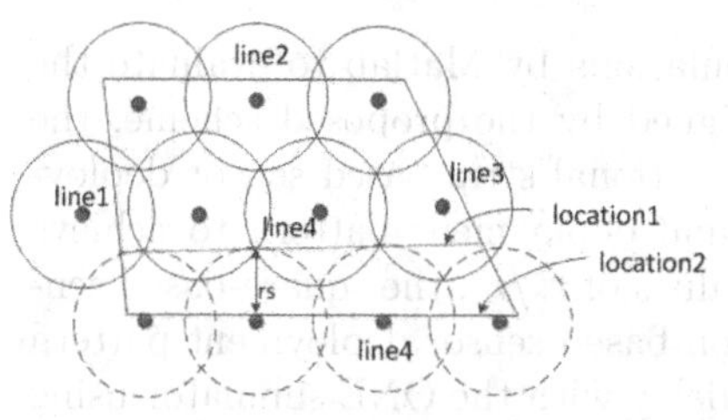

Fig. 7. The coverage hole generated along the horizontal direction

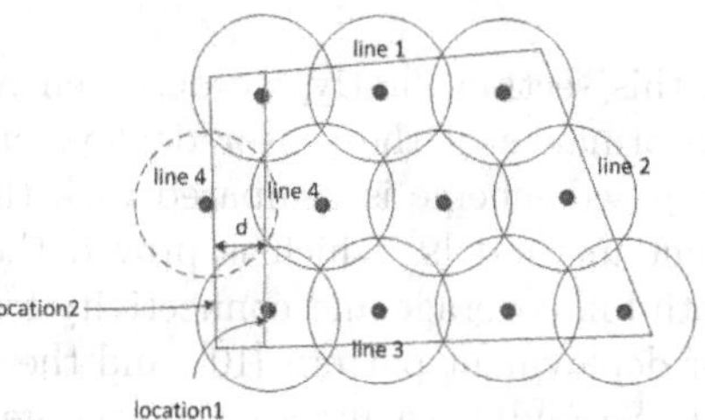

Fig. 8. The coverage hole generated along the vertical direction

Observation. *The largest diameter of the coverage hole is* r_s.

The above result can be easily derived in the following way. The generation of the coverage hole can be divided into two cases:

Case 1: As shown in Fig. 7, the solid circle donates the sensor that already deployed, the dashed circle donates the sensors that have not been deployed. The target region is composed of line1, line2, line3 and line4. If line4 is located at location1, then the sensors deployed can coverage the target region without any coverage hole. But when line4 moves down, then the coverage hole appears. When line4 moves to location2, the size of the coverage hole is the largest, and if line4 continues to move down, then the center of the dashed circle is inside the target region, so these locations will be chosen by algorithm MSSDB to deploy sensors, then the target area can be totally covered without any coverage hole. So the largest diameter of the coverage hole is r_s.

Case 2: As shown in Fig. 8, the solid circle donates the sensor that already deployed, the dashed circle donates the sensors that have not been deployed. The target region is composed of line1, line2, line3 and line4. If line4 is located at location1, then the sensors deployed can coverage the target region without any coverage hole. But when line4 moves to the left, then the coverage hole appears. When line4 moves to location2, the size of the coverage hole is the largest, and if line4 continues to move to the left, then the center of the dashed circle is inside the target region, so these locations will be chosen by algorithm MSSDB to deploy sensors, then the target area can be totally covered without any coverage hole. So the largest diameter of the coverage hole is d, is given by:

$$d = \begin{cases} r_c & \text{if } r_c < \sqrt{3}r_s \\ \frac{\sqrt{3}r_s}{2} & \text{if } r_c > \sqrt{3}r_s \end{cases}$$

Combining Case 1 and Case 2, we can figure out that the largest diameter of the coverage hole is r_s.

6 Simulations

In this section, firstly, we do extensive simulations by Matlab to evaluate the performance of the sensor deployment designed by the proposed scheme, the proposed scheme is compared with the conventional strip-based sensor deployment method [9] which is proved the optimal deployment pattern to achieve both full coverage and connectivity for all values of r_c/r_s, the square-based sensor deployment pattern [10], and the hexagon-based sensor deployment pattern [11]. Secondly, we analyse the average time delay with the ONE simulator using the representative algorithm Epidemic Routing [20].

Fig. 9. Map: A 3700 m by 4000 m area of Helsinki in Finland (map data provided by OpenStreetMap)

6.1 Simulation Scenarios

The simulation area is based on real map about 3700 m by 4000 m area of Helsinki, Finland, as shown in Fig. 9. There are 50 to 100 buses traveling on the roads, the location of the BS is set to (2397,3147) which is converted from the KKJ (Finnish National Coordinate System) [21]. The simulation is simulated for 12 h and the locations of 14 stationary nodes are selected. The bus which is within the communication range of the stationary nodes can collect the data from the subregion which is connected by the stationary nodes. The message is created by the stationary nodes every 25 to 35 s and the TTL of message is set to be 300 min.

The simulation parameters are set in Table 2. We assume that the buses are powerful and can support communication devices with a large transmission range.

Table 2. Table of simulation parameters

Environmental parameters	Used values
Experimental area	3700 × 4000 m
Sensing radius of ordinary sensors	15, 15, 15, 15 (m)
Communication radius of ordinary sensors	10, 15, 20, 30 (m)
Sensing radius of vehicle-mounted sensors	50 m
Communication radius of vehicle-mounted sensors	100, 150 (m)
Velocity of bus	7~15 m/s
Number of bus line	8
Number of bus on each line	4, 6, 8, 10, 12
Size of messages	500 kb~1 M
Size of buffer space on the bus	50 M
Width of the road	50 m

6.2 Simulation Analysis for Sensor Deployment

In this section, we present the performance evaluation of the sensor deployment designed by the proposed scheme. As shown in Fig. 8, the shape of the sensing field is irregular, we consider four cases: $(r_s,r_c) = (15,10)$, (15,15), (15,20) and (15,30) to reflect the relationships of $r_s > r_c$, $r_s = r_c$, $r_s < r_c \leq r_s$, and $r_c > \sqrt{3}r_s$, respectively. For each case, we run the simulation for 5 times and report the average.

Figure 10 presents the performance of the conventional strip-based sensor deployment algorithm and the modified strip-based sensor deployment with boundaries MSSDB, the square-based sensor deployment algorithm and the hexagon-based sensor deployment in terms of the number of deployed sensors

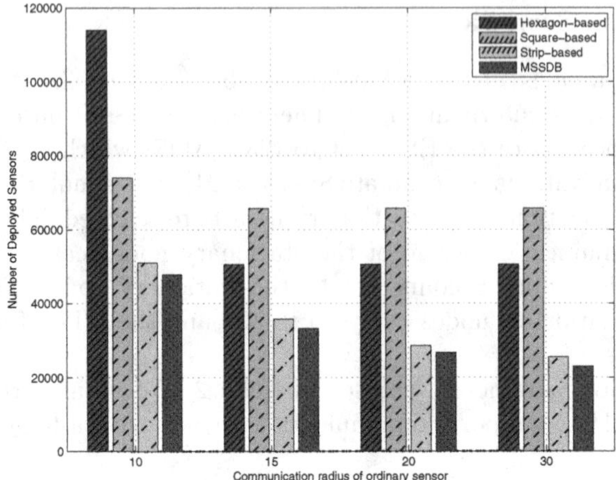

Fig. 10. Number of sensors needed by different deployment patterns. r_s is 15 m.

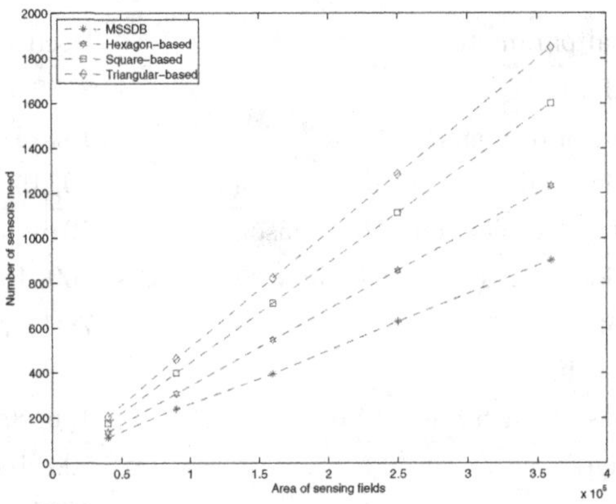

Fig. 11. Comparison of the growing rate of sensor numbers when the area of sensing field increases

versus the relationship of the sensing radius and communication radius. We can see that MSSDB outperforms the other three sensor deployment methods on the whole. With the value of r_c/r_s increasing, the number of deployed sensors decrease in all the four algorithms. Our scheme needs less sensors than the conventional strip-based method about 2000 in average without considering the relationship of the sensing radius and communication radius. The other two conventional deployment method need more sensors, when r_c is 10 m, square-based deployment method need less sensors than hexagon-based deployment method, but when r_c is increasing, square-based deployment method need more sensors than hexagon-based deployment method. Our scheme saves about 20 thousand in average compared with square-based and hexagon-based sensor deployment method.

Figure 11 illustrates that, when the sensing field expands from approximate 40000 m^2 to 360000 m^2, the number of sensors required in MSSDB, hexagon-based, square-based and triangular-based increases by about 800, 1100, 1400 and 1600 respectively. The lowest growing rate of node numbers required by MSSDB proves a better adaptability to the changing of network scale.

6.3 Simulation Analysis for Average Delay

This section analyse the average delivery delay between the ordinary nodes and the BS. Due to that the data transmission speed in WSNs is far greater than the data propagation speed in BANETs, which mean that they are not on an order of magnitude, we only consider the delivery delay in BANETs in this paper.

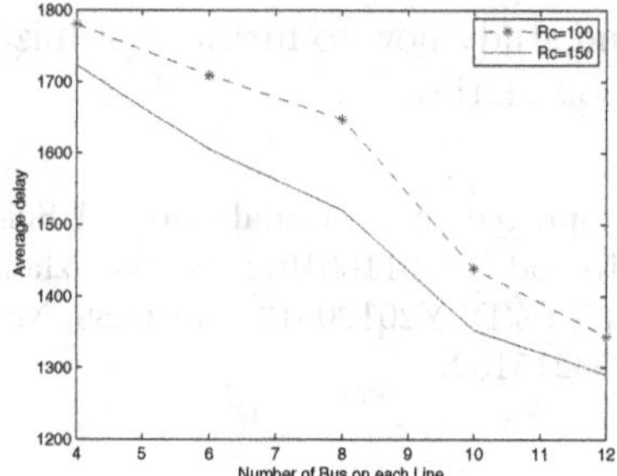

Fig. 12. Average delay for different number of bus on each line and different R_c.

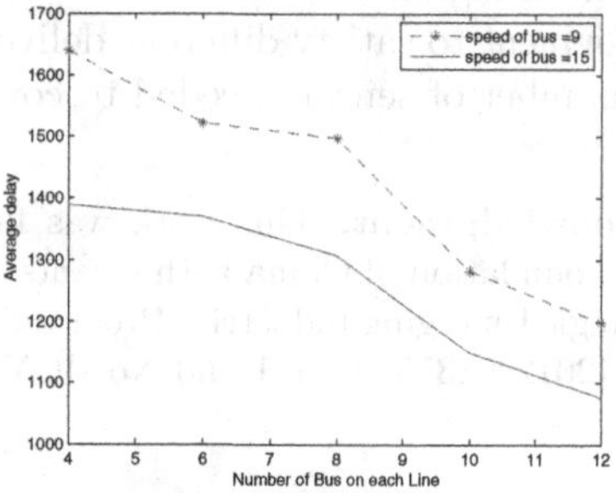

Fig. 13. Average delay for different number of bus on each line and different speed of bus.

Figure 12 presents the average delay versus number of bus on each line when the speed of the bus is set to be 7 to 10. We can see that with the number of bus on each line increasing, the average delay decreased at the same time. At the beginning, the delivery latency is high, it means that buses running on the road are not enough to encounter with each other, so maybe only when certain bus reaches to the BS, the BS can obtain all the data. But with the number of bus increasing, the chance that two buses encounter with each other is high, then the messages can be delivered through multi-hop rather than carried by the bus all the time, so the delivery latency becomes less. Figure 12 also show that transmission range have an important influence on the average delay, when the transmission range is high, the average delay is less.

Figure 13 presents the average delay versus the number of bus on each line when the communication radius is set to be 150. We can see that the speed of the buses running on the road also influence the average delay, when the speed is high, the delay is less.

In summary, the average delay is influenced by three important factors: the number of bus, the communication radius of the bus, and the speed of the bus running on the road. The delay can be less to a great extent with more buses, higher communications and higher speed. Generally speaking, The average delay can be controlled in about 25 min.

7 Conclusion and Future Works

In this paper, we introduce how to utilize WSNs and BANETs to monitor the city area cooperatively and analyse that using the hybrid network framework can reduces the sensors deployed considerably. We propose an effective algorithm to deploy sensors in irregular sensing fields. Simulations show that our method outperforms the conventional strip-based sensor deployment algorithm, the square-based sensor deployment pattern and the hexagon-based sensor deployment pattern. And the messages can be delivered to the BS in time, which is valuable and meaningful. For the future work, we will study how to adjust the sensor deployment to satisfy different delivery delay and study how to further optimize the number of sensors needed to coverage the boundaries.

Acknowledgment. This work was financially supported by National Natural Science Foundation of China with Grants No. 61370216 and No. 61100191, and Shenzhen Strategic Emerging Industries Program with Grants No. ZDSY20120613125016389, No. JCYJ20120613151201451 and No. JCYJ20130329153215152.

References

1. Keränen, A., Ott, J., Kärkkäinen, T.: The ONE simulator for DTN protocol evaluation. In: Proceedings of the 2nd International Conference on Simulation Tools and Techniques, Brussels, Article No. 55 (2009)
2. Wu, L., Du, H., Wu, W., Li, D., Lv, J., Lee, W.: Approximations for minimum connected sensor cover. In: IEEE INFOCOM (2013)
3. Bai, X., Yun, Z., Xuan, D., Jia, W., Zhao, W.: Pattern mutation in wireless sensor deployment. In: IEEE INFOCOM, pp. 1–9 (2010)
4. Luo, H., Du, H., Kim, D., Ye, Q., Zhu, R., Zhang, J.: Imperfection better than perfection: beyond optimal lifetime barrier coverage in wireless sensor networks. In: 10th International Conference on Mobile Ad-hoc and Sensor Networks (MSN), December 2014
5. Chong, C.Y., Kumar, S.: Sensor networks: evolutions, opportunities, challenges. Proc. IEEE **91**(8), 1247–1256 (2003)
6. Kershiner, R.: The number of circles covering a set. Am. J. Math. **61**, 665–671 (1939)
7. Zhang, H., Hou, J.: Maintaining sensing coverage and connectivity in large sensor networks. In: NSF International Workshop on Theoretical and Algorithmic Aspects of Sensor, Ad Hoc Wireless, and Peer-to-Peer Networks (2004)
8. Iyengar, R., Kar, K., Banerjee, S.: Low-coordination topologies for redundancy in sensor networks. In: Proceedings of the 6th ACM International Symposium on Mobile Ad Hoc Networking and Computing (Mobihoc) (2005)

9. Bai, X., Kumar, S., Xuan, D., Yun, Z., Lai, T.H.: Deploying wireless sensors to achieve both coverage and connectivity. In: Proceedings of the 7th ACM International Symposium on Mobile Ad Hoc Networking and Computing (Mobihoc) (2006)
10. Bai, X., Yun, Z., Xuan, D., Lai, T.H., Jia, W.: Deploying four-connectivity and full-coverage wireless sensor networks. In: Proceedings 27th IEEE International Conference on Computer Communications, IEEE INFOCOM (2008)
11. Bai, X., Xuan, D., Yun, Z., Lai, T.H., Jia, W.: Complete optimal deployment patterns for full-coverage and k-connectivity ($k \leq 6$) wireless sensor networks. In: Proceedings of the 9th ACM International Symposium on Mobile Ad Hoc Networking and Computing (Mobihoc), pp. 401–410 (2008)
12. Ishizuka, M., Aida, M.: Performance study of node placement in sensor networks. In: Proceedings of 24th International Conference on Distributed Computing Systems Workshops, pp. 598–603 (2004)
13. Miller, L.E.: Distribution of link distances in a wireless network. J. Res. Nat. Inst. Stand. Technol. **106**, 401–412 (2001)
14. Cheng, X., Du, D., Wang, L., Xu, B.: Relay sensor placement in wireless sensor networks. Wireless Netw. **14**, 347–355 (2008)
15. Lin, G., Xue, G.: Steiner ree problem with minimum number of Steiner points and bounded edge-length. J. Inf. Process. Lett. **69**, 53–57 (1999)
16. Chen, D., Du, D.Z., Hu, X.D., Lin, G., Wang, L., Xue, G.: Approximations for Steiner trees with minimum number of Steiner points. J. Global Optim. **18**, 17–33 (2000)
17. Kashyap, A., Khuller, S., Shayman, M.: Relay placement for higher order connectivity in wireless sensor networks. In: Proceedings 25th IEEE International Conference on Computer Communications (INFOCOM), pp. 1–12 (2006)
18. Zhang, W., Xue, G., Misra, S.: Fault-tolerant relay node placement in wireless sensor networks: problems and algorithms. In: Proceedings of 26th IEEE International Conference on Computer Communications (INFOCOM 2007), pp. 1649–1657 (2007)
19. Lloyd, E.L., Xue, G.: Relay node placement in wireless sensor networks. IEEE Trans. Comput. **56**, 134–138 (2007)
20. Vahdat, A., Becker, D.: Epidemic routing for partially connected ad hoc networks. Technical report CS-200006, Duke University, April 2000
21. http://www.kolumbus.fi/eino.uikkanen/geodocsgb/ficoords.html

New Insight into 2-Community Structures in Graphs with Applications in Social Networks

Cristina Bazgan[1,2], Janka Chlebíková[3], and Thomas Pontoizeau[1(✉)]

[1] PSL, Université Paris-Dauphine, LAMSADE UMR CNRS 7243, Paris, France
{bazgan,thomas.pontoizeau}@lamsade.dauphine.fr
[2] Institut Universitaire de France, Paris, France
[3] School of Computing, University of Portsmouth, Portsmouth, UK
janka.chlebikova@port.ac.uk

Abstract. We investigate the structural and algorithmic properties of 2-community structure in graphs introduced by Olsen [13]. A 2-community structure is a partition of vertex set into two parts such that for each vertex of the graph number of neighbours in/outside own part is in correlation with sizes of parts. We show that every 3-regular graph has a 2-community structure which can be found in polynomial time, even if the subgraphs induced by each partition must be connected. We introduce a concept of a 2-weak community and prove that it is NP-complete to find a balanced 2-weak community structure in general graphs even with additional request of connectivity for both parts. On the other hand, the problem can be solved in polynomial time in graphs of degree at most 3.

Keywords: Graph theory · Complexity · Graph partitioning · Community structure · Clustering · Social networks

1 Introduction

The problematic around community structures is closely related to the well established research areas of clustering and graph partitioning where similar problems have been studied from different aspects. A good introduction and overview of clustering and partitioning results can be found in [6,11,12,14]. The problems associated with communities are well motivated by current research in social networks such as Facebook, Linkedin, see also Sect. 2 in [5] for more details about various applications. A standard abstract model for such networks are graphs, in which a community in a graph intuitively corresponds to a dense subgraph. More formally, a community structure is a partition of vertices with some additional constrains such as number of edges between parts or general constrains as connectivity for each part. Therefore the new results for communities may find applications in the areas similar to a graph partition such as parallel-computing, VLSI-circuit design, route planning [8] and divide-and-conquer algorithms [15].

There are several definitions proposed for a community structure in the literature together with some structural and complexity results [3,4,9,13]. The results

Z. Lu et al. (Eds.): COCOA 2015, LNCS 9486, pp. 236–250, 2015.
DOI: 10.1007/978-3-319-26626-8_18

in this paper are based on the definition of community structure introduced recently by Olsen [13] which seems to be a natural model for a community in undirected connected graphs. We introduce the concept of a weak community structure in which every member of a community considers itself as a part of the community. We investigate the structural properties of the members of communities for fixed number of two communities in the graphs of maximum degree 3 and present some algorithmic results. The results are further extended for a weak community together with additional constrains such as connectivity or the same size for both parts (balanced partition).

The partition of graphs are intensively studied in the literature with various measures to evaluate their optimality, see for example [1,6]. In the balanced partition problem, which can be seen as a generalisation of the bisection problem to any given number of parts, the goal is to minimise the number of edges between partitions. It is known that the problem cannot be approximated within any finite factor in polynomial time in general graphs and it remains APX-hard even if the maximum degree of the tree is constant [10]. It demonstrates that some graph partitions problems which are related to e.g. balanced communities are hard to solve even for restricted graph classes. This indicates hardness of different problems related to a community structure too, hence any positive results in community structure problems are important to get better understanding of differences between community and partition problems.

The paper is structured as follows. In Sect. 2 we introduce formally some notation and definitions of problems we study. In Sect. 3 we show that every 3-regular graph has a 2-community structure which can be found in polynomial time, even if the subgraphs induced by each partition must be connected. In Sect. 4 we prove that every graph of maximum degree 3 has a balanced weak 2-community structure that can be found in polynomial time, while the problem is NP-complete in general graphs even when both parts should remain connected. Conclusions and open questions are provided in Sect. 5. Due to the space constraints, some proofs are deferred to a full version.

2 Preliminaries

In the paper, all considered graphs are undirected and connected. For any subgraph C of the graph $G = (V, E)$ and a vertex $v \in V$ let $N_C(v)$ (resp. $N_C[v]$) be the set of the neighbours (resp. closed neighbours) of v in C, $d(v)$ be the degree of the vertex v in G. For a partition of V into two parts and $v \in V$ let $d_{in}(v)$ (resp. $d_{out}(v)$) be the number of neighbours in its own part (resp. out of its part). A partition $\{C_1, C_2\}$ of V is connected if the subgraphs induced by C_1 and C_2 are connected. A partition $\{C_1, C_2\}$ of V is balanced if the sizes C_1, C_2 differ by at most 1. A graph is k-regular if every vertex is of degree k, $k \geq 2$. A pendant vertex of G is any vertex of degree 1. A star is a complete bipartite graph $K_{1,\ell}$ for any $\ell \geq 1$.

Now we introduce Olsen's definition of a k-community structure from [13].

Definition 1. *A k-community structure for a connected graph $G = (V, E)$ is a partition $\Pi = \{C_1, \ldots, C_k\}$ of V, $k \geq 2$, such that $\forall i \in \{1, \ldots, k\}, |C_i| \geq 2$, and $\forall v \in C_i, \forall C_j \in \Pi$, $j \neq i$, the following holds*

$$\frac{|N_{C_i}(v)|}{|C_i| - 1} \geq \frac{|N_{C_j}(v)|}{|C_j|} \tag{1}$$

For a weak k-community structure, (1) is replaced by a "weaker" condition

$$\frac{|N_{C_i}[v]|}{|C_i|} \geq \frac{|N_{C_j}(v)|}{|C_j|}. \tag{2}$$

Notice that a k-community structure is obviously a weak k-community structure since $\frac{|N_{C_i}[v]|}{|C_i|} = \frac{|N_{C_i}(v)|+1}{|C_i|-1+1} \geq \frac{|N_{C_i}(v)|}{|C_i|-1}$, but the opposite is not true (see Fig. 1).

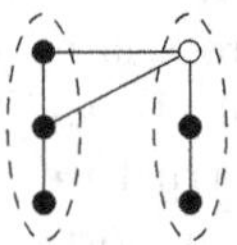

Fig. 1. A 2-weak-community structure of a graph in which the blank vertex does not satisfy condition (1) but satisfies condition (2).

If we remove the restriction k on the number of communities from Definition 1, but (1) is still true for each vertex, we obtain a concept of *a community structure* introduced by Olsen [13]. Olsen proved that a community structure without any restriction on the number of communities can be found in polynomial time for any graphs (with at least 4 vertices) except the star graphs. He also proved that it was NP-complete to find a community structure in a graph in which a given set of vertices is included in a part [13].

In this paper we investigate the problems of a community structure for fixed number of two communities:

2-Community

Input: A graph G.

Question: Does G have a 2-community structure?

Obviously, if G has a 2-community structure, it must have at least 4 vertices and not be isomorphic to a star which we assume in the paper.

In the Weak 2-Community problem we are looking for a weak 2-community structure in a graph. Adding the balanced condition to the 2-Community problem (for graphs with even number of vertices), we obtain the Balanced 2-Community problem introduced by Estivill-Castro et al. [9]. Similarly we can define the Balanced Weak 2-community problem.

The additional constrain which asks for subgraphs induced by each part of the partition to be connected is a natural constrain useful for the problems related to the connectedness. The CONNECTED 2-COMMUNITY problem is to decide if a graph has a connected 2-community structure, i.e. a 2-community structure $\{C_1, C_2\}$ such that the subgraphs induced by C_1, C_2 are connected. We can define analogous problems for weak and balanced versions.

The following overview summarises the classes of graphs in which a 2-community structure (hence also weak 2-community) always exists. Depending on the case the results can be extended to connected or balanced communities. All these results are either easy observations or contained as the main results in this paper.

In this way, there always exists a 2-community structure which can be found in polynomial time in (considering graphs with at least 4 vertices):

- 3-regular graphs, even a connected 2-community structure (Sect. 3).
- graphs of bounded tree-width (except stars), even a balanced 2-community structure. (The results follow from [2], see Sect. 4 for more details.)
- graphs with minimum degree $\lceil \frac{(c-1).|V|}{c} \rceil$ where c is the size of an inclusion-wise maximal clique in G. Denote by C such a clique in G and consider the partition $\{C, V \setminus C\}$. Then the condition (1) is satisfied for all vertices in C (the left part of the inequality is 1). The condition (1) is also satisfied for all neighbours $x \in V \setminus C$ of vertices in C since $\frac{d_{in}(x)}{|V|-c-1} \geq \frac{\frac{(c-1).|V|}{c}-(c-1)}{|V|-c-1} \geq \frac{c-1}{c} \geq \frac{d_{out}(x)}{c}$. Finally, the other vertices in $V \setminus C$ trivially satisfy condition (1) since the right part of the inequality is 0.
- graphs of maximum degree 2 (hence either a cycle or a path). Indeed, any balanced connected partition is a connected 2-community structure (an easy exercise).
- complete graphs, since any partition in which each part has at least 2 vertices is an example of a connected 2-community structure (an easy exercise).

Moreover, there always exists a balanced weak 2-community structure in graphs of maximum degree 3, but this is not true for 2-community structure, see Sect. 4.

Estivill-Castro et al. [9] proved that the problem of finding a balanced 2-community structure is NP-complete. In Sect. 4 we show that the same result also holds for a weak community, even with additional constrain of connectivity for both parts. We also present a shorter proof of the known NP-complete result for a balanced 2-community in general graphs based on an alternative definition of community structure [3], which also implies NP-completeness for a connected balanced 2-community. On the other hand, we prove that every graph of degree at most 3 has a weak balanced 2-community structure which can be found in a polynomial time.

3 2-Community Structure in 3-Regular Graphs

In this section we show that any 3-regular graph has a 2-community structure (even connected) computable in polynomial time.

First, the restrictions on the size of partitions are discussed to ensure the vertices fulfil the condition (1) in case of a 2-community structure.

Lemma 1. *Let G be a 3-regular graph of size n. Let $\{C_1, C_2\}$ be a partition of G such that $\lceil \frac{n-1}{3} \rceil \leq |C_1| \leq n - \lceil \frac{n-1}{3} \rceil$. Then each vertex of G which has at most one neighbour out of its own part fulfils the condition (1) of a 2-community structure.*

Lemma 2. *Let G be a 3-regular graph of size n. Let $\{C_1, C_2\}$ be a 2-partition of G such that $|C_1| = \lceil \frac{n}{3} \rceil$ or $|C_1| = \lfloor \frac{n}{3} \rfloor$. Then each vertex of degree 3 from C_1 which has two neighbours in C_2 fulfils the condition (1).*

Theorem 1. *Every 3-regular graph has a 2-community structure. Moreover it can be found in polynomial time.*

Proof. Let $G = (V, E)$ be a 3-regular graph of size n. The algorithm runs in two stages.

Stage 1: The algorithm finds a partition $\{C_1, C_2\}$ of V such that $|C_1| = \lceil \frac{n-1}{3} \rceil$ and at most two vertices from C_1 have more than one neighbour in C_2.

Stage 2: The algorithm moves some vertices between C_1 and C_2 until $\lceil \frac{n-1}{3} \rceil \leq |C_1| \leq n - \lceil \frac{n-1}{3} \rceil$ and each vertex of G has a restricted number of neighbours out of its own part in a such way that Lemmas 1 or 2 can be applied.

Stage 1: Let $u, v \in V$ be such that $(u, v) \in E$ and put $C_1 = \{u, v\}$. Now repeat the following steps (S1) and (S2) until $|C_1| = \lceil \frac{n-1}{3} \rceil$:

(S1) Let w be a neighbour of u (or v) which is not in C_1, put $C_1 := C_1 \cup \{w\}$, $u := w$ (or $v := w$).

(S2) If there is no such vertex w, the degree of each vertex in the subgraph induced by C_1 is 2 or 3. In such a case let u be any vertex of degree 2 in the subgraph induced by C_1.

It is clear that at the end of the first stage the algorithm finishes with a set C_1 such that $|C_1| = \lceil \frac{n-1}{3} \rceil$. If it is not possible to apply (S1) and (S2) and $|C_1| < \lceil \frac{n-1}{3} \rceil$ then all vertices in C_1 must have all neighbours in C_1 which means that G is not connected.

Furthermore at most two vertices (u and v) from C_1 may have more than one neighbour outside C_1 and the subgraph induced by C_1 is connected. Define $C_2 = V \setminus C_1$.

Stage 2: We distinguish two major cases:

Case 1: If $\forall w \in C_2$, $d_{out}(w) \leq 1$ then all vertices in G except u and v have at most one neighbour out of its part. Using Lemma 1, these vertices fulfil the condition (1). Moreover, $\lceil \frac{n-1}{3} \rceil$ equals $\lfloor \frac{n}{3} \rfloor$ or $\lceil \frac{n}{3} \rceil$ and according to Lemma 2, (1) is also true if u, v have two vertices out of C_1. Hence, $\{C_1, C_2\}$ is a 2-community structure.

Case 2: There exists a vertex $w \in C_2$, such that $d_{out}(w) \geq 2$.

Now we distinguish several subcases:

(A) $\forall x \in C_1, d_{out}(x) \leq 1$
(B) All vertices from C_1 which have more than one neighbor outside C_1 are adjacent to w (only $u, v \in C_1$ are possible candidates).
(C) No vertex from C_1 which has more than one neighbor outside C_1 is adjacent to w (only $u, v \in C_1$ are possible candidates).
(D) Both vertices $u, v \in C_1$ have more than one neighbor outside C_1, but only one of them is adjacent to w.

Case 2(A): Repeat the update step until it is possible:
if $\exists z \in C_2$, $d_{out}(z) \geq 2$, update C_1 and C_2 as follows: $C_1 := C_1 \cup \{z\}, C_2 := C_2 \setminus \{z\}$.
If no update step is possible, then return $\{C_1, C_2\}$.

After each update step, two neighbours of z in C_1 have degree three. After repeating the update step k times, C_1 has $\lceil \frac{n-1}{3} \rceil + k$ vertices and at least $2k$ vertices in C_1 have degree three in the subgraph induced by C_1. Hence, we can repeat the update step at most $\lceil \frac{n-1}{3} \rceil - 1$ times. Otherwise the degree of each vertex in C_1 is three which implies G is not connected. Thus $|C_1| \leq 2\lceil \frac{n-1}{3} \rceil - 1 \leq n - \lceil \frac{n-1}{3} \rceil$. Note that now every vertex in G has at most one neighbour out of its part and thus applying Lemma 1, $\{C_1, C_2\}$ is a 2-community structure.

Case 2(B): Wlog we suppose that $d_{out}(u) = 2$ and u is adjacent to w. Furthermore, if $d_{out}(v) = 2$ then also v is adjacent to w.

Now using C_1, C_2 we define a 2-community partition. After the initial update: $C_1 := C_1 \cup \{w\}, C_2 := C_2 \setminus \{w\}$, all vertices in C_1 have at most one neighbor in C_2 and $|C_1| = \lceil \frac{n-1}{3} \rceil + 1$. Then repeat the update step:

- if $\exists z \in C_2, d_{out}(z) \geq 2$, define $C_1 := C_1 \cup \{z\}, C_2 := C_2 \setminus \{z\}$, until $|C_1| = 2\lceil \frac{n-1}{3} \rceil - 1$ or if there is no such a vertex z to update C_1.

There are two possible scenarios:

(i) if $|C_1| \leq 2\lceil \frac{n-1}{3} \rceil - 1$ and there is no such vertex $z \in C_2$ such that $d_{out}(z) \geq 2$, then each vertex in G has at most one neighbour out of its own part. Obviously, $|C_1| \leq n - \lceil \frac{n-1}{3} \rceil$ and due to Lemma 1, $\{C_1, C_2\}$ is a 2-community structure.
(ii) $|C_1| = 2\lceil \frac{n-1}{3} \rceil - 1$ and $\exists z \in C_2$ such that $d_{out}(z) \geq 2$: the update step has been repeated $\lceil \frac{n-1}{3} \rceil - 2$ times and in each step the number of vertices $x \in C_1$ with $d_{out}(x) = 1$ is decreased by at least 2 (the neighbours of z in C_1). It means every vertices in V has all neighbours in its own part, except at most three vertices, each having one neighbour out of its part.

- If $n \equiv 2 \mod 3$, then the size of $|C_2| = n - (2\lceil \frac{n-1}{3} \rceil - 1) = \lceil \frac{n}{3} \rceil$. Because $|C_1| \leq n - \lceil \frac{n-1}{3} \rceil$, due to Lemma 1, all vertices with at most one neighbor out of its own part fulfil the condition (1). If a vertex of C_2 is adjacent to exactly two vertices from C_1 then the condition (1) is true according to Lemma 2. A vertex of C_2 cannot be adjacent to all three vertices in C_1, otherwise $C_1 \cup \{w\}$ is a disconnected part of G. Hence, $\{C_1, C_2\}$ is a 2-community partition.

- If $n \equiv 0 \mod 3$ or $n \equiv 1 \mod 3$, then define the last update: $C_1 := C_1 \cup \{z\}, C_2 := C_2 \setminus \{z\}$. Now only one vertex of C_1 has one neighbour in C_2. Because $|C_1| = 2\lceil\frac{n-1}{3}\rceil \leq n - \lceil\frac{n-1}{3}\rceil$, the condition (1) is true for all vertices of G because of Lemma 1. Hence the updated partition $\{C_1, C_2\}$ is a 2-community structure.

Case 2(C): Wlog we suppose that $d_{out}(u) = 2$. Update $C_1 := C_1 \cup \{w\} \setminus \{u\}, C_2 := V \setminus C_1$.

Notice that after the update, $|C_1| = \lceil\frac{n-1}{3}\rceil$ and there may be at most two vertices in C_1 which have two neighbours in C_2. Hence we are again in one of the cases (A)-(D) of second stage, but each time we apply this update the size of the cut between C_1 and C_2 is decreases by two. Therefore the process is finite.

Case 2(D): Wlog suppose that u is adjacent to w and $d_{out}(u) = 2$. Update $C_1 := C_1 \cup \{w\} \setminus \{v\}$, $C_2 := V \setminus C_1$.

Notice that after the update, $|C_1| = \lceil\frac{n-1}{3}\rceil$. Moreover, u has two neighbours inside C_1 since u is obviously not adjacent to v. Hence we are in one of the previous cases of stage 2, since there is at most one vertex in C_1 (the neighbour of v) which could have two neighbours in C_2. □

Now we investigate the problem of finding a 2-community structure with additional condition of connectivity for each part. Using the algorithm from Theorem 1 as a tool we extend the result for a connected 2-community structure in 3-regular graphs, but with many fine details in the proof.

Lemma 3. *Let G be a 3-regular graph and $\{C_1, C_2\}$ a connected 2-partition of G with $\lceil\frac{n-1}{3}\rceil \leq |C_1| \leq n - \lceil\frac{n-1}{3}\rceil$ such that each part has at most one vertex with two neighbours out of its own part. Then G has a connected 2-community structure which can be found in polynomial time.*

Proof. The main idea is to move selected vertices between two parts in such a way that it preserves connectivity and offers the options to use Lemmas 1 or 2.

We discuss four cases depending on which vertices have two neighbours out of its own part. Notice that transferring a vertex which has two neighbours out of its part does not compromise the connectivity of the partition.

(a) If there is no vertex in C_1 and C_2 with two neighbours out of its own part, then using Lemma 1 the partition $\{C_1, C_2\}$ is already a connected 2-community structure.
(b) If the only vertex with two neighbours out of its own part is in C_2, then update C_1, C_2 using the following loop:
    while $|C_1| < n - \lceil\frac{n-1}{3}\rceil$ and there exists a vertex z in C_2 which has two neighbours in C_1, update $C_2 := C_2 \setminus \{z\}$, $C_1 := C_1 \cup \{z\}$.
    Obviously after each run of the while loop both parts of the partition remains connected. At the end of the while loop
    - if $|C_1| < n - \lceil\frac{n-1}{3}\rceil$, then all vertices in G have at most one neighbour out of their parts and satisfy the properties of 2-community structure due to Lemma 1,

  - if $|C_1| = n - \lceil\frac{n-1}{3}\rceil$ then $|C_2| = \lceil\frac{n-1}{3}\rceil$ and hence all vertices in C_2 with two neighbours out of the own part satisfy the properties of 2-community structure due to Lemma 2, the rest of vertices satisfy Lemma 1.

(c) The only vertex with two neighbours out of its own part is in C_1. Then the case is similar to (b) by symmetry swapping the roles between C_1 and C_2.

(d) There are two vertices:
  - $v_1 \in C_1$ with two neighbours in C_2 and let v_1^0 be the neighbour of $v_1 \in C_1$;
  - $v_2 \in C_2$ with two neighbours in C_1 and let v_2^0 be the neighbour of $v_2 \in C_2$.

  Now we need to distinguish two cases:

  (i) If $(v_1, v_2) \in E$, then we update the partition as follows. If $|C_1| < n - \lceil\frac{n-1}{3}\rceil$ then define a new partition $C_1 := C_1 \cup \{v_2\}$; $C_2 := C_2 \setminus \{v_2\}$, otherwise $C_1 := C_1 \setminus \{v_1\}$; $C_2 := C_2 \cup \{v_1\}$. Obviously, $\{C_1, C_2\}$ is a connected partition which fulfil initial conditions of lemma, so we can apply case (a), (b), (c) or (d) again. Notice that the case (d) can be repeated only finite number of times since the cut size between C_1 and C_2 decreases each time the case is applied.

  (ii) If $(v_1, v_2) \notin E$ we define the following update $C_1 := C_1 \cup \{v_2\} \setminus \{v_1\}$, $C_2 := C_2 \cup \{v_1\} \setminus \{v_2\}$. The new partition is a connected partition with no change in sizes and the following options are possible:
    - If $d_{out}(v_1^0) = 0$ or $d_{out}(v_2^0) = 0$ before the update, then update removes at least one of the vertices with two outgoing edges. Now we can again apply one of cases (a), (b) or (c) which leads to a connected 2-community structure.
    - If $d_{out}(v_1^0) > 0$ and $d_{out}(v_2^0) > 0$ then we can apply case (d) again. This process is finite because each time the size of the cut size between C_1 and C_2 is decreased by 2.

Obviously, the whole procedure can be run in polynomial time. □

Theorem 2. *Every* 3*-regular graph has a connected* 2*-community structure. Moreover it can be found in polynomial time.*

Proof. The algorithm runs in two stages similarly to the algorithm in Theorem 1.

Stage 1: The algorithm finds either a connected partition $\{C_1, C_2\}$ such that $|C_1| = \lceil\frac{n-1}{3}\rceil$ and at most two vertices from C_1 have two neighbours in C_2 or ends up with a connected 2-community structure.

Stage 1: Apply directly Stage 2 from Theorem 1.

The difference to the approach from Theorem 1 is that C_2 remains connected until the end of the first stage, where C_1 is connected in both approaches.

Then we apply the second stage of the algorithm from Theorem 1. Since moving a vertex which has 2 neighbours in the other part never disconnect any part and all transfers only affect such vertices, the final partition $\{C_1, C_2\}$ remains connected at the end of the second stage.

Stage 1: (for a connected partition)
Choose any vertices $u, v \in V$ such that $\{u, v\} \in E$ and the subgraph induced by $V \backslash \{u, v\}$ is connected. Label the vertices u, v and define $C_1 := \{u, v\}$, $C_2 := V \backslash \{u, v\}$.

The initial construction: While $|C_1| < \lceil \frac{n-1}{3} \rceil$ and one of the updates (S1), (S2) (in this order) can be applied do:

(S1) If there exists a vertex $x \in C_2$ such that $d_{out}(x) = 2$, then update $C_1 := C_1 \cup \{x\}$, $C_2 := C_2 \backslash \{x\}$. If all labelled vertices have three neighbours in C_1, then removes all labels and label one vertex in C_1 which has one neighbour in C_2.

(S2) If there exists a vertex $x \in C_2$ such that x is a neighbour of a labelled vertex w in C_1 and $C_2 \backslash \{x\}$ remains connected then update $C_1 := C_1 \cup \{x\}$, $C_2 := C_2 \backslash \{x\}$, label the vertex x and remove label from w. If all labelled vertices have three neighbours in C_1, then removes all labels and label one vertex in C_1 which has one neighbour in C_2.

Obviously after each update we can have at most two labelled vertices in C_1.

Now there are two possibilities how the initial construction can finish:

(1) The algorithm finishes with $|C_1| = \lceil \frac{n-1}{3} \rceil$. Due to the properties of the construction, the partition $\{C_1, C_2\}$ is connected and at most two vertices from C_1 may have two neighbours in C_2. In such a case we can move directly to Stage 2.
(2) If none of the updates (S1), (S2) can be applied and $|C_1| < \lceil \frac{n-1}{3} \rceil$ then we redefine the partition $\{C_1, C_2\}$ using the major update construction to obtain a new partition which leads to a connected 2-community structure.

The major update construction:

Step A: Let $N \subseteq C_2$ be the set of neighbours of all labelled vertices from C_1, hence $1 \leq |N| \leq 4$. In the first part we define the subsets Q, Q', Z of C_2 and the vertices q, $q' \in N$ (q, q' are not necessarily distinct) such that $\{Q, Q', Z, \{q\}, \{q'\}\}$ is a connected partition of the graph induced by C_2 and each vertex from $Q \cup Q'$ has at most one neighbour in C_1. Furthermore, the entire set Q (resp. Q') has exactly one neighbour outside Q (resp. Q') in C_2 and this is the vertex $q \in N$ (resp. $q' \in N$).

We show that such Q, Q', Z exist and are always connected. Consider a vertex $v_1 \in N$. The vertex v_1 has necessarily two neighbours in C_2 and the subgraph induced by $C_2 \setminus \{v_1\}$ is disconnected (otherwise (S1) or (S2) from Stage 1 could be applied). Let C_2^1 and C_2^2 be the two connected components of $C_2 \setminus \{v_1\}$.

Define $Q := C_2^1$, $Q' := C_2^2$ and $q = q' = v_1$. Moreover, $Z = \emptyset$ is trivially connected. Hence in case $|N| = 1$ we can now move directly to Step B.

If $|N| > 1$, select another vertex $v_2 \in N$. Such vertex must be in Q or Q' defined above. Consider wlog that $v_2 \in Q$. If $\{v_1, v_2\} \in E$, then update $Q := Q \backslash \{v_2\}$, $q := v_2$, $q' := v_1$ and $Z := \emptyset$. Notice that Q is still connected since v_2 has only one neighbour in Q. If $\{v_1, v_2\} \notin E$, $Q \setminus \{v_2\}$ must be disconnected into two part Q^1 and Q^2. Name Q^2 the set which contains a neighbour of v_1 (Q^1

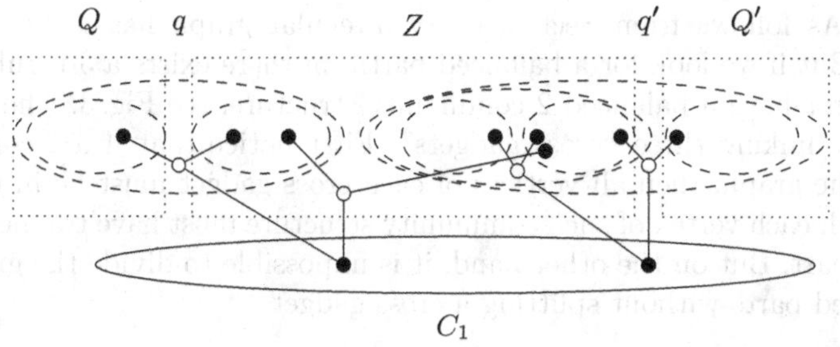

Fig. 2. Splitting C_2 when $|N| = 4$ (vertices in N are blank)

obviously cannot have a neighbour of v_1, since the other two neighbours are in Q' and C_1, respectively). Update $Q := Q^1$, $q := v_2$ and $Z := Q^2$. Hence in case $|N| = 2$, the construction in Step A is over and we can continue directly with Step B. Indeed, Q has only $v_2 \in N$ as a neighbour in $C_2 \setminus Q$ and similarly for Q' and $v_1 \in N$.

Suppose now that $|N| \geq 3$. Then select another vertex $v_3 \in N$, the vertex v_3 must be in Q, Q' or Z defined above. If $v_3 \in Z$, then Q, Q', Z, q and q' already satisfy the properties. Otherwise, $v_3 \in Q \cup Q'$ and wlog we suppose $v_3 \in Q$. If $\{v_3, v_2\} \in E$, then update $Q := Q \setminus \{v_3\}$, $q := v_3$ and $Z := Z \cup \{v_2\}$ which is trivially connected. Otherwise, $Q \setminus \{v_3\}$ must be disconnected, let Q^1 and Q^2 be its connected parts. Denote Q^2 the set which has a neighbour of v_2 (then Q^1 cannot have a neighbour of v_2). Update $Q := Q^1$, $q := v_3$, $Z := Z \cup Q^2 \cup \{v_2\}$. Again, if $|N| = 3$, the construction is over. Indeed, Q has only $v_3 \in N$ as a neighbour in $C_2 \setminus Q$ and there are no changes for Q'. Moreover, Z is connected. Hence in case $|N| = 3$ we can move to Step B. If $|N| = 4$, the construction is similar to the discussion for $|N| = 3$ (see Fig. 2 as an example for $|N| = 4$).

It is also important to notice that if $|N| > 2$ then following the construction $N \setminus \{q, q'\} \subseteq Z$.

Step B: Using the sets Q, Q', Z and vertices q and q' we define a new partition of V. Then the step B only consists in looking at the size of the sets $Z, Q, Q', Z \cup Q \cup \{q\}, Z \cup Q' \cup \{q'\}$ and update $\{C_1, C_2\}$ depending on the size of those sets which are known to be connected.

The detail of Step B is based on awkward discussion of several cases. □

4 Balanced 2-Community Structure

In this section we first prove that every graph of maximum degree 3 has a balanced weak 2-community structure that can be found in polynomial time. On the other hand, we show that the BALANCED WEAK 2-COMMUNITY problem is NP-complete in general graphs similarly to the BALANCED 2-COMMUNITY problem. The latter result was presented as the main result in [9] and here an alternative shorter proof is given. Both NP-complete results are valid if connectivity of both parts is required.

Remark. As follows from Sect. 3, every 3-regular graph has a 2-community structure. But if we look for a balanced partition there exists a 3-regular graph which doesn't have a balanced 2-community structure, see Fig. 3. The graph is obtained by linking three "cross gadgets". First notice that if a 2-community exists for the graph, then all vertices of each cross gadget must be in the same part. Indeed, each vertex of such community structure must have two neighbours in its own part. But on the other hand, it is impossible to divide the graph into two balanced parts without splitting a cross gadget.

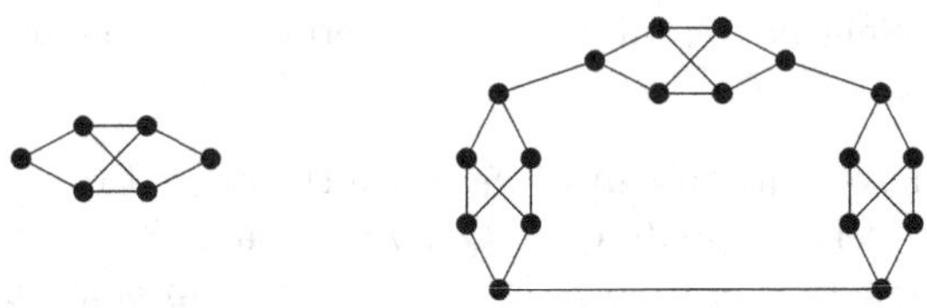

Fig. 3. A cross gadget and a 3-regular graph with no balanced 2-community structure.

Nevertheless, if we focus on a weak community, we can prove that BALANCED WEAK 2-COMMUNITY has always a solution in graphs of maximum degree 3. Moreover, a balanced weak 2-community can be found in polynomial time in such graphs.

Theorem 3. *Any graph of maximum degree* 3 *with at least* 4 *vertices has a balanced weak* 2*-community structure. Moreover, such a community structure can be found in polynomial time.*

Proof. Let $G = (V, E)$ be a connected graph of maximum degree 3. First notice that in any balanced partition, any vertex of degree 1 fulfils condition (2), even if its neighbour is not in its own part. Then, the only vertices which may not satisfy this condition are vertices of degree 2 or 3 which have no neighbour in their own part.

Moreover, for any balanced partition of G, every vertex of degree 2 or 3, which has at least one neighbour in its own part, satisfies condition (2).

Choose any balanced partition $\{C_1, C_2\}$ of G. Then repeat the following steps until the partition is a weak 2-community structure:

1. If every vertex of degree 2 or 3 has at least one neighbour in its own part, then return $\{C_1, C_2\}$ as a weak 2-community structure.
2. If there exists one vertex of degree 2 or 3 in both parts which has no neighbour in its own part, then swap these two vertices.
3. If there is only one partition that contains a vertex v of degree 2 or 3 which has no neighbour in its own part (wlog suppose $v \in C_1$), then choose a vertex $w \in C_2$ such that w has at least one neighbor in C_1 and update: $C_1 := C_1 \cup \{w\} \backslash \{v\}, C_2 := C_2 \cup \{v\} \backslash \{w\}$.

First notice that if case 3 occurs, such vertex w always exists since the graph is connected. Moreover, in cases 2 and 3, notice that the partition is still balanced. Besides, the size of the cut between partitions C_1, C_2 always decreases (by at least 4 in case 2, by at least 2 in case 3) so after finite numbers of iteration (bounded trivially by $O(|V|^2)$, only case 1 remains. Hence at the end of the loop, the algorithm returns a balanced weak 2-community structure. □

It can be observed that the BALANCED 2-COMMUNITY problem (hence also BALANCED WEAK 2-COMMUNITY) is polynomially solvable for graphs with bounded tree-width. Such result follows directly from [2] where problems closely related to communities where studied. Indeed, BALANCED 2-COMMUNITY corresponds to a t-Decomposition defined in [2] where the function $t = n/2$, the functions a, b are equal and for each vertex v in G let $a(v) = b(v) = \frac{n/2-1}{n-1}d(v)$, where n is the order of the graph. Since this problem was proved to be polynomial-time solvable for bounded tree-width in [2], we can conclude the same result for the BALANCED 2-COMMUNITY problem.

Now we focus on the problem of 2-communities in general graphs. In [7] it has been proved that to find a connected balanced partition without any additional constrains is an NP-complete problem in general graphs. We prove similar results for BALANCED WEAK 2-COMMUNITY and BALANCED 2-COMMUNITY and their connected variants. To show that BALANCED WEAK 2-COMMUNITY is NP-complete, we use the BALANCED CO-SATISFACTORY PARTITION problem which was proved NP-complete by Bazgan et al. [4]. The problems is defined as follow:

BALANCED CO-SATISFACTORY PARTITION

Input: A graph $G = (V, E)$ on an even number of vertices.

Question: Is there a balanced partition $\{C_1, C_2\}$ of V such that for every $v \in V$, $d_{in}(v) \leq d_{out}(v)$?

Theorem 4. BALANCED WEAK 2-COMMUNITY *is NP-complete.*

Proof. Clearly this problem is in NP. We reduce BALANCED CO-SATISFACTORY PARTITION to BALANCED WEAK 2-COMMUNITY. Let G be a graph on $2n$ vertices as an instance of BALANCED CO-SATISFACTORY PARTITION, and let G', the complement of G, be an instance of BALANCED WEAK 2-COMMUNITY. If G admits a balanced co-satisfactory partition $\{C_1, C_2\}$ then $\{C_1, C_2\}$ is also a weak 2-community. Indeed, for every vertex $v \in V$, then $d_{in}(v) \leq d_{out}(v)$ in G. Thus, in G' we have $d_{in}^{G'}(v) = 2n - 1 - d_{in}(v) \geq 2n - 1 - d_{out}(v) = d_{out}^{G'}(v)$. Conversely, any balanced weak 2-community in G' is a balanced co-satisfactory partition in G. □

Due to the construction of G' in Theorem 4 and the reduction from [4] we can conclude the following:

Theorem 5. CONNECTED BALANCED WEAK 2-COMMUNITY *is NP-complete.*

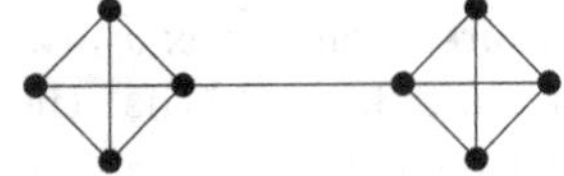

Fig. 4. A graph in which all 2-community structures are balanced

Estivill-Castro et al. [9] have shown that BALANCED 2-COMMUNITY is NP-complete by constructing a reduction from a variant of the CLIQUE problem. We propose a shorter alternative proof which is also valid for the CONNECTED BALANCED 2-COMMUNITY problem. The proof is based on the NP-complete problem BALANCED SATISFACTORY PARTITION which was introduced by Bazgan et al. [3] as follows:

BALANCED SATISFACTORY PARTITION

Input: A graph $G = (V, E)$ on an even number of vertices.

Question: Is there a balanced partition $\{C_1, C_2\}$ of V such that for every $v \in V$, $d_{in}(v) \geq \frac{d(v)}{2}$?

It can be proved that these two problems are in fact equivalent when the number of vertices is even.

Lemma 4. *Let $G = (V, E)$ be a graph with n vertices. Consider a partition $\{C_1, C_2\}$ of V and $v \in C_1$. Then the following assertions are equivalent:*

1. $\frac{d_{in}(v)}{|C_1|-1} \geq \frac{d(v)}{n-1}$
2. $\frac{d_{out}(v)}{|C_2|} \leq \frac{d(v)}{n-1}$
3. $\frac{d_{in}(v)}{|C_1|-1} \geq \frac{d_{out}(v)}{|C_2|}$

Remark. Notice that the third assertion in Lemma 4 is the condition (1) of a 2-community structure.

Lemma 5. *Let $G = (V, E)$ be a graph with an even number n of vertices. Consider a balanced partition $\{C_1, C_2\}$ of V. Then for any vertex $v \in V$, $d_{in}(v) = \frac{|C_1|-1}{n-1} d(v)$ if and only if $d(v) = n - 1$.*

Remark. Let $\{C_1, C_2\}$ be a balanced partition of G and $v \in C_1$ be a vertex of degree $n - 1$. Since v has $\frac{n}{2} - 1$ neighbors in its own part and $\frac{n}{2}$ in other part, v does not satisfy the condition of BALANCED SATISFACTORY PARTITION. However, v satisfies the BALANCED 2-COMMUNITY condition since $\frac{d_{in}(v)}{|C_1|-1} = 1$.

Proposition 1. *Let $G = (V, E)$ be a graph with n vertices without vertices of degree $n - 1$. Then* BALANCED SATISFACTORY PARTITION *and* BALANCED 2-COMMUNITY *are equivalent on G.*

Proof. Suppose that G is a yes-instance of BALANCED SATISFACTORY PARTITION, that is there exists a balanced partition $\{C_1, C_2\}$ of V such that any vertex $v \in V$ satisfies the condition $d_{in}(v) \geq \frac{1}{2}d(v)$, which implies that $d_{in}(v) \geq \frac{|C_1|-1}{2|C_1|-1}d(v) = \frac{|C_1|-1}{n-1}d(v)$. Thus, G is a yes-instance of BALANCED 2-COMMUNITY.

Suppose now that G is a yes-instance of BALANCED 2-COMMUNITY, that is there exists a balanced partition $\{C_1, C_2\}$ of V such that any vertex $v \in V$ satisfies the condition $d_{in}(v) \geq \frac{|C_1|-1}{|C_2|}d_{out}(v)$ that is equivalent to $d_{in}(v) \geq \frac{|C_1|-1}{n-1}d(v)$ using Lemma 4. According to Lemma 5, there is no vertex v such that $d_{in}(v) = \frac{|C_1|-1}{n-1}d(v)$.

We have to show that for every vertex $v \in V$, $d_{in}(v) \geq \frac{1}{2}d(v)$. Suppose by contradiction that there exists a vertex $v \in V$ that does not satisfy the inequality that is

$$\frac{|C_1|-1}{n-1}d(v) < d_{in}(v) < \frac{1}{2}d(v)$$

First, notice that $\frac{1}{2}d(v) - \frac{|C_1|-1}{n-1}d(v) = \frac{1}{2(n-1)}d(v) < 1$, which means that there is at most one whole number between $\frac{|C_1|-1}{n-1}d(v)$ and $\frac{1}{2}d(v)$.

Moreover, $d(v)$ cannot be even, since otherwise $d(v)/2$ would be a whole number and thus $d_{in}(v)$ could not be a whole number. Then $d(v)$ is odd and let $d(v) = 2p + 1$, $p \in \mathbb{N}$. We will arrive to a contradiction by showing that $p < d_{in}(v) < p + 1/2$. Notice that $d(v) < n - 1 \Rightarrow \frac{d(v)-1}{2} < \frac{|C_1|-1}{n-1}d(v)$ that implies $p < \frac{|C_1|-1}{n-1}d(v) < d_{in}(v)$. Then, we have necessarily $d_{in}(v) \geq \frac{1}{2}d(v)$ for every vertex $v \in V$, that is G is a yes-instance of BALANCED SATISFACTORY PARTITION. □

BALANCED SATISFACTORY PARTITION has already been proved NP-complete in [3], even if both parts are required to be connected. Moreover, the reduction which is used to prove it does not construct a graph with vertices of degree $n-1$. Thus we obtain a similar result as in [9] (the authors mentioned along the lines that the proof also works in connected case).

Theorem 6. CONNECTED BALANCED 2-COMMUNITY *is NP-complete.*

On the other hand, it is interesting to notice that there exist graphs in which all 2-community structures are balanced (see Fig. 4).

5 Conclusion and Open Problems

We studied the problems of existence and determination of a 2-community structure and its variants in graphs. We showed that every 3-regular graph has a 2-community structure and such a structure can be found in polynomial time. This remains true even if connectivity of the partitions is required. The interesting open question is to determine if a graph of order at least 4 (except stars) has always a 2-community structure, even connected one. BALANCED 2-COMMUNITY is NP-complete in general graphs, but the complexity of determining a balanced

2-community structure in 3-regular graphs remains open. This last problem is equivalent to finding a cut whose edges forms a matching in 3-regular graphs.

For the weak version the situation is slightly different since any graph of maximum degree 3 has even a balanced weak 2-community structure that can be found in polynomial time. Furthermore, we proved that Balanced weak 2-Community is NP-complete on general graphs, even for connected communities. It remains open if any graph of order at least 4 has a weak 2-community structure, even connected one (except stars).

References

1. Andreev, K., Racke, H.: Balanced graph partitioning. Theory Comput. Syst. **39**(6), 929–939 (2006)
2. Bazgan, C., Tuza, Z., Vanderpooten, D.: Degree-constrained decompositions of graphs: bounded treewidth and planarity. Theoret. Comput. Sci. **355**(3), 389–395 (2006)
3. Bazgan, C., Tuza, Z., Vanderpooten, D.: The satisfactory partition problem. Discrete Appl. Math. **154**(8), 1236–1245 (2006)
4. Bazgan, C., Tuza, Z., Vanderpooten, D.: Approximation of satisfactory bisection problems. J. Comput. Syst. Sci. **74**(5), 875–883 (2008)
5. Bazgan, C., Tuza, Z., Vanderpooten, D.: Satisfactory graph partition, variants, and generalizations. Europ. J. Operat. Res. **206**(2), 271–280 (2010)
6. Buluç, A., Meyerhenke, H., Safro, I., Sanders, P., Schulz, C.: Recent advances in graph partitioning. arXiv:1311.3144
7. Chlebíková, J.: Approximating the maximally balanced connected partition problem in graphs. Inf. Process. Lett. **60**(5), 223–230 (1996)
8. Delling, D., Goldberg, A.V., Pajor, T., Werneck, R.F.: Customizable route planning. In: Pardalos, P.M., Rebennack, S. (eds.) SEA 2011. LNCS, vol. 6630, pp. 376–387. Springer, Heidelberg (2011)
9. Estivill-Castro, V., Parsa, M.: On connected two communities. In Proceedings of the 36th Australasian Computer Science Conference (ACSC), pp. 23–30 (2013)
10. Feldmann, A.E., Foschini, L.: Balanced partitions of trees and applications. Algorithmica **71**(2), 354–376 (2015)
11. Fortunato, S.: Community detection in graphs. Phys. Rep. **486**, 75–174 (2010)
12. Newman, M.E.J.: Detecting community structure in networks. Europ. Phys. J. B-Condens. Matter Complex Syst. **38**(2), 321–330 (2004)
13. Olsen, M.: A general view on computing communities. Math. Soc. Sci. **66**(3), 331–336 (2013)
14. Schaeffer, S.E.: Graph clustering. Comput. Sci. Rev. **1**, 27–64 (2007)
15. Shmoys, D.B.: Cut problems and their application to divide-and-conquer. In: Approximation Algorithms for NP-Hard Problems, pp. 192–235. PWS Publishing (1996)

WDCS: A Weight-Based Distributed Coordinate System

Yaning Liu[1], Hongwei Du[1(✉)], and Qiang Ye[2]

[1] Shenzhen Key Laboratory of Internet Information Collaboration of Computer Science and Technology, Harbin Institute of Technology Shenzhen Graduate School, Shenzhen, China
yn.liuhit@gmail.com, hwdu@hitsz.edu.cn

[2] Department of Computer Science and Information Technology, University of Prince Edward Island, Charlottetown, Canada
qye@upei.ca

Abstract. Many Internet-based applications, such as Content Distribution Networks (CDNs), BitTorrent and Network Game Accelerator, require the knowledge of the distance between each pair of network hosts. Due to the large number of hosts in these applications, measuring all of the pairwise distances is too time-consuming and traffic-inefficient. Network Coordinate System (NCS) is an innovative approach to obtain the pairwise distances by measuring only part of the distances and thereafter utilizing the collected information to predict the remaining pairwise distances. However, due to Triangle Inequality Violation (TIV) and measurement errors, the accuracy of the existing NCS schemes is low. In this paper, we propose a novel distributed NCS scheme, WDCS, that uses a well-designed algorithm to select an appropriate set of landmarks and employs the W_RUN_PACE method to eliminate the impact of measurement errors or outliers. Through extensive experiments, we found that WDCS outperforms the state-of-the-art NCS schemes in terms of prediction accuracy and convergence speed.

Keywords: Network Coordinate System · Distributed system · Landmark · High accuracy

1 Introduction

In the past, Internet applications tended to use the client-server model, in which a client communicates with a single fixed server. With the development of networking technologies and low-cost hardware, most Internet applications are using multiple servers to serve clients. For example, in a Content Distribution Network (CDN) [1], a client desire to download web objects from a particular mirror site which has the highest bandwidth. We refer to latency and bandwidth as distance. These distributed applications such as BitTorrent [2], Content Distribution Network (CDN) and Multi-player Online Games [3] provide better service by knowing the distance between each pair of network nodes. For small-scale

Z. Lu et al. (Eds.): COCOA 2015, LNCS 9486, pp. 251–260, 2015.
DOI: 10.1007/978-3-319-26626-8_19

network, we can use the direct measurement method to get distances among all nodes. For example, in a network contains N nodes, the direct measurement method needs $N * (N - 1)$ explicit measurements. The problem is, this method is not scaled well with the increase of network nodes. The measurement time complexity is $O(N^2)$. In addition, direct measurement increases the burden of clients. To solve this problem, Network coordinate system is proposed to predict the distance between each pair of network nodes through measurements. The measurement time complexity of NCS is $O(N)$. Moreover, NCS depletes the burden of clients.

Existing NCS models can be divided into two types: Euclidean Distance Model [4] and Matrix Factorization Model [5,6]. With Euclidean Distance Model, the pairwise distance is the euclidean distance between two hosts which are calculated through their coordinates. This model has two disadvantages. Firstly, the distance between each pair nodes is symmetrical. But in fact, the distances are asymmetric. Secondly, in Euclidean space, the distance between arbitrary three nodes must satisfy triangle inequality. While in a real network, Triangle Inequality Vilation (TIV) [7] is existed extensively. With Matrix Factorization Model, a large distance matrix can be approximately factorized into two small matrices. The distance between each pair of nodes is the dot product of two predefined vectors. The existing factorization algorithms have Singular Value Decomposition (SVD) and Non-negative Matrix Factorization (NMF) [10]. Since Matrix Factorization Model does not suffer from the problems that are associated with Euclidean Distance Model, it has been widely adopted in varied NCS schemes. The proposed method, Weighted Distributed Coordinate System (WDCS), is also based on Matrix Factorization Model.

Despite the advantages of Matrix Factorization Model, there is still a problem with many existing NCS schemes based on Matrix Factorization Model. In a variety of different real-life applications, the distance matrix D is not completely low-rank due to outliers and measurement errors. But these algorithms attempt to use a completely low-rank matrix to approximate D, which will lead to low accuracy. Different than these algorithms, RNC [9] takes outliers and errors into consideration to arrive at high precision. Technically, RNC first splits distance matrix D into two sub-matrix L and S. Note that L is a completely low-rank matrix and it consistent with the non-negativity characteristic; S is the matrix which corresponds to outliers and measurement errors.

The proposed NCS scheme, WDCS, is based on RNC. But WDCS attempts to improve the precision of RNC from the following two aspects. First of all, RNC is a landmark-based approach, which randomly selects 32 nodes as landmarks. Random selection is a low-complexity approach, which contributes to the fast computation of RNC. However, random selection could lead to non-optimal landmarks, which could degrade the precision performance of RNC. WDCS uses a clustering-based technique, K_SEL, to choose the appropriate landmark nodes in order to avoid the problem with the random approach. With K_SEL, each landmark node is also assigned a weight. Secondly, the specific algorithm used by RNC to split D into L and S is called RUN_PACE [9].

With the weights obtained during the landmark selection process, WDCS uses a weighted algorithm, W_RUN_PACE, to distinguish the distances associated with different landmarks, ultimately arriving at high prediction precision. The detailed contributions of this paper are summarized as follows:

1. We propose a novel NCS scheme, WDCS, that uses an innovative landmark selection method based on k-means clustering, K_SEL. K_SEL is capable of selecting appropriate landmarks in a large-scale distributed system.
2. With WDCS, a weighted algorithm, W_RUN_PACE is used to split the distance matrix D into two components. Since the distances associated with different landmarks are associated with different weights, WDCS leads to high prediction precision.
3. WDCS is a distributed NCS scheme, which leads to outstanding scalability performance.

The rest of this paper is organized as follows. Section 2 presents the problem formulation and Sect. 3 describes the details of WDCS. Our experimental results are discussed in Sect. 4. Finally, we conclude the paper in Sect. 5.

2 Problem Formulation

In a network which contains N hosts. Every host is assigned a global ID in the range of 1 to N. The distance between these hosts constitute an $N \times N$ matrix D. With Matrix Factorization Model, the distance between host i and host j can be calculated using Eq. (1):

$$D(i, j) = X_i \cdot Y_j^T \tag{1}$$

Where $D(i, j)$ represents the distance from i to j. X_i is the outgoing vector of host i and Y_j^T is the incoming vector of host j. Outgoing vectors and incoming vectors of all nodes form two matrixes U and V. U, V can be called the outgoing matrix and the incoming matrix. The distance matrix D is obtained via UV^T. Mathematically, we have:

$$D = U \cdot V^T \tag{2}$$

In many real-life applications, due to outliers and measurement errors, D is not completely low-rank. But a variety of different existing NCS schemes tend to use a low-rank matrix $\hat{D}$ to approximate D, which results in low prediction accuracy. WDCS adopts a weighted algorithm, W_RUN_PACE, to solve this problem. Technically, W_RUN_PACE decomposes D into two matrices of the same dimension L and S. Note that $D = L + S$. L is the low-rank component and nonnegative. S corresponds to outliers and measurement errors. Once the decomposition is completed, only L is used to predict distances. Namely, the estimated coordinates of hosts only need to match the low-rank component L according to Eq. (3):

$$L(i, j) = X_i \cdot Y_j^T \tag{3}$$

Similarly, L can be calculated with U and V according to Eq. (4):

$$L = U \cdot V^T \tag{4}$$

3 WDCS: A High-Precision Scheme

In this section, we present the details of WDCS. Technically, WDCS maps each host to a pair of r-dimensional vectors, namely the incoming vector and the outgoing vector. In our research, these two vectors are referred to as the coordinates of a host. The predicted distance from host i to host j is simply the dot product of i's outgoing vector and j's incoming vector.

With WDCS, the process of distance prediction is composed of two phases. In the first phase, an appropriate set of landmark nodes are selected with K_SEL. K_SEL also leads to a set of weights for the selected landmarks. With the distances between landmarks, the RUN_PACE method used in RNC and the Non-negative Matrix Factorization (NMF) [10] method are utilized to determine the coordinates of landmarks. In the second phase, each node in the distributed system uses W_RUN_PACE and the Non-Negative Least Squares (NNLS) method to calculate its own coordinates in a distributed manner. Once the coordinates are available, the distance between any pair of nodes can be predicted.

3.1 Overview Of WDCS

As mentioned previously, WDCS is composed of two phases. The details of the first phase are presented as follows:

Step 1 In a distributed system, 128 hosts are randomly selected as landmark candidates. The distance between each pair of these hosts is measured.

Step 2 K_SEL, a landmark selection algorithm based on the k-means method [11], is used to choose 32 landmarks among the 128 candidates. Technically, the k-means method is a widely-used clustering technique to divide a set of nodes into multiple clusters. To use the k-means method, we need to map each host to a point in a vector space. In our research, we used the Lipschitz embedding method. Namely, each host i is represented as a 127-dimensional vector whose elements are the distances from host i to the other landmark candidates. With the k-means method, the 128 candidates are divided into 32 clusters. The host that is closest to the centroid of each cluster is selected as a landmark.

Step 3 A weight is assigned to a landmark according to the size of the corresponding cluster.

Step 4 Based on the distances between landmarks, RUN_PACE is used to determine the low-rank component.

Step 5 Based on the low-rank component, NMF is utilized to determine the coordinates of landmarks.

During the second phase, each host whose coordinates are not available uses W_RUN_PACE and NNLS to determine its own coordinates in a distributed manner. The details of the second phase are described as follows:

Step 1 Each host i measures its distances to the landmarks.

Step 2 With the distances to the landmarks, each host i uses W_RUN_PACE to calculate the low-rank component.

Step 3 With the low-rank component, each host i utilizes NNLS to determine its own coordinates.

Step 4 With the coordinates of all nodes in the distributed system, the distance between any pair of nodes can be easily predicted.

3.2 Details of WDCS

In this section, we present the details of the two phases of WDCS. During the first phase, the distances between the 128 landmark candidates are stored in a distance matrix D_{lc}. With the K_SEL method that is based on k-means clustering, 32 clusters are obtained and thereafter 32 landmark nodes are selected. In our research, we use S_c and S_l to denote the set of clusters and the set of landmarks respectively. The elements in S_c (i.e. the clusters) are represented as c_1, c_2, ..., c_{32}. The size of these clusters are denoted as s_1, s_2, ..., s_{32}. The maximum size is represented as max_s. The elements in S_l (i.e. the selected landmarks) are denoted as l_1, l_2, ..., l_{32}. Once S_l is available, the distance matrix corresponding to the landmarks, D_l, can be easily acquired by retrieving the relevant entries from D_{lc} using the EXTRACT_DL method.

With S_c and S_l, a weight calculation algorithm, CAL_WEI, counts the number of hosts in each cluster and assigns a weight w_i to the landmark l_i using Eq. (5):

$$w_i = s_i/max_s \tag{5}$$

In our research, we use W_l to store the weights assigned to the landmarks.

Once the distance matrix for landmark hosts, D_l, is available, RUN_PACE is used to calculate the low-rank component of D_l, L_l. Mathematically, L_l can be determined using Eq. (6):

$$\begin{aligned} &\min_{L_l,S_l} \ ||L_l||_* + \lambda||S_l||_1, \\ &subject\ to\ D_l = L_l + S_l, \\ &L_l \geq 0 \end{aligned} \tag{6}$$

where $||\cdot||_*$ denotes the nuclear norm of a matrix, $||\cdot||_1$ represents the sum of the absolute value of the matrix entries, and λ is a tuning parameter.

With the low-rank component L_l, WDCS uses NMF to calculate the coordinates of landmark hosts. In our research, we use U_l and V_l to store the outgoing and incoming vectors of landmark hosts. Mathematically, U_l and V_l can be determined by solving the following minimization problem:

$$\begin{aligned} &\min_{U_l,V_l} \ \|D_l - U_l \cdot V_l^T\|_F^2 \\ &\text{subject to} \ \ U_l \geq 0, V_l \geq 0 \end{aligned} \tag{7}$$

where $\|\cdot\|_F$ denotes the Frobenius norm of a matrix; $U_l \geq 0$ and $V_l \geq 0$ are the constraints that ensure that each element of U_l and V_l is greater than or

Algorithm 1. Phase 1 of WDCS

Input: D_{lc}
1: $(S_c, S_l) = K\_SEL(D_{lc})$
2: $D_l = EXTRACT\_DL(D_{lc}, S_l)$
3: $W_l = CAL\_WEI(S_c, S_l)$
4: $L_l = RUN\_PACE(D_l)$
5: $(U_l, V_l) = NMF(L_l)$
Output: U_l, V_l, S_l, W_l

equal to 0. Once the minimization problem is solved, the coordinates of landmark hosts are available. The steps involved in the first phase of WDCS are summarized in Algorithm 1.

During the second phase of WDCS, each host i measures the distances between itself and the landmarks. The measured distances and the distances between landmarks are stored in a 33×33 matrix D_i (note that there are 32 landmarks). Furthermore, WDCS assigns a weight $W_i(m, n)$ to each entry of D_i, $D_i(m, n)$, using the WEI-ASSIGN method. Obviously, $D_i(m, n)$ corresponds to one of three types of distances: a distance between two landmarks, a distance between host i and a landmark, and a distance from i to i. As mentioned previously, each landmark is assigned a weight. If $D_i(m, n)$ corresponds to the first type of distance, WDCS checks the weights of the involved landmarks and assigns the higher weight to $D_i(m, n)$. If $D_i(m, n)$ corresponds to the second type of distance, WDCS assigns the weight of the relevant landmark to $D_i(m, n)$. If $D_i(m, n)$ corresponds to the third type of distance, a zero weight is assigned to $D_i(m, n)$.

With D_i and W_i, WDCS uses W_RUN_PACE to extract the low-rank component L_i by solving the following minimization problem:

$$\begin{aligned} \underset{L_i, S_i}{min} \quad & ||L_i||_* + \lambda ||W_i \cdot S_i||_1, \\ subject\ to\ & D_i = L_i + S_i, \\ & L_i \geq 0 \end{aligned} \tag{8}$$

Because W_i is incorporated into the minimization problem, WDCS prefers to match the distance involving a landmark with a higher weight over that involving a landmark with a lower weight. As a result, the distances associated with large clusters are matched better, ultimately leading to high prediction precision.

After L_i is acquired, WDCS uses NNLS to calculate i's coordinates (X_i, Y_i). Once the coordinates of all hosts are available, the distance between any pair of hosts can be easily predicted. Formally, the steps in the second phase of WDCS are summarized using Algorithm 2.

4 Experimental Results

In this section, we present the performance of WDCS and two state-of-the-art NCS schemes, Phoenix [12] and Vivaldi [13]. To illustrate the impact of the

Algorithm 2. Phase 2 of WDCS

Input: D_i, U_l, V_l, W_l
$W_i = WEI\_ASSIGN(D_i, W_l, U_l, V_l)$
$L_i = W\_RUN\_PACE(D_i, W_i)$
$(X_i, Y_i) = NNLS(L_i)$
Output: X_i, Y_i

weights acquired during the landmark selection process, we devised a variant of WDCS, Non-Weighted Distributed Coordinate System (NWDCS). NWDCS uses the same landmark selection process as WDCS. However, the second phase of NWDCS is similar to that of RNC (i.e. it does not utilize the weights). In our experiments, we implemented these schemes using Matlab. Our experiments were carried out with a 64-bit windows 7 machine with Intel Core i7 CPU and 8 GB memory.

Phoenix is a popular NCS scheme based on Matrix Factorization Model. Its prediction accuracy is relative high and its computing time is small. In addition, Phoenix first uses edge-weight thought. Vivaldi is an NCS scheme based on Euclidean Distance Model which first adopts the physical mass-spring system. Therefore, we chose these two schemes in our experiments. For Pheonix and Vivaldi, each host selects $K = 32$ 1-hop reference hosts. For WDCS and NWDCS, 32 landmark hosts are selected. The dimension of coordinates r was set to 8.

In our experiments, we use two real world data sets PL and P2PSim. The distance in these two matrix is RTTs between pairwise network nodes. PL [14] data set contain RTTs of 169 PlanetLab hosts deployed all over the world. P2PSim [15] data set contains latencies among 1740 DNS servers measured using King method [16].

4.1 Evaluation Metrics

The evaluation focuses on two main aspects of network coordinate system. The first aspect is accuracy which is the main metric in evaluation. The second aspect is convergence which represents the speed of coordinate convergence process. There are many metrics that could be used to evaluate network coordinate system. In this paper, the following metrics are used:

Relative Error: Relative error is the main metrics in evaluation. It is used to compare the performance of different network coordinate systems. Suppose a network contain N hosts, distances among these hosts compose a $N \times N$ matrix D. D is the real distance matrix. $\hat{D}$ represent the prediction matrix. The definition of relative error is presented using Eq. (9):

$$RE = \frac{||D - \hat{D}||}{D} \tag{9}$$

Median Relative Error: Median relative error, for short MRE, is median value of relative error.

Cumulative Distribution Function: For short CDF, is the cumulative distribution function of relative error.

The Ninetieth Percentile Relative Error(NPRE): It is the ninetieth error of relative error.

4.2 Accuracy

We use PL data and King data to evaluate the prediction accuracy of different schemes. Figure 1 includes the CDF of relative error of WDCS, NWDCS, Phoenix and Vivaldi. In Fig. 1, We can see WDCS and NWDCS outperform Phoenix and Vivaldi in all data sets. Due the use of the weights, WDCS outperforms NWDCS.

In Fig. 1, the prediction accuracy of all the schemes in King data is better than that in the case of PL data. The reason is that the distances in the PL data are not symmetric, which is more difficult to reconstruct the matrix. In these two data sets, WDCS and NWDCS are the more accurate schemes. The reason is that when there are some abnormal points, WDCS and NWDCS can well eliminate their influence. In addition, WDCS outperforms NWDCS.

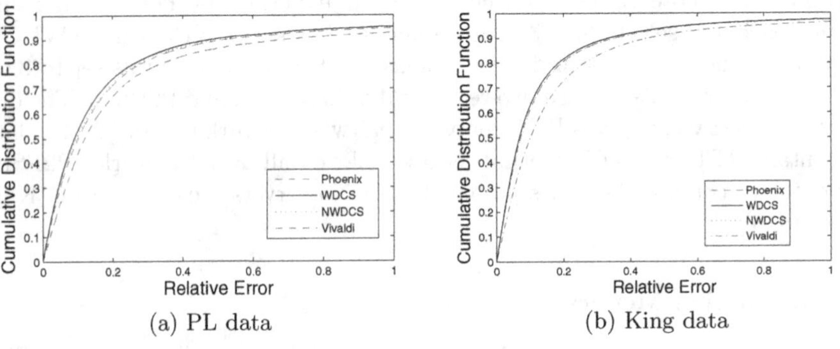

(a) PL data (b) King data

Fig. 1. CDF of relative error.

4.3 Convergence Behaviour

In this section, We evaluate the convergence performance of WDCS, NWDCS, Phoenix and Vivaldi. The number of iterations is set to 30. We compare their performances using NPRE and MRE. Figure 2 shows that WDCS and NWDCS converge at the fastest speed in terms of NPRE. For all of the data sets, the convergence process of WDCS and NWDCS can be completed within 1 time slot. Figure 3 shows the convergence performance in terms of MRE. Similarly, WDCS and NWDCS lead to the fastest convergence speed and use 1 time slot to finish the convergence process.

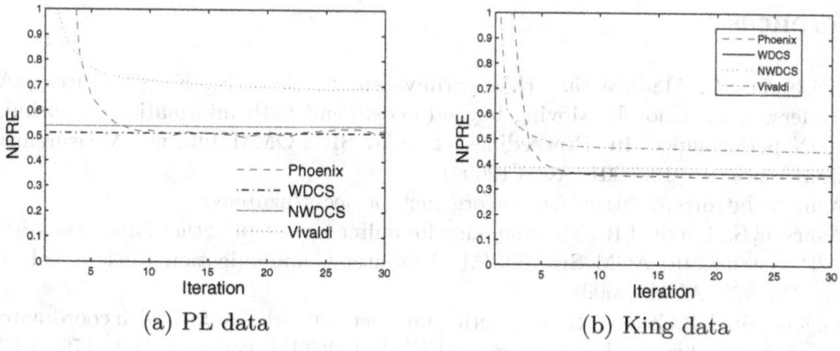

(a) PL data (b) King data

Fig. 2. NPRE comparison of Phoenix, Vivaldi and WDCS

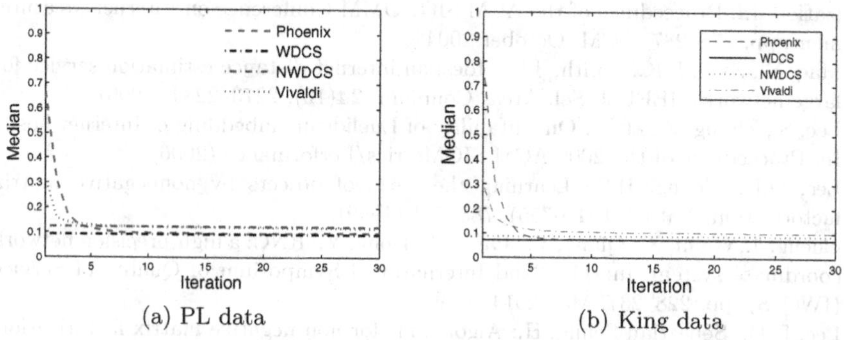

(a) PL data (b) King data

Fig. 3. MRE comparison of Phoenix, Vivaldi and WDCS

5 Conclusion

In this paper, we propose WDCS, a weight-based distributed NCS scheme. Technically, WDCS completes distance prediction using two phases. In the first phase, 32 landmark hosts are selected and thereafter associated with different weights. With the pairwise distances between these landmark hosts, the coordinates of these hosts are determined. In the second phase, each non-landmark host measures the distances between itself and the landmark hosts. The measured distances, the coordinates and the weights of landmark hosts are utilized to calculate the coordinates of the non-landmark host. Once the coordinates of every host are available, the distance between any pair of hosts can be easily predicted using the dot product of the outgoing vector of one host and the incoming vector of the other host. Our experimental results indicate that WDCS outperforms the state-of-the-art NCS schemes in terms of prediction accuracy and convergence speed.

Acknowledgement. This work was financially supported by National Natural Science Foundation of China with Grants No. 61370216 and No. 61100191, and Shenzhen Strategic Emerging Industries Program with Grants No. ZDSY20120613125016389, No. JCYJ20120613151201451 and No. JCYJ20130329153215152.

References

1. Krishnan, R., Madhyastha, H.V., Srinivasan, S., Jain, S., Krishnamurthy, A., Anderson, T., Gao, J.: Moving beyond end-to-end path information to optimize CDN performance. In: Proceedings of ACM SIGCOMM Internet Measurement Conference, pp. 190–201. ACM (2009)
2. Azureus bittorrent. http://sourceforge.net/projects/azureus/
3. Agarwal, S., Lorch, J.R.: Matchmaking for online games and other latency-sensitive P2P systems. In: ACM SIGCOMM Computer Communication Review, vol. 39, pp. 315–326. ACM (2009)
4. Eugene Ng, T.S., Zhang, H.: Predicting internet network distance with coordinates-based approaches. In: Proceedings of IEEE INFOCOM, vol. 1, pp. 170–179. IEEE, April 2002
5. Mao, Y., Saul, L.K.: Modeling distances in large-scale networks by matrix factorization. In: Proceedings of the ACM SIGCOMM Conference on Internet Measurement, pp. 278–287. ACM, October 2004
6. Mao, Y., Saul, L.K., Smith, J.M.: Ides: an internet distance estimation service for large networks. IEEE J. Sel. Areas Commun. **24**(12), 2273–2284 (2006)
7. Lee, S., Zhang, Z., et al.: On suitability of Euclidean embedding of Internet hosts. In: Proceedings of the 2006 ACM SIGMetrics/Performance (2006)
8. Lee, D.D., Seung, H.S.: Learning the parts of objects by nonnegative matrix factorization. Nature **401**(6755), 788–791 (1999)
9. Cheng, J., Guan, X., Qiang, Y., Jiang, H., Dong, Y.: RNC: a high-precision network coordinate system. In: The 22nd International Symposium of Quality of Service (IWQoS), pp. 228–237, May 2014
10. Lee, D.D., Sebastian Seung, H.: Algorithms for non-negative matrix factorization
11. Bishop, C.M.: Neural Networks for Pattern Recognition. Oxford University Press, Oxford (1995)
12. Chen, Y., Wang, X., Shi, C., Lua, E.K., Fu, X., Deng, B., Li, X.: Phoenix: a weight-based network coordinate system using matrix factorization. IEEE Trans. Netw. Serv. Manage. **8**(4), 334–347 (2011)
13. Dabek, F., Cox, R., Kaashoek, F., Morris, R.: Vivaldi: a decentralized network coordinate system. In: Proceedings of the ACM SIGCOMM, pp. 15–26. ACM Press, New York (2004)
14. Stribling, J.: All pairs of ping data for planetlab. http://www.pdos.lcs.mit.edu/~strib/pl_app
15. The P2PSim project. http://pdos.csail.mit.edu/p2psim/
16. Gummadi, K.P., Saroiu, S., Gribble, S.D.: King: estimating latency between arbitrary internet end hosts. In: Proceedings of ACM SIGCOMM Internet Measurement Workshop, pp. 5–18. ACM, November 2002

Adaptive Scheduling Over a Wireless Channel Under Constrained Jamming

Antonio Fernández Anta[1], Chryssis Georgiou[2], and Elli Zavou[1,3](✉)

[1] IMDEA Networks Institute, Madrid, Spain
[2] University of Cyprus, Nicosia, Cyprus
[3] Universidad Carlos III de Madrid, Madrid, Spain
elli.zavou@imdea.org

Abstract. We consider a wireless channel between a single pair of stations (sender and receiver) that is being "watched" and disrupted by a malicious, adversarial jammer. The sender's objective is to transmit as much useful data as possible, over the channel, despite the jams that are caused by the adversary. The data is transmitted as the payload of packets, and becomes useless if the packet is jammed. In this work, we develop deterministic scheduling algorithms that decide the lengths of the packets to be sent, in order to maximize the total payload successfully transmitted over period T in the presence of up to f packet jams, *useful payload*.

We first consider the case where all packets must be of the same length and compute the optimal packet length that leads to the best possible useful payload. Then, we consider adaptive algorithms; ones that change the packet length based on the feedback on jammed packets received. We propose an optimal scheduling algorithm that is essentially a recursive algorithm that calculates the length of the next packet to transmit based on the packet errors that have occurred up to that point. We make a thorough non trivial analysis for the algorithm and discuss how our solutions could be used to solve a more general problem than the one we consider.

Keywords: Packet scheduling · Wireless channel · Unreliable communication · Adversarial jamming

1 Introduction

Motivation and Prior Work. Transmitting data over wireless media is becoming increasingly popular, especially with the dramatic increase of the use of mobile devices (e.g., smart phones). A major challenge that needs to be addressed is to cope with disruptions of the communication over such media, especially when these disruptions are caused intentionally, e.g., by jamming devices. Several research efforts have been made in addressing this challenge under different assumptions and constraints (e.g., [1–6, 9–12]).

This research was supported in part by Ministerio de Economía y Competitividad grant TEC2014-55713-R, Regional Government of Madrid (CM) grant Cloud4BigData (S2013/ICE-2894, cofunded by FSE &FEDER), and grant FPU12/00505 from MECD.

Z. Lu et al. (Eds.): COCOA 2015, LNCS 9486, pp. 261–278, 2015.
DOI: 10.1007/978-3-319-26626-8_20

In a recent work [2], we have initiated the investigation of the following problem: We consider a wireless channel between a single pair of stations (sender and receiver) that is being "watched" and disrupted by a malicious, adversarial jammer. The sender's objective is to transmit as much useful data as possible, over the channel, despite the jams that are caused by the adversary. The data is transmitted as the payload of *packets*, and becomes useless if the packet is jammed. The adversary has complete knowledge of the packet scheduling algorithm and it decides on how to jam the channel dynamically. However, the jamming power of the adversary is constrained by two parameters, ρ and σ, whose values depend on technological aspects. The parameter σ represents the maximum number of "error tokens" available for the adversary to use at any point in time, and ρ represents the rate at which new error tokens become available (one at a time). Each error token models the ability of the adversary to jam one packet. This adversarial model could represent a jamming entity with limited resource of rechargeable energy, e.g., malicious mobile devices or battery-operated military drones. In these cases, σ represents the capacity of the battery (in packets that can be jammed) and ρ the rate at which the battery can be recharged (for instance, with solar cells). To evaluate scheduling algorithms, two efficiency measures are used: the *transmission time*, to completely send a fixed pre-defined amount of data, and the *goodput ratio* (successful transmission rate) achieved to do so, which intuitively are reversely proportional.

Under this model, we first showed in [2] upper and lower bounds on the transmission time and goodput when the sender sends packets of the same length throughout the execution (uniform case); in this case the scheduling policy does not take into account the history of jams. Then, considering the case $\sigma = 1$, we proposed an adaptive scheduling algorithm that changes the packet length based on the feedback on jammed packets received, and showed that it can achieve better goodput and transmission time with respect to the uniform case, for most values of ρ. However, the analysis technique used for the case $\sigma = 1$ turned out not to be easily generalized for cases where $\sigma > 1$.

In order to better understand the above problem and lay the groundwork for obtaining its optimal solutions, in this work we consider a simpler, more "static" version of the problem. In particular, we focus on a specific time interval of length T, and instead of assuming that new error tokens are continuously arriving we assume a fixed number of error tokens f. As before, the sender's objective is to correctly transmit the maximum amount of data, in the form of packets, under the jamming of the adversary. Now, the adversary is constrained only by parameter f, which is the maximum number of errors (packet jams) it can introduce in the corresponding interval T and are available from the very beginning of the interval. Therefore, given T and f as input, we would like to maximize the total *useful payload* transmitted within the interval of interest.[1] (Our modeling assumptions are detailed in Sect. 2.)

[1] As we assume that the transmission time of each packet is equal to its length, it follows that T is an absolute upper bound on the useful payload transmitted..

We plan to use the results obtained for this problem to derive solutions of the continuous version of [2], but we believe that this static problem is a fundamental and challenging problem and hence of interest by itself.

Contributions. We provide a comprehensive solution to the abovementioned problem (static version). Specifically:

- We first consider the case where the scheduling algorithm is restricted in sending packets of the same length (uniform case); this could be due to limitations in the communication protocol or the sender's specification (Sect. 3). Given a period of length T and up to f packet jams, we show that the optimal packet length is $p^* \approx \sqrt{T/f}$ that leads to optimal useful payload $T + f - 2\sqrt{T/f}$. In the case the adversary does not cause any packet error in the interval of time T, we show that the useful payload achieved by uniform packets of length p^* is $T - \sqrt{T/f}$.
- Then, we devise adaptive scheduling algorithms, that is, algorithms that change the packet length based on the feedback on jammed packets received. We start by first considering the case of $f = 1$ (Sect. 4). We devise algorithm ADP$(T, 1)$ and prove its optimality. We show that the algorithm achieves optimal useful payload of $\frac{i-1}{i}T - \frac{i+1}{2} + \frac{1}{i}$, where i is the integer such that $T \in \left[\frac{(i-1)i}{2} + 1, \frac{i(i+1)}{2} + 1\right)$.

  Algorithm ADP$(T, 1)$ chooses the length p of the first packet to be transmitted as a function of T. If the packet is jammed then it transmits a second packet of length $T - p$ which is now guaranteed not to be jammed. If the first packet goes through, then the algorithm is invoked recursively as ADP$(T - p, 1)$.
- Next, we generalize algorithm ADP$(T, 1)$ into algorithm ADP(T, f) and show that it obtains optimal useful payload for any f (Sect. 5). Algorithm ADP(T, f) is essentially a recursive algorithm that also begins by choosing length p of the first packet to be transmitted as a function of T (a different function from that of ADP$(T, 1)$). If the packet is jammed, the adversary (unlike in the case of $f = 1$) still has error tokens that it can use. Therefore, instead of sending a packet that spans the rest of the interval, ADP(T, f) makes the recursive call ADP$(T - p, f - 1)$. If the packet is not jammed, then it makes a recursive call to ADP$(T - p, f)$.

  Although the above algorithmic approach is natural, the choice of the length p of the packet to be sent as well as the algorithm's analysis of optimality, are nontrivial.
- Finally, we discuss and compare the version of the problem considered in this work (static) with the one of [2] (continuous) and draw interesting conclusions (Sect. 6).

Related Work. Several studies have investigated the effect of jamming in wireless channels. For example, Thuente et al. [12] studied the effects of different jamming techniques in wireless networks and the trade-off with their energy efficiency. Their study includes from trivial/continuous to periodic and intelligent jamming (taking into consideration the size of packets being transmitted).

Pelechrinis et al. [6] present a detailed survey of the Denial of Service attacks in wireless networks. They present the various techniques used to achieve malicious behaviors and describe methodologies for their detection as well as for the network's protection from the jamming attacks. Dolev et al. [4] present a survey of several existing results in adversarial interference environments in the unlicensed bands of the radio spectrum, discussing their vulnerability.

Awerbuch et al. [3] designed a medium access (MAC) protocol for single-hop wireless networks that is robust against adaptive adversarial jamming (the adversary knows the protocol and its history and decides to jam the channel at any time) and requires only limited knowledge about the adversary (an estimate of the number of nodes, n, and an approximation of a time threshold T). One of the differences with our work is that the adversary they consider is allowed to jam $(1-\varepsilon)$-fraction of the time steps. On a later work [10], Richa et al. explored the design of a robust MAC protocol that takes into consideration the signal to interference plus noise ratio (SINR) at the receiver end. In [11] they extended their work to the case of multiple co-existing networks; they proposed a randomized MAC protocol which guarantees fairness between the different networks and efficient use of the bandwidth. In [9], Richa et al. considered an adaptive adversarial jammer that is also reactive: it is allowed to make a jamming decision based on the actions of the nodes at the current step; this is similar to the adversary we consider in this work. However, they consider a different constrain on jamming, following the previously mentioned works: given a time period of length T, the adversary can jam at most $(1-\varepsilon)T$ of the time steps in that period. In our case, the adversary, within a time period T, can cause f channel jams, where f does not correspond to a fraction of time, but on the number of packets it can corrupt. Other differences is that they consider n nodes transmitting over the channel (hence, they deal with transmission collisions) and their objective is to optimize throughput over the non-jammed time periods.

Finally, Gilbert et al. [5] investigated the impact on the communication delay between two honest nodes that a third malicious, energy-constraint node can have. In particular, the three nodes share a time-slotted single-hop wireless radio channel and the two honest nodes begin with a value to communicate. The malicious node wishes to prevent them from communicating for as long as it can, by broadcasting messages. However, it is allowed to broadcast up to β messages. This is similar to the restriction we impose in our work, by allowing the adversary to cause up to f packet errors. The setting and objectives of the work [5], though, are different. First they show a tight bound on the number of rounds that the malicious node can delay the communication between the two honest entities: $2\beta + \Theta(\log |V|)$ rounds, where V is the set of possible values the two honest nodes may communicate. Then, they study the implication of this bound on more general n-node problems, such as reliable broadcast and leader election.

2 Model

We now present in detail the model considered in this paper.

Network Setting. We consider an unreliable wireless channel that connects two end stations: a sender and a receiver. The sender has data to be transmitted to the receiver over a fixed time interval T, which is sent as the payload of the packets scheduled.[2] However, the channel is prone to instantaneous jams, that cause any packet in transit to be corrupted/destroyed. Hence, the sender needs to decide the length of the packets to be sent in the time interval, taking or not into consideration the history of transmissions and jams that already occurred. This is done by using an *online scheduling algorithm* [7,8]. The objective of the sender is to provide the receiver with as much data as possible over the period of time T, despite the channel jams.

As in [2], each packet sent across the channel consists of a *header* of a fixed predefined size h and a *payload* of length l chosen by the algorithm; the total length of the packet is $p = h + l$. For simplicity we assume that $h = 1$, i.e., $p = l + 1$ (we assume l to be a real number). Recall that the payload corresponds to the useful data sent across the channel. In addition, we assume that the transmission time of each packet is equal to its length; the channel has a constant transmission rate. (Therefore, T is an absolute upper bound on the useful payload transmitted.)

Packet Jamming. We assume that the jams occurring in the channel are orchestrated by an omniscient and adaptive adversary. That is, the adversary has complete knowledge of the packet scheduling and transmission algorithm it decides to cause the jams during the course of the computation. However, it has a constrained number of jams it can cause in a given period. Specifically, we consider adversary (T, f)-$\mathcal{A}$, that for a time interval of length T, $T \geq 1$, it can cause up to f packet jams. As in [2], for a worst case analysis, we assume that the adversary jams some bit in the header of the packets in order to ensure their destruction. So given T, the adversary defines the *error pattern* E as a set of up to f jamming events on the channel over that period, each given by a time instant in the period. We will sometimes use the special error pattern $E = \emptyset$ that corresponds to the case in which the adversary causes no jamming. For a given T, we assume that f is known to the scheduling algorithm.

Efficiency Measures. As in [2], we consider two efficiency measures, *useful payload* and *goodput rate*. The useful payload, is the sum of payloads of the successfully transmitted packets within a time interval of length T, under any f-size error pattern E. The goodput rate, is the corresponding ratio of the useful payload sent during the interval and is of interest mostly for the continuous version of the model presented in [2].

More formally, we denote by $\mathrm{UP}_A(T, f, E)$ the useful payload (payload correctly received) when using scheduling algorithm A in an interval of length T

[2] We assume that the sender has data with total payload greater than T.

against an adversary of power f that uses error pattern E. For a fixed algorithm A, its useful payload is then simply $\mathrm{UP}_A(T, f) = \min_{E \in \mathcal{E}(f)} \mathrm{UP}_A(T, f, E)$, where $\mathcal{E}(f)$ is the set of all possible error patterns with at most f jams. From this, we define the optimal useful payload as $\mathrm{UP}^*(T, f) = \max_A \mathrm{UP}_A(T, f)$.

The goodput metric is defined similarly, by simply dividing the useful payload by the length of the interval. More precisely, when using scheduling algorithm A in an interval of length T against an adversary of power f that uses error pattern E, the goodput rate is $G_A(T, f, E) = \mathrm{UP}_A(T, f, E)/T$, the goodput of algorithm A is $G_A(T, f) = \mathrm{UP}_A(T, f)/T$, and the optimal goodput is $G^*(T, f) = \mathrm{UP}^*(T, f)/T$.

Feedback Mechanism. Following [1,2], we consider instantaneous feedback. In particular, at the time a packet is successfully received by the receiver, a notification/acknowledgement message is immediately received by the sender. If such a message is not received by the sender, then it considers the packet to be jammed. We assume that the notification packets cannot be jammed by the errors in the channel because of their relatively small size.
Remark: Observe that if $T \leq f$, then the adversary can jam all packets sent in the interval and no useful data will be received. Therefore, from this point onward we focus only in time periods that are initially of length $T > f$.

3 Uniform Packets

We first consider the case in which all the packets scheduled are of the same length. Having to use uniform packets may be a requirement due to limitations in the communication protocol, or the sender's specifications. In this case, the following result gives the uniform packet length that has to be used in order to maximize the minimum useful payload. (Note that the approximations below are due to floors and ceilings; these approximations get closer to equality as Tf grows.)

Theorem 1. *Let $U(p)$ denote the algorithm that only uses uniform packets of length p. In an interval of length T and maximum number of errors f, the optimal packet length for these algorithms is $p^* \approx \sqrt{T/f}$ that achieves useful payload $UP_{U(p^*)}(T, f) \approx T + f - 2\sqrt{Tf}$. When the adversary causes no jamming, the useful payload achieved by $U(p^*)$ is $UP^*_{U(p^*)}(T, f, \emptyset) \approx T - \sqrt{Tf}$.*

Proof. Let us denote by n the number of uniform packets of length $p = \frac{T}{n}$ sent in an interval of length T when the adversary has f error tokens available. Hence, we will be using $U(n)$ and $U(p)$ to denote the same algorithm. In the worst case, the adversary will use its error tokens to jam f packets in the interval, and hence there will be at least $n - f$ successfully received packets by the receiver by the end of the interval. The useful payload of the uniform algorithm using n (and hence p) will thus be $\mathrm{UP}_{U(n)}(T, f) = (n - f)\left(\frac{T}{n} - 1\right)$ (recall that each packet consists of the payload and a unit-size header).

Deriving this expression with respect to n, we get

$$\frac{\partial \mathrm{UP}_{U(n)}(T,f)}{\partial n} = \frac{fT}{n^2} - 1,$$

which implies that $\mathrm{UP}_{U(n)}(T,f)$ is maximized for $n = \sqrt{Tf}$. Moreover, the derivative is positive for $n < \sqrt{Tf}$ and negative for $n > \sqrt{Tf}$, which implies that the useful payload is strictly increasing on the left of $n = \sqrt{Tf}$ and strictly decreasing on the right. From this, we get that (1) there is no other n that maximizes the useful payload, and (2) since the number of packets has to be an integer value, the only two candidates for the optimal number of packets n^* are $\lfloor\sqrt{Tf}\rfloor$ and $\lceil\sqrt{Tf}\rceil$. Hence the value of these two that maximizes the useful payload is the optimal number n^*. From this, and the fact that $p = \frac{T}{n}$, we get that $p^* \approx \sqrt{T/f}$, as claimed.

Then, the optimal number of packets n^* gives optimal useful payload $\mathrm{UP}_{U(n^*)}(T,f) = (n^* - f)(\frac{T}{n^*} - 1) \approx T + f - 2\sqrt{Tf}$, as claimed. Finally, when n^* packets are used, and no packet is jammed by the adversary, the useful payload is maximized reaching $\mathrm{UP}_{U(n^*)}(T,f,\emptyset) = n^*(\frac{T}{n^*} - 1) \approx T - \sqrt{Tf}$, as desired. □

Corollary 1. *The optimal achievable goodput rate is* $G_{U(p^*)}(T,f) \approx \left(1 - \sqrt{f/T}\right)^2$.

4 Optimal Algorithm for $f = 1$

In this section, we turn our focus on the case of a single error token available to the adversary for an interval of length T. We give an adaptive algorithm, named ADP$(T,1)$, and prove its optimality. By doing so, we hope to give an intuition to the reader for how the general optimal algorithm, for any number of error tokens, works.

Algorithm 1. ADP$(T,1)$

If $T \in [1,2)$ then
 Send packet with length $p = T$
else
 Let i be the integer such that $T \in \left[\frac{(i-1)i}{2} + 1, \frac{i(i+1)}{2} + 1\right)$
 Let $\alpha = i - 2$, and $\beta = \frac{(i-1)i}{2} - 1$
 Send packet π with length $p = \frac{T+\beta}{\alpha+2} = \frac{T-1}{i} + \frac{i-1}{2}$
 If packet π is jammed then
  Send packet with length $p' = T - p$
 else
  Call ADP$(T - p, 1)$

Algorithm ADP$(T,1)$ is used in a time recursive fashion, with respect to the length of the interval of interest, T. Its scheduling policy is as follows: It chooses the length p of the first packet to be transmitted as a function of T. If the packet is jammed then it transmits a second packet of length $T-p$ which is guaranteed not to be jammed. If the first packet goes through, then the algorithm is invoked recursively as ADP$(T-p,1)$.

The detailed description of the algorithm is given as Algorithm 1. Let us fix the interval length $T \geq 1$, and let i be the integer such that $T \in \left[\frac{(i-1)i}{2}+1, \frac{i(i+1)}{2}+1\right)$, as described in the above pseudocode. Let us also define parameters $\alpha = i-2$ and $\beta = \frac{(i-1)i}{2}-1$, packet length $p = \frac{T+\beta}{\alpha+2}$, and interval length $T' = T-p$. We first present the following two lemmas that are used to show the optimality of Algorithm ADP$(T,1)$.

Lemma 1. *Interval length $T' = T-p$ is such that $T' \in \left[\frac{(j-1)j}{2}+1, \frac{j(j+1)}{2}+1\right)$ for $j = i-1$, where i is an integer such that $i \geq 1$.*

Proof. Replacing the values of α and β in the calculation of $T' = T-p$,

$$T' = \frac{(\alpha+1)T-\beta}{\alpha+2} = \frac{(i-2+1)T - \left(\frac{(i-1)i}{2}-1\right)}{i-2+2} = \frac{(i-1)T - \frac{(i-1)i}{2}+1}{i}.$$

Now, using the fact that $T \geq \frac{(i-1)i}{2}+1$, we have

$$T' \geq \frac{(i-1)\left(1+\frac{(i-1)i}{2}\right) - \frac{(i-1)i}{2}+1}{i} = \cdots = \frac{(i-1)(i-2)}{2}+1.$$

Similarly, using the fact that $T < \frac{i(i+1)}{2}+1$, we have

$$T' < \frac{(i-1)\left(1+\frac{i(i+1)}{2}\right) - \frac{(i-1)i}{2}+1}{i} = \cdots = \frac{(i-1)i}{2}+1.$$

Setting $j = i-1$ in both cases, we have $T' \in \left[\frac{(j-1)j}{2}+1, \frac{j(j+1)}{2}+1\right)$ as claimed. □

Lemma 2. *Let $T \geq 2$ and assume that $UP_{ADP}(T',1) = \frac{\alpha T'-\beta}{\alpha+1}$, where $T' = T-p$. Then, Algorithm ADP$(T,1)$ achieves useful payload $UP_{ADP}(T,1) = \frac{(\alpha+1)T-(\beta+\alpha+2)}{\alpha+2}$.*

Proof. Since $T \geq 2$, that Algorithm ADP$(T,1)$ schedules first a packet π with length $p = \frac{T+\beta}{\alpha+2}$. If π is jammed, then a packet of length equal to the rest of the interval, i.e., $T' = T-p$, can be sent successfully, and hence the useful payload will be $\text{UP}_{\text{ADP}}(T,1) = T - \frac{T+\beta}{\alpha+2} - 1 = \frac{(\alpha+1)T-(\beta+\alpha+2)}{\alpha+2}$.

Otherwise, if π is not jammed, the useful payload is obtained as $\text{UP}_{\text{ADP}}(T,1)$ $= p-1+\text{UP}_{\text{ADP}}(T',1) = p-1+\frac{\alpha T'-\beta}{\alpha+1} = p-1+\frac{\alpha(T-p)-\beta}{\alpha+1} = \frac{(\alpha+1)T-(\beta+\alpha+2)}{\alpha+2}$. In both cases, the useful payload is as claimed, which completes the proof. □

Theorem 2. *Given an interval of length $T \geq 1$, Algorithm ADP$(T,1)$ achieves optimal useful payload UP*$^*(T,1) = \frac{i-1}{i}T - \frac{i+1}{2} + \frac{1}{i}$*, where i is the integer such that $T \in \left[\frac{(i-1)i}{2}+1, \frac{i(i+1)}{2}+1\right)$.*

Proof. The proof is by induction on T. The base case is when $T \in [1,2)$, which implies that $i = 1$. In this case only one packet is sent by ADP$(T,1)$, which spans the whole interval and can be jammed by the adversary. Observe that in this case at most one packet can in fact be sent in the interval. This matches the claim that ADP$(T,1)$ achieves optimal useful payload UP$^*(T,1) = 0$ in this case.

Let us now consider any interval length $T \geq 2$, which implies $i \geq 2$. Then, from Lemma 1, interval length $T' = T - p \in \left[\frac{(j-1)j}{2}+1, \frac{j(j+1)}{2}+1\right)$ for $j = i-1$. By induction hypothesis, $\text{UP}_{\text{ADP}}(T',1) = \text{UP}^*(T',1) = \frac{j-1}{j}T - \frac{j+1}{2} + \frac{1}{j} = \frac{\alpha T' - \beta}{\alpha+1}$, and from Lemma 2 we have that $\text{UP}_{\text{ADP}}(T,1) = \frac{(\alpha+1)T-(\beta+\alpha+2)}{\alpha+2} = \frac{i-1}{i}T - \frac{i+1}{2} + \frac{1}{i}$.

To show that the useful payload achieved by ADP is optimal for this case $T \geq 2$, consider an algorithm A that follows one of the following approaches:

(a) First sends a packet π' of length $p' > \frac{T+\beta}{\alpha+2}$. We assume then that the adversary jams π'. The length of the rest of the interval is $T - p' < T - \frac{T+\beta}{\alpha+2}$. Hence the useful payload will be

$$\text{UP}_A(T,1) < T - \frac{T+\beta}{\alpha+2} - 1 = \frac{(\alpha+1)T - (\beta+\alpha+2)}{\alpha+2} = \text{UP}_{\text{ADP}}(T,1).$$

(b) First sends a packet π' of length $p' < \frac{T+\beta}{\alpha+2}$, $p' \geq 1$. Then the adversary does not jam π'. The rest of the interval has length $T - p' = T' + (p - p') > T'$. We consider two cases (from Lemma 1 no other case is possible):

Case (b).1: $T - p' = T' + (p-p') \in \left[\frac{(j-1)j}{2}+1, \frac{j(j+1)}{2}+1\right)$ for $j = i-1$. Then, by induction hypothesis, $\text{UP}^*(T'+(p-p'),1) = \frac{j-1}{j}(T'+(p-p')) - \frac{j+1}{2} + \frac{1}{j} < \frac{j-1}{j}T' - \frac{j+1}{2} + \frac{1}{j} + (p-p') = \text{UP}^*(T',1) + (p-p')$. Hence,

$$\begin{aligned}\text{UP}_A(T,1) &\leq p' - 1 + \text{UP}^*(T'+(p-p'),1) < p' - 1 + \text{UP}^*(T',1) + (p-p')\\ &= p - 1 + \text{UP}^*(T',1) = \text{UP}_{\text{ADP}}(T,1).\end{aligned}$$

Case (b).2: $T - p' = T' + (p-p') \in \left[\frac{(i-1)i}{2}+1, \frac{i(i+1)}{2}+1\right)$. In this case,

$$\begin{aligned}\text{UP}_A(T,1) &\leq p' - 1 + \text{UP}^*(T-p',1) = p' - 1 + \frac{i-1}{i}(T-p') - \frac{i+1}{2} + \frac{1}{i}\\ &< \frac{i-1}{i}T - \frac{i+1}{2} + \frac{1}{i} = \text{UP}_{\text{ADP}}(T,1),\end{aligned}$$

where the first equality follows from induction hypothesis, and the second inequality follows from the fact that $p' < i$ (derived from $p' < \frac{T+\beta}{\alpha+2}$, the definition of α and β, and the fact that $T < \frac{i(i+1)}{2}+1$).

Hence, in none of the two cases, neither (a) nor (b), Algorithm A was able to achieve a higher useful payload than ADP, which implies that the latter achieves optimality. □

5 Optimal Algorithm for ANY $f > 1$

We now turn our focus on the case of any number of error tokens $f > 1$ available to the adversary for an interval of length T. We present the general adaptive algorithm ADP(T, f) for $f > 1$ as Algorithm 2, and prove its optimality in the rest of the section. The pseudocode of ADP(T, f) for $f > 1$ is similar to that of ADP$(T, 1)$, with a couple of differences. First, in this case it is not possible to explicitly give the length p of the first packet π sent (values of α, β, and γ) when $T \geq f + 1$ (see Theorem 3). Second, if π is jammed, the adversary still has some error tokens that it can use. Hence, instead of sending a packet that spans the rest of the interval, ADP(T, f) makes the call ADP$(T - p, f - 1)$, which could be recursive if $f > 2$, or a call to the algorithm ADP$(T - p, 1)$ (see Algorithm 1), if $f = 2$. It will not be surprising then that the proof of optimality of the algorithm ADP(T, f) will use induction on f.

Algorithm 2. ADP(T, f), for $f > 1$

If $T < f + 1$ then
  Send packet with length $p = T$
else
  Send packet π with length $p = \frac{\alpha T + \beta}{\gamma}$ // α, β and γ depend on T; see Theorem 3
  If packet π is jammed then
    Call ADP$(T - p, f - 1)$
  else
    Call ADP$(T - p, f)$

Let us first prove some observations that hold for any optimal algorithm OPT, to be used later in the analysis of Algorithm ADP(T, f).

Observation 1. *The useful payload of an optimal algorithm OPT, follows a non-decreasing function with respect to the length of the interval of interest, T, when there are $f \geq 0$ available errors, i.e., $UP^*(T, f) \leq UP^*(T + \delta, f)$, for $\delta > 0$.*

Proof. Let us consider an optimal algorithm OPT that achieves optimal useful payload $\mathrm{UP}^*(T, f) = \alpha$, for an interval of length T and f error tokens available within the interval. Now let us construct an algorithm A, that for interval length $T + \delta$ initially uses the exact same approach as OPT for T; choosing the same packet lengths OPT does during the initial T time of the interval. This means that it has at least the same useful payload as OPT for T, i.e., $\mathrm{UP}_A(T + \delta, f) \geq \alpha$.

Since OPT is the optimal algorithm, it must achieve at least the same useful payload as A for the interval of length $T+\delta$, i.e., $\mathrm{UP}^*(T+\delta, f) \geq \mathrm{UP}_A(T+\delta, f)$. Hence, $\mathrm{UP}^*(T, f) \leq \mathrm{UP}^*(T+\delta, f)$ as claimed. □

Observation 2. *The useful payload of an optimal algorithm OPT, follows a non-increasing function with respect to the number of available errors in an interval of length T, i.e., $UP^*(T, f) \leq UP^*(T, f-1)$, where $f \geq 1$.*

Proof. Let us consider an optimal algorithm OPT, with a useful payload $\mathrm{UP}^*(T, f) = \beta$ for an interval length T with f errors available. Then, let us construct an algorithm A, that for $f-1$ error tokens during the same interval length T, uses the exact approach as OPT for f errors; choosing the same packet lengths until $f-1$ error tokens are used by the adversary. Then, it schedules one packet equal to the size of the remaining interval. This means that it has at least the same useful payload as OPT does for f errors, $\mathrm{UP}_A(T, f-1) \geq \beta$. And since OPT is the optimal algorithm, it must achieve at least the same useful payload for the same interval and $f-1$ errors, i.e., $\mathrm{UP}^*(T, f-1) \geq \mathrm{UP}_A(T, f-1)$. Hence, $\mathrm{UP}^*(T, f) \leq \mathrm{UP}^*(T, f-1)$ as claimed. □

Lemma 3. *There is an optimal algorithm OPT that is work-conserving, i.e., for each T and for each f, there is an optimal work-conserving strategy deciding the packet lengths.*

Proof. Assume by contradiction that there is some combination of interval and number of error tokens (T, f), for which no work-conserving scheduling strategy is optimal. We choose the smallest such T and consider the following:

(1) There is an optimal strategy for this pair of T and f that does not send any packet during the interval. Hence the optimal useful payload is zero, $\mathrm{UP}^*(T, f) = 0$. In this case, sending one packet that spans the whole interval will lead to the same payload.

(2) There is a strategy that waits for Δ time at the beginning of the interval before sending a packet of length p. This packet can be jammed. Therefore,

$$\begin{aligned}\mathrm{UP}^*(T, f) &= \min\{\mathrm{UP}^*(T-\Delta-p, f-1), p-1+\mathrm{UP}^*(T-\Delta-p, f)\}\\ &\leq \min\{\mathrm{UP}^*(T-p, f-1), p-1+\mathrm{UP}^*(T-p, f)\}.\end{aligned}$$

where the inequality follows from Observation 1. The right side of the inequality is the useful payload obtained by the strategy that does not wait the Δ period, but instead schedules the packet of length p at the beginning of the interval (which is work-conserving). Since both cases lead to a contradiction, the claim follows. □

Lemma 4. *The optimal useful payload is a continuous function with respect to the length of the interval, T, when there are $f \geq 1$ errors available.*

Proof. Assume by contradiction that the optimal useful payload is not a continuous function. This means that there is an interval length T for which the following holds: $\lim_{\epsilon \to 0} \text{UP}^*(T-\epsilon, f) < \text{UP}^*(T, f)$. Let us fix parameter $\epsilon > 0$, and observe the behavior of a work-conserving optimal algorithm OPT for interval lengths T and $T - \epsilon$ (such an algorithm exists by Lemma 3). Let us then denote by p_O and p_ϵ the lengths of the first packet scheduled by OPT in each case respectively. These packets can be jammed or not. We observe:

$$\text{UP}^*(T-\epsilon, f) = \min\{\text{UP}^*(T-\epsilon-p_\epsilon, f-1), p_\epsilon - 1 + \text{UP}^*(T-\epsilon-p_\epsilon, f)\} \quad (1)$$

$$\text{UP}^*(T, f) = \min\{\text{UP}^*(T-p_O, f-1), p_O - 1 + \text{UP}^*(T-p_O, f)\} \quad (2)$$

However, if we construct an alternative algorithm A that chooses a packet of length $p'' = p_O - \epsilon$ in the case of interval of length $T - \epsilon$, and works as OPT for smaller intervals, then

$$\text{UP}_A(T-\epsilon, f) = \min\{\text{UP}^*(T-p_O, f-1), p_O - \epsilon - 1 + \text{UP}^*(T-p_O, f)\} \geq \text{UP}^*(T, f) - \epsilon.$$

Since $\text{UP}^*(T-\epsilon, f) \geq \text{UP}_A(T-\epsilon, f)$, it is then trivial to conclude that $\lim_{\epsilon \to 0} \text{UP}^*(T - \epsilon, f) = \text{UP}^*(T, f)$, which is a contradiction. Hence the optimal useful payload is a continuous function with respect to the length of the interval, as claimed. □

We will now show how Algorithm ADP(T, f) computes the packet length p of the packet π sent when $T \geq f + 1$. The computation assumes that it is possible to recursively call ADP(T', j) for any $T' < T$ and $j \leq f$, and that the useful payload of each of these recursive calls is the optimal value $\text{UP}^*(T', j)$. Then, ADP(T, f) chooses as length of packet π the smallest value $p \in [1, T]$ that satisfies the equality $\text{UP}^*(T-p, f-1) = p - 1 + \text{UP}^*(T-p, f)$. Table 1 shows the values of p chosen for some interval lengths T when $f = 2$. It also shows the useful payload achieved by the algorithm using these values of p.

Table 1. Values of packet length p and optimal useful payload $\text{UP}^*(T, 2)$ achieved with Algorithm ADP($T, 2$).

T	$[1, 3)$	$[3, 9/2)$	$[9/2, 17/3)$	$[17/3, 19/3)$	$[19/3, 70/9)$	$[70/9, 308/36)$
p	T	$\frac{T}{3}$	$\frac{T+6}{7}$	$\frac{3T+3}{12}$	$\frac{5T+16}{26}$	$\frac{6T+42}{42}$
$\text{UP}^*(T, 2)$	0	$\frac{T-3}{3}$	$\frac{3T-10}{7}$	$\frac{6T-22}{12}$	$\frac{14T-54}{26}$	$\frac{24T-98}{42}$

We now prove that the described process to make the choice leads to optimality.

Theorem 3. *Given an interval of length $T \geq f + 1$, Algorithm ADP(T, f) achieves optimal useful payload by choosing the smallest value $p \in [1, T]$ that satisfies the equality*

$$UP^*(T-p, f-1) = p - 1 + UP^*(T-p, f).$$

Moreover, there are constants α_l, β_l, γ_l, α_k, β_k, and γ_k such that $UP^*(T-p,f) = \frac{\alpha_l(T-p)-\beta_l}{\gamma_l}$ and $UP^*(T-p,f-1) = \frac{\alpha_k(T-p)-\beta_k}{\gamma_k}$, and hence

$$p = \frac{(\alpha_k\gamma_l - \gamma_k\alpha_l)T + \gamma_k\gamma_l + \gamma_k\beta_l - \beta_k\gamma_l}{\gamma_k\gamma_l + \alpha_k\gamma_l - \gamma_k\alpha_l}.$$

(Observe that the parameters used in Algorithm 2 are hence $\alpha = \alpha_k\gamma_l - \gamma_k\alpha_l$, $\beta = \gamma_k\gamma_l + \gamma_k\beta_l - \beta_k\gamma_l$, and $\gamma = \gamma_k\gamma_l + \alpha_k\gamma_l - \gamma_k\alpha_l$.) The optimal useful payload obtained is then

$$UP^*(T,f) = \frac{\alpha_k\gamma_l T - (\alpha_k\gamma_l + \alpha_k\beta_l + \beta_k\gamma_l - \beta_k\alpha_l)}{\gamma_k\gamma_l + \alpha_k\gamma_l - \gamma_k\alpha_l}.$$

Proof. We prove by a double induction on the number of error tokens f and the length of the interval T, that the approach followed by Algorithm ADP(T,f) gives the optimal useful payload.

Base Cases. We have as base case of the induction on the number of error tokens the fact that (1) when $f = 0$ the optimal strategy is to send a single packet of length T that spans the whole interval, leading to $\text{UP}^*(T,0) = T-1$, and (2) that the algorithm ADP$(T,1)$ presented in Sect. 4 is optimal for any T, which covers the case $f = 1$.

For a given $f > 1$, we also use induction in the length of the interval T. In this case the base case is when $T < f+1$, which has optimal payload $\text{UP}^*(T,f) = 0$, since the adversary can jam each of the up to f packets that can be sent.

Induction Hypotheses. We first inductively assume that ADP(T,j) is optimal for any number of tokens $j < f$ available to the adversary at the beginning of the interval and any interval length $T > j$. In particular, for any $j < f$ and any $T > j$, there is a known range of lengths $R_{ij} = [a_{ij}, b_{ij})$ such that $T \in R_{ij}$, $b_{ij} = a_{(i+1)j}$ and the optimal useful payload is known to be $\text{UP}^*(T,j) = \frac{\alpha_{ij}T - \beta_{ij}}{\gamma_{ij}}$. Parameters α_{ij}, β_{ij} and γ_{ij} are known positive integers, such that $\beta_{ij} > \gamma_{ij} > \alpha_{ij}$.

We inductively also assume that for f error tokens, there are m known ranges of lengths $R_{if} = [c_{if}, d_{if})$ for $i = 1, 2, \ldots, m$, such that $\bigcup_{i=1}^{m} R_{if} = [1, d_{mf})$, $c_{1f} = 1$, and $d_{if} = c_{(i+1)f}, \forall 1 \le i < m$. Also, for any interval length T such that $T < d_{mf}$ and $T \in R_{if} = [c_{if}, d_{if})$, the optimal useful payload is known to be $\text{UP}^*(T,f) = \frac{\alpha_{if}T - \beta_{if}}{\gamma_{if}}$. Parameters α_{if}, β_{if} and γ_{if} are known positive integers such that (1) $\beta_{if} > \gamma_{if} > \alpha_{if}$, and for any $R_{\ell f}$, R_{rf} where $1 \le \ell \le r \le m$, it holds that (2) $\frac{\beta_{rf}}{\gamma_{rf}} \ge \frac{\beta_{\ell f}}{\gamma_{\ell f}}$ and (3) $\frac{\alpha_{rf}}{\gamma_{rf}} \ge \frac{\alpha_{\ell f}}{\gamma_{\ell f}}$.

Inductive Step. For interval length $T \in [d_{mf}, d_{mf}+1)$, the algorithm ADP(T,f) chooses the *smallest* packet length $p \in [1,T]$ that satisfies the following condition

$$\text{UP}^*(T-p, f-1) = p - 1 + \text{UP}^*(T-p, f). \tag{3}$$

Claim. There is at least one packet length $p \in [1,T]$ that satisfies Eq. 3.

Proof of Claim. Observe that, when $p = 1$, from Observation 2 we have that $\text{UP}^*(T-p, f-1) \ge p - 1 + \text{UP}^*(T-p, f)$. On the other hand, when $p = T$, we have that $\mathrm{UP}^*(T-p, f-1) = 0 \leq p-1+\mathrm{UP}^*(T-p, f) = T-1$. Hence, taking into consideration the continuity of the useful payload function of both $f-1$ and f error tokens (Lemma 4) and the Mean Value Theorem, there always exists a packet size $p \in [1, T]$ such that $\mathrm{UP}^*(T-p, f-1) = p-1+\mathrm{UP}^*(T-p, f)$. $\square_{Claim}$

Now, let p be the packet length chosen, and let $T-p \in R_{kj} = [a_{kj}, b_{kj})$ and $T-p \in R_{lf} = [c_{lf}, d_{lf})$. Note that R_{kj} and R_{lf} are among the known ranges from the induction hypothesis. Then, by induction hypothesis $\mathrm{UP}^*(T-p, f) = \frac{\alpha_{lf}(T-p)-\beta_{lf}}{\gamma_{lf}}$ and $\mathrm{UP}^*(T-p, f-1) = \frac{\alpha_{kj}(T-p)-\beta_{kj}}{\gamma_{kj}}$. Then, solving Eq. 3 for p, the packet length is

$$p = \frac{(\alpha_{kj}\gamma_{lf} - \gamma_{kj}\alpha_{lf})T + \gamma_{kj}\gamma_{lf} + \gamma_{kj}\beta_{lf} - \beta_{kj}\gamma_{lf}}{\gamma_{kj}\gamma_{lf} + \alpha_{kj}\gamma_{lf} - \gamma_{kj}\alpha_{lf}},$$

and the useful payload obtained is

$$\mathrm{UP}_{\mathrm{ADP}}(T, f) = \mathrm{UP}^*(T-p, f-1) = p-1+\mathrm{UP}^*(T-p, f) = \frac{\alpha_{kj}(T-p)-\beta_{kj}}{\gamma_{kj}}$$

$$= \frac{\alpha_{kj}\gamma_{lf}T - (\alpha_{kj}\gamma_{lf} + \alpha_{kj}\beta_{lf} + \beta_{kj}\gamma_{lf} - \beta_{kj}\alpha_{lf})}{\gamma_{kj}\gamma_{lf} + \alpha_{kj}\gamma_{lf} - \gamma_{kj}\alpha_{lf}},$$

as claimed. To complete the induction step, we define $\alpha = \alpha_{kj}\gamma_{lf}$, $\beta = \alpha_{kj}\gamma_{lf} + \alpha_{kj}\beta_{lf} + \beta_{kj}\gamma_{lf} - \beta_{kj}\alpha_{lf}$ and $\gamma = \gamma_{kj}\gamma_{lf} + \alpha_{kj}\gamma_{lf} - \gamma_{kj}\alpha_{lf}$. Then, we show the following three properties (1) $\beta > \gamma > \alpha$, (2) $\frac{\beta}{\gamma} \geq \frac{\beta_{lf}}{\gamma_{lf}}$, and (3) $\frac{\alpha}{\gamma} \geq \frac{\alpha_{lf}}{\gamma_{lf}}$ as follows.

Property 1. For the new parameters $\alpha = \alpha_{kj}\gamma_{lf}$, $\beta = \alpha_{kj}\gamma_{lf} + \alpha_{kj}\beta_{lf} + \beta_{kj}\gamma_{lf} - \beta_{kj}\alpha_{lf}$ and $\gamma = \gamma_{kj}\gamma_{lf} + \alpha_{kj}\gamma_{lf} - \gamma_{kj}\alpha_{lf}$, it holds that $\beta > \gamma > \alpha$.

Proof of Property 1. First, from the *induction hypotheses*, recall the definition of parameters α_{ij}, β_{ij} and γ_{ij}, being known positive integers such that $\beta_{ij} > \gamma_{ij} > \alpha_{ij}$. Looking now at the current parameters α, β and γ individually, we have the following:

(a) $\alpha = \alpha_{kj}\gamma_{lf}$.
(b) $\beta = \alpha_{kj}\gamma_{lf} + \alpha_{kj}\beta_{lf} + \beta_{kj}\gamma_{lf} - \beta_{kj}\alpha_{lf} = \alpha_{kj}(\gamma_{lf} + \beta_{lf}) + \beta_{kj}(\gamma_{lf} - \alpha_{lf})$.
(c) $\gamma = \gamma_{kj}\gamma_{lf} + \alpha_{kj}\gamma_{lf} - \gamma_{kj}\alpha_{lf} = \gamma_{kj}(\gamma_{lf} - \alpha_{lf}) + \alpha_{kj}\gamma_{lf}$.

Observe that $\gamma_{kj}(\gamma_{lf} - \alpha_{lf}) + \alpha_{kj}\gamma_{lf} > \alpha_{kj}\gamma_{lf}$, since $\gamma_{kj} > 0$ and $\gamma_{lf} - \alpha_{lf} > 0$ by induction hypothesis. Hence, from (a) and (c) $\gamma > \alpha$. Also, $\alpha_{kj}(\gamma_{lf} + \beta_{lf}) + \beta_{kj}(\gamma_{lf} - \alpha_{lf}) > \gamma_{kj}(\gamma_{lf} - \alpha_{lf}) + \alpha_{kj}\gamma_{lf}$, since by induction hypothesis $\beta_{kj} > \gamma_{kj}$, $\gamma_{lf} - \alpha_{lf} > 0$, and all parameters are positive. Hence, from (b) and (c) $\beta > \gamma$ holds as well. This completes the proof of the claim. $\square_{Property1}$

Property 2. For the new parameters $\beta = \alpha_{kj}\gamma_{lf} + \alpha_{kj}\beta_{lf} + \beta_{kj}\gamma_{lf} - \beta_{kj}\alpha_{lf}$ and $\gamma = \gamma_{kj}\gamma_{lf} + \alpha_{kj}\gamma_{lf} - \gamma_{kj}\alpha_{lf}$, it holds that $\frac{\beta}{\gamma} > \frac{\beta_{lf}}{\gamma_{lf}}$.

Proof of Property 2. For this proof observe first, that since $\beta > \gamma$ (as shown in Property 1), we can safely use the fact that $\frac{\beta}{\gamma} > \frac{\beta - c}{\gamma - c}$, where c is positive. Also by induction hypothesis we have that $\gamma_{lf} - \alpha_{lf} > 0$ and $\beta_{kj} - \gamma_{kj} > 0$. We therefore use some fraction inequality properties as follows:

$$\frac{\beta}{\gamma} = \frac{\alpha_{kj}\gamma_{lf} + \alpha_{kj}\beta_{lf} + \beta_{kj}\gamma_{lf} - \beta_{kj}\alpha_{lf}}{\gamma_{kj}\gamma_{lf} + \alpha_{kj}\gamma_{lf} - \gamma_{kj}\alpha_{lf}} = \frac{\alpha_{kj}(\gamma_{lf} + \beta_{lf}) + \beta_{kj}(\gamma_{lf} - \alpha_{lf})}{\gamma_{kj}(\gamma_{lf} - \alpha_{lf}) + \alpha_{kj}\gamma_{lf}}$$
$$> \frac{\alpha_{kj}(\gamma_{lf} + \beta_{lf}) + (\beta_{kj} - \gamma_{kj})(\gamma_{lf} - \alpha_{lf})}{\alpha_{kj}\gamma_{lf}} > \frac{\alpha_{kj}\gamma_{lf} + \alpha_{kj}\beta_{lf}}{\alpha_{kj}\gamma_{lf}} = 1 + \frac{\beta_{lf}}{\gamma_{lf}} > \frac{\beta_{lf}}{\gamma_{lf}},$$

which completes the proof. $\square_{Property 2}$

Property 3. For the new parameters $\alpha = \alpha_{kj}\gamma_{lf}$ and $\gamma = \gamma_{kj}\gamma_{lf} + \alpha_{kj}\gamma_{lf} - \gamma_{kj}\alpha_{lf}$, it holds that $\frac{\alpha}{\gamma} > \frac{\alpha_{lf}}{\gamma_{lf}}$.

Proof of Property 3. For this proof observe first, that since $\gamma > \alpha$ (as shown in Property 1), we can safely use the fact that $\frac{\alpha}{\gamma} > \frac{\beta+c}{\gamma+c}$, where c is positive. Also by induction hypothesis we have that $\gamma_{lf} - \alpha_{lf} > 0$. We therefore use some fraction inequality properties as follows:

$$\frac{\alpha}{\gamma} = \frac{\alpha_{kj}\gamma_{lf}}{\gamma_{kj}\gamma_{lf} + \alpha_{kj}\gamma_{lf} - \gamma_{kj}\alpha_{lf}} = \frac{\alpha_{kj}\gamma_{lf} + \gamma_{kj}\alpha_{lf}}{\alpha_{kj}\gamma_{lf} + \gamma_{kj}\gamma_{lf}}$$
$$= \frac{\alpha_{kj}\alpha_{lf} + \alpha_{kj}(\gamma_{lf} - \alpha_{lf}) + \gamma_{kj}\alpha_{lf}}{\gamma_{lf}(\alpha_{kj} + \gamma_{kj})} = \frac{\alpha_{lf}(\alpha_{kj} + \gamma_{kj})}{\gamma_{lf}(\alpha_{kj} + \gamma_{kj})} + \frac{\alpha_{kj}(\gamma_{lf} - \alpha_{lf})}{\gamma_{lf}(\alpha_{kj} + \gamma_{kj})} > \frac{\alpha_{lf}}{\gamma_{lf}},$$

which completes the proof. $\square_{Property 3}$

Observe that the above proof holds for all Ts in the interval $[d_{mf}, d_{mf}+1)$; for each one of these, the algorithm would compute the smaller p that satisfies Eq. 3 and the computation of the parameters α, β, γ is done analogously. Therefore, the known ranges of lengths are extended in this interval.

We must now show that the useful payload is in fact optimal. Let us assume by contradiction that an algorithm A is able to achieve a larger useful payload for the pair (T, f) by sending first a different packet length $p' \neq p$. We consider the following cases.

(a) Algorithm A chooses a packet π' of length $p' > p$. Then, we assume that the adversary will jam the packet π'. Hence, the useful payload achieved by A will be upper bounded as $\mathrm{UP}_A(T, f) \leq \mathrm{UP}^*(T - p', f - 1)$ which by Observation 1 is smaller than $\mathrm{UP}^*(T - p, f - 1) = \mathrm{UP}_{\mathrm{ADP}}(T, f)$, since $T - p' < T - p$.

(b) Algorithm A chooses a packet π' of length $p' < p$. Observe that p' does not satisfy Eq. 3, since p is the smallest length that does. Then the adversary does not jam π'. Then, $\mathrm{UP}_A(T, f) \leq p' - 1 + \mathrm{UP}^*(T - p', f)$. We show now that this value is no larger than $p - 1 + \mathrm{UP}^*(T - p, f) = \mathrm{UP}_{\mathrm{ADP}}(T, f)$. Let us assume that $T - p' \in R_{rf}$, where $r \geq l$. Then, $\mathrm{UP}^*(T - p', f) = \frac{\alpha_{rf}(T-p') - \beta_{rf}}{\gamma_{rf}} \leq \frac{\alpha_{rf}}{\gamma_{rf}}(T - p') - \frac{\beta_{lf}}{\gamma_{lf}}$, since $\frac{\beta_{rf}}{\gamma_{rf}} \geq \frac{\beta_{lf}}{\gamma_{lf}}$ as shown by Property 2. Similarly, $\mathrm{UP}^*(T - p, f) = \frac{\alpha_{lf}(T-p) - \beta_{lf}}{\gamma_{lf}} \geq \frac{\alpha_{rf}}{\gamma_{rf}}(T - p) - \frac{\beta_{lf}}{\gamma_{lf}}$, since $\frac{\alpha_{rf}}{\gamma_{rf}} \geq \frac{\alpha_{lf}}{\gamma_{lf}}$ as shown by Property 3. Finally, combining these bounds and the fact that

$\frac{\alpha_{rf}}{\gamma_{rf}} < 1$ (see Property 1), we get that

$$\begin{aligned}\mathrm{UP}_A(T,f) &\le p' - 1 + \mathrm{UP}^*(T-p',f) \le p' - 1 + \frac{\alpha_{rf}}{\gamma_{rf}}(T-p') - \frac{\beta_{lf}}{\gamma_{lf}} \\ &\le p' - 1 + \frac{\alpha_{rf}}{\gamma_{rf}}(T-p') - \frac{\beta_{lf}}{\gamma_{lf}} + (p-p') - \frac{\alpha_{rf}}{\gamma_{rf}}(p-p') \\ &= p - 1 + \frac{\alpha_{rf}}{\gamma_{rf}}(T-p) - \frac{\beta_{lf}}{\gamma_{lf}} \le \mathrm{UP}_{\mathrm{ADP}}(T,f).\end{aligned}$$

In all cases the resulting useful payload is smaller than the one achieved by choosing the smallest packet size p such that $\mathrm{UP}^*(T-p, f-1) = p - 1 + \mathrm{UP}^*(T-p, f)$. Hence the packet size calculated by $\mathrm{ADP}(T, f)$ is optimal. □

6 Discussion

Recall that the problem we considered up to this point in the paper is a "static" version of the problem we considered in [2] (continuous version). In this section we discuss the use of our proposed algorithms when applied to the continuous version of the problem. (Recall from Sect. 1 the definitions of ρ and σ.)

We begin with the following observation: If we divide the time interval of the continuous version of the problem into successive intervals of length $1/\rho$, and σ error tokens are available at the beginning of each interval, then each of these intervals can be considered an instance of the static version of the problem, where $T = 1/\rho$ and $f = \sigma$.

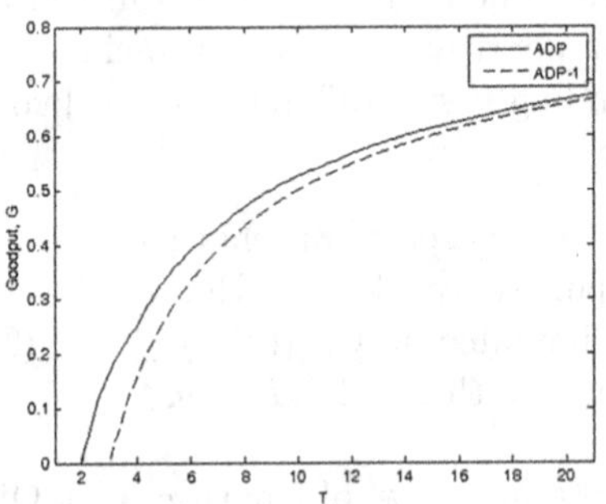

Fig. 1. The goodput rate of algorithms ADP-1 [2] and $\mathrm{ADP}(T, 1)$ (Sect. 4) for $T = 1 \ldots 22$

Therefore, by running algorithm $\mathrm{ADP}(1/\rho, \sigma)$ in each of these intervals we obtain a solution to the continuous version of the problem. However, this solution is possibly not the best possible, as we make the pessimistic assumption that at the beginning of each interval, the adversary has all σ error tokens available to use; this is true for the first interval, but in successive intervals this might not be the case (with the exception of the case $\sigma = 1$, which we discuss below). Based on the model defined in [2], a new error token will be arriving at the beginning of each interval. If there are already σ tokens, then a token is lost (σ represents, for example, the capacity of the battery of a jamming device – this cannot be exceeded). If in this interval, the adversary performs, say, three packet jams, then at the beginning of the next interval it will have $\sigma - 2$ available tokens. If the scheduling algorithm keeps track of this, then in this interval it should use ADP$(1/\rho, \sigma - 2)$ instead of ADP$(1/\rho, \sigma)$. So, in order to produce more efficient solutions, the scheduling algorithm needs to keep track (using the feedback mechanism) how many jams took place in the previous interval, and using its knowledge of $1/\rho$, run the appropriate version of ADP(). Although there are other subtle issues that also need to be considered, the proposed approach can be used as the basis for obtaining an optimal solution to the continuous version of the problem. We plan to pursue this direction in future research.

Regarding the case of $f = \sigma = 1$, as demonstrated in Fig. 1, algorithm ADP$(1/\rho, 1)$ obtains better results than the solution developed in [2] (called Algorithm ADP-1). In [2], for $\sigma = 1$ it was shown that the goodput rate of Algorithm ADP-1 is $1 - \frac{\rho}{2}\left(1 + \sqrt{1 + \frac{8}{\rho}}\right)$. Figure 1 depicts this goodput rate and the goodput rate of algorithm ADP$(1/\rho, 1)$ as obtained from our analysis in Sect. 4, for $T = 1 \ldots 22$. Since in the case of $\sigma = 1$ it is best for the adversary to use the error token (otherwise it will lose it), our improved goodput demonstrates the promise of the abovementioned approach.

References

1. Fernández Anta, A., Georgiou, C., Kowalski, D.R., Widmer, J., Zavou, E.: Measuring the impact of adversarial errors on packet scheduling strategies. In: Moscibroda, T., Rescigno, A.A. (eds.) SIROCCO 2013. LNCS, vol. 8179, pp. 261–273. Springer, Heidelberg (2013)
2. Fernández Anta, A., Georgiou, C., Zavou, E.: Packet scheduling over a wireless channel: AQT-based constrained jamming. In: Proceedings of NETYS (2015)
3. Awerbuch, B., Richa, A., Scheideler, C.: A jamming-resistant mac protocol for single-hop wireless networks. In: Proceedings of PODC, pp. 45–54 (2008)
4. Dolev, S., Gilbert, S., Guerraoui, R., Kowalski, D., Newport, C., Kohn, F., Lynch, N.: Reliable distributed computing on unreliable radio channels. In: Proceedings of MobiHoc S^3 workshop, pp. 1–4 (2009)
5. Gilbert, S., Guerraoui, R., Newport, C.: Of malicious motes and suspicious sensors: on the efficiency of malicious interference in wireless networks. Theor. Comput. Sci. **410**(6), 546–569 (2009)
6. Pelechrinis, K., Iliofotou, M., Krishnamurthy, S.V.: Denial of service attacks in wireless networks: the case of jammers. IEEE Commun. Surv. Tutor. **13**(2), 245–257 (2011)
7. Pruhs, K.: Competitive online scheduling for server systems. ACM SIGMETRICS Perform. Eval. Rev. **34**(4), 52–58 (2007)
8. Pruhs, K., Sgall, J., Torng, E.: Online scheduling. In: Handbook of Scheduling: Algorithms, Models, and Performance Analysis, pp. 15-1–15-41 (2004)

9. Richa, A., Scheideler, C., Schmid, S., Zhang, J.: Competitive and fair medium access despite reactive jamming. In: Proceedings of ICDCS, pp. 507–516 (2011)
10. Richa, A., Scheideler, C., Schmid, S., Zhang, J.: Towards jamming-resistant and competitive medium access in the sinr model. In: Proceedings of the 3rd ACM Workshop on Wireless of the Students, by the Students, for the Students, pp. 33–36 (2011)
11. Richa, A., Scheideler, C., Schmid, S., Zhang, J.: Competitive and fair throughput for co-existing networks under adversarial interference. In: Proceedings of PODC, pp. 291–300 (2012)
12. Thuente, D., Acharya, M.: Intelligent jamming in wireless networks with applications to 802.11 b and other networks. In: Proceedings of MILCOM, vol. 6 (2006)

Metric and Distributed On-Line Algorithm for Minimizing Routing Interference in Wireless Sensor Networks

Kejia Zhang[1], Qilong Han[1], Zhipeng Cai[1,2](✉), Guisheng Yin[1], and Junyu Lin[1]

[1] College of Computer Science and Technology, Harbin Engineering University, Harbin, China
[2] Department of Computer Science, Georgia State University, Atlanta, USA
zcai@gsu.edu

Abstract. First, this paper gives a metric to quantify interference level among multiple routing paths in wireless sensor networks. Unlike the existing metrics, the proposed one can precisely measure interference level among any set of paths, even though these paths have some nodes or links in common. Based on the proposed metric, the Minimizing-Interference-for-Multiple-Paths (MIMP) problem is considered: Given k routing requests $\{s_1, t_1\}, \ldots, \{s_k, t_k\}$, find k paths connecting these routing requests with minimum interference. The MIMP problem is NP-hard even for $k = 2$. This paper proposes heuristic distributed on-line algorithm for the MIMP problem. The efficiency of the proposed algorithm is verified by simulation results.

1 Introduction

A typical wireless sensor network (WSN) consists of hundreds or thousands of sensor nodes deployed in the monitoring area. Usually, these sensor nodes are battery-powered devices with simple communication components, which makes their communication distance relatively short. If two sensor nodes want to exchange their data, the data usually needs to be forwarded by many relaying nodes, so routing in WSNs is a crucial problem. Let s and t denote a routing request, i.e., two sensor nodes who want to communicate with each other. The main objective of routing is to optimally find paths connecting s and t (i.e., $s \sim t$ paths).

Because of the large scale and frequent data exchange of WSNs, there are usually more than one routing requests need to be met at the same time. Even if there is only one routing request in the network, sometimes the users may want to use multiple paths, say k paths, for routing to increase throughput or load balance. In this case, we can see the multi-path routing request as k individual routing requests with the same s and t. To get a better routing performance like throughput or energy efficiency, network designers always want to use independent paths to satisfy all the routing requests. Here, "independent" means that these routing paths do not interfere with each other.

Z. Lu et al. (Eds.): COCOA 2015, LNCS 9486, pp. 279–292, 2015.
DOI: 10.1007/978-3-319-26626-8_21

There are many works [1–13,15–23] study how to construct multiple independent paths in WSNs and other wireless ad hoc networks. Some works [7,9,23] borrow the thoughts in wired networks and use node-disjoint paths as independent paths. A set of paths are node-disjoint if these paths do not have any nodes in common. Since sensor nodes share the same communication channel in most of WSNs, when node v is transmitting data to node u, the transmission will block all v's neighbors except u from receiving data. This feature of WSNs is called *wireless interference*. Node-disjoint paths are not independent enough for routing since wireless interference. It has been shown in [22] that the wireless interference among multiple paths will dramatically decrease the throughput of routing even though the routing paths are node-disjoint. And [21] claims that for two node-disjoint paths, the more wireless interference between them, the larger the average end-to-end delay for both paths. Therefore, the concept of non-interference paths is proposed by [19]. A set of paths are non-interference if the nodes in one path do not interfered by (are not neighbors of) any node in the other paths.

Although using non-interference paths can greatly improve the performance of routing, sometimes, especially when there are many routing requests have to be satisfied at the same time, it is impossible to find non-interference paths to satisfy all the routing requests. In these cases, the best of way to meet all the routing requests is relaxing the requirement for non-interference and using a set of paths with the lowest level of interference. Therefore, we need a metric to quantify interference level among multiple routing paths and a method to build routing paths with minimum interference. There already exist some works [1,11,13,17,18,21] to quantify interference level, but they all have apparent drawbacks. Some of them can only quantify the interference level between two paths. Some of them require all the routing paths are node-disjoint, which is impossible in some cases.

In this paper, we consider the Minimizing-Interference-for-Multiple-Paths (MIMP) problem, which is defined as: Given k routing requests $\{s_1, t_1\}, \ldots, \{s_k, t_k\}$, find k paths connecting these routing requests with minimum interference. We deeply study how the transmission in different routing paths will interfere one another, especially when some of these paths share common nodes. We give a metric to measure the interference level of multiple paths. Compared with the existing interference metrics, our metric can more precisely measure the interference level and deal with the situation that some of the paths share common nodes or links. Base on the proposed metric, we design a distributed on-line algorithm to find multiple paths with minimum interference. Therefore, the main contributions of this paper are: (i) Giving a metric that can precisely measure the interference level among any set of routing paths and (ii) Designing an distributed on-line algorithm for the MIMP problem based on the given metric.

2 Related Works

[17] proposes an interference-aware multi-path routing protocol IM2PR for event reporting. Upon the detection of an interesting event, the source node uses mul-

tiple node-disjoint paths to report data to the sink. These paths are constructed one by one. During the path construction, each node chooses the next hop with higher delivery probability, lower interference level and more remaining energy. The interference level of node v is measured by how many neighbors of v are in the other active paths. Unfortunately, this paper does not give a clear metric to measure the interference level among multiple paths. Furthermore, IM2PR requires these multiple paths connecting the source node and the sink are node-disjoint, which is a too strong requirement for many cases.

[18] assumes that the interference range of each node is twice the transmission range and uses the conflict graph proposed by [13] to measure the interference level of multiple paths. It proposes protocol I2MR to construct three non-interference paths (two for data transmission and one for backup) connecting the given source and destination. I2MR builds the shortest path from source to destination at first, then marks all the one-hop and two-hop neighbors of the first path as in the interference zone so that they can not participate in other paths. Next, I2MR finds two paths along each side of the interference zone. I2MR is only suitable for constructing interference-free paths. In many cases, it impossible to find interference-free paths to meet all the routing requests. Besides, I2MR has an assumption which may be not true in practice, i.e., two nodes are in the twice of the transmission range from each other have at least one common neighbor.

[13] proposes the idea of using conflict graph to measure interference of multiple paths. The conflict graph does not consider the situation that two or more routing paths have some links in common, which is inevitable if we try to route some certain node pairs simultaneously. [21] uses the criteria called *correlation factor* to measure the interference level of multiple routing paths. The correlation factor of two node-disjoint paths is defined as the number of the links connecting the two paths. The total correlation factor of a set of multiple paths is defined as the sum of the correlation factor of each pair of the paths. This correlation factor criteria does not consider the situation that some routing paths have nodes or links in common either.

[8] studies the problem of finding multiple non-interference or interference-minimized paths from a set of sources to a receiver. It proposes a metric to measure the interference level of a set of routing paths. The metric is not precise enough since it cannot distinguish the situation that one path is interfered by another path once and the situation that one path is interfered by another path many times. [16] gives the metric to measure the interference level between two paths. However, the proposed metric is not symmetrical, i.e., for two paths P_1 and P_2, the interference for P_1 w.r.t. P_2 is not equal to the interference for P_2 w.r.t. P_1. [1,11] also propose criteria to measure how badly one routing path is interfered by other concurrent routing paths. These works do not consider the situation that some of the routing paths share common nodes or links either.

[19,20] defines that two node-disjoint paths are collision-free (non-interference) if there is no link between any two nodes on different paths. They gives algorithms to find two collision-free paths connecting a given source and destination node

pair. [15] states that even though some paths are not non-interference, we can still overcome these interference by appropriate schedule.

So far, all the existing metrics to measure interference level among multiple paths can not deal with the situation that some paths have nodes or links in common. When there are many routing requests have to be satisfied at the same time or for some certain routing requests, it is inevitable that we meet these routing requests with paths having common nodes or links. Moreover, even if we can use a set of node-disjoint paths to satisfy all the routing request, wireless interference may make the routing performance of these paths very poor. In such case, routing with certain set of paths sharing common nodes may be a much better solution.

3 Interference Metric

In this section, we give our metric to measure the interference level among a set of routing paths. This metric can deal with the situation that some paths have common nodes or links very well. We assume that only the nodes who have wireless links to v will be interfered by v's transmission. Hereinafter, if not otherwise specified, "neighbor" means one-hop neighbor.

3.1 Network Model

We model the given network as an undirected graph $G = (V, E)$, where $V = \{v \mid v$ is a sensor node$\}$ and $E = \{(u, v) \mid$ there is a wireless link between $u \in V$ and $v \in V\}$. $n = |V|$ is the number of nodes and $m = |E|$ is the number of wireless links in the given network. Here, we assume that all the links are bidirectional, *i.e.*, if v can receive the signal from u then u can receive the signal from v. $N(v)$ is the set including all v's neighbors. Hereinafter, we do not distinguish the terms "network" and "graph" and the terms "edge" and "wireless link".

A path P in G is a subgraph which can be expressed as a sequence of distinct nodes $(v_1, \ldots, v_m)$, where each $v_i \in V$ and each $(v_i, v_{i+1}) \in E$. Path $P = (v_1, \ldots, v_m)$ can be called a $v_1 \sim v_m$ path. We say a path P is the *acyclic path* if there is no edge connecting two non-adjacent nodes in P. Since the paths we construct are all the shortest paths w.r.t. some criteria, "path" means acyclic path in the rest of this paper.

Suppose that k routing paths $P_1, \ldots, P_k$ are being used in the given network. Next, we discuss how to measure the interference level among these paths. We say a edge is a *cross-path edge* if its two end nodes belong to different paths of $P_1, \ldots, P_k$. We use the function $C : V \to \mathbb{N}^0$ to define how many paths each node is in.

3.2 Quantifying Interference Level

By our assumption of the nodes' interference scope, if $P_1, \ldots, P_k$ are node-disjoint, it is reasonable to use the number of cross-path edges to measure their interference level as was done in [21]. In Example 1 with two node-disjoint paths as shown in Fig. 1, the cross-path edge (b, e) corresponds to the interference between node b in P_1 and node e in P_2. When b is sending messages for P_1, e cannot receive messages in P_2. When e is sending messages for P_2, b cannot receive messages in P_1. Therefore, for a set of node-disjoint paths, each cross-path edge corresponds to such a two-way send-and-receive interference between two nodes in different paths.

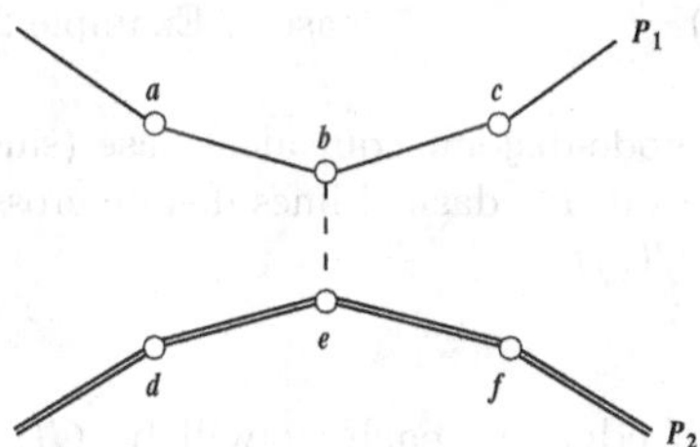

Fig. 1. Example 1 with two node-disjoint paths (single lines denote edges in P_1, double lines denote edges in P_2, dashed lines denote cross-path edges)

As we mentioned before, in many cases, it is impossible to find a set of node-disjoint path to meet all the routing requests. Sometimes, because of wireless interference, routing with paths sharing common nodes may be a better solution than routing with node-disjoint paths. Therefore, the proposed interference metric must be able to deal with the situation that the given routing paths $P_1, \ldots, P_k$ have some nodes or edges in common.

In Example 2 with two paths sharing a common node as shown in Fig. 2(a), let us consider how interference will happen. P_1 and P_2 have a common node b. The edge (b, d) can be seen as a cross-path edge which corresponds to the two-way send-and-receive interference between b in P_1 and d in P_2. Similarly, the edges (b, f), (b, a) and (b, c) can be seen as cross-path edges too. When b is sending messages for P_1, it cannot receive messages in P_2. When b is sending messages for P_2, it cannot receive messages in P_1 either. Thus, there is a two-way send-and-receive interference between b in P_1 and b in P_2. To get a equivalent case with two node-disjoint paths, we can split b into two nodes b^1 and b^2 as shown in Fig. 2(b). b^1 is the b in P_1 and b^2 is the b in P_2. Since there is a two-way send-and-receive interference between b in P_1 and b in P_2, we add a cross-path edge (b^1, b^2). From this equivalent case, we can easily see that the interference level between P_1 and P_2 (the number of cross-path edges) is 5.

We extend the above analysis to measure the interference level of k paths $P_1, \ldots, P_k$. Suppose u and v are neighbors of each other, $C(u) > 1$ and $C(v) > 1$. To get the node-disjoint equivalent case of $P_1, \ldots, P_k$, we split u into $C(u)$ new nodes $u^1, \ldots, u^{C(u)}$ and split v into $C(v)$ new nodes $v^1, \ldots, v^{C(v)}$. There is an edge connecting each u^i $(i = 1, \ldots, C(u))$ and each v^j $(j = 1, \ldots, C(v))$. And all these $u^1, \ldots, u^{C(u)}$ form a clique, so do $v^1, \ldots, v^{C(v)}$. Therefore, in the node-disjoint equivalent case, if we only count the edges connecting the nodes in

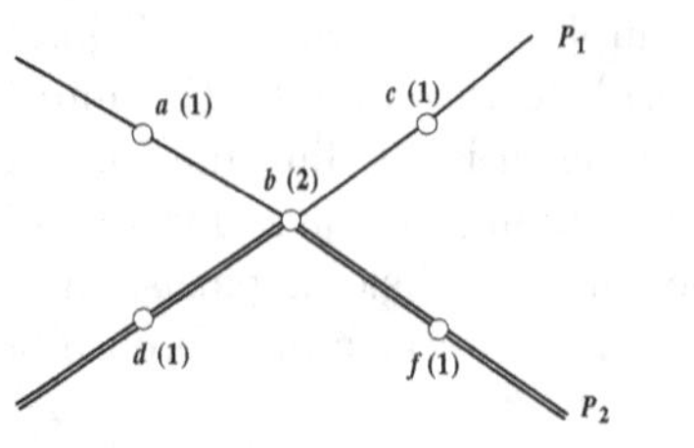

(a) Example 2 (two paths sharing a common node)

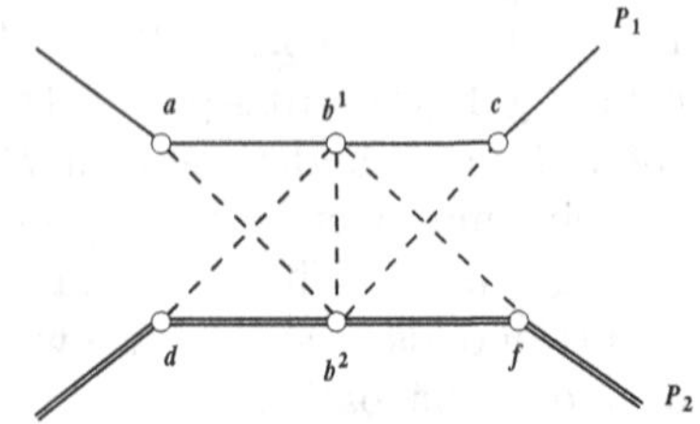

(b) The node-disjoint equivalent case of Example 2

Fig. 2. Example 2 and its node-disjoint equivalent case (single lines denote edges in P_1, double lines denote edges in P_2, dashed lines denote cross-path edges, numbers in the brackets denote nodes' $C(v)$)

$P_1, \ldots, P_k$, the number of edges at each v^j will be $C(v) - 1 + \sum_{u \in N(v)} C(u)$. In those edges, except the edges connecting v^j to its previous hop and next hop in the same path, all the other edges are cross-path edges. Therefore, the number of cross-path edges at each v^j is $W(v) - 3$ where $W : V \rightarrow \mathbb{N}^0$ is defined as $W(v) = C(v) + \sum_{u \in N(v)} C(u)$. The total number of cross-path edges at all these $v^1, \ldots, v^{C(v)}$ is $C(v)\big(W(v) - 3\big)$. The total number of cross-path edges in the node-disjoint equivalent case of $P_1, \ldots, P_k$, i.e., the interference level among these paths can be given by Definition 1.

Definition 1. *For a set of paths $P_1, \ldots, P_k$, their interference level is defined as*

$$IN(P_1, \ldots, P_k) = \frac{\sum_{v \in V} C(v)\big(W(v) - 3\big)}{2}$$

where $C(v)$ is the number of paths v is in and $W(v) = C(v) + \sum_{u \in N(v)} C(u)$.

In Example 2. We have $W(b) = 6, W(a) = W(c) = W(e) = W(f) = 4$ (Although the other neighbor of a is not shown in the figure, we suppose that its $C(v)$ is 1). The number of cross-path edges at b^1 / b^2 in the node-disjoint equivalent case is $W(b) - 3 = 3$. The number of cross-path edges at $a/b/e/f$ is $4 - 3 = 1$. The total number of cross-path edges in the node-disjoint equivalent case is $(2 * 3 + 1 + 1 + 1 + 1)/2 = 5$.

3.3 Analysis of Interference Metric

Suppose that $P_1, \ldots, P_k$ are constructed one by one in the order of their suffixes. Define $C_i(v)$ $(i = 1, \ldots, k)$ as the number of paths v is in after deploying $P_1, \ldots, P_i$ and $C_0(v) = 0$ for each $v \in V$. Accordingly, $W_i(v) = C_i(v) + \sum_{u \in N(v)} C_i(u)$. Since $P_1, \ldots, P_k$ are acyclic paths, we have the following Proposition.

Proposition 1. *For each intermediate node v in P_{i+1}, $W_{i+1}(v) = W_i(v) + 3$. And*

$$\sum_{v\in P_{j+1}} W_j(v) = \sum_{v\in P_{j+1}} (W_{j+1}(v) - 3)$$

.

Now, let us consider the change of interference level before and after deploying path P_{i+1}. For simplicity, we use IN_i to denote $IN(P_1, \ldots, P_i)$, and use C_i and W_i to denote $C_i(v)$ and $W_i(v)$ respectively. By the definition of interference level we have

$$2(IN_{i+1} - IN_i) = \sum_{v\in V} (C_{i+1}(W_{i+1} - 3) - C_i(W_i - 3)) \tag{1}$$

$$\begin{aligned} &= \sum_{v\in P_{i+1}} (C_{i+1}(W_{i+1} - 3) - C_i(W_i - 3)) \\ &\quad + \sum_{v\in V-P_{i+1}} (C_{i+1}(W_{i+1} - 3) - C_i(W_i - 3)) \end{aligned} \tag{2}$$

$$\begin{aligned} &= \sum_{v\in P_{i+1}} ((C_i + 1)(W_{i+1} - 3) - C_i(W_i - 3)) \\ &\quad + \sum_{v\in V-P_{i+1}} (C_i(W_{i+1} - 3) - C_i(W_i - 3)) \end{aligned} \tag{3}$$

$$= \sum_{v\in P_{i+1}} (W_{i+1} - 3) + \sum_{v\in P_{i+1}} C_i(W_{i+1} - W_i) + \sum_{v\in V-P_{i+1}} C_i(W_{i+1} - W_i) \tag{4}$$

$$= \sum_{v\in P_{i+1}} W_i + \sum_{v\in V} C_i(W_{i+1} - W_i) \tag{5}$$

Equation 3 can be get from Eq. 2 because $C_{i+1}(v) = C_i(v) + 1$ for each $v \in P_{i+1}$ and $C_{i+1}(v) = C_i(v)$ for each $v \in V - P_{j+1}$. By Proposition 1, we can get Eq. 5 from Eq. 4.

By the definition of function W, we have

$$\sum_{v\in P_{i+1}} W_i(v) = \sum_{v\in P_{i+1}} C_i(v) + \sum_{v\in P_{i+1}} \sum_{u\in N(v)} C_i(u) \tag{6}$$

$$= \sum_{v\in P_{i+1}} C_i(v) + \sum_{\substack{v\in P_{i+1}\\(u,v)\in E}} C_i(u) \tag{7}$$

and

$$\sum_{v\in V} C_i(W_{i+1}(v) - W_i(v))$$

$$= \sum_{v\in V} C_i(v)(C_{i+1}(v) - C_i(v)) + \sum_{v\in V} \Big(C_i(v) \sum_{u\in N(v)} (C_{i+1}(u) - C_i(u))\Big) \tag{8}$$

$$= \sum_{v\in P_{j+1}} C_i(v) + \sum_{\substack{u\in P_{j+1}\\(u,v)\in E}} C_i(v) \tag{9}$$

We can get Eq. 9 from Eq. 8 because $C_{i+1}(v) - C_i(v) = 1$ for $v \in P_{j+1}$ and $C_{i+1}(v) - C_i(v) = 0$ for $v \in V - P_{j+1}$.

Combining Eqs. 5, 7 and 9, we have Lemma 1 and Corollary 1.

Lemma 1. *For $i = 0, 1, \ldots k-1$, the interference level among paths satisfies*

$$IN(P_1, \ldots, P_{i+1}) - IN(P_1, \ldots, P_i) = \sum_{v \in P_{i+1}} W_i(v)$$

Corollary 1. *The interference level among paths $\{P_1, \ldots, P_k\}$ can be also computed by*

$$IN(P_1, \ldots, P_k) = \sum_{v \in P_1} W_0(v) + \sum_{v \in P_2} W_1(v) + \cdots + \sum_{v \in P_k} W_{k-1}(v)$$

Corollary 1 gives us the idea of designing on-line algorithm for MIMP. We satisfy the given routing requests $\{s_1, t_1\}, \ldots, \{s_k, t_k\}$ in order. For each $\{s_i, t_i\}$, we find the shortest $s_i \sim t_i$ path w.r.t. function W_{i-1}. Here, the length of path P_i w.r.t. W_{i-1} is defined as $\sum_{v \in P_i} W_{i-1}(v)$.

4 Distributed On-Line Algorithm

In this paper, we consider the MIMP problem, which is defined as: Given k routing requests $\{s_1, t_1\}, \ldots, \{s_k, t_k\}$, finding k paths $P_1, \ldots, P_k$ satisfying these routing requests, i.e., P_i is a $s_i \sim t_i$ path for $i = 1, \ldots, k$. The optimizing goal is to minimize interference among these paths, i.e., $IN(P_1, \ldots, P_k)$.

According to [14], given two node pairs $\{s_1, t_1\}$ and $\{s_2, t_2\}$ in a graph, it is NP-complete to determine whether there exist a pair of non-interference (there is no cross-path edge) $s_1 \sim t_1$ path and $s_2 \sim t_2$ path. Therefore, we have Theorem 1.

Theorem 1. *MIMP is NP-hard even for $k = 2$.*

MIMP is very hard to solve or to approximate. In this paper, we design a heuristic on-line algorithm named DOAMI (short for Distributed On-line-Algorithm for Minimizing Interference) for the MIMP problem. DOAMI is executed in a totally distributed way, i.e., each node participates by only exchanging short messages with its neighbors and the exchange of messages does not rely on any pre-defined structure. DOAMI is also an on-line algorithm, which means that DOAMI deploys paths to satisfy the given routing requests one by one. After DOAMI receives a routing request $\{s_i, t_i\}$, it immediately finds a $s_i \sim t_i$ path to satisfy it without waiting for the coming of the next routing request.

4.1 Description of DOAMI

The design DOAMI is motivated by Corollary 1. DOAMI uses two phases to find path for each routing request $\{s_i, t_i\}$. In Phase 1, starting from s_i, by exchanging FIND-SWP messages, each node v finds the shortest path from s_i to itself w.r.t. function W_{i-1}. In Phase 2, starting from t_i, the shortest path from s_i to t_i w.r.t. W_{i-1} is traced and confirmed. Meanwhile, each node computes its $W_i(v)$. DOAMI is given by Algorithm 1. We assume that the ID of the sender of each message is contained in the message's header, so the receiver can always know who is the sender of the received message. In the following, "SWP" means shortest path w.r.t. function W_{i-1}.

Algorithm 1. Distributed On-line Algorithm for Minimizing Interference (DOAMI)

Input: k routing requests $\{s_1, t_1\}, \ldots, \{s_k, t_k\}$
Output: k paths $P_1, \ldots, P_k$ that P_i is a $s_i \sim t_i$ path for $i = 1, \ldots, k$
/*codes to deal with each $\{s_i, t_i\}$ */
//Phase 1:
s_i sets $W\_len(s_i) = W(s_i)$ and broadcasts a FIND-SWP message $\{t_i, W\_len(s_i)\}$;
while $v \neq s_i$ receives a FIND-SWP message $\{t_i, W\_len(u)\}$ from u **do**
    if $W\_len(v) > W\_len(u) + W(v)$ **then**
        $W\_len(v) \leftarrow W\_len(u) + W(v)$;
        $SWP\_LP(v) \leftarrow u$;
        if $v = t_i$ **then**
            Go to Phase 2;
        else
            Broadcast a FIND-SWP message $\{t_i, W\_len(v)\}$;
//Phase 2:
t_i waits for the end of FIND-SWP message exchanging then sends a TRACE-SWP message to $SWP\_LP(t_i)$;
while v receives a TRACE-SWP message from u **do**
    $SWP\_NP(v) \leftarrow u$;
    $W(v) \leftarrow W(v) + 1$;
    Broadcast a PATH-CONF message;
    Send a TRACE-SWP message to $SWP\_LP(v)$;
while v receives a PATH-CONF message **do**
    $W(v) \leftarrow W(v) + 1$;

Each node v maintains four variables $W(v), W\_len(v), SWP\_LP(v), SWP\_NP(v)$. While executing routing request $\{s_i, t_i\}$, $W(v)$ records $W_{i-1}(v)$ in Phase 1 and is updated to $W_i(v)$ in Phase 2. $W\_len(v)$ records the length of the SWP from s_i to v found by now. Let P_i be the SWP from s_i to t_i found by DOAMI at the end of Phase 1. If v is a node in P_i, variables $SWP\_LP(v)$ and $SWP\_NP(v)$ record v's last and next hop in P_i respectively.

In Phase 1, by exchanging FIND-SWP messages, each node v finds the SWP from s_i to itself. The data field of each FIND-SWP message (let the sender be u) contains two variables: the destination node's ID t_i and the length of the SWP from s_i to u, i.e., $W\_len(u)$. At first, the source node s_i broadcasts a FIND-SWP message $\{t_i, W\_len(s_i)\}$ to start Phase 1 (Line 1 in Algorithm 1). While a node $v \neq s_i$ receives a FIND-SWP message $\{t_i, W\_len(u)\}$ from u, it finds a new path from s_i to u then to itself, whose W_{i-1}-length is $W\_len(u) + W(v)$. If the new path is shorter (w.r.t. W_{i-1}) than the current SWP from s_i to v, it should be the new SWP from s_i to v. In this case, v updates $W\_len(v)$ to $W\_len(u)+W(v)$ and updates $SWP\_LP(v)$ to u, then broadcasts a FIND-SWP message $\{t_i, W\_len(v)\}$ (Line 2-5, 8, 9 in Algorithm 1). From receiving the first FIND-SWP message, t_i waits for the end of Phase 1 then broadcasts a TRACE-SWP message to start Phase 2 (Line 6, 7, 10 in Algorithm 1).

Let P_i be the SWP from s_i to t_i found by DOAMI in Phase 1. In Phase 2, starting from t_i, each node in P_i sends a TRACE-SWAP message to its last hop in P_i. In this way, P_i is confirmed and established. Each TRACE-SWP message has an empty data field. In the process of confirming P_i, each node in the network should update their $W(v)$. In Phase 1, $W(v)$ records $W_{i-1}(v)$. At the end of Phase 2, $W(v)$ should be updated to $W_i(v)$. While node v receives a TRACE-SWP message from u, it sets u as its next hop in P_i (Line 12 in Algorithm 1). Since v joins into P_i, $C(v)$ should be increased by one and all the neighbors of v should add one to their W variable. Therefore, we let v add 1 to $W(v)$ (Line 13 in Algorithm 1) and broadcast a PATH-CONF message with empty data field to inform its neighbors. Receiving a PATH-CONF message, each node adds one to their W variable (Line 16, 17 in Algorithm 1).

We refer the execution time for each routing request $\{s_i, t_i\}$ as a "round". We assume that for each node v, $W\_len(v)$ is positive infinity at the beginning of each round. This can be implemented by involving a round number i in each FIND-SWP message. While v receives a FIND-SWP message $\{t_i, W\_len(u)\}$ with a larger round number, no matter what value $W\_len(v)$ is, it updates $W\_len(v)$ to $W\_len(u) + W(v)$.

5 Simulation Results

We use a simulator written in C++ to evaluate the proposed algorithm. We set the simulation environment as follows: 2500 sensor nodes are randomly deployed in a 1500 m × 1500 m area. The effective transmitting range R of each sensor node is set to 50 m. Let v be the sender of any data packet. There are two kinds of interference caused by v's transmission: one-hop interference and two-hop interference. The nodes (besides the receiver) whose distance to v is less than or equal to 50 m will detect v's signal and be interfered by the transmission for sure. On the other hand, the nodes whose distance to v is in (50 m, 100 m) has a certain probability to be interfered by v's signal. We call this probability the Interference Probability (IP). Two-hop interference cannot be detected by the interfered node.

To simulate routing, we randomly generate different number of routing requests, i.e., $\{s, t\}$ pairs. Given $\{s_1, t_1\}, \ldots, \{s_k, t_k\}$, we use routing algorithm to built paths to satisfy all the routing requests. Next, simultaneously, for each

$\{s_i, t_i\}$, we let s_i generate a new data packet in every 3 time slots and send the packets to t_i along the built path. Each routing request has to transmit 100 data packets. The data field of each packet contains 50 bytes. The efficiency of routing is measured at four aspects: (1) Transmission Delay. It is the number of time slots we use to complete the transmission of data packets for all the routing requests. This figure indicates the throughput of routing. (2) Routing Energy. It is the energy we use to transmit the data packets for all the routing request.

We compare the proposed algorithm DOAMI with three other algorithms: (1) IM2PR proposed by [17]. IM2PR requires the built paths are node-disjoint. (2) I2MR proposed by [18]. Any two nodes in different paths built by I2MR are more than two hops away from each other. (3) Naive. It is the algorithm without considering interference. For each routing request $\{s_i, t_i\}$, Naive simply finds the shortest paths (with smallest number of hops) from s_i to t_i.

Table 1. The largest number of satisfied routing requests

DOAMI	IM2PR	I2MR	Naive
∞	13	5	∞

At first, we compare the largest number of routing requests each algorithm can satisfy. From Table 1, we can see that DOAMI and Naive can satisfy all the given k routing requests no matter what value k is. In contrast, IM2PR can build paths for at most 13 routing requests. Since IM2PR requires that all the built paths are node-disjoint, when k becomes larger, it is hard to find a path for the latter routing request to get through the "barrier" formed by the former built paths. I2MR gives the worst performance in this aspect because it has the strictest requirement of the built paths. The paths built by I2MR have to be more than two hops away from each other. In the following, we only compare the algorithms when they build the same number of routing paths.

In the first group of simulation, we set the number of routing requests as $k = 5$ and compare all the four algorithms. The comparison is done at three different values of IP, i.e., in what probability the nodes in $(R, 2R]$ from sender will be interfered by the transmission.

As shown in Fig. 3(a), I2MR has the best performance in the aspect of Transmission Delay. Since I2MR has the strictest requirement of the routing paths, the transmissions in different paths barely interfere with each other. DOAMI outperforms IM2PR a little in this aspect when IP becomes larger. The difference among I2MR, DOAMI and IM2PR is very limited. Since Naive does not consider the interference while building routing paths, it performs worst in this aspect.

As shown in Fig. 3(d), I2MR also has the best performance in the aspect of Routing Energy. Since I2MR almost avoids all one-hop and two-hop interference among different paths, when IP become larger, the advantage of I2MR gets more and more apparent. However, when $k > 5$, I2MR cannot satisfy all the routing

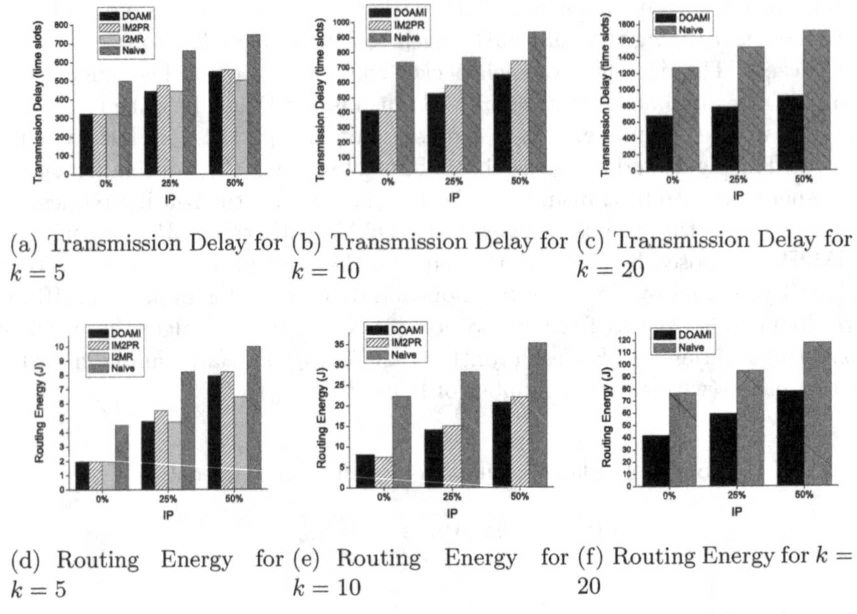

(a) Transmission Delay for $k = 5$

(b) Transmission Delay for $k = 10$

(c) Transmission Delay for $k = 20$

(d) Routing Energy for $k = 5$

(e) Routing Energy for $k = 10$

(f) Routing Energy for $k = 20$

Fig. 3. Simulation Results

requests. In this aspect, DOAMI and IM2PR almost have the same performance and DOAMI outperforms IM2PR a little. Naive still gives the worst performance.

In the second group of simulation, we set $k = 10$ to compare DAOMI, IM2PR and Naive. As shown in Fig. 3(b) and (e), DOAMI has the best performance in both aspects of Transmission Delay and Routing Energy. When IP gets larger, DOAMI outperforms the other two algorithms more and more apparently. Naive always gives the worst performance since it does not consider interference.

In the third group of simulation, we set $k = 20$ to compare DAOMI and Naive. As shown in Fig. 3(c) and (f), DOAMI apparently outperforms Naive in both aspects.

Let us summary the simulation results. For both routing throughput and energy efficiency, I2MR is a little better than DOAMI when k is small. However, I2MR cannot deal with the cases with a relative large k. DOAMI is a little better than IM2PR and much better than Naive in both aspects of routing throughput and energy efficiency. Compared with IM2PR, DOAMI has the ability to deal with larger k whereas IM2PR cannot.

6 Conclusion

In this paper, we give a metric to measure the interference level among multiple routing paths in wireless sensor networks. Unlike the existing metrics, the proposed metric can deal with the situation that the routing paths have some nodes or links in common. Based on the metric, we propose a distributed on-line algorithm DOAMI to build paths with minimum interference for multiple routing requests. The efficiency of DOAMI is confirmed by our simulation results. For the future works, maybe we will consider how to design an algorithm for the same problem with guaranteed bound.

Acknowledgement. This work is supported by National Natural Science Foundation of China (Grant No. 61300207, Grant No. 61272186, Grant No. 61370084), Fundamental Research Funds for the Central Universities (Grant No. HEUCF100610, Grant No. HEUCF100609).

References

1. Bidai, Z., Maimour, M.: Interference-aware multipath routing protocol for video transmission over zigbee wireless sensor networks. In: International Conference on Multimedia Computing and Systems (ICMCS), pp. 837–842 (2014)
2. Cai, Z., Chen, Z.Z., Lin, G.: A 3.4713-approximation algorithm for the capacitated multicast tree routing problem. Theoret. Comput. Sci. **410**(52), 5415–5424 (2008)
3. Cai, Z., Chen, Z.-Z., Lin, G., Wang, L.: An improved approximation algorithm for the capacitated multicast tree routing problem. In: Yang, B., Du, D.-Z., Wang, C.A. (eds.) COCOA 2008. LNCS, vol. 5165, pp. 286–295. Springer, Heidelberg (2008)
4. Cai, Z., Goebel, R., Lin, G.: Size-constrained tree partitioning: approximating the multicast k-tree routing problem. Theoret. Comput. Sci. **412**(3), 240–245 (2011)
5. Cai, Z., Lin, G., Xue, G.: Improved approximation algorithms for the capacitated multicast routing problem. In: Wang, L. (ed.) COCOON 2005. LNCS, vol. 3595, pp. 136–145. Springer, Heidelberg (2005)
6. Cheng, S., Cai, Z., Li, J.: Curve query processing in wireless sensor networks. IEEE Trans. Veh. Technol. **PP**(99), 1 (2015). doi:10.1109/TVT.2014.2375330
7. Deb, B., Bhatnagar, S., Nath, B.: Reinform: reliable information forwarding using multiple paths in sensor networks. In: Proceedings of the 28th Annual IEEE International Conference on Local Computer Networks (2003)
8. Ding, Y., Yang, Y., Xiao, L.: Multi-path routing and rate allocation for multi-source video on-demand streaming in wireless mesh networks. In: Proceedings of IEEE INFOCOM, pp. 2051–2059 (2011)
9. Ganesan, D., Govindan, R., Shenker, S., Estrin, D.: Highly-resilient, energy-efficient multipath routing in wireless sensor networks. In: Proceedings of the 2nd ACM International Symposium on Mobile Ad Hoc Networking & Computing (MobiHoc) (2001)
10. Guo, L., Li, Y., Cai, Z.: Minimum-latency aggregation scheduling in wireless sensor network. J. Comb. Optim. 1–32 (2014). doi:10.1007/s10878-014-9748-7
11. Guoqiang, Y., Weijun, D., Chao, M., Liang, H.: Minimize interference while using multipath transportation in wireless multimedia sensor networks. In: Qian, Z., Cao, L., Su, W., Wang, T., Yang, H. (eds.) Recent Advances in CSIE 2011. LNEE, vol. 127, pp. 239–244. Springer, Heidelberg (2012)
12. He, Z., Cai, Z., Cheng, S., Wang, X.: Approximate aggregation for tracking quantiles in wireless sensor networks. In: Zhang, Z., Wu, L., Xu, W., Du, D.-Z. (eds.) COCOA 2014. LNCS, vol. 8881, pp. 161–172. Springer, Heidelberg (2014). doi:10.1007/978-3-319-12691-3_13

13. Jain, K., Padhye, J., Padmanabhan, V.N., Qiu, L.: Impact of interference on multi-hop wireless network performance. In: Proceedings of the 9th Annual International Conference on Mobile Computing and Networking, MobiCom 2003, pp. 66–80 (2003)
14. Kobayashi, Y.: Induced disjoint paths problem in a planar digraph. Discrete Appl. Math. **157**(15), 3231–3238 (2009)
15. Mohanoor, A., Radhakrishnan, S., Sarangan, V.: Interference aware multi-path routing in wireless networks. In: 5th IEEE International Conference on Mobile Ad Hoc and Sensor Systems (MASS), pp. 516–518, September 2008
16. Pearlman, M., Haas, Z., Sholander, P., Tabrizi, S.: On the impact of alternate path routing for load balancing in mobile ad hoc networks. In: First Annual Workshop on Mobile and Ad Hoc Networking and Computing (MobiHOC), pp. 3–10 (2000)
17. Radi, M., Dezfouli, B., Bakar, K., Razak, S., Hwee-Pink, T.: IM2PR: interference-minimized multipath routing protocol for wireless sensor networks. Wirel. Netw. **20**(7), 1807–1823 (2014)
18. Teo, J.Y., Ha, Y., Tham, C.K.: Interference-minimized multipath routing with congestion control in wireless sensor network for high-rate streaming. IEEE Trans. Mob. Comput. **7**(9), 1124–1137 (2008)
19. Voigt, T., Dunkels, A., Braun, T.: On-demand construction of non-interfering multiple paths in wireless sensor networks. In: Proceedings of the 2nd Workshop on Sensor Networks at Informatik (2005)
20. Wang, Z., Bulut, E., Szymanski, B.: Energy efficient collision aware multipath routing for wireless sensor networks. In: IEEE International Conference on Communications (ICC), pp. 1–5, June 2009
21. Wu, K., Harms, J.: Performance study of a multipath routing method for wireless mobile ad hoc networks. In: Proceedings of the 9th International Symposium on Modeling, Analysis and Simulation of Computer and Telecommunication Systems, pp. 99–107 (2001)
22. Zhang, J., Jeong, C.K., Lee, G.Y., Kim, H.J.: Cluster-based multi-path routing algorithm for multi-hop wireless network. Future Gener. Commun. Network. **1**, 67–75 (2007)
23. Zhang, K., Han, Q., Yin, G., Pan, H.: OFDP: a distributed algorithm for finding disjoint paths with minimum total length in wireless sensor networks. J. Comb. Optim., 1–19 (2015)

Distributed Algorithm for Mending Barrier Gaps via Sensor Rotation in Wireless Sensor Networks

Yueshi Wu and Mihaela Cardei(✉)

Department of Computer and Electrical Engineering and Computer Science, Florida Atlantic University, Boca Raton, FL 33431, USA
{wuy2013,mcardei}@fau.edu

Abstract. When deploying sensors to monitor boundaries of battlefields or country borders, sensors are usually dispersed from an aircraft following a predetermined path. In such scenarios sensing gaps are usually unavoidable. We consider a wireless sensor network consisting of directional sensors deployed using the line-based sensor deployment model. In this paper we proposed a distributed algorithm for weak barrier coverage that allows sensors to determine their orientation such that the total number of gaps and the total gap length are minimized. We use simulations to analyze the performance of our algorithm and to compare it with two related work algorithms.

Keywords: Wireless sensor networks · Weak barrier coverage · Directional sensors · Distributed algorithm

1 Introduction and Related Works

A major application of Wireless Sensor Networks (WSNs) is area monitoring. Examples of area monitoring are intrusion detection and border surveillance. Unlike full area coverage problem where every point of a region has to be covered, these applications aim to detect an intruder attempting to enter or exit the border of a certain region.

Deploying a set of sensor nodes on a region of interest where sensors form barriers for intruders is often referred to as the *barrier coverage* [6]. When deploying sensors to monitor boundaries of battlefields or country borders, sensors are usually dispersed from an aircraft following a predetermined path [9]. Therefore sensing gaps are usually unavoidable. In this paper, we aim to address the barrier coverage problem by mending gaps via a distributed manner as well as to minimize the total length of non-mendable gaps.

Many studies on WSNs barrier coverage consider constructing barriers with stationary sensors. [10] studies the scenario where sensors are airdropped along a straight line. They investigate how different sensor deployment strategies impact the barrier coverage of a WSN and provide a tight lower bound on the existence

Z. Lu et al. (Eds.): COCOA 2015, LNCS 9486, pp. 293–304, 2015.
DOI: 10.1007/978-3-319-26626-8_22

of barrier coverage under the assumption that the offset of each sensor's actual landing point follows a normal distribution.

However, [10] does not address the issue of mending the barrier gaps formed after the initial deployment. When deploying sensors to monitor boundaries of battlefields or country borders, sensors are usually dispersed from an aircraft following a predetermined path. Therefore sensing gaps may occur.

In [3], the authors proposed a measure of the goodness of sensors called *path exposure*, which measures the likelihood of detecting a target traversing the region using a given path. They determine the number of sensors that have to be deployed for target detection under the assumption that sensors are randomly deployed in the region of interest. When using stationary sensors, the full barrier coverage is difficult to achieve when sensors are randomly deployed [7].

There are also works investigating the use of mobile sensors to mend barrier gaps, for example [4,5]. A chain reaction algorithm is proposed by [5] to move sensors automatically and cooperatively such that to form a barrier coverage. [4] considers a hybrid network which consists of both stationary and mobile sensors. Mobile sensors can be moved in order to mend gaps. They provide a scheme to minimize the maximal energy consumed by moving sensors and another scheme to maximize the lifetime of barrier coverage after mending all the gaps.

All of the related works mentioned above are based on the isotropic sensing model of the sensors. However, in many practical applications using infrared sensors, radar sensors, audio sensors, or camera sensors, they often have a directional sensing model [1]. Barrier coverage using directional sensors is an important research topic. One application is providing surveillance of country borders using barrier coverage with sensors equipped with cameras.

[2] proposes two algorithms for weak barrier coverage: the Simple Rotation Algorithm(SRA) and the Chain Reaction-based rotation Algorithm(CRA). These algorithms aim to mend barrier gaps via sensor rotation for a line-based sensor deployment. SRA uses two critical sensors: the right most sensor P in the jth sub-barrier B_j and the left most sensor Q in the $(j+1)$th sub-barrier B_{j+1} in order to mend the gap between B_j and B_{j+1}.

SRA first rotates P clockwise to cover the gap while maintaining the coverage connection with its left neighbor. If the gap cannot be mended by rotating only P, then SRA rotates Q anti-clockwise to cover the gap while maintaining the coverage connectivity with the right neighbor. If rotating P cannot mend the barrier, SRA rotates P to cover the maximum on the right without creating new gaps, and then rotates Q to cover the gap. SRA fails if rotating both P and Q cannot mend the gap.

CRA first rotates the sensor nodes in B_j in a chain reaction manner. Node P is first rotated until the gap is mended. If a new gap is created by rotating P, then the node $P-1$ is rotated in the same manner. The first step of CRA fails when a new barrier gap always exists even after all sensors in B_j have been rotated. Step 2 of CRA rotates Q anti-clockwise only to see if the gap can be mended, and if not, then step 3 is performed. In step 3 CRA rotates P only to reduce the gap length without creating new gaps, then rotates sensors in B_j similar to the step 1.

The algorithms proposed in [2] are centralized. Centralized approaches are not suitable since they require a central coordinator and have large delays for a large WSN. In this paper we consider a line-based sensor deployment model with directional sensors, similar to [2]. Our objective is to design a distributed mechanism to mend gaps and to minimize the total length of non-mendable gaps. The performance of our distributed algorithm is compared with SRA and CRA.

2 Network Model and Problem Definition

In this section we start by specifying the network model, notations, and terminologies used in this paper. When sensors are remotely deployed, e.g. sprayed from an aircraft, a larger amount of sensors have to deployed to ensure the barrier coverage. Besides the uniform deployment, the *line-based sensor deployment* model has been used recently for barrier coverage.

We assume a line-based stationary sensor deployment similar to [2,11]. A total of n sensor nodes $\{s_1, s_2, ..., s_n\}$ are deployed in a rectangular field of length L and width H, see Fig. 1. In the line-based sensor deployment, sensors are evenly deployed on a horizontal-line (e.g. $y = 0$). Such an example is illustrated in Fig. 1a by the "target" positions, where the coordinate of the ith node is computed as $(2i - 1)L/2N$. This is hard to achieve in practice for remotely deployed sensors, and their "actual" positions have random offsets. We follow the model in [2] where the random offset distances have a Gaussian distribution. We assume that each sensor node knows its location using GPS or other localization mechanisms. A sensor node s_i's location is denoted by (x_i, y_i).

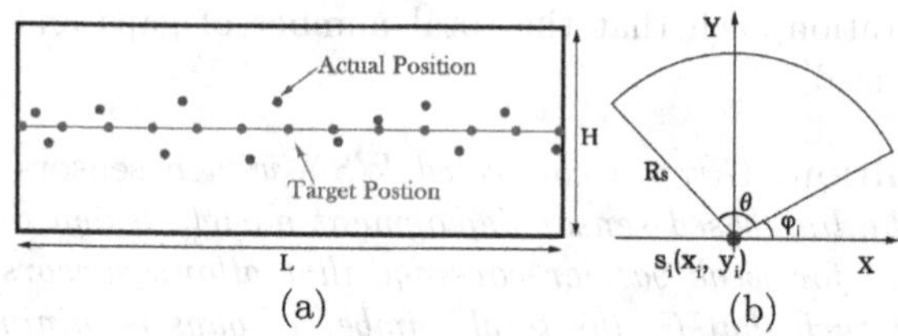

Fig. 1. (a) Sensor deployment; (b) Sensor coverage model.

In a WSN designed for barrier coverage, sensors have the ability to exchange messages with each other during initialization, to exchange information, or in response to an intruder detection. We assume an omnidirectional communication model, where the transmission range of each sensor is R_t. In addition, we assume directional sensor nodes which have a finite view angle. Different from isotropic sensors, they cannot sense the whole circular area.

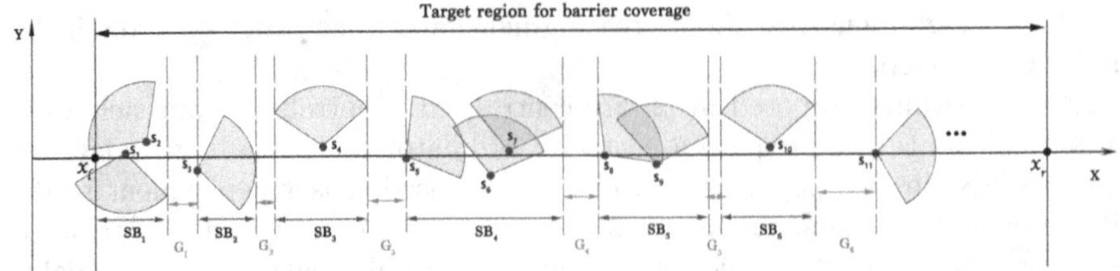

Fig. 2. An example of initial sensor deployment for weak barrier coverage with sub-barriers (SB) and gaps (G)

Directional sensing can be modeled as 2D sector-shaped sensing [8]. We describe a sensor s_i using the five tuple $< (x_i, y_i), \theta, \varphi_i, R_s, R_t >$, see Fig. 1b, where (x_i, y_i) are the Cartesian coordinates of s_i, θ is the view angle, φ_i is the orientation angle, R_s is the sensing range, and R_t is the communication range. We assume that all sensors in the network have the same view angle, sensing range, and communication range.

We assume that intruders move north-south. In the barrier coverage problem, the objective is to construct a barrier such that any north-south path falls into the coverage area of at least one sensor. For the *weak barrier coverage*, the barrier of sensors has to provide coverage when intruders move along vertical traversing paths.

Due to the random sensor deployment, a full weak barrier coverage may be impossible to attain, and in this case the target region is divided into sub-barriers (where coverage is provided) and gaps. Figure 2 shows an initial sensor deployment scenario with sub-barriers (SB) and gaps (G). Intruders crossing through a SB region are detected, while those crossing though a G region are not detected.

In this paper we seek to design a distributed algorithm where each sensor decides its orientation such that the total number of gaps and the total gap length are minimized.

Problem Definition: *Given a connected WSN with n* sensors $\{s_1, s_2, ..., s_n\}$ *deployed using the line-based sensor deployment model, design an efficient distributed algorithm for weak barrier coverage that allows sensors to determine their orientation such that (i) the total number of gaps is minimized, and (ii) the total gap length is minimized.*

3 Distributed Algorithm for Weak Barrier Coverage

In this section, we present our distributed algorithm to mend barrier gaps and minimize the total length of non-mendable gaps. Our algorithm has three phases. In phase 1, sensor nodes perform neighbor discovery and exchange location information with neighboring sensors.

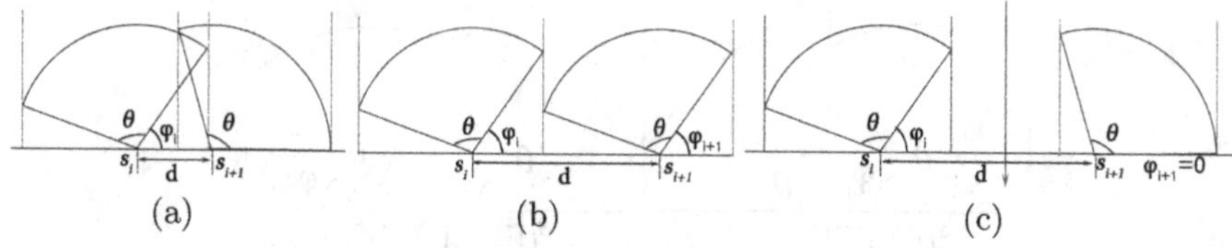

Fig. 3. Cases for orientation decision when $\theta > 90°$.

We propose an optimal orientation algorithm in phase 2. Starting from the left boundary node, each node decides its own orientation angle such that the total number of sensing gaps throughout the network is minimized. In phase 3, we minimize the total gap length by maximizing the total sensing coverage.

3.1 Phase 1: Neighbor Discovery

During the phase 1, each sensor node broadcasts a $HELLO$ message including its ID and location. A sensor node s_i receiving a $HELLO(s_j, (x_j, y_j))$ message stores s_j in its inSensingRangeNeighbor list if $\sqrt{(x_i - x_j)^2 + (y_i - y_j)^2} \leq R_s$.

For the *weak barrier coverage problem*, we work with sensor projections on the x-axis. In this case s_i stores s_j in its inSensingRangeNeighbor list if $|x_i - x_j| \leq R_s$. Each sensor node s_i computes its closest right neighbor N_i^r and its closest left neighbor N_i^l from the nodes in the inSensingRangeNeighbor list. For example, in Fig. 2, sensor s_6 sets $N_6^r = s_7$ and $N_6^l = s_5$.

3.2 Phase 2: Optimal Orientation

We assume that sensors have random rotations when they are deployed initially. Phase 2 is started by the left boundary sensor, e.g. s_1 in Fig. 2, which computes $\varphi_1 = \pi - \theta - arccos(\frac{x_1 - x_l}{R_s})$, where x_l is the x coordinate of the left boundary of the target region. s_1 then broadcasts the message $POS(s_1, pos_1)$, where $pos_i =< s_i, x_i, y_i, \varphi_i >$

A sensor node s_i waits to receive a POS message from its closest left neighbor N_i^l, then computes its orientation angle φ_i, and broadcasts a message $POS(s_i$, pos_1, pos_2, ..., $pos_i)$. The first field is the sender ID, and the other fields are the positioning information for all prior sensors in the same sub-barrier. s_i appends pos_i at the end of the list received from s_{i-1}. For example in Fig. 2, node s_2 receives the POS message from s_1, computes φ_2 accordingly, then broadcasts $POS(s_2, pos_1, pos_2)$.

Next we describe the mechanism used by a sensor s_{i+1} to set up its orientation angle φ_{i+1} after it has received a POS message from its closest left neighbor, denoted s_i. We distinguish two main cases to be addressed individually: $\theta > 90°$ and $\theta \leq 90°$.

Since we are concerned with the weak barrier coverage, we work with sensor projections on the x-axis. Let d be the horizontal distance between s_i and s_{i+1}.

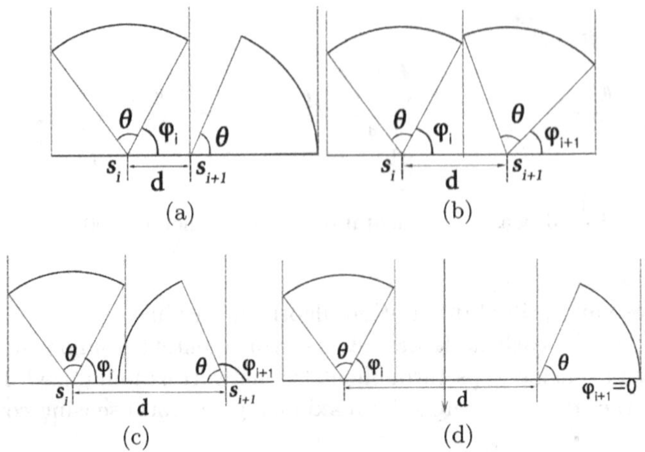

Fig. 4. Cases for orientation decision when $\theta \leq 90°$ and $\varphi_i < 90°$.

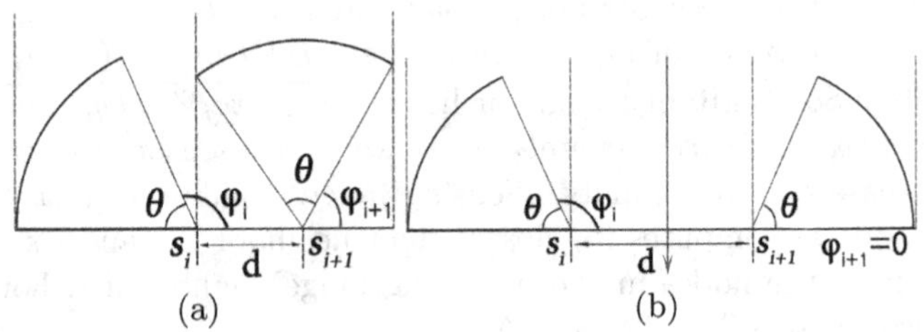

Fig. 5. Cases for orientation decision when $\theta \leq 90°$ and $\varphi_i \geq 90°$.

For a view angle $\theta > 90°$, we distinguish three cases, see Fig. 3. Let us assume that s_{i+1} receives a *POS* message from its closest left neighbor s_i and let d be the horizontal distance between s_i and s_{i+1}.

Case 1: $d \leq R_s cos\varphi_i + R_s cos(\pi - \theta)$
In this case the node s_{i+1} sets $\varphi_{i+1} = 0$. This provides sensing overlapping with s_i on the left and maximum coverage on the right. Note that rotating s_{i+1} anticlockwise (e.g. $\varphi_{i+1} > 0$) does not increase the coverage on the left of s_{i+1}, but it decreases the coverage on the right. Such an example is illustrated in Fig. 3a.

Case 2: $R_s cos\varphi_i + R_s cos(\pi - \theta) < d \leq R_s + R_s cos\varphi_i$
In this case s_{i+1} calculates its orientation angle φ_{i+1} such that to maintain the sensing coverage connection with s_i and to maximize the coverage on the right. s_{i+1} computes $\varphi_{i+1} = \pi - \theta - arccos(\frac{d - R_s cos\varphi_i}{R_s})$. An example is illustrated in Fig. 3b.

Case 3: $d > R_s + R_s cos\varphi_i$
In this case s_{i+1} is not able to mend the gap between s_i and s_{i+1}. The node s_{i+1} sets $\varphi_{i+1} = 0$ to give maximum coverage on the right. An example is illustrated in Fig. 3c.

When $\theta \leq 90°$, the orientation angle φ_i of the node s_i is set up either as $0 \leq \varphi_i < 90°$ or $\varphi_i = \pi - \theta$. Note that $\varphi_i \geq 90°$ provides no coverage on s_i's right side and setting $\varphi_i = \pi - \theta$ maximizes the coverage on s_i's left side. We first discuss the case when $0 \leq \varphi_i < 90°$, see Fig. 4.

Case 1: $0 \leq d \leq R_s cos\varphi_i$
The node s_{i+1} sets $\varphi_{i+1} = 0$, see Fig. 4a.

Case 2: $R_s cos\varphi_i < d \leq R_s cos\varphi_i + R_s cos(\frac{\pi}{2} - \theta)$
The node s_{i+1} calculates its orientation angle φ_{i+1} such that to maintain sensing connection with s_i and to maximize the coverage on the right. φ_{i+1} is computed as $\varphi_{i+1} = \pi - \theta - arccos(\frac{d - R_s cos\varphi_i}{R_s})$, see Fig. 4b.

Case 3: $R_s cos\varphi_i + R_s cos(\frac{\pi}{2} - \theta) < d \leq R_s cos\varphi_i + R_s$
Node s_{i+1} sets $\varphi_{i+1} = \pi - \theta$, see Fig. 4c.

Case 4: $d > R_s cos\varphi_i + R_s$
In this case there will be a gap between s_i and s_{i+1}. The node s_{i+1} sets $\varphi_{i+1} = 0$ to maximize the coverage on the right, see Fig. 4d.

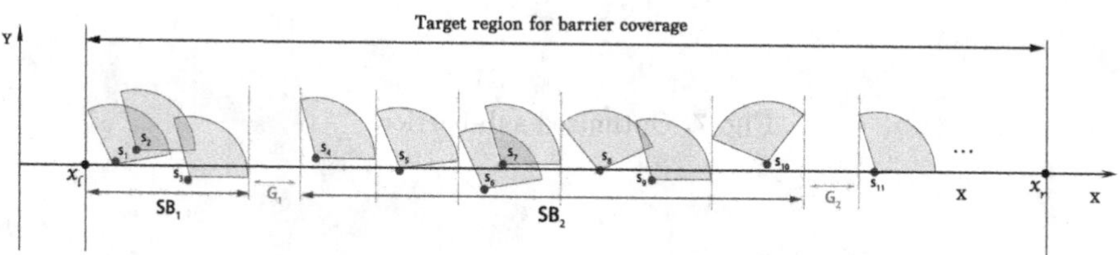

Fig. 6. Nodes positioning after phase 2.

For $\varphi_i \geq 90°$ (e.g. $\varphi_i = \pi - \theta$), if $0 \leq d \leq R_s cos(\frac{\pi}{2} - \theta)$, then s_{i+1} sets $\varphi_{i+1} = \pi - \theta - arccos(\frac{d}{R_s})$, see Fig. 5a, otherwise s_{i+1} sets $\varphi_{i+1} = 0$, see Fig. 5b. Note that there is a gap between s_i and s_{i+1} when $d > R_s$.

Figure 2 shows the initial sensors deployment and Fig. 6 shows nodes orientation at the end of phase 2. A sensor node identifies itself as a left boundary node of a sub-barrier if it is in the case 3 for $\theta > 90°$ (e.g. Figure 3c) or in the case 4 for $\theta \leq 90°$ (e.g. Figure 4d). A left-boundary node of a sub-barrier will reset the *pos* list in the POS message. For example, in Fig. 6, s_3 broadcasts $POS(s_3, pos_1, pos_2, pos_3)$. s_4 is the left-boundary node of SB_2 and it broadcasts $POS(s_4, pos_4)$. s_3 overhears the POS message from its closest right neighbor and since the *pos* list was reset, it concludes that it is the right-boundary node of its sub-barrier SB_1.

Note that the right boundary of a sub-barrier will have positioning information of all the nodes in the same sub-barrier. In Fig. 6, s_3 has positioning information of all the nodes in SB_1, s_{10} has positioning information of all the nodes in SB_2, so on.

3.3 Phase 3: Minimizing the Gap Length

We distinguish 2 types of sub-barriers: to be optimized and optimized. If the right boundary node of a sub-barrier has the orientation angle $\pi - \theta$, then the sub-barrier is optimized. An example is shown in Fig. 7.

If the orientation angle of the right boundary node of SB_j is different than $\pi-\theta$ and is between 0 and 90°, then SB_j is a "to be optimized" sub-barrier. Phase 3 has the objective to optimize sensor orientation in SB_j such that to minimize the gap length. An example of optimizing a sub-barrier SB_j is illustrated in Fig. 8. By sequentially rotating sensors anti-clockwise, the gap G_{j-1} is decreased by $\Delta x^-_{G_{j-1}}$, while the gap G_j is increased by $\Delta x^+_{G_j}$.

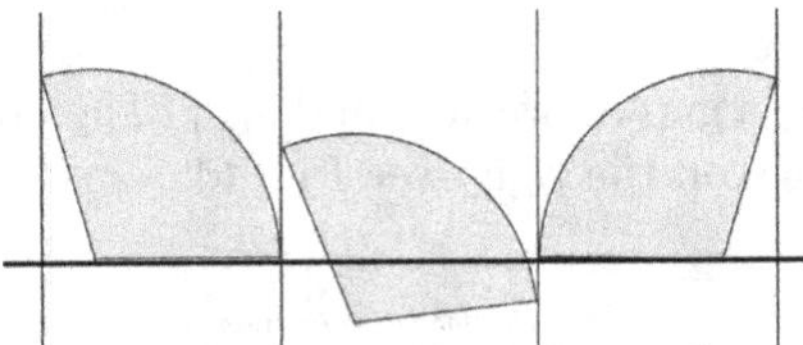

Fig. 7. Optimized sub-barrier.

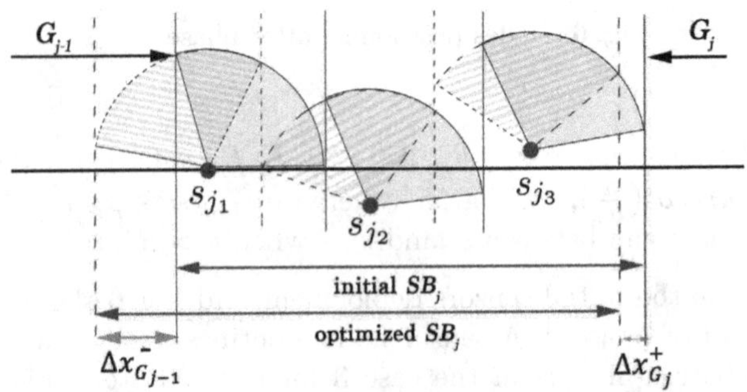

Fig. 8. Optimizing a sub-barrier SB_j.

In Fig. 8 the solid sectors indicate the original sensor positioning after phase 2. By rotating sensors anti-clockwise, see the dashed line sectors, the decreased

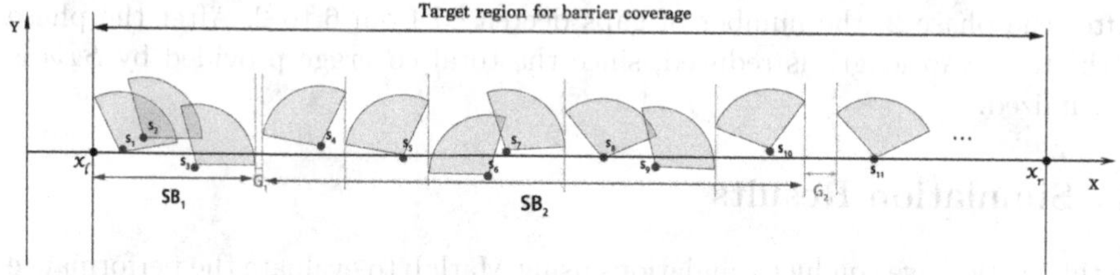

Fig. 9. Nodes positioning after phase 3.

length of the gap G_{j-1} is larger than the increased length of the gap G_j, thus the total gap length is reduced.

Consider a sub-barrier SB_j with sensors $s_{j_1}, s_{j_2}, ..., s_{j_n}$ in order from left to right. The sensor s_{j_1} is the left most node and s_{j_n} is the right most node. Let us assume that s_{j_n}'s orientation angle $\varphi_{j_n} \neq \pi - \theta$ and $0 \leq \varphi_{j_n} \leq 90°$, that is SB_j is a "to be optimized" sub-barrier.

The objective of phase 3 is to minimize $\sum_{j \geq 1}$ length(G_j) while maintain the coverage connectivity, e.g. no new gaps are formed. This gap length minimization problem is equivalent to maximizing the coverage of each sub-barrier at the left-most and right-most nodes. s_{j_n} computes the orientation of all the sensors in SB_j by formulating a nonlinear optimization problem:

Maximize:

$$max\{0, R_s cos(\pi - \theta - \varphi_{j_1})\} + R_s cos(\varphi_{j_n})$$

Subject to:

$max\{0, R_s cos(\varphi_{j_i})\} + max\{0, R_s cos(\pi - \theta - \varphi_{j_{i+1}})\} \geq dist(s_{j_i}, s_{j_{i+1}})$ for $i = 1, 2, 3, ..., n-1$

$0 \leq \varphi_{j_i} \leq \pi - \theta$ for $i = 1, 2, ..., n$

The objective function maximizes the coverage of the left most and right most sensors of the sub-barrier SB_j. The first constraint makes sure that no new gap is formed between consecutive sensors in SB_j. The variable constraint specifies the range of the orientation angle for each sensor in SB_j.

The nonlinear optimization of the sub-barrier SB_j is computed by s_{j_n}. Let us denote $\Phi^* = (\varphi^*_{j_1}, \varphi^*_{j_2} ... \varphi^*_{j_n})$ the orientation angles of all nodes in SB_j when the global maximum of the objective function is achieved. The node s_{j_n} then sends the Φ^* values to all other nodes in SB_j using a message $SetOrientation(s_{j_n}, \Phi^*)$. When a node receives the Φ^* values from its closest right neighbor, it switches to its optimal orientation accordingly. Then it waits a random delay and re-transmits the *SetOrientation* message by replacing the first field with its own ID.

Figure 2 shows the initial line-based sensor deployment, Fig. 6 shows sensor orientations after phase 2, and Fig. 9 shows sensor orientation after phase 3.

After the phase 2, the number of gaps decreased from 6 to 2. After the phase 3 the total gap length is reduced, since the total coverage provided by SB_2 is maximized.

4 Simulation Results

In this section, we conduct simulations using Matlab to evaluate the performance of our proposed distributed algorithm for mending barrier gaps. The sensor field is a belt region of length L $= 500m$ and width H $= 100m$. The initial sensor orientation angle follows a uniform distribution in the range $[0, 2\pi]$. The sensing range is $R_s = 15m$.

As explained in Sect. 2, the actual sensor positions have random offsets following a Gaussian distribution with mean 0 and variance σ. If we denote σ_i^x and σ_i^y the offset distance of the sensor node s_i in the horizontal and vertical directions, then we have $\sigma_i^x, \sigma_i^y \sim N(0, \sigma^2)$. In our simulations we set $\sigma_x = \sigma_y = 5$.

We generate 100 different sensor deployments. Each data point in our simulation results is an average of 100 experiments. We compare the performance of our distributed gap mending algorithm with SRA and CRA algorithms proposed in [2]. SRA and CRA [2] are briefly described in Sect. 1.

In the first experiment in Fig. 10a, we measure the number of unmended gaps before and after running gap mending algorithms. The total number of sensors deployed in the field varies between 20 and 150 with a step size of 5. Each sensor has a view angle of $\theta = 60°$.

Figure 10a shows that deploying a larger number of sensors leads to a smaller number of unmended gaps. The CRA algorithm mends more gaps than the SRA algorithm, while our distributed algorithm mends the most number of gaps among the three algorithms.

Figure 10b compares the total gap length of unmended gaps. Our distributed gap mending algorithm has a smallest total gap length. We can also observe the impact of phase 3 of our algorithm in reducing the total gap length.

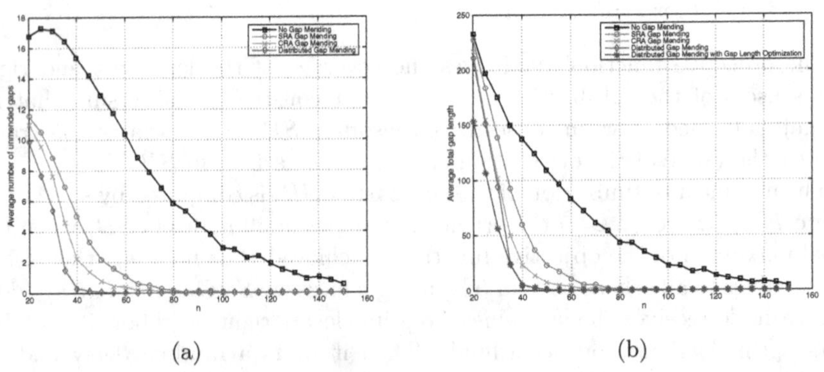

Fig. 10. (a)Average number of unmended gaps when $\theta = 60°$; (b)Average total gap length when $\theta = 60°$.

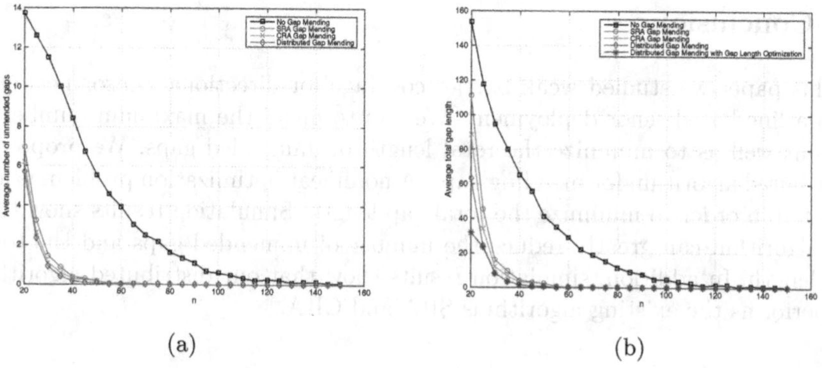

Fig. 11. (a)Average number of unmended gaps when $\theta = 120°$; (b)Average total gap length when $\theta = 120°$.

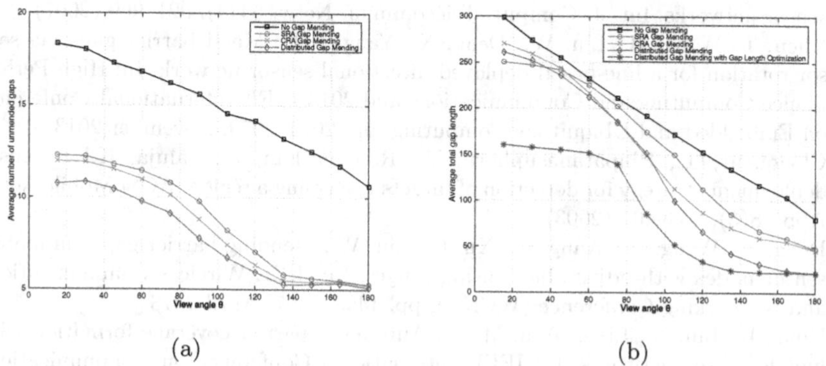

Fig. 12. (a)Average number of unmended gaps when we vary the view angle; (b)Average total gap length when we vary the view angle.

In the second experiment, we measure the number of unmended gaps and total gap length when each sensor has a view angle of $\theta = 120°$. The results are shown in Fig. 11. Our algorithm mend more gaps then SRA and CRA. We also provide a smaller total gap length.

In the third experiment in Fig. 12, we fix the total number of sensors to 20 and we vary the view angle between 15° and 180° with a step size of 15°. The picture shows that a larger view angle leads to a lower number of unmended gaps, since each sensor provider a larger coverage. Our distributed algorithm achieves the least number of unmended gaps for all view angles. In addition, our algorithm with gap length optimization phase provides a smaller total gap length compared to SRA and CRA.

5 Conclusions

In this paper we studied weak barrier coverage for directional sensor networks with a line-based senor deployment. We aim to mend the maximum number of gaps as well as to minimize the total length of unmended gaps. We propose a distributed algorithm for mending gaps. A nonlinear optimization problem is formulated in order to minimize the total gap length. Simulation results show that our algorithm can greatly reduce the number of unmended gaps and the total gap length. In addition, simulation results show that our distributed algorithm outperforms the existing algorithms SRA and CRA.

References

1. Akyildiz, I.F., Melodia, T., Chowdhury, K.R.: A survey on wireless multimedia sensor networks. Int. J. Comput. Telecommun. Netw. **51**(4), 921–960 (2007)
2. Chen, J., Wang, B., Liu, W., Deng, X., Yang, L.T.: Mend barrier gaps via sensor rotation for a line-based deployed directional sensor network. In: High Performance Computing and Communications and 2013 IEEE International Conference on Embedded and Ubiquitous Computing, pp. 2074–2079, November 2013
3. Clouqueur, T., Phipatanasuphorn, V., Ramanathan, P., Saluja, K.K.: Sensor deployment strategy for detection of targets traversing a region. ACM Mobile Netw. Appl. **8**(3), 453–461 (2003)
4. Deng, X., Wang, B., Wang, C., Xu, H., Liu, W.: Mending barrier gaps via mobile sensor nodes with adjustable sensing ranges. In: IEEE Wireless Communications and Networking Conference (WCNC), pp. 1493–1497, April 2013
5. Kong, L., Liu, X., Li, Z., Wu, M. Y.: Automatic barrier coverage formation with mobile sensor networks. In: IEEE International Conference on Communications (ICC), pp. 1–5, May 2010
6. Kumar, S., Lai, T. H., Arora, A.: Barrier coverage with wireless sensors. In: Proceedings ACM MobiCom, pp. 284–298, 2005
7. Liu, B., Dousse, O., Wang, J., Saipulla, A.: Strong barrier coverage of wireless sensor networks. In: Proceedings of The ACM International Symposium on Mobile Ad Hoc Networking and Computing (MobiHoc), pp. 411–420 (2008)
8. Ma, H., Liu, Y.: On coverage problems of directional sensor networks. In: Jia, X., Wu, J., He, Y. (eds.) MSN 2005. LNCS, vol. 3794, pp. 721–731. Springer, Heidelberg (2005)
9. Saipulla, A., Liu, B., Wang, J.: Barrier coverage with airdropped sensors. In: Proceedings of IEEE International Conference for Military Communications (MilCom), pp. 1–7, November 2008
10. Saipulla, A., Westphal, C., Liu, B., Wang, J.: Barrier coverage of line-based deployed wireless sensor networks. In Proceedings of IEEE Conference on Computer Communications (InfoCom), pp. 127–135, April 2009
11. Shih, K.P., Chou, C.M., Liu, I.H., Li, C.C.: On barrier coverage in wireless camera sensor networks. In: 24th IEEE International Conference on Advanced Information Networking and Applications (AINA), pp. 873–879, April 2010

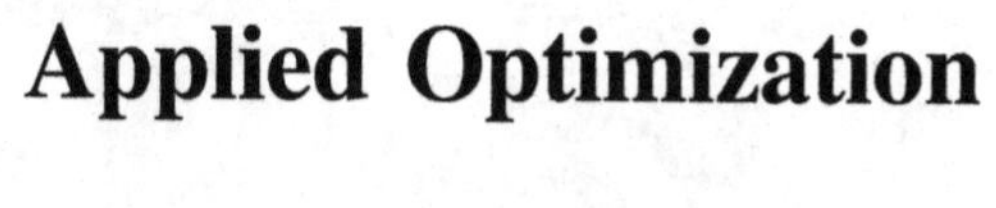
Applied Optimization

A Hybrid Large Neighborhood Search for Dynamic Vehicle Routing Problem with Time Deadline

Dan Yang[1,2], Xiaohan He[1,2], Liang Song[1,2], Hejiao Huang[1,2](✉), and Hongwei Du[1,2]

[1] Harbin Institute of Technology Shenzhen Graduate School, Shenzhen, China
{yangdanhsc,hexiaohan818}@gmail.com
[2] Shenzhen Key Laboratory of Internet Information Collaboration, Shenzhen, China
{songliang,hjhuang,hwdu}@hitsz.edu.cn
http://carc.hitsz.edu.cn/MICCRC/index.html

Abstract. In order to consider the customer's real-time requests and service time deadline, this paper studies the dynamic vehicle routing problem with time deadline (DVRPTD), and proposes a hybrid large neighborhood search (HLNS) algorithm. This strategy divides the working interval into multiple equal time slices, so DVRPTD is decomposed into a sequence of static problems. We use an insert heuristic to incorporate new requests into the current solution and a large neighborhood search algorithm (LNS) with various remove and re-insert strategies to optimize at the end of each time slice. The computational results of 22 benchmarks ranging from 50 up to 385 customers prove the superiority of our algorithm comparing with those existing ones .

Keywords: Time deadline · DVRPTD · Insert heuristic · LNS

1 Introduction

Vehicle routing problem (VRP) mentioned in [1] is a combinatorial optimization problem, and it plays an important role in the transportation and logistics fields. There are some dispatching systems that used heuristic algorithms to plan routes, but most of them developed as static ones so they cannot satisfy the real-time request. It is meaningful to focus on the dynamic vehicle routing problems (DVRP).

The earliest research on DVRP was in [2], it solved a Dial-and-Ride problem with only one vehicle. Then Psaraftis introduced the concept of "immediate request" in [3] and defined the DVRP as planning routes to meet the requests of the real-time customers in [4]. So far, DVRP mostly relies on heuristic algorithms

H. Huang—This work was financially supported by National Natural Science Foundation of China with Grant No. 11371004 and Shenzhen Strategic Emerging Industries Program with Grant No. ZDSY20120613125016389.

Z. Lu et al. (Eds.): COCOA 2015, LNCS 9486, pp. 307–318, 2015.
DOI: 10.1007/978-3-319-26626-8_23

as they can quickly find the optimal solution for the current state. Khoadjia et al. [5] used the Particle Swarm Algorithm; Gendreau et al. mentioned a Neighborhood Search for dynamic Pickup and Delivery Problems [6]; Beaudry et al. gave an Insertion Method and Tabu Search for the dynamic patients' transportation problem in hospital [7]; Francesco et al. adopted Tabu Search and a real-time control approach for urgent delivery of goods [8]. Moreover, the extensional version of DVRP, for example, DVRP with time windows (DVRPTW) has been studied, Alvarenga et al. proposed a Genetic and Set Partitioning approach [9]; Meidan et al. studied an Insertion Heuristic and GIS buffer analysis [10]; Lianxi [11] introduced an improved LNS which performs very well.

In real life, it is unpractical to provide all customers with service within the driver's working interval. In this paper, the DVRPTD is taken into consideration, which refers that the available service time equals working interval, all vehicles start from depot to serve yesterday's left customers after opening time and must return to the depot by the deadline. The similar problem was first introduced in [12]. A cut-off time T_{co} was put forward to show that new requests received before T_{co} will be committed to vehicles, and these who appears after T_{co} will be postponed to the next day to be served as a-priori. Then Montemanni et al. proposed an Ant Colony System (ACS) and an events manager [13]; Messaoud et al. studied a Genetic Algorithm (GA) [14] and an ACOLNS algorithm [15]. In this paper, we presented an HLNS heuristic to solve the DVRPTD which is better than algorithms above in some ways.

The remaining sections of this paper are organized as follows. Section 2 gives a formulation of DVRPTD and Sect. 3 proposes our HLNS algorithm. Section 4 shows the computational results and analysis on benchmarks. Finally, Sect. 5 gives a summary.

2 Problem Formulation

The DVRPTD is described as an undirected graph $G = (V, E)$, $V = \{0\} \cup M$ is the set of vertices including depot node (denoted as 0) and customer nodes (denoted as M), E is a set of edges between vertices. The customer nodes M include static nodes (denoted as N) which indicate the rest requests of yesterday and dynamic nodes (denoted as $M \setminus N$) which appear with time. The homogeneous vehicles are denoted as K, each with a same capacity Q. For each customer $i \in M$, it has a demand $q_i(q_i < Q)$, an appearance moment a_i and a service duration s_i. At the same time, d_{ij} and t_{ij} represent the Euclidean distance and travel time of edge $(i, j) \in E$, respectively. Therefore, for each customer $i \in M$, the vehicle's arriving time which is denoted as b_i should satisfy $b_i = b_{i-1} + t_{(i-1)i} + s_{i-1}$.

The driver's working interval denoted as $[W_1, W_2]$ is divided into multiple slices $\{T_1, T_2, \ldots, T_n\}$ with equal length t_d, we serve the static customers N in T_1, and dynamic ones $M \setminus N$ in $\{T_2, T_3, \ldots, T_n\}$. Some of the other useful variables are defined as follows:

Q_{kl} indicates the total demand of vehicle k at the end of T_{l-1}; c_{kl} indicates the last request of vehicle k during T_{l-1}; t'_{kl} indicates the end moment of serving the last request by vehicle k during T_{l-1}; C_l indicates the non-served requests at the beginning of T_l; D_l indicates the already served requests during T_l. It is assumed that $Q_{kl} = 0, t'_{kl} = 0, C_l = N$ when $l = 1$. Finally two decision variables x_{ijk} and y_{ki} are defined as follows:

$x_{ijk} = 1$ if node j is visited after node i by vehicle k and 0 otherwise.

$y_{ki} = 1$ if node i is visited by vehicle k and 0 otherwise.

The general mathematical model for DVRPTD is described as follows:

$$\min \sum_{i,j \in N \cup \{0\}} \sum_{k \in K} d_{ij} x_{ijk} \tag{1}$$

s.t.

$$\sum_{i \in N} q_i y_{ki} \le Q, \qquad \forall k \in K \tag{2}$$

$$W_1 + \sum_{i,j \in N \cup \{0\}} x_{ijk} t_{ij} + \sum_{i \in N} s_i y_{ki} \le W_2, \qquad \forall k \in K \tag{3}$$

The dynamic model for each time slice is described as follows:

$$\min \sum_{i,j \in C_l \cup \{0, c_{kl}\}} \sum_{k \in K} d_{ij} x_{ijk} \tag{4}$$

s.t.

$$\sum_{j \in C_l} x_{ijk} \le 1, \qquad \forall k \in K, i \in \{0, c_{kl}\} \tag{5}$$

$$\sum_{i \in C_l} x_{i0k} \le 1, \qquad \forall k \in K \tag{6}$$

$$Q_{kl} + \sum_{i \in C_l} q_i y_{ki} \le Q, \qquad \forall k \in K \tag{7}$$

$$t'_{kl} + \sum_{i,j \in C_l \cup \{0, c_{kl}\}} x_{ijk} t_{ij} + \sum_{i \in C_l} s_i y_{ki} \le W_2, \qquad \forall k \in K \tag{8}$$

$$D_l = \{i \mid b_i + s_i \le l * t_d, \quad \forall i \in C_l\} \tag{9}$$

Any other constraints are:

$$\sum_{k \in K} y_{ki} = 1, \qquad \forall i \in M \tag{10}$$

$$\sum_{i \in M} x_{ijk} = \sum_{i \in M} x_{jik} \le 1, \qquad \forall j \in M, \forall k \in K \tag{11}$$

$$\sum_{i \in M} \sum_{k \in K} x_{0ik} = \sum_{i \in M} \sum_{k \in K} x_{i0k} \le |K| \tag{12}$$

$$W_1 \le s_i \le W_2, \qquad \forall i \in M \tag{13}$$

$$M \setminus N = \{i \mid W_1 \le a_i < T_{co}\} \quad \forall i \in M \tag{14}$$

The objective function (1) aims at minimizing the total travel distance for all customers in a static environment, and the expressions (2) and (3) are the initial capacity and deadline constraints for static customers, respectively. The objective function (4) seeks the optimization solution for the current state at the end of each time slice. The expressions (5) and (6) show the requirement for the start position, namely, the vehicles can start from the depot or the last customer and finally must return to the depot, and the expressions (7) and (8) are the capacity and time deadline constraints for adding new customers, respectively. The expression (9) defines the served requests within each time slice. The expressions (10)–(12) aim at all customers, while the expressions (10) and (11) ensure that each customer is served only once by only one vehicle, and the expression (12) ensures that all vehicles start from depot during initialization and finally return to the depot, meanwhile the number of vehicles put in use cannot exceed the total quantity at depot. The expression (13) ensures that the time of service should be within work interval, and the expression (14) ensures the served customers' arrival time.

3 Hybrid Large Neighborhood Search

The DVRPTD is a dynamic deterministic problem according to [16], and we deal with it by using the division idea in [13] which belongs to the periodic optimization strategy. Based on the ideas of the periodic optimization, we can convert the DVRPTD into a series of static snapshots, at each static snapshot, vehicles start from their current position or depot to serve all non-served requests including the unfinished and the new ones (denoted as C_{new}). If there are no working vehicles suitable for new customers, we need to choose another one from depot.

This paper proposes an HLNS algorithm to solve the DVRPTD which includes two important parts. One is an insert heuristic for incorporating new requests into the current solution at the end of each time slice; the other is a LNS algorithm for statically optimizing the solution. Since LNS has a high requirement on the initial solutions, so we adopt the famous saving algorithm [17] which can provide a relatively good initial solution. At the same time, our HLNS algorithm studies the population thought of genetic algorithm which means that a group of the current feasible solutions is saved, and at each iteration we randomly select one feasible solution to run LNS optimization respectively, then we can improve the overall quality of the whole group. From what has been discussed above, the overall framework of our HLNS is shown in Fig. 1.

Let the symbol S be denoted as the current complete solution, P_S be denoted as the set of the current complete solutions, S_{best} be denoted as the best current complete solution, S_{worst} be denoted as the worst current complete solution, Φ be denoted as the removing node set, S_{Φ} be denoted as the residual solution by deleting Φ from S, S' be denoted as the solution after reinserting Φ into S_{Φ}, S_{new} be denoted as the new solution by incorporating C_{new}, S_x be denoted as the residual solution by deleting node x and connecting $x-1$ to $x+1$ of S, $f(S)$ be denoted as the objective function value of the solution S.

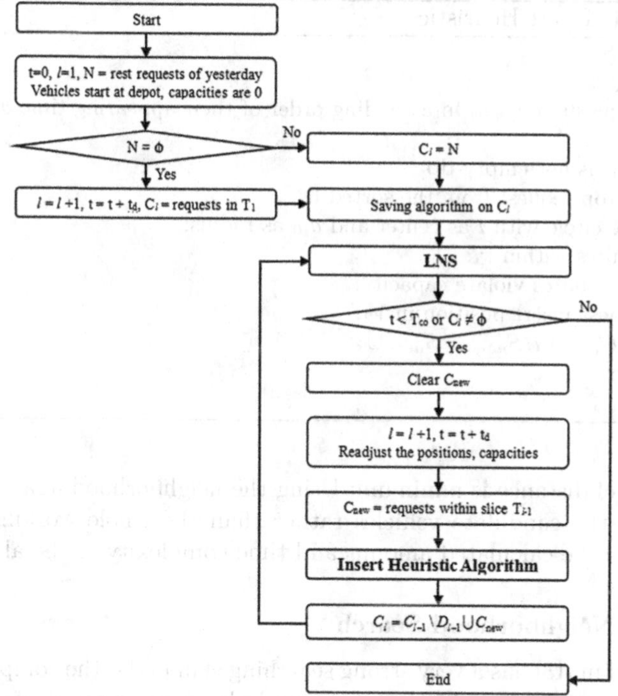

Fig. 1. The framework of HLNS.

3.1 Insert Heuristic

New requests appear constantly with the time passing, then how to re-plan the current solution by taking new requests into consideration is the core problem. To find the appropriate solution as soon as possible, this paper adopts an insert heuristic at the end of each time slice. It is described in Algorithm 1.

1. Sort all the new requests in ascending order of their appearance moment, and select an earliest request to insert in turn until all requests have been inserted;
2. For each inserting request, use neighborhood analysis to select the suitable candidate vehicles. One's neighborhood area π_i is defined as a circle with the inserting request as center and the distance between depot and it as radius, thus vehicles in this circle are saved as the candidate set V_1;
3. Select such vehicles from V_1 which have enough capacity for the serving request and save as V_2;
4. Find the best position in V_2 for this inserting request by greedy selection.

This strategy is desirable, because the vehicles within neighborhood area are closer to the inserting request than these who are not within. Inserting it into neighborhood area will obviously shorten the distance increment so that

Algorithm 1. Insert Heuristic

Input: C_{new}, S
Output: S_{new}
Sort all requests in C_{new} in ascending order of their appearing time a_i;
$S_{new} = S$;
while C_{new} is not empty **do**
  $i \leftarrow$ the top request from the sorted list;
  $\pi_i \leftarrow$ the circle with i as center and d_{i0} as radius;
  $V_1 \leftarrow$ routes within π_i;
  $V_2 \leftarrow V_1 \setminus$ routes violate capacity;
  get i 's best insert position in V_2;
  $C_{new} = C_{new} \setminus i, S_{new} = S_{new} \cup i$;
end while
Return S_{new}.

the total travel distance is minimum; Using the neighborhood analysis to insert requests into the candidate vehicles rather than the whole working fleet, can greatly reduce the calculated amount and time complexity of the algorithm.

3.2 Large Neighborhood Search

LNS proposed in [18] has a very strong searching skill under the complicated constraints. It optimizes the solution continuously by removing part of nodes of initial solution and re-inserting them to obtain some better solutions. The improvement of LNS in this paper is the multiple remove strategies which take more elements into consideration to remove nodes. This section will give a detailed description on remove and re-insert strategies of LNS, and first we give a process of LNS in Algorithm 2.

Algorithm 2. LNS Algorithm

Input: P_S
Output: better P_S and S_{best}
while not satisfy the max iteration I_1 **do**
  Randomly select a solution S in P_S;
  Let $S_1 = S$;
  $S_\Phi \leftarrow$ **Remove Algorithm** on S_1;
  $S' \leftarrow$ **Reinsert Algorithm** on S_Φ;
  if $f(S') < f(S_{worst})$ **then**
    $P_S = P_S \setminus S_{worst} \cup S'$;
    if $f(S') < f(S_{best})$ **then**
      $S_{best} = S'$;
    end if
  end if
end while
Return P_S and its S_{best}.

Remove Algorithm. Remove algorithm is an important part of LNS, which removes some nodes from the current solution to obtain new results. To improve the effectiveness, we use three strategies, that is, shaw-relatedness remove, saving remove and mean remove strategies, respectively. They are chosen randomly in each iteration. Many different remove strategies instead of only one strategy can improve robustness of our heuristic and avoid the local optimum. In addition, a tabu list L is put in use to save the deleting nodes in case we remove same nodes repeatedly.

(1) Shaw-Relatedness Remove

This strategy mentioned in [18,19] is based on the relatedness among all nodes, namely, Φ is made up of related nodes. In the first step, Φ incorporates a node from S randomly; then adds the node i which has the max relatedness $R(i,\Phi)$. $R(i,\Phi)$ is the relatedness of node i and set Φ which equals the total relatedness between the node i and all nodes in Φ. It is defined as follows:

$$R(i,\Phi) = \sum_{j\in\Phi} r(i,j) \qquad i \in S_{\Phi} \tag{15}$$

Where $r(i,j)$ is the relatedness of nodes i and j, it includes 3 terms: the distance between nodes, the demand between nodes and nodes are whether in a same car. It is defined as follow:

$$r(i,j) = \frac{1}{\lambda_1 d_{ij} + \lambda_2 \mid a_i - a_j \mid + \lambda_3(1 - x_{ijk})} \qquad \forall k \in K, i \in S_{\Phi}, j \in \Phi \tag{16}$$

$\lambda_1, \lambda_2, \lambda_3$ are the controlled parameters.

(2) Saving Remove

There are some relatively far nodes which may greatly increase the travel distance, so it is necessary to remove these far nodes to effectively reduce the total distance, another saving value RSA is proposed, and the node with maximum RSA is added to Φ in each iteration.

$$RSA_x = f(S) - f(S_x) \qquad \forall x \in S \tag{17}$$

Where RSA_x is the saving value of node x, which means the possible saving cost in distance when deleting node x in S. It is described in the following equation:

$$RSA_x = d_{(x-1)x} + d_{x(x+1)} - d_{(x-1)(x+1)} \qquad \forall x \in S \tag{18}$$

(3) Mean Remove

In consideration of some routes with little nodes but long distances, a mean distance is proposed. It equals the total distance divided by nodes' quantity. In order to shorten the average distance, we should remove the fewest routes with long distance. This strategy tries to shorten the mean distance so that the total distance is shortened. First select the route with maximal mean distance. If this route has little nodes, delete this whole route, otherwise randomly delete one or two nodes of this route, then recalculate the mean distance and repeat the process above until there are a certain number of nodes in Φ.

Re-insert Algorithm. Re-insert algorithm aims to re-insert Φ into the solution to generate a more optimal solution. To make the re-insertion more feasible, this paper adopts the greedy re-insert strategy. This strategy inserts only one node at minimum cost in each iteration, so how to choose the inserting node becomes important. Because each node i in Φ has many cost values (same as objective function differentials, denoted as $\Delta f_{loc_{i1}}, \Delta f_{loc_{i2}}, \Delta f_{loc_{i3}} \dots$) by inserting into all positions in S, where the position with the minimum cost value of node i is denoted as $besti$, its corresponding cost value is denoted as Δf_{best_i}. So in each iteration, we select the node i with the minimum Δf_{best_i} (denoted as i_{best}) from Φ and re-insert it into its best position $best_{i_{best}}$, till Φ is empty.

4 Computational Results

This section shows the computational results on 22 benchmarks under a static and a dynamic circumstance, respectively. In order to prove the effectiveness of our approach, a comparison of HLNS and other algorithms is also provided.

Table 1. Static results and comparisons

Instances	NC	BR	BTD	RE	Distribution
c50	50	524.61	538.74	2.69 %	Uniform
c75	75	835.26	859.43	2.89 %	Uniform
c100	100	826.14	861.09	4.23 %	Uniform
c150	150	1028.42	1085.28	5.53 %	Uniform
c199	199	1291.45	1371.13	6.17 %	Uniform
t100d	100	1581.25	1643.47	3.93 %	Uniform
c100b	100	819.56	837.80	2.23 %	Clustered
c120	120	1042.11	1069.96	2.67 %	Clustered
f71	71	241.97	242.63	0.27 %	Clustered
t75b	75	1344.64	1355.93	0.84 %	Clustered
t75d	75	1365.42	1374.35	0.65 %	Clustered
t100c	100	1407.44	1435.17	1.97 %	Clustered
t150a	150	3055.23	3117.63	2.04 %	Clustered
t150b	150	2727.99	2787.83	2.19 %	Clustered
t150c	150	2362.79	2418.21	2.35 %	Clustered
t150d	150	2655.67	2709.23	2.01 %	Clustered
t385	385	24435.50	25500.10	4.36 %	Clustered
f134	134	1162.96	1204.74	3.59 %	Mixed
t75a	75	1618.36	1659.53	2.54 %	Mixed
t75c	75	1291.01	1317.91	2.08 %	Mixed
t100a	100	2047.90	2109.81	3.02 %	Mixed
t100b	100	1940.61	2007.45	3.44 %	Mixed

Table 2. Best dynamic results and comparisons

Instances	Best results				
	GRASP	ACS	ACOLNS	GA	HLNS
c50	696.92	631.30	601.78	602.75	598.93
c75	1066.59	1009.38	1003.20	962.79	1011.07
c100	1080.33	973.26	987.65	1000.98	1018.40
c100b	978.39	944.23	932.35	899.05	891.79
c120	1546.50	1416.45	1272.65	1328.54	1292.66
c150	1468.36	1345.73	1370.33	1412.03	1386.37
c199	1774.33	1771.04	1717.31	1778.56	1782.45
f71	359.16	311.18	311.33	304.52	300.49
f134	15433.84	15135.51	15557.82	16063.65	15934.22
t75a	1911.48	1843.08	1832.84	1822.38	1835.13
t75b	1582.24	1535.43	1456.97	1433.98	1458.24
t75c	1596.17	1574.98	1612.10	1505.06	1573.29
t75d	1545.21	1472.35	1470.52	1434.18	1456.51
t100a	2427.07	2375.92	2257.05	2223.04	2251.65
t100b	2302.95	2283.97	2203.63	2221.58	2210.36
t100c	1599.19	1562.30	1660.48	1518.08	1532.43
t100d	1973.03	2008.13	1952.15	1870.50	1957.80
t150a	3787.53	3644.78	3436.40	3508.09	3504.83
t150b	3313.03	3166.88	3060.02	3019.90	3013.74
t150c	3110.10	2811.48	2735.39	2959.58	2721.50
t150d	3159.21	3058.87	3138.70	3008.30	3049.23
t385	-	-	33062.06	40238	38325.17

The experiments are coded in Pathon and execute on a MacBook Inter (R) Core (TM) i5-3337U CPU @1.80 GHz, 4 GB RAM Memory and Window 7 operation system.

4.1 Static Results and Analysis

In this subsection, we test our LNS algorithm on the 22 famous static benchmarks which are proposed in [20–22] and include ranging from 50 up to 385 customers. There are three iterative parameters for LNS, which indicate the max iterative times I for the whole experiment, the max iterative times I_1 for LNS and the max non-improving iterative times I_2 to avoid the invalid loops, respectively. Through many experiments on different parameters, we find that the results fluctuate at a relatively stable state when $I = 50$, $I_1 = 1000$, $I_2 = 50$.

Table 3. Average dynamic results and comparisons

Instances	Average results				
	GRASP	ACS	ACOLNS	GA	HLNS
c50	719.56	681.86	623.09	618.86	647.22
c75	1079.16	1042.39	1013.47	1027.08	1043.86
c100	1119.06	1066.16	1012.30	1013.03	1071.13
c100b	1022.12	1023.60	943.05	931.35	928.25
c120	1643.15	1525.15	1451.60	1418.13	1409.48
c150	1501.35	1455.50	1394.77	1461.55	1452.21
c199	1898.20	1844.82	1757.02	1843.06	1859.13
f71	376.66	348.69	320.00	323.91	318.56
f134	16458.47	16083.56	16030.53	16671.17	16547.28
t75a	2005.44	1945.20	1880.87	1871.46	1938.36
t75b	1758.88	1704.06	1477.15	1533.63	1527.69
t75c	1674.37	1653.58	1692.00	1558.70	1631.10
t75d	1588.73	1529.00	1491.84	1458.93	1486.53
t100a	2510.29	2428.38	2331.28	2290.05	2347.20
t100b	2512.27	2347.90	2317.30	2263.46	2360.47
t100c	1704.40	1655.91	1717.61	1541.25	1592.69
t100d	2087.55	2060.72	2087.96	2004.78	2089.45
t150a	3899.16	3840.18	3595.40	3570.51	3633.04
t150b	3485.79	3327.47	3095.61	3120.57	3116.64
t150c	3219.27	3016.14	2840.69	3065.73	2984.07
t150d	3298.76	3203.75	3233.39	3175.37	3226.40
t385	-	-	35188.99	41319.39	40550.81

The computational results with above parameters are shown in Table 1 according to the customer's distribution.

There are six items, namely, the instances, the number of customers (NC), the best known results (BR), the best total travel distance (BTD) by using our approach on 100 runs, the relative error (RE) (i.e. $RE = (BTD - BR)/BR * 100\,\%$) and the distribution of customers(Distribution).

From Table 1 we can see that our best results are acceptable, because the relative errors are less than 7 % for all instances, especially less than 5 % for all most instances. Combined with the customer's distribution, the results imply that our algorithm may do better in the aspect of clustered and mixed areas. A conclusion is that, the effective LNS algorithm we proposed can quickly get optimal results, it lays a foundation for the dynamic tests.

4.2 Dynamic Results and Analysis

The dynamic data sets are derived from above benchmarks by Kilby [12]. According to the simulation in [13], the length of the working interval is set 1500 s, T_{co} is set to 1500/2, the suitable number of time slices is 25. Tables 2 and 3 show the best computational results by our HLNS algorithm and the average results over 20 runs, respectively.

There are other best and average results in Tables 2 and 3 provided by some heuristic algorithms, in which GRASP and ACS by [13], ACOLNS by [15], GA by [14].

From Table 2, we can see that our best results win GRASP out in 19 instances, win ACS out in 16 instances, win ACOLNS out in 10 instances and wins GA out in 11 instances. From Table 3, our average results win GRASP out in almost all instances, win ACS out in 14 instances, win ACOLNS out in 7 instances and win GA out in 9 instances. All the results and comparisons show our approach is competitive, and it is effective for the DVRPTD in some ways, especially for clustered customers. The reasons for better optimization of our approach are that, the saving algorithm in the first step can provide a good initial solution, and the insert heuristic is used to select the relatively legal positions.

5 Conclusion

This paper studies the DVRPTD, which refers that part or all of the requests appear dynamically so that it requires a combination of the vehicles' current state and the requests' current information to re-plan routes. The time deadline is also taken into account. In this paper, we adopt the time-division strategy and present an HLNS algorithm containing two parts, the insert heuristic for incorporating new requests and the LNS algorithm for improving the quality of the solution at the end of each time slice. In addition, we use the saving algorithm to get an initial solution of all requests known a-priori at the first time slice. The computational results show that the HLNS can do well and find satisfactory solutions under the dynamic circumstance, that prove the effectiveness of the algorithm.

References

1. Dantzig, G.B., Ramser, J.H.: The truck dispatching problem. Manag. Sci. **6**(1), 80–91 (1959)
2. Wilson, N.H.M., Colvin, N.J.: Computer control of the Rochester dial-a-ride system. Massachusetts Institute of Technology. Center for Transportation Studies (1977)
3. Psaraftis, H.N.: A dynamic programming solution to the single vehicle many-to-many immediate request dial-a-ride problem. Transp. Sci. **14**(2), 130–154 (1980)
4. Psaraftis, H.N.: Dynamic vehicle routing: status and prospects. Ann. Oper. Res. **61**(1), 143–164 (1995)

5. Khouadjia, M.R., Alba, E., Jourdan, L., Talbi, E.-G.: Multi-swarm optimization for dynamic combinatorial problems: a case study on dynamic vehicle routing problem. In: Dorigo, M., Birattari, M., Di Caro, G.A., Doursat, R., Engelbrecht, A.P., Floreano, D., Gambardella, L.M., Groß, R., Şahin, E., Sayama, H., Stützle, T. (eds.) ANTS 2010. LNCS, vol. 6234, pp. 227–238. Springer, Heidelberg (2010)
6. Gendreau, M., Guertin, F., Potvin, J.Y., et al.: Neighborhood search heuristics for a dynamic vehicle dispatching problem with pick-ups and deliveries. Transp. Res. Part C, Emerg. Technol. **14**(3), 157–174 (2006)
7. Beaudry, A., Laporte, G., Melo, T., et al.: Dynamic transportation of patients in hospitals. OR Spectrum **32**(1), 77–107 (2010)
8. Ferrucci, F., Bock, S., Gendreau, M.: A pro-active real-time control approach for dynamic vehicle routing problems dealing with the delivery of urgent goods. Eur. J. Oper. Res. **225**(1), 130–141 (2013)
9. Alvarenga, G.B., De Abreu, Silva, R.M., Mateus, G.R.: A hybrid approach for the dynamic vehicle routing problem with time windows. In: Fifth International Conference on Hybrid Intelligent Systems, HIS 2005, 7 p. IEEE (2005)
10. Meidan, L., Yehua, S., Jing, W., et al.: Insertion heuristic algorithm for dynamic vehicle routing problem with time window. In: 2010 2nd International Conference on Information Science and Engineering (ICISE), pp. 3789–3792. IEEE (2010)
11. Hong, L.: An improved LNS algorithm for real-time vehicle routing problem with time windows. Comp. Oper. Res. **39**(2), 151–163 (2012)
12. Kilby, P., Prosser, P., Shaw, P.: Dynamic VRPs: a study of scenarios. University of Strathclyde Technical report, pp. 1–11 (1998)
13. Montemanni, R., Gambardella, L.M., Rizzoli, A.E., et al.: Ant colony system for a dynamic vehicle routing problem. Comb. Optim. **10**(4), 327–343 (2005)
14. Elhassania, M., Jaouad, B., Ahmed, E.A.: Solving the dynamic vehicle routing problem using genetic algorithms. In: 2014 International Conference on Logistics and Operations Management (GOL), pp. 62–69. IEEE (2014)
15. Elhassania, M.J., Jaouad, B., Ahmed, E.A.: A new hybrid algorithm to solve the vehicle routing problem in the dynamic environment. Int. J. Soft Comput. **8**(5), 327–334 (2013)
16. Pillac, V., Gendreau, M., Guret, C., et al.: A review of dynamic vehicle routing problems. Eur. J. Oper. Res. **225**(1), 1–11 (2013)
17. Clarke, G., Wright, J.W.: Scheduling of vehicles from a central depot to a number of delivery points. Oper. Res. **12**(4), 568–581 (1964)
18. Shaw, P.: Using constraint programming and local search methods to solve vehicle routing problems. In: Maher, M.J., Puget, J.-F. (eds.) CP 1998. LNCS, vol. 1520, pp. 417–431. Springer, Heidelberg (1998)
19. Ropke, S., Pisinger, D.: An adaptive large neighborhood search heuristic for the pickup and delivery problem with time windows. Transp. Sci. **40**(4), 455–472 (2006)
20. Fisher, M.L., Jaikumar, R.: A generalized assignment heuristic for vehicle routing. Netw. **11**(2), 109–124 (1981)
21. Christofides, N., Beasley, J.E.: The period routing problem. Netw. **14**(2), 237–256 (1984)
22. Taillard, É.: Parallel iterative search methods for vehicle routing problems. Netw. **23**(8), 661–673 (1993)

Indoor Localization via Candidate Fingerprints and Genetic Algorithm

Zeqi Song, Hongwei Du(✉), Hejiao Huang, and Chuang Liu

Shenzhen Key Laboratory of Internet Information Collaboration,
Department of Computer Science and Technology,
Harbin Institute of Technology Shenzhen Graduate School, Shenzhen, China
{zeqisong1990,chuangliuhit}@gmail.com, {hwdu,hjhuang}@hitsz.edu.cn

Abstract. WiFi-based indoor localization was proposed to be a practical method to locate WiFi-enabled devices due to the popularity of WiFi networks. However, it suffers from large localization errors ($6 \sim 10$ m). In this paper, we propose a novel localization scheme: indoor localization using candidate fingerprints (CFs) and genetic algorithm (GA). We come up with candidate fingerprints (CFs) selection to increase the probability of obtaining the best location estimations of indoor devices. Furthermore the GA are used to search for the optimal combination of CFs of each device using the relative distance constraint information. In addition, we provide an analytical model for selecting CFs to predict the probability of CFs could cover their true location of target device. The experimental results on realistic data set indicate that our method can reduce the 50 % and 80 % errors to 1.6 m and 2.4 m respectively. And typical running times for our simulations are only within a few seconds (less than 5 s).

Keywords: Indoor localization · Candidate fingerprints · Genetic algorithm · Analytical model

1 Introduction

WiFi-based indoor localization approach [1,13] have attracted much attention, due to their low deployment costs and potential for reasonable accuracy. WiFi-based localization [1,3,13] leverage prevalent wireless access points, thus being easily applied to mobile devices. In addition, acoustic ranging-based techniques [7,12] that relies on range measurement between proximate devices has been motivated in recent years. Mobile localization schemes [17,18] utilize the traces of mobile devices to get estimate their location. While these methods have been shown to achieve promising localization accuracy (below 3 m at 80 % tile), significant errors ($6 \sim 10$ m) always exist. Those achieving high accuracy usually require special hardware [5,11] not readily available on most mobile devices. There has also been work on combining WiFi and acoustic localization [4]. While hybrid schemes have greatly improved the positioning accuracy, the systems seem

Z. Lu et al. (Eds.): COCOA 2015, LNCS 9486, pp. 319–333, 2015.
DOI: 10.1007/978-3-319-26626-8_24

to be much more complex. For PLWLS [4], the system needs to detect the presence of large errors, only devices with small initial location errors can serve as peers. Centaur [19] takes much time on Bayesian inference as the number of devices growing larger. Therefore, accurate indoor localization on mobile devices is still an open problem.

For WiFi-based localization, the closest matching WiFi signature may not be the best location estimation, which has been reported by *Hongbo Liu's* [4]: *the root cause is the existence of distinct locations with similar signatures.* On the other hand, we observed that the modern devices are equipped with both WiFi interfaces and speakers/microphones, which are amenable to WiFi measurements and acoustic ranging [6]. Our idea is leveraging all the information available to achieve accurate localization. For one thing, WiFi signal measurements performed by mobile devices provide an initial probability distribution of possible location of them. For another, the relative distances of pairwise devices could be used as physical constrains on their possible locations. Our consideration is producing highly accurate location estimation by fusing initial WiFi RSS probability distributions and geometric constrains.

In summary, we make the following major contributions:

1. We propose a novel localization scheme: indoor localization using candidate fingerprints (CFs) and genetic algorithm (GA). Using initial WiFi fingerprints incorporate relative distance constrains, our method has greatly improved the localization accuracy.
2. We come up with candidate fingerprints (CFs) which are more likely to include a closer position to the real position of indoor devices. We design a genetic algorithm to search for the optimal combination of CFs of each device. In addition, we provide an analytical model for selecting CFs to predict the probability of candidate fingerprints could cover the true location of target device.
3. We carried out our method on realistic data set collected from an office building. The experimental results indicate that our method outperforms state-of-the-art localization scheme in terms of localization errors. And typical running times for our simulations are within a few seconds.

The rest of this paper is organized as follows. In Sect. 2, we review the related works. In Sect. 3, we formulate the problem of indoor localization mathematically. Sections 4 and 5 described the details of our method of selecting candidate fingerprints and searching closest fingerprints using GA. In Sect. 6, we evaluate our approach using both analytical model and simulations. Finally, Sect. 7 concludes this paper.

2 Related Work

Indoor localization has been an active area of research for the past two decades. Here we provide a brief overview of some key researchs focusing on the WiFi and acoustic ranging techniques most closely related to ours.

WiFi-Based Localization: WiFi-based localization can be categorized into fingerprint based and model based. Fingerprint based techniques work in two phases. The off-line phase is to fingerprint each location in the space of interest with a vector of received signal strength (RSSI) measurement from various access points (APs), where the fingerprint could either be expressed in a deterministic form (e.g., RADAR [1]) or in probabilistic form (e.g., Horus [13]). In on-line phase, the measured RSSI values are compared with the fingerprints recorded previously to find the closest match. Model based techniques use a RF propagation model to derive RSS at various location. The log distance path loss (LDPL) model is a widely used model. EZ [3] and EZ-Perfect [19] use existing WiFi infrastructure to localize mobile devices without collecting prior knowledge about WiFi APs. While WiFi based techniques have been used to many applications, they sometimes suffer from the lack of localization accuracy.

Acoustic Ranging-Based Localization: The earliest kind of this scheme usually relied on deploying specialized infrastructure. For example, Active Bat [12] measures the times-of-flight of the ultrasound pulse to the mounted receivers on the ceiling. The speed of sound in air is used to calculate the distances from the Bat to each receiver. The Cricket location support system [7] uses a combination of RF and ultrasound technologies to provide a location-support service to users and applications. In addition, BeepBeep [6] introduced a novel way to do acoustic ranging, without requiring the device to be synchronized. Acoustic ranging-based techniques may suffer from a couple of problems: the direct acoustic signal is strong while the line-of-sight acoustic could be blocked by obstructions within indoor environment; it is assumed that the devices includes special purpose ultrasound hardware or speakers and microphones.

Hybrid Indoor Localization Scheme: There has also been work on combining WiFi and acoustic localization. PLWLS [4] proposed a peer assisted localization approach to eliminate such large errors. It obtains accurate acoustic ranging estimates among peer phones, then maps their locations jointly against WiFi signature map subjecting to ranging constraints. Centaur [19] draws inspirations from Bayesian graphical model for indoor localization, but the specific Bayesian graphical models it employs incorporate the geometric constraints arising from acoustic range measurements, in addition to WiFi constrains. While these hybrid schemes have greatly improved the positioning accuracy, the systems seem to be very complex. For PLWLS, the system needs to detect the presence of large errors, only devices with small initial can serve as peers. Centaur takes much time on Bayesian inference as the number of devices growing larger.

3 Problem Formulation

Consider the indoor environments such as office buildings or shopping malls which are overlaid on a WLAN. There are M mobile devices, each of them is assigned a global ID in the range of 1 to M. We assume that there are N access points (APs) in the area and they are all visible to all devices throughout the area. In addition, these devices are equipped with speakers/microphones, thus are able to implement acoustic ranging. Our problem is locating these mobile devices using the received signal strength from APs as well as the relative distances between devices obtained by acoustic ranging.

Construct Indoor *Radio Map.* We use Horus [13] method to construct *radio map* in this paper. Horus captures the RSS characteristics of the indoor apace through a probability distribution, $P(r = r_j|x_i)$, which is the probability of seeing an RSS value of r from access point j at location x_i. Since the signal strength values received from different access points are independent, the joint probability of observing $R = \{r_1, r_2, \ldots, r_n\}$ at location x_i is computed as:

$$P(R|x_i) = \prod_j P(r = r_j|x_i) \tag{1}$$

Acoustic Ranging. BeepBeep [6] method is used in this paper to measured the relative distances between mobile devices. BeepBeep work as follows: devices A and B transmit sound pulses to each other and then using the sampling rate of the sound-card to keep the measure of the time. Given the sampling frequency F of the sound card, BeepBeep computes the propagation delay Δt_{AB} in seconds as:

$$\Delta t_{AB} = \frac{\Delta N}{F} = \frac{1}{2F}[(N_B^A - N_A^A) - (N_B^B - N_A^B)] \tag{2}$$

where N_X^Y represents the sample number at which device $y's$ microphone detected the chirp emitted by device $X's$ speaker.

4 Pick the CFs

For WiFi-based localization, the closest matching WiFi signature may not be the best location estimation. In this paper, we come up with the concept of candidate fingerprints (CFs) which will be used throughout this paper. Instead of selecting one closest fingerprint as the exactly location estimation, we select several CFs as the alternative positions, which are more likely to include a closer position to the real position of indoor devices. To select the CFs, we should get the initial WiFi location estimations of devices firstly. As previously expressed, there are many WiFi fingerprints based indoor position models, which can further divided into deterministic techniques and probabilistic techniques. Moustafa [8] have proved that deterministic technique are not optimal as this method does not store any information about the signal strength distribution. According to Moustafa's theory, the Horus system outperforms the Radar system since the expected error for the former is less than the later. In this paper, we choose k closest location as CFs.

During training phase, the radio-map stores a Gaussian distribution that fits the signal strength received from an access point for every training point. Given a signal strength vector $R_j = \{r_1^j, r_2^j, \ldots, r_N^j\}$ of device j, we want to find the probability $P(l_i|R_j)$, which represent the probability that the device j is located in training points l_i. This probability is given by [13]. Then we select the front k location as CFs from $P(l_i|R_j), i = 1, 2, \ldots, s$, which are sorted in descending order. The algorithm is detailed in Algorithm 1.

Algorithm 1. $CF = CF\_Selecting(R, l, RM)$

Input:

$\{R_j\}$: measured signal strength vector, $j = 1, 2, \ldots, M$, where $R_j = \{r_1^j, r_2^j, \ldots, r_N^j\}$;

$\{l_i\}$: the training points, $i = 1, 2, \ldots, s$;

RM: radio-map, where $RM[a][l_i](q)$ represents the *pdf* of signal strength from access point a at location $l_i \in \mathbb{X}$.

Output:

$\{CF_j\}$: the CFs of device j, $j = 1, 2, \ldots, M$, where $CF_j = \{cf_1^j, cf_2^j, \ldots, cf_k^j\}$.

for $j = 1$ to M **do**
  for $i = 1$ to s and $l_i \in \mathbb{X}$ **do**
    $P(l_i|R_j) \leftarrow \prod_{n=1}^{N} \int_{r_n^j - 0.5}^{r_n^j + 0.5} RM[n][l_i](q)dq$
  end for
  Sort $\{P(l_i|R_j)\}$ in descending order;
  Chose the front k locations from the sorted set: $\{P(l_{i_1}|R_j), P(l_{i_2}|R_j), \ldots, P(l_{i_s}|R_j)\}$, where $P(l_{i_1}|R_j) \geqslant P(l_{i_2}|R_j) \geqslant \ldots \geqslant P(l_{i_s}|R_j)$;
  $CF_j \leftarrow \{cf_1^j, cf_2^j, \ldots, cf_k^j\}$, where cf_m^j corresponding to a location l_{i_m}.
end for

By introducing the CFs, we increase the probability of obtaining the true locations of indoor devices. Another problem is how to choose a value k, for the number of CFs so as to cover the true location as well as small enough. Note that if the value of k being very large, much time will be taken to make further computation and the method will become inefficient. Further discussion about the parameter k will be given in Sect. 4. Table 1 shows that each devices corresponding to a set of fingerprints.

Table 1. CFs for each devices

Devices	CFs
$dev_1 = (x_1, y_1)$	$CF_1 = \{cf_1^1, cf_2^1, \ldots, cf_k^1\}$
$dev_2 = (x_2, y_2)$	$CF_2 = \{cf_1^2, cf_2^2, \ldots, cf_k^2\}$
$\vdots$	$\vdots$
$dev_M = (x_M, y_M)$	$CF_M = \{cf_1^M, cf_2^M, \ldots, cf_k^M\}$

5 Search Closest Fingerprint Using GA

After selecting CFs, each device is assigned to a location set in which an entry is the closest position to its real location. The next step is to get the new location estimation: searching for the optimal combination of CFs of each device using other constraint information. By exploiting acoustic ranging, a device can use peer devices as reference points and obtain its relative position to them. This imposes unique physical constraints on the possible location of the devices, thus reducing the uncertainty and improving the accuracy.

5.1 Example Scenario

Before providing a physical intuition to the working of our algorithm, we start with the example depicted in Fig. 1. Device A and B are located within audible distance of each other at unknown location x_A and x_B. Using its WiFi fingerprint measurements, device A computes $CF_A = \{CF_{A1}, CF_{A2}, CF_{A3}\}$ - the CFs of device A. For case of exposition in this example, we shall set the number of CF as $k = 3$. But more formally, the value of k may be greater according your realities of situation. Similarly, based WiFi fingerprint localization, device B's CFs is denoted as $CF_B = \{CF_{B1}, CF_{B2}, CF_{B3}\}$. By performing acoustic ranging, the devices learn that the distance between them $d_{AB} = 5\,\text{m}$. This distance constraint further limit the choosing of the closer location from their CFs. Through calculation, the distance between device A and B's CFs are shown in Table 2. As is shown in Table 2, the distance between CF_{A2} and CF_{B2} is $d_{22} = 5.2\,\text{m}$, which is the nearest distance value to the distance measured by acoustic ranging. Where $\|d_{22} - d_{AB}\| \leqslant \|d_{ij} - d_{AB}\|, i, j \in \{1, 2, 3\}$ ($\|d_{ij} - d'_{ij}\|$ represent the Euclidean distance of d_{ij} and d'_{ij}). Then the locations of device A and B are more likely to be CF_{A2} and CF_{B2} rather than other combinations.

The basic idea of above example can be easily extend to several devices on a large floor as depicted in Fig. 2. Initially, each device has a location estimation (vertices in the dash-line graph) from WiFi fingerprint localization. Compared

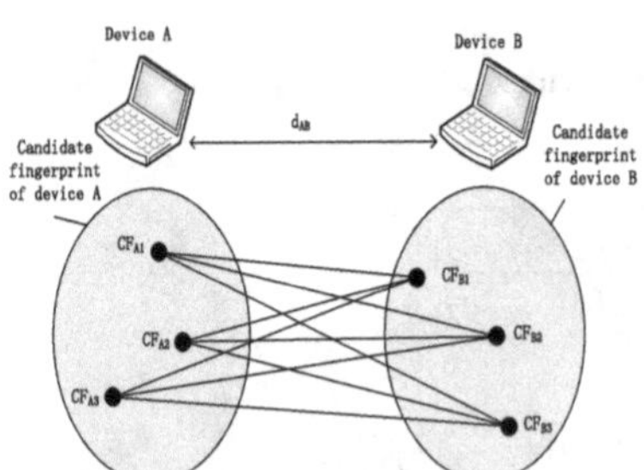

Fig. 1. A sample of distance constrain between two devices

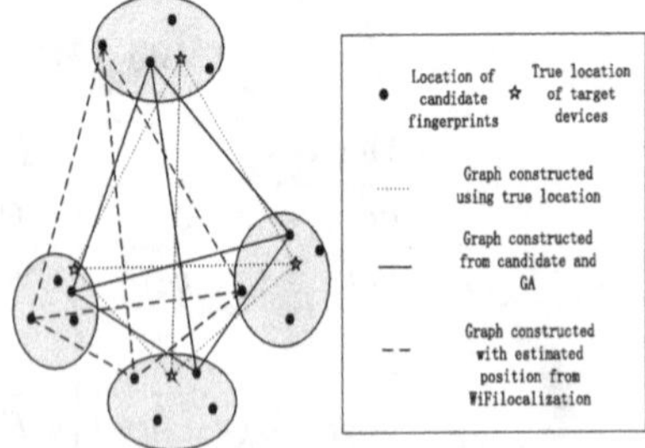

Fig. 2. A sample of using CFs and physical constrains to perform localization

Table 2. Calculated distances between CFs

Distances(m)	CF_{B1}	CF_{B2}	CF_{B3}
CF_{A1}	4.2	5.8	6.5
CF_{A2}	3.8	5.2	5.5
CF_{A3}	5.4	6.8	7.2

with WiFi signal strength within complex indoor environment, the relative distance between devices, which are measured by acoustic ranging, is much more reliable. The relative distances of vertices in the solid-line is another combination of CFs, which is more conform to the ground truth (the dotted-line graph). Apparently, the new combination of CFs is a better choice compared with the initial estimation. Such additional relative distance constraints "force" the new CFs combination more closer to their real locations, thus refining the location estimation obtained from initial WiFi localization.

The intuition of our algorithm is searching a combination of CFs among all the CFs to find a optimal one which is most satisfied with the distance constraints. In the algorithm, the objective function are designed as follow:

$$J(G) = \underset{i,j\in\{1,2,\ldots,M\}}{\operatorname{argmin}} \frac{1}{n^2}\sum_{i,j}\sqrt{(d_{ij}-d'_{ij})^2} \tag{3}$$

where G is a set of location estimations of devices, n denotes the number of pair distances measured by acoustic ranging. In ideal condition, the number of pair distances is $n = M^2$, where M is the number of device within the area. Generally, due to communication distance, non-line-of-sight in acoustic ranging, we may miss some pairwise distances with a large probability. The number of pair distances, in some cases, is less than M^2 which will be discussed in Sect. 5. Equation 3 produces a set of locations $G = \{l_i\}, i = 1, 2, \ldots, M$ from the CF database $l_i \in \mathbb{X}$, so as to minimize the distance differences between our location estimation d_{ij} and distances from acoustic ranging d'_{ij}.

It should be noted that the distances between some pairs of devices may not determine the shape of a graph. For example, a square is flexible since its vertices can rotate against each other and form a family of rhombi while preserving the edge lengths, whereas the shape of a triangle is "rigid" given the lengths of the three edges. The graph rigidity theory [14] describes under what conditions a graph is rigid: A graph is called generically rigid (or called rigid) if one cannot continuously deform the graph embedding in the plane while preserving the distance constraints. A graph is generically globally rigid (or called globally rigid) if there is a unique realization in the plane. *Jackson et al.* [15] prove that a graph is globally rigid if and only if it is 3-connected and redundantly rigid. A graph is redundantly rigid if the removal of any edge results in a graph that is still rigid. PLWLS [4] proposed a acoustic signal design, detection and scheduling techniques that satisfy the requirements of concurrent multi-peer ranging, which is robust to noise and having minimum impact on user's regular activities. So we can believe that the concurrent ranging among multiple devices and the resulting pairwise distances give a complete, thus rigid graph.

5.2 The Genetic Algorithm

An alternative is exhaustion method: searching over all possible combinations of the CFs. Assume that the number of device is M with k CFs for each device. The complexity of the problem is $O(k^M)$, which would be very time-consuming. While genetic algorithm (GA) can search the solution space efficiently, it can miss local minima that might provide a reasonably good solution.

Table 3. Chromosome encoding and initial generation

Chromosome encoding	Combination of CFs
$chro_1 = (c_1^1, c_2^1, \ldots, c_M^1)$	$CCF_1 = \{cf_{c_1^1}^1, cf_{c_2^1}^2, \ldots, cf_{c_M^1}^M\}$
$chro_2 = (c_1^2, c_2^2, \ldots, c_M^2)$	$CCF_2 = \{cf_{c_1^2}^1, cf_{c_2^2}^2, \ldots, cf_{c_M^2}^M\}$
$\vdots$	$\vdots$
$chro_L = (c_1^L, c_2^L, \ldots, c_M^L)$	$CCF_L = \{cf_{c_1^L}^1, cf_{c_2^L}^2, \ldots, cf_{c_M^L}^M\}$

The GA start by selecting an initial set of solutions(initial generation) randomly. As is shown in Table 3, the initial chromosome generation consists of a vector of values $chro_i$, which correspond a combination of CFs CCF_i. While L represent the number of chromosome individuals included in the initial generation; M, with the above meaning, denotes the number of devices within the area. $c_j^i \in \{1, 2, \ldots, k\}$, where k denote the number of CFs. Algorithm 2 gives the details of the genetic algorithm. During each iteration, each chromosome is evaluated by a fitness function (line 2). According to the optimal function J, the fitness function is designed as:

$$f(chro^*) = \frac{1}{n^2} \sum_{i,j} \sqrt{(d_{ij}^* - d_{ij}')^2} \quad (4)$$

where d_{ij}^* is the distance between estimate location of device i and j. For a chromosome individual $chro* = \{c_1^*, c_2^*, \ldots, c_M^*\}$, we can get a feasible solution of CFs $CCF^* = \{cf_1^*, cf_2^*, \ldots, cf_M^*\}$ by chromosome decoding. As every CFs combination is corresponding a set of location $G^* = \{l_1^*, l_2^*, \ldots, l_M^*\}$, then d_{ij}^* represent the distance between l_i^* and l_j^*, which comes from:

$$\begin{aligned} d_{ij}^* &= \sqrt{(x_i^* - x_j^*)^2 + (y_i^* - y_j^*)^2}, \\ l_i^* &= (x_i^*, y_i^*), l_j^* = (x_j^*, y_j^*) \end{aligned} \quad (5)$$

$P(t+1)$ is from previous generation $P_c(t)$ and crossover it while $P_m(t)$ is the chromosome population after mutating. For the "select" operation, consecutive generation of chromosomes are generated in the following manner:

Algorithm 2. $G^* = GA\_Searching(CF, MG, L)$

Input:

$CF = \{CF_1, CF_2, \ldots, CF_M\}$: The CFs of each target device;
MG: Maximum number of generations;
L: The size of population.

Output:

$G^* = \{l_1^*, l_2^*, \ldots, l_M^*\}$: The new location estimations of all the target devices.

Random generate CF combinations: $CCF \leftarrow \{CCF_1, CCF_2, \ldots, CCF_L\}$
Chromosome encoding: $chro \leftarrow \{chro_1, chro_2, \ldots, chro_L\}$
$t \leftarrow 0$
Initialize $P(t)$: $P(t) \leftarrow \{chro_1(t), chro_2(t), \ldots, chro_L(t)\}$
Evaluate $P(t)$: $J(P(t)) \leftarrow \{J(chro_1(t)), J(chro_2(t)), \ldots, J(chro_L(t))\}$
while $t \leqslant MG$ **do**
  $P_c(t) \leftarrow crossover\{P(t)\}$
  $P_m(t) \leftarrow mutation\{P(t)\}$;
  Evaluate $P_c(t)$ and $P_m(t)$
  $P(t+1) \leftarrow select(P_c(t) \cup P_m(t))$
  $t \leftarrow t + 1$
end while
Select the fittest chromosome individual: $chro^* \leftarrow best(P(t))$
Chromosome decoding: $CCF^* \leftarrow chro^*$
Location mapping: $G^* \leftarrow CCF^*$

1. 20 % of the solutions with the highest fitness are retained.
2. 20 % of the solutions are randomly generated for every generation.
3. 60 % of the solution are generated by picking two solutions $P_c(t)$ and $P_m(t)$ from the previous generation and mix them using a random convex liner combination.

$$P(t+1) = a \bullet P_c(t) + (1-a) \bullet P_m(t) \tag{6}$$

where a is a random vector with each element independently randomly draw from $(0,1)$ and $\bullet$ represents a vector dot product. $P_c(t)$ is from previous generation $P(t)$ and crossover it while $P_m(t)$ is the chromosome population after mutating. a is a random vector with each element independently randomly draw from $(0,1)$ and $\bullet$ represents a vector dot product.

As generations evolve, solutions with higher fitness are discovered. The GA terminates when solutions do not improve for 3 consecutive generations or it comes to the max iteration steps.

6 Performance Evaluation

Our experiments use the dataset of received signal strength indication (RSSI) collected from within an indoor office building Contributed by *Kevin Bauer etc.* [2]. The office building environment is a single storey building measuring roughly 50 × 70 m. The interior consists of small offices, cubicles, long hallways, and large warehouse-like rooms. A floor plan of this environment with measurement points and the passive monitors' locations labeled is provided in Fig. 3. This data captures RSSI behavior when 802.11 frames are transmitted using:a stock omnidirectional antenna. Omnidirectional RSSI measurements are collected from roughly 180 distinct physical locations throughout a large office building. To quantify how the transmitter's received signal strength varies with their physical location in the building, the experiment transmit 500 packets from each of the 180 physical positions.

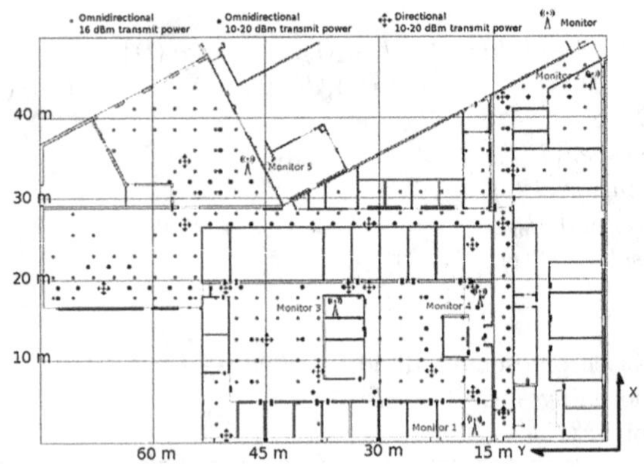

Fig. 3. Floor plan of the office space where the RSSI were collected

6.1 Analytical Model for Selecting CFs

Consider a grid system with two grid points x_i and x_j shown in Fig. 4, and a device is at the location of x_i with the signal strength $\tilde{R}_i$ (for the convenience of the analysis, the device's location is limited to the point on the grid). Assume that the distribution of signal strength at location x_i is $R_i \sim N(\mu_i, \sigma_i)$ (the red curve), and the distribution at x_j is $R_j \sim N(\mu_j, \sigma_j)$ (the blue curve). We can get the *pairwise error probability(PEP)* [10] that we have an incorrect estimate of the device as follows:

$$\begin{aligned} PEP(x_i, x_j) &= P(P(x_i|\tilde{R}_i) < P(x_j|\tilde{R}_i)) \\ &= \int_{p=r_{ij}}^{x=\infty} \frac{1}{\sqrt{2\pi\sigma_i^2}} e^{\frac{-(p-\mu_i)^2}{2\sigma_i^2}} dp \\ &= \Phi(\frac{r_{ij} - u_i}{2\sigma_i}) \end{aligned} \tag{7}$$

where r_{ij} represent the signal strength where the probability of locating device at x_i equals to the probability of locating the target device at x_j. That is the intersection of signal strength distribution of x_i and x_j. For a given signal strength, the system select the location that has maximum probability. When the target device' signal strength $\tilde{R}_i$ is larger than r_{ij}, the system will choose x_j as the location estimation rather than x_i. In Fig. 4, red area represent the probability of having a wrong location estimation. When considering only two location fingerprints, the probability of return the correct location or *pairwise correct probability (PCP)* can be computed: $PCP(x_i, x_j) = 1 - PEP(x_i, x_j)$.

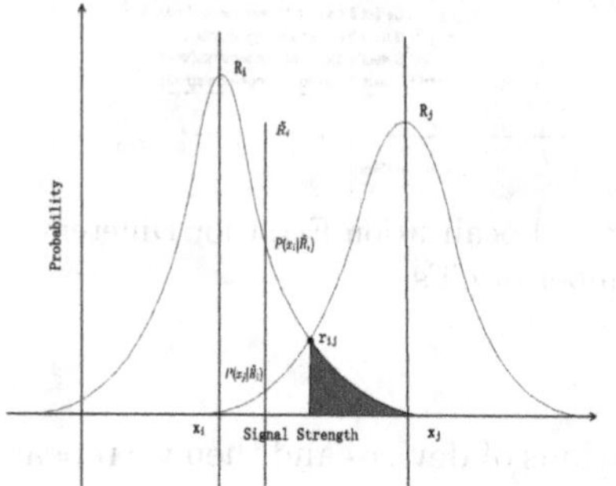

Fig. 4. *Pairwise error probability(PEP)* from a set of two (Color figure online).

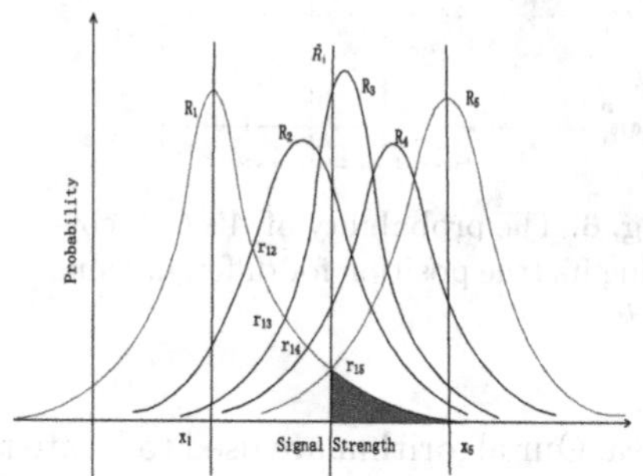

Fig. 5. *Pairwise error probability(PEP)* of CFs missing its correct position (Color figure online).

For our method, the radio-map contains several entries and fingerprints. As shown in Fig. 5, $\{x_1, x_2, x_3, x_4, x_5\}$ is a set of training points. We have fit a set of distribution for their RSS: $R_i \sim N(\mu_i, \sigma_i), i = \{1, 2, 3, 4, 5\}$. R_1 and $\{R_2, R_3, R_4, R_5\}$ intersect at $\{r_{12}, r_{13}, r_{14}, r_{15}\}$ respectively. Assume the target device is located at x_1, so its RSS obey the distribution $\tilde{R_1} \sim N(\mu_1, \sigma_1)$. If $\tilde{R_1} > r_{12}$, we can get $P(x_1|\tilde{R_1}) > P(x_2|\tilde{R_1}) > P(x_3|\tilde{R_1}) > P(x_4|\tilde{R_1}) > P(x_5|\tilde{R_1})$. If we choose the top 4 ($k = 4$) as the CFs and $\tilde{R_1} > r_{15}$, so we can get the CFs $CF_1 = \{R_1, R_2, R_3, R_4\}$ where its true location is included in the CFs. On the other hand, if $\tilde{R_1} < r_{15}$, we can get $P(x_3|\tilde{R_1}) > P(x_2|\tilde{R_1}) > P(x_3|\tilde{R_1}) > P(x_5|\tilde{R_1}) > P(x_1|\tilde{R_1})$. In this case, x_1 will not be selected and the CFs is $CF_A = \{R_2, R_3, R_4, R_5\}$. According to this model, we can get the probability of missing its correct location from $PEP(R_1, R_5) = \Phi(\frac{r_{15} - u_1}{2\sigma_1})$ (red area in Fig. 5).

Given a RSS from device A within the area, Fig. 6 shows the probability of A's CFs covering its true position for different value of k calculated using Eq. 7. From the figure we see that when $k = 6$ the probability comes to as high as 90 %, which will lead to increased accuracy.

6.2 The Effect of CFs Selecting

To investigate the performance of our method, we carry out intensive simulation based on the test bed described in this Section. In particular, we select 10 target devices locations within the area. The RSSI of each device is randomly selected from the training RSS dataset. The device's location are random distributed in this

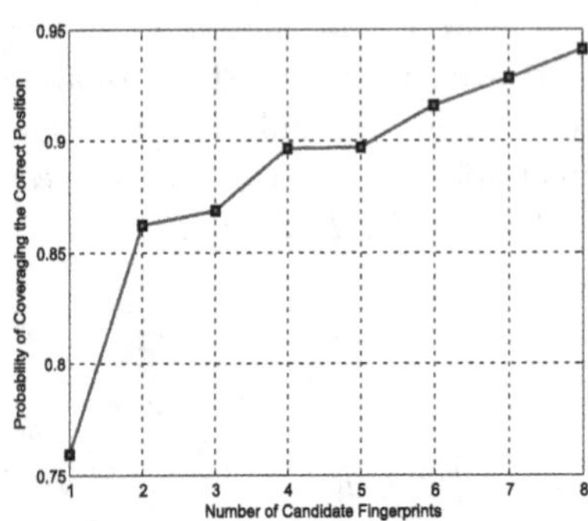

Fig. 6. The probability of A's CFs covering its true position for different value of k.

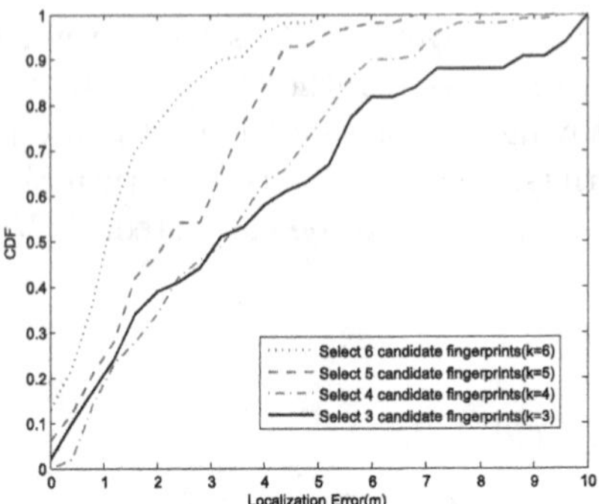

Fig. 7. Localization Error for Different Number of CFs.

area. Our algorithm are used to locate the locations of devices and then we repeat this process for 10 times to measure the location accuracy and time consumption. In total we have 100 location estimation of target devices for our simulation.

Firstly, we carry out the simulation introduced above for different value of k ($k = \{3, 4, 5, 6\}$). Figure 7 shows the effect of the parameter k on the performance of our method. The cumulation distribution function of localization errors are obtained for the four schemes. As is shown in the figure, $k = 6$ performs the best with 50th and 80th percentile error of 1.1 m and 2.3 m respectively. Followed by $k = 5$ with the 50th and 80th percentile error of 2.2 m and 3.8 m respectively. As the value k (number of CFs) increases, the accuracy of our method should increase correspondingly.

Table 4. Running time for various experiments

No of Cfs	No of Devices		
	$M = 5$	$M = 10$	$M = 15$
$k = 3$	0.6 s	1.3 s	1.9 s
$k = 4$	0.6 s	1.5 s	2.2 s
$k = 5$	1.3 s	2.4 s	3.0 s
$k = 6$	1.5 s	3.6 s	7.2 s

Note that if the value of k being very large, much time will be taken to make further computation and the method will become inefficient. Table 4 provide the time consumption by our program in running these four schemes. We can see from the table that the time consumption increfases as the number of CFs increase. Typical running times for our simulations are within a few seconds (less than 5 s when the number of devices is less than 10). These are reasonable times to locate devices in an indoor environment.

6.3 The Effect of Communication Distance

As mentioned in Sect. 4, due to non-line-of-sight, communication distance and outliers in acoustic ranging, we may miss some pairwise distances with a large probability. Figure 8 shows the effect of different communication distances for the number of pairwise distances measured. As shown in the graph, The longer the communication distance, the more measured distances we can get. In Fig. 8, we select 10 ($M = 10$) indoor devices to measure their distances and repeat this process for 10 times. Without missing, we will get $10 * 10 * 10$ pairwise distances altogether, but we get at most about 780 pairwise distances in practice when the communication distance is equal to 10. Our method is robust to missing pairwise distances of devices. In the graph, we experiment communication distance $c = \{4, 6, 8, 10\}$ to investigated the influence of communication distances on localization error. As shown in Fig. 9, The longer the communication distance, the more measured distances we can get, thus provide more distance constrains, our method will perform better.

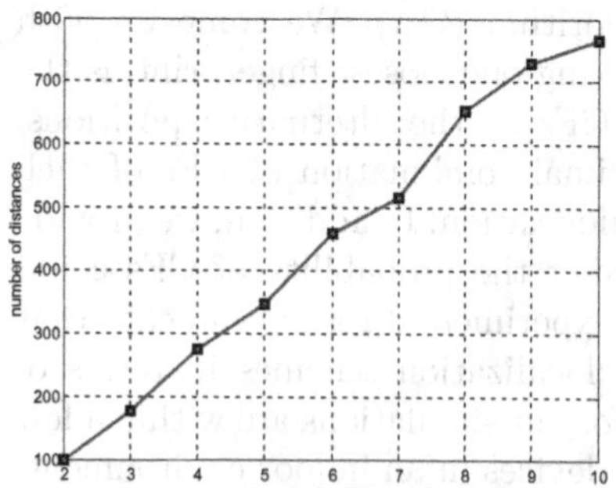

Fig. 8. Number of measured distances for different communication radius

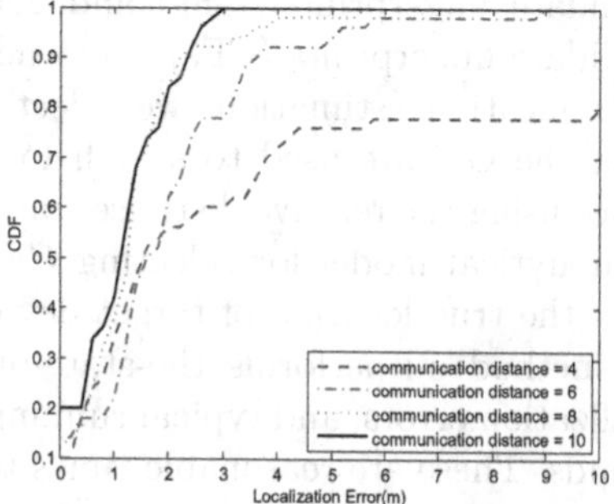

Fig. 9. Localization error for different communication distances.

6.4 Localization Errors

We compare the performance of our method with the following existing localization schemes: Radar, Horus and PLWLS. Figure 10 shows the cumulation distribution function of localization errors for the three schemes. Figure 11 summarizes the comparison results. From Fig. 10, we observe that WiFi only localization (Radar and Horus) can result in significant error ($6 \sim 10$ m). Adding distance constrains, whether PLWLS and our method helps regin in these outliers ($5 \sim 7$ m). Comparing the our method to the PLWLS system shows that the average error is decreased by 0.3 m. These results show the effectiveness of the proposed techniques. Figure 11 shows that the mean error is decreased by more than 48 % for Horus and 14 % for PLWLS. The 50 % error is decreased by more than 38 % for the Horus system and 30 % for PLWLS. The 80 % error is decreased by more than 36 % for the Horus system and 12 % for PLWLS respectively.

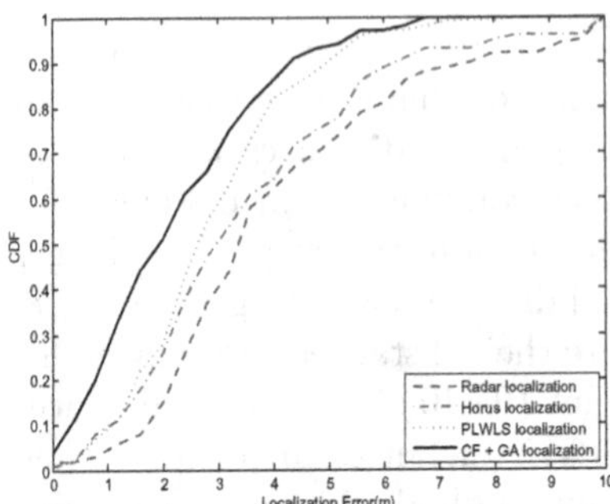

Fig. 10. Localization error of different localization schemes.

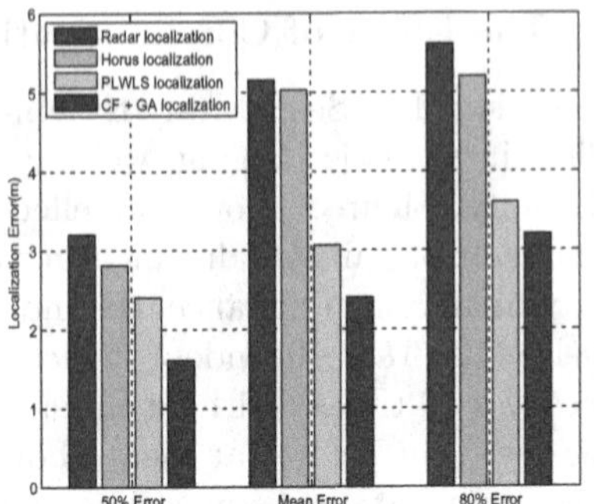

Fig. 11. Comparison of different localization schemes.

7 Conclusion

In this paper, we propose a novel localization scheme: indoor localization using candidate fingerprints (CFs) and genetic algorithm (GA). We come up with candidate fingerprints (CFs) - instead of selecting one closest fingerprint as the exact location estimation, we select several CFs as the alternative positions. Then the GA are used to search for the optimal combination of CFs of each device using the relative distance constraint information. In addition, we provide an analytical model for selecting CFs to predict the probability of CFs could cover the true location of target device. The experimental results indicate that our method outperforms the-state-of-the-art localization schemes in terms of localization errors, and typical running times for our simulations are within a few seconds. These are reasonable times to locate devices in an indoor environment.

Acknowledgments. This work was financially supported by National Natural Science Foundation of China with Grants No. 61370216 and No. 61100191, and Shenzhen Strategic Emerging Industries Program with Grants No. ZDSY20120613125016389, No. JCYJ20120613151201451 and No. JCYJ20130329153215152.

References

1. Bahl, P., Padmanabhan, V.: Radar: an in-building RF-based user location and tracking system. In: Proceedings of the IEEE Nineteenth Annual Joint Conference of the IEEE Computer and Communications Societies, INFOCOM 2000, vol. 2, pp. 775–784 (2000)
2. Bauer, K., McCoy, D., Grunwald, D., Sicker, D.C.: CRAWDAD data set cu/rssi (v. 2009–05-28), May 2009. http://crawdad.org/cu/rssi/
3. Chintalapudi, K.K., Iyer, A.P., Padmanabhan, V.: Indoor localization without the pain. In: Mobicom. Association for Computing Machinery Inc., September 2010
4. Liu, H., Gan, Y., Yang, J., Sidhom, S., Wang, Y., Chen, Y., Ye, F.: Push the limit of wifi based localization for smartphones. In: Proceedings of the 18th Annual International Conference on Mobile Computing and Networking, Mobicom 2012, pp. 305–316. ACM, New York (2012)

5. Ni, L., Liu, Y., Lau, Y.C., Patil, A.: Landmarc: indoor location sensing using active RFID. In: Proceedings of the First IEEE International Conference on Pervasive Computing and Communications (PerCom 2003), pp. 407–415, March 2003
6. Peng, C., Shen, G., Han, Z., Zhang, Y., Li, Y., Tan, K.: Demo abstract: a beepbeep ranging system on mobile phones, November 2007
7. Priyantha, N.B., Chakraborty, A., Balakrishnan, H.: The cricket location-support system. In: Proceedings of the 6th Annual International Conference on Mobile Computing and Networking, MobiCom 2000, pp. 32–43. ACM, New York (2000)
8. Rehim, Y.A.: HORUS: A WLAN-based Indoor Location Determination System-date. Chapter 4, pp. 44–46 (2004)
9. Roos, T., Myllymäki, P., Tirri, H., Misikangas, P., Sievänen, J.: A probabilistic approach to wlan user location estimation. Int. J. Wirel. Inf. Netw. **9**(3), 155–164 (2002)
10. Swangmuang, N., Krishnamurthy, P.: An effective location fingerprint model for wireless indoor localization. Pervasive Mob. Comput. **4**(6), 836–850 (2008). PerCom 2008
11. Want, R., Hopper, A., Falcao, V., Gibbons, J.: The active badge location system. ACM Trans. Inf. Syst. **10**(1), 91–102 (1992)
12. Ward, A., Jones, A., Hopper, A.: A new location technique for the active office. IEEE Pers. Commun. **4**(5), 42–47 (1997)
13. Youssef, M., Agrawala, A.: The horus WLAN location determination system. In: Proceedings of the 3rd International Conference on Mobile Systems, Applications, and Services, MobiSys 2005, pp. 205–218. ACM, New York (2005)
14. Jack Graver, B.S., Servatius, H.: Combinatorial Rigidity. American Mathematical Society (1993)
15. Jackson, B., Jordán, T.: Connected rigidity matroids and unique realizations of graphs. J. Combin. Theor. Ser. B **94**(1), 1–29 (2005)
16. Wang, X., Qiu, J., Ye, S., Dai, G.: An advanced fingerprint-based indoor localization scheme for WSNs. In: 2014 IEEE 9th Conference on Industrial Electronics and Applications (ICIEA), pp. 2164–2169, 9–11 June 2014. doi:10.1109/ICIEA.2014.6931530
17. Cheng, J., Ye, Q., Du, H., Liu, C.: DISCO: a distributed localization scheme for mobile networks. In: 2015 IEEE 35th International Conference on Distributed Computing Systems (ICDCS), pp. 527–536 (2015). doi:10.1109/ICDCS.2015.60
18. Ye, Q., Cheng, J., Du, H., Jia, X., Zhang, J.: A matrix-completion approach to mobile network localization. In: Proceedings of the 15th ACM International Symposium on Mobile Ad Hoc Networking and Computing, pp. 327–336. ACM (2014)
19. Nandakumar, R., Chintalapudi, K.K., Padmanabhan, V.: Centaur: Locating devices in an office environment. In: Mobicom (August 2012)

On Clustering Without Replication in Combinatorial Circuits

Zola Donovan[1(✉)], Vahan Mkrtchyan[2], and K. Subramani[2]

[1] Department of Mathematics, West Virginia University, Morgantown, WV, USA
zdonovan@mix.wvu.edu
[2] LDCSEE, West Virginia University, Morgantown, WV, USA
vahanmkrtchyan2002@ysu.am, ksmani@csee.wvu.edu

Abstract. In this paper, we consider the problem of clustering combinatorial circuits for delay minimization, when logic replication is not allowed (**CN**). The problem of delay minimization when logic replication is allowed (**CA**) has been well studied, and is known to be solvable in polynomial-time [8]. However, unbounded logic replication can be quite expensive. Thus, **CN** is an important problem. We show that selected variants of **CN** are **NP-hard**. We also obtain approximability and inapproximability results for these problems.

Keywords: Clustering without replication · Computational complexity · **NP-completeness** · Approximation · Inapproximability

1 Introduction

In this paper, we consider the problem of clustering combinatorial circuits for delay minimization when logic replication is not allowed (**CN**). Combinatorial circuits implement Boolean functions, and produce a unique output for every combination of input signals [14]. The gates and their interconnections in the circuit represent implementations of one or more Boolean function(s). The Boolean functions are realized by the assignment of the gates to chips.

Due to manufacturing process and capacity constraints, it is generally not possible to place all of the circuit elements in one chip. Consequently, the circuit must be partitioned into clusters, where each cluster represents a chip in the overall circuit design. The circuit elements are assigned to clusters, while satisfying certain design constraints (e.g., area capacity) [8].

Gates and their interconnections usually have delays. The delays of the interconnections are determined by the way the circuit is clustered. Intra-cluster delays are associated with the interconnections between gates in the same cluster. Inter-cluster delays are associated with the interconnections between gates

V. Mkrtchyan and K. Subramani—The author is supported, in part, by the Air Force of Scientific Research through Award FA9550-12-1-0199.
K. Subramani—The author is supported by the National Science Foundation through Award CCF-1305054.

Z. Lu et al. (Eds.): COCOA 2015, LNCS 9486, pp. 334–347, 2015.
DOI: 10.1007/978-3-319-26626-8_25

in different clusters. The delay along a path from an input to an output is the sum of the delays of the gates and interconnections on the respective paths. The delay of the overall circuit, with respect to its clustering, is the maximum delay among all paths that connect an input to any output in the circuit.

The problem of clustering combinatorial circuits for delay minimization, when logic replication is allowed (**CA**), is well studied. It arises frequently in VLSI design. In **CA**, the goal is to find a clustering of a circuit for which the delay of the overall circuit is minimized. **CA** has been shown to be solvable in polynomial-time [8]. However, unbounded replication can be quite expensive. As systems become increasingly more complex, the need for clustering without logic replication is crucial. It follows that **CN** is an important problem in VLSI design.

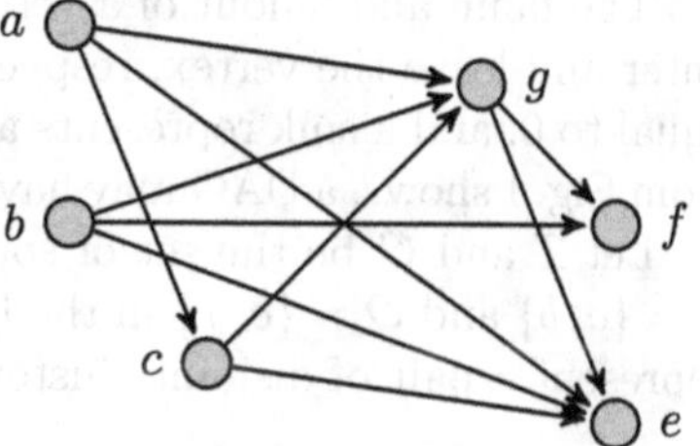

Fig. 1. A DAG representing a combinatorial network with two sources and two sinks.

In this paper, we consider several variants of **CN**. We prove **NP-hardness** results for these variants. We design an approximation algorithm for one of them. We also obtain an inapproximability result.

The rest of this paper is organized as follows: We give the necessary graph-theoretic definitions in Sect. 2. The problem is formally described in Sect. 3. We then examine related work in Sect. 4. In Sect. 5, we give some hardness results for the clustering problem. We also show that one of the hardness results implies that the problem is inapproximable. In Sect. 6, we propose an approximation algorithm for solving the clustering problem, when the gates are unweighted and the cluster capacity, M, is 2. We conclude the paper with Sect. 7 by summarizing our main results, and identifying avenues for future work.

2 Graph Preliminaries

In this section, we define the main graph-theoretic concepts that are used in the paper.

Graphs considered in this paper do not contain loops or parallel edges. The degree of a vertex v of a graph G is the number of edges of G incident with v.

A path of G is a sequence $P = v_0, e_1, v_1, ..., e_l, v_l$, where $v_0, v_1, ..., v_l$ are vertices of G, $e_1, ..., e_l$ are edges of G, and $e_j = (v_{j-1}, v_j)$, $1 \leq j \leq l$. l is called the length of the path P, and sometimes we say that P is an l-path of G. The edge $e_{\lceil \frac{l}{2} \rceil}$ is called a central edge of P. G is connected if any two vertices of G are joined by a path of G. P is said to be a cycle, if $v_0 = v_l$.

A directed path of a directed graph D is a sequence $Q = v_0, e_1, v_1, ..., e_l, v_l$, where $v_0, v_1, ..., v_l$ are vertices of D, $e_1, ..., e_l$ are edges of D, and $e_j = (v_{j-1}, v_j)$, $1 \leq j \leq l$. l is called the length of the path Q, and sometimes we say that Q is a directed l-path of D. If $v_0 = v_l$, then Q is called a directed cycle. D is said to be a directed acyclic graph (DAG), if it contains no directed cycles.

A cluster is defined as a subset of the vertices of a graph. If C is a cluster in a graph, then an edge is said to be a cut-edge if it connects a vertex of C to a vertex from $V \setminus C$. The degree of C is the number of cut-edges incident with a vertex in C.

The fanin and fanout of a vertex are the number of arcs belonging to E that enter and leave the vertex, respectively. A source represents a vertex with fanin equal to 0, and a sink represents a vertex with fanout equal to 0. As the example from Fig. 1 shows, a DAG may have more than one source and more than one sink.

Let $\mathcal{I}$ and $\mathcal{O}$ be the set of sources and sinks of G, respectively. Notice that $\mathcal{I} = \{a, b\}$ and $\mathcal{O} = \{e, f\}$ in the DAG in Fig. 1; $C_1 = \{a, c, g\}$ and $C_2 = \{b, e, f\}$ represent a pair of disjoint clusters.

3 Statement of Problems

In this section, we formally describe the problem studied in this paper. We start with formulating the problem using the language of combinatorial circuits. Then, we represent such circuits as directed acyclic graphs and formulate the main problem using graph-theoretic terminology.

In general, each gate in a circuit has an associated delay [7]. In the model that we consider in this paper, each interconnection has one of the following types of delays: (1) an intra-cluster delay, d, when there is an interconnection between two gates in the same cluster, or (2) an inter-cluster delay, D, when there is an interconnection between two gates in different clusters.

Note that $D >> d$, so inter-cluster delays typically dominate in all delay calculations.

The delay along a path from an input to an output is the sum of the delays of the gates and interconnections that lie on the path. The delay of the overall circuit is the maximum delay among all source to sink paths in the circuit.

Technology and design paradigms impose a number of constraints on the clustering of a circuit. So, a clustering is feasible if all clusters obey the imposed constraints. Constraints can be either monotone or non-monotone. In [5], the following definition is given:

Definition 1. *A constraint is said to be monotone if and only if any connected subset of gates in a feasible cluster is also feasible.*

Typical constraints include: capacity (a monotone constraint), which is a fixed constant M, denoting an upper-bound on the number of gates allowed in a cluster; and, pin-limitations (a non-monotone constraint), which is a fixed constant P, denoting an upper-bound on the total degree of a cluster.

In **CN**, a clustering partitions the circuits into disjoint subsets.

A clustering algorithm tries to achieve one or both of the following goals, subject to one or more constraints:

(1) Delay minimization through the circuit [1,5,7,8].
(2) Minimize the total number of cut-edges [2–4,6,9,10].

In this paper, we study **CN** under the delay model described as follows:

1. Associated with every gate v of the circuit, there is a delay $\delta(v)$ and a size $w(v)$.
2. The delay of an interconnection between two gates within a single cluster is d.
3. The delay of an interconnection between two gates in different clusters is D, where $D >> d$.

The size of a cluster is the sum of the sizes of the gates in the cluster.

The precise formulation of the problem is as follows:

CN: Given a combinatorial circuit, with each gate having a size and a delay, intra- and inter-cluster delays d and D, respectively, and a positive integer M called cluster capacity, the goal is to partition the circuit into clusters such that

1. The size of each cluster is bounded by M,
2. The delay of the circuit is minimized.

A combinatorial circuit can be represented as a directed graph $G = (V, E)$, with vertex-set V and edge-set E, such that G has no directed cycles. In G, each vertex $v \in V$ represents a gate, and each edge $(u, v) \in E$ represents an interconnection between gates u and v.

Given a clustering of the combinatorial circuit, the delays on the interconnections between gates induce an edge-length function $l : E(G) \to \{d, D\}$ of G. The weight of a cluster is the sum of the weights of the vertices in the cluster.

In the rest of the paper, we focus on a graph-theoretic formulation of **CN**. We employ the following notations and concepts: The length of a path P in G is calculated as the sum of all delays of vertices and edge-lengths of edges of P. X can be either W, which means that the vertices are weighted, or N, which means that the vertices are unweighted. M is the cluster capacity. Δ is the maximum number of arcs entering or leaving any vertex of the DAG.

CN is formulated (graph-theoretically) as follows:

CN$\langle X, M, \Delta \rangle$: Given a DAG $G = (V, E)$, with vertex-weight function $w : V \to \mathbb{N}$, delay function $\delta : V \to \mathbb{N}$, constants d and D, and a cluster capacity M, the goal is to partition V into clusters such that

1. The weight of each cluster is bounded by M,
2. The maximum length of any path from a source to a sink of G is minimized.

A clustering of G, such that the weight of each cluster is bounded by M, is called feasible. Given a feasible clustering of G, one can consider the corresponding edge-length function $l : E(G) \to \{d, D\}$ of G. A maximum length path (with respect to l) from a source to a sink of G is called an optimal path. A clustering of G is optimal, if the length of an optimal path is the smallest. An optimal path with respect to an optimal clustering is called a critical path.

In Fig. 2, we consider a simple example of a clustering of a combinatorial circuit represented by a DAG, where logic replication is not allowed. In this example, the weights and delays of all vertices are equal to 1 (i.e., $\delta(v) = 1$ and $w(v) = 1$ for all vertices v in the DAG); the upper bound for the weight of the cluster is $M = 2$; the intra-cluster delay is $d = 1$; and, the inter-cluster delay is $D = 2$. It can be easily seen that the partition $\Sigma = \{\{s, a\}, \{b, e\}, \{c, t\}\}$ forms a feasible clustering such that the length of the optimal path is 9. Moreover, it can be checked that this clustering is optimal.

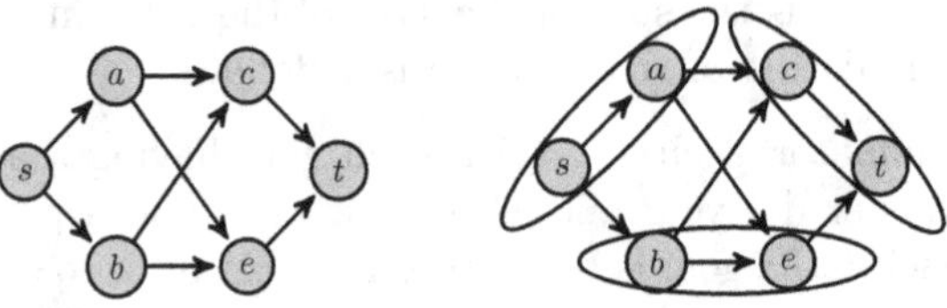

Fig. 2. An example of a DAG and its clustering.

In this paper, we focus on a restriction of $\mathbf{CN}\langle X, M, \Delta\rangle$, when $\delta(v) = 0$ for any vertex v of G.

The main contributions of this paper are as follows:

1. We show that $\mathbf{CN}\langle W, M, \Delta\rangle$ and several of its variants are **NP-hard** (Sect. 5).
2. For $\mathbf{CN}\langle N, 2, \Delta\rangle$, we present a 2-approximation algorithm (Sect. 6).
3. We prove that $\mathbf{CN}\langle W, M, \Delta\rangle$ does not admit a $(2 - \varepsilon)$-approximation algorithm for every $\varepsilon > 0$, unless $\mathbf{P} = \mathbf{NP}$ (Sect. 5).

4 Related Work

In this section, we describe some related work in the literature.

In [5], the authors present an exact polynomial-time algorithm for **CA**. The problem is solved under the so-called unit delay model [5].

A more general delay model is presented in [7]. The problem of disjoint clustering for minimum delay under the area or pin constraint is shown to be intractable in [7]. To minimize the delay, the authors propose an algorithm which constructs a clustering. This algorithm achieves the optimal delay under specific conditions.

In [8], **CA** is considered under the more general delay model proposed in [7]. However, [8] presents a different polynomial-time algorithm. Their heuristic is shown to always find an optimal clustering under any monotone clustering constraint.

Similar to [7], the problem of disjoint clustering for minimum delay under the area or pin constraint is also shown in [11] to be intractable. However, an improved heuristic is proposed in [11]. The authors also share comparative experimental results which show that a decrease in clusters generally leads to an increase in maximum delay.

In [10], the authors propose an efficient network-flow based algorithm which determines an optimal partitioning of the circuit. Using the least amount of replication, the optimal partitioning separates the nodes of the circuit into two subsets with the smallest cut size. The algorithm presented in [10] is also applicable to size-constrained partitioning.

[12,13] explore the advantage of evolutionary algorithms aimed at reducing the delay and area in partitioning and floorplanning. In turn, this would reduce the wirelength. A hybrid of the evolutionary algorithms are used to find optimal solutions to VLSI physical design problems.

5 Computational Complexity of CN

In this section, we obtain the main results that deal with the computational complexity of **CN**. We prove two theorems that establish the **NP-completeness** of some variants of **CN**. One of our reductions implies that **CN** is inapproximable within a certain factor.

In order to formulate the results, we consider **CNWD**, which is formulated as follows:

CNWD: Given a DAG $G = (V, E)$, with vertex-weight function $w : V \to \mathbb{N}$, delay function $\delta : V \to \mathbb{N}$, constants d and D, cluster capacity M and a positive integer k, partition V into clusters such that

1. The weight of each cluster is bounded by M,
2. The length of an optimal path of G is at most k.

It is not hard to see that **CNWD** is the decision version of $\mathbf{CN}\langle W, M, \Delta\rangle$. We make this correspondence explicit by writing **CNWD** as $\mathbf{CNWD}\langle W, M, \Delta\rangle$. We use the same notation for restrictions of $\mathbf{CN}\langle W, M, \Delta\rangle$.

Note that $\mathbf{CNWD}\langle W, M, \Delta\rangle$ is in **NP**. This follows from the well-known fact that a maximum weighted path in an edge-weighted DAG can be found in polynomial time.

If A is a subset of positive integers, then we denote by $\mathbf{CNWD}\langle A, M, \Delta\rangle$, the restriction of $\mathbf{CNWD}\langle W, M, \Delta\rangle$, when the weights of verticies of the input DAG are from A.

We recall the partition problem:

Partition: Given a set $S = \{a_1, a_2, \ldots, a_n\}$, the goal is to check whether there is a set $S_1 \subset S$, such that $\sum_{x \in S_1} x = \sum_{x \in S - S_1} x$.

Without loss of generality, we assume that $B = \sum_{i \in S} a_i$ is even, otherwise the problem is trivial.

The following theorem establishes the **NP-completeness** of **CNWD**$\langle W, M, \Delta\rangle$. Clearly, this means that **CN**$\langle W, M, \Delta\rangle$ is **NP-hard**.

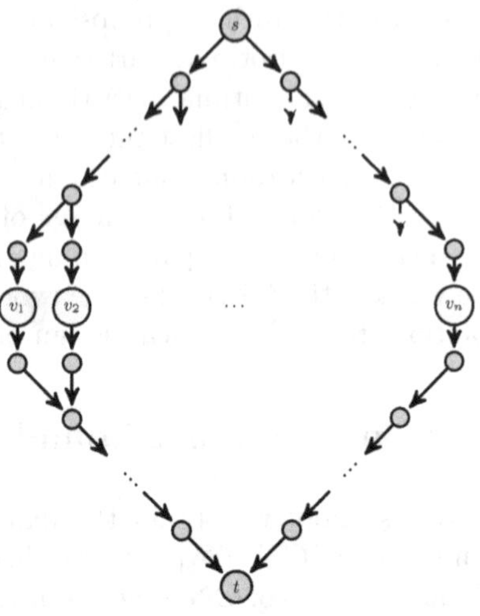

Fig. 3. Reduction from the partition problem to **CNWD**$\langle W, M, 3\rangle$.

Theorem 1. CNWD$\langle W, M, 3\rangle$ *is* **NP-complete.**

Proof. In order to prove that **CNWD**$\langle W, M, 3\rangle$ is **NP-complete**, we present a reduction from **Partition**.

We construct a new instance I' of **CNWD**$\langle W, M, 3\rangle$ as shown in Fig. 3.

For each $i \in \{1, \ldots, n\}$, there is a path connecting the source s to the sink t, through a vertex v_i. Let V denote the set of all v_i vertices. Let S denote the set of all vertices that are predecessors to the vertices in V, and let T denote the set of all vertices that are successors to the vertices in V. Since $|S| = |T|$, let m denote the size of S and T. No pair of vertices in V are connected. Each vertex $v_i \in V$ has a weight of a_i. Every vertex in S and T has weight 1. So, the sum of the weights of all vertices in S is equal to m, and the sum of the weights of all vertices in T is equal to m. We set $D = 1$ and $d = 0$. Every vertex is given a delay of 0. The cluster capacity M is set to $(\frac{B}{2} + m)$, and we take $k = 1$. The description of I' is complete.

Observe that I' can be constructed from I in polynomial time. In order to complete the proof of the theorem, we show that I is a "yes" instance of **Partition**, if and only if I' is a "yes" instance of **CNWD**$\langle W, M, 3\rangle$.

Assume that I is a "yes" instance of **Partition**. This means that there exists a partition of A into A_1 and A_2, such that $\sum_{x \in A_1} x = \sum_{x \in A_2} x = \frac{B}{2}$. Group the vertices corresponding to the elements in A_1 with S, and the remaining vertices with T. Observe that the cluster capacity constraint is met. Moreover, the length of the optimal path from a source to a sink is 1. This means that I' is a "yes" instance of **CNWD**$\langle W, M, 3\rangle$.

Conversely, assume that I' is a "yes" instance of **CNWD**$\langle W, M, 3\rangle$. This means that there is a way of packing the vertices of the DAG in Fig. 3 into clusters, such that the cluster capacity is not exceeded, and the length of the optimal path from s to t is 1.

Observe that if there are two vertices of S which belong to different clusters, then the $s-t$ path(s) going though one of them will have length at least 2. This means that any two vertices of S must be in the same cluster. The same is true for T.

If there is a vertex $v_i \in V$ which is not packed with either S or T, then the $s-t$ path going through that vertex will have length at least 2. Therefore, V cannot be partitioned into more than two sets.

Let V_S and V_T denote the subset of vertices $v_i \in V$ that are packed with S and T, respectively. Observe that $V_S \cup V_T = V$. Moreover, the length of the path from s to any vertex in V_S must be 0, and the length of the path from any vertex in V_T to t must also be 0.

Let $w(S)$ denote the sum of the weights of all vertices in S, and $w(T)$ denote the sum of the weights of all vertices in T. Notice that $w(S) = w(T) = m$. Let $w(V_S)$ and $w(V_T)$ denote the sum of the weights of all vertices in V_S and V_T, respectively.

Notice that,

$$w(S) + w(V_S) + w(T) + w(V_T) = B + 2 \cdot m.$$

Since

$$w(S) + w(V_S) \leq \left(\frac{B}{2} + m\right) \text{ and}$$

$$w(T) + w(V_T) \leq \left(\frac{B}{2} + m\right),$$

then

$$w(V_S) \leq \frac{B}{2} \text{ and } w(V_T) \leq \frac{B}{2}.$$

This implies that

$$w(V_S) = \frac{B}{2} \text{ and } w(V_T) = \frac{B}{2}.$$

Thus, we have obtained the desired partition of A. Hence, I is a "yes" instance of **Partition**.

The proof of the theorem is complete. ∎

The proof of Theorem 1 implies an inapproximability result for $\mathbf{CN}\langle W, M, 3\rangle$.

Corollary 1. $\mathbf{CN}\langle W, M, 3\rangle$ *does not admit a* $(2 - \varepsilon)$*-approximation algorithm for each* $\varepsilon > 0$, *unless* $\mathbf{P} = \mathbf{NP}$.

Proof. Consider the reduction from **Partition** described in the proof of Theorem 1. Observe that in any approximate solution of the clustering problem, there must exist at least one vertex which is not packed with either s or t. This means that s and t do not belong to the same cluster. Hence, the approximation algorithm can be used to solve the partition problem exactly.

The proof of the corollary is complete. ∎

In the proof of the following theorem, we use a **3SAT** reduction modeled after the one presented in [11]. For that purpose, we recall the definition of **3SAT**:

3SAT: Given a 3-CNF formula ϕ with n variables $x_1, \ldots, x_n$ and m clauses $C_1, \ldots, C_m$, the goal is to check whether ϕ has a satisfying assignment.

Without loss of generality, for all $i \in \{1, \ldots, n\}$ we assume that each variable x_i in ϕ appears at most 3 times and each literal at most twice. (Any **3SAT** instance can be transformed to satisfy these properties in polynomial time [15].)

Theorem 2. **CNWD**$\langle\{1,2,3\},3,3\rangle$ *is* **NP-complete.**

Proof. In order to prove the completeness of **CNWD**$\langle\{1,2,3\},3,3\rangle$ for **NP**, we present a reduction from **3SAT**.

Let each variable x_i $(1 \leq i \leq n)$, be represented by a variable gadget as shown in Fig. 4(a). Let each clause C_j $(1 \leq j \leq m)$, be represented by a clause gadget as shown in Fig. 4(b). If a variable x_i or its complement $\bar{x}_i$ is the 1st, 2nd, or 3rd literal of a clause C_j, then the corresponding vertex labeled x_i (or $\bar{x}_i$) is connected to a sink labeled C_j through a pair of vertices labeled y_{j1} and z_{j1}, y_{j2} and z_{j2}, or y_{j3} and z_{j2}, respectively.

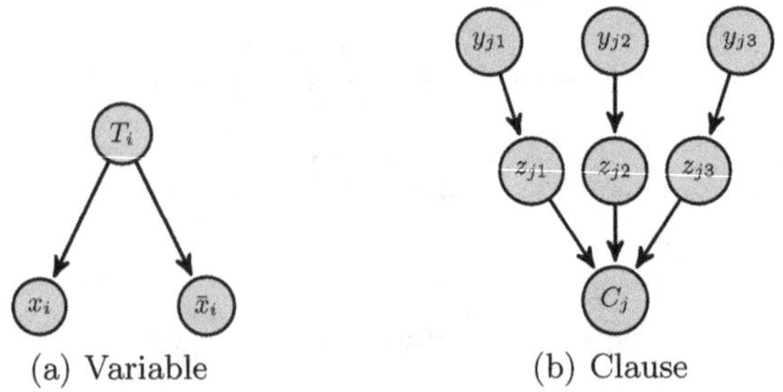

Fig. 4. Gadgets used to represent variables and clauses.

We now construct an instance I' of **CNWD**$\langle\{1,2,3\},3,3\rangle$ as shown in Fig. 5.

The resulting DAG G represents a combinatorial circuit. Let V denote the set of all vertices labeled x_i or $\bar{x}_i$ $(1 \leq i \leq n)$. There are n sources T_i $(1 \leq i \leq n)$ connected to m sinks C_j $(1 \leq j \leq m)$ through some vertices in V and $3m$ pairs of vertices labeled y_{jp} and z_{jp} $(1 \leq j \leq m, 1 \leq p \leq 3)$. Each y_{jp} is connected to exactly one vertex gadget, and for fixed j, no two vertices in $\{y_{j1}, y_{j2}, y_{j3}\}$ are adjacent to the vertices x_i and $\bar{x}_i$ belonging to the same vertex gadget. In other words, both x_i and $\bar{x}_i$ cannot both be connected to the same clause gadget. Every T_i, z_{jp}, and C_j has a weight of 1, every $x_i, \bar{x}_i \in V$ has a weight of 2, and every y_{jp} has a weight of 3. We set $D = 1$ and $d = 0$. All vertices are given a delay of 0. The cluster capacity M is set to 3, and we take $k = 3$. The description of I' is complete.

Observe that I' can be constructed from I in polynomial time. In order to complete the proof of the theorem, we show that I is a "yes" instance of **3SAT**, if and only if I' is a "yes" instance of **CNWD**$\langle\{1,2,3\},3,3\rangle$.

Suppose that I is a "yes" instance of **3SAT**. This means that there exists an assignment of ϕ such that every clause has at least one true literal. If a literal is set to **true**, then the corresponding vertex x_i (or $\bar{x}_i$) should be clustered with T_i, but if it is set to **false**, then the corresponding vertex is clustered alone. Notice that every y_{jp} must be clustered alone. Since each clause C_j has at least one true literal, the vertex z_{jp} corresponding to that literal should be clustered alone. This means that the source to sink path going through vertices y_{jp} and

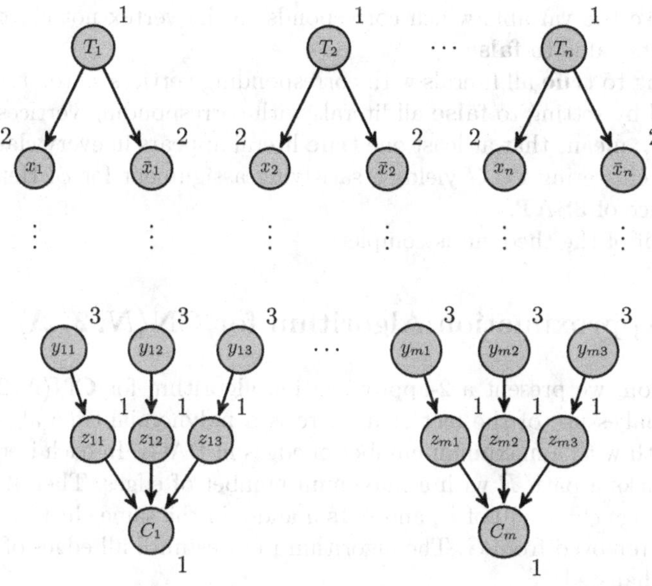

Fig. 5. Reduction from the 3SAT problem to **CNWD**$\langle\{1, 2, 3\}, 3, 3\rangle$.

z_{jp} corresponding to **true** literals have length 3. If either of the other two z_{jp} vertices belonging to the respective clause gadget corresponds to literals which are set to **false**, they should be clustered with C_j. Otherwise, they may also be clustered alone. Note that clustering two z_{jp} vertices with C_j, even if they both correspond to **true** literals, leads to paths of length $2 < 3 = k$. Observe that the cluster capacity constraint is met, and the length of the optimal path from any source T_i to any sink C_j is 3. This means that I' is a "yes" instance of **CNWD**$\langle\{1, 2, 3\}, 3, 3\rangle$.

Conversely, suppose that I' is a "yes" instance of **CNWD**$\langle\{1, 2, 3\}, 3, 3\rangle$. This means that there is a way of packing the vertices in G into clusters of capacity $M = 3$, such that the length of the optimal path from source to sink is 3.

Since $M = 3$, again notice that every y_{jp} must be clustered alone. Each vertex C_j may be clustered with at most 2 of the z_{jp} vertices. So, at least one z_{jp} is clustered alone. However, notice that any source to sink path with a vertex z_{jp} clustered alone, has length at least 3. In order to satisfy the optimal path constraint, each T_i must be clustered with the vertex x_i (or $\bar{x}_i$) which corresponds to a vertex z_{jp} clustered alone. Otherwise, the length of the path would be $4 > k = 3$. To avoid exceeding the cluster capacity, either x_i or $\bar{x}_i$ (but not both) may be clustered with T_i. Finally, notice that z_{jp} vertices along paths where their corresponding literals are considered **false**, must be clustered with their respective sinks C_j. Otherwise, the length of such a path with all internal vertices belonging to single vertex clusters would have length $4 > 3 = k$. Take the variable which corresponds to the vertex clustered with T_i, and set its value to **true**. Take the variable which corresponds to the vertex not clustered with T_i, and set its value to **false**.

By setting to **true** all literals with corresponding vertices x_i (or $\bar{x}_i$) clustered with T_i, and by setting to **false** all literals with corresponding vertices not clustered with T_i, means that at least one **true** literal appears in every clause. Thus, a satisfying clustering for G yields a satisfying assignment for ϕ. Hence, I is a "yes" instance of **3SAT**.

The proof of the theorem is complete. ■

6 A 2-Approximation Algorithm for CN$\langle N, 2, \Delta\rangle$

In this section, we present a 2-approximation algorithm for **CN**$\langle N, 2, \Delta\rangle$. Our algorithm makes use of the fact that there is a polynomial-time algorithm for finding a path with a maximum number of edges in DAGs. In each iteration, the algorithm picks a path P with a maximum number of edges. Then it considers the central edge $e = (u, v)$ of P, and puts u and v in the same cluster. After that u and v are removed from G. The algorithm iterates until all edges of the input DAG are exhausted.

Theorem 3. *Algorithm 1 is a 2-approximation algorithm for* **CN**$\langle N, 2, \Delta\rangle$.

Algorithm 1. A 2-approximation algorithm for the clustering problem.

```
Input: a DAG G;
Output: a clustering of vertices of G;
Take a longest path P in the DAG G;
Declare the central edge e_⌈l/2⌉ of P as a d-edge, where l denotes the length of P,
  and e_1, ··· , e_l are the edges of P.
The edges adjacent to e_⌈l/2⌉ should be declared as D-edges.
Remove edge e_⌈l/2⌉ together with its adjacent edges.
Continue this process until all edges of G are exhausted.
```

Proof. For a path P, let $l(P)$ be the length of P (i.e., the number of edges of P). Moreover, let

$$l = \max_{P} l(P).$$

So, l denotes the length of a longest path of G.

The following shows a lower bound for OPT, where OPT is the delay of the optimal clustering of G when $M = 2$.

$$OPT \geq \lceil \frac{l(P)}{2} \rceil \cdot d + \lfloor \frac{l(P)}{2} \rfloor \cdot D.$$

Since P represents any path, then the above inequality must also be true for the longest path. Thus,

$$OPT \geq \lceil \frac{l}{2} \rceil \cdot d + \lfloor \frac{l}{2} \rfloor \cdot D.$$

Now, let us estimate ALG, where ALG is the delay of the clustering found by the algorithm. Let Q be an optimal path with respect to this clustering. We consider three cases.

Case 1: When $l(Q) < l$, then the following chain of inequalities holds:

$$\begin{aligned} ALG &\leq (l-1) \cdot D \\ &\leq 2 \cdot (\lceil \frac{l}{2} \rceil \cdot d + \lfloor \frac{l}{2} \rfloor \cdot D) \\ &\leq 2 \cdot OPT. \end{aligned}$$

Case 2: When $l(Q) = l$ and Q contains at least one d-edge, then the following chain of inequalities holds:

$$\begin{aligned} ALG &\leq d + (l-1) \cdot D \\ &\leq 2 \cdot (\lceil \frac{l}{2} \rceil \cdot d + \lfloor \frac{l}{2} \rfloor \cdot D) \\ &\leq 2 \cdot OPT. \end{aligned}$$

Case 3: When $l(Q) = l$ and all edges of Q are D-edges, we consider two sub-cases of this case.

Sub-case (3a): When l is even, then the following chain of inequalities holds:

$$\begin{aligned} ALG &= l \cdot D \\ &\leq 2 \cdot (\lceil \frac{l}{2} \rceil \cdot d + \lfloor \frac{l}{2} \rfloor \cdot D) \\ &\leq 2 \cdot OPT. \end{aligned}$$

Sub-case (3b): When l is odd, assume that $l = 2 \cdot k + 1$. In this case, we show that:

$$OPT \geq \lfloor \frac{l}{2} \rfloor \cdot d + \lceil \frac{l}{2} \rceil \cdot D.$$

Clearly, we must rule out the case when

$$\begin{aligned} OPT &= \lceil \frac{l}{2} \rceil \cdot d + \lfloor \frac{l}{2} \rfloor \cdot D \\ &= (k+1) \cdot d + k \cdot D, \end{aligned}$$

which is equivalent to saying that in any optimal path of odd length l, the edges with odd numbered indices, are d-edges. In particular, this means that all paths of length l in G are vertex disjoint.

Since we are in case 3, all edges of Q are D-edges, specifically the central, $(k+1)$-edge $e = (u, v)$ of Q. Now, since this edge is also a D-edge, it means that there is an edge e' adjacent to e, such that the algorithm has declared e' as a d-edge. Clearly, this means that e' is a central edge of some other path Q' of length l, contradicting our conclusion as stated above.

Finally, observe that the following chain of inequalities holds:

$$\begin{aligned} ALG &= l \cdot D \\ &\leq 2 \cdot (\lfloor \frac{l}{2} \rfloor \cdot d + \lceil \frac{l}{2} \rceil \cdot D) \\ &\leq 2 \cdot OPT. \end{aligned}$$

The proof of the theorem is complete. ■

Figure 6 shows an example of a DAG for which the algorithm achieves an approximation factor of 2.

Fig. 6. A DAG which obtains a factor 2 approximation.

Observe that in this example, $OPT = 2 \cdot d + D$. However, the algorithm picks up the central edge first, then the delay of the resulting clustering would be $ALG = 2 \cdot D + d$, which shows that the bound of at most $2 \cdot OPT$ cannot be improved for this algorithm. Finally observe that one can construct an infinite sequence of DAGs, such that the algorithm achieves a bound of at most $2 \cdot OPT$ (simply take vertex disjoint copies of 3-paths).

7 Conclusion

In this paper, we studied the problem of clustering combinatorial networks for delay minimization when logic replication is not allowed (**CN**$\langle X, M, \Delta \rangle$). We showed that several versions of **CN**$\langle W, M, \Delta \rangle$ are **NP-hard**. The strategy developed for the proofs allowed us to prove that the problem does not admit a $(2-\varepsilon)$-approximation algorithm for any $\varepsilon > 0$, unless **P** = **NP**. On the positive side, there exists a 2-approximation algorithm for **CN**$\langle N, 2, \Delta \rangle$.

We are interested in the following open problems:

1. Finding an approximation algorithm for **CN**$\langle N, 2, \Delta \rangle$ whose performance ratio is smaller than 2. There may exist a combinatorial approximation algorithm for **CN**$\langle N, 2, \Delta \rangle$ with smaller performance ratio. The following idea may be helpful in the design of such an algorithm. Take a longest path in the input DAG. Put the first two vertices in one cluster, the second two in another cluster, and so on. Remove all the vertices that are clustered with some other vertex. Iterate until all edges of the DAG are exhausted.
2. Finding inapproximabilility results for **CN**$\langle N, M, \Delta \rangle$. It might be the case that **CN**$\langle N, M, \Delta \rangle$ is **NP-hard**. Moreover, it may not admit an FPTAS.

References

1. Cong, J., Ding, Y.: FlowMap: an optimal technology mapping algorithm for delay optimization in lookup-table based FPGA designs. IEEE Trans. Comput. Aided Des. Integr. Circuits Syst. **13**(1), 1–12 (1994)
2. Barke, E., Behrens, D., Hebrich, K.: Heirarchical partitioning. In: Proceedings of the IEEE International Conference on CAD, pp. 470–477 (1997)
3. Hwang, L.J., Gamal, A.E.: Min-cut replication in partitioned networks. IEEE Trans. Comput. Aided Des. Integr. Circuits Syst. **14**, 96–106 (1995)
4. Cheng, C.-K., Liu, L.T., Kuo, M.-T.: A replication cut for two-way partitioning. IEEE Trans. Comput. Aided Des. Integr. Circuits Syst. **14**, 623–630 (1995)
5. Lawler, E.L., Levitt, K.N., Turner, J.: Module clustering to minimize delay in digital networks. IEEE Trans. Comput. **C–18**(1), 47–57 (1966)
6. Kuh, E.S., Shih, M.: Circuit partitioning under capacity and I/O constraints. In: IEEE Custom Integrated Circuits Conference, vol. 28 (1994)
7. Brayton, R.K., Murgai, R., Sangiovanni-Vincentelli, A.: On clustering for minimum delay/area. In: International Conference on Computer Aided Design, pp. 6–9, November 1991
8. Rajaraman, R., Wong, D.F.: Optimal clustering for delay minimization. In: Proceedings of the 30th ACM/IEEE Design Automation Conference, pp. 309–314. IEEE Computer Society, Washington, DC, June 1993
9. Deng, W., Dutt, S.: VLSI circuit partitioning by cluster-removal using iterative improvement techniques. In: Proceedings of the IEEE International Conference on CAD, pp. 194–200 (1996)
10. Wong, D.F., Mak, W.-K.: Minimum replication min-cut partitioning. In: Proceedings of the IEEE International Conference on CAD, pp. 205–210 (1996)
11. Kagaris, D.: On minimum delay clustering without replication. Integr. VLSI J. **36**, 27–39 (2003)
12. Shanavas, I.H., Gnanamurthy, R.K.: Wirelength minimization in partitioning and floorplanning using evolutionary algorithms. VLSI Design **2011**, Article ID 896241, 9 (2011). doi:10.1155/2011/896241
13. Shanavas, I.H., Gnanamurthy, R.K.: Optimal solution for VLSI physical design automation using hybrid genetic algorithm. Math. Probl. Eng. **2014**, Article ID 809642, 15 (2014). doi:10.1155/2014/809642
14. Koshy, T.: Boolean Algebra and Combinatorial Circuits, Discrete Mathematics with Applications, pp. 803–865. Elsevier Academic Press (2004)
15. Papadimitriou, C.M.: Computational Complexity. Addison-Wesley, Redaing (1994)

On Replica Placement in High-Availability Storage Under Correlated Failure

K. Alex Mills(✉), R. Chandrasekaran, and Neeraj Mittal

Department of Computer Science,
The University of Texas at Dallas, Richardson, TX, USA
{k.alex.mills,chandra,neerajm}@utdallas.edu

Abstract. A new model describing dependencies among system components as a directed graph is presented and used to solve a novel replica placement problem in data centers. A criterion for optimizing replica placements is formalized and explained. In this work, the optimization goal is to choose placements in which correlated failure events disable as few replicas as possible. A fast optimization algorithm is given for dependency models represented by trees. The main contribution of the paper is an $O(n + \rho \log \rho)$ dynamic programming algorithm for placing ρ replicas on a tree with n vertices.

1 Introduction

With the surge towards the cloud, our websites, services and data are increasingly being hosted by third-party data centers. These data centers are often contractually obligated to ensure that data is rarely, if ever unavailable. One cause of unavailability is co-occurring component failures, which can result in outages that can affect millions of websites, and can cost millions of dollars in profits. An extensive one-year study of availability in Google's cloud storage infrastructure showed that such failures are relatively harmful. Their study emphasizes that "correlation among node failure dwarfs all other contributions to unavailability in our production environment" [4].

We believe that the correlation found among failure events arises due to dependencies among system components. Much effort has been made in the literature to produce quality statistical models of this correlation. But in using such models researchers do not make use of the fact that these dependencies can be explicitly modeled, since they are known to the system designers. In contrast, we propose a model wherein such dependencies are included, and demonstrate how an algorithm may make use of this information to optimize placement of data replicas within a data center.

To achieve high availability, data centers typically store multiple replicas of data to tolerate the potential failure of system components. This gives rise to a *replica placement problem*, which, broadly speaking, involves determining which

This work was supported, in part, by the National Science Foundation (NSF) under grant number CNS-1115733.

Z. Lu et al. (Eds.): COCOA 2015, LNCS 9486, pp. 348–363, 2015.
DOI: 10.1007/978-3-319-26626-8_26

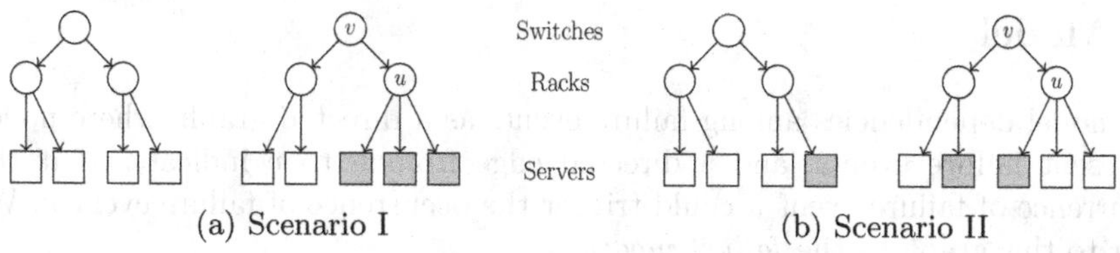

(a) Scenario I (b) Scenario II

Fig. 1. Two scenarios represented by directed trees.

subset of nodes in the system should store a copy of a given file so as to maximize a given objective function (e.g., reliability, communication cost, response time, or access time). While our focus is on replica placements, our model could also be used to place replicas of other entities that require high-availability, such as virtual machines or mission-critical tasks.

In this work, we present a new model for causal dependencies among failures, and a novel algorithm for optimal replica placement in our model. See Fig. 1 for an example model, in which three identical replicas of the same block of data are distributed on servers in a data center. As can be seen from in Fig. 1(a), a failure in the power supply unit (PSU) on a single rack could result in a situation where every replica of a data block is completely unavailable, whereas in Fig. 1(b) three PSUs would need to fail in order to achieve the same result. Best practices can avoid Scenario I by ensuring that each replica is housed on a separate rack. However, this simple heuristic can be suboptimal in some cases. For instance, failures that occur higher in the tree can impact the availability of every data replica stored on adjacent racks.

While many approaches for replica placement have been proposed, our approach of modeling causal dependencies among failure events appears to be new. Other work on reliability in storage area networks has focused on objectives such as mean time to data loss [1,3,4,8,10,12]. These exemplify an approach towards correlated failure which we term "measure-and-conquer". In measure-and-conquer approaches, a measured degree of correlation is given as a parameter to the model. In contrast, we model explicit causal relations among failure events which we believe give rise to the correlation seen in practice. More recently, Pezoa and Hayat [6] have presented a model in which spatially correlated failures are explicitly modeled. However, they consider the problem of task allocation, whereas we are focused on replica placement. In the databases community, work on replica placement primarily focuses on finding optimal placements in storage-area networks with regard to a particular distributed access model or mutual exclusion protocol [5,11,13]. In general, much of the work from this community focuses on system models and goals which are substantially different from our own. Recently, there has been a surge of interest in computer science concerning cascading failure in networks. While our model is conceptually related to this work, it does not appear to directly follow from any published model. Moreover, current work in this area is focused on fault-tolerant network design [2] modeling cascading failure [7], and developing techniques for adversarial analysis [14]. To our knowledge, no one has yet considered the problem of replica placement in such models.

2 Model

We model dependencies among failure events as a directed graph, where nodes represent failure events, and a directed edge from u to v indicates that the occurrence of failure event u could trigger the occurrence of failure event v. We refer to this graph as the *failure model*.

Given such a graph as input, we consider the problem of selecting nodes on which to store data replicas. Roughly, we define a *placement problem* as the problem of selecting a subset of these vertices, hereafter referred to as a *placement*, from the failure model so as to satisfy some safety criterion. In our application, only those vertices which represent storage servers are candidates to be part of a placement. We refer to such vertices as *placement candidates*. Note that the graph also contains vertices representing other types of failure events, which may correspond to real-world hardware unsuitable for storage (e.g., a ToR switch), or even to abstract events which have no associated physical component.

More formally, let E denote the set of failure events, and C denote the set of placement candidates. We are interested in finding a *placement* of size ρ, which is defined to be a set $P \subseteq C$, with $|P| = \rho$. Throughout this paper we will use P to denote a placement, and ρ to denote its size. We consistently use C to denote the set of placement candidates, and E to denote the set of failure events.

Let $G = (V, A)$ be a directed graph with vertices in V and edges in A. The vertices represent both events in E and candidates in C, so let $V = E \cup C$. A directed edge from event e_1 to event e_2 indicates that the occurrence of failure event e_1 can trigger the occurrence of failure event e_2. A directed edge from event e to candidate c indicates that the occurrence of event e could compromise candidate c. We will assume failure to act transitively (pessimistic model). That is, if a failure event occurs, all failure events reachable from it in G also occur.

Definition 1 (failure number). *Given an event $e \in E$ and a placement P, the failure number of e with respect to P, denoted $f(e, P)$ is defined as the number of candidates in P whose correct operation could be compromised by occurrence of event e. In particular,*

$$f(e, P) = |\{p \in P \mid p \text{ is reachable from } e \text{ in } G\}|.$$

As an example, node u in Fig. 1 has failure number 3 in Scenario I, and failure number 1 in Scenario II. The following property can be easily verified.

Property 1. For any placement P of replicas in tree T, if node u has descendant v, then $f(v, P) \leq f(u, P)$.

The failure number captures a conservative criterion for a safe placement. Intuitively, we consider the worst case scenario, in which every candidate which *could* fail due to an occurring event *does* fail. Our goal is to find a placement which does not induce large failure numbers in any event. To aggregate this idea across all events, we define the *failure aggregate*, a measure that accounts for the failure number of every event in the model.

Definition 2 (failure aggregate). *The* failure aggregate *of a placement P is a vector in $\mathbb{N}^{\rho+1}$, denoted $\boldsymbol{f}(P)$, where $\boldsymbol{f}(P) := \langle p_\rho, \ldots, p_1, p_0 \rangle$, and each $p_i := \left|\{e \in E \mid f(e, P) = i\}\right|$.*

Note that nodes in C are not counted as part of the failure aggregate. In Fig. 1, Scenario I has failure aggregate of $\langle 2, 0, 0, 4 \rangle$ and Scenario II has failure aggregate of $\langle 0, 1, 4, 1 \rangle$ in Scenario II.

In all of the problems considered in this paper, we are interested in optimizing $\boldsymbol{f}(P)$. When optimizing a vector quantity, we must choose a meaningful way to totally order the vectors. In the context of our problem, we find that ordering the vectors with regard to the *lexicographic order* is both meaningful and convenient. The lexicographic order $\leq_L$ between $\boldsymbol{f}(P) = \langle p_\rho, \ldots, p_1, p_0 \rangle$ and $\boldsymbol{f}(P') = \langle p'_\rho, \ldots, p'_1, p'_0 \rangle$ is defined via the following formula:

$$\boldsymbol{f}(P) \leq_L \boldsymbol{f}(P') \iff \exists\, m > 0,\ \forall\, i < m\left[p_i = p'_i \land p_m \leq p'_m\right].$$

To see why this is desirable, consider a placement P which lexico-minimizes $\boldsymbol{f}(P)$ among all possible placements. Such a placement is guaranteed to minimize p_ρ, i.e. the number of nodes which compromise *all* of the entities in our placement. Further, among all solutions minimizing p_ρ, P also minimizes $p_{\rho-1}$, the number of nodes compromising *all but one* of the entities in P, and so on for $p_{\rho-2}, p_{\rho-3}, \ldots, p_0$. Clearly, the lexicographic order nicely prioritizes minimizing the entries of the vector in an appealing manner.

Throughout the paper, any time a vector quantity is maximized or minimized, we are referring to the maximum or minimum value in the lexicographic order. We will also use $\boldsymbol{f}(P)$ to denote the failure aggregate, and p_i to refer to the ith component of $\boldsymbol{f}(P)$, where P can be inferred from context. Variables i, j are reserved to index sequences, and u, v, w are reserved to refer to vertices in a graph. Their intended meaning is context-dependent. We will also use the phrase "place a replica on node u" to mean, "set $P \leftarrow P \cup \{u\}$".

In the most general case, we could consider the following problem.

Problem 1 (Replica Placement). Given graph $G = (V, A)$ with $V = C \cup E$, and positive integer ρ with $\rho < |C|$, find a placement $P \subseteq C$ with $|P| = \rho$ such that $\boldsymbol{f}(P)$ is lexico-minimum.

The replica placement problem, as defined above, is NP-hard to solve, even in the case where G is a bipartite graph. In particular, we can reduce an instance of the independent set problem to an instance of the replica placement problem. The reduction can be found in the full paper [9]. The problem is, however, tractable for special classes of graphs, one of which is the case wherein the graph forms a directed, rooted tree with leaf set L and $C = L$. Our main contribution in this paper is a fast algorithm for solving the replica placement problem in such a case. We briefly mention a naïve greedy algorithm which solves the problem in $O(n^2\rho)$ time, where $n = |V|$. However, since $n \gg \rho$ in practice our result of an $O(n + \rho \log \rho)$ algorithm is much preferred.

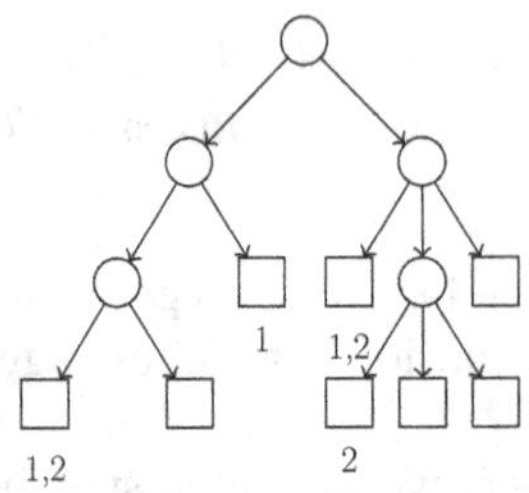

Fig. 2. Two round-robin placements with different objective function values.

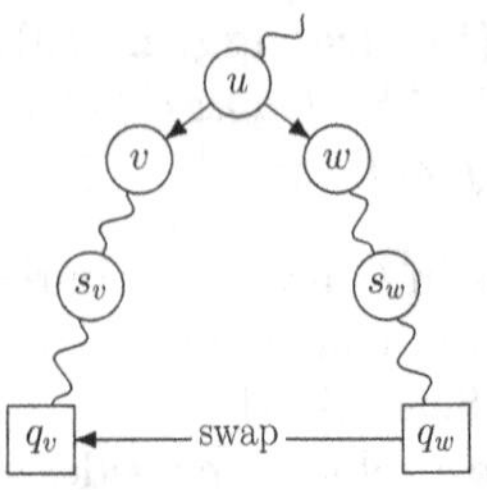

Fig. 3. Illustration of swap to reduce imbalance.

3 An $O(n^2\rho)$ Greedy Algorithm

The greedy solution to this problem forms a partial placement P', to which new replicas are added one at a time, until ρ replicas have been placed overall. P' starts out empty, and at each step, the leaf u which lexico-minimizes $\boldsymbol{f}(P' \cup \{u\})$ is added to P'. That this naïve greedy algorithm works correctly is not immediately obvious. It can be shown via an exchange argument that each partial placement found by the greedy algorithm is a subset of some *optimal* placement. The correctness proof can be found in the full version of the paper [9].

The running time of the above greedy algorithm is $O(n^2\rho)$ for a tree in general. Each iteration requires visiting $O(|L|)$ leaves for inclusion. For each leaf q which is checked, every node on a path from q to the root must have its failure number computed. Both the length of a leaf-root path and the number of leaves can be bounded by $O(n)$ in the worst case, yielding the result.

4 Balanced Placements

Consider a round-robin placement in which the set of replicas placed at each node is distributed among its children, one replica per child, until all replicas have been placed. This process is then continued recursively at the children. Throughout the process, no child is given more replicas than its subtree has leaf nodes. This method has intuitive appeal, but it does not compute an optimal placement exactly as can be seen from Fig. 2. Let placements P_1 and P_2 consist of the nodes labeled by 1 and 2 in Fig. 2 respectively. Note that both outcomes are round-robin placements. A quick computation reveals that $\boldsymbol{f}(P_1) = \langle 1, 1, 2, 1\rangle \neq \langle 1, 1, 3, 0\rangle = \boldsymbol{f}(P_2)$. Since the placements have different failure aggregates, round-robin placement alone cannot guarantee optimality.

For any node u, let ℓ_u refer to the number of leaves on node u. When a sequence of nodes is indexed, (i.e. $c_1, ..., c_i, ...$) we will use ℓ_i to refer to the number of leaves on the ith node. Similarly, r_u refers to the number of replicas in placement P which can be reached from node u, and r_i is defined analogously to ℓ_i.

Key to our algorithm is the observation that any placement which lexico-minimizes $\boldsymbol{f}(P)$ must be *balanced*. If we imagine each child c_i of u as a bin of capacity ℓ_i, balanced nodes are those in which all unfilled children are approximately "level", and no child is filled while children of smaller capacity remain unfilled. These ideas are formalized in the following definitions.

Definition 3. *Let node u have children* $c_1, ..., c_i, ..., c_k$. *A node for which* $\ell_i - r_i = 0$ *is said to be* filled. *A node for which* $\ell_i - r_i > 0$ *is said to be* unfilled.

Definition 4 (balanced placement). *Node u is said to be* balanced *in placement P iff:*

$$\ell_i - r_i > 0 \implies \forall j \in \{1, \ldots, k\} \; (r_i \geq r_j - 1).$$

Placement P is said to be balanced *if all nodes* $v \in V$ *are balanced.*

Consider Fig. 4. In the figure, Fig. 4(a) shows a hierarchical failure mode, Fig. 4(b) and (c) show two different placements P_1 and P_2 of three replicas on leaf nodes. Upon computing $\boldsymbol{f}(P_1)$ and $\boldsymbol{f}(P_2)$, we find that $\boldsymbol{f}(P_1) = \langle 2, 1, 0, 2 \rangle \geq_L \langle 1, 1, 1, 2 \rangle = \boldsymbol{f}(P_2)$. Note that for placement P_1, the root of the tree is unbalanced, therefore P_1 is unbalanced. Note also, that P_2 is balanced, since each of its nodes are balanced. We invite the reader to verify that P_2 is an optimal solution for this tree.

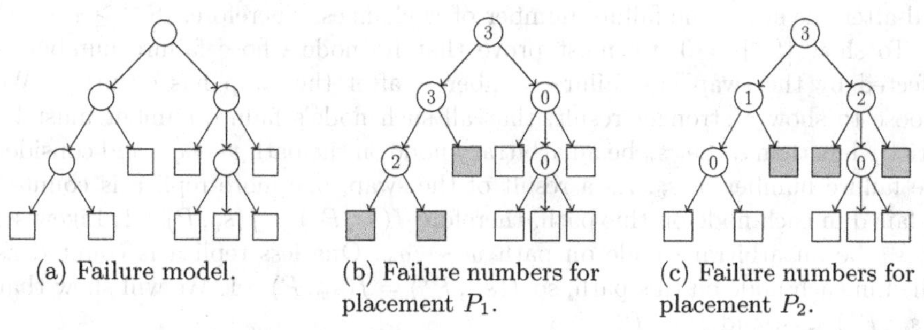

(a) Failure model. (b) Failure numbers for placement P_1. (c) Failure numbers for placement P_2.

Fig. 4. Two different placements. Shaded leaf nodes denote nodes with replicas.

Our main result is that it is *necessary* for an optimal placement to be balanced. However, the balanced property alone is not sufficient to guarantee optimality. To see this, consider the two placements in Fig. 2. By definition, both placements are balanced, yet they have different failure aggregates. Therefore, balancing alone is insufficient to guarantee optimality. Despite this, we can use Theorem 1 to justify discarding unbalanced solutions as suboptimal. We exploit this property of optimal placements in our algorithm.

Theorem 1. *Any placement P in which* $\boldsymbol{f}(P)$ *is lexico-minimum among all placements for a given tree must be balanced.*

Proof. Suppose P is not balanced, yet $\boldsymbol{f}(P)$ is lexico-minimum among all placements P. We proceed to a contradiction, as follows.

Let u be an unbalanced node in T. Let v be an unfilled child of u, and let w be a child of u with at least one replica such that $r_v < r_w - 1$. Since v is unfilled, we can take one of the replicas placed on w and place it on v. Let q_w be the leaf node from which this replica is taken, and let q_v be the leaf node on which this replica is placed (see Fig. 3). Let $P^* := (P - \{q_w\}) \cup \{q_v\}$. We aim to show that P^* is more optimal than P, contradicting P as a lexico-minimum.

Let $\boldsymbol{f}(P) := \langle p_\rho, \ldots, p_0 \rangle$, and $\boldsymbol{f}(P^*) := \langle p^*_\rho, \ldots, p^*_0 \rangle$. For convenience, we let $f(w, P) = \alpha$. To show that $\boldsymbol{f}(P^*) <_L \boldsymbol{f}(P)$, we aim to prove that $p^*_\alpha < p_\alpha$, and that for any i with $\rho \geq i > \alpha$, that $p^*_i = p_i$. We will concentrate on proving the former, and afterwards show that the latter follows easily.

To prove $p^*_\alpha < p_\alpha$, observe that as a result of the swap, some nodes change failure number. These nodes all lie on the paths $v \rightsquigarrow q_v$ and $w \rightsquigarrow q_w$. Let S^- (resp. S^+) be the set of nodes whose failure numbers change to α (resp. change from α), as a result of the swap. Formally, we define

$$S^- := \{x \in V \mid f(x, P) = \alpha, f(x, P^*) \neq \alpha\},$$
$$S^+ := \{x \in V \mid f(x, P) \neq \alpha, f(x, P^*) = \alpha\}.$$

By definition, $p^*_\alpha = p_\alpha - |S^-| + |S^+|$. We claim that $|S^-| \geq 1$ and $|S^+| = 0$, which yields $p^*_\alpha < p_\alpha$. To show $|S^-| \geq 1$, note that $f(w, P) = \alpha$ by definition, and after the swap, the failure number of w changes. Therefore, $|S^-| \geq 1$.

To show $|S^+| = 0$, we must prove that no node whose failure number is affected by the swap has failure number α after the swap has occurred. We choose to show a stronger result, that all such node's failure number must be strictly less than α. Let s_v be an arbitrary node on the path $v \rightsquigarrow q_v$, and consider the failure number of s_v. As a result of the swap, one more replica is counted as failed in each node on this path, therefore $f(s_v, P^*) = f(s_v, P) + 1$. Likewise, let s_w be an arbitrary node on path $w \rightsquigarrow q_w$. One less replica is counted as failed in each node on this path, so $f(s_w, P^*) = f(s_w, P) - 1$. We will show that $f(s_w, P^*) < \alpha$, and $f(s_v, P^*) < \alpha$.

First, note that for any s_w, by Property 1, $f(s_w, P^*) \leq f(w, P^*) = \alpha - 1 < \alpha$. Therefore, $f(s_w, P^*) < \alpha$, as desired.

To show $f(s_v, P^*) < \alpha$, note that by supposition $r_w - 1 > r_v$, and from this we immediately obtain $f(w, P) - 1 > f(v, P)$ by the definition of failure number. Now consider the nodes s_v, for which

$$f(s_v, P) \leq f(v, P) < f(w, P) - 1 = \alpha - 1 \implies f(s_v, P^*) - 1 < \alpha - 1,$$

where the first inequality is an application of Property 1, and the implication follows by substitution. Therefore $f(s_v, P^*) < \alpha$ as desired.

Therefore, among all nodes in P^* whose failure numbers change as a result of the swap, no node has failure number α, so $|S^+| = 0$ as claimed. Moreover, since $f(s, P^*) < \alpha$ for any node s whose failure number changes as a result of the swap, we also have proven that $p_i = p^*_i$ for all i where $\rho \geq i > \alpha$. This completes the proof. $\square$

5 An $O(n + \rho \log \rho)$ Algorithm

Our algorithm executes in three consecutive phases, each named *divide*, *transform* and *combine* respectively. Each phase consists of a post-order traversal of the nodes of the tree starting from the root. Before the divide phase begins, a preprocessing step gathers the following information about each node u: values of the leaf capacity of u, denoted ℓ_u, and the least-depth leaf which is a descendant of u. It is easy to see that this information can be gathered by a post-order traversal in $O(n)$ time. During the *divide* phase, values of $r(u)$ are computed for each node u, where $r(u)$ is the maximum number of replicas which can be placed on u in any balanced placement. This phase also computes the set of unfilled children of u. Since u itself must be balanced, each unfilled child c will have either $r(c)$ or $r(c) - 1$ replicas placed upon it. The value $r(c)$ is computed for each unfilled child of u during the divide phase. The *transform* phase is performed between the divide and combine phases to remove *degenerate chains* from the tree. These chains result in wasted processing during the combine phase. Removing them allows us to obtain our running time of $O(n + \rho \log \rho)$. More details are given in Sect. 5.3. The *combine* phase recursively computes two optimal placements of size $r(u)$ and $r(u) - 1$ at each unfilled node u. To do so optimal placements of size $r(c)$ and $r(c) - 1$ are computed for each unfilled child of u. An optimal combination of these child placements is then made. We show that at each node, an optimal combination of child placements can be made which yields an optimal placement for u. By an easy structural induction on the tree, it can be shown that the placement of size ρ returned at the root is optimal.

We present the divide and combine phases first, then the transform phase.

5.1 Divide Phase

The divide phase computes the value of $r(u)$ for each node u, along with the set of unfilled children of node u. At the root, $r(u) = \rho$. Given a value $r(u)$ for node u, we describe how to compute $r(c_i)$ for each of u's k children, c_i, where $1 \leq i \leq k$. Let child c_i have capacity ℓ_i. We first determine which children are filled, and which are unfilled.

The set of unfilled children can be determined (without sorting) in an iterative manner using an $O(k)$ time algorithm similar to that for the Fractional Knapsack problem. The main idea of the algorithm is as follows: in each iteration, at least one-half of the children whose status is currently unknown are assigned a filled/unfilled status. To determine which half, the median capacity child (with capacity ℓ_{med}) is found using the selection algorithm. Based upon the number of replicas that have not been assigned to the filled nodes, either (a) the set of children c_i with $\ell_i \geq \ell_{med}$ are labeled as "unfilled" or (b) the set of children c_i with $\ell_i \leq \ell_{med}$ are labeled as "filled". The algorithm recurses on the remaining unlabeled children. Pseudocode can be found in Algorithm 1. Algorithm 1 is only invoked if the sum of the capacities of all the children is strictly greater than r implying that at least one child will be labeled as unfilled. The proof of correctness of the labeling algorithm can be found in the full paper [9].

Algorithm 1. Determine filled and unfilled nodes

Function Label-Children$\{c_1, c_2, \ldots, c_k\}$, r**begin**
  $F \leftarrow \emptyset$; $U \leftarrow \emptyset$; // F := filled children U := unfilled children
  $M \leftarrow \{c_1, c_2, \ldots, c_k\}$; // M := unassigned children
  $s \leftarrow r$; // s := number of replicas not yet assigned to filled children
  while $M \neq \emptyset$ **do**
    $\ell_{med} \leftarrow$ median capacity of children in M;
    $M_\ell \leftarrow \{c_i \in M \mid \ell_i < \ell_{med}\}$;
    $M_e \leftarrow \{c_i \in M \mid \ell_i = \ell_{med}\}$;
    $M_g \leftarrow \{c_i \in M \mid \ell_i > \ell_{med}\}$;
    $x \leftarrow s - \sum_{c_i \in M_\ell} \ell_i$;
    if $x \leq (\ell_{med} - 1) \cdot (|U| + |M_e| + |M_g|)$ **then** // $M_e \cup M_g$ guaranteed unfilled
      $U \leftarrow U \cup M_e \cup M_g$; $M \leftarrow M - (M_e \cup M_g)$;
    else if $x > (\ell_{med}) \cdot (|U| + |M_e| + |M_g|)$ **then** // $M_\ell \cup M_e$ guaranteed filled
      $F \leftarrow F \cup M_\ell \cup M_e$; $M \leftarrow M - (M_\ell \cup M_e)$;
      $s \leftarrow x - \sum_{c_i \in M_e} \ell_i$;
    else // M_ℓ filled, $M_e \cup M_g$ unfilled
      $U \leftarrow U \cup M_e \cup M_g$; $F \leftarrow F \cup M_\ell$; $M \leftarrow \emptyset$;
  return (F, U); // return filled and unfilled children

Let U be the set of unfilled children found by Algorithm 1. Suppose we know that we only need to find placements of size $r(u)$ and $r(u) - 1$ for node u. Moreover, we know that in an optimal placement of size $r(u)$, each child c_i only needs to accommodate either $r(c_i)$ or $r(c_i) - 1$ replicas. We also have optimal placements of size $r(c_i)$ and $r(c_i) - 1$ available for each child c_i. Theorem 2 shows that optimal child placements having these two sizes are all that is required to compute optimal placements of size $r(u)$ *and* $r(u) - 1$.

Theorem 2. *In any case where* $r(u)$ *or* $r(u)-1$ *replicas must be balanced among* k *unfilled children, it suffices to consider placing either* $\lceil \frac{|U|}{k} \rceil$ *or* $\lfloor \frac{|U|-1}{k} \rfloor$ *children at each unfilled child.*

Proof (Proof of Theorem 2). Suppose $|U| \bmod k = 0$. If $r(u)$ replicas are placed at u, then all unfilled children receive exactly $\frac{|U|}{k}$ $(= \lceil \frac{|U|}{k} \rceil)$ replicas. If $r(u) - 1$ replicas are placed at u, one child gets $\frac{|U|}{k} - 1 = \lfloor \frac{|U|-1}{k} \rfloor$ replicas. If instead $|U| \bmod k > 0$, then the average number of replicas on each unfilled child is $\frac{|U|}{k} \notin \mathbb{Z}$. To attain this average using integer values, values both above and below $\frac{|U|}{k}$ are needed. However, since the unfilled children must be balanced, whatever values selected must have absolute difference at most 1. The only two integer values satisfying these requirements are $\lceil \frac{|U|}{k} \rceil$ and $\lfloor \frac{|U|}{k} \rfloor$. But $\lfloor \frac{|U|}{k} \rfloor = \lfloor \frac{|U|-1}{k} \rfloor$ when $|U| \bmod k > 0$. □

5.2 Combine Phase

The combine phase selects $|U| \bmod k$ of the unfilled children on which to place $\lceil \frac{|U|}{k} \rceil$ replicas, and places $\lfloor \frac{|U|-1}{k} \rfloor$ replicas on the remaining unfilled children. This selection must be made so as to yield two optimal placements at node u, one with $r(u)$ replicas, and another with only $r(u)-1$. We show how to obtain a solution in the $r(u)-1$ case, the $r(u)$ case is easily obtained thereafter.

Let $\boldsymbol{a}_i$ (respectively $\boldsymbol{b}_i$) represent the lexico-minimum value of $\boldsymbol{f}(P)$ where P is any placement of $\lfloor \frac{|U|-1}{k} \rfloor$ (respectively $\lceil \frac{|U|}{k} \rceil$) replicas on child i. Recall that $\boldsymbol{a}_i, \boldsymbol{b}_i \in \mathbb{N}^{\rho+1}$, and are available as the result of the recursive call. We formulate an optimization problem by encoding the decision to take $\boldsymbol{b}_i$ over $\boldsymbol{a}_i$ as a decision variable $x_i \in \{0,1\}$, for which either $x_i = 0$ if $\boldsymbol{a}_i$ is selected, or $x_i = 1$ if $\boldsymbol{b}_i$ is selected. The problem can then be described as an assignment of values to x_i according to the following system of constraints, in which all arithmetic operations are performed component-wise.

$$\min \sum_i \boldsymbol{a}_i + (\boldsymbol{b}_i - \boldsymbol{a}_i)x_i, \quad \text{subj. to: } \sum_i x_i = |U| \bmod k. \tag{1}$$

An assignment of x_i which satisfies the requirements in (1) can be found by computing $\boldsymbol{b}_i - \boldsymbol{a}_i$ for all i, and greedily assigning $x_i = 1$ to those i which have the $|U| \bmod k$ smallest values of $\boldsymbol{b}_i - \boldsymbol{a}_i$. That this assignment is correct is not immediately obvious, see Theorem 3 below for a proof. The required greedy solution can be quickly made by finding the unfilled child having the $(|U| \bmod k)^{th}$ largest value of $\boldsymbol{b}_i - \boldsymbol{a}_i$ using linear-time selection, and the partition procedure from quicksort can be used to find those children having values below the selected child. For ease of exposition, we assume that the unfilled children c_i are assigned indices in increasing order of $\boldsymbol{b}_i - \boldsymbol{a}_i$ even though the algorithm performs no such sorting.

At the end of the combine phase, we compute and return the sum

$$\sum_{i < |U| \bmod k} \boldsymbol{b}_i + \sum_{i \geq |U| \bmod k} \boldsymbol{a}_i + \sum_{j:\text{filled}} \boldsymbol{f}(P_j) + \mathbf{1}_{r(u)}, \tag{2}$$

where P_j is the placement of ℓ_j replicas on child j and $\mathbf{1}_{r(u)}$ is a vector of length $\rho+1$ having a one in entry $r(u)$ and zeroes everywhere else. The term $\mathbf{1}_{r(u)}$ accounts for the failure number of u. This sum gives the value of an optimal placement of size $r(u)-1$.

It remains to show that the greedy assignment of values x_i to system (1) is optimal. This is the content of Theorem 3.

Theorem 3. *Let $\pi = (\pi_1, \pi_2, \ldots, \pi_k)$ be a permutation of $\{1, 2, \ldots, k\}$ such that:*

$$\boldsymbol{b}_{\pi_1} - \boldsymbol{a}_{\pi_1} \leq_L \boldsymbol{b}_{\pi_2} - \boldsymbol{a}_{\pi_2} \leq_L \ldots \leq_L \boldsymbol{b}_{\pi_k} - \boldsymbol{a}_{\pi_k}.$$

If vector $\boldsymbol{x} = \langle x_1, \ldots, x_k \rangle$ is defined according to the following rules: set $x_{\pi_i} = 1$ iff $i < |U| \bmod k$, else $x_{\pi_i} = 0$, then $\boldsymbol{x}$ is an optimal solution to (1).

The proof of Theorem 3 is greatly simplified through use of an algebraic property of addition on $\mathbb{Z}^n$ under lexicographic order. Recall that a *group* is a pair $\langle S, \cdot \rangle$, where S is a set, and $\cdot$ is a binary operation which is, closed for S, is associative, and has both an identity and inverses. A *linearly-ordered group* is a group $\langle S, \cdot \rangle$, along with a linear-order $\leq$ on S in which for all $x, y, z \in S$, $x \leq y \implies x \cdot z \leq y \cdot z$, i.e. the linear-order on $\langle S, \cdot \rangle$ is *translation-invariant.*

Lemma 1. *$\langle \mathbb{Z}^n, + \rangle$ forms a linearly-ordered group under $\leq_L$. In particular, for any $\boldsymbol{x}, \boldsymbol{y}, \boldsymbol{z} \in \mathbb{Z}^n, \boldsymbol{x} \leq_L \boldsymbol{y} \implies \boldsymbol{x} + \boldsymbol{z} \leq_L \boldsymbol{y} + \boldsymbol{z}$.*

The proof of the above lemma can be found in the full paper [9].

Proof (Proof of Theorem 3*).* First, notice that a solution to (1) which minimizes the quantity $\sum_i (\boldsymbol{b}_i - \boldsymbol{a}_i)x_i$ also minimizes the quantity $\sum_i \boldsymbol{a}_i + (\boldsymbol{b}_i - \boldsymbol{a}_i)x_i$. It suffices to minimize the former quantity, which can be done by considering only those values of $(\boldsymbol{b}_i - \boldsymbol{a}_i)$ for which $x_i = 1$. For convenience, we consider $\boldsymbol{x}$ to be the characteristic vector of a set $S \subseteq \{1, \ldots, k\}$. We show that no other set S' can yield a characteristic vector $\boldsymbol{x}'$ which is strictly better than $\boldsymbol{x}$ as follows.

Let $\beta = |U|$ mod k, and let $S = \{\pi_1, \ldots, \pi_{\beta-1}\}$ be the first $\beta - 1$ entries of π taken as a set. Suppose that there is some S' which represents a feasible assignment of variables to $\boldsymbol{x}'$ for which $\boldsymbol{x}'$ is a strictly better solution than $\boldsymbol{x}$. $S' \subseteq \{1, \ldots, k\}$, such that $|S'| = \beta - 1$, and $S' \neq S$. Since $S' \neq S$, and $|S'| = |S|$ we have that $S - S' \neq \emptyset$ and $S' - S \neq \emptyset$. Let $i \in S - S'$ and $j \in S' - S$. We claim that we can form a better placement, $S^* = (S' - \{j\}) \cup \{i\}$. Specifically,

$$\sum_{\ell \in S^*} (\boldsymbol{b}_\ell - \boldsymbol{a}_\ell) \leq_L \sum_{m \in S'} (\boldsymbol{b}_m - \boldsymbol{a}_m). \tag{3}$$

which implies that replacing a single element in S' with one from S does not cause the quantity minimized in (1) to increase.

To prove (3) note that $j \notin S$ and $i \in S \implies (\boldsymbol{b}_i - \boldsymbol{a}_i) \leq_L (\boldsymbol{b}_j - \boldsymbol{a}_j)$. We now apply Lemma 1, setting $\boldsymbol{x} = (\boldsymbol{b}_i - \boldsymbol{a}_i)$, $\boldsymbol{y} = (\boldsymbol{b}_j - \boldsymbol{a}_j)$, and $\boldsymbol{z} = \sum_{\ell \in (S^* - \{i\})} (\boldsymbol{b}_\ell - \boldsymbol{a}_\ell)$. This yields

$$\sum_{\ell \in (S^* - \{i\})} (\boldsymbol{b}_\ell - \boldsymbol{a}_\ell) + (\boldsymbol{b}_i - \boldsymbol{a}_i) \leq_L \sum_{\ell \in (S^* - \{i\})} (\boldsymbol{b}_\ell - \boldsymbol{a}_\ell) + (\boldsymbol{b}_j - \boldsymbol{a}_j).$$

But since $S^* - \{i\} = S' - \{j\}$, we have that

$$\sum_{\ell \in (S^* - \{i\})} (\boldsymbol{b}_\ell - \boldsymbol{a}_\ell) + (\boldsymbol{b}_i - \boldsymbol{a}_i) \leq_L \sum_{m \in (S' - \{j\})} (\boldsymbol{b}_m - \boldsymbol{a}_m) + (\boldsymbol{b}_j - \boldsymbol{a}_j). \tag{4}$$

Clearly, (4) $\implies$ (3), thereby proving (3). This shows that any solution which is not S can be modified to swap in one extra member of S without increasing the quantity minimized in (1). By induction, it is possible to include every element from S, until S itself is reached. Therefore, $\boldsymbol{x}$ is an optimal solution to (1). □

Naïve implementation of the combine phase yields a $O(n\rho)$ algorithm. But observe that the maximum failure number returned from child c_i is $r(c_i)$. This along with Property 1 implies that the vector returned from c_i will have a zero in indices $\rho, \rho - 1, \ldots, r(c_i) + 1$. To avoid wasting time, we modify the algorithm to return the non-zero suffix of this vector, which has length at most $r(c_i)$. At each node, we then compute (2) by summing the vectors in increasing order of their component index. Specifically, to compute $\boldsymbol{v}_1 + \boldsymbol{v}_2 + \ldots + \boldsymbol{v}_k$, where each vector $\boldsymbol{v}_i$ has length $r(c_i)$, we first allocate an empty vector $\boldsymbol{w}$, of size $r(u)$, to store the result of the sum. Then, for each vector $\boldsymbol{v}_i$, we set $\boldsymbol{w}[j] \leftarrow \boldsymbol{w}[j] + \boldsymbol{v}_i[j]$ for indices j from 0 up to $r(c_i)$. After all vectors have been processed, $\boldsymbol{w} = \boldsymbol{v}_1 + \ldots + \boldsymbol{v}_k$. This algorithm takes $r(c_1) + \ldots + r(c_k) = O(r(u))$ time. Using smaller vectors also implies that the $(|U| \bmod k)^{th}$ best child is found in $O(r(u))$ time, since each unfilled child returns a vector of size at most $O(\frac{r(u)}{k})$, and there are only k unfilled children to compare. With these modifications the entire combine phase takes only $O(r(u))$ time at node u.

5.3 Transform Phase

Note that in any placement, nodes at the same depth have ρ replicas placed on them in total. We can therefore achieve an $O(\rho \log \rho)$ combine phase overall by ensuring the combine phase only needs to occur in at most $O(\log \rho)$ levels of the tree. To do this, we observe that when $r(u) = 1$, any leaf with minimum depth forms an optimal placement. The combine phase can therefore be stopped once $r(u) = 1$. To ensure that there are only $O(\log \rho)$ levels, we transform the tree to guarantee that as the conquer phase proceeds down the tree, $r(u)$ decreases by at least a factor of two at each level. Balancing ensures this will automatically occur when there are two or more unfilled children at each node. Problems can therefore only occur when a tree has a *degenerate chain*, a path of nodes each of which have a single unfilled child. By removing degenerate chains we can achieve an $O(\rho \log \rho)$ combine phase.

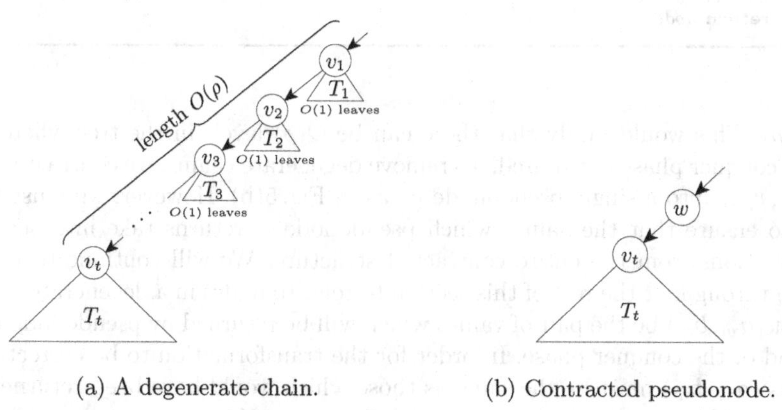

(a) A degenerate chain. (b) Contracted pseudonode.

Fig. 5. Illustration of a degenerate chain.

Figure 5(a) illustrates a degenerate chain. In this figure, each T_i with $1 \leq i \leq t-1$ is the set of all descendant nodes of v_i which are filled. Thus, $v_1, \ldots, v_{t-1}$ each have only a single unfilled child (since each v_i has v_{i+1} as an child). In contrast, node v_t has at least two unfilled children. It is easy to see that if the number of leaves in each T_i is $O(1)$ then the length of the chain, can be as large

Algorithm 2. Transform phase

```
Function Transform(u, chain, s)begin
    if u has two or more unfilled children then                  // not chain node
        foreach child c_i unfilled do
            (−, −, −, x) ←Transform(c_i, false, ⊥);
            if c_i ≠ x then c_i ← x;                  // replace c_i with pseudonode
        if chain = false then return (⊥, ⊥, ⊥, u);
        else return (0_{s+1}, 0_{s+1}, 0_{s+1}, u);               // last node of chain
    if u has one unfilled child, v then                             // chain node
        if chain = false then                                // first node of chain
            (a, b, f, x) ← Transform(v, true, r(v));               // pass down r(v)
        else (a, b, f, x) ← Transform(v, true, r);
        foreach filled child c_i do Filledc_i, f;                   // O(n_i) time
        k ← Σ_{i:i filled} ℓ_i + r(v) − 1;            // k is min failure number of u
        a[k + 1] ← a[k + 1] + 1;    b[k] ← b[k] + 1;              // update a and b
        if chain = false then x ← Make-Pseudonode(a, b, f, x);
        return (a, b, f, x);

Function Filledu, fbegin
    if u is a leaf then
        f[0] ← f[0] + 1;
        return;
    foreach child c_i do
        Filledc_i, f
    a ← Σ_i ℓ_i;
    f[a] ← f[a] + 1;
    return;

Function Make-Pseudonode(a, b, f, x)begin
    allocate a new node node;
    node.a ← a + f;
    node.b ← b + f;
    node.child ← x;
    return node;
```

as $O(\rho)$. This would imply that there can be $O(\rho)$ levels in the tree where the entire conquer phase is required. To remove degenerate chains, we contract nodes $v_1, \ldots, v_{t-1}$ into a single pseudonode w, as in Fig. 5(b). However, we must take care to ensure that the values which pseudonode w returns take into account contributions from the entire contracted structure. We will continue to use v_i and T_i throughout the rest of this section to refer to nodes in a degenerate chain.

Let $(\boldsymbol{a_w}, \boldsymbol{b_w})$ be the pair of values which will be returned by pseudonode w at the end of the conquer phase. In order for the transformation to be correct, the vectors $(\boldsymbol{a_w}, \boldsymbol{b_w})$ must be the same as those which would have been returned at node v_1 had no transformation occurred. To ensure this, we must consider and include the contribution of each node in the set $T_1 \cup \ldots \cup T_{t-1} \cup \{v_1, \ldots, v_{t-1}\}$. It is easy to see that the failure numbers of nodes in $\{v_1, \ldots, v_{t-1}\}$ depend only upon whether $r(v_t)$ or $r(v_t) - 1$ replicas are placed on node v_t, while the filled nodes in sets $T_1, \ldots, T_{t-1}$ have no such dependency. Since values of $r(v_i)$ are available at each node after the divide phase, enough information is present to contract the degenerate chain.

Pseudocode for the transform phase is given in Algorithm 2. Let $S_w := T_1 \cup \ldots \cup T_{t-1} \cup \{v_1, \ldots, v_{t-1}\}$, and let the contribution of nodes in S_w to $\boldsymbol{a_w}$ and $\boldsymbol{b_w}$

be given by vectors $\boldsymbol{a}$ and $\boldsymbol{b}$ respectively. The transform phase is started at the root of the tree by invoking `Transform`($root, false, \rho$). `Transform` is a modified recursive breadth-first search, which returns a 4-tuple $(\boldsymbol{a}, \boldsymbol{b}, \boldsymbol{f}, x)$, where x is the node v_t which ends the degenerate chain. As the recursion proceeds down the tree, each node is tested to see if it is part of a degenerate chain (lines 2 and 8). If a node is not part of a degenerate chain, the call continues on all unfilled children (line 3). The first node (v_1) in a degenerate chain is marked (by passing down $chain \leftarrow true$ at lines 10 and 11). Once the bottom of the chain (node v_t) has been reached, the algorithm allocates memory for three vectors, $\boldsymbol{a}, \boldsymbol{b}$ and $\boldsymbol{f}$, each of size $s + 1$ (line 7). The value of $r(v_1)$ is passed down to the bottom of the chain at lines 10 and 11. These vectors are then passed up through the entire degenerate chain (cf. lines 7 and 16), along with node u (at line 7), whose use will be explained later. When a node u in a degenerate chain receives $\boldsymbol{a}, \boldsymbol{b}$, and $\boldsymbol{f}$, u adds its contribution to each vector (lines 12–14). The contribution of node u consists of two parts. First, the contribution of the filled nodes is added to $\boldsymbol{f}$ by invoking a special Filled subroutine which computes the sum of the failure aggregates of each filled child of u (line 12). Note that Filled uses pass-by-reference semantics when passing in the value of $\boldsymbol{f}$. Then, the contribution of node u itself is added, by summing the number of leaves in all of the filled children, and the number of replicas on the single unfilled child, v (lines 13–14). By the time that the recursion reaches the start of the chain on the way back up, all nodes have added their contribution, and the pseudonode is created and returned (line 15).

`Make-Pseudonode` runs in $O(1)$ time. It is easy to see that Filled runs in $O(n_j)$ time, where n_j is the number of nodes in the subtree rooted at child c_j. `Transform` therefore takes $O(|T_i|)$ time to process a single node v_i. The time needed for `Transform` to process an entire degenerate chain is therefore $O(|S_w|) + 3 \cdot O(r(v_1))$, where the $3 \cdot O(r(v_1))$ term arises from allocating memory for vectors $\boldsymbol{a}$, $\boldsymbol{b}$ and $\boldsymbol{f}$ at the last node of the chain.

When we sum this time over all degenerate chains, we obtain a running time of $O(n + \rho \log \rho)$ for the transform phase. To reach this result, we examine the sum of $r(v_1)$ for all pseudonodes having the same depth. Since there are at most ρ replicas among such pseudonodes, this sum can be at most $O(\rho)$ at any depth. After degenerate chains have been contracted, there are only $O(\log \rho)$ levels where $r(u) > 1$, thus, pseudonodes can be only be present in the first $O(\log \rho)$ levels of the final tree. Therefore the $3 \cdot O(r(v_1))$ term sums to $O(\rho \log \rho)$ overall. Since $|S_w|$ clearly sums to $O(n)$ overall, the transform phase takes at most $O(n + \rho \log \rho)$ time.

Including the transform phase implies that there are only $O(\log \rho)$ levels where the conquer phase needs to be run in its entirety. Therefore, the conquer phase takes $O(\rho \log \rho)$ time overall. When combined with the $O(n)$ divide phase and the $O(n + \rho \log \rho)$ transform phase, this yields an $O(n + \rho \log \rho)$ algorithm for solving replica placement in a tree.

6 Conclusion and Future Work

In this paper, we formulate the replica placement problem and show that it can be solved by a greedy algorithm in $O(n^2\rho)$ time. In search of a faster algorithm, we prove that any optimal placement in a tree must be balanced. We then exploit this property to derive a near-optimal $O(n + \rho \log \rho)$ algorithm. In future work we plan to consider replica placement on additional classes of graphs, such as special cases of bipartite graphs, and also design good approximation algorithms for general graphs.

Acknowledgments. We would like to acknowledge insightful comments from S. Venkatesan and Balaji Raghavachari during meetings discussing this work, as well as comments from Conner Davis on a draft version of this paper.

References

1. Bakkaloglu, M., Wylie, J.J., Wang, C., et al.: On Correlated Failures in Survivable Storage Systems. Tech. rep. CMU-CS- 02–129, Carnegie Mellon University (2002)
2. Blume, L., Easley, D., Kleinberg, J., Kleinberg, R., Tardos, E.: Which networks are least susceptible to cascading failures? In: Proceedings of the 52nd Annual Symposium on Foundations of Computer Science (FOCS), October 2011
3. Chen, M., Chen, W., Liu, L., Zheng, Z.: An analytical framework and its applications for studying brick storage reliability. In: Proceedings of 26th International Symposium on Reliable Distributed Systems (SRDS) (2007)
4. Ford, D., Labelle, F., Popovici, F., et al.: Availability in globally distributed storage systems. In: Proceedings of the 9th USENIX Symposium on Operating Systems Design and Implementation (OSDI) (2010)
5. Hu, X.D., Jia, X.H., Du, D.Z., et al.: Placement of data replicas for optimal data availability in ring networks. J. Parallel Distrib. Comput. (JPDC) **61**(10), 1412–1424 (2001)
6. Pezoa, J.E., Hayat, M.M.: Reliability of heterogeneous distributed computing systems in the presence of correlated failures. IEEE Trans. Parallel Distrib. Comput. **25**(4), 1034–1043 (2014)
7. Kim, J., Dobson, I.: Approximating a loading-dependent cascading failure model with a branching process. IEEE Trans. Reliab. **59**(4), 691–699 (2010)
8. Lian, Q., Chen, W., Zhang, Z.: On the impact of replica placement to the reliability of distributed brick storage systems. In: Proceedings of the International Conference on Distributed Computing Systems (ICDCS) (2005)
9. Mills, K.A., Chandrasekaran, R., Mittal, N.: Algorithms for replica placement in high-availability storage (2015). arXiv:1503.02654
10. Nath, S., Yu, H., Gibbons, P.B., Seshan, S.: Subtleties in tolerating correlated failures in wide-area storage systems. In: Proceedings of the 3rd USENIX Symposium on Networked Systems Design and Implementation (NSDI) (2006)
11. Shekhar, S., Wu, W.: Optimal placement of data replicas in distributed database with majority voting protocol. Theoret. Comput. Sci. **258**(1–2), 555–571 (2001)
12. Weatherspoon, H., Moscovitz, T., Kubiatowicz, J.: Introspective failure analysis: avoiding correlated failures in peer-to-peer systems. In: Proceedings of the 21st Symposium on Reliable Distributed Systems (SRDS) (2002)

13. Zhang, Z., Wu, W., Shekhar, S.: Optimal placements of replicas in a ring network with majority voting protocol. J. Parallel Distribut. Comput. (JPDC) **69**(5), 461–469 (2009)
14. Zhu, Y., Yan, J., Sun, Y., et al.: Revealing cascading failure vulnerability in power grids using risk-graph. IEEE Trans. Parallel Distribut. Syst. (TPDS) **25**(12), 1 (2014)

Observing the State of a Smart Grid Using Bilevel Programming

Sonia Toubaline(✉), Pierre-Louis Poirion, Claudia D'Ambrosio, and Leo Liberti

CNRS LIX, Ecole Polytechnique, 91128 Palaiseau, France
{toubaline,poirion,dambrosio,liberti}@lix.polytechnique.fr

Abstract. Monitoring an electrical network is an important and challenging task. Phasor measurement units are measurement devices that can be used for a state estimation of this network. In this paper we consider a PMU placement problem without conventional measurements and with zero injection nodes for a full observability of the network. We propose two new approaches to model this problem, which take into account a propagation rule based on Ohm's and Kirchoff's law. The natural binary linear programming description models an iterative observability process. We remove the iteration by reformulating its fixed point conditions to a bilevel program, which we then further reformulate to a single-level mixed-integer linear program. We also present a bilevel algorithm to solve directly the proposed bilevel model. We implemented and tested our models and algorithm: the results show that the bilevel algorithm is better in terms of running time and size of instances which can be solved.

Keywords: Bilevel program · Mixed integer linear program · Monitoring electrical network · PMU placement problem

1 Introduction

One of the properties making a grid *smart* is that its state is continuously monitored. The term *state* is an abstract concept which may be represented in many ways. We consider that a state of a grid is the set of values of all the branch currents and node voltages. Monitoring the state of a grid can be achieved using tools of measurement and control. A piece of equipment which can be used is Phasor Measurement Unit (PMU). PMUs are monitoring devices that provide time synchronized phasor measurement (a phasor is a complex number that represents both the magnitude and phase angle of the sine waves found in electricity). A PMU placed at a (sub)station measures the voltage and phase angle of this (sub)station and branch current phasor of all transmission line emerging from it [11]. PMUs are synchronized via global positioning systems (GPS) and send large bursts of data to a system monitoring centre. Due to the relatively high cost of PMUs, their optimal placement constitutes an important challenge.

Modelling the network by a graph where nodes correspond to (sub)stations and edges to transmission lines, the optimal location problem of PMUs, called

Z. Lu et al. (Eds.): COCOA 2015, LNCS 9486, pp. 364–376, 2015.
DOI: 10.1007/978-3-319-26626-8_27

PMU PLACEMENT PROBLEM, consists of determining the minimum number of PMUs to place on the nodes, in order to ensure a full observability of the graph. A graph is said fully observed if the voltage is known at each node, and the current known on each edge. In [2], the observability of a graph is defined by two rules: (i) if a node has a PMU then this node and all its neighbours are observed; (ii) if an observed node has all its neighbours observed, except one, then this latter is observed. The PMU PLACEMENT PROBLEM is also known as POWER DOMINATING SET (PDS) problem [8]. This problem has been largely studied in literature. The PDS is shown to be *NP*-complete even for bipartite, chordal graphs [8] and planar bipartite graphs [2], and polynomial for trees and grids [4]. Some approximation results are presented in [1]. Different solution methods have been proposed to solve the PMU PLACEMENT PROBLEM [11,12]. In all these approaches, the propagation rule has been considered for the zero injection nodes with a limited depth.

The PMU PLACEMENT PROBLEM has also been studied for PMUs with limited channels ℓ where only a limited number of incident transmission lines can be observed [9,10]. Korkali and Abur propose a binary linear program considering for each node of the graph all the possible combinations of ℓ edges among incident edges of that node [9]. The number of combinations can be exponential. Kumar and Rao propose a new method to solve PMU PLACEMENT PROBLEM based on node connectivity and edge selectivity matrices where the number of channels are less than the minimum degree of the graph [10]. For PMUs with one channel, only one incident line can be observed. The placement of PMUs is then no more on nodes but on edges. The first rule of observability is then: if an edge has a PMU then its extremity nodes are observed. Emami et al. propose a binary linear program for this variant of the problem [5,6], which turns out to be equivalent to the minimum edge cover problem. They consider the second rule of observability defined in [2] for zero injection nodes with only one depth.

We propose in this paper a new approach to model the optimal location of PMUs with one channel. We place PMUs on edges (next to one of the adjacent nodes) and take into account both rules of observability. We call this particular variant the POWER EDGE SET (PES) problem. More specifically, we consider the case without conventional measurements (measures provided by non synchronized sensors) and all nodes are zero injection nodes (no current is injected in the network at those nodes). We present two mathematical formulations for this problem: iterative and bilevel models; the latter can be formally inferred from the former by means of a fixed-point argument. The former is a Binary Linear Program. We show that the latter can be reformulated exactly to a Mixed-Integer Linear Program (MILP) with binary variables, and also propose a cutting-plane algorithm to solve it in its native bilevel formulation. We benchmarked our solution methods on standard IEEE bus-systems [14] and randomly generated graphs.

2 Problem Statement

Let $G = (V, E)$ be a graph modelling the electrical network where $V = \{1, \ldots, n\}$ is the set of nodes representing the (sub)stations and E the set of edges corresponding to transmission lines. For $i \in V$, $\Gamma(i) = \{j : \{i, j\} \in E\}$ is the set of neighbours (adjacent nodes) of i. For graph-theoretical notions, see [7,13].

In this paper, we are interested in the optimal placement of PMUs with one channel, so as to ensure a full observability of G. We consider that no conventional measures exist, which would reduce the number of PMUs to install.

A PMU is placed on an edge $\{i, j\}$ close to node i, for $i \in V$ and $\{i, j\} \in E$.

We remark that the fact that PMU placement occurs closer to an adjacent node than the other is relevant for physical reasons, but irrelevant for our abstract modelling purposes. Henceforth, we shall simply assume that placement occurs on an edge $\{i, j\} \in E$. A graph is said to be observable if all node voltages and current edges are known either measured by a PMU or estimated using electrical laws. The problem is defined as follows:

Power Edge Set **(PES) problem**
Input: A graph $G = (V, E)$.
Output: A PMU placement $\Pi \subset E$ of minimum cardinality such that G is fully observable.

2.1 Observability of a Graph

Let Π be a given PMU placement on G and Ω the set of observed nodes. The observability of G is defined by the two following rules based on electrical laws explained below.

*R*1: If a PMU is placed on an edge $\{i, j\}$, then nodes i and j are observed

$$\{i, j\} \in \Pi \Rightarrow i, j \in \Omega$$

*R*2: If an observed node i has all its neighbors observed, save one, then this node is observed

$$i \in \Omega \text{ and } |\Gamma(i) \setminus \Omega| \leq 1 \Rightarrow \Gamma(i) \subseteq \Omega$$

By rule *R*1, the PMU placed at $\{i, j\}$ measures the voltage at i and the current on $\{i, j\}$. Using Ohm's law, we can deduce the voltage on j. Then i and j are observed. By rule *R*2, if a node i and all its neighbors $k \in \Gamma(i)$ are observed, except a single node j, then using Ohm's law we can determine the current on $\{i, k\}$ for $k \in \Gamma(i) \setminus \{j\}$; knowing the currents on all $\{i, k\}$ (for $k \neq j$) we can deduce the current on $\{i, j\}$ using Kirchoff's law. Then, knowing the voltage at i and the current on $\{i, j\}$, we determine the voltage on j using Ohm's law. Hence, j is observed.

3 Mathematical Modelling

We present two Mathematical Programming (MP) models for PES problem: an iterative based model, and a bilevel one.

3.1 Iterative Model

The model is based on the iterative process of observability given by rules $R1$ and $R2$. Assuming the problem instance to be a feasible one, the set of observed nodes can be found in at most $n - 2$ steps: this is because, in the worst case, only one PMU is placed on an edge (this observes the adjacent nodes), and one more node is observed at each iteration. Let $T = n - 2$ be the maximum number of steps. The iterative model (P_{IT}) is as follows.

Variables. We define the following set of variables:

$$\forall i \in V, j \in \Gamma(i), \quad s_{ij} = \begin{cases} 1 \text{ if a PMU is installed on } \{i,j\} \\ 0 \text{ otherwise} \end{cases}$$

$$\forall i \in V, t = 0, \ldots, T, \quad \omega_{it} = \begin{cases} 1 \text{ if the node } i \text{ is observed at step } t \\ 0 \text{ otherwise} \end{cases}$$

and $\forall i \in V, j \in \Gamma(i), t = 0, \ldots, T-1,$

$$y_{ijt} = \begin{cases} 1 \text{ if R2 is used to observe j using the observed node i at step t} \\ 0 \text{ otherwise.} \end{cases}$$

Constraints. The set of constraints is the following:

- All nodes must be observed at step T

$$\forall i \in V, \quad \omega_{iT} = 1$$

- If a node i is observed at step 0 then at least one PMU is placed at $\{i,j\}$ or $\{j,i\}$ for a given neighbour j of i

$$\forall i \in V, \quad \omega_{i0} \leq \sum_{j \in \Gamma(i)} (s_{ij} + s_{ji})$$

- The set of constraints corresponding to rule $R2$ is the following:
  - If i not observed at step t is observed at step $t+1$ then at least one neighbour observed node has been used to observe i

$$\forall i \in V, t = 0, \ldots, T-1, \quad \omega_{i(t+1)} \leq \omega_{it} + \sum_{\ell : i \in \Gamma(\ell)} y_{\ell it}$$

  - If an observed node i is used at step t to observe a neighbour node j and j is observed at step $t+1$ then j is not observed at step t

$$\forall i \in V, j \in \Gamma(i), t = 0, \ldots, T-1, \quad \omega_{j(t+1)} + y_{ijt} \leq \omega_{jt} + \omega_{it} + 1$$

- If an observed node i is used at step t to observe a neighbour node j and j is observed at step $t+1$ then j is not observed at step t and all the other neighbour nodes k of i are observed at step t

$$\forall i \in V, j, k \in \Gamma(i), k \neq j, t = 0, \dots, T-1, \quad \omega_{j(t+1)} + y_{ijt} \leq \omega_{jt} + \omega_{kt} + 1$$

– If a node i is observed at step t then i is observed from step t to step T

$$\forall i \in V, t = 0, \dots, T-1, \quad \omega_{it} \leq \omega_{i(t+1)}.$$

Objective Function. The aim of PES problem is to minimize the number of PMUs to install and that allow a full observability of G. Hence the objective function is given by

$$\min \sum_{i \in V} \sum_{j \in \Gamma(i)} s_{ij}.$$

3.2 From Iterative to Bilevel Model

We show in this subsection how to deduce a bilevel model from the iterative one using a fixed-point method.

Let $\omega^t = (\omega_{it} \mid i \in V)$ be the characteristic vector describing the observability of nodes at step t. The iterative model computes the vector values for $t \in \{1, \dots, T\}$. Let $\omega = (\omega_i \mid i \in V)$ be the characteristic vector of Ω:

$$\forall i \in V, \quad \omega_i = \begin{cases} 1 \text{ if node } i \text{ is observed} \\ 0 \text{ otherwise.} \end{cases}$$

We have that $\omega = \omega^T$. We show now how to obtain a non iterative model where ω are the only variables that model the observability of the graph.

Let $t \leq T$ and i a node in V. The recursive relation that allows to express $\omega_{i(t+1)}$ in function of ω_t is:

$$\omega_{i(t+1)} = \max \left(\omega_{it}, \max_{j \in \Gamma(i)} \left(1 - |\Gamma(j)| + \omega_{jt} + \sum_{k \in \Gamma(j), k \neq i} \omega_{kt} \right) \right)$$

meaning that a node i is observed at step $t+1$ if it was already observed at step t or if there exists a neighbour j of i such that all the other neighbours $k \neq i$ of j are observed.

Let $\theta : \{0,1\}^n \mapsto \{0,1\}^n$ be a function where

$$\forall i \in V, \theta_i(x) = \max \left(x_i, \max_{j \in \Gamma(i)} \left(1 - |\Gamma(j)| + x_j + \sum_{k \in \Gamma(j), k \neq i} x_k \right) \right).$$

with $x = (x_i \mid i \in V)$ and $(\theta(x))_i = \theta_i(x)$.

By definition we have that:

$$\forall t \in \{1, \ldots, T-1\} \quad \omega^{(t+1)} = \theta(\omega^t).$$

Recursive Computation of ω: The vector ω is determined recursively as follows:

- Based on $R1$, for $t = 0$ we have:

$$\forall i \in V,\ \omega_{i0} = \max(\max_{j \in \Gamma(i)} s_{ij},\ \max_{j \in \Gamma(i)} s_{ji})$$

- Knowing ω^t, we can compute $\omega^{(t+1)}$ by looking for an optimal solution of the linear program:

$$(*)\begin{cases} \min\limits_{\omega^{t+1} \in \{0,1\}^n} & \sum\limits_{i=1}^{n} \omega_{i(t+1)} \\ \forall i \in V & \omega_{i(t+1)} \geq \omega_{it} \\ \forall i \in V,\ j \in \Gamma(i) & \omega_{i(t+1)} \geq 1 - |\Gamma(j)| + \omega_{jt} + \sum\limits_{\substack{k \in \Gamma(j) \\ k \neq i}} \omega_{kt}. \end{cases}$$

Theorem 1. *We have that ω is the smallest fixed point of θ.*

Proof. By the definition of the function θ, we have that $\theta_i(\omega) \geq \omega_i, \forall i \in V$. Suppose that $\exists i \in V,\ \theta_i(\omega) > \omega_i$ i.e. $\omega_i = 0$ and $\theta_i(\omega) = 1$. Then $i \in \Omega$ by an application of $R2$ which implies $\omega_i = 1 > 0 = \omega_i$, contradiction. Hence $\theta(\omega) = \omega$.

Assume now that $\exists \omega' < \omega : \omega' = \theta(\omega')$. This means that $R2$ cannot be used to observed more nodes. Hence the number of nodes observed in ω' is less then the one in ω, i.e. $\sum_{i \in V} \omega'_i < \sum_{i \in V} \omega_i$ which contradict the optimality of $(*)$.

Therefore, ω is the smallest fixed point of θ and correspond to the optimal solution of the following linear program:

$$\begin{cases} \min\limits_{\omega \in \{0,1\}^n} \sum_{i=1}^{n} \omega_i & \\ \omega_i \geq s_{ij} + s_{ji} & \forall i \in V,\ j \in \Gamma(i) \\ \omega_i - \omega_j - \sum\limits_{k \in \Gamma(j), k \neq i} \omega_k \geq 1 - |\Gamma(j)| & \forall i \in V,\ j \in \Gamma(i) \end{cases}$$

□

3.3 Bilevel Model

We describe in this subsection the bilevel program proposed to model the PES problem. We also show how it can be reformulated to a MILP.

The formulation

$$(\dagger)\quad \begin{cases} \min\limits_{s} \sum\limits_{i\in V}\sum\limits_{j\in\Gamma(i)} s_{ij} \\ s_{ij}\in\{0,1\} \qquad \forall i\in V, j\in\Gamma(i) \\ f(s)\geq n \\ f(s)=\begin{cases} \min\limits_{\omega} \sum\limits_{i\in V}\omega_i \\ \omega_i \geq s_{ij}+s_{ji} & \forall i\in V, j\in\Gamma(i) \\ \omega_i-\omega_j-\sum\limits_{k\in\Gamma(j),k\neq i}\omega_k \geq 1-|\Gamma(j)| & \forall i\in V, j\in\Gamma(i) \\ \omega_i\in\{0,1\} & \forall i\in V \end{cases} \end{cases}$$

In the upper level problem, the objective is to minimize the number of PMUs to install such that the number of observed nodes given fy the function $f(s)$ is at least n. The function f corresponds to the optimal value of the lower level problem described below and s is the vector representing s_{ij}.

In the lower level problem, the objective is to minimize the number of nodes observed. The first set of constraints says that if a PMU is placed at $\{i,j\}$ or $\{j,i\}$ then i and j are observed. The second expresses the propagation rule $R2$: if a non observed node i has an observed neighbour j that has all its others node neighbours $k (k\neq i)$ observed then i is observed.

MILP Reformulation. The integrality of variables ω_i can be relaxed in the lower level problem.

Lemma 1. *For each $i\in V$, the constraint $\omega_i\in\{0,1\}$ can be replaced by $\omega_i\geq 0$.*

Proof. Let $\bar{\omega}$ be an optimal solution of the slave problem and consider a certain configuration of installed PMUs in the graph.

By the first constraint

$$\omega_i \geq s_{ij}+s_{ji},\ \forall i\in V, j\in\Gamma(i) \tag{1}$$

we have that $\exists S\subseteq V, \forall i\in S, \bar{\omega_i}=1$. If we rewrite the second constraint of the slave problem as

$$\forall i\in V, j\in\Gamma(i),\quad \omega_i\geq\omega_j+\sum_{k\in\Gamma(j),k\neq i}\omega_k-|\Gamma(j)|+1$$

we have that the right hand side $r(\bar{\omega})\in[1-|\Gamma(j)|,1]$.

If $r(\bar{\omega})\in]0,1[$ then $\exists z\in\Gamma(j)\cup\{j\}:\bar{\omega_z}\in]0,1[$. Hence, $\exists Z\subseteq V$, $\forall z\in Z$: $\bar{\omega_z}\in]0,1[$. Also, $\forall z\in Z$, z is not constrained by (1) otherwise $\bar{\omega_z}=1$. By the objective function direction, $\forall z\in Z$ we can set $\bar{\omega_z}=0$ and still be feasible, which contradict the optimality of $\bar{\omega}$. Therefore, we can relax the integrity of variables ω to $[0,1]^n$.

Similarly, we can prove that $\forall i\in V, \omega_i\geq 0$. □

Hence, by replacing the lower-level problem by its KarushKuhnTucker (KKT) conditions [3], we obtain the following MILP:

$$(P)\begin{cases} \min_s \sum_{i\in V}\sum_{j\in\Gamma(i)} s_{ij} & \\ s_{ij}\in\{0,1\} & \forall i\in V, j\in\Gamma(i) \\ \sum_{i\in V}\sum_{j\in\Gamma(i)} s_{ij}\mu_{ij}+s_{ji}\mu_{ij}+(1-|\Gamma(j)|)\lambda_{ij}\geq n & \\ \sum_{j\in\Gamma(i)}(\mu_{ij}+\lambda_{ij}-\lambda_{ji}-\sum_{k\in\Gamma(j),k\neq i}\lambda_{kj})\leq 1 & \forall i\in V \\ \lambda_{ij},\mu_{ij}\geq 0 & \forall i\in V, j\in\Gamma(i) \end{cases}$$

We now prove that the dual variables μ_{ij} are bounded, $\forall i\in V, j\in\Gamma(i)$.

Proposition 1. $\forall i\in V,\ j\in\Gamma(i),\ \exists M>0:\ \mu_{ij}\leq M.$

Proof. Let (s^*,μ^*,λ^*) be an optimal solution of (P) and (s^*,ω^*) be the corresponding optimal solution of the bilevel formulation. In particular, we consider (s^*,μ^*,λ^*) such that (μ^*,λ^*) is a basis solution of the dual program of the linear program that defines f:

$$\begin{cases} \max_s \sum_{i\in V}\sum_{j\in\Gamma(i)} s_{ij}\mu_{ij}+s_{ji}\mu_{ij}+(1-|\Gamma(j)|)\lambda_{ij} & \\ \sum_{j\in\Gamma(i)}(\mu_{ij}+\lambda_{ij}-\lambda_{ji}-\sum_{k\in\Gamma(j),k\neq i}\lambda_{kj})\leq 1 & \forall i\in V \\ \lambda_{ij},\mu_{ij}\geq 0 & \forall i\in V, j\in\Gamma(i) \end{cases}$$

Necessarily at most n dual variables are non-zero. Let $I=\{(i,j)\mid \mu_{ij}\neq 0\}$ and $J=\{(i,j)\mid \lambda_{ij}\neq 0\}$. We have $|I|+|J|\leq n$.

Let $i\in\{1,...,n\}$ such that $\omega_i^*=1$. By complementary slackness conditions we have

$$\sum_{j\in\Gamma(i)}(\mu_{ij}^*+\lambda_{ij}^*-\lambda_{ji}^*-\sum_{k\in\Gamma(j),k\neq i}\lambda_{kj}^*)=1 \tag{2}$$

Let $A_B\in\mathbb{R}^{n\times n}$ be the basis matrix corresponding to the optimal solution (μ^*,λ^*). By Eq. (2), $v=(\mu^*,\lambda^*,\beta^*)$ is a solution of the system $A_B\,v=e$, where β^* denotes the slack variables used to write the above dual program in standard form, e is a vector in $\mathbb{R}^n$ where each component is one, and all elements of A_B are in $\{-1,0,1\}$.

Since $A_B^{-1}=\frac{\operatorname{adj}(A_B)}{\det(A_B)}$, where $\operatorname{adj}(A_B)$ is the adjugate matrix of A_B and $det(A_B)$ is the determinant of A_B, using Hadamard inequality for determinant, we obtain that the dual variables μ_{ij} are all bounded by $M=n^{\frac{n}{2}}$, $\forall i\in V, j\in\Gamma(i)$. □

By Proposition 1, we can linearize the program (P) by replacing the variable products by $\forall i \in V, j \in \Gamma(i)$, $p_{ij} = s_{ij}\mu_{ij}$ and $q_{ij} = s_{ji}\mu_{ij}$. Therefore, we obtain the MILP (P_{MILP}).

$$(P_{\mathsf{MILP}}) \quad \begin{cases} \min \sum\limits_{i\in V}\sum\limits_{j\in\Gamma(i)} s_{ij} & \\ s_{i,j} \in \{0,1\} & \forall i \in V, j \in \Gamma(i) \\ \sum\limits_{i\in V}\sum\limits_{j\in\Gamma(i)} p_{ij} + q_{ij} + (1 - |\Gamma(j)|)\lambda_{ij} \geq n & \\ \sum\limits_{j\in\Gamma(i)} (\mu_{ij} + \lambda_{ij} - \lambda_{ji} - \sum\limits_{k\in\Gamma(j),k\neq i} \lambda_{kj}) \leq 1 & \forall i \in V \\ p_{ij} \leq M\ s_{ij} & \forall i \in V, j \in \Gamma(i) \\ p_{ij} \leq \mu_{ij} & \forall i \in V, j \in \Gamma(i) \\ p_{ij} \geq \mu_{ij} - M(1 - s_{ij}) & \forall i \in V, j \in \Gamma(i) \\ q_{ij} \leq M\ s_{ji} & \forall i \in V, j \in \Gamma(i) \\ q_{ij} \leq \mu_{ij} & \forall i \in V, j \in \Gamma(i) \\ q_{ij} \geq \mu_{ij} - M(1 - s_{ji}) & \forall i \in V, j \in \Gamma(i) \\ \lambda_{ij}, \mu_{ij} \geq 0 & \forall i \in V, j \in \Gamma(i) \end{cases}$$

4 An Algorithm for the Bilevel Problem

We propose a cutting plane algorithm BILEVELSOLVE to solve the bilevel program (†) directly. BILEVELSOLVE iteratively solves a modified version of the upper level problem as a *master* MILP, adding a new cut at each iteration. The cuts are generated by means of the combinatorial procedure GENERATECUT on the lower level *slave* problem.

Consider the following MILP P^k:

$$[P^k] \quad \begin{cases} \min\limits_{s\in\{0,1\}^{|E|}} \sum\limits_{i\in V}\sum\limits_{j\in\Gamma(i)} s_{ij} \\ \forall h \leq k \qquad \alpha^h s \geq 1, \end{cases} \tag{3}$$

where $\alpha^h \in \{0,1\}^{|E|}$ for each $h \leq k$, and k is the main algorithm iteration counter: at iteration k, P^k has k linear covering constraint, starting with $\alpha^1 = (1, \ldots, 1)$.

Although BILEVELSOLVE needs exponentially many cuts in the worst case, we found it to perform very well empirically.

5 Computational Results

All the experimentations presented here were performed on a 2.70 GHz computer with 8.0 GB RAM. The models (P_{IT}), (P_{MILP}) and the bilevel algorithm were implemented using IBM ILOG CPLEX 12.6. We considered as instances a 5-bus system and standard IEEE n-bus systems, with $n \in \{7, 14, 30, 57, 118\}$ [14]. We also generate randomly graphs with n nodes and $m = 1.4 \times n$ for $n = \{5 \times i, i = 1, \ldots, 10\}$ where 1.4 is the average rate of edges over nodes in standard IEEE

Algorithm 1. BILEVELSOLVE

```
k = 1
termination ← 0
while termination = 0 do
    s ← MILPSOLVE(P^k)
    k ← k + 1
    α^k = GENERATECUT(s, termination)
    P^k ← [P^{k−1} s.t. α^k s ≥ 1]
end while
```

Algorithm 2. GENERATECUT(s, termination)

```
termination ← 0
// observe nodes according to PMUs in s
place PMUs in G in all edges in the support of s
apply rules R1 and R2 to G, to obtain Ω ⊆ V (observed nodes)
if Ω = V then
    // if PMUs in s suffice to observe all nodes, terminate
    termination ← 1
    α ← (0, ..., 0)
else
    // otherwise, apply more PMUs and aim to observe all nodes
    Θ ← Ω
    while Ω ⊊ V do
        choose v ∈ VsetminusΩ and {u, v} ∈ E
        place PMU in {u, v} and apply R1, R2 to update Ω
        if Ω ≠ V then
            Θ ← Ω
        end if
    end while
    // generate cut on edges not induced by nodes observed
    // at R2 application step before full observability
    let F be the set of edges induced by Θ
    let α be the support of E \ F
end if
return (α, termination)
```

bus systems. The instances can be forests and no node is isolated. For each value of n, 10 different instances were generated and tested. The results obtained are reported in Table 1 where each given value for the randomly generated graphs is the average over 10 instances. We limited the running time to 2 h. For any instance which is not solved optimally within the time limit, the running time is set to this limit. We reported: (i) the *Gap*, expressed as a percentage, that is the average over ratios $\frac{UB - BS}{UB}$ computed on all instances returning at least one feasible solution, where UB is the final best upper bound and BS is the best solution value found; and (ii) the number of instances #opt solved optimally, and the number of instances that run out of memory (mof, for memory overflow).

Table 1. Computational results

Networks/graphs	n	m	Iterative			MILP			Bilevel algo.		
			Time (s)	Gap (%)	#opt (mof)	Time (s)	Gap (%)	#opt (mof)	Time (s)	Gap (%)	#opt (mof)
IEEE bus system	5	6	1.30	0	1	**1.29**	0	1	2.01	0	1
	7	8	6.22	0	1	**1.33**	0	1	2.14	0	1
	14	20	19.30	0	1	**1.38**	0	1	2.17	0	1
	30	41	7200	100	0	40.93	0	1	**2.37**	0	1
	57	80	7200	100	0	7200	63.51	0	**4.26**	0	1
	118	176	7200	100	0	7200	100	0	**247.41**	0	1
Rand. gen. graphs	5	7	**1.30**	0	10	6.42	0	10	2.09	0	10
	10	14	2.20	0	10	**1.37**	0	10	2.16	0	10
	15	21	297.60	0	10	**1.48**	0	10	2.18	0	10
	20	28	7200	69.89	0	2.34	0	10	**2.26**	0	10
	25	35	7200	96.33	0	29.16	0	10	**2.64**	0	10
	30	42	7200	95.71	0	1820.58	5.31	8	**4.86**	0	10
	35	49	*7200*	*98.14*	0(1)	3789.77	16.81	6	**15.26**	0	10
	40	56	7200	93.11	0	*6316.36*	*33.56*	1(5)	**24.34**	0	10
	45	62	7200	98.57	0	7200	47.14	0	**148.58**	0	9
	50	70	7200	93.16	0	*7200*	*50.36*	0(4)	**414.24**	0	7

italics: average over instances that did not run out of memory

We note that the iterative model cannot be used to solve medium and larger size instances. The MILP model can solve instances with more larger size than the iterative one but cannot solve large size instances. The bilevel algorithm can solve almost all the instances considered in few seconds. It did not solve only 4 instances of the random generated graphs considered within the time limit. For small instances, MILP performs a little better than the bilevel. This is due to the choice of the solution selected at each iteration to generate the cutting plane in the bilevel algorithm. Hence some iterations may be needed to converge to the optimal solution in the bilevel algorithm while in the MILP model, having a small number of variables and constraints for those instances, the model converges in few seconds.

Therefore the bilevel algorithm is better in terms of running time and size of instances that can be solved.

Remark 1. We assumed here that the installation cost is the same for every PMU location at a node along an edge. If not, the problem consists then in finding the placement of PMUs that ensures a full observability of the graph and minimize the total installation cost. Let, $\forall i \in V, j \in \Gamma(i)$, c_{ij} be the cost of installing a PMU on $\{i, j\}$ at i. The new objective function is then given by:

$$\min \sum_{i \in V} \sum_{j \in \Gamma(i)} c_{ij}\, s_{ij}$$

Remark 2. The models proposed for PMU PLACEMENT PROBLEM, where PMUs are with unlimited number of channels and hence the placement is done on nodes, do not consider the propagation rule too. Our proposed models can easily be adapted to this node version.

6 Conclusions

We presented a new approach to model PES problem using a propagation rule based on Ohm's and Kirchoff's laws to reduce the number of PMUs to place. We proposed two mathematical models: an iterative model and a bilevel one. The iterative model is based on the observability propagation process and is given by a binary linear program. The bilevel model is deduced from the iterative one using fixed point method. We showed that we can transform the bilevel model to a MILP. We proposed also an algorithm to solve the bilevel model. We implemented these models and algorithm for the bilevel program and we performed tests on different IEEE bus systems and randomly generated graphs. The results showed that: the iterative model cannot be used for medium and large instances; the MILP model can solve instances with more large size than the iterative one but cannot solve large size instances; and the bilevel algorithm can solve instances with large sizes. Therefore, the bilevel algorithm is better in terms of running time and size of instances that can be solved. Further future work could be to model the case of conventional measures. We can also consider the case of line outage and single contingency of PMUs. Another further future work would be to generalize our models for the case of PMUs with limited channels ℓ. Also, due to maintenance or repairing works the electrical network topology is not fixed. Hence, another interesting perspective is to study the PMU PLACEMENT PROBLEM under these conditions by proposing a robust model and a solution method to solve it.

Acknowledgments. This work was carried out as part of the SOGRID project (www.so-grid.com), co-funded by the French agency for Environment and Energy Management (ADEME) and developed in collaboration between participating academic and industrial partners.

References

1. Aazami, A., Stilp, M.D.: Approximation algorithms and hardness for domination with propagation. In: Charikar, M., Jansen, K., Reingold, O., Rolim, J.D.P. (eds.) RANDOM 2007 and APPROX 2007. LNCS, vol. 4627, pp. 1–15. Springer, Heidelberg (2007)
2. Brueni, D.J., Heath, L.: The PMU placement problem. SIAM J. Discrete Math. **19**(3), 744–761 (2005)
3. Boyd, S., Vandenberghe, L.: Convex Optimization. Cambridge University Press, Cambridge (2004)
4. Dorfling, M., Henning, M.A.: A note on power domination in grid graphs. Discrete Appl. Math. **154**, 1023–1027 (2006)

5. Emami, R., Abur, A.: Robust measurement design by placing synchronized phasor measurements on network branches. IEEE Trans. Power Syst. **25**(1), 38–43 (2010)
6. Emami, R., Abur, A., Galvan, F.: Optimal placement of phasor measurements for enhanced state estimate: a case study. In: Proceedings of the 16th Power Systems Computation Conference, Glasgow, Scotland, pp. 1–6, 14–18 July 2008
7. Gross, J.L., Yellen, J.: Handbook of Graph Theory. Discrete Mathematics and Its Applications, vol. 25. CRC Press, Boca Raton (2003)
8. Haynes, T.W., Hedetniemi, S.M., Hedetniemi, S.T., Henning, M.A.: Domination in graphs applied to electric power networks. SIAM J. Discrete Math. **15**(4), 519–529 (2002)
9. Korkali, M., Abur, A.: Placement of PMUs with channel limits. In: IEEE Power and Energy Society General Meeting (2009)
10. Kumar, R., Rao, V.S.: Optimal placement of PMUs with limited number of channels. In: Proceedings of North American Power Symposium (NAPS), Boston, MA, pp. 1–7, 4–6 August 2011
11. Manousakis, N.M., Korres, G.N., Georgilakis, P.S.: Optimal placement of phasor measurement units: a literature review. In: Proceedings of the 16th International Conference on Intelligent System Application to Power Systems, ISAP 2011, Hersonissos, Greece, pp. 1–6, 25–28 September 2011
12. Manousakis, N.M., Korres, G.N., Georgilakis, P.S.: Taxonomy of PMU placement methodologies. IEEE Trans. Power Syst. **27**(2), 1070–1077 (2012)
13. West, D.B.: Introduction to Graph Theory, 2nd edn. Prentice Hall, Englewood Cliffs (2000)
14. Zimmerman, R.D., Murillo-Sánchez, C.E.: MATPOWER. http://www.pserc.cornell.edu/matpower/

Optimizing Static and Adaptive Probing Schedules for Rapid Event Detection

Ahmad Mahmoody(✉), Evgenios M. Kornaropoulos, and Eli Upfal

Department of Computer Science, Brown University, Providence, USA
{ahmad,evgenios,eli}@cs.brown.edu

Abstract. We formulate and study a fundamental search and detection problem, *Schedule Optimization*, motivated by a variety of real-world applications, ranging from monitoring content changes on the web, social networks, and user activities to detecting failure on large systems with many individual machines.

We consider a large system consists of many nodes, where each node has its own rate of generating new events, or items. A monitoring application can probe a small number of nodes at each step, and our goal is to compute a probing schedule that minimizes the expected number of undiscovered items at the system, or equivalently, minimizes the expected time to discover a new item in the system.

We study the Schedule Optimization problem both for deterministic and randomized memoryless algorithms. We provide lower bounds on the cost of an optimal schedule and construct close to optimal schedules with rigorous mathematical guarantees. Finally, we present an adaptive algorithm that starts with no prior information on the system and converges to the optimal memoryless algorithms by adapting to observed data.

1 Introduction

We introduce and study a fundamental stochastic search and detection problem, *Schedule Optimization*, that captures a variety of practical applications, ranging from monitoring content changes on the web, social networks, and user activities to detecting failure on large systems with many individual machines.

Our optimization problem consists of a large set of units, or *nodes*, that generate events, or *items*, according to a random process with known or unknown parameters. A detection algorithm can discover new items in the system by probing a small number of nodes in each step. This setting defines a discrete, infinite time process, and the goal of the stochastic optimization problem is to construct a probing schedule that minimizes the long term expected number of undiscovered items in the system, or equivalently, minimizes the expected time to discover a new item in the system.

We outline several important applications of this schedule optimization problem:

NSF grants IIS-1016648 and IIS-1247581

Z. Lu et al. (Eds.): COCOA 2015, LNCS 9486, pp. 377–391, 2015.
DOI: 10.1007/978-3-319-26626-8_28

News and Feed Aggregators. To provide up to date summary of the news, news aggregator sites need to constantly browse the Web, and often also the blogosphere and social networks, for new items. Scanning a site for new items requires significant communication and computation resources, thus the news aggregator can scan only a few sites simultaneously. The frequency of visiting a site has to depend on the likelihood of finding new items in that site. [1,9,21]

Algorithmic Trading on Data. An emerging trend in algorithmic stock trading is the use of automatic search through the Web, the blogosphere, and social networks for relevant information that can be used in fast trading, before it appears in the more popular news sites [4,5,8,11,14,16,17]. The critical issue in this application is the speed of discovering new events, but again there is a resource limit on the number of sites that the search algorithm can scan simultaneously.

Detecting Anomaly and Machine Malfunction. In large server farm or any other large collection of semi-autonomous machines a central controller needs to identify and contain anomalies and malefactions as soon as possible, before they spread in the system. To minimize interference with the system's operation the controller must probe only a small number of machines in each step.

1.1 Our Contribution

We consider an infinite, discrete time process in which n nodes generate new items according to a stochastic process which is governed by a generating vector π (see Sect. 3 for details). An algorithm can probe up to c nodes per step to discover all new items in these nodes. The goal is to minimize the *cost* of the algorithm (or the probing schedule), which we define as the long term (steady state) expected number of undiscovered items in the system.

We first show that the obvious approach of probing at each step the nodes with maximum expected number of undiscovered items at that step is not optimal. In fact, the cost of such a schedule can be arbitrary far from the optimal.

Our first result toward the study of efficient schedules is a lower bound on the cost of any deterministic or random schedule as a function of the generating vector π.

Next we assume that the generating vector π is known and study explicit constructions of deterministic and random schedules. We construct a deterministic schedule whose cost is within a factor of $(3 + (c - 1)/c)$ of the optimal cost, and a very simple, memoryless random schedule with cost that is within a factor of $(2 + (c - 1)/c)$ from optimal, where c is the maximum number of probes at each step.

Finally, we address the more realistic scenario in which the generating vector, π, is not known to the algorithm and may change in time. We construct an adaptive scheduling algorithm that learns from probing the nodes and converges to the optimal memoryless random schedule.

2 Related Work

The *News and Feed Aggregation* problem is a very well-studied topic, in which the general goal is to obtain the updates of news websites (e.g. by RSS feeds).

Among many introduced objectives [1,9,19] in studying this problem, the most similar one to our cost function is the *delay* function presented by [21]. In [21] it is assumed that the rates of the news publication does not change, where in our setting these rates may change and our algorithm (`Adaptive`) can adapt itself to the new setting. Also, we assume at any given time the number of probes is fixed (or bounded) regarding the limited computational power for simultaneous probes, but [21] uses a relaxed assumption by fixing the number of probes over a *time window* of a fixed length which may result in high number of probes at a single time step. Finally, [21] introduces a deterministic algorithm in which the number of probes to each feed is obtained by applying the Lagrange multipliers method (very similar result to Theorem 3), but they loose the guarantee on optimality of their solution, by rounding the estimated number of probes to integers. In contrast, our solution provides theoretical guarantee on optimality of our output schedule.

Web-crawling is another related topic, where a web-crawler aims to obtain the most recent snapshots of the web. However, it differs from our model substantially: in web-crawling algorithm data get *updated*, so missing some intermediate snapshot would not affect the quality of the algorithm, where in our model data are generated and they all need to be processed [3,22].

There has been an extensive work on *Outbreak Detection* (motivated in part by the "Battle of Water Sensors Network" challenge [20]) using statistic or mobile sensor in physical domains, and regarding a variety of objectives [7,10,13]. Our model deviates from the Outbreak Detection problem as it is geared to detection in virtual networks such as the Web or social networks embedded in the Internet, where a monitor can reach (almost) any node at about the same cost.

Another related problem is the *Emerging Topic Detection* problem, where the goal is to identify emergent topics in a social network, assuming full access to the stream of all postings. Besides having different objectives, our model differs mainly in this accessibility assumption: the social network providers have an immediate access to all tweets or postings as they are submitted to their servers, whereas in our model we consider an outside observer who needs an efficient mechanism to monitor changes, without having such full access privilege [2,15].

In the next section, we formally define our model and the Schedule Optimization problem.

3 Model and Problem Definition

We study an infinite, discrete time process in which a set of n *nodes*, indexed by $1, \ldots, n$, generate new *items* according to a random generating process. The generating process at a given time step is characterized by a *generating vector* $\pi = (\pi_1, \ldots, \pi_n)$, where π_i is the expected number of new items generated at node i at that step (by either a Bernoulli or a Poisson process). The generation processes in different nodes are independent.

We focus first on a *static generating process* in which the generating vector does not change in time. We then extend our results to adapt to generating vectors that change in time.

Our goal is to detect new events as fast as possible by probing in each step a small number of nodes. In particular, we consider probing schedules that can probe up to c nodes per step.

Definition 1 (Schedule). *A* c-schedule *is a function* $\mathcal{S} : \mathbb{N} \to \{1, \dots, n\}^c$ *specifying a set of* c *nodes to be probed at any time* $t \in \mathbb{N}$*. A deterministic function* $\mathcal{S}$ *defines a* deterministic *schedule, otherwise the schedule is* random.

Definition 2 (Memoryless Schedule). *A random schedule is* memoryless *if it is defined by a vector* $p = (p_1, \dots, p_n)$ *such that at any step the schedule probes a set* C *of* c *items with probability* $\prod_{j \in C} p_i$ *independent of any other event. In that case we use the notation* $\mathcal{S} = p$.

Definition 3 (Cyclic Schedule). *A schedule,* $\mathcal{S}$*, is* ℓ-cyclic *if there is a finite time* t_0 *such that from time* t_0 *on, the schedule repeats itself every period of* ℓ *steps. A schedule is cyclic if it is* ℓ*-cyclic for some positive integer* ℓ.

The quality of a probing schedule is measured by the speed in which it discovers new items in the system. When a schedule probes a node i at a time t, all items that were generated at that node by time $t - 1$ are discovered (thus, each item is not discovered in at least one step). We define the *cost* of a probing schedule as the long term expected number of undiscovered items in the system.

Definition 4 (Cost). *The* cost *of schedule* $\mathcal{S}$ *in a system of* n *nodes with generating vector* π *is*

$$cost(\mathcal{S}, \pi) = \lim_{t \to \infty} \frac{1}{t} \sum_{t'=1}^{t} \mathbb{E}\left[Q^{\mathcal{S}}(t')\right] = \lim_{t \to \infty} \frac{1}{t} \sum_{t'=1}^{t} \sum_{i=1}^{n} \mathbb{E}\left[Q_i^{\mathcal{S}}(t')\right],$$

where $Q_i^{\mathcal{S}}(t')$ *is the number of undiscovered items at node* i *and at time* t'*, and* $Q^{\mathcal{S}}(t') = \sum_{i=1}^{n} Q_i^{\mathcal{S}}(t')$*. The expectation is taken over the distribution of the generating system and the probing schedule.*

While the cost can be unbounded for some schedules, the cost of the optimal schedule is always bounded. To see that, consider a round-robin schedule, $\mathcal{S}$, that probes each node every n steps. Clearly no item is undiscovered in this schedule for more than n steps, and the expected number of items generated in an interval of n steps is $n \sum_{i=1}^{n} \pi_i$. Thus, $Q^{\mathcal{S}}(t) \leq n \sum_{i=1}^{n} \pi_i$, which implies $\text{cost}\,(\mathcal{S}, \pi) \leq n \sum_{i=1}^{n} \pi_i$. Therefore, without loss of generality we can restrict our discussion to bounded cost schedules. Also, note that when the sequence $\left\{\mathbb{E}\left[Q^{\mathcal{S}}(t)\right]\right\}_{t \in \mathbb{N}}$ converges we have $\text{cost}\,(\mathcal{S}, \pi) = \lim_{t \to \infty} \mathbb{E}\left[Q^{\mathcal{S}}(t)\right]$ (Cesaro Means [6]).

One can equivalently define the cost of a schedule in terms of the expected time that an item is in the system until it is discovered.

Lemma 1. *Let* $\omega_i^{\mathcal{S}}$ *be the expected waiting time of an item generated at node* i *until node* i *is probed by schedule* $\mathcal{S}$*. Then*

$$cost(\mathcal{S}, \pi) = \sum_{i=1}^{n} \pi_i \omega_i^{\mathcal{S}}.$$

Proof. Following the definition of the cost function we have

$$\text{cost}(\mathcal{S},\pi) = \lim_{t\to\infty}\frac{1}{t}\sum_{t'=1}^{t}\sum_{i=1}^{n}\mathbb{E}\left[Q_i^{\mathcal{S}}(t')\right] = \sum_{i=1}^{n}\left[\lim_{t\to\infty}\frac{\sum_{t'=1}^{t}\mathbb{E}\left[Q_i^{\mathcal{S}}(t')\right]}{t}\right]$$
$$= \sum_{i=1}^{n}\pi_i\omega_i^{\mathcal{S}},$$

where the last eqaulity is obtained by applying Little's Law [12]. □

Corollary 1. *A schedule that minimizes the expected number of undiscovered items in the system simultaneously minimizes the expected time that an item is undiscovered.*

Corollary 2. *For any schedule $\mathcal{S}$, cost$(\mathcal{S},\pi) \geq \sum_{i=1}^{n}\pi_i$.*

Proof. As mentioned above, when we probe a node i at time t we discover only the items that have been generated by time $t-1$. Therefore, $\omega_i^{\mathcal{S}} \geq 1$, and by Lemma 1 the proof is complete. □

Now, our main problem is defined as the following:

Definition 5 (Schedule Optimization). *Given a generating vector π and a positive integer c, find a c-schedule with minimum cost.*

When the generating vector is not known a priori to the algorithm the goal is to design a schedule that *converges* to an optimal one. For that we need the following definition:

Definition 6 (Convergence). *We say schedule $\mathcal{S}$* converges *to schedule $\mathcal{S}'$, if for any generating vector π, $\lim_{t\to\infty}\left|\mathbb{E}\left[Q^{\mathcal{S}}(t)\right] - \mathbb{E}\left[Q^{\mathcal{S}'}(t)\right]\right| = 0$.*

4 Results

We start this section by, first, showing that the obvious approach of maximizing the expected number of detections at each step is far from optimal. We then prove a lower bound on the cost of any schedule, and provide deterministic and memoryless c-schedules that are within a factor of $(3+(c-1)/c)$ and $(2+(c-1)/c)$, respectively, from the optimal. Finally, we introduce an algorithm, `Adaptive` , which outputs a schedule $\mathcal{A}$ that converges to the optimal *memoryless* 1-schedule when the generating vector π is not known in advance. We also show that `Adaptive` can be used to obtain a c-schedule $\mathcal{A}^c$ whose cost is within $(2+(c-1)/c)$ factor of any optimal c-schedule.

Throughout this section, by $\tau_i^{\mathcal{S}}(t)$ we mean the number of steps from the last time that node i was probed until time t, while executing schedule $\mathcal{S}$; if i has not been probed so far, we let $\tau_i^{\mathcal{S}}(t) = t$. Using the definition, it is easy to see that

$$\mathbb{E}\left[Q_i^{\mathcal{S}}(t)\right] = \pi_i\mathbb{E}\left[\tau_i^{\mathcal{S}}(t)\right], \tag{1}$$

when the expectations are over the randomness of *both* $\mathcal{S}$ and π. Therefore, if the expectation is over *only* the randomness of π we have

$$\mathbb{E}\left[Q_i^{\mathcal{S}}(t)\right] = \pi_i\tau_i^{\mathcal{S}}(t). \tag{2}$$

4.1 On Maximizing Immediate Gain

Let $\mathcal{S}$ be a 1-schedule that at each step, probes the node with the maximum expected number of undetected items. By (2), the expected number of undetected items at node i and at time t is $\pi_i \tau_i^{\mathcal{S}}(t)$, and thus, $\mathcal{S}(t) = \arg\max_i \pi_i \tau_i^{\mathcal{S}}(t)$.

Now, suppose $\pi_i = 2^{-i}$, for $1 \leq i \leq n$. Since the probability that node 1 has an undetected item in each step is at least 1/2, node i is probed no more than once in each 2^{i-1} steps. Thus, the expected number of time steps that an item at node i will stay undetected is at least $\frac{1}{2^{i-1}}(1+\ldots+2^{i-1}) = \frac{2^{i-1}+1}{2} > 2^{i-2}$. Using Lemma 1, the cost of this schedule is at least $\sum_{i=1}^{n} \pi_i \omega_i > \sum_{i=1}^{n} 2^{-i} 2^{i-2} = \Omega(n)$. Now, consider an alternative schedule that probes node i in each step with probability $2^{-i/2}/Z$, where $Z = \sum_{j=1}^{n} 2^{-j/2}$. The expected number of steps between two probes of i is $Z/2^{-i/2}$, and the cost of this schedule is

$$\sum_{i=1}^{n} 2^{-i} \left(\frac{2^{-i/2}}{\sum_{j=1}^{n} 2^{-j/2}} \right)^{-1} = \left(\sum_{j=1}^{n} 2^{-j/2} \right)^{2} = O(1).$$

Thus, optimizing immediate gain is not optimal in this problem.

4.2 Lower Bound on Optimal Cost

In this section we provide a lower bound on the optimal cost, i.e., the cost of an optimal schedule.

Theorem 1. *For any c-schedule $\mathcal{O}$ with finite cost we have*

$$cost(\mathcal{O}, \pi) \geq \max \left\{ \sum_{i=1}^{n} \pi_i, \frac{1}{2c} \left(\sum_{i=1}^{n} \sqrt{\pi_i} \right)^{2} \right\}.$$

Proof. First, by Corollary 2, cost $(\mathcal{O}, \pi) \geq \sum_{i=1}^{n} \pi_i$. Now we show cost $(\mathcal{O}, \pi) \geq \frac{1}{2c} \left(\sum_{i=1}^{n} \sqrt{\pi_i} \right)^2$. Fix a positive integer $t > 0$, and suppose during the time interval $[0, t]$, $\mathcal{O}$ probes node i at steps $t_1, t_2, \ldots, t_{n_i}$. Let $t_0 = 0$ and $t_{n_i+1} = t$. So, the sequence $t_0, \ldots, t_{n_i+1}$ partition the interval $[0, t]$ into $n_i + 1$ intervals $I_i(j) = [t_j + 1, t_{j+1}]$, for $0 \leq j \leq n_i$, and the length of $I_i(j)$ is $\ell_i(j) = t_{j+1} - t_j$. Applying the Cauchy-Schwartz inequality we have:

$$\sum_{j=0}^{n_i} \ell_i(j)^2 \sum_{j=0}^{n_i} 1 \geq \left(\sum_{j=0}^{n_i} \ell_i(j) \right)^2$$

$$\Longrightarrow \sum_{j=0}^{n_i} \ell_i(j)^2 \geq \frac{1}{n_i + 1} \left(\sum_{j=0}^{n_i} \ell_i(j) \right)^2 = \frac{t^2}{n_i + 1} = \frac{t^2}{n_i} \left(1 - \frac{1}{n_i + 1} \right).$$

For $t' \in I_i(j)$, $Q_i^{\mathcal{O}}(t')$ is a Poisson random variable with parameter $\pi_i(t' - t_j)$. Therefore,

$$\sum_{t=1}^{t} \mathbb{E}\left[Q_i^{\mathcal{O}}(t')\right] = \sum_{j=0}^{n_i} \sum_{t' \in I_i(j)} \mathbb{E}\left[Q_i^{\mathcal{O}}(t')\right] = \pi_i \sum_{j=0}^{n_i} (1 + \ldots + \ell_i(j))$$
$$= \pi_i \sum_{j=0}^{n_i} \frac{\ell_i(j)(\ell_i(j)+1)}{2} \geq \frac{\pi_i}{2} \sum_{j=0}^{n_i} \ell_i(j)^2 \geq \frac{\pi_i}{2} \frac{t^2}{n_i}\left(1 - \frac{1}{n_i+1}\right).$$

By summing over all nodes and averaging over t, we have

$$\sum_{i=1}^{n} \sum_{t'=1}^{t} \frac{1}{t} \mathbb{E}\left[Q_i^{\mathcal{O}}(t')\right] \geq \sum_{i=1}^{n} \frac{1}{t} \frac{\pi_i}{2} \frac{t^2}{n_i}\left(1 - \frac{1}{n_i+1}\right)$$
$$= \sum_{i=1}^{n} \frac{\pi_i}{2} \frac{t}{n_i}\left(1 - \frac{1}{n_i+1}\right) \tag{3}$$
$$\geq \frac{1}{c}\left(\sum_{i=1}^{n} \frac{n_i}{t}\right)\left(\sum_{i=1}^{n} \frac{\pi_i}{2} \frac{t}{n_i}\left(1 - \frac{1}{n_i+1}\right)\right)$$
$$\geq \frac{1}{2c}\left(\sum_{i=1}^{n} \sqrt{\pi_i} \sqrt{\left(1 - \frac{1}{n_i+1}\right)}\right)^2, \tag{4}$$

where in the second line we use the fact that if the schedule executed c probes in each step then $\sum_{i=1}^{n} \frac{n_i}{t} \leq c$, and the third line is obtained by applying the Cauchy-Schwartz inequality.

It remains to show that for any schedule with finite cost, and any i such that $\pi_i > 0$, $\lim_{t \to \infty} n_i = \infty$. For sake of contradiction assume that there is a time s such that the node i is never probed by $\mathcal{O}$ at time $t > s$. So, $\mathbb{E}\left[Q_i^{\mathcal{O}}(t)\right] = \pi(t-s)$ and we have $\text{cost}(\mathcal{O}, \pi) \geq \frac{1}{t} \sum_{t'=s}^{t} \mathbb{E}\left[Q_i^{\mathcal{O}}\right](t) = \frac{\pi_i}{t} \frac{(t-s)(t-s-1)}{2}$ which converges to ∞ as $t \to \infty$, which is a contradiction. Hence, for all i, $\lim_{t \to \infty} n_i = \infty$, and using (4.2) we obtain

$$\text{cost}(\mathcal{O}, \pi) \geq \lim_{t \to \infty} \frac{1}{2c}\left(\sum_{i=1}^{n} \sqrt{\pi_i} \sqrt{\left(1 - \frac{1}{n_i+1}\right)}\right)^2 = \frac{1}{2c}\left(\sum_{i=1}^{n} \sqrt{\pi_i}\right)^2,$$

which completes the proof. □

4.3 Deterministic $(3 + (c-1)/c)$-Approximation Schedule

We construct a deterministic 1-schedule in which each node i is probed approximately every $n_i = \frac{\sum_{j=1}^{n} \sqrt{\pi_j}}{\sqrt{\pi_i}}$ steps, and using that, present our $(3 + (c-1)/c)$-approximation schedule. For each i let r_i be a nonnegative integer such that $2^{r_i} \geq n_i > 2^{r_i - 1}$, and let $\rho = \max_i r_i$.

Lemma 2. *There is a 2^ρ-cyclic 1-schedule $\mathcal{D}$ such that node i is probed exactly every 2^{r_i} steps.*

Proof. Without loss of generality assume $\sum_{i=1}^n 2^{-r_i} = 1$, otherwise we can add auxiliary nodes to complete the sum to 1, with the powers (r_i's) associated with the auxiliary nodes all bounded by ρ.

We prove the lemma by induction on ρ. If $\rho = 0$, then there is only one node, and the schedule is 1-cyclic. Now, assume the statement holds for all $\rho' < \rho$. Since the smallest frequency is $2^{-\rho}$, and the sum of the frequencies is 1, there must be two nodes, v and u, with same frequency $2^{-\rho}$. Join the two nodes to a new node w with frequency $2^{-\rho+1}$. Repeat this process for all nodes with frequency $2^{-\rho}$. We are left with a collection of nodes all with frequencies $> 2^{-\rho}$. By the inductive hypothesis there is a $\left(2^{\rho-1}\right)$-cyclic schedule $\mathcal{D}'$ such that each node i is probed exactly each 2^{r_i} steps. In particular a node w that replaced u and v is probed exactly each $2^{-\rho+1}$ steps.

Now, we create an 2^ρ-schedule, $\mathcal{D}$, whose cycle is obtained by repeating the cycle of $\mathcal{D}'$ two times. For each probe to w that replaced a pair u, v, in the first cycle we probe u and in the second cycle we probe v. Thus, u and v are probed exactly every 2^ρ steps, and the new schedule does not change the frequency of probing nodes with frequency larger than $2^{-\rho}$. □

Theorem 2. *The cost of the deterministic 1-schedule $\mathcal{D}$ is no more than 3 times of the optimal cost.*

Proof. By Lemma 2 each node i is probed exactly every 2^{r_i} steps. Using $2^{r_i-1} < \frac{\sum_{j=1}^n \sqrt{\pi_j}}{\sqrt{\pi_i}}$ we have $2^{r_i} + 1 \le \frac{2\cdot\sum_{j=1}^n \sqrt{\pi_j}}{\sqrt{\pi_i}} + 1$, and therefore

$$
\begin{aligned}
\lim_{t\to\infty} \frac{1}{t} \sum_{t'=1}^{t} \mathbb{E}\left[Q_i^{\mathcal{D}}(t')\right] &= \lim_{t\to\infty} \frac{1}{t}\frac{t}{2^{r_i}} \sum_{t'=1}^{2^{r_i}} \mathbb{E}\left[Q_i^{\mathcal{D}}(t')\right] = \frac{1}{2^{r_i}} \sum_{t'=1}^{2^{r_i}} \pi_i t' \\
&= \frac{\pi_i}{2^{r_i}} \frac{2^{r_i}(2^{r_i}+1)}{2} \le \frac{\pi_i}{2} \left(\frac{2\sum_{j=1}^n \sqrt{\pi_j}}{\sqrt{\pi_i}} + 1 \right) \\
&= \sqrt{\pi_i} \cdot \sum_{j=1}^{n} \sqrt{\pi_j} + \frac{\pi_i}{2}.
\end{aligned}
$$

Thus by Theorem 1, we have

$$
\begin{aligned}
\mathrm{cost}\left(\mathcal{D}, \pi\right) &= \lim_{t\to\infty} \frac{1}{t} \sum_{i=1}^{n} \sum_{t'=1}^{t} \mathbb{E}\left[Q_i^{\mathcal{D}}(t')\right] \le \sum_{i=1}^{n} \left(\sqrt{\pi_i} \cdot \sum_{j=1}^{n} \sqrt{\pi_j} \right) + \frac{1}{2} \sum_{i=1}^{n} \pi_i \\
&= \left(\sum_{j=1}^{n} \sqrt{\pi_j} \right)^2 + \frac{1}{2} \sum_{j=1}^{n} \pi_j \le 3 \cdot \mathrm{cost}\left(\mathcal{O}, \pi\right),
\end{aligned}
$$

where $\mathrm{cost}\left(\mathcal{O}, \pi\right)$ is the optimal cost. □

Using the previous deterministic 1-schedule, the following corollary provides a c-schedule whose cost is within $(3 + (c-1)/c)$ factor of the optimal cost.

Corollary 3. *There is a deterministic c-schedule $\mathcal{D}^c$ whose cost is at most $(3 + (c-1)/c)$ times of the optimal cost.*

Proof. Consider the execution of the deterministic 1-schedule $\mathcal{D}$ constructed in Theorem 2 on generating vector $\frac{1}{c}\pi$. Let $\mathcal{D}^c$ be a deterministic c-schedule obtained by grouping c consecutive probes of $\mathcal{D}$ into one step. Suppose $\mathcal{O}$ is an optimal c-schedule. Applying the equation in Lemma 1,

$$\begin{aligned}\text{cost}\,(\mathcal{D}^c, \pi) &= \sum_{i=1}^{n} \pi_i \omega_i^{\mathcal{D}^c} = \sum_{i=1}^{n} \frac{\pi_i}{c} c\omega_i^{\mathcal{D}^c} \leq \sum_{i=1}^{n} \frac{\pi_i}{c}(\omega_i^{\mathcal{D}} + c - 1)\\ &= \text{cost}\,(\mathcal{D}, \pi) + \sum_{i=1}^{n} \frac{(c-1)\pi_i}{c} \leq 3\left(\sum_{i=1}^{n} \sqrt{\frac{\pi_i}{c}}\right)^2 + \sum_{i=1}^{n} \frac{(c-1)\pi_i}{c}\\ &= \frac{3}{2c}\left(\sum_{i=1}^{n} \sqrt{\pi_i}\right)^2 + \sum_{i=1}^{n} \frac{(c-1)\pi_i}{c} \leq \left(3 + \frac{c-1}{c}\right) \text{cost}\,(\mathcal{O}, \pi),\end{aligned}$$

where the first inequality holds because some items could be detected in less than c steps in the 1-schedule but are counted in one full step of the c-schedule. The last inequality is obtained by applying Theorem 1. □

4.4 On Optimal Memoryless Schedule

Here, we consider memoryless schedules, and show that the memoryless 1-schedule with minimum cost can be easily computed. We call a memoryless schedule with minimum cost among memoryless schedules, an optimal memoryless schedule. We also provide an upper bound on the minimum cost of a memoryless c-schedule.

Theorem 3. *Let $\mathcal{R} = (p_1, \ldots, p_n)$ be a memoryless 1-schedule. Then we have $cost(\mathcal{R}, \pi) \geq \left(\sum_{i=1}^{n} \sqrt{\pi_i}\right)^2$, and the equality holds if and only if $p_i = \frac{\sqrt{\pi_i}}{\sum_{j=1}^{n} \sqrt{\pi_j}}$, for all i.*

Proof. Since probing each node i is a geometric distribution with parameter p_i, the expected time until an item generated in node i is discovered, is $\omega_i^{\mathcal{R}} = 1/p_i$. Therefore, by Lemma 1, we have $\text{cost}\,(\mathcal{R}, \pi) = \sum_{i=1}^{n} \frac{\pi_i}{p_i}$. We find $p^* = \text{argmin}_{\mathcal{S}=p}\text{cost}\,(\mathcal{R}, \pi)$, using the Lagrange multipliers:

$$\frac{\partial}{\partial p_j}\left(\sum_{i=1}^{n} \frac{\pi_i}{p_i} + \lambda \sum_{i=1}^{n} p_i\right) = 0 \Longrightarrow p_j \propto \sqrt{\pi_j}.$$

Therefore, cost $(\mathcal{R}, \pi)$ is minimized if $p_i = \frac{\sqrt{\pi_i}}{\sum_{j=1}^{n} \sqrt{\pi_j}}$, and in this case the (minimized) cost will be

$$\text{cost}\,(\mathcal{R}, \pi) = \sum_{i=1}^{n} \left(\sqrt{\pi_i} \cdot \sum_{j=1}^{n} \sqrt{\pi_j} \right) = \left(\sum_{i=1}^{n} \sqrt{\pi_i} \right)^2 .$$

□

Corollary 4. *The cost of the optimal memoryless 1-schedule is within a factor of 2 of the cost of any optimal 1-schedule.*

Proof. The cost of the schedule $\mathcal{R}$ in Theorem 3 is $\left(\sum_{i=1}^{n} \sqrt{\pi_i}\right)^2$, which is bounded by 2. cost $(\mathcal{O}, \pi)$ for an optimal 1-schedule $\mathcal{O}$ using Theorem 1. □

Corollary 5. *There is memoryless c-schedule, $\mathcal{R}^c$, whose cost is within a factor of $(2 + (c-1)/c)$ of any optimal c-schedule.*

Proof. Suppose $\mathcal{R}^c$ is a memoryless c-schedule obtained by choosing c probes in each step, each chosen according to the optimal memoryless 1-schedule, $\mathcal{R}$, computed in Theorem 3. Using the same argument as in the proof of Corollary 3 we have

$$\text{cost}\,(\mathcal{R}^c, \pi) \leq \frac{1}{c} \left(\sum_{i=1}^{n} \sqrt{\pi_i} \right)^2 + \frac{c-1}{c} \sum_{i=1}^{n} \pi_i \leq \left(2 + \frac{c-1}{c} \right) \text{cost}\,(\mathcal{O}, \pi) ,$$

for an optimal c-schedule $\mathcal{O}$. □

4.5 On Adaptive Algorithm for Memoryless Schedules

Assume now that the scheduling algorithm starts with no information on the generating vector π (or that the vector has changed). We design and analyze an adaptive algorithm, `Adaptive`, that outputs a schedule $\mathcal{A}$ convergent to the optimal memoryless algorithm $\mathcal{R}$ (see Sect. 4.4) by gradually learning the vector π by observing the system. To simplify the presentation we present and analyze a 1-schedule algorithm. The results easily scale up to any integer $c > 1$, where the adaptive algorithm outputs a c-schedule convergent to $\mathcal{R}^c$ (as in Sect. 4.4).

Each iteration of the algorithm `Adaptive` starts with $\tilde{\pi} = (\tilde{\pi}_1, \ldots, \tilde{\pi}_n)$ as an estimate of the unknown generating vector $\pi = (\pi_1, \ldots, \pi_n)$. Based on this estimate the algorithm chooses to probe node i with probability $p_i(t) = \frac{\sqrt{\tilde{\pi}_i}}{\sum_{j=1}^{n} \sqrt{\tilde{\pi}_j}}$ (which is the optimal memoryless schedule if $t\pi$ was the correct estimate). If nodes i is probed at time t, the estimate of π_i is updated to $\tilde{\pi}_{i_0} \leftarrow \frac{\max(1, c_{i_0})}{t}$, where c_{i_0} is the total number of new items discovered in that node since time 0.

Algorithm 1: `Adaptive`

```
Outputs: A(t), for t = 1,2,....
begin
    (c_1,...,c_n) ← (0,...,0);
    (π̃_1,...,π̃_n) ← (1,...,1);
    for t = 1,2,... do
        for i ∈ {1,...,n} do
            p_i(t) ← √π̃_i / Σ_{j=1}^n √π̃_j ;
        A(t) ~ p(t);
        output A(t);
        c' ← number of new items caught at i_0;
        c_{i_0} ← c_{i_0} + c';
        π̃_{i_0} ← max(1, c_{i_0}) / t ;
end
```

We denote the output of `Adaptive` schedule by $\mathcal{A}$ and the optimal memoryless 1-schedule by $\mathcal{R} = p^* = (p_1^*, \dots, p_n^*)$; see Sect. 4.4. Our main result of this section is the following theorem.

Theorem 4. *The schedule $\mathcal{A}$ converges to $\mathcal{R}$, and thus, $cost(\mathcal{A}, \pi) = cost(\mathcal{R}, \pi)$.*

To prove Theorem 4 we need the following lemmas.

Lemma 3. *For any time t and $i \in [n]$ we have $p_i(t) \geq \frac{1}{n\sqrt{t}}$.*

Proof. It is easy to see that $p_i(t)$ will reach its lowest value at time t only if for $j \neq i$ we have $\tilde{\pi}_j = 1$ and $\tilde{\pi}_i = \frac{1}{t-1}$ (which requires i to be probed at time $t-1$). Therefore, $p_i(t) \geq \frac{1/\sqrt{t-1}}{1/\sqrt{t-1}+n-1} = \frac{1}{1+(n-1)\sqrt{t-1}} \geq \frac{1}{n\sqrt{t}}$. □

Define $\delta(t) = 4n\exp\left(-\frac{\pi_* t^{1/3}}{6}\right)$ and let N_0 be the smallest integer t such that $\exp\left(-\frac{\sqrt{t}}{2n}\right) \leq 2\exp\left(-\frac{\pi_* t^{1/3}}{6}\right)$. Note that one can choose $\delta(t) = 4ne^{-\frac{\pi_* t^{1/2-\epsilon}}{6}}$ for any $\epsilon \in (0, 1/2)$, and for convenience we chose $\epsilon = 1/6$.

Lemma 4. *For any time $t \geq N_0$, with probability $\geq 1 - \delta(t)/2$, all the nodes are probed during the time interval $[t/2, t)$.*

Proof. By Lemma 3, the probability of not probing i during the time interval $[t/2, t)$ is at most

$$\prod_{t'=t/2}^{t-1} (1 - p_i(t')) \leq \left(1 - \frac{1}{n\sqrt{t}}\right)^{t/2} \leq e^{-\frac{t}{2n\sqrt{t}}} \leq 2\exp\left(-\frac{\pi_* t^{1/3}}{6}\right) = \frac{\delta(t)}{2n}.$$

A union bound over all the nodes completes the proof. □

Lemma 5. *Suppose node i is probed at a time $t' > t/2$. Then,*

$$Pr\left[|\tilde{\pi}_i(t') - \pi_i| > t^{-\frac{1}{3}}\pi_i\right] < \frac{\delta(t)}{2n}.$$

Proof. We estimate π from $t' > t/2$ steps, each with π_i expected number of new items. Applying a Chernoff bound [18] for the sum of t' independent random variables with either Bernulli or Poisson distribution we have

$$\Pr\left[|\tilde{\pi}_i(t') - \pi_i| > t^{-\frac{1}{3}}\pi_i\right] < 2\exp\left(-\frac{t^{-\frac{2}{3}}\pi_i t'}{3}\right) \leq 2\exp\left(-\frac{t^{-\frac{2}{3}}\pi_* t}{6}\right) = \frac{\delta(t)}{2n}.$$

□

Note that by union bound, Lemma 5 holds, with probability at least $1 - \delta(t)/2$, for all the nodes that are probed after $t/2$.

Lemma 6. *Suppose $t \geq N_0$. With probability at least $1 - \delta(t)$ we have for all $i \in [n]$,*

$$\left(1 - \frac{1}{t^{1/3}+1}\right)p_i^* \leq \sqrt{\frac{1-t^{-1/3}}{1+t^{-1/3}}}p_i^* \leq p_i(t) \leq \sqrt{\frac{1+t^{-1/3}}{1-t^{-1/3}}}p_i^* \leq \left(1 + \frac{1}{t^{1/3}-1}\right)p_i^*$$

Proof. Applying Lemmas 4, 5 and a union bound, with probability $1 - \delta(t)$ all the nodes are probed during the time $[t/2, t)$ and $|\tilde{\pi}_i(t) - \pi_i| \leq t^{-1/3}\pi_i$ for all $i \in [n]$. Since $p_i(t) = \frac{\sqrt{\tilde{\pi}_i(t)}}{\sum_j \sqrt{\tilde{\pi}_j(t)}}$, we obtain

$$p_i(t) \geq \frac{\sqrt{(1-t^{-1/3})\pi_i}}{\sum_j \sqrt{(1+t^{-1/3})\pi_j}} = \sqrt{\frac{1-t^{-1/3}}{1+t^{-1/3}}}\frac{\sqrt{\pi_i}}{\sum_j\sqrt{\pi_j}} = \sqrt{\frac{1-t^{-1/3}}{1+t^{-1/3}}}p_i^* \geq (1 - \frac{1}{t^{1/3}+1})p_i^*$$

where the last inequality uses the Taylor series of $\sqrt{1+x}$. The upper bound is obtained by a similar argument. □

Corollary 6. *The variation distance between the distribution used by algorithm* `Adaptive` *at time $t \geq N_0$, $p(t) = (p_1(t), \ldots, p_n(t))$, and the distribution $p^* = (p_1^*, \ldots, p_n^*)$ used by the optimal memoryless algorithm satisfy*

$$\| p(t) - p^* \| = \frac{1}{2}\sum_{i=1}^{n} |p_i(t) - p_i^*| \leq \frac{n}{t^{1/3}-1} + \delta(t) \stackrel{t\to\infty}{\longrightarrow} 0.$$

Finally, we present our proof for Theorem 4.

Proof of Theorem 4. Recall that we defined $\tau_i^S(t)$ as the number of steps from the last time that node i was probed until time t in an execution of an schedule S, and $\mathbb{E}\left[Q_i^S\right] = \pi_i \mathbb{E}\left[\tau_i^S(t)\right]$.

Let $F(t)$ indicate the event that the inequalities in Lemma 6 are held for $\forall t' \in [t/2, t)$. Therefore, $\Pr[F(t)] < 1 - \frac{\delta(t/2)\cdot t}{2}$ by applying union bound over all $t' \in [t/2, t)$, and using the fact that $\delta(t') \leq \delta(t/2)$. Therefore,

$$\begin{aligned}\left|\mathbb{E}\left[Q^{\mathcal{A}}(t)\right] - \mathbb{E}\left[Q^{\mathcal{R}}(t)\right]\right| &= \left|\sum_{i=1}^{n} \pi_i \mathbb{E}\left[\tau_i^{\mathcal{A}}(t)\right] - \sum_{i=1}^{n} \pi_i \mathbb{E}\left[\tau_i^{\mathcal{R}}(t)\right]\right| \\ &\leq \sum_{i=1}^{n} \pi_i \left|\mathbb{E}\left[\tau_i^{\mathcal{A}}(t)\right] - \mathbb{E}\left[\tau_i^{\mathcal{R}}(t)\right]\right| \\ &\leq \sum_{i=1}^{n} \pi_i \left|\mathbb{E}\left[\tau_i^{\mathcal{A}}(t)\right] - \mathbb{E}\left[\tau_i^{\mathcal{A}}(t) \mid F(t)\right]\right| \\ &+ \sum_{i=1}^{n} \pi_i \left|\mathbb{E}\left[\tau_i^{\mathcal{A}}(t) \mid F(t)\right] - \mathbb{E}\left[\tau_i^{\mathcal{R}}(t)\right]\right|,\end{aligned}$$

where we used the triangle inequality for both inequalities. So, it suffices to show that for every i,

$$\lim_{t\to\infty} \left|\mathbb{E}\left[\tau_i^{\mathcal{A}}(t)\right] - \mathbb{E}\left[\tau_i^{\mathcal{A}}(t) \mid F(t)\right]\right| = \lim_{t\to\infty} \left|\mathbb{E}\left[\tau_i^{\mathcal{A}}(t) \mid F(t)\right] - \mathbb{E}\left[\tau_i^{\mathcal{R}}(t)\right]\right| = 0.$$

Obviously, $\tau_i^{\mathcal{A}}(t) \leq t$. Now by letting $t \geq 2N_0$ we have,

$$\begin{aligned}\mathbb{E}\left[\tau_i^{\mathcal{A}}(t)\right] &= \Pr[F(t)]\,\mathbb{E}\left[\tau_i^{\mathcal{A}}(t) \mid F(t)\right] + \Pr[\neg F(t)]\,\mathbb{E}\left[\tau_i^{\mathcal{A}}(t) \mid \neg F(t)\right] \\ &\leq \mathbb{E}\left[\tau_i^{\mathcal{A}}(t) \mid F(t)\right] + \frac{\delta(t/2)t}{2} t = \mathbb{E}\left[\tau_i^{\mathcal{A}}(t) \mid F(t)\right] + \frac{\delta(t/2)t^2}{2}. \qquad (5)\end{aligned}$$

We also get

$$\begin{aligned}\mathbb{E}\left[\tau_i^{\mathcal{A}}(t)\right] &\geq \left(1 - \frac{\delta(t/2)t}{2}\right) \mathbb{E}\left[\tau_i^{\mathcal{A}}(t) \mid F(t)\right] \qquad (6) \\ &= \mathbb{E}\left[\tau_i^{\mathcal{A}}(t) \mid F(t)\right] - \frac{\delta(t/2)t}{2} \mathbb{E}\left[\tau_i^{\mathcal{A}}(t) \mid F(t)\right] \geq \mathbb{E}\left[\tau_i^{\mathcal{A}}(t) \mid X\right] - \frac{\delta(t)t^2}{2}.\end{aligned}$$

Note that $\lim_{t\to\infty} \frac{\delta(t/2)t^2}{2} = \lim_{t\to\infty} 4ne^{-\frac{\pi_* t^{1/3}}{6\sqrt[3]{2}}} t^2 = 0$, and thus by (5) and (6) we have

$$\lim_{t\to\infty} \mathbb{E}\left[\tau_i^{\mathcal{A}}(t)\right] - \mathbb{E}\left[\tau_i^{\mathcal{A}}(t) \mid F(t)\right] = 0 \Rightarrow \lim_{t\to\infty} \left|\mathbb{E}\left[\tau_i^{\mathcal{A}}(t)\right] - \mathbb{E}\left[\tau_i^{\mathcal{A}}(i) \mid F(t)\right]\right| = 0. \qquad (7)$$

Now, we show that $\lim_{t\to\infty} \left|\mathbb{E}\left[\tau_i^{\mathcal{A}}(t) \mid F(t)\right] - \mathbb{E}\left[\tau_i^{\mathcal{R}}(t)\right]\right| = 0$. So here, we assume $F(t)$ holds. So for every $i \in [n]$, node i is probed in $[t/2, t)$, and for all $t' \in [t/2, t)$ we have

(i) $p_i(t') \geq \left(1 - \frac{1}{t'^{1/3}+1}\right) p_i^* \geq \left(1 - \frac{1}{(t/2)^{1/3}+1}\right) p_i^*$. So,

$$\mathbb{E}\left[\tau_i^{\mathcal{A}}(t) \mid F(t)\right] \leq \left(1 - \frac{1}{(t/2)^{1/3}+1}\right)^{-1} \frac{1}{p_i^*} = \left(1 + (t/2)^{-1/3}\right) \frac{1}{p_i^*}.$$

(ii) $p_i(t') \leq \left(1 + \frac{1}{t'^{1/3}-1}\right) p_i^* \leq \left(1 + \frac{1}{(t/2)^{1/3}-1}\right) p_i^*$. Hence,

$$\mathbb{E}\left[\tau_i^{\mathcal{A}}(t) \mid F(t)\right] \geq \left(1 + \frac{1}{(t/2)^{1/3}-1}\right)^{-1} \frac{1}{p_i^*} = \left(1 - (t/2)^{-1/3}\right) \frac{1}{p_i^*}.$$

Obviously $\mathbb{E}\left[\tau_i^{\mathcal{R}}(t)\right] = \frac{1}{p_i^*}$, since probing node i by $\mathcal{R}$ can be viewed as a geometric distribution with parameter p_i^*, and since $\delta(t) \to 0$ as $t \to \infty$ we have

$$\begin{aligned}\mathbb{E}\left[\tau_i^{\mathcal{R}}(t)\right] &= \frac{1}{p_i^*} = \lim_{t\to\infty}\left(1 - (t/2)^{-1/3}\right)\frac{1}{p_i^*} \leq \lim_{t\to\infty} \mathbb{E}\left[\tau_i^{\mathcal{A}}(t) \mid F(t)\right] \\ &\leq \lim_{t\to\infty}\left(1 + (t/2)^{-1/3}\right)\frac{1}{p_i^*} = \frac{1}{p_i^*} = \mathbb{E}\left[\tau_i^{\mathcal{R}}(t)\right].\end{aligned}$$

Therefore,

$$\lim_{t\to\infty} \left|\mathbb{E}\left[\tau_i^{\mathcal{A}}(t) \mid F(t)\right] - \mathbb{E}\left[\tau_i^{\mathcal{R}}(t)\right]\right| = 0. \tag{8}$$

Thus, by (7) and (8) we have $\lim_{t\to\infty} \left|\mathbb{E}\left[Q^{\mathcal{A}}(t)\right] - \mathbb{E}\left[Q^{\mathcal{R}}(t)\right]\right| = 0$, and $\mathcal{A}$ converges to $\mathcal{S}$, and since $\lim_{t\to\infty} \mathbb{E}\left[Q^{\mathcal{S}}(t)\right] = \left(\sum_{i=1}^n \sqrt{\pi_i}\right)^2$, it implies that $\lim_{t\to\infty} \mathbb{E}\left[Q^{\mathcal{A}}(t)\right] = \left(\sum_{i=1}^n \sqrt{\pi_i}\right)^2 = \text{cost}(\mathcal{A}, \pi)$ (by Cesaro Mean [6]). □

Note that one can obtain an adaptive schedule $\mathcal{A}^c$ by choosing c probes in each step, at each round of `Adaptive`, and using similar argument as in Sect. 4.4 (and similar to Corollary 3), it is easy to see that $\mathcal{A}^c$ converges to $\mathcal{R}^c$.

Finally, if π changes, the `Adaptive` algorithm converges to the new optimal memoryless algorithm, as the change in the rate of generating new items is observed by `Adaptive`.

References

1. Bright, L., Gal, A., Raschid, L.: Adaptive pull-based policies for wide area data delivery. ACM Trans. Database Syst. (TODS) **31**(2), 631–671 (2006)
2. Cataldi, M., Di Caro, L., Schifanella, C.: Emerging topic detection on twitter based on temporal and social terms evaluation. In: Proceedings of the Tenth International Workshop on Multimedia Data Mining, MDMKDD 2010, pp. 4:1–4:10. ACM, New York (2010)
3. Dasgupta, A., Ghosh, A., Kumar, R., Olston, C., Pandey, S., Tomkins, A.: The discoverability of the web. In: Proceedings of the 16th international conference on World Wide Web, pp. 421–430. ACM (2007)
4. Delaney, A.: The growing role of news in trading automation, October 2009., http://www.machinereadablenews.com/images/dl/Machine_Readable_News_and_Algorithmic_Trading.pdf
5. Group, D.B.: Alphaflash trader automated trading based on economic events (2015). http://www.alphaflash.com/product-info/alphaflash-trader

6. Hardy, G.H.: Divergent series. American Mathematical Soc. 334 (1991)
7. Hart, W., Murray, R.: Review of sensor placement strategies for contamination warning systems in drinking water distribution systems. J. Water Resour. Plan. Manage. **136**(6), 611–619 (2010)
8. Hope, B.: How computers trawl a sea of data for stock picks. The Wall Street Journal, April 2015. http://www.wsj.com/articles/how-computers-trawl-a-sea-of-data-for-stock-picks-1427941801?KEYWORDS=computers+trawl+sea
9. Horincar, R., Amann, B., Artières, T.: Online refresh strategies for content based feed aggregation. World Wide Web **18**(4), 1–35 (2014)
10. Krause, A., Leskovec, J., Guestrin, C., Van Briesen, J., Faloutsos, C.: Efficient sensor placement optimization for securing large water distribution networks. J. Water Resour. Plan. Manage. **134**(6), 516–526 (2008)
11. Latar, N.L.: The robot journalist in the age of social physics: the end of human journalism? In: Einav, G. (ed.) The New World of Transitioned Media, pp. 65–80. Springer, Switzerland (2015)
12. Leon-Garcia, A.: Probability, Statistics, and Random Processes for Electrical Engineering, 3rd edn. Prentice Hall, Upper Saddle River (2008)
13. Leskovec, J., Krause, A., Guestrin, C., Faloutsos, C., VanBriesen, J.M., Glance, N.S.: Cost-effective outbreak detection in networks. In: Berkhin, P., Caruana, R., Wu, X. (eds.) KDD, pp. 420–429. ACM (2007)
14. Eagle Alpha Ltd.: Discovering the web's hidden alpha, June 2014. http://www.eaglealpha.com/whitepaper_pdf
15. Mathioudakis, M., Koudas, N.: Twittermonitor: Trend detection over the twitter stream. In: Proceedings of the ACM SIGMOD International Conference on Management of Data, SIGMOD 2010, pp. 1155–1158. ACM, New York (2010)
16. McKinney, W.: Structured Data Challenges in Finance and Statistics (November 2011), http://www.slideshare.net/wesm/structured-data-challenges-in-finance-and-statistics
17. Mitra, G., Mitra, L.: The handbook of news analytics in finance, vol. 596. John Wiley & Sons (2011)
18. Mitzenmacher, M., Upfal, E.: Probability and computing: Randomized algorithms and probabilistic analysis. Cambridge University Press (2005)
19. Oita, M., Senellart, P.: Deriving dynamics of web pages: A survey. In: TWAW (Temporal Workshop on Web Archiving) (2011)
20. Ostfeld, A., et al.: The battle of the water sensor networks (BWSN): a design challenge for engineers and algorithms. J. Water Resour. Plan. Manage. **134**(6), 556–568 (2008)
21. Sia, K.C., Cho, J., Cho, H.K.: Efficient monitoring algorithm for fast news alerts. IEEE Trans. Knowl. Data Eng. **19**(7), 950–961 (2007)
22. Wolf, J.L., Squillante, M.S., Yu, P., Sethuraman, J., Ozsen, L.: Optimal crawling strategies for web search engines. In: Proceedings of the 11th international conference on World Wide Web, pp. 136–147. ACM (2002)

Complexity and Game

Vertex Cover in Conflict Graphs: Complexity and a Near Optimal Approximation

Dongjing Miao(✉), Jianzhong Li, Xianmin Liu, and Hong Gao

Harbin Institute of Technology, Harbin 150001, Heilongjiang, China
{miaodongjing,lijzh,liuxianmin,honggao}@hit.edu.cn

Abstract. Given finite number of forests of complete multipartite graph, *conflict graph* is a sum graph of them. Graph of this class can model many natural problems, such as in database application and others. We show that this property is non-trivial if limiting the number of forests of complete multipartite graph, then study the problem of vertex cover on *conflict graph* in this paper. The complexity results list as follow,

- If the number of forests of complete multipartite graph is fixed, *conflict graph* is non-trivial property, but finding 1.36-approximation algorithms is NP-*hard*.
- Given 2 forests of complete multipartite graph and maximum degree less than 7, vertex cover problem of *conflict graph* is NP-*complete*. Without the degree restriction, it is shown to be NP-*hard* to find an algorithm for vertex cover of *conflict graph* within $\frac{17}{16} - \varepsilon$, for any $\varepsilon > 0$.

Given *conflict graph* consists of r forests of complete multipartite graph, we design an approximation algorithm and show that the approximation ratio can be bounded by $2 - \frac{1}{2^r}$. Furthermore, under the assumption of UGC, the approximation algorithm is shown to be near optimal by proving that, it is hard to improve the ratio with a factor independent of the size (number of vertex) of *conflict graph*.

Keywords: Approximation algorithm · Vertex cover · Complete multipartite graph

1 Introduction

Vetex Cover is one of the classical problems in graph theory: given a graph $G(V, E)$, find a vertex set of vertices in V, say V', such that for each edge (u, v) of E, at least one of u and v belongs to V', and V' has the minimize size.

We focus on the vertex cover problem in *conflict graph* which is a restricted class of graph in this paper.

This work was supported in part by the National Grand Fundamental Research 973 Program of China under grant 2012CB316200, the Major Program of National Natural Science Foundation of China under grant 61190115, Project 61502121 supported by National Natural Science Foundation of China.

Z. Lu et al. (Eds.): COCOA 2015, LNCS 9486, pp. 395–408, 2015.
DOI: 10.1007/978-3-319-26626-8_29

Definition 1. *Conflict Graph. Given an vertex set V and r forests of complete multipartite graph, namely F_1, ..., F_r defined on V, such that each connected component is a multipartite graph in each F_i. The conflict graph $G(V, E)$ is an undirected graph where edge $(u, v) \in E$ if there is an F_i such that edge $(u, v) \in F_i$.*

1.1 Database Application

Managing inconsistent data is a core problem in the area of data quality management. Inconsistent data indicates that there is conflicted information in the data, which can be formalized as the violations of given semantic constraints. For relational data, usually, dependencies are utilized to capture the inconsistencies of data. There are many kinds of dependencies in database theory. Given a relational database schema $\mathbf{R} = \{R_1, \ldots, R_c\}$, a formal form of general dependency can be specified as a first-order logic sentence as follows.

$$\forall x_1 \ldots \forall x_n[\varphi(x_1, \ldots, x_n) \rightarrow \exists z_1 \ldots \exists z_k \psi(y_1, \ldots, y_m)]$$

Here, the variables in ψ are taken from $\{x_i\}$ and $\{z_j\}$, both φ and ψ are conjunctions of *relation atoms* of the form $R_l(w_1, \ldots, w_h)$ and *equality atoms* of the form $w = w'$, where R_l is a relation in $\mathbf{R}$ and each of the w, w', w_1, ..., w_h is a variable appearing in the sentence. Furthermore, the dependencies can be classified from following aspects. (i) *Full versus embedded.* A *full* dependency is a dependency that has no existential quantifiers. (ii) *Tuple generating versus equality generating.* A *tuple generating dependency* (tgd) is a dependency in which no equality atoms occur, while an *equality generating dependency* (egd) is a dependency for which the right-hand formula is a single equality atom. (iii) *Typed versus untyped.* A dependency is *typed* if there is an assignment of variables to column positions such that variables in relation atoms occur only in their assigned position, and each equality atom involves a pair of variables assigned to the same position. (iv) *k-ary.* A *k-ary* dependency contains k relation atoms in φ.[1] Several kinds of dependencies have been successfully applied in the area of inconsistent data management, especially *2-ary* dependencies, such as functional dependency (FD for short) [1] and so on. *2-ary* dependencies is a typical example of such depdencies utilized in capturing inconsistencies, and the reason is easy to understand after observing that the "inconsistency" is usually explained to be the conflict between 2 tuples when some conditions are satisfied.

Example. An FD constraint $AB \rightarrow C$ defined over relation R says, "for two tuples in relation R, if their have equal values on attributes A and B, they must have equal values on attribute C also". Given the above FD constraint, it is expected that the constraint is valid on all possible data instances. Then, if two tuples $t_1 = \{A : x, B : abc, C : m\}$ and $t_2 = \{A : x, B : abc, C : n\}$ are found in the real instance, it is easy to verify that t_1 and t_2 violate the given constraint. That is, *inconsistent data* are found according to the given semantic constraints.

[1] A more strict definition is based on the assumption that φ is a conjunction of relation atoms, and the assumption will not affect the expressability of dependencies.

An important task focusing on managing inconsistent data is to evaluate how inconsistent the data is. A natural model is using the vertex cover size to show this value based on the input dependencies. Concretely, for an inconsistent database instance with respect to several given FDs, we model each tuple as a vertex in graph, and each tuple pair has an edge between them if they are conflict with each other. Then we can see that each FD will generate a forest of complete multipartite graph, because if a tuple has the same value on the left attributes with some other tuples, it must be conflict with other tuples with different right attributes' values, i.e., it has edges with all of them. Finally, if we want to remove the minimum tuples to make it satisfying all the given FDs, we must compute the vertex cover of the sum graph of those generated by single FD, here the sum graph is just a *conflict graph.*

Therefore, in this scenario, vertex cover problem in conflict graph is the basis of evaluating data inconsistency and provide several theoretical analysis results. Since inconsistent data is defined under the help of dependencies, a specific kind of dependency FD is choosed when introducing our results in the following parts. In fact, all results in this paper can be extended to model the *full*, *equality-generating*, *typed* and *2-ary* dependencies [1].

1.2 Literature Review of the Vertex Cover Problem

The related work is vast, however, we only mention those very related. For any $\varepsilon > 0$, minimum vertex cover might be hard to be approximated within $2-\varepsilon$ [8], this is its UGC-hardness. For any $\varepsilon > 0$, $7/6-\varepsilon$ inapproximability for minimum vertex cover has been proved by Hastad et al. [6], and this factor was improved by Dinur and Safra [4] to 1.36, this is its NP-hardness. A practical 2-approximation algorithm of minimum vertex cover was provided by Gavril et al. [5], it is to find a maximal matching of a graph and output all vertices in the matching, since the size of any maximal matching of a graph is always a 2-approximation of its minimum vertex cover. For complete multipartite graph, there are few works about computing its vertex cover or maximal matching. A formula of computing maximum matching size has been developed in [12].

On the other side, the vertex cover problem on general graph is approximable within $2-\frac{\log\log|V|}{2\log|V|}$ [10] and $2-\frac{\ln\ln|V|}{2\ln|V|}(1-o(1))$ [3]. Karakostas reduce the approximation factor to $2-\Theta\left(\frac{1}{\sqrt{\log|V|}}\right)$ [7] instead of the previous $2-\frac{\log\log|V|}{2\log|V|}$. We mention that several simple 2-approximation algorithms are known. Half-integral solution [11] is a well-known approach to obtain a better approximation. The algorithm uses a k-coloring of the subgraph induced by the half value vertices as input to reduce approximation ratio into $2-\frac{k}{2}$. For graphs with degree bounded by d, based on Brook's theorem, this directly leads to a $(2-\frac{2}{d})$-approximation. As referred above, the basic approach has been improved to $2-\Theta\left(\frac{1}{\sqrt{\log|V|}}\right)$ [7] by replacing the LP relaxation to SDP relaxation. We also mentioned the recent work on local biclique coloring given by F. Kuhn [9].

They generalize the result on bounded degree obtained by SDP relaxation to bounded local chromatic number. It can be guaranteed that any valid coloring with k colors also is a local k-coloring. They proved the following theorem [9],

Theorem 1. *Assuming the UGC, it is NP-*hard *to approximate the vertex cover problem in graphs for which a* $(\Delta + 1)$*-local coloring is given as input, within any constant factor better than* $2 - \frac{2}{\Delta+1}$. *If the given coloring is also a biclique coloring, there will be a randomized polynomial-time algorithm with approximation ratio* $2 - \Omega(1)\frac{\ln \ln \Delta}{\ln \Delta}$.

In this paper, based on the above theorem, we show that it is hard to give a bounded local (biclique) coloring with size independent of $|V|$ to improve the ratio with use of the approach they gave and our approach seems good enough to solve vertex cover problem on *conflict graph*.

1.3 Our Contribution

We first prove that *conflict graph* is a non-trivial property when the number of forests of complete multipartite graph is fixed, i.e., if we fix the number r, it is always possible to find a graph can not be represented by sum of r forests of complete multipartite graph. However, we proved that (i) If the number of forests of complete multipartite graph is fixed, finding 1.36-approximation is also NP-*hard*. (ii) Given 2 forests of complete multipartite graph and maximum degree less than 7, vertex cover problem of *conflict graph* is NP-*complete*. Without the degree restriction, it is shown to be still NP-*hard* to find an algorithm for vertex cover of *conflict graph* within $\frac{17}{16} - \varepsilon$, for any $\varepsilon > 0$. By directly using previous results on vertex cover problem, it is shown that it is NP-*hard* to obtain a $(2-\varepsilon)$-approximation for any $\varepsilon > 0$ unless the UGC(unique gaming conjecture) does not hold.

Given *conflict graph* consists of r forests of complete multipartite graph, we design an deterministic approximation algorithm and show that the approximation ratio can be bounded by $2 - \frac{1}{2^r}$ which is a bound does not depend on $|V|$ but only r usually a small constant in real applications. Also, this bound is not a expected ratio of a random algorithm, but it cannot be improved by applying the related SDP approach [9]. Furthermore, the approximation algorithm is shown to be near optimal by proving that it is impossible to improve the ratio with any constant factor under the assumption of UGC. Actually, it is an approximation algorithm for vertex cover on a special graph class, namely *conflict graph*. Comparing with the previous results, the results in this paper are obtained based on the characteristic of *conflict graph*.

2 Complexity and Inapproximation

In the following, we study the complexity of vertex cover problem on *conflict graph*. First, a trivial case is the vertex cover on only one forest of complete multipartite graph.

Proposition 1. *Minimum vertex cover in a forest of complete multipartite graph can be found polynomially.*

Proof. Due to the definition of *conflict graph*, each two connected component are disjoint and there is no edge between them. Given a *conflict graph* G, for each connected component (a complete multipartite graph), take all vertices of it into cover C excepting those of the largest part. At last, C is a minimum vertex cover of G.

Lemma 1. *Given any graph G, it's not always possible to find a collection of subgraphs such that*

(a) each subgraph is a complete k-partite graph for some k,
(b) each edge in occurs in at least one of the subgraphs, and
(c) each vertex in occurs in at most a constant number of the subgraphs.

Moreover, the probability is close to 1.

Proof. Given n, let's take a random bipartite graph $G = ([n], [n], E)$ where $Pr[(v_i, v_j) \in E] = \frac{1}{2}$ for each pair (v_i, v_j). (The answer is no with probability close to 1.)

Since G is bipartite, any complete k-partite subgraph has to be bipartite. First, we claim that with probability $1 - o(1)$, every complete bipartite subgraph in G has at most $4n$ edges. For any pair of subsets $L \subseteq [n]$ and $R \subseteq [n]$ with $|L \times R| \geq 4n$, the probability that "the complete bipartite subgraph with edge set $L \times R$ is in G" is $2^{-|L \times R|}$, and it is no more than 2^{-4n}. There are fewer than $2^n \times 2^n = 4^n$ such pairs L and R, thus, by the naive union bound, the probability that any of the corresponding subgraphs is present in G is at most $4^n 2^{-4n}$, and it is no more than 2^{-2n}.

Also, with probability $1 - o(1)$, the graph G has at least $\frac{n^2}{4}$ edges.

Thus, with probability $1 - o(1)$, every complete bipartite subgraph in G has at most $4n$ edges, and G has at least $\frac{n^2}{4}$ edges. Assume this happens.

Now suppose for contradiction that a collection of subgraphs with the desired properties exists. Each of the subgraphs has edge set $L \times R$ for some pair of subsets L and R. In G, direct all the edges in $L \times R$ from the larger side to the smaller side (to the left if $|L| \leq |R|$, and to the right otherwise). Since $|L \times R| \leq 4n$, the smaller of L or R must have size at most $2\sqrt{n}$.

Since each vertex is in $O(1)$ of the subgraphs, each vertex now has $O(\sqrt{n})$ edges directed out of it. But all edges are directed one way or the other, so the number of edges in G is at most the number of vertices times the maximum out-degree of any vertex, that is, at most $O(n\sqrt{n})$. This contradicts the graph having at least $\frac{n^2}{4}$ edges.

Lemma 1 shows that *conflict graph* is a non-trivial graph class under the constant number restriction. Inspired by this, we next have the following theorem on the complexity for a more restricted condition.

Theorem 2. *Given* 2 *forests of complete multipartite graph and maximum degree no more than* 6, *decision version of vertex cover problem on conflict graph is NP-*complete.

Proof. Thus the problem is in NP obviously. We give the proof of the lower bound as follow.

NP-*hardness*. The lower bound is established by a reduction from 3-SAT problem. An instance of 3-SAT problem includes a set U of n variables $x_1, ..., x_n$ and a collection S of m clauses $s_1, ..., s_m$, while in each clause $s_i = \alpha_{i1} + \alpha_{i2} + \alpha_{i3}$, each α_{ij} $(1 \leq j \leq 3)$ is the j-*th* literal of s_i. Given an instance of 3-SAT problem, it is to decide whether there is a satisfying truth assignment for S. The 3-SAT problem is NP-*complete*, and it remains NP-*complete* even if for each $x_i \in U$, there are at most 5 clauses in S that contain either x_i or $\overline{x}_i$.
A polynomial reduction from 3-SAT can be constructed as follows.

Given an instance of 3-SAT with n variables and m clauses, let *conflict graph* $G = F_1 + F_2$ and $k = n + 2m$. Then we introduce 2 forests of complete multipartite graph F_1 and F_2 share a vertex set with $2n + 3m$ vertices,

F_1: Set edge $(2i-1, 2i)$ for each $i \in [n]$, and edges $(3i-2, 3i-1)$, $(3i-2, 3i)$, $(3i-2, 3i-1)$ (i.e., a triangle for each clause) for each $i \in [m]$. That is, F_1 consists of n complete bipartite subgraphs and m complete tripartite subgraphs;
F_2: For each variable x_i of 3-SAT instance, build a complete bipartite subgraph $L_i \times R_i$, let L_i includes vertex $2i - 1$ and all vertices corresponding to the positive literals of x_i, and R_i includes vertex $2i$ and all vertices corresponding to the negative literals of x_i;

Note that, each vertex has a degree at most 7 assuming each variable occurs at most 5 clauses.

Suppose the 3-SAT instance is satisfiable, i.e., there is an satisfying truth assignment $\rho : U \rightarrow \{0, 1\}^n$ for S, then there is a vertex cover VC of G such that its size is at most $n + 2m$. Concretely, it can be computed as follows, for each variable x_i, (1) if $\rho(x_i) = 1$, delete vertex $2i$ from G. And for each clause s_j, if α_{jq} is a positive literal of x_i and vertices $2n + 3j - 2, 2n + 3j - 1, 2n + 3j$ is currently in G, delete vertex $2n + 3(j - 1) + q$ from G; (2) if $\rho(x_i) = 0$, delete vertex $2i + 1$ from G. And for each clause s_j, if α_{jq} is a negative literal of x_i and vertices $2n + 3j - 2, 2n + 3j - 1, 2n + 3j$ is currently in G, delete vertex $2n + 3(j - 1) + q$ from G. We have that for each i, either $2i$ or $2i + 1$ is deleted from G, and for each j, either of $\{2n + 3j - 2, 2n + 3j - 1, 2n + 3j\}$ is deleted from G for each j. This is because in each clause, there is at least one literal that is made *true* by assignment ρ. Therefore, there is a set VC of the rest tuples such that it is a cover and has a size no more than $n + 2m = k$.

To see the converse, let VC is the cover such that $|VC| \leq k = n + 2m$. To cover F_1, either $2i - 1$ or $2i$ should be included in VC for each $i \in [n]$, and at least two of $2n + 3j - 2$, $2n + 3j - 1$, $2n + 3j$ should be included in VC. That is, the size of VC is at least $n+2m$. Thus, $|VC| = n+2m$. After covering F_2, at most one literal of each variable rests in $G - VC$. Then, there is a satisfying truth assignment τ for S such that, for each $i \in [n]$,

$$\tau(x_i) = \begin{cases} 0, \text{ if } 2i - 1 \in V(G) - VC, \\ 1, \text{ if } \textit{otherwise} \end{cases} \tag{1}$$

It is sure that τ will make all clauses *true*. Therefore, it is NP-*complete*, even if given only 2 forest of complete multipartite graph and vertex with at most 7 degree.

We next prove that vertex cover problem on *conflict graph* is *Max-SNP-hard* if without the degree restriction. To analyze the lower bound of approximation of vertex cover, we next use the same reduction in Theorem 2 above, but from MAX-E3SAT [6] to this problem if given 2 forest of complete multipartite graph. We know that unless P≠NP, there is no polynomial-time algorithm approximates MAX-E3SAT with $\frac{7}{8} + \varepsilon$ [6] (we use the ratio notion less than 1 for *maximizing* problem)for any $\varepsilon > 0$. That is, there is no guarantee that "more clause satisfied, more vertex preserved". This is really because three *free* tuples are built for each clause in the reduction. Concretely, in that reduction, there may exist two an assignments τ_p and τ_q $(p > q)$where they makes p and q clauses *true*. However, in its corresponding instance, there may be p' and q' tuples can be preserved respectively where $p' < q'$.

Therefore, we give a linear reduction carefully designed by modifying the one used in Theorem 2.

Theorem 3. *Vertex cover problem on conflict graph can not be approximated in* $\frac{17}{16}$ *if given* 2 *forest of complete multipartite graph.*

Proof. This lower bound is established by a reduction from MAX-E3SAT whose instance includes a set U of n variables and a collection S of m disjunctive clauses of *exactly* 3 literals. Given an instance of MAX-E3SAT problem, it is to find a satisfying truth assignment maximizing the number of clauses satisfied by it.
We build the reduction the same in Theorem 2 and also use the assignment function τ such that

$$\tau(x_i) = \begin{cases} 0 \text{ if } 2i - 1 \in V(G) - VC, \\ 1 \text{ if } \textit{otherwise} \end{cases} \tag{2}$$

Let $\#\tau$ is the number of clauses satisfied by any assignment τ, let $\#\tau_{\max}$ refer to the optimal assignment $\tau_{\max}$which maximizing the number of clauses satisfied in the MAX-E3SAT instance. Let $|V(G)|$ is the number of vertex in G.

We first consider $VC_{\min}$ the minimum vertex cover of G. We claim that any pair of variable vertices cannot occurs in $VC_{\min}$, because if not, there must be a smaller $VC'_{\min}$ obtained by returning redundant variable vertices from $VC_{\min}$ into $V(G) - VC_{\min}$ without any edge left. Based on this, given the optimal assignment $\tau_{\max}$ maximizing the number of clauses satisfied in the MAX-E3SAT instance, we have

$$\#\tau_{\max} = |V(G)| - |VC_{\min}| - n \tag{3}$$

And for any solution VC of $V(G)$, we have

$$\#\tau_{vc} \geq |V(G)| - |VC| - n \tag{4}$$

Additionally, we have the fact that

$$|V(G)| = 2n + 3m, m > 0 \tag{5}$$

Now, let $\hat{VC}$ is an r-approximation ($r > 1$) of minimum culprit $VC_{\min}$ such that $|\hat{VC}| \leq r \cdot |VC_{\min}|$. We have

$$\frac{\#\tau_{\hat{vc}}}{\#\tau_{\max}} > \frac{|V(G)| - |\hat{VC}| - n}{|V(G)| - |VC_{\min}| - n} > 1 + \frac{(1-r) \cdot |VC_{\min}|}{|V(G)| - |VC_{\min}| - n} \tag{6}$$

Since each clause has exactly 3 literals, we have

$$|VC_{\min}| \geq n + 2 \cdot \frac{|V(G)| - 2n}{3} \tag{7}$$

Apply this fact in the right hand of inequality 6, it is

$$\frac{|VC_{\min}|}{|V(G)| - |VC_{\min}| - n} \geq \frac{2|V(G)| - n}{|V(G)| - 2n} = 2 + \frac{3}{\frac{|V(G)|}{n} - 2} \tag{8}$$

Since $|V(G)| > 2n$, therefore we get

$$\frac{|VC_{\min}|}{|V(G)| - |VC_{\min}| - n} > 2 \tag{9}$$

Then, apply this into inequality 6, then

$$\frac{\#\tau_{\hat{vc}}}{\#\tau_{\max}} > 3 - 2r \tag{10}$$

That is, if minimum culprit can be approximated within ration r, then MAX-E3SAT can be approximated within $3-2r$. However, if the former can be approximated within $\frac{17}{16}$, then the later will be approximated better than $\frac{7}{8}$, and this contraries to the result of [6].

We next show that for the fixed number of subgraph, minimum vertex cover has a global inapproximability bound. The observation is that sum of fixed number of subgraphs is able to represent any bounded degree graph.

Theorem 4. *For sufficiently large fixed number of forests of complete multipartite graph, it is NP-hard to approximate vertex cover on conflict graph within 1.3606.*

Proof. Consider any input instance is a graph $G\langle V, E\rangle$ and each vertex has a degree constraint d ($d > 0$), where each vertex in V has a degree no more than d. It is easy to build d forests of complete multipartite graph to represent G.

For each vertex i, we can easily distribute its each edge into d forests of complete multipartite graph one by one. For sufficiently large bounded degree d, Dinur and Safra [4] has proved that it is NP-*hard* to approximate minimum vertex cover within any constant factor smaller than 1.3606. The number of forests of complete multipartite graph are $O(d)$ in the reduction herein, this concludes the theorem.

The theorem gives an approximation lower bound for the case with fixed number of forest of complete multipartite graph. For the *non-fixed* case, it is able to encode arbitrary graphs with unbounded degree, just set $|E|$ subgraphs. B!eyond the NP-*hardness*of 1.3606, Khot et. al [8] had proved that, for any $\varepsilon > 0$, there is no $(2 - \varepsilon)$-polynomial approximation, unless the *unique games conjecture* is false. For the sake of clarity, we summary the hardness results of vertex cover problem as follows.

Cases	**Complexity/Inapproximation**
1 subgraph	PTIME
2 subgraphs (with degree ≤ 7)	$\frac{17}{16}$ (NP-*complete*)
Fixed number	1.3606
Non-fixed number	1.3606(NP-*hardness*), $2 - \varepsilon$(UGC-*hardness* [8])

3 A Near Optimal $(2 - \frac{1}{2^r})$-approximation

In this section, we use the a linear based $(2 - \frac{1}{2^C})$-approximation for our input graph where C is the number of connected components in the input graph, then improve the approximation.

We start by the following observations on the chromatic number of the input graph G, given r forests of complete multipartite graph, namely $F_1, ..., F_r$. We have the following useful observation of a forest of complete multipartite graph.

Property 1. If a graph G is a sum of C complete multipartite graphs, k_i-partite graph respectively, then it has a chromatic number $\prod k_i$. Moreover, this chromatic number is tight.

Proof. The graph G is $\prod k_i$-colorable. We can do this in linear time by assigning a distinct color for each partite of each complete k_i-partite graph. A k-partite graph (assuming non-empty parts) is k-colorable (color all of the vertices in one part with the same color). A complete k-partite graph has chromatic number k (clearly two vertices in two different parts require different colors). In fact, it has a maximal set of edges such that the graph has chromatic number (adding an edge within a part will force another color). If the graphs g_1 and g_2 are k_1 and k_2 colorable respectively, the sum $G = g_1 + g_2$ is $k_1 k_2$ colorable by the product coloring. It follows that each g_i is k_i colorable, and that the sum, G is $\prod k_i$-colorable.

This property is tight if we know nothing more about the graphs. For arbitrary n, let $p_1 p_2 \cdots p_k$ be the prime factorization of n (where repeated primes are listed repeatedly). If G is the sum of k correctly-chosen complete p_i-partite graphs, $G = K_n$. To choose G_1, let the parts be the residues modulo p_1. For G_2, let i be in a part according to the residue of $\lfloor \frac{1}{p_1} \rfloor$ modulo p_2. Continuing this for each k makes the right subgraphs. The best way to imagine this is to express i has a number where the 1 s place goes up to p_1, the p_1s place (the next digit over) goes up to p_2, etc. Then G is complete because if $i \neq j$ then The representations of i and j are different.

3.1 A Basic Approximation Algorithm

We next combine this property and linear program relaxation to give a better approximation. Recall the classical linear program of vertex cover.

$$minimize \sum_{i \in V} x_i \tag{11}$$

subject to:

$$x_i + x_j \geq 1, \forall (i,j) \in E, \tag{12}$$

$$x_i \geq 0, i \in V \tag{13}$$

We can relax it with condition: $x_i \in \{0, \frac{1}{2}, 1\}$, and we have that $OPT^{relax} \leq OPT$. Then, we use a coloring on the input graph which can be found trivially based on Property 1 to improve the approximation. We use a solution similar with [11] to improve the approximation as follow.

Algorithm 1

```
Solve the linear programming relaxation to obtain a solution x' such
  that x' ∈ {0, 1/2, 1} for all i ∈ V.
Let P_j is the set of vertices of color j
j ← arg max_j |{x'_i | i ∈ P_j ∧ x'_i = 1/2}|
for each i ∈ V do
    if x'_i = 1 or (x'_i = 1/2 and i ∉ P_j) then
        add v_i into ṼC
return ṼC
```

First, OPT^{relax} can be returned in polynomial time as shown in [13], and it is easy to see that $\tilde{VC}$ is a cover. Actually, if an $x'_i = \frac{1}{2}$ and it is not chosen into $\tilde{VC}$, then its neighbor must be added into $\tilde{VC}$ since its neighbor is not in P_j, and the sum of two adjacent variables is at least 1.

Second, we claim that the approximation ratio is $2\left(1 - \frac{1}{\prod k_i}\right)$.

Lemma 2. *Algorithm 1 returns a* $2\left(1-\frac{1}{\prod k_i}\right)$*-approximation.*

Proof. Let S_1 be the set $\{x'_i | i \in V, x'_i = 1\}$, $S_{\frac{1}{2}}$ be the set $\{x'_i |\, i \in V, x'_i = \frac{1}{2}\}$, and S_{P_j} be the set $\{x'_i |\, i \in P_j, x'_i = \frac{1}{2}\}$. Obviously, $OPT^{relax} = \sum_{i \in S_1 \cup S_{\frac{1}{2}}} x'_i \leq OPT$. We have,

$$\left|\tilde{VC}\right| \leq |S_1| + \left|S_{\frac{1}{2}}\right| - \left|S_{P_j}\right| \tag{14}$$

$$= \sum_{i \in S_1} x'_i + 2 \sum_{i \in S_{\frac{1}{2}}} x'_i - 2 \sum_{i \in S_{P_j}} x'_i \tag{15}$$

$$\leq \sum_{i \in S_1} x'_i + 2 \sum_{i \in S_{\frac{1}{2}}} x'_i - 2 \cdot \frac{1}{\prod k_i} \sum_{i \in S_{\frac{1}{2}}} x'_i \tag{16}$$

$$\leq \sum_{i \in S_1} x'_i + (2 - \frac{2}{\prod k_i}) \sum_{i \in S_{\frac{1}{2}}} x'_i \tag{17}$$

$$\leq (2 - \frac{2}{\prod k_i}) \sum_{i \in S_1 \cup S_{\frac{1}{2}}} x'_i \tag{18}$$

$$\leq (2 - \frac{2}{\prod k_i}) OPT \tag{19}$$

3.2 Improve the Approximation by Triangle Eliminating

To improve the approximation, it is necessary to decrease the chromatic number of the input graph. We next use triangle eliminating technique to decrease the chromatic number. An improved algorithm is shown as follow.

Algorithm 2

```
for each forest of complete multipartite graph F_i do
    while there is a triangle do
        include the three vertices of a triangle into ṼC_1
        remove them and their adjacent edges in G
run Algorithm 1 on the residual graph to get a cover ṼC_2
return ṼC_1 ∪ ṼC_2
```

We formally state that the algorithm above will return a $(2-\frac{1}{2C})$-approximation as the following theorem.

Theorem 5. *Algorithm 2 returns a* $(2-\frac{1}{2C})$*-approximation.*

Proof. First, Algorithm 2 does find a cover, because the edges connected triangles removed and the residual graph are covered by $\tilde{VC}_1$, meanwhile, Algorithm 1 guarantees $\tilde{VC}_2$ is a cover of the residual graph.

Second, as it known [2], in a graph G, if v, u and w form a triangle, then we can include all three vertices in a vertex cover for a $\frac{3}{2}$-approximation. This is because at least two vertices of a triangle is needed in order to cover it. Therefore, $\tilde{VC_1}$ is a $\frac{3}{2}$-approximation for the subgraph induced by all the triangles removed.

Third, for each connected component, if all the triangles are removed, it is at most a complete bipartite graph, therefore it has a chromatic number at most 2. Then the residual graph has a chromatic number at most 2^C. Due to Lemma 2, $\tilde{C}_2$ is an $\left(2 - \frac{1}{2^C}\right)$-approximation on the residual graph. Then combine the two covers will obtain a $\max\left\{\frac{3}{2}, 2 - \frac{1}{2^C}\right\}$-approximation. Without loss of generality, we have $C \geq 1$, then Algorithm 2 returns a $\left(2 - \frac{1}{2^C}\right)$-approximation.

Remark. In fact, after removing all the triangles, the residual graph has a chromatic number 2^r which is far less than C. Moreover, there is also a simple polynomial coloring algorithm as follow.

```
Algorithm Coloring
Removing all the triangles from each forest of complete multipartite
graph
for each forest of complete multipartite graph F_i do
    color each connected component in F_i with colors col_{2i-1} and clo_{2i},
    such that this subgraph is colored with only two colors.
for each vertex in G do
    color it with col (col_{j_1}, col_{j_2}, ..., col_{j_r}), where graph col_{j_i} is its color in
    F_i
```

Using the coloring algorithm after triangles elimination, we can reduce the approximation ratio to $2 - \frac{1}{2^r}$. Theoretically, this ration does not depend on size of input graph. Practically, when the number of forest of complete multipartite graph is a fixed small constant, it is a good approximation ratio.

Corollary 1. *Algorithm 2 returns a* $\left(2 - \frac{1}{2^r}\right)$*-approximation.*

3.3 Near Optimality

In the following, we show that **Algorithm** 2 is a near optimal approximation based on the following two definitions in [9].

Local Coloring. Let Δ be a positive integer. A Δ-local coloring of a graph is a valid vertex coloring such that for every vertex $u \in V$, all the neighbors of u are colored by at most $\Delta - 1$ colors.

Biclique Coloring. A coloring of a graph is called a biclique coloring if for any two colors i and j, the subgraph induced by " the vertices with color j that have a neighbor with color i" and "the vertices with color i that have a neighbor with color j" is either empty or a complete bipartite graph.

Remark. On one side, the best approximation for vertex cover of general graph is $2 - \Theta\left(\frac{1}{\sqrt{\log|V|}}\right)$ [7]. More strictly, it is a randomized algorithm and the ratio is parameterized by size of vertex set $|V|$. While **Algorithm** 2 is deterministic and returns an approximation independent with $|V|$ but only depends on $|\Sigma|$ always small.

On the other side, as proven in [9] that it is UGC-*hard* to improve the approximation ratio of $2-\frac{2}{\Delta+1}$ by any constant factor if a $(\Delta+1)$-local coloring is given as input and it is not a biclique coloring. In general, computing a local biclique coloring is a hard problem for general graphs. Actually, due to the reduction in the proof of 4, the instance of our problem can simulate arbitrary graph, that is, finding a local biclique coloring for the input conflict graph is also hard. However, **Algorithm** *Coloring* provides a 2^r-local coloring, this is to say that if assume UGC, then it is hard to improve any constant factor better than $\left(2-\frac{1}{2^r}\right)$.

Moreover, we also claim that **Algorithm** *Coloring* will return a 2^r-local coloring and it is tight but not a biclique coloring. Actually, This property is tight if we know nothing more about the graphs. For arbitrary n, let $p_1p_2\cdots p_k$ be the prime factorization of n (where repeated primes are listed repeatedly). If G' is the sum of k correctly-chosen complete p_i-partite graphs, $G' = K_n$. To choose G_1, let the parts be the residues modulo p_1. For G_2, let i be in a part according to the residue of $\lfloor\frac{1}{p_1}\rfloor$ modulo p_2. Continuing this for each k makes the right subgraphs. The best way to imagine this is to express i has a number where the 1 s place goes up to p_1, the p_1s place (the next digit over) goes up to p_2, etc. Then G' is complete because if $i \neq j$ then The representations of i and j are different.

Proposition 2. *The biclique local coloring of conflict graph depends on the number of vertices, even if the number of forest of complete multipartite graph is fixed ($r \geq 2$).*

Proof. We can build a valid conflict graph with $4n$ vertices and its biclique coloring of size $\Omega(\sqrt{n})$. We show the conflict graph as the summation of two simple subgraphs which can be easily expressed as a database instance and two FDs by the reduction in the proof of Theorem 4.

Construct two subgraphs, a complete bipartite graph F_1 $(V \bigcup U, E_1)$ where $V = \{v_1, \cdots, v_{2n}\}$, $U = \{u_1, \cdots, u_{2n}\}$ and $E_1 = V \times U$, and a bipartite graph F_2 $(V \bigcup U, E_2)$ where $E_2 = \{v_1v_2, v_3v_4, \cdots, v_{2n-1}v_{2n}, u_1u_2, u_3u_4, \cdots, u_{2n-1}u_{2n}\}$. Let the conflict graph G is $F_1 + F_2$ as follow example (dash edges from E_2, others from E_1),

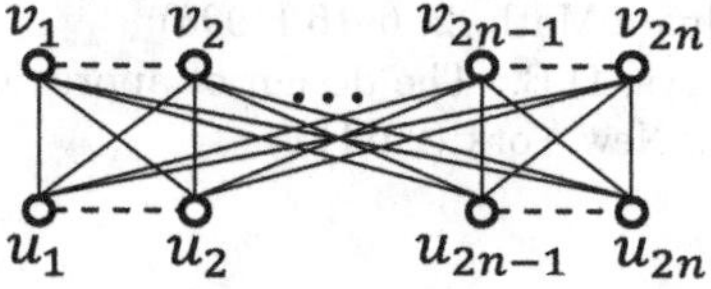

For each odd i, we see that v_i (u_i) is not adjacent with other vertices of V (U) except $v_{i+1}(u_{i+1})$, therefore, in order to keep "biclique coloring" property, we have the following conditions,

(a) $v_i(u_i)$ should be colored different from $v_{i+1}(u_{i+1})$;
(b) the color pair of v_i and v_{i+1} (u_i and u_{i+1}) can not be the same as any other color pair of v_j and v_{j+1} (u_j and u_{j+1}) for any other odd j.
(c) all the colors of vertices in V should be different from those in U.

Based on this, if a biclique local coloring has a size f, then it must satisfy that

$$\binom{2}{\frac{f}{2}} \geq \frac{n}{2} \tag{20}$$

Therefore, the biclique local coloring depends on the number of vertices.

Corollary 2. *Algorithm 2 is near optimal.*

References

1. Abiteboul, S., Hull, R., Vianu, V.: Foundations of Databases. Addison-Wesley, Boston (1995)
2. Ásgeirsson, E.I., Stein, C.: Divide-and-conquer approximation algorithm for vertex cover. SIAM J. Discret. Math. **23**(3), 1261–1280 (2009)
3. Bar-Yehuda, R., Even, S.: A local-ratio theorem for approximating the weighted vertex cover problem. Ann. Discrete Math. **25**, 27–46 (1985)
4. Dinur, I., Safra, S.: On the hardness of approximating minimum vertex cover. Ann. Math. **162**, 439–485 (2005)
5. Gavril, F.: Algorithms for minimum coloring, maximum clique, minimum covering by cliques, and maximum independent set of a chordal graph. SIAM J. Comput. **1**(2), 180–187 (1972)
6. Håstad, J.: Some optimal inapproximability results. J. ACM (JACM) **48**(4), 798–859 (2001)
7. Karakostas, G.: A better approximation ratio for the vertex cover problem. ACM Trans. Algorithms **5**(4), 41:1–41:8 (2009)
8. Khot, S., Regev, O.: Vertex cover might be hard to approximate to within 2 - ϵ. J. Comput. Syst. Sci. **74**(3), 335–349 (2008)
9. Kuhn, F., Mastrolilli, M.: Vertex cover in graphs with locally few colors. In: Aceto, L., Henzinger, M., Sgall, J. (eds.) ICALP 2011, Part I. LNCS, vol. 6755, pp. 498–509. Springer, Heidelberg (2011)
10. Monien, B., Speckenmeyer, E.: Ramsey numbers and an approximation algorithm for the vertex cover problem. Acta Inform. **22**(1), 115–123 (1985)
11. Nemhauser, G.L., Trotter Jr, L.E.: Vertex packings: structural properties and algorithms. Math. Program. **8**(1), 232–248 (1975)
12. Sitton, D.: Maximum matchings in complete multipartite graphs. Furman Univ. Electron. J. Undergraduate Math. **2**, 6–16 (1996)
13. Williamson, D.P., Shmoys, D.B.: The design of approximation algorithms. Cambridge University Press, New York (2011)

On the Complexity of Scaffolding Problems: From Cliques to Sparse Graphs

Mathias Weller[1,2], Annie Chateau[1,2](✉), and Rodolphe Giroudeau[1]

[1] LIRMM - CNRS UMR 5506, Montpellier, France
{mathias.weller,annie.chateau,rodolphe.giroudeau}@lirmm.fr
[2] IBC, Montpellier, France

Abstract. This paper is devoted to new results about the scaffolding problem, an integral problem of genome inference in bioinformatics. The problem consists of finding a collection of disjoint cycles and paths covering a particular graph called the "scaffold graph". We examine the difficulty and the approximability of the scaffolding problem in special classes of graphs, either close to trees, or very dense. We propose negative and positive results, exploring the frontier between difficulty and tractability of computing and/or approximating a solution to the problem.

1 Introduction

The scaffolding of genomes is one of the operations performed when producing a genomic sequence from the real molecule. After sequencing and assembly, we end up with a certain amount of sequences (words of various length over the alphabet $\{A, T, G, C\}$) called contigs. To complete the whole genome sequence, those contigs must be relatively ordered and oriented. In previous work on scaffolding, this problem has been modeled as a combinatorial problem on graphs which is, unfortunately, computationally hard [4]. Some methods use heuristic ways to simplify the graph [9], others use a decomposition of the problem into two separate steps (orienting and ordering), whose difficulty could be bypassed under certain restrictions [7]. A good presentation of the mainly used recent methods can be found in [12].

The following work is based on a simple formulation of the problem. We consider a scaffold graph, that is, an undirected graph for which a perfect matching is given. Edges in the matching represent the contigs, whereas other edges represent witnesses for the relative locations of the contigs. These latter edges are weighted by a flexible confidence measure that can be read from the sequencing data or mixed with, for example, ancestral support in a phylogenetic context. Then, the scaffolding problem consists in finding at most a number of σ_{p} paths and σ_{c} cycles that, together, cover all matching edges (contigs). We formally describe this problem in Sect. 2.

In previous works, we stated that the problem is $\mathcal{NP}$-complete, even in bipartite and planar graphs, and initiated the quest to the frontier between polynomial-time solvability and $\mathcal{NP}$-completeness [3,4]. Exploring the structure of the scaffold graphs on real instances, we noted that many vertices of

Z. Lu et al. (Eds.): COCOA 2015, LNCS 9486, pp. 409–423, 2015.
DOI: 10.1007/978-3-319-26626-8_30

Table 1. Complexity results for SCAFFOLDING on various graph classes depending on $\omega_{\max}$, σ_{p}, and σ_{c}.

G	$\omega_{\max} = 1$			$\omega_{\max} = 0$		
	$\sigma_{\mathrm{p}}, \sigma_{\mathrm{c}} > 0$	$\sigma_{\mathrm{c}} = 0$	$\sigma_{\mathrm{p}} = 0$	$\sigma_{\mathrm{p}}, \sigma_{\mathrm{c}} > 0$	$\sigma_{\mathrm{c}} = 0$	$\sigma_{\mathrm{p}} = 0$
bipartite	$\mathcal{NP}$c (Theorem 3 & [4])					
co-bipartite	$\mathcal{NP}$c (Corollary 1), no $2^{o(m)}$-time algorithm (Corollary 2)			$\mathcal{P}$ (Theorem 1)		
split				open		
quasi forest	$\mathcal{NP}$c, $\mathcal{W}[1]$-h wrt. k (Thm. 5) no $2^{o(m)}$- or $n^{o(k)}$-time algorithm (Corollary 5)		$\mathcal{P}$ (Cor. 8)	open		$\mathcal{P}$ (Cor. 8)

Table 2. Complexity to approximate SCAFFOLDING.

G	max	min
clique, complete bipartite	2-approx. (Theorem 2)	$\notin \mathcal{APX}$ (Corollary 3 & [4]), $\omega_{\max}/\omega_{\min}$-approximation
co-bipartite, split	open	
quasi forest	no $n^{\frac{1}{2}-\epsilon}$-approx. (Cor. 6)	$\mathcal{APX}$-hard (Corollary 7)

the scaffold graph have small degrees, leading to overall sparsity [15,16]. We aim to exploit this property to design algorithms tuned to instances occurring in practice, by exploring different classes of graphs, described in Sect. 2. Indeed, we focus on two types of graphs: First, in Sect. 3, we consider dense graphs who we know are susceptible to polynomial-time approximation algorithms [3,4]. We focus on dense graphs which are not entirely complete, yet allow encoding some structure, namely co-bipartite and split graphs. On co-bipartite graphs, the unweighted version of the scaffolding problem becomes polynomial-time solvable, which is a first step towards designing algorithms for the general problem on these graphs. We consider a slightly relaxed version of the problem to improve the known approximation algorithm on complete graphs [4] to a ratio of two.

Second, in Sect. 4, we focus on classes of graphs that resemble real instances. Since SCAFFOLDING can be solved in polynomial time on graphs that are close to trees by measure of "treewidth" [15], we are interested in other distance measures to trees. To this end, we consider the class of graphs that can be turned into a (linear) forest by removing the edges of the given perfect matching M^* from it ("quasi forest"). In Sect. 4, we consider SCAFFOLDING on graphs G such that $G - M^*$ is a linear forest, a forest, a tree, or a path and show that the problem remains $\mathcal{NP}$-hard even for very restricted inputs. We reduce the $\mathcal{NP}$-complete WEIGHTED 2-SAT problem to it, allowing the inheritance of various hardness results of this problem.

The complexity and approximation results are respectively summarized in Tables 1 and 2. Next section is devoted to formal description of problems.

2 Notation and Problem Description

Let $G = (V, E)$ be a graph. For a vertex set $V' \subseteq V$, let $G[V']$ denote the subgraph of G induced by V' and let $G-V' := G[V \setminus V']$. Further, for any $S \subseteq E$, we define $\mathrm{Gr}(S) := (\bigcup_{e \in S} e, S)$ and $G - S := (V, E \setminus S)$. An edge-set M^* of a graph is called *matching* if no two of its edges intersect, that is, $e_1 \cap e_2 = \emptyset$ for all distinct $e_1, e_2 \in M^*$. A pair (G, M^*) where M^* is a perfect matching on G is called a *scaffold graph*. For a matching M^* and a vertex u, we define $M^*(u)$ as the unique vertex v with $uv \in M^*$ if such a v exists, and $M^*(u) = \bot$, otherwise. We abbreviate $X - \{x\} =: X - x$ for any set X of elements of the same type as x. Slightly abusing notation, we identify a path with the set of its edges. A path p is *alternating* with respect to a matching M^* if, for all vertices u of p, also $M^*(u)$ is a vertex of p. Thus, alternating paths have an even number of vertices. If M^* is clear from context, we do not mention it explicitly. For a function $\omega : E \to \mathbb{N}$ and a set $S \subseteq E$, we abbreviate $\sum_{e \in S} \omega(e) =: \omega(S)$ and we let $\omega_{\max} := \max_{e \in E} \omega(e)$. Thus, $\omega_{\max} = 1$ (resp. $= 0$) means that the weights can take only two values (resp. one value). The center of this work is the following problem.

SCAFFOLDING (SCA)
Input: $G = (V, E)$, $\omega : E \to \mathbb{N}$, perfect matching M^* in G, $\sigma_{\mathrm{p}}, \sigma_{\mathrm{c}}, k \in \mathbb{N}$
Question: Is there an $S \subseteq E \setminus M^*$ such that $\mathrm{Gr}(S \cup M^*)$ is a collection of $\leq \sigma_{\mathrm{p}}$ paths and $\leq \sigma_{\mathrm{c}}$ cycles and $\omega(S) \geq k$?

If ω is uniform, that is, all edges have the same weight, then we call the problem *unweighted* SCAFFOLDING. The variant of the problem that asks for *exactly* σ_{p} paths and *exactly* σ_{c} cycles is called STRICT SCAFFOLDING (SSCA). If we are looking for paths and cycles of *fixed lengths* ℓ_{p} and ℓ_{c}, we replace σ_{p} and σ_{c} by pairs $(\sigma_{\mathrm{p}}, \ell_{\mathrm{p}})$ and $(\sigma_{\mathrm{c}}, \ell_{\mathrm{c}})$ (length means the number of edges). We refer to the optimization variants of SCAFFOLDING that ask to minimize or maximize $\omega(S)$ as MIN SCAFFOLDING and MAX SCAFFOLDING, respectively.

Classes of Graphs. A graph is *bipartite* if it does not contain an odd cycle or, equivalently, if it admits a proper vertex two-coloring. A graph is *co-bipartite* if its complement is bipartite. Thus, a co-bipartite graph can also be considered as a pair of disjoint cliques, with some edges between them. For disjoint I and C, a graph $G = (I \cup C, E)$ such that I is an independent set, and C induces a clique in G, is called *split graph.* A scaffold graph (G, M^*) is called *quasi-forest* (resp. *quasi-tree* or *quasi-path*) if $G - M^*$ is a forest (resp. tree or path).

3 Dense Graphs

3.1 Good News

We start by showing that SCAFFOLDING can be solved in polynomial time on some co-bipartite graphs, so let $G = (V_1 \uplus V_2, E)$ be co-bipartite. We suppose that $G_1 := G[V_1]$ and $G_2 := G[V_2]$ are cliques, both G_1 and G_2 contain at least one matching edge, and that $n_1, n_2 \geq 2\sigma_p + 4\sigma_c$, where $n_1 := |V_1|$ and $n_2 := |V_2|$. We call *big* co-bipartite graphs the co-bipartite graphs fulfilling the above conditions. Let $B \subseteq E$ denote the set of edges with one endpoint in V_1 and one endpoint in V_2.

A *bridge* between G_1 and G_2 is either an edge of $B \cap M^*$, or a set of three edges of the form $uv \in M^* \cap G_1, vw \in B \setminus M^*, wx \in M^* \cap G_2$. A *round-trip* between G_1 and G_2 is a set of three edges of the form $uv \in B \cap M^*, vw \in B \setminus M^*$ and $wz \in G_1 \cap M^*$ or $G_2 \cap M^*$. We define necessary and sufficient conditions to guarantee a feasible solution to SCAFFOLDING in big co-bipartite graphs.

Lemma 1. *Unweighted* STRICT SCAFFOLDING *admits a feasible solution in big co-bipartite graphs if and only if the following properties hold:*

$(\sigma_p, \sigma_c) = (1, 0)$: $B \neq \emptyset$ (1)

$(\sigma_p, \sigma_c) = (0, 1)$: $\exists B' \subseteq B : B' \neq \emptyset, B \cap M^* \subseteq B'$ *and* B' *contains an even number of edges of disjoint bridges, and eventually round-trips.* (2)

$(\sigma_p, \sigma_c) = (0, \geq 2)$: $\exists B' \subseteq B : B \cap M^* \subseteq B'$ *and* B' *contains an even number of edges of disjoint bridges, and eventually round-trips.* (3)

Other cases : *A feasible solution always exists.* (4)

The proof is omitted, due to lack of place[1].

Since every condition appearing in Lemma 1 is checkable in polynomial time, we conclude the following.

Theorem 1. *Unweighted* STRICT SCAFFOLDING *can be solved in polynomial time in big co-bipartite graphs.*

Unfortunately, Theorem 1 holds only for unweighted instances. As we will see in Sect. 3.2, SCAFFOLDING is $\mathcal{NP}$-hard if we allow weights to be 0 or 1. However, we can still show a simple factor-2 approximation[2] for MAX SCAFFOLDING in case G is a complete graph or a complete bipartite graph.

Algorithm 1 starts with a maximal-cardinality maximum-weight matching S of $G - M^*$, implying that $\text{Gr}(S \cup M^*)$ is a collection of cycles. Then, it merges cycles, two at a time. Finally, it turns cycles into paths until the correct numbers of paths and cycles are reached.

Lemma 2. *If G is a complete graph, Algorithm 1 produces a solution whose weight is at least half the optimum.*

Proof. Let S^{org} denote the set S as computed in line 1 and let $\tilde{S}$ denote the set S returned in line 14. First, we show that $\tilde{S}$ is a solution. To this end, note

[1] available on http://www.lirmm.fr/~chateau/proof_cocoa_2015.pdf.

[2] That is, Algorithm 1 produces a solution of weight at least half the optimum weight.

Algorithm 1. A 2-approximation for MAX SCAFFOLDING on complete bipartite graphs.

1 $S \leftarrow$ a maximal-cardinality maximum-weight matching in $G - M^*$;
2 $\mathcal{C} \leftarrow$ the set of cycles in $\mathrm{Gr}(S \cup M^*)$;
3 $X \leftarrow \bigcup_{C \in \mathcal{C}} \operatorname{argmin}\{\omega(uv) \mid uv \in C \setminus M^*\}$;
4 **while** $|X| > \sigma_c + \sigma_p$ **do**
5   $e, e' \leftarrow \operatorname{argmin}\{\omega(e), \omega(e') \mid e, e' \in X \wedge e \neq e'\}$;
6   $Y \leftarrow$ a maximum-weight 4-cycle containing e and e' in G;
7   $S \leftarrow S \Delta Y$;
8   $e^* \leftarrow \operatorname{argmin}\{\omega(e^*) \mid e^* \in S \cap Y\}$;
9   $X \leftarrow (X \setminus \{e, e'\}) + e^*$;
10 **while** $|X| > \sigma_c$ **do**
11   $e \leftarrow \operatorname{argmin}\{\omega(e) \mid e \in X\}$;
12   $S \leftarrow S - e$;
13   $X \leftarrow X - e$;
14 **return** S;

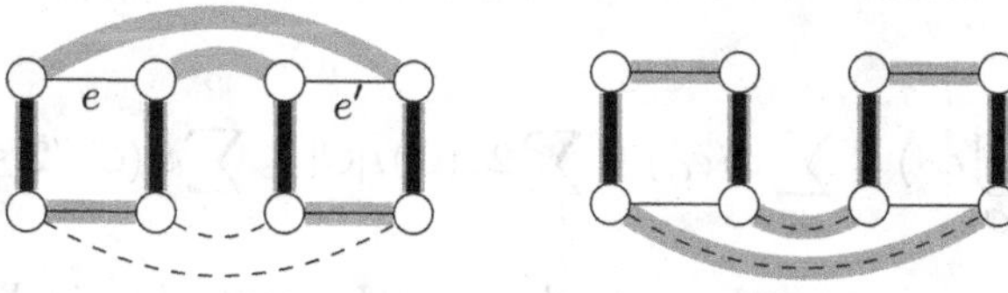

Fig. 1. An example with $\sigma_c = 1$ for which Algorithm 1 gives a solution of half optimal weight. Drawn edges (solid and dashed) have weight 1, all other edges have weight 0. The solid edges are a maximal-cardinality maximum-weight matching. Left: Algorithm 1 replaces e and e' to form the highlighted solution of weight 2. Right: an optimal solution of weight 4.

that S^{org} is a matching in $G - M^*$ and $\mathrm{Gr}(S^{\mathrm{org}} \cup M^*)$ is a collection of cycles since S^{org} is maximal-cardinality (and, thus, perfect). Since the only times S changes is when its symmetric difference with a 4-cycles is formed (line 7) or when edges are removed from S (line 12), the set $\tilde{S}$ is a matching in $G - M^*$. Thus, $\mathrm{Gr}(\tilde{S} \cup M^*)$ has maximum degree two.

Further, note that "$X \subseteq S$" and "$\mathrm{Gr}(S \cup M^*)$ is a collection of cycles" are invariants of the first while loop. Since, in line 9, we know that $\mathrm{Gr}(S \cup M^*)$ has at most $\sigma_p + \sigma_c$ connected components, all of which are cycles, we conclude that $\mathrm{Gr}(\tilde{S} \cup M^*)$ is a collection of at most σ_p paths and at most σ_c cycles.

Next, we show that the weight of the set S returned in line 14 is at least half the weight of a maximum matching in $G - M^*$, which is an upper bound on the solution weight and which is equal to $\omega(S^{\mathrm{org}})$. To this end, note that for all cycles C of $\mathrm{Gr}(S^{\mathrm{org}} \cup M^*)$, we selected a minimum-weight edge e_C of C into X in line 3. Thus, $\omega(C) \geq |C|/2 \cdot \omega(e_C)$ for each cycle C in $\mathrm{Gr}(S^{\mathrm{org}} \cup M^*)$.

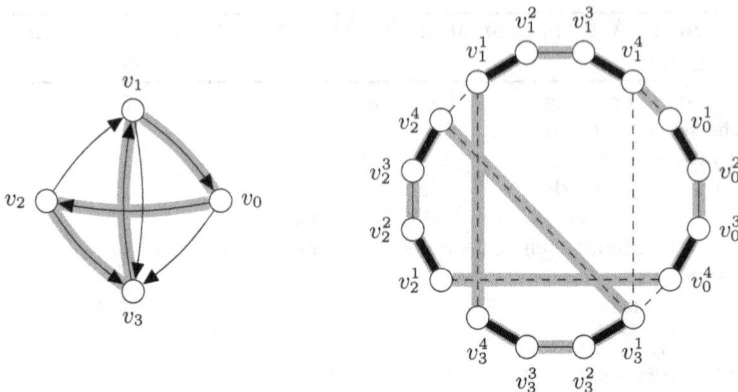

Fig. 2. Example of Construction 1, transforming the left instance of DIRECTED HAMILTONIAN CYCLE to the right graph with edges of M^* in bold and edges of the form $v_i^4 v_j^1$ dashed. A corresponding solution for both instances is highlighted.

Finally, let X^{org} denote the set X as computed in line 3. Then, since $|C| \geq 4$ for each C,

$$\omega(X^{\text{org}}) = \omega(\bigcup_C e_C) \leq \sum_C \omega(e_C) \leq \sum_C 2\omega(C)/|C| \leq \sum_C \omega(C)/2 \leq \frac{1}{2}\omega(S^{\text{org}}).$$

Since Algorithm 1 never touches any edge of S^{org} except edges in X^{org}, we know that $S^{\text{org}} \subseteq \tilde{S} \cup X^{\text{org}}$ and, thus, $\omega(\tilde{S}) \geq \omega(S^{\text{org}}) - \omega(X^{\text{org}}) \geq \omega(S^{\text{org}})/2$. □

Note that all arguments remain valid for complete bipartite graphs. Furthermore, Fig. 1 gives an example of a configuration in which Algorithm 1 gives a solution of weight half the optimum, implying that the bound of two is tight.

Theorem 2. *If G is a complete bipartite graph or a clique, then* MAX SCAFFOLDING *can be approximated to within a factor* 2 *in asymptotically the same time as it takes to compute a bipartite matching in G (currently $O(|V|^3)$). This factor is tight.*

3.2 Bad News

In the following, we show that SCAFFOLDING is $\mathcal{NP}$-hard on unweighted bipartite graphs and weighted co-bipartite and split graphs by reducing the $\mathcal{NP}$-complete DIRECTED HAMILTONIAN CYCLE problem [10] to SCAFFOLDING.

DIRECTED HAMILTONIAN CYCLE (DHC)
Input: A directed graph G
Question: Does G contain a simple cycle visiting all vertices?

Construction 1. *Let $G = (V = \{v_1, v_2, \ldots, v_n\}, A)$ be an instance of DHC. We construct $G' = (V_1 \uplus V_2, E)$ as follows (see Fig. 2).*

$$V_1 := \{v_i^1, v_i^3 \mid v_i \in V\} \qquad V_2 := \{v_i^2, v_i^4 \mid v_i \in V\}$$

$$E := \{v_i^1 v_i^2, v_i^2 v_i^3, v_i^3 v_i^4 \mid v_i \in V\} \cup \{v_i^4 v_j^1 \mid v_i v_j \in A\}.$$

Finally, let $M^* := \{v_i^1 v_i^2, v_i^3 v_i^4 \mid v_i \in V\}$, let $\omega : E \to \{0\}$, and let $k := 0$.

Theorem 3. *Unweighted* SCAFFOLDING *is $\mathcal{NP}$-complete on bipartite graphs, even if $\sigma_p = 0$ and $\sigma_c = 1$.*

Proof. The problem is clearly in $\mathcal{NP}$. We show that it is also $\mathcal{NP}$-hard by proving that G has a Hamiltonian cycle if and only if G' has an alternating Hamiltonian cycle with respect to M^*.

"$\Rightarrow$": Let $\mathcal{C}$ be a Hamiltonian cycle in G. Then, $S := \{v_i^2 v_i^3 \mid v_i \in V\} \cup \{v_i^4 v_j^1 \mid v_i v_j \in \mathcal{C}\}$ is a feasible solution, and $\mathrm{Gr}(S \cup M^*)$ is an alternating Hamiltonian cycle with respect to M^*.

"$\Leftarrow$": Let S' be a matching in $G' - M^*$ such that $\mathcal{C}' := \mathrm{Gr}(S \cup M^*)$ is an alternating Hamiltonian cycle in G' with respect to M^*. Since $\mathcal{C}'$ contains $v_i^3 v_i^4$ for each $v_i \in V$ (because they are all in M^*) and $\mathcal{C}'$ is a cycle, we know that S' contains n edges of the form $v_i^4 v_j^1$ for $v_i v_j \in A$. Since S' is a matching in $G - M^*$, no two of these edges are adjacent. Now, $\mathcal{C} := \{v_i v_j \mid v_i^4 v_j^1 \in S'\}$ is a collection of cycles covering all vertices of V in G. However, if $\mathcal{C}$ induces more than one cycle in G, then so does $S' \cup M^*$ in G'. Therefore, $\mathcal{C}$ is a Hamiltonian cycle in G. □

Note that we can change the construction such that $\omega : E \to \{1\}$ and $k := 4n$. This further enables us to add any number of edges of weight 0 to Construction 1 without affecting the correctness argument. This implies that SCAFFOLDING is $\mathcal{NP}$-complete on, for example, split graphs and co-bipartite graphs.

Corollary 1. *Let $\mathfrak{G}$ be a class of graphs such that, for each bipartite graph G there is a supergraph of G in $\mathfrak{G}$.* SCAFFOLDING *is $\mathcal{NP}$-complete on $\mathfrak{G}$, even if $\sigma_p = 0$ and $\sigma_c = 1$ and $\omega_{max} = 1$.*

Construction 1 also implies subexponential lower bounds for our problems based on the widely believed complexity-theoretic hypothesis known as the "Exponential-Time Hypothesis"[3] (ETH, see [13,17]). In fact, the lower bound is established directly from the fact that (planar) DIRECTED HAMILTONIAN CYCLE does not admit a $O(2^{o(|E(G)|)})$-time algorithm [14, Theorem 3.5] and that Construction 1 only linearly blows up the instance size.

Corollary 2. *Let $\mathfrak{G}$ be a class of graphs such that, for each bipartite graph G there is a supergraph of G in $\mathfrak{G}$. Assuming ETH, there is no $2^{o(|E(G)|)}$-time algorithm for* SCAFFOLDING *on $\mathfrak{G}$, even if $\sigma_p = 0$ and $\sigma_c = 1$ and $\omega_{max} = 1$.*

---

[3] The ETH states: there is a constant $c > 1$ such that no $O(c^n)$-time algorithm for n-variable 3-SAT exists.

Furthermore, we derive inapproximability of SCAFFOLDING from Construction 1.

Corollary 3. *Let $\mathfrak{G}$ be a class of graphs such that, for each bipartite graph G there is a supergraph of G in $\mathfrak{G}$. For all $\rho \in \mathbb{N}$, MIN SCAFFOLDING on $\mathfrak{G}$ is $\mathcal{NP}$-hard to approximate to within a factor of ρ, even if $\sigma_p = 0$ and $\sigma_c = 1$ and $\omega_{\max} = 1$.*

Proof. Suppose that there is a polynomial-time approximation algorithm $\mathfrak{A}$ for this problem with approximation ratio $\rho > 1$. Let $G = (V, E)$ be an instance of DIRECTED HAMILTONIAN CYCLE with $|V| = n$. We use Construction 1 to construct a bipartite graph G' with matching M^*. Then, we let $\omega : E' \to \mathbb{N}$ such that $\omega(E_1) = 0$ and set $k := 0$. Then, we can add any number of edges of weight 1 and no solution computed by $\mathfrak{A}$ can contain any of these edges. Then, replacing $4n$ by 0 in the proof of Theorem 1 yields a proof for Corollary 3. Indeed, if G has a Hamiltonian cycle, then $\mathfrak{A}$ finds a solution of weight $\rho \cdot 0 = 0$. Conversely, if G does not have a Hamiltonian cycle, then at least one edge of weight 1 must be taken in a solution produced by $\mathfrak{A}$. Thus, $\mathfrak{A}$ decides the $\mathcal{NP}$-complete DIRECTED HAMILTONIAN CYCLE problem in polynomial time. □

While there is little hope of finding a constant-factor polynomial-time approximation algorithm for MIN SCAFFOLDING, there is a linear-time algorithm with approximation ratio $\frac{\omega_{\max}}{w_{\min}}$ (where $\omega_{\max}$ and $\omega_{\min}$ denote the respective maximum and minimum edge weights) on complete bipartite graphs with $\sigma_p = 0$ and $\sigma_c = 1$. This algorithm repeatedly chooses the lowest weighted edge that does not close the cycle.

4 Sparse Graphs

4.1 Bad News

The hardness of SCAFFOLDING for dense graphs proved by Theorem 3 motivates the search for tractable cases among classes of sparse graphs. It is known that SCAFFOLDING is polynomial-time solvable on graphs that are close to being a forest (constant treewidth) [15], so we consider a different sparsity measure here. We investigate whether SCAFFOLDING becomes polynomial-time solvable if the result of removing the given perfect matching M^* from G forms a forest. We call this class of graphs "quasi forests". Remark that scaffold graphs originating from real data are not always quasi-forest, however this is a first step towards their structure. We start off by modifying Construction 1 to make the resulting graph a quasi tree. Unfortunately, this requires fixing the length of the sought Hamiltonian cycle. To circumvent this, we present another construction, reducing the $\mathcal{NP}$-complete WEIGHTED 2-SAT to SCAFFOLDING, that does not require fixing the lengths.

Construction 2. *Let $G = (V, A)$ be an instance of DHC, with $V = \{v_1, v_2, \ldots, v_n\}$. We construct $G' = (V', E)$ from G as follows:*

- *For each* $u \in V$*, we construct a "vertex-path"* $P_{4,u} = (u_1, u_2, u_3, u_4)$*, and we call* $\{u_2, u_3\}$ *an* inner edge. *The set of all such paths is denoted by* $P_4 = \bigcup_{u \in V} P_{4,u}$
- *For each* $(u, v) \in A$*, we construct an "edge-path"* $PE_{4,uv} = (uv^1, uv^2, uv^3, uv^4)$ *and the two edges* $\{u_4, uv^1\}, \{uv^4, v\}$.
- *We add a path* $Z = (z_1, z_2, z_3, z_4)$ *plus the edges* $\{z_1, z_3\}$ *and* $\{z_2, z_4\}$.
- *For each vertex* $u \in V$*, we add the edges* $\{u_1, z_1\}, \{u_2, z_1\}, \{u_4, z_1\}$. *For each* $(u, v) \in A$*, we add the edge* $\{uv^2, z_1\}$.
- *We let the perfect matching*

$$M^* := \{\{u_1, u_2\}, \{u_3, u_4\} \mid u \in V\} \cup \{\{uv^1, uv^2\}, \{uv^3, uv^4\} \mid (u, v) \in A\} \cup \{\{z_1, z_2\}, \{z_3, z_4\}\}.$$

Lemma 3. *$G' - M^*$ is a tree.*

Proof. First, $G' - M^*$ is a connected graph since all vertices, except z_1 are connected to z_1.

To show that $G' - M^*$ has no cycles, we count the number of edges in $G' - M^*$. For each $v \in V$, we have the edges $\{z_1, v_1\}$, $\{v_2, v_3\}$, $\{z_1, v_2\}$, and $\{z_1, v_4\}$, and for each $(u, v) \in A$ we have the edges $\{u_4, uv^1\}$, $\{z_1, uv^2\}$, $\{z_1, uv^3\}$, $\{uv^4, v_1\}$. Finally, we have the edges $\{z_1, z_3\}, \{z_2, z_4\}, \{z_2, z_3\}$. Therefore, $|E| = 4n + 4|A| + 3 = |V'| - 1$. So $G' - M^*$ is a tree. □

Theorem 4. Scaffolding *with* $(\sigma_p, \ell_p) = (|E| - n + 1, 3)$ *and* $(\sigma_c, \ell_c) = (1, 8n)$ *is* $\mathcal{NP}$*-complete on quasi-trees.*

Proof. Clearly, the problem is in $\mathcal{NP}$. We show that Construction 2 is correct, that is, G has a Hamiltonian cycle if and only if (G', M^*) can be covered by $|E| - n + 1$ paths of length 3 and 1 cycle of length $8n$.

"$\Rightarrow$": Let $\mathcal{C} = (v_1, v_2, \ldots, v_n)$ be a directed Hamiltonian cycle in G. We construct a solution for the Scaffolding-instance as follows: The alternating cycle of length $8n$ consists of the vertex-paths P_{4,v_i} for all $i \leq n$ and the edge-paths $PE_{4,uv}$ with $(u, v) \in \mathcal{C}$. A path of length 3 is given by Z, the remaining paths are given by the paths-edges $PE_{4,uv}$ such that $(u, v) \notin \mathcal{C}$.

"$\Leftarrow$": Suppose there is a set of one alternating cycle $\mathcal{C}$ of length $8n$ and $|E| - n + 1$ alternating paths of length 3. Clearly, the vertices of Z are not in $\mathcal{C}$ and therefore, they are included in a path of length 3. Thus, the edges $\{z_1, x\}$ for each $x \in V' \setminus \{z_2, z_3, z_4\}$ cannot be used in the covering. By construction, vertex-paths (resp. edge-paths) cannot appear consequently in $\mathcal{C}$. Since each vertex- and edge- path contains exactly four vertices and $\mathcal{C}$ has $8n$ vertices, $\mathcal{C}$ alternately contains n vertex-paths and n edge-paths. Then, a Hamiltonian cycle in G is given by the order of the vertex-paths in $\mathcal{C}$. □

Note that we had to fix the lengths of the paths and cycles we are looking for in the Scaffolding-instance. To show that Scaffolding is also hard on quasi-forests without restricting the lengths, we give another reduction. We reduce the $\mathcal{NP}$-complete Weighted 2-SAT problem (see [1]) to Scaffolding.

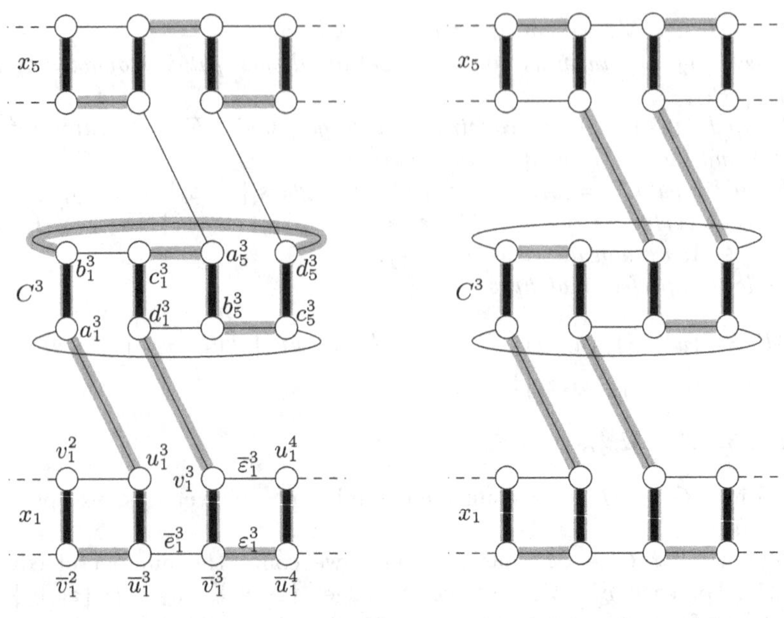

(a) $x_1 \vee \overline{x}_5$ satisfied by $x_1 = 1$. (b) $x_1 \vee \overline{x}_5$ satisfied by $x_1 = 1$ and $x_5 = 0$.

Fig. 3. Example of Construction 3 for the clause $x_1 \vee \overline{x}_5$. Bold edges are in M^*. Gray paths are solution paths corresponding to the respective assignments.

WEIGHTED 2-SAT
Input: n variables x_i with weights $w_i \geq 0$, m size-two clauses, $k \in \mathbb{N}$
Question: Is there a truth assignment β s.t., $\sum_{i \mid \beta(x_i)=1} w_i \geq k$?

The optimization variants of WEIGHTED 2-SAT that ask to find a satisfying assignment β that minimizes or maximizes $\sum_{i \mid \beta(x_i)=1} w_i$ are called MIN W2SAT and MAX W2SAT, respectively.

Construction 3. *Let (φ, k) be an instance of* WEIGHTED 2-SAT *with n variables $x_0, x_1, \ldots, x_{n-1}$ and m clauses $C^0, C^1, \ldots, C^{m-1}$. We produce the following instance $(G, \omega, M^*, n, 0, k)$ of* SCAFFOLDING *(see Fig. 3). For each variable x_i and for each $0 \leq j \leq m$, introduce*

- *vertices u_i^j, $\overline{u}_i^j$, v_i^{j-1},$\overline{v}_i^{j-1}$,*
- *edges $u_i^j\overline{u}_i^j$,$v_i^{j-1}\overline{v}_i^{j-1}$ that are also added to M^*,*
- *edges $\overline{\varepsilon}_i^{j-1} := v_i^{j-1}u_i^j$, and $\varepsilon_i^{j-1} := \overline{v}_i^{j-1}\overline{u}_i^j$.*
- *for $j < m$, if C^j contains $\overline{x}_i$, the edge $e_i^j := u_i^j v_i^j$, otherwise, $\overline{e}_i^j := \overline{u}_i^j\overline{v}_i^j$.*

For each clause C^j on the variables x_{ℓ_0} and x_{ℓ_1}, introduce

- *for each $i \in \{\ell_0, \ell_1\}$,*

  - *vertices* a_i^j, b_i^j, c_i^j, d_i^j
  - *edges* $a_i^j b_i^j$ *and* $c_i^j d_i^j$ *that are added to* M^* *and* $b_i^j c_i^j$,
  - *if* C^j *contains* $\overline{x}_i$, *edges* $\overline{u}_i^j a_i^j$, $\overline{v}_i^j d_i^j$, *otherwise,* $u_i^j a_i^j$, $v_i^j d_i^j$,
- *edges* $a_{\ell_0}^j c_{\ell_1}^j$, $c_{\ell_0}^j a_{\ell_1}^j$, $b_{\ell_0}^j d_{\ell_1}^j$, *and* $d_{\ell_0}^j b_{\ell_0}^j$.

Finally, set $\omega(\varepsilon_i^{m-1}) := 1$ *for each variable* x_i *and set the weights of all other edges to* 0.

Lemma 4. *Construction 3 is correct, that is,* φ *has a satisfying assignment of weight* k *if and only if* $(G, \omega, M^*, n, 0, k)$ *is a yes-instance of* SCAFFOLDING.

Proof. "$\Rightarrow$": Let β denote a solution for (φ, k). Then, we construct a solution S for $(G, \omega, M^*, n, 0, k)$ as follows. For each variable x_i and each $0 \leq j \leq m$, if $\beta(x_i) = 1$ then include $\{e_i^j, \varepsilon_i^j\} \cap E(G)$ in S, otherwise include $\{\overline{e}_i^j, \overline{\varepsilon}_i^j\} \cap E(G)$ in S.

For all clauses C^j, if exactly one of its literals is true, include edges according to Fig. 3a, if both its literals are true, include edges according to Fig. 3b in S. Then, $S \cup M^*$ contains exactly 1 alternating path for each of the n variables and, since $\varepsilon_i^{m-1} \in S$ for each x_i with $\beta(x_i) = 1$, the weight of S equals the weight of β, which is at least k.

"$\Leftarrow$": Let S be a solution for $(G, \omega, M^*, n, 0, k)$ and note that $S \cup M^*$ contains at most n paths and no cycles. Since $\mathrm{Gr}(S \cup M^*)$ does not contain cycles, for each $i < n$ and each $j \leq m$ we have $\varepsilon_i^j \notin S$ or $\overline{\varepsilon}_i^j \notin S$. This directly implies that, for each $i < n$ there is a path ending at u_i^m or $\overline{u}_i^m$ and there is a path ending at v_i^{-1} or $\overline{v}_i^{-1}$. Since there are at most n paths in $\mathrm{Gr}(S \cup M^*)$, the "or" above are exclusive and all other vertices have degree exactly two in $\mathrm{Gr}(S \cup M^*)$, implying that

$$\text{all other vertices are incident to exactly one edge in } S. \tag{5}$$

Next, we show for all $i < n$ and $j < m$ that

$$u_i^j a_i^j \in S \iff v_i^j d_i^j \in S. \tag{6}$$

To show $u_i^j a_i^j \in S \Rightarrow v_i^j d_i^j \in S$, assume $u_i^j a_i^j \in S$ and $v_i^j d_i^j \notin S$. Then, either $b_i^j c_i^j \in S$ or $b_i^j d_\ell^j \in S$ for some $\ell \neq i$. In the first case, we have $d_i^j b_\ell^j \in S$ and, thus, c_ℓ^j cannot have an incident edge in S without violating (5). In the second case, note that the only edges incident to c_i^j and d_i^j that could be in S without violating (5) are $c_i^j a_\ell^j$ and $d_i^j b_\ell^j$, respectively. However, if both are in S, then $\mathrm{Gr}(S \cup M^*)$ contains a forbidden cycle. The direction $v_i^j d_i^j \in S \Rightarrow u_i^j a_i^j \in S$ can be shown analogously.

Next, we show for each $i < n$ and $j \leq m$, that $\varepsilon_i^j \in S$ or $\overline{\varepsilon}_i^j \in S$, implying

$$\overline{\varepsilon}_i^j \in S \iff \varepsilon_i^j \notin S. \tag{7}$$

This is easy to see for $j = m$ since one of u_i^m and $\overline{u}_i^m$ has degree 2 in $\mathrm{Gr}(S \cup M^*)$. So let the claim hold for $j+1$ but not for j, that is, $\varepsilon_i^j, \overline{\varepsilon}_i^j \notin S$. If x_i is not contained in C^j, this means that both e_i^{j+1} and $\overline{e}_i^{j+1}$ are in S, forming a forbidden cycle.

Thus, by symmetry, let C^j contain x_i non-negated. Then, S contains both $\overline{e}_i^{j+1}$ and $u_i^{j+1}a_i^{j+1}$ and, by (6), also $v_i^{j+1}d_i^{j+1}$. Then, by (5), none of ε_i^{j+1} and $\overline{\varepsilon}_i^{j+1}$ are in S, contradicting that the claim holds for $j+1$. Thus, (7) holds by induction.

Next, we show for each $i < n$ and $j < m$ that

$$\varepsilon_i^j \in S \iff \varepsilon_i^{j-1} \in S. \tag{8}$$

Note that, by (7) it is sufficient to prove $\varepsilon_i^j \in S \Rightarrow \varepsilon_i^{j-1} \in S$ and $\overline{\varepsilon}_i^j \in S \Rightarrow \overline{\varepsilon}_i^{j-1} \in S$. Consider some $i < n$ and $j < m$ such that $\varepsilon_i^j \in S$. Then, by (5), we have $\overline{v}_i^j d_i^j \notin S$ and $\overline{e}_i^j \notin S$. By (6), it follows that $\overline{u}_i^j a_i^j \notin S$ and, thus, by (5), $\varepsilon_i^{j-1} \in S$. Note that $\overline{\varepsilon}_i^j \in S \Rightarrow \overline{\varepsilon}_i^{j-1} \in S$ can be shown analogously.

Finally, we define the assignment β for φ as $\beta(x_i) = 1 \iff \varepsilon_i^{m-1} \in S$. Then, since $\omega(\varepsilon_i^{m-1}) = 1$ for all $i < n$, we know that β assigns 1 to at most k variables. It remains to show that β satisfies φ. To this end, assume that a clause C^j is not satisfied and let x_i and x_ℓ denote the variables occuring in C^j. Note that at least one of the edges $u_i^j a_i^j$, $\overline{u}_i^j$, $u_\ell^j a_\ell^j$, and $\overline{u}_\ell^j a_\ell^j$ is in S since, otherwise, none of the n paths ending in the variable gadgets can visit the clause gadget of C^j. Since the prove is symmetric in all four cases, let us assume $u_i^j a_i^j \in S$. Then, C^j contains x_i non-negated. By (5), we have $\overline{\varepsilon}_i^{j-1} \notin S$, which, by (7) implies $\varepsilon_i^{j-1} \in S$ and, by (8), we arrive at $\varepsilon_i^{m-1} \in S$. Thus, $\beta(x_i) = 1$ and, thus, C^j is satisfied by β. □

Since WEIGHTED 2-SAT is known to be $\mathcal{W}[1]$-hard with respect to k (that is, an algorithm that is exponential only in k is unlikely to exist [8]), by Lemma 4, so is SCAFFOLDING.

Theorem 5. SCAFFOLDING *is $\mathcal{NP}$-hard and $\mathcal{W}[1]$-hard with respect to k, even on bipartite graphs G with $G - M^*$ being a linear forest, $\omega_{\max} = 1$ and $\sigma_c = 0$.*

Construction 3 can be modified to restrict the problem even further: consider two paths $p = (v_0, v_1, \ldots)$ and $q := (u_0, u_1, \ldots)$ in $G - M^*$. We can add new vertices α_j, β_j, γ_j with $j \in \{u, v\}$ with matching edges $\alpha_u\alpha_v$, $\beta_u\gamma_u$, $\beta_v\gamma_v$ and non-matching edges $\gamma_u\gamma_v$, $v_0\alpha_v$, $u_0\alpha_u$ of weight 0 and non-matching edges $\alpha_u\beta_u$, $\alpha_v\beta_v$ of weight $n+1$. Finally, we ask for a solution of weight $2n(n+1)+k$ containing $\sigma_p := 2n$ paths. Then, since all solutions have to contain the heavy edges $\alpha_u\beta_u$ and $\alpha_v\beta_v$, no solution can contain either $u_0\alpha_u$ or $v_0\alpha_v$ and, thus, any solution contains a solution for the original instance.

Corollary 4. SCAFFOLDING *is $\mathcal{NP}$-hard and $\mathcal{W}[1]$-hard with respect to k, even on bipartite graphs G with $G - M^*$ being a path, ω being tristate, and $\sigma_c = 0$.*

In analogy with Corollary 2, Construction 3 implies subexponential-time lower bounds for exact algorithms.

Corollary 5.
- *Assuming ETH, there is no $2^{o(|E(G)|)}$-time algorithm for* SCAFFOLDING, *and,*
- *assuming $\mathcal{W}[1] \neq \mathcal{FPT}$, there is no $n^{o(k)}$-time algorithm for* SCAFFOLDING, *even if $\sigma_c = 0$, $\omega_{\max} = 1$, and $G - M^*$ is a linear forest.*

Proof. We modify Construction 3 slightly such that the gadget for each variable x_i contains a "module" (subgraph induced by u_i^j, $\overline{u}_i^j$, v_i^j, and $\overline{v}_i^j$) *only for the clauses it is actually contained in.* Thus, the number of vertices and edges in the produced instance can be bounded linearly in the number of clauses of the WEIGHTED 2-SAT instance. Then, since INDEPENDENT SET does not have a $2^{o(m)}$-time algorithm [13] (with m denoting the number of edges), SCAFFOLDING does not have a $2^{o(m)}$-time algorithm (unless the ETH fails).

Furthermore, note that k-INDEPENDENT SET $\equiv$ k-WEIGHTED 2-SAT and k-WEIGHTED 2-SAT reduces to k-SCAFFOLDING by Construction 3. Thus, since INDEPENDENT SET does not have an $n^{o(k)}$-time algorithm [5], SCAFFOLDING does not have an $n^{o(k)}$-time algorithm (unless $\mathcal{W}[1] = \mathcal{FTP}$).

Since MAX W2SAT is $\mathcal{NP}$-hard to approximate to within a factor of $n^{1-\epsilon}$ for any $\epsilon > 0$ [1,11] and the number of vertices in the instance produced by Construction 3 is bounded in the number of variables, we conclude that, in contrast to the factor-3 approximation for SCAFFOLDING in complete graphs [4] (and the factor-2 approximation presented in Sect. 3), the problem is hard to approximate in the restricted class described above.

Corollary 6. MAX SCAFFOLDING *is $\mathcal{NP}$-hard to approximate to within a factor of $n^{\frac{1}{2}-\epsilon}$ for any $\epsilon > 0$, even on bipartite graphs G with $G - M^*$ being a linear forest, $\omega_{\max} = 1$ and $\sigma_c = 0$.*

For the minimization version, MIN SCAFFOLDING, we derive approximation hardness as well. To see this, note that Construction 3 is an S-reduction (see [6]) and MIN W2SAT is $\mathcal{APX}$-complete [1]. Thus, MIN SCAFFOLDING is $\mathcal{APX}$-hard.

Corollary 7. MIN SCAFFOLDING *is $\mathcal{APX}$-hard even on bipartite graphs G with $G - M^*$ being a linear forest, $\omega_{\max} = 1$ and $\sigma_c = 0$.*

Curiously, the approximation hardness result for MIN SCAFFOLDING is weaker than that for MAX SCAFFOLDING, which contrasts earlier observations on general graphs [4]. Thus, we suspect that Corollary 7 can be strengthened to at least the same hardness-level as we have for MAX SCAFFOLDING (Corollary 6). To this end, we conjecture existence of a gap-preserving reduction from an $\mathcal{NP}$-complete problem Π to MIN SCAFFOLDING with a non-constant gap.

Note that all results in this section hold for any numbers $\sigma_p \geq n$ and $\sigma_c \geq 0$ since we can add more paths artificially by adding isolated matching edges and we can add more cycles by adding new 4-cycles. Clearly, the isolated matching edges must constitute isolated paths. Further, if any isolated 4-cycle is covered by two paths, there are only $(n-2)$ paths and a cycle to cover all of the u_i^m, $\overline{u}_i^m$, v_i^{-1}, $\overline{v}_i^{-1}$, which can be seen to be impossible.

4.2 Good News

We show that, if G is a quasi forest and $\sigma_p = 0$, then SCAFFOLDING, and even STRICT SCAFFOLDING, can be solved in linear time. To this end, we employ the following reduction rule.

Rule 1. *Let u be a leaf in $G - M^*$ such that the parent v of u in $G - M^*$ is not a leaf. Then, delete all edges incident with v in $G - M^*$ that are not uv.*

Proof (Correctness of Rule 1). The proof is based on the argument that any solution S for G is a perfect matching (that is, $\mathrm{Gr}(S \cup M^*)$ has no degree-1 vertices). Since uv is the only edge of $G - M^*$ incident with u, it is apparent that $uv \in S$ and, thus, no other edge incident with v is in S. □

If we maintain a list of leaves on each edge-deletion, we can apply Rule 1 exhaustively in linear time. Moreover, if it is no longer applicable to $G - M^*$, then $G - M^*$ is a matching and checking whether G has the correct number of cycles can be done in linear time. Finally, we can extend this idea to work for any σ_p and σ_c by guessing all $2\sigma_p$ end points of paths in the solution and deleting the non-matching edges incident with them. Clearly, the result of this operation remains a quasi-forest and all vertices having a parent in $G-M^*$ have degree two in the solution, so the correctness of Rule 1 remains valid.

Corollary 8. STRICT SCAFFOLDING *can be solved in $O(n^{2\sigma_p+1})$ time on quasi forests.*

Corollary 8 raises the interesting question of whether (STRICT) SCAFFOLDING on quasi forests is fixed-parameter tractable with respect to σ_p, that is, whether it has an algorithm that is exponential only in σ_p.

5 Conclusion

In this article, we focus on the SCAFFOLDING problem and its minimization and maximization versions in the framework of complexity and approximation with performance guarantee. In such context, we consider several types of scaffold graphs, some being dense, some being sparse (can be turned into a tree, path, or forests by removing the edges of the perfect matching). We prove several negative results for these classes of graphs according to complexity hypotheses (see Tables 1 and 2 for a summary). On the positive side, we design a polynomial-time approximation algorithm with a ratio at most two (resp. $\frac{\omega_{\max}}{\omega_{\min}}$), for a variant of MAX SCAFFOLDING (resp. MIN SCAFFOLDING). These algorithms could be useful for the genome scaffolding problem in bioinformatics — an implementation would have to be tested and included in the available scaffolding tools. A further interesting task would be to design an $\mathcal{FPT}$-algorithm with respect to σ_p, or a polynomial-time algorithm for the unweighted case, under the assumption that $G-M^*$ is a forest. We may also consider extensions of this case to $G-M^*$ being of bounded tree-width. Finally, on the side of approximation theory, some more questions remain towards the goal of understanding the SCAFFOLDING problem. For instance, can the 2-approximation for MAX SCAFFOLDING on dense graphs be improved to a PTAS? Moreover, it would be interesting to extend our polynomial-time algorithms for unweighted graphs to approximation or parameterized algorithms for the weighted case. Finally, we are still interested in extending our results to other classes of graphs generalizing cliques, split and co-bipartite graphs, for instance (r, l)-graphs, which are graphs which are decomposeable into r independent sets and l cliques [2].

References

1. Alimonti, P., Ausiello, G., Giovaniello, L., Protasi, M.: On the complexity of approximating weighted satisfiability problems. Technical report, Università degli Studi di Roma La Sapienza. Rapporto Tecnico RAP 38.97 (1997)
2. Brandstädt, A.: Partitions of graphs into one or two independent sets and cliques. Discrete Math. **152**(1–3), 47–54 (1996)
3. Chateau, A., Giroudeau, R.: Complexity and polynomial-time approximation algorithms around the scaffolding problem. In: Dediu, A.-H., Martín-Vide, C., Truthe, B. (eds.) AlCoB 2014. LNCS, vol. 8542, pp. 47–58. Springer, Heidelberg (2014)
4. Chateau, A., Giroudeau, R.: A complexity and approximation framework for the maximization scaffolding problem. Theor. Comput. Sci. **595**, 92–106 (2015)
5. Chen, J., Chor, B., Fellows, M., Huang, X., Juedes, D.W., Kanj, I.A., Xia, G.: Tight lower bounds for certain parameterized NP-hard problems. Inf. Comput. **201**(2), 216–231 (2005)
6. Crescenzi, P.: A short guide to approximation preserving reductions. In: Proceedings of the Twelfth Annual IEEE Conference on Computational Complexity, Ulm, Germany, June 24–27, 1997, pp. 262–273 (1997)
7. Dayarian, A., Michael, T., Sengupta, A.: SOPRA: scaffolding algorithm for paired reads via statistical optimization. BMC Bioinform. **11**, 345 (2010)
8. Downey, R.G., Fellows, M.R.: Fundamentals of Parameterized Complexity. Texts in Computer Science. Springer, London (2013)
9. Gao, S., Sung, W.-K., Nagarajan, N.: Opera: reconstructing optimal genomic scaffolds with high-throughput paired-end sequences. J. Comput. Biol. **18**(11), 1681–1691 (2011)
10. Garey, M.R., Johnson, D.S.: Computers and intractability: a guide to the theory of NP-completeness (1979)
11. Håstad, J.: Clique is hard to approximate within $n^{1-\varepsilon}$. Electron. Colloquium Comput. Complex. (ECCC) **4**(38) (1997)
12. Hunt, M., Newbold, C., Berriman, M., Otto, T.: A comprehensive evaluation of assembly scaffolding tools. Genome Biol. **15**(3), R42 (2014)
13. Impagliazzo, R., Paturi, R., Zane, F.: Which problems have strongly exponential complexity? J. Comput. Syst. Sci. **63**(4), 512–530 (2001)
14. Lokshtanov, D., Marx, D., Saurabh, S.: Lower bounds based on the exponential time hypothesis. Bull. EATCS **105**, 41–72 (2011)
15. Weller, M., Chateau, A., Giroudeau, R.: Exact approaches for scaffolding. Accepted in RECOMBCG 2015, to appear in BMC Bioinformatics 2015a
16. Weller, M., Chateau, A., Giroudeau, R.: On the implementation of polynomial-time approximation algorithms for scaffold problems. Technical report (2015)
17. Woeginger, G.: Exact algorithms for NP-hard problems: a survey. In: Jünger, M., Reinelt, G., Rinaldi, G. (eds.) Combinatorial Optimization — Eureka, You Shrink!. LNCS, vol. 2570, pp. 185–207. Springer, Heidelberg (2003)

Parameterized Lower Bound and NP-Completeness of Some H-Free Edge Deletion Problems

N.R. Aravind[1], R.B. Sandeep[1(✉)], and Naveen Sivadasan[2]

[1] Department of Computer Science & Engineering, Indian Institute of Technology Hyderabad, Hyderabad, India
{aravind,cs12p0001}@iith.ac.in
[2] TCS Innovation Labs, Hyderabad, India
naveen@atc.tcs.com

Abstract. For a graph H, the H-free Edge Deletion problem asks whether there exist at most k edges whose deletion from the input graph G results in a graph without any induced copy of H. We prove that H-free Edge Deletion is NP-complete if H is a graph with at least two edges and H has a component with maximum number of vertices which is a tree or a regular graph. Furthermore, we obtain that these NP-complete problems cannot be solved in parameterized subexponential time, i.e., in time $2^{o(k)} \cdot |G|^{O(1)}$, unless Exponential Time Hypothesis fails.

1 Introduction

Graph modification problems ask whether we can obtain a graph G' from an input graph G by at most k number of *modifications* on G such that G' satisfies some properties. Modifications could be any kind of operations on vertices or edges. For a graph property Π, the Π Edge Deletion problem is to check whether there exist at most k edges whose deletion from the input graph results in a graph with property Π. Π Edge Completion and Π Edge Editing are defined similarly, where Completion allows only adding (completing) edges and Editing allows both completion and deletion. Another graph modification problem is Π Vertex Deletion, where at most k vertex deletions are allowed. The focus of this paper is on H-free Edge Deletion. It asks whether there exist at most k edges whose removal from the input graph G results in a graph G' without any induced copy of H. The corresponding Completion problem H-free Edge Completion is equivalent to $\overline{H}$-free Edge Deletion where $\overline{H}$ is the complement graph of H. Hence the results we obtain on H-free Edge Deletion translate to that of $\overline{H}$-free Edge Completion.

Graph modifications problems have been studied rigorously from 1970s onward. Initially, the studies were focused on proving that a modification problem is NP-complete or solvable in polynomial time. These studies resulted a good

R.B. Sandeep — Supported by TCS Research Scholarship.

Z. Lu et al. (Eds.): COCOA 2015, LNCS 9486, pp. 424–438, 2015.
DOI: 10.1007/978-3-319-26626-8_31

yield for vertex deletion problems: Lewis and Yannakakis proved [13] that Π VERTEX DELETION is NP-complete if Π is non-trivial and hereditary on induced subgraphs. In other words, Π VERTEX DELETION is NP-complete if Π is defined by a finite set of forbidden induced subgraphs. Interestingly, researchers could not find a dichotomy result for Π EDGE DELETION similar to that of Π VERTEX DELETION. The scarcity of hardness results for Π EDGE DELETION is mentioned in many papers in the last four decades. For examples, see [7,16]. It is a folklore result that H-FREE EDGE DELETION can be solved in polynomial time if H is a graph with at most one edge. Only these H-FREE EDGE DELETION problems are known to have polynomial time algorithms. Cai and Cai proved that H-FREE EDGE DELETION is incompressible if H is 3-connected but not complete, and H-FREE EDGE COMPLETION is incompressible if H is 3-connected and has at least two non-edges, unless NP$\subseteq$coNP/poly [3]. Further, under the same assumption, it is proved that H-FREE EDGE DELETION and H-FREE EDGE COMPLETION are incompressible if H is a tree on at least 7 vertices, which is not a star graph and H-FREE EDGE DELETION is incompressible if H is the star graph $K_{1,s}$, where $s \geq 10$ [4]. They use polynomial parameter transformations for the reductions. This implies that these problems are NP-complete. The H-FREE EDGE DELETION problems are NP-complete where H is C_ℓ for any fixed $\ell \geq 3$, claw ($K_{1,3}$) [16], P_ℓ for any fixed $\ell \geq 3$ [8], $2K_2$ [6] and diamond ($K_4 - e$) [9]. In this paper, we prove that H-FREE EDGE DELETION is NP-complete if H has at least two edges and has a component with maximum number of vertices which is a tree or a regular graph. For every such graph H, to obtain that H-FREE EDGE DELETION is NP-complete, we compose a series of polynomial time reductions starting from the reductions from one of the four base problems: P_3-FREE EDGE DELETION, P_4-FREE EDGE DELETION, K_3-FREE EDGE DELETION and $2K_2$-FREE EDGE DELETION (see Fig. 1). We believe that this technique can be extended to obtain a dichotomy result - H-FREE EDGE DELETION is NP-complete if and only if H has at least two edges. The evidence for this belief is discussed in the concluding section.

Another active area of research is to give parameterized lower bounds for graph modification problems. For example, to prove that a problem cannot be solved in parameterized subexponential time, i.e., in time $2^{o(k)} \cdot |G|^{O(1)}$, under some complexity theoretic assumption, where the parameter k is the size of the solution being sought. For this, the technique used is a linear parameterized reduction - a polynomial time reduction where the parameter blow up is only linear - from a problem which is already known to have no parameterized subexponential time algorithm under the Exponential Time Hypothesis (ETH). ETH is a widely believed complexity theoretic assumption that 3-SAT cannot be solved in subexponential time, i.e., in time $2^{o(n)}$, where n is the number of variables in the 3-SAT instance. Sparsification Lemma [11] implies that, under ETH, there exist no algorithm to solve 3-SAT in time $2^{o(n+m)} \cdot (n+m)^{O(1)}$, where m is the number of clauses in the 3-SAT instance. Sparsification Lemma considerably helps to obtain linear parameterized reductions from 3-SAT as it is allowed to have a parameter k such that $k = O(m+n)$ in the reduced problem instance. It is known that the base problems mentioned in the last paragraph cannot be solved in parameterized subexponential time, unless ETH fails. Since all the reductions we introduce here are compositions of linear parameterized reductions from the base problems, we obtain that H-FREE EDGE DELETION cannot solved in parameterized subexponential time, unless ETH fails, if H is a graph with at least two edges and has a component with maximum number of vertices which is a tree or a regular graph.

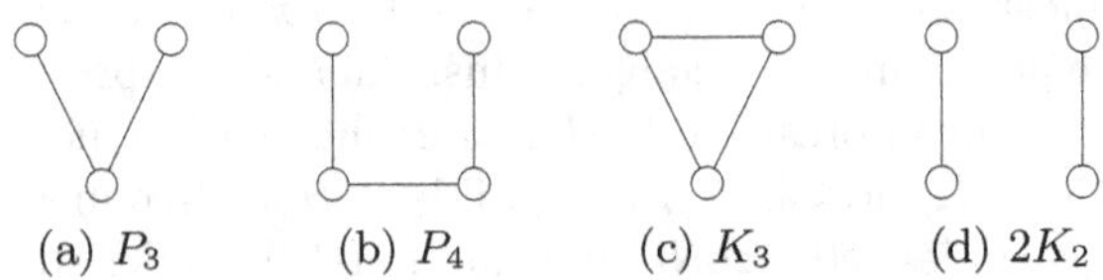

(a) P_3 (b) P_4 (c) K_3 (d) $2K_2$

Fig. 1. The four base problems are P_3-FREE EDGE DELETION, P_4-FREE EDGE DELETION, K_3-FREE EDGE DELETION and $2K_2$-FREE EDGE DELETION.

Graph modification problems have applications in DNA physical mapping [2,10], numerical algebra [14], circuit design [8] and machine learning [1].

Outline of the Paper: Section 2 gives the notations and terminology used in the paper. It also introduces two constructions which are used for the reductions. Section 3 proves that for any tree T with at least two edges, T-FREE EDGE DELETION is NP-complete and cannot be solved in parameterized subexponential time, unless ETH fails. Section 4 proves that for any connected regular graph R with at least two edges, R-FREE EDGE DELETION is NP-complete and cannot be solved in parameterized subexponential time, unless ETH fails. Section 5 combines the results from Sects. 3 and 4 to prove that for any graph H with at least two edges such that H has a component with maximum number of vertices which is a tree or a regular graph, H-FREE EDGE DELETION is NP-complete and cannot be solved in parameterized subexponential time, unless ETH fails. As a consequence of the equivalence between H-FREE EDGE DELETION and $\overline{H}$-FREE EDGE COMPLETION, we obtain the same results for $\overline{H}$-FREE EDGE COMPLETION.

2 Preliminaries and Basic Tools

Graphs: We consider simple, finite and undirected graphs. The vertex set and the edge set of a graph G is denoted by $V(G)$ and $E(G)$ respectively. G is represented by the tuple $(V(G), E(G))$. A simple path on ℓ vertices is denoted by P_ℓ. For a vertex set $V' \subseteq V(G)$, $G[V']$ denotes the graph induced by V' in G. $G - V'$ denotes the graph obtained by deleting all the vertices in V' and the edges incident to them from G. For an edge set $E' \subseteq E(G)$, $G - E'$ denotes the graph $(V(G), E(G) \setminus E')$. The diameter of a graph G, denoted by $\text{diam}(G)$, is the number of edges in the longest induced path in G. An r-regular graph is a graph in which every vertex has degree r. A regular graph is an r-regular graph for some non-negative integer r. A dominating set of a graph G is a set of vertices $V' \subseteq V(G)$ such that every vertex in G is either in V' or adjacent to at least one vertex in V'. For a graph G, the disjoint union of t copies of G is denoted by tG. A component of a graph G is a maximal connected subgraph of G. A largest component of a graph is a component with maximum number of vertices. We denote $|V(G)| + |E(G)|$ by $|G|$. We follow [15] for further notations and terminology.

Technique for Proving Parameterized Lower Bounds: Exponential Time Hypothesis (ETH) is the assumption that 3-SAT cannot be solved in time $2^{o(n)}$, where n is the number of variables in the 3-SAT instance. Sparsification Lemma [11] implies that there exists no algorithm for 3-SAT running in time $2^{o(n+m)} \cdot (n+m)^{O(1)}$, unless ETH fails, where n and m are the number of variables and the number of clauses respectively of the 3-SAT instance. A linear parameterized reduction is a polynomial time reduction from a parameterized problem A to a parameterized problem A' such that for every instance (G, k) of A, the reduction gives an instance (G', k') of B such that $k' = O(k)$.

Proposition 2.1 ([5]). *If there is a linear parameterized reduction from a parameterized problem A to a parameterized problem B and if A does not admit a parameterized subexponential time algorithm, then B does not admit a parameterized subexponential time algorithm.*

We refer the book [5] for an excellent exposition on this and other aspects of parameterized algorithms and complexity.

Proposition 2.2. *The following problems are* NP-complete. *Furthermore, they cannot be solved in time* $2^{o(k)} \cdot |G|^{O(1)}$, *unless ETH fails.*

(i) P_3-FREE EDGE DELETION *[12]*
(ii) P_4-FREE EDGE DELETION *[6]*
(iii) C_ℓ-FREE EDGE DELETION *for any fixed* $\ell \geq 3$ *[16]*[1]
(iv) $2K_2$-FREE EDGE DELETION *[6]*

For any fixed graph H, the H-FREE EDGE DELETION problem trivially belongs to NP. Hence, we may state that an H-FREE EDGE DELETION problem is NP-complete by proving that it is NP-hard.

[1] Yannakakis gives a polynomial time reduction from VERTEX COVER to C_ℓ-FREE EDGE DELETION, for any fixed $\ell \geq 3$ [16]. If $\ell \neq 3$, the reduction he gives is a linear parameterized reduction. When $\ell = 3$, the reduction is not a linear parameterized reduction as it gives an instance with a parameter $k' = O(|E(G)| + k)$, where (G, k) is the input VERTEX COVER instance. But, it is straight-forward to verify that composing the standard 3-SAT to VERTEX COVER reduction (which is a linear parameterized reduction and gives a graph with $O(n+m)$ edges) with this reduction gives a linear parameterized reduction from 3-SAT to $K_3(C_3)$-FREE EDGE DELETION.

2.1 Basic Tools

We introduce two constructions which will be used for the polynomial time reductions in the upcoming sections.

Construction 1. *Let (G', k, H, V') be an input to the construction, where G' and H are graphs, k is a positive integer and V' is a subset of vertices of H. Label the vertices of H such that every vertex get a unique label. Let the labelling be ℓ_H. For every subgraph (not necessarily induced) C with a vertex set $V(C)$ and an edge set $E(C)$ in G' such that C is isomorphic to $H[V']$, do the following:*

- *Give a labelling ℓ_C for the vertices in C such that there is an isomorphism f between C and $H[V']$ which maps every vertex v in C to a vertex v' in $H[V']$ such that $\ell_C(v) = \ell_H(v')$, i.e., $f(v) = v'$ if and only if $\ell_C(v) = \ell_H(v')$.*
- *Introduce $k+1$ sets of vertices $V_1, V_2, \ldots, V_{k+1}$, each of size $|V(H) \setminus V'|$.*
- *For each set V_i, introduce an edge set E_i of size $|E(H) \setminus E(H[V'])|$ among $V_i \cup V(C)$ such that there is an isomorphism h between H and $(V(C) \cup V_i, E(C) \cup E_i)$ which preserves f, i.e., for every vertex $v \in V(C)$, $h(v) = f(v)$.*

This completes the construction. Let the constructed graph be G.

An example of the construction is shown in Fig. 2. Let C be a copy of $H[V']$ in G'. Then, C is called a *base* in G'. Let $\{V_i\}$ be the $k+1$ sets of vertices introduced in the construction for the base C. Then, each V_i is called a *branch* of C and the vertices in V_i are called the *branch vertices* of C. C is called the *base* of V_i for $1 \leq i \leq k+1$. The vertex set of G' in G is denoted by $V_{G'}$.

Since H is a fixed graph, the construction runs in polynomial time. In the construction, for every base C in G', we introduce new vertices and edges such that there exist $k+1$ copies of H in G and C is the common intersection of every pair of them. This enforces that every solution of an instance (G, k) of H-FREE EDGE DELETION is a solution of an instance (G', k) of H'-FREE EDGE DELETION, where H' is $H[V']$. This is proved in the following lemma.

Lemma 2.3. *Let G be obtained by Construction 1 on the input (G', k, H, V'), where G' and H are graphs, k is a positive integer and $V' \subseteq V(H)$. Then, if (G, k) is a yes-instance of H-FREE EDGE DELETION, then (G', k) is a yes-instance of H'-FREE EDGE DELETION, where H' is $H[V']$.*

Proof. Let F be a solution of size at most k of (G, k). For a contradiction, assume that $G' - F$ has an induced H' with a vertex set U. Hence there is a base C in G' isomorphic to H' with the vertex set $V(C) = U$. Since there are $k+1$ copies of H in G, where each pair of copies of H has the intersection C, and $|F| \leq k$, deleting F cannot kill all the copies of H associated with C. Therefore, since U induces an H' in $G' - F$, there exists a branch V_i of C such that $U \cup V_i$ induces H in $G - F$, which is a contradiction. □

Now we introduce a simple construction, which is used in the next section. This construction attaches a clique of $k+1$ vertices to each vertex in the input graph of the construction.

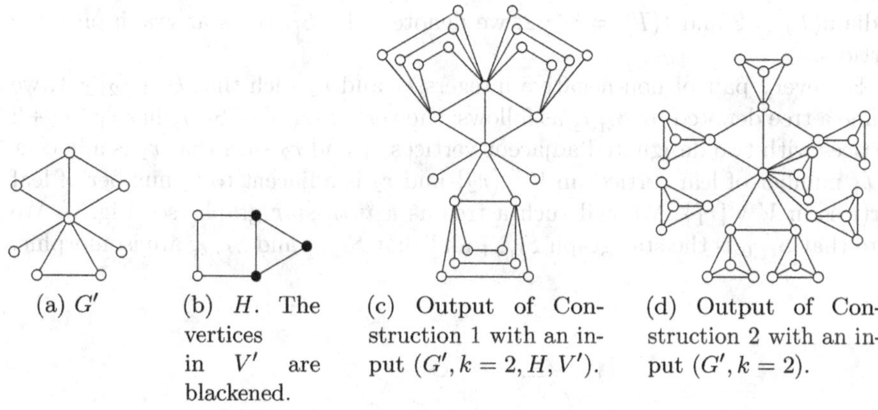

(a) G' (b) H. The vertices in V' are blackened. (c) Output of Construction 1 with an input $(G', k = 2, H, V')$. (d) Output of Construction 2 with an input $(G', k = 2)$.

Fig. 2. Examples showing Constructions 1 and 2.

Construction 2. *Let (G', k) be an input to the construction, where G' is a graph and k is a positive integer. For every vertex v_i in G', introduce a set of $k+1$ vertices V_i and make every pair of vertices in $V_i \cup \{v_i\}$ adjacent. This completes the construction. Let the resultant graph be G.*

An example of the construction is shown in Fig. 2. Here, we call all the newly introduced vertices as *branch vertices.*

3 T-FREE EDGE DELETION

Let T be any tree with at least two edges. We use induction on the diameter of T to prove that T-FREE EDGE DELETION is NP-complete. The base cases are when $\text{diam}(T) = 2$ or 3. To prove the base cases, we use polynomial time reductions from P_3-FREE EDGE DELETION and P_4-FREE EDGE DELETION. For any T with $\text{diam}(T) > 3$, we give polynomial time reduction from T'-FREE EDGE DELETION to T-FREE EDGE DELETION, where T' is a subtree of T such that $\text{diam}(T') = \text{diam}(T) - 2$. To prove each of the base cases, we apply induction on the number of leaf vertices. All our reductions are linear parameterized reductions and hence from the non-existence of parameterized subexponential algorithms for P_3-FREE EDGE DELETION and P_4-FREE EDGE DELETION, we obtain that there exists no parameterized subexponential time algorithm for T-FREE EDGE DELETION, unless ETH fails.

3.1 Base Cases

As mentioned above, the base cases are when $\text{diam}(T) = 2$ or 3. By $\ell(T)$, we denote the number of leaf vertices of T. We call the vertices in T with degree one as *leaf vertices* and the vertices with degree more than one as *internal vertices.*

If $\mathrm{diam}(T) = 2$ and $\ell(T) = \ell \geq 2$, we denote T by S_ℓ, the star graph on $\ell + 1$ vertices.

For every pair of non-negative integers ℓ_1 and ℓ_2 such that $\ell_1 + \ell_2 \geq 1$, we define a tree denoted by S_{ℓ_1,ℓ_2} as follows: the vertex set V of S_{ℓ_1,ℓ_2} has $\ell_1+\ell_2+2$ vertices with two designated adjacent vertices r_1 and r_2 such that r_1 is adjacent to ℓ_1 number of leaf vertices in $V \setminus \{r_2\}$ and r_2 is adjacent to ℓ_2 number of leaf vertices in $V \setminus \{r_1\}$. We call such a tree as a *twin-star* graph (see Fig. 3). We note that $S_{\ell_1,0}$ is the star graph S_{ℓ_1+1} and that S_{ℓ_1,ℓ_2} and S_{ℓ_2,ℓ_1} are isomorphic.

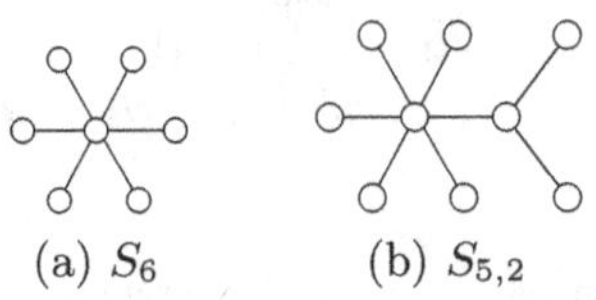

Fig. 3. A star graph and a twin-star graph

Lemma 3.1. *Let $\ell > 2$. Then, there is a linear parameterized reduction from $S_{\ell-1}$-FREE EDGE DELETION to S_ℓ-FREE EDGE DELETION.*

Proof. Let (G', k) be an instance of $S_{\ell-1}$-FREE EDGE DELETION. Apply Construction 2 on (G', k) to obtain G. We claim that (G', k) is a yes-instance of $S_{\ell-1}$-FREE EDGE DELETION if and only if (G, k) is a yes-instance of S_ℓ-FREE EDGE DELETION.

Let (G', k) be a yes-instance of $S_{\ell-1}$-FREE EDGE DELETION. Let F' be a solution of size at most k of (G', k). For a contradiction, assume that $G - F'$ has an induced S_ℓ with a vertex set U. Let r be the internal vertex of the S_ℓ induced by U in $G - F'$. Now there are two cases and in both the cases we obtain contradictions.

- r is a branch vertex: Since the neighborhood of any branch vertex in $G - F'$ is a clique, r cannot be the internal vertex, which is a contradiction.
- r is a vertex in $V_{G'}$: Since the branch vertices in the neighborhood of r in $G - F'$ induce a clique, at most one branch neighbor u of r is present in U (as a leaf vertex). Hence, the remaining leaf vertices of the S_ℓ induced by U in $G - F'$ belong to $V_{G'}$. This implies that $U \setminus \{u\}$ induces $S_{\ell-1}$ in $G' - F'$, which is a contradiction.

Conversely, let (G, k) be a yes-instance of S_ℓ-FREE EDGE DELETION. Let F be a solution of size at most k of (G, k). For a contradiction, assume that $G' - F$ has an induced $S_{\ell-1}$ with a vertex set U. Let r be the internal vertex of $S_{\ell-1}$ induced by U in $G' - F$. Since $|F| \leq k$ and $k + 1$ branch vertices are adjacent to r in G, there is at least one branch vertex u adjacent to r in $G - F$. Hence, $U \cup \{u\}$ induces an S_ℓ in $G - F$, which is a contradiction. □

Theorem 3.2. *For every integer $\ell \geq 2$, S_ℓ-*FREE EDGE DELETION *is* NP-complete. *Furthermore, S_ℓ-*FREE EDGE DELETION *is not solvable in time $2^{o(k)} \cdot |G|^{O(1)}$, unless ETH fails.*

Proof. The proof is by induction on ℓ. When $\ell = 2$, S_ℓ is the graph P_3. Hence, Proposition 2.2(i) proves this case. Assume that the statements are true for $S_{\ell-1}$-FREE EDGE DELETION, if $\ell - 1 \geq 2$. Now the statements follow from Lemma 3.1. □

We apply a similar technique to prove the NP-completeness and parameterized lower bound for T-FREE EDGE DELETION when $\text{diam}(T) = 3$. As described before, we denote these graphs by S_{ℓ_1,ℓ_2}, the twin-star graph having $\ell_1 \geq 1$ leaf vertices adjacent to an internal vertex r_1 and $\ell_2 \geq 1$ leaf vertices adjacent to another internal vertex r_2.

Lemma 3.3. *For any pair of integers ℓ_1 and ℓ_2 such that $\ell_1, \ell_2 \geq 1$ and $\ell_1 + \ell_2 \geq 3$, there is a linear parameterized reduction from S_{ℓ_1-1,ℓ_2-1}-*FREE EDGE DELETION *to S_{ℓ_1,ℓ_2}-*FREE EDGE DELETION.

Proof. Let (G', k) be an instance of S_{ℓ_1-1,ℓ_2-1}-FREE EDGE DELETION. Apply Construction 2 on (G', k) to obtain G. We claim that (G', k) is a yes-instance of S_{ℓ_1-1,ℓ_2-1}-FREE EDGE DELETION if and only if (G, k) is a yes-instance of S_{ℓ_1,ℓ_2}-FREE EDGE DELETION.

Let (G', k) be a yes-instance of S_{ℓ_1-1,ℓ_2-1}-FREE EDGE DELETION. Let F' be a solution of size at most k of (G', k). For a contradiction, assume that $G - F'$ has an induced copy of S_{ℓ_1,ℓ_2} with a vertex set U. Let r_1 and r_2 be the two internal vertices of the S_{ℓ_1,ℓ_2} induced by U in $G - F'$. Now, there are the following cases and in each case, we obtain a contradiction.

- Either r_1 or r_2 is a branch vertex: This is not possible as the neighborhood of every branch vertex induces a clique in $G - F'$.
- Both r_1 and r_2 are in $V_{G'}$: Since the branch vertices adjacent to r_1 forms a clique in $G - F'$, at most one branch vertex u_1 can be a leaf vertex adjacent to r_1 in the S_{ℓ_1,ℓ_2} induced by U in $G - F'$. Similarly, at most one branch vertex u_2 can be a leaf vertex adjacent to r_2 in the S_{ℓ_1,ℓ_2} induced by U in $G - F'$. The remaining vertices of U belong to $V_{G'}$. Hence $U \setminus \{u_1, u_2\}$ induces S_{ℓ_1-1,ℓ_2-1} in $G' - F'$, which is a contradiction.

Conversely, let (G, k) be a yes-instance of S_{ℓ_1,ℓ_2}-FREE EDGE DELETION. Let F be a solution of size at most k of (G, k). For a contradiction, assume that $G' - F$ has an induced S_{ℓ_1-1,ℓ_2-1} with a vertex set U. Since $\ell_1 + \ell_2 \geq 3$, there exists at least one internal vertex, say r_1, in the S_{ℓ_1-1,ℓ_2-1} induced by U in $G' - F$. If there is no other internal vertex r_2 in the S_{ℓ_1-1,ℓ_2-1}, then let r_2 be any leaf vertex of the S_{ℓ_1-1,ℓ_2-1}. Let V_1 and V_2 be the set of branch vertices introduced in the construction such that every vertex in V_1 is adjacent to r_1 and every vertex in V_2 is adjacent to r_2. Since $|F| \leq k$ and $|V_1|, |V_2| = k + 1$, there exist a vertex $v_1 \in V_1$ adjacent to r_1 and a vertex $v_2 \in V_2$ adjacent to r_2 in $G - F$. Hence, $U \cup \{v_1, v_2\}$ induces an S_{ℓ_1,ℓ_2} in $G - F$, which is a contradiction. □

Theorem 3.4. *For every pair of integers ℓ_1 and ℓ_2 such that $\ell_1, \ell_2 \geq 0$ and $\ell_1+\ell_2 \geq 1$, S_{ℓ_1,ℓ_2}-FREE EDGE DELETION is* NP-complete *and S_{ℓ_1,ℓ_2}-FREE EDGE DELETION is not solvable in time $2^{o(k)} \cdot |G|^{O(1)}$, unless ETH fails.*

Proof. The proof is by induction on $\ell_1 + \ell_2$. The base cases are:

- $\ell_1 = 0$ ($\ell_2 = 0$): This is the case when the tree is S_{ℓ_2+1} (S_{ℓ_1+1}), the case handled by Theorem 3.2.
- $\ell_1 = \ell_2 = 1$: Here the tree is a P_4 and hence the statements follow from Proposition 2.2(ii).

Assume that the statements holds true for the integers $\ell_1 - 1, \ell_2 - 1$ such that $\ell_1 - 1, \ell_2 - 1 \geq 0$ and $(\ell_1 - 1) + (\ell_2 - 1) \geq 1$. Now, the statements follow from Lemma 3.3. □

3.2 Induction

In the previous subsection, we proved the base cases of the inductive proof for the NP-completeness and parameterized lower bound of T-FREE EDGE DELETION. The base cases were diam(T) = 2 (star graph) and diam(T) = 3 (twin-star graph). Before concluding the proof, we give a lemma which is stronger than what we require and the further implications of this lemma will be discussed in the concluding section.

Lemma 3.5. *Let H be any graph and d be any integer. Let V' be the set of all vertices in H with degree more than d. Let H' be $H[V']$. Then, there is a linear parameterized reduction from H'-FREE EDGE DELETION to H-FREE EDGE DELETION.*

Proof. Let (G', k) be an instance of H'-FREE EDGE DELETION. Obtain G by applying Construction 1 on (G', k, H, V'). We claim that (G', k) is a yes-instance of H'-FREE EDGE DELETION if and only if (G, k) is a yes-instance of H-FREE EDGE DELETION.

Let (G', k) be a yes-instance of H'-FREE EDGE DELETION. Let F' be a solution of size at most k of (G', k). For a contradiction, assume that $G - F'$ has an induced H with a vertex set U. Let U' be the set of all vertices in U such that every vertex in U' has degree more than d in $(G - F')[U]$. Since every branch vertex in G has degree at most d, every vertex in U' must be in $V_{G'}$. Hence U' induces an H' in $G' - F'$, which is a contradiction. Lemma 2.3 proves the converse. □

Corollary 3.6 is obtained by invoking Lemma 3.5 with $H = T$ and $d = 1$.

Corollary 3.6. *Let T be any tree with* diam(T) > 3. *Let T' be obtained from T by deleting all leaf vertices. Then, there exists a linear parameterized reduction from T'-FREE EDGE DELETION to T-FREE EDGE DELETION.*

Theorem 3.7. *Let T be any tree with at least two edges. Then, T-FREE EDGE DELETION is* NP-complete. *Furthermore, T-FREE EDGE DELETION is not solvable in time $2^{o(k)} \cdot |G|^{O(1)}$, unless ETH fails.*

Proof. We apply induction on the diameter of T. Theorems 3.2 and 3.4 prove the statements when $\text{diam}(T) = 2$ and $\text{diam}(T) = 3$ respectively. Let the statements be true when $\text{diam}(T) = t'$ for all t' such that $2 \leq t' \leq t$ for some $t \geq 3$. Assume that T has diameter $t+1$. Deleting all leaf vertices from T gives a graph T' with diameter $t+1-2 = t-1 \geq 2$. Now the statements follow from Corollary 3.6. □

4 R-FREE EDGE DELETION

In this section, for any connected r-regular graph R, where $r > 2$, we give a direct reduction either from P_3-FREE EDGE DELETION or from K_3-FREE EDGE DELETION to R-FREE EDGE DELETION. The following three observations are used to prove the reduction which is given in Lemma 4.4.

Observation 4.1. *Let R be an r-regular graph for some $r > 2$. Let $V' \subseteq V(R)$ be such that $|V'| = 3$. Then, $V \setminus V'$ is a dominating set in R.*

Proof. To prove that $V \setminus V'$ is a dominating set of R, we need to prove that for every vertex $v \in V(R)$, either v is in $V \setminus V'$ or v is adjacent to a vertex in $V \setminus V'$. If $v \notin V \setminus V'$, then $v \in V'$. Since $|V'| = 3$ and v has degree $r \geq 3$, v must have at least one edge to a vertex in $V \setminus V'$. □

Observation 4.2. *Let G be a graph and $r > 0$ be an integer. Let $W \subseteq V(G)$ be such that every vertex in W has degree r in G and $G[W]$ is connected. Let R be any r-regular graph and G has an induced copy of R on a vertex set W' containing at least one vertex in W. Then $W \subseteq W'$.*

Proof. Let W'' be $W \setminus W'$. For a contradiction, assume that W'' is non-empty. It is given that $W \cap W'$ is non-empty, i.e., $W \setminus W''$ is non-empty. Therefore, since $G[W]$ is connected, there exists a vertex $v \in W''$ such that v is adjacent to a vertex $u \in W \setminus W''$. Since $u \in W'$ and $G[W']$ induces an r-regular graph and u has degree r in G, we obtain that every neighbor of u must be in W'. This is a contradiction as v is a neighbor of u and is not in W'. Hence $W \subseteq W'$. □

Observation 4.3. *Let G and G' be two graphs such that $|V(G)| = |V(G')| = 3$ and $|E(G)| = |E(G')|$. Then G and G' are isomorphic.*

Proof. If a graph has exactly three vertices, the graph is completely defined by its number of edges e: If $e = 0$, the graph is a null graph, if $e = 1$, the graph is $K_1 \cup K_2$, if $e = 2$, the graph is a P_3 and if $e = 3$, the graph is a K_3. □

Lemma 4.4. *Let R be any connected r-regular graph for any $r > 2$. Assume that there exists a set of vertices $V' \subseteq V(R)$ such that $R[V']$ is a P_3 or a K_3 and $R - V'$ is connected. Let $R[V']$ be H'. Then, there is a linear parameterized reduction from H'-FREE EDGE DELETION to R-FREE EDGE DELETION.*

Proof. Let (G', k) be an instance of H'-FREE EDGE DELETION. We apply Construction 1 on $(G', k, H = R, V')$ to obtain G. We claim that (G', k) is a yes-instance of H'-FREE EDGE DELETION if and only if (G, k) is a yes-instance of R-FREE EDGE DELETION.

Let F' be a solution of size at most k of (G', k). We claim that F' is a solution of (G, k). Let G'' be $G - F'$. Assume that the claim is false. Then, there is a set of vertices $U \subseteq V(G'')$ which induces R in G''. Since $R \setminus V'$ is connected, there is a set of vertices $U' \subseteq U$ which induces H' in G'' such that $G''[U \setminus U']$ is a connected graph. Since $G' - F'$ is H'-free, at least one vertex $v \in U'$ must be from a branch V_j. Since $R \setminus V'$ is connected, by the construction, V_j induces a connected graph in G and hence in G''. Furthermore, every vertex in V_j has degree r in G''. Now, by Observation 4.2 (invoked with $G = G''$, $W = V_j$ and $W' = U$), every vertex in V_j is in U. Since $|V'| = 3$, by the construction, $|V_j| = |U| - 3$. Hence, by Observation 4.1 (invoked with $V' = U \setminus V_j$), V_j is a dominating set in $G''[U]$. Therefore, $U = V_j \cup B_j$ where B_j is the set of base vertices of V_j in G. Since every vertex in V_j has degree r and $G''[U]$ induces an r-regular graph, every edge incident to the vertices in V_j is in $G''[U]$, i.e., $E_j \subseteq E(G''[U])$, where E_j is the edge set introduced along with V_j in Construction 1. Now, by an edge counting argument, $E(G''[B_j])$ must have $|E(H')|$ number of edges. Therefore, since $|B_j| = 3$, by Observation 4.3, B_j induces H' in $G' - F'$, which is a contradiction. Lemma 2.3 proves the converse. □

Observation 4.5. *Let G be a connected graph with at least $d \geq 1$ vertices. Then, there is a set of vertices $V' \subseteq V(G)$ such that $|V'| = d$ and $G[V']$ is connected.*

Proof. Let v be any vertex in G. Do a breadth first search starting from v until d number of vertices are visited. Let V' be the set of visited vertices. Clearly, $G[V']$ is connected. □

The following lemma may be of independent interest. The assumption in Lemma 4.4 comes as a special case of it.

Lemma 4.6. *Let H be any connected graph with minimum degree d for any $d > 2$. Then, there exists $V' \subseteq V(H)$ such that $|V'| = d$, $H[V']$ is connected and $H \setminus V'$ is connected.*

Proof. Let $\mathcal{H}$ be the set of all connected graphs with d number of vertices. Since the minimum degree of H is d, H has at least $d + 1$ vertices. Hence, by Observation 4.5, there exists at least one $H' \in \mathcal{H}$ as an induced subgraph of H. For a contradiction, assume that for every $V' \subseteq V(H)$ which induces any $H' \in \mathcal{H}$ in H, $H \setminus V'$ is disconnected. Among all such sets of vertices, consider a set of vertices $V' \subseteq V(H)$ which induces any $H' \in \mathcal{H}$ in H such that $H - V'$ leaves a component with maximum number of vertices. Let the $t > 1$ components of $H \setminus V'$ be composed of sets of vertices $V_1, V_2, \ldots, V_t$. Without loss of generality, assume that $H[V_1]$ is a component with maximum number of vertices. Every other component has at most $d - 1$ vertices. Otherwise, by Observation 4.5, there will be a connected induced subgraph of d vertices in that component deleting which we get a larger component composed of $V_1 \cup V'$. Consider V_j for any j such that $2 \leq j \leq t$. We obtained that $|V_j| \leq d-1$. Hence, the degree of any vertex $v \in V_j$ is at most $d-2$ in $H[V_j]$. Since the minimum degree of H is d, there is at least 2 edges from v to V'. Let the neighbourhood of v in V' be V''. If none of the vertices in V'' is adjacent to V_1, then v and any of its $d-1$ neighbours induces a connected graph deleting which gives a larger component. If one of the vertices in V'' is adjacent to V_1, excluding that we get $d-1$ neighbours of v which along with v induce a connected subgraph and deleting which gives a larger component. This is a contradiction. □

Corollary 4.7. *Let H be a connected graph with minimum degree 3. Then there exists an induced P_3 or K_3 with a vertex set V' in H such that $H \backslash V'$ is connected.*

Theorem 4.8. *Let R be a connected regular graph with at least two edges. Then, R-*FREE EDGE DELETION *is* NP-complete. *Furthermore, R-*FREE EDGE DELETION *is not solvable in time $2^{o(k)} \cdot |G|^{O(1)}$, unless ETH fails.*

Proof. Let R be an r-regular graph. Since R is connected and has at least 2 edges, $r > 1$. If $r = 2$ then R is a cycle and the statements follow from Proposition 2.2(iii). Assume that $r \geq 3$. By Corollary 4.7, there exists an induced P_3 or K_3 with a vertex set V' in R such that $R - V'$ is connected. Now the statements follow from Lemma 4.4, Proposition 2.2(i) and (iii). □

The complement graph of a regular graph with at least two non-edges is a regular graph with at least two edges. Thus, we obtain the following corollary.

Corollary 4.9. *Let R be a regular graph with at least two non-edges. Then, R-*FREE EDGE COMPLETION *is* NP-complete. *Furthermore, R-*FREE EDGE COMPLETION *is not solvable in time $2^{o(k)} \cdot |G|^{O(1)}$, unless ETH fails.*

5 Handling Disconnected Graphs

We have seen in Sects. 3 and 4 that for any tree or connected regular graph H with at least two edges, H-FREE EDGE DELETION is NP-complete and does not admit parameterized subexponential time algorithm unless ETH fails. In this section, we extend these results to any H with at least two edges such that H has a largest component which is a tree or a regular graph.

Lemma 5.1. *Let H be a graph with $t \geq 1$ components. Let H_1 be a component of H with maximum number of vertices. Let H' be the disjoint union of all components of H isomorphic to H_1. Then, there is a linear parameterized reduction from H'-*FREE EDGE DELETION *to H-*FREE EDGE DELETION.

Proof. Let $V' \subseteq V(H)$ be the vertex set which induces H' in H. Let (G', k) be an instance of H'-FREE EDGE DELETION. We apply Construction 1 on (G', k, H, V') to obtain G. We claim that (G', k) is a yes-instance of H'-FREE EDGE DELETION if and only if (G, k) is a yes-instance of H-FREE EDGE DELETION.

Let F' be a solution of size at most k of (G', k). For a contradiction, assume that $G - F'$ has an induced H with a vertex set U. Hence there is a vertex set $U' \subseteq U$ such that U' induces H' in $G - F'$. It is straightforward to verify that a branch vertex can never be part of an induced H' in $G - F'$. Hence U' does not contain a branch vertex and hence U' induces an H' in $G' - F'$, which is a contradiction. Lemma 2.3 proves the converse. □

Lemma 5.2 handles the case of disjoint union of isomorphic connected graphs.

Lemma 5.2. *Let H be any connected graph. For every pair of integers t, s such that $t \geq s \geq 1$, there is a linear parameterized reduction from sH-FREE EDGE DELETION to tH-FREE EDGE DELETION.*

Proof. The proof is by induction on t. The base case when $t = s$ is trivial. Assume that the statement is true for $t - 1$, if $t - 1 \geq s$. Now, we give a linear parameterized reduction from $(t-1)H$-FREE EDGE DELETION to tH-FREE EDGE DELETION.

Let (G', k) be an instance of $(t-1)H$-FREE EDGE DELETION. Let G'' be a disjoint union of $k + 1$ copies of H. Make every pair of vertices (v_i, v_j) adjacent in G'' such that $v_i \in V(H_i)$ and $v_j \in V(H_j)$ where H_i and H_j are two different copies of H in G''. Let the resultant graph be $\hat{G}$. Let G be the disjoint union of G' and $\hat{G}$. We need to prove that (G', k) is a yes-instance of $(t-1)H$-FREE EDGE DELETION if and only if (G, k) is a yes-instance of tH-FREE EDGE DELETION.

Let F' be a solution of size at most k of (G', k). It is straightforward to verify that $\hat{G}$ is $2H$-free. Hence, if $G - F'$ has an induced tH then $G' - F'$ has an induced $(t-1)H$, which is a contradiction. Conversely, let (G, k) be a yes-instance of tH-FREE EDGE DELETION. Let F be a solution of size at most k of (G, k). For a contradiction, assume that $G' - F$ has an induced $(t-1)H$ with a vertex set U. Since $|F| \leq k$, F cannot kill all the induced Hs in $\hat{G}$. Hence, let $U' \subseteq V(\hat{G})$ induces an H in $G - F$. Therefore, $U \cup U'$ induces tH in $G - F$, which is a contradiction. □

Corollary 5.3 is obtained by invoking Lemma 5.2 with $s = 1$. Lemma 5.4 follows from Lemma 5.1 and Corollary 5.3.

Corollary 5.3. *Let H be any connected graph. For every integer $t \geq 1$, there is a linear parameterized reduction from H-FREE EDGE DELETION to tH-FREE EDGE DELETION.*

Lemma 5.4. *Let H be a graph such that H has a component with at least two edges. Let H_1 be a component of H with maximum number of vertices. Then there is a linear parameterized reduction from H_1-FREE EDGE DELETION to H-FREE EDGE DELETION.*

Proof. Let H' be the disjoint union of the components of H which are isomorphic to H_1. By Lemma 5.1, there is a linear parameterized reduction from H'-FREE EDGE DELETION to H-FREE EDGE DELETION. Then, by Corollary 5.3, there is a linear parameterized reduction from H_1-FREE EDGE DELETION to H'-FREE EDGE DELETION. Composing these two reductions will give a linear parameterized reduction from H_1-FREE EDGE DELETION to H-FREE EDGE DELETION. □

Theorem 5.5. *For every $t > 1$, tK_2-FREE EDGE DELETION is* NP-complete. *Furthermore, tK_2-FREE EDGE DELETION is not solvable in time $2^{o(k)} \cdot |G|^{O(1)}$, unless ETH fails.*

Proof. Follows from Proposition 2.2(iv) and Lemma 5.2 (invoked with $s = 2$). □

Theorem 5.6. *Let H be any graph with at least two edges such that a largest component of H is a tree or a regular graph. Then H-FREE EDGE DELETION is* NP-complete. *Furthermore, H-FREE EDGE DELETION is not solvable in time $2^{o(k)} \cdot |G|^{O(1)}$, unless ETH fails.*

Proof. Let H be $tK_2 \bigcup t'K_1$, for some $t' \geq 0$. Since H has at least two edges, $t > 1$. Then the statements follow from Theorem 5.5 and Lemma 5.1. If H is not $tK_2 \bigcup t'K_1$, let H_1 be a largest component which is a tree or a regular graph. Clearly, H_1 has at least two edges. Then, Lemma 5.4 gives a linear parameterized reduction from H_1-FREE EDGE DELETION to H-FREE EDGE DELETION. Now, the theorem follows from Theorems 3.7 and 4.8. □

Since H-FREE EDGE DELETION is equivalent to $\overline{H}$-FREE EDGE COMPLETION, we obtain the following corollary.

Corollary 5.7. *Let $\mathcal{H}$ be the set of all graphs H with at least two edges such that H has a largest component which is either a tree or a regular graph. Let $\overline{\mathcal{H}}$ be the set of graphs such that a graph is in $\overline{\mathcal{H}}$ if and only if its complement is in $\mathcal{H}$. Then, for every $H \in \overline{\mathcal{H}}$, H-FREE EDGE COMPLETION is* NP-complete. *Furthermore, H-FREE EDGE COMPLETION is not solvable in time $2^{o(k)} \cdot |G|^{O(1)}$, unless ETH fails.*

6 Concluding Remarks

We proved that H-FREE EDGE DELETION is NP-complete if H is a graph with at least two edges and a largest component of H is a tree or a regular graph. We also proved that, for these graphs H, H-FREE EDGE DELETION cannot be solved in parameterized subexponential time, unless Exponential Time Hypothesis fails. The same results apply for $\overline{H}$-FREE EDGE COMPLETION.

Assume that we obtain a graph H' from H by deleting every vertex with degree $\delta(H)$, the minimum degree of H. Also assume that H'-FREE EDGE DELETION is NP-complete. Then by Lemma 3.5, we obtain that H-FREE EDGE DELETION is NP-complete. The reduction in Lemma 3.5 is not useful if H' is a graph with at most one edge, as for this H'-FREE EDGE DELETION is polynomial time solvable. Hence we believe that, if we can prove the NP-completeness of H'-FREE EDGE DELETION where H' is a graph in which the set of vertices with degree more than $\delta(G)$ induces a graph with at most one edge, we can prove that H-FREE EDGE DELETION is NP-complete if and only if H has at least two edges.

References

1. Bansal, N., Blum, A., Chawla, S.: Correlation clustering. Mach. Learn. **56**(1–3), 89–113 (2004)
2. Bodlaender, H.L., de Fluiter, B.: On intervalizing k-colored graphs for DNA physical mapping. Discrete Appl. Math. **71**(1), 55–77 (1996)
3. Cai, L., Cai, Y.: Incompressibility of H-free edge modification problems. Algorithmica **71**(3), 731–757 (2015)
4. Cai, Y.: Polynomial kernelisation of H-free edge modification problems. M.Phil. thesis, Department of Computer Science and Engineering, The Chinese University of Hong Kong, Hong Kong SAR, China (2012)
5. Cygan, M., Fomin, F.V., Kowalik, Ł., Lokshtanov, D., Marx, D., Pilipczuk, M., Pilipczuk, M., Saurabh, S.: Parameterized Algorithms. Springer International Publishing, Heidelberg (2016)
6. Drange, P.G., Fomin, F.V., Pilipczuk, M., Villanger, Y.: Exploring subexponential parameterized complexity of completion problems. In: STACS (2014)
7. Drange, P.G., Pilipczuk, M.: A polynomial kernel for trivially perfect editing. In: Bansal, N., Finocchi, I. (eds.) ESA 2015. LNCS, vol. 9294, pp. 424–436. Springer, Heidelberg (2015)
8. El-Mallah, E.S., Colbourn, C.J.: The complexity of some edge deletion problems. IEEE Trans. Circuits Syst. **35**(3), 354–362 (1988)
9. Fellows, M.R., Guo, J., Komusiewicz, C., Niedermeier, R., Uhlmann, J.: Graph-based data clustering with overlaps. Discrete Optim. **8**(1), 2–17 (2011)
10. Goldberg, P.W., Golumbic, M.C., Kaplan, H., Shamir, R.: Four strikes against physical mapping of DNA. J. Comput. Biol. **2**(1), 139–152 (1995)
11. Impagliazzo, R., Paturi, R., Zane, F.: Which problems have strongly exponential complexity? In: Proceedings of the 39th Annual Symposium on Foundations of Computer Science, pp. 653–662. IEEE (1998)
12. Komusiewicz, C., Uhlmann, J.: Cluster editing with locally bounded modifications. Discrete App. Math. **160**(15), 2259–2270 (2012)
13. Lewis, J.M., Yannakakis, M.: The node-deletion problem for hereditary properties is NP-complete. J. Comput. Syst. Sci. **20**(2), 219–230 (1980)
14. Rose, D.J.: A graph-theoretic study of the numerical solution of sparse positive definite systems of linear equations. Graph Theor. Comput., 183–217 (1972)
15. West, D.B.: Introduction to Graph Theory, 2nd edn. Prentice Hall Inc., Upper Saddle River (2001)
16. Yannakakis, M.: Edge-deletion problems. SIAM J. Comput. **10**(2), 297–309 (1981)

Multicast Network Design Game on a Ring

Akaki Mamageishvili[1](✉) and Matúš Mihalák[2]

[1] Department of Computer Science, ETH Zurich, Zurich, Switzerland
akaki@inf.ethz.ch
[2] Department of Knowledge Engineering, Maastricht University, Maastricht, The Netherlands

Abstract. In this paper we study quality measures of different solution concepts for the multicast network design game on a ring topology. We recall from the literature a lower bound of $\frac{4}{3}$ and prove a matching upper bound for the price of stability, which is the ratio of the social costs of a best Nash equilibrium and of a general optimum. Therefore, we answer an open question posed by Fanelli et al. in [12]. We prove an upper bound of 2 for the ratio of the costs of a potential optimizer and of an optimum, provide a construction of a lower bound, and give a computer-assisted argument that it reaches 2 for any precision. We then turn our attention to players arriving one by one and playing myopically their best response. We provide matching lower and upper bounds of 2 for the myopic sequential price of anarchy (achieved for a worst-case order of the arrival of the players). We then initiate the study of myopic sequential price of stability and for the multicast game on the ring we construct a lower bound of $\frac{4}{3}$, and provide an upper bound of $\frac{26}{19}$. To the end, we conjecture and argue that the right answer is $\frac{4}{3}$.

Keywords: Network design game · Nash equilibrium · Price of stability/anarchy · Ring topology · Myopic sequential price of stability/anarchy · Potential-optimum price of stability/anarchy

1 Introduction

Network design game is played by n players on an edge-weighted graph. Each player i, $i = 0, \ldots, n-1$, connects her terminal vertices s_i and t_i by selecting an s_i-t_i path P_i. Using an edge e costs c_e and all players using it share the cost equally. In total, player i's cost for using path P_i is the sum of all shares towards the edges of P_i.

Network design game belongs to the broader class of congestion games. It is a special congestion game in that increasing the congestion on a resource makes it cheaper to use (in contrast to the more established and studied games with monotone increasing cost functions). Finite congestion games are exact potential games, i.e., games for which a potential function exists, i.e., a function $\Phi(P_0, \ldots, P_{n-1}) \to \mathbb{R}$ that exactly reflects the difference in any player's

Z. Lu et al. (Eds.): COCOA 2015, LNCS 9486, pp. 439–451, 2015.
DOI: 10.1007/978-3-319-26626-8_32

cost, if this unilaterally changes her path from P_i to P'_i. It is well-known that exact potential games always possess a pure Nash equilibrium, for example the vector $(P_0, \ldots, P_{n-1})$ minimizing the potential function Φ. The price of anarchy is the ratio of the worst Nash equilibrium cost and the general optimum cost, and can be as large as n. The price of stability, which is the ratio of the best Nash equilibrium cost and the general optimum cost, of network design games is well understood for directed graphs – it is at most H_n [3], where $H_n = 1 + 1/2 + 1/3 + \cdots + 1/n$ is the n-th harmonic number (equal asymptotically to $\log n$) and the matching lower bound example has also been constructed. The price of stability of the game is much less understood for undirected graphs. While it is known to be strictly smaller than H_n [11,19], namely, at most $H_{n/2}$ [19], the largest known example has price of stability equal to roughly 2.245 [6]. Closing this gap is a major open problem in the field of congestion games and in the computational game theory in general.

For the special type of the game where all players have the same target vertex t, better bound on the price of stability has been proven [15]. If additionally each vertex of a graph is a source vertex of some player, a series of papers improved the upper bound [8,13,17], where the latest result of Bilò et al. [8] shows that price of stability is $O(1)$ in this class of games. In many results, the potential function, and the equilibria minimizing it, play an important role. Actually, equilibria minimizing potential function are regarded as stable (against noise) by Asadpour and Saberi [4], and accordingly, some authors studied the price of stability restricted to these kind of equilibria [16,19], the so-called *potential-optimum price of stability.*

One of the motivations to study best Nash equilibria is that they can be regarded as outcomes of the game if a little coordination is present – an authority that suggests the players the strategies P_i. Then, players have no incentive to unilaterally deviate from the suggested strategy profile. It is questionable whether such an authority exists – it would need to be very strong, both computationally and imperatively. To address this applicability issue of equilibrium concepts, sequential versions of the game were studied: the players arrive one by one, and upon arrival, player i chooses *myopically* the best path P_i as if this was the end of the game (i.e., no further players would arrive). Chekuri et al. [10] show that the total cost achieved by a worst-case permutation of the arriving players is at most $O(\sqrt{n}\log n)$ times the optimum cost. Subsequently, Charikar et al. [9] improved this bound to $O(\log^2 n)$ (the original version [9] is erroneous, but the authors provide corrected arguments upon request). The worst-case approach to the order in which the players arrive naturally models the complete lack of coordination. In this paper, we suggest to study also the best-case order in which players arrive. This is motivated by the presence of an authority that can control the access to the resources over time (and thus decide an order of the arriving players). Such an authority is arguably weaker than the one mentioned above, as it does not impose any decision upon the players, and it leaves them to decide their strategies freely upon arriving. Biló et al. [7] studied a version of a cost sharing scheme for multicast network design game, in which each player only knows strategies of some other players, and pays fair share of edge costs that she uses based only on her information. Sequential versions described above can be modeled with this cost sharing scheme.

In this paper, we focus on one specific network topology: the ring. This is a fundamental topology in networking and communications. It is the edge-minimal topology that is resistant against a single link fault. From the decentralized point of view, *call control* comes close in spirit to network design games, in that the connecting s_i-t_i paths needs to be chosen to obey given capacities on the links [1]. The study of approximation algorithms is the counterpart to bounding the prices of anarchy and stability. Rings have also been intensively studied in the distributed setting, e.g., among plenty of others, in the context of the fundamental leader election problem [5].

Network design games on rings has previously been studied by Fanelli et al. [12], which show a tight bound of 3/2 on the price of stability for the general setting. In this paper we restrict ourselves to the *multicast* version in which all players share the same target vertex $t = t_i, i = 0, \dots, n-1$ and answer the open question asked by Fanelli et al. [12] about tight bounds of the price of stability for multicast game on a ring. We study various solutions concepts and analyze their quality compared to an optimum network (with respect to the social cost). In most cases, we are able to provide tight bounds. Furthermore, we also study the myopic sequential price of stability in general multicast network design games, and give a simpler proof of an upper bound of 4 for this class of games compared to a more general proof in [7] (cf. this with the upper bound of $\log^2 n$ on the myopic sequential price of anarchy for multicast games).

2 Preliminaries

Network design game is a strategic game of n players played on an edge-weighted graph $G = (V, E)$ with non-negative edge costs c_e, $e \in E$. Each player i, $i = 0, \dots, n-1$, has a dedicated *source* node s_i and a *target* node t_i. In the multicast game all t_i's are the same and we denote it by t throughout a paper. All s_i-t_i paths form the set $\mathcal{P}_i$ of the *strategies* of player i. Each player i chooses one path $p_i \in \mathcal{P}_i$, the union of which creates a network in which all s_i-t_i pairs are connected. The cost for player i is $\sum_{e \in p_i} \frac{c_e}{n(e)}$ where $n(e)$ is the number of players that use the edge e in their chosen paths. A strategy profile p is a vector of strategies for all players, $p = (p_0, \dots, p_{n-1})$. A strategy profile is Nash equilibrium if no player can unilaterally change her strategy p_i to p_i' and improve her cost. The (social) cost of a strategy profile p is the cost of the created network, i.e., the sum of the costs of all edges in the created network, which is, in turn, the sum of all players' costs. An optimum network is a network of minimum social cost in which all s_i-t_i pairs are connected. An optimum network can be equivalently described by a strategy profile p^*, and then we refer to p^* as an optimum strategy profile. Note that an optimum strategy profile is not, in general, a Nash equilibrium. Observe also that in a multicast game an optimum network forms a Steiner tree on the terminals s_i and t for $i = 0, \dots, n-1$. If an underlying graph G is a ring, then there are only 2 possible strategies for each player.

In this paper, we focus on the multicast game on rings. We can assume, without loss of generality, that every node but the target t is a source of exactly one player. Otherwise, we can modify the topology by the following two operations. If there are $l > 1$ players sharing the same node x of the ring as a source vertex, we make l copies of this vertex, add $l - 1$ consecutive edges of cost 0 between them to make a path of length $l - 1$, replace x in the ring with this path in a natural way, and associate each vertex with a unique source (copy of x). If there is a node x in the ring which is not a target nor a source of any player, we delete x from the ring, and connect its two neighbors by an edge of cost $c_e + c_{e'}$, where e, e' are the two adjacent edges of x. A repetitive application of these two operations preserve the cost of optimum and Nash equilibrium strategy profiles, and also preserves the equilibrium properties of strategy profiles (if the strategies are expressed in the form "go clockwise/counterclockwise to s_i").

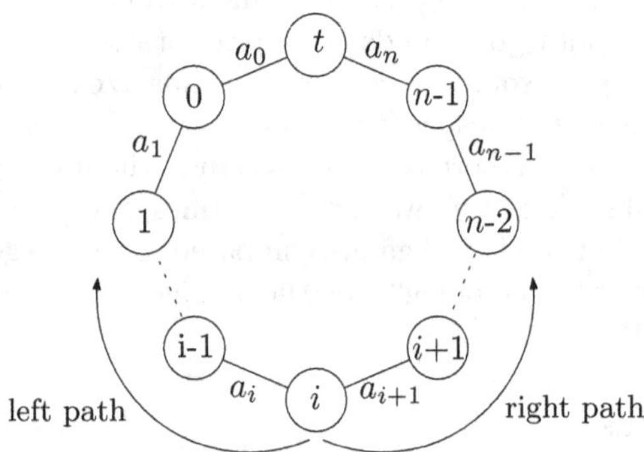

Fig. 1. Multicast game on rings.

We label the sources (players) and the edges connecting them in a counterclockwise order as in Fig. 1, where a_i denotes the cost of the i-th edge. Player i has exactly 2 strategies, one is to go *left*, i.e., *clockwise*, taking edges $i, i-1, \ldots 0$, or to go *right*, i.e., *counterclockwise*, taking edges $i+1, \ldots n$. Observe that the optimum strategy profile is the one which uses all edges except the most expensive edge. Let o denote the most expensive edge. Then the (social) cost of an optimum network is $\sum_{i \neq o} a_i$.

Price of anarchy of a network design game is the ratio of the costs of a worst Nash equilibrium and of an optimum strategy profile. *Price of stability* is the ratio of the costs of a best Nash equilibrium and of an optimum strategy profile. *Potential optimum* is a strategy profile p that minimizes the potential function $\Phi = \sum_e \sum_{i=1}^{n(e)} \frac{c_e}{i}$. *Potential-optimum price of anarchy/stability* is the ratio of the costs of the worst/best potential-optimum profile and of an optimum strategy profile. *Myopic sequential price of anarchy/stability* is the worst-case/best-case ratio of the costs of a strategy profile that can be obtained by ordering the players as in a permutation π and letting player $\pi(i)$ choose the best-response

$p_{\pi(i)}$ in the game induced by the first i players $\pi(1), \pi(2), \ldots, \pi(i)$ and of an optimum profile.

Note on related concepts. The term *sequential price of anarchy* has been used [2,18] to express a different, yet still closely related, concept compared to the notion of the myopic sequential price of anarchy/stability. In the sequential price of anarchy, players also come one by one, and decide their strategy upon arrival, but the stability of the outcome is measured in terms of Nash equilibria again. In some sense, the game resembles extensive games. Observe that profiles p that get compared to optima in the myopic sequential price of anarchy/stability are in general no Nash equilibria.

3 Price of Anarchy/Stability for Multicast on Rings

It is known that the price of anarchy on general graphs is at most n, and that this bound is tight. The tight example actually is a multicast game on a ring, and the general analysis of the price of anarchy thus carries over to our multicast game on rings. For completeness, we show the example in Fig. 2.

Theorem 1 ([3]). *Price of anarchy for mutlicast games on rings is at most n. This is tight.*

We now turn our attention to the price of stability. The example from Fig. 3, due to Anshelevich et al. [3], shows that the price of stability can be as high as 4/3 (observe that the game possesses a unique Nash equilibrium where both players use the direct edge to get connected to t). We now show that the price of stability cannot get larger than that for multicast games on rings, and therefore answer the open question asked by Fanelli et al. [12].

Theorem 2. *Price of stability in the multicast game on rings is at most $\frac{4}{3}$.*

In the proof of the theorem we will use the following lemma.

Lemma 1. *If a strategy profile p in which an edge i is not used is not Nash equilibrium, then either player i or player $i-1$ can improve her cost by changing her strategy.*

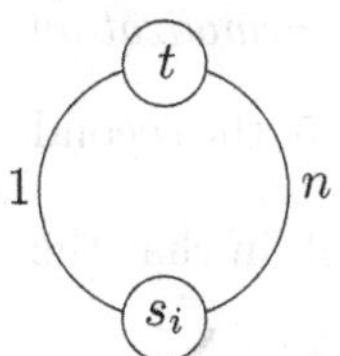

Fig. 2. In the worst equilibrium, all players use the edge of cost n.

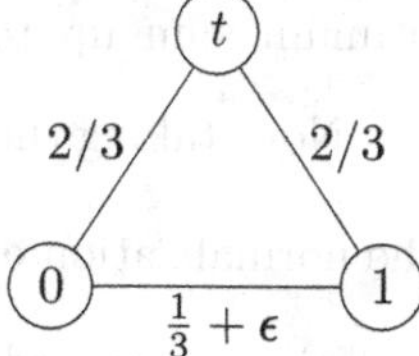

Fig. 3. Example of a lower bound 4/3.

Proof. Since the strategy profile p is not a Nash equilibrium, there exists a player k that can change her strategy and improve the cost. Assume, without loss of generality, that $k < i - 1$. Since edge i is not used in p, it follows that player k uses the left path to get to t. The cost of k in p is thus $\sum_{l=0}^{k} \frac{a_l}{i-l}$, which is, by our assumption, bigger than the cost of k if she switches to the right path, i.e., bigger than $\sum_{l=k+1}^{i-1} \frac{a_l}{i-l+1} + \sum_{l=i}^{n} \frac{a_l}{l-i+1}$. It follows that player $i-1$ also uses the left path in p, and thus her cost is at least the cost of player k, whereas the alternative cost of $i-1$ if she switches to the right path is at most the alternative cost of player k. Hence, the alternative cost of player $i-1$ is smaller than her cost in p, and player $i-1$ thus improves her cost as well. □

Proof (of Theorem 2). Consider an optimum strategy profile and let o be the edge that is not used in it. If optimum is also Nash equilibrium, then price of stability is 1 and the claim follows. Otherwise, optimum is not a Nash equilibrium and, by Lemma 1, one of the endpoints of the edge o can improve its cost. Assume, without loss of generality, that player $o-1$ can improve. We now consider the following best-response dynamics: let $o-1$ improve; then, edge $o-1$ is not used, and in case we have not reached Nash equilibrium, let player $o-2$ improve (the player $o-2$ must be able to improve by Lemma 1), and so on, until some player $o-k$ cannot improve anymore (this happens at the latest for player 0), and we reach a Nash equilibrium.

We will show that the social cost of a Nash equilibrium that is reached by this best response dynamics is maximized for $k = 1$, i.e., for the strategy profile reached after one step of the dynamics. We then show that the cost of such a profile is at most 4/3 times the cost of the optimum, which proves the theorem.

Let us first show the second part. Assume therefore that player $o-1$ switches to improve her cost, and the resulting profile is an equilibrium. In particular, we have that player $o-2$ does not want to switch. This can be expressed by the following two inequalities: $\sum_{l=o}^{l=n} \frac{a_l}{l-o+1} \leq \sum_{l=0}^{l=o-1} \frac{a_l}{o-l}$, and $\sum_{l=0}^{l=o-2} \frac{a_l}{o-1-l} \leq \sum_{l=o-1}^{l=n} \frac{a_l}{l-o+2}$. We further introduce a normalization of the edge costs so that the edges in the optimum sum up to 1. Thus, we obtain the *normalization* equation $\sum_{i=0, i\neq o}^{n} a_i = 1$. Now, taking the first inequality with weight 5, the second with weight 1, and the normalization equality with weight 6, we obtain that the cost of the Nash equilibrium where edge $o-1$ is not used has cost $\sum_{i=0, i\neq o-1}^{i=n} a_i$ at most $\frac{4}{3}$.

We can proceed in the same way for every other value of $k = 2, 3, \ldots$ for which the reached Nash equilibrium does not use edge $o-k$. For every k, we get for each of the players $o-k-1, o-k, \ldots, o-1$ an inequality stating that the player did not want, respectively wanted to swap her strategy. For all values of $k = 1, 2, 3, 4, 5, 6, 7$, we provide in the appendix the coefficients with which we need to take the inequalities and to obtain the upper bound of at most 4/3 on the cost of the Nash equilibrium.

If the length of the best-response dynamics is 8 or more, it follows that we do not need to add further inequalities, and the 7 inequalities obtained for the first 7 deviating players are enough to show the upper bound of 4/3 on the cost of the reached Nash equilibrium. □

4 Potential-Optimum Price of Anarchy for Multicast on Rings

The potential-optimum price of anarchy/stability has been first studied, in the context of the network design games, by Kawase and Makino [16]. Besides other results, they proved that for multicast network design games, the two values collide. Therefore, in the following, we only study the potential-optimum price of anarchy (POPoA for short), and we show that it is at most two for rings, and provide an infinite family of examples with increasing POPoA, which we conjecture converges to two, but leave the formal analysis as an open problem. We have analyzed one such game from the family which shows that POPoA can be as large as 1.99992.

Theorem 3. *POPoA is at most* 2 *in the multicast game on rings.*

Proof. Consider an optimal strategy profile O and let o be the edge that is not used in it. Consider a potential optimum strategy profile P and let p be the edge in it that is not used by any player. Assume, without loss of generality, that $p < o$.

By the definition of P, we have, for any strategy profile Q, $\Phi(P) \le \Phi(Q)$, and in particular $\Phi(P) \le \Phi(O)$, i.e.,

$$\sum_{i=0}^{p-1} a_i \cdot H_{p-i} + \sum_{i=p+1}^{n} a_i \cdot H_{i-p} \le \sum_{i=0}^{o-1} a_i \cdot H_{o-i} + \sum_{i=o+1}^{n} a_i \cdot H_{i-o}. \quad (1)$$

We now concentrate on a_o and show that a_o is at most the cost of optimum, i.e., at most $\sum_{i \neq o} a_i$. This then shows that any strategy profile (and, in particular, P) has cost at most twice the cost of optimum.

Isolate in the second sum of the left hand side (LHS for short) of Eq. (1) the term with a_o and put the rest of the sum to the right hand side (RHS). This rest will dominate the second sum on the RHS, and by neglecting the resulting negative number, we get that $\sum_{i=0}^{p-1} a_i \cdot H_{p-i} + a_o \cdot H_{o-p} \le \sum_{i=0}^{o-1} a_i \cdot H_{o-i}$, or, equivalently, that $a_o \le \frac{\sum_{i=0}^{o-1} a_i \cdot H_{o-i} - \sum_{i=0}^{p-1} a_i \cdot H_{p-i}}{H_{o-p}}$. Analyzing the influence of p on the RHS, one can show that the RHS is maximized for $p = 1$. Thus, we obtain that $a_o \le \frac{\sum_{i=0}^{o-1} a_i \cdot H_{o-i} - a_0}{H_{o-1}}$. Then, since $H_o - 1 \le H_{o-1}$ we get that $a_o \le \sum_{i=0}^{o-1} a_i \le \sum_{i \ne o} a_i$, which proves the claim and thus the theorem. □

We now provide a construction of a game which shows that POPoA is at least 1.99992. We conjecture that the construction can be used to prove an asymptotic lower bound of 2 on POPoA.

Consider non-zero numbers $a_0, \ldots, a_{2 \cdot l}$ that sum up to 1, and where l is constant, $o = n$, $p = l - 1$ and where a_n is equal to $\frac{H_n - a}{H_n}$, for some constant a. Compare the potentials of the strategy profiles which do not use edge i for $i = 0, \ldots, i = 2 \cdot l$ to the potential of P (the strategy profile minimizing Φ) that does not include the p-th edge. Note that after canceling the coefficients on both sides, the coefficient in front of a_n is a sum of a constant number of terms converging to 0 for n tending to infinity, so these terms can be neglected. The potential of the strategy profiles which do not use edge i for $\frac{n}{2} > i > 2 \cdot l$ is increasing when i is increasing and decreasing towards n. We solved the resulting system of linear equations and obtained a lower bound for POPoA converging to 1.99992 for $l = 1000$ and n tending to infinity. Thus, we have the following proposition.

Proposition 1. *There are games that have POPoA* 1.99992.

We leave it as an open problem to analyze the convergence of the POPoA of the above construction, and conjecture that it converges to two.

Conjecture 1. There are games that have POPoA arbitrarily close to 2.

5 Myopic Sequential Prices of Anarchy/Stability

In this section we study the myopic sequential price of anarchy and the myopic sequential price of stability.

5.1 Sequential Price of Anarchy in Multicast Game on Rings

Lemma 2. *The myopic sequential price of anarchy is at most* 2 *in the multicast games on rings.*

Proof. Consider an optimal strategy profile and let o be the edge that is not used. Consider any permutation (order) π of the players. If any player $\pi(i)$, $i < o$, decides to take a path containing edge o for the first time then it means that $a_o \le \sum_{l=0}^{l=i} a_l$ which is bounded by the cost of optimum. Therefore, the whole cost of the ring is bounded by 2 times the cost of optimum. □

The presented upper bounds is tight, as shows the example in Fig. 4, where $\pi = \{0, 1, 3, 2\}$ results in myopic sequential price of anarchy equal to 2.

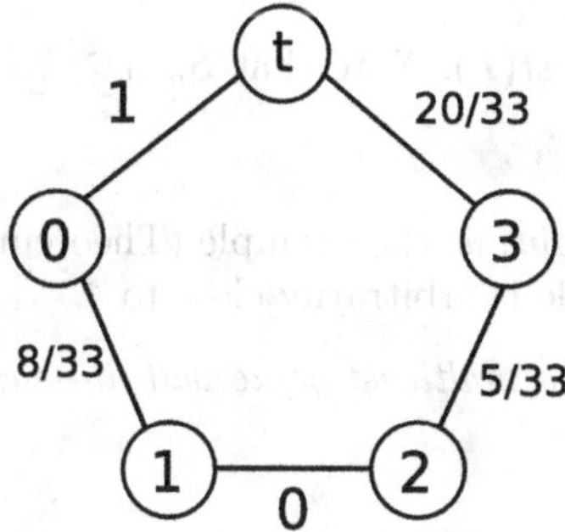

Fig. 4. Lower bound example for the sequential price of anarchy

5.2 Myopic Sequential Price of Stability in Multicast Game

In the myopic sequential price of stability we consider the best permutation of players, with respect to the resulting network cost. In [7] authors prove that when the social knowledge network graph is directed acyclic then the price of anarchy is bounded by 4 (Theorem 8). If we consider that in the social knowledge graph each incoming player knows all the previous players then the result can be directly translated into our setting, but we give a different (simpler than the proof of general result in [7]) proof for our setting:

Theorem 4. *The myopic sequential price of stability in multicast games on arbitrary graphs is at most* 4.

Proof. Since there is a common target vertex t, any optimum strategy profile forms a Steiner tree T on terminals s_i, $i = 0, \ldots, n-1$ and t. Consider a permutation of the vertices that corresponds to a depth-first search of the tree T, and make it the identity permutation $(0, 1, \ldots n-1)$. Let the players enter the game in this order, and make the myopic best responses. Denote by B_i the cost of the edges that player i uses alone in her strategy at the moment she enters the game, and let S_i be the overall cost of player i when she enters. Then the cost of the resulting network is $\sum_{i=0}^{i=n-1} B_i$. Since every player optimizes her cost when she enters the game, we have the following chain of inequalities: $S_i \leq d_T(s_i, s_{i-1}) + S_{i-1} - \frac{1}{2}B_{i-1}$, for $i = 1, \ldots, n-1$, where $d_T(u, v)$ is the distance between nodes u and v using only the edges of the tree T. Each player i has the following alternative strategy: first travel to the source (vertex) s_{i-1} using the edges of T, and then follow the strategy of player $i-1$. Note that in this alternative strategy, player i saves at least half of the cost of the edges that player $i-1$ takes alone when she enters the game. For the first player, we have the following inequality $S_0 \leq d_T(s_0, t)$, because when she enters the game, one of the possible strategies is to take a direct path from s_0 to t using only the edges of T. By summing up all inequalities given above, we get that

$\frac{1}{2}\sum_{i=0}^{i=n-2} B_i + S_{n-1} \leq 2 \cdot cost(T)$. Note that $S_{n-1} \geq \frac{1}{2}B_{n-1}$, which results into the upper bound of 4. □

This upper bound is tight, as the example (Theorem 5) from [7] shows, ratio in the lower bound example is arbitrarily close to 4.

Proposition 2. *There is a multicast game with the myopic sequential price of anarchy arbitrarily close to 4.*

5.3 Myopic Sequential Price of Stability on Rings

In this section we consider the myopic sequential price of stability of the multicast games on rings. The example from Fig. 3 shows that it can be as high as $\frac{4}{3}$. We prove the following upper bound.

Theorem 5. *The myopic sequential price of stability in the multicast games on rings is at most $\frac{26}{19}$.*

Proof. Assume that the optimum strategy profile does not include the edge of cost a_o, and without loss of generality $\sum_{i=0}^{i=o-1} a_i \geq \sum_{i=o+1}^{i=n} a_i$. Consider the permutation $\pi = \{n-1, \ldots, o, 0, 1, \ldots, o-1\}$. First $n-o$ players clearly take the right path, by our assumption. Consider the remaining players. If there is no player which, upon arrival, prefers the right path over the left path, then only edges of an optimum strategy profile are included into the resulting network which means that the myopic sequential price of stability is 1. If the very first player 0 prefers the right path, then all other players necessarily prefer the right path as well, and the resulting network consists of all edges except for that of weight a_0. But then a_0 is at least as large as a_o, resulting again the myopic sequential price of stability equal to 1. Suppose that there exists i such that every player $l \leq i$ prefers to take the left path, and only the player (vertex) $i+1$ prefers to take the right path. This implies the following inequalities:

$$\sum_{k=0}^{k=i} \frac{a_k}{i-k+1} \leq \sum_{k=i+1}^{k=n} a_k, \text{ and} \tag{2}$$

$$\sum_{k=i+2}^{k=o} a_k \leq \sum_{k=0}^{k=i+1} \frac{a_k}{i+2-k}, \tag{3}$$

where the first inequality (2) indicates that the i-th player prefers the left path, and the second inequality (3) indicates that the $i+1$-th player prefers the right. Our goal is to investigate the maximum possible cost c of the resulting network, where $c = a_0 + \cdots + a_i + a_{i+2} + \cdots + a_n$. Take the first inequality (2) with weight $\frac{2}{19}$, the second inequality (3) with weight $\frac{24}{19}$, and the normalization equation $a_0 + \cdots + a_{o-1} + a_{o+1} + \cdots + a_n = 1$ with weight $\frac{26}{19}$. We obtain that the sum on the left hand side s satisfies $c \leq s \leq \frac{26}{19}$, which gives that $c \leq \frac{26}{19} \approx 1.368$. □

The permutation from the proof of Theorem 5 cannot be used to provide a better bound, as there exists an example of a game, where the permutation results in a network of cost $\frac{26}{19}$ times larger than the cost of optimum. The example consists of 3 players and edges have weights $\frac{6}{19}$, $\frac{10}{19}$, $\frac{3}{19}$ and $\frac{10}{19}$ in the counter-clockwise order. Players who come in the game according to the permutation $\{0, 1, 2, 3\}$ take all edges except for the 3-rd edge of weight $\frac{3}{19}$, resulting into a network of cost $\frac{26}{19}$, while the optimum network cost is 1. Note that if players come according to the "opposite" permutation $(n - 1, \ldots, 0)$, then the resulting network has the same cost as the optimum network. We have experimentally played with these two permutations, and for all inputs we tried, one of the two permutations resulted in networks of cost no more than the 4/3 of the optimum cost. Actually, we have checked that there is no instance of at most 1000 players where the better of the two permutations fails in that respect.

Conjecture 2. The myopic sequential price of stability in the multicast game on rings is at most $\frac{4}{3}$.

6 Conclusions

We have analyzed several solution concepts for the multicast network design games on rings, and demonstrated that they differ in terms of quality. Some of the derived bounds are not shown to be tight, and we leave it for future work to make them tight.

We have also initiated the study of the myopic sequential price of stability, and analyzed it for the multicast network design game on a ring. It is certainly an interesting challenge to provide better bounds on this concept for general (not multicast) network design games.

Acknowledgements. This work has been supported by the Swiss National Science Foundation (SNF) under the grant number 200021_143323/1. We used the CGAL linear and quadratic programming solver [14] for solving all linear programs described in this article. We thank anonymous reviewer for insightful comments and especially for pointing out the relation between myopic sequential price of stability and graphical multicast cost sharing games.

A Weights for Inequalities from the Proof of Theorem 2

In this appendix we provide the multiplicative weights of the inequalities using a dual to a linear program that was solved to upper bound the price of stability in the multicast game on rings. The first inequality is the normalization inequality, therefore its weight is the upper bound on the price of stability. The next k inequalities indicate that the first $k \leq 7$ players left of edge e prefer to deviate, i.e., prefer to choose the right path instead of the left path, and the last inequality indicates that we have a Nash equilibrium, i.e., the last player considered in the best-response dynamics prefers to stick with the left path than to switch to the right path. The objective of the linear program is to minimize the sum of the edge costs without the edge that is not used by the Nash equilibrium achieved via the best response dynamics. The coefficients (weights) are as follows:

- $k = 1$ (0: 4/3; 1: 10/9; 2: 2/9)
- $k = 2$ (0: 22/17; 1: 252/323; 2: 202/323; 3: 90/323)
- $k = 3$ (0: 29/23; 1: 2976/4025; 2: 1206/4025; 3: 2256/4025; 4: 1224/4025)
- $k = 4$ (0: 1.243533565; 1: 0.722076586; 2: 446160/1659763; 3: 0.268809463; 4: 0.528169383; 5: 0.329251827)
- $k = 5$ (0: 1.229596836; 1: 0.711037768; 2: 0.257115234; 3: 0.201170436; 4: 0.199302216; 5: 0.50797093; 6: 0.348431623)
- $k = 6$ (0: 1.217310111; 1: 0.702648246; 2: 0.250967669; 3: 0.189905238; 4: 0.168566505; 5: 0.179311025; 6: 0.494134279; 7: 0.362553601)
- $k = 7$ (0: 1.206536915; 1: 0.69586637; 2: 0.247111078; 3: 0.184286036; 4: 0.157438535; 5: 0.148587957; 6: 0.165607593; 7: 0.484007846; 8: 0.373384452)

For $k > 7$, we take only the first 7 inequalities indicating that the first 7 players prefer to take the right path than to stick to the left path. This is enough to prove an upper bound of 1.33081 for the price of stability. In the following, we list the weights of the inequalities of the dual to our linear program (index k : denotes the weight of the inequality to player k): (0: 1.330802428; 1: 0.750587484; 2: 0.246845878; 3: 0.168106752; 4: 0.12615003; 5: 0.096800836; 6: 0.072578056; 7: 0.048719834).

References

1. Adamy, U., Ambühl, C., Anand, R.S., Erlebach, T.: Call control in rings. Algorithmica **47**(3), 217–238 (2007)
2. Angelucci, A., Bilo, V., Flammini, M., Moscardelli, L.: On the sequential price of anarchy of isolation games. J. Comb. Optim. **29**(1), 165–181 (2015)
3. Anshelevich, E., Dasgupta, A., Kleinberg, J.M., Tardos, É., Wexler, T., Roughgarden, T.: The price of stability for network design with fair cost allocation. In: FOCS, pp. 295–304 (2004)
4. Asadpour, A., Saberi, A.: On the inefficiency ratio of stable equilibria in congestion games. In: Leonardi, S. (ed.) WINE 2009. LNCS, vol. 5929, pp. 545–552. Springer, Heidelberg (2009)
5. Attiya, H., Snir, M., Warmuth, M.K.: Computing on an anonymous ring. J. ACM **35**(4), 845–875 (1988)
6. Bilò, V., Caragiannis, I., Fanelli, A., Monaco, G.: Improved lower bounds on the price of stability of undirected network design games. Theory Comput. Syst. **52**(4), 668–686 (2013)
7. Bilò, V., Fanelli, A., Flammini, M., Moscardelli, L.: When ignorance helps: graphical multicast cost sharing games. Theor. Comput. Sci. **411**(3), 660–671 (2010). http://dx.doi.org/10.1016/j.tcs.2009.10.007
8. Bilò, V., Flammini, M., Moscardelli, L.: The price of stability for undirected broadcast network design with fair cost allocation is constant. In: FOCS, pp. 638–647 (2013)

9. Charikar, M., Karloff, H.J., Mathieu, C., Naor, J., Saks, M.E.: Online multicast with egalitarian cost sharing. In: SPAA 2008: Proceedings of the 20th Annual ACM Symposium on Parallelism in Algorithms and Architectures, Munich, Germany, 14–16 June 2008, pp. 70–76 (2008)
10. Chekuri, C., Chuzhoy, J., Lewin-Eytan, L., Naor, J., Orda, A.: Non-cooperative multicast and facility location games. In: Proceedings of the 7th ACM Conference on Electronic Commerce (EC 2006), Ann Arbor, Michigan, USA, 11–15 June 2006, pp. 72–81 (2006)
11. Disser, Y., Feldmann, A.E., Klimm, M., Mihalák, M.: Improving the H_k-bound on the price of stability in undirected shapley network design games. In: Spirakis, P.G., Serna, M. (eds.) CIAC 2013. LNCS, vol. 7878, pp. 158–169. Springer, Heidelberg (2013)
12. Fanelli, A., Leniowski, D., Monaco, G., Sankowski, P.: The ring design game with fair cost allocation. Theor. Comput. Sci. **562**, 90–100 (2015). http://dx.doi.org/10.1016/j.tcs.2014.09.035
13. Fiat, A., Kaplan, H., Levy, M., Olonetsky, S., Shabo, R.: On the price of stability for designing undirected networks with fair cost allocations. In: Bugliesi, M., Preneel, B., Sassone, V., Wegener, I. (eds.) ICALP 2006. LNCS, vol. 4051, pp. 608–618. Springer, Heidelberg (2006)
14. Fischer, K., Gärtner, B., Schönherr, S., Wessendorp, F.: Linear and quadratic programming solver. In: CGAL User and Reference Manual, 4.6 edn. CGAL Editorial Board (2015). http://doc.cgal.org/4.6/Manual/packages.htmlPkgQPSolverSummary
15. Jian, L.: An upper bound on the price of stability for undirected shapley network design games. Inf. Process. Lett. **109**, 876–878 (2009)
16. Kawase, Y., Makino, K.: Nash equilibria with minimum potential in undirected broadcast games. Theor. Comput. Sci. **482**, 33–47 (2013)
17. Lee, E., Ligett, K.: Improved bounds on the price of stability in network cost sharing games. In: EC, pp. 607–620 (2013)
18. Leme, R.P., Syrgkanis, V., Tardos, É.: The curse of simultaneity. In: Innovations in Theoretical Computer Science 2012, Cambridge, MA, USA, 8–10 January 2012, pp. 60–67 (2012)
19. Mamageishvili, A., Mihalák, M., Montemezzani, S.: An $H_{n/2}$ upper bound on the price of stability of undirected network design games. In: Csuhaj-Varjú, E., Dietzfelbinger, M., Ésik, Z. (eds.) MFCS 2014, Part II. LNCS, vol. 8635, pp. 541–552. Springer, Heidelberg (2014)

Extreme Witnesses and Their Applications

Andrzej Lingas[1(✉)] and Mia Persson[2]

[1] Department of Computer Science, Lund University, Lund, Sweden
Andrzej.Lingas@cs.lth.se
[2] Department of Computer Science, Malmö University, Malmö, Sweden
mia.persson@mah.se

Abstract. We study the problem of computing the so called minimum and maximum witnesses for Boolean vector convolution. We also consider a generalization of the problem which is to determine for each positive coordinate of the convolution vector, q smallest (or, largest) witnesses, where q is the minimum of a parameter k and the number of witnesses for this coordinate. We term this problem the smallest k-witness problem or the largest k-witness problem, respectively. We also study the corresponding smallest and largest k-witness problems for Boolean matrix product. In both cases, we provide algorithmic solutions and applications to the aforementioned witness problems, among other things in string matching and computing the (min, +) vector convolution.

Keywords: Boolean vector convolution · Boolean matrix product · String matching · Witnesses · Minimum and maximum witnesses · Time complexity · Lightest triangles

1 Introduction

For a potential alignment of a pattern string with a text string over the same alphabet, a position in the alignment where the pattern symbol is different from the text symbol is a *witness* to the symbol *mismatch* while a position where the pattern and text symbol are equal is a witness to the symbol *match*.

Similarly, if A and B are two $n \times n$ Boolean matrices and C is their Boolean matrix product then for any entry $C[i,j] = 1$ of C, a witness is an index m such that $A[i,m] \wedge B[m,j] = 1$. The smallest (or, largest) possible witness is called the *minimum witness* (or, *maximum witness*, respectively).

The problems of finding "witnesses" have been extensively studied for several decades, at the beginning independently within stringology and graph algorithms relying on matrix computations. In string matching, witnesses for symbol mismatches or matches in potential alignments of two strings are sought [3,9,17] while in graph algorithms, witnesses for the Boolean matrix product are typically sought, originally in order to solve shortest path problems in graphs [2,4]. In both cases, highly non-trivial efficient algorithmic solutions have been presented [2–4,17].

Z. Lu et al. (Eds.): COCOA 2015, LNCS 9486, pp. 452–466, 2015.
DOI: 10.1007/978-3-319-26626-8_33

Also in both areas, useful generalizations and/or specializations of the problems of finding witnesses have been studied. A natural generalization introduced for string matching in [17] is to request up to k witnesses instead of a single one. It has been efficiently solved by using concepts from group testing in [3] and conveyed to Boolean matrix product in [3,16]. A natural specialization is to request minimum or maximum witnesses. This specialization has been introduced and efficiently solved in [10] in the context of finding lowest common ancestors in directed acyclic graphs and it found many other applications since then (cf. [8,18,20]).

In analogy to witnesses for Boolean matrix product, if a and b are two n-dimensional Boolean vectors and c is their Boolean convolution then for any coordinate $c_i = 1$ of c, a *witness* is an index l such that $a_l \wedge b_{i-l} = 1$. In contrast to string matching and Boolean matrix product, the problem of computing the witnesses of Boolean vector convolution does not seem to be explicitly studied in the literature. On the other hand, Boolean vector convolution is very much related to string matching [12], and hence the algorithms for reporting witness or more generally up to k witnesses can be easily conveyed from stringology to Boolean vector convolution (see Propositions 1, 2 in Sect. 3).

In this paper, we study the problem of computing minimum and maximum witnesses for Boolean vector convolution. We also consider a generalization of the problem which is to determine for each positive coordinate of the convolution vector, q smallest (or, largest) witnesses, where q is the minimum of a parameter k and the number of witnesses for this coordinate. We term this problem the smallest k-witness problem or the largest k-witness problem, respectively. We also study the corresponding generalization for Boolean matrix product.

Let $\omega(1, r, 1)$ denote the exponent of fast arithmetic multiplication of an $n \times n^r$ matrix by an $n^r \times n$ matrix. In particular, $\omega(1, 1, 1)$ denoted by ω is known to not exceed 2.373 [14,21]. Next, let the notation $\tilde{O}(\ )$ suppress polylogarithmic in n factors. Our main contributions are as follows:

- an $\tilde{O}(n^{1.5})$-time algorithm for reporting minimum and maximum witnesses for the Boolean convolution of two n-dimensional vectors, and more generally, an $\tilde{O}(n^{1.5}k^{0.5})$-time algorithm for the smallest or largest k-witness problem for the convolution;
- as corollaries, $\tilde{O}(n^{1.5}k^{0.5})$ time bounds for the smallest or largest k-witness problems in string matching;
- in part as corollaries, several upper time bounds on computing the (min, +) integer vector convolution in restricted cases, summarized in Table 1;
- an $O(n^{2+\lambda}k)$-time algorithm for the smallest or largest k-witness problem for the Boolean matrix product of two $n \times n$ Boolean matrices, where λ is a solution to the equation $\omega(1, \lambda, 1) = 1 + 2\lambda + \log_n k$;
- as a corollary, an $O(n^{2+\lambda}k)$ time bound for the problem of reporting for each edge of a vertex-weighted graph k lightest (or, heaviest) triangles containing it, where λ satisfies the aforementioned equation; also, an $O(\min\{n^{\omega}k + n^{2+o(1)}k, n^{2+\lambda}k\})$ time bound for the problem of reporting k lightest (or, heaviest) triangles in the input vertex-weighted graph.

Table 1. Our upper time bounds for computing the convolution of two n-dimensional integer vectors either with coordinates having a bounded number of different values, or decomposable into a number of non-decreasing or non-increasing subsequences, or just monotone subsequences.

vector a/vector b	c_b dif. values	m_b non-decr. subs.	m_b non-incr. subs.	m_b mon. subs.
c_a different values	$\tilde{O}(c_a c_b n)$	$\tilde{O}(c_a m_b n^{1.5})$	$\tilde{O}(c_a m_b n^{1.5})$	$\tilde{O}(c_a m_b n^{1.5})$
m_a non-decr. subs.	$\tilde{O}(m_a c_b n^{1.5})$	$\tilde{O}(m_a m_b n^{1.5})$	?	?
m_a non-incr. subs.	$\tilde{O}(m_a c_b n^{1.5})$	?	$\tilde{O}(m_a m_b n^{1.5})$	?
m_a mon. subs.	$\tilde{O}(m_a c_b n^{1.5})$	?	?	?
Arbitrary	$\tilde{O}(c_b n^{1.844})$	?	?	?

2 Preliminaries

For two n-dimensional vectors $a = (a_0, ..., a_{n-1})$ and $b = (b_0, ..., b_{n-1})$ over a semi-ring $(\mathbb{U}, \oplus, \odot)$, their convolution over the semi-ring is a vector $c = (c_0, ..., c_{2n-2})$, where $c_i = \bigoplus_{l=\max\{i-n+1,0\}}^{\min\{i,n-1\}} a_l \odot b_{i-l}$ for $i = 0, ..., 2n-2$. Similarly, for a $p \times q$ matrix A and a $q \times r$ matrix B over the semi-ring, their matrix product over the semi-ring is a $p \times r$ matrix C such that $C[i,j] = \bigoplus_{m=1}^{q} A[i,m] \odot B[m,j]$ for $1 \le i \le p$ and $1 \le j \le r$. In particular, for the semi-rings $(\mathbb{Z}, +, \times)$, $(\mathbb{Z}, \min, +)$, $(\mathbb{Z}, \max, +)$, and $(\{0,1\}, \vee, \wedge)$, we obtain the arithmetic, $(\min, +)$, $(\max, +)$, and the Boolean convolutions or matrix products, respectively.

We shall use the unit-cost RAM computational model [1] with computer word of length logarithmic in the maximum of the size of the input and the value of the largest input integer.

The following fact is well known (cf. [12]).

Fact 1. *Let p and q be two n-dimensional integer vectors. The arithmetic convolution of p and q can be computed in $\tilde{O}(n)$ time. Hence, also the Boolean convolution of two n-dimensional vectors can be computed in $\tilde{O}(n)$ time.*

Fact 2 *[13, 19]. A sequence of n integers can be decomposed into a number of monotone subsequences within $O(\log n)$ of the minimum in $O(n^{1.5} \log n)$ time.*

Fact 3 *[5]. The problem of computing the convolution of two n-dimensional vectors over a semi-ring can be reduced to computing $O(\sqrt{n})$ products of two $O(\sqrt{n}) \times O(\sqrt{n})$ matrices over the semi-ring. Importantly, the matrices can be constructed in $O(n^{1.5})$ time in total and their entries are appropriately filled with the coordinates of the vectors.*

Fact 4 *(see Theorem 3.2 in [7]). Let A and B be two $n \times n$ integer matrices where the entries of A range over at most c different integers. The $(\min, +)$ matrix product of A and B can be computed in $O(cn^{2.688})$ time.*

Fact 5 *[11]. A lightest (or, heaviest) triangle in an undirected vertex weighted graph on n vertices can be found in $O(n^{\omega} + n^{2+o(1)})$ time.*

3 Extreme Witnesses for Boolean Convolution

Let $c = (c_0, ..., c_{2n-2})$ be the Boolean convolution of two n-dimensional Boolean vectors a and b. A *witness* of $c_i = 1$ is any $l \in [\max\{i-n+1, 0\}, \min\{i, n-1\}]$ for which $a_l \wedge b_{i-l} = 1$. A *minimum witness* (or *maximum witness*) of $c_i = 1$ is the smallest (or, the largest, respectively) witness of c_i. The *witnesses problem* (or *minimum witness problem*, or *maximum witness problem*) for the Boolean convolution of two n-dimensional Boolean vectors is to determine witnesses (or, the minimum witnesses or the maximum witnesses, respectively) for all non-zero coordinates of the Boolean convolution of the vectors. The *k-witness problem* (or, *the smallest k-witness problem* or *the largest k-witness problem*) for the Boolean convolution of two n-dimensional Boolean vectors is to determine for each non-zero coordinate of the convolution q witnesses (or, q smallest witnesses or q largest witnesses, respectively), where q is the minimum of k and the number of witnesses for this coordinate.

The Boolean vector convolution is very much related to string matching problems [12]. The corresponding problems of reporting a symbol mismatch or match, or up to k such mismatches or matches for each potential alignments of the pattern with the text have been studied in the so called non-standard stringology [3,17]. Also, the focus of this paper is on extreme witnesses. For these reasons and on the other hand, for the completeness sake, we just state two propositions on standard witnesses for Boolean vector convolution that can be obtained analogously as the well known corresponding facts on string matching or Boolean matrix product.

Proposition 1 *(Analogous to [4]). The witnesses problem for Boolean convolution of two n-dimensional vectors can be solved in $\tilde{O}(n)$ time.*

Proof. Sketch. The witnesses for the Boolean convolution c of two n-dimensional vectors a and b can be computed analogously as the witnesses for the Boolean matrix product [4]. The first observation is that for all coordinates of c that have a single witness, their witnesses can be obtained by computing the arithmetic convolution of a with the vector b' resulting from replacing each 1 in b with the number of the respective coordinate. The next idea is to dilute the other vector b gradually so the number of witnesses for each positive coordinate of c decreases finally to zero but in most cases passing through 1 first. For instance, if c_i has l witnesses and in each phase each coordinate of b is set to 0 with probability $\frac{1}{2}$ then after a logarithmic number of such phases there is a positive probability that exactly one witness will remain. By iterating the process a logarithmic number of times witnesses for all positive coordinates of c can be determined with high probability.

In order to remove the randomness, we can use small c-wise ϵ-bias sample spaces analogously as Alon and Naor in their deterministic algorithm for witnesses of Boolean matrix product [4].

The algorithm, its analysis and derandomization are totally analogous to those of the algorithm of Alon and Naor for witnesses of Boolean matrix product [4].

We refer the reader for the technical details to their paper, it is sufficient to replace matrices with vectors, entries with coordinates and Boolean matrix product with Boolean vector convolution in their proof. □

Proposition 2 *(Analogous to [3] and [16]). The k-witness problem for Boolean convolution of two n-dimensional vectors can be solved in $\tilde{O}(nk)$ time.*

With a moderate technical effort, the minimum or maximum witness problem for Boolean convolution could be solved by combining the known $O(n^{2.575})$-time algorithm for the corresponding problem of minimum or maximum witnesses of Boolean matrix product [10] with the known reduction of vector convolution over an arbitrary semi-ring to matrix product over the semi-ring described in Fact 3 [5]. The combination results in an $O(n^{1.787})$-time solution to the witness problem for Boolean convolution. We shall show that a substantially more efficient solution can be obtained directly.

Theorem 6. *The minimum witness problem (maximum witness problem, respectively) for Boolean convolution of two n-dimensional vectors can be solved in $\tilde{O}(n^{1.5})$ time.*

Proof. Let a and b be two n-dimensional vectors. Let r be an integer parameter between 1 and n. For $p = 1, ..., \lceil n/r \rceil$, let a^p be the Boolean n-dimensional vector resulting from setting to zero all coordinates of a with indices not exceeding $(p-1)r$ and those with indices greater that pr. We compute, for each $p = 1, ..., \lceil n/r \rceil$, the Boolean convolution c^p of a^p and b. Next, for each $i = 0, ..., 2n-2$, we determine the smallest p such that $c_i^p = 1$. Then, if such a p exists, we determine the interval of the implicants $a_l \wedge b_{i-l}$ of c_i^p that potentially can have a non-zero value, i.e., where $l \in ((p-1)r, pr]$, and return the smallest l in the interval for which $a_l \wedge b_{i-l} = 1$. The $\lceil n/r \rceil$ computations of Boolean convolutions c^p takes $\tilde{O}(n^2/r)$ time. The total time taken by the determination of the smallest p is $O(n \times n/r)$. To determine the smallest l for a given i and p requires examining the value of $O(r)$ implicants and hence it takes $O(nr)$ time in total. By setting $r = \lceil \sqrt{n} \rceil$, we obtain the claimed time complexity. □

The method of Theorem 6 can be generalized to include the smallest k-witness problem and the largest k-witness problem.

Theorem 7. *The smallest k-witness problem as well as the largest k-witness problem for Boolean convolution of two n-dimensional vectors can be solved in $\tilde{O}(n^{1.5}k^{0.5})$ time.*

Proof. Let a and b be two input n-dimensional vectors. Let r be an integer parameter between 1 and n. Analogously as in the proof of Theorem 6, for $p = 1, ..., \lceil n/r \rceil$, we let a^l to denote the Boolean n-dimensional vector resulting from setting to zero all coordinates of a with indices not exceeding $(p-1)r$ and those with indices greater that pr. Next, we compute for each $p = 1, ..., \lceil n/r \rceil$, the arithmetic convolution w^p of a^l and b by interpreting these vectors as $\{0, 1\}$ ones. The arithmetic convolutions provide us with the number of witnesses in

Input: two n-dimensional Boolean vectors $a = (a_0, ..., a_{n-1})$ and $b = (b_0, ..., b_{n-1})$, and an integer parameter $k \in [1, n]$.
Output: for each coordinate c_i of the Boolean convolution $c = (c_0, ..., c_{2n-2})$ of a and b, q smallest witnesses of c_i, where q is the minimum of k and the number of witnesses of c_i.

1: **for** $i = 0$ **to** $2n - 2$ **do**
2: $\quad w_i \leftarrow 0$
3: $\quad witset(c_i) \leftarrow \emptyset$
4: **end for**
5: $r \leftarrow \lceil \sqrt{\frac{n}{k}} \rceil$
6: **for** $p = 1$ **to** $\lceil n/r \rceil$ **do**
7: $\quad$ set a^p to an n-dimensional Boolean vector obtained from a by zeroing all coordinates of a outside the interval $((p-1)r, pr]$
8: **end for**
9: **for** $p = 1$ **to** $\lceil n/r \rceil$ **do**
10: $\quad$ compute the arithmetic convolution w^p of a^p and b treated as $\{0, 1\}$ vectors
11: $\quad$ **for** $i = 0$ **to** $2n - 2$ **do**
12: $\quad\quad w_i \leftarrow w_i + w_i^p$
13: $\quad$ **end for**
14: **end for**
15: **for** $p = 1$ **to** $\lceil n/r \rceil$ **do**
16: $\quad$ **for** $i = 0$ **to** $2n - 2$ **do**
17: $\quad\quad$ **if** $\#witset(c_i) < \min\{w_i, k\} \wedge w_i^p \geq 1$ **then** extend $witset(c_i)$ by $\min\{w_i^p, \min\{w_i, k\} - \#witset(c_i)\}$ smallest witnesses of c_i in the interval $((p-1)r, pr]$.
18: $\quad$ **end for**
19: **end for**
20: $witset \leftarrow (witset(c_0), ..., witset((c_{2n-2}))$
21: **return** $witset$

Fig. 1. An algorithm for the smallest k-witness problem for the Boolean convolution of two n-dimensional vectors a and b.

each interval $((p-1)r, pr]$ for each coordinate c_i of the Boolean convolution c of a and b. Their coordinate-wise sum provides us with the total number of witnesses for each coordinate of c. In order to solve the smallest k-witness problem, for $p = 1, ..., \lceil n/r \rceil$, and for $i = 0, ..., 2n - 2$, whenever $w_i^p > 0$ and the number of witnesses for c_i found so far is less than the minimum of k and the number of witnesses of c_i, we search through the interval $((p-1)r, pr]$ from the left to the right for further witnesses. For details see the algorithm depicted in Fig. 1. In the worst case, for each $i = 0, ..., 2n - 2$, we need to search through k of such intervals. The total cost of the searches becomes $O(n \times \frac{n}{r} + n \times k \times r)$, see lines 15-19 in the algorithm depicted in Fig. 1. On the other hand, the $\lceil n/r \rceil$ computations of the arithmetic convolutions w^p takes $\tilde{O}(n^2/r)$ time. By setting $r = \lceil \sqrt{\frac{n}{k}} \rceil$, we obtain the claimed time complexity for the smallest k-witness problem.

The largest k-witness problem can be solved analogously in the same asymptotic time by considering the intervals in the opposite order and searching them from the right to the left instead. □

3.1 String Matching

Fisher and Patterson showed already in 1974 [12] that several string matching problems can be efficiently reduced to Boolean vector convolution.

Suppose we are given two strings $\tau = \tau_{m-1}\tau_{m-2}...\tau_0$ and $\rho = \rho_0\rho_1...\rho_{n-1}$, where $m < n$, over a finite alphabet Σ. Following [12], for $\gamma \in \Sigma$, let $H_\gamma(\ )$ be a predicate defined on Σ which is true only for γ. If $i + m \leq n$, the question of whether $\tau_{m-1}\tau_{m-2}...\tau_0$ matches $\rho_i\rho_{i+1}...\rho_{i+m-1}$ is equivalent to a conjunction of the negations of terms $\bigvee_{l=0}^{m-1} H_\alpha(\rho_{i+l}) \wedge H_\beta(\tau_{m-1-l})$, where $\alpha, \beta \in \Sigma$ and $\alpha \neq \beta$. Note that whenever such a term is true, the matching cannot take place as at some position α clashes with β. In this way, the standard string matching problem for τ and ρ easily reduces to $O(|\Sigma|^2)$ Boolean convolutions of two Boolean vectors of length at most n.

Observe now that witnesses for the aforementioned Boolean convolutions yield positions of the clashes, in other words, symbol mismatches. If we modify the terms to $\bigvee_{l=0}^{m-1} H_\alpha(\rho_{i+l}) \wedge H_\alpha(\tau_{m-1-l})$, for $\alpha \in \Sigma$, the witnesses for the $O(|\Sigma|)$ Boolean convolutions yield positions of two sided matches with $\alpha \in \Sigma$. Hence, we obtain the following theorem as a corollary from Theorem 7.

Theorem 8. *Consider the string matching problem for a text string of length n and a pattern string of length $m < n$, both over a finite alphabet. For each potential alignment of the pattern with the text, we can provide locations of the k earliest symbol mismatches and/or the k earliest symbol matches as well as locations of the k latest symbol mismatches and the k latest symbol matches in the alignments in $\tilde{O}(n^{1.5}k^{0.5})$ time in total. In particular, we can also provide positions of the earliest and/or latest two-side symbol matches with a given alphabet symbol (cf. ones problem in [17]) in the potential alignments in $\tilde{O}(n^{1.5}k^{0.5})$ time in total.*

3.2 (min, +) Convolution

Our original motivation has been an extension of the $O(n^{1.859})$-time algorithm due to Chan and Levenstein for the (min, +) convolution of two n-dimensional vectors with integer coordinates of size $O(n)$ forming monotone sequences [6] to include the case where the vectors are decomposable into relatively few monotone subsequences. The major difficulty here is that a completion of the subsequences to full monotone sequences can affect the result. We can avoid this difficulty when the coordinates of one of the vectors range over relatively few different values (see Fig. 2) or all the subsequences are simultaneously either non-decreasing or non-increasing. The idea is to use our algorithm for minimum and maximum witnesses of Boolean convolution.

The correctness of the algorithm depicted in Fig. 2 relies on the following straightforward lemma.

Input: two n-dimensional vectors $a = (a_0, ..., a_{n-1})$ and $b = (b_0, ..., b_{n-1})$ with integer coordinates such that the coordinates of a range over c_a different values and the sequence of consecutive coordinates of b is decomposable into m_b monotone subsequences.

Output: the convolution $c = (c_0, ..., c_{2n-2})$ of a and b

1: Decompose the sequence $a_0, ..., a_{n-1}$ into a minimum number of constant subsequences $a^i = a^i_1, ..., a^i_{l_i}$
2: **for each** a^i **do**
3: form an n-dimensional Boolean vector $char(a^i)$ with n coordinates indicating with ones the coordinates of a covered by a^i
4: **end for**
5: Decompose the sequence $b_0, ..., b_{n-1}$ into a number of monotone subsequences $b^j = b^j_1, ..., b^j_{l_p}$ within $O(\log n)$ of the minimum
6: **for each** b^j **do**
7: form an n-dimensional Boolean vector $char(b^j)$ with n coordinates indicating with ones the coordinates of b covered by b^j
8: **end for**
9: **for each** pair a^i and b^j **do**
10: **if** b^j is non-decreasing **then** compute the minimum witnesses $wit(d_0)$,...,$wit(d_{2n-2})$ of the Boolean convolution $d = (d_0, ..., d_{2n-2})$ of $char(a^i)$ and $char(b^j)$
11: **if** b^j is non-increasing **then** compute the maximum witnesses $wit(d_0)$,...,$wit(d_{2n-2})$ of the Boolean convolution $d = (d_0, ..., d_{2n-2})$ of $char(a^i)$ and $char(b^j)$
12: **for** $k = 0$ **to** $2n - 2$ **do**
13: **if** $d_k \neq 0$ **then** $c_k \leftarrow \min\{a^i_{wit(d_k)} + b^j_{k-wit(d_k)}, c_k\}$
14: **end for**
15: **end for**
16: $c \leftarrow (c_0, ..., c_{2n-2})$
17: **return** c

Fig. 2. An algorithm for computing the convolution c of two n-dimensional integer vectors a and b, where the coordinates of a range over c_a different values and the sequence of consecutive coordinates of b is decomposable into m_b monotone subsequences.

Lemma 1. *In the algorithm depicted in Fig. 2, the following equivalence holds:* $d_k \neq 0$ *in line 13 if and only if* $c'_k = \min\{a_l + b_m | l + m = k \wedge a_l \in a^i \wedge b_m \in b^j\}$.

Theorem 9. *Let a and b be two n-dimensional integer vectors such that the coordinates of a range over at most c_a different values while the sequence of the consecutive coordinates of b can be decomposed into m_b monotone subsequences. The algorithm depicted in Fig. 2 computes their* $(\min, +)$ *convolution in* $\tilde{O}(c_a m_b n^{1.5})$ *steps.*

Proof. By Lemma 1 and line 13 in the algorithm, none of the coordinates of the output vector has a lower value than the corresponding coordinate of the $(\min, +)$ convolution of a and b. Conversely, if the k-th coordinate of the $(\min, +)$ convolution of a and b equals $a_l + b_m$, where $l + m = k$, then there exist i, j such that $a_l \in a^i$ and $b_m \in b^j$. Hence again by Lemma 1 and line 13 in the algorithm, the k-th coordinate in the output vector has value not larger than the k-th coordinate of the $(\min, +)$ convolution of a and b.

The decomposition of the vector a into c_a constant subsequences in line 1 trivially takes $O(n)$ time. Next, the decomposition of the vector b into $\tilde{O}(m_a)$ monotone subsequences in line 5 takes $O(n^{1.5} \log n)$ time by Fact 2. The forming of the vectors $char(a^i)$ in lines 2-3 and $char(b^j)$ in lines 6-7 take $\tilde{O}(c_a n + m_b n)$ time in total. The $\tilde{O}(c_a m_b)$ computations of the minimum and maximum witnesses of the Boolean convolution d in lines 9-11 take $\tilde{O}(c_a m_b n^{1.5})$ time in total by Theorem 6. Finally, the line 13 is executed $\tilde{O}(c_a m_b n)$ times. The bound $\tilde{O}(c_a m_b n^{1.5})$ follows. □

If we are given decompositions of the two input n-dimensional vectors a and b into monotone subsequences that are either all non-decreasing or all non-increasing then we can use an algorithm analogous to that depicted in Fig. 2. in order to compute the $(\min, +)$ convolution of a and b. Thus, first for each subsequence a^i of a and each subsequence b^j of b, we compute the Boolean vectors $char(a^i)$ and $char(b^j)$ indicating with ones the coordinates of a or b covered by a^i or b^j, respectively. Next, depending if the subsequences are non-decreasing or non-increasing, for each pair of such subsequences a^i and b^j, we compute the minimum witnesses of the Boolean convolution of $char(a^i)$ and $char(b^j)$ or the maximum witnesses of this convolution, respectively. We use the extreme witnesses to update the current coordinates of the computed $(\min, +)$ convolution analogously as in the algorithm depicted in Fig. 2. Hence, we obtain the following theorem.

Theorem 10. *Let a and b be two n-dimensional integer vectors given with the decompositions of the sequences of their consecutive coordinates into m_a and m_b monotone subsequences respectively such that all the subsequences are either non-decreasing or non-increasing. The $(\min, +)$ convolution of a and b can be computed in $\tilde{O}(m_a m_b n^{1.5})$ time.*

Proof. The proof of the correctness of the algorithm proposed in the discussion preceding the theorem is analogous to that of the algorithm depicted in Fig. 2. The time complexity analysis of the algorithm is also similar to that of the previous algorithm. The main difference is that the decompositions of a and b into subsequences are given and that the $O(n^{1.5})$-time algorithm for minimum or maximum witnesses of Boolean convolution is run $m_a m_b$ times instead of $c_a m_b$ times. □

By combining Fact 3 with Fact 4, we obtain also the following bound.

Theorem 11. *Let a and b be two n-dimensional integer vectors such that the coordinates of a range over at most c_a different values. The $(\min, +)$ convolution of a and b can be computed in $\tilde{O}(c_a n^{1.844})$ time.*

We can also consider the problem of computing the $(\min, +)$ integer vector convolution of the input vectors a and b, when their coordinates range over c_a and c_b different integers, respectively. Again, we can use an algorithm analogous to that depicted in Fig. 2. The first difference is that the subsequences b^j on the side of b are also constant. It follows that for any pair of such constant subsequences a^i and b^j, the value $sum(i,j)$ of the sum of any element from a^i with any element from b^j is constant and it can be trivially computed a priori. For this reason, it is sufficient to compute the Boolean convolution d of $char(a^i)$ and $char(b^j)$ for any such pair a^i and b^j. Then, for any non-zero coordinate of d, we need to update the corresponding coordinate of the computed $(\min,+)$ convolution of a and b by taking the minimum of the coordinate and $sum(i,j)$. By Fact 1, we obtain the following theorem.

Theorem 12. *Let a and b be two n-dimensional integer vectors such that their coordinates range over at most c_a or c_b different values, respectively. Their $(\min,+)$ convolution can be computed in $\tilde{O}(c_a c_b n)$ time.*

Proof. The discussed algorithm can be easily implemented in $\tilde{O}(c_a c_b n)$ time by running $c_a c_b$ times the known $\tilde{O}(n)$-time algorithm for Boolean convolution of two n-dimensional Boolean vectors, see Fact 1. □

4 Extreme Witnesses for Boolean Matrix Product

For two $n \times n$ Boolean matrices A and B, a witness of a $C[i,j]$ entry of the Boolean matrix product of A and B is any index m such that $A[i,m] \wedge B[m,j] = 1$. Next, the minimum witness and maximum witness for an entry of C as well as the witness problem, the minimum and maximum witness problems, the k-witness problem, and the smallest k-witness and largest k-witness problems for Boolean matrix product of A and B are defined analogously as those for Boolean vector convolution.

In this section, we shall present a generalization of the algorithm for minimum and maximum witnesses for Boolean matrix product from [10] to include the smallest and largest k-witness problems.

Let ℓ be a positive integer smaller than n. We may assume w.l.o.g that n is divisible by ℓ. Partition the matrix A into $n \times \ell$ sub-matrices A_p and the matrix B into $\ell \times n$ sub-matrices B_p, such that $1 \le p \le n/\ell$, and the sub-matrix A_p covers the columns $(p-1)\,\ell+1$ through $p\,\ell$ of A whereas the sub-matrix B_p covers the rows $(p-1)\,\ell+1$ through $p\,\ell$ of B.

For $p = 1,\ldots,n/\ell$, let W_p be the arithmetic product of A_p and B_p treated as $\{0,1\}$ matrices. On the other hand, let C denote the Boolean matrix product of A and B. Then, $W_p[i,j] = q$ iff there are exactly q witnesses of $C[i,j]$ in the interval $((p-1)\,\ell, p\,\ell]$. Consequently, the total number of witnesses of $C[i,j]$ is given by $\sum_{p=1}^{n/\ell} W_p[i,j]$. Therefore, the following lemma follows.

Lemma 2. *Suppose that a $C[i,j]$ entry of the Boolean product C of A and B is positive. Let q be the minimum of k and the total number of witnesses of $C[i,j]$. Next, let p' be the minimum value of p such that $\sum_{u=1}^{p} W_u[i,j]$ is not less than q. The smallest q witnesses of $C[i,j]$ belong to the interval $[1, p'\,\ell]$.*

Input: two $n \times n$ Boolean matrices A and B, and integer parameters $\ell, k \in [1, n]$, where n is assumed to be divisible by ℓ.
Output: for each entry $C[i,j]$ of the Boolean matrix product C of A and B, q smallest witnesses of $C[i,j]$, where q is the minimum of k and the number of witnesses of $C[i,j]$.

```
for i = 1 to n do
   for j = 1 to n do
      W[i,j] ← 0
      witset(C[i,j]) ← ∅
   end for
end for
for p = 1 to n/ℓ do
   set A_p to an n × n Boolean matrix that is a restriction of A to columns with
   indices in ((p − 1)ℓ, pℓ]
end for
for p = 1 to n/ℓ do
   set B_p to an n × n Boolean matrix that is a restriction of B to rows with indices
   in ((p − 1)ℓ, pℓ]
end for
for p = 1 to n/ℓ do
   compute the arithmetic matrix product W_p of A_p and B_p treated as {0,1}
   matrices
   for i = 1 to n do
      for j = 1 to n do
         W[i,j] ← W[i,j] + W_p[i,j]
      end for
   end for
end for
for p = 1 to n/ℓ do
   for i = 1 to n do
      for j = 1 to n do
         if #witset(C[i,j]) < min{W[i,j], k} ∧ W_p[i,j] ≥ 1 then extend
         witset(C[i,j]) by min{W_p[i,j], min{W[i,j], k} − #witset(C[i,j])} smallest
         witnesses of C[i,j] in the interval ((p − 1)ℓ, pℓ]
      end for
   end for
end for
for i = 1 to n do
   for j = 1 to n do
      return witset(C[i,j])
   end for
end for
```

Fig. 3. An algorithm for the smallest k-witness problem for the Boolean matrix product of two $n \times n$ Boolean matrices A and B.

By this lemma, after computing all the matrix products $W_p = A_p \cdot B_p$, $1 \le p \le n/\ell$, we need $O(n/\ell + k\ell)$ time per positive entry of C to find up to k smallest witnesses: $O(n/\ell)$ time to determine p' and then $O(k\ell)$ time to locate the up to k smallest witnesses. See Fig. 3 for our algorithm for the smallest k-witness problem.

Recall that $\omega(1,r,1)$ denotes the exponent of the multiplication of an $n \times n^r$ matrix by an $n^r \times n$ matrix. It follows that the total time taken by our algorithm for the smallest k-witness problem is

$$O((n/\ell) \cdot n^{\omega(1,\log_n \ell,1)} + n^3/\ell + n^2 k\ell) .$$

By setting r to $\log_n \ell$ and z to $\log_n k$, our upper bound transforms to $O(n^{1-r+\omega(1,r,1)} + n^{3-r} + n^{2+r+z})$. Note that by assuming $r \geq \frac{1}{2} - \frac{z}{2}$, we can get rid of the additive n^{3-r} term. See Fig. 5 in [22] for the graph of the function $1-r+\omega(1,r,1)$. By solving the equation $1-\lambda+\omega(1,\lambda,1) = 2+z+\lambda$ implying $\lambda \geq \frac{1}{2} - \frac{z}{2}$ by $\omega(1,\lambda,1) \geq 2$, we obtain our main result.

Theorem 13. *Let λ be such that $\omega(1,\lambda,1) = 1 + 2\lambda + \log_n k$. The smallest k-witness problem as well as the largest k-witness problem for the Boolean matrix product of two $n \times n$ Boolean matrices can be solved in $O(n^{2+\lambda}k)$ time.*

Le Gall has recently substantially improved upper time bounds on rectangular matrix multiplication in [15]. In consequence, he could show that for the equation $\omega(1,\mu,1) = 1 + 2\mu$, $\mu < 0.5302$. This in particular improves the upper time bound for the minimum and maximum witness problems from $O(n^{2.575})$ to $O(n^{2.5302})$. It follows that for $k >> 1$, λ in Theorem 13 is substantially smaller than 0.5302.

4.1 Lightest Triangles

By generalizing the reduction of the problem of reporting for each edge of a vertex-weighted graph a heaviest triangle containing it to the maximum witness problem for Boolean matrix product from [20] to include reporting k heaviest triangles and the largest k-witness problem, we obtain the following theorem as a corollary from Theorem 13.

Theorem 14. *Let G be an undirected vertex weighted graph on n vertices and let k be a natural number not exceeding n. Next, let λ be such that $\omega(1,\lambda,1) = 1 + 2\lambda + \log_n k$. We can list for each edge $\{u,v\}$ of G, q_e lightest (or, heaviest) triangles $\{u,v,w\}$ in G, where q_e is the minimum of k and the number of triangles $\{u,v,w\}$ in G, in $O(n^{2+\lambda}k)$ time.*

Proof. Number the vertices of G in non-decreasing vertex-weight order. Next, solve the smallest (or, largest) k-witness problem for the Boolean matrix product C of the adjacency matrix of G with itself. For each edge $e = \{i,j\}$ of G, the up to k smallest (or, largest) witnesses of $C[i,j]$ yield the q_e lightest (or, heaviest, respectively) triangles in G including e. Theorem 13 yields the claimed upper bound. □

As for the problem of finding k lightest (or, heaviest) triangles in a vertex-weighted graph, iterating the $O(n^{\omega}+n^{2+o(1)})$-time algorithm for finding a lightest or heaviest triangle described in Fact 5 seems to be a better choice for up to moderate values of k. Before each next iteration, we remove the three vertices of the lastly reported triangle. After k iterations, we stop and find among the reported triangles and no more than $3(k-1)n^2$ other triangles incident to the removed vertices, the k lightest (or, heaviest) triangles if possible. The method takes $O(n^{\omega}k+n^{2+o(1)}k+n^2k)$, i.e., $O(n^{\omega}k+n^{2+o(1)}k)$ time.

Theorem 15. *Let G be an undirected vertex weighted graph on n vertices and let k be a natural number $\leq n$. Next, let λ be such that $\omega(1,\lambda,1)=1+2\lambda+\log_n k$. We can list q lightest (or, heaviest) triangles in G, where q is the minimum of k and the number of triangles in G, in $O(\min\{n^{\omega}k+n^{2+o(1)}k, n^{2+\lambda}k\})$ time.*

5 Final Remarks

It is an interesting open problem if any of our upper time bounds on minimum and maximum witnesses for Boolean vector convolution and the extreme k-witness problems both for Boolean vector convolution and Boolean matrix product can be substantially improved? Note here that so far the $O(n^{2+\lambda})$ time bound on minimum and maximum witnesses of Boolean matrix product established about one decade ago [10] couldn't be improved (see also [8]).

The problems of Boolean vector convolution and Boolean matrix product seem to be similar but there are some substantial differences between them. The former problem admits almost a linear in the input size algorithm while for the latter problem the current upper time bound is substantially non-linear [14,21]. There is a moderately efficient reduction of vector convolution to matrix product described in Fact 3 while such a reverse reduction is not known. Our upper time bounds for minimum and maximum witnesses of Boolean vector convolution show that a direct approach to Boolean vector convolution can yield better upper time bounds than those obtained by conveying known upper time bounds for Boolean matrix product via Fact 3 to Boolean vector convolution.

The extreme k-witness problems for Boolean matrix product presumably admit several other applications often corresponding to generalizations of the applications for minimum and maximum witnesses of Boolean matrix product [18,20] and/or the applications of the k-witness problem for Boolean matrix product [3], e.g., the all-pairs k-bottleneck paths.

Acknowledgments. We thank Mirosław Kowaluk for valuable comments.

References

1. Aho, A. V., Hopcroft, J. E., Ullman, J.: The Design and Analysis of Computer Algorithms. Addison-Wesley Publishing Company, Reading (1974)

2. Alon, N., Galil, Z., Margalit, O., Naor, M.: Witnesses for Boolean matrix multiplication and for shortest paths. In: Proceedings 33rd Symposium on Foundations of Computer Science (FOCS), pp. 417–426 (1992)
3. Aumann, Y., Lewenstein, M., Lewenstein, N., Tsur, D.: Finding witnesses by peeling. In: Ma, B., Zhang, K. (eds.) CPM 2007. LNCS, vol. 4580, pp. 28–39. Springer, Heidelberg (2007)
4. Alon, N., Naor, M.: Derandomization, witnesses for Boolean matrix multiplication and construction of perfect hash functions. Algorithmica **16**, 434–449 (1996)
5. Bremner, D., Chan, T.M., Demaine, E.D., Erickson, J., Hurtado, F., Iacono, J., Langerman, S., Taslakian, P.: Necklaces, convolutions, and $X + Y$. In: Azar, Y., Erlebach, T. (eds.) ESA 2006. LNCS, vol. 4168, pp. 160–171. Springer, Heidelberg (2006)
6. Chan, T. M., Lewenstein, M.: Clustered integer 3SUM vis additive combinatorics. In: Proceedings 47th ACM Symposium on Theory of Computing (STOC 2015)
7. Chan, T.M.: More algorithms for all-pairs shortest paths in weighted graphs. SIAM J. Comput. **39**(5), 2025–2089 (2010)
8. Cohen, K., Yuster, R.: On minimum witnesses for Boolean matrix multiplication. Algorithmica **69**(2), 431–442 (2014)
9. Crochemore, M., Rytter, W.: Text Algorithms. Oxford University Press, New York (1994)
10. Czumaj, A., Kowaluk, M., Lingas, A.: Faster algorithms for finding lowest common ancestors in directed acyclic graphs. Theor. Comput. Sci. **380**(1–2), 37–46 (2007). The special ICALP 2005 issue
11. Czumaj, A., Lingas, A.: Finding a heaviest vertex-weighted triangle is not harder than matrix multiplication. SIAM J. Comput. **39**(2), 431–444 (2009)
12. Fisher, M. J., Paterson, M. S.: String-matching and other products. In: Proceedings 7th SIAM-AMS Complexity of Computation, pp. 113–125 (1974)
13. Fomin, F.V., Kratsch, D., Novelli, J.: Approximating minimum cocolorings. Inf. P. Lett. **84**(5), 285–290 (2002)
14. Le Gall, F.: Powers of tensors and fast matrix multiplication. In: Proceedings 39th International Symposium on Symbolic and Algebraic Computation, pp. 296–303 (2014)
15. Le Gall, F.: Faster algorithms for rectangular matrix multiplication. In: Proceedings 53rd Symposium on Foundations of Computer Science (FOCS), pp. 514–523 (2012)
16. Gasieniec, L., Kowaluk, M., Lingas, A.: Faster multi-witnesses for Boolean matrix product. Inf. Process. Lett. **109**, 242–247 (2009)
17. Muthukrishnan, S.: New results and open problems related to non-standard stringology. In: Galil, Z., Ukkonen, E. (eds.) CPM 1995. LNCS, vol. 937, pp. 298–317. Springer, Heidelberg (1995)
18. Shapira, A., Yuster, R., Zwick, U.: All-pairs bottleneck paths in vertex weighted graphs. Algorithmica **59**(4), 621–633 (2011)
19. Yang, B., Chen, J., Lu, E., Zheng, S.Q.: A comparative study of efficient algorithms for partitioning a sequence into monotone subsequences. In: Cai, J.-Y., Cooper, S.B., Zhu, H. (eds.) TAMC 2007. LNCS, vol. 4484, pp. 46–57. Springer, Heidelberg (2007)

20. Vassilevska, V., Williams, R., Yuster, R.: Finding heaviest H-subgraphs in real weighted graphs, with applications. ACM Trans. Algorithms **6**(3), 44:1–44:23 (2010)
21. Williams, V. V.: Multiplying matrices faster than Coppersmith-Winograd. In: Proceedings 44th Annual ACM Symposium on Theory of Computing (STOC), pp. 887–898 (2012)
22. Zwick, U.: All pairs shortest paths using bridging sets and rectangular matrix multiplication. J. ACM **49**(3), 289–317 (2002)

Orbital Independence in Symmetric Mathematical Programs

Gustavo Dias(✉) and Leo Liberti

CNRS LIX, Ecole Polytechnique, 91128 Palaiseau, France
{dias,liberti}@lix.polytechnique.fr

Abstract. It is well known that symmetric mathematical programs are harder to solve to global optimality using Branch-and-Bound type algorithms, since the solution symmetry is reflected in the size of the Branch-and-Bound tree. It is also well known that some of the solution symmetries are usually evident in the formulation, making it possible to attempt to deal with symmetries as a preprocessing step. One of the easiest approaches is to "break" symmetries by adjoining some symmetry-breaking constraints to the formulation, thereby removing some symmetric global optima, then solve the reformulation with a generic solver. Sets of such constraints can be generated from each orbit of the action of the symmetries on the variable index set. It is unclear, however, whether and how it is possible to choose two or more separate orbits to generate symmetry-breaking constraints which are compatible with each other (in the sense that they do not make all global optima infeasible). In this paper we discuss a new concept of orbit independence which clarifies this issue.

1 Introduction

In this paper we address an important issue which arises when breaking symmetries of Mathematical Programs (MP) in view of solving them using Branch-and-Bound (BB) type algorithms. Symmetry-breaking devices are usually derived from orbits of the action of the formulation group on the decision variables. However, one cannot simply use such devices for all orbits: some orbits depend on each other, in a very precise mathematical sense, and hence it may be impossible to use more than one orbit for symmetry-breaking purposes. Below, we discuss a notion of orbit independence which permits to break symmetries from different orbits concurrently.

2 Previous Work and Notation

2.1 Mathematical Programming

An MP is a formulation which formally describes an optimization problem in terms of known parameters (input), decision variables (output), an objective

Z. Lu et al. (Eds.): COCOA 2015, LNCS 9486, pp. 467–480, 2015.
DOI: 10.1007/978-3-319-26626-8_34

function, and some constraints. We consider MPs of the following general form:

$$\left.\begin{array}{r} \min_x \ f(x) \\ \forall i \leq m \ g_i(x) \leq 0 \\ x \in D. \end{array}\right\} \quad (1)$$

In Eq. (1), $f, g_i : \mathbb{R}^n \to \mathbb{R}$ are functions for which we have closed form expressions f, g_i for each $i \leq m$. The expressions are written in terms of a formal language $\mathcal{L}$ based on an alphabet $\mathcal{A}$ consisting of a finite number of operators (e.g. sum, difference, product, fractions, powers, square roots, basic transcendental functions such as logarithm and exponentials, and possibly more complicated operators depending on the application at hand), a countable supply of variable symbols $\mathsf{x}_1, \ldots, \mathsf{x}_n$ representing the decision variables $x_1, \ldots, x_n$, and the rational numbers. The set D might contain non-functional constraints such as ranges $[x^L, x^U]$ for the decision variables, and/or integrality constraints, encoded as an index set $Z \subseteq X = \{1, \ldots, n\}$ such that $x_j \in \mathbb{Z}$ for each $j \in Z$. This modelling paradigm contains Linear Programming (LP), Nonlinear Programming (NLP), Mixed-Integer Linear Programming (MILP), Mixed-Integer Nonlinear Programming (MINLP) and Semidefinite Programming (SDP) if $x_1, \ldots, x_n$ are matrices.

2.2 Symmetry Detection

We emphasize that Eq. (1) subsumes the description of *two* mathematical entities: the MP itself, denoted by P, and its formal description P in the language $\mathcal{L}$, which we obtain when replacing x, f, g by their representing symbols $\mathsf{x}, \mathsf{f}, \mathsf{g}$. It is well known that P can be parsed into a Directed Acyclic Graph (DAG) data structure T (an elementary graph contraction of the well-known *parsing tree*) using a fairly simple context-free grammar [2]. The leaf nodes of T are labelled by constants or decision variable symbols, whereas the other nodes of T are labelled by operator symbols. The incidence structure of T encodes the parent-child relationships between operators, variables and constants. In practice, we can write P using a modelling language such as AMPL [3] and use an unpublished but effective AMPL API to derive T [4]. Since T is a labelled graph, we know how to compute the group $\mathcal{G}$ of its label-invariant isomorphisms (which must also respect a few other properties, such as non-commutativity of certain operators) [15,16]. In practice, we can use the software codes Nauty or Traces [16] to obtain $\mathcal{G}$ and the set Θ of the orbits of its action on the nodes $V(T)$ of the DAG.

2.3 Formulation and Solution Groups

It was shown in [8] that: (a) the action of $\mathcal{G}$ can be projected to the leaf nodes of $V(T)$, which represent decision variables; (b) this projection induces a group homomorphism ϕ mapping $\mathcal{G}$ to a certain group image G_P; (c) G_P is a group of permutations of the indices of the variable symbols $\mathsf{x}_1, \ldots, \mathsf{x}_n$; (d) G_P is precisely the group of variable permutations of P which keeps $f(x)$ and $\{g_i(x) \mid i \leq m\}$

invariant. In other words, [8] provides a practical methodology for computing the *formulation group* of a MP given as in Eq. (1). Since it is not hard to show that G_P is a subgroup of the *solution group* of P, meant as the group of permutations which keeps the set $\mathscr{G}(P)$ of global optima of P invariant, this methodology can be used to extract symmetries from P prior to solving it.

2.4 Symmetry Breaking Constraints

So much for detecting (some) symmetries. Once these are known, their most efficient exploitation appears to be their usage within the BB algorithm itself [12,13,17,18]. Such approaches are, unfortunately, difficult to implement, as each solver code must be addressed separately. Their simplest exploitation is *static symmetry breaking* [14, Sect. 8.2] which, simply put, consists in adjoining some Symmetry-Breaking Constraints (SBCs) to the original formulation Eq. (1) in the hope of making all but one of the symmetric global optima infeasible. Following the usual trade-off between efficiency and generality, approaches which offer provable guarantees of removing symmetric optima are limited to special structures [6], whereas approaches which hold for any MP in the large class Eq. (1) are mostly common-sense constraints designed to work in general [9]. The consensus seems to be that sets of SBCs are derived from each orbit of the action of G_P on X (though this is not the only possibility: SBCs can also be derived from cyclic subgroups of G_P or single permutations).

2.5 Orbits

We recall that an orbit is an equivalence class of the quotient set $X/\sim$, where $i \sim j$ if there is $g \in G_P$ such that $g(i) = j$. This way, G_P partitions X into a set Ω_{G_P} of orbits $\omega_1, \ldots, \omega_p$, each of which can be used to generate SBCs. The projection homomorphism ϕ defined above for $\mathcal{G}$ and the leaf nodes of the parsing tree can be restricted to act on G_P and generalized to project its action to any subset $Y \subseteq X$ as follows: for each $\pi \in G_P$ let $\phi(\pi)$ be the product of the cycles of π having all components in Y. We denote by ϕ_Y this generalized action projection homomorphism. The image of ϕ_Y, when Y is some orbit $\omega \in \Omega_{G_P}$, is a group $G_P[\omega]$ called the *transitive constituent* of ω (a group action is *transitive* on a set S if $s \sim t$ for each $s, t \in S$).

2.6 Strong and Weak SBCs

We borrow the square bracket notation to localize vectors: if $x^* \in \mathscr{G}(P)$ is a global optimum of P, then $x^*[\omega]$ is a projection of x^* on the coordinates indexed by ω. If $G_P[\omega]$ is the full symmetric group $\mathsf{Sym}(\omega)$ on the orbit, it means that $\mathscr{G}(P)$ contains vectors which, when projected onto ω, yield every possible order of $x^*[\omega]$. This implies that we can arbitrarily choose one order, e.g.:

$$\forall \ell < |\omega| \quad x_{\omega(\ell)} \leq x_{\omega(\ell+1)}, \tag{2}$$

where $\omega(\ell)$ is the ℓ-th element of ω (stored as a list), enforce this order by means of SBCs, and still be sure that at least one global optimum remains feasible. The SBCs in Eq. (2) are called *strong SBCs*. If $G_P[\omega]$ has any other structure, we observe that, by transitivity of the transitive constituents, at least one permutation in $G_P[\omega]$ will map the component having minimum value in $x^*[\omega]$ to the first component (the choice if minimum value and first components are arbitrary — alternative SBC sets can occur by choosing maximum and/or any other component). This yields the *weak SBCs*:

$$\forall \ell \in \omega \setminus \{\omega(1)\} \quad x_{\omega(1)} \leq x_{\omega(\ell)}. \tag{3}$$

Strong SBCs select one order out of $|\omega|!$ many, and hence are able to break all symmetries in $G_P[\omega]$. Weak SBCs are unlikely to be able to achieve that. We let $g(x[B]) \leq 0$ denote SBCs involving only variables x_j with j in a given set B.

2.7 Stabilizers

Let $Y \subseteq X$. We recall that the pointwise stabilizer of Y w.r.t. G_P (or any group G) is defined as the subgroup of elements of G_P fixing each element of Y, i.e., $G^Y = \{g \in G_P \mid \forall y \in Y \ (gy = y)\}$. The setwise stabilizer of Y w.r.t. G_P is the subgroup of those elements of G_P under which Y is invariant, i.e., $\mathsf{stab}(Y, G_P) = \{g \in G_P \mid \forall y \in Y \ (gy \in Y)\}$. By definition, if Y is an orbit of G_P, then G^Y is the kernel of ϕ_Y and $\mathsf{stab}(Y, G_P) = G_P$.

3 Orbital Independence Notions

In this section we introduce our main results regarding orbit independence (OI). First we illustrate how SBCs built from different orbits may cut global optima from a MP; then we recall the conditions of OI originally introduced in [8], and finally we present a new concept of OI based on pointwise stabilizers.

3.1 Incompatible SBCs

In general, one may only adjoin to P the SBCs from *one* orbit. Adjoining SBCs from two or more orbits chosen arbitrarily may result in all global optima being infeasible, as Example 1 shows.

Example 1. Let P be the following MILP:

$$\min_{x \in \{0,1\}^4} \quad x_1 + x_2 + 2x_3 + 2x_4$$

$$\begin{pmatrix} 1 & 1 & 0 & 0 \\ 0 & 0 & 1 & 1 \\ -1 & 0 & -1 & 0 \\ 0 & -1 & 0 & -1 \end{pmatrix} \begin{pmatrix} x_1 \\ x_2 \\ x_3 \\ x_4 \end{pmatrix} \leq \begin{pmatrix} 1 \\ 1 \\ -1 \\ -1 \end{pmatrix}.$$

It has formulation group $G_P = \langle(1\,2)(3\,4)\rangle$, optima $\mathcal{G}(P) = \{(0,1,1,0),(1,0,0,1)\}$ and orbits $\Omega_{G_P} = \{\omega_1,\omega_2\} = \{\{1,2\},\{3,4\}\}$. Valid SBCs for ω_1 (resp. ω_2) are $x_1 \le x_2$ (resp. $x_3 \le x_4$). Whereas adjoining either of the two SBCs yields a valid narrowing, adjoining both simultaneously leads to an infeasible problem.

Yet, breaking symmetries from only one orbit does not generally make a strong computational impact in MPs of the form Eq. (1). In what follows, we explore the concept of "orbital independence" meant as sufficient conditions to guarantee that SBCs from many orbits preserve at least one global optimum of P feasible.

3.2 Some Existing OI Conditions

In order to concurrently combine sets of SBCs generated by two orbits $\omega,\theta \in \Omega_{G_P}$ into a valid narrowing (i.e. a reformulation guaranteed to keep at least one global optimum [7]) of a MINLP, two sufficient conditions were provided in [8]:

- there is a subgroup $H \le G_P[\omega \cup \theta]$ such that $H[\omega] \cong C_{|\omega|}$ and $H[\theta] \cong C_{|\theta|}$;
- $\mathsf{gcd}(|\omega|,|\theta|) = 1$.

Two orbits with these properties are called *coprime*. Coprime orbits occur relatively rarely in practice [8].

Another set of conditions for OI was hinted at in [11], by means of the following iterative procedure. Initially, one sets $G \leftarrow G_P$ and picks an orbit $\omega \in \Omega_{G_P}$; then adjoins SBCs for ω to P, and then replaces G by G^ω. Termination occurs when G is the trivial group. At each iteration, the SBCs from different orbits can be concurrently adjoined to P. On the other hand, the orbits refer to the action of different groups: G_P initially, then the groups in a normal chain of pointwise stabilizers. In the following, we expand on this idea.

3.3 New Conditions for OI

Our goal now is to introduce the concept of *independent* set of orbits and provide conditions that will help us to identify such sets. These new necessary conditions for OI will be established based on pointwise stabilizers.

First, let $\omega,\theta \in \Omega_{G_P}$. We look at what happens to θ when ω is pointwise stabilized: either G^ω fixes θ, or a subset of θ, or it does not fix any element of θ at all. So three cases follow:

(a) for any subset $\sigma \subseteq \theta$, $\sigma \notin \Omega_{G^\omega}$;
(b) there is a subset $\sigma \subsetneq \theta$ such that $\sigma \in \Omega_{G^\omega}$;
(c) $\theta \in \Omega_{G^\omega}$.

We can thus state the following binary *dependence* relations on the set Ω_{G_P}.

Definition 1. *The orbit θ is dependent of ω, denoted by $\theta \to \omega$, if θ is stabilized when ω is stabilized (case (a) above).*

Definition 2. *The orbit θ is semi-dependent of ω, denoted by $\theta \rightsquigarrow \omega$, if θ splits when ω is stabilized (case (b) above).*

Definition 3. *The orbit θ is independent of ω, denoted by $\theta \looparrowright \omega$, if θ is not stabilized when ω is stabilized (case (c) above).*

Next, let Γ^ω be the set of permutations of G_P which move elements of the orbit ω nontrivially. By definition, Γ^ω does not contain the identity permutation e of G_P and thus it is not itself a group. Moreover, the following properties hold: $G^\omega \cap \Gamma^\omega = \varnothing$, $\mathsf{stab}(\omega, G_P) = G^\omega \cup \Gamma^\omega = G_P$ and $\phi_\omega(\Gamma^\omega) = G_P[\omega] \smallsetminus e$.

Theorem 1 establishes the dependence relation between two orbits $\omega, \theta \in \Omega_{G_P}$ by comparing the sets Γ^ω and Γ^θ.

Theorem 1. *The following statements are true:*

(1) If $\Gamma^\theta = \Gamma^\omega$ then $\theta \to \omega$ and $\omega \to \theta$;
(2) If $\Gamma^\theta \subset \Gamma^\omega$ then $\theta \to \omega$ and either $\omega \looparrowright \theta$ or $\omega \rightsquigarrow \theta$;
(3) If $\Gamma^\theta \cap \Gamma^\omega \neq \varnothing$ then $(\theta \looparrowright \omega$ or $\theta \rightsquigarrow \omega)$ and $(\omega \looparrowright \theta$ or $\omega \rightsquigarrow \theta)$;
(4) If $\Gamma^\theta \cap \Gamma^\omega = \varnothing$ then $\theta \looparrowright \omega$ and $\omega \looparrowright \theta$.

Proof. (1) Assume $\Gamma^\theta = \Gamma^\omega$ and consider ω. Then $G^\omega = G_P \smallsetminus \Gamma^\omega \Rightarrow G^\omega \cap \Gamma^\theta = \varnothing \Rightarrow \theta \notin \Omega_{G^\omega}$ and $\theta \to \omega$. Since the same argument holds if we consider θ, we also have $\omega \to \theta$.

(2) Assume $\Gamma^\theta \subset \Gamma^\omega$ and consider ω. Then $G^\omega = G_P \smallsetminus \Gamma^\omega \Rightarrow G^\omega \cap \Gamma^\theta = \varnothing \Rightarrow \theta \notin \Omega_{G^\omega}$ and $\theta \to \omega$. Considering θ, we have that $G^\theta = G_P \smallsetminus \Gamma^\theta \Rightarrow G^\theta \cap \Gamma^\omega \neq \varnothing$. If the action of G^θ is transitive on ω, we have $\omega \looparrowright \theta$. Otherwise, we have $\omega \rightsquigarrow \theta$.

(3) Assume $\Gamma^\theta \cap \Gamma^\omega \neq \varnothing$ but neither set is wholly contained in the other, and consider ω. Then $G^\omega = G_P \smallsetminus \Gamma^\omega \Rightarrow G^\omega \cap \Gamma^\theta \neq \varnothing$. If the action of G^ω is transitive on θ, we have $\theta \looparrowright \omega$. Otherwise, we have $\theta \rightsquigarrow \omega$. The same argument holds if we consider θ.

(4) Assume $\Gamma^\theta \cap \Gamma^\omega = \varnothing$ and consider ω. Then $G^\omega = G_P \smallsetminus \Gamma^\omega \Rightarrow G^\omega \supset \Gamma^\theta \Rightarrow \theta \in \Omega_{G^\omega}$ and $\theta \looparrowright \omega$. The argument is similar if we consider θ, thus $\omega \looparrowright \theta$. □

Lemma 1. *The premise $\Gamma^\theta \cap \Gamma^\omega = \varnothing$ to condition (4) in Theorem 1 never holds.*

Proof. Let Δ be the set of generators of G_P. If there is $g \in \Delta$ such that $g[\omega]$ and $g[\theta]$ are nontrivial, then $g \in \Gamma^\theta \cap \Gamma^\omega$. Otherwise, let $\Delta^\theta = \{g \in \Delta \mid g[\omega] = e\}$ and $\Delta^\omega = \{g \in \Delta \mid g[\theta] = e\}$. Because every element of G_P can be expressed as the combination (under the group operation) of finitely many elements of Δ, there is $g \in G_P$ such that $g = g_\omega g_\theta$ where $g_\omega \in \Delta^\omega$ and $g_\theta \in \Delta^\theta$. Thus $g \in \Gamma^\theta \cap \Gamma^\omega$. □

Based on the above definitions and results, the following lemmata hold.

Lemma 2. *The relation $\to$ is reflexive and the relations $\rightsquigarrow$ and $\looparrowright$ are irreflexive.*

Lemma 3. *The relation $\rightarrow$ is symmetric iff $\Gamma^\theta = \Gamma^\omega$ and asymmetric iff $\Gamma^\theta \subset \Gamma^\omega$.*

Lemma 4. *The relation $\rightarrow$ is transitive.*

Proof. Let $\theta, \omega, \tau \in \Omega_{G_P}$ be distinct orbits satisfying $\theta \rightarrow \omega$ and $\omega \rightarrow \tau$. From Theorem 1, $\theta \rightarrow \omega$ implies that either $\Gamma^\theta = \Gamma^\omega$ or $\Gamma^\theta \subset \Gamma^\omega$. Similarly, $\omega \rightarrow \tau$ implies that either $\Gamma^\omega = \Gamma^\tau$ or $\Gamma^\omega \subset \Gamma^\tau$. Then:

(i) $\Gamma^\theta = \Gamma^\omega \wedge \Gamma^\omega = \Gamma^\tau \Rightarrow \Gamma^\theta = \Gamma^\tau \Rightarrow \theta \rightarrow \tau$;
(ii) $\Gamma^\theta = \Gamma^\omega \wedge \Gamma^\omega \subset \Gamma^\tau \Rightarrow \Gamma^\theta \subset \Gamma^\tau \Rightarrow \theta \rightarrow \tau$;
(iii) $\Gamma^\theta \subset \Gamma^\omega \wedge \Gamma^\omega = \Gamma^\tau \Rightarrow \Gamma^\theta \subset \Gamma^\tau \Rightarrow \theta \rightarrow \tau$;
(iv) $\Gamma^\theta \subset \Gamma^\omega \wedge \Gamma^\omega \subset \Gamma^\tau \Rightarrow \Gamma^\theta \subset \Gamma^\tau \Rightarrow \theta \rightarrow \tau$. □

Whenever the dependence relations are symmetric, we write $\omega \leftrightarrow \theta$ or $\omega \leftrightsquigarrow \theta$ or $\omega \not\leftrightarrow \theta$. Using this notation, we set forth that:

Definition 4. *Two orbits $\omega, \theta \in \Omega_{G_P}$ are dependent if $\omega \leftrightarrow \theta$, semi-dependent if $\omega \leftrightsquigarrow \theta$ and independent if $\omega \not\leftrightarrow \theta$.*

Following, we extend the dependence relations presented above to sets of orbits. In this sense, consider a set $\Omega \subseteq \Omega_{G_P}$ and let $\Omega^\omega = \Omega \setminus \omega$ for $\omega \in \Omega$. We look at what happens to ω when the set Ω^ω is pointwise stabilized, i.e., when all of the orbits in Ω^ω are (simultaneously) pointwise stabilized. Similar cases to (a)–(c) may occur and suitable definitions can be stated.

Definition 5. *The orbit ω is dependent of Ω^ω, denoted by $\omega \hookrightarrow \Omega^\omega$, if ω is stabilized when all orbits of Ω^ω are stabilized.*

Definition 6. *The orbit ω is semi-dependent of Ω^ω, denoted by $\omega \rightsquigarrow \Omega^\omega$, if ω splits when all orbits of Ω^ω are stabilized.*

Definition 7. *The orbit ω is independent of Ω^ω, denoted by $\omega \not\hookrightarrow \Omega^\omega$, if ω is not stabilized when all orbits of Ω^ω are stabilized.*

Lemma 5 establishes necessary conditions to have $\omega \not\hookrightarrow \Omega^\omega$. The pointwise stabilizer of a set Ω of orbits is denoted as G^Ω hereafter.

Lemma 5. *If $\omega \not\hookrightarrow \Omega^\omega$, then $\omega \not\rightarrow \theta$ for all $\theta \in \Omega^\omega$.*

Proof. By definition, $\omega \not\hookrightarrow \Omega^\omega$ implies that the action of G^{Ω^ω} on ω is transitive. Since G^{Ω^ω} is a subgroup of G^θ for every $\theta \in \Omega^\omega$, G^θ also acts transitively on ω and thus $\omega \not\rightarrow \theta$. □

Finally we can define an independent set of orbits. We remark that, although we do not state them explicitly, corresponding definitions can be laid down concerning the concepts of dependent and semi-dependent sets of orbits.

Definition 8. *A set Ω is said to be independent if $\omega \not\hookrightarrow \Omega^\omega$ for all $\omega \in \Omega$.*

Corollary 1 provides necessary conditions so as to a set Ω be independent.

Corollary 1. *If the set Ω is independent, then $\omega \not\leftrightarrow \theta$ for all $\omega, \theta \in \Omega$.*

Proof. By Definition 8 and Lemma 5. □

3.4 SBCs from Independent Sets

Let Ω_I denote an independent set of orbits. Similarly to the results presented in [8], the following propositions set appropriate conditions to build weak and strong SBCs, respectively, from independent sets of orbtis.

Proposition 1. *The constraints* (3) *are SBCs for P and $G^{\Omega_I^\omega}$ with respect to $\omega \in \Omega_I$.*

Proof. Let $y \in \mathcal{G}(P)$. Since $G^{\Omega_I^\omega}$ acts transitively on ω, there exists $\pi \in G^{\Omega_I^\omega}$ mapping $\min y[\omega]$ to $y_{\omega(1)}$. □

Proposition 2. *Provided that $G^{\Omega_I^\omega}[\omega] = \mathsf{Sym}(\omega)$, the constraints* (2) *are SBCs for P and $G^{\Omega_I^\omega}$ with respect to $\omega \in \Omega_I$.*

Proof. Let $y \in \mathcal{G}(P)$. Since $G^{\Omega_I^\omega}[\omega] = \mathsf{Sym}(\omega)$, there exists $\pi \in G^{\Omega_I^\omega}$ such that $(\pi y)[\omega]$ is ordered by $\leq$. Therefore πy is feasible w.r.t. the contraints (2). □

4 Orbital Independence Algorithm

In this section we describe the methodology used to find an independent set of orbits of a mathematical program. We present how to model and solve the problem of finding such a set by means of a classical combinatorial optimization problem. Moreover, we describe in details the algorithm proposed to build SBCs from all orbits contained in an independent set.

4.1 Independence Graph

Our interest relies in finding the largest $\Omega_I \subseteq \Omega_{G_P}$. Nevertheless, so far we do not have theoretical results providing sufficient conditions to find such a set. Yet we can use the necessary conditions provided by Corollary 1 and search for the largest set $\Omega_K \subseteq \Omega_{G_p}$ whose elements are pairwise independent. Having obtained Ω_K, we can then search for the largest $\Omega_I \subseteq \Omega_K$.

Hence we propose to find Ω_K by encoding the independence relation between orbits of G_P as an undirected graph $G_I = (V, E)$, as of now called the *independence* graph of P, where $V = \Omega_{G_P}$ and E is the set of pairs of independent orbits in Ω_{G_P}. In this manner we reduce the problem of finding Ω_K to the problem of finding the maximum clique in G_I.

4.2 Orbital Independence Reformulations

We expect that the larger the number of SBCs adjoined to the original formulation, the stronger their computational impact. Particularly, the larger the number of strong SBCs, the better. We thus propose two different reformulations based on the concept of OI: the first prioritizing the total number of SBCs generated and the second prioritizing the total number of strong SBCs generated. In this sense, we look for cliques in G_I that either involve large orbits or involve mostly orbits which may satisfy the conditions to build strong SBCs.

In order to find such cliques, we associate a weight function $d : V \to W$ to $G_I = (V, E, d)$ and solve the Maximum Weight Clique problem (MWCP) for G_I using the MP formulation described in [1]. In the first reformulation, which we call *orbital independence narrowing*, we have $W = \{|\omega_1|, \ldots, |\omega_{|V|}|\}$ and $d(\omega_i) = |\omega_i|$ for all $\omega_i \in V$. In the second, which we call *strong orbital independence narrowing*, $W = \{d_1, d_2\}$. It is worth pointing out that the strong orbital independence narrowing prioritizes cliques having mostly orbits which satisfy $G_P[\omega] = \mathsf{Sym}(\omega)$; this is a necessary condition to have $G^{\Omega_I^\omega}[\omega] = \mathsf{Sym}(\omega)$ since $G^{\Omega_I^\omega}[\omega]$ is a subgroup of $G_P[\omega]$ for every $\omega \in \Omega_I$.

4.3 Algorithm Description

Algorithm 1. Orbital Independence SBC generator

```
Require: nontrivial G_P and reformulation strategy ς
Let C = ∅ and Ω_I = ∅
Let Ω_{G_P} = computeOrbits(G_P)
if |Ω_{G_P}| > 1 then
   for ω ∈ Ω_{G_P} do
      Let G^ω = computePointStab(ω)
      for θ ∈ Ω_{G_P} such that pos(θ) > pos(ω) do
         Let G^θ = computePointStab(θ)
         if isTransitiveAction(G^ω, θ) ∧ isTransitiveAction(G^θ, ω) then
            Let E = E ∪ {{ω, θ}, {θ, ω}}
         end if
      end for
   end for
   if |E| ≥ 2 then
      Let G_I = buildGraph(Ω_{G_P}, E, ς)
      Let Ω_K = Ω_I = solveMWCP(G_I)
      for ω ∈ Ω_K do
         if not isTransitiveAction(G^{Ω_I^ω}, ω) then
            Let Ω_I = Ω_I ∖ ω
         end if
      end for
      for ω ∈ Ω_I do
         Let g(x[ω]) ≤ 0 be some SBCs for P and G^{Ω_I^ω} w.r.t. ω
         Let C = C ∪ {g(x[ω]) ≤ 0}
      end for
   end if
end if
return C
```

The Algorithm 1 generates a set C containing SBCs derived from the largest independent set of orbits of P. It takes as inputs a nontrivial formulation group (parameter G_P) and a reformulation strategy (parameter ς). Some functions simplify the pseudocode of Algorithm 1: $\mathsf{computeOrbits}(G_P)$ returns the orbits of the group G_P; $\mathsf{computePointStab}(\omega)$ returns the pointwise stabilizer of ω; $\mathsf{pos}(\omega)$ returns the position of orbit ω in the list Ω_{G_P}; $\mathsf{isTransitiveAction}(G, \omega)$ returns true if the action of the group G is transitive on the orbit ω and false otherwise; $\mathsf{buildGraph}(V, E, \varsigma)$ returns a graph with vertices V, edges E and weights appropriate to the strategy ς; $\mathsf{solveMWCP}(G_I)$ returns a solution of the MWCP for the graph G_I.

If G_P has more than one orbit ($|\Omega_{G_P}| > 1$), the algorithm first iteratively looks for all pairs of independent orbits in order to build the set E. Because the Condition (3) in Theorem 1 is not sufficient to ascertain whether two orbits $\omega, \theta \in \Omega_{G_P}$ satisfy $\omega \dashv\!\vdash \theta$, ultimately we must check if the action of the stabilizers G^ω and G^θ is transitive on θ and ω, respectively. Thus the algorithm does not compare the sets Γ^ω and Γ^θ but rather directly checks whether the actions are transitive. Following the first loop, if at least one pair of independent orbits is found ($|E| \geq 2$), the algorithm builds the independence graph G_I according to the reformulation strategy ς and calls a third party MILP solver to solve the MWCP for G_I. Once Ω_K is known, the algorithm converges to a set Ω_I by iteratively removing (from a copy of Ω_K stored as Ω_I) the orbits that do not satisfy $\omega \dashv\!\vdash \Omega_I^\omega$. We remark that our approach here is not optimal in the sense that the resulting Ω_I may not be the largest one; evaluating all possible $\Omega_I \subseteq \Omega_k$ would most likely require a large computational effort due to many stabilizer computations. Then, for each orbit in the set Ω_I, the algorithm builds and adds SBCs to the set C. We remark that if $|\Omega_{G_P}| = 1$ (unique orbit) or $|E| = 0$ (no pair of independent orbits in Ω_{G_P}), no reformulation is carried out.

Theorem 2. *The constraint set* $C_{\Omega_I} = \{g(x[\omega_k]) \leq 0 \mid \omega_k \in \Omega_I\}$ *is an SBC system for* P.

Proof. If P is infeasible then adjoining the constraints in C_{Ω_I} to P does not change its infeasibility, so assume P is feasible. Since $g(x[\omega_k]) \leq 0$ are SBCs for P and $G^{\Omega_I^{\omega_k}}$ w.r.t. ω_k, there exist $y \in \mathcal{G}(P)$ and $\pi_{\omega_k} \in G^{\Omega_I^{\omega_k}}$ such that $\pi_{\omega_k} y$ satisfies $g((\pi_{\omega_k} y)[\omega_k]) \leq 0$. But $\pi_{\omega_k} \in G_P$ for all $\omega_k \in \Omega_I$ and, due to the closure of the group operation, there exists $\pi \in G_P$ such that $\pi = \prod \pi_{\omega_k}$. So $\pi y \in \mathcal{G}(P)$. But $\pi[\omega_k] = \pi_{\omega_k}[\omega_k]$ since $\pi_{\omega_{k'}}$ stabilizes ω_k pointwise for every $k' \neq k$ and thus $(\pi y)[\omega_k] = (\pi_{\omega_k} y)[\omega_k]$. Therefore πy satisfies $g((\pi y)[\omega_k]) \leq 0$ for all $\omega_k \in \Omega_I$. □

5 Computational Experiments

In this section we show the computational impact on the resolution of MILPs when adjoining SBCs arising from different orbits simultaneously. We describe the computational environment involved (machinery, solvers, instances) and analyze the results obtained from the conducted experiments.

5.1 Environment

Our test set consists of symmetric MPs taken from the library MIPLIB2010. The reformulations were obtained on a quad-CPU Intel Xeon at 2.66 GHz with

24 Gb RAM. Automatic group detection is carried out using the ROSE reformulator [10] and the Traces software [16]. Other group computations are carried out using GAP v. 4.7.4 [19]. The MP results were obtained on a 24-CPU Intel Xeon at 2.53 GHz with 48 Gb RAM. All problems were solved under the AMPL [3] environment using CPLEX 12.6 [5]. The execution time was limited to 1800 s of user cpu time. In order to try and provide a fair assessment of our methodology, we disabled the symmetry handling methods built into CPLEX. We also ran CPLEX in single thread mode to impose its sequential (and deterministic) behaviour and increase the chances of measuring performance differences.

5.2 Results

We first comment the results regarding the reformulation process. Table 1 reports, per instance, the number of variables (n) and orbits ($|\Omega_{G_P}|$) of the original formulation, and the total number of variables indexed by the orbits Ω_{G_P} (#svar); for each OI narrowing type, the table reports the size of the maximum clique ($|\Omega_K|$), the size of the largest independent set ($|\Omega_I|$), the total number of variables indexed by all of the orbits in Ω_I (#var), and the number of weak (#wea) and strong (#str) SBCs generated.

We would like to remark that both reformulation strategies yielded the same narrowings for the most part of the instances. In these cases, we do not present results regarding the strong orbital independence reformulation. Additionally, we also point out that the size of the maximum cliques is equal to the size of the largest independent sets for all instances.

Apart from the structure of the group G_P, intuitively, the ratio ν =(#svar/n) may also indicate how symmetric a formulation P is. Similarly, the ratios $\rho = (|\Omega_I|/|\Omega_{G_P}|)$ and υ =(#var/#svar) may indicate how extensively we have exploited the symmetries of P. All together, we expect SBCs to make a strong computational impact whenever the triplet (ν, ρ, υ) tends to $(1, 1, 1)$. Table 1 shows that the symmetric instances tested so far have, in general, two low ratios, which suggests that the impact of the SBCs may not be too significative.

Table 2 reports the optimization results. Per instance and for each formulation, the table exhibits the best solution found, the user cpu time (in seconds), the gap (%) and the solver status at termination (opt = optimum found, lim = time limit reached, inf = infeasible instance). Best values are emphasized in boldface. Some instances from Table 1 do not appear in Table 2 because no method performed better than the other.

As expected, we do not observe cases of infeasible narrowings due to the usage of SBCs derived from different orbits simultaneously. Moreover, we also observe consistent improvements in favor of the orbital independence narrowings. In 22 out of 48 instances, the SBCs slightly helped to improve the performance of the solver. On the other hand, in 14 cases the SBCs were harmful and, in 12 other instances, they made no difference at all. Although they provided good results, the few soi-narrowings did not achieve outstanding performances. Interestingly,

Table 1. OI narrowings of symmetric instances from MIPLIB2010.

	Original			Oi-narrowing				
Instance	n	$\lvert\Omega_{G_P}\rvert$	#svar	$\lvert\Omega_K\rvert$	$\lvert\Omega_I\rvert$	#var	#wea	#str
bab5	21600	1936	3872	4	4	8	0	4
blp-ar98	16017	2	4	2	2	4	0	2
blp-ic97	8445	2	4	2	2	4	0	2
core4872-1529	24605	505	1046	46	46	96	0	50
gmu-35-40	842	40	111	4	4	13	0	9
gmu-35-50	1177	40	111	4	4	13	0	9
gmut-75-50	36164	64	242	6	6	19	0	13
gmut-77-40	13140	70	280	7	7	26	0	19
iis-bupa-cov	345	2	7	2	2	7	0	5
lectsched-4-obj	3513	267	557	17	17	36	0	19
map06	46015	107	245	10	10	20	0	10
map10	46015	107	245	10	10	20	0	10
map14	46015	107	245	10	10	20	0	10
map18	46015	107	245	10	10	20	0	10
map20	46015	107	245	10	10	20	0	10
mcsched	1669	45	90	15	15	30	0	15
mzzv11	10240	155	310	16	16	32	0	16
neos-1311124	1092	52	1092	4	4	84	0	80
neos-1426635	520	52	520	4	4	40	0	36
neos-1426662	832	52	832	4	4	64	0	60
neos-1436709	676	52	676	4	4	52	0	48
neos-1440460	468	52	468	4	4	36	0	32
neos-1442119	728	52	728	4	4	56	0	52
neos-1442657	624	52	624	4	4	48	0	44
neos-911880	888	259	888	7	7	24	0	17
neos-952987	31329	37	81	4	4	8	0	4
neos18	963	53	248	5	5	26	0	21
ns1631475	22696	105	210	11	11	22	0	11
ns2081729	661	300	600	3	3	6	0	3
ns2122603	18052	36	72	18	18	36	0	18
p2m2p1m1p0n100	100	25	92	3	3	12	0	9
protfold	1835	558	1800	2	2	4	0	2
rocII-4-11	3409	2	27	2	2	27	0	25
rococoC10-001000	2566	41	82	4	4	8	0	4
rvb-sub	33765	113	226	12	12	24	0	12
satellites1-25	9013	200	400	20	20	40	0	20
seymour-disj-10	1209	49	106	5	5	12	0	7
seymour	1255	55	156	5	5	41	29	7
swath	6404	21	163	2	2	8	0	6
transportmoment	9099	85	189	17	17	38	0	21
uc-case3	36921	2687	5374	2	2	4	0	2
uct-subprob	2236	136	306	7	7	14	0	7

	Original			Oi-narrowing					Soi-narrowing				
Instance	n	$\lvert\Omega_{G_P}\rvert$	#svar	$\lvert\Omega_K\rvert$	$\lvert\Omega_I\rvert$	#var	#wea	#str	$\lvert\Omega_K\rvert$	$\lvert\Omega_I\rvert$	#var	#wea	#str
core2536-691	15288	88	187	12	12	29	3	14	12	12	27	0	15
macrophage	2260	251	566	18	18	42	5	19	18	18	39	0	21
neos-555424	3815	132	3810	8	8	190	107	75	8	8	145	58	79
neos-826841	5516	156	5436	3	3	200	191	6	4	4	46	0	42
neos-849702	1737	128	1737	2	2	36	34	0	2	2	9	0	7
toll-like	2883	386	1091	26	26	91	44	21	26	26	59	0	33

the SBCs were harmful to all instances of the family map#. We shall investigate why this happens in order to get more insights on the impact of SBCs.

Overall, we understand that the results are few and at most reasonable, but they support our motivation and encourage a more extensive experimental evaluation against a larger set of instances that exhibit nontrivial symmetries.

Table 2. MIPLIB2010 results obtained with CPLEX 12.6.

	Original formulation				Oi-narrowing			
Instance	Best	Time (s)	Gap (%)	St.	Best	Time (s)	Gap (%)	St.
bab5	-106412	1800.10	0.16	lim	-106412	849.51	0	**opt**
blp-ar98	6205.21	1293.84	0	**opt**	6205.6	1800.12	0.12	lim
blp-ic97	4032.94	1800.13	0.69	lim	4025.02	1800.08	**0.39**	lim
core4872-1529	1479	1800.10	3.07	lim	1472	1800.11	**2.56**	lim
gmu-35-40	-2406600	52.83	0	opt	-2406600	**52.51**	0	opt
gmut-75-50	-14176700	1800.48	0.03	lim	-14178800	1800.41	**0.01**	lim
gmut-77-40	-14166700	1800.29	0.04	lim	-14167400	1800.30	**0.03**	lim
iis-bupa-cov	36	1800.04	7.90	lim	36	1800.05	**7.51**	lim
lectsched-4-obj	4	9.05	0	opt	4	**8.15**	0	opt
map06	-289	**683.60**	0	opt	-289	808.28	0	opt
map10	-495	**581.88**	0	opt	-495	719.42	0	opt
map14	-674	**642.28**	0	opt	-674	678.25	0	opt
map18	-847	**271.84**	0	opt	-847	323.00	0	opt
map20	-922	**148.02**	0	opt	-922	167.72	0	opt
mcsched	211913	**320.23**	0	opt	211913	361.66	0	opt
mzzv11	-21718	**20.43**	0	opt	-21718	34.13	0	opt
neos-1426635	-176	1800.83	1.14	lim	-176	1800.19	**0.57**	lim
neos-1426662	-44	1800.41	14.74	lim	-44	1800.35	**13.59**	lim
neos-911880	54.76	7.57	0	opt	54.76	**7.05**	0	opt
neos18	13	25.24	0	opt	13	**15.97**	0	opt
ns2081729	9	**394.59**	0	opt	9	812.08	0	opt
protfold	-25	1800.02	46.34	lim	-27	1800.02	**35.14**	lim
rocII-4-11	-5.65564	**389.22**	0	opt	-5.65564	397.36	0	opt
rvb-sub	27.51	1800.36	58.90	lim	27.4683	1800.31	**58.83**	lim
satellites1-25	-5	**195.57**	0	opt	-5	422.41	0	opt
seymour-disj-10	288	1800.04	1.91	lim	288	1800.04	**1.85**	lim
seymour	307	1800.06	**2.00**	lim	307	1800.06	2.06	lim
swath	467.408	1800.34	11.95	lim	467.408	1800.53	**11.27**	lim
transportmoment	∞	2.69	∞	inf	∞	**2.52**	∞	inf
uc-case3	6931.73	1800.39	0.12	lim	6931.39	1800.40	**0.11**	lim
uct-subprob	315	1800.06	**4.87**	lim	317	1800.09	7.29	lim

	Original formulation				Oi-narrowing				Soi-narrowing			
Instance	Best	Time (s)	Gap (%)	St.	Best	Time (s)	Gap (%)	St.	Best	Time (s)	Gap (%)	St.
core2536-691	683	56.47	0	opt	683	64.99	0	opt	683	**50.04**	0	opt
macrophage	374	755.91	0	opt	374	372.96	0	opt	374	**224.31**	0	opt
neos-555424	1286800	**5.71**	0	opt	1286800	6.68	0	opt	1286800	6.52	0	opt
neos-849702	0	717.74	0	opt	0	**8.68**	0	opt	0	89.52	0	opt
toll-like	617	1800.03	21.81	lim	611	1800.04	**20.33**	lim	617	1800.04	20.38	lim

6 Conclusions

In this paper we discussed the notion of orbital independence by presenting theoretical results that establish sufficient conditions to break symmetries from different orbits of MPs concurrently. These conditions allowed us to design an algorithm that efficiently generates SBCs to the largest independent set of orbits of MPs. We evaluated the impact of our methodology by conducting experiments with symmetric instances taken from MIPLIB2010. The results were at most reasonable but encouraging; we aim to extend our computational tests to a larger set of symmetric instances, either taken from public libraries such as MINLPLIB2 or generated so as to contain formulation groups with specific structures.

Acknowledgments. The first author (GD) is financially supported by a CNPq Ph.D. thesis award.

References

1. Bomze, I., Budinich, M., Pardalos, P., Pelillo, M.: The maximum clique problem. In: Du, D.Z., Pardalos, P. (eds.) Handbook of Combinatorial Optimization, pp. 1–74. Kluwer Academic Publishers, Dordrecht (1998)

2. Costa, A., Hansen, P., Liberti, L.: Formulation symmetries in circle packing. In: Mahjoub, R. (ed.) Proceedings of the International Symposium on Combinatorial Optimization. Electronic Notes in Discrete Mathematics, vol. 36, pp. 1303–1310. Elsevier, Amsterdam (2010)
3. Fourer, R., Gay, D.: The AMPL Book. Duxbury Press, Pacific Grove (2002)
4. Galli, S.: Parsing AMPL internal format for linear and non-linear expressions (2004), B.Sc. dissertation, DEI, Politecnico di Milano, Italy
5. IBM: ILOG CPLEX 12.6 User's Manual. IBM (2014)
6. Kaibel, V., Pfetsch, M.: Packing and partitioning orbitopes. Math. Program. **114**(1), 1–36 (2008)
7. Liberti, L.: Reformulations in mathematical programming: definitions and systematics. RAIRO-RO **43**(1), 55–86 (2009)
8. Liberti, L.: Reformulations in mathematical programming: automatic symmetry detection and exploitation. Math. Program. A **131**, 273–304 (2012)
9. Liberti, L.: Symmetry in mathematical programming. In: Lee, J., Leyffer, S. (eds.) Mixed Integer Nonlinear Programming, IMA, vol. 154, pp. 263–286. Springer, New York (2012)
10. Liberti, L., Cafieri, S., Savourey, D.: The reformulation-optimization software engine. In: Fukuda, K., Hoeven, J., Joswig, M., Takayama, N. (eds.) ICMS 2010. LNCS, vol. 6327, pp. 303–314. Springer, Heidelberg (2010)
11. Liberti, L., Ostrowski, J.: Stabilizer-based symmetry breaking constraints for mathematical programs. J. Global Optim. **60**, 183–194 (2014)
12. Margot, F.: Pruning by isomorphism in branch-and-cut. Math. Program. **94**, 71–90 (2002)
13. Margot, F.: Exploiting orbits in symmetric ILP. Math. Program. B **98**, 3–21 (2003)
14. Margot, F.: Symmetry in integer linear programming. In: Jünger, M., Liebling, T., Naddef, D., Nemhauser, G., Pulleyblank, W., Reinelt, G., Rinaldi, G., Wolsey, L. (eds.) 50 Years of Integer Programming, pp. 647–681. Springer, Berlin (2010)
15. McKay, B.: Practical graph isomorphism. Congressus Numerantium **30**, 45–87 (1981)
16. McKay, B., Piperno, A.: Practical graph isomorphism. II. Journal of Symbolic Computation **60**, 94–112 (2014)
17. Ostrowski, J.P., Linderoth, J., Rossi, F., Smriglio, S.: Constraint Orbital Branching. In: Lodi, A., Panconesi, A., Rinaldi, G. (eds.) IPCO 2008. LNCS, vol. 5035, pp. 225–239. Springer, Heidelberg (2008)
18. Ostrowski, J., Linderoth, J., Rossi, F., Smriglio, S.: Orbital branching. Math. Program. **126**, 147–178 (2011)
19. The GAP Group: GAP - Groups, Algorithms and Programming. Version 4.7.4 (2014)

Symbolic Model Checking for Alternating Projection Temporal Logic

Haiyang Wang, Zhenhua Duan(✉), and Cong Tian(✉)

ICTT and ISN Laboratory, Xidian University,
Xi'an 710071, People's Republic of China
{zhhduan,ctian}@mail.xidian.edu.cn

Abstract. This paper presents a symbolic model checking approach for Alternating Projection Temporal Logic (APTL). In our approach, the model of a system to be verified is specified by an Interpreted System IS, and a property of the system is expressed by an APTL formula ϕ. To check whether ϕ is valid on IS or not: first, the system IS is symbolically represented and $\neg\phi$ is transformed into its normal form. Then, the set $Sat(\neg\phi)$, containing all the states from which there exists at least one computation such that $\neg\phi$ holds, is computed. Finally, whether the property is valid on the system is equivalently evaluated by checking the emptiness of the intersection of the set of initial states in the system and $Sat(\neg\phi)$. Supporting tool is also developed to show how the proposed approach works in practice.

Keywords: Alternating projection temporal logic · Interpreted system · Symbolic model checking · OBDD · Verification

1 Introduction

In the past three decades, model checking [1–3] approach has been intensively developed. At the very time of its introduction in the early 1980's, the prevailing paradigm for verification was a manual theoretic proof or reasoning, using formal axioms and inference rules towards sequential programs. With the development of computer programs and systems, model checking has pervaded into concurrent programs. Model checking is now widely used in many areas such as control systems, resource schedulers, security protocols, auctions and election mechanisms, etc. With model checking, the system is modeled as a state-transition structure and the desired properties are specified in temporal logic formulas [4,5]. Temporal logics provide key inspiration for model checking. In 1977, Pnueli introduced Linear Temporal Logic (LTL) [2] which is a linear logic to specify and verify reactive systems. The universal quantification over all computations is implicit in the LTL semantics. In 1980, Clarke and Emerson introduced Computation Tree Logic (CTL) [2], which is a branching temporal logic and allows the expression of properties of some or all computations of a system. Interval Temporal

This research is supported by the NSFC Grant Nos. 61133001, 61272117, 61322202, 61420106004, and 91418201.

Z. Lu et al. (Eds.): COCOA 2015, LNCS 9486, pp. 481–495, 2015.
DOI: 10.1007/978-3-319-26626-8_35

Logic (ITL) [6] is also a useful formalism for specifying and verifying concurrent systems. Projection Temporal Logic (PTL) [7] is an extention of ITL. Within ITL, $(P; Q)$ holds over an interval if and only if either the interval can be splitted into two parts, P holds on the first part, Q holds on the second part (the interval can be finite or infinite), or P holds over the interval and the interval is infinite; while, within PTL $(P; Q)$ is true only for the first case. In Propositional PTL (PPTL), chop and projection constructs are useful for specifying properties of sequential and iterative behaviors, respectively.

Recently, alternative approaches have been involved and extended to logics such as Multi-agent system (MAS) logics which make them possible to verify a range of MAS against temporal logics and modalities. For example, Alur et al. introduces Alternating-time Temporal Logic (ATL) [8], which offers selective quantification over those paths that are possible out-comes of games. Alternating Interval Temporal Logic (AITL) [9–11] and Alternating Projection Temporal Logic (APTL) [9,12,13] are the extensions of Propositional ITL and PPTL, respectively. AITL and APTL are also logics indispensable for the specification and verification of MAS.

Model checkers for MAS based on OBDD have also been implemented, such as Mocha [14] on CTL and ATL, MCMAS [15] on CTL, ATL and epistemic logics. Model checking can easily lead to state space explosion when the number of agents is large. Symbolic model checking [16,17] can some what conquer the state space explosion problem. Therefore, we investigate a symbolic model checking approach for APTL. In our approach, the system to be verified is modeled as an interpreted system [18,19] $IS = <(L_i, Act_i, P_i, t_i)_{i\in\Sigma\cup\{E\}}, S_0, h>$, while a property of the system is specified by an APTL formula ϕ. Firstly, $\neg\phi$ is transformed to its normal form, and then the sets of states in IS satisfying the sub-formulas of $\neg\phi$ can be recursively computed according to the evolution function of the IS. Thus, whether ϕ is satisfied on the interpreted system IS equals to whether the intersection of $Sat(\neg\phi)$ and S_0 is empty. Since the model IS is represented symbolically by boolean functions, the checking procedure is implemented as an efficient graph algorithm operated on OBDDs [20]. We have develop an APTL model checker named MCMAS_APTL and show how it works with a case study.

The rest of the paper is organized as follows. The next section introduces the syntax, semantics and normal forms of APTL formulas. The definition and the symbolic representation of interpreted systems are given in Sect. 3. In Sect. 4, the symbolic model checking algorithms for APTL are studied and the supporting tool is presented with a case study to show how it works. Finally, conclusions are drawn in Sect. 5.

2 Alternating Projection Temporal Logic

In this section, the syntax and semantics of APTL are presented. The normal forms of APTL are defined and an example to illustrate how an APTL formula is translated to its normal form is given.

2.1 APTL Syntax

Let $\mathcal{P}$ be a finite set of atomic propositions and $\mathcal{A}$ a finite set of agents. The formulas of APTL are given as follows:

$$P ::= p \mid \neg P \mid P \vee Q \mid \bigcirc_{\ll A \gg} P \mid (P_1, \cdots, P_m) prj_{\ll A \gg} Q$$

where $p \in \mathcal{P}$, $A \subseteq \mathcal{A}$, $P_1, \cdots, P_m$, P and Q are well-formed APTL formulas. $\bigcirc_{\ll A \gg}$ (next) and $prj_{\ll A \gg}$ (projection) are basic temporal operators. An APTL formula is called a state formula if it contains no temporal operators, otherwise a temporal formula. The abbreviations *true*, *false*, $\vee$, $\rightarrow$ and $\leftrightarrow$ are defined as usual. Moreover, we have the following derived formulas:

$$\varepsilon \stackrel{\text{def}}{=} \neg \bigcirc_{\ll \emptyset \gg} true \qquad more \stackrel{\text{def}}{=} \bigcirc_{\ll \emptyset \gg} true$$

$$\bigcirc^0_{\ll A \gg} P \stackrel{\text{def}}{=} P \qquad \bigcirc^n_{\ll A \gg} P \stackrel{\text{def}}{=} \bigcirc_{\ll A \gg}(\bigcirc^{n-1}_{\ll A \gg} P)$$

$$len(n) \stackrel{\text{def}}{=} \bigcirc^n_{\ll \emptyset \gg} \varepsilon \qquad skip \stackrel{\text{def}}{=} len(1)$$

$$\Diamond_{\ll A \gg} P \stackrel{\text{def}}{=} true ;_{\ll A \gg} P \qquad \odot_{\ll A \gg} P \stackrel{\text{def}}{=} \varepsilon \vee \bigcirc_{\ll A \gg} P$$

$$fin(P) \stackrel{\text{def}}{=} \Box_{\ll \emptyset \gg}(\varepsilon \rightarrow P) \qquad keep(P) \stackrel{\text{def}}{=} \Box_{\ll \emptyset \gg}(\neg\varepsilon \rightarrow P)$$

$$halt(P) \stackrel{\text{def}}{=} \Box_{\ll \emptyset \gg}(\varepsilon \leftrightarrow P) \qquad P ;_{\ll A \gg} Q \stackrel{\text{def}}{=} (P, Q) prj_{\ll A \gg} \varepsilon$$

where $\odot_{\ll A \gg}$ (weak next), $\Box_{\ll A \gg}$ (always), $;_{\ll A \gg}$ (chop), $\Diamond_{\ll A \gg}$ (sometimes) are derived temporal operators; ε (*empty*) denotes an interval with zero length, *more* represents that the current state is not the final one over a finite computation, $halt(p)$ holds over a finite computation if and only if p is true at the final state, $fin(p)$ holds as long as p is true at the final state, and $keep(p)$ holds over a finite computation as long as p holds at all states ignoring the last one.

2.2 APTL Semantics

Traditionally, the sematics of APTL formulas are given in terms of Concurrent Game Structures (CGSs) [8,9]. A CGS is a tuple $C = <\mathcal{P}, \mathcal{A}, S, S_0, l, \Delta, \tau>$ where

- $\mathcal{P}$ is a finite nonempty set of atomic propositions;
- $\mathcal{A}$ is a finite set of agents;
- S is a finite nonempty set of states;
- S_0 is a finite nonempty set of initial states;
- $l : S \rightarrow 2^{\mathcal{P}}$ is a labeling function that decorates each state with a subset of the atomic propositions;
- $\Delta^a(s)$ is a nonempty set of possible decisions for an agent $a \in \mathcal{A}$ at state s; $\Delta^A(s) = \Delta^{a_1}(s) \times \ldots \times \Delta^{a_k}(s)$ is a nonempty set of decision vectors for the set of agents $A = \{a_1, \ldots, a_k\} \in 2^{\mathcal{A}}$ at state s; accordingly, $\Delta^{\mathcal{A}}(s)$ is simplified as $\Delta(s)$ and denotes the decisions of all agents in $\mathcal{A}$; and for a decision $d \in \Delta(s)$, d_a denotes the decision of agent a within d, and d_A denotes the decision of the set of agents $A \subseteq \mathcal{A}$ within d;

- For each state $s \in S$, $d \in \Delta(s)$, $\tau(s, d)$ maps s and a decision d of the agents in $\mathcal{A}$ to a new state in S. Note that in a CGS, for a state s , each transition is made by a decision $d \in \Delta(s)$ of all agents in $\mathcal{A}$. In some cases, if we just concern with the decisions of $A \subseteq \mathcal{A}$ without caring about the ones of other agents, notation d_A is used. Particularly, if A is a singleton, d_a is adopted.

A computation $\lambda = s_0, s_1, \ldots$ is a nonempty sequence of states, which can be finite or infinite. The length $|\lambda|$ of λ is ω if λ is infinite, and the number of states minus 1 if λ is finite. To have a uniform notation for both finite and infinite intervals, we use extended integers as indices. That is, we consider the set N_0 of non-negative integers and ω , $N_\omega = N_0 \cup \omega$, and extend the comparison operators, $=, <, \leq$, to N_ω by considering $\omega = \omega$, and for arbitrary $i \in N_0$, $i < \omega$. Moreover, we define $\preceq$ as $\leq - \{(\omega, \omega)\}$. Let Γ denote the set of all computations. For any computation $\lambda \in \Gamma$ and $0 \leq i \leq j \preceq |\lambda|$, we use $\lambda[i]$, $\lambda[0, i]$, $\lambda[i, |\lambda|]$ and $\lambda[i, j]$ to denote the i-th state in λ, the finite prefix $s_0, s_1, \ldots, s_i$ of λ, the suffix $s_i, s_{i+1}, \ldots$ of λ, and an interval $s_i, \ldots, s_j$ of λ respectively. A strategy for an agent $a \in \mathcal{A}$ is a function f_a that maps a nonempty finite state sequence $\lambda \in S^+$ to a state in S by $f_a(\lambda) = \tau(s, d_a)$ if the last state s in λ is not a dead state. Similarly, given a set $A \subseteq \mathcal{A}$ of agents, a strategy for the agents in A is a function f_A that maps a nonempty finite state sequence $\lambda \in S^+$ to a state in S by $f_A(\lambda) = \tau(s, d_A)$ if the last state s in λ is not a dead state.

Following the definition of CGS, we define a state s over $\mathcal{P}$ to be a mapping from $\mathcal{P}$ to $B = \{true, false\}$, $s : \mathcal{P} \to B$. A computation $\lambda(s)$ starting from a state s in a CGS satisfies the APTL formula P, denoted by $\lambda(s) \models P$. A CGS C satisfies an APTL formula P iff all of the computations starting from initial states of the CGS satisfy the APTL formula P, denoted by $C \models P$.

Let $\lambda = s_0, s_1, \ldots$ be a computation, and $r_1, \ldots, r_k$ be integers ($h \geq 1$) such that $0 = r_1 \leq \ldots \leq r_h \preceq |\lambda|$. The projection of λ onto $r_1, \ldots, r_h$ is the computation, $\lambda \downarrow (r_1, \ldots, r_h) = s_{t_1}, s_{t_2}, \ldots, s_{t_l}$ where $t_1, \ldots, t_l$ are obtained from $r_1, \ldots, r_h$ by deleting all duplicates. $t_1, \ldots, t_l$ is the longest strictly increasing subsequence of $r_1, \ldots, r_h$. For example, $s_0, s_1, s_2, s_3, s_4 \downarrow (0, 0, 2, 2, 2, 3) = s_0, s_2, s_3$.

The relation $\models$ is inductively defined as follows:

- $\lambda(s) \models p$ for propositions $p \in \mathcal{P}$, iff $p \in l(s)$.
- $\lambda(s) \models \neg P$, iff $\lambda(s) \not\models P$.
- $\lambda(s) \models P \vee Q$, iff $\lambda(s) \models P$ or $\lambda(s) \models Q$.
- $\lambda(s) \models \bigcirc_{\ll A \gg} P$ iff $|\lambda(s)| \geq 2$, and there exists a strategy f_A for the agents in A, such that $\lambda(s) \in out(s, f_A)$, and $\lambda(s)[1, |\lambda|] \models P$.
- $\lambda(s) \models (P_1, \ldots, P_m) prj_{\ll A \gg} Q$ iff there exists a strategy f_A for the agents in A, and $\lambda(s) \in out(s, f_A)$, and integers $0 = r_0 \leq r_1 \leq \ldots \leq r_m \leq |\lambda(s)|$ such that $\lambda(s)[r_{i-1}, r_i] \models P_i$, $0 < i \leq m$ and $\lambda \models Q$ for one of the following λ:
(a) $r_m < |\lambda(s)|$ and $\lambda = \lambda(s) \downarrow (r_0, \ldots, r_m) \cdot \lambda(s)[r_m + 1, \ldots, |\lambda(s)|]$ or
(b) $r_m = |\lambda(s)|$ and $\lambda = \lambda(s) \downarrow (r_0, \ldots, r_m)$ for some $0 \leq h \leq m$.

2.3 Normal Form of APTL

Normal forms are requisite in model checking of APTL formulas and illustrated in the following.

Definition 1 (Normal Form, NF). *Let Q_P be the set of atomic propositions appearing in the APTL formula Q. Normal form of Q can be defined as $Q \equiv Q_e \wedge \varepsilon \vee \bigvee_{i=0}^{n} (Q_{ci} \wedge Q_i)$ where $Q_i \equiv \bigwedge_{j=1}^{r} \bigcirc_{\ll A_{ij} \gg} Q_{ij}$, Q_e is a state formula, $Q_{ci} \equiv \wedge_{k=1}^{l} \dot{q}_k$, $q_k \in Q_p$, $\dot{q}_k$ denotes q_k or $\neg q_k$, and $Q_{ci} \neq Q_{cj}$ if $i \neq j$; each Q_{ij} is an arbitrary APTL formula.*

The normal form of an APTL formula contains two parts: the present part $Q_e \wedge \varepsilon$ indicates that Q_e is satisfied over a computation with a single state s_0, and the future part $Q_{ci} \wedge Q_i$ where $Q_i \equiv \wedge_{j=1}^{r} \bigcirc_{\ll A_{ij} \gg} Q_{ij}$ means that Q_{ci} is satisfied by the current state s'_0 of a computation λ ($|\lambda| > 1$) and Q_{ij} are satisfied over the sub-computation $\lambda[1, |\lambda|]$.

Definition 2 (Complete Normal Form, CNF). *Let Q_P be the set of atomic propositions appearing in an APTL formula Q. The complete normal form of Q is defined as $Q \equiv Q_e \wedge \varepsilon \vee \bigvee_{i=0}^{n} (m_i \wedge M_i)$ where $M_i \equiv \bigwedge_{j=1}^{r} \bigcirc_{\ll A_{ij} \gg} M_{ij}$, Q_e is a state formula, $m_i \equiv \wedge_{k=1}^{l} \dot{q}_k$, $q_k \in Q_p$, $\dot{q}_k$ denotes q_k or $\neg q_k$, $\vee_{i=0}^{n} m_i \equiv true$ and $\bigvee_{i \neq j} (m_i \wedge m_j) \equiv false$, m_i is the min-products of the atomic propositions, and if there are n atomic propositions then we have 2^n min-products, m_0, m_1, ..., m_{2^n-1}; M_{ij} is the max-sums of the atomic propositions, and we also have 2^n max-sums.*

If we have transformed a formula Q to its normal form, then we can further transform it to its complete normal form. If Q is transformed into complete normal form $Q \equiv Q_e \wedge \varepsilon \vee \bigvee_{i=0}^{n} (m_i \wedge M_i)$ then $\neg Q$ can be transformed into its normal form as $\neg Q \equiv \neg Q_e \wedge \varepsilon \vee \bigvee_{i=0}^{n} (m_i \wedge \neg M_i)$.

When we rewrite an APTL formula P into its normal form, we first make preprocesses to eliminate all implications, equivalence, double negations, *skip*, *len*(n), $\Diamond$ in P by replacing all sub-formulas of the form $P_1 \rightarrow P_2$ by $\neg P_1 \vee P_2$, $P_1 \leftrightarrow P_2$ by $\neg P_1 \wedge \neg P_2 \vee P_1 \wedge P_2$, $\neg\neg P_1$ by P_1, *skip* by $\bigcirc_{\ll \emptyset \gg} \varepsilon$, *len*($n$) by $\bigcirc^n_{\ll \emptyset \gg} \varepsilon$ and $\Diamond_{\ll A \gg} P$ by $true;_{\ll A \gg} P$.

We propose the algorithms for transforming APTL formulas into normal forms and implement them. The following example is used to illustrate how an APTL formula can be transformed to its normal form.

Example 1. Transform formula $\Box_{\ll A_1 \gg} p \wedge \Diamond_{\ll A_2 \gg} q;_{\ll A_3 \gg} \Box_{\ll A_4 \gg} r$ to its normal form.

$$\begin{aligned} &\text{NF}(\Box_{\ll A_1 \gg} p \wedge \Diamond_{\ll A_2 \gg} q;_{\ll A_3 \gg} \Box_{\ll A_4 \gg} r) \\ &\equiv \Box_{\ll A_1 \gg} p \wedge (true;_{\ll A_2 \gg} q);_{\ll A_3 \gg} \Box_{\ll A_4 \gg} r \end{aligned}$$

$$\equiv (p \wedge \varepsilon \vee p \wedge \bigcirc_{\ll A_1 \gg} \Box_{\ll A_1 \gg} p) \wedge (true;_{\ll A_2 \gg} q);_{\ll A_3 \gg} \Box_{\ll A_4 \gg} r$$
$$\equiv (p \wedge \varepsilon \vee p \wedge \bigcirc_{\ll A_1 \gg} \Box_{\ll A_1 \gg} p) \wedge (\varepsilon \vee \bigcirc_{\ll \emptyset \gg} true;_{\ll A_2 \gg} q);$$
$$_{\ll A_3 \gg} \Box_{\ll A_4 \gg} r$$
$$\equiv (p \wedge \varepsilon \vee p \wedge \bigcirc_{\ll A_1 \gg} \Box_{\ll A_1 \gg} p) \wedge ((\varepsilon;_{\ll A_2 \gg} q)$$
$$\vee (\bigcirc_{\ll \emptyset \gg} true;_{\ll A_2 \gg} q));_{\ll A_3 \gg} \Box_{\ll A_4 \gg} r$$
$$\equiv (p \wedge \varepsilon \vee p \wedge \bigcirc_{\ll A_1 \gg} \Box_{\ll A_1 \gg} p) \wedge (q \wedge \varepsilon \vee q \wedge \bigcirc_{\ll \emptyset \gg} true \vee$$
$$\bigcirc_{\ll \emptyset \gg} (true;_{\ll A_2 \gg} q));_{\ll A_3 \gg} \Box_{\ll A_4 \gg} r$$
$$\equiv p \wedge q \wedge \varepsilon \vee p \wedge q \wedge \bigcirc_{\ll A_1 \gg} \Box_{\ll A_1 \gg} p \wedge \bigcirc_{\ll \emptyset \gg} true \vee p$$
$$\wedge \bigcirc_{\ll \emptyset \gg} (true;_{\ll A_2 \gg} q);_{\ll A_3 \gg} \Box_{\ll A_4 \gg} r$$
$$\equiv p \wedge q \wedge \varepsilon \vee p \wedge q \wedge \bigcirc_{\ll A_1 \gg} \Box_{\ll A_1 \gg} p \vee p \wedge \bigcirc_{\ll A_1 \gg} \Box_{\ll A_1 \gg} p$$
$$\wedge \bigcirc_{\ll \emptyset \gg} (true;_{\ll A_2 \gg} q);_{\ll A_3 \gg} \Box_{\ll A_4 \gg} r$$
$$\equiv (p \wedge q \wedge \varepsilon;_{\ll A_3 \gg} \Box_{\ll A_4 \gg} r) \vee (p \wedge q \wedge \bigcirc_{\ll A_1 \gg} \Box_{\ll A_1 \gg} p;_{\ll A_3 \gg}$$
$$\Box_{\ll A_4 \gg} r) \vee (p \wedge \bigcirc_{\ll A_1 \gg} \Box_{\ll A_1 \gg} p \wedge \bigcirc_{\ll \emptyset \gg} (true;_{\ll A_2 \gg} q);$$
$$_{\ll A_3 \gg} \Box_{\ll A_4 \gg} r$$
$$\equiv p \wedge q \wedge \Box_{\ll A_4 \gg} r \vee p \wedge q \wedge (\bigcirc_{\ll A_1 \gg} \Box_{\ll A_1 \gg} p;_{\ll A_3 \gg} \Box_{\ll A_4 \gg} r)$$
$$\vee p \wedge (\bigcirc_{\ll A_1 \gg} \Box_{\ll A_1 \gg} p \wedge \bigcirc_{\ll \emptyset \gg} (true;_{\ll A_2 \gg} q);_{\ll A_3 \gg} \Box_{\ll A_4 \gg} r)$$
$$\equiv p \wedge q \wedge (r \wedge \varepsilon \vee r \wedge \bigcirc_{\ll A_4 \gg} \Box_{\ll A_4 \gg} r) \vee p \wedge q \wedge \bigcirc_{\ll A_1 \gg} (\Box_{\ll A_1 \gg} p;$$
$$_{\ll A_3 \gg} \Box_{\ll A_4 \gg} r) \vee p \wedge (\bigcirc_{\ll A_1 \gg} \Box_{\ll A_1 \gg} p;_{\ll A_3 \gg} \Box_{\ll A_4 \gg} r) \wedge$$
$$(\bigcirc_{\ll \emptyset \gg} (true;_{\ll A_2 \gg} q);_{\ll A_3 \gg} \Box_{\ll A_4 \gg} r)$$
$$\equiv p \wedge q \wedge r \wedge \varepsilon \vee p \wedge q \wedge r \wedge \bigcirc_{\ll A_4 \gg} \Box_{\ll A_4 \gg} r \vee p \wedge q \wedge \bigcirc_{\ll A_1 \gg}$$
$$(\Box_{\ll A_1 \gg} p;_{\ll A_3 \gg} \Box_{\ll A_4 \gg} r) \vee p \wedge \bigcirc_{\ll A_1 \gg} (\Box_{\ll A_1 \gg} p;$$
$$_{\ll A_3 \gg} \Box_{\ll A_4 \gg} r) \wedge \bigcirc_{\ll \emptyset \gg} ((true;_{\ll A_2 \gg} q);_{\ll A_3 \gg} \Box_{\ll A_4 \gg} r)$$

3 Interpreted Systems and Symbolization

In this section, we describe the formalism of interpreted systems to model multi-agent systems, and compare the semantics of interpreted systems with CGSs. The symbolic representation of interpreted systems is also presented in this section.

3.1 Interpreted Systems

Formally, an interpreted system IS is a tuple $IS = <(L_i, Act_i, P_i, t_i)_{i \in \Sigma \cup \{E\}}, S_0, h>$ where

- Each agent $i (i \in \{1, ..., n\})$ in the system is characterised by a finite set of local states L_i and a finite set of actions Act_i;
- Actions are performed in compliance with a protocol $P_i : L_i \rightarrow 2^{Act_i}$, specifying which actions may be performed at a given state;
- The environment is modelled by means of a special agent E, a set of local states L_E, a set of actions Act_E, and a protocol P_E associated with E;
- A tuple $g = (l_1, ..., l_n, l_E) \in L_1 \times ... \times L_n \times L_E$, where $l_i \in L_i$ for each i and $l_E \in L_E$, is called a global state which gives a description of the system at a particular instant of time;

- The evolution of the agents' local states is described by a function $t_i : L_i \times L_E \times Act_1 \times ... \times Act_n \times Act_E \rightarrow L_i$. It means that a local state for agent i is returned for a "current" local state of agent i, a "current" local state of the environment, and all the agents' actions. For the environment agent E, the evolution of its local states is described by a function $t_E : L_E \times Act_1 \times ... \times Act_n \times Act_E \rightarrow L_E$;
- It is assumed that, at every state, all agents evolve simultaneously. The evolution of the global states of the system may be described by a function $t : G \times Act \rightarrow G$, where $G \subseteq (L_1 \times \cdots \times L_n \times L_E)$ denotes the set of reachable global states, and $Act = Act_1 \times ... \times Act_n \times Act_E$ denotes the set of "joint" actions. Function t is the composition of all the functions t_i, and it is defined by $t(g, a) = g'$ iff $\forall i, t_i(l_i(g), l_E(g), a) = l_i(g')$. Here $l_i(g)$ denotes the local state of agent i in global state g and $a \in Act$;
- $S_0 \subseteq G$: a set of initial global states;
- To complete the description of an interpreted system, a set of atomic propositions AP is introduced together with a valuation relation $h \subseteq AP \times S$.

Interpreted systems and CGSs are closely related. Considering an interpreted system $IS = <(L_i, Act_i, P_i, t_i)_{i \in \Sigma \cup \{E\}}, S_0, h>$ and a CGS $C = <\mathcal{P}, \mathcal{A}, S, S_0, l, \Delta, \tau>$:

- both structures comprise a set of agents Σ and $\mathcal{A}$ respectively, and a set of states, called global states in IS;
- the function Δ, which returns the number of decisions available to a player in a state of a C, intuitively corresponds to the protocols P_i in IS;
- the evolution function τ of C is an "accessibility" relation between states, while the evolution function t of IS is defined in terms of the evolution functions t_i;
- both the valuation functions l of C and h of IS label states with propositions.

Indeed, C assumes that every player has perfect information, i.e., every player is fully aware of the state $s \in S$ at every time. Conversely, in interpreted systems an agent is aware of its private local state only.

Difference between interpreted systems and CGSs lies in their own definitions of strategies. In CGSs, a strategy is defined as a function from sequences of states to an action. In interpreted systems, instead, a strategy for agent i is a function from a local state to an action of agent i. Strategies and protocols in interpreted systems are closely relative because they associate actions to states. Particularly, in deterministic interpreted systems, strategies and protocols are the same mathematical object.

Interpreted systems have been proven a suitable formalism for reasoning about temporal properties of agents.

3.2 Symbolic Representation of System Models

The key idea about symbolic model checking is expressing a problem in a form, where all the objects are represented as boolean functions which can then be efficiently manipulated by ROBDDs. In the following, we review ROBDDs and present the method for the symbolic representation of interpreted systems.

Reduced Ordered Binary Decision Diagrams. A Binary Decision Diagram (BDD) represents a boolean function as a rooted, directed acyclic graph, where internal vertices corresponding to the variables over which the function is defined and terminal vertices being labeled by 0 and 1. For example, Fig. 1 represents the BDD of the boolean function $f(a, b, c) = a \vee (b \wedge \bar{c})$.

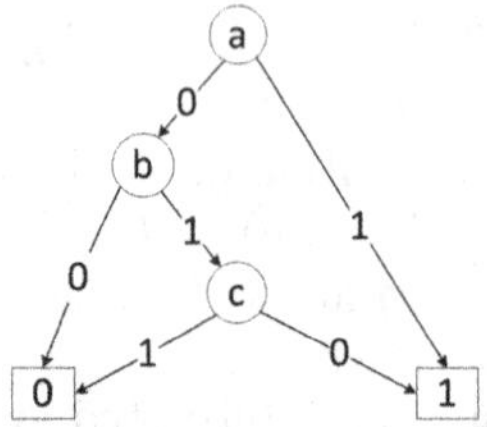

Fig. 1. The BDD of boolean function $f(a, b, c) = a \vee (b \wedge \bar{c})$

A boolean function can be represented by many different BDDs. To ensure that each boolean function has a unique representation, two restrictions must be placed on the form of BDDs: (1) A total ordering is given for the variables in the boolean function. For instance, the variable ordering in Fig. 1 is $a < b < c$. (2) There should be no isomorphic subtrees or redundant vertices in the BDD. This can be achieved by repeatedly applying three transformation rules in [20].

Symbolic Representation of Interpreted Systems. We express how to represent an interpreted system by boolean functions in detail. Given an interpreted system $IS = <(L_i, Act_i, P_i, t_i)_{i \in \Sigma \cup \{E\}}, S_0, h>$:

- The number of boolean variables $nv(i)(i \in \mathcal{N})$ required to encode the local states of an agent i is $nv(i) = \lceil log_2 |L_i| \rceil$. A global state g can be encoded as a boolean vector $\bar{v} = (v_1, ..., v_N)$, where $N = \sum_i nv(i)$;
- To encode an agent's action, the number $na(i)$ of boolean variables w_i required is $na(i) = \lceil log_2 |Act_i| \rceil$. A joint action a can be encoded as a boolean vector $\bar{w} = (w_1, ..., w_M)$, where $M = \sum_i na(i)$.
- Protocols can be encoded by implications between boolean formulas representing local states and actions. We use $P(\bar{v}, \bar{w})$ to denote the boolean formula obtained by taking the conjunction of the boolean formulas encoding the protocols for all the agents.
- The evolution functions can be translated into boolean formulas, and we denote $t(\bar{v}, \bar{w}, \bar{v'})$ as a boolean function by taking the conjunction of all the boolean formulas encoding the evolution functions for the agents.
- The set of initial states is easily translated into a boolean function $S_{0\bar{v}}$. So as the evaluation function h.

Boolean function $R_t(g, g')$ represents the temporal transitions. It can be obtained from the evolution functions t_i by quantifying over actions. This quantification can be translated into a propositional formula using a disjunction: $R_{t(\bar{v},\bar{v'})} = \bigvee_{\bar{w} \in Act}(t(\bar{v}, \bar{w}, \bar{v'}) \wedge P(\bar{v}, \bar{w}))$. The formula provides a boolean relation between global states that can be used in the evaluation of temporal operators. The set of reachable states is also needed in the algorithm: the set G of reachable global states can be expressed symbolically by a boolean formula, and it can be computed as the fix-point of the operator $\tau(Q) = (S_{0\bar{v}} \vee \exists(\bar{v'})(R_{t(\bar{v'},\bar{v})} \wedge Q_{\bar{v'}}))$, where Q is the set of states of the system. Intuitively, $\tau(Q)$ computes the sets of states that are reachable from Q in a single step. The fix-point of τ can be computed by iterating from $\tau(\emptyset)$ as usual (see [1]).

4 Symbolic Model Checking for APTL

In this section, symbolic model checking algorithms for APTL are presented and a model checker named MCMAS_APTL is introduced with a case study.

4.1 Symbolic Model Checking Algorithm for APTL

Our approach employs interpreted system $IS = <(L_i, Act_i, P_i, t_i)_{i \in \Sigma \cup \{E\}}, S_0, h>$ to model a system and an APTL formula ϕ to specify the property. The notation $IS, \lambda \models \phi$ means that ϕ holds along a computation λ in IS. We simply write $\lambda \models \phi$ when IS is unambiguous in the context. $IS \models \phi$ iff all of the computations which departing from the initial states in the system satisfy ϕ. Accordingly, we give the definition of $Sat(\phi)$ in the following.

Definition 3. *Let ϕ be an APTL formula, $IS = <(L_i, Act_i, P_i, t_i)_{i \in \Sigma \cup \{E\}}, S_0, h>$ an interpreted system. A computation λ is a nonempty sequence of states of the system IS, which can be finite or infinite. The set $Sat(\phi) \subseteq G$ includes all the states departing from which there exists at least one computation satisfying ϕ:*

$$Sat(\phi) = \{g \in G | \exists \lambda \in Computations(G) \models \phi \text{ and } \lambda[0] = g\}.$$

Theorem 1. *Let $IS = <(L_i, Act_i, P_i, t_i)_{i \in \Sigma \cup \{E\}}, S_0, h>$ be an interpreted system and ϕ be a property formula in APTL. $IS \models \phi$ iff $Sat(\neg\phi) \cap S_0$ is empty.*

Proof. $\Rightarrow$*: If $IS \models \phi$, then $Sat(\neg\phi) \cap S_0$ is empty.*

If $IS \models \phi$, then ϕ is true over all computations in $Computations(S_0)$. Therefore, there exists no computation over which $\neg\phi$ is true. In order to determine the emptiness of the intersection of $Sat(\neg\phi)$ and the set of the initial states S_0, we take two cases about $SubComputations(G)$ into account:

1. *Suppose that for each subcomputation $\lambda' \in SubComputations(G)$, $\lambda' \not\models \neg\phi$. By Definition 3, it is easy to see that $Sat(\neg\phi) = \emptyset$. The consequence $Sat(\neg\phi) \cap S_0 = \emptyset$ is straightforward.*

2. *Suppose that there exists a subcomputation* $\lambda' \in SubComputations(G)$ *such that* $\lambda' \models \neg\phi$. *By Definition 3, it follows that for the starting global state of* λ', $g' \in Sat(\neg\phi)$. *If* $g' \in S_0$ *then* $\lambda' \in Computation(S_0)$ *and* $\lambda' \models \neg\phi$. *This leads to a conflict with our hypothesis. Consequently, we can infer that* $Sat(\neg\phi) \cap S_0 = \emptyset$.

$\Leftarrow$: *If* $Sat(\neg\phi) \cap S_0$ *is empty, then* $IS \models \phi$.

By Definition 3, if $Sat(\neg\phi) \cap S_0 = \emptyset$, *then* $Sat(\neg\phi)$ *can be either be an empty set or a nonempty set involving some state* $g' \in G$ *starting a subcomputation* $\lambda' \in SubComputations(g')$ *such that* $\lambda' \models \neg\phi$.

1. *If* $Sat(\neg\phi) = \emptyset$, *then there exists no subcomputation* $\lambda' \in \neg\phi$. *In other words,* ϕ *holds along all the subcomputations in* $SubComputations(G)$. *For all* $Computations(S_0) \subseteq SubComputations(G)$, *we can conclude that* $IS \models \phi$.
2. *If* $Sat(\neg\phi)$ *is a nonempty set, then for each global state* $g' \in Sat(\neg\phi)$, $g' \notin S_0$. *Thus, there exists no computation* $\lambda \in Computations(S_0)$ *such that* $\lambda \models \neg\phi$. *This means that* ϕ *holds along all computations in* IS.

Let $IS = <(L_i, Act_i, P_i, t_i)_{i \in \Sigma \cup \{E\}}, S_0, h>$ be an interpreted system. Given a set $A \subseteq \Sigma \cup \{E\}$ of agents, a set $G_1 \subseteq G$ and a set $t' \subseteq t$, function PRE returns a set of states. Here G is the global states and t is the evolution function of the global states. The function PRE is defined below:

$$\text{PRE}(A, G_1, t') = \{g \in G_1 | \exists g' \in G_1.t'(g, P_A) = g'\}.$$

We present algorithm CHARFUN for checking whether an APTL formula ϕ is satisfied on an interpreted system $IS = <(L_i, Act_i, P_i, t_i)_{i \in \Sigma \cup \{E\}}, S_0, h>$ or not: Firstly, we invoke CHARFUN to calculate the characteristic function $Sat_{\bar{v}}(\neg\phi)$ for $Sat(\neg\phi)$. Secondly, if $Sat_{\bar{v}}(\neg\phi) \cdot S_{0\bar{v}}$ equals to 0, it follows that the intersection of $Sat(\neg\phi)$ and S_0 is empty and there exists no computation $\lambda \in Computations(g_0)(g_0 \in S_0)$ such that $\lambda \models \neg\phi$. Conversely, ϕ is true over all computations in IS. Further, if $Sat_{\bar{v}}(\neg\phi) \cdot S_{0\bar{v}} \neq 0$, then we can always find a computation λ_{ce} in IS starting from some state $g \in Sat(\neg\phi) \cap S_0$ such that $\lambda_{ce} \models \neg\phi$. The counterexample λ_{ce} is returned when a bug is found.

The pseudo code of algorithm CHARFUN is demonstrated in Table 1. Note that, "+" and "·" are logic operators "*OR*" and "*AND*" respectively. $Sat_{\bar{v}}(\phi)$ is the boolean function of $Sat(\phi)$. $NF(\phi)$ is the procedure for constructing the normal form of ϕ. $\text{PRE}(A, Sat_{\bar{v}}(\varphi), R_{t(\bar{v},\bar{v'})})$ is used to compute the preimage of $Sat_{\bar{v}}(\varphi)$ based on the transition function $R_{t(\bar{v},\bar{v'})}$ and the agents set A. $\text{FIXPOINT}(\mathfrak{c}(Sat_{\bar{v}}(\phi_{ir_e}), R_{t(\bar{v},\bar{v'})}))$ calculates the fixed point of $\mathfrak{c}(Sat_{\bar{v}}(\phi_{ir_e}), R_{t(\bar{v},\bar{v'})})$ by invoking the algorithm shown in Table 2.

We develop the algorithm CHARFUN by first rewriting ϕ into normal form NF_ϕ, and then considering each disjunct ψ_i of NF_ϕ:

1. If $\psi_i \equiv \phi_e \wedge \varepsilon$, according to the definition of normal form, ϕ_e is a state formula and those states are final states in finite computations.

Table 1. Algorithm for computing the boolean function of $Sat(\phi)$

```
Algorithm CharFun(φ, R_t(v̄,v̄'))
/*precondition: φ is an APTL formula, R_t(v̄,v̄') is the boolean function for transitions
        of the system IS*/
/*postcondition: CharFun returns Sat_v̄(φ)*/
mark[φ] = 0, Sat_v̄(φ)=0;
NF_φ=NF(φ)= ∨_{i=1}^n ψ_i; /* ψ_i ≡ φ_e ∧ ε or ψ_i ≡ φ_i' ∧ ∧_{j=1}^m ○≪A_i'j≫φ_i'j */
mark[φ_i'j] = 0;
for i = 1 to n
    case
     ψ_i ≡ φ_e ∧ ε: Sat_v̄(ψ_i) = Sat_v̄(φ_e)·Pre(∅, Sat_v̄(ε), R_t(v̄,v̄'));
     ψ_i ≡ φ_i ∧ ∧_{j=1}^m ○≪A_ij≫φ_ij and for some formula φ_ik,
          1 ≤ k ≤ m, mark[φ_ik] == 0 and for some formula φ_ir,
          1 ≤ r ≤ m, mark[φ_ir] == 1:
       for all φ_ik make mark[φ_ik] = 1;
       Sat_v̄(ψ_i) = Sat_v̄(φ_i) · ∏_k(Pre(A_ik,CharFun(φ_ik, R_t(v̄,v̄'))), R_t(v̄,v̄'))·
       ∏_r(Pre(A_ir,FixPoint(c(Sat_v̄(φ_ir_e), R_t(v̄,v̄'))), R_t(v̄,v̄')));
       c(X, R) is computed with the update value of X.
           /*φ_ir ≡ φ_ir_e ∧ ∧_{x=1}^y ○≪A_irex≫φ_irex */
    end case
end for
Sat_v̄(φ) = Sat_v̄(ψ_1) + Sat_v̄(ψ_2) + ... + Sat_v̄(ψ_n);
return Sat_v̄(φ)
```

2. If $\psi_i \equiv \phi_i \wedge \bigwedge_{j=1}^{m} \bigcirc\!\ll\! A_{ij} \!\gg\! \phi_{ij}$, we first compute $Sat_{\bar{v}}(\phi_i)$ with CharFun. Then for the sub-formulas $\phi_{ik}(1 \le k \le m)$ which never appear during the transformation of ϕ, we calculate $\text{Pre}(A_{ik}, \text{CharFun}(\phi_{ik}, R_{t(\bar{v},\bar{v}')}), R_{t(\bar{v},\bar{v}')})$. Moreover, for the sub-formulas $\phi_{ir}(1 \le r \le m)$ which have been encountered during transforming ϕ into its normal form, the procedure of computing $Sat_{\bar{v}}(\phi_{ir})$ becomes more complicated since the involvement of infinite computation. This property can easily be characterized by fixpoint theory. We can obtain $Sat_{\bar{v}}(\phi_{ir})$ by $\text{FixPoint}(\mathfrak{c}(Sat_{\bar{v}}(\phi_{ir_e}), R_{t(\bar{v},\bar{v}')}))$, and then calculate preimage by $\text{Pre}(A_{ir}, \text{FixPoint}(\mathfrak{c}(Sat_{\bar{v}}(\phi_{ir_e}), R_{t(\bar{v},\bar{v}')})), R_{t(\bar{v},\bar{v}')})$. $Sat_{\bar{v}}(\psi_i)$ can be obtained by exerting logic "*AND*" operation on these boolean functions calculated above.

The algorithm FixPoint shown in Table 2 consists of a main loop which converges when the value of intermediate variable $Sat_{\bar{v}}(old)$ coincides on two successive visits. Initially, we set $Sat_{\bar{v}}(old)$ with $Sat_{\bar{v}}(\phi_e)$ and assign $\mathfrak{c}(Sat_{\bar{v}}(old), R_{t(\bar{v},\bar{v}')})$ to $Sat_{\bar{v}}(new)$ where $\mathfrak{c}(Sat_{\bar{v}}(old), R_{t(\bar{v},\bar{v}')})$ is computed with the update value of $Sat_{\bar{v}}(old)$. This iterating procedure will not terminate until $Sat_{\bar{v}}(old) = \mathfrak{c}(Sat_{\bar{v}}(old), R_{t(\bar{v},\bar{v}')})$. Actually, $Sat_{\bar{v}}(\phi_e)$ is the final value of $Sat_{\bar{v}}(old)$ with $Sat_{\bar{v}}(old) = Sat_{\bar{v}}(new)$ holds.

Finally, we can obtain the boolean function $Sat_{\bar{v}}(\phi)$ by exerting logic "*OR*" operation on all the $Sat_{\bar{v}}(\phi)$, $Sat_{\bar{v}}(\phi) = Sat_{\bar{v}}(\psi_1) + Sat_{\bar{v}}(\psi_2) + \ldots + Sat_{\bar{v}}(\psi_n)$.

Table 2. Algorithm for computing fixed point

Algorithm FIXPOINT($\mathfrak{c}(Sat_{\bar{v}}(\phi_e), R_{t(\bar{v},\bar{v'})})$)
/*precondition: $\mathfrak{c} : \bigcup_{G' \in \mathcal{P}(G)} G'_{\bar{v}} \rightarrow \bigcup_{G' \in \mathcal{P}(G)} G'_{\bar{v}}$*/
/*postcondition: FIXPOINT($\mathfrak{c}(Sat_{\bar{v}}(\phi_e), R_{t(\bar{v},\bar{v'})})$) computes the fix point of $\mathfrak{c}$*/

$Sat_{\bar{v}}(old) = Sat_{\bar{v}}(\phi_e)$, $Sat_{\bar{v}}(new) = \mathfrak{c}(Sat_{\bar{v}}(old), R_{t(\bar{v},\bar{v'})})$;
while($Sat_{\bar{v}}(old) \neq Sat_{\bar{v}}(new)$)
$Sat_{\bar{v}}(old) = Sat_{\bar{v}}(new)$;
$Sat_{\bar{v}}(new) = \mathfrak{c}(Sat_{\bar{v}}(old), R_{t(\bar{v},\bar{v'})})$;
end while
return $Sat_{\bar{v}}(old)$

Table 3. Algorithm for checking whether system IS satisfies APTL formula ϕ or not

Algorithm CHECKAPTL($\phi, R_{t(\bar{v},\bar{v'})}$)
/*precondition: ϕ is an APTL formula, $R_{t(\bar{v},\bar{v'})}$ is the boolean function for transitions of the system IS*/
/*postcondition: CHECKAPTL($\phi, R_{t(\bar{v},\bar{v'})}$) checks whether $IS \models \phi$ or not.*/

$Sat_{\bar{v}}(\neg\phi)$ =CHARFUN($\neg\phi, R_{t(\bar{v},\bar{v'})}$)
if $Sat_{\bar{v}}(\neg\phi) \cdot S_{0\bar{v}} == 0$
return $IS \models \phi$
else
return $IS \not\models \phi$

In the following, we present algorithm CHECKAPTL in Table 3 for checking whether an interpreted system satisfies an APTL formula ϕ. Firstly, we invoke algorithm CHARFUN to calculate the characteristic function $Sat_{\bar{v}}(\neg\phi)$ for $Sat(\neg\phi)$. Secondly, if $Sat_{\bar{v}}(\neg\phi) \cdot S_{0\bar{v}} = 0$, it means that the intersection of $Sat(\neg\phi)$ and S_0 is empty and there exists no computation that satisfies $\neg\phi$, i.e. ϕ is true over all computations in IS. Further, if $Sat_{\bar{v}}(\neg\phi) \cdot S_{0\bar{v}} \neq 0$, then we can find a computation λ_{ce} in IS starting from state $g \in Sat(\neg\phi) \cap S_0$ such that $\lambda_{ce} \models \neg\phi$. The counterexample λ_{ce} is shown when a bug is found.

4.2 Model Checker MCMAS_APTL

We have developed a model checker named MCMAS_APTL based on the proposed algorithms in C++. The structure of the tool is shown in Fig. 2. The tool mainly contains three parts: representing interpreted systems by boolean functions; translating APTL formulas to normal forms; and checking whether APTL formulas hold on the interpreted systems or not. The first part symbolically represent interpreted systems by employing the corresponding part of MCMAS. In the second part, we translate APTL formulas to normal forms since all of the well-formed APTL formulas can be translated into normal forms. In the last part, we check whether an APTL formula is satisfied on an interpreted system based on the algorithms introduced in Sect. 4.1.

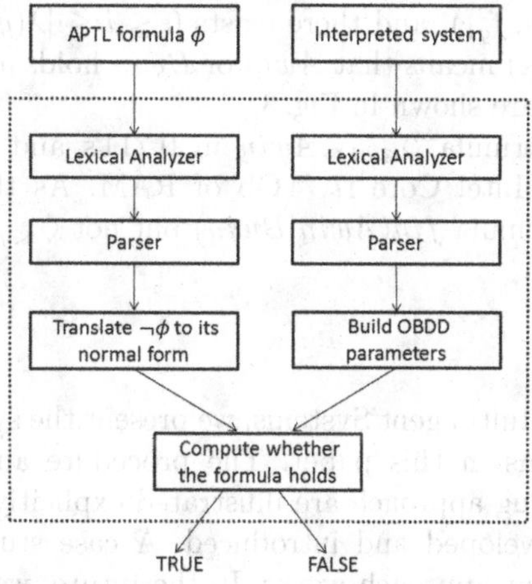

Fig. 2. The structure of MCMAS_APTL.

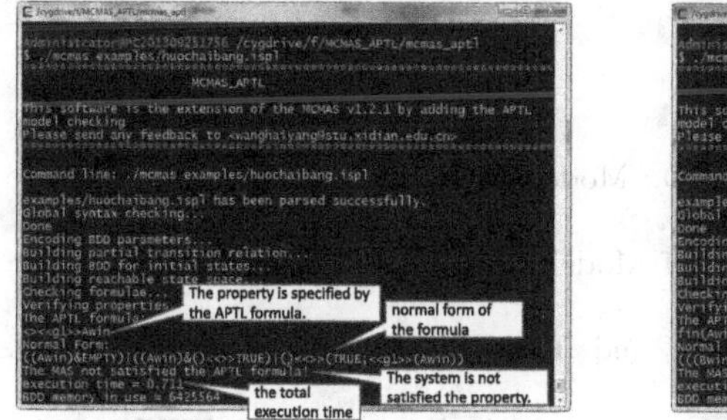

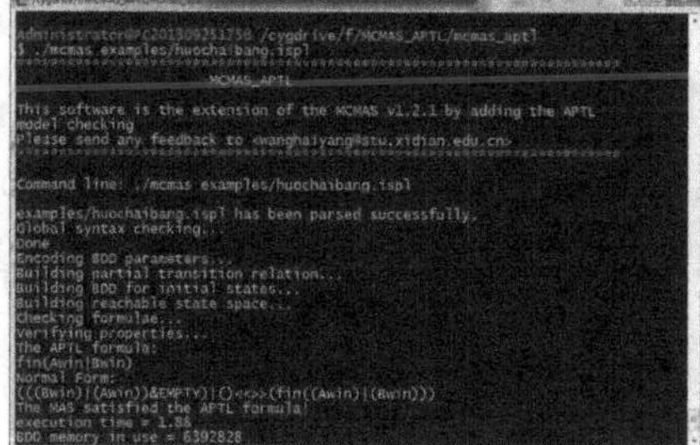

Fig. 3. The result of checking $\Diamond_{\ll g1 \gg} Awin$ and $fin(Awin|Bwin)$.

4.3 A Case Study

We present a match game where two players take out matches from a pile of matches. We assume that there are five matches and both of the players take at least one match in each step. The player who takes the last one loses and the other one wins. We encode the game rule in the formalism of the interpreted systems and impose that only one agent makes a move at each step. The interpreted system has three agents, i.e. *Environment*, A and B. The agent *Environment* contains two variables i and $childA$ respectively representing the number of the remanent matches and the number of matches taken by A in each step. The agent $A(B)$ contains one local variable *state* representing state of agent $A(B)$, and the agent B is similar. The set of atomic properties is $AP = \{Awin, Bwin\}$. $Awin$ represents that A wins and $Bwin$ means B wins. We assume that the system has a group $g1 = \{Environment, A\}$. We give two APTL formulas $\Diamond_{\ll g1 \gg} Awin$ and $fin(Awin|Bwin)$ to verify whether they are satisfied on the system or not. $\Diamond_{\ll g1 \gg} Awin$ means that there exists a strategy f_{g1} for the agents in group $g1$, such that $\lambda(g) \in out(s, f_{g1})$, and there exists $0 \leq i \leq |\lambda(g)|$, and $\lambda(g)[i, |\lambda|] \models Awin$. $fin(Awin|Bwin)$ means that $Awin$ or $Bwin$ holds at the final state. The corresponding results are shown in Fig. 3.

We could verify formula $\Diamond_{\ll g1 \gg} Awin$ in 0.711 s and $fin(Awin|Bwin)$ in 1.88 s on a 2.93 GHZ Intel Core i7, 4 Gb of RAM. As the results show, the system satisfies the formula $fin(Awin|Bwin)$ but not $\Diamond_{\ll g1 \gg} Awin$.

5 Conclusions

To specify and verify Multi-agent Systems, we present the symbolic model checking for APTL formulas in this paper. The procedure and algorithms of the symbolic model checking approach are illustrated explicitly. The model checker MCMAS_APTL is developed and introduced. A case study is also presented to demonstrate how our approach works. In the future, we will further investigate how the proposed approach can be utilized in the verification of real-world multi-agent systems.

References

1. Clarke, E.M., Grumberg, O., Peled, D.: Model Checking. MIT Press, Cambridge (1999)
2. Baier, C., Katoen, J.P.: Principles of Model Checking. MIT Press, Cambridge (2008)
3. Katoen, J.P.: Concepts, Algorithms, and Tools for Model Checking. IMMD, Erlangen (1999)
4. Pnueli, A.: The temporal logic of programs. In: Proceedings of 18th IEEE Symposium Foundation of Computer Science, pp. 46–57 (1977)
5. Wang, H., Xu, Q.: Temporal logics over infinite intervals. Technical report 158, UNU/IIST, Macau (1999)
6. Cau, A., Moszkowski, B., Zedan, H.: Interval Temporal Logic (2006). http://www.cms.dmu.ac.uk/~cau/itlhomepage/itlhomepage.html
7. Duan, Z.: Temporal Logic and Temporal Logic Programming. Science Press, Beijing (2005)
8. Alur, R., Henzinger, T.A., Kupferman, O.: Alternating-time temporal logic. J. ACM **49**, 672–713 (2002)
9. Tian, C., Duan, Z.: Alternating interval based temporal logics. In: Dong, J.S., Zhu, H. (eds.) ICFEM 2010. LNCS, vol. 6447, pp. 694–709. Springer, Heidelberg (2010)
10. Moszkowski, B., Manna, Z., Kozen, D.: Reasoning in interval temporal logic. In: Clarke, E., Kozen, D. (eds.) Logics of Programs. LNCS, vol. 164, pp. 371–382. Springer, Heidelberg (1984)
11. Moszkowski, B.: A complete axiom system for propositional interval temporal logic with infinite time. ArXiv Prepr (2012). ArXiv1207.3816
12. Duan, Z., Tian, C., Zhang, L.: A decision procedure for propositional projection temporal logic with infinite models. Acta Inform. **45**, 43–78 (2008)
13. Duan, Z., Koutny, M., Holt, C.: Projection in temporal logic programming. In: Pfenning, F. (ed.) LPAR 1994. LNCS, vol. 822, pp. 333–344. Springer, Heidelberg (1994)

14. Alur, R., Henzinger, T.A., Mang, F.Y.C., Qadeer, S., Rajamani, R.K., Tasiran, S.: MOCHA: modularity in model checking. In: Hu, A.J., Vardi, M.Y. (eds.) CAV 1998. LNCS, vol. 1427, pp. 521–525. Springer, Heidelberg (1998)
15. Lomuscio, A., Qu, H., Raimondi, F.: MCMAS: a model checker for the verification of multi-agent systems. In: Bouajjani, A., Maler, O. (eds.) CAV 2009. LNCS, vol. 5643, pp. 682–688. Springer, Heidelberg (2009)
16. Burch, J.R., Clarke, E.M., McMillan, K.L., et al.: Symbolic model checking: 10^{20} states and beyond. In: Proceedings of Fifth Annual IEEE Symposium on Logic in Computer Science, LICS 1990, pp. 428–439. IEEE (1990)
17. Pang, T., Duan, Z., Tian, C.: Symbolic model checking for propositional projection temporal logic. In: Sixth International Symposium on Theoretical Aspects of Software Engineering (TASE 2012). IEEE, pp. 9–16 (2012)
18. Lomuscio, A., Raimondi, F.: Model checking knowledge, strategies, and games in multi-agent systems. In: Proceedings of the Fifth International Joint Conference on Autonomous Agents and Multiagent Systems, pp. 161–168. ACM (2006)
19. Halpern, J.Y., Moses, Y., Vardi, M.Y.: Reasoning About Knowledge. MIT Press, Cambridge (1995)
20. Bryant, R.E.: Graph-based algorithms for boolean function manipulation. IEEE Trans. Comput. **100**(8), 677–691 (1986)

Optimization in Graphs

An I/O Efficient Algorithm for Minimum Spanning Trees

Alka Bhushan[1,2] and Gopalan Sajith[1](✉)

[1] Department of Computer Science and Engineering, Indian Institute of Technology Guwahati, Guwahati 781039, India
sajith@iitg.ac.in

[2] GISE Lab, Department of Computer Science and Engineering, Indian Institute of Technology Bombay, Mumbai 400076, India
abhushan@iitb.ac.in

Abstract. An $O(\text{Sort}(E) \cdot \log\log_{E/V} B)$ I/Os algorithm for computing a minimum spanning tree of a graph $G = (V, E)$ is presented, where $\text{Sort}(E) = (E/B)\log_{M/B}(E/B)$, M is the main memory size, and B is the block size. This improves on the previous bound of $O(\text{Sort}(E) \cdot \log\log(VB/E))$ I/Os by Arge et al. for all values of V, E and B, for which I/O optimality is still open. In particular, our algorithm matches the lowerbound $\Omega(E/V \cdot \text{Sort}(V))$, when $E/V \geq B^{\epsilon}$ for a constant $\epsilon > 0$, an $O(\log\log B)$ factor improvement over the algorithm of Arge et al. Our algorithm can compute the connected components too, for the same number of I/Os, which is an improvement on the best known upper bound.

Keywords: External memory algorithms · Minimum spanning trees · Graph algorithms

1 Introduction

The *minimum spanning tree (MST) problem* on an input undirected graph $G = (V, E)$, where each edge is assigned a real-valued weight, is to compute a spanning forest (a spanning tree for each connected component) of G so that the total weight of the edges in the spanning forest is a minimum. In this paper we consider the problem on the I/O model of Aggarwal and Vitter [3]. This model has been used to design algorithms intended to work on large data sets that do not fit in the main memory.

In the I/O model, M is the size of the main memory and B is the size of a disk block. It is assumed that $2B < M < V, E$. In an I/O operation one block of data is transferred between the disk and the internal memory. The measure of performance of an algorithm on this model is the number of I/Os it performs. The number of I/Os needed to read (write) N contiguous items from (to) the disk is $\text{Scan}(N) = \Theta(N/B)$. The number of I/Os required to sort N items is $\text{Sort}(N) = \Theta((N/B) \cdot \log_{M/B}(N/B))$ [3]. For all realistic values of N, B, and M, $\text{Scan}(N) < \text{Sort}(N) \ll N$.

Z. Lu et al. (Eds.): COCOA 2015, LNCS 9486, pp. 499–509, 2015.
DOI: 10.1007/978-3-319-26626-8_36

I/O-efficient graph algorithms have been studied for a host of problems; for a review see [15,17]. For the MST problem, a lower bound of $\Omega(E/V \cdot \mathrm{Sort}(V))$ on the number of I/Os is known [16].

Most I/O efficient MST algorithms are based on Borůvka phases [6]. Each phase selects the lightest edge incident to each vertex v, and outputs as part of the MST. At the end of the phase, the selected edges are contracted; that is, each set of vertices connected by the selected edges is fused into a supervertex. Proofs of the correctness of this approach can be found in [6–8,10,14,16].

(We use set theoretic notations for supervertices. Every vertex of the input graph G is a singleton supervertex. The size $|v|$ of a supervertex v is the number of vertices it contains. For supervertices u and v, we say that u participates in v, if $u \subseteq v$. An edge (u, v) is an internal edge of supervertex w, if $u, v \subseteq w$. Edges (u', v') and (u'', v'') are multiple edges, if there are supervertices u and v such that $u', u'' \subseteq u$ and $v', v'' \subseteq v$.)

At the end of a Borůvka phase, algorithms typically remove the internal edges, and for each set of multiple edges retain only the lightest among them. After the i-th phase, the size of every supervertex is at least 2^i. After $O(\log(V/M))$ phases, $O(M)$ supervertices remain, and these fit in the main memory. The MST can be now computed in one scan of the sorted edge set using the disjoint set data structure and Kruskal's algorithm [11]. As a Borůvka phase takes $O(\mathrm{Sort}(E))$ I/Os [4,7,14,16], this results in an $O(\mathrm{Sort}(E)\log(V/M))$ algorithm. Kumar and Schwabe [14] observe that after $\Theta(\log B)$ phases, with the number of vertices reduced to $O(V/B)$, a contraction phase can be performed more efficiently. They obtain an $O(\mathrm{Sort}(E)\log B + \mathrm{Scan}(E)\log V)$ I/Os algorithm.

Arge et al. [4] present an improved $O(\mathrm{Sort}(E)\log\log(VB/E))$ I/Os algorithm. This is the best known algorithm for MST, and is optimal when $E = \Omega(VB)$. Arge et al. use the fact that after $\Theta(\log(VB/E))$ phases, with number of vertices reduced to E/B, a modified version [4] of Prim's algorithm [11] can be used to construct an MST in the remaining graph. They divide the $\Theta(\log(VB/E))$ phases into $\Theta(\log\log(VB/E))$ *superphases* requiring $O(\mathrm{Sort}(E))$ I/Os each. The i-th superphase consists of $\lceil \log \sqrt{N_i} \rceil$ phases, where $N_i = 2^{(3/2)^i}$. The phases in superphase i work only on a subset E_i of edges. This subset contains the $\lceil \sqrt{N_i} \rceil$ lightest edges incident to each vertex v. These edges are sufficient to perform $\lceil \log \sqrt{N_i} \rceil$ phases as proved in [4,9,16]. Using this subset of edges, each superphase is performed in $O(\mathrm{Sort}(E_i)) = O(\mathrm{Sort}(E))$ I/Os.

The modified Prim's algorithm [4] works as follows: Initialize an external memory priority queue (EMPQ) with the edges of a particular vertex. In each step, add the lightest *border edge* (which connects a vertex that is in the MST to one that is not) to the MST, and add all the edges incident to the newly captured vertex to the EMPQ, except for the edge through which it is captured. The EMPQ stores all the current border edges and some internal edges (i.e., edges between vertices in the MST). A deletemin operation on the EMPQ produces the lightest edge in it; it is an internal edge iff there are two copies of it in the EMPQ; discard it if it is an internal edge. The algorithm performs $\Theta(E)$ EMPQ operations, which take $O(\mathrm{Sort}(E))$ I/Os [1,5,12,14]; it also needs one I/O per vertex. Thus, its I/O complexity is $O(V + \text{Sort}(E))$. When $V = E/B$, this is $O(\text{Sort}(E))$.

If $E = O(V)$ in the input graph $G = (V, E)$, presented as an edgelist, an adjacency list representation of G on its non-isolated vertices can be obtained in $O(\text{Sort}(E))$ I/Os. Then we run Borůvka phases until, for the resultant graph $G' = (V', E')$, $E' = \emptyset$, or $E' = \omega(V')$. The latter is possible, if isolated supervertices are removed from the graph as soon as they form. The first of those phases would run in $O(\text{Sort}(E))$ I/Os [4]. The I/O cost of the subsequent phases will fall geometrically. Thus, the total I/O cost too will be $O(\text{Sort}(E)) = O(E/V \cdot \text{Sort}(V))$.

Therefore, without loss of generality, we assume that $E = \omega(V)$. Let α denote E/V. Then $VB/E = B/\alpha$.

Thus, the problem is open only on sparse graphs with $E = o(VB)$, and $E = \omega(V)$. Computing of connected components (CC) and MSTs are related. The best known upper bound for CC is also $O(\text{Sort}(E) \cdot \log\log(VB/E))$ [16]. However, an optimal randomized algorithm is known for both CC and MST [2,7].

We propose a new MST algorithm that has an I/O complexity of $O(\text{Sort}(E) \cdot \log\log_{E/V} B)$. Our algorithm can compute the connected components too, for the same number of I/Os, which is an improvement on the best known the upper bound [16].

For $\alpha = E/V$, $\log_\alpha B > \log \frac{B}{\alpha}$ iff either $\log\alpha > (\log\sqrt{B})(1+(1-4/\log B)^{0.5})$, or $\log\alpha < (\log\sqrt{B})(1-(1-4/\log B)^{0.5})$. In the former case, for $B > 16$, $\log\alpha > \log\sqrt{B}$, and therefore, our algorithm matches the lowerbound. In the latter case, since $(\log\sqrt{B})(1-(1-4/\log B)^{0.5})$ has a maximum of 2 at $B = 16$, $\log\alpha < 2$, and therefore $E = O(V)$. Hence, our algorithm uses asymptotically fewer I/Os than that of Arge et al. [4] for all values of V, E and B for which I/O optimality is still open. In particular, when $\alpha = B^\epsilon$, for a constant $\epsilon > 0$, our algorithm has an I/O complexity of $O(\text{Sort}(E))$, an $O(\log\log B)$ factor improvement over the algorithm of Arge et al.

The rest of the paper is organized as follows. The high level description of the algorithm is presented in Sect. 2. Section 3 describes a stage. The correctness and I/O complexity of the algorithm are discussed in Sects. 4 and 5.

2 The Stages

The structure of our algorithm is similar to that of Arge et al. [4]: reduce the number of vertices from V to E/B using $\log(VB/E)$ Borůvka phases, and then apply the modified Prim's algorithm to the resultant graph. Our algorithm differs from that of Arge et al. [4] in how we schedule the phases.

Let α denote E/V. Then $VB/E = B/\alpha$.

Our algorithm schedules its $\log(B/\alpha)$ Borůvka phases in a number of stages. The j-th stage, $j \geq 0$, executes $2^j \log\alpha$ phases. The number of supervertices at the start of the j-th stage is at most $V/2^{(2^j-1)\log\alpha} = E/\alpha^{2^j}$. Thus, $\log\log_\alpha B$ stages are required to reduce the number of vertices to E/B. We implement each stage in $O(\text{Sort}(E))$ I/Os, for a total of $O(\text{Sort}(E)\log\log_\alpha B)$ I/Os.

For $j > 0$, let $G_j = (V_j, E_j)$ denote the output of the $(j-1)$-st stage; G_j is also the input to the j-th stage. The input to the 0-th stage, $G_0 = (V_0, E_0)$, is G. Let $g(j) = j + \log\log\alpha$. For $v \in V_j$, let $deg_j(v)$ denote the degree of v in G_j. From E_j we construct $g(j)+2$ buckets: $B_0, \ldots, B_{g(j)+1}$. Each bucket is a set of edges, and is maintained as a sorted array with the composite $\langle$source vertex, edge-weight$\rangle$ as the sort key. In bucket B_k, $0 \leq k \leq g(j)$, we store, for each vertex $v \in V_j$, the $\min\{deg_j(v), 2^{2^k} - 1\}$ lightest edges incident to v. Clearly, $B_k \subseteq B_{k+1}$. Set $B_{g(j)+1} = E_j$.

For $0 \leq k \leq g(j)$, bucket B_k is of size at most $V_j(2^{2^k} - 1)$. The total space used by all the buckets of the j-th stage is, therefore, at most $O(E_j) + \sum_{k=0}^{g(j)} V_j(2^{2^k} - 1) = O(E_j) + O(V_j \alpha^{2^j}) = O(E)$.

In the j-th stage, first we form bucket $B_{g(j)+1}$ by sorting E_j on the composite key $\langle$source vertex, edgeweight$\rangle$. Next we form buckets $B_{g(j)}, \ldots, B_0$ in that order. For $g(j) \geq k \geq 0$, bucket B_k can be formed by scanning B_{k+1} and choosing for each vertex $v \in V_j$, the $\min\{deg_j(v), 2^{2^k} - 1\}$ lightest edges incident to v. Clearly, this involves scanning each bucket twice, once for write and once for read. We do not attempt to align bucket and block boundaries. As soon as the last record of bucket B_{k+1} is written, we start the writing of B_k; but this requires us to start reading from the beginning of B_{k+1}; if we retain a copy of the block that contains the beginning of B_{k+1} in the main memory, we can do this without performing an additional I/O. The total I/O cost of buckets formation is $O(\text{Sort}(E_j))$.

We now define a threshold value $h_k(v)$ for every $v \in V_j$ and bucket B_k as follows: if $deg_j(v) \geq 2^{2^k}$, then let $h_k(v)$ be the weight of the 2^{2^k}-th lightest edge incident to v; otherwise, let $h_k(v)$ be ∞. Note that when $h_k(v)$ is finite, it is the weight of the lightest edge of v not in B_k; $h_k(v) = \infty$ implies that every edge of v is in B_k. We store $h_k(v)$ at the end of v's edgelist in B_k.

After constructing the buckets from E_j, the algorithm performs $2^j \log\alpha$ Borůvka phases. These phases are described in the next section. A phase includes, in addition to the hook and contract operations, the clean up of an appropriate bucket and some bookkeeping.

3 A Phase

In this section, we describe the i-th phase of the j-th stage; $1 \leq i \leq 2^j \log\alpha$. For $i > 1$, let $G_{j,i} = (V_{j,i}, E_{j,i})$ be the graph output by the $(i-1)$-st phase; $G_{j,i}$ is also the input to the i-th phase. The input to the first phase is $G_{j,1} = (V_{j,1}, E_{j,1}) = G_j$. For $i \geq 1$, every $v \in V_{j,i+1}$ is an i-supervertex. Every $v \in V_j$ is a 0-supervertex.

For each $(i-1)$-supervertex v, if v is not overgrown for phase i of stage j (i.e., $|v| < \alpha^{2^j - 1} \cdot 2^{i+1}$), then B_0 contains the lightest external edge of v. (See Sect. 4.) There can be at most one edge of v in B_0, because $2^{2^0} - 1 = 1$.

Let $z(i)$ denote the number of trailing 0's in the binary representation of i.

At the start of the i-th phase, buckets $B_{z(i)}, \ldots, B_1$ are "empty", in that they contain star graphs that represent supervertices formed in some of the past phases. But bucket $B_{z(i)+1}$ is "full", in that it contains the edges of G_j placed in it earlier. (See Sect. 4.)

First we describe the three steps of a phase numbered $i < 2^j \log \alpha = 2^{g(j)}$. The last phase of a stage is special, and will be described later.

Step 1: (Hook & Contract) Let H_i be the graph induced by the edges in $B_{z(i)} \cup \ldots \cup B_1 \cup B_0$. Compute the connected components in H_i and select one representative vertex for each component. These representatives form the set $V_{j,i+1}$. Construct a star graph for each component, with the representative as the root, and the other vertices of the component pointing to it. Let F_i denote the union of these star graphs.

Details of Step 1: Each component of $H_i = (V_h, E_h)$ is a pseudo tree, a connected directed graph with exactly one cycle. Moreover, each cycle of H_i has a size of two.

For each edge (u, v) of H_i, we call v the parent $p(u)$ of u. Concurrently read E_h (which is sorted on source) and a copy C of E_h that is sorted on destination, and for each $(v, w) \in E_h$ and each $(u, v) \in C$, add (u, w) into C'. For each u such that $(u, u) \in C'$ and $u < p(u)$, delete $(u, p(u))$ from H_i and mark u as a root in H_i. Now H_i is a forest. Let H_i' be the underlying undirected tree of H_i. Form an adjacency list representation of H_i' with twin pointers.

Find an Euler tour U of H_i' by simulating the $O(1)$ time Exclusive Read Exclusive Write (EREW) Parallel Random Access Machine (PRAM) algorithm [13] on the external memory model [7]. For each root node r of H_i, delete from U the first edge with r as the source. Now U is a collection of disjoint linked lists. For each element (u, r) without a successor in U, set $\text{rank}(u, r) = r$, and for every other element (u, v), set $\text{rank}(u, v) = 0$. Now invoke list ranking on U. The result gives us the connected components of U, and therefore of H_i'. Each edge and therefore each vertex of H_i' now holds a pointer to the root of the tree to which it belongs.

Each connected component of H_i' forms a supervertex. Its root shall be its representative. Thus, we have a star graph for each supervertex. Therefore, F_i is a list of edges (u, u_r), where u_r is the representative of the supervertex in which u participates.

Remark: Every edge of $B_{z(i)+1}$ is between two vertices of H_i. (See Sect. 4.)

Step 2: (Clean-up) For each i-supervertex v, let $r_i(v)$ denote $\min\{h_{z(i)+1}(x) \mid x \in V_h \text{ and } v \text{ represents the component of } F_i \text{ that contains } x\}$. Clean the edgelists of bucket $B_{z(i)+1}$ of internal and multiple edges. Initialize $X_i = \phi$. For each i-supervertex v, copy from $B_{z(i)+1}$ into X_i all the edges of v with weight less than $r_i(v)$. Store the edges of F_i in $B_{z(i)+1}$.

Details of Step 2: A scan of $B_{z(i)+1}$, and a sort-and-scan of the edges of F_i are enough to find $r_i(v)$ for each i-supervertex v.

Rename the edges of $B_{z(i)+1}$ as follows: sort F_i and $B_{z(i)+1}$ on source, and then read them concurrently; replace each $(u, v) \in B_{z(i)+1}$ with (u_r, v); sort

$B_{z(i)+1}$ on destination; read F_i and $B_{z(i)+1}$ concurrently; replace each $(u, v) \in B_{z(i)+1}$ with (u, v_r); sort $B_{z(i)+1}$ on the composite ⟨source, destination, weight⟩.

Remove the internal edges, which are of the form (u, u), from $B_{z(i)+1}$. For each $(u, v) \in B_{z(i)+1}$, delete all copies of (u, v), except the lightest, from $B_{z(i)+1}$.

Step 3: (Prepare the buckets) For $k = z(i)$ to 0, and for each i-supervertex v, if X_i has at least 2^{2^k} edges with v as the source, then copy into B_k the $(2^{2^k} - 1)$ lightest of them, and set $h_k(v)$ to the weight of the 2^{2^k}-th lightest of them; else copy into B_k all the edges in X_i with v as the source, and set $h_k(v)$ to $r_i(v)$.

Remark: Now the buckets are ready for the next phase. In particular, for each i-supervertex u, if $|u| < \alpha^{2^j - 1} \cdot 2^{i+2}$, then B_0 contains the lightest external edge of u.

The Last Phase. In the $2^{g(j)}$-th phase, execute Step 1 as in a regular phase. Clean the edgelists of bucket $B_{g(j)+1}$ of internal and multiple edges, as in Step 2. This leaves us with a clean graph $G_{j+1} = (V_{j+1}, E_{j+1})$ with which to begin the next stage.

4 A Proof of Correctness

In any stage, the buckets are repeatedly filled and emptied over the phases. We call a bucket "full" when it contains edges, and "empty" when it contains star graphs. The i-th phase (i) uses up the edges in B_0 for hooking, thereby emptying B_0, (ii) fills B_k, for all $k < z(i)+1$, from $B_{z(i)+1}$, and finally (iii) empties $B_{z(i)+1}$. A simple induction, therefore, proves:

Lemma 1. *For $i \geq 1$ and $k \geq 1$, bucket B_k is full at the end of the i-th phase, if and only if the k-th bit (with the least significant bit counted as the first) in the binary representation of i is 0.*

The emptying of $B_{z(i)+1}$ and filling of $B_{z(i)}, \ldots, B_1$ is analogous to summing 1 and $(i - 1)$ using binary representations. $B_{z(i)+1}$ is filled in the $(i - 2^{z(i)})$-th phase, emptied in the i-th phase and filled again in the $(i + 2^{z(i)})$-th phase, and is never accessed in the phases in between. For $1 \leq k \leq g(j)$, B_k is alternately filled and emptied at intervals of 2^{k-1} phases. In particular, B_1 is filled and emptied in alternate phases. B_0 fulfills the role of a buffer that holds the lightest edge incident to every vertex that is not overgrown. It is filled in every step.

The clean up of $B_{z(i)+1}$ that is done in phase i needs the star graphs formed since the last filling of $B_{z(i)+1}$, which was in phase $(i - 2^{z(i)})$. The following proposition is helpful:

Proposition 1. *For $1 \leq i < 2^{g(j)}$, and $i < l \leq i + 2^{z(i)}$, the star graphs formed in phases numbered $i - 2^{z(i)} + 1, \ldots, i$ are present in bucket $B_{z(i)+1}$ at the start of the l-th phase.*

Proof: This can be proved by induction as follows. The case for $i = 1$ forms the basis: the first phase uses up B_0, fills it from B_1, empties B_1 and then stores the newly found star graphs in B_1. Now consider $i > 1$. Inductively hypothesize that the lemma is true for all smaller values of i.

If $z(i) = 0$, then $l = i+1$. The i-th phase uses up B_0, fills it from B_1, empties B_1 and then stores the newly found star graphs in B_1.

Suppose $z(i) > 0$. For $0 \le q < z(i)$, let $r = i - 2^q$; then $z(r) = q$ and $r \le i - 1 < r + 2^{z(r)}$; therefore, by the hypothesis, the star graphs formed in phases numbered $i - 2^{q+1} + 1, \ldots, i - 2^q$ are present in bucket B_{q+1} at the start of the i-th phase. That is, the star graphs formed since the last time $B_{z(i)+1}$ was filled (which was in the $(i - 2^{z(i)})$-th phase) till the start of the i-th phase are available in buckets $B_{z(i)}$ through B_1. These, along with the MST edges found in the i-th phase, form graph H_i (Step 1, Sect. 3), and summarise all the hooks done in phases $(i - 2^{z(i)} + 1)$ through i. At the end of the i-th phase, F_i (the set of star graphs obtained from H_i) is stored in $B_{z(i)+1}$, which is not accessed again till the $(i + 2^{z(i)})$-th phase. Hence the induction holds. □

Corollary 1. *For $1 < i \le 2^{g(j)}$, at the start of the i-th phase, (i) for $0 \le q < z(i)$, the star graphs formed in phases numbered $i - 2^{q+1} + 1, \ldots, i - 2^q$ are present in bucket B_{q+1}, (ii) $B_{z(i)} \cup \ldots \cup B_1 \cup S$ summarises all the hooks performed in phases $i - 2^{z(i)} + 1, \ldots, i$, and (iii) every edge of $B_{z(i)+1}$ is between two vertices of H_i, which is a graph on $(i - 2^{z(i)})$-supervertices.*

Definition: We say that a set P of edges is a minset of supervertex v, if for any two external edges e_1 and e_2 of v in G_j with $\text{wt}(e_1) < \text{wt}(e_2)$, $e_2 \in P$ implies that $e_1 \in P$.

Lemma 2. *For $0 \le i < 2^{g(j)}$, $1 \le k \le z(i)$, and for every i-supervertex v, at the start of the $(i+1)$-st phase, B_k is a minset of v, and $h_k(v)$ is a lower bound on the weight of the lightest external edge of v not in B_k.*

Proof: The proof is by induction. The case of $i = 0$ forms the basis. Hypothesize that for every $(i - 2^{z(i)})$-supervertex x, at the start of the $(i - 2^{z(i)} + 1)$-st phase, $B_{z(i)+1}$ is a minset of x, and $h_{z(i)+1}(x)$ is a lower bound on the weight of the lightest external edge of x not in $B_{z(i)+1}$.

We claim that for each i-supervertex v, X_i forms a minset of v. Suppose it does not. Then, among the edges of G_j there must exist an external edge e of v such that $\text{wt}(e) < r_i(v)$ and $e \notin X_i$. Then e is an external edge of some $(i - 2^{z(i)})$-supervertex y that participates in v. Clearly, $\text{wt}(e) < r_i(v) \le h_{z(i)+1}(y)$. Of the external edges of y in $B_{z(i)+1}$, exactly those of weight less than $r_i(v)$ are copied into X_i. Therefore, $e \notin B_{z(i)+1}$. That is, $B_{z(i)+1}$ is not a minset of y. Contradiction. Hence the claim.

For each i-supervertex v, since X_i is a minset of v, each B_k constructed out of X_i in Step 3 is a minset of v too. Also, $h_k(v)$ is a lower bound on the weight of the lightest external edge of v not in B_k. □

Definition: For a t-supervertex v and an $(i - 2^{z(i)})$-supervertex x, where $t \in [i - 2^{z(i)}, i + 2^{z(i)}]$, x is an $(i - 2^{z(i)})$-seed of v if x has the smallest $h_{z(i)+1}$ threshold among all the $(i - 2^{z(i)})$-supervertices that participate in v.

Note that $r_i(v) = h_{z(i)+1}(x)$, where x is an $(i - 2^{z(i)})$-seed of v.

Next we want to show that an i-supervertex v that does not have an external edge in B_k at the start of the $(i + 1)$-st phase has grown to a size of at least $\alpha^{2^j-1} \cdot 2^{f(k,i)}$, where $f(k, i) = i + 2^k$. First we prove this for buckets filled at the beginning of the stage.

For $s \geq 0$, let t_s denote $\lfloor \frac{t}{2^s} \rfloor 2^s$. We defined $z(i)$ as the number of trailing 0's in the binary representation of i, for $i \geq 1$. Let $z(0)$ be defined as $g(j)$. A supervertex v is fully grown, if it has no external edge in G_j.

Lemma 3. *For* $0 \leq k \leq z(0) = g(j)$, $0 \leq t < 2^k$, *for every* t*-supervertex* v, *if* v *is not fully grown and does not have an external edge in* B_k *at the start of the* $(t + 1)$*-st phase, then* $|v| \geq \alpha^{2^j-1} \cdot 2^{f(k,0)}$.

Proof. By Lemma 2, every 0-supervertex has an external edge in B_k, for $0 \leq k \leq g(j)$. Hence the case of $t = 0$ is vacuously true. When $0 < t < 2^k$, $t_k = 0$; B_k was filled before the first phase, and, by Lemma 1, will be filled again in the 2^k-th phase. Let x be a 0-seed of v; as B_k does not contain an external edge of v, by Lemma 2, the weight of the lightest external edge of v is at least $h_k(x)$. But $h_k(x) \neq \infty$, because otherwise, for every 0-supervertex $x' \subseteq v$, $h_k(x') = \infty$, implying that v is fully grown. Hence $2^{2^k} - 1$ edges of x were included in B_k before the first phase. All those edges are internal to v. So $|v| \geq \alpha^{2^j-1} \cdot 2^{2^k} = \alpha^{2^j-1} \cdot 2^{f(k,0)}$.

Lemma 4. *For* $0 \leq i < 2^{g(j)}$, $0 \leq k \leq z(i)$, $i \leq t < i + 2^k$, *for every* t*-supervertex* v, *if* v *is not fully grown and does not have an external edge in* B_k *at the start of the* $(t + 1)$*-st phase, then* $|v| \geq \alpha^{2^j-1} \cdot 2^{f(k,i)}$.

Proof. The proof is by induction on i. The case of $i = 0$, proved in Lemma 3 forms the basis. Consider $i > 0$. Hypothesize that the lemma is true for all smaller values of i. In particular, for all $p < i$, for every p-supervertex v, if v is not fully grown and does not have an external edge in B_0 at the start of the $(p + 1)$-st phase, then $|v| \geq \alpha^{2^j-1} \cdot 2^{p+1}$. It follows that for all $p \leq i$, for every p-supervertex v, $|v| \geq \alpha^{2^j-1} \cdot 2^p$.

For $t \in [i, i + 2^k)$, let v be a t-supervertex that is not fully grown and does not have an external edge in B_k at the start of the $(t + 1)$-st phase. Let x be an i-seed of v, and y an $(i - 2^{z(i)})$-seed of x; recall, the edges filled into B_k in the i-th phase have all come from $B_{z(i)+1}$, which (by Lemma 1) was last filled in phase $(i - 2^{z(i)})$, and $(i - 2^{z(i)}) < i \leq t < i + 2^k < (i + 2^{z(i)})$.

If in the i-th phase, $2^{2^k} - 1$ edges of x were included in B_k, when it was filled from $B_{z(i)+1}$, then all those edges are internal to v. Therefore, at least 2^{2^k} i-supervertices participate in v. That is, $|v| \geq \alpha^{2^j-1} \cdot 2^i \cdot 2^{2^k} = \alpha^{2^j-1} \cdot 2^{f(k,i)}$.

Otherwise, $h_k(x) = h_{z(i)+1}(y)$. By Lemma 2, the weight of the lightest external edge of v is at least $h_k(x) = h_{z(i)+1}(y)$. So, y is an $(i - 2^{z(i)})$-seed of v. That is, v does not have an external edge in $B_{z(i)+1}$ either. Thus, v qualifies for an application of the hypothesis with $i \leftarrow i - 2^{z(i)}$ and $k \leftarrow z(i)+1$. Therefore, $|v| \geq \alpha^{2^j-1} \cdot 2^{f(z(i)+1,i-2^{z(i)})} \geq \alpha^{2^j-1} \cdot 2^{f(k,i)}$.

Corollary 2. *For $0 \leq i < 2^{g(j)}$, for every i-supervertex v, if v is not fully grown and $|v| < \alpha^{2^j-1} \cdot 2^{i+1}$, then the lightest external edge of v is present in B_0 at the start of the $(i+1)$-st phase.*

There can be at most one edge of v in B_0, because $2^{2^0} - 1 = 1$. That together with Corollary 2, completes the correctness proof.

5 The I/O Complexity

We discuss below the number of I/Os taken by each step of phase i in stage j:

Step 1: One step of a PRAM that uses N processors and $O(N)$ space can be simulated on the external memory model in $O(\text{Sort}(N))$ I/Os [7]. The Euler tour of a tree of N vertices given in adjacency list representation with twin pointers can be formed in $O(1)$ time with $O(N)$ processors on an EREW PRAM [13].

If Y is a permutation of an array X of n elements, and if each element in X knows its position in Y, then any $O(1)$ amount of information that each element of X holds can be copied into the corresponding element of Y and vice versa in $O(1)$ time using n processors on an EREW PRAM. Therefore, if each element of X holds a pointer to another element of X, then these pointers can be replicated in Y in $O(1)$ time using n processors on an EREW PRAM, and hence in $O(\text{Sort}(N))$ I/Os on the external memory model.

The list ranking algorithm of [7] when invoked on a list of size n takes $O(\text{Sort}(n))$ I/Os.

Let $b_j(k,i) = V_{j,i} \cdot 2^{2^k}$; this an upper bound on the size of bucket B_k in phase i of stage j. Clearly, for any k, $b_j(k,i) \geq \Sigma_{l=0}^{k-1} b_j(l,i)$. Also, $b_j(l,i) = 2b_j(l,i+1)$. Thus, H has a size of at most $b_j(z(i)+1, i-2^{z(i)})$. $B_{z(i)+1}$ has a size of at most $b_j(z(i)+1,i)$. Therefore, the total I/O requirement of Step 2 is $\text{Sort}(b_j(z(i)+1, i-2^{z(i)}))$.

Step 2: The cost of this step is clearly dominated by that of Step 1. See the Details of Step 2 in Sect. 3.

Step 3: Once $B_{z(i)+1}$ has been cleaned up, for $z(i) \geq k \geq 0$, bucket B_k can be formed by scanning B_{k+1} and choosing for each vertex $v \in V_{j,i}$, the $(2^{2^k} - 1)$ lightest edges incident to v. Clearly, this involves scanning each bucket twice, once for write and once for read, and can be done in $O(\text{Scan}(b_j(z(i)+1,i)))$ I/Os.

The total number of I/Os executed by the i-th phase of the j-th stage is therefore $O(\text{Sort}(b_j(z(i)+1, i-2^{z(i)})))$. Therefore, the total I/O cost of phases $1, \ldots, 2^{g(j)}-1$ is

$$\sum_{i=1}^{2^{g(j)}-1} \text{Sort}(b_j(z(i)+1, i-2^{z(i)})) \leq \sum_{k=1}^{g(j)} \sum_{r=0}^{\infty} O(\text{Sort}(b_j(k, r.2^k)))$$

which is $\sum_{k=1}^{g(j)} O(\text{Sort}(b_j(k,0)) = \sum_{k=1}^{g(j)} O(\text{Sort}(V_j(2^{2^k}-1))) = O(\text{Sort}(E))$. The total I/O cost of the bucket formation at the start of the stage, as well as the $2^{g(j)}$-th phase is only $O(\text{Sort}(E_j))$ I/Os. Therefore, the total I/O cost of the j-th stage is $O(\text{Sort}(E))$.

Thus, we reduce the minimum spanning tree problem of an undirected graph $G = (V, E)$ to the same problem on a graph with $O(E/B)$ vertices and $O(E)$ edges. On this new graph, the external memory version of Prim's algorithm can compute a minimum spanning tree in $O(E/B + \text{Sort}(E))$ I/Os. From the MST of the reduced graph, an MST of the original graph can be constructed; the I/O complexity of this will be dominated by the one of the reduction.

Putting everything together, therefore,

Theorem 3. *The minimum spanning forest of an undirected graph* $G = (V, E)$ *can be computed in* $O(\text{Sort}(E) \log\log_{E/V} B)$ *I/Os.*

Acknowledgements. We wish to thank anonymous reviewers for their comments on an earlier version of this paper.

References

1. Arge, L.: The buffer tree: a technique for designing batched external data structures. Algorithmica **37**(1), 1–24 (2003)
2. Abello, J., Buchsbaum, A.L., Westbrook, J.R.: A functional approach to external graph algorithms. Algorithmica **32**, 437–458 (2002)
3. Aggarwal, A., Vitter, J.S.: Complexity of sorting and related problems. Commun. ACM **31**(9), 1116–1127 (1988)
4. Arge, L., Brodal, G.S., Toma, L.: On external-memory MST, SSSP, and muti-way planar graph separation. J. Algorithms **53**, 186–206 (2004)
5. Bhushan, A., Gopalan, S.: External memory soft heap, and hard heap, a meldable priority queue. In: Gudmundsson, J., Mestre, J., Viglas, T. (eds.) COCOON 2012. LNCS, vol. 7434, pp. 360–371. Springer, Heidelberg (2012)
6. Borůvka, O.: O jistém problému minimálním. Práca Moravské Přírodovědecké Společnpsti **3**, 37–58 (1926)
7. Chiang, Y.J., Goodrich, M.T., Grove, E.F., Tamassia, R., Vengroff, D.E., Vitter, J.S.: External-memory graph algorithms. In: Proceedings of the ACM-SIAM Symposium on Discrete Algorithms, pp. 139–149 (1995)
8. Chin, F., Lam, J., Chen, I.: Efficient parallel algorithms for some graph problems. Commun. ACM **25**, 659–665 (1982)
9. Chong, K.W., Han, Y., Igarashi, Y., Lam, T.W.: Improving the efficiency of parallel minimum spanning tree algorithms. Discrete Appl. Math. **126**, 33–54 (2003)
10. Cole, R., Vishkin, U.: Approximate parallel scheduling, II. Applications to logarithmic-time optimal parallel algorithms. Inf. Comput. **92**(1), 1–47 (1991)
11. Cormen, T.H., Leiserson, C.E., Rivest, R.L.: Introduction to Algorithms. MIT Press, Cambridge (1990)
12. Fadel, R., Jakobsen, K.V., Katajainen, J., Teuhola, J.: Heaps and heapsort on secondary storage. Theoret. Comput. Sci. **220**, 345–362 (1999)
13. JáJá, J.F.: Introduction to Parallel Algorithms. Addison-Wesley, Reading (1992)

14. Kumar, V., Schwabe, E.: Improved algorithms and data structures for solving graph problems in external memory. In: Proceedings of the IEEE Symposium on Parallel and Distributed Processing, pp. 169–177 (1996)
15. Meyer, U., Sanders, P., Sibeyn, J.F. (eds.): Algorithms for Memory Hierarchies. LNCS, vol. 2625. Springer, Heidelberg (2003)
16. Mungala, K., Ranade, A.: I/O-complexity of graph algorithms. In: Proceedings of the ACM-SIAM Symposium on Discrete Algorithms, pp. 687–694 (1999)
17. Vitter, J.S.: Algorithms and Data Structures for External Memory, Series on Foundations and Trends in Theoretical Computer Science. Now Publishers, Hanover (2008)

The Connected p-Centdian Problem on Block Graphs

Liying Kang[1], Jianjie Zhou[1], and Erfang Shan[1,2(✉)]

[1] Department of Mathematics, Shanghai University, Shanghai 200444, People's Republic of China
lykang@shu.edu.cn, 15565791770@126.com, efshan@i.shu.edu.cn

[2] School of Management, Shanghai University, Shanghai 200444, People's Republic of China

Abstract. In this paper, we consider the problems of locating p-vertex X_p on block graphs such that the induced subgraph of the selected p vertices is connected. Two problems are proposed: one problem is to minimizes the sum of its weighted distances from all vertices to X_p, another problem is to minimize the maximum distance from each vertex in $V - X_p$ to X_p and at the same time to minimize the sum of its distances from all vertices. We prove that the first problem is linearly solvable on block graphs with unit edge length. For the second problem, it is shown that the set of Pareto-optimal solutions of the two criteria has cardinality not greater than n, and can be obtained in $O(n^2)$ time, where n is the number of vertices of the block graph.

Keywords: Connected p-center · Median · Centdian · Block graphs

1 Introduction

In network location theory, two main criteria that are often used on a network for locating a facility are: the maximal distance between the facility and a customer and the average distance between the facility and the customers. However, neither of the two above criteria alone capture all essential elements of a location problem. The sum of the distances criterion alone may result in solutions which are unacceptable from the point of view of the service level for the clients who are located far away from the facilities. On the other hand, the criterion of the minimization of the maximum distance, if used alone, may lead to very costly service systems. To capture more real-word problems and provide good ways to trade-off minisum and minimax approaches, Halpern [5] introduce the centdian criteria which combine the minimax and minisum objective functions.

Problems of locating a facility at a point of a network with combinations of the two criteria are investigated and efficient algorithms for them are developed

Research was partially supported by the National Nature Science Foundation of China (Nos. 11471210, 11571222).

Z. Lu et al. (Eds.): COCOA 2015, LNCS 9486, pp. 510–520, 2015.
DOI: 10.1007/978-3-319-26626-8_37

in [6–8]. Averbakh and Berman [1] considered problems of finding the optimal location of a path (of unrestricted length) on a tree, using different combinations of the minisum and minimax criteria. Becker et al. [2] considered the first two problems introduced in [1] with an additional constraint, namely that the two optimal paths must have length (or cost) bounded by a fixed constant. Tamir et al. [9] studied the problem finding the optimal location of a tree shaped facility of a specified size in a tree network, using the centdian criterion, and developed an $O(nlogn)$ algorithm.

Yen [11] studied the connected p-center problem on block graphs. In this paper, we consider problems of finding the optimal location of connected p-median and connected p-centdian on a block graph. Shan et al. [10] considered the connected p-center and connected p-median problems on interval and circular-arc graphs and shown that all the problems can by solved in polynomial time.

The paper is organized as follows. In the next section we formally introduce the notation and the problems that we study in this paper. In Sect. 3, we study the connected p-median problem on block graphs, we prove that the connected p-median problem is linearly solvable on block graphs with unit edge length. Section 4 studies the connected p-centdian problem on unweighted block graphs. We prove some properties of the Pareto-optimal solutions and shown that there are at most n Pareto-optimal solutions. Then two algorithms are proposed to obtain all the Pareto-optimal solutions. In the last section, we describe an example to illustrate the whole process.

2 Problem Formulation

Let $G = (V, E, w, l)$ be a finite, connected, undirected graph with n-vertex-set V and m-edge-set E, where each vertex $v \in V$ is associated with a nonnegative weight $w(v)$ and each edge $(v_i, v_j) \in E$ is associated with a certain cost or length $l(v_i, v_j)$. For convenience, we denote $G = (V, E)$ as the unweighted graph that $w(v) = 1$ for all vertices and $l(e) = 1$ for all edges. Given a graph G, a vertex u is called a cut vertex of G if $\kappa(G - \{u\}) > \kappa(G)$, where $\kappa(G)$ denotes the number of components of G. A connected subgraph H of G is called a block of G if H is maximal and it contains no cut vertices. A graph G is a block graph if all blocks of G are cliques and any two distinct blocks B_1 and B_2 have at most one common vertex [3]. In this paper, we study the location problems on block graph, we denote G as the block graph in the following.

For any two vertices of $u, v \in G$, let $P(u, v)$ denote the shortest path between u and v. For any p-vertex set $X_p = \{x_1, \cdots, x_p\}$ of G, let $\langle X_p \rangle$ denote the subgraph induced by X_p and $d(v, X_p) = min_{1\leq j\leq p}\{d(v, x_i)\}$ denote the distance between vertex v and vertex set X_p where $d(v, x_i)$ is the length of a shortest path in G between v and x_i. A p-vertex set X_p is called connected p-vertex if $\langle X_p \rangle$ is a connected subgraph of G. Let Φ be the set of all connected p-vertex of the graph G.

The minimax objective seeks a connected p-vertex X_p on the graph G that minimize the maximum weighted distances from each vertices to X_p:

$$min_{X_p \in \Phi} F_1(X_p) = max_{v \in V}\{w(v)d(v, X_p)\}.$$

This problem is known as connected p-center (CpC) problem and studied in [11]. Yen [11] shown that the CpC problem is NP-hard on block graphs G when $w(v) = 1$ for all vertices v and $l(e) = \{1, 2\}$ for all edges e. We consider the following problems on block graph G.

Problem 1: Find a connected p-vertex X_p on the graph G that minimize the sum of its distances from all vertices: $min_{X_p \in \Phi} F_2(X_p) = \sum_{v \in V} w(v)d(v, X_p)$. This problem is known as connected p-median (CpM) problem.

Problem 2: Find the set of all Pareto-optimal solutions Π, $X_p \in \Phi$ of the bi-objective problem:

$$\{\min_{X_p \in \Phi} F_1(X_p) = \max_{v \in V}\{w(v)d(v, X_p)\},\ \min_{X_p \in \Phi} F_2(X_p) = \sum_{v \in V} w(v)d(v, X_p)\},$$

where a connected p-vertex $X_p \in \Phi$ is called Pareto-optimal, if there is no connected p-vertex $X'_p \in \Phi$ such that $F_1(X'_p) \leq F_1(X_p)$ and $F_2(X'_p) \leq F_2(X_p)$ and at least one is satisfied as strict inequality. This problem is known as connected p-centdian problem.

Given a block graph $G = (V, E)$, let r be a vertex of G, we consider the rooted block graph $G(r)$. Each vertex v of $G(r)$ is associated with a label $L(v)$, called the level of v which can computed by the BFS traversal in $O(n)$ time. The parent of v, denoted by $par(v)$, is the vertex u such that $(v, u) \in E$ and $L(v) = L(u)+1$. Note that $par(v) = NULL$ if $v = r$. The children set of v, denoted by $chi(v)$, is defined as $chi(v) = \{u|(v, u) \in E \ and \ L(v) = L(u)-1\}$. The descendent set of v, denoted by $des(v)$, is defined as $des(v) = \{u|L(v) < L(u) \ and \ v \in P(u, r)\}$. $G(v)$ denotes the subgraph of $G(r)$ induced by $\{v\} \cup des(v)$. We define $W(v) = \sum_{u \in G(v)} w(u)$ which can computed bottom-up by the following formula:

$$W(v) = w(v) + \sum_{u \in chi(v)} W(u). \tag{1}$$

When we consider unweighted block graph, $W(v)$ is the number of vertices in subgraph $G(v)$. For each vertex v of $G(r)$, let $cs(v) = \{u|(u, v) \in E \ and \ L(u) = L(v)\}$. Note that $cs(v)$ may be empty under this definition.

For a vertex v of the rooted block graph $G(r)$, let $f(v), g(v)$ be the sum of the weighted distance from vertices of $G(v)$ to v and vertices of $G(r) - G(v)$ to v, respectively. Then

$$\begin{aligned} f(v) &= \sum_{u \in G(v)} w(u)d(u, v) \\ &= \sum_{z \in chi(v)} f(z) + \sum_{z \in chi(v)} W(z) \\ &= \sum_{u \in G(v)-v} W(u), \end{aligned} \tag{2}$$

and

$$\begin{aligned} g(v) &= \sum_{u \in G(r)-G(v)} w(u)d(u,v) \\ &= \sum_{u \in G(r)-G(par(v))} w(u)d(u,v) + \sum_{u \in G(par(v))-G(v)} w(u)d(u,v) \\ &= g(par(v)) + W(r) - W(par(v)) + f(par(v)) - (f(v) + W(v)) \\ &\quad +(W(par(v)) - W(v) - \sum_{z \in cs(v)} W(z)) \\ &= g(par(v)) + f(par(v)) - f(v) + W(r) - 2W(v) - \sum_{z \in cs(v)} W(z). \quad (3) \end{aligned}$$

Note that $g(r) = 0$ and $F_2(v) = f(v) + g(v)$ for each vertex v of G.

3 The CpM Problem on Block Graph with Unit Edge Length

Suppose that m is the 1-median of G and m belongs to a block $B_j = \{v_1 = m, \cdots, v_t\}$ of G. When we delete all the edges in B_j, we obtain a collection of connected components $G_1, \cdots, G_t$ with $v_i \in G_i$, $1 \leq i \leq t$. Let $W(G_i) = \sum_{v \in G_i} w(v)$. For each vertex $v \in G_i$, $d(v,m) = d(v,v_i) + d(v_i,m)$. Then, we have

$$\begin{aligned} F_2(m) &= \sum_{v \in G} w(v)d(v,m) \\ &= \sum_{i=1}^{t} \sum_{v \in G_i} w(v)d(v,v_i) + \sum_{i=1}^{t} \Big(\sum_{v \in G_i} w(v) \Big) d(v_i,m) \\ &= \sum_{i=1}^{t} \sum_{v \in G_i} w(v)d(v,v_i) + \sum_{i=1}^{t} W(G_i) - W(G_1). \end{aligned}$$

Thus we have the following lemma

Lemma 1. *If the* 1*-median vertex m of G belongs to a block B_j, then m is the vertex of B_j with maximum value $W(G_i)$.*

Chen et al. [4] introduce an extension version of Goldman's algorithm which (in linear time) either finds the 1-median of G or finds the single block of G which contains the 1-median of G. An immediate consequence is

Lemma 2. *We can find the* 1*-median of G in linear time.*

Lemma 3. *Let m be the* 1*-median of a block graph $G = (V, E, w)$. There exist a connected p-median X_p^* of G containing the vertex m.*

Proof. By contradiction, suppose that X_p^* is a connected p-median of G such that $d(m, X_p^*)$ is minimized and $m \notin X_p^*$. Assume x is a vertex in X_p^* such that $d(m,x) = d(m, X_p^*)$ and x' is a vertex in path $P(m,x)$ which is adjacent to vertex x. We assume that the 1-median vertex m of G belongs to a block B_j of G. Delete all the edges in B_j, we obtain connected components of $G-E(B_j)$. Assume G_m is the component of $G-E(B_j)$ containing m. Since G is a block graph, then there exists a vertex $y \in X_p$ such that $y \neq x$ and $\langle X_p - y\rangle$ is connected. Let $V_x = \{v|d(v,x) = d(v,X_p), v \in V\}$, $V_y = \{v|d(v,y) = d(v,X_p), v \in V - V_x\}$. Set $X_p' = X_p^* - \{y\} + \{x'\}$. Obviously, $\langle X_p'\rangle$ is connected and $d(m, X_p') < d(m, X_p^*)$.

If $X_p^* \subseteq G - G_m$. Then, we have

$$F_2(X_p') - F_2(X_p^*) \leq \sum_{v \in V_y} w(v) - \sum_{v \in G_m} w(v).$$

Since m is a 1-median vertex of G, V_y belongs to a component of $G - E(B_j) - G_m$. By Lemma 1, $\sum_{v \in V_y} w(v) \leq \sum_{v \in G_m} w(v)$. Thus $F_2(X_p') \leq F_2(X_p^*)$. This contradicts the assumption.

If $X_p^* \subseteq G_m$. Then, we have

$$F_2(X_p') - F_2(X_p^*) \leq \sum_{v \in V_y} w(v) - \sum_{v \in G - G_m + m} w(v).$$

Since $X_p \subseteq G_m$, there exists another block $B_k \neq B_j$ such that $m \in B_k$ and y is contained in a component of $G - E(B_k)$. Denote the component contains y as G_y. Since $\langle X_p\rangle$ is connected, in view of the choices of x and y, $V_y \subseteq V(G_y)$. By Lemma 1, $\sum_{v \in V_y} w(v) \leq \sum_{v \in G - G_m + m} w(v)$. Thus $F_2(X_p') \leq F_2(X_p^*)$. We get a contradiction. The result follows. □

Lemma 4. *Let m be the 1-median of a block graph $G = (V, E, w)$. For the rooted graph $G(m)$, let $X_p^* = \{v_1, \cdots, v_p\}$ be a p-vertex set of G such that $W(v_1), \cdots, W(v_p)$ are the first p largest numbers among $\{W(v)|v \in G(m)\}$ and $L(v_1), \cdots, L(v_p)$ as small as possible. Then, X_p^* is a connected p-median of G.*

Proof. Obviously, for each non-root vertex v, $W(v) \leq W(par(v))$. This implies that $m \in X_p^*$. Next we show that X_p^* is connected. We will prove this statement by induction on p. Without loss of a generality, assume that $W(v_1) \geq W(v_2) \geq \cdots \geq W(v_n)$, and $X_i^* = \{v_1, \cdots, v_i\}$, $1 \leq i \leq p$.

It is trivial that $\langle X_1^*\rangle$ is connected. Assume that $\langle X_k^*\rangle$ is connected for $1 \leq k < p$. The choice of v_{k+1} implies that $v_{k+1} \in \{y|y \in chi(X_k^*) - X_k^*\}$. Then $\langle X_{k+1}^*\rangle$ is connected.

It is easily seen that

$$\begin{aligned} F_2(X_p^*) &= F_2(X_{p-1}^*) - W(x_p) \\ &= F_2(m) - \sum_{v \in X_p^* - m} W(v). \end{aligned}$$

By Lemma 3 and the assumption of the lemma, for any connected p-vertex set $X'_p \neq X^*_p$ of G, we have

$$\sum_{v \in X^*_p} W(v) \geq \sum_{v \in X'_p} W(v).$$

Thus $F_2(X^*_p) \leq F_2(X'_p)$, X^*_p is an optimal solution. □

Theorem 1. *For a given block graph $G = (V, E, w)$ with n vertices, the CpM problem can be solved in $O(n)$ time.*

4 The Connected p-Centdian Problem on Unweighted Block Graphs

In [11], Yen gave an algorithm to find a diameter path $P^S(u^*, v^*)$ of an unweighted block graph G in $O(n + m)$ time. Obviously, the 1-center vertex c is the middle vertex of diameter path $P^S(u^*, v^*)$ and $F_1(c) = \left\lceil \frac{L(P^S(u^*,v^*))}{2} \right\rceil$. Hence, we can find the 1-center vertex of an unweighted block graph in $O(n+m)$ time. As shown in Sect. 3, we can find the 1-median of G in linear time.

Lemma 5. *Let c be the 1-center vertex and m be the 1-median vertex of a block graph $G = (V, E)$. For any connected p-vertex $X_p \in \Pi$, we have $X_p \cap P(c, m) \neq \emptyset$.*

Proof. By contradiction, we suppose that $X_p \in \Pi$, $X_p \cap P(c, m) = \emptyset$ and $d(P(c, m), X_p)$ is minimized. Let x be a vertex in X_p such that $d(P(c, m), x) = d(P(c, m), X_p)$. Assume x' is adjacent to x and x' is a vertex of the shortest path from x to $P(c, m)$. Since G is a block graph, then there exists a vertex $y \in X_p$ such that $y \neq x$ and $\langle X_p - y \rangle$ is connected. Let $V_x = \{v | d(v, x) = d(v, X_p), v \in V\}$, $V_y = \{v | d(v, y) = d(v, X_p), v \in V - V_x\}$. Set $X'_p = X_p - \{y\} + \{x'\}$. Obviously, $\langle X'_p \rangle$ is connected, $d(P(c, m), X'_p) = d(P(c, m), X_p) - 1$ and $d(c, X'_p) = d(c, X_p) - 1$.

Since c lies on the diameter path $P^S(u^*.v^*)$, c is a cut vertex. Let G_{u^*} and G_{v^*} be the subgraphs of $G - \{c\}$ containing u^* and v^*, respectively. Since $c \notin X_p$ and $\langle X_p \rangle$ is connected, X_p cannot lie within two distinct subgraphs G_{u^*} and G_{v^*}. We have

$$\begin{aligned} F_1(X_p) &= max\{d(u^*, X_p), d(v^*, X_p)\} \\ &= \left\lceil \frac{L(P^S(u^*, v^*))}{2} \right\rceil + d(c, X_p) \text{ or } \left\lfloor \frac{L(P^S(u^*, v^*))}{2} \right\rfloor + d(c, X_p). \end{aligned}$$

Similarly, $F_1(X'_p) = \left\lceil \frac{L(P^S(u^*,v^*))}{2} \right\rceil + d(c, X'_p)$ or $\left\lfloor \frac{L(P^S(u^*,v^*))}{2} \right\rfloor + d(c, X'_p)$. Obviously, X_p and X'_p lies in the same component of $G - \{c\}$. The first part of $F_1(X_p)$ and $F_1(X'_p)$ are the same. Then $F_1(X'_p) < F_1(X_p)$.

We assume that the 1-median vertex m of G belongs to a block B_j of G. Deleting all the edges in B_j, we obtain a collection of connected components. Suppose G_m is the component containing m. If $X_p \subseteq G - G_m$, we have

$$F_2(X'_p) - F_2(X_p) \leq \sum_{v \in V_y} w(v) - \sum_{v \in G_m} w(v).$$

Since $\langle X_p \rangle$ is connected, in view of the choices of x and y, V_y belongs to a component of $G - E(B_j) - G_m$. By Lemma 1, $\sum_{v \in V_y} w(v) \leq \sum_{v \in G_m} w(v)$. Thus $F_2(X'_p) \leq F_2(X_p)$. If $X_p \subseteq G_m$, we have

$$F_2(X'_p) - F_2(X_p) \leq \sum_{v \in V_y} w(v) - \sum_{v \in G - G_m + m} w(v).$$

Since $X_p \subseteq G_m$, there exists another block $B_k \neq B_j$ such that $m \in B_k$ and y is contained in a component of $G - E(B_k)$. Denote the component containing y as G_y. Since $\langle X_p \rangle$ is connected, in view of the choices of x and y, $V_y \subseteq V(G_y)$. By Lemma 1, $\sum_{v \in V_y} w(v) \leq \sum_{v \in G - G_m + m} w(v)$. Thus $F_2(X'_p) \leq F_2(X_p)$. This contradicts the assumption that X_p is Pareto-optimal. □

Lemma 6. *Suppose $X_p \in \Pi$, the 1-center vertex $c \notin X_p$ and x is a vertex of X_p such that $d(x,c)$ is minimum. Consider the rooted graph $G(c)$. Let $x_2, \cdots, x_p$ be the vertices such that $W(x_2), \cdots, W(x_p)$ are the first $p-1$ largest numbers among $\{W(v) | v \in G(x) \cup_{z \in cs(x)} G(z) - \{x\}\}$ and $L(x_2), \cdots, L(x_p)$ as small as possible, then $X_p = \{x, x_2, \cdots, x_p\}$.*

Proof. Using the similar method as in Lemma 4, we can shown that $X_p \in \Phi$. Since x is a vertex of X_p such that $d(x,c)$ is minimum, X_p only contain vertices of $G(x) \cup_{z \in cs(x)} G(z)$. Then we have

$$\begin{aligned} F_1(X_p) &= max_{v \in V}\{d(v, X_p)\} \\ &= \left\lceil \frac{L(P^S(u^*, v^*))}{2} \right\rceil + d(x,c) \text{ or } \left\lfloor \frac{L(P^S(u^*, v^*))}{2} \right\rfloor + d(x,c), \end{aligned}$$

and

$$\begin{aligned} F_2(X_p) &= \sum_{v \in G(c)} d(v, X_p) \\ &= \sum_{v \in G(c) - G(x)} d(v, X_p) + \sum_{v \in G(x)} d(v, X_p) \\ &= g(x) - \sum_{x' \in X_p \cap (\cup_{z \in cs(x)} G(z))} W(x') + f(x) - \sum_{x' \in X_p \cap G(x) - \{x\}} W(x') \\ &= F_2(x) - \sum_{x' \in X_p - x} W(x'). \end{aligned}$$

So for any $X'_p \in \Phi$ satisfying the conditions of the lemma, we have $\sum_{v \in X_p} W(v) \geq \sum_{v \in X'_p} W(v)$. Thus $F_2(X_p) \leq F_2(X'_p)$, X_p is Pareto-optimal. □

We choose 1-center vertex c as the root of G and compute $W(v)$ for $v \in G(c)$ by formula (1). Find the optimal solution Q of G by the algorithm in [11] and set $\beta_1 = F_1(Q)$. For each vertex v of $G(c)$, define $\mu(v) = max\{d(y,v)|y \in G(v)\}$. Then we have:

$$\mu(v) = \begin{cases} 0 & \text{if } chi(v) = \emptyset, \\ max\{\mu(u)|u \in chi(v)\} + 1 & \text{if } chi(v) \neq \emptyset. \end{cases} \tag{4}$$

For integer k with $\beta_1 \leq k \leq F_1(c))$, let $Y_k = \{x|x \in G(c), \mu(v) \geq k\}$. Then Y_k is the minimum set of connected vertices of G such that $F_1(Y_k) = k$ and $|Y_k| \leq p$. It is easy to check that Y_k can be found in $O(n)$ time.

The definition of Y_k implies that $Y_k \subseteq X_p$. Using the similar method as in Lemma 6, we can prove the following lemma.

Lemma 7. *Suppose $X_p \in \Pi$, $c \in X_p$ and $F_1(X_p) \leq k$, where $\beta_1 \leq k \leq F_1(c)$. Then $X_p = Y_k \cup \{x_1, \cdots, x_{p-|Y_k|}\}$, where $W(x_1), \cdots, W(x_{p-|Y_k|})$ are the first $p - |Y_k|$ largest numbers among $\{W(v)|v \in G(c) - Y_k\}$ and $L(x_1), \cdots, L(x_{p-|Y_k|})$ as small as possible.*

We give the following algorithm to find a set containing all Pareto-optimal solutions

---

Algorithm 1.

Set $\Psi = \emptyset$, $F_1(c) = \left\lceil \frac{L(P^S(u^*,v^*))}{2} \right\rceil$, $P(m,c) = \{y_1 = m, y_2, \cdots, y_t = c\}$;
for $i = 1$ to t **do**
  Compute $F_2(y_i) = f(y_i) + g(y_i)$ by formula (2) and (3),
end for
for $i = 1$ to $t - 1$ **do**
  Find $x_2, \cdots, x_p$ such that $W(x_2), \cdots, W(x_p)$ are the first $p-1$ largest numbers among $\{W(v)|v \in G(y_i) \cup_{u \in cs(y_i)} G(u) - \{y_i\}\}$ and $L(x_2), \cdots, L(x_p)$ as small as possible,
  set $X_p^{y_i} = \{y_i, x_2, \cdots, x_p\}$ and $\Psi = \Psi \cup X_p^{y_i}$,
  compute $F_1(X_p^{y_i})$ and $F_2(X_p^{y_i}) = F_2(y_i) - \sum_{x' \in X_p^{y_i} - y_i} W(x')$,
end for
for $k = F_1(c)$ to β_1 **do**
  Find minimum set connected vertices Y_k that $F_1(Y_k) = k$,
  find $x_1, \cdots, x_{p-|Y_k|}$ such that $W(x_1), \cdots, W(x_{p-|Y_k|})$ are the first $p - |Y_k|$ largest numbers among $\{W(v)|v \in G(c) - Y_k\}$ and $L(x_1), \cdots, L(x_{p-|Y_k|})$ as small as possible,
  set $X_p^k = Y_k \cup \{x_1, \cdots, x_{p-|Y_k|}\}$ and $\Psi = \Psi \cup X_p^k$,
  compute $F_1(X_p^k)$ and $F_2(X_p^k) = F_2(c) - \sum_{x \in X_p - c} W(x)$,
end for

Lemma 8. *Algorithm 1 obtains a set Ψ such that $\Pi \subseteq \Psi \subseteq \Phi$ in $O(n^2)$ time, furthermore we get $|\Pi| \leq n$.*

Proof. Obviously, all $X_p^{y_i}$ and X_p^k are connected p-vertex, i.e. $\Psi \subseteq \Phi$. By Lemma 5, for any connected p-vertex $X_p \in \Pi$, we have $X_p \cap P(c,m) \neq \emptyset$. By Lemma 6, Steps 5–9 in Algorithm 1 find all $X_p \in \Pi$ with $c \notin X_p$. By Lemma 7, Steps 10–15 in Algorithm 1 find all $X_p \in \Pi$ with $c \in X_p$. Note that some $X_p \in \Psi$ may not be Pareto-optimal. Thus $\Pi \subseteq \Psi$. We have:

$$|\Pi| \leq |\Psi| \leq L(P(m,c)) + F_1(c) - \beta_1 \leq \frac{n}{2} + \frac{n}{2} = n.$$

In Steps 2–4, the values $F_2(y_i)$ $(1 \leq i \leq t)$ can be computed in $O(n)$ time. The for loop Steps 5–9 take $t-1$ times and each loop takes $O(n)$ time. The Steps 10–15 take $F_1(c) - \beta_1$ times and each loop takes $O(n)$ time. Thus, the total computational complexity is $O(n^2)$. □

Based on Algorithm 1, we give the following algorithm to find all Pareto-optimal solutions.

Algorithm 2.

---

for each $X_p \in \Psi$ **do**
  Two functions $h_1, h_2 : \Psi \to \{1, 2, \cdots, |\Psi|\}$ are given such that for any different connected p-vertex $X_p, X'_p \in \Psi$:
  $h_1(X_p) < h_1(X'_p)$ iff $F_1(X_p) < F_1(X'_p)$ or $F_1(X_p) = F_1(X'_p)$ and $F_2(X_p) < F_2(X'_p)$,
  $h_2(X_p) < h_2(X'_p)$ iff $F_2(X_p) < F_2(X'_p)$ or $F_2(X_p) = F_2(X'_p)$ and $F_1(X_p) < F_1(X'_p)$,
end for
for all $X'_p \in \Psi$ **do**
  if there exists X_p such that $h_1(X_p) < h_1(X'_p)$ and $h_2(X_p) < h_2(X'_p)$ **then**
    set $\Psi = \Psi - X'_p$,
  end if
end for
Set $\Pi = \Psi$.

Given Ψ, Algorithm 2 obtains the Pareto-optimal set Π of the connected p-centdian problem in $O(nlogn)$ time. Thus we have:

Theorem 2. *The Pareto-optimal set Π of the connected p-centdian problem can be obtained in $O(n^2)$ time.*

5 An Example for the Connected p-Centdian Problem

Suppose the block graph shown in Fig. 1 is unweighted and $p = 5$. Applying Algorithm 1, we get:

1. The diameter path $P^S(u^*, v^*)$ and the middle vertex c;
2. $W(v)$ for all $v \in G$ is presented in Table 1;

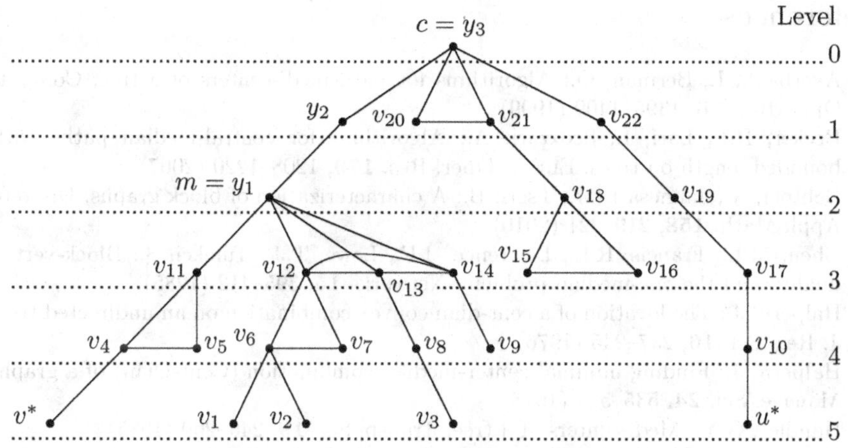

Fig. 1. Illustration for the example.

Table 1. $W(v)$ for all $v \in G$

	v^*	v_1	v_2	v_3	u^*	v_4	v_5	v_6	v_7	v_8	v_9	v_{10}	v_{11}	v_{12}	v_{13}	v_{14}	v_{15}	v_{16}	v_{17}	y_1	v_{18}	v_{19}	y_2
$W(\cdot)$	1	1	1	1	1	2	1	3	1	2	1	2	4	5	3	2	1	1	3	15	3	4	16
	v_{20}	v_{21}	v_{22}	y_3																			
$W(\cdot)$	1	4	5	27																			

3. The connected 5-center $Q = \{y_1, y_2, y_3, v_{22}, v_{19}\}$, and $\beta_1 = F_1(Q) = 3$;
4. The 1-median m;
5. $F_1(c) = 5$ and $P(m, c) = \{y_1, y_2, y_3\}$;
6. $F_2(y_1) = 76$, $F_2(y_2) = 79$, $F_2(y_3) = 84$;
7. $X_5^{y_1} = \{y_1, v_{11}, v_{12}, v_{13}, v_6\}$, $F_1(X_5^{y_1}) = 7$, $F_2(X_5^{y_1}) = 61$,
   $X_5^{y_2} = \{y_2, y_1, v_{11}, v_{12}, v_{13}\}$, $F_1(X_5^{y_2}) = 6$, $F_2(X_5^{y_2}) = 52$;
8. $Y_5 = \{y_3\}$, $X_5^5 = \{y_3, y_2, y_1, v_{12}, v_{22}\}$, $F_1(X_5^5) = 4$, $F_2(X_5^5) = 43$,
   $Y_4 = \{y_2, y_3, v_{22}\}$, $X_5^4 = \{y_3, y_2, y_1, v_{12}, v_{22}\}$, $F_1(X_5^4) = 4$, $F_2(X_5^4) = 43$,
   $Y_3 = \{y_1, y_2, y_3, v_{19}, v_{22}\}$, $X_5^3 = \{y_3, y_2, y_1, v_{19}, v_{22}\}$, $F_1(X_5^3) = 3$,
   $F_2(X_5^4) = 44$;
9. $\Psi = \{X_5^{y_1}, X_5^{y_2}, X_5^4, X_5^3\}$.

Applying Algorithm 2, we get:

1. $a(X_5^3) = 1$, $a(X_5^4) = 2$, $a(X_5^{y_2}) = 3$, $a(X_5^{y_1}) = 4$,
   $b(X_5^4) = 1$, $b(X_5^3) = 2$, $b(X_5^{y_2}) = 3$, $b(X_5^{y_1}) = 4$;
2. $\Pi = \{X_5^4, X_5^3\}$.

References

1. Averbakh, I., Berman, O.: Algorithms for path medi-centers of a tree. Comput. Oper. Res. **26**, 1395–1409 (1999)
2. Becker, R.I., Lari, I., Scozzari, A.: Algorithms for central-median paths with bounded length on trees. Eur. J. Oper. Res. **179**, 1208–1220 (2007)
3. Behtoei, A., Jannesari, M., Tseri, B.: A characterization of block graphs. Discrete Appl. Math. **158**, 219–221 (2010)
4. Chen, M.L., Francis, R.L., Lawrence, J.F., Lowe, T.J., Tufekci, S.: Block-vertex duality and the one-median problem. Networks **15**, 395–412 (1985)
5. Halpern, J.: The location of a cent-dian convex combination on an undirected tree. J. Reg. Sci. **16**, 237–245 (1976)
6. Halpern, J.: Finding minimal center-median combination (Cent-Dian) of a graph. Manage. Sci. **24**, 535–544 (1978)
7. Handler, G.Y.: Medi-centers of a tree. Transp. Sci. **19**, 246–260 (1985)
8. Hansen, P., LabbeH, M., Thisse, J.-F.: From the median to the generalized center. Oper. Res. **25**, 73–86 (1991)
9. Tamir, A., Puerto, J., Pérez-Brito, D.: The centdian subtree on tree networks. Discrete Appl. Math. **118**, 263–278 (2002)
10. Shan, E., Zhou, J., Kang, L.: The connected p-center and p-median problems on interval and circular-arc graphs. Acta Mathematicae Applicatae Sinica (Accepted)
11. Yen, W.C.-K.: The connected p-center problem on block graphs with forbidden vertices. Theor. Comput. Sci. **426**, 13–24 (2012)

Searching for (near) Optimal Codes

Xueliang Li[1(✉)], Yaping Mao[2], Meiqin Wei[1], and Ruihu Li[3]

[1] Center for Combinatorics and LPMC-TJKLC, Nankai University, Tianjin 300071, China
lxl@nankai.edu.cn, weimeiqin8912@163.com

[2] Department of Mathematics, Qinghai Normal University, Xining 810008, Qinghai, China
maoyaping@ymail.com

[3] Institute of Science, The Air Force Engineering University, Xi'an 710051, China
liruihu@aliyun.com

Abstract. Formally self-dual (FSD) codes are interesting codes and have received an enormous research effort due to their importance in mathematics and computer science. Danielsen and Parker proved that every self-dual additive code over $GF(4)$ is equivalent to a graph codes in 2006, and hence graph is an important tool for searching (near) optimal codes. In this paper, we introduce a new method of searching (near) optimal binary (formally self-dual) linear codes and additive codes from circulant graphs.

Keywords: Graph code · FSD code · Additive code · Optimal code · Circulant graph

1 Introduction

Let F_2 be the binary field, and let F_2^n denote the n-dimensional binary vector space. A k-dimensional linear subspace $\mathcal{C}$ of F_2^n is called an $[n, k]$ *linear code* and vectors in $\mathcal{C}$ are called *codewords*. Define $GF(4) = \{0, 1, \omega, \omega^2\}$, where $\omega^2 = 1 + \omega$. An additive code $\mathcal{C}$ over $GF(4)$ of length n is an additive subgroup of $GF(4)$. It is clear that $\mathcal{C}$ contains codewords for some $0 \leq k \leq 2n$, and can be defined by a $k \times n$ generator matrix, with entries from $GF(4)$, whose rows span $\mathcal{C}$ additively. We call $\mathcal{C}$ an $(n, 2^k)$ *additive code*. The *Hamming weight* of a vector $x = (x_1, \cdots, x_n)$, denoted by $wt(x)$, is the number of its nonzero coordinates, the *Hamming distance* between two vectors x, y is equal to the Hamming weight $wt(x - y)$. The *minimum distance* d of a code is defined as the smallest possible distance between pairs of distinct codewords. An $[n, k]$ linear code with minimum distance d is denoted as an $[n, k, d]$ code and an $(n, 2^k)$ additive code with minimum distance d is denoted as an $(n, 2^k, d)$ code. The *weight distribution* of a code $\mathcal{C}$ is the sequence $(A_0, A_1, \cdots, A_n)$, where A_i is the number of codewords of weight i in $\mathcal{C}$. The *weight enumerator* of the

Supported by "973" program No. 2013CB834204.

Z. Lu et al. (Eds.): COCOA 2015, LNCS 9486, pp. 521–536, 2015.
DOI: 10.1007/978-3-319-26626-8_38

code is the polynomial $W(z) = \sum_{i=0}^{n} A_i z^i$. The *inner product* of two vectors $x, y \in F_2^n$ is defined as $(x, y) = \sum_{i=1}^{n} x_i y_i$. The *dual code* of an $[n, k, d]$ code $\mathcal{C}$ is defined as $\mathcal{C}^{\perp} = \{x \in F_2^n \mid (x, y) = 0, \text{for all } y \in \mathcal{C}\}$. A binary code with the same weight distribution as its dual code is called *formally self-dual* (FSD). The *conjugation* of $x \in GF(4)$ is defined by $\bar{x} = x^2$, and the *trace map* is defined by $Tr(x) = x + \bar{x}$. The *Hermitian trace inner product* of $u = (u_1, \cdots, u_n)$ and $v = (v_1, \cdots, v_n)$. where $u, v \in GF(4)$, is defined as $u * v = Tr(u \cdot \bar{v}) = \sum_{i=1}^{n} Tr(u_i \bar{v_i}) = \sum_{i=1}^{n} (u_i v_i^2 + u_i^2 v_i)$. We define the *dual of the additive code* $\mathcal{C}$ with respect to the Hermitian trace inner product as $\mathcal{C}^{\perp} = \{u \in GF(4) \mid u * c = 0 \; for \; all \; c \in \mathcal{C}\}$. Then $\mathcal{C}$ is *self-orthogonal* if $\mathcal{C} \subseteq \mathcal{C}^{\perp}$, and $\mathcal{C}$ is *self-dual* if $\mathcal{C} = \mathcal{C}^{\perp}$.

In 2002, Tonchev [20] set up a relationship between an undirected graph and a binary linear code. Given a graph Γ_n on n vertices with adjacency matrix A_n, one can define a binary linear code with generator matrix $G = (I; A_n)$, where I is the identity matrix. Such a code is called the *linear graph code* of Γ_n, and a linear graph code is a $[2n, n]$ FSD code. In 2006, Danielsen and Parker [10] proved that every self-dual additive code over $GF(4)$ is equivalent to a graph code. Among additive codes over $GF(4)$, a graph code is an additive code over $GF(4)$ that has a generator matrix of form $\mathcal{C} = \Gamma + \omega I$, where Γ is the adjacency matrix of a simple undirected graph. An $[n, k, d]$ (or $(n, 2^k, d)$) code is *optimal* if there is no $[n, k, d+1]$ (or $(n, 2^k, d+1)$) code, and *near optimal* if there is no $[n, k, d+2]$ (or $(n, 2^k, d+2)$) code. An $[n, k, d]$ (or $(n, 2^k, d)$) code is *known best* if d attains the highest known minimum distance for $[n, k]$ (or $(n, 2^k)$) codes. For parameters of optimal codes, or lower and upper bounds on minimum distances of optimal codes, see [12].

Formally self-dual codes are important class of codes, they have connections to other mathematical structures such as block designs, lattices, modular forms, and sphere packing. Many works have been done on the FSD codes, and optimal FSD codes of length $n \leq 28$ have been classified, see [1,2,13,14]. In 2002, Tonchev [20] set up a relationship between an FSD code and the adjacency matrix of an undirected graph, and showed that some interesting codes can be obtained from graphs with high degree of symmetry, such as strongly regular graphs. In 2006, Danielsen and Parker [10] proved that every self-dual additive code over $GF(4)$ is equivalent to a graph code. In 2012, Danielsen [6] focused his attention on additive codes over $GF(9)$ and transformed the problem of code equivalence into a problem of graph isomorphism. By an extension technique, they classified all optimal codes of lengths 11 and 12. In fact, computer searching reveals that circulant graph codes usually contain many strong codes, and some of these codes have highly regular graph representations, see [21]. In [6], Danielsen obtained some optimal additive codes from circulant graphs in 2005. Later, Varbanov investigated additive circulant graph codes over $GF(4)$, see [21]. Recently, finding optimal codes from graphs has received a wide attention of many researchers, see [6–11,15,20,21]. Inspired by these works, we discuss the construction of (near) optimal FSD codes and additive codes from undirected circulant graphs in this paper.

The paper is organized as follows. Section 2 recalls some concepts in graph theory. In Sect. 3, we propose a new method to find (near) optimal binary linear codes from circulant graphs, and construct some (near) optimal or known best binary linear codes by using this method. In Sect. 4, we propose a new method to find additive optimal codes from circulant graphs.

2 Preliminaries

We introduce some concepts of graph theory for latter use in this paper, for more details please see [5]. An *undirected graph* $\Gamma = (V, E)$ is a set $V(\Gamma) = \{v_1, v_2, \cdots, v_n\}$ of vertices together with a collection $E(\Gamma)$ of edges, where each edge is an unordered pair of vertices. The vertices v_i and v_j are *adjacent* if $\{v_i, v_j\}$ is an edge. Then v_j is a *neighbour* of v_i. All the neighbours of a vertex v_i in Γ form the *neighbourhood* of v_i, and it is denoted by $N_\Gamma(v_i)$. The *degree of a vertex* v is the number of vertices adjacent to v. A graph is *regular* of degree k if all vertices have the same degree k. For a graph $\Gamma = (V, E)$, suppose that V' is a nonempty subset of V. The subgraph of Γ whose vertex set is V' and whose edge set is the set of those edges of Γ that have both ends in V' is called the *subgraph of* Γ *induced by* V', denoted by $\Gamma[V']$. We say that $\Gamma[V']$ is an induced subgraph of Γ. The *adjacency matrix* $A = (a_{ij})$ of $\Gamma = (V, E)$ is a symmetric $(0,1)$-matrix defined as follows: $a_{i,j} = 1$ if the i-th and j-th vertices are adjacent, and $a_{i,j} = 0$ otherwise.

Circulant graphs and their various applications are the objects of intensive study in computer science and discrete mathematics, see [3,4,16,18]. Recently, Monakhova published a survey paper on this subject, see [17]. Let $S = \{a_1, a_2, \cdots, a_k\}$ be a set of integers such that $0 < a_1 < \cdots < a_k < \frac{n+1}{2}$, and let the vertices of an n-vertex graph be labelled as $0, 1, 2, \cdots, n-1$. Then the *circulant graph* $C(n, S)$ has $i \pm a_1, i \pm a_2, \cdots, i \pm a_k \pmod n$ adjacent to each vertex i. A *circulant matrix* is obtained by taking an arbitrary first row, and shifting it cyclically one position to the right in order to obtain successive rows. We say that a circulant matrix is generated by its first row. Formally, if the first row of an n-by-n circulant matrix is $a_0, a_1, \cdots, a_{n-1}$, then the $(i,j)^{th}$ element is a_{j-i}, where subscripts are taken modulo n. The term circulant graph arises from the fact that the adjacency matrix for such a graph is a circulant matrix.

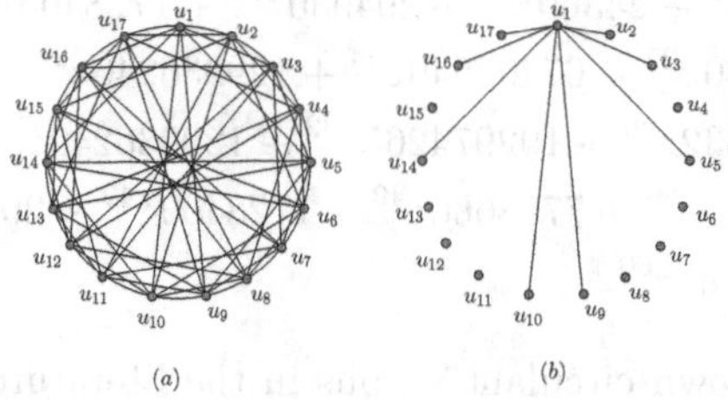

Fig. 1. (*a*) The (4, 4)-Ramsey graph Γ; (*b*) the edge-induced graph $\Gamma(E_1)$.

For example, the (4, 4)-*Ramsey graph* Γ (see Fig. 1) is a famous circulant graph, which can be obtained by regarding the vertices as elements of the field of integers modulo 17, and joining two vertices if their difference is a quadratic residue of 17 (either 1, 2, 4, 8, 9, 13, 15 or 16). For the vertex u_1, we have $E_1 = \{u_1u_2, u_1u_3, u_1u_5, u_1u_9, u_1u_{10}, u_1u_{14}, u_1u_{16}, u_1u_{17}\} \subseteq E(\Gamma)$ and a vector $\alpha_{17} = (0,1,1,0,1,0,0,0,1,1,0,0,0,1,0,1,1)$. It is clear that the adjacency matrix A_{17} of the (4, 4)-Ramsey graph is generated by α_{17}, where

$$A_{17} = \begin{pmatrix} 01101000110001011 \\ 10110100011000101 \\ 11011010001100010 \\ \cdots\cdots\cdots\cdots\cdots \end{pmatrix}.$$

3 New Binary Linear Codes Searching from Circulant Graphs

In this section, we discuss the construction of binary linear codes from circulant graphs. Since a circulant graph Γ_n can be uniquely determined by its adjacency matrix A_n, or by the vector α_n corresponding to E_1 ($E_1 \subseteq E(\Gamma_n)$), whose elements are incident with the vertex $u_1 \in V(\Gamma_n)$. We will make no difference of α_n, A_n or a circulant graph Γ_n, and simply say a circulant graph Γ_n with vector α_n or a vector α_n of a circulant graph Γ_n. And in this section, we denote the binary linear code with generator matrix $G = (I, A_n)$ by $\mathcal{C}_n$. According to the relation between a graph code and the adjacency matrix of an undirected graph introduced by [20], we can get a [34, 17, 8] optimal FSD code from the matrix $(I; A_{17})$, where A_{17} is the adjacency matrix of the (4, 4)-Ramsey graph.

In [6], Danielsen got some optimal additive codes. One of them is the optimal additive code $(30, 2^{30}, 12)$ obtained from the vector $\beta_{30} = (\omega,0,1,1,0,0,0,0,1,1,0,1,1,1,1,1,1,1,1,1,0,1,1,0,0,0,0,1,1,0)$, which corresponds to a circulant graph of order 30 with vector $\alpha_{30} = (0,0,1,1,0,0,0,0,1,1,0,1,1,1,1,1,1,1,1,1,0,1,1,0,0,0,0,1,1,0)$. The graph code $\mathcal{C}_{30}$ with generator matrix $G = (I; A_{30})$ is a [60, 30, 12] code, where A_{30} is the circulant matrix generated by α_{30}. The code $\mathcal{C}_{30}$ is a known best FSD code. The weight enumerator of $\mathcal{C}_{30}$ is

$$\begin{aligned} W_{\mathcal{C}_{30}}(z) = 1 &+ 4060z^{12} + 24360z^{14} + 294930z^{16} + 1728400z^{18} + 7758400z^{20} \\ &+ 26336640z^{22} + 67403540z^{24} + 129936240z^{26} + 192974265z^{28} \\ &+ 220819632z^{30} + 192974265z^{32} + 129936240z^{34} + 67403540z^{36} \\ &+ 26336640z^{38} + 7758660z^{40} + 1728400z^{42} + 294930z^{44} + 24360z^{46} \\ &+ 4060z^{48} + z^{60}. \end{aligned}$$

However, using known circulant graphs in the literature, we only get a few binary linear codes that are optimal, near optimal or known best. So, we need to study a new method for designing good binary linear codes from circulant graphs. To achieve this goal, we give some notations first.

Denote by L_n the highest known minimum distance of all $[2n, n]$ codes. For a graph Γ_n with vector α_n, we let $\mathcal{C}_n$ be its graph code and d_n be the distance of $\mathcal{C}_n$. If $d_n \geq L_n$, then $\mathcal{C}_n$ is a good FSD code. If $d_n < L_n$, then $\mathcal{C}_n$ is a "poor" code and α_n is a "poor" vector. For a graph Γ_n with "poor" vector α_n, we manage to find a new vector α'_n that gives a binary linear code $\mathcal{C}'_n = [2n, n, d'_n]$ such that $\mathcal{C}'_n$ is better than the code $\mathcal{C}_n$. The idea of finding a new vector α'_n from a "poor" vector $\alpha_n = (b_1, b_2, \cdots, b_n)$ is introduced as follows:

(1) Let A_n be the circulant matrix generated by α_n and denote $G_n = (I; A_n)$. Then the j-th row of G_n is

$$\begin{aligned} g_j &= (e_j \mid a_j) \\ &= (0, \cdots, 0, 1, 0, \cdots, 0 \mid a_{j,1}, a_{j,2}, \cdots, a_{j,j-1}, a_{j,j}, a_{j,j+1}, \cdots, a_{j,n}) \\ &= (0, \cdots, 0, 1, 0, \cdots, 0 \mid b_{n-j+2}, b_{n-j+3}, \cdots, b_n, b_1, b_2, \cdots, b_{n-j+1}). \end{aligned}$$

(2) We call codeword $\beta \in \mathcal{C}_n$ "bad" if $wt(\beta) < L_n$. Find "bad" codewords $\beta_1, \beta_2, \cdots, \beta_m$ (if exist) such that their weights are $d_n, d_n+1, \cdots, d_n+(m-1)$, where $m = L_n - d_n$. If there is no codeword with weight $d_n + i$ $(0 \leq i \leq m-1)$, then β_{i-1} is not under consideration.

(3) Let $\beta_i = g_{j_1} + g_{j_2} + \cdots + g_{j_r} = (u_i \mid v_i)$, where $u_i, v_i \in F_2^n$ and $j_1, j_2, \cdots, j_r \in \{1, 2, \cdots, n\}$.

(4) From $g_{j_l} = (e_{j_l} \mid a_{j_l})$, one can define a column vector set $\{\Lambda_i : i = 1, 2, \cdots, n\}$, where $\Lambda_i = (\lambda_{i,j_1}, \lambda_{i,j_2}, \cdots, \lambda_{i,j_r})$ and $\lambda_{i,j_1} = \sum_{l=1}^{r} a_{j_\ell, j_1+(i-1)}$, $\lambda_{i,j_2} = \sum_{l=1}^{r} a_{j_\ell, j_2+(i-1)}$, $\cdots$, $\lambda_{i,j_r} = \sum_{l=1}^{r} a_{j_\ell, j_r+(i-1)}$. Note that all the additions are proceeding modular n.

(5) Using these Λ_i to set up a standard for adjusting a "poor" vector $\alpha_n = (b_1, b_2, \cdots, b_n)$ to α'_n.

The above method can be realized by the following Algorithm 1:

Step 1. Give a circulant graph Γ_n with vector $\alpha_n = (b_1, b_2, \cdots, b_n)$, and generate $G_n = (I; A_n)$.

Step 2. Calculate the distance d_n of binary linear code $\mathcal{C}_n$ (Algorithm 1-1). If $d_n \geq L_n$, then $\mathcal{C}_n$ is a good binary linear code, and stop. Or else, go to Step 3.

Step 3. Do adjustments of the elements of the "poor" vector α_n as follows.

Step 3.1. Find "bad" codewords $\beta_1, \beta_2, \cdots, \beta_m$ such that their weights are $d_n, d_n+1, \cdots, d_n+(m-1)$ by Algorithm 1-2, where $m = L_n - d_n$. If there is no codeword with weight $d_n + i$ $(0 \leq i \leq m-1)$, then β_{i-1} is not under consideration.

Step 3.2. For each β_i $(0 \leq i \leq m-1)$, we can find a combination of β_i by Algorithm 1-2. Suppose $\beta_i = g_{j_1} + g_{j_2} + \cdots + g_{j_r} = (u_i \mid v_i)$, where $u_i, v_i \in F_2^n$ and $j_1, j_2, \cdots, j_r \in \{1, 2, \cdots, n\}$ and

$$\begin{aligned} g_{j_k} &= (e_{j_k} \mid a_{j_k}) \\ &= (0, \cdots, 0, 1, 0, \cdots, 0 \mid a_{j_k,1}, a_{j_k,2}, \cdots, a_{j_k,j_k-1}, a_{j_k,j_k}, a_{j_k,j_k+1}, \cdots, a_{j_k,n}) \\ &= (0, \cdots, 0, 1, 0, \cdots, 0 \mid b_{n-j_k+2}, b_{n-j_k+3}, \cdots, b_n, b_1, b_2, \cdots, b_{n-j_k+1}). \end{aligned}$$

Step 3.3. Determine whether each element 1 of the generator vertex α_n is a "bad" element in the following way (since $b_1 = 0$, we begin with element b_2). If $b_2 = 1$, then $a_{j_1,j_1+1} = a_{j_2,j_2+1} = \cdots = a_{j_r,j_r+1} = b_2 = 1$. We calculate the exact value $\lambda_{2,j_1} = \sum_{\ell=1}^{r} a_{j_\ell,j_1+1}, \lambda_{2,j_2} = \sum_{\ell=1}^{r} a_{j_\ell,j_2+1}, \cdots, \lambda_{2,j_r} = \sum_{\ell=1}^{r} a_{j_\ell,j_r+1}$. Note that $\lambda_{2,j_k} = 0$ or $\lambda_{2,j_k} = 1$ $(1 \le k \le r)$. Consider the set $\Lambda_2 = \{\lambda_{2,j_1}, \lambda_{2,j_2}, \cdots, \lambda_{2,j_r}\}$. If the number of elements with value "0" in Λ_2 is larger than the number of elements with value "1", then the element b_2 is called a *"bad" element* of the generator vector α_n. If b_2 is a "bad" element, then we change $b_2 = 1$ into $b_2' = 0$ and obtain a new vector $\alpha_n' = (b_1, b_2', \cdots, b_n)$. Then we return to Step 1. If b_2 is not a "bad" element or $b_2 = 0$, then we consider b_3 and continue to determining whether b_3 is a "bad" element. The procedure terminates till b_n has been considered.

Algorithm 1-1. Minimum distance of a binary linear code

Input: The value of n, the generator vector α_n of a binary linear code C_n
Objective: The minimum distance of binary linear code C_n
1. Input the value of n, the generator vector $\alpha_n = (b_1, b_2, \cdots, b_n)$;
2. Obtain the generator matrix $G = (I; A_n)$ of the binary linear code C_n;
3. Get the minimum distance of the binary linear code C_n.

For example, let $n = 19$ and $\alpha_n = (0,1,1,0,1,0,0,0,0,1,1,0,0,0,0,1,0,1,1)$. The algorithm details are stated as follows:

Program:

```
n = 19;
a = [0,1,1,0,1,0,0,0,0,1,1,0,0,0,0,1,0,1,1]
m = matrix(GF(2), [[a[(i - k)%n] for i in [0..(n - 1)]] for k in [0..n - 1]]);
f = lambda s : sum(map(lambda x : m[x], s));
s = [];
for k in [1..8]:
    t = min([list(i).count(1) for i in Subsets(range(n), k).map(f)]);
    s+ = [t];
    print k, t;
```

Output: s_i : 1 2 3 4 5 6 7 8
s_i' : 8 6 4 2 2 4 2 2

Result: The elements of the first row $(s_1, s_2, \cdots, s_8)$ are the contribution of the matrix I for the weight of a codeword. The elements of the second row $(s_1', s_2', \cdots, s_8')$ are the contribution of the matrix A_{19} for the weight of a codeword. The value of $\min\{s_i + s_i' \mid 1 \le i \le 8\} = 6$ is the minimum weight of the code C_{19} and then the minimum distance of the code C_{19} is also 6.

Algorithm 1–2. "Bad" codewords and their combinations

Input: The value of n, the generator vector α_n of a binary linear code $\mathcal{C}_n$
Objective: "Bad" codewords and their combinations
1. Input the value of n, the generator vector $\alpha_n = (b_1, b_2, \cdots, b_n)$;
2. Obtain the generator matrix $G = (I; A_n)$ of the binary linear code $\mathcal{C}_n$;
3. Get "bad" codewords $\beta_1, \beta_2, \cdots, \beta_m$ and a combination of each β_i $(1 \le i \le m)$.

For example, let $n = 19$ and $\alpha_n = (0, 1, 1, 0, 1, 0, 0, 0, 0, 1, 1, 0, 0, 0, 0, 1, 0, 1, 1)$.
The algorithm details are stated as follows:

Program:

```
n = 19;
a = [0,1,1,0,1,0,0,0,0,1,1,0,0,0,0,1,0,1,1]
m = matrix(GF(2), [[a[(i - k)%n] for i in [0..(n - 1)]] for k in [0..n - 1]]);
f = lambda s : sum(map(lambda x : m[x], s));
g = lambda s : str(sorted(map(lambda x : x + 1, s))).replace('[','').replace(']','');
s = [];
for k in [1..8]:
    t = min([(i, list(f(i)).count(1)) for i in Subsets(range(n), k)],
    key = lambda x : x[-1]);
    s+ = [t];
    print k, t[1], g(t[0]);
```

Output:

s_i	s'_i	$\{j_1, j_2, \cdots, j_r\}$ (as defined in Step 3.2)
1	8	{1}
2	6	{1, 9}
3	4	{1, 4, 12}
4	2	{1, 2, 6, 16}
5	2	{1, 2, 8, 11, 14}
6	4	{1, 2, 3, 4, 6, 13}
7	2	{1, 2, 4, 6, 7, 10, 17}
8	2	{1, 2, 3, 6, 7, 8, 12, 16}

Now we show how to use our Algorithm 1 for searching $[38, 19, 8]$ codes.

Step 1. Among all graphs with 19 vertices, we consider the graph Γ_{19}, which can be generated by the edge set $E_1 = \{u_1u_2, u_1u_3, u_1u_5, u_1u_{10}, u_1u_{11}, u_1u_{16}, u_1u_{18}, u_1u_{19}\}$, and then $\alpha_{19} = (0, 1, 1, 0, 1, 0, 0, 0, 0, 1, 1, 0, 0, 0, 0, 1, 0, 1, 1)$. Obviously, $b_2 = b_3 = b_5 = b_{10} = b_{11} = b_{16} = b_{18} = b_{19} = 1$.

Step 2. From the Code Tables, we know that the lower bound of the minimum distance of linear code $[38, 19]$ over $GF(2)$ is 8, that is, $L_{19} = 8$. By Algorithm 1-1, we obtain that the minimum distance d_{19} of the code $(I; A_{19})$ is just 6, that is, $d_{19} = 6$. Clearly, $6 = d_{19} < L_{19} = 8$.

Step 3. Obviously, $m = L_{19} - d_{19} = 2$.

Step 3.1. From Algorithm 1–2, we find two "bad" codewords

$$\beta_1 = (1,1,0,0,0,1,0,0,0,0,0,0,0,0,0,1,0,0,0 \\ |\,0,0,0,0,0,0,0,1,0,0,0,0,0,1,0,0,0,0,0),$$
$$\beta_2 = (1,0,0,1,0,0,0,0,0,0,0,1,0,0,0,0,0,0,0 \\ |\,0,1,1,0,0,1,0,0,0,0,0,0,0,0,0,0,0,1,0).$$

such that their weights are 6 and 7, that is, $wt(\beta_1) = 6$ and $wt(\beta_2) = 7$.

Step 3.2. For β_1, we can find a combination of $\beta_1 = \alpha_{19,1} + \alpha_{19,2} + \alpha_{19,6} + \alpha_{19,16}$ by Algorithm 2-2, where

$$\alpha_{19,1} = (1,0,0,0,0,0,0,0,0,0,0,0,0,0,0,0,0,0,0 \\ |\,0,1,1,0,1,0,0,0,0,1,1,0,0,0,0,1,0,1,1),$$
$$\alpha_{19,2} = (0,1,0,0,0,0,0,0,0,0,0,0,0,0,0,0,0,0,0 \\ |\,1,0,1,1,0,1,0,0,0,0,1,1,0,0,0,0,1,0,1),$$
$$\alpha_{19,6} = (0,0,0,0,0,1,0,0,0,0,0,0,0,0,0,0,0,0,0 \\ |\,0,1,0,1,1,0,1,1,0,1,0,0,0,0,1,1,0,0,0),$$
$$\alpha_{19,16} = (0,0,0,0,0,0,0,0,0,0,0,0,0,0,0,1,0,0,0 \\ |\,1,0,0,0,0,1,1,0,0,0,0,1,0,1,1,0,1,1,0).$$

Note that $r = 4$, $j_1 = 1$, $j_2 = 2$, $j_3 = 6$ and $j_4 = 16$.
For β_2, we can find a combination of $\beta_2 = \alpha_{19,1} + \alpha_{19,4} + \alpha_{19,12}$ by Algorithm 2-2, where

$$\alpha_{19,1} = (1,0,0,0,0,0,0,0,0,0,0,0,0,0,0,0,0,0,0 \\ |\,0,1,1,0,1,0,0,0,0,1,1,0,0,0,0,1,0,1,1),$$
$$\alpha_{19,4} = (0,0,0,1,0,0,0,0,0,0,0,0,0,0,0,0,0,0,0 \\ |\,0,1,1,0,1,1,0,1,0,0,0,0,1,1,0,0,0,0,1),$$
$$\alpha_{19,12} = (0,0,0,0,0,0,0,0,0,0,0,1,0,0,0,0,0,0,0 \\ |\,0,1,1,0,0,0,0,1,0,1,1,0,1,1,0,1,0,0,0).$$

Note that $r = 3$, $j_1 = 1$, $j_2 = 4$ and $j_3 = 12$.

Step 3.3. Recall that $\alpha_{19} = (0,1,1,0,1,0,0,0,0,1,1,0,0,0,0,1,0,1,1)$ and $b_2 = b_3 = b_5 = b_{10} = b_{11} = b_{16} = b_{18} = b_{19} = 1$. Since $b_2 = 1$, we consider whether b_2 is a "bad" element in α_{19}.

For β_1, since $r = 4$, $j_1 = 1$, $j_2 = 2$, $j_3 = 6$ and $j_4 = 16$, we have $a_{1,2} = a_{2,3} = a_{6,7} = a_{16,17} = b_2 = 1$, and

$$\lambda_{2,j_1} = \sum_{\ell=1}^{r} a_{j_\ell,j_1+1} = a_{1,2} + a_{2,2} + a_{6,2} + a_{16,2} = 0,$$
$$\lambda_{2,j_2} = \sum_{\ell=1}^{r} a_{j_\ell,j_2+1} = a_{1,3} + a_{2,3} + a_{6,3} + a_{16,3} = 0,$$
$$\lambda_{2,j_3} = \sum_{\ell=1}^{r} a_{j_\ell,j_3+1} = a_{1,7} + a_{2,7} + a_{6,7} + a_{16,7} = 0,$$
$$\lambda_{2,j_4} = \sum_{\ell=1}^{r} a_{j_\ell,j_4+1} = a_{1,17} + a_{2,17} + a_{6,17} + a_{16,17} = 0.$$

For β_2, since $r = 3$, $j_1 = 1$, $j_2 = 4$ and $j_3 = 12$, we have $a_{1,2} = a_{4,5} = a_{12,13} = b_2 = 1$. Then

$$\lambda_{2,j_1} = \sum_{\ell=1}^{r} a_{j_\ell, j_1+1} = a_{1,2} + a_{4,2} + a_{12,2} = 1,$$
$$\lambda_{2,j_2} = \sum_{\ell=1}^{r} a_{j_\ell, j_2+1} = a_{1,5} + a_{4,5} + a_{12,5} = 0,$$
$$\lambda_{2,j_3} = \sum_{\ell=1}^{r} a_{j_\ell, j_3+1} = a_{1,13} + a_{4,13} + a_{12,13} = 0.$$

It is clear that the number of elements with value "0" in Λ_2's is larger than the number of elements with value "1", then the element b_2 is called a *"bad" element* of the generator vector α_n. We change $b_2 = 1$ into $b_2' = 0$ and obtain a new vector $\alpha_{19}' = (0,0,1,0,1,0,0,0,0,1,1,0,0,0,0,1,0,1,1)$. Then we return to Step 1.

Let us now investigate the linear code C_{19}' with generator matrix $G = (I; A_{19}')$. By Algorithm 1-1, we get that the minimum distance d_{19}' of the linear code C_{19}' is 8. The code C_{19}' is a near optimal and also known best binary linear code over $GF(2)$. The weight enumerator of the code C_{19}' is

$$W_{C_{19}'}(z) = 1 + 133z^8 + 2052z^{10} + 10108z^{12} + 36575z^{14} + 85595z^{16} + 127680z^{18} + 127680z^{20} + 85595z^{22} + 36575z^{24} + 10108z^{26} + 2052z^{28} + 133z^{30} + z^{38}.$$

With the above approach and algorithms, we can also find three other near optimal binary linear $[38, 19, 8]$ codes by the generator matrices $G = (I; A_{19}'')$, $G = (I; A_{19}''')$ and $G = (I; A_{19}'''')$. The circulant matrices A_{19}'', A_{19}''' and A_{19}'''' are separately generated by

$$\alpha_{19}'' = (0,1,1,0,1,0,0,0,0,0,1,0,0,0,0,1,0,1,1),$$
$$\alpha_{19}''' = (0,1,1,0,1,0,0,0,0,1,0,0,0,0,0,1,0,1,1),$$
$$\alpha_{19}'''' = (0,1,1,0,1,0,0,0,0,1,1,0,0,0,0,1,0,1,0).$$

The weight enumerator of the code C_{19}'' is

$$W_{C_{19}''}(z) = 1 + 190z^8 + 1767z^{10} + 10507z^{12} + 36860z^{14} + 84341z^{16} + 128478z^{18} + 128478z^{20} + 84341z^{22} + 36860z^{24} + 10507z^{26} + 1767z^{28} + 190z^{30} + z^{38}.$$

One can also check that the weight enumerators of the codes C_{19}''' and C_{19}'''' are equal to the ones of C_{19}'' and C_{19}', respectively.

At the end of this section, we list some more binary linear codes constructed from known circulant graphs by applying Algorithm 1.

1. In [6], Danielsen got a $(15, 2^{15})$ additive code from the vector $(\omega, 0, 1, 1, 1, 0, 0, 1, 1, 0, 0, 1, 1, 1, 0)$, which corresponds to a circulant graph of order 15 with vector $\alpha_{15} = (0,0,1,1,1,0,0,1,1,0,0,1,1,1,0)$. Its graph code $C_{15} = [30, 15, 6]$ is a poor code. Applying Algorithm 1, we obtain a new vector $\alpha_{15}' =$

$(0,0,1,1,1,0,0,1,1,0,0,1,1,0,0)$. The code $\mathcal{C}'_{15} = [30,15,8]$ generated by α'_{15} is an optimal binary linear code. The weight enumerator of this code $\mathcal{C}'_{15}$ is

$$W_{\mathcal{C}'_{15}}(z) = 1 + 450z^8 + 1848z^{10} + 5040z^{12} + 9045z^{14} + 9045z^{16} \\ +5040z^{18} + 1848z^{20} + 450z^{22} + z^{30}.$$

2. Extending vector $\alpha_{17} = (0,1,1,0,1,0,0,0,1,1,0,0,0,1,0,1,1)$ of the $(4,4)$-Ramsey graph, one can obtain a vector of length 19. For example, let $\alpha_{19} = (0,1,1,0,1,0,0,0,0,1,1,0,0,0,0,1,0,1,1)$, which corresponds to a circulant graph of order 19. It corresponds to the linear graph code $\mathcal{C}_{19} = [38,19,6]$, which is a poor code. By Algorithm 1, we obtain four new vectors as follows:

$\alpha'_{19} = (0,0,1,0,1,0,0,0,0,1,1,0,0,0,0,1,0,1,1),$
$\alpha''_{19} = (0,1,1,0,1,0,0,0,0,1,1,0,0,0,0,1,0,1,0),$
$\alpha'''_{19} = (0,1,1,0,1,0,0,0,0,1,0,0,0,0,0,1,0,1,1),$
$\alpha''''_{19} = (0,1,1,0,1,0,0,0,0,0,1,0,0,0,0,1,0,1,1).$

The codes $\mathcal{C}'_{19}$ and $\mathcal{C}''_{19}$ are $[38,19,8]$ codes with the same weight enumerators. The weight enumerator of the code $\mathcal{C}'_{19}$ is

$$W_{\mathcal{C}'_{19}}(z) = 1 + 133z^8 + 2052z^{10} + 10108z^{12} + 36575z^{14} + 85595z^{16} \\ +127680z^{18} + 127680z^{20} + 85595z^{22} + 36575z^{24} + 10108z^{26} \\ +2052z^{28} + 133z^{30} + z^{38}.$$

The codes $\mathcal{C}'''_{19}$ and $\mathcal{C}''''_{19}$ are $[38,19,8]$ codes with the same weight enumerators. The weight enumerator of the code $\mathcal{C}'''_{19}$ is

$$W_{\mathcal{C}'''_{19}}(z) = 1 + 190z^8 + 1767z^{10} + 10507z^{12} + 36860z^{14} + 84341z^{16} \\ +128478z^{18} + 128478z^{20} + 84341z^{22} + 36860z^{24} + 10507z^{26} \\ +1767z^{28} + 190z^{30} + z^{38}.$$

These four binary linear codes are all near optimal and known best.

3. In addition, we consider graphs with large number of vertices by similar approach. Let $\mathcal{C}_{25,1}$ and $\mathcal{C}_{25,2}$ be the two codes generated by vectors $\alpha^1_{25} = (0,1,1,0,1,0,1,0,0,0,0,1,1,0,0,0,0,1,0,1,0,1,0,1,1)$ and $\alpha^2_{25} = (0,1,1,0,1,0,1,0,1,0,0,0,0,1,1,0,0,0,0,1,0,1,0,1,1)$, respectively. These two codes are both poor codes. Then by Algorithm 1, we find $\alpha'_{25} = (0,1,0,0,1,0,1,0,0,0,0,1,1,0,0,0,0,1,0,1,0,1,0,1,1)$, and $\alpha''_{25} = (0,1,1,0,1,0,1,0,1,0,0,0,0,1,1,0,0,0,0,1,0,1,0,0,1)$. Both α'_{25} and α''_{25} give binary linear codes known best. Let $\mathcal{C}'_{25}$ and $\mathcal{C}''_{25}$ be codes of α'_{25} and α''_{25}, respectively. It is easy to check that $\mathcal{C}'_{25}$ and $\mathcal{C}''_{25}$ are $[50,25,10]$ codes, and the weight enumerator of the code $\mathcal{C}'_{25}$ and $\mathcal{C}''_{25}$ is

$$\begin{aligned}W_{C'_{25}}(z) &= 1 + 225z^{10} + 1250z^{11} + 3825z^{12} + 11525z^{13} + 28050z^{14} + 64005z^{15} \\ &+ 147075z^{16} + 294975z^{17} + 535075z^{18} + 9111100z^{19} + 1409205z^{20} \\ &+ 1999925z^{21} + 2642200z^{22} + 3219675z^{23} + 3623325z^{24} + 377243z^{25} \\ &+ 3621975z^{26} + 3216050z^{27} + 2643475z^{28} + 2009175z^{29} + 1408010z^{30} \\ &+ 904475z^{31} + 535400z^{32} + 292725z^{33} + 147525z^{34} + 68880z^{35} \\ &+ 27975z^{36} + 9775z^{37} + 3500z^{38} + 1125z^{39} + 375z^{40} + 125z^{41} \\ &= W_{C''_{25}}(z).\end{aligned}$$

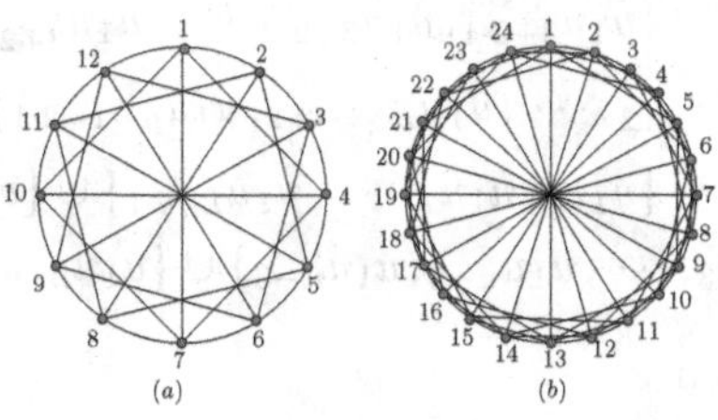

Fig. 2. (*a*) A 5-valent graph Γ_{12}; (*b*) Circulant graph Γ_{24}.

4 New Additive Codes Searching from Circulant Graphs

In fact, Glynn et al. [11] obtained an optimal code from a circulant graph, called a 5-valent graph. Recall that Γ_{12} is a circulant graph of order 12; see Fig. 2(*a*). Let $V(\Gamma_{12}) = \{u_1, u_2, \cdots, u_{12}\}$. For the vertex u_1, we let $E_1 = \{v_1v_2, v_1v_4, v_1v_7, v_1v_{10}, v_1v_{12}\} \subseteq E(\Gamma_{12})$. For the vertex u_2, we just rotate the above vertices and edges, that is, we only permit the existence of the edge set $E_2 = \{v_2v_3, v_2v_4, v_2v_6, v_2v_{10}, v_2u_{12}\} \subseteq E(G)$. For each vertex $u_i \in V(\Gamma) \setminus \{u_1, u_2\} = \{u_3, u_4, \cdots, u_{12}\}$, we can also obtain an edge set E_i $(3 \leq i \leq 17)$. Observe that $E(\Gamma_{12}) = \bigcup_{i=1}^{12} E_i$. The adjacency matrix of the graph Γ_{12} is the following circulant matrix

$$A_{12} = \begin{pmatrix} 010100100101 \\ 101010010010 \\ 010101001001 \\ \cdots\cdots\cdots \end{pmatrix}.$$

The above matrix can also be obtained by vector $\alpha_{12} = (0, 1, 0, 1, 0, 0, 1, 0, 0, 1, 0, 1)$. Observe that this vector just corresponds to the set E_1 of edges, which is an expression of the adjacency relation for the vertex u_1. We conclude that a 5-valent graph can be determined by the edge set E_1, and the adjacency matrix of this graph is determined by the above vector α_{12}. Furthermore, the matrix $A'_{12} = A_{12} + \omega I$ is also a circulant matrix, which can be obtained by vector $\alpha'_{12} = (\omega, 1, 0, 1, 0, 0, 1, 0, 0, 1, 0, 1)$. From the matrix $A_{12} + \omega I$, we can get a graph code $\mathcal{C}_{12}$. From the Code Tables, we know that $\mathcal{C}_{12}$ is an optimal $(12, 2^{12}, 6)$ additive code over $GF(4)$. The above statement suggests the following method for finding (near) optimal additive codes.

Step 1. Given an even integer n. Denote by L_n the lower bound of the additive code $(n,2^n)$ over $GF(4)$. From the Code Tables, we find the exact value of L_n for given n. We now construct a circulant graph Γ_n by a set E_1 of edges as follows.

Step 1.1. Arrange the vertices from $V(\Gamma_n) = \{u_1, u_2, \cdots, u_n\}$ in a circular order.

Step 1.2. Determine the set E_1 of edges satisfying $|E_1| = L_n - 1$ or $|E_1| = L_n + 1$, where $E_1 = \{u_i u_1 \mid u_i \in N_{\Gamma_n}(u_1)\}$. If $|E_1| = L_n + 1$, then

$$\begin{aligned}E_1 &= \{u_1u_2, u_1u_3\} \cup \{u_1u_{3+2\cdot 1}, u_1u_{3+2\cdot 2}, \cdots, u_1u_{3+2\cdot\frac{L_n-4}{2}},\} \cup \{u_1u_{\frac{n}{2}+1}\}\\ &\quad \cup \{u_1u_{n-1-2\cdot\frac{L_n-4}{2}}, \cdots, u_1u_{n-1-2\cdot 2}, u_1u_{n-1-2\cdot 1}\} \cup \{u_1u_n, u_1u_{n-1}\}\\ &= \{u_1u_2, u_1u_3\} \cup \{u_1u_5, u_1u_7, \cdots, u_1u_{L_n-1}\} \cup \{u_1u_{\frac{n}{2}+1}\}\\ &\quad \cup \{u_1u_{n-L_n+3}, \cdots, u_1u_{n-5}, u_1u_{n-3}\} \cup \{u_1u_n, u_1u_{n-1}\}.\end{aligned}$$

If $|E_1| = L_n - 1$, then

$$\begin{aligned}E_1 &= \{u_1u_2, u_1u_3\} \cup \{u_1u_{3+2\cdot 1}, u_1u_{3+2\cdot 2}, \cdots, u_1u_{3+2\cdot\frac{L_n-6}{2}},\} \cup \{u_1u_{\frac{n}{2}+1}\}\\ &\quad \cup \{u_1u_{n-1-2\cdot\frac{L_n-6}{2}}, \cdots, u_1u_{n-1-2\cdot 2}, u_1u_{n-1-2\cdot 1}\} \cup \{u_1u_n, u_1u_{n-1}\}\\ &= \{u_1u_2, u_1u_3\} \cup \{u_1u_5, u_1u_7, \cdots, u_1u_{L_n-3}\} \cup \{u_1u_{\frac{n}{2}+1}\}\\ &\quad \cup \{u_1u_{n-L_n+5}, \cdots, u_1u_{n-5}, u_1u_{n-3}\} \cup \{u_1u_n, u_1u_{n-1}\}.\end{aligned}$$

Step 2. By the edge set E_1, we write the vector α_n corresponding to E_1. If $|E_1| = L_n + 1$, then

$$\alpha_n = (0\ \overset{u_2}{1}\ \overset{u_3}{1}\ 0\ \overset{u_5}{1} \cdots 0\ \overset{u_{L_n-1}}{1}\ 0\ 0 \cdots 0\ \overset{u_{\frac{n}{2}+1}}{1}\ 0\ 0 \cdots 0\ \overset{u_{n-L_n+3}}{1}\ 0 \cdots \overset{u_{n-3}}{1}\ 0\ \overset{u_{n-1}}{1}\ \overset{u_n}{1}).$$

If $|E_1| = L_n - 1$, then

$$\alpha_n = (0\ \overset{u_2}{1}\ \overset{u_3}{1}\ 0\ \overset{u_5}{1} \cdots 0\ \overset{u_{L_n-3}}{1}\ 0\ 0 \cdots 0\ \overset{u_{\frac{n}{2}+1}}{1}\ 0\ 0 \cdots 0\ \overset{u_{n-L_n+5}}{1}\ 0 \cdots \overset{u_{n-3}}{1}\ 0\ \overset{u_{n-1}}{1}\ \overset{u_n}{1}).$$

Step 3. Change the first component of the vector α_n into ω. Denote by α'_n the new vector. We generate a circulant matrix A'_n from α'_n,
$\alpha'_n = (\omega, 1, 1, 0, 1, 0, 1, \cdots, 0, 1, 0, 0, \cdots, 0, 1, 0, 0, \cdots, 0, 1, 0, 1, 0, \cdots, 1, 0, 1, 1)$.

Step 4. By Algorithm 2, we obtain the minimum distance d_n of the additive code $\mathcal{C}_n$ and determine whether $d_n = L_n$. If so, the code $\mathcal{C}_n$ is a (near) optimal or at least known best additive code.
Below is an algorithm (running in SAGE). For more details, we refer to [19].

Algorithm 2. Minimum distance of a circulant graph code

Input: The value of n, the generator vector α_n of a circulant graph code $\mathcal{C}_n$
Objective: The minimum distance of the circulant graph code $\mathcal{C}_n$

1. Input the value of n, the generator vector $\alpha_n = (b_1, b_2, \cdots, b_n)$;
2. Obtain the generator matrix G of the circulant graph code $\mathcal{C}_n$;
3. Get the minimum distance of the circulant graph code $\mathcal{C}_n$.

For example, let $n = 24$ and $\alpha_n = (\omega, 1, 1, 0, 1, 0, 0, 0, 0, 0, 0, 0, 1, 0, 0, 0, 0, 0, 0, 0, 1, 0, 1, 1)$. The algorithm details are stated as follows:

Program:

```
F.<x> = GF(4,'x')
n = 24;
a = [x,1,1,0,1,0,0,0,0,0,0,0,1,0,0,0,0,0,0,0,1,0,1,1]
m = matrix(F,n,n,[[a[(i - k)%n] for i in [0..(n - 1)]] for k in [0..n - 1]]);
f = lambda s : sum(map(lambda x : m[x],s));
s = [ ];
for k in [1..8]:
    t = min([n - list(i).count(0) for i in Subsets (range(n),k).map(f)]);
    s+ = [t];
    print s;
```

Output:

```
[8]
[8, 8]
[8, 8, 10]
[8, 8, 10, 8]
[8, 8, 10, 8, 8]
[8, 8, 10, 8, 8, 8]
[8, 8, 10, 8, 8, 8, 8]
[8, 8, 10, 8, 8, 8, 8, 10]
```

Result: The minimum element of the last array is the minimum distance d_{24} of the code $\mathcal{C}_{24}$, that is, $d_{24} = 8$.

Inspired by the graph code $\mathcal{C}_{12}$ which corresponds to the 5-valent graph, we hope to find out some other optimal additive codes for $n = 24$.

Step 1. Recall that L_{24} is the lower bound of the additive code $(24, 2^{24})$ over $GF(4)$. From the Code Tables, we find that the exact value of L_{24} is 8, that is, $L_{24} = 8$. We now construct a circulant graph Γ_{24} by a set E_1 of edges as follows.

Step 1.1. Arrange the vertices from $V(\Gamma_n) = \{u_1, u_2, \cdots, u_{24}\}$ in a circular order.

Step 1.2. Determine the set E_1 of edges satisfying $|E_1| = L_{24} - 1 = 7$ or $|E_1| = L_{24}+1 = 9$, where $E_1 = \{u_i u_1 \mid u_i \in N_{\Gamma_n}(u_1)\}$. If $|E_1| = 9$, then $E_1 = \{u_1u_2, u_1u_3, u_1u_5, u_1u_7, u_1u_{13}, u_1u_{19}, u_1u_{21}, u_1u_{23}, u_1u_{24}\}$. If $|E_1| = 7$, then $E_1 = \{u_1u_2, u_1u_3, u_1u_5, u_1u_{13}, u_1u_{21}, u_1u_{23}, u_1u_{24}\}$. In this case, the circulant graph Γ_{24} can be found out; see Fig. 2(b).

Step 2. By the edge set E_1, we write the vector α_{24} corresponding to E_1. If $|E_1| = 9$, then $\alpha_{24} = (0, 1, 1, 0, 1, 0, 1, 0, 0, 0, 0, 0, 1, 0, 0, 0, 0, 0, 1, 0, 1, 0, 1, 1)$. If $|E_1| = 7$, then $\alpha_{24} = (0, 1, 1, 0, 1, 0, 0, 0, 0, 0, 0, 0, 1, 0, 0, 0, 0, 0, 0, 0, 1, 0, 1, 1)$.

Step 3. Change the first component of the vector α_{24} into ω. Denote by α'_{24} the new vector. We generate a circulant matrix A'_{24} (A''_{24}) from α'_{24} (α''_{24}):

$$\alpha'_{24} = (\omega, 1, 1, 0, 1, 0, 1, 0, 0, 0, 0, 0, 1, 0, 0, 0, 0, 0, 1, 0, 1, 0, 1, 1),$$
$$\alpha''_{24} = (\omega, 1, 1, 0, 1, 0, 0, 0, 0, 0, 0, 0, 1, 0, 0, 0, 0, 0, 0, 0, 1, 0, 1, 1).$$

Step 4. By Algorithm 2, we obtain the minimum distance $d'_{24} = d''_{24} = 8$. Therefore, both C'_{24} and C''_{24} are known best additive codes over $GF(4)$. The weight enumerators of the codes C'_{24} and C''_{24} are

$$W_{C'_{24}}(z) = 1 + 528z^8 + 13992z^{10} + 171276z^{12} + 1118040z^{14} + 3773517z^{16} + 6218520z^{18} + 4413948z^{20} + 1034088z^{22} + 33306z^{24},$$
$$W_{C''_{24}}(z) = 1 + 648z^8 + 13032z^{10} + 174636z^{12} + 1111320z^{14} + 3781917z^{16} + 6211800z^{18} + 4417308z^{20} + 1033128z^{22} + 33426z^{24}.$$

Applying the above method, we can obtain (near) optimal additive codes over $GF(4)$ from the first two generator vectors.

n	d	L_n	First row of generator matrix	about the code
8	4	4	$(\omega, 1, 1, 0, 1, 0, 1, 1)$	optimal
8	4	4	$(\omega, 1, 0, 0, 1, 0, 0, 1)$	optimal
10	4	4	$(\omega, 1, 1, 0, 0, 1, 0, 0, 1, 1)$	optimal
10	4	4	$(\omega, 1, 0, 0, 0, 1, 0, 0, 0, 1)$	optimal
16	6	6	$(\omega, 1, 1, 0, 1, 0, 0, 0, 1, 0, 0, 0, 1, 0, 1, 1)$	optimal
22	8	8	$(\omega, 1, 1, 0, 1, 0, 1, 0, 0, 0, 1, 0, 0, 0, 1, 0, 1, 0, 1, 1)$	optimal
24	8	8	$(\omega, 1, 1, 0, 1, 0, 1, 0, 0, 0, 0, 0, 1, 0, 0, 0, 0, 0, 1, 0, 1, 0, 1, 1)$	known best
24	8	8	$(\omega, 1, 1, 0, 1, 0, 0, 0, 0, 0, 0, 0, 1, 0, 0, 0, 0, 0, 0, 0, 1, 0, 1, 1)$	known best

From the above analysis, we see that the circulant graphs under our consideration are all relatively sparse. So one may think that only sparse circulant graphs produce optimal graph codes. However, the following fact gives it a negative answer. Danielsen [6] obtained an optimal additive code $(30, 2^{30}, 12)$ from the vector $\alpha_{30} = (\omega, 0, 1, 1, 0, 0, 0, 0, 1, 1, 0, 1, 1, 1, 1, 1, 1, 1, 1, 0, 1, 1, 0, 0, 0, 0, 1, 1, 0)$, which corresponds to a circulant graph of order 30 such that its adjacency matrix A_{30} is generated by the first row $\alpha_{30} = (0, 0, 1, 1, 0, 0, 0, 0, 1, 1, 0, 1, 1, 1, 1, 1, 1, 1, 1, 1, 0, 1, 1, 0, 0, 0, 0, 1, 1, 0)$. It is clear that the circulant graph is 17-regular. Since the order of this graph is 30, it follows that the degree of each vertex is relatively large, i.e., it is a relatively dense graph. Let $\alpha_n = (\omega, 0, 1, 1, 0, 0, 0, 0, 1, 1, 0, \overbrace{1, 1, \ldots, 1, 1}^{n-21}, 0, 1, 1, 0, 0, 0, 0, 1, 1, 0)$. For $n = 32$ and $n = 34$, we have the following vectors:

$$\alpha_{32} = (\omega, 0, 1, 1, 0, 0, 0, 0, 1, 1, 0, 1, 1, 1, 1, 1, 1, 1, 1, 1, 1, 1, 0, 1, 1, 0, 0, 0, 0, 1, 1, 0),$$
$$\alpha_{34} = (\omega, 0, 1, 1, 0, 0, 0, 0, 1, 1, 0, 1, 1, 1, 1, 1, 1, 1, 1, 1, 1, 1, 1, 1, 0, 1, 1, 0, 0, 0, 0, 1, 1, 0).$$

By Algorithm 2, we know that C_{32} and C_{34} are $(32, 2^{32}, 10)$ and $(34, 2^{34}, 10)$ additive codes, respectively, and both are known best additive codes over $GF(4)$.

The weight enumerators of the codes $\mathcal{C}_{32}$ and $\mathcal{C}_{34}$ are

$$\begin{aligned}W_{\mathcal{C}_{32}}(z) = {} & 1 + 1325z^{10} + 41973z^{12} + 745155z^{14} + 8030541z^{16} + 53150370z^{18} \\ & + 213875634z^{20} + 510617670z^{22} + 691665390z^{24} \\ & + 491629473z^{26} + 159600905z^{28} + 17838471z^{30} + 286740z^{32}, \\ W_{\mathcal{C}_{34}}(z) = {} & 1 + 492z^{10} + 14373z^{12} + 291849z^{14} + 3494061z^{16} + 26279603z^{18} \\ & + 123536402z^{20} + 357928154z^{22} + 620714798z^{24} + 614055698z^{26} \\ & + 319190777z^{28} + 75747789z^{30} + 6157556z^{32} + 72095z^{34}.\end{aligned}$$

5 Concluding Remarks

In recent years, graphs have been used to construct codes. But, usually they cannot produce good codes. Proper graphs have to be chosen in order to construct good codes. Because the structure of a circulant graph is very symmetric, the row vectors of its adjacency matrix may span a subspace with the property that the minimum Hamming distance among the vectors of the subspace is comparatively large. Our paper uses this possibility to develop two algorithms for searching for good binary linear codes and additive codes. Starting from a circulant graph, the algorithms modify the generator vector of the circulant graph successively, and get a good code at some step, or fail when all "1"'s of the generator vector have been checked. Therefore, sometimes our algorithms can find good codes, but sometimes they would fail to produce any good code. In general, the running time of our algorithms is exponential in the order n of the circulant graph, since we have to generate all the vectors of the subspace spanned by the row vectors of an $n \times n$ or $n \times 2n$ generator matrix, and this complexity cannot be lower down generally. However, as is seen in the above sections, the row vectors of a circulant are generated by a single vector–its first row vector. This gives us some reasonable hope to use the property to lower down the complexity of generating the subspace spanned by the row vectors of a circulant, which will be left for us to further study.

References

1. Betsumiya, K., Harada, M.: Binary optimal odd formally self-dual codes. Des. Codes Cryptogr. **23**, 11–21 (2001)
2. Betsumiya, K., Harada, M.: Classification of formally self-dual even codes of lengths up to 16. Des. Codes Cryptogr. **23**, 325–332 (2001)
3. Bermond, J.C., Comellas, F., Hsu, D.F.: Distributed loop computer networks: a survey. J. Parallel Distribut. Comput. **24**, 2–10 (1995)
4. Boesch, F.T., Wang, J.F.: Reliable circulant networks with minimum transmission delay. IEEE Trans. Circuits Syst. **32**, 1286–1291 (1985)
5. Bondy, J.A., Murty, U.S.R.: Graph Theory. Graduate Texts in Mathematics, vol. 244. Springer, London (2008)

6. Danielsen, L.E.: On Self-Dual Quantum Codes, Graphs and Boolean Functions. University of Bergen, Norway (2005)
7. Danielsen, L.E.: Graph-based classification of self-dual additive codes over finite field. Adv. Math. Commun. **3**(4), 329–348 (2009)
8. Danielsen, L.E.: On the classification of Hermitian self-dual additive codes over GF(9). IEEE Trans. Inform. Theory **58**(8), 5500–5511 (2012)
9. Danielsen, L.E., Parker, M.G.: Directed graph representation of half-rate additive codes over GF(4). Des. Codes Cryptogr. **59**, 119–130 (2011)
10. Danielsen, L.E., Parker, M.G.: On the classification of all self-dual additive codes over $GF(4)$ of length up to 12. J. Combin. Theory Series **113**, 1351–1367 (2006)
11. Glynn, D.G., Gulliver, T.A., Marks, J.G., Gupta, M.K.: The Geometry of Additive Quantum Codes, Preface. Springer, Berlin (2006)
12. Grassl, M.: Bounds on the minimum distance of linear codes. http://www.codetables.de. Accessed 31 October 2013
13. Gulliver, T.A., Östergård, P.R.J.: Binary optimal linear rate 1/2 codesraphs. Discrete Math. **283**, 255–261 (2004)
14. Han, S., Lee, H., Lee, Y.: Binary formally self-dual odd codes. Des. Codes Cryptogr. **61**, 141–150 (2011)
15. Huffman, W.C., Pless, V.: Fundamentals of Error-Correcting Codes. Cambridge University, Cambridge (2003)
16. Mans, B., Pappalardi, F., Shparlinski, I.: On the spectral Adam property for circulant graphs. Discrete Math. **254**(1–3), 309–329 (2002)
17. Monakhova, E.A.: A survey on undirected circulant graphs. Discrete Math. Algor. Appl. **4**(1), 1250002 (2012)
18. Muzychuk, M.E., Tinhofer, G.: Recognizing circulant graphs of prime order in polynomial time. Electron. J. Combin. **5**(1), 501–528 (1998)
19. Stein, W.A., et al.: Sage Mathematics Software (Version 6.1.1), The Sage Development Team (2014). http://www.sagemath.org
20. Tonchev, V.: Error-correcting codes from graphs. Discrete Math. **257**, 549–557 (2002)
21. Varbanov, Z.: Additive circulant graph codes over GF(4). Math. Maced. **6**, 73–79 (2008)

Dynamic Single-Source Shortest Paths in Erdös-Rényi Random Graphs

Wei Ding[1](✉) and Ke Qiu[2]

[1] Zhejiang University of Water Resources and Electric Power, Hangzhou 310018, Zhejiang, China
dingweicumt@163.com

[2] Department of Computer Science, Brock University, St. Catharines, Canada
kqiu@brocku.ca

Abstract. This paper studies the *dynamic* single-source shortest paths (SSSP) in Erdös-Rényi random graphs generated by $G(n, p)$ model. In 2014, Ding and Lin (AAIM 2014, LNCS 8546, 197–207) first considered the dynamic SSSP in general digraphs with *arbitrary* positive weights, and devised a nontrivial local search algorithm named DSPI which takes at most $O(n \cdot \max\{1, n \log n/m\})$ expected update time to handle a single weight increase, where n is the number of nodes and m is the number of edges in the digraph. DSPI also works on undirected graphs. This paper analyzes the expected update time of DSPI dealing with edge weight increases or edge deletions in Erdös-Rényi (a.k.a., $G(n, p)$) random graphs. For weighted $G(n, p)$ random graphs with arbitrary positive edge weights, DSPI takes at most $O(h(T_s))$ expected update time to deal with a single edge weight increase as well as $O(pn^2 h(T_s))$ total update time, where $h(T_s)$ is the height of input SSSP tree T_s. For $G(n, p)$ random graphs, DSPI takes $O(\ln n)$ expected update time to handle a single edge deletion as well as $O(pn^2 \ln n)$ total update time when $20 \ln n/n \leq p < \sqrt{2 \ln n/n}$, and $O(1)$ expected update time to handle a single edge deletion as well as $O(pn^2)$ total update time when $p > \sqrt{2 \ln n/n}$. Specifically, DSPI takes the least total update time of $O(n \ln n h(T_s))$ for weighted $G(n, p)$ random graphs with $p = c \ln n/n, c > 1$ as well as $O(n^{3/2}(\ln n)^{1/2})$ for $G(n, p)$ random graphs with $p = c\sqrt{\ln n/n}, c > \sqrt{2}$.

Keywords: Dynamic SSSP · Weight increase · Edge deletion · Random graph

1 Introduction

Given a weighted digraph, a *single-source shortest paths* (SSSP) algorithm computes the shortest paths from a given source node to all the other nodes. When dynamic changes occur to the digraph, a *dynamic* SSSP algorithm maintains the shortest paths. One could recompute the shortest paths using the *static* algorithms [8,15], but a truly dynamic algorithm seeks for updating operations using fundamental properties of the shortest paths, and is expected to run faster than recomputation by the static algorithms.

Z. Lu et al. (Eds.): COCOA 2015, LNCS 9486, pp. 537–550, 2015.
DOI: 10.1007/978-3-319-26626-8_39

Dynamic edge changes of a digraph include *weight update* and *topology update*. Weight update includes *weight increase* and *decrease*, and topology update includes *edge deletions* and *insertions*. Topology update can be realized by weight update. When dealing with only weight updates, an algorithm is said to be *fully dynamic* if it can handle both weight increases and decreases, is *incremental* if it only can handle weight increases but not decreases, and is *decremental* if it only can handle weight decreases but not increases [1,7,13,23,27,28]. When dealing with only topology updates, an algorithm is said to be *fully dynamic* if it can handle both edge insertions and deletions, is *incremental* if it can handle only edge insertions but not deletions, and is *decremental* if it can handle only deletions but not insertions [2–4,12,18–21,24,25]. Incremental and decremental algorithms are sometimes collectively called *partially dynamic*.

1.1 Previous Work

The *dynamic SSSP problem* with topology update (in particular, edge deletions) has been studied extensively. For *undirected unweighted* graphs, Even and Shiloach [12] presented an $O(mn)$ total update time decremental algorithm, which has been the fastest known algorithm for three decades. Roditty and Zwick [25] proved that the decremental and incremental SSSP problems on unweighted digraphs are at least as hard as the Boolean matrix multiplication problem. This hardness result hints that it will be difficult to improve Even and Shiloach's algorithm. Recently, a number of papers have beaten the barrier in approximation and/or probability sense. Bernstein and Roditty [4] devised an $O(n^{2+O(1/\sqrt{\log n})})$ total update time decremental algorithm to maintain $(1+\epsilon)$-approximate SSSP. This is the first algorithm that breaks the long-standing $O(mn)$ update time barrier on decremental SSSP problem, in not-too-sparse graphs. Very recently, Henzinger *et al.* [18] presented a faster $(1+\epsilon)$-approximation decremental SSSP algorithm which improves the total update time to $O(n^{1.8+O(1/\sqrt{\log n})} + m^{1+O(1/\sqrt{\log n})})$, and further reduced the expected total update time to $O(m^{1+O(\sqrt{\log\log n/\log n})})$ [20]. For *undirected weighted* graphs where the edge weights are integers from 1 to W, Henzinger *et al.* [20] proposed $(1+\epsilon)$-approximation decremental SSSP algorithm with an $O(m^{1+O(\sqrt{\log\log n/\log n})}\log W)$ expected total update time. Henzinger and King [17] discovered that Even and Shiloach's algorithm can be easily adapted to digraphs. For *weighted digraphs* where edge weights are integers from 1 to W, King [21] devised a decremental SSSP algorithm with $O(mnW)$ total update time. The combination of the techniques used in Bernstein's random algorithms [2,3] and Madry's algorithm [22] yielded a $(1+\epsilon)$-approximation randomized decremental SSSP algorithm with $\widetilde{O}(mn\log W)$ total update time, where the $\widetilde{O}(\cdot)$ notation hides polylog$n$ factors. More recently, Henzinger *et al.* [19] first proposed an improved $(1+\epsilon)$-approximation randomized decremental SSSP algorithm with $\widetilde{O}(mn^{0.986})$ total update time for unweighted digraphs, and further extended it to obtain an $O(mn^{0.986}\log^2 W)$ total update time for weighted digraphs. This is the first algorithm that breaks the $O(mn)$ total update time barrier on decremental SSSP problem in digraphs.

Until now, few of literatures dealt with the dynamic SSSP problem with weight updates. Fakcharoemphol and Rao [13] studied the fully dynamic variant in planar digraphs, and devised an $O(n^{\frac{4}{5}} \log^{\frac{13}{5}} n)$ amortized time algorithm. Demetrescu and Italiano [7] proposed the open problem on whether or not we can solve efficiently (*i.e.*, better than recomputation) fully dynamic SSSP problem (with both weight increases and decreases) in general graphs. In [9], Ding and Lin studied the incremental SSSP problem in weighted digraphs with arbitrary positive arc weights. They examined some properties of the shortest paths and presented a nontrivial local search algorithm named DSPI which can handle a single arc weight increase in at most $O(n \cdot \max\{1, n \log n/m\})$ expected time. That is, the total update time is at most $O(\max\{mn, n^2 \log n\})$. Specifically, when $m = \Omega(n \log n)$, DSPI is a linear-time incremental SSSP algorithm. To the best of our knowledge, DSPI is the first incremental SSSP algorithm for handling weight increases in weighted digraphs with arbitrary positive arc weights.

1.2 Our Results

We observe that Ding and Lin's local search algorithm DSPI [9] can work on undirected graphs. In this paper, we rewrite DSPI using pointer approach, and analyze the expected update time of DSPI handling edge weight increases or edge deletions in Erdös-Rényi (a.k.a. $G(n,p)$) random graphs [6,10,11]. This paper also considers the weighted Erdös-Rényi random graphs, denoted by $G_w(n,p)$, generated by applying $G(n,p)$ model to $\binom{n}{2}$ possible weighted edges with arbitrary positive weights. Note that Erdös-Rényi is abbreviated to ER throughout this paper, and all the results shown below are obtained in probability sense.

For $G_w(n,p)$ random graphs with arbitrary positive edge weights, DSPI takes at most $O(h(T_s))$ expected update time to handle a single edge weight increase as well as $O(pn^2 h(T_s))$ total update time, where $h(T_s)$ is the height of input SSSP tree T_s. Specifically, when $p = c \ln n/n$ with a constant $c > 1$, DSPI takes the least total update time of $O(n \ln n h(T_s))$. For $G(n,p)$ random graphs, DSPI takes the same update time as that on $G_w(n,p)$ random graphs when $\ln n/n < p < 20 \ln n/n$, $O(\ln n)$ expected update time to handle a single edge deletion as well as $O(pn^2 \ln n)$ total update time when $20 \ln n/n \le p < \sqrt{2 \ln n/n}$, and $O(1)$ expected update time to handle a single edge deletion as well as $O(pn^2)$ total update time when $p > \sqrt{2 \ln n/n}$. We observe $\ln n/n$, $20 \ln n/n$ and $\sqrt{2 \ln n/n}$ are three threshold values of p, and DSPI takes less (*resp.* more) expected update time in denser (*resp.* sparser) random graphs to deal with a single edge deletion in general. Specifically, when $p = c\sqrt{\ln n/n}$ with a constant $c > \sqrt{2}$, DSPI takes the least total update time of $O(n^{3/2}(\ln n)^{1/2})$. Thus, DSPI is a best decremental SSSP algorithm for $G(n,p)$ random graphs with $p > \sqrt{2 \ln n/n}$, that is, the best for most of large-scale Erdös-Rényi random graphs.

Organization. The rest of the paper is organized as follows. In Sect. 2, we recall Ding and Lin's local search algorithm named DSPI and rewrite it using pointer approach. In Sect. 3, we analyze the expected update time of DSPI handling a single edge weight increase as well as the total update time in weighted ER random graphs with arbitrary positive weights. In Sect. 4, we analyze the expected update time of DSPI handling a single edge deletion as well as the total update time in ER random graphs. Finally, we offer concluding remarks in Sect. 5. Due to page limit, most proofs are omitted in this paper.

2 Ding and Lin's Local Search Algorithm

In this section, we recall Ding and Lin's local search algorithm named DSPI and rewrite it using pointer approach.

2.1 Preliminaries

Let $D = (V, A, w, s)$ be a weighted digraph, where V is the node set, A is the arc set, s is a designated node called *source*, and $w(\cdot)$ is a weight function $w : A \to \mathbb{R}^+$. Suppose the weight of an arc a increases from $w(a)$ up to $w'(a)$ while all the other arc weights remain unchanged, and $w'(a)$ always remains positive in this paper. The resulting digraph is denoted by $D' = (V, A, w', s)$. Let $d^*_D(s,v)$ (*resp.* $d^*_{D'}(s,v)$) denote the shortest path distance from s to node v in D (*resp.* D'), and T_s (*resp.* T'_s) denote the *single-source shortest paths tree* in D (*resp.* D'). The **partially dynamic SSSP problem with a single weight increase**, abbreviated to the **incremental SSSP problem**, aims to update T_s to T'_s when a single arc weight increase occurs.

Obviously, T_s and T'_s are both a rooted tree at s. Let T_u (*resp.* T'_u) be the subtree of T_s (*resp.* T'_s) rooted at $u \in V$. We use $\pi_{T_s}(s,v)$ to denote the s-to-v simple path in T_s, and use $d_{T_s}(s,v)$ to denote the length of $\pi_{T_s}(s,v)$ which is equal to the sum of all arc weights on $\pi_{T_s}(s,v)$. Clearly, $d_{T_s}(s,v) = d^*_D(s,v)$ and $d_{T'_s}(s,v) = d^*_{D'}(s,v)$ for all $v \in V$. Let $V(\cdot)$ and $A(\cdot)$ be the node set and arc set of one digraph or subgraph, respectively. For any subset $U \subseteq V$, we use $S(D,U)$ to denote the subgraph of D induced by U.

Let $a = (u,v)$ be the arc in D from u to v, and $w(u,v)$ or $w(a)$ denote its weight. We call u the *tail* of a and denote u by t_a, and call v the *head* of a and denote v by h_a. So, $a = (t_a, h_a)$. We call a an *outgoing arc* from t_a and an *incoming arc* to h_a. For any $v \in V$, we use $\mathcal{I}(v)$ and $\mathcal{O}(v)$ to denote the set of tails of all the incoming arcs to v and the set of heads of all the outgoing arcs from v, respectively.

When a weight increase occurs to $a \in A$, an auxiliary graph $D_a = (V_a, A_a, w_a)$ is constructed from D based on a in the following way. Let

$$b(v') = \arg \min_{u \in \mathcal{I}(v') - V(T_{h_a})} \{d_T(s,u) + w(u,v')\}, \quad \forall v' \in V(T_{h_a}), \tag{1}$$

and call $b(v')$ a *bridge node* of v'. Let $b(v') = \text{null}$ and $(b(v'), v') = \text{null}$ if $b(v')$ does not exist in D'. In addition, we use an arc $\epsilon(v')$ to represent the path comprising $\pi_{T_s}(s, b(v'))$ and arc $(b(v'), v')$, for any $v' \in V(T_{h_a})$. Let $V_a = V(T_{h_a}) \cup \{s\}$, and

$$A_a = A(S(D, V(T_{h_a}))) \cup \{\epsilon(v) : v \in V(T_{h_a})\}, \tag{2}$$

and

$$w_a(r) = \begin{cases} d_{T_s}(s, b(v)) + w(b(v), v) & \text{if} \quad r = \epsilon(v), \\ w(r) & \text{otherwise,} \end{cases} \quad \forall r \in A_a, \tag{3}$$

where $w(b(v), v) = \infty$ if $(b(v), v) = \text{null}$.

Ding and Lin examine several fundamental properties of the shortest paths [9], based on which the local search algorithm is devised.

Theorem 1 (see [9]**).** *For any $a \notin T_s$, it always holds that $d^*_{D'}(s, v) = d_{T_s}(s, v)$ for all $v \in V$ no matter how much $w(a)$ increases by to $w'(a)$.*

Theorem 2 (see [9]**).** *For any $a \in T_s$, regardless of how much $w(a)$ increases by to $w'(a)$, it always holds that $d^*_{D'}(s, v) = d_{T_s}(s, v)$ for all $v \in V - V(T_{h_a})$.*

Theorem 3 (see [9]**).** *For any $a \in T_s$, regardless of how much $w(a)$ increases by to $w'(a)$, it always holds that $d^*_{D'}(s, v) = d^*_{D_a}(s, v)$ for all $v \in V(T_{h_a})$.*

2.2 Local Search Algorithm

By Theorems 1, 2, 3, we need to discuss the situation of the arc a with its weight increased, *i.e.*, discuss whether $a \notin T_s$ or $a \in T_s$. When $a \notin T_s$, we claim by Theorem 1 that $\pi_{T_s}(s, v)$ is also an s-to-v shortest path in D', for any $v \in V$. So, T_s is also a single-source shortest paths tree in D' with s as origin. When $a \in T_s$, we conclude from Theorem 2 that $\pi_{T_s}(s, u)$ is also an s-to-u shortest path in D' for any $u \in V - V(T_{h_a})$, and from Theorem 3 that one is an s-to-v shortest path in D' iff it is an s-to-v shortest path in D_a for any $v \in V(T_{h_a})$. Therefore, our main task is to update the s-to-v shortest path for all $v \in V(T_{h_a})$.

Let $f(v)$ be the *head pointer* of v which records the parent node of v, for any node v of a rooted tree $\mathcal{T}$. All the head pointers form a set

$$\mathcal{F} = \{f(v) : v \in V(\mathcal{T})\}. \tag{4}$$

Any rooted tree corresponds to a unique set of head pointers. We can easily obtain all the children of each nonleaf node in $\mathcal{T}$ by traversing the head pointer set of $\mathcal{T}$, and thus can tour $\mathcal{T}$ in $O(|V(\mathcal{T})|)$ time, *i.e.*, a linear time, based on its head pointer set.

Let $\mathcal{F}_s$ (*resp.* $\mathcal{F}'_s$) be the head pointer set of T_s (*resp.* T'_s) as defined in Eq. (4), which takes the place of T_s (*resp.* T'_s). Let $f(v)$ (*resp.* $f'(v)$) denote the head pointer of v in T_s (*resp.* T'_s). Note that $f(s) = \text{null}$ and $f'(s) = \text{null}$. The essence of updating T_s to T'_s is to update $\mathcal{F}_s$ to $\mathcal{F}'_s$. Based on above analysis, we claim that $f(v)$ stays unchanged, for any $v \in V$ when $a \notin T_s$ and for any $v \in V - V(T_{h_a})$ when $a \in T_s$. Therefore, our main task is actually to update $f(v)$ to $f'(v)$, for any $v \in V(T_{h_a})$ when $a \in T_s$. As a fact that $f'(v), v \in V(T_{h_a})$, is either $b(v)$ or one node in $\mathcal{I}(v) \cap V(T_{h_a})$, both $b(v)$ and all the nodes in $\mathcal{I}(v) \cap V(T_{h_a})$ are candidates of $f'(v)$.

By Eq. (1), we need to visit all the nodes in $\mathcal{I}(v)$ so as to find $b(v)$ for any $v \in V(T_{h_a})$, judging whether or not they are in T_{h_a}. For this purpose, we define a 0-1 variable $J(v)$. Specifically, $J(v) = 0$ means v is not in T_{h_a} while $J(v) = 1$ means v is in T_{h_a}. Initially, set $J(u) = 0$ for all $u \in V$. In addition, we use $\text{dist}(v)$ (*resp.* $\text{dist}'(v)$) to record the shortest path distance in D (*resp.* D') from s to any $v \in V$. Obviously, the value of $\text{dist}(v)$ and $\text{dist}'(v)$ is equal to $d_{T_s}(s, v)$ and $d_{T'_s}(s, v)$, respectively.

Above discussions leads to a *local search algorithm*, named DSPI. In order to facilitate implementing DSPI, we use appropriate linked lists to store digraphs and SSSP trees. Initially, $\mathcal{F}'_s$ is equal to $\mathcal{F}_s$, and $\text{dist}'(\cdot)$ is equal to $\text{dist}(\cdot)$. In Step_0, DSPI judges whether or not $a \in T_s$. DSPI goes to Step_1 if $a \in T_s$ and returns if $a \notin T_s$. In Step_1, DSPI initializes $J(v)$ to zero for all $v \in V$, builds a linked list to store $\mathcal{F}_s$, and then uses DFS to tour T_{h_a} with h_a as origin. During the tour, a series of update works are done every time a new node is visited. For example, when $v \in V(T_{h_a})$ is visited, DFS updates $J(v)$ to one, finds $b(v)$ by Eq. (1), replaces $f(v)$ with $b(v)$ and updates $\text{dist}(v)$ to $\text{dist}(b(v)) + w(b(v), v)$. When DFS terminates, DSPI goes to Step_2. In this step, DSPI replaces $\mathcal{O}(s)$ with $V(T_{h_a})$ into D', and then uses Dijkstra's algorithm with s as origin in the updated D' to compute the single-source shortest paths from s to the nodes with $J(\cdot) = 1$. Finally, $\text{dist}(\cdot)$ and $f(\cdot)$ are updated accordingly.

Algorithm DSPI:
Input: $D = (V, A, w, s)$, $w'(a)$, $\mathcal{F}_s$ and $\text{dist}(\cdot)$ of D; **Output:** $\mathcal{F}'_s$ and $\text{dist}'(\cdot)$ of D'
Step_0: If $a \notin T_s$, then return; if $a \in T_s$, then goto Step_1; **Step_1**: $J(v) \leftarrow 0, \forall v \in V$; build a linked list to store T_s based on $\mathcal{F}_s$; use DFS to tour T_{h_a} with h_a as origin, and in the meanwhile visit $\mathcal{I}(v)$ and do the following works for all $v \in V(T_{h_a})$: (1) $J(v) \leftarrow 1$; (2) find $b(v)$ using Eq. (1); (3) $f(v) \leftarrow b(v)$; $\text{dist}(v) \leftarrow \text{dist}(b(v)) + w(b(v), v)$; **Step_2**: Take $\mathcal{O}(s) \leftarrow V(T_{h_a})$ into D'; use Dijkstra's algorithm with s as origin in updated D' to compute the single-source shortest paths from s to the nodes with $J(\cdot) = 1$; accordingly update $\text{dist}(\cdot)$ and $f(\cdot)$;

Suppose that a single arc weight increase in $D = (V, A, w, s)$ occurs at random from the *uniform distribution* on A. Theorems 4 and 5 show the expected update time of DSPI handling a single arc weight increase in a weighted digraph with arbitrary positive weights.

Theorem 4 (see [9]). *Given any $D = (V, A, w, s)$, the expected update time of DSPI dealing with a single arc weight increase is*

$$O(\frac{1}{|A|}\sum_{u\in V}\sum_{v\in V(T_u)}|\mathcal{I}(v)| + \frac{1}{|A|}\sum_{u\in V}|V(T_u)|\log|V(T_u)| + \frac{1}{|A|}\sum_{u\in V}|\mathcal{I}(u)|). \quad (5)$$

Theorem 5 (see [9]). *Given any $D = (V, A, w, s)$ with n nodes and m arcs, the expected update time of DSPI dealing with a single arc weight increase is at most $O(n \cdot \max\{1, n\log n/m\})$.*

Since the topology of D stays unchanged when a series of arc weight increases occur to D, the *total update time* of DSPI is actually equal to the sum of time costs required by applying DSPI to all m arcs of D. Therefore, the corollary below follows directly from Theorem 5.

Corollary 1. *The total update time of DSPI in any $D = (V, A, w, s)$ with n nodes and m arcs is at most $O(\max\{mn, n^2\log n\})$.*

We observe that DSPI can be applied to undirected graphs. Each undirected weighted edge is associated with two arcs with a same weight as the edge. So, each weighted undirected graph can be transformed into a bi-directed weight-symmetric digraph. The occurrence of a single edge weight increase in an undirected graph means a same weight increase occurs to two arcs associated with this edge in the bi-directed weight-symmetric digraph. So, we can apply DSPI to handle a single edge weight increase in an undirected graph by using DSPI twice to handle two arc weight increases in the corresponding bi-directed weight-symmetric digraph. As a fact that at least one arc is not in T_s, the time cost of applying DSPI to an undirected graph is almost the same as on a digraph.

3 Weight Increase in Weighted ER Random Graphs

In this section, we first present several properties of ER random graphs, and then analyze the expected update time of DSPI dealing with a single edge weight increase in weighted ER random graphs.

3.1 ER and Weighted ER Random Graphs

The Erdös-Rényi random graph, denoted by the $G(n, p)$ random graph, is referred to as the random graph generated by the $G(n, p)$ *model*, which was first introduced by Erdös-Rényi [10,11] in 1959 and also independently proposed by Solomonoff and Rapoport [26] and Gilbert [16]. See [6] for more details. Erdös-Rényi random graphs have been applied in a wide range of theoretical studies [6,14].

The $G(n, p)$ model can be formally stated as follows: given a set $\mathcal{N}$ of n nodes, the edge between each pair of nodes is selected with a uniform probability p, independent of all other edges. There are $\binom{n}{2}$ possible edges in total. One $G(n, p)$

random graph contains a subset of edges, and thus the $G(n,p)$ model can produce $2^{\binom{n}{2}}$ random graph [6,16]. This paper also considers the *weighted* ER random graph, denoted by the $G_w(n,p)$ random graph, which is generated by applying the $G(n,p)$ model to $\binom{n}{2}$ possible edges with arbitrary positive weights. It is easy to verify that $G(n,p)$ and $G_w(n,p)$ random graphs have the same properties in terms of random topologies.

The node set of $G(n,p)$ random graph is $\mathcal{N}$ and its edge set is denoted by $E(G(n,p))$. For any $v \in \mathcal{N}$, we use $in(v)$ to denote the set of adjacent edges to v in $G(n,p)$. Clearly, $|in(v)|$ and $|E(G(n,p))|$ are both a random variable. Let $N = \frac{1}{2}n(n-1)$. We know that $|E(G(n,p))|$ subjects to the *binomial distribution* $B(N,p)$ [6]. Obviously, $\frac{|in(v)|}{|E(G(n,p))|}$ is also a random variable, and its possible values are

$$\frac{|in(v)|}{|E(G(n,p))|} = \frac{x}{y}, \quad 1 \leq y \leq N, 0 \leq x \leq \min\{y, n-1\}. \tag{6}$$

Note that the fraction is not simplified in the paper. For instance, we differentiate $\frac{2}{4}$ with $\frac{1}{2}$. Lemma 2 shows the distribution law of $\frac{|in(v)|}{|E(G(n,p))|}$.

The issue studied in this section can be stated as follows: given an original $G_w(n,p)$ random graph, a designated source node s and an SSSP tree T_s of the original graph, we are asked to update T_s to T_s' when one edge weight of the original graph is increased. Obviously, DSPI can be applied to solve this problem. The rough analysis on its expected update time for this problem is shown in Theorem 6. First of all, we need to consider whether or not the random graphs generated by $G(n,p)$ model is connected. Lemma 1 shows a property of $G(n,p)$ random graphs.

Lemma 1 (see Theorem 4.6 in [5]). *Given $p = c\ln n/n$, $G(n,p)$ has no isolated node if $c > 1$ and has at least one isolated node if $c < 1$.*

Lemma 1 means the random $G(n,p)$ graph is connected when $p > \ln n/n$. In the rest of this paper, we focus on the case of $p > \ln n/n$.

Theorem 6. *Given $p > \ln n/n$, DSPI takes at most $O(n)$ expected update time to deal with a single edge weight increase in $G_w(n,p)$, and at most $O(pn^3)$ total update time.*

Proof. The theorem follows by taking $m = \binom{n}{2}p > \frac{1}{2}(n-1)\log n$ into Theorem 5 and Corollary 1. □

3.2 Fundamental Lemmas

In this section, we show several properties of $G(n,p)$ random graphs, which will play an important role in the detailed analysis on the expected update time of DSPI handling a single edge weight increase in $G_w(n,p)$ random graphs.

Lemma 2. *The $G(n,p)$ random graph satisfies, for any $v \in \mathcal{N}$,*

$$\Pr[\frac{|in(v)|}{|E(G(n,p))|} = \frac{x}{y}] = \binom{n-1}{x}\binom{N-(n-1)}{y-x} p^y (1-p)^{N-y}, \quad \forall 1 \le y \le N, 0 \le x \le \min\{y, n-1\}. \tag{7}$$

Lemma 3. $\mathbb{E}[\frac{|in(v)|}{|E(G(n,p))|}] \le O(\frac{1}{n}), \forall v \in \mathcal{N}$.

Similarly, $\frac{1}{|E(G(n,p))|}$ is also a random variable since $|E(G(n,p))|$ is a random variable which subjects to $B(N,p)$. Lemma 4 estimates the upper bound on the expectation of $\frac{1}{|E(G(n,p))|}$.

Lemma 4. $\mathbb{E}[\frac{1}{|E(G(n,p))|}] \le \frac{4}{p(n^2-n+2)}$.

3.3 Expected Update Time in Weighted ER Random Graphs

Given any $G_w(n,p)$ random graph and any designated source s, we also use T_s to denote the SSSP tree in $G_w(n,p)$ with s as origin, T_u to denote its subtree rooted at u, and $h(T_s)$ to denote the height of T_s. Let TIME denote the expected time of DSPI dealing with a single edge weight increase in $G_w(n,p)$. By Theorem 4, we replace m with $|E(G_w(n,p))|$ into Eq. (5) to obtain

$$\begin{aligned}\text{TIME} = O(\frac{1}{|E(G_w(n,p))|}(&\sum_{u\in V}\sum_{v\in V(T_u)}|in(v)| \\ &+\sum_{u\in V}|V(T_u)|\log|V(T_u)| + \sum_{u\in V}|in(u)|)).\end{aligned} \tag{8}$$

From the fact that $|E(G_w(n,p))|$ as well as all of $|in(u)|, |V(T_u)|, \forall u \in V$, are random variables, we know TIME is also a random variable. The theorem below shows the upper bound on the expected update time of DSPI dealing with a single edge weight increase in $G_w(n,p)$.

Theorem 7. *Given any $G_w(n,p)$ random graph with $p > \ln n/n$, the expected update time of DSPI handling a single edge weight increase in $G_w(n,p)$ is*

$$\mathbb{E}[\text{TIME}] = O(\max\{\frac{1}{n}, \frac{4\log n}{p(n^2-n+2)}\}\sum_{u\in V}|V(T_u)|). \tag{9}$$

Given any $G_w(n,p)$ random graph, Theorem 7 implies that the upper bound on the expected update time of DSPI handling a single edge weight increase in graph depends on the value of $\sum_{u\in V}|V(T_u)|$. It is easy to see that this sum depends on the topology of the SSSP tree T_s in graph, and the tree topology is determined by the topology and edge weights of graph. As a consequence, we consider the worst-case SSSP tree topology and analyze the upper bound on this sum. Lemma 5 shows the both upper bound and lower bound on the value of $\sum_{u\in V}|V(T_u)|$.

Lemma 5. *Given any edge-weighted connected graph $G = (V, E, w)$ with n nodes and any designated source s, the SSSP tree T_s in G satisfies*

$$2n + \frac{1}{2}h^2(T_s) - \frac{3}{2}h(T_s) \le \sum_{u \in V} |V(T_u)| \le nh(T_s) - \frac{1}{2}h^2(T_s) + \frac{1}{2}h(T_s). \quad (10)$$

Theorem 8. *The expected update time of DSPI handling a single edge weight increase in $G_w(n,p)$ with $p > \frac{\ln n}{n}$ is $O(h(T_s)\max\{1, \frac{4n\log n}{p(n^2-n+2)}\})$.*

Proof. From Theorem 7 and Lemma 5, it follows directly that

$$\begin{aligned}\mathbb{E}[\text{TIME}] &\le O(\max\{\frac{1}{n}, \frac{4\log n}{p(n^2-n+2)}\}[nh(T_s) - \frac{1}{2}h^2(T_s) + \frac{1}{2}h(T_s)]) \\ &\le O(\max\{\frac{1}{n}, \frac{4\log n}{p(n^2-n+2)}\}nh(T_s)) \\ &= O(h(T_s)\max\{1, \frac{4n\log n}{p(n^2-n+2)}\}).\end{aligned}$$

□

Corollary 2. *Given $p > \ln n/n$, DSPI takes at most $O(h(T_s))$ expected update time to deal with a single edge weight increase in $G_w(n,p)$, as well as at most $O(pn^2h(T_s))$ total update time. Specifically, when $p = c\ln n/n$ where c is a constant larger than 1, the total update time of DSPI is $O(n\ln nh(T_s))$.*

4 Edge Deletion in ER Random Graphs

An unweighted graph can be viewed as a weighted graph with a uniform edge weight. As mentioned in Sect. 1, to delete one edge can be realized by replacing the weight of this edge with positive infinity. So, DSPI also can be applied to handle a single edge deletion on $G(n,p)$ random graphs. The scenario can be stated as follows: given an original $G(n,p)$ random graph, a designated source s and an SSSP tree T_s of the original graph, we are asked to update T_s to T'_s when one edge of the original graph is deleted. In this section, we analyze the expected update time of DSPI in the scenario.

For any pair of nodes u and v of $G(n,p)$ random graphs, the u-to-v shortest path, also named the *minimum hop path*, is also referred to as a simple path on graph with the least number of edges. The longest one among all-pairs shortest paths is called the *diameter* of $G(n,p)$ random graphs and denoted by diam$(G(n,p))$ [6]. Obviously, the height of SSSP tree with any node as its origin is no more than diam$(G(n,p))$. So,

$$h(T_s) \le \text{diam}(G(n,p)), \quad \forall s \in \mathcal{N}. \quad (11)$$

According to Theorem 4.17 in [5] and its proof therein, we present an important property of diam$(G(n,p))$ in Lemma 6.

Lemma 6 (see Theorem 4.17 in [5]). *Given $p = c \ln n/n, c \geq 20$, the diameter of $G(n,p)$ random graph is $O(\ln n)$.*

The combination of Corollary 2 and Lemma 6 yields Theorem 9.

Theorem 9. *The expected update time of DSPI handling a single edge deletion in $G(n,p)$ is $O(\ln n)$ when $p \geq 20 \ln n/n$, and is $O(h(T_s))$ when $\ln n/n < p < 20 \ln n/n$.*

Lemma 7 shows another important property of $\text{diam}(G(n,p))$.

Lemma 7 (see Theorem 4.5 in [5]). *Given $p = c\sqrt{\ln n}/\sqrt{n}$, if $c > \sqrt{2}$ then $\text{diam}(G(n,p))$ is at most two and if $c < \sqrt{2}$ then $\text{diam}(G(n,p))$ is greater than two.*

Theorem 10. *Given $p > \sqrt{2 \ln n/n}$, the expected update time of DSPI dealing with a single edge deletion in $G(n,p)$ is $O(1)$.*

In fact, a series of edge deletions can be taken as a series of edge weight replacements from their original values to positive infinity. Therefore, the topology of the given $G(n,p)$ random graph remains unchanged when DSPI deals with a series of edge deletions. By combining $|E(G(n,p))| = \binom{n}{2}p$ with Theorems 9 and 10, respectively, we directly obtain the corollary below. Specifically, when $p = c\sqrt{\ln n/n}$ with a constant c larger than $\sqrt{2}$, we conclude from Theorem 10 that the total update time is

$$p\binom{n}{2}O(1) = O(pn^2) = O(cn^2\sqrt{\ln n/n}) = O(n^{3/2}(\ln n)^{1/2}).$$

Corollary 3. *The total update time of DSPI in $G(n,p)$ is $O(pn^2 h(T_s))$ when $\ln n/n < p < 20 \ln n/n$, is $O(pn^2 \ln n)$ when $20 \ln n/n \leq p < \sqrt{2 \ln n/n}$, is $O(n^2 p)$ when $p > \sqrt{2 \ln n/n}$, and is $O(n^{3/2}(\ln n)^{1/2})$ when $p = c\sqrt{\ln n/n}$ where c is a constant larger than $\sqrt{2}$.*

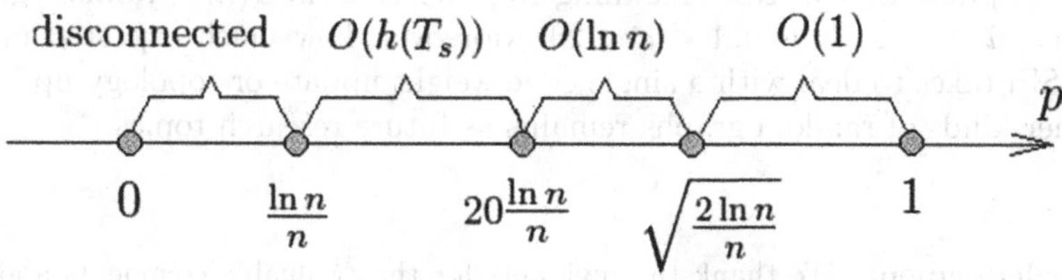

Fig. 1. Expected update time of DSPI varies with the value of p.

We know from Lemma 1 that the $G(n,p)$ random graph is disconnected when $p < \ln n/n$, and from Theorems 9, 10 and Corollary 3 that DSPI takes at most $O(h(T_s))$ expected update time to handle a single edge deletion as well as $O(pn^2h(T_s))$ total update time when $\ln n/n < p < 20\ln n/n$, $O(\ln n)$ expected update time to handle a single edge deletion as well as $O(pn^2 \ln n)$ total update time when $20\ln n/n \leq p < \sqrt{2\ln n/n}$, and $O(1)$ expected update time to handle a single edge deletion as well as $O(n^2p)$ total update time when $p > \sqrt{2\ln n/n}$. See Fig. 1. Specifically, DSPI takes the least total update time, $O(n^{3/2}(\ln n)^{1/2})$, when $p = c\sqrt{\ln n/n}$ with a constant $c > \sqrt{2}$. In general, DSPI has less expected updated time for denser random graphs while more expected update time for sparser random graphs. Clearly, $\ln n/n$, $20\ln n/n$ and $\sqrt{2\ln n/n}$ are three threshold values of p. Since $\sqrt{2\ln n/n}$ converges to zero when n tends to infinity, DSPI takes only $O(1)$ expected update time as well as $O(pn^2)$ total update time in large-scale Erdös-Rényi random graphs generated by $G(n,p)$ model even using a very small positive number p.

5 Concluding Remarks

In this paper, we have analyzed the expected update time of Ding and Lin's algorithm named DSPI [9] handling edge weight increases or edge deletions in Erdös-Rényi random graphs. For $G_w(n,p)$ random graphs with arbitrary positive edge weights, DSPI takes at most $O(h(T_s))$ expected update time to handle a single edge weight increase as well as $O(pn^2h(T_s))$ total update time, where $h(T_s)$ is the height of input SSSP tree T_s. Specifically, when $p = c\ln n/n$ with a constant $c > 1$, DSPI takes the least total update time of $O(n\ln nh(T_s))$.

For $G(n,p)$ random graphs, DSPI takes at most the same updated time as that on $G_w(n,p)$ random graphs when $\ln n/n < p < 20\ln n/n$, $O(\ln n)$ expected update time to handle a single edge deletion as well as $O(pn^2\ln n)$ total update time when $20\ln n/n \leq p < \sqrt{2\ln n/n}$, and $O(1)$ expected update time to handle a single edge deletion as well as $O(pn^2)$ total update time when $p > \sqrt{2\ln n/n}$. Specifically, when $p = c\sqrt{\ln n/n}$ with a constant $c > \sqrt{2}$, DSPI takes the least total update time of $O(n^{3/2}(\ln n)^{1/2})$. We think that DSPI is a best decremental SSSP algorithm for $G(n,p)$ random graphs with $p > \sqrt{2\ln n/n}$, that is, the best for most of large-scale Erdös-Rényi random graphs.

It is of interest to tighten the value of c in Lemma 6 and further analyze the expected update time of DSPI handling edge deletions in $G(n,p)$ random graphs with $\ln n/n < p < c\ln n/n, 1 < c < 20$. Moreover, how much expected update time DSPI takes to deal with a single edge weight update or topology update in the other kinds of random graphs remains as future research topics.

Acknowledgement. We thank the reviewers for their valuable comments and suggestions.

References

1. Ausiello, G., Italiano, G.F., Marchetti-Spaccamela, A., Nanni, U.: Incremental algorithms for minimal length paths. J. Algorithms **12**, 615–638 (1991)

2. Bernstein, A.: Fully dynamic $(2+\epsilon)$ approximate all-pairs shortest paths with fast query and close to linear update time. In: Proceedings of the 50th FOCS, pp. 693–702 (2009)
3. Bernstein, A.: Maintaining shortest paths under deletions in weighted directed graphs. In: Proceedings of the 45th STOC, pp. 725–734 (2013)
4. Bernstein, A., Roditty, L.: Improved dynamic algorithms for maintaining approximate shortest paths under deletions. In: Proceedings of the 22th SODA, pp. 1355–1365 (2011)
5. Blum, A., Hopcroft, J., Kannan, R.: Foundation of Data Science, Manuscript (14 May 2015). http://www.cs.cornell.edu/jeh/bookMay2015.pdf
6. Bollobás, B.: Random Graphs. Cambridge University Press, Cambridge (2001)
7. Demetrescu, C., Italiano, G.F.: A new approach to dynamic all pairs shortest paths. J. ACM **51**, 968–992 (2004)
8. Dijkstra, E.W.: A note on two problems in connection with graphs. Numer. Math. **1**, 269–271 (1959)
9. Ding, W., Lin, G.: Partially dynamic single-source shortest paths on digraphs with positive weights. In: Gu, Q., Hell, P., Yang, B. (eds.) AAIM 2014. LNCS, vol. 8546, pp. 197–207. Springer, Heidelberg (2014)
10. Erdös, P., Rényi, A.: On random graphs-I. Publicationes Mathematicae (Debrecen) **6**, 290–297 (1959)
11. Erdös, P., Rényi, A.: On the Evolution of Random Graphs. Akad. Kiado, Budapest (1960)
12. Even, S., Shiloach, Y.: An on-line edge-deletion problem. J. ACM **28**, 1–4 (1981)
13. Fakcharoemphol, J., Rao, S.: Planar graphs, negative weight edges, shortest paths, and near linear time. In: Proceedings of the 42nd FOCS, pp. 232–241 (2001)
14. Feller, W.: An Introduction to Probability Theory and Its Applications. Wiley, New York (1968)
15. Fredman, M.L., Tarjan, R.E.: Fibonacci heaps and their uses in improved network optimization algorithms. J. ACM **34**(3), 596–615 (1987)
16. Gilbert, E.N.: Random graphs. Ann. Math. Stat. **30**(4), 1141–1144 (1959)
17. Henzinger, M., King, V.: Fully dynamic biconnectivity and transitive closure. In: Proceedings of the 36th FOCS, pp, 664–672 (1995)
18. Henzinger, M., Krinninger, S., Nanongkai, D.: A subquadratic-time algorithm for dynamic single-source shortest paths. In: Proceedings of the 25th SODA, pp, 1053–1072 (2014)
19. Henzinger, M., Krinninger, S., Nanongkai, D.: Sublinear-time decremental algorithms for single-source reachability and shortest paths on directed graphs. In: Proceedings of the 46th STOC, pp. 674–683 (2014)
20. Henzinger, M., Krinninger, S., Nanongkai, D.: Decremental single-source shortest paths on undirected graphs in near-linear total update time. In: Proceedings of the 55th FOCS, pp. 146–155 (2014)
21. King, V.: Fully dynamic algorithms for maintaining all-pairs shortest paths and transitive closure in digraphs. In: Proceedings of the 40th FOCS, pp. 81–99 (1999)
22. Madry, A.: Faster approximation schemes for fractional multicommodity flow problems via dynamic graph algorithms. In: Proceedings of the 42th STOC, pp. 121–130 (2010)
23. Peres, Y., Sotnikov, D., Sudakov, B., Zwick, U.: All-pairs shortest paths in $O(n^2)$ time with high probability. In: Proceedings of the 51th FOCS, pp. 663–672 (2010)
24. Roditty, L., Zwick, U.: Dynamic approximate all-pairs shortest paths in undirected graphs. In: Proceedings of the 45th FOCS, pp. 499–508 (2004)

25. Roditty, L., Zwick, U.: On dynamic shortest paths problems. In: Albers, S., Radzik, T. (eds.) ESA 2004. LNCS, vol. 3221, pp. 580–591. Springer, Heidelberg (2004)
26. Solomonoff, R., Rapoport, A.: Connectivity of random nets. Bull. Math. Biol. **13**(2), 107–117 (1951)
27. Thorup, M.: Fully-dynamic all-pairs shortest paths: faster and allowing negative cycles. In: Hagerup, T., Katajainen, J. (eds.) SWAT 2004. LNCS, vol. 3111, pp. 384–396. Springer, Heidelberg (2004)
28. Thorup, M.: Worst-case update times for fully-dynamic all-pairs shortest paths. In: Proceedings of the 37th STOC, pp. 112–119 (2005)

Trees, Paths, Stars, Caterpillars and Spiders

Minghui Jiang(✉)

Department of Computer Science, Utah State University, Logan, UT 84322, USA
mjiang@cc.usu.edu

Abstract. For any $k \geq 2$, deciding whether the linear arboricity, star arboricity, caterpillar arboricity, and spider arboricity, respectively, of a bipartite graph are at most k are all NP-complete.

Keywords: Arboricity · Edge partition · NP-hardness

1 Introduction

The *arboricity* of a graph is the minimum number of parts in an edge partition of the graph such that each part induces a forest. The trees in the forests may be restricted to various subclasses of trees, leading to several related concepts of arboricity. For example, the *linear arboricity* (respectively, *star arboricity*, *caterpillar arboricity*, and *spider arboricity*) of a graph is the minimum number of parts in an edge partition of the graph such that each part induces a disjoint union of paths (respectively, stars, caterpillars, and spiders).

Recall that a *path* is a tree with maximum degree at most two. A *star* is a tree with diameter at most two, or, equivalently, a tree with at most one vertex of degree more than one. A *caterpillar* is a tree whose internal vertices induce a path. A *spider* is a tree with at most one vertex of degree more than two. Intuitively, a caterpillar is a path of stars, and a spider is a star of paths.

The concept of arboricity and its several variants such as star/caterpillar/spider arboricities are closely related to thickness, another classic concept in graph theory and particularly in graph drawing. Recall that the *thickness* of a graph is the minimum number of planar graphs into which the edges of the graph can be partitioned. Since any planar graph with n vertices has at most $3n - 6$ edges, the formula of Nash-Williams [14] implies that its arboricity is at most 3. It is also known that the star arboricity (hence the spider arboricity too) of any planar graph is at most 5 [9], and the caterpillar arboricity of any planar graph is at most 4 [7]. On the other hand, any forest is clearly planar. Thus for any graph, these arboricities differ from thickness by at most a constant factor. Also, linear arboricity and caterpillar arboricity are important for characterizing certain variants of rectangle visibility graphs that naturally arise in two-layer routing problems in VLSI design and are useful for labeled graph layouts [16].

While the arboricity of any graph can be computed in polynomial time [5], the arboricities with restrictions on the trees often turn out to be hard to compute [9,16]. In this paper, we show that for any $k \geq 2$, deciding whether a graph

Z. Lu et al. (Eds.): COCOA 2015, LNCS 9486, pp. 551–565, 2015.
DOI: 10.1007/978-3-319-26626-8_40

has linear arboricity, star arboricity, caterpillar arboricity, and spider arboricity, respectively, at most k are all hard, even if the graph is bipartite.

Peroche [15] first proved that deciding whether a multigraph (which may contain multiple parallel edges) has linear arboricity at most 2 is NP-complete. This result was later strengthened by Shermer [16], who proved that deciding whether a graph has linear arboricity at most 2 is NP-complete too. Our first theorem extends the NP-completeness of linear arboricity k to all $k \geq 2$ even for the restricted class of bipartite graphs.

Theorem 1. *For any $k \geq 2$, deciding whether a bipartite graph has linear arboricity at most k is NP-complete.*

Denote by $\mathrm{la}(G)$ and $\Delta(G)$, respectively, the linear arboricity and the maximum degree of a graph G. Our proof of Theorem 1 implies that for any $k \geq 2$, deciding whether $\mathrm{la}(G) = k$ for a bipartite graph G with $\Delta(G) = 2k$ and $k \leq \mathrm{la}(G) \leq k + 1$ is NP-complete. Note that for any graph G, $\mathrm{la}(G)$ would be exactly $(\Delta(G)+1)/2$ when Δ is odd, and either $\Delta(G)/2$ or $\Delta(G)/2+1$ when Δ is even, if the linear arboricity conjecture of Akiyama, Exoo, and Harary [1] is true; see also [2].

Hakimi et al. [9] proved that deciding whether a graph has star arboricity at most 2 is NP-complete. Gonçalves and Ochem [8] strengthened this result by showing that the problem remains NP-complete even for a very restricted graph class. More precisely, they proved [8, Theorem 2] that for any $g \geq 3$, deciding whether a bipartite planar graph with girth at least g and maximum degree 3 has star arboricity at most 2 is NP-complete. They also proved [8, Theorem 5] that deciding whether a 2-degenerate bipartite planar graph has star arboricity at most 3 is NP-complete. Our next theorem extends the NP-completeness of star arboricity k to all $k \geq 2$ for bipartite graphs.

Theorem 2. *For any $k \geq 2$, deciding whether a bipartite graph has star arboricity at most k is NP-complete.*

Shermer [16] also proved that deciding whether a graph has caterpillar arboricity at most 2 is NP-complete. Gonçalves and Ochem [8, Theorem 6] extended this result and proved that deciding whether a 2-degenerate bipartite planar graph has caterpillar arboricity at most 2 and respectively at most 3 are both NP-complete. Our next theorem extends the NP-completeness of caterpillar arboricity k to all $k \geq 2$ for bipartite graphs.

Theorem 3. *For any $k \geq 2$, deciding whether a bipartite graph has caterpillar arboricity at most k is NP-complete.*

To our best knowledge, the complexity of determining the spider arboricity of a graph was open, although spiders play an important role in algorithms for certain network design problems [11]. Our last theorem fills this gap in the literature.

Theorem 4. *For any $k \geq 2$, deciding whether a bipartite graph has spider arboricity at most k is NP-complete.*

All these arboricity problems are clearly in NP. In Sects. 2–5, we prove their NP-hardness. Due to space constraints, we defer the proofs of some lemmas for Theorems 2 and 3 to the full version of this paper.

2 Linear Arboricity of Bipartite Graphs

In this section we prove Theorem 1. We show that for any $k \geq 2$, deciding whether a bipartite graph has linear arboricity at most k is NP-hard. Our proof consists of two parts: $k = 2$ and $k \geq 3$.

The first part of our proof shows that deciding whether a *bipartite graph* has linear arboricity at most 2 is NP-hard. This strengthens two previous results [15,16]. Peroche [15] proved, by a reduction from 3-SAT, that deciding whether a *multigraph* has linear arboricity at most 2 is NP-hard. Shermer [16] extended this result and proved that deciding whether a *graph* has linear arboricity at most 2 is NP-hard. The core element in Shermer's proof is a local transformation that replaces each doubled edge uv in a multigraph G' by four edges ua, ub, vc, vd between u, v and the four vertices a, b, c, d from a distinct copy of K_4, which results in a graph G with linear arboricity at most 2 if and only if G' has linear arboricity at most 2. This graph is not triangle-free, however, and hence not bipartite either. Indeed to obtain a bipartite graph with the desired property seems difficult using only local transformations.

Since any graph can be turned into a bipartite graph simplyby subdividing all edges, one may speculate that this would yield a straightforward reduction from the linear arboricity problem for general graphs to the linear arboricity problem for bipartite graphs. But this approach does not work because the linear arboricity of a graph may change after its edges are subdivided.

Take the complete graph K_5 for example. Clearly, $\mathrm{la}(K_5) > 2$: in any partition of the edges of K_5 into two linear forests, each linear forest must include exactly two edges incident to each vertex, and hence must be a cycle and not a forest at all. Let K_5' be the graph obtained from K_5 by subdividing each edge, and let K_5'' be the graph obtained from K_5' by subdividing each edge. Then K_5' has $\binom{5}{2} \cdot 4$ edges, with each pair of degree-4 vertices connected by a path of length 4. We next show that $\mathrm{la}(K_5'') = 2$. For each degree-4 vertex, color the four incident edges arbitrarily using two colors, with two edges of each color. Then in the path of length 4 between any pair of degree-4 vertices, the two outer edges are colored, and the two inner edges remain to be colored. If the two outer edges have the same color, then color both inner edges with the other color; otherwise, color the two inner edges with different colors such that the colors of the four edges alternate along the path. Then the edges of each color form a linear forest in which every path has length at most two. We have shown that $\mathrm{la}(K_5) > 2 = \mathrm{la}(K_5'')$. It follows that either $\mathrm{la}(K_5) \neq \mathrm{la}(K_5')$ or $\mathrm{la}(K_5') \neq \mathrm{la}(K_5'')$.

2.1 $K_{2k-1,2k}$ and T_c

The central component of our construction, for both $k = 2$ and $k \geq 3$, is the complete bipartite graph $K_{2k-1,2k}$.

Lemma 1. *For any $k \geq 2$, $K_{2k-1,2k}$ has linear arboricity k. Moreover, for any edge partition of $K_{2k-1,2k}$ into k forests, the k forests must be k spanning trees. In particular, for any edge partition of $K_{2k-1,2k}$ into k linear forests, the k forests must be k spanning paths.*

Proof. It is well-known that for any $k \geq 2$, $K_{2k-1,2k}$ has linear arboricity at most k; see for example [10, Fig. 4]. For any edge partition of $K_{2k-1,2k}$ into k forests, each forest must be a spanning tree because the average number of edges of the k forests is $(2k-1)2k/k = 4k-2$, while the maximum number of edges of any forest is $(2k-1+2k)-1 = 4k-2$, which is attained only when the forest is a spanning tree. For linear forests, the spanning trees are spanning paths. □

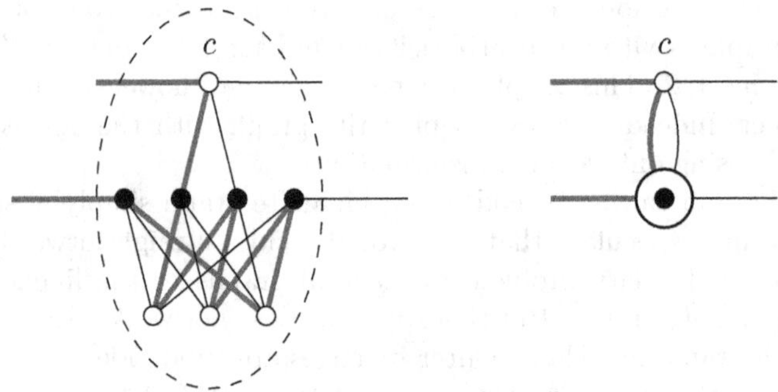

Fig. 1. Left: The graph T_c, enclosed in the oval, consists of a vertex c connected to k of the $2k$ vertices of degree $2k-1$ in $K_{2k-1,2k}$. The edges crossing the oval connect c and free vertices of T_c to outside vertices. Right: A schematic representation of T_c, where $K_{2k-1,2k}$ is represented by a large circle around a black dot (Color figure online).

Let T_c be the graph consisting of a vertex c connected to any k of the $2k$ vertices of degree $2k-1$ in $K_{2k-1,2k}$ (the degrees of these k vertices are $2k-1$ in $K_{2k-1,2k}$ and $2k$ in T_c). We call the other k vertices of degree $2k-1$ in $K_{2k-1,2k}$ that are not connected to c the *free vertices* of T_c. Refer to Fig. 1 for the $k = 2$ case.

Lemma 2. *For any graph that contains T_c as a vertex-induced subgraph, and for any k-color assignment to the edges of this graph that partitions it into k linear forests, the following three properties hold:*

(i) all edges connecting c to vertices outside T_c must have different colors;
(ii) all edges connecting a free vertex of T_c to a vertex outside T_c must have different colors.

(iii) any two edges of the same color that connect c and a free vertex of T_c *respectively to two vertices outside* T_c *must extend the two opposite ends of a monochromatic spanning path in* T_c *of this color.*

Proof. Consider any k-color assignment to the edges of $K_{2k-1,2k}$ that partitions it into k linear forests. By Lemma 1, these k linear forests must be k spanning paths: each of the $2k-1$ vertices of degree $2k$ is incident to exactly two edges of each color, and each of the $2k$ vertices of degree $2k-1$ is one of the $2k$ ends of k monochromatic paths. Any edge connecting one of these $2k$ ends to a vertex outside $K_{2k-1,2k}$ must have the same color as the corresponding monochromatic path in $K_{2k-1,2k}$, because this end is already incident to two edges of any other color in $K_{2k-1,2k}$. If both ends of a monochromatic path were connected to the same outside vertex, then this path would close into a monochromatic cycle, a contradiction. Thus the k edges incident to c in T_c must extend k different monochromatic paths in $K_{2k-1,2k}$ and hence have k different colors. Then the three properties easily follow. □

2.2 Linear Arboricity 2

Our proof for $k = 2$ is based on a reduction from the NP-complete problem 2-Colorability of 3-Uniform Hypergraphs [12]. Recall that a *q-uniform hypergraph* consists of a set V of vertices and a set E of hyperedges, where each hyperedge in E is a size-q subset of the vertices in V, and that a hypergraph is *k-colorable* if the vertices can be colored with k colors such that in each hyperedge, not all of its vertices have the same color. Given a 3-uniform hypergraph G, we will construct a bipartite graph G_2 with maximum degree 4 such that G has a 2-coloring if and only if G_2 admits an edge partition into two linear forests.

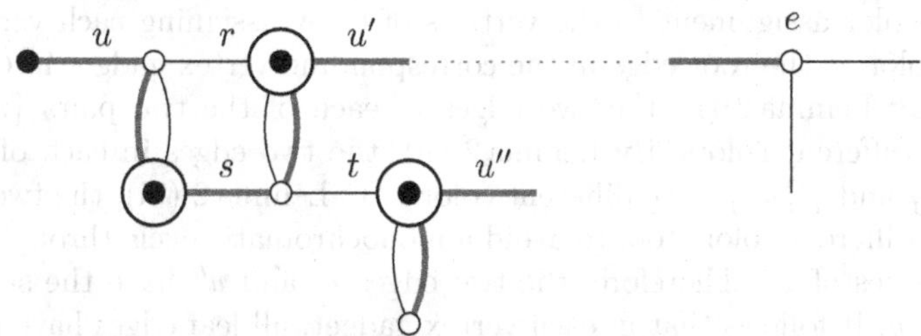

Fig. 2. Left: A duplicator forks an edge u to two edges u' and u'' via three copies of the graph T_c inside a vertex gadget. Right: Three leaf edges from three vertex gadgets join at a vertex e to form a hyperedge gadget (Color figure online).

For each vertex v of degree $\deg(v)$ in G, we construct a vertex gadget in G_2 as follows. If $\deg(v) = 1$, then the vertex gadget is just a single edge. Now suppose that $\deg(v) \geq 2$. Take an arbitrary binary tree with $\deg(v)$ leaves and $\deg(v) - 1$ internal nodes, then connect its root to an additional vertex which becomes the new root of the tree, so that every internal node of this rooted tree except the new root has degree exactly 3. We call the deg(v) edges incident to the deg(v) leaves of this rooted tree the *leaf edges*, and call the single edge incident to the root the *root edge*. Replace each internal node of degree 3 in this rooted tree, which is connected to its parent by some edge u and to its two children by two edges u' and u'', by a duplicator consisting of three disjoint copies of the graph T_c connected by three additional edges r, s, and t, as illustrated in Fig. 2 left.

For each hyperedge e in G, we construct a hyperedge gadget in G_2 which consists of a single vertex e joining three distinct leaf edges from three vertex gadgets, respectively, corresponding to the three vertices of the hyperedge. Refer to Fig. 2 right for an illustration. This completes the construction of G_2.

Lemma 3. *The 3-uniform hypergraph G is 2-colorable if and only if the graph G_2 admits an edge partition into two linear forests.*

Proof. We first prove the direct implication. Suppose that G is 2-colorable. We obtain an edge partition of G_2 into two linear forests as follows. Refer to Fig. 1. For the internal edges of all copies of T_c in G_2, we can clearly find a 2-color assignment, by Lemma 1, that partitions these edges into two linear forests. Refer to Fig. 2. For the other edges in G_2, we simply assign the root edge of each vertex gadget in G_2 the same color as the corresponding vertex in G, then propagate the colors along the duplicators as illustrated, such that all leaf edges have the same color as the root edge. In each hyperedge e in G, not all three vertices have the same color, by our assumption that G is 2-colorable. Correspondingly, in each hyperedge gadget in G_2, not all three leaf edges incident to the vertex e have the same color. It is straightforward to verify that the edges of each color in G_2 form a linear forest.

We next prove the reverse implication. Suppose that G_2 admits a 2-color assignment to its edges that partitions it into two linear forests. Then we can obtain a 2-color assignment to the vertices of G by assigning each vertex in G the same color as the root edge in the corresponding vertex gadget in G_2. Refer to Fig. 2. By Lemma 2 (i), the two edges in each of the two pairs $\{u, r\}$ and $\{s, t\}$ have different colors. By Lemma 2 (ii), the two edges in each of the two pairs $\{r, u'\}$ and $\{t, u''\}$ have different colors. By Lemma 2 (iii), the two edges r and s have different colors too, to avoid a monochromatic cycle through the two adjacent copies of T_c. Therefore, the two edges u' and u'' have the same color as the edge u. It follows that in each vertex gadget, all leaf edges have the same color as the root edge. Moreover, since the two forests in the edge partition of G_2 are both linear, not all three edges have the same color in each hyperedge gadget in G_2. Correspondingly, not all three vertices have the same color in each hyperedge in G. □

The graph G_2 is bipartite as indicated by the black and white dots that represent its vertices. The reduction is clearly polynomial. We have proved that deciding whether a bipartite graph has linear arboricity at most 2 is NP-hard.

2.3 Linear Arboricity k for $k \geq 3$

Our proof for $k \geq 3$ is based on a reduction from the problem k-Colorability, which asks whether a given graph G admits a vertex coloring with k colors. Maffray and Preissmann [13] proved that for any $k \geq 3$, the problem k-Colorability is NP-hard even for triangle-free graphs. Note that k-Colorability for $k = 2$ is just bipartite testing which is well-known to be solvable in polynomial time. Thus we had to prove the $k = 2$ case by reduction from a different problem.

Given a graph G, we will construct a bipartite graph G_k with maximum degree $2k$ such that G admits a vertex coloring with k colors if and only if G_k admits an edge partition into k linear forests.

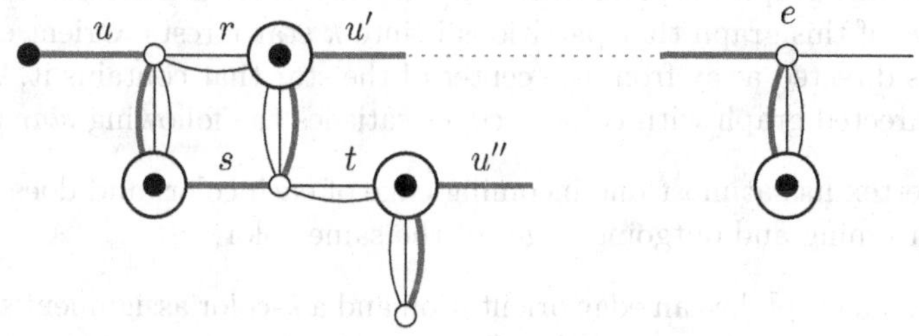

Fig. 3. Left: A duplicator forks an edge u to two edges u' and u'' via three copies of the graph T_c inside a vertex gadget. Right: Two leaf edges from two vertex gadgets join at the vertex e of T_e to form an edge gadget (Color figure online).

The vertex gadget for $k \geq 3$ is a straightforward generalization of the vertex gadget for $k = 2$. Refer to Fig. 3 left for an illustration of duplicator for the $k = 3$ case. Note that each of the two labels r and t now corresponds to $k - 1$ edges between two adjacent copies of T_c.

For each edge e in G, we construct an edge gadget in G_k which consists of a copy T_e of the graph T_c, with the vertex e in T_e joining two distinct leaf edges from two vertex gadgets, respectively, corresponding to the two vertices of the edge. Refer to Fig. 3 right for an illustration.

Following a similar argument as in proof of Lemma 3, it is easy to verify that G admits a vertex coloring with k colors if and only if G_k admits an edge partition into k linear forests. Again, the graph G_k is bipartite as indicated by the black and white dots, and the reduction is clearly polynomial. We have proved that for any $k \geq 3$, deciding whether a bipartite graph has linear arboricity at most k is NP-hard. This completes the proof of Theorem 1.

Remark. Note that the graph G_2 in our reduction for $k = 2$ has maximum degree 4, and that the graph G_k in our reduction for $k \geq 3$ has maximum degree $2k$. So $\Delta(G_k) = 2k$ and hence $\mathrm{la}(G_k) \geq \lceil \Delta(G_k)/2 \rceil = k$ for $k \geq 2$. On the other hand, from the proof of the direct implication of Lemma 3 for $k = 2$ and its analog for $k \geq 3$, the linear arboricities of G_k excluding an arbitrary leaf edge from each hyperedge gadget (Fig. 2 right) and each edge gadget (Fig. 3 right), respectively, are at most k, and hence it is easy to show that $\mathrm{la}(G_k) \le k+1$ for $k \ge 2$. Thus our proof of Theorem 1 implies that for any $k \ge 2$, deciding whether $\mathrm{la}(G_k) = k$ for a bipartite graph G_k with $\Delta(G_k) = 2k$ and $k \le \mathrm{la}(G_k) \le k+1$ is NP-complete.

3 Star Arboricity of Bipartite Graphs

In this section we prove Theorem 2. The two cases $k = 2$ and $k = 3$ of the theorem have been proved by Gonçalves and Ochem [8]. We show that for any $k \ge 3$, deciding whether a bipartite graph has star arboricity at most k is NP-hard.

Consider any graph with star arboricity k. Then there is a k-color assignment to the edges of this graph that partitions it into k star forests. Orient each edge so that it is directed away from the center of the star that contains it. Then the resulting directed graph with colored edges satisfies the following *star property*:

($\star$) Each vertex has at most one incoming edge of each color, and does not have both incoming and outgoing edges of the same color.

Conversely, if a graph has an edge orientation and a k-color assignment satisfying this star property, which we call a *k-color star orientation*, then the graph admits an edge partition into k star forests, with stars centered at the vertices with outgoing edges. Therefore, a graph has star arboricity k if and only if it has a k-color star orientation.

Our proof is based on a reduction from k-Colorability [13]. Given a graph G, we will construct a bipartite graph G_X such that G admits a vertex coloring with k colors if and only if G_X has a k-color star orientation.

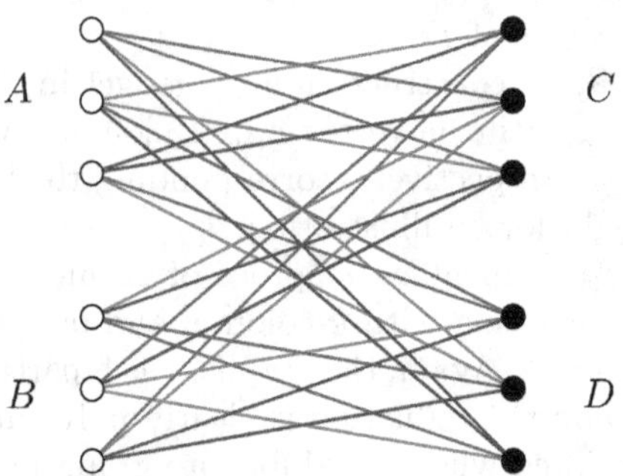

Fig. 4. The graph X for $k = 3$ (Color figure online).

Refer to Fig. 4. The central component of our construction is a bipartite graph X with vertices $A \cup B \cup C \cup D$, where $|A| = |B| = |C| = |D| = k$. Write $A = \{a_1, \dots, a_k\}$, $B = \{b_1, \dots, b_k\}$, $C = \{c_1, \dots, c_k\}$, and $D = \{d_1, \dots, d_k\}$.

The edges of X include all possible edges $a_i c_j, a_i d_j, b_i c_j, b_i d_j$ between $A \cup B$ and $C \cup D$ except the edges $a_i c_i, a_i d_i, b_i c_i, b_i d_i$ between vertices of equal indices. There are $4k$ vertices and $4k^2 - 4k$ edges in X.

Lemma 4. *X has a k-color star orientation.*

Lemma 5. *For any graph that contains X as a vertex-induced subgraph, and for any k-color star orientation of this graph, each edge between a vertex v in X and a vertex outside X must be outgoing from v, and moreover all outgoing edges from v have the same color.*

To obtain G_X from G, we simply identify each vertex v in G with an arbitrary vertex in a distinct copy of X, then subdivide each edge $e = uv$ in G into two edges ue and ve with an intermediate vertex e.

Lemma 6. *G admits a vertex coloring with k colors if and only if G_X admits a k-color star orientation.*

Clearly, the graph G_X is bipartite, and the reduction is polynomial. We have proved that for any $k \geq 3$, deciding whether a bipartite graph has star arboricity at most k is NP-hard. This completes the proof of Theorem 2.

4 Caterpillar Arboricity of Bipartite Graphs

In this section we prove Theorem 3. The two cases $k = 2$ and $k = 3$ of the theorem have been proved by Gonçalves and Ochem [8]. We show that for any $k \geq 3$, deciding whether a bipartite graph has caterpillar arboricity at most k is NP-hard.

Let S_n be the tree obtained by subdividing each edge of an n-star $K_{1,n}$ into two edges. We call S_n an *n-superstar* although it is just a spider with n legs of length two. Gonçalves and Ochem [8] observed that a tree is a caterpillar if and only if it is S_3-free, and consequently the caterpillar arboricity of a graph is just the $n = 3$ case of the more general concept of S_n-free arboricity, where every forest in the edge partition must not contain S_n as a vertex-induced subgraph. Following their framework, we prove a more general result than Theorem 3:

Theorem 5. *For any $n \geq 2$ and any $k \geq 3$, deciding whether a bipartite graph has S_n-free arboricity at most k is NP-complete.*

This problem is clearly in NP. We next prove its NP-hardness by a reduction from k-Colorability [13]. Given a graph G, we will construct a bipartite graph G_Y such that G admits a vertex coloring with k colors if and only if G_Y has S_n-free arboricity k.

Refer to Fig. 5. The central component of our construction is a bipartite graph Y with vertices $A \cup B \cup C$, where $|A| = |B| = k^2(n-1)$ and $|C| = k$. Write $A = \{a_i : 0 \leq i < k^2(n-1)\}$, $B = \{b_i : 0 \leq i < k^2(n-1)\}$, and $C = \{c_j : 0 \leq j < k\}$. The edges of Y include all edges $a_i b_i$ between vertices of equal indices in A and B, and all possible edges $b_i c_j$ between B and C. There are $2k^2(n-1) + k$ vertices and $k^2(n-1) + k^3(n-1)$ edges in Y.

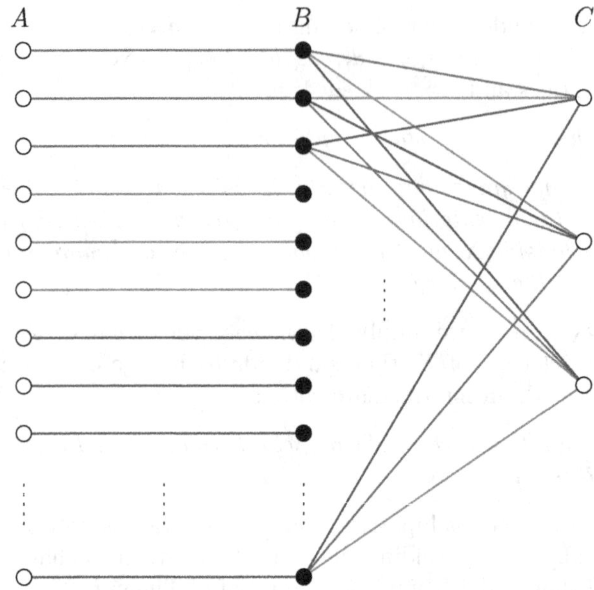

Fig. 5. The graph Y for $k = 3$ (Color figure online).

Lemma 7. *Y has S_n-free arboricity at most k.*

Lemma 8. *For any graph that contains Y as a vertex-induced subgraph and contains another vertex v outside Y that is adjacent to $k-1$ distinct vertices of C in Y, and for any k-color assignment to the edges of this graph that partitions it into k S_n-free forests, the $k-1$ edges between Y and v must have $k-1$ distinct colors, and all other edges incident to v must have the only remaining color.*

To obtain G_Y from G, we simply connect each vertex v in G to $k-1$ distinct vertices of C in a distinct copy of Y, then replace each edge uv in G by a cycle of four edges $u\bar{u}, \bar{u}v, v\bar{v}, \bar{v}u$.

Lemma 9. *G admits a vertex coloring with k colors if and only if G_Y admits an edge partition into k S_n-free forests.*

Clearly, the graph G_Y is bipartite, and the reduction is polynomial. We have proved that for any $n \geq 2$ and any $k \geq 3$, deciding whether a bipartite graph has S_n-free arboricity at most k is NP-hard. This completes the proof of Theorem 3.

5 Spider Arboricity of Bipartite Graphs

In this section we prove Theorem 4. We show that for any $k \geq 2$, deciding whether a bipartite graph has spider arboricity at most k is NP-hard. Our proof consists of two parts: $k = 2$ and $k \geq 3$. For $k \geq 3$, the reduction is from k-Colorability [13]. For $k = 2$, the reduction is from 2-Subcolorability of triangle-free graphs. A *subcoloring* of a graph is a partition of its vertices such that each part induces a disjoint union of cliques. For a triangle-free graph, a 2-subcoloring corresponds to a vertex partition into two forests with maximum degree one. The problem 2-Subcolorability is NP-complete even for triangle-free planar graphs with maximum degree 4 [4,6].

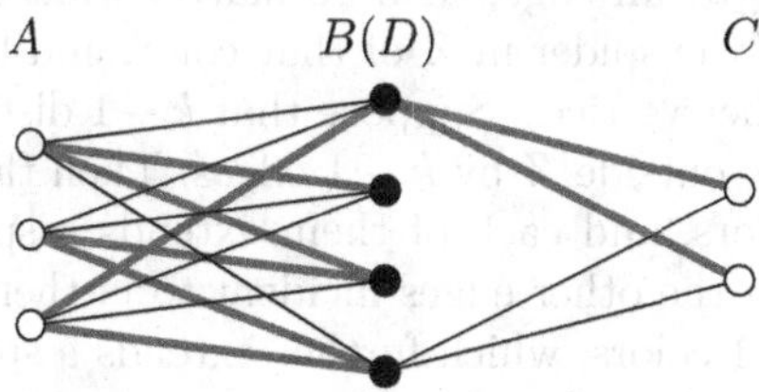

Fig. 6. The graph Z for $k = 2$ (Color figure online).

Refer to Fig. 6. The central component of our construction, for both $k = 2$ and $k \geq 3$, is a bipartite graph Z with vertices $A \cup B \cup C$, where $|A| = 2k - 1$, $|B| = 2k$, and $|C| = k$. Let D be a subset of k vertices in B. The edges of Z include all possible edges between A and B and between C and D. There are $5k - 1$ vertices and $(2k - 1)(2k) + k^2$ edges in Z.

Lemma 10. *Z has spider arboricity at most k.*

Proof. We obtain a k-color assignment to the edges of Z that partitions it into k spider forests (indeed k spiders) as follows. First color the edges between A and B to partition this complete bipartite graph $K_{2k-1,2k}$ into k monochromatic paths such that the $2k$ ends of the k paths are the $2k$ vertices in B (recall Lemmas 1 and 2) and moreover each path has one end at a vertex in D and the other end at a vertex in $B \setminus D$, next color each edge between C and D with the same color as the incident monochromatic path. Then the edges of Z are partitioned into k spiders centered at the vertices in D. □

Lemma 11. *For any graph that contains Z as a vertex-induced subgraph and contains another vertex v outside Z that is adjacent to $k - 1$ distinct vertices of C in Z, and for any k-color assignment to the edges of this graph that partitions it into k spider forests, the $k - 1$ edges between Z and v must have $k - 1$ distinct colors. Moreover, among all the other edges incident to v, there can be at most one edge of each of these $k - 1$ colors. Except these edges, all other edges incident to v must have the only remaining color.*

Proof. By Lemma 1, the spider forest of each color in Z, when restricted to the $(2k - 1)2k$ edges between A and B, must be a spanning tree of A and B. Thus the k spider forests in Z must indeed be k spiders. Morever, the k edges incident to each vertex in C must have k distinct colors, because any two edges of the same color would create a cycle together with the spanning tree of A and B of that color. Also, since each vertex in D has degree $2k-1+k \geq 2k+1$ for $k \geq 2$, and hence has degree at least 3 in one of the k spiders, the k vertices in D must be the centers of the k spiders in Z.

These two properties, (i) that the k edges incident to each vertex in C have k distinct colors, and (ii) that the k vertices in D are the centers of the k spiders in Z, together have the following implications. If any vertex in C is connected to a vertex v outside Z by an edge, then no matter what color this edge has, it must extend one leg of the spider in Z of that color, and this leg must not fork as v is connected to other vertices. Suppose that $k-1$ distinct vertices in C are connected to a vertex v outside Z by $k-1$ edges. Then these $k-1$ edges must have $k-1$ distinct colors, and each of them extends a spider leg of a distinct color. Moreover, among the other edges incident to v, there can be at most one edge of each of the $k-1$ colors, which further extends a spider leg of that color. Except these edges, all other edges incident to v must have the only remaining color. □

To obtain G_Z from G, we first connect each vertex v in G to $k-1$ distinct vertices of C in a distinct copy of Z. Then we proceed differently for the two cases: if $k \geq 3$, replace each edge $e = uv$ in G by $2k-1$ pairs of edges $\{ue_i, ve_i\}$ for $1 \leq i \leq 2k-1$; if $k = 2$, first connect each vertex v in G to two distinct dummy vertices, then replace each edge $e = uv$ in G by two pairs of edges $\{ue_u, ve_u\}$ and $\{ue_v, ve_v\}$.

Lemma 12. *For $k \geq 3$, G admits a vertex coloring with k colors if and only if G_Z admits an edge partition into k spider forests.*

Proof. We first prove the direct implication. Suppose that G admits a vertex coloring with k colors. We obtain a k-color assignment to the edges of G_Z that partitions it into k spider forests as follows. For each vertex v of color κ_v in G, first color the edges in the corresponding copy of Z as in the proof of Lemma 10, next color the $k-1$ edges between v and C with $k-1$ distinct colors except κ_v, and finally color all other edges incident to v with the same color κ_v.

We next prove the reverse implication. Suppose there is a k-color assignment to the edges of G_Z that partitions it into k spider forests. It follows by Lemma 11 that for each vertex v in G, the $k-1$ edges in G_Z between v and the corresponding copy of Z must have $k-1$ distinct colors. Let the remaining color be the color κ_v of v. We claim that this yields a vertex coloring with k colors in G.

We prove this claim by contradiction. Suppose that for some edge $e = uv$ in G, the two vertices u and v received the same color $\kappa_u = \kappa_v = \kappa$. By Lemma 11, among the edges incident to u, except those between u and the corresponding copy of Z, there are at most $k-1$ edges not colored κ. In particular, among the $2k-1$ edges ue_i, there are at most $k-1$ edges not colored κ. Similarly, among the $2k-1$ edges ve_i, there are at most $k-1$ edges not colored κ. Thus at least $(2k-1) - 2(k-1) = 1$ pair of edges ue_i and ve_i have the same color κ. Also, among the $2k-1$ edges ue_i (respectively, ve_i), u (respectively, v) must be incident to at least $(2k-1)-(k-1)=k\geq 3$ edges of color κ. Thus we have a monochromatic connected subgraph containing at least two vertices with degree at least three, which is not a spider. This contradicts the assumption that the k-color assignment to the edges of G_Z partitions it into k spider forests. □

Lemma 13. *For $k=2$, G admits a vertex subcoloring with two colors if and only if G_Z admits an edge partition into two spider forests.*

Proof. We first prove the direct implication. Suppose that G admits a vertex subcoloring with two colors. We obtain a two-color assignment to the edges of G_Z that partitions it into two spider forests as follows. For each vertex v of color κ_v in G, first color the edges in the corresponding copy of Z as in Lemma 10, such that the $k-1=1$ edge between v and C is not colored κ_v. Then proceed to color the two edges between v and the corresponding dummy vertices, and the two edges ve_v and ve_u for each edge $e=uv$ in G. If v has no neighbor of the same color κ_v by the subcoloring in G, color all these edges with κ_v. Otherwise if v has one neighbor u of the same color by the subcoloring, where $uv=e$, color all these edges with κ_v, except the edge ve_u with the other color.

We next prove the reverse implication. Suppose there is a two-color assignment to the edges of G_Z that partitions it into two spider forests. Then for each vertex v in G, the $k-1=1$ edge in G_Z between v and the corresponding copy of Z takes one color. Let the other color be the color κ_v of v. We claim that this yields a vertex subcoloring with two colors in G.

We prove this claim by contradiction. Suppose that for two edges $e=uv$ and $f=uw$ in G, the three vertices u,v,w received the same color $\kappa_u=\kappa_v=\kappa_w=\kappa$. By Lemma 11, among the edges incident to u, except the one between u and the corresponding copy of Z, there is at most $k-1=1$ edge not colored κ. In particular, at most 1 of the 4 edges ue_u, ue_v, uf_u, uf_w incident to u is not colored κ. Similarly, at most 1 of the 2 edges ve_v, ve_u (respectively, wf_w, wf_u) incident to v (respectively, w) is not colored κ. Thus at least $4-1-1-1=1$ of the 4 pairs of edges $\{ue_u, ve_u\}, \{ue_v, ve_v\}, \{uf_u, wf_u\}, \{uf_w, wf_w\}$ have the same color κ. Also, among the 2 edges to the dummy vertices and the 4 edges to e_u, e_v, f_u, f_w, u must be incident to at least $2+4-1=5>3$ edges of color κ, and, among the 2 edges to the dummy vertices and the 2 edges to e_u, e_v (respectively, f_u, f_w), v (respectively, w) must be incident to at least $2+2-1=3$ edges of color κ. Thus we have a monochromatic connected subgraph containing at least two vertices with degree at least three, which is not a spider. This contradicts the assumption that the two-color assignment to the edges of G_Z partitions it into two spider forests. □

Clearly, for both $k\geq 3$ and $k=2$, the graph G_Z is bipartite, and the reduction is polynomial. We have proved that for any $k\geq 2$, deciding whether a bipartite graph has spider arboricity at most k is NP-hard. This completes the proof of Theorem 4.

6 Concluding Remarks

The *track number* of a graph is the minimum number of parts in an edge partition of the graph such that each part induces an interval graph. For a triangle-free graph, the track number is equal to the caterpillar arboricity. Because of this connection, the aforementioned result of Gonçalves and Ochem [8, Theorem 6] on caterpillar arboricity actually implies that deciding whether a triangle-free graph has track number at most 2 is NP-complete. Similarly, our theorem on caterpillar arboricity implies the following corollary.

Corollary 1. *For any $k \geq 2$, deciding whether a bipartite (hence also triangle-free) graph has track number at most k is NP-complete.*

The *unit track number* of a graph is the minimum number of parts in an edge partition of the graph such that each part induces a unit interval graph. For a triangle-free graph, the unit track number is equal to the linear arboricity. The author proved in [10] that for any $k \geq 2$, deciding whether a graph has unit track number at most k (equivalently, recognizing unit k-track interval graphs) is NP-complete. Our theorem on linear arboricity implies that this holds even for bipartite graphs.

Corollary 2. *For any $k \geq 2$, deciding whether a bipartite (hence also triangle-free) graph has unit track number at most k is NP-complete.*

Recall that our result on linear arboricity holds for bipartite graphs with maximum degree $2k$. Indeed our other results for star/caterpillar/spider arboricities also hold for bipartite graphs with maximum degree bounded by some function of k (note that in our reductions for $k \geq 3$, the maximum degree of the graph G for k-Colorability [13] is bounded by some function of k).

In light of the previous NP-hardness results [8,9,16] for small values of k, some of our results for all $k \geq 2$ may not seem surprising, but still we cannot take them for granted. Indeed for certain graph classes, these restricted arboricity problems may be hard for $k = 2$ but not for all $k \geq 2$. For example, besides bipartite graphs, planar graphs are another interesting class of graphs whose arboricities are often studied. Gonçalves and Ochem [8] proved that deciding whether a bipartite planar graph has star/caterpillar arboricity at most $k = 2, 3$ are all NP-hard. Analogous results showing the NP-hardness of deciding whether a bipartite planar (or just planar) graph has linear/star/caterpillar/spider arboricity at most k for all $k \geq 2$ are not possible, however, unless P = NP. This is because the linear arboricity of any planar graph with maximum degree Δ is known to be exactly $\lceil \Delta/2 \rceil$ for $\Delta \geq 9$ [3] and is at most $\Delta + 1 \leq 9$ for $\Delta < 9$ by Vizing's theorem, the star arboricity (hence the spider arboricity too) of any planar graph is at most 5 [9], and the caterpillar arboricity of any planar graph is at most 4 [7]. Hence it is trivial to decide whether the various restricted arboricities of a planar graph are at most k when k exceeds certain constant thresholds. Note in particular the sharp dichotomy implied by the results in [7,8]: deciding whether a planar (or bipartite planar) graph has caterpillar arboricity at most k is NP-hard for $k = 2, 3$ but is in P for any $k \geq 4$.

Question 1. For what constant values of k is it NP-hard to decide whether a planar (or bipartite planar) graph has linear/star/spider arboricity at most k?

Question 2. Rather than constructing ad hoc proofs for all values of k, can we establish a meta-theorem for all graph-partitioning problems, such that the NP-hardness of a problem with k equal to some constant c implies its NP-hardness for all $k \geq c$, as long as the source and target graph classes of the partition satisfy certain properties?

References

1. Akiyama, J., Exoo, G., Harary, F.: Covering and packing in graphs III: cyclic and acyclic invariants. Math. Slovoca **30**, 405–417 (1980)
2. Alon, N.: The linear arboricity of graphs. Israel J. Math. **62**, 311–325 (1988)
3. Cygan, M., Hou, J., Kowalik, Ł., Lužar, B., Wu, J.: A planar linear arboricity conjecture. J. Graph Theor. **69**, 403–425 (2012)
4. Fiala, J., Jansen, K., Le, V.B., Seidel, E.: Graph subcolorings: complexity and algorithms. SIAM J. Discrete Math. **16**, 635–650 (2003)
5. Gabow, H.N., Westermann, H.H.: Forests, frames, and games: algorithms for matroid sums and applications. Algorithmica **7**, 465–497 (1992)
6. Gimbel, J., Hartman, C.: Subcolorings and the subchromatic number of a graph. Discrete Math. **272**, 139–154 (2003)
7. Gonçalves, D.: Caterpillar arboricity of planar graphs. Discrete Math. **307**, 2112–2121 (2007)
8. Gonçalves, D., Ochem, P.: On star and caterpillar arboricity. Discrete Math. **309**, 3694–3702 (2009)
9. Hakimi, S.L., Mitchem, J., Schmeichel, E.: Star arboricity of graphs. Discrete Math. **149**, 93–98 (1996)
10. Jiang, M.: Recognizing d-interval graphs and d-track interval graphs. Algorithmica **66**, 541–563 (2013)
11. Klein, P., Ravi, R.: A nearly best-possible approximation algorithm for node-weighted Steiner trees. J. Algorithms **19**, 104–115 (1995)
12. Lovász, L.: Coverings and colorings of hypergraphs. In: Proceedings of the 4th Southeastern Conference on Combinatorics, Graph Theory and Computing, pp. 3–12. Utilitas Mathematica Publishing, Winnipeg (1973)
13. Maffray, F., Preissmann, M.: On the NP-completeness of the k-colorability problem for triangle-free graphs. Discrete Math. **162**, 313–317 (1996)
14. Nash-Williams, C.S.J.A.: Decomposition of finite graphs into forests. J. Lond. Math. Soc. **39**, 12 (1964)
15. Peroche, B.: Complexité de l'arboricité linéaire d'un graphe. RAIRO Recherche Operationelle **16**, 125–129 (1982)
16. Shermer, T.C.: On rectangle visibility graphs III: external visibility and complexity. In: Proceedings of the 8th Canadian Conference on Computational Geometry (CCCG 1996), pp. 234–239 (1996)

Algorithms for the Densest Subgraph with at Least k Vertices and with a Specified Subset

Wenbin Chen(✉), Lingxi Peng, Jianxiong Wang, Fufang Li, and Maobin Tang

Department of Computer Science, Guangzhou University, Guangzhou, People's Republic of China
cwb2011@gzhu.edu.cn

Abstract. The density of a subgraph in an undirected graph is the sum of the subgraph's edge weights divided by the number of the subgraph's vertices. Finding an induced subgraph of maximum density among all subgraphs with *at least* k vertices is called as the densest *at-least-k*-subgraph problem (DalkS).

In this paper, we first present a polynomial time algorithms for DalkS when k is bounded by some constant c. For a graph of n vertices and m edges, our algorithm is of time complexity $O(n^{c+3}\log n)$, which improve previous best time complexity $O(n^c(n+m)^{4.5})$.

Second, we give a greedy approximation algorithm for the Densest Subgraph with a Specified Subset Problem. We show that the greedy algorithm is of approximation ratio $2\cdot(1+\frac{k}{3})$, where k is the element number of the specified subset.

1 Introduction

In many graphs, such as graphs arising from search auctions, from links of blogs, or from protein-protein interactions, it is often required to find the densest subgraph with the large or small size. Based on these requirement, the densest *at-least-k*-subgraph problem (DalkS) and the densest *at-most-k*-subgraph problem (DamkS) are introduced by Andersen in [1].

These problems are closely related to the densest subgraph problem [2] and the dense k-subgraph problem [5]. Based on the maximum flow problem, the densest subgraph of a given graph can be found in a polynomial time [6,7]. Charikar gave a 2-approximation algorithm with linear time in [2]. The dense k-subgraph problem is to find the densest induced subgraph of exactly k vertices. It is shown that the dense k-subgraph problem is NP-hard [4]. There are several approximation algorithms for it [4,5,9,11,14].

For DamkS, Andersen showed that DamkS is NP-complete, but no approximation algorithms for DamkS were provided. Recently, an approximation algorithm for DamkS is given in [3].

For DalkS, Andersen a 2-approximation algorithm based on the parametric flow algorithm [1,6]. In [10], Samir Khuller and Barna Saha proved that DalkS is

Z. Lu et al. (Eds.): COCOA 2015, LNCS 9486, pp. 566–573, 2015.
DOI: 10.1007/978-3-319-26626-8_41

NP-hard and also proposed 2-approximation algorithms based on the max-flow algorithm and linear programming. They also studied the densest subgraph problem for directed graphs. For the densest subgraph problem of directed graphs, Khuller and Saha proposed a polynomial time algorithm based on max-flow technology, as well as a greedy 2-approximation algorithm.

Though DalkS is NP-hard, it is shown in [3] that DalkS is still polynomial time solvable when k is bounded by some constant c. In [3], when k is bounded by some constant c, an algorithm of time complexity $O(n^c(n+m)^{4.5})$ for the DalkS problem is proposed based on Linear Programming method.

In some cases, it is required to find the densest subgraph with a specified subset, such as gene annotation graphs [8]. To meet this requirement, Saha et al. introduced the densest subgraph with a specified subset problem [13].

In [13], Saha et al. give an $O(n^4 \log n)$-time algorithm for the densest subgraph with a specified subset problem based on the maximum flow (min-cut) problem. Later, Wenbin et al. propose a LP-based polynomial time algorithm for this problem with less time $O(n^{3.5})$ [3].

It is an open problem whether there exist any less time exact algorithms or approximation algorithms for the densest subgraph with a specified subset problem.

In this paper, firstly, we propose a polynomial time algorithm for DalkS when k is bounded by some constant c based on the algorithm for the minimum cut with at least k vertices. Our algorithm is of time complexity $O(n^{c+3} \log n)$, which improve previous best time complexity $O(n^c(n+m)^{4.5})$.

Second, we propose an approximation algorithm for finding the densest subgraph with a specified subset problem in time $O(n^2)$. We show that the greedy algorithm is of approximation ratio $2 \cdot (1 + \frac{k}{3})$, where k is the element number of the specified subset.

The remainder of the paper is organized as follows. Section 2 reviews some basic definitions. In Sect. 3, we give an algorithm for solving DalkS based on a minimum s-t cut with at least k vertices problem. In Sect. 4, we design a greedy approximation algorithm for finding the densest subgraph with a specified subset. Finally, Sect. 5 give some open problems.

2 Definitions

In this section, we give some basic definitions. Let $G = (V, E)$ be an undirected graph in which every edge is assigned a positive weight by a weight function $w : E \rightarrow \mathbb{R}^+$. Also, define the weighted degree of a vertex v in G, $w(v, G)$, to be the sum of the weights of the edges incident with v, and let the total weight of G, $W(G)$, be the sum of the weights of all of the edges in G.

Definition 1. *For any induced subgraph H of G, we define the density of H to be $d(H) = \frac{W(H)}{|H|}$.*

Definition 2. *For an undirected graph G, we define the quantity $dal(G, k)$ = the maximum density of an induced subgraph on at least k vertices.*

Definition 3. *The densest at-least-k-subgraph problem (DalkS) is the problem of finding an induced subgraph of at least k vertices with density $dal(G, k)$.*

In order to design our algorithm, we still need the following definition of a minimum cut with at least k vertices and some facts about it in [3].

Definition 4. *Given a graph $G = (V, E)$, a cut is a partition of the vertex set V into two sets V_1 and V_2. A cut edge set $C = \{(v_1, v_2) \in E : v_1 \in V_1, v_2 \in V_2\}$ is associated with every cut. The capacity of a cut is the sum of the capacities (i.e. weights) of the cut edges.*

Definition 5. *Given vertices $s, t \in V$, an s-t cut is a cut (V_1, V_2) such that $s \in V_1$ and $t \in V_2$. An s-t cut with* **at least** *k vertices is an s-t cut (V_1, V_2) such that $|V_1 \setminus \{s\}| \geq k$.*

In [3], it is shown that finding a minimum cut with at least k vertices is polynomial time solvable when k is bounded by some constant c.

Lemma 1. *When k is bounded by a constant value c, finding a minimum cut with at least k vertices is solvable in time $O(n^{(c+3)})$.*

Definition 6. *Given an undirected graph G and a vertices subset $v_F = \{v_{f_1}, \ldots v_{f_k}\}$, we define the following quantity*
$mdv(G, v_F) :=$ the maximum density of an induced subgraph containing v_F.

Definition 7. *The densest subgraph with a specified subset problem (DSS) is the problem of finding an induced subgraph containing this specified subset v_F with density $mdv(G, v_F)$.*

Now, we give a formal definition for an approximation algorithm.

Definition 8. *An algorithm $A(G, k)$ is a γ-approximation algorithm for the densest subgraph with a specified subset problem if for any graph G and a specified subset v_F, it returns an induced subgraph H containing v_F that $\frac{mdv(G, v_F)}{d(H)} \leq \gamma$.*

3 An Algorithm for DalkS Based on the Minimum *s-t* Cut with at Least *k* Vertices

In [7], Goldberg proposes a polynomial time algorithm for the densest subgraph problem based on the minimum *s-t* cut problem. In this section, we generalize Goldberg's algorithm to find the densest subgraph with at least k vertices based on the algorithm of finding a minimum *s-t* cut with at least k vertices.

The basic idea of the algorithm is as follows. At each stage, we guess a value g for $dal(G, k)$. Then, we construct a new graph and compute a minimum *s-t* cut with at least k vertices that enables us to decide whether $g \leq dal(G, k)$ or $g > dal(G, k)$. Thus, we search for $dal(G, k)$ in a binary search fashion.

As in [7], we first consider the simplest case, in which G is an unweighted, undirected graph—this is equivalent to a weighted graph in which all weights are either 1 or 0.

Let $G = (V, E)$ be an unweighted, undirected graph, and let $n = |V|$ and $m = |E|$. Let d_i be the degree of vertex i of G. Given a "guess" g, we construct another graph $N = (V_N, E_N)$ as follows [7].

We add source s and sink t to the set of vertices of G; connect source s to every node i of G by an edge of capacity m; and connect every node i of G to the sink t by an edge of capacity $(m + 2g - d_i)$. Each edge of G is assigned capacity 1. Since $d_i \leq m$ for all i, all capacities are positive. Whereas Goldberg constructs a directed graph, we construct an undirected graph. This change makes no difference, though.

Let (S, T) be an s-t cut in N, and let $V_1 = S \setminus \{s\}$ and $V_2 = T \setminus \{t\}$. In [7], Goldberg proved the following result:

Lemma 2. *The capacity of the cut* $(S, T) = m|V| + 2|V_1|(g - D_1)$, *where* D_1 *is the density of the subgraph of* G *generated by* V_1.

In order to decide whether $g \leq dal(G, k)$ or $g > dal(G, k)$, we need the following result, which is implicit in the Theorem 4.1 of [3].

Lemma 3. *Let* R *be the value of a minimum s-t cut* (S, T) *with at least* k *vertices. If* $R \leq m|V|$, *then* $g \leq dal(G, k)$. *If* $R > m|V|$, *then* $g > dal(G, k)$.

As shown in [7], the smallest distance between two different possible subgraph densities for undirected graphs is $\frac{1}{n(n-1)}$. For completeness, we explain the reason as follows. Since any density of one subgraph can be denoted as $\frac{p}{q}$, where $0 \leq p \leq m, 1 \leq q \leq n$. So the difference Δd between two subgraphs' density can be denoted as $\frac{p_1}{q_1} - \frac{p_2}{q_2} = \frac{p_1 q_2 - p_2 q_1}{q_1 q_2}$. When $q_1 \neq q_2$, $q_1 q_2 \leq n(n-1)$, $|\Delta d| \geq \frac{1}{n(n-1)}$. When $q_1 = q_2$, $|\Delta d| \geq 1/n \geq \frac{1}{n(n-1)}$. Thus, we have the following theorem.

Theorem 1. *If* H *is a subgraph of* G *with at least* k *vertices, and there exists no subgraph* H' *with at least* k *vertices such that* $d(H') \geq d(H) + \frac{1}{n(n-1)}$, *then* $d(H) = dal(G, k)$.

In the following, we design an algorithm for $dal(G, k)$:

In the following, we show that below algorithm is correct and is of time complexity $O(n^{c+3} \log n)$.

Theorem 2. *Algorithm 1 is correct and is of time complexity* $O(n^{c+3} \log n)$ *when* k *is bounded by a constant* c.

Proof. In the process of the algorithm, the value of u or ℓ changes in each loop and the value of $u - \ell$ decreases gradually. Thus, at the beginning of each loop, the value of g also changes, which influence the capacities of edges in the graph N. In particular, V_1 doesn't influence the construction of N.

In above algorithm, when $R \leq m|V|$, the value of ℓ is changed and V_1 is set $S \setminus \{s\}$. Thus, by the step 7 in the algorithm, V_1 contains at least k vertices of a

Input: A unweighted graph $G = (V, E)$ with n vertices, m edges and an integer k
Output: An induced subgraph with at least k vertices

$\ell = 0$;
$u = m$;
$V_1 = V$;
while $u - \ell \geq \frac{1}{n(n-1)}$ **do**
    $g = \frac{u+\ell}{2}$;
    Construct $N = (V_N, E_N)$;
    Find a minimum s-t cut (S, T) with at least k vertices and compute its cut value R;
    if $R > m|V|$ **then**
        $u = g$;
    else
        $\ell = g$;
        $V_1 = S \setminus \{s\}$;
    end
end
return *the subgraph of* G *induced by* V_1

Algorithm 1. The algorithm for $dal(G, k)$

subgraph of G. By the Lemma 3, V_1 has density at least ℓ. When the algorithm stops, we know that there is no subgraph of at least k vertices with density $\ell + \frac{1}{n(n-1)}$ or greater, so the subgraph returned is a maximum density subgraph with at least k vertices by the result of Theorem 1 So, above algorithm is correct.

If $T(n)$ be the time of finding a minimum s-t cut with at least k vertices in a graph of n vertices and m edges, the time of Algorithm 1 is $O(T(n) \log n)$. Thus, we get above algorithm is of time complexity conclusion $O(n^{c+3} \log n)$ by Lemma 1. Hence, we get the theorem.

Using Megiddo's technique [12], Goldberg generalizes his algorithm to weighted graphs. Similarly, our algorithm can be generalized to weighted graphs. Details can be found [7], we omit them here.

4 A Greedy Approximation Algorithm for *DSS*

Given $G = (V, E)$ and $v_F = \{v_{f_1}, \ldots v_{f_k}\} \subseteq V$, a greedy approximation algorithm is designed as follows: we repeatedly remove the vertex of $V \setminus v_F$ with the minimum weighted degree in all non-v_F vertices until the subgraph has only those v_F vertices. Let G_i denote each subgraph of i vertices formed in the process, where $i \in \{n, \ldots, k\}$. We then output the subgraph with the maximum density from $\{G_n, \ldots, G_k\}$.

Assume that $|V| = n$ and $|E| = m$. It is not difficult to know that the greedy algorithm can be implemented in time $O(n^2)$.

In the following, we show that the greedy approximation algorithm for DSS has approximation ratio $2 \cdot (1 + \frac{k}{3})$. First, we prove two lemmas.

Lemma 4. *Suppose H is a subgraph of G and $v \in H$. Let $w(H,v)$ denote the sum of the weights of the edges incident with v in H. Let $H' = H \setminus \{v\}$. Then $d(H') = d(H) + \frac{d(H)-w(H,v)}{|H|-1}$.*

Proof. Note that $W(H') + w(H,v) = W(H)$. So $(|H|-1)d(H') + deg(H,v) = |H|d(H)$. Hence, $d(H') = d(H) + \frac{d(H)-w(H,v)}{|H|-1}$.

Lemma 5. *If H is an optimum solution of DSS and $v \in H$, then $w(G,v) \geq mdv(G, v_F)$.*

Proof. Let $H' = H \setminus \{v\}$. Since $d(H) \geq d(H')$, $d(H) \geq d(H) + \frac{d(H)-w(H,v)}{h-1}$ by Lemma 4. Thus, we get $w(H,v) \geq d(H)$. So $w(H,v) \geq mdv(G, v_F)$. Thus, $w(G,v) \geq w(H,v) \geq mdv(G, v_F)$.

Theorem 3. *The greedy approximation algorithm for DSS is of approximation ratio $2 \cdot (1 + \frac{k}{3})$.*

Proof. We prove the claim by induction on the vertex number n of G. Basis: when $n = k+1, k+2$, it is easy to verify the conclusion.

Inductive Hypothesis: We suppose the conclusion holds when $n = s-1$, for some s such that $s - k \geq 3$.

We will show that the conclusion also holds for $n = s$.

Let v_1 be the first vertex that is removed; i.e., v_1 is the vertex of the minimum weighted degree out of all non-v_F vertices. So, $G_{s-1} = G \setminus \{v_1\}$. There are two cases for v_1: either v_1 does not appear in the vertex set of the optimum solution of DSS or v_1 does appear in the optimum solution of DSS.

When v_1 does not appear in the vertex set of the optimum solution of DSS, we get $mdv(G_{s-1}, v_F) = mdv(G, v_F)$. Let $A(G)$ be the output solution of greedy approximation algorithm for G. Let $A(G_{s-1})$ be the solution produced by the greedy approximation algorithm for G_{s-1}.

By the induction hypothesis, $mdv(G_{s-1}, v_F)/A(G_{s-1}) \leq 2 \cdot (1 + \frac{k}{3})$. Since $A(G) = \max\{d(G), A(G_{s-1})\}$, we get $mdv(G, v_F)/A(G) \leq mdv(G_{s-1}, v_F)/A(G_{s-1}) \leq 2 \cdot (1 + \frac{k}{3})$.

Suppose v_1 does appear in the vertex set of the optimum solution of $DS(G, v_F)$. Then $w(G,v) \geq mdv(G, v_F)$ by Lemma 5. Since v_1 has the minimum weighted degree in all non-v_F vertices in G and there are $s-k$ non-v_F vertices in G, the sum of all $w(G,v)$ ($v \notin v_F$) is at least $(s-k)w(G, v_1)$. Thus, the sum of edges weights in G is at least $\frac{(s-k)w(G,v_1)}{2}$; i.e., $W(G) \geq \frac{(s-k)w(G,v_1)}{2}$.

Hence, $d(G) = \frac{W(G)}{s} \geq \frac{(s-k)w(G,v_1)}{2s} \geq \frac{mdv(G,v_F)}{2} \cdot \frac{s-k}{s}$. Since $s - k \geq 3$, $\frac{s-k}{s} \geq \frac{3}{k+3}$. Thus, $d(G) \geq \frac{mdv(G,v_F)}{2} \cdot \frac{3}{k+3}$. Hence, $\frac{mdv(G,v_F)}{d(G)} \leq \frac{2(k+3)}{3}$. So $mdv(G, v_F)/A(G) \leq mdv(G, v_F)/d(G) \leq \frac{2(k+3)}{3} = 2 \cdot (1 + \frac{k}{3})$.

Hence, the conclusion holds for $n = s$. By induction, the conclusion holds for all $n \geq k$. Thus, the greedy approximation algorithm for DSS is of approximation ratio $2 \cdot (1 + \frac{k}{3})$.

5 Conclusion

In this paper, we have presented a $O(n^{c+3}\log n)$ time algorithm for the densest *at-least-k*-subgraph problem (DalkS) when k is bounded by a constant c. Second, we design an approximation algorithm with time $O(n^2)$ for finding the densest subgraph with a specified subset problem. We prove that the approximation ratio of the approximation algorithm is $2 \cdot (1 + \frac{k}{3})$, where k is the element number of the specified subset.

One open problem is whether DalkS is still solvable in polynomial time when k is not bounded by a constant (particularly as k approaches $\frac{n}{2}$). Another open problem is to propose approximation algorithms with better approximation better or exact algorithms with less time.

Acknowledgments. We would like to thank the anonymous referees for their careful readings of the manuscripts and many useful suggestions.

Wenbin Chen's research has been supported by the National Science Foundation of China (NSFC) under Grant No. 11271097. Lingxi Peng's research has been partly supported by the Funding Program for Research Development in Institutions of Higher Learning Under the Jurisdiction of Guangzhou Municipality under Grant No. 2012A077. Jianxiong Wang's research was partially supported under Foundation for Distinguished Young Talents in Higher Education of Guangdong (2012WYM0105 and 2012LYM0105) and Funding Program for Research Development in Institutions of Higher Learning Under the Jurisdiction of Guangzhou Municipality (2012A143). FuFang Li's work had been co-financed by: Natural Science Foundation of China under Grant No. 61472092; Guangdong Provincial Science and Technology Plan Project under Grant No. 2013B010401037; and GuangZhou Municipal High School Science Research Fund under grant No. 1201421317. Maobin Tang's research has been supported under Guangdong Province's Science and Technology Projects under Grant No. 2012A020602065 and the research project of Guangzhou education bureau under Grant No. 2012A075.

References

1. Andersen, R., Chellapilla, K.: Finding dense subgraphs with size bounds. In: Avrachenkov, K., Donato, D., Litvak, N. (eds.) WAW 2009. LNCS, vol. 5427, pp. 25–37. Springer, Heidelberg (2009)
2. Charikar, M.: Greedy approximation algorithms for finding dense components in a graph. In: Jansen, K., Khuller, S. (eds.) APPROX 2000. LNCS, vol. 1913, pp. 84–95. Springer, Heidelberg (2000)
3. Chen, W., Samatova, N.F., Stallmann, M.F., Hendrix, W.: On size-constrained minimum *s-t* cut problems and size-constrained dense subgraph problems, submitted to Theoretical Computer Science, under review
4. Feige, U., Seltser, M.: On the densest k-subgraph problems. Technical report: CS97-16, Department of Applied Mathematics and Computer Science (1997)
5. Feige, U., Kortsarz, G., Peleg, D.: The dense *k*-subgraph problem. Algorithmica **29**, 410–421 (2001)
6. Gallo, G., Grigoriadis, M., Tarjan, R.: A fast parametric maximum flow algorithm and applications. SIAM J. Comput. **18**(1), 30–55 (1989)

7. Goldberg, A.: Finding a maximum density subgraph, Technical report UCB/CSB 84/171, Department of Electrical Engineering and Computer Science, University of California, Berkeley (1984)
8. Hu, H., Yan, X., Huang, Y., et al.: Mining coherent dense subgraphs across massive biological networks for functional discovery. Bioinformatics **21**, 213–221 (2005)
9. Han, Q.M., Ye, Y.Y., Zhang, J.W.: Approximation of Dense-k subgraph, http://citeseerx.ist.psu.edu/viewdoc/summary?doi=10.1.1.41.1899
10. Khuller, S., Saha, B.: On finding dense subgraphs. In: Albers, S., Marchetti-Spaccamela, A., Matias, Y., Nikoletseas, S., Thomas, W. (eds.) ICALP 2009, Part I. LNCS, vol. 5555, pp. 597–608. Springer, Heidelberg (2009)
11. Kortsarz, G., Peleg, D.: On choosing a dense subgraph. In: Proceedings of the 34th Annual IEEE Symposium on Foundations of Computer Science, pp. 692–701 (1993)
12. Megiddo, N.: Combinatorial optimization with rational objective function. Math. Operat. Res. **4**(4), 414–424 (1979)
13. Saha, B., Hoch, A., Khuller, S., Raschid, L., Zhang, X.-N.: Dense subgraphs with restrictions and applications to gene annotation graphs. In: Berger, B. (ed.) RECOMB 2010. LNCS, vol. 6044, pp. 456–472. Springer, Heidelberg (2010)
14. Srivastav, A., Wolf, K.: Finding dense subgraphs with semidefinite programming. In: Jansen, K., Rolim, J.D.P. (eds.) APPROX 1998. LNCS, vol. 1444, pp. 181–191. Springer, Heidelberg (1998)

Deleting Edges to Restrict the Size of an Epidemic: A New Application for Treewidth

Jessica Enright[1](✉) and Kitty Meeks[2]

[1] Computing Science and Mathematics, University of Stirling, Stirling, UK
jae@cs.stir.ac.uk
[2] School of Mathematics and Statistics, University of Glasgow, Glasgow, UK
kitty.meeks@glasgow.ac.uk

Abstract. Motivated by applications in network epidemiology, we consider the problem of determining whether it is possible to delete at most k edges from a given input graph (of small treewidth) so that the maximum component size in the resulting graph is at most h. While this problem is NP-complete in general, we provide evidence that many of the real-world networks of interest are likely to have small treewidth, and we describe an algorithm which solves the problem in time $O((wh)^{2w}n)$ on an input graph having n vertices and whose treewidth is bounded by a fixed constant w.

1 Introduction

Network epidemiology seeks to understand the dynamics of disease spreading over a network or graph, and is an increasingly popular method of modelling real-world disease. The rise of network epidemiology corresponds to a rapid increase in the availability of contact network datasets that can be encoded as networks or graphs: typically, the vertices of the graph represent agents that can be infected and infectious, such as individual humans or animals, or appropriate groupings of these, such as cities, households, or farms. The edges are then the potentially infectious contacts between those agents. Considering the contacts within a population as the edges of a graph can give a large improvement in disease modelling accuracy over mass action models, which assume that a population is homogeneously mixing. For example, if we consider a sexual contact network in which the vertices are people and the edges are sexual contacts, the heterogeneity in contacts is very important for explaining the pattern and magnitude of an AIDS epidemic [1].

Our work has been especially motivated by the idea of controlling diseases of livestock by preventing disease spread over livestock trading networks. As required by European law, individual cattle movements between agricultural holdings in Great Britain are recorded by the British Cattle Movement Service (BCMS) [20]; in early 2014, this dataset contained just under 300 million trades and just over 133,000 agricultural holdings. For modelling disease spread across

Z. Lu et al. (Eds.): COCOA 2015, LNCS 9486, pp. 574–585, 2015.
DOI: 10.1007/978-3-319-26626-8_42

the British cattle industry, it is common to create vertices from farms, and edges from trades of cattle between those farms: a disease incursion starting at a single farm could spread across this graph through animal trades, as is thought to have happened during the economically-damaging 2001 British foot-and-mouth disease crisis [15].

We are interested in controlling or limiting the spread of disease on this sort of network, and so have focussed our attention on edge deletion, which might correspond to forbidden trade patterns or, more reasonably, extra vaccination or disease surveillance along certain trade routes. Introducing extra controls of this kind is costly, so it is important to ensure that this is done as effectively as possible. Our target graph class is also informed by our disease motivation: when a contagion spreads over the edges of a graph, the maximum component size is an upper bound on the maximum number of vertices infected from a single initially infected vertex. To this end, we consider the problem of determining whether a given graph can be modified, using only up to k edge-deletion operations, so that the resulting graph has maximum component size at most h. We also discuss a number of relevant extensions:

- assigning different weights to different vertices (e.g. corresponding to the number of animals in a particular animal holding), and seeking to bound the total weight of each connected component;
- associating different costs with the deletion of different edges;
- imposing different limits on the size of components containing individual vertices (for example, we might want to enforce a smaller size limit for components containing certain vertices considered to be particularly high risk).

This problem is intractable in general, so in order to develop useful algorithms for real-world applications we need to exploit structural properties of the input network. In Sect. 2 we provide evidence that many animal trade networks of interest are likely to have small treewidth. In Sect. 3 we then go on to describe an algorithm to solve the problem whose running time on an n-vertex graph of treewidth w is bounded by $O((wh)^{2w}n)$; this algorithm is easily adapted to output an optimal solution. Many problems that are thought to be intractable in general are known to admit polynomial-time algorithms when restricted to graphs of bounded treewidth, often by means of a dynamic programming strategy similar to that used to attack the problem considered here; however, to the best of the authors' knowledge, the usefulness of such algorithms for solving real-world network problems has yet to be investigated thoroughly.

In reality, policy decisions about where to introduce controls are likely to be influenced by a range of factors, which cannot all be captured adequately in a network model. Thus, the main application of our algorithm will be in comparing any proposed strategy with the theoretical optimum: a policy-maker can determine whether there is a solution with the same total cost that results in a smaller maximum component size; extensive experiments on real animal movement data are left as a task for future work.

In the remainder of this section, we begin by reviewing previous related work in Sect. 1.1 before introducing some important notation in Sect. 1.2 and reviewing the key features of tree decompositions in Sect. 1.3. A discussion of the treewidth of real-world networks is given in Sect. 2, and our algorithm is described in Sect. 3.

1.1 Review of Previous Work

The problem of modifying a graph to bound the maximum component size has previously been studied both in the setting of epidemiology [18] and in the study of network vulnerability [7,13]. The edge-modification version we consider here appears in the literature under various names, including the *component order edge connectivity problem* [13] and the *minimum worst contamination problem* [18]. Li and Tang [18] show that it is NP-hard to approximate the minimisation version of the problem to within $2 - \epsilon$, while Gross et al. [13] describe a polynomial-time algorithm to solve the problem when the input graph is a tree.

From a combinatorial perspective, this problem belongs to the more general family of *edge-deletion problems.* An edge-deletion problem asks if there is a set of at most k edges that can be deleted from an input graph to produce a graph in some target class. In contrast to the related well-characterised vertex-deletion problems [17], there is not yet a complete characterisation of the hardness of edge-deletion problems by target graph class.

Yannakakis [24] gave early results in edge-deletion problems, showing that edge-deletion to planar graphs, outer-planar graphs, line graphs, and transitive digraphs is NP-complete. Subsequently, Watanabe et al. [23] showed that edge-deletion problems are NP-complete if the target graph class can be finitely characterised by 3-connected graphs. There are a number of further hardness results known for edge-deletion to well-studied graph classes, including for interval and unit interval graphs [11], cographs [8], and threshold graphs [19] and, as noted in [21], hardness of edge-deletion to bipartite graphs follows from the hardness of a MAX-CUT problem. Natanzon et al. [21] further showed NP-completeness of edge-deletion to disjoint unions of cliques, and perfect, chain, chordal, split, and asteroidal-triple-tree graphs, but also give polynomial-time algorithms, in the special case of the input graph having bounded degree, for edge-deletion to chain, split, and threshold graphs.

Given the large number of hardness results in the literature, it is natural to consider the parameterised complexity of these problems. Cai [5] initiated this investigation, showing that edge-deletion to a graph class characterisable by a finite set of forbidden induced subgraphs is fixed-parameter tractable when parameterised by k (the number of edges to delete): he gave an algorithm to solve the problem in time $O(d^{2k} \cdot n^{d+1})$, where n is the number of vertices in the input graph and d is the maximum number of vertices in a forbidden induced subgraph. Further fpt-algorithms have been obtained for edge-deletion to split graphs [10] and to chain, split, threshold, and co-trivially perfect graphs [14].

Considering the problem of deleting edges to obtain a graph with restricted maximum componet size, restricted to graphs of small treewidth, the algorithm we describe in this paper represents a significant improvement on Cai's result [5] above, which implies the existence of an algorithm running in time $O(h^{2k} \cdot n^h)$

(on arbitrary input graphs). While the fixed parameter tractability of this problem (parameterised by the maximum component size h) restricted to graphs of bounded treewidth does follow from the optimization version of Courcelle's Theorem [3,6], this does not lead to a practical algorithm for addressing real-world problems.

1.2 Notation and Problem Definition

Unless otherwise stated, all graphs are simple, undirected, and loopless. For graph $G = (V, E)$, $V = V(G)$ is the vertex set of G, and $E = E(G)$ the edge set of G. We denote the sizes of the edge and vertex sets of G as $e(G) = |E(G)|$ and $v(G) = |V(G)|$. For further general graph notation, we direct the reader to [12].

A *partition* $\mathcal{P}$ of a set X is a collection of disjoint, non-empty sets whose union is X. We call each set in the partition a *block* of the partition, and every partition corresponds to a unique equivalence relation on X where $x \sim y$ if and only if x and y belong to the same block of X.

In this paper, we consider the following problem, were $\mathcal{C}_h$ is the set of all connected graphs on h vertices.

$\mathcal{C}_h$-FREE EDGE DELETION
Input: A Graph $G = (V, E)$ and an integer k.
Question: Does there exist $E' \subseteq E$ with $|E'| = k$ such that $G \setminus E'$ does not contain any $H \in \mathcal{C}_h$ as an induced subgraph?

This problem is NP-complete even for $h = 4$: in [9] we outline an easy proof of this result, by means of a reduction from PERFECT TRIANGLE COVER, which relies on the observation that the maximum number of edges in a graph having maximum component size h is obtained if the graph is a disjoint union of h-cliques. In particular, this indicates that parameterisation by h alone will not be sufficient to give an fpt-algorithm.

1.3 Tree Decompositions

In this section we review the concept of a tree decomposition (introduced by Robertson and Seymour in [22]) and introduce some of the key notation we will use throughout the rest of the paper.

Given any tree T, we will assume that it contains some distinguished vertex $r(T)$, which we will call the *root* of T. For any vertex $v \in V(T) \setminus r(T)$, the *parent* of v is the neighbour of v on the unique path from v to $r(T)$; the set of *children* of v is the set of all vertices $u \in V(T)$ such that v is the parent of u. The *leaves* of T are the vertices of T whose set of children is empty. We say that a vertex u is a *descendant* of the vertex v if v lies somewhere on the unique path from u to $r(T)$ (note therefore that every vertex is a descendant of the root). Additionally, for any vertex v, we will denote by T_v the subtree induced by v together with the descendants of v.

We say that $(T, \mathcal{D})$ is a *tree decomposition* of G if T is a tree and $\mathcal{D} = \{\mathcal{D}(t) : t \in V(T)\}$ is a collection of non-empty subsets of $V(G)$ (or *bags*), indexed by the nodes of T, satisfying:

1. $V(G) = \bigcup_{t \in V(T)} \mathcal{D}(t)$,
2. for every $e = uv \in E(G)$, there exists $t \in V(T)$ such that $u, v \in \mathcal{D}(t)$,
3. for every $v \in V(G)$, if $T(v)$ is defined to be the subgraph of T induced by nodes t with $v \in \mathcal{D}(t)$, then $T(v)$ is connected.

The *width* of the tree decomposition $(T, \mathcal{D})$ is defined to be $\max_{t \in V(T)} |\mathcal{D}(t)| - 1$, and the *treewidth* of G is the minimum width over all tree decompositions of G.

We will denote by V_t the set of vertices in G that occur in bags indexed by the descendants of t in T. Thus, $V_t = \bigcup_{t' \in V(T_t)} \mathcal{D}(t')$.

Although it is NP-hard to determine the treewidth of an arbitrary graph [2], it is shown in [4] that the problem of determining whether a graph has treewidth at most w, and if so computing a tree-decomposition of width at most w, can be solved in linear time for any constant w.

Theorem 1 ([4]). *For each $w \in N$, there exists a linear time algorithm, that tests whether a given graph $G = (V, E)$ has treewidth at most w, and if so, outputs a tree decomposition of G with treewidth at most w.*

A special kind of tree decomposition, known as a *nice tree decomposition*, was introduced in [16]. The nodes in such a decomposition can be partitioned into four types (examples in Fig. 1):

Leaf nodes: t is a leaf in T.
Introduce nodes: t has one child t', such that $\mathcal{D}(t') \subset \mathcal{D}(t)$ and $|\mathcal{D}(t)| = |\mathcal{D}(t')| + 1$.
Forget nodes: t has one child t', such that $\mathcal{D}(t') \supset \mathcal{D}(t)$ and $|\mathcal{D}(t)| = |\mathcal{D}(t')| - 1$.
Join nodes: t has two children, t_1 and t_2, with $\mathcal{D}(t_1) = \mathcal{D}(t_2) = \mathcal{D}(t)$.

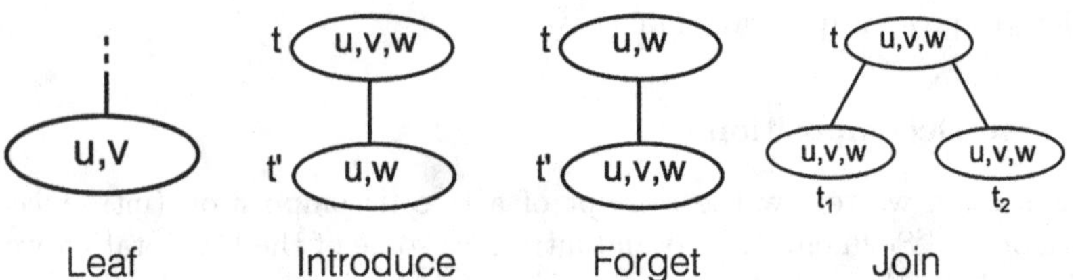

Fig. 1. The four types of node in a nice tree decomposition. From left to right: a leaf, an introduce node, a forget node, and a join node.

Any tree decomposition can be transformed into a nice tree decomposition in linear time:

Lemma 1 ([16]). *For constant k, given a tree decomposition of a graph G of width w and $O(n)$ nodes, where n is the number of vertices of G, one can find a nice tree decomposition of G of width w and with at most $4n$ nodes in $O(n)$ time.*

2 Treewidth of Real Networks

While the overall graph of cattle trades in Great Britain from 2001 to 2014 is fairly dense, many of the edges are repeated or parallel trades: that is, a farm sending animals over time to the same place, or many individual animals being moved at the same time; when we restrict our attention to a limited time frame, and ignore movements that would generate multiple edges (that is, we require our graph to be simple), the graph is quite sparse. When considering an epidemic, it is much more relevant only to consider trades occurring within some restricted time frame (whose precise duration depends on the disease under consideration).

Moreover, the networks that are obtained by considering shorter time frames typically have an approximately hub-and-spoke or tree-like structure, which results in small treewidth. This can be explained to some extent by considering the structure of the industry and the directionality of farm management styles. For example, beef cattle are likely to flow through dealers or markets, and lead quite short lives, which is likely to result in a hub-and-spoke network. Additionally, farms can sometimes be characterised by "type", with breeders producing calves who then might be grown at one or two other farms before eventual slaughter: this means that cycles are unlikely to occur frequently in the network.

These anecdotal observations about the treewidth of livestock trade networks have been supported by computational calculations on some examples of real cattle trading graphs. First of all, for each year from 2009 to 2014, we generated a graph from a type of persistent trade link recorded by BCMS in Scotland. The largest of these is derived from the trades in 2013, and includes approximately 7,000 nodes and 6,000 edges (this lower density is typical when considering only persistent trade links, or trades over a restricted time period). None of these six graphs has treewidth more than four.

Secondly, in addition to these persistent trade links, we have computed an upper bound of the treewidth of the largest component of an aggregated, undirected version of the overall network of cattle trades in Scotland in 2009 over a variety of time windows Fig. 2. The treewidths of these components remains low even for large time windows: for an aggregation of all movements in a 200-day window the treewidth is below 10, and for all movements over the year it is below 18. It is unlikely to be necessary to include a full year of movements in the analysis of any single epidemic, as the time scale of most exotic epidemics is much shorter.

While we have by no means completed an exhaustive study of the structural properties of real-world livestock trade networks, the evidence given here seems sufficient to suggest that algorithms which achieve a good running time on graphs of bounded treewidth will be useful for this application in practice.

3 The Algorithm

In this section, we describe an algorithm which, given a graph G together with a nice tree decomposition $(T, \mathcal{D})$ of G of width at most w, determines whether or

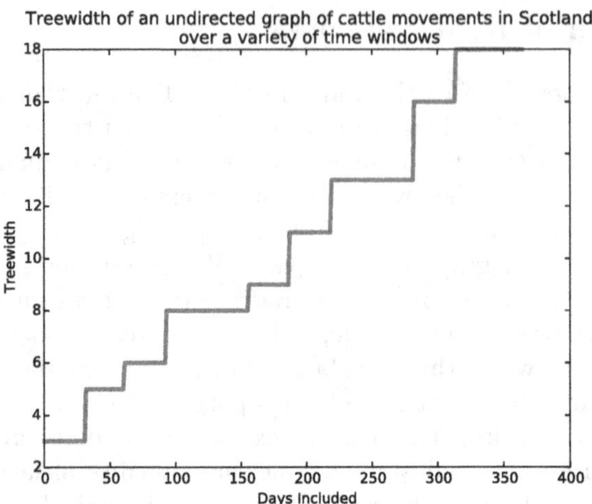

Fig. 2. A plot of an upper bound treewidth of the largest component in an undirected version of the cattle movement graph in Scotland in 2009 over a number of different days included: all day sets start on January 1, 2009. Treewidths below eight are exact, treewidths over eight are upper bounds of the true treewidth.

not it is possible to delete at most k edges from G so that the resulting graph has no component on more than h vertices. Since there exist linear-time algorithms both to compute a tree-decomposition of any graph G of fixed treewidth w, and to transform an arbitrary tree-decomposition into a nice tree decomposition, this in fact gives an algorithm which takes as input just a graph G of treewidth at most w. Thus, we prove the following theorem.

Theorem 2. *There exists an algorithm to solve C_h-*Free Edge Deletion *in time $O((wh)^{2w}n)$ on an input graph with n vertices whose treewidth is at most w.*

As with many algorithms that use tree decompositions, our algorithm works by recursively carrying out computations for each node of the tree, using the results of the same computation carried out on any children of the node in question. In this case, we recursively compute the *signature* of each node: we define the signature of a node in Sect. 3.1. It is then possible to determine whether we have a yes- or no-instance to the problem by examining the signature of the root of T.

The techniques used to calculate each node's signature from those of its children (which differ slightly depending on which of the four types of node in the nice tree decomposition is being considered) are fairly standard in the literature, and are omitted here due to space constraints. Full details of the algorithm, together with a mathematical proof of its correctness, are given in [9]. The running time of the algorithm is justified in Sect. 3.2, and several extensions are also discussed.

3.1 The Signature of a Node

In this section, we describe the information we compute for each node, and define the *signature* of a node.

Throughout the algorithm, we need to record the possible states corresponding to a given bag. A valid *state* of a bag $\mathcal{D}(t)$ is a triple consisting of:

1. a partition $\mathcal{P}$ of $\mathcal{D}(t)$ into disjoint, non-empty subsets or *blocks* of size at most h, and
2. a function $c : \mathcal{P} \rightarrow [h]$ such that, for each $X \in \mathcal{P}$, $|X| \leq c(X)$.

We will write $u \sim_{\mathcal{P}} v$ to indicate that u and v belong to the same block of $\mathcal{P}$.

Intuitively, $\mathcal{P}$ tells us which vertices are allowed to belong to the same component of the graph we obtain after deleting edges and c tells us the maximum number of vertices which are permitted in components corresponding to a given block of the partition.

For any bag $\mathcal{D}(t)$, we denote by $\text{st}(t)$ the set of possible states of $\mathcal{D}(t)$. Note that there are at most B_w partitions of a set of size w (where B_w is the wth Bell number) and at most h^w functions from a set of size at most w to $[h]$; thus the total number of valid states for $\mathcal{D}(t)$ is at most $B_w h^w < (wh)^w$ (although not all possible combinations of a partition and a function will give rise to a valid state).

For any given state $\sigma = (\mathcal{P}, c) \in \text{st}(t)$, we set $\mathcal{E}(t,\sigma)$ to be the set of edge-sets $E' \subset E(G[V_t])$ such that $\widetilde{G_t} = G[V_t] \setminus E'$ has the following properties:

1. for each connected component C of $\widetilde{G_t}$:
   (a) $|V(C)| \leq h$, and
   (b) if $C_t = V(C) \cap \mathcal{D}(t) \neq \emptyset$, then C_t is contained in a single block X_C of $\mathcal{P}$,
2. for each block X in $\mathcal{P}$, the total number of vertices in connected components of $\widetilde{G_t}$ that intersect X is at most $c(X)$.

Note that, whenever σ is a valid state for t, the set $\mathcal{E}(t,\sigma)$ will be non-empty: setting $E' = E(G[V_t])$ will always satisfy both conditions. Since we are interested in determining whether it is possible to delete at most k edges to obtain a graph with maximum component size h, we will primarily be interested in a subset of $\mathcal{E}(t,\sigma)$: for any node t and $\sigma \in \text{st}(t)$ we define this subset as

$$\mathcal{E}_k(t,\sigma) = \{E' \in \mathcal{E}(t,\sigma) : |E'| \leq k\}.$$

We then define

$$\text{del}_k(t,\sigma) = \min_{E' \in \mathcal{E}_k(t,\sigma)} |E'|,$$

adopting the convention that the minimum, taken over an empty set, is equal to infinity. To simplify notation, given any $a, b \in \mathbb{N}$, we define $[a]_{\leq b}$ to be equal to a if $a \leq b$, and equal to ∞ otherwise. Finally, we define the *signature* of a node t to be the function $\text{sig}_t : \text{st}(t) \rightarrow \{0, 1, \ldots, k, \infty\}$ such that $\text{sig}_t(\sigma) = \text{del}_k(t,\sigma)$.

Observe that, with this definition, our input graph is a yes-instance to $\mathcal{C}_h$-Free Edge Deletion if and only if there exists some $\sigma \in \text{st}(r)$ such that $\text{sig}_r(\sigma) \leq k$, where r is the root of the tree indexing the decomposition.

3.2 Running Time and Extensions

At each of the $O(n)$ nodes of the nice tree decomposition, we will generate, and then iterate over, fewer than $(wh)^w$ states for that node. For each of those states, we will need to consider a collection of *inherited states* for the node's children; there are at most $(wh)^w$ such states (or pairs of states, in the case of a join node) that need to be considered for each state of the parent node. In the algorithm, we first generate each of the states for a given node, and the corresponding set of inherited states for its children, then iterate over each relevant combination of states, performing various constant-time operations. Thus, at each of $O(n)$ nodes we do $O\left((wh)^{2w}\right)$ work, giving an overall time complexity of $O((wh)^{2w}n)$.

For simplicity, we have only described the most basic version of the algorithm; however, it is straightforward to extend it to deal with more complicated situations, involving any or all of the following.

Deleting edges so that the sum of weights of vertices in any component is at most h, where a weight function $w : V(G) \to \mathbb{N}$ is given: change condition 1(a) in the definition of $\mathcal{E}(t,\sigma)$ to $\sum_{v \in V(C)} w(v) \leq h$, and add to the definition of the set of valid states for a node the condition that, for each block X of $\mathcal{P}$, we have $\sum_{v \in X} w(X) \leq c(X)$.

Determining if it is possible to delete a set of edges whose total cost is at most k, where a cost function $f : E(G) \to \mathbb{N}$ is given: define $\text{del}_k(t,\sigma)$ to be $\min_{E' \in \mathcal{E}(\sigma,t)} \sum_{e \in E'} f(e)$.

Deleting edges so that each vertex v belongs to a component containing at most $\ell(v)$ vertices, where a limit function $\ell : V(G) \to \mathbb{N}$ is given: change condition 1(a) in the definition of $\mathcal{E}(t,\sigma)$ to $|V(C)| \leq \min_{v \in V(C)} \ell(v)$, and add to the definition of the set of valid states for a node the condition that, for each block X of $\mathcal{P}$, we have $c(X) \leq \min_{v \in X} \ell(v)$.

None of these adaptations changes the asymptotic running time of the algorithm.

Additionally, if we wish to output an optimal set of edges to delete in any of the variants (note that in general there may be many such optimal sets), we can simply record, for each node t and each state $\sigma \in \text{st}(t)$, a set of edges $E' \in \mathcal{E}_k(t,\sigma)$ such that $|E'| = \text{del}_k(t,\sigma)$; computing such a set from the relevant sets for the node's children requires only basic set operations. An element of $\mathcal{E}(r,\sigma)$, where r is the root of the tree decomposition and $\text{del}_k(r,\sigma) = \min_{\sigma \in \text{st}(r)}$ is then an optimal solution for the problem.

4 Conclusions and Open Problems

We have investigated the relevance of the well-studied graph parameter treewidth to the structure of real-world animal trade networks, and have provided evidence that this parameter is likely to be small for many networks of interest for epidemiological applications. Motivated by this observation, we have derived an algorithm to solve $\mathcal{C}_h$-Free Edge Deletion on input graphs having n vertices and treewidth bounded by some fixed constant w in time $O((wh)^{2w}n)$. It is straightforward to adapt this algorithm to deal with more complicated situations likely to arise in the application.

An implementation of our approach and its application to real livestock data sets will be one of our next steps; this presents an unusual opportunity to apply a treewidth-based optimisation algorithm to a real-life problem.

Many open questions remain concerning the complexity of this problem more generally, as we are far from having a complete complexity classification. We know that useful structure in the input graph is required to give an fpt-algorithm: we demonstrated that it is not sufficient to parameterise by the maximum component size h alone (unless P=NP). However, it remains open whether the problem might belong to FPT when parameterised only by the treewidth w; we conjecture that treewidth alone is *not* enough, and that the problem is W[1]-hard with respect to this parameterisation. Considering other potentially useful structural properties of input graphs, one question of particular relevance to epidemiology would be the complexity of the problem on planar graphs: this would be relevant for considering the spread of a disease based on the geographic location of animal holdings (in situations where a disease is likely to be transmitted between animals in adjacent fields).

Furthermore, animal movement networks can capture more information on real-world activity when considered as *directed* graphs, and the natural generalisation of the problem to directed graphs in this context would be to consider whether it is possible to delete at most k edges from a given directed graph so that the maximum number of vertices *reachable* from any given starting vertex is at most h. Exploiting information on the direction of movements might allow more efficient algorithms for this problem when the underlying undirected graph does not have very low treewidth; a natural first question would be to consider whether there exists an efficient algorithm to solve this problem on directed acyclic graphs.

Finally, based on our investigation of the treewidth of real-world animal trade networks, it is natural to ask what other relevant problems can be solved efficiently on graphs of bounded treewidth. For example, we might wish to delete edges to achieve membership in a more complicated graph class (for example, some class of graphs on which intervention strategies used in the event of a disease outbreak are likely to be effective); alternatively, if deleting edges to achieve a small component size is too costly to be practical in some situations, we might wish to consider more relaxed criteria that nevertheless retain some desirable properties.

Acknowledgements. We are very grateful to Ivaylo Valkov for his assistance in implementing this algorithm as part of a summer research project, and to EPIC: Scotland's Centre of Expertise on Animal Disease Outbreaks, which supported JE for part of her work on this project.

References

1. Anderson, R.M., Gupta, S., Ng, W.: The significance of sexual partner contact networks for the transmission dynamics of HIV. J. Acquir. Immune Defic. Syndr. Hum. Retrovirology **3**, 417–429 (1990)
2. Arnborg, S., Corneil, D.G., Proskurowski, A.: Complexity of finding embeddings in a k-tree. SIAM J. Alg. Disc. Meth. **8**, 277–284 (1987)
3. Arnborg, S., Lagergren, J., Sesse, D.: Easy problems for tree-decomposable graphs. J. Algorithms **12**, 308–340 (1991)
4. Bodlaender, H.L.: A linear time algorithm for finding tree-decompositions of small treewidth. In: Proceedings of the Twenty-fifth Annual ACM Symposium on Theory of Computing, STOC 1993, pp. 226–234. ACM, New York, NY, USA (1993)
5. Cai, L.: Fixed-parameter tractability of graph modification problems for hereditary properties. Inf. Process. Lett. **58**(4), 171–176 (1996)
6. Courcelle, B., Mosbah, M.: Monadic second-order evaluations on tree-decomposable graphs. Theor. Comput. Sci. **109**(1–2), 49–82 (1993)
7. Drange, P.G., Dregi, M.S., van't Hof, P.: On the computational complexity of vertex integrity and component order connectivity. In: Ahn, H.-K., Shin, C.-S. (eds.) ISAAC 2014. LNCS, vol. 8889, pp. 285–297. Springer, Heidelberg (2014)
8. El-Mallah, E.S., Colbourn, C.J.: The complexity of some edge deletion problems. IEEE Trans. Circ. Syst. **3**, 354–362 (1988)
9. Enright, J., Meeks, K.: Deleting edges to restrict the size of an epidemic (2015). arXiv:1504.05773 [cs.DS]
10. Ghosh, E., Kolay, S., Kumar, M., Misra, P., Panolan, F., Rai, A., Ramanujan, M.S.: Faster parameterized algorithms for deletion to split graphs. In: Fomin, F.V., Kaski, P. (eds.) SWAT 2012. LNCS, vol. 7357, pp. 107–118. Springer, Heidelberg (2012)
11. Goldberg, P.W., Golumbic, M.C., Kaplan, H., Shamir, R.: Four strikes against physical mapping of DNA. J. Comput. Biol. **2**, 139–152 (1993)
12. Golumbic, M.C.: Algorithmic Graph Theory and Perfect Graphs, vol. 57. Elsevier, Amsterdam (2004)
13. Gross, D., Heinig, M., Iswara, L., Kazmierczak, L.W., Luttrell, K., Saccoman, J.T., Suffel, C.: A survey of component order connectivity models of graph theoretic networks. SWEAS Trans. Math. **12**(9), 895–910 (2013)
14. Guo, J.: Problem Kernels for NP-complete edge deletion problems: split and related graphs. In: Tokuyama, T. (ed.) ISAAC 2007. LNCS, vol. 4835, pp. 915–926. Springer, Heidelberg (2007)
15. Kao, R.R., Green, D.M., Johnson, J., Kiss, I.Z.: Disease dynamics over very different time-scales: foot-and-mouth disease and scrapie on the network of livestock movements in the UK. J. R. Soc. Interface **4**(16), 907–916 (2007)
16. Kloks, T.: Treewidth: Computations and Approximations. LNCS, vol. 842. Springer, Heidelberg (1994)
17. Lewis, J.M., Yannakakis, M.: The node-deletion problem for hereditary properties is NP-complete. J. Comput. Syst. Sci. **20**(2), 219–230 (1980)
18. Li, A., Tang, L.: The complexity and approximability of minimum contamination problems. In: Ogihara, M., Tarui, J. (eds.) TAMC 2011. LNCS, vol. 6648, pp. 298–307. Springer, Heidelberg (2011)
19. Margot, F.: Some complexity results about threshold graphs. Discrete Appl. Math. **49**(1:3), 299–308 (1994). (Special Volume Viewpoints on Optimization)

20. Mitchell, A., Bourn, D., Mawdsley, J., Wint, W., Clifton-Hadley, R., Gilbert, M.: Characteristics of cattle movements in britain: an analysis of records from the cattle tracing system. Anim. Sci. **80**, 265–273 (2005)
21. Natanzon, A., Shamir, R., Sharan, R.: Complexity classification of some edge modification problems. Discrete Appl. Math. **113**(1), 109–128 (2001). (Selected Papers: 12th Workshop on Graph-Theoretic Concepts in Computer Science)
22. Robertson, N., Seymour, P.D.: Graph Minors. III. Planar Tree-Width. J. Comb. Theory Ser. B **36**(1), 49–64 (1984)
23. Watanabe, T., Ae, T., Nakamura, A.: On the NP-hardness of edge-deletion and -contraction problems. Discrete Appl. Math. **6**(1), 63–78 (1983)
24. Yannakakis, M.: Node-and edge-deletion NP-complete problems. In: Proceedings of the Tenth Annual ACM Symposium on Theory of Computing, STOC 1978, pp. 253–264. ACM, New York, NY, USA (1978)

Optimal Approximation Algorithms for Maximum Distance-Bounded Subgraph Problems

Yuichi Asahiro[1], Yuya Doi[2], Eiji Miyano[2(✉)], and Hirotaka Shimizu[2]

[1] Department of Information Science, Kyushu Sangyo University, Fukuoka 813-8503, Japan
asahiro@is.kyusan-u.ac.jp

[2] Department of Systems Design and Informatics, Kyushu Institute of Technology, Fukuoka 820-8502, Japan
miyano@ces.kyutech.ac.jp

Abstract. A *d-clique* in a graph $G = (V, E)$ is a subset $S \subseteq V$ of vertices such that for pairs of vertices $u, v \in S$, the distance between u and v is at most d in G. A *d-club* in a graph $G = (V, E)$ is a subset $S' \subseteq V$ of vertices that induces a subgraph of G of diameter at most d. Given a graph G with n vertices, the goal of MAX d-CLIQUE (MAX d-CLUB, resp.) is to find a d-clique (d-club, resp.) of maximum cardinality in G. MAX 1-CLIQUE and MAX 1-CLUB cannot be efficiently approximated within a factor of $n^{1-\varepsilon}$ for any $\varepsilon > 0$ unless $\mathcal{P} = \mathcal{NP}$ since they are identical to MAX CLIQUE [14,21]. Also, it is known [3] that it is $\mathcal{NP}$-hard to approximate MAX d-CLUB to within a factor of $n^{1/2-\varepsilon}$ for any fixed $d \geq 2$ and for any $\varepsilon > 0$. As for approximability of MAX d-CLUB, there exists a polynomial-time algorithm which achieves an optimal approximation ratio of $O(n^{1/2})$ for any even $d \geq 2$ [3]. For any odd $d \geq 3$, however, there still remains a gap between the $O(n^{2/3})$-approximability and the $\Omega(n^{1/2-\varepsilon})$-inapproximability for MAX d-CLUB [3]. In this paper, we first strengthen the approximability result for MAX d-CLUB; we design a polynomial-time algorithm which achieves an *optimal* approximation ratio of $O(n^{1/2})$ for MAX d-CLUB for any odd $d \geq 3$. Then, by using the similar ideas, we show the $O(n^{1/2})$-approximation algorithm for MAX d-CLIQUE on general graphs for any $d \geq 2$. This is the best possible in polynomial time unless $\mathcal{P} = \mathcal{NP}$, as we can prove the $\Omega(n^{1/2-\varepsilon})$-inapproximability. Furthermore, we study the tractability of MAX d-CLIQUE and MAX d-CLUB on subclasses of graphs.

1 Introduction

Let $G = (V, E)$ be an unweighted graph, where V and E denote the set of vertices and the set of edges, respectively. $V(G)$ and $E(G)$ also denote the vertex set and the edge set of G, respectively. A *clique* in a graph G is a subset $Q \subseteq V(G)$ of pairwise adjacent vertices, i.e., the diameter of the subgraph induced by Q is one. In graph theory and theoretical computer science, one of the most important

Z. Lu et al. (Eds.): COCOA 2015, LNCS 9486, pp. 586–600, 2015.
DOI: 10.1007/978-3-319-26626-8_43

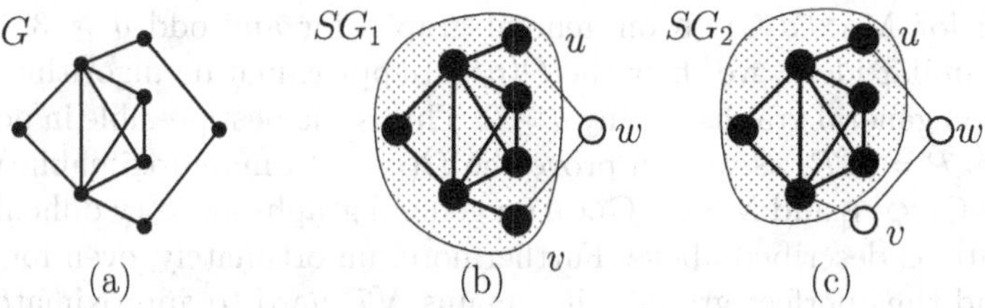

Fig. 1. (a) Graph G, (b) maximum 2-clique SG_1, and (c) maximum 2-club SG_2

and most investigated computational problems is the maximum clique problem (Max Clique): Given a graph G, the goal of Max Clique is to find a clique of maximum cardinality in G. There are a huge number of its applications in diverse fields; many different problems have been modeled using cliques.

In this paper, we consider two types of generalizations of Max Clique, the maximum distance-d clique problem (Max d-Clique) and the maximum diameter-d club problem (Max d-Club) for a positive integer d [2,18]: A *distance-d clique* (*d-clique* for short) in a graph G is a subset $S \subseteq V(G)$ of vertices such that for pairs of vertices $u, v \in S$, the distance between u and v is at most d in G. A *diameter-d club* (*d-club* for short) in a graph G is a subset $S' \subseteq V(G)$ of vertices that induces a subgraph of G of diameter at most d. Given a graph G, the goal of Max d-Clique (Max d-Club, resp.) is to find a d-clique (d-club, resp.) of maximum cardinality in G. For $d = 1$, Max 1-Club is the same problem as Max 1-Clique, i.e., Max 1-Club is simply Max Clique. For $d \geq 2$, however, Max d-Club and Max d-Clique are quite different: Take a look at a graph G having eight vertices illustrated in Fig. 1(a). The 2-clique of maximum cardinality in G is SG_1 induced by seven black vertices in Fig. 1(b), and the 2-club of maximum cardinality is SG_2 induced by only six black vertices in Fig. 1(c). Note that the distance between u and v is only two in G since those vertices are connected through the "outside" vertex w of SG_1. Also, note that the diameter of SG_1 itself is three, but the diameter of SG_2 is two.

In the following we assume that the input graph $G = (V, E)$ has $|V(G)| = n$ vertices and $|E(G)| = m$ edges. Since Max 1-Clique and Max 1-Club are identical to Max Clique, they cannot be efficiently approximated within a factor of $n^{1-\varepsilon}$ for any $\varepsilon > 0$ unless $\mathcal{P} = \mathcal{NP}$ [14,21]. For any $\varepsilon > 0$ and a fixed $d \geq 2$ it can be shown that it is $\mathcal{NP}$-hard to approximate Max d-Club to within a factor of $n^{1/3-\varepsilon}$ by using the gap preserving reduction provided by Marinček and Mohar [17] (although they assumed that $\mathcal{NP} \neq \mathcal{ZPP}$ in their original proof). Then, Asahiro et al. [3] improve the inapproximability from $\Omega(n^{1/3-\varepsilon})$ to $\Omega(n^{1/2-\varepsilon})$ for general graphs. As for the approximability, they present a polynomial-time algorithm which achieves an optimal approximation ratio of $O(n^{1/2})$ for any even $d \geq 2$. For any odd $d \geq 3$, however, there still remains a gap between the $O(n^{2/3})$-approximability and the $\Omega(n^{1/2-\varepsilon})$-inapproximability for Max d-Club on general graphs [3]. In this paper, we first strengthen the approximability result for Max d-Club; we design a simple polynomial-time algorithm which achieves an *optimal* approximation ratio of $O(n^{1/2})$ for MAX d-CLUB on general graphs for any odd $d \geq 3$. Then, by using the similar ideas, we show the $O(n^{1/2})$-approximation algorithm for MAX d-CLIQUE on general graphs for any $d \geq 2$. This is the best possible in polynomial time unless $\mathcal{P} = \mathcal{NP}$, as we can prove the $\Omega(n^{1/2-\varepsilon})$-inapproximability.

MAX d-CLIQUE and MAX d-CLUB on general graphs are very difficult even to approximate as described above. Furthermore, unfortunately, even for bipartite graphs (and thus perfect graphs), it remains $\mathcal{NP}$-hard to approximate MAX d-CLUB to within a factor of $n^{1/3-\varepsilon}$ for any $\varepsilon > 0$ and for any fixed integer $d \geq 3$. For every fixed *even* integer $d \geq 2$, it remains $\mathcal{NP}$-hard to approximate MAX d-CLUB to within a factor of $n^{1/3-\varepsilon}$ for any $\varepsilon > 0$ if the input graph is chordal. A graph G is *chordal* if every cycle of length at least four in G has at least one chord, which is an edge joining non-consecutive vertices in the cycle.

To cope with such hard situations, we investigate a graph structure on the length of cycles for MAX d-CLUB, motivated by the fact that a bipartite graph has only cycles of even length, and a chordal graph does not have long induced cycles. More precisely, we show that if every cycle of length at least $d+3$ in the input graph G has at least one chord, i.e., G has no induced cycle of length at least $d+3$, then a simple algorithm exactly solves MAX d-CLUB for G. As its direct consequences with simple observations, it is shown that for every fixed integer $d \geq 1$, MAX d-CLIQUE and MAX d-CLUB can be solved in polynomial time for (i) *interval* graphs, (ii) *trapezoid* graphs, and (iii) *strongly chordal* graphs. Furthermore, we show that for every fixed *odd* integer $d \geq 1$, MAX d-CLIQUE and MAX d-CLUB can be solved in polynomial time for (iv) *weakly chordal* graphs and thus (v) *chordal* graphs. The formal definitions of those graphs, and inclusion relations among the graph classes are given in Sect. 4.1, however the graphs, (i), (ii), and (v), are typical *intersection graphs*, and (i) through (v) do not include any induced cycle of length at least five. Note that although the tractability results (i) and (v) have been already stated in [3], their proofs contained errors which are fixed in this paper. Also, note that Golovach et al. [13] have independently shown the tractability of MAX d-CLUB on (i) interval, (iii) strongly chordal graphs for any $d \geq 2$ and MAX d-CLUB on (iv) weakly chordal and (v) chordal graphs for any odd $d \geq 1$ by proving the so-called *d-clique-power* properties of graphs without induced cycle of length at least five.

Although this paper focuses on the polynomial-time solvability and the approximability of the problems, here we would like to cite more theoretical results: The decision version d-CLUB of MAX d-CLUB asks whether there exists a d-club of size at least k in a given graph. Schäfer et al. [20] develop fixed parameter algorithms for the problem, whose running times are $O((k-2)^k \cdot k! \cdot kn + nm)$ and $O(2^{n-k} \cdot nm)$. Chang et al. [7] propose an $O^*(1.62^n)$ time algorithm for MAX d-CLUB.

2 Problems and Previous Results

2.1 Definitions

Let $G = (V, E)$ be a connected undirected graph. We denote an edge with endpoints u and v by (u, v). The maximum and the minimum degree among all vertices in a graph G is denoted by $\Delta(G)$ and $\delta(G)$, respectively. For a vertex v, the set of vertices adjacent to v in G, i.e., the open neighborhood of v is denoted by $N_G(v)$, and $N_G^+(v)$ denotes the closed neighborhood of v, i.e., $N_G(v) \cup \{v\}$.

A graph S is a subgraph of a graph $G = (V, E)$ if $V(S) \subseteq V$ and $E(S) \subseteq E$. For a subset of vertices $U \subseteq V$, let $G[U]$ be the subgraph of G induced by U. For a subgraph S of G, if $E(S) = V(S) \times V(S)$, then S (or $G[V(S)]$) and $V(S)$ are called a *clique* and a *clique set*, respectively.

A path P_{v_0,v_i,v_ℓ} of length ℓ from a vertex v_0 to a vertex v_ℓ, which passes through a vertex v_i, is represented as a sequence of vertices such that $P_{v_0,v_i,v_\ell} = \langle v_0, v_1, \cdots, v_i, \cdots, v_\ell \rangle$. The length of a path P is denoted by $|P|$. A cycle C of length ℓ is similarly written as $C = \langle v_0, v_1, \cdots, v_{\ell-1}, v_0 \rangle$. In this paper, we deal with simple paths and simple cycles only, that is, $v_i \neq v_j$ for any v_i and v_j in the sequences of vertices. For a path P, $\overline{P}$ represents a path of the reverse order. Consider two paths $P_1 = \langle v_0, \cdots, v_i \rangle$ and $P_2 = \langle v_i, \cdots, v_\ell \rangle$, where the end vertex of P_1 and the start vertex of P_2 are identical. Then, a path $P_{v_0,v_i,v_\ell} = \langle v_0, \cdots, v_i, \cdots, v_\ell \rangle$ constituted by P_1 and P_2 is denoted by $P_1 \oplus P_2$. For a path P passing through two vertices u and v, $P(u, v)$ is the subpath of P from u to v. For a pair of vertices u and v in G, the length of a shortest path from u to v, i.e., the *distance* between u and v is denoted by $dist_G(u, v)$, and the *diameter* of G is defined as $diam(G) = \max_{u,v \in V(G)}\{dist_G(u, v)\}$.

For a positive integer $d \geq 1$ and a graph G, the *d-th power of* G, denoted by $G^d = (V(G), (E(G))^d)$, is the graph formed from $V(G)$, where all pairs of vertices $u, v \in G$ such that $dist_G(u, v) \leq d$ are connected by an edge (u, v). Note that $E(G) \subseteq (E(G))^d$, i.e., the original edges in $E(G)$ are retained.

Definition 1. *A subgraph S (or a vertex set $V(S)$) of G is a* d-clique *(or a* d-clique set*) if $dist_G(u, v) \leq d$ holds for every pair of $u, v \in V(S)$.*

The diameter of a d-clique may be greater than d. A d-clique with diameter at most d is called d-club:

Definition 2. *A subgraph S of G (or a vertex set $V(S)$) is a* d-club *(or a* d-club set*) if $dist_S(u, v) \leq d$ holds for every pair of $u, v \in V(S)$, i.e., $diam(S) \leq d$.*

Although in [18], a d-club (or a d-clique) is defined as a maximal subgraph, i.e., no super set of vertices forms a d-club (or d-clique), we do not restrict our attention to maximal ones in this paper; it is known to be $\mathcal{NP}$-complete to answer whether a given d-club is maximal or not [19]. Clearly, a d-club of G is also a d-clique of G from their definitions. The sizes of the maximum clique, the maximum d-clique, and the maximum d-club in G are denoted by $\#clique(G)$, $\#clique(G, d)$, and $\#club(G, d)$, respectively.

For the maximization problems, an algorithm `ALG` is called a σ-approximation algorithm and `ALG`'s approximation ratio is σ if $OPT(G)/ALG(G) \leq \sigma$ holds for every input G, where $ALG(G)$ and $OPT(G)$ are the numbers of vertices of obtained subgraphs by `ALG` and an optimal algorithm, respectively.

The maximum d-clique problem (MAX d-CLIQUE) and the maximum d-club problem (MAX d-CLUB) are formulated as follows: Given a connected undirected graph G, the goal of MAX d-CLIQUE (MAX d-CLUB, resp.) is to find a d-clique (d-club, resp.) of maximum cardinality in G. In the case $d = 1$, the above two problems are the same; the problems MAX 1-CLIQUE and MAX 1-CLUB are identical to MAX CLIQUE, and hence they are generalizations of MAX CLIQUE in terms of the distance between vertices and the diameter of output subgraphs. Here we observe that it holds that $\#clique(G) = \#clique(G, 1) = \#club(G, 1)$. Since Erdös et al. [9] show that the diameter of a connected undirected graph G having n vertices is at most $3n/(\delta(G) + 1) + O(1)$, the above two problems are defined for the range $1 \leq d \leq 3n/(\delta(G) + 1) + O(1)$.

2.2 Optimal Approximation Algorithm for MAX d-CLUB with Even d in [3]

In this subsection, we give a brief explanation on the optimal approximation algorithm for MAX d-CLUB with even d, proposed in [3]; its basic idea is to find a 2-clique in the $\lfloor d/2 \rfloor$-th power of the original G: The algorithm `FindStar` finds a vertex v having the maximum degree $\Delta(G)$ in the input graph G and then outputs the subgraph $G[N_G^+(v)]$ induced by the vertex v and its adjacent $\Delta(G)$ vertices. This output subgraph is clearly a 2-club and hence a d-club for $d \geq 2$ of size $\Delta(G) + 1$. `FindStar` is rather simple and runs in linear time. Its approximation ratio is given as follows:

Lemma 3. *Given an n-vertex graph and a fixed integer $d \geq 2$, `FindStar` achieves an approximation ratio of $O(n^{1-1/d})$ for MAX d-CLUB.*

By the above Lemma 3, one can see that the approximation ratio of `FindStar` is $O(n^{1/2})$ for MAX 2-CLUB. This approximation ratio of `FindStar` is the best possible for MAX 2-CLUB in the sense that the lower bound of the approximation ratio of the problem is $\Omega(n^{1/2-\varepsilon})$ for any $\varepsilon > 0$ as shown also in [3].

A simple polynomial time algorithm `PowerOfGraph` is designed to construct the d-th power of a given connected undirected graph $G = (V, E)$ and an integer d: `PowerOfGraph` first computes $dist_G(u, v)$ for any pair of vertices $u, v \in V$, and then adds an edge (u, v) if $dist_G(u, v) \leq d$. By combining `PowerOfGraph` and `FindStar`, the following polynomial-time algorithm $\texttt{ByFindStar}_d$ was proposed for MAX d-CLUB, in which the $\lfloor d/2 \rfloor$-th power of the input graph is constructed.

Algorithm $\texttt{ByFindStar}_d$

Input: A connected undirected graph $G = (V, E)$
Output: A subgraph S of G
Step 1. Obtain the $\lfloor d/2 \rfloor$-th power $G^{\lfloor d/2 \rfloor}$ of the graph G by applying `PowerOfGraph`$(G, \lfloor d/2 \rfloor)$.
Step 2. Apply `FindStar` to $G^{\lfloor d/2 \rfloor}$, and then obtain a largest star T.
Step 3. Output $S = G[V(T)]$.

The following lemma states that the output of $\texttt{ByFindStar}_d$ is a d-club.

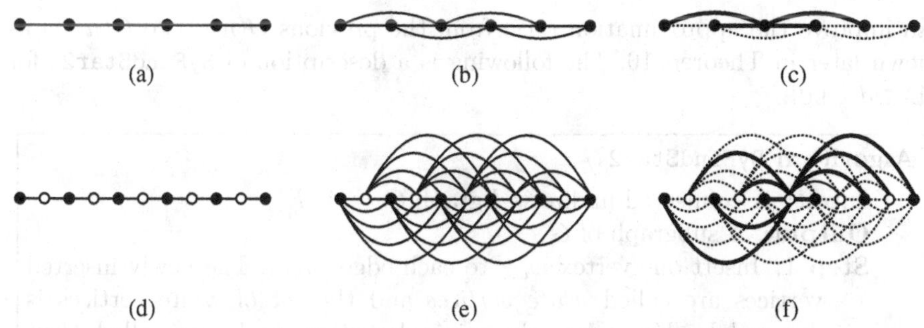

Fig. 2. (a) Example graph G for $d = 5$; (b) G^2; (c) an output of $\texttt{ByFindStar}_5$ (in solid line); (d) the graph H; (e) H^d; (f) an output by $\texttt{ByFindStar2}_5$ (in solid line)

Lemma 4. *The output of* $\texttt{ByFindStar}_d$ *is a* $2 \cdot \lfloor d/2 \rfloor$*-club.*

The approximation ratio of $\texttt{ByFindStar}_d$ for even d is as follows.

Theorem 5. *For an n-vertex graph and a fixed even integer $d \geq 2$,* $\texttt{ByFindStar}_d$ *is a polynomial time $O(n^{1/2})$-approximation algorithm for* Max *d*-Club.

Unfortunately, however, for odd d's, the approximation ratio of $\texttt{ByFindStar}_d$ becomes worse, which is one of the motivations of this paper:

Theorem 6. *For an n-vertex graph and a fixed odd integer $d \geq 3$,* $\texttt{ByFindStar}_d$ *is a polynomial time $O(n^{2/3})$-approximation algorithm for* Max *d*-Club.

3 Optimal Approximation Algorithm for Max *d*-Club with Odd *d*

As described in Sect. 2.2 the performance of $\texttt{ByFindStar}_d$ for odd d's is worse than that for even d's. This reason can be seen in Lemma 4: In the worst case, although the optimal solution is a subgraph of diameter exactly d (i.e., d-club), $\texttt{ByFindStar}_d$ outputs a subgraph of diameter $2 \cdot \lfloor d/2 \rfloor$, which is $d-1$ if d is odd. For example, take a look at Fig. 2. Now suppose that $d = 5$. One bad example for the previous algorithm $\texttt{ByFindStar}_5$ is shown in Fig. 2(a), which is a path of length five, i.e., it is a 5-club and thus the whole graph is the optimal solution for Max 5-Club. Given the graph G as input, however, $\texttt{ByFindStar}_5$ first constructs G^2 in (b), and then finds a star depicted by solid lines in (c), which has five vertices. That is, the output is a 4-club (path of length 4) in G.

The above observation leads us to a new algorithm $\texttt{ByFindStar2}_d$ below; $\texttt{ByFindStar2}_d$ first inserts an extra vertex into each edge as in Fig. 2(d), and then constructs its d-th power graph, while the previous $\texttt{ByFindStar}_d$ constructs the $\lfloor d/2 \rfloor$-th power graph. By this modification, edges connected to a newly inserted vertex can represent distance of $d/2$ for odd d, and the obtained star in Fig. 2(f) corresponds to a d-club in G. Although the ideas are quite simple, they can improve the approximation ratio from the previous $O(n^{2/3})$ to $O(n^{1/2})$ as shown later in Theorem 10. The following is a description of $\texttt{ByFindStar2}_d$ for Max d-Club.

Algorithm $\texttt{ByFindStar2}_d$

Input: A connected undirected graph $G = (V, E)$

Output: A subgraph of G

Step 1. Insert one vertex $w_{u,v}$ to each edge (u, v). The newly inserted vertices are called *white vertices* and the set of white vertices is denoted by V_W, while the original vertices in V are called *black vertices*. The obtained graph is denoted by $H = (V \cup V_W, E')$, where $E' = \{(u, w_{u,v}), (v, w_{u,v}) \mid (u, v) \in E\}$.

Step 2. Obtain the d-th power H^d of the graph H by applying $\texttt{PowerOfGraph}(H, d)$.

Step 3. For H^d, find a star $T = (V', E')$ having the maximum number of black vertices. (We call this procedure `FindStar2`)

Step 4. Output $G[V' \cap V]$, i.e., the subgraph of G induced by the black vertices in V'.

For a set B of black vertices in a graph H constructed in Step 1, let $W(B) = \{w_{u,v} \mid u, v \in B \text{ and } (u, v) \in E(G)\}$ in H, i.e., $W(B)$ includes every white vertex whose two neighbors (black vertices) both belong to B. The next lemma relates the distance between (black) vertices in G to the distance between them in H.

Lemma 7. *For any pair of two black vertices u, v in a d-club set S of G, $dist_{G[S]}(u, v) \leq d$ if and only if $dist_{H[S \cup W(S)]}(u, v) \leq 2d$. As a result, $G[S]$ is a d-club of G if and only if $H[S \cup W(S)]$ is a $2d$-club of H.*

Proof. ($\Rightarrow$) Consider a shortest path P from u to v in $G[S]$. Its corresponding path P' in $H[S \cup W(S)]$ includes the set of black vertices, say, $V(P)$ and the set of white vertices $W(V(P))$ determined by $V(P)$. Thus, the length $|P'|$ is twice of $|P|$, namely, if $dist_{G[S]}(u, v) \leq d$, then $dist_{H[S \cup W(S)]}(u, v) \leq 2d$.

($\Leftarrow$) For simplicity, let $H' = H[S \cup W(S)]$. For any pair of white vertices w_1 and w_2 in H', w_1 and w_2 are not adjacent, based on the construction of H. Similarly, any two black vertices b_1 and b_2 in H' are not adjacent either. Let P' be a shortest path from u to v in H'. The length of the path P in G corresponding to P' in H' is half of $|P'|$. Note that the black vertices in P' are included in S and so P exists inside $G[S]$. Thus, if $dist_{H[S \cup W(S)]}(u, v) \leq 2d$, then $dist_{G[S]}(u, v) \leq d$. □

Next we examine how short the d-th power operation makes the original distance between any pair of two black vertices in a $2d$-club set in H:

Lemma 8. *If S is a $2d$-club set in H, then for any two black vertices $u, v \in S$, $dist_{H^d[S]}(u, v) \leq 2$ holds, i.e., S is a 2-club set in H^d.*

Proof. Consider a path P of length at most $2d$ from u to v in $H[S]$. Let x be the center vertex of P. That is, $dist_H(u,x) \leq d$ and $dist_H(x,v) \leq d$ hold. These imply that $H^d[S]$ has two edges (u,x) and (x,v) which are incident with $x \in S$, i.e., $dist_{H^d[S]}(u,v) \leq 2$. □

The next lemma guarantees the output of $\texttt{ByFindStar2}_d$ is a feasible solution for Max d-Club.

Lemma 9. *The output of ByFindStar2$_d$ is a d-club in the input G.*

Proof. Let r be the root of the star $T = (V', E')$ in H^d, obtained in Step 3. Consider a vertex $v \in V'$ such that $v \neq r$. Since $(r,v) \in E'$, there is a path P of length at most d from r to v in H. Notice that the vertices in P are also adjacent to r in H^d because the distance from r to each of them is also at most d. This implies that these vertices are also included in V'. Therefore, if both u and v belong to V' such that $r \notin \{u,v\}$, that is, there are two edges (r,u) and (r,v) in T, then $dist_{H[V']}(u,v) \leq 2d$ because $H[V']$ includes two paths of length at most d from r to u, and from r to v. The distance between r and the another vertex $w \in V(T)$ in $H[V']$ is at most d. In summary, $diam(H[V']) \leq 2d$, i.e., $H[V']$ is a $2d$-club. Here, $V' \cap V$ is a set of black vertices in V'. From Lemma 7 and the fact that $H[V']$ is a $2d$-club, for any two black vertices $u, v \in V' \cap V$, $dist_{G[V' \cap V]} \leq d$, i.e., $G[V' \cap V]$ is a d-club. □

Finally we can show the approximation ratio of $\texttt{ByFindStar2}_d$ for Max d-Club with odd d's.

Theorem 10. *ByFindStar2$_d$ is a polynomial-time $\lceil n^{1/2} \rceil$-approximation algorithm for* Max *d-*Club *with odd d.*

Proof. Since FindStar2 in Step 3 runs in linear time in the size of H^d, the total running time of $\texttt{ByFindStar2}_d$ is polynomial.

From Lemmas 7 and 8, a d-club of G corresponds to a 2-club of H^d. What we want to do here is to find a 2-club in H^d having the maximum number of "black" vertices (original vertices in G), which may be different from the maximum 2-club in H^d. So we consider a variant of the problem, finding a 2-club having maximum "black" vertices in H^d, where the solution size is defined as the number of black vertices. While the previous procedure FindStar finds a 2-club having maximum vertices, the new FindStar2 finds a (possibly different) 2-club having maximum "black" vertices. One can observe that an optimal solution (2-club) in H^d for this new problem may not give any d-club in the input G, however, Lemma 9 guarantees that the output of $\texttt{ByFindStar2}_d$ surely forms a d-club in G. In this proof, the *blackdegree* of a vertex is the number of black vertices adjacent to v. Let Δ_b be the maximum blackdegree of the vertices in H^d.

Let S^* be a 2-club having maximum black vertices in H^d. We show an upper bound of the number of black vertices in S^*. If S^* has only black vertices, then the number of vertices in S^* is bounded above by $1+\Delta_b+(\Delta_b-1)^2 = (\Delta_b)^2-\Delta_b+2$, which is an upper bound on the number of black vertices reachable from a black vertex by a path of length two. As a next case, assume that S^* contains a white vertex w. Let B_1 be the set of black vertices in $N_1(w)$. The size $|B_1|$ is at most Δ_b, and at most $(\Delta_b)^2$ black vertices in $N_2(w)$ are adjacent to the vertices of B_1. Let B_2 be the set of those black vertices in $N_2(w)$, where $|B_2| \leq (\Delta_b)^2$. There may exist white vertices in $N_1(w)$ and they may be adjacent to black vertices in $N_2(w)$. Here we observe that the black vertices in $N_2(w)$ adjacent to white vertices in $N_1(w)$ are all included in B_2: Let x be a white vertex in $N_1(w)$, and let y be one of its adjacent black vertices in $N_2(w)$. In H, $dist_H(w, x) \leq d - 1$ because the distance between two white vertices is even. Then, there is a black vertex z which is adjacent to x and is on a path of length at most d from x to y in H. Here, $dist_H(w, z) \leq d$ and $dist_H(z, y) \leq d-1$ hold and so there are two edges (w, z) and (z, y) in H^d. Since z is a black vertex in $N_1(w)$, the vertex y is already enumerated as a member of B_2. Namely, such a new black vertex in $N_2(w)$ does not appear as a neighborhood of white vertices in $N_1(w)$. Thus we can bound the number of black vertices in S^* above by $\max\{(\Delta_b)^2 - \Delta_b + 2, |B_1| + |B_2|\} = \max\{(\Delta_b)^2 - \Delta_b + 2, (\Delta_b)^2 + \Delta_b\}$. Since $\Delta_b \geq 1$, $(\Delta_b)^2 + \Delta_b$ is the upper bound.

The size $FindStar2(H^d)$ of the output by `FindStar2` for H^d is at least Δ_b, the number of black leaves of the obtained star, where the root vertex may be white. Then the size $\#club_b(H^d, 2)$ of the optimal solution for the new problem is bounded by $\min\{n, (\Delta_b)^2 + \Delta_b\}$ from the above discussion. Note that H^d has $n + m$ vertices in total, but since the number of black vertices is n, one upper bound of the optimal size is n here. Hence, if $\Delta_b \geq \lceil n^{1/2} \rceil$, then

$$\frac{\#club_b(H^d, 2)}{FindStar2(H^d)} \leq \frac{n}{\Delta_b} \leq \frac{n}{\lceil n^{1/2} \rceil} \leq n^{1/2}.$$

Conversely, if $\Delta_b \leq \lceil n^{1/2} \rceil - 1$ (note that Δ_b is an integer), then

$$\frac{\#club_b(H^d, 2)}{FindStar2(H^d)} \leq \frac{(\Delta_b)^2 + \Delta_b}{\Delta_b} = \Delta_b + 1 \leq \left\lceil n^{1/2} \right\rceil.$$

It holds that $\#club_b(H^d, 2) \geq \#club(H, 2d) = \#club(G, d)$ based on Lemmas 7 and 8. Finally, the fact $FindStar2(H^d) = ByFindStar2(G)$ derives that

$$\frac{\#club(G, d)}{ByFindStar2(G)} \leq \frac{\#club_b(H^d, 2)}{FindStar2(H^d)} \leq \left\lceil n^{1/2} \right\rceil,$$

where $ByFindStar2(G)$ is the output size of $\texttt{ByFindStar2}_d$ for the input G. □

Remark 11. The above proof is done under the assumption that d is odd in its third paragraph. However, the upper bound $(\Delta_b)^2 + \Delta_b$ of the number of black vertices can be shown also for the case where d is even, by which we can show $\texttt{ByFindStar2}_d$ is also an $\lceil n^{1/2} \rceil$-approximation algorithm for even d's.

4 Polynomial-Time Algorithms of Max *d*-Club for Graph Classes

4.1 Graph Classes

Let $\mathcal{S} = \{S_1, S_2, \cdots, S_n\}$ be any family of nonempty sets. The *intersection graph* of $\mathcal{S}$, denoted by $G_{\mathcal{S}}$ is the graph having $\mathcal{S}$ as vertex set with S_i adjacent to S_j

if and only if $i \neq j$ and $S_i \cap S_j \neq \emptyset$. Here, $\mathcal{S}$ is called a *set representation* of $G_{\mathcal{S}}$. When $\mathcal{S}$ is allowed to be an arbitrary family of sets, any graph can be represented as an intersection graph, but typically, some important special classes of graphs may be defined by the types of sets that are used to form a set representation of them. We consider the following intersection graphs (see, e.g., [6] for details).

A graph G is *chordal* if every cycle in G of length at least four has at least one chord. *Interval graphs* are one of the most important subclasses of chordal graphs: A graph $G = (V, E)$ is an *interval graph* if the following two conditions are satisfied for a family $\mathcal{I}$ of intervals on the real line: (i) There is a one-to-one correspondence between V and $\mathcal{I}$, and (ii) for a pair of vertices $u, v \in V$ and their corresponding two intervals $I_u, I_v \in \mathcal{I}$, $I_u \cap I_v \neq \emptyset$ if and only if $(u, v) \in E$.

Other related two classes to chordal graphs are *weakly chordal* and *strongly chordal* graphs: A graph G is *weakly chordal* if every cycle of length at least five in G and its complement $\overline{G}$ has at least one chord. A graph G is *strongly chordal* if it is chordal, and every even cycle of length at least six has at least one odd chord which is a chord connecting two vertices of odd distance on the cycle. The following inclusion relations among the graph classes are known [6]:

Proposition 12. *(i) Interval graphs are a subclass of strongly chordal graphs. (ii) Strongly chordal graphs are a subclass of chordal graphs. (iii) Chordal graphs are a subclass of weakly chordal graphs.*

Another way to generalize interval graphs is to make the intervals higher dimensional. Suppose $\mathcal{I}$ and $\mathcal{J}$ be two parallel lines of which $\mathcal{I} = \{I_1, \cdots, I_n\}$ and $\mathcal{J} = \{J_1, \cdots, J_n\}$ are families of intervals. Then, each $i \in \{1, \cdots, n\}$ determines a trapezoid having parallel sides I_i and J_i, respectively (allowing degenerate trapezoids with either I_i or J_i a single point). Let $\mathcal{T} = \{T_1, T_2, \cdots, T_n\}$ be a family of such n trapezoids: A graph $G = (V, E)$ is a *trapezoid graph* if the following two conditions are satisfied for a family $\mathcal{T}$ of trapezoids between the two parallel lines $\mathcal{I}$ and $\mathcal{J}$: (i) There is a one-to-one correspondence between V and $\mathcal{T}$, and (ii) for a pair of vertices $u, v \in V$ and their corresponding two trapezoids $T_u, T_v \in \mathcal{T}$, $T_u \cap T_v \neq \emptyset$ if and only if $(u, v) \in E$.

Proposition 13. *(i) Trapezoid graphs are a subclass of weakly chordal graphs [8]. (ii) Interval graphs are a subclass of trapezoid graphs.*

Note that interval, chordal, weakly chordal, strongly chordal, and trapezoids are subclasses of perfect graphs [6].

4.2 Algorithms

Let `FindClique` be an algorithm which can obtain optimal solutions for MAX CLIQUE (or, MAX 1-CLUB). Here, the running time of `FindClique` might be exponential. By using `FindClique` and `PowerOfGraph` described in Sect. 2.2, we can design a simple algorithm called $\mathtt{ByFindClique}_d$ [3]. This algorithm outputs an optimal solution for MAX d-CLIQUE for $d \geq 2$. On the other hand, for MAX d-CLUB, the output of this algorithm is not guaranteed to be optimal for general graphs. However, we will show that for some special classes of graphs this algorithm obtains an optimal solution. The following is a description of the algorithm $\texttt{ByFindClique}_d$:

Algorithm $\texttt{ByFindClique}_d$

Input: A connected undirected graph $G = (V, E)$
Output: A subgraph S of G
Step 1. Obtain the d-th power G^d of G by $\texttt{PowerOfGraph}(G, d)$.
Step 2. Apply $\texttt{FindClique}$ to G^d, and then obtain a maximum clique set Q in G^d.
Step 3. Output $S = G[Q]$.

In the following, Lemma 14 plays a key role to provide polynomial-time algorithms. More precisely, it shows the optimality of the algorithm $\texttt{ByFindClique}_d$ for MAX d-CLUB on the graph such that every cycle of length at least $d+3$ has at least one chord in the graph. We assume that $d \geq 2$ in this section (since MAX 1-CLUB is equivalent to MAX CLIQUE).

Lemma 14. *If every cycle of length at least $d+3$ in the input graph has at least one chord, then the diameter of the graph induced by the maximum clique set in the d-th power G^d of the input graph G must be at most d.*

Proof. To prove the lemma by a contradiction, we assume that for a subset $S \subseteq V(G)$ $(= V(G^d))$ of vertices, $G^d[S]$ is a maximum clique and $diam(G[S]) \geq d+1$. From the assumption that $diam(G[S]) \geq d+1$, there exists a pair of vertices $\ell, r \in S$ such that $dist_{G[S]}(\ell, r) \geq d+1$. Since $G^d[S]$ is a clique, there exists an edge (ℓ, r) in $G^d[S]$, which implies that $dist_G(\ell, r) \leq d$ in the original graph G. Namely, there must be another vertex u such that $2 \leq dist_G(\ell, u) + dist_G(u, r) \leq d$ and $u \notin S$. Without loss of generality, we assume u is on the shortest path $P_{\ell,u,r}$ between ℓ and r in G and among such shortest paths $P_{\ell,u,r}$ includes the minimum number of vertices not in S. The fact $u \notin S$ and the maximality of the clique $G^d[S]$ derives the existence of a vertex $w \in S$ such that $dist_G(u, w) \geq d+1$.

Let the shortest paths from u to ℓ and u to r in G be $P_{u,\ell} = \langle u, \cdots, \ell\rangle$ and $P_{u,r} = \langle u, \cdots, r\rangle$, respectively. Similarly, two shortest paths from ℓ to w and r to w in $G[S]$ are denoted by $P_{\ell,w}$ and $P_{r,w}$, respectively. Here, a subpath from w' to w for a vertex w' may be shared among $P_{\ell,w}$ and $P_{r,w}$. Consider a path $P'_{\ell,r}$ from ℓ to r such that $P'_{\ell,r} = P_{\ell,w}(\ell, w') \oplus \overline{P_{r,w}}(w', r)$, where $w' = w$ if such a w' does not exist. Let $P'_{\ell,r} = \langle \ell_0(= \ell), \ell_1, \ell_2, \cdots, \ell_{p-1}, w'(= \ell_p = r_q), r_{q-1}, \cdots, r_1, r_0(= r)\rangle$, i.e., the distance between ℓ (or r) and w' in this path is p (or q) for some integer $1 \leq p \leq d$ (or $1 \leq q \leq d$). Since $dist_{G[S]}(\ell, r) \geq d+1$, the length $|P'_{\ell,r}| = p+q$ is also at least $d+1$.

(Case 1) First, for simplicity, suppose that the length $|P'_{\ell,r}| = d+1$. This implies that for $v_1, v_2 \in V(P'_{\ell,r})$ such that $(v_1, v_2) \notin E(P'_{\ell,r})$, there is no edge between v_1 and v_2, since otherwise $dist_{G[S]}(\ell, r) \leq d$ which contradicts the assumption $dist_{G[S]}(\ell, r) \geq d+1$. We will refer this fact by writing "there is no shortcut edge for $P'_{\ell,r}$" in the below. In addition, there is no edge connecting

u and the vertices of $P'_{\ell,r}$ except for ℓ and r, since if such an edge, e.g., (u, ℓ_i), exists, the length of the path $\langle u, \ell_i \rangle \oplus P_{\ell,w}(\ell_i, w)$ is at most d, i.e., $dist_G(u, w) \leq d$, which again contradicts the assumption $dist_G(u, w) \geq d + 1$.

(Case 1-i) Assume that $|P_{u,\ell}| = |P_{u,r}| = 1$, i.e., (u, ℓ) and (u, r) are in $E(G)$. Consider a cycle $C_1 = P_{u,\ell} \oplus P'_{\ell,r} \oplus \overline{P_{u,r}}$. The length of this cycle C_1 is $d + 3$. Moreover, since there is no shortcut edge for $P'_{\ell,r}$ and u is not adjacent to any vertex in $P'_{\ell,r}(\ell_1, r_1)$, we observe that C_1 has no chord. This contradicts the assumption that every cycle of length at least $d + 3$ has at least one chord.

(Case 1-ii) Without loss of generality, assume that $|P_{u,\ell}| \geq 2$. Let $P_{u,\ell} = \langle u(= x_0), x_1, x_2, \ldots, x_s, \ell \rangle$. The vertex x_i can be connected only to $\ell_1, \ldots, \ell_i$ and $r_1, \ldots, r_i$ in $P'_{\ell,r}$, because if x_i is adjacent to another vertex z in $P'_{\ell,r}$, $dist_G(u, w) \leq d$ holds, again contracting the assumption $dist_G(u, w) \geq d + 1$: If $z = \ell_j$ (or r_j) for $j > i$, the length of the path $\langle u, x_1, \ldots, x_i, z(= \ell_j) \rangle \oplus P_{l,w}(\ell_j, w)$ (or $\langle u, x_1, \ldots, x_i, z(= r_j) \rangle \oplus P_{r,w}(r_j, w)$) is at most $|P_{\ell,w}| \leq d$ (or $|P_{r,w}| \leq d$) since $j > i$. In addition, if x_i is adjacent to $z \in \{r_1, \ldots, r_i\}$, it also contradicts the assumption on $P_{\ell,u,r}$ that u is on a shortest path between ℓ and r in G which includes the minimum number of vertices not in S; the path $\overline{P_{u,\ell}}(\ell, x_i) \oplus \langle x_i, z \rangle \oplus P'_{\ell,r}(z, r)$ must be a shortest path between ℓ and r and its number of vertices not in S is smaller than that of the path $P_{\ell,u,r}$. Namely, x_i is connected only to $\ell_1, \ldots, \ell_i$. For another path $P_{u,r} = \langle u(= y_0), y_1, y_2, \ldots, y_t, r \rangle$, a similar argument can be done; y_i is connected only to $r_1, \ldots, r_i$.

Let $i = \arg\min_k \{(x_k, z) \in E(G) \mid z \in V(P'_{\ell,r})\}$, and analogously $j = \arg\min_k \{(y_k, z) \in E(G) \mid z \in V(P'_{\ell,r})\}$. Let z_i (or z_j) represents the vertex in $P'_{\ell,r}$ such that the edge (x_i, z_i) (or (y_j, z_j)) exists and z_i (or z_j) is the closest to w' among such vertices. As discussed above, z_i (or z_j) is a vertex belonging to $\{\ell_1, \ldots, \ell_i\}$ (or $\{r_1, \ldots, r_j\}$). Consider a cycle $C_2 = P_{u,\ell}(u, x_i) \oplus \langle x_i, z_i \rangle \oplus P'_{\ell,r}(z_i, z_j) \oplus \langle z_j, y_j \rangle \oplus \overline{P_{u,r}}(y_j, u)$. Since $|P_{u,\ell}(u, x_i)| = i$, $|P'_{\ell,r}(z_i, z_j)| \geq |P'_{\ell,r}| - i - j = d + 1 - i - j$, and $|\overline{P_{u,r}}(y_j, u)| = j$, the length of the cycle C_2 is at least $d + 3$. Furthermore, C_2 has no chord based on the fact that $u, x_1, \ldots, x_{i-1}$, $y_1, \ldots, y_{j-1}$ are not adjacent to any vertex in $P'_{\ell,r}$, and there is no shortcut edge for $P'_{\ell,r}$. This contradicts the assumption that every cycle of length at least $d + 3$ has at least one chord.

(Case 2) The remaining case we have to consider is that $|P'_{\ell,r}| \geq d + 2$. In this case, there may be edges connecting ℓ_i and r_j for some i and j, where no edge connects two ℓ_i's or two r_j's because $P_{\ell,w}$ and $P_{r,w}$ are the shortest paths from ℓ to w and r to w, respectively. Let $P''_{\ell,r}$ be the shortest path from ℓ and r defined on the vertices of $P'_{\ell,r}$. Since $dist_{G[S]}(\ell, r) \geq d + 1$, $|P''_{\ell,r}| \geq d + 1$ also holds. As in the above Case 1-ii we set $i = \arg\min_k \{(x_k, z) \in E(G) \mid z \in V(P''_{\ell,r})\}$ and $j = \arg\min_k \{(y_k, z) \in E(G) \mid z \in V(P''_{\ell,r})\}$. Then, we consider a cycle $C_3 = P_{u,\ell}(u, x_i) \oplus \langle x_i, z_i \rangle \oplus P''_{\ell,r}(z_i, z_j) \oplus \langle z_j, y_j \rangle \oplus \overline{P_{u,r}}(y_j, u)$. We observe that the length of this cycle C_3 is at least $d + 3$ as well and it has no chord, which is a contradiction. This completes the proof of the lemma. □

From Lemma 14, if every cycle of length at least $d + 3$ has at least one chord, then an optimal solution of MAX d-CLUB must be the maximum clique of G^d:

Theorem 15. *For a fixed integer d, if every cycle of length at least $d+3$ has at least one chord in G, ByFindClique$_d$ exactly solves* MAX d-CLUB *for G.*

Therefore, `ByFindClique`$_d$ solves MAX d-CLUB in polynomial time if we restrict the input to graphs having such property and a maximum clique of its d-th power can be found in polynomial time. In the following subsections, we give several polynomial-time solvabilities of MAX d-CLUB according to this fact.

4.3 Chordal and Weakly Chordal Graphs for Odd d

If the input is restricted to chordal graphs, then the next corollary holds from Theorem 15:

Corollary 16. *For chordal graphs, ByFindClique$_d$ finds an optimal solution for* MAX d-CLUB *for $d \geq 1$.*

Proof. In chordal graphs, every cycle of length at least four always has a chord by definition, which means that every cycle of length at least $d+3$ always has a chord for any $d \geq 1$. Then, for chordal graphs, `ByFindClique`$_d$ finds an optimal solution for MAX d-CLUB. □

Furthermore, if d is odd, the next theorem can be shown:

Theorem 17. *If the input is a chordal graph, then* MAX d-CLUB *can be solved in polynomial time by ByFindClique$_d$ for odd d.*

Proof. For chordal graphs, polynomial time algorithms to find a maximum clique are known [12]. If d is odd and $d \geq 3$, the d-th power G^d of a chordal graph G is also chordal [1,4]. Thus, adopting one of the algorithms in [12] as the procedure `FindClique`, the algorithm `ByFindClique`$_d$ solves MAX d-CLUB in polynomial time for odd d if the input is a chordal graph. □

From the definition, there exists no induced cycle of length at least five in weakly chordal graphs. Similarly to the case of chordal graphs, polynomial time algorithms to find a maximum clique in weakly chordal graphs are known [15]. In addition, the d-th power G^d of a weakly chordal graph G is also weakly chordal if d is odd [5]. Hence the following theorem holds:

Theorem 18. *If the input is a weakly chordal graph, then* MAX d-CLUB *can be solved in polynomial time by ByFindClique$_d$ for odd d.*

4.4 Strongly Chordal, Trapezoid, and Interval Graphs

As stated in Proposition 12(ii), the class of strongly chordal graphs is a subclass of chordal graphs. Since for every integer $d \geq 2$ the d-th power G^d of a strongly chordal graph G is also strongly chordal [16], we have:

Theorem 19. *If the input is a strongly chordal graph, then* MAX d-CLUB *can be solved in polynomial time by ByFindClique$_d$.*

Since the class of trapezoid graphs is a subclass of weakly chordal graphs from Proposition 13(i), every cycle of length at least five always has a chord in trapezoid graphs. For trapezoid graphs, Max Clique can be solved in polynomial time [10], and the d-th power G^d of a trapezoid graph G is also trapezoid [11]. As a result, the following theorem holds:

Theorem 20. *If the input is a trapezoid graph, then* Max d-Club *can be solved in polynomial time by* $\mathtt{ByFindClique}_d$.

Proposition 13(ii) and Theorem 20 derive the following theorem:

Theorem 21. *If the input is an interval graph, then* Max d-Club *can be solved in polynomial time by* $\mathtt{ByFindClique}_d$.

5 Max d-Clique

This section summarizes the obtained results for Max d-Clique. (Due to the space limitation, all the proofs are omitted in this section.) The first result shows the approximation ratio of $\mathtt{ByFindStar2}_d$ for Max d-Clique.

Theorem 22. *For an n-vertex graph and a fixed integer d,* $\mathtt{ByFindStar2}_d$ *is a polynomial-time $n^{1/2}$-approximation algorithm for* Max d-Clique.

As for inapproximability, we can obtain the following theorem:

Theorem 23. *For any $\varepsilon > 0$, is NP-hard to approximate* Max d-Clique *to within a factor of $n^{1/2-\varepsilon}$.*

We can show that Max d-Clique can be solved in polynomial time for the following special graph classes based on similar discussions in Sect. 4:

Theorem 24. *If the input graph is interval, trapezoid, or strongly chordal, then* Max d-Clique *can be solved in polynomial time by* $\mathtt{ByFindClique}_d$. *If the input graph is chordal or weakly chordal, then* Max d-Clique *can be solved in polynomial time by* $\mathtt{ByFindClique}_d$ *for odd d.*

Acknowledgments. This work is partially supported by KAKENHI grant numbers 25330018 and 26330017.

References

1. Agnarsson, G., Greenlaw, R., Halldórsson, M.M.: On powers of chordal graphs and their colorings. Congr. Numer. **144**, 41–65 (2000)
2. Alba, R.: A graph-theoretic definition of a sociometric clique. J. Math. Sociol. **3**, 113–126 (1973)
3. Asahiro, Y., Miyano, E., Samizo, K.: Approximating maximum diameter-bounded subgraphs. In: López-Ortiz, A. (ed.) LATIN 2010. LNCS, vol. 6034, pp. 615–626. Springer, Heidelberg (2010)

4. Balakrishnan, R., Paulraja, P.: Powers of chordal graphs. Aust. J. Math. Ser. A **35**, 211–217 (1983)
5. Brandstädt, A., Dragan, F.F., Xiang, Y., Yan, C.: Generalized powers of graphs and their algorithmic use. In: Arge, L., Freivalds, R. (eds.) SWAT 2006. LNCS, vol. 4059, pp. 423–434. Springer, Heidelberg (2006)
6. Brandstädt, A., Le, V.B., Spinrad, J.P.: Graph classes: a survey. In: SIAM (1999)
7. Chang, M.-S., Hung, L.-J., Lin, C.-R., Su, P.-C.: Finding large k-clubs in undirected graphs. Computing **95**, 739–758 (2013)
8. Corneil, D.G., Kamula, P.A.: Extensions of permutation and interval graphs. Congr. Number. **58**, 267–275 (1987)
9. Erdös, P., Pach, J., Pollack, R., Tuza, Z.: Radius, diameter, and minimum degree. J. Combin. Theor. Ser. B **47**, 73–79 (1989)
10. Felsner, S., Müller, R., Wernisch, L.: Trapezoid graphs and generalization, geometry and algorithms. Discrete Appl. Math. **74**(1), 13–32 (1997)
11. Flotow, C.: On powers of m-trapezoid graphs. Discrete Appl. Math. **63**(2), 187–192 (1995)
12. Gavril, F.: Algorithms for minimum coloring, maximum clique, minimum covering by cliques, and maximum independent set of a chordal graph. SIAM J. Comput. **1**(2), 180–187 (1972)
13. Golovach, P.A., Heggernes, P., Kratsch, D., Rafiey, A.: Finding clubs in graph classes. Discrete Appl. Math. **174**, 57–65 (2014)
14. Håstad, J.: Clique is hard to approximate within $n^{1-\varepsilon}$. Acta Math. **182**(1), 105–142 (1999)
15. Hayward, R., Hoáng, C., Maffray, F.: Optimizing weakly triangulated graphs. Graphs Comb. **5**, 339–349 (1989)
16. Lubiw, A.: Γ-free matrices. Masters thesis, Department of Combinatorics and Optimization, University of Waterloo, Canada (1982)
17. Marinček, J., Mohar, B.: On approximating the maximum diameter ratio of graphs. Discrete Math. **244**, 323–330 (2002)
18. Mokken, R.J.: Cliques, clubs and clans. Qual. Quant. **13**, 161–173 (1979)
19. Pajouh, F.M., Balasundaram, B.: On inclusionwise maximal and maximum cardinality k-clubs in graphs. Discrete Optim. **9**, 84–97 (2012)
20. Schäfer, A., Komusiewicz, C., Moser, H., Niedermeier, R.: Parameterized computational complexity of finding small-diameter subgraphs. Optim. Lett. **6**(5), 883–891 (2012)
21. Zuckerman, D.: Linear degree extractors and the inapproximability of max clique and chromatic number. Theor. Comput. **3**, 103–128 (2007)

The Influence of Preprocessing on Steiner Tree Approximations

Stephan Beyer(✉) and Markus Chimani

Institute of Computer Science, University of Osnabrück, Osnabrück, Germany
{stephan.beyer,markus.chimani}@uni-osnabrueck.de

Abstract. Given an edge-weighted graph G and a node subset R, the Steiner tree problem asks for an R-spanning tree of minimum weight. There are several strong approximation algorithms for this NP-hard problem, but research on their practicality is still in its early stages.

In this study, we investigate how the behavior of approximation algorithms changes when applying preprocessing routines first. In particular, the shrunken instances allow us to consider algorithm parameterizations that have been impractical before, shedding new light on the algorithms' respective drawbacks and benefits.

1 Introduction

Given a connected graph $G = (V, E)$ with edge costs $c\colon E \to \mathbb{R}_{\geq 0}$ and a subset $R \subseteq V$ of required nodes called *terminals*, the *Minimum Steiner Tree Problem in Graphs (STP)* asks for a minimum-cost terminal-spanning subtree $T = (V_T, E_T)$ in G, that is, a tree with $R \subseteq V_T \subseteq V$ and minimum $c(T) := c(E_T) := \sum_{e \in E_T} c(e)$. The STP is long known to be NP-hard and even APX-hard [15].

While 2-approximations can be found easily, breaking this barrier has been a theoretically challenging but fruitful field, resulting in a series of ever-decreasing approximation ratios (see Sect. 2). The currently strongest known algorithms guarantee a ratio of $1.39 + \varepsilon$ for general instances. All these below-2 approximation algorithms share a common trait that makes them usually worrisome in practice: ε depends on a parameter k, assumed to be constant. The algorithm's running time, however, is exponentially dependent on this k, as we have to consider all trees spanning any subset of up to k terminals (see below for details). Hence, to achieve low ε, we have to live with increasingly high running times.

Research on the practicality of below-2 Steiner tree approximations started only very recently. In a series of papers [1,2,4][1], it was shown that the algorithms' dependency on k seems to be an insurmountable stumbling block in practice. Typically, only values of $k = 3$ and probably, for small graphs, $k = 4$

Funded by project CH 897/1-1 of the German Research Foundation (DFG).

[1] Article [2] is an extended version of both [1,4], with several implementation improvements; when referring to these studies in the following, we will only cite [2].

Z. Lu et al. (Eds.): COCOA 2015, LNCS 9486, pp. 601–616, 2015.
DOI: 10.1007/978-3-319-26626-8_44

seemed worthwhile. These findings were validated in [5] in the context of a further approximation algorithm not considered in [2].[2]

It is certainly beneficial to the field that the STP has garnered enough interest (in particular also from the communities of exact and (meta)heuristic approaches) that SteinLib [11], a wide collection of generated and real-world instances, has been established. We use this well-established testbed as the basis of our study. Some SteinLib instances have special properties (like being Euclidean, planar, or quasi-bipartite), and various special-purpose algorithms have been developed for such settings. We, however, refrain from explicitly exploiting such properties since we are interested in the approximation algorithms for general graphs.

Contribution. Preprocessing means to shrink a given input by simplifying parts that are 'easy' to solve, favorably in theoretically and practically small time. Our considered research questions can be summarized as follows:

- *Interplay between preprocessing and approximations.* Neither of the above mentioned studies on approximation algorithms consider preprocessing. This was in order to concentrate on the algorithmic behavior of the approximations and to rule out any shadowing influences arising from preprocessed data. In this study, we want to analyze the influence of preprocessing on the approximations. Clearly, the approximations become faster when run only on reduced instances. But it is already unclear if preprocessed instances help the algorithms to find significantly better solutions.
- *Approximation behavior for larger k.* As noted above, the approximation algorithms are typically restricted to $k = 3$ when run on the original SteinLib instances. As we will see, the preprocessing routines are strong enough to obtain graphs so small that larger values of k become feasible. So now, for the first time, we can investigate the approximation algorithms' behavior for such values. This is in particular interesting due to a central finding in [2]: There, the oldest below-2 approximations (AC3 and RC3 in the description below) typically outperform the newer, formally stronger approximations — not only w.r.t. running time but also often w.r.t. solution quality. This can be attributed to the fact that, for too small values of k, the actual approximation guarantees of the latter are still worse than AC3's 11/6. Only for larger k, their ratios become (at least formally) strictly better. The question is whether these theoretical results carry over to practical benefits.

We are explicitly *not* arguing that our considered algorithms would be particularly good in practice. In fact, we will see in the last section that state-of-the-art heuristics and exact approaches still outperform all theoretically strong approximations. We are interested in learning more about practical stumbling blocks and possible achievements for the class of strong approximation algorithms, in the hope that such a deeper understanding will at some point help to find practically relevant strong approximation algorithms, at least for certain circumstances.

[2] In [5], an implementation of [3] is considered; in [2], the improved variant [9] is investigated instead, as it gives the same guarantee with a smaller runtime complexity.

Outline. The following two sections summarize the approximation algorithms and preprocessing routines we use for our study, respectively. Section 4 contains the actual experimental evaluation.

Preliminaries. For any graph H, we denote its nodes and edges by V_H and E_H, respectively. When referring to the input graph G, we omit the subscript. Considering edge costs, we use the terms *cost* (expensive, cheap) and *length* (long, short) interchangeably. Let $d(u,v)$ be the *distance* between nodes $u, v \in V$, i.e., the cost of the cheapest path between u, v. Let the *neighborhood* $\mathcal{N}(v)$ of $v \in V$ be the set of nodes adjacent to v, and the *Voronoi region* $\mathcal{V}(r) = \{r\} \cup \{v \in V \setminus R \mid d(r,v) \leq d(s,v)\ \forall s \in R\}$ of a terminal $r \in R$ be the set of nodes that is nearer to r than to any other terminal. If $d(r,v) = d(s,v)$ for $s \neq r$, the node v is arbitrarily assigned to either $\mathcal{V}(r)$ or $\mathcal{V}(s)$. Hence the set of Voronoi regions of each terminal is a partition of V. We define $\text{base}(v) := r$ for all $v \in \mathcal{V}(r)$ and $r \in R$. By $\text{MST}(H)$ we denote a minimum spanning tree in H.

2 Approximation Algorithms and Their Engineering

We briefly summarize the algorithms considered for our experimental evaluation. They are identical to the ones considered in [2]. This allows direct comparisons w.r.t. the additional influence of the preprocessing routines. The only additional algorithmic modification for the approximations is the addition of a Dreyfus-Wagner type generation of full components (see below).

Algorithm TM. One of the simplest and most efficient approximation algorithms is the 2-approximation algorithm by Takahashi and Matsuyama [21]. We start with a single terminal node as T_0. In the i-th iteration, we construct T_i from T_{i-1} by adding the shortest path between T_{i-1} and t, where $t \in R \setminus V_{T_{i-1}}$ is the nearest terminal to T_{i-1}. The output of the algorithm is $T_{|R|-1}$. Well-implemented, it has proven to be the most efficient 2-approximation (among the algorithms by Kou et al. [12] and Mehlhorn [14]) in terms of time and solution quality [2,16].

Full Components. Given a Steiner tree T, we can assume that all its leaves are terminals. We can split all its non-leaf terminals to obtain connected components whose leaves are exactly their terminals. Such components are called *full components* of T. They are *k-restricted* if they contain at most k terminals.

Let $k \geq 3$ be constant and $\mathcal{C}_k$ be the set of all possible k-restricted full components. All below-2 approximation algorithms evaluate the elements of $\mathcal{C}_k$ to find a good subset $\mathcal{S} \subseteq \mathcal{C}_k$, such that the components in $\mathcal{S}$ together yield a feasible Steiner tree. The evaluation method differs per approximation algorithm.

In [2], several methods to construct $\mathcal{C}_k$ are evaluated. All methods are based on shortest path algorithms, and the first beneficial idea is to use a variant of shortest path algorithms that, in case of a tie, prefers a path over terminals. Such paths can be discarded for construction of full components. For the special case of $k = 3$, one can also choose between multiple calls to a single-source shortest path algorithm (SSSP) in overall $\mathcal{O}(|R| \cdot |V|^2)$ time and a single call to an all-pair shortest path algorithm (APSP) in time $\mathcal{O}(|V|^3)$. When applied to SteinLib, SSSP should be chosen if and only if the graph's *density* $|E|/\binom{|V|}{2}$ is at most 0.25.

For $k \geq 4$, we have to use a shortest-path matrix between all nodes, so APSP is the only viable choice. However, the success rates of computing $\mathcal{C}_k$ using the enumeration scheme given in [2] are rather bad. In [5], it is proposed to compute $\mathcal{C}_k$ essentially by running the first k iterations of the Dreyfus-Wagner algorithm [7]; we call this method *DW*. It is based on the simple observation that in order to compute a minimum-cost tree spanning k terminals, the Dreyfus-Wagner algorithm also computes all minimum-cost trees spanning at most k terminals. Hence one call of *DW* yields all k-restricted full components.

Algorithms ACk *and* RCk. Zelikovsky [22] was the first one to exploit the idea of k-restricted full components for a below-2 approximation algorithm. He summarized the ideas as the *greedy contraction framework* [25]: we first compute a 2-approximation T_0. Given T_{i-1}, we seek the full component $C \in \mathcal{C}_k$ that promises the maximum improvement over T_{i-1} according to some given *win function*. T_i is constructed by contracting C in T_{i-1} and removing the most expensive edge in each arising cycle, that is, $T_i = \mathrm{MST}(T_{i-1}/C)$. This is repeated until no promising component is found. The contracted components and the remaining edges in the last constructed tree together form the approximative solution to the original Steiner tree problem.

In this setting, ACk is the algorithm that uses an *absolute* win function that describes the actual cost reduction of T_{i-1} when including a component C. It provides an approximation ratio of $11/6 \approx 1.83$ [22–24] for $k \geq 3$. RCk uses a *relative* win function [25] that relates the saved cost by contracting C in T_{i-1} to the cost of C. For $k = 3$, it only achieves ratio 1.97; however, the ratio decreases for higher k, down to 1.69 for $k \to \infty$. We say the approximation ratio is $1.69+\varepsilon$.

Generating $\mathcal{C}_3$ can be accelerated by using an on-demand generation [23] for AC3, or a Voronoi-based approach [24] for RC3. The latter is also valid for $k \geq 4$, but the general enumeration approach (even without *DW*) is favorable [2].

We also use the beneficial implementation details mentioned in [2]: a strategy that allows to discard non-promising components early (and hence reduces the number of full components) and a data structure for lowest common ancestor queries on static auxiliary trees to efficiently obtain values for the win functions.

Algorithms LCk. Robins and Zelikovsky [20] achieved a ratio $1.55 + \varepsilon$ by using *loss-contractions* (instead of simple contractions) of C in the greedy contraction framework. The *loss* of C is a sub-forest in C such that each nonterminal is connected to exactly one terminal. A loss-contraction performs a contraction of the loss edges only (instead of all edges). The used win function represents the concept of a relative win function transfered to loss-contractions.

For the generation of full components, we have to use the general enumeration approach for every $k \geq 3$. It is beneficial to compute the loss on components containing edges representing shortest paths instead of the original edges [2].

Algorithms LPk. The algorithm by Goemans et al. [9] is the most recent Steiner tree approximation and achieves a ratio of $1.39 + \varepsilon$, similar to the algorithm by Byrka et al. [3]. Both algorithms are based on linear programming techniques. The corresponding LP formulations have exponentially many constraints but can be solved in polynomial time using a separation oracle that involves a linear number of maximum flow computations. However, the approach by Goemans et al. is favorable since the resulting LP formulation has less variables (by a factor proportional to k) and the LP relaxation is solved only once.

Each variable in the formulation represents a full component in $\mathcal{C}_k$. The algorithm first solves the LP and then applies a sophisticated randomized rounding scheme to obtain a provably good integral solution. There, the idea is to iteratively contract a (fractionally chosen) component, and make the solution feasible for the contracted problem again (which involves removal and splitting of other components). In order to minimize the cost of the resulting feasible solution, we estimate $w(C)$, the maximum cost of the edges that are to be removed in order to re-establish feasibility after contracting C. Any component C with $c(C) \leq w(C)$ can be chosen. (It is guaranteed that there always exists such a component.) We note that finding such a maximum-cost set of edges to be removed reduces to a minimum-cost flow computation in an auxiliary network.

For the generation of $\mathcal{C}_k$, the general enumeration algorithm has to be used. It is beneficial to augment the maximum-flow-based separation oracle with further preconditional tests; to choose 'leaf components' before actually invoking the approximation algorithm on the fractional solution; and to use edges representing shortest paths (instead of original edges) in the auxiliary network [2].

Exact Algorithms BC *and* BCS. We compare the approximations to two exact algorithms. Algorithm BC is a branch-and-cut approach based on the bidirected cut LP formulation [13]; it is our own straight-forward, simple, and short implementation using a standard branch-and-cut framework (ABACUS [10]), and therefore arguably easier to implement than, e.g., the sophisticated strong approximation algorithms. Furthermore, we consider the highly tuned, more sophisticated version of BC that has been presented in [8]. It has been one of the winners of the *11th DIMACS Implementation Challenge* on Steiner tree problems [6], and we denote it by BCS.

3 Preprocessing Techniques

We use reduction tests to preprocess STP instances. These tests check for conditions that imply the inclusion or exclusion of nodes and edges in a Steiner minimum tree (SMT)[3]. For our experimental evaluation, we only consider efficient tests, i.e., with running time bounded by $\mathcal{O}(|V|^2)$. Hence, when applied iteratively, the total running time of the preprocessing process is $\mathcal{O}(|E| \cdot |V|^2)$.

We give a brief overview over the used reduction tests. A more detailed decription (including proofs) is given, e.g., by Polzin and Vahdati-Daneshmand [18].

[3] This word order and abbreviation is historically common, to avoid conflicts with the established abbreviation 'MST' for minimum *spanning* tree.

Trivial Reductions. Test D_1 deals with nodes of degree 1. Let $v \in V$ with $\deg(v) = 1$ and $e = \{v, w\}$. If $v \in R$ and $|R| \geq 2$, edge e must be in every SMT. We can hence contract e and let the resulting node be a terminal. If $v \in V \setminus R$, remove v (and e) since it is not part of any SMT. Test NTD_2 considers nonterminals of degree 2 and replaces its two incident edges by a single edge. Test S removes self-loops and parallel edges (keeping the edge of minimum cost). Another trivial test is to remove connected components not containing any terminals (NTC). We summarize the former tests as trivial test set T.

Consider the *distance network* on the terminals R, i. e., the complete graph on R with edge costs resembling the length of the shortest paths in G between the respective terminals. Here and in the remainder of this paper, let M denote an MST in this distance network. It can be computed in time $\mathcal{O}(|E| + |V| \log |V|)$ using Voronoi regions [14]. The simple test CTD (cost vs. terminal distance) is to remove every edge e with cost larger than the shortest path between two terminals, that is, if $c(e) > \max_{f \in E_M}\{c(f)\}$.

Classical Inclusion Tests Using Voronoi Regions. Let $s \in R$. Test NV (nearest vertex) is as follows: Let $\{s, u\}$ and $\{s, w\}$ be the shortest and second shortest edge incident to s, respectively. The edge $\{s, u\}$ belongs to at least one SMT if

$$c(\{s, w\}) \geq \begin{cases} \mathrm{rdist}_u(s) & \text{if } u \in \mathcal{V}(s), \\ c(\{s, u\}) + d(u, \mathrm{base}(u)) & \text{otherwise}, \end{cases}$$

where $\mathrm{rdist}_u(s)$ is the shortest distance between s and any $t \in R$, $t \neq s$, over $\{s, u\}$.

Test SL (short links) is as follows: Let $e_1 = \{v_1, w_1\}$ and $e_2 = \{v_2, w_2\}$ with $v_i \in \mathcal{V}(s), w_i \notin \mathcal{V}(s), i = 1, 2$, be the shortest and second shortest edge, respectively, that leaves $\mathcal{V}(s)$. Then e_1 belongs to at least one SMT if $c(e_2) \geq d(s, v_1) + c(e_1) + d(w_1, \mathrm{base}(w_1))$.

Tests Based on the Steiner Bottleneck Distance. A path P in G can be decomposed into a sequence of *full paths*, i. e., full components with exactly two terminals each. Let $L(P)$ be the longest full path of P. The *bottleneck Steiner distance* $b(u, v)$ is the minimum $c(L(P))$ over all paths P between u and v. We use an upper bound on $b(u, v)$ as described in [18]: if u and v are terminals, we query the largest cost on the unique u-v-path in M (which is a lowest common ancestor query in a weight-tree); if u and/or v is not a terminal, we restrict ourselves to paths over their respective κ nearest terminals, for some constant κ.

Test PT (paths with many terminals) removes every edge $\{u, v\}$ with $c(\{u, v\}) > b(u, v)$. Here, PT should be followed by NTC to remove nonterminal components.

Let $d \geq 3$ be constant. Test NTD_d (nonterminals of degree d) checks for every $v \in V \setminus R$ with $\deg(v) \leq d$ whether it satisfies

$$\sum_{w \in N} c(\{v, w\}) \geq c(\mathrm{MST}((N, N \times N, b))) \qquad \forall N \subseteq \mathcal{N}(v), |N| \geq 3.$$

Here $(N, N \times N, b)$ is the complete graph over N where the costs are given by (upper bounds on) the bottleneck Steiner distances. If the above universally quantified property holds, v has $\deg_T(v) \leq 2$ in at least one SMT T. Consequently, v can be removed, and edges $\{u, w\}$ with cost $c(\{u, v\}) + c(\{v, w\})$ are introduced for (all) $u, w \in \mathcal{N}(v)$.

After a pilot study, we chose $\kappa = 5$ and $d = 10$ as the best practical values.

Bound-Based Tests Using Voronoi Regions. Bound-based tests compute a lower bound for the SMT under the assumption that a nonterminal or an edge is included in the SMT. If that lower bound exceeds an upper bound U, the assumption must be false and the nonterminal or edge can be removed.

We can use Voronoi regions to compute lower bounds. Let $\text{exit}(s), s \in R$, be the shortest distance to any $v \in V \setminus \mathcal{V}(s)$. Let X be the sum of all these values except the largest two, and let $d_i(v)$ be the distance to the i-th nearest terminal of $v \in V$. Test *LBE* (lower bound on edges) removes $e = \{u, v\}$ if $c(e) + d_1(u) + d_1(v) + X > U$. Let $G' := (R, E', c')$ with

$$E' := \{\{s,t\} \mid \{u,v\} \in E, u \in \mathcal{V}(s), v \in \mathcal{V}(t)\} \text{ and}$$
$$c'(\{s,t\}) := \min\{c(e) + \min\{d_1(u), d_1(v)\} \mid e = \{u,v\} \in E, u \in \mathcal{V}(s), v \in \mathcal{V}(t)\}.$$

Test *LBN* (lower bound on nodes) removes $v \in V \setminus R$ if $d_1(v) + d_2(v) + \min\{X, c'(\text{MST}(G')) - \max_{e \in E'} c'(e)\} > U$.

4 Experimental Evaluation

In the following experimental evaluation, we use an Intel Xeon E5-2430 v2, 2.50 GHz running Debian 8. The binaries are compiled in 64bit with g++ 4.9.0 and `-O3` optimization flag. All algorithms are implemented as part of the free C++ Open Graph Drawing Framework (OGDF), the used LP solver is CPLEX 12.6. We apply our algorithms on all 1200 connected instances from SteinLib.

As in [2,8], we say that an algorithm *fails* for a specific instance if it exceeds one hour of computation time or needs more than 16 GB of memory; otherwise it *succeeds*. The *success rate* is the percentage of instances that succeed.

By χ we denote the ratio of edges remaining after the preprocessing.

We evaluate the solution quality of a solved instance by computing a *gap* as $\frac{c(T)}{c(T^*)} - 1$, the relative discrepancy between the cost of the found tree T and the cost of the optimal Steiner tree T^*, usually given in thousandths (‰). When no optimal solution values are known, we use the currently best known upper bounds from the 11th DIMACS Challenge [6].

Preliminaries. Before we go into the depth of our main research questions, we first have two discuss two relevant preliminary issues.

Component Generation via DW. Fig. 1 illustrates the success rates of different component enumeration methods: the 'smart' one from [2], and *DW*. The results

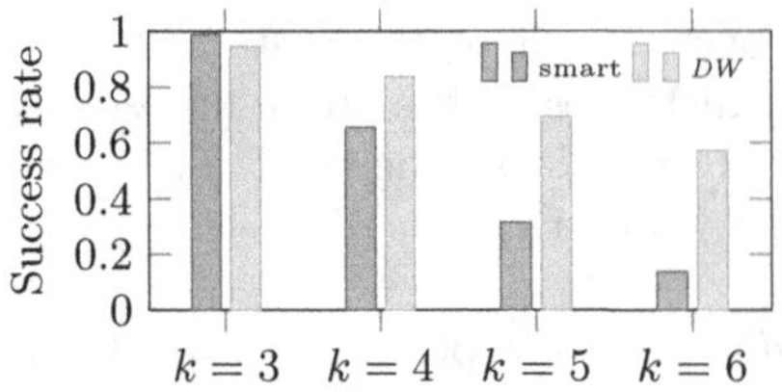

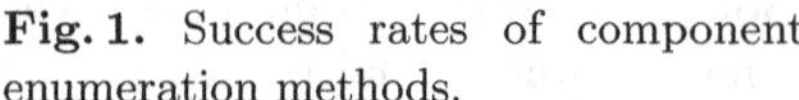

Fig. 1. Success rates of component enumeration methods.

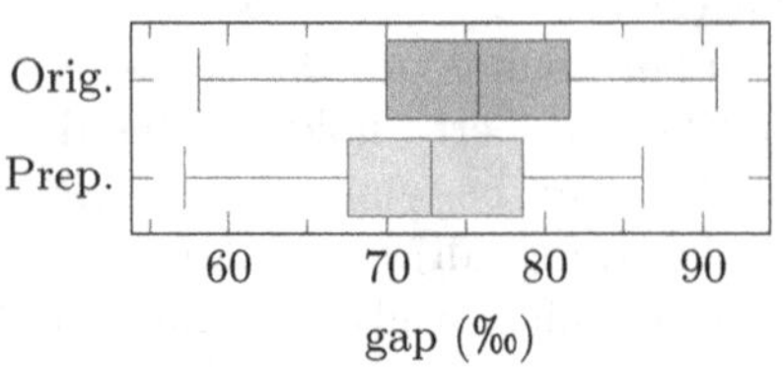

Fig. 2. Average variation in solution gaps due to different shufflings.

are clear and consistent with theory: While the former is better for $k = 3$, the latter outperforms it for $k \geq 4$. Hence, for $k \geq 4$ we will from now on use DW. For $k = 3$, we can still use the the best applicable approaches [2]: the direct approach for AC3, the Voronoi-based approach for RC3, the 'smart' approach for LC3 and LP3.

Preprocessing Sequence. The described reduction tests can be applied in various permutations and repetition schemes. The application of one reduction may help or hinder another reduction to be applicable. Since the used preprocessing is always faster than the subsequent algorithms, we focus on the reduction quality.

Using only *T* already gives an average $\chi = 90.6\,\%$. We considered all $7! = 5040$ possible permutations of the preprocessing steps that start with *T*. For each sequence, we iterate until no further reduction is achieved. The sequence ⟨*T*, *CTD*, *LBE*, *PT*, *NTD*, *SL*, *LBN*, *NV*⟩ —using *T* in between whenever necessary— gives the best average χ of 75.49 %. However, all sequences resulted in comparable reduction ratios and running times; the worst permutation gives average $\chi = 75.57\,\%$. The percentage of instances solved purely via preprocessing is 5.25 %, and 9.75 % of the instances are not susceptible to any of our considered reductions.

The average runtime of our preprocessing is negligible for the subsequent strong approximation algorithms (0.28 s on average). The only notable outliers are `alue7080` taking 15 s and `es10000fst01` taking 35 s.

Effect of Preprocessing on Solution Quality. In this subsection we temporarily disregard instances that are not susceptible to preprocessing at all. Our first attempt to compare the solution quality with and without preprocessing was surprisingly hazy. While the gaps slightly decreased *on average*, the situation was muddled on the instance level: for many instances the gap indeed shrunk, but it also increased for many. The reason seems to be primarily that all algorithms have to rely on some tie-breaking when choosing between multiple equally long shortest paths, multiple components with the same win-value, etc. This tie-breaking is implicitly done based on the internal order of the nodes and edges in the graph data structure. To investigate the interplay between these orders and the preprocessing, we *shuffle* all instances 50 times, i. e., we

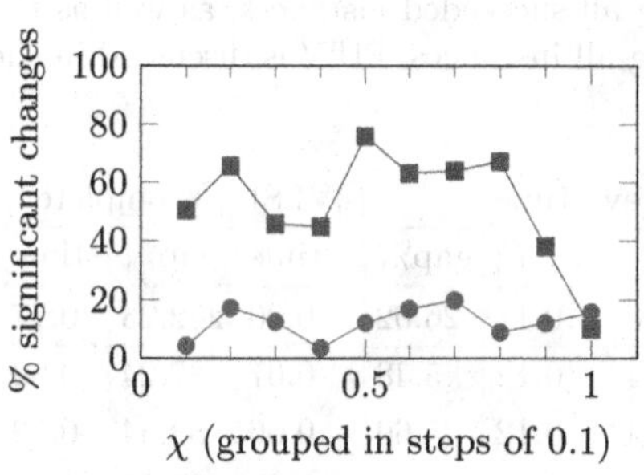

Fig. 3. Preprocessing: percentage of solution quality changes that are significant. Green squares are improvements, red circles are degradations.

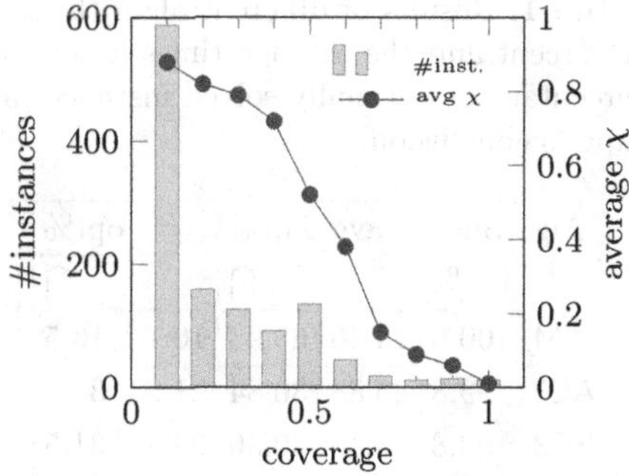

Fig. 4. Average χ (ratio of edges remaining after preproc.) over all graphs.

have 50 versions of the *same* instances differing only in internal storage order, and report on TM without and with preprocessing. Our results for the other algorithms, albeit only on a random sample of the instances, are analogous.

Probably despite one's first intuition, for any given instance, the so-generated 50 solutions with (or without) preprocessing do *not* resemble a normal distribution; in fact they are very far away from doing so (see Appendix A for illustrative examples). The central limit theorem does not apply since choices of edges (and hence solution costs) are not independent. In other words, we cannot apply the standard (parametric) significance tests in our statistical evaluation.

Figure 2 gives a comparison of the average distributions for the gaps obtained by our original and preprocessed instances. We observe that the median gap improves, and the whole range as well as the quartile ranges shrink and move to the left. This indicates a favorable (though statistically insignificant) effect of preprocessing on the gaps in general. Interestingly, for VLSI and Euclidean instances we even see a slight increase of the gap, see Appendix B.

Our first hypothesis is that the (arbitrary) order of the elements in the graph data structure has a larger influence on the solution quality than the preprocessing. Consider some instance. We are interested in the probability p that its solution will improve by preprocessing. If $p = 0.5$, any improvements are only due to chance based on different shufflings; if $p = 1$, preprocessing will always improve the solution independent of the shuffling. We compute these probabilities based on our 50 shufflings, and obtain an average of only $p = 0.58$ over all 1083 instances, i.e., we accept the hypothesis.

Our second hypothesis is that, despite this overshadowing effect, preprocessing does significantly improve (or at least does not significantly decrease) solutions. Due to the lack of a normal distribution, we resort to the Wilcoxon-Mann-Whitney test, and choose a significance level of 5 % (cf. Fig. 3). We observe a significant improvement for 35.5 % (55.7 %, 60.5 %) of the instances with $\chi < 1$ ($\chi \leq 0.9$, $\chi \leq 0.8$), and significant degradation only for 14.1 % (12.7 %, 12.8 %, resp.).

Table 1. Results of different algorithms. Per algorithm, we provide the success rates in percent and the average times in *sec* among all succeeded instances, as well as the percentage of optimally solved instances among all instances. PUW is discussed in the paper's conclusion.

Alg	Succ	avg gap‰		opt%		avg time		⋂VLSI		⋂Complete	
	%		⋂		⋂		⋂	gap‰	time	gap‰	time
TM	100.0	74.46	65.93	10.7	16.5	0.26	0.11	26.62	0.06	262.23	0.77
AC3	99.8	35.88	30.84	21.2	31.8	2.44	0.12	5.48	0.07	87.24	0.82
RC3	99.8	35.56	30.36	20.8	31.5	3.64	0.12	5.60	0.06	88.64	0.82
LC3	99.2	41.06	36.57	18.1	27.1	20.62	0.14	22.11	0.07	87.87	1.00
LP3	92.8	36.38	34.36	21.5	32.1	29.22	0.18	6.39	0.08	99.73	1.21
AC4	90.5	23.17	20.63	25.8	38.1	66.35	3.63	3.31	19.53	53.72	2.43
RC4	90.5	22.46	19.46	25.2	36.8	66.86	3.61	3.28	19.57	54.07	2.34
LC4	90.5	29.15	27.40	20.2	30.3	67.78	3.64	19.42	19.37	53.69	2.89
LP4	86.1	21.98	21.32	33.1	46.6	67.90	3.71	1.60	19.42	67.10	2.63
AC5	79.5	17.28	15.29	26.9	40.6	170.85	16.14	4.09	60.21	36.72	11.86
RC5	79.5	16.60	14.62	27.0	40.3	168.60	16.37	2.86	60.74	36.79	11.82
LC5	79.0	25.50	24.24	20.9	31.1	158.86	17.35	18.86	60.75	36.70	12.62
LP5	74.9	16.43	15.49	38.1	54.4	104.55	17.04	1.03	61.55	46.26	13.75
AC6	67.4	13.50	12.11	27.7	43.5	197.90	154.59	4.12	339.96	23.77	93.84
RC6	67.5	13.19	11.76	27.2	42.7	197.36	156.66	2.55	346.44	24.09	92.57
LC6	67.5	23.66	22.23	21.0	33.1	199.67	155.93	18.92	343.81	24.70	96.86
LP6	64.8	13.63	12.95	39.8	62.1	172.32	159.63	0.40	347.23	35.86	99.02
BCS	88.2	0.00	0.00	88.2	100.0	94.78	44.40	0.00	92.33	0.00	94.25
PUW$_0$	100.0	16.04	15.04	23.3	35.3	(0.06)	(0.03)	6.71	0.01	9.53	0.25
PUW$_2$	100.0	6.27	4.61	41.3	59.7	(0.22)	(0.09)	1.64	0.03	5.01	0.62
PUW$_8$	100.0	0.54	0.07	84.7	98.0	(16.08)	(5.57)	0.02	1.73	0.01	38.87

Overall, we have seen that the (rather unpredicable) effect of different storage orders dominates the effect of preprocessing on the solution quality. Preprocessing achieves (slight) improvements on average, but the evidence is not strong enough to conclude that preprocessing improves the gaps significantly. We can, however, accept the hypothesis that, overall, preprocessing does not significantly degrade the expected solution quality. Hence, preprocessing can and should be applied.

Higher *k*. We have run experiments on all SteinLib instances using all algorithms described in Sect. 2 with $k \leq 6$ and preprocessing enabled.

Table 1 summarizes the results. For a simpler comparison between different algorithms and values for k, we also report the respective numbers on the instance subset that was solved by *every* considered algorithm in the table. This subset ('⋂') contains only 63.25 % of the SteinLib, with average $\chi = 70.22\,\%$.

Be aware that it contains, essentially, instances that are relatively 'simple' for the approximation algorithms.

The overall success rate of the exact BCS is 88.2 % and an average successful run takes approximately 95 s. Note that the much simpler BC, obtains an overall success rate of 79.1 % with 90 s running time on average in the successful cases.

Almost all algorithms (except for LP5) could be considered practical for up to $k = 5$ where they achieve a success rate comparable to BC, trading solution quality for speed. For $k = 6$, however, the success rates drop harshly below BC's. Considering BCS, the approximation algorithms' success rates are only comparable up to $k = 4$. For any given k, success rates and times are comparable among all algorithms. In other words, the time to generate the set of all k-restricted full components dominates the running time.

We observe that higher k yield smaller gaps (as suggested in theory for RC, LC and LP) in practice. There are interesting things that can be observed:

- RC is the best, and LC is, by far, the worst choice among all strong approximation algorithms. This is remarkable since, according to theoretical bounds, RCk is the worst algorithm for $k \leq 18$, and LCk is the best for $6 \leq k \leq 14$.
- LP has the highest chances to find the optimum. Despite this, it has a worse average gap and a much worse success rate than RC, i. e., if it fails to find the optimum, its solution is either quite weak or non-existent at all.
- AC's gaps decrease with increasing k and are comparable to those of RC and LP. However, for AC we only know the approximation ratio 11/6 for $k \geq 3$. According to theoretical bounds, it should be the best one for $k \leq 5$.

Dependency on Instance Classes. Most of our observations on general algorithmic behavior is surprisingly consistent over the different instance groups of SteinLib. There are a couple of notable exceptions and observations, however.

When an instance has high *coverage* $|R|/|V|$, the preprocessing is typically much stronger, cf. Fig. 4. This consequently leads to better average gaps and running times. Generally, the instance class and the obtained χ are very strongly correlated. While SteinLib's rectilinear instances have high coverage, the VLSI and wire-routing instances, for example, have low.

Most interestingly, the density of an instance has a big influence on the algorithms' behaviors. On average, the LC algorithms give gaps 1.5–2 times larger than those of the other strong approximations. For sparse instances ($|E| \leq 2 \cdot |V|$) this worsens to factors 2–3. On sparse instances and for larger k, LP's gaps become at least as strong as RC's. VLSI instances are particularly extreme: LP's gaps are the by far strongest, while LC's gaps are over 10-fold. For complete instances, however, the picture changes drastically: we observe that LC becomes comparable to RC and AC, while LP's gaps deteriorate. Table 1 shows the average gaps and running times of VLSI ($\chi = 89.81\,\%$) and complete ($\chi = 74.80\,\%$) instances, restricted to the '$\bigcap$'-instances.

5 Conclusion

We showed that the efficient Steiner tree preprocessings do not degrade but also not significantly improve the observable solution quality. Using preprocessing —and the efficient Dreyfus-Wagner type enumeration due to [5]—allowed us to raise the bar of considered values for k. We may now say that $k \leq 5$ is applicable in practice. The probably biggest surprises for the larger values of k are that (a) the (comparably old) RCk continues to perform best in practice (this was previously only established for the special case $k = 3$); (b) ACk's solution quality improves in lockstep with RCk and LPk, despite the fact that its proven approximation ratio does not change for larger k; and (c) for identical k, LCk gives consistently clearly weaker results than any of the other approximations.

Although not the focus of this research, it remains to briefly discuss the overall practicality of the strong approximations. Therefore, we can compare to the currently practically strongest algorithms as presented in the recent DIMACS challenge [6][4]. In Table 1 we also report on the strong and fast heuristic PUW$_i$ [17], where the number of iterations is 2^i. We see that the heuristic clearly dominates all approximations by orders of magnitudes. For exact approaches, we have considered a simple branch-and-cut approach BC, as well as its more sophisticated variant BCS from [8]. We may also consider the exact results from [19]. For them, we only have detailed data over a subset of 434 SteinLib instances that have been considered particularly interesting. Over those, BCS achieves a success rate of 75.1 % and average running time of 122 s, while [19] achieve 89.4 % and 9.5 s, respectively. We can hence deduce that the current theoretically strongest approximation algorithms are neither competitive to state-of-the-art heuristics nor to exact approaches.

Acknowledgements. We thank Mihai Popa for implementations of reduction tests, and Google for funding him through the Google Summer of Code 2014 program. We also thank the authors of [8] for making their exact solver available, and Renato Werneck for detailed logs of their experiments in [17].

A Distribution of Solution Values

For each of the three specific instances below, we generated 2000 shufflings. The plots below show the distribution of the corresponding TM solution values.

[4] Those results were obtained on different machines, with different preprocessing and memory limits. The machines are compared via the DIMACS benchmark [6]; higher is faster. We: 375, 16 GB limit. PUW [17]: 389, no (relevant) limit. [19]: 307, no limit. All success rates are reported with respect to a 1-hour time limit.

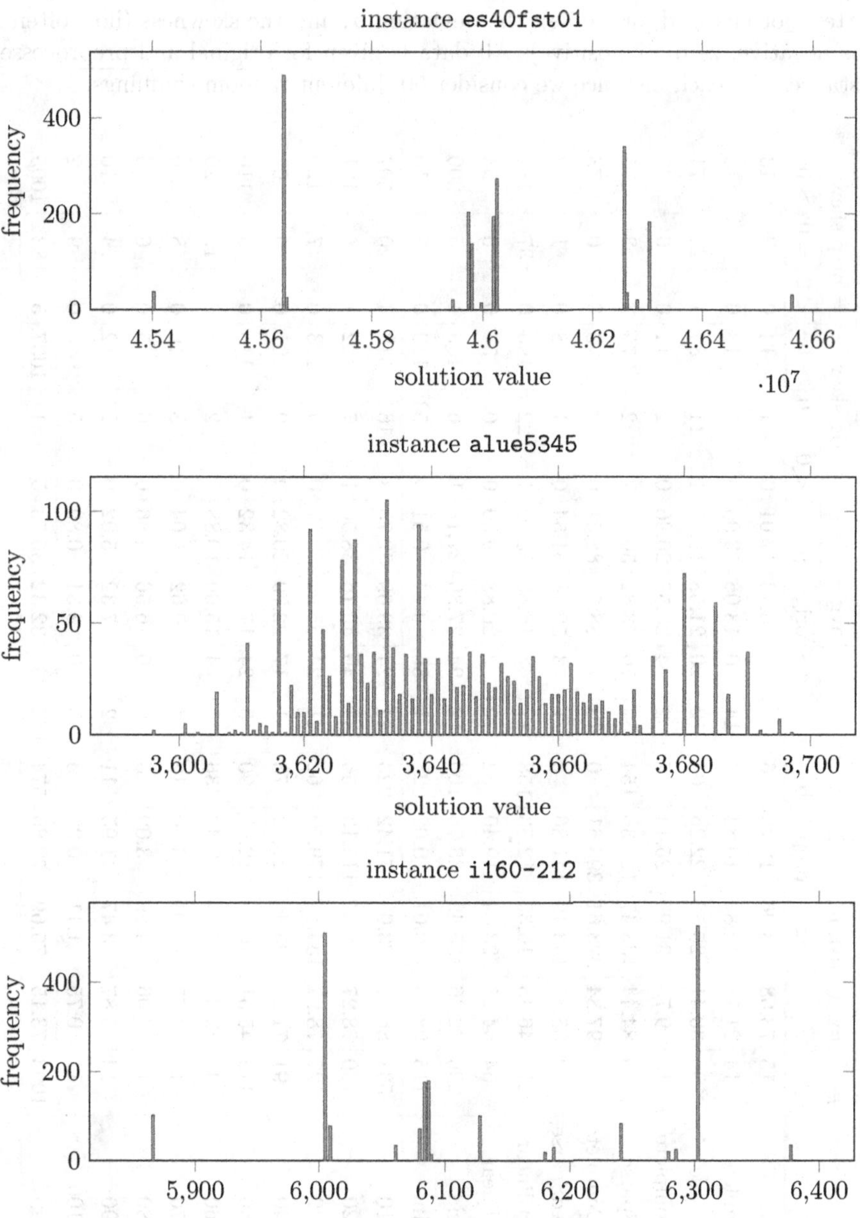

B More Detailed Table for Influence of Preprocessing

The table below shows detailed information (number of instances, average χ in %, and statistical data on the gaps) averaged over the instance groupings proposed in [2]. Statistical data is: the mean μ of the gaps (averaged and how often it was better, not changed, or worse); the deviation σ, and the skewness (how often it was negative, zero, or positive). All data is given for original and preprocessed instances; for each instance we consider 50 different random shufflings.

Group	#	avgχ	avg μ		# μ			avg σ		# orig skew			# prep skew		
			orig	prep	b	n	w	orig	prep	< 0	= 0	> 0	< 0	= 0	> 0
EuclidSparse	15	73.68	13.06	13.73	6	1	8	15.43	17.01	0	1	14	0	2	13
EuclidComplete	14	21.11	10.84	10.84	0	14	0	13.06	13.06	0	1	13	0	1	13
RandomSparse	96	30.41	25.52	22.26	64	12	20	21.58	19.00	1	11	84	1	24	71
RandomComplete	13	9.75	31.90	25.14	7	2	4	28.53	20.86	0	2	11	0	6	7
IncidenceSparse	280	84.44	133.47	127.36	154	50	76	53.82	50.68	2	2	276	3	2	275
IncidenceComplete	73	97.84	393.61	393.61	0	73	0	88.51	88.51	0	0	73	0	0	73
ConstructedSparse	9	85.57	143.19	141.86	5	1	3	50.48	51.54	0	1	8	0	1	8
SimpleRectilinear	218	46.45	16.35	12.53	158	17	43	16.86	13.90	1	22	195	0	37	181
HardRectilinear	54	64.27	23.04	19.40	52	0	2	20.88	19.30	0	0	54	0	0	54
VLSI / Grid	206	92.36	35.49	35.76	100	12	94	27.36	27.13	1	6	199	1	5	200
WireRouting	115	98.48	0.01	0.02	28	1	86	0.44	0.51	0	5	110	0	5	110
Coverage 10	531	86.77	73.09	71.42	223	84	224	33.69	32.75	3	16	512	4	20	507
Coverage 20	130	78.27	118.35	115.13	56	34	40	40.16	38.50	1	7	122	1	8	121
Coverage 30	127	78.92	184.86	179.50	62	41	24	55.90	52.67	0	4	123	0	7	120
Coverage 40	91	71.20	25.46	21.50	73	1	17	22.79	20.82	0	0	91	0	0	91
Coverage 50	119	45.51	16.15	12.46	90	5	24	17.17	14.32	0	1	118	0	9	110
Coverage 60	41	32.33	13.61	9.44	36	1	4	15.09	10.88	1	2	38	0	12	29
Coverage 70	18	14.77	8.01	3.60	12	4	2	9.62	6.04	0	8	10	0	8	10
Coverage 80	11	9.06	4.89	3.09	5	6	0	6.56	4.56	0	6	5	0	6	5
Coverage 90	14	5.87	3.47	1.95	11	2	1	7.32	5.02	0	2	12	0	4	10
Coverage 100	11	0.75	1.11	0.22	6	5	0	3.31	0.89	0	5	6	0	9	2
All	1093	73.15	75.69	72.86	574	183	336	32.32	30.53	5	51	1037	5	83	1005

References

1. Beyer, S., Chimani, M.: Steiner tree 1.39-approximation in practice. In: Hliněný, P., Dvořák, Z., Jaroš, J., Kofroň, J., Kořenek, J., Matula, P., Pala, K. (eds.) MEMICS 2014. LNCS, vol. 8934, pp. 60–72. Springer, Heidelberg (2014)
2. Beyer, S., Chimani, M.: Strong Steiner Tree Approximations inPractice. Submitted to Journal (2014). http://arxiv.org/abs/1409.8318
3. Byrka, J., Grandoni, F., Rothvoß, T., Sanità, L.: Steiner tree approximation via iterative randomized rounding. J. ACM **60**(1), 6:1–6:33 (2013)
4. Chimani, M., Woste, M.: Contraction-based Steiner tree approximations in practice. In: Asano, T., Nakano, S., Okamoto, Y., Watanabe, O. (eds.) ISAAC 2011. LNCS, vol. 7074, pp. 40–49. Springer, Heidelberg (2011)
5. Ciebiera, K., Godlewski, P., Sankowski, P., Wygocki, P.: Approximation Algorithms for Steiner Tree Problems Based on Universal Solution Frameworks. arXiv abs/1410.7534 (2014). http://arxiv.org/abs/1410.7534
6. 11th DIMACS Challenge. http://dimacs11.cs.princeton.edu. Bounds 12 September 2014
7. Dreyfus, S.E., Wagner, R.A.: The Steiner problem in graphs. Networks **1**(3), 195–207 (1971)
8. Fischetti, M., Leitner, M., Ljubic, I., Luipersbeck, M., Monaci, M., Resch, M., Salvagnin, D., Sinnl, M.: Thinning out Steiner trees: a node-based model for uniform edge costs. In: 11th DIMACS Challenge (2014)
9. Goemans, M.X., Olver, N., Rothvoß, T., Zenklusen, R.: Matroids and integrality gaps for hypergraphic steiner tree relaxations. In: STOC 2012, pp. 1161–1176. ACM (2012)
10. Junger, M., Thienel, S.: The ABACUS system for branch-and-cut-and-price algorithms in integer programming and combinatorial optimization. Softw.: Pract. Exp. **30**, 1325–1352 (2000)
11. Koch, T., Martin, A., Voß, S.: SteinLib: An Updated Library on Steiner Tree Problems in Graphs. ZIB-Report 00–37 (2000). http://steinlib.zib.de
12. Kou, L.T., Markowsky, G., Berman, L.: A fast algorithm for Steiner trees. Acta Informatica **15**, 141–145 (1981)
13. Ljubic, I., Weiskircher, R., Pferschy, U., Klau, G.W., Mutzel, P., Fischetti, M.: An algorithmic framework for the exact solution of the prize-collecting Steiner tree problem. Math. Program. **105**(2–3), 427–449 (2006)
14. Mehlhorn, K.: A faster approximation algorithm for the steiner problem in graphs. Inf. Process. Lett. **27**(3), 125–128 (1988)
15. Papadimitriou, C.H., Yannakakis, M.: Optimization, approximation, and complexity classes. In: STOC 1988, 229–234. ACM (1988)
16. Poggi de Aragão, M., Ribeiro, C.C., Uchoa, E., Werneck, R.F.: Hybrid local search for the steiner problem in graphs. In: MIC 2001 (2001)
17. Pajor, T., Uchoa, E., Werneck, R.F.: A robust and scalable algorithm for the Steiner problem in graphs. In: 11th DIMACS Challenge (2014)
18. Polzin, T., Vahdati Daneshmand, S.: Improved algorithms for the Steiner problem in networks. Discrete Appl. Math. **112**(1–3), 263–300 (2001)
19. Polzin, T., Vahdati Daneshmand, S.: The Steiner tree challenge: an updated study. In: 11th DIMACS Challenge (2014)
20. Robins, G., Zelikovsky, A.: Tighter bounds for graph Steiner tree approximation. SIAM J. Discrete Math. **19**(1), 122–134 (2005)

21. Takahashi, H., Matsuyama, A.: An approximate solution for the Steiner problem in graphs. Math. Jpn. **24**, 573–577 (1980)
22. Zelikovsky, A.: An 11/6-approximation algorithm for the Steiner problem on graphs. Ann. Discrete Math. **51**, 351–354 (1992)
23. Zelikovsky, A.: A faster approximation algorithm for the Steiner tree problem in graphs. Inf. Process. Lett. **46**(2), 79–83 (1993)
24. Zelikovsky, A.: An 11/6-approximation algorithm for the network Steiner problem. Algorithmica **9**(5), 463–470 (1993)
25. Zelikovsky, A.: Better Approximation Bounds for the Network and Euclidean Steiner Tree Problems. Technical report. CS-96-06, University of Virginia (1995)

Legally $(\Delta + 2)$-Coloring Bipartite Outerplanar Graphs in Cubic Time

Danjun Huang[1](✉), Ko-Wei Lih[2], and Weifan Wang[1]

[1] Department of Mathematics, Zhejiang Normal University, Jinhua 321004, China
{hdanjun,wwf}@zjnu.cn

[2] Institute of Mathematics, Academia Sinica, Taipei 10617, Taiwan
makwlih@sinica.edu.tw

Abstract. The 2-distance vertex-distinguishing index $\chi'_{d2}(G)$ of a graph G is the least number of colors required for a proper edge coloring of G such that any pair of vertices at distance 2 have distinct sets of colors on their incident edges. Let G be a bipartite outerplanar graph of order n with maximum degree Δ. We give an algorithm of time complexity $O(n^3)$ to show that $\chi'_{d2}(G) \leq \Delta + 2$.

Keywords: Algorithm · Edge coloring · 2-distance vertex-distinguishing index · Outerplanar graph · Bipartite graph

1 Introduction

Only simple and finite graphs are considered in this paper. Let G be a graph with vertex set $V(G)$, edge set $E(G)$, maximum degree $\Delta(G)$, and minimum degree $\delta(G)$. For a vertex v, we use $E(v)$ to denote the set of edges incident to v. Let $d_G(v) = |E(v)|$ denote the *degree* of v in G. A vertex of degree d is also called a d-vertex. The *distance* between two vertices u and v, denoted by $d_G(u, v)$, is the length of a shortest path connecting them if there is any. Otherwise, $d_G(u, v) = \infty$ by convention. If $d_G(u, v) = r$ for $u, v \in V(G)$, then u is called an r*-distance vertex* or an r*-neighbor* of v, and vice versa. Moreover, we use $N^r_G(v)$ to denote the set of r-neighbors of v in the graph G. In particular, we simply call a 1-neighbor of v a neighbor of v and abbreviate $N^1_G(v)$ to $N_G(v)$. If no ambiguity arises, $\Delta(G), d_G(u, v), N^r_G(v)$, and $N_G(v)$ are written as $\Delta, d(u, v), N^r(v)$, and $N(v)$, respectively. A graph G is called *normal* if it contains no isolated edges.

A *proper edge* k*-coloring* of a graph G is a mapping $\phi : E(G) \to C = \{1, 2, \ldots, k\}$ such that $\phi(e) \neq \phi(e')$ for any two adjacent edges e and e'. For a vertex $v \in V(G)$, let $C_\phi(v)$ denote the set of colors assigned to the edges in $E(v)$, i.e., $C_\phi(v) = \{\phi(uv) \mid uv \in E(G)\}$. Let $r \geq 1$ be an integer. The r*-strong edge chromatic number* $\chi'_s(G, r)$ of a graph G is the minimum number of colors

D. Huang—Research supported by NSFC (No.11301486) and ZJNSFC (No.LQ13A 010009).

W. Wang—Research supported by NSFC (No.11371328).

Z. Lu et al. (Eds.): COCOA 2015, LNCS 9486, pp. 617–632, 2015.
DOI: 10.1007/978-3-319-26626-8_45

required for a proper edge coloring of G such that any two vertices u and v with $d(u,v) \leq r$ have $C_\phi(u) \neq C_\phi(v)$. Note that a graph G has an r-strong edge coloring if and only if G is normal. The parameter $\chi'_s(G,r)$ was introduced by Akbari et al. [1], and independently by Zhang et al. [15].

Let diam(G) denote the *diameter* of a connected graph G, i.e., the maximum of distances between any pair of distinct vertices in G. Clearly, $\chi'_s(G,r) = \chi'_s(G)$ for $r \geq \text{diam}(G)$, where $\chi'_s(G)$ is the *strong edge chromatic number* of G and has been extensively investigated ([3–5]).

The *adjacent vertex distinguishing edge chromatic number* $\chi'_a(G)$ is precisely $\chi'_s(G,1)$. Zhang, Liu and Wang [14] first introduced this concept (*adjacent strong edge coloring* in their terminology). Among other things, they proposed the following challenging conjecture, in which C_5 denotes the cycle on five vertices.

Conjecture 1. If G is a normal graph and $G \neq C_5$, then $\chi'_a(G) \leq \Delta + 2$.

Conjecture 1 was confirmed for bipartite graphs and subcubic graphs [2]. Using probabilistic method, Hatami [9] showed that every graph G with $\Delta > 10^{20}$ satisfies $\chi'_a(G) \leq \Delta + 300$. Wang et al. [13] showed that every graph G satisfies $\chi'_a(G) \leq 2.5\Delta$ and every semi-regular graph G satisfies $\chi'_a(G) \leq \frac{5}{3}\Delta + \frac{13}{3}$. A graph G is said to be *semi-regular* if each edge of G is incident to at least one Δ-vertex. If G is a planar graph, then it is shown in [10] that $\chi'_a(G) \leq \Delta + 2$ when $\Delta \geq 12$.

More recently, Wang et al. [11] defined the 2-distance vertex-distinguishing edge coloring of graphs, which can be regarded as a relaxed form of the 2-strong edge coloring. Given a proper edge coloring ϕ, we say that two vertices u and v at distance 2 are *in conflict* with each other if $C_\phi(u) = C_\phi(v)$. The coloring ϕ is called *2-distance vertex-distinguishing* (or *2-neighbor-distinguishing*) if $C_\phi(u) \neq C_\phi(v)$ for any pair of vertices u and v at distance 2. The *2-distance vertex-distinguishing index* (or *2-neighbor-distinguishing index*) $\chi'_{d2}(G)$ of a graph G is the smallest integer k such that G has a 2-distance vertex-distinguishing edge k-coloring. Clearly, $\Delta \leq \chi'(G) \leq \chi'_{d2}(G) \leq \chi'_s(G,2)$.

In [11], the authors obtained results on the 2-distance vertex-distinguishing indices for cycles, paths, trees, complete graphs, complete bipartite graphs, unicycle graphs, Halin graphs, etc. They proposed the following conjecture.

Conjecture 2. For any graph G, $\chi'_{d2}(G) \leq \Delta + 2$.

In this paper, we are going to show that Conjecture 2 holds for bipartite outerplanar graphs.

Before proving our main results, we need to introduce some notions and preliminary results. Let K_n be the complete graph on n vertices. A graph is called *bipartite* if its vertices can be partitioned into two subsets V_1 and V_2 such that no edge has both endpoints in the same subset. A *complete bipartite graph* is a special kind of bipartite graph where every vertex of V_1 is connected to every vertex of V_2. A complete bipartite graph with partitions of size $|V_1| = m$ and $|V_2| = n$, is denoted by $K_{m,n}$. A *planar* graph is a graph that can be embedded in the Euclidean plane, i.e., it can be drawn on the plane in such a way that its edges intersect only at their endpoints. Such a drawing is called a *plane* graph or *planar embedding* of the graph. A planar graph is called *outerplanar* if it has an embedding in the Euclidean plane such that all vertices belong to the unbounded face. An *outerplane graph* is a particular planar embedding of an outerplanar graph. Given an outerplane graph G, we use $F(G)$ to denote the set of faces in G. For a face $f \in F(G)$, let $b(f)$ denote the boundary walk of f and write $f = [u_1 u_2 \cdots u_n]$ if $u_1, u_2, \ldots, u_n$ are the vertices of $b(f)$ in a cyclic order. A *leaf* is a 1-vertex. If a 3-vertex v is adjacent to a leaf v', then we call v' the *handle* of v.

Suppose that G is an outerplane graph. Then the following properties (P1)-(P3) hold. Note that (P3) follows from (P2) easily.

(P1) $\delta(G) \leq 2$.
(P2) G does not contain a subdivision of K_4 or $K_{2,3}$ as a subgraph ([7]).
(P3) G does not contain a separating cycle C (i.e., each of the interior and the exterior of C contains at least one vertex.).

2 Bipartite Outerplanar Graphs with $\Delta = 3$

This section is devoted to the study of the 2-distance vertex-distinguishing index of bipartite outerplanar graphs of maximum degree at most 3. We first establish a structural lemma that will be applied in the proof of Theorem 1.

Lemma 1. *Let G be a connected bipartite outerplane graph with maximum degree $\Delta \leq 3$. Then G contains either (B1) or (B2) of the following configurations.*

(B1) *A vertex v that is adjacent to at least $d_G(v) - 1$ leaves.*
(B2) *A path $x_1 x_2 x_3 x_4$ such that each of x_2 and x_3 is either a 2-vertex or a 3-vertex with a handle.*

Proof. Assume to the contrary that G contains none of (B1) and (B2). Since G has no (B1), no 2-vertex is adjacent to a leaf and every 3-vertex is adjacent to at most one leaf. Let H be the graph obtained by removing all leaves of G. Then H is a connected bipartite outerplane graph. It is easy to check that $\delta(H) = 2$. So H contains an end face $f = [u_1 u_2 \cdots u_n]$, $n \geq 4$, such that $d_H(u_1) \leq 3$, $d_H(u_n) \leq 3$, and $d_H(u_i) = 2$ for all $i = 2, 3, \ldots, n-1$. It is easy to see that G contains (B2), a contradiction. □

In the sequel, a 2-distance vertex-distinguishing edge k-coloring of a graph G is called a 2DVDE k-coloring for short. An edge uv is said to be *legally colored* if the color assigned to it differs from the colors of all adjacent edges and no pair of 2-distance vertices that are in conflict with each other are produced. Let H be a subgraph of G. A proper edge coloring ϕ of H is called a 2DVDE *partial coloring* of G on H if $C_\phi(u) \neq C_\phi(v)$ for each pair of vertices $u, v \in V(H)$ satisfying $d_H(u) = d_G(u)$, $d_H(v) = d_G(v)$, and $d_H(u, v) = 2$.

Theorem 1. *If G is a bipartite outerplane graph with maximum degree $\Delta \leq 3$, then $\chi'_{d2}(G) \leq 5$.*

Proof. The proof proceeds by induction on $|T(G)| := |V(G) \cup E(G)|$. If $|T(G)| \leq 5$, the theorem holds trivially. Suppose that G is a bipartite outerplane graph with $\Delta \leq 3$ and $|T(G)| \geq 6$. We may assume that G is connected since $\chi'_{d2}(G) = \max\{\chi'_{d2}(H) \mid H \text{ is a component of } G\}$. By the induction hypothesis, any bipartite outerplane graph H with $\Delta(H) \leq 3$ and $|T(H)| < |T(G)|$ has a 2DVDE 5-coloring. Let $C = \{1, 2, \ldots, 5\}$ be the set of five colors.

If G is a tree or an even cycle, then it has been shown in [11] that $\chi'_{d2}(G) \leq \Delta + 1 \leq 4$. Now, assume that G is neither a tree nor an even cycle. By Lemma 1, G contains configurations (B1) or (B2). Each case will be dealt with separately.

(B1) G contains a vertex v adjacent to at least $d_G(v) - 1$ leaves.

Let $v_1, v_2, \ldots, v_k$, where $k = d_G(v)$, denote all the neighbors of v with $d_G(v_1) = \cdots = d_G(v_{k-1}) = 1$ and $d_G(v_k) \geq 1$. Let $H = G - \{v_1, v_2, \ldots, v_{k-1}\}$. By the induction hypothesis, H has a 2DVDE 5-coloring ϕ using the color set C. It is easy to check that $vv_1, vv_2, \ldots, vv_{k-1}$ can be legally colored with different colors in $C \setminus C_\phi(v_k)$. The extended coloring is a 2DVDE 5-coloring of G.

(B2) G contains a path $x_1x_2x_3x_4$ such that each of x_2 and x_3 is either a 2-vertex or a 3-vertex with a handle.

Clearly, $x_1 \neq x_4$, for otherwise G contain a 3-cycle, making G non-bipartite. In view of the proof of Case (B1), we may assume that $d_G(x_1) \geq 2$ and $d_G(x_4) \geq 2$. Let y_1 and y_2 be the neighbors of x_1 other than x_2 if $d_G(x_1) = 3$; z_1 and z_2 be the neighbors of x_4 other than x_3 if $d_G(x_4) = 3$. For $i \in \{2, 3\}$, let x'_i be a leaf adjacent to x_i provided x_i is a 3-vertex with a handle. Then the proof splits into the following three cases by symmetry.

Case 1. $d_G(x_2) = d_G(x_3) = 3$.

Let $H = G - x'_3$. By the induction hypothesis, H admits a 2DVDE 5-coloring ϕ using the color set C such that $\phi(x_2x_3) = 1$ and $\phi(x_3x_4) = 2$. If $x_3x'_3$ cannot be legally colored, then it follows easily that $d_G(x_1) = d_G(x_4) = 3$. We may further assume that $C_\phi(x_1) = \{1, 2, 3\}$, $C_\phi(z_1) = \{1, 2, 4\}$, and $C_\phi(z_2) = \{1, 2, 5\}$. Since $C_\phi(x_1) = \{1, 2, 3\}$ and $\phi(x_1y_i) \notin \{\phi(x_1x_2), 4, 5\}$, it implies that $C_\phi(y_i) \neq \{\phi(x_1x_2), 4, 5\}$ for $i = 1, 2$. Then we need to handle the following two cases.

- $\phi(x_1x_2) = 3$. Color or recolor $x_2x_3, x_2x'_2, x_3x'_3$ with 4, 5, 3, respectively, to form a 2DVDE 5-coloring of G.
- $\phi(x_1x_2) = 2$. If $C_\phi(x_4) \neq \{2, 4, 5\}$, then we color or recolor $x_2x_3, x_2x'_2, x_3x'_3$ with 4, 5, 3, respectively. If $C_\phi(x_4) = \{2, 4, 5\}$, then $\phi(x_4z_1) = 4$ and $\phi(x_4z_2) = 5$. So the edges in $(E(z_1) \cup E(z_2)) \setminus \{x_4z_1, x_4z_2\}$ are colored with 1 or 2. We can recolor x_3x_4 with 3 so that the resultant coloring is also a 2DVDE 5-coloring of H. Then it suffices to color or recolor $x_2x_3, x_2x'_2, x_3x'_3$ with 4, 5, 2, respectively.

Case 2. $d_G(x_2) = 2$ and $d_G(x_3) = 3$.

Let $H = G - x'_3$. By the induction hypothesis, H admits a 2DVDE 5-coloring ϕ using the color set C such that $\phi(x_2x_3) = 1$ and $\phi(x_3x_4) = 2$. If $x_3x'_3$ cannot be legally colored, then it follows easily that $d_G(x_1) = d_G(x_4) = 3$. We may further assume that $C_\phi(x_1) = \{1, 2, 3\}$ and $\{\{1, 2, 4\}, \{1, 2, 5\}\} = \{C_\phi(z_1), C_\phi(z_2)\}$. Recolor x_2x_3 with 4 and color x_3x_3' with 5 to form a 2DVDE 5-coloring of G.

Case 3. $d_G(x_2) = d_G(x_3) = 2$.

Let $H = G - x_2x_3$. By the induction hypothesis, H has a 2DVDE 5-coloring ϕ using the color set C with $\phi(x_1x_2) = 1$ and $\phi(x_3x_4) = \alpha$.

- $\alpha \neq 1$, say $\alpha = 2$. If x_2x_3 cannot be legally colored, then it is easy to see that at least one of x_1 and x_4, say x_1, is a 3-vertex. Thus, we may further assume that $C_\phi(y_1) = \{1, 3\}$, $C_\phi(y_2) = \{1, 4\}$, and $\{2, 5\} \in \Omega = \{C_\phi(z_1), C_\phi(z_2)\}$. This implies that $\phi(x_1y_1) = 3$ and $\phi(x_1y_2) = 4$. Since at least one of $\{2, 3\}$ and $\{2, 4\}$, say $\{2, 3\}$, is not in Ω, we recolor x_1x_2 with 5 and color x_2x_3 with 3. The resultant coloring is a 2DVDE 5-coloring of G.
- $\alpha = 1$. If at least one of x_1 and x_4, say $d_G(x_4) = 2$, is of degree 2, then let x_5 be the second neighbor of x_4. Without loss of generality, we may assume that $\phi(x_4x_5) = 2$. Delete the color of x_3x_4, and then recolor legally x_2x_3 with some color in $\{3, 4, 5\}$ to produce a 2DVDE 5-coloring of $H' = G - x_3x_4$. The proof is reduced to the previous case. So suppose that $d_G(x_1) = d_G(x_4) = 3$. If x_2x_3 cannot be legally colored, then we may assume that $C_\phi(y_1) = \{1, 2\}$, $C_\phi(y_2) = \{1, 3\}$, $C_\phi(z_1) = \{1, 4\}$, and $C_\phi(z_2) = \{1, 5\}$. This implies that $\phi(x_1y_1) = 2$ and $\phi(x_1y_2) = 3$. We recolor x_1x_2 with 5 and color x_2x_3 with 3. The resultant coloring is a 2DVDE 5-coloring of G. □

3 Bipartite Outerplanar Graphs with $\Delta \geq 4$

In this section, we construct an algorithm of cubic time to legally color the edges of a bipartite outerplanar graph with maximum degree $\Delta \geq 4$ using at most $\Delta+2$ colors.

3.1 Ordered Breadth First Search

A *rooted tree* T is a tree with a particular vertex r designated as its root. The vertices of a rooted tree can be arranged in layers, with vertices at distances i to the root r forming layer i. Hence, layer 0 consists of the root only. For a vertex v in layer $i \geq 1$, the neighbor of v in layer $i - 1$ is called its *parent* and all the neighbors of v in layer $i+1$ are called its *children*. Vertices in layer i are ordered from left to right with labels $v_1^i, v_2^i, \ldots, v_{l_i}^i$ so that, for any j, either v_j^i and v_{j+1}^i have the same parent, or the parent of v_j^i is to the left side of the parent of v_{j+1}^i.

Let G be a connected outerplane graph. Beginning with a chosen vertex r, we order all vertices clockwise. Calamoneri and Petreschi [6] constructed an algorithm OBFT for G. It is a breadth first search starting from r in such a way that vertices coming first in the cyclic ordering are visited first. Using OBFT, G can be edge-partitioned into a spanning tree T rooted at r and a subgraph H with $\Delta(H) \leq 4$, i.e., $E(G) = E(T) \cup E(H)$ and $E(T) \cap E(H) = \emptyset$. Edges in $E(T)$

and $E(H)$ are called *tree-edges* and *non-tree-edges* of G, respectively. This edge-partition is called an OBFT partition. Calamoneri and Petreschi [6] used OBFT to determined the $L(h, 1)$-labeling number of G. This edge-partition technique was also successfully employed in [12] to study the surviving rate of outerplanar graphs.

Lemma 2. [6] *Every OBFT partition $T \cup H$ of a connected outerplane graph G satisfies the following properties.*

1. *If v_j^i is adjacent to v_k^i with $j < k$, then $v_j^i v_k^i$ is a non-tree-edge and $k = j+1$.*
2. *If $v_j^i v_k^{i-1} \in E(H)$ and v_j^i is a child of v_r^{i-1}, then $k = r+1$ and v_j^i is the rightmost child of v_r^{i-1}.*

To give an example of an OBFT partition, we consider the bipartite outerplane graph G^* depicted in Fig. 1. In Fig. 2, vertex 1 is the root of the tree T produced by OBFT and solid (broken) lines denote tree-edges (non-tree-edges).

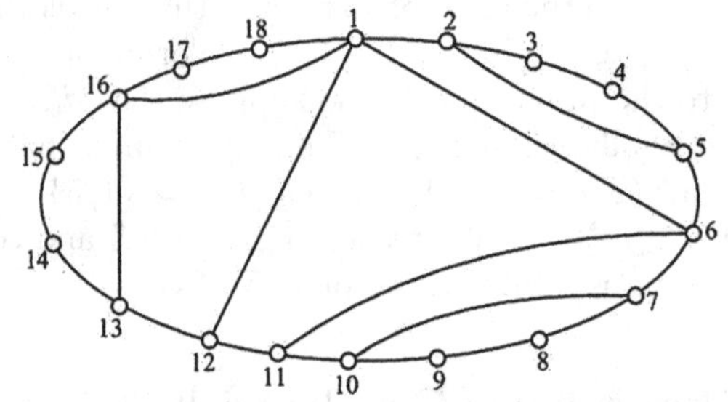

Fig. 1. A bipartite outerplane graph G^* on 18 vertices.

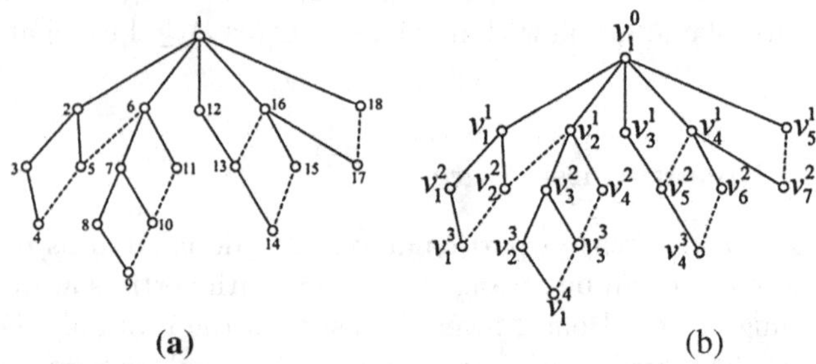

Fig. 2. (a) An OBFT partition of G^*; (b) the same edge-partition with vertices renamed.

3.2 A Legal $(\Delta + 2)$-coloring Algorithm

Let G be a connected bipartite outerplane graph. Let $T \cup H$ be the OBFT partition of G such that the root r of T satisfies $d_G(r) = \Delta$ and $\Delta(H) \leq 4$. Recall that v_l^i denotes the l-th vertex in layer i.

Claim 1. *v_l^i has no neighbor in layer i.*

Proof. Assume to the contrary that v_l^i has a neighbor x in layer i. By Lemma 2, we may assume that $x = v_{l+1}^i$. Let Q_1 be the path in T connecting r and v_l^i and Q_2 be the path in T connecting r and v_{l+1}^i. Then Q_1 and Q_2 have the same length by the structure of T. It follows that $Q_1 \cup Q_2 \cup \{v_l^i v_{l+1}^i\}$ contains an odd cycle, making G non-bipartite. □

It follows from Lemma 2 and Claim 1 that $\Delta(H) \leq 2$ and every non-root vertex v_l^i has at most two neighbors in layer $i-1$, no neighbor in layer i, at most $d_G(v_l^i) - 1$ neighbors in layer $i+1$, and no 2-neighbor in layer $i+1$ and $i-1$.

Theorem 2. *If G is a bipartite outerplane graph, then $\chi'_{d2}(G) \leq \Delta + 2$.*

Proof. We may assume that $\Delta \geq 4$ in view of Theorem 1. We will construct an algorithm to find a 2DVDE $(\Delta + 2)$-coloring of G using the color set $C = \{1, 2, \ldots, \Delta + 2\}$. Let u_j, $1 \leq j \leq |V(G)|$, be the j-th vertex visited in OBFT, where $u_1 = r$ is the root. Let $E_j = E(u_1) \cup E(u_2) \cup \cdots \cup E(u_j)$ and $G_j = G[E_j]$.

At Step 1, we color $u_1u_2, u_1u_3, \ldots, u_1u_{\Delta+1}$ with $1, 2, \ldots, \Delta$, respectively. By Claim 1, u_j has one parent u_1 and $d_G(u_j) - 1$ children when $2 \leq j \leq \Delta + 1$.

At Step j, $j = 2, 3, \ldots, \Delta + 1$, we properly color the edges in $E(u_j) \setminus \{u_ju_1\}$ from left to right in the following manner.

- If j is even, use the first $d_G(u_j) - 1$ colors in the set $\{\Delta + 1, j, j+1, \ldots, \Delta, 1, 2, \ldots, j-2\}$.
- If j is odd, use the first $d_G(u_j) - 1$ colors in the set $\{\Delta + 2, j, j+1, \ldots, \Delta, 1, 2, \ldots, j-2\}$.

Let $j \geq \Delta + 2$. Suppose that E_{j-1} has been colored such that a 2DVDE partial coloring ϕ_{j-1} of G on G_{j-1} was established. At Step j, we will color $E(u_j)$ to form a 2DVDE partial coloring ϕ_j of G on G_j. When Step j is finished, we go to Step $j+1$. Once all $E(u_j)$'s, $1 \leq j \leq |V(G)|$, have been colored, we obtain a 2DVDE coloring of G using at most $\Delta + 2$ colors.

A vertex $z \in V(G)$ is called *full* if all the edges in $E(z)$ have been colored in the foregoing steps. We use $\mathcal{F}(z)$ to denote the set of full vertices that are in conflict with z.

Let $u_j = v_l^i$. Then $i \geq 2$. Let v_k^{i-1} be the parent of v_l^i and v_h^{i-2} be the parent of v_k^{i-1}. Let $d_{G_j}(v_l^i) = p$, $d_{G_{j-1}}(v_l^i) = q$, and $t = p - q$. Then $q \leq 2$ by Claim 1.

Claim 2. *v_l^i has at most two 2-neighbors in layer $i-2$.*

Proof. If $v_l^i v_{k+1}^{i-1} \notin E(G)$, then only v_h^{i-2} and v_{h+1}^{i-2} may be 2-neighbors of v_l^i, and hence the claim holds. Assume that $v_l^i v_{k+1}^{i-1} \in E(G)$. By the outerplanarity of G, the parent of v_{k+1}^{i-1} is v_h^{i-2} or v_{h+1}^{i-2}. For the former, at most v_h^{i-2} and v_{h+1}^{i-2} are 2-neighbors of v_l^i in layer $i-2$. For the latter, it is easy to derive that $v_{k+1}^{i-1} v_{h+2}^{i-2} \notin E(G)$, for otherwise G will contain a separating cycle with v_{h+1}^{i-2} as an internal vertex, contradicting (P3). Thus, in layer $i-2$, only v_h^{i-2} and v_{h+1}^{i-2} can be the 2-neighbors of v_l^i. □

It suffices to discuss the following cases, depending on the size of t.

Case 1. $t = 0$.

Since all edges in $E(v_l^i)$ have been colored, we are done.

Case 2. $t = 1$.

Remark 1. No child z of v_{k+1}^{i-1} is in $\mathcal{F}(v_l^i)$.

In fact, if $z \in \mathcal{F}(v_l^i)$, then it is easy to derive that $v_l^i v_{k+1}^{i+1} \in E(H)$ and hence $d_{G_j}(v_l^i) \geq 3$. However, z is a leaf or a 2-vertex in G_j.

Let x be the neighbor of v_l^i not in G_{j-1}. By Claim 1, x is in layer $i+1$. There are two subcases.

Case 2.1. $xv_l^i \in E(T)$.

Then xv_l^i has exactly q forbidden colors. In G_j, the child of v_l^i is a leaf. It follows that each 2-neighbor of v_l^i is adjacent to v_k^{i-1} or v_{k+1}^{i-1}. Therefore, by Claim 2 and Remark 1, we have

$$|\mathcal{F}(v_l^i)| \leq |(N_G(v_k^{i-1}) \setminus \{v_l^i\}) \cup \{v_{h+1}^{i-2}\}| \leq d_G(v_k^{i-1}). \tag{1}$$

We need to consider certain subcases as follows.

- $q = 1$ or $|\mathcal{F}(v_l^i)| \leq \Delta - 1$. Since $|C| - (|\mathcal{F}(v_l^i)| + q) \geq |C| - (\Delta + 1) \geq 1$, we can legally color xv_l^i with a color in C to establish a 2DVDE partial coloring ϕ_j on G_j.
- $q = 2$ and $|\mathcal{F}(v_l^i)| = \Delta \geq 4$. Then $v_l^i v_{k+1}^{i-1} \in E(H)$ and v_l^i is the rightmost child of v_k^{i-1}. By inequality (1), we have $d_G(v_k^{i-1}) = \Delta$ and $v_{h+1}^{i-2} v_k^{i-1} \notin E(G)$. It follows that v_k^{i-1} has $\Delta - 1$ neighbors in layer i by Claim 1. Let $y_1, y_2, \ldots, y_{\Delta-1}$ be the neighbors of v_k^{i-1} in layer i from left to right, where $y_{\Delta-1} = v_l^i$. Suppose that xv_l^i cannot be legally colored. Then we may, without loss of generality, assume that $\phi_{j-1}(v_l^i v_k^{i-1}) = 1$, $\phi_{j-1}(v_l^i v_{k+1}^{i-1}) = 2$, $C_{\phi_{j-1}}(v_{h+1}^{i-2}) = \{1, 2, 3\}$, $C_{\phi_{j-1}}(v_h^{i-2}) = \{1, 2, 4\}$, and $C_{\phi_{j-1}}(y_s) = \{1, 2, s+4\}$ for $s \in \{1, 2, \ldots, \Delta - 2\}$. Notice that $\Delta \geq 4$, and $y_{\Delta-2}$ is the middle neighbor of v_k^{i-1} in layer i. Hence, every 2-neighbor of $y_{\Delta-2}$ in G_j is adjacent to v_k^{i-1}. Let $zy_{\Delta-2} \in E(G)$ with $\phi_{j-1}(zy_{\Delta-2}) = 1$. If $d_{G_j}(z) = 1$ or $d_G(z) > d_{G_j}(z)$, then we recolor $zy_{\Delta-2}$ with 3 and color xv_l^i with $\Delta+2$. Otherwise, it follows that $d_G(z) = 2$, $zy_{\Delta-2} \in E(H)$, $zy_{\Delta-3} \in E(T)$, and $|\mathcal{F}(z)| \leq 1$. Let $\mathcal{F}(z) = \{z^*\}$ when $|\mathcal{F}(z)| = 1$. Then $d_G(z^*) = d_G(z) = 2$ and $z^* y_{\Delta-3} \in E(H)$. Moreover $\phi_{j-1}(z^* y_{\Delta-3}) = 1$ since $C_{\phi_{j-1}}(y_{\Delta-3}) = \{1, 2, \Delta + 1\}$ and $\phi_{j-1}(zy_{\Delta-2}) = 1$. Again, we recolor $zy_{\Delta-2}$ with 3 and color xv_l^i with $\Delta + 2$. The induced coloring is a 2DVDE partial coloring ϕ_j on G_j.

Case 2.2. $xv_l^i \in E(H)$.

By Lemma 2, v_{l-1}^i is the parent of x. Note that xv_l^i has at most $q+1$ forbidden colors, i.e., the colors of xv_{l-1}^i, $v_l^i v_k^{i-1}$, and $v_l^i v_{k+1}^{i-1}$ (if it exists).

Claim 3. If $d_G(x) = d_{G_j}(x)$, then $|\mathcal{F}(x)| \leq 2$ and $|\mathcal{F}(x) \setminus \{v_k^{i-1}, v_{k+1}^{i-1}\}| \leq 1$.

Proof. Each 2-neighbor of x is adjacent to v_l^i or v_{l-1}^i. Each child of v_{l-1}^i other than x is a leaf in G_j. If v_{l-1}^i has a neighbor in layer $i+1$ which is not a child, let y^* be such a vertex. Let z be the parent of v_{l-1}^i. By Lemma 2, $z = v_k^{i-1}$ or $z = v_{k-1}^{i-1}$. For the former, we have $\mathcal{F}(x) \subseteq \{v_{k+1}^{i-1}, y^*\}$. For the latter, we have $v_l^i v_{k+1}^{i-1} \notin E(G)$, for otherwise G has a separating cycle with v_k^{i-1} as an internal vertex, contradicting (P3). So $\mathcal{F}(x) \subseteq \{v_{k-1}^{i-1}, v_k^{i-1}, y^*\}$. However, if y^* exists, then v_{l-2}^i is the parent of y^* by Lemma 2. Further, v_{k-1}^{i-1} is the parent of v_{l-2}^i, for otherwise G has a separating cycle with v_{k-1}^{i-1} as an internal vertex. Hence $d_G(v_{k-1}^{i-1}) \geq 3$ that implies $v_{k-1}^{i-1} \notin \mathcal{F}(x)$. Therefore, $|\mathcal{F}(x)| \leq 2$. □

Case 2.2.1. v_l^i is the unique child of v_k^{i-1}.

Then the parent of v_{l-1}^i is v_{k-1}^{i-1} by Lemma 2. Thus $v_l^i v_{k+1}^{i-1} \notin E(G)$, or else G contains a separating cycle with v_k^{i-1} as an internal vertex. So $q = 1$. By Claim 2, v_l^i are in conflict with at most two full vertices in layer $i-2$. As v_{l-1}^i is a 2-neighbor of v_l^i, we see $|\mathcal{F}(v_l^i)| \leq 3$. If $d_G(x) > d_{G_j}(x)$, then we can legally color xv_l^i since $|C| - (|\mathcal{F}(v_l^i)| + q + 1) \geq 1$. Next suppose that $d_G(x) = d_{G_j}(x) = 2$. Then $\mathcal{F}(v_l^i) \subseteq \{v_h^{i-2}, v_{h+1}^{i-2}, v_{l-1}^i\}$ and $\mathcal{F}(x) \subseteq \{v_k^{i-1}, v_{k-1}^{i-1}, y^*\}$, where $y^* v_{l-1}^i \in E(H)$ and y^* is in layer $i+1$. Recall that at most one of y^* and v_{k-1}^{i-1} belongs to $\mathcal{F}(x)$ by the proof of Claim 3. Moreover, if $\mathcal{F}(v_l^i) = \{v_h^{i-2}, v_{h+1}^{i-2}, v_{l-1}^i\}$, then $d_G(v_k^{i-1}) \geq 3$, and so $\mathcal{F}(x) \subseteq \{v_{k-1}^{i-1}, y^*\}$. Therefore, $|\mathcal{F}(x)| + |\mathcal{F}(v_l^i)| \leq 4$. If $\phi_{j-1}(xv_{l-1}^i) = \phi_{j-1}(v_l^i v_k^{i-1})$, then $|C| - (|\mathcal{F}(x)| + |\mathcal{F}(v_l^i)| + q) \geq 1$. If $\phi_{j-1}(xv_{l-1}^i) \neq \phi_{j-1}(v_l^i v_k^{i-1})$, then v_l^i and v_{l-1}^i will admit different color sets whenever xv_l^i is properly colored. Hence, $|\mathcal{F}(x)| + |\mathcal{F}(v_l^i)| \leq 3$, and then $|C| - (|\mathcal{F}(x)| + |\mathcal{F}(v_l^i)| + q + 1) \geq 1$. So we always can legally color xv_l^i with a color in C to establish a 2DVDE partial coloring ϕ_j on G_j.

Case 2.2.2. $d_G(v_k^{i-1}) \geq 3$ and v_l^i is not the rightmost child of v_k^{i-1}.

Then $q = 1$, $d_G(v_l^i) = 2$, and $v_l^i v_{k+1}^{i-1} \notin E(G)$. Moreover, $|\mathcal{F}(x)| \leq 1$ by Claim 3. Note that every vertex in $\mathcal{F}(v_l^i)$ is v_{l-1}^i or is adjacent to v_k^{i-1}. Let z^* be the rightmost child of v_k^{i-1}. Then $1 \leq d_{G_j}(z^*) \leq 2$. Now the proof splits into the following two cases.

Case 2.2.2.1. $d_{G_j}(z^*) = 1$.

Then $|\mathcal{F}(v_l^i)| \leq |(N_G(v_k^{i-1}) \setminus \{v_l^i, z^*\}) \cup \{v_{l-1}^i\}| \leq \Delta - 1$ since $z^* \notin \mathcal{F}(v_l^i)$. If $\phi_{j-1}(xv_{l-1}^i) = \phi_{j-1}(v_l^i v_k^{i-1})$, then $|C| - (|\mathcal{F}(x)| + |\mathcal{F}(v_l^i)| + q) \geq 1$. Otherwise, v_l^i and v_{l-1}^i have different color sets whenever xv_l^i is properly colored. This implies that $|\mathcal{F}(v_l^i)| \leq |N_G(v_k^{i-1}) \setminus \{v_l^i, z^*\}| \leq \Delta - 2$, and hence $|C| - (|\mathcal{F}(x)| + |\mathcal{F}(v_l^i)| + q + 1) \geq 1$. So we can legally color xv_l^i with a color in C.

Case 2.2.2.2. $d_{G_j}(z^*) = 2$.

Then $z^* v_{k+1}^{i-1} \in E(H)$. We have to consider two possibilities.

- Assume that $v_k^{i-1} v_{h+1}^{i-2} \in E(G)$. Clearly, $v_k^{i-1} v_{h+1}^{i-2}$ is a non-tree-edge. Then v_{h+1}^{i-2} is the parent of v_{k+1}^{i-1} by the outerplanarity of G, and so $d_G(v_{h+1}^{i-2}) \geq 3$.

Then $|\mathcal{F}(v_l^i)| \leq |(N_G(v_k^{i-1}) \setminus \{v_l^i, v_{h+1}^{i-2}\}) \cup \{v_{l-1}^i\}| \leq \Delta - 1$. If $\phi_{j-1}(xv_{l-1}^i) = \phi_{j-1}(v_l^i v_k^{i-1})$, then $|C| - (|\mathcal{F}(x)| + |\mathcal{F}(v_l^i)| + q) \geq 1$. Otherwise, v_l^i and v_{l-1}^i have different color sets whenever xv_l^i is properly colored. This implies that $|\mathcal{F}(v_l^i)| \leq |N_G(v_k^{i-1}) \setminus \{v_l^i, v_{h+1}^{i-2}\}| \leq \Delta - 2$, and hence $|C| - (|\mathcal{F}(x)| + |\mathcal{F}(v_l^i)| + q + 1) \geq 1$. Hence we can legally color xv_l^i with certain color in C.

- Assume that $v_k^{i-1} v_{h+1}^{i-2} \notin E(G)$. If $v_{l-1}^i \in N_G(v_k^{i-1})$, then $|\mathcal{F}(v_l^i)| \leq |N_G(v_k^{i-1}) \setminus \{v_l^i\}| \leq \Delta - 1$. If $v_{l-1}^i \notin N_G(v_k^{i-1})$ and $d_G(v_k^{i-1}) \leq \Delta - 1$, then $|\mathcal{F}(v_l^i)| \leq |(N_G(v_k^{i-1}) \setminus \{v_l^i\}) \cup \{v_{l-1}^i\}| \leq d_G(v_k^{i-1}) \leq \Delta - 1$. Otherwise, $v_{l-1}^i \notin N_G(v_k^{i-1})$ and $d_G(v_k^{i-1}) = \Delta \geq 4$, then v_k^{i-1} has one child z' with $d_{G_j}(z') = 1$ since v_k^{i-1} has $\Delta - 1 \geq 3$ children in layer i. Hence, we also have $|\mathcal{F}(v_l^i)| \leq |(N_G(v_k^{i-1}) \setminus \{v_l^i, z'\}) \cup \{v_{l-1}^i\}| \leq \Delta - 1$. Arguing similarly as before, we can establish a 2DVDE partial coloring ϕ_j on G_j.

Case 2.2.3. $d_G(v_k^{i-1}) \geq 3$ and v_l^i is the rightmost child of v_k^{i-1}.

Then v_k^{i-1} is the parent of v_{l-1}^i by Case 2.2.1. The proof splits into the following subcases, depending on the size of q.

Case 2.2.3.1. $q = 1$.

Then $d_G(v_l^i) = 2$, $v_l^i v_{k+1}^{i-1} \notin E(G)$, and each vertex in $\mathcal{F}(v_l^i)$ is adjacent to v_k^{i-1}. So $|\mathcal{F}(v_l^i)| \leq |N_G(v_k^{i-1}) \setminus \{v_l^i\}| \leq \Delta - 1$. Moreover, $|\mathcal{F}(x)| \leq 1$ by Claim 3. If $\phi_{j-1}(xv_{l-1}^i) = \phi_{j-1}(v_l^i v_k^{i-1})$, then $|C| - (|\mathcal{F}(x)| + |\mathcal{F}(v_l^i)| + q) \geq 1$. Otherwise, whenever xv_l^i is properly colored, v_l^i and v_{l-1}^i have different color sets implying $|\mathcal{F}(v_l^i)| \leq |N_G(v_k^{i-1}) \setminus \{v_l^i, v_{l-1}^i\}| \leq \Delta - 2$, and hence $|C| - (|\mathcal{F}(x)| + |\mathcal{F}(v_l^i)| + q + 1) \geq 1$. Consequently, xv_l^i can be legally colored with some color in C.

Case 2.2.3.2. $q = 2$.

Then $v_l^l v_{k+1}^{i-1} \in E(H)$ and $d_G(v_l^i) = 3$. Let $\phi_{j-1}(v_l^i v_k^{i-1}) = 1$, $\phi_{j-1}(v_l^i v_{k+1}^{i-1}) = 2$, and $\phi_{j-1}(xv_{l-1}^i) = \alpha$. Without loss of generality, we may assume that $\alpha \in \{1, 2, 3\}$. By Claim 2 and Remark 1, we have

$$|\mathcal{F}(v_l^i)| \leq |(N_G(v_k^{i-1}) \setminus \{v_l^i\}) \cup \{v_{h+1}^{i-2}\}|. \tag{2}$$

Moreover, if v_k^{i-1} has m neighbors in layer i, then we use $y_1, y_2, \ldots, y_m$ to denote these neighbors ordered from left to right, where $y_m = v_l^i$. For $w \in \{x, v_l^i\}$, we define $\mathcal{C}(w) = \{c \in C \mid \text{if } \phi_j(xv_l^i) = c, \text{ then there is } v \in \mathcal{F}(v_l^i) \text{ such that } C_{\phi_j}(w) = C_{\phi_j}(v)\}$.

Obviously, $|\mathcal{C}(v_l^i)| \leq |\mathcal{F}(v_l^i)|$ and $|\mathcal{C}(x)| \leq |\mathcal{F}(x)|$. It remains to consider the following two cases.

Case 2.2.3.2.1. $\alpha \in \{1, 2\}$.

Case 2.2.3.2.1(a). $d_G(x) > d_{G_j}(x)$.

If $|\mathcal{F}(v_l^i)| \leq \Delta - 1$, then $|C| - (|\mathcal{F}(v_l^i)| + q) \geq |C| - ((\Delta - 1) + 2) \geq 1$, xv_l^i can be legally colored. Thus assume that $|\mathcal{F}(v_l^i)| \geq \Delta$. By (2), $d_G(v_k^{i-1}) = \Delta$ and $v_{h+1}^{i-2} \notin N_G(v_k^{i-1})$ that implies $m = \Delta - 1$ by Claim 1. If xv_l^i cannot be legally colored, then we may suppose that $C_{\phi_{j-1}}(v_{h+1}^{i-2}) = \{1,2,3\}$, $C_{\phi_{j-1}}(v_h^{i-2}) = \{1,2,4\}$, and $C_{\phi_{j-1}}(y_s) = \{1,2,s+4\}$ for $s \in \{1,2,\ldots,\Delta-2\}$. Since $\Delta \geq 4$, all 2-neighbors of $y_{\Delta-2}$ in G_j are adjacent to v_k^{i-1}. It suffices to recolor $xy_{\Delta-2}$ with 3, and then color xv_l^i with $\Delta+2$.

Case 2.2.3.2.1(b). $d_G(x) = d_{G_j}(x) = 2$.

Let y^* denote a possible vertex in layer $i+1$ such that $y^*v_{l-1}^i \in E(H)$. Note that y^* may not exist. Once y^* exists, its parent must be v_{l-2}^i by Lemma 2. Our argument splits into the following two cases.

(b1) $d_G(v_{k+1}^{i-1}) \geq 3$.

By Claim 3, $|\mathcal{F}(x)| \leq |\{y^*\}| = 1$. There are two possibilities to consider.

(b1.1) $|\mathcal{F}(x)| = 0$.

Equivalently, if y^* exists, then $C_{\phi_{j-1}}(y^*) \neq \{\alpha,\beta\}$ for any $\beta \in \{3,4,\ldots,\Delta+2\}$. If $|\mathcal{F}(v_l^i)| \leq \Delta-1$, then $|C| - (|\mathcal{F}(v_l^i)| + q) \geq |C| - ((\Delta-1)+2) \geq 1$, we can legally color xv_l^i. So assume that $|\mathcal{F}(v_l^i)| \geq \Delta$. By (2), $d_G(v_k^{i-1}) = \Delta$ and $v_{h+1}^{i-2} \notin N_G(v_k^{i-1})$. It follows that $m = \Delta-1$ by Claim 1. Similar to the previous argument, we can legally color xv_l^i through a necessary recoloring for $xy_{\Delta-2}$.

(b1.2) $|\mathcal{F}(x)| = 1$.

In this case, y^* exists and $C_{\phi_{j-1}}(y^*) = \{\alpha,\beta\}$ for some $\beta \in \{3,4,\ldots,\Delta+2\}$, say $\beta = 3$. Then $3 \in C_{\phi_{j-1}}(v_{l-1}^i)$. Note that if xv_l^i is colored with any color in $\{4,5,\ldots,\Delta+2\}$, then v_l^i and v_{l-1}^i will have different color sets. Hence, $|\mathcal{F}(v_l^i) \setminus \{v_{l-1}^i\}| \leq |(N_G(v_k^{i-1}) \setminus \{v_l^i, v_{l-1}^i\}) \cup \{v_{h+1}^{i-2}\}| \leq d_G(v_k^{i-1}) - 1$. If $|\mathcal{F}(v_l^i) \setminus \{v_{l-1}^i\}| \leq \Delta-2$, then $|\{4,5,\ldots,\Delta+2\}| - |\mathcal{F}(v_l^i) \setminus \{v_{l-1}^i\}| \geq 1$, we can legally color xv_l^i with any color in $\{4,5,\ldots,\Delta+2\}$. Otherwise, $|\mathcal{F}(v_l^i) \setminus \{v_{l-1}^i\}| = \Delta-1$. It turns out that $d_G(v_k^{i-1}) = \Delta$ and $v_{h+1}^{i-2} \notin N_G(v_k^{i-1})$. It follows that $m = \Delta-1$ by Claim 1. If xv_l^i cannot be legally colored, then $C_{\phi_{j-1}}(v_{h+1}^{i-2}) = \{1,2,4\}$, $C_{\phi_{j-1}}(v_h^{i-2}) = \{1,2,5\}$, and $C_{\phi_{j-1}}(y_s) = \{1,2,s+5\}$ for $s \in \{1,2,\ldots,\Delta-3\}$. Since $\Delta \geq 4$, all 2-neighbors of $y_{\Delta-2}$ in G_j are adjacent to v_k^{i-1}. We recolor $xy_{\Delta-2}$ with some color in $\{\alpha,3,4,5,\ldots,\Delta+2\} \setminus C_{\phi_{j-1}}(y_{\Delta-2})$ and color xv_l^i with 3.

(b2) $d_G(v_{k+1}^{i-1}) = 2$.

Note that the parent of v_{k+1}^{i-1} is v_h^{i-2} or v_{h+1}^{i-2} by Lemma 2 and $|\mathcal{F}(x)| \leq |\{y^*, v_{k+1}^{i-1}\}| = 2$ by Claim 3. We need to consider the following two subcases.

(b2.1) v_{k+1}^{i-1} is the child of v_h^{i-2}.

Then $|\mathcal{F}(v_l^i)| \leq \Delta-1$ by (2) since $v_{h+1}^{i-2} \notin \mathcal{F}(v_l^i)$. Note that if $\phi_{j-1}(v_h^{i-2}v_{k+1}^{i-1}) = a \in \{3,4,\ldots,\Delta+2\}$, then we can color xv_l^i with some color in $\{3,4,\ldots,\Delta+2\} \setminus \{a\}$ such that v_l^i and v_h^{i-2} have different color sets; x and v_{k+1}^{i-1} have different color sets. Hence, $|\mathcal{C}(x) \cup \mathcal{C}(v_l^i)| \leq |\mathcal{F}(x)| + |\mathcal{F}(v_l^i)| - 1$. Note that, if $|\mathcal{F}(x)| \leq 1$, then $|\{3,4,\ldots,\Delta+2\}| - |\mathcal{C}(x) \cup \mathcal{C}(v_l^i)| \geq \Delta - (|\mathcal{F}(x)| + |\mathcal{F}(v_l^i)| - 1) \geq 1$; or if $|\mathcal{F}(x)| = 2$ and $C_{\phi_{j-1}}(y^*) \neq \{\alpha,\beta\}$ for any $\beta \in \{3,4,\ldots,\Delta+2\}$, then

$|\{3,4,\ldots,\Delta+2\}|-|\mathcal{C}(x)\cup\mathcal{C}(v_l^i)|\geq\Delta-(|\mathcal{F}(x)\setminus\{y^*\}|+|\mathcal{F}(v_l^i)|-1)\geq 1$. In these two cases, we can legally color xv_l^i. Otherwise, we may assume that $|\mathcal{F}(x)|=2$ and $C_{\phi_{j-1}}(y^*)=\{\alpha,3\}$. Then $3\in C_{\phi_{j-1}}(v_{l-1}^i)$. Note that $|\mathcal{F}(v_l^i)\setminus\{v_{l-1}^i\}|\leq\Delta-2$ by (2). Since $|\{4,5,\ldots,\Delta+2\}|-|\mathcal{C}(x)\cup\mathcal{C}(v_l^i)|\geq(\Delta-1)-(|\mathcal{F}(x)\setminus\{y^*\}|+|\mathcal{F}(v_l^i)\setminus\{v_{l-1}^i\}|-1)\geq(\Delta-1)-(1+(\Delta-2)-1)\geq 1$, we can legally color xv_l^i with a color in C to establish a 2DVDE partial coloring ϕ_j on G_j.

(b2.2) v_{k+1}^{i-1} is the child of v_{h+1}^{i-2}.

Note that if $\phi_{j-1}(v_{h+1}^{i-2}v_{k+1}^{i-1})=a\in\{3,4,\ldots,\Delta+2\}$, then xv_l^i can be colored with any color in $\{3,4,\ldots,\Delta+2\}\setminus\{a\}$ with v_l^i and v_{h+1}^{i-2} have different color sets; x and v_{k+1}^{i-1} have different color sets. Hence, $|\mathcal{C}(x)\cup\mathcal{C}(v_l^i)|\leq|\mathcal{F}(x)|+|\mathcal{F}(v_l^i)|-1$.

First assume that either $|\mathcal{F}(x)|\leq 1$ or $|\mathcal{F}(x)|=2$ but $C_{\phi_{j-1}}(y^*)\neq\{\alpha,\beta\}$ for any $\beta\in\{3,4,\ldots,\Delta+2\}$. If xv_l^i cannot be legally colored, then $|\mathcal{C}(x)\cup\mathcal{C}(v_l^i)|=\Delta$. Therefore, $|\mathcal{F}(v_l^i)|=\Delta$. By (2), $d_G(v_k^{i-1})=\Delta$ and $v_{h+1}^{i-2}\notin N_G(v_k^{i-1})$. It follows that $m=\Delta-1$ by Claim 1. Similar to the previous argument, we can legally color xv_l^i through a necessary recoloring.

Next assume that $|\mathcal{F}(x)|=2$ with $C_{\phi_{j-1}}(y^*)=\{\alpha,3\}$. Then $3\in C_{\phi_{j-1}}(v_{l-1}^i)$. If xv_l^i cannot be legally colored with a color in $\{4,5,\ldots,\Delta+2\}$, then it follows that $\Delta-1=|(\mathcal{C}(x)\cup\mathcal{C}(v_l^i))\setminus\{3\}|\leq|\mathcal{F}(x)\setminus\{y^*\}|+|\mathcal{F}(v_l^i)\setminus\{v_{l-1}^i\}|-1$, and hence $|\mathcal{F}(v_l^i)\setminus\{v_{l-1}^i\}|\geq\Delta-1$. By (2), $d_G(v_k^{i-1})=\Delta$, $v_{h+1}^{i-2}\notin N_G(v_k^{i-1})$, and consequently $m=\Delta-1$. Hence, $C_{\phi_{j-1}}(v_{h+1}^{i-2})=\{1,2,4\}$ or $C_{\phi_{j-1}}(v_{k+1}^{i-1})=\{2,4\}$, $C_{\phi_{j-1}}(v_h^{i-2})=\{1,2,5\}$, and $C_{\phi_{j-1}}(y_s)=\{1,2,s+5\}$ for $s\in\{1,2,\ldots,\Delta-3\}$. Since $\Delta\geq 4$, all 2-neighbors of $y_{\Delta-2}$ in G_j are adjacent to v_k^{i-1}. Recolor $xy_{\Delta-2}$ with some color in $\{\alpha,3,4,\ldots,\Delta+2\}\setminus\{C_{\phi_{j-1}}(v_{l-1}^i)\}$, and color xv_l^i with 3. A 2DVDE partial coloring ϕ_j on G_j is constructed.

Case 2.2.3.2.2. $\alpha=3$.

It is easy to observe that when xv_l^i is colored with a color in $\{4,5,\ldots,\Delta+2\}$, we have $C_{\phi_j}(x)\neq C_{\phi_j}(v_{k+1}^{i-1})$ and $C_{\phi_j}(v_l^i)\neq C_{\phi_j}(v_{l-1}^i)$. Hence,

$$|\mathcal{F}(v_l^i)|\leq|(N_G(v_k^{i-1})\setminus\{v_l^i,v_{l-1}^i\})\cup\{v_{h+1}^{i-2}\}|\leq d_G(v_k^{i-1})-1. \tag{3}$$

By Claim 3, $|\mathcal{F}(x)|\leq 1$. In addition, if $|\mathcal{F}(x)|=1$, we set $\mathcal{F}(x)=\{y^*\}$, where y^* lies in layer $i+1$ such that $y^*v_{l-1}^i\in E(H)$. By Lemma 2, v_{l-2}^i is the parent of y^*. Hence, if y^* exists, then v_k^{i-1} is the parent of v_{l-2}^i, for otherwise G will contain a separating cycle with v_k^{i-1} as internal vertex, contradicting (P3).

Case 2.2.3.2.2(a). $d_G(x)>d_{G_j}(x)$.

If $|\mathcal{F}(v_l^i)|\leq\Delta-2$, then $|C|-(|\mathcal{F}(v_l^i)|+q+1)\geq|C|-((\Delta-2)+3)\geq 1$ and we can legally color xv_l^i. So assume that $|\mathcal{F}(v_l^i)|\geq\Delta-1$. By (3), we have $d_G(v_k^{i-1})=\Delta$ and $v_{h+1}^{i-2}\notin N_G(v_k^{i-1})$, and hence $m=\Delta-1$. If xv_l^i cannot be legally colored, then we may assume that $C_{\phi_{j-1}}(v_{h+1}^{i-2})=\{1,2,4\}$, $C_{\phi_{j-1}}(v_h^{i-2})=\{1,2,5\}$, and $C_{\phi_{j-1}}(y_s)=\{1,2,s+5\}$ for $s\in\{1,2,\ldots,\Delta-3\}$. Since $\Delta\geq 4$, all 2-neighbors of $y_{\Delta-2}$ in G_j are adjacent to v_k^{i-1}. Recolor $xy_{\Delta-2}$ with some color

$\gamma \in C \setminus C_{\phi_{j-1}}(y_{\Delta-2})$. If $\gamma \in \{1,2\}$, then it is reduced to Case 2.2.3.2.1(a). If $\gamma \notin \{1,2\}$, then we color xv_l^i with 3.

Case 2.2.3.2.2(b). $d_G(x) = d_{G_j}(x) = 2$.

If $|\mathcal{F}(x)| = 0$ or y^* exists such that $C_{\phi_{j-1}}(y^*) \neq \{3,\beta\}$ for any $\beta \in \{3,4,\ldots,\Delta+2\}$, then we can legally color xv_l^i analogously to the previous proof. Otherwise, $|\mathcal{F}(x)| = 1$, we may assume that y^* exists such that $C_{\phi_{j-1}}(y^*) = \{3,4\}$. Hence $\phi_{j-1}(y^*v_{l-1}^i) = 4$ and $\phi_{j-1}(y^*v_{l-2}^i) = 3$. Note that xv_l^i can be colored with any color in $\{5,6,\ldots,\Delta+2\}$ such that $C_{\phi_j}(v_l^i) \neq C_{\phi_j}(v_{l-1}^i)$, $C_{\phi_j}(v_l^i) \neq C_{\phi_j}(v_{l-2}^i)$, and $C_{\phi_j}(x) \neq C_{\phi_j}(y^*)$. By Claim 2 and the fact that $v_{l-2}^i \in N_G(v_k^{i-1})$, we have $|\mathcal{F}(v_l^i)| \leq |(N_G(v_k^{i-1}) \setminus \{v_l^i, v_{l-1}^i, v_{l-2}^i\}) \cup \{v_{h+1}^{i-2}\}| \leq d_G(v_k^{i-1}) - 2$. If xv_l^i cannot be legally colored with any color in $\{5,6,\ldots,\Delta+2\}$, then $|\mathcal{F}(v_l^i)| = \Delta - 2$. Therefore, $d_G(v_k^{i-1}) = \Delta$, $v_{h+1}^{i-2} \notin N_G(v_k^{i-1})$, and hence $m = \Delta - 1$. Assume that $C_{\phi_{j-1}}(v_{h+1}^{i-2}) = \{1,2,5\}$, $C_{\phi_{j-1}}(v_h^{i-2}) = \{1,2,6\}$, and $C_{\phi_{j-1}}(y_s) = \{1,2,s+6\}$ for $s \in \{1,2,\ldots,\Delta-4\}$. Since $\Delta \geq 4$, all 2-neighbors of $y_{\Delta-2}$ in G_j are adjacent to v_k^{i-1}. We recolor $xy_{\Delta-2}$ with some color γ in $\{1,3,4,\ldots,\Delta+2\} \setminus C_{\phi_{j-1}}(y_{\Delta-2})$. If $\gamma = 1$, then it is reduced to Case 2.2.3.2.1(b). Otherwise, we color xv_l^i with 4.

Case 3. $t \geq 2$.

By Claim 1, all the neighbors of v_l^i that are not in G_{j-1}, say $x_1, x_2, \ldots, x_t$, lie in layer $i+1$. Assume that $x_1, x_2, \ldots, x_t$ are ordered from left to right. Note that $d_{G_j}(x_i) = 1$ for all $i = 2,3,\ldots,t$ and $1 \leq d_{G_j}(x_1) \leq 2$.

Case 3.1. $d_{G_j}(x_1) = 1$.

By Claim 2 and Remark 1, $|\mathcal{F}(v_l^i)| \leq |(N_G(v_k^{i-1}) \setminus \{v_l^i\}) \cup \{v_{h+1}^{i-2}\}| \leq \Delta$. Note that $2 \leq t \leq \Delta - q$ and $\binom{\Delta+2-q}{t} \geq \binom{\Delta+2-q}{2} \geq \binom{\Delta}{2} > \Delta$. There exists a subset C' of available colors in C with $|C'| = t$ such that $x_1, x_2, \ldots, x_t$ can be legally colored with C'.

Case 3.2. $d_{G_j}(x_1) = 2$.

We see that $x_1v_l^i \in E(H)$ and the parent of x_1 is v_{l-1}^i by Lemma 2. Let $\phi_{j-1}(v_l^i v_k^{i-1}) = c_1$, $\phi_{j-1}(x_1 v_{l-1}^i) = c_2$, and $\phi_{j-1}(v_l^i v_{k+1}^{i-1}) = c_3$ if $v_l^i v_{k+1}^{i-1} \in E(G)$. Now the proof splits into the following three cases.

Case 3.2.1. $d_G(v_k^{i-1}) = 2$.

Note that the parent of v_{l-1}^i is v_{k-1}^{i-1}. If $q = 2$, then G will contain a separating cycle with v_k^{i-1} as an internal vertex, contradicting (P3). Thus, $q = 1$, i.e., $v_l^i v_{k+1}^{i-1} \notin E(G)$. Furthermore, $|\mathcal{F}(v_l^i)| \leq |\{v_h^{i-2}, v_{l-1}^i\}| = 2$ by Claim 2 and $|\mathcal{F}(x_1)| \leq 2$ by Claim 3. Since $|C \setminus \{c_1, c_2\}| \geq 6 - 2 = 4$, we can properly color $x_1 v_l^i$ with some color $c^* \in C \setminus \{c_1, c_2\}$ such that x_1 is not in conflict with its 2-neighbors. Since $2 \leq t \leq \Delta - q$ and $\binom{\Delta+2-q-1}{t-1} = \binom{\Delta}{t-1} \geq \Delta > 2$, we can find a subset C' of available colors in $C \setminus \{c^*, c_1\}$ with $|C'| = t - 1$ such that $x_2 v_l^i, x_3 v_l^i, \ldots, x_t v_l^i$ can be legally colored with C'.

Case 3.2.2. $d_G(v_k^{i-1}) \geq 3$ and v_l^i is not the rightmost child of v_k^{i-1}.

It is easy to see that $v_l^i v_{k+1}^{i-1} \notin E(G)$, and henceforth $q = 1$. In this case, every 2-neighbor of v_l^i is v_{l-1}^i or is adjacent to v_k^{i-1}. Every child v_s^i of v_k^{i-1} with $s > l$ is of degree at most 2 in G_j and $d_{G_j}(v_l^i) \geq t+1 \geq 3$. These facts imply immediately that $|\mathcal{F}(v_l^i)| \leq \Delta - 1$. By Claim 3, $|\mathcal{F}(x_1)| \leq 1$. Since $|C \setminus \{c_1, c_2\}| \geq 6-2 = 4$, we can first color $x_1 v_l^i$ with some color $c^* \in C \setminus \{c_1, c_2\}$ such that x_1 is not in conflict with its 2-neighbor. Since $2 \leq t \leq \Delta - q$ and $\binom{\Delta+2-q-1}{t-1} = \binom{\Delta}{t-1} \geq \Delta > \Delta - 1$, we can find a subset C' of available colors in $C \setminus \{c^*, c_1\}$ with $|C'| = t-1$ such that $x_2 v_l^i, x_3 v_l^i, \ldots, x_t v_l^i$ can be legally colored with C'.

Case 3.2.3. $d_G(v_k^{i-1}) \geq 3$ and v_l^i is the rightmost child of v_k^{i-1}.

Then $q \leq 2$, $|\mathcal{F}(v_l^i)| \leq |(N_G(v_k^{i-1}) \setminus \{v_l^i\}) \cup \{v_{h+1}^{i-2}\}| \leq \Delta$ by Claim 2 and Remark 1; $|\mathcal{F}(x_1)| \leq 2$ by Claim 3. Since $2 \leq t \leq \Delta - q$ and $\binom{\Delta+2-q}{t} \geq \binom{\Delta+2-q}{2} \geq \binom{\Delta}{2} \geq \Delta + 2$, there exist two subsets $C_1', C_2' \subseteq C \setminus \{c_1, c_3\}$ such that (i) $C_1' \neq C_2'$; (ii) $|C_1'| = |C_2'| = t$; and (iii) for $i = 1$ and 2, $x_1 v_l^i, x_2 v_l^i, \ldots, x_t v_l^i$ can be properly colored with C_i' such that v_l^i is not in conflict with its 2-neighbors. Since $t \geq 2$, $|C_1' \cup C_2'| \geq 3$. Let us now construct a 2DVDE partial coloring ϕ_j on G_j as follows. If $c_2 \in \{c_1, c_3\}$, then we first color $x_1 v_l^i$ with a color $c^* \in C_1' \cup C_2'$, say $c^* \in C_1'$, such that x_1 is not in conflict with its 2-neighbors. Then we color $x_2 v_l^i, \ldots, x_t v_l^i$ with $C_1' \setminus \{c^*\}$. If $c_2 \notin \{c_1, c_3\}$, then x_1 and v_{k+1}^{i-1} will have different color sets if $x_1 v_l^i$ is colored properly. Thus, $|\mathcal{F}(x_1)| \leq 1$. We color $x_1 v_l^i$ with a color $c^* \in (C_1' \cup C_2') \setminus \{c_2\}$, say $c^* \in C_1'$, such that x_1 is not in conflict with its 2-neighbors. Then we color $x_2 v_l^i, \ldots, x_t v_l^i$ with $C_1' \setminus \{c^*\}$. In view of the choices of C_1' and C_2', the resultant coloring is a 2DVDE partial coloring ϕ_j on G_j. □

Based on Theorem 2, we give the following algorithm for finding a 2DVDE $(\Delta+2)$-coloring of a bipartite outerplanar graph G when $\Delta \geq 4$.

ALGORITHM: 2DVDE-Color Bipartite Outerplanar Graphs.
INPUT: A connected bipartite outerplanar graph G with $\Delta \geq 4$.
OUTPUT: A 2DVDE $(\Delta+2)$-coloring of G.

Begin

1. Choose a Δ-vertex u_1 as the root to run the OBFT algorithm.
2. Coloring $E(u_1)$ with the colors in $C = \{1, 2, \ldots, \Delta\}$.
3. Coloring $E(u_j) \setminus \{u_j u_1\}$, for $j = 2, 3, \ldots, \Delta+1$, from left to right as follows.
   - If j is even, use the first $d_G(u_j) - 1$ colors in the set $\{\Delta+1, j, j+1, \ldots, \Delta, 1, 2, \ldots, j-2\}$,
   - Else j is odd, use the first $d_G(u_j) - 1$ colors in the set $\{\Delta+2, j, j+1, \ldots, \Delta, 1, 2, \ldots, j-2\}$.
4. From top to bottom and from left to right, for each vertex u_j with $j \geq \Delta+2$, color the edges in $E(u_j) \setminus (E(u_1) \cup E(u_2) \cup \cdots \cup E(u_{j-1}))$ according to Theorem 2.

End

Theorem 3. *Let G be a connected bipartite outerplanar graph with $n \geq 2$ vertices. The algorithm* **2DVDE-Color Bipartite Outerplanar Graphs** *runs in $O(n^3)$ time.*

Proof. First, it requires $O(n)$ time to run OBFT to get a spanning tree T (cf. [8]). Next, the algorithm performed in Theorem 2 is iterated n times. At each iteration, we legally color the edge subset $E_j^* := E(u_j) \setminus E_{j-1}$. The time complexity to consider in this part includes performing the following four tasks.

(i) Compute the total number τ_1 of forbidden colors for the edges in E_j^*.
(ii) Compute the total number τ_2 of full vertices that are in conflict with at most two fixed vertices.
(iii) Select a subset C' of available colors from C.
(iv) Color E_j^* properly with color set C'.

For (i), it follows from the proof of Theorem 2 that $\tau_1 \leq 3|E(u_j)| \leq 3\Delta$. For (ii), Claims 1 to 3 showed that each vertex x is in conflict with at most Δ full vertices in G_{j-1}. Thus, $\tau_2 \leq 2\Delta$. For (iii), we need to eliminate at most Δ color subsets from C, whereas every color subset consists of at most Δ elements. For (iv), at most Δ edges are required for a legal coloring. The above analysis shows that the total running time of our algorithm is at most $n(3\Delta + 2\Delta + \Delta^2 + \Delta) = O(n\Delta^2)$. Since $\Delta \leq n - 1$, our algorithm runs in $O(n^3)$ time. □

References

1. Akbari, S., Bidkhori, H., Nosrati, N.: r-Strong edge colorings of graphs. Discrete Math. **306**, 3005–3010 (2006)
2. Balister, P.N., Győri, E., Lehel, J., Schelp, R.H.: Adjacent vertex distinguishing edge-colorings. SIAM J. Discrete Math. **21**, 237–250 (2007)
3. Bazgan, C., Harkat-Benhamdine, A.H., Li, H., Woźniak, M.: On the vertex-distinguishing proper edge-colorings of graphs. J. Combin. Theory Ser. B **75**, 288–301 (1999)
4. Burris A.C.: Vertex-distinguishing edge-colorings. Ph.D. Dissertation, Memphis State University (1993)
5. Burris, A.C., Schelp, R.H.: Vertex-distinguishing proper edge-colorings. J. Graph Theory **26**, 73–82 (1997)
6. Calamoneri, T., Petreschi, R.: $L(h, 1)$-labeling subclasses of planar graphs. J. Parallel Distrib. Comput. **64**, 414–426 (2004)
7. Chartrand, G., Harary, F.: Planar permutation graphs. Ann. Inst. H. Poincaré Sect. B (N. S.) **3**, 433–438 (1967)
8. Cormen, T.H., Leiserson, C.E., Rivest, R.L., Stein, C.: Introduction to Algorithms, 3rd edn. The MIT Press, Cambridge (2009)
9. Hatami, H.: Δ+300 is a bound on the the adjacent vertex distinguishing edge chromatic number. J. Combin. Theory Ser. B **95**, 246–256 (2005)
10. Horňák, M., Huang, D., Wang, W.: On neighbor-distinguishing index of planar graphs. J. Graph Theory **76**, 262–278 (2014)
11. Wang, W., Wang, Y., Huang, D., Wang, Y.: 2-Distance vertex-distinguishing edge coloring of graphs (submitted, 2015)

12. Wang, W., Yue, X., Zhu, X.: The surviving rate of an outerplanar graph for the firefighter problem. Theoret. Comput. Sci. **412**, 913–921 (2011)
13. Wang, Y., Wang, W., Huo, J.: Some bounds on the neighbor-distinguishing index of graphs. Discrete Math. **338**, 2006–2013 (2015)
14. Zhang, Z., Liu, L., Wang, J.: Adjacent strong edge coloring of graphs. Appl. Math. Lett. **15**, 623–626 (2002)
15. Zhang, Z., Li, J., Chen, X., Cheng, H., Yao, B.: $D(\beta)$-vertex-distinguishing proper edge-coloring of graphs. Acta Math. Sin. (Chin. Ser.) **49**, 703–708 (2006)

Maximum Independent Set on B_1-VPG Graphs

Abhiruk Lahiri[1], Joydeep Mukherjee[2(✉)], and C.R. Subramanian[2]

[1] Department of Computer Science and Automation, Indian Institute of Science, Bangalore, India
abhiruk.lahiri@csa.iisc.ernet.in
[2] Theoretical Computer Science, The Institute of Mathematical Sciences, Chennai, India
{joydeepm,crs}@imsc.res.in

Abstract. We present two approximation algorithms for the maximum independent set (MIS) problem over the class of B_1-VPG graphs and also for the subclass, equilateral B_1-VPG graphs. The first algorithm is shown to have an approximation guarantee of $O((\log n)^2)$ whereas the second one is shown to have an approximation guarantee of $O(\log d)$ where d denotes the ratio d_{max}/d_{min} and d_{max} and d_{min} denote respectively the maximum and minimum length of of any arm in the input L-representation of the graph. No approximation algorithms have been known for the MIS problem for these graph classes before. Also, the NP-completeness of the decision version restricted to unit length equilateral B_1-VPG graphs is established.

1 Introduction

The problem of computing a maximum independent set (*MIS*) in an arbitrary graph is notoriously hard, even if we aim only for a good approximation to an optimum solution. It is known that, for every fixed $\epsilon > 0$, MIS cannot be approximated within a multiplicative factor of $n^{1-\epsilon}$ for a general graph, unless $NP = ZPP$ [Hås97]. Throughout, n stands for the number of vertices in the input graph. Naturally, there have been algorithmic studies of this problem on special classes of graphs like: (*i*) efficient and exact algorithms for perfect graphs, (*ii*) linear time exact algorithms for chordal graphs and interval graphs, (*iii*) $O(n^2)$ time exact algorithms for comparability and co-comparability graphs, (*iv*) PTAS's (polynomial time approximation schemes) for planar graphs [Bak94] and unit disk graphs [HMR+98], (*v*) efficient $(\frac{k}{2} + \epsilon)$-approximation algorithms for $(k + 1)$ claw-free graphs [Hal95].

MIS has also been studied on classes formed by intersection graphs of geometric objects. These classes are interesting not only from an applications point of view but also from a purely algorithmic point of view. One such graph class is denoted by B_1-VPG. Before describing the class B_1-VPG, we give a brief introduction to the class VPG of graphs.

Vertex intersection graphs of Paths on Grid (or, in short, VPG graphs) was first introduced by Golumbic et al. [ACG+12]. For a member of this class of

Z. Lu et al. (Eds.): COCOA 2015, LNCS 9486, pp. 633–646, 2015.
DOI: 10.1007/978-3-319-26626-8_46

graphs, its vertices represent paths joining grid-points on a rectangular grid and two such vertices are adjacent if and only if the corresponding paths intersect. The study of MIS for this class of graphs is motivated by a corresponding problem in VLSI circuit design. Since each path on a grid corresponds to wires in a VLSI circuit and since intersection between wires is to be minimized, the MIS problem represents a finding a large collection of mutually non-intersecting paths on the grid. We embed paths in an independent set in one layer of the circuit.

We denote a VPG representation of a graph $G = (V, E)$ by $\{P_v\}_{v \in V(G)}$, where P_v is the path corresponding to vertex v. Also, let $|V| = n$. A *bend* is a right-angle turn of a path at some grid point. The *bend-number* ($b(G)$) of a graph G is the minimum integer k such that G has a VPG representation where none of the paths has more than k bends. "B_k-VPG graphs" denotes the collection of VPG graphs each of which has bend-number k. In particular, B_1-VPG graphs denotes the class of intersection graphs of paths on a grid where each path has one of the following shapes: $\llcorner$, $\ulcorner$, $\urcorner$ and $\lrcorner$. By an "arm" of a $\llcorner$, we mean either a horizontal or a vertical line segment associated with $\llcorner$. We often refer to a B_1-VPG graph, representable with only paths of type $\llcorner$ as an L-graph. An L-graph in vertices are represented by $\llcorner$-paths each having equal length arms, is called an *equilateral L-graph.* It is possible that two vertices correspond to paths of different arm lengths. We often refer to a L-shape as a'l' for the sake of brevity.

A number of mathematical results on VPG graphs have been obtained recently [ACG+12, CU13, BD15, FKMU14, CGTW15, CKU13, CJKV12, CCS11]. Relationships between other known graph classes and VPG graphs have also been studied in [ACG+12]. In [CU13], it has been shown that planar graphs form a subset of B_2-VPG graphs. Recently, this result has been further tightened by Therese Biedl and Martin Derka. They have shown that planar graphs form a subset of 1-string B_2-VPG graphs [BD15] which is a subclass of B_2-VPG graphs. In [FKMU14], authors have shown that any full subdivision of any planar graph is an L-graph. By a full subdivision of a graph G, we mean a graph H obtained by replacing every edge of G by a path of length two or more with every newly added vertex being part of exactly one path. They have also shown that every co-planar graph (complement of a planar graph) is a B_{19}-VPG graph. A relationship between poset dimension and VPG bend-number has also been obtained in [CGTW15]. Contact representation of L-graphs has been studied in [CKU13]. In this work, the authors have studied the problems of characterizing and recognizing contact L-graphs and have also shown that every contact L-representation has an equivalent equilateral contact L-representation. By a *contact* L-representation, we mean a more restricted intersection, namely, that two vertices are adjacent if and only if they just touch. Recognizing VPG graphs is shown to be NP-complete in [CJKV12]. In the same work, it is also shown that recognizing if a given B_{k+1}-VPG graph is a B_k-VPG graph is NP-complete even if we are given a B_{k+1}-VPG representation of the input. The recognition problem has also been looked at for some subclasses of B_0-VPG graphs in [CCS11].

VPG graphs are a special type of *string graphs*, which are intersection graphs of curves in the plane [ACG+12]. The decision version of maximum independent set problem on string graphs is known to be NP-complete. It follows from the fact that planar graph forms a subclass of string graphs [CGO10] and it is well known that decision version of MIS is NP-complete over planar graphs [GJ77]. Also, PTAS (polynomial time approximation schemes) have been obtained for planar graphs [Bak94]. The best known algorithm for MIS on string graphs has an approximation factor n^{ϵ}, for some $\epsilon > 0$ [FP11]. In this context, it would be interesting to know if there are subclasses of string graphs (other than planar graphs) where better approximation guarantees have been obtained.

Our Results: In this paper, we present new approximation algorithms for the class of B_1-VPG graphs which is a sub-class of string graphs. We also present new approximation results over equivaleteral B_1-VPG graphs. To the best of our knowledge, no previous work on approximately solving MIS on B_1-VPG graphs has been published. Precisely, we obtain the following results.

Theorem 1. *There exists an efficient $O((\log n)^2)$-approximation algorithm for MIS restricted to B_1-VPG graphs.*

Theorem 2. *There exists an efficient $O(\log d)$-approximation algorithm for MIS restricted to equilateral B_1-VPG graphs. Here, d denotes the ratio d_{max}/d_{min} where d_{max} and d_{min} denote respectively the maximum and minimum length of any arm of any L-shape in the input instance.*

As a consequence, we obtain an $O(1)$-approximation algorithm for those equilateral B_1-VPG graphs with a bounded value of d. We assume that the input graph is presented as a set of ls each l is specified as a 4-tuple as described in Sect. 2. We have used combinatorial techniques on the L-representation of B_1-VPG graphs to obtain the approximations. For B_1-VPG graphs, we employ the divide and conquer paradigm by splitting the given input instance into few smaller sub-instances and recurse on them. The recursion stops when the input is either a graph of constant size or if it forms a special type of graph. The latter case corresponds to a co-comparability graph and hence can be solved optimally in polynomial time. For equilateral B_1-VPG graphs, we reduce the given instance into solving the problem on few sub-instances and then compute the MIS for each of those sub-instances and report the one with maximum size. Here, we use a combinatorial characterization of the sub-instances which helps us to solve MIS optimally for the sub-instances.

We introduce the notations in Sect. 2. In Sect. 3, we present the approximation algorithm for B_1-VPG graphs. In Sect. 4, we present the approximation algorithm for equilateral B_1-VPG graphs and analyse it in Sect. 5. NP-Completeness proof for the decision version of MIS over unit B_1-VPG graphs is provided in Sect. 6. Finally, we conclude with some remarks in Sect. 7.

2 Preliminaries

We work with geometric objects in the shape of "L" and the three shapes obtained by rotating "L" by 90, 180 and 270 degrees in the clockwise direction around the common point on the two arms of L. Thus we get four distinct shapes. For ease of further discussion, we refer to them as follows. L_1 refers to the shape "L", L_2 refers to the shape when "L" is rotated by 90 degrees clockwise, L_3 refers to the shape when "L" is rotated by 180 degrees clockwise and L_4 refers to the shape when "L" is rotated by 270 degrees clockwise. Henceforth, we use l to denote a geometric object with one of the four shapes L_1, L_2, L_3 and L_4. Initially, we confine our discussion only to ls of shape L_1. The other shapes can be treated similarly. The intersection point of the two sides of an l is defined as the *corner* of the l and is denoted by c_l, the tip of the horizontal arm is denoted by h_l and that of the vertical arm is denoted by v_l. For an object l, we use (cx, cy, hx, vy) to denote respectively the x- and y- coordinates of c_l, the x-coordinate of h_l and the y- coordinate of v_l. This 4-tuple completely describes l. The set of points constituting l is denoted by P_l and is given by

$$P_l = \{(x, cy) \ : \ cx \leq x \leq hx\} \cup \{(cx, y) \ : \ cy \leq y \leq vy\}.$$

We say that two distinct objects l_1 and l_2 intersect if $P_{l_1} \cap P_{l_2} \neq \emptyset$. l_1 and l_2 are said to be *independent* if and only if they do not intersect. A set of l's such that no two of them forms an intersecting pair is said to be an *independent* set. Suppose two objects l_1 and l_2 are such that $l_1.cx < l_2.cx$ and $l_1.cy < l_2.cy$. Then we say that $c_{l_1} < c_{l_2}$. When the length of vertical side of an l is equal to the horizontal side of an l we say that it is equilateral. Since for equilateral l's the length of the horizontal side is equal to that of the vertical side, we simply use $le(l)$ to denote the length of the horizontal side as well as the vertical side. All logarithms used below are with respect to base 2. We denote a set $\{1, 2, \ldots, n\}$ by $[n]$.

3 Approximation for B_1-VPG

MAXIMUM INDEPENDENT SET IN B_1- VPG
Input: A set $\mathcal{S}$ of l's
Output: a set $I \subset \mathcal{S}$ such that I is independent and $\lvert I \rvert$ is maximized.

The decision version of this problem is NP-complete (see Theorem 5). Below, we present approximation algorithms for this problem.

Define $S_{L_1} = \{l \in \mathcal{S} \mid l \text{ is of type } L_1\}$. Similarly, we define $S_{L_2}, S_{L_3}, S_{L_4}$. We reduce approximately solving for S to approximately solving for each of S_{L_i}, $i = 1, 2, 3, 4$. By symmetry, solving for any of these sub-instances is similar to solving for S_{L_1}. Hence, we focus on approximately solving for inputs S where each l is of type L_1. Below, we present an approximation algorithm with this assumption. Before proceeding further, we introduce an assumption which is stated in the following claim and which can be justified easily.

Claim. Without loss of generality, we can assume that

(i) $l_1.cx \neq l_2.cx$ and $l_1.cy \neq l_2.cy$ for any pair of distinct $l_1, l_2 \in S$ provided l_1, l_2 are of same type;
(ii) the number $n = |S|$ of objects is even.

Our approach is broadly divide and conquer. We sort the objects in S in increasing order of their cx values. Define x_{med} to be the middle point between the cx values of the $n/2$-th object and the $(n/2)+1$-th object in this order. Then, we compute the sets S_1, S_2 and S_{12} defined as follows.

$S_1 := \{l \in S \ : \ l.hx < x_{med}\}$.
$S_2 := \{l \in S \ : \ l.cx > x_{med}\}$.
$S_{12} := \{l \in S \ : \ l.cx \leq x_{med} \leq l.hx\}$.

The sets S_1, S_2 and S_{12} form a partition of S. Also, any pair of $l_1 \in S_1, l_2 \in S_2$ are independent. The problem is solved by applying the recursive Algorithm *IndSet*1. This algorithm (on input S) computes the partition $S = S_1 \cup S_2 \cup S_{12}$. Then, it recursively computes an approximately optimal solution for each of S_1 and S_2 and computes their disjoint union. This is one candidate approximate solution. Then, it computes an approximate solution to the instance with S_{12} as its input using Algorithm *IndSet*2, which is also a recursive procedure. This is another candidate approximate solution. *IndSet*1 then compares the two candidate solutions and outputs the one of larger size.

Now we give an outline of how Algorithm *IndSet*2 works. This algorithm (on input T satisfying the required assumption) computes a partition $T = T_1 \cup T_2 \cup T_{12}$ defined as before for S_1, S_2, S_{12} except that we use the cy values of the l's in T for calculating the median and the sets and for partitioning it into T_1, T_2, T_{12} we use vy values. It is shown (in Lemma 1) that the intersection graph of T_{12} is a co-comparability graph and hence a maximum independent set can be computed efficiently. Approximate independent sets are computed recursively for each of the two sub-instances specified by T_1 and T_2 and their disjoint union is also computed. As before, we compare the two candidate solutions and output the better one.

Algorithm 1. IndSet1

Require: A non-empty set S of l's of type L_1.
if $|S| \leq 3$ **then**
  return Compute and return a maximum independent set I_S of S
else
  Compute x_{med} and also the partition $S = S_1 \cup S_2 \cup S_{12}$.
  Compute $IndSet1(S_1) \cup IndSet1(S_2)$ and also $IndSet2(S_{12})$.
  Return I_S defined as the larger of the two sets computed before.
end if

We state and prove the following lemma in terms of l's of type L_1. But it holds for l's of each of the other three types as well.

Lemma 1. *Suppose S' is a set of l's, each being of type L_1. Suppose there exist a horizontal line $y = b$ and a vertical line $x = a$ such that each $l \in S'$ intersects both $y = b$ and $x = a$. Then, the intersection graph of members of S' is a co-comparability graph.*

Algorithm 2. IndSet2

Require: A non-empty set Y of l's of type L_1 satisfying: for some vertical line $x = a$, each member of T intersects $x = a$.

```
if |Y| ≤ 3 then
    return Compute and return a maximum independent set I_Y of Y.
else
    Compute y_med and also the partition Y = Y_1 ∪ Y_2 ∪ Y_12.
    Compute J_union = IndSet2(Y_1) ∪ IndSet2(Y_2) and also
    Compute a maximum independent set J*_12 of Y_12.
    Return J_Y defined as the larger of the two sets computed before.
end if
```

Proof. We begin with the following claim.

Claim. A pair $l_1, l_2 \in S'$ is independent if and only if $c_{l_1} < c_{l_2}$ or vice versa.

Proof. (of Claim) It is easy to see that if either $c_{l_1} < c_{l_2}$ or $c_{l_2} < c_{l_1}$, then l_1 and l_2 are independent. To prove the converse: Assume that l_1 and l_2 are independent. We also know that $l_1.cx \neq l_2.cx$ and $l_1.cy \neq l_2.cy$ by Claim 3. Suppose that neither $c_{l_1} < c_{l_2}$ holds nor $c_{l_2} < c_{l_1}$ holds. As a consequence, we have one of the following two scenarios: (1) $l_1.cx < l_2.cx$ and $l_1.cy > l_2.cy$ **or** (2) $l_1.cx > l_2.cx$ and $l_1.cy < l_2.cy$. For Case (1), we have $(l_2.cx, l_1.cy) \in P_{l_1} \cap P_{l_2}$. For Case (2), we have $(l_1.cx, l_2.cy) \in P_{l_1} \cap P_{l_2}$. In both cases, we have used our assumption that both l_1 and l_2 intersect the lines $y = b$ and $x = a$. In either case, l_1 and l_2 intersect and hence are not independent, a contradiction to our assumption. □

Consider the complement of the intersection graph formed by members of S'. Its vertices are members of S' and there is an edge between two members if and only if they do not intersect. We denote this graph by G^C. We orient each edge (l_1, l_2) as follows: it is oriented as $l_1 \rightarrow l_2$ if $c_{l_1} < c_{l_2}$ and as $l_2 \rightarrow l_1$ otherwise. To prove that G^C is a co-comparability graph, it suffices to show that $\forall l_i, l_j, l_k \in S'$, we have $l_i \rightarrow l_j, l_j \rightarrow l_k \Rightarrow l_i \rightarrow l_k$. But by the above claim $l_i \rightarrow l_j \Rightarrow c_{l_i} < c_{l_j}$ and $l_j \rightarrow l_k \Rightarrow c_{l_j} < c_{l_k}$. It then follows that $c_{l_i} < c_{l_k}$. This implies that (l_i, l_k) is oriented as $l_i \rightarrow l_k$ by the above claim. This establishes the transitivity of the orientation and hence G is a co-comparability graph. This completes the proof of Lemma 1. □

4 Analysis of IndSet1 and IndSet2

Denote by I^* any maximum independent set of S. Similarly, denote by I_1^*, I_2^* and I_{12}^* any maximum independent set of S_1, S_2 and S_{12} respectively. Denote by I, I_1, I_2 and I_{12} the independent set produced by $IndSet1$ when provided with S, S_1, S_2 and S_{12} as input respectively.

Lemma 2. $|I_{12}| \geq \frac{|I_{12}^*|}{\log |S_{12}|}$.

Proof. We use Y to denote the set S_{12}. Let $|Y| = m$. Let Y_1, Y_2, Y_{12} denote the partition of Y computed in Step 4 of $IndSet(S_{12})$. It follows that $|Y_1| \leq \frac{m}{2}$, $|Y_2| \leq \frac{m}{2}$ and $|Y_{12}| \leq \frac{m}{2}$ by our assumption stated in (i) of Claim 3. We prove the lemma by induction on m.

The base case is when $|Y| \leq 3$ or when $Y = Y_{12}$. For this case, we can solve the instance optimally since $|Y|$ is either small or its intersection graph is a co-comparability graph. This takes care of the base case.

Let J_1^*, J_2^* and , J_{12}^* denote respectively a maximum independent set of Y_1, Y_2 and Y_{12}. Let J_1, J_2 and J_{12} denote respectively the solutions returned by $IndSet2$ when the input is Y_1, Y_2 and Y_{12}. Since Y_{12} induces a co-comparability intersection graph, we have $|J_{12}| = |J_{12}^*|$. Recall that I_{12}^* denote the maximum independent set of S_{12}. By induction, $|J_1| \geq \frac{|J_1^*|}{\log(m/2)} \geq \frac{|I_{12}^* \cap Y_1|}{\log m - 1}$, $|J_2| \geq \frac{|I_{12}^* \cap Y_2|}{\log m - 1}$. Thus,

$$\begin{aligned} I_{12} &= \max\{|J_{12}|, |J_1| + |J_2|\} \\ &\geq \max\{|I_{12}^* \cap Y_{12}|, \frac{|I_{12}^* \cap Y_1| + |I_{12}^* \cap Y_2|}{\log m - 1}\} \\ &\geq \max\{|I_{12}^* \cap Y_{12}|, \frac{|I_{12}^*| - |I_{12}^* \cap Y_{12}|}{\log m - 1}\} \end{aligned}$$

If $|I_{12}^* \cap Y_{12}| \geq \frac{|I_{12}^*|}{\log |S_{12}|}$ we are done. Otherwise,

$$\frac{|I_{12}^*| - |I_{12}^* \cap Y_{12}|}{\log m - 1} \geq \frac{|I_{12}^*| - |I_{12}^*|/\log m}{\log m - 1} = \frac{|I_{12}^*|}{\log |S_{12}|}.$$

This establishes the induction step, thereby completing the inductive proof. □

Recall that $|S| = n$.

Lemma 3. $I \geq \frac{|I^*|}{\log^2 n}$.

Proof. Due to our assumption stated in (i) of Claim 3, we have

$$|S_1| \leq \frac{n}{2}, \ |S_2| \leq \frac{n}{2}, \ |S_{12}| \leq \frac{n}{2}.$$

Again the proof is based on induction on n. We have the following.

$$I_1 \geq \frac{|I_1^*|}{\log^2(n/2)} \geq \frac{|I^* \cap S_1|}{(\log n - 1)^2} \tag{1}$$

$$I_2 \geq \frac{|I_2^*|}{\log^2(n/2)} \geq \frac{|I^* \cap S_2|}{(\log n - 1)^2} \tag{2}$$

From Lemma 2, we have

$$I_{12} \geq \frac{I_{12}^*}{\log |S_{12}|} \geq \frac{|I^* \cap S_{12}|}{\log |S_{12}|} \tag{3}$$

$$\text{Also, } |I| = \max\{|I_{12}|, |I_1| + |I_2|\} \quad (4)$$
$$\geq \max\{\frac{|I^* \cap S_{12}|}{\log |S_{12}|}, \frac{|I^*| - |I^* \cap S_{12}|}{(\log n - 1)^2}\}$$

The last inequality follows from applying Inequalities (1), (2) and (3).

The base case corresponding to $n \leq 3$ follows, since we can find a maximum independent set in constant time.

For an arbitrary $n > 3$, the inductive argument is as follows: If $\frac{|I^* \cap S_{12}|}{\log |S_{12}|} \geq \frac{|I^*|}{\log^2 n}$, the the induction step is proved. Otherwise, we have $|I^* \cap S_{12}| < \frac{|I^*| \log |S_{12}|}{\log^2 n}$. Thus,

$$\frac{|I^*| - |I^* \cap S_{12}|}{(\log n - 1)^2} \geq \frac{|I^*| - \frac{|I^*| \log |S_{12}|}{\log^2 n}}{(\log n - 1)^2}$$
$$\geq \frac{|I^*| - \frac{|I^*|}{\log n}}{(\log n - 1)^2}$$
$$\geq \frac{|I^*|}{\log^2 n}$$

This proves the induction step for the case when $\frac{|I^* \cap S_{12}|}{\log |S_{12}|} < \frac{|I^*|}{\log^2 n}$. Hence the proof. □

Lemma 3 obtains an upper bound of $(\log^2 n)$ on the approximation factor of $IndSet1$ assuming that each member of S is of type L_1, where n denotes the size of S. By symmetry, we get $(\log |S_{L_2}|)^2$, $(\log |S_{L_3}|)^2$, $(\log |S_{L_4}|)^2$ approximation for the sets $S_{L_2}, S_{L_3}, S_{L_4}$. Thus given a set $\mathcal{S}$ we compute I_{L_1}, I_{L_2}, I_{L_3} and I_{L_4} where the above sets denote the independent sets computed by the $IndSet1$ for sets $S_{L_1}, S_{L_2}, S_{L_3}, S_{L_4}$ respectively. Then we return the one with maximum cardinality among I_{L_1}, I_{L_2}, I_{L_3} and I_{L_4}. This leads us to the following theorem on approximating a maximum independent set in $\mathcal{S}$.

Theorem 3. *There exists an efficient algorithm which, given a set of $\mathcal{S}$ of l's such that $|S| = n$, outputs an independent set of size at least $\frac{1}{4(\log^2 n)}$ of that of an optimum solution of $\mathcal{S}$.*

4.1 Analysis of Running Time

Let $s(m)$ denote the running time of $IndSet2(Y)$ on an input Y of size m. We have $s(m) = O(1)$ if $m \leq 3$. If Y induces a co-comparability graph, then $s(m) = O(m^2)$. Otherwise, $s(m) \leq 2s(m/2) + O(m^2)$. Unravelling the recursion, we deduce that $s(m) = O(m^2(\log m))$.

Let $t(n)$ denote the running time of $IndSet1(S)$ on an input S of size n. We have $t(n) = O(1)$ if $n \leq 3$. Otherwise, $t(n) \leq 2t(n/2) + s(n/2) \leq 2t(n/2) + O(n^2(\log n))$. Unravelling the recursion, we deduce that $t(n) = O(n^2(\log n)^2)$. Thus, $IndSet1(S)$ runs in time $O(n^2(\log n)^2)$ on an input of size n.

5 Approximation for Equilateral B_1-VPG:

MAXIMUM INDEPENDENT SET IN EQUILATERAL B_1-VPG
Input: A set S of equilateral l's such that $le(l) \in [2, d]\ \forall l \in S$.
Output: An independent set $I \subseteq S$ such that $\|I\|$ is maximized.

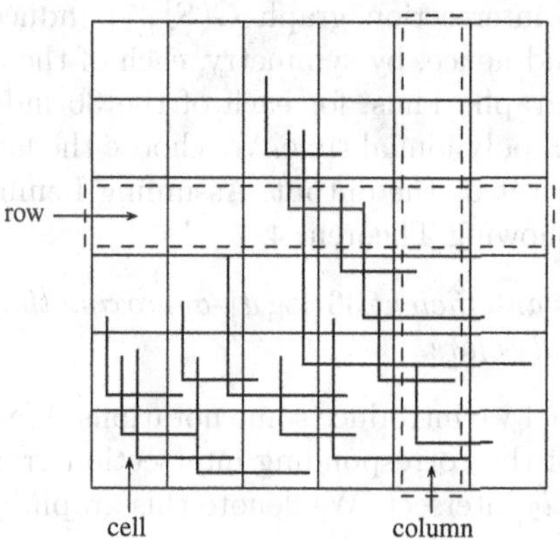

Fig. 1. The grid is for l's of type 1 whose length varies within the range 2^i to 2^{i+1}

We call the above problem as *MISL*. We call an equilateral l a unit l if $le(l) = 1$. In Theorem 5, we establish that the decision version of MISL restricted to unit ls (and denoted by $MIS1$) is NP-Complete. As a consequence, it follows that the decision version of MISL is also NP-complete. In rest of this section, we present a new approximate algorithm for MISL. Before that, we present a claim which can be justified easily.

Claim. Without loss of generality, assume that the input to MISL satisfies $d_{min} = 2$.

Proof. (sketch:) Rescale the the coordinates of x-axis and y-axis by stretching both of them by a multiplicative factor of $2/d_{min}$. □

The algorithm begins by dividing the input set S into disjoint sets $S_1, S_2, \ldots, S_{\lfloor \log d \rfloor}$ where $S_i = \{l \in S \mid 2^i \leq le(l) < 2^{i+1}\}$, $\forall i \in [\lfloor \log d \rfloor]$. This split is to exploit the fact that $d \leq 2$ when the input is restricted to only members of S_i, for any i. For the ith set, we do the following. We place a sufficiently large but finite grid structure on the plane covering all members of S_i. The grid is chosen in such a way so that grid-length in each of the x and y directions is 2^i. What we get is a rectangular array of square boxes of side length 2^i each. We number the rows of boxes from bottom and the columns of boxes from left.

We denote a *box* by (r', c') if it is in the intersection of r'^{th} row and c'^{th} column. We say l is inside a box if its corner c_l lies inside the box, but not on its top horizontal edge or its right vertical edge. If l lies inside a box (r', c') we denote it by $l \in (r', c')$.

For every $m \in [4], k_r, k_c \in [3]$, define

$$S^m_{i,k_r,k_c} = \{l \in (r', c') \mid r' = k_r \mod 3, c' = k_c \mod 3, \text{ l is of type m}\}.$$

Here, for purposes of simplicity, we use 3 in place of 0 in (mod 3) arithmetic.

Thus, we partition input S into $36(\lfloor \log d \rfloor)$ subsets S^m_{i,k_r,k_c}. In Lemma 4, we establish that the intersection graph $G(S^1_{i,1,1})$ induced by $S^1_{i,1,1}$ is a co-comparability graph and hence, by symmetry, each of the 36 induced subgraphs is a co-comparability graph. Thus, for each of the 36 induced subgraphs, MIS can be solved exactly in polynomial time. We choose the largest of these 36 independent sets and return it as the output. Assuming Lemma 4 (which we prove below), we have the following Theorem 4.

Theorem 4. *There is an efficient* $36\lfloor \log d \rfloor$*-approximation for MISL where* $d = d_{max}/d_{min}$ *is as defined before.*

For proving Lemma 4 we introduce some notations. We consider the set $S^1_{i,1,1}$ and the complement of the corresponding intersection graph. We draw an edge between l_1, l_2 if l_1 and l_2 intersect. We denote this graph by $G(S^1_{i,1,1})$. Below we prove the following lemma.

Lemma 4. $G(S^1_{i,1,1})^C$ *is a comparability graph.*

Proof. Note that all members of $s^1_{i,1,1}$ lie in boxes which are in the intersection of rows and columns both numbered from $\{1, 4, 7, \ldots\}$. We prove the claim by showing that there exist a transitive orientation of the edges of this graph. We describe the orientation in two steps. First, we orient those edges which connect two l's whose corner lie in the same box. In the second step, we orient those edges which connect two l's located in two different boxes. For the first step, we employ the following claim which is an immediate consequence of Lemma 1.

Claim. 5 Suppose l_1 and l_2 are two members such that $|l_1.cx - l_2.cx| \leq 2^i$ and $|l_1.cy - l_2.cy| \leq 2^i$. Then, l_1, l_2 are independent if and only if $c_{l_1} < c_{l_2}$ or vice versa.

Orientation: Let l_1 and l_2 be two arbitrary members of $S^1_{i,1,1}$ joined by an edge in $G(S^1_{i,1,1})^C$.

(i) If l_1 and l_2 are lying in a common box, we employ Claim 5 and orient it from l_1 to l_2 if $c_{l_1} < c_{l_2}$ and from l_2 to l_1 otherwise.

(ii) Suppose l_1 and l_2 lie in different boxes in the same row and let $l_1.cx < l_2.cx$ without loss of generality. We orient the edge from l_1 to l_2.

(iii) Suppose l_1 and l_2 lie in different rows and let $l_1.cy < l_2.cy$ without loss of generality. We orient the edge from l_1 to l_2.

If the orientation of an edge (l_1, l_2) is from l_1 to l_2, we denote it by $\overrightarrow{(l_1, l_2)}$.

We prove that this orientation is transitive. We prove it by performing a case analysis. For an edge $\overrightarrow{(l_1, l_2)}$, we call it *h-oriented* if vertices l_1 and l_2 lie in the same row and we call it *v-oriented* if l_1 lies in a row which is below the row in which l_2 is present. We denote by "case h,v", the case of 3 vertices l_1, l_2, l_3 such that (l_1, l_2) is h-oriented, (l_2, l_3) is v-oriented. Then we prove that there exist an edge $\overrightarrow{(l_1, l_3)}$. Similarly, the other cases "h,h", "v,v" and "v,h" are defined and handled.

Case h, h: In this case we have three sub-cases. They are (1) l_1, l_2, l_3 are in the same box, (2) Two of the three vertices are in the same box different from the box of the other, (3) All three are in different boxes.

First, we handle the sub-case (1). l_1, l_2 are in the same box and they are independent. In view of Claim 5, this implies that $c_{l_1} < c_{l_2}$. Similarly, we infer that $c_{l_2} < c_{l_3}$. Hence, it follows that $c_{l_1} < c_{l_3}$ and hence (l_1, l_3) is oriented from l_1 to l_3. Thus it is transitive.

Now, for the sub-case (2): either l_1, l_2 will be in the same box or l_2, l_3 will be in the same box. In both the cases l_1, l_3 will be in different boxes. Since any two points in different boxes of the same row differ in their x-coordinates by at least 2^{i+1}, the edge (l_1, l_3) exists and is directed l_1 to l_3, thereby proving the required transitivity.

The sub-case (3): Since l's lie in different boxes in the same row, $l_3.cx - l_1.cx \geq 2^{i+2}$ and hence the edge (l_1, l_3) exists and is directed l_1 to l_3, thereby proving the required transitivity. This completes the proof of Case h, h.

Case h, v: Since l_1 and l_3 are in different rows, we have $l_3.cy - l_1.cy \geq 2^{i+1}$, the edge (l_1, l_3) exists and is directed l_1 to l_3, thereby proving the required transitivity.

Case v, h: In this case l_3 is in a box above that of l_1 by our hypothesis. As before, $l_3.cy - l_1.cy \geq 2^{i+1}$ and hence the edge (l_1, l_3) exists and is directed l_1 to l_3, thereby proving the required transitivity.

Case v, v: By our hypothesis l_3 is in a box above that of l_1. Hence, $l_3.cy - l_1.cy \geq 2^{i+2}$ and transitivity is established. □

6 Hardness of MIS on Unit L-Graphs

Theorem 5. *The decision version of* MAXIMUM INDEPENDENT SET *on unit L-graphs is NP-complete.*

Proof. By *MIS*, we mean the decision version of the Maximum Independent Set problem on a generic class graphs. It is known that MIS is NP-complete [GJ79] on planar graphs with maximum degree four. We exhibit a polynomial time reduction between this problem and the MIS problem on the class of unit L-graphs. The reduction is as follows. Given a planar $G = (V, E)$ (with maximum degree four), we construct a unit L-graph $G' = (V', E')$ (of polynomial size) from G such that for any k, G has an independent set of size k if and only if G' has an independent set of size $k + f(G')$ for some polynomial time computable quantity $f(G')$. This establishes our claim. Our proof is motivated by [KN90].

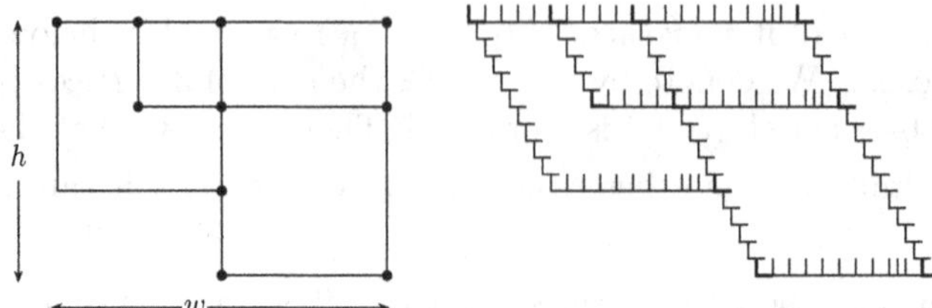

Fig. 2. Planar graph with maximum degree four and its unit L VPG representation.

It is known that every planar graph of degree at most four can be drawn on a grid of linear size such that the vertices are mapped to points of the grid and the edges to piecewise linear curves made up of horizontal and vertical line segments whose endpoints are also points of the grid [Sch90]. It is reasonable to assume that a path between two vertices of G, if exists, use horizontal and vertical segments which has length more than one on the grid (otherwise it is possible to consider fine enough grid such that this property holds). Let $R(w, h)$ be the rectangular grid where the graph G has been drawn. We denote the width and height of the grid by w and h respectively. Let us consider $\delta = 1/2h$. Now for each vertex of the graph G, draw an unit length L whose corner point has co-ordinates $(x - \delta y, y)$, where in the grid $R(w, h)$ the vertex is positioned at (x, y). Let P_e be the path on the grid corresponding to edge e. Also let $|P_e|$ denote the number of intermediate grid vertices on the path P_e. Now for every path P_e, where $e = (u, v) \in E(G)$, if $|P_e|$ is even then for every intermediate grid vertex (x, y) on the path P_e draw a unit length L whose corner lie on $(x - \delta y, y)$. If $|P_e|$ is odd then for every intermediate grid vertex (x, y) except last one on the path P_e draw an unit length L whose corner lie on $(x - \delta y, y)$. If the last intermediate grid vertex (x, y) on the path is on a vertical segment of P_e then draw two L's as follows one L has corner point at $(x - \delta y, y - \epsilon)$ and other L has corner at $(x - \delta y, y)$ where $\epsilon > 0$ is a small number. If the last intermediate grid vertex (x, y) on the path is on a horizontal segment of P_e then draw two L's as follows one L has corner point at $(x - \delta y, y)$ and other L has corner at $(x - \delta y - \epsilon, y)$ where $\epsilon > 0$ is a small number. We denote this graph as G'. From the construction it is clear that it is an intersection graph of unit L' s.

Clearly G' is obtained from G by subdividing every edge $e \in E$ with an even number of new vertices. We refer to this fact by saying that G' is an (*even subdivision*) of G. Let us denote the set of new vertices introduced to subdivide e by V_e. Clearly $V' = V \bigcup \cup_{e \in E(G)} V_e$. The correctness of the reduction follows from the following claim.

Claim. Let H denote a graph and H' be an even subdivision of H. Then, H has an independent set of size k if and only if H' has an independent set of size $k + \sum_{e \in E(H)} |V_e|/2$.

Proof. The *only if* part is easy to verify. We focus on proving the *if* part. Suppose H' has an independent set I of size $k + \sum_{e \in E(H)} |V_e|/2$. Notice that $|V_e|$ is even for each $e \in E(H)$. Also, any independent set of H' contains at most half of the vertices of V_e, for each $e \in E(H)$. Hence $|I - I \cap V(H)| \leq \sum_{e \in E(H)} |V_e|/2$. Removing all new vertices, we get an independent set of size at least k in H. Hence the claim. □

The above claim implies that $\alpha(G') = \alpha(G) + \sum_{e \in E(G)} |V_e|/2$. This completes the proof. □

7 Conclusion and Remarks

We presented new approximation algorithms for Maximum Independent Set problem on B_1-VPG graphs and for equilateral B_1-VPG graphs. This leads us to the following interesting questions. A natural problem is to improve the approximation guarantee ratio for both problems. Another interesting direction would be to obtain some conditional in-approximability results for MIS by obtaining lower bounds on the approximation ratios for these graph classes.

References

[ACG+12] Asinowski, A., Cohen, E., Golumbic, M.C., Limouzy, V., Lipshteyn, M., Stern, M.: Vertex intersection graphs of paths on a grid. J. Graph Algorithms Appl. **16**(2), 129–150 (2012)

[Bak94] Baker, B.S.: Approximation algorithms for NP-complete problems on planar graphs. J. ACM (JACM) **41**(1), 153–180 (1994)

[BD15] Biedl, T.C., Derka, M.: 1-string b_2-VPG representation of planar graphs. In: 31st International Symposium on Computational Geometry, SoCG 2015, 22–25 June 2015, Eindhoven, The Netherlands, pp. 141–155 (2015)

[CCS11] Chaplick, S., Cohen, E., Stacho, J.: Recognizing some subclasses of vertex intersection graphs of 0-bend paths in a grid. In: Kolman, P., Kratochvíl, J. (eds.) WG 2011. LNCS, vol. 6986, pp. 319–330. Springer, Heidelberg (2011)

[CGO10] Chalopin, J., Gonçalves, D., Ochem, P.: Planar graphs have 1-string representations. Discrete Comput. Geom. **43**(3), 626–647 (2010)

[CGTW15] Cohen, E., Golumbic, M.C., Trotter, W.T., Wang, R.: Posets and VPG graphs. Order, pp. 1–11 (2015)

[CJKV12] Chaplick, S., Jelínek, V., Kratochvíl, J., Vyskočil, T.: Bend-bounded path intersection graphs: sausages, noodles, and waffles on a grill. In: Golumbic, M.C., Stern, M., Levy, A., Morgenstern, G. (eds.) WG 2012. LNCS, vol. 7551, pp. 274–285. Springer, Heidelberg (2012)

[CKU13] Chaplick, S., Kobourov, S.G., Ueckerdt, T.: Equilateral L-contact graphs. In: Brandstädt, A., Jansen, K., Reischuk, R. (eds.) WG 2013. LNCS, vol. 8165, pp. 139–151. Springer, Heidelberg (2013)

[CU13] Chaplick, S., Ueckerdt, T.: Planar graphs as VPG-graphs. J. Graph Algorithms Appl. **17**(4), 475–494 (2013)

[FKMU14] Felsner, S., Knauer, K., Mertzios, G.B., Ueckerdt, T.: Intersection graphs of L-shapes and segments in the plane. In: Csuhaj-Varjú, E., Dietzfelbinger, M., Ésik, Z. (eds.) MFCS 2014, Part II. LNCS, vol. 8635, pp. 299–310. Springer, Heidelberg (2014)

[FP11] Fox, J., Pach, J.: Computing the independence number of intersection graphs. In: Proceedings of the Twenty-Second Annual ACM-SIAM Symposium on Discrete Algorithms, SODA 2011, 23–25 January 2011, San Francisco, California, USA, pp. 1161–1165 (2011)

[GJ77] Garey, M.R., Johnson, D.S.: The rectilinear steiner tree problem is NP-complete. SIAM J. Appl. Math. **32**(4), 826–834 (1977)

[GJ79] Garey, M.R., Johnson, D.S.: Computers and Intractability: A Guide to the Theory of NP-Completeness. A Series of Books in the Mathematical Sciences. W. H Freeman and Company, San Francisco (1979)

[Hal95] Halldórsson, M.M.: Approximating discrete collections via local improvements. In: SODA, vol. 95, pp. 160–169 (1995)

[Hås97] Håstad, J.: Clique is hard to approximate within $n^{1-\varepsilon}$. Acta Mathematica **182**, 105–142 (1997)

[HMR+98] Hunt, H.B., Marathe, M.V., Radhakrishnan, V., Ravi, S.S., Rosenkrantz, D.J., Stearns, R.E.: NC-approximation schemes for NP-and PSPACE-hard problems for geometric graphs. J. Algorithms **26**(2), 238–274 (1998)

[KN90] Kratochvíl, J., Nesetril, J.: INDEPENDENT SET and CLIQUE problems in intersection-defined classes of graphs. Commentationes Mathematicae Universitatis Carolinae **031**(1), 85–93 (1990)

[Sch90] Schnyder, W.: Embedding planar graphs on the grid. In: Proceedings of the First Annual ACM-SIAM Symposium on Discrete Algorithms, 22–24 January 1990, San Francisco, California, pp. 138–148 (1990)

Approximating the Restricted 1-Center in Graphs

Wei Ding[1(✉)] and Ke Qiu[2]

[1] Zhejiang University of Water Resources and Electric Power, Hangzhou 310018, Zhejiang, China
dingweicumt@163.com

[2] Department of Computer Science, Brock University, St. Catharines, Canada
kqiu@brocku.ca

Abstract. This paper studies the *restricted vertex 1-center problem* (RV1CP) and *restricted absolute 1-center problem* (RA1CP) in general undirected graphs with each edge having two weights, cost and delay. First, we devise a simple FPTAS for RV1CP with $O(mn^3(\frac{1}{\epsilon}+\log\log n))$ running time, based on FPTAS proposed by Lorenz and Raz (Oper. Res. Lett. 28(1999), 213–219) for computing end-to-end *restricted shortest path* (RSP). During the computation of the FPTAS for RV1CP, we derive a RSP distance matrix. Next, we discuss RA1CP in such graphs where the delay is a *separable* (e.g., linear) function of the cost on edge. We investigate an important property that the FPTAS for RV1CP can find a $(1+\epsilon)$-approximation of RA1CP when the RSP distance matrix has a *saddle point.* In addition, we show that it is harder to find an approximation of RA1CP when the matrix has no saddle point. This paper develops a scaling algorithm with at most $O(mn^3K(\frac{\log K}{\eta}+\log\log n))$ running time where K is a step-size parameter and η is a given positive number, to find a $(1+\eta)$-approximation of RA1CP.

Keywords: Restricted 1-center · Restricted shortest path · Saddle point · FPTAS

1 Introduction

Given an undirected graph $G=(V,E,w)$ with n nodes, m edges and every edge $e\in E$ having a weight $w(e)>0$, the *vertex 1-center* (V1C) of G is a node such that the longest distance with the node as origin is minimized. The problem of finding a V1C of G, called the *vertex 1-center problem* (V1CP), is tractable [2], e.g., using the $O(mn+n^2\log n)$-time algorithm proposed by Fredman *et al.* [4] or the $O(m^*n+n^2\log n)$-time algorithm proposed by Karger *et al.* [9] where m^* is the number of edges used by shortest paths or Pettie's $O(mn+n^2\log\log n)$-time algorithm [12] to find *all-pairs shortest paths* (APSP) and then determining V1C.

A point on an edge of G is called *absolute 1-center* (A1C) of G if it minimizes the longest distance with the point as origin. Hakimi first proposed the concept of A1C, and examined an important property that A1C must be at one

Z. Lu et al. (Eds.): COCOA 2015, LNCS 9486, pp. 647–659, 2015.
DOI: 10.1007/978-3-319-26626-8_47

of $\frac{1}{2}n(n-1)$ break points or at one node of graph [6]. The problem of finding A1C of G, called *absolute 1-center problem* (A1CP), is thus tractable. When G is a vertex-weighted graph, Hakimi *et al.* [7] devised an $O(mn^2 \log n)$-time algorithm to find an A1C of G based on Hakimi's property, and later Kariv and Hakimi [10] developed an $O(mn^3)$-time basic algorithm and an $O(mn \log n)$-time improved algorithm. When G is a vertex-unweighted graph, A1CP admitted an $O(mn \log n)$-time algorithm [7] and an $O(mn)$-time improved algorithm [10]. Specifically, Kariv and Hakimi [10] also presented an $O(n \log n)$-time algorithm and an $O(n)$-time algorithm for finding a V1C or A1C of a vertex-weighted tree and vertex-unweighted tree, respectively. We refer readers to [14] or [2] and literatures listed therein for more results on A1CP.

The distance involved in V1CP and A1CP is referred to as the length of the *shortest path* (SP) in G between two distinct nodes. However, given an undirected graph $G = (V, E, c, d)$ with two edge weights, we are frequently concerned with the length of the *restricted shortest path* (RSP) between two distinct nodes in many practical settings. For instance, one city intends to build an emergent facility with an aim of getting to each of a group of important locations via a shortest path within a uniform (or nonuniform) delay bound. This problem can be modeled as the *restricted absolute 1-center problem* (RA1CP). Specifically, it can be modeled as the *restricted vertex 1-center problem* (RV1CP) if the facility is restricted to be located at one of a group of locations. Both RA1CP and RV1CP are defined formally in Sect. 2.

Obviously, both RA1CP and RV1CP could involve the computation of RSP, which is NP-hard [5]. So far, a number of *fully polynomial time approximation schemes* (FPTAS) have been devised for computing end-to-end RSP [1,3,8,11, 13,15]. Let ϵ be an arbitrary small positive number. In [1], given an undirected graph, Bernstein's randomized algorithm can find a $(1+\epsilon)$-approximation of RSP with delay at most $1+\epsilon$ times the given bound in nearly linear time. Since we cannot ensure that the properties of RA1CP shown in Sect. 4 always hold for RSP with $1+\epsilon$ times bound, we cannot employ Bernstein's algorithm here. All the other algorithms listed above can compute a $(1+\epsilon)$-approximation of RSP with strict delay bound in different running times. Both Warburton's FPTAS [15] and its improved version obtained by Ergun *et al.* [3] only work in acyclic graphs and thus cannot be generalized to undirected graphs. Therefore, this paper employs the improved FPTAS with $O(mn(\log \log n + \frac{1}{\epsilon}))$ running time, derived by Lorenz and Raz [11] from Hassin's FPTAS [8], since they apply to general digraphs and thus can be generalized to general undirected graphs.

This paper focuses on RA1CP in a general undirected graph $G = (V, E, c, d)$ where every edge has a *separable* (e.g., linear) delay function of its cost. We first study RV1CP, in which two node subsets $S \subseteq V$ and $T \subseteq V$ are involved and we aim to find a restricted vertex 1-center in T that minimizes the cost of the most costly RSP from it to S. We devise a simple FPTAS for RV1CP with $O(mn^3(\frac{1}{\epsilon} + \log \log n))$ running time, based on the FPTAS for RSP designed by Lorenz and Raz [11]. In RA1CP, a continuum set $\mathcal{P}$ of points on a collection of edges in G and a node subset $S \subseteq V$ are involved, and we are asked to find a restricted absolute 1-center in $\mathcal{P}$ such that the cost of the most costly RSP from it to S is minimized. If we let a subset $T' \subseteq V$ be $\mathcal{P}$, we have an instance J of RA1CP which is an instance I' of RV1CP. We can derive a RSP distance matrix by applying the FPTAS to I' for any given $\epsilon > 0$. We examine an important property that the FPTAS for RV1CP also can find a $(1+\epsilon)$-approximation of RA1CP when the matrix has a *saddle point*. However, the solution produced by this FPTAS may be arbitrarily bad in the worst case when the matrix has no saddle point. For this case, we further design a scaling algorithm based on this property, with at most $O(mn^3 K(\frac{\log K}{\eta} + \log\log n))$ running time where K is a size-step parameter, to find a $(1+\eta)$-approximation of RA1CP, for any $\eta > 0$.

The rest of this paper is organized as follows. In Sect. 2, we define RV1CP and RA1CP formally and discuss their NP-hardness. In Sect. 3, we propose a simple FPTAS for RV1CP. In Sect. 4, we first discuss some important properties and then present approximation algorithms for RA1CP. In Sect. 5, we conclude this paper. Due to page limit, most proofs are omitted in this paper.

2 Problem Definitions and Intractability

Let $G = (V, E, c, d)$ be an undirected graph where V is the node set and E is the edge set. Every edge $e \in E$ has two nonnegative integer weights, $c(e) \geq 0$ representing the cost of e and $d(e) \geq 0$ representing the delay of e. Without otherwise specified, we always use G to denote such an undirected graph. Let π be a simple path on G. The cost of π, denoted by $c(\pi)$, is equal to the sum of all the costs on its edges, i.e., $c(\pi) = \sum_{e\in\pi} c(e)$. Let $d(\pi)$ denote the delay of π. Similarly, $d(\pi) = \sum_{e\in\pi} d(e)$. In this paper, for any pair of nodes v and u of G, a *restricted shortest path* (RSP) connecting v and u is referred to as a minimum cost v-to-u path with delay bounded by a constant $\mathbb{D} \geq 0$, denoted by $\pi^*[v, u; \mathbb{D}]$ and also called a v-to-u *restricted cheapest path with delay bounded by* $\mathbb{D} \geq 0$ (abbreviated to $\mathbb{D}$-RCP).

This paper focuses on the *restricted vertex 1-center problem* defined in Problem 1 and the *restricted absolute 1-center problem* defined in Problem 2.

Problem 1. Given $G = (V, E, c, d)$, two node subsets $S \subseteq V$ and $T \subseteq V$, and a constant $\mathbb{D} \geq 0$, the **restricted vertex 1-center problem (RV1CP)** aims to find a node in T such that the cost of the most costly $\mathbb{D}$-RCP from it to S is minimized.

Let $C(v, u; \mathbb{D})$ denote the cost of $\pi^*[v, u; \mathbb{D}]$. Clearly, $C(v, u; \mathbb{D}) = c(\pi^*[v, u; \mathbb{D}])$. For any node $v \in T$, we use $f(v, S; \mathbb{D})$ to denote the cost of the most costly $\mathbb{D}$-RCP from v to S, i.e.,

$$f(v, S; \mathbb{D}) = \max_{u \in S\setminus\{v\}} C(v, u; \mathbb{D}), \quad \forall v \in T. \tag{1}$$

Let v^* be the optimal solution of Problem 1. We have

$$f(v^*, S; \mathbb{D}) = \min_{v\in T} f(v, S; \mathbb{D}). \tag{2}$$

Let $\mathcal{P}(e)$ and $\mathcal{P}(G)$ denote the continuum set of points on e and those on all the edges of G respectively. For any $x \in \mathcal{P}(G)$, we also use $\pi^*[x, u; \mathbb{D}]$ to denote an x-to-u $\mathbb{D}$-RCP. Note that, for any point $x \in \mathcal{P}(e)$, the computation of the cost or delay from x to two endpoints of e is given in Sect. 4.

Problem 2. Given $G = (V, E, c, d)$, a node subset $S \subseteq V$ and a point subset $\mathcal{P} \subseteq \mathcal{P}(G)$, and a constant $\mathbb{D} \geq 0$, the **restricted absolute 1-center problem (RA1CP)** asks to find a point in $\mathcal{P}$ such that the cost of the most costly $\mathbb{D}$-RCP from it to S is minimized.

Similarly, we also use $C(x, u; \mathbb{D})$ to denote the cost of $\pi^*[x, u; \mathbb{D}]$ and have $C(x, u; \mathbb{D}) = c(\pi^*[x, u; \mathbb{D}])$, and also use $f(x, S; \mathbb{D})$ to denote the cost of the most costly $\mathbb{D}$-RCP from x to S, i.e.,

$$f(x, S; \mathbb{D}) = \max_{u \in S \setminus \{x\}} C(x, u; \mathbb{D}), \quad \forall x \in \mathcal{P}. \tag{3}$$

Let x^* be the optimal solution of Problem 2. We have

$$f(x^*, S; \mathbb{D}) = \min_{x \in \mathcal{P}} f(x, S; \mathbb{D}). \tag{4}$$

As we all know, V1CP and A1CP are both tractable in general graphs [2]. However, we find that the hardness of RV1CP (*resp.* RA1CP) is much higher than that of V1CP (*resp.* A1CP). The proofs of the NP-hardness of RV1CP and RA1CP are shown in Theorems 1 and 2, respectively.

The decision versions of RV1CP and RA1CP are defined as follows.

Problem 3. Given $G = (V, E, c, d)$, two node subsets $S \subseteq V$ and $T \subseteq V$, and two constants $\mathbb{D} \geq 0$ and $\mathcal{W} \geq 0$, is there a node $\widehat{v}$ in T such that the cost of the most costly $\mathbb{D}$-RCP from $\widehat{v}$ to S does not exceed $\mathcal{W}$?

Problem 4. Given $G = (V, E, c, d)$, a node subset $S \subseteq V$ and a point subset $\mathcal{P} \subseteq \mathcal{P}(G)$, and two constants $\mathbb{D} \geq 0$ and $\mathcal{W} \geq 0$, is there a point $\widehat{x}$ in $\mathcal{P}$ such that the cost of the most costly $\mathbb{D}$-RCP from $\widehat{x}$ to S does not exceed $\mathcal{W}$?

Theorem 1. *RV1CP is NP-hard.*

Proof. We consider the special case of RV1CP where both S and T have a single node. Suppose $S = \{u\}$ and $T = \{v\}$. The decision version of the special case of RV1CP over $S = \{u\}$ and $T = \{v\}$ can be described as: "Given $G = (V, E, c, d)$, two nodes u and v, and two constants $\mathbb{D} \geq 0$ and $\mathcal{W} \geq 0$, is there a v-to-u path on G with delay bounded by $\mathbb{D}$ and cost bounded by $\mathcal{W}$?", which is the decision version of the restricted shortest path problem (RSPP). Since RSPP is NP-hard [5], the decision version of RV1CP is also NP-hard. This completes the proof. □

Similarly, we can extend Theorem 1 to Theorem 2 by letting $S = \{u\}$ and $\mathcal{P} = \{v\}$.

Theorem 2. *RA1CP is NP-hard.*

3 An FPTAS for RV1CP

In this section, we study RV1CP on undirected graphs, and develop an FPTAS named AlgRV1CP for RV1CP based on an FPTAS for RSP.

We use the FPTAS proposed by Lorenz and Raz [11], called RSP, to compute a v-to-u $\mathbb{D}$-RCP for any pair of nodes $v \in T$ and $u \in S$. Given any $\epsilon > 0$, the solution path produced by RSP is a $(1+\epsilon)$-approximation of v-to-u $\mathbb{D}$-RCP, denoted by $\pi^\epsilon[v, u; \mathbb{D}]$ and called a v-to-u $(1+\epsilon)$-*approximation restricted cheapest path with delay bounded by* $\mathbb{D}$ (abbreviated to $(\mathbb{D}, \epsilon)$-ARCP). Let $C^\epsilon(v, u; \mathbb{D})$ denote the cost of $\pi^\epsilon[v, u; \mathbb{D}]$, i.e., $C^\epsilon(v, u; \mathbb{D}) = c(\pi^\epsilon[v, u; \mathbb{D}])$. Since RSP is an FPTAS, it follows that for any $\epsilon > 0$

$$C(v, u; \mathbb{D}) \le C^\epsilon(v, u; \mathbb{D}) \le (1+\epsilon)C(v, u; \mathbb{D}). \tag{5}$$

Let I represent an input instance of RV1CP, including an undirected graph $G = (V, E, c, d)$, two node subsets $S \subseteq V$ and $T \subseteq V$, and a constant $\mathbb{D} \ge 0$. Let $\mathrm{OPT}(I)$ denote an optimal solution to I and $\mathrm{A}(I, \epsilon)$ denote an algorithm solution produced by AlgRV1CP with input $\epsilon > 0$. For any $v \in T$, we let $g(v, S; \mathbb{D}, \epsilon)$ denote the cost of the most costly $(\mathbb{D}, \epsilon)$-ARCP from v to S, i.e.,

$$g(v, S; \mathbb{D}, \epsilon) = \max_{u \in S \setminus \{v\}} C^\epsilon(v, u; \mathbb{D}), \quad \forall v \in T. \tag{6}$$

Let v^ϵ be a node in T that minimizes $g(v, S; \mathbb{D}, \epsilon)$. We have

$$g(v^\epsilon, S; \mathbb{D}, \epsilon) = \min_{v \in T} g(v, S; \mathbb{D}, \epsilon). \tag{7}$$

For any node $v \in T$, we use RSP to compute $C^\epsilon(v, u; \mathbb{D})$ for each node u in S other than v (specifically, $C^\epsilon(v, v; \mathbb{D}) = 0$), and then determine $g(v, S; \mathbb{D}, \epsilon)$ by Eq. (6). We pick out the node such that $g(v, S; \mathbb{D}, \epsilon)$ is minimized. This idea forms a very simple algorithm for solving RV1CP, called AlgRV1CP. Theorem 3 shows that AlgRV1CP is an FPTAS and its time complexity.

```
Algorithm AlgRV1CP(I, ε):
Input: an instance I, and a real constant ε > 0;
Output: A(I, ε)
for each node v ∈ T do
    for each node u ∈ S \ {v} do
        Use RSP to compute C^ε(v, u; D);
    endfor
    g(v, S; D, ε) := max_{u∈S\{v}} C^ε(v, u; D);
endfor
A(I, ε) := arg min_{v∈T} g(v, S; D, ε)
```

First of all, we give a property frequently used in the rest of this paper, whose proof is straightforward and omitted here.

Lemma 1. *Given two q-dimensional (q is either a constant or ∞) arrays $\{X_j\}_q$ and $\{Y_j\}_q$ satisfying $X_j \leq Y_j$ for all j, we have*

$$\min\{X_1, X_2, \ldots, X_q\} \leq \min\{Y_1, Y_2, \ldots, Y_q\}. \tag{8}$$

and

$$\max\{X_1, X_2, \ldots, X_q\} \leq \max\{Y_1, Y_2, \ldots, Y_q\}. \tag{9}$$

Theorem 3. *Given an input I in which G has n nodes and m edges, and a real constant $\epsilon > 0$, if there is a restricted vertex 1-center in G, then AlgRV1CP can find a $(1+\epsilon)$-approximation of RV1CP in $O(mn^3(\frac{1}{\epsilon} + \log\log n))$ time.*

4 Approximation Algorithms for RA1CP

In this section, we study RA1CP on $G = (V, E, c, d)$ with $\mathcal{P}$ over a collection of selected edges in E, e.g., $\mathcal{P} = \{\mathcal{P}(e) | e \in E' \subseteq E\}$, and show AlgRV1CP is not only an FPTAS for RV1CP but also for RA1CP in some cases. Suppose G has m edges and n nodes. All m edges are labelled by $i = 1, 2, \ldots, m$ in order and the edge with label i is denoted by e_i. Let M be the set of indices of edges in E', and T' be the set of endpoints of edges in E'. For any $i \in M$, e_i has cost c_i and delay d_i.

Given two points $x', x'' \in \mathcal{P}(e_i), i \in M$, the segment of e_i between x' and x'' is denoted by $[x', x'']_i$. For RA1CP we assume that the delay over $[x', x'']_i$ is a *separable* function of the cost over it, denoted by $d = \lambda_i(c), 0 \leq c \leq c_i$, which satisfies

$$\lambda_i(c' + c'') = \lambda_i(c') + \lambda_i(c''), \quad \forall c', c'' \geq 0, c' + c'' \leq c_i. \tag{10}$$

Obviously, λ_i is a monotonic increasing function. For example, for a linear function $\lambda_i(c) = \ell_i \cdot c, \ell_i > 0$ defined on e_i, the delay on $[x', x'']_i$ is $\ell_i \cdot c_0$ when the cost on $[x', x'']_i$ is c_0.

4.1 Fundamental Properties

For each $i \in M$, we use t_i^1 and t_i^2 to denote the endpoints of e_i. So, $T' = \{t_i^1, t_i^2 | i \in M\}$. Let x_i be a point in $\mathcal{P}(e_i)$. Let $c_i(x_i)$ denote the cost of $[t_i^1, x_i]_i$ and then the delay of $[t_i^1, x_i]_i$ is equal to $\lambda_i(c_i(x_i))$, for any $x_i \in \mathcal{P}(e_i)$. We observe every x_i-to-u $\mathbb{D}$-RCP must pass through t_i^1 or t_i^2, for any $u \in S \backslash \{x_i\}$. The cheapest one in all x_i-to-u $\mathbb{D}$-RCPs passing through t_i^1 (*resp.* t_i^2) is denoted by $\pi_1^*[x_i, u; \mathbb{D}]$ (*resp.* $\pi_2^*[x_i, u; \mathbb{D}]$). Let $C_1(x_i, u; \mathbb{D})$ and $C_2(x_i, u; \mathbb{D})$ denote the cost of $\pi_1^*[x_i, u; \mathbb{D}]$ and $\pi_2^*[x_i, u; \mathbb{D}]$, respectively.

Lemma 2. *Given any $i \in M$ and a point $x_i \in \mathcal{P}(e_i)$, it follows that*

1. *for any $u \in S \setminus \{x_i\}$ and a point $x' \in [t_i^1, x_i]_i$, we have*

$$C_1(x_i, u; \mathbb{D}) = C_1(x', u; \mathbb{D} - \lambda_i(c_i(x_i) - c_i(x'))) + c_i(x_i) - c_i(x'), \tag{11}$$

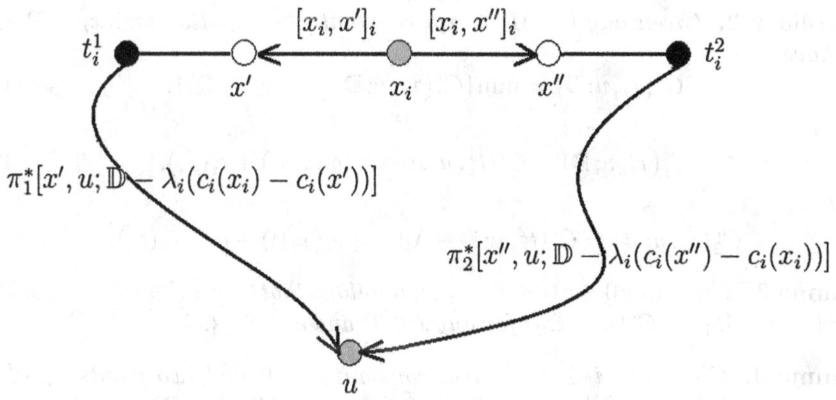

Fig. 1. Illustration for Lemma 2.

2. for any $u \in S \setminus \{x_i\}$ and a point $x'' \in [t_i^2, x_i]_i$, we have

$$C_2(x_i, u; \mathbb{D}) = C_2(x'', u; \mathbb{D} - \lambda_i(c_i(x'') - c_i(x_i))) + c_i(x'') - c_i(x_i). \quad (12)$$

It is clear that t_i^1 is a point satisfying Case 1 of Lemma 2 and t_i^2 is a point satisfying Case 2, for any $x_i \in \mathcal{P}(e_i), i \in M$. Specifically, we substitute $x' = t_i^1$ into Eq. (11) and $x'' = t_i^2$ into Eq. (12). From the fact that $c_i(t_i^1) = 0$ and $c_i(t_i^2) = c_i$, we obtain Eqs. (14) and (15), respectively. Since any x_i-to-u $\mathbb{D}$-RCP must pass through either t_i^1 or t_i^2, we obtain Eq. (13). As a result, we derive the following corollary.

Corollary 1. *Given any $i \in M$ and a point $x_i \in \mathcal{P}(e_i)$, we infer that*

$$C(x_i, u; \mathbb{D}) = \min\{C_1(x_i, u; \mathbb{D}),\ C_2(x_i, u; \mathbb{D})\}, \quad \forall u \in S \setminus \{x_i\}, \quad (13)$$

where

$$C_1(x_i, u; \mathbb{D}) = C(t_i^1, u; \mathbb{D} - \lambda_i(c_i(x_i))) + c_i(x_i), \quad (14)$$

and

$$C_2(x_i, u; \mathbb{D}) = C(t_i^2, u; \mathbb{D} - \lambda_i(c_i - c_i(x_i))) + c_i - c_i(x_i). \quad (15)$$

□

Given any $i \in M$, we construct two new x_i-to-u paths with delay bounded by $\mathbb{D}$, for any $x_i \in \mathcal{P}(e_i)$ and $u \in S \setminus \{x_i\}$. The one composed of $[x_i, t_i^1]_i$ and $\pi^\epsilon[t_i^1, u; \mathbb{D} - \lambda_i(c_i(x_i))]$ is denoted by $\Pi_1^\epsilon[x_i, u; \mathbb{D}]$, and the other composed of $[x_i, t_i^2]_i$ and $\pi^\epsilon[t_i^2, u; \mathbb{D} - \lambda_i(c_i - c_i(x_i))]$ is denoted by $\Pi_2^\epsilon[x_i, u; \mathbb{D}]$. Let $\Pi^\epsilon[x_i, u; \mathbb{D}]$ be the cheaper one between $\Pi_1^\epsilon[x_i, u; \mathbb{D}]$ and $\Pi_2^\epsilon[x_i, u; \mathbb{D}]$. We use $C_1^\epsilon(x_i, u; \mathbb{D})$, $C_2^\epsilon(x_i, u; \mathbb{D})$ and $C^\epsilon(x_i, u; \mathbb{D})$ to denote the cost of $\Pi_1^\epsilon[x_i, u; \mathbb{D}]$, $\Pi_2^\epsilon[x_i, u; \mathbb{D}]$ and $\Pi^\epsilon[x_i, u; \mathbb{D}]$, respectively.

By Corollary 1, we can propose one method to compute $C_1^\epsilon(x_i, u; \mathbb{D})$, $C_2^\epsilon(x_i, u; \mathbb{D})$ and $C^\epsilon(x_i, u; \mathbb{D})$, shown in Corollary 2.

Corollary 2. *Given any $i \in M$, a real constant $\epsilon > 0$ and a point $x_i \in \mathcal{P}(e_i)$, we have*

$$C^\epsilon(x_i, u; \mathbb{D}) = \min\{C_1^\epsilon(x_i, u; \mathbb{D}),\ C_2^\epsilon(x_i, u; \mathbb{D})\}, \tag{16}$$

where

$$C_1^\epsilon(x_i, u; \mathbb{D}) = C^\epsilon(t_i^1, u; \mathbb{D} - \lambda_i(c_i(x_i))) + c_i(x_i), \tag{17}$$

and

$$C_2^\epsilon(x_i, u; \mathbb{D}) = C^\epsilon(t_i^2, u; \mathbb{D} - \lambda_i(c_i - c_i(x_i))) + c_i - c_i(x_i). \tag{18}$$

Lemma 3. *Given any $0 \le \mathbb{D}_1 \le \mathbb{D}_2 \le \mathbb{D}$, it follows that $C(x, u; \mathbb{D}_1) \ge C(x, u; \mathbb{D}_2)$ and $C^\epsilon(x, u; \mathbb{D}_1) \ge C^\epsilon(x, u; \mathbb{D}_2)$ for any $x \in \mathcal{P}$ and $u \in S \setminus \{x\}$.*

Lemma 4. *Given any $i \in M$, a real constant $\epsilon > 0$ and two points $x_i', x_i'' \in \mathcal{P}(e_i)$, provided that $c_i(x_i') < c_i(x_i'')$, it follows that $C_1^\epsilon(x_i', u; \mathbb{D}) < C_1^\epsilon(x_i'', u; \mathbb{D})$ and $C_2^\epsilon(x_i', u; \mathbb{D}) > C_2^\epsilon(x_i'', u; \mathbb{D})$ for any $u \in S \setminus \{x_i\}$.*

Obviously, $\pi^*[x_i, u; \mathbb{D}]$ is a simple path on G. We note that $\Pi_1^\epsilon[x_i, u; \mathbb{D}]$ is not a simple path when $\pi^\epsilon[t_i^1, u; \mathbb{D} - \lambda_i(c_i(x_i))]$ contains e_i. This is also for $\Pi_2^\epsilon[x_i, u; \mathbb{D}]$. However, Theorem 4 shows $\Pi^\epsilon[x_i, u; \mathbb{D}]$ is surely a simple path.

Theorem 4. *Given any $i \in M$ and a real constant $\epsilon > 0$, $\Pi^\epsilon[x_i, u; \mathbb{D}]$ must be a simple path, for any $x_i \in \mathcal{P}(e_i)$ other than endpoints and any $u \in S \setminus \{x_i\}$.*

Theorem 4 shows $\Pi^\epsilon[x_i, u; \mathbb{D}]$ is a simple path. Moreover, we further conclude that $C^\epsilon(x_i, u; \mathbb{D}) \le (1 + \epsilon)C(x_i, u; \mathbb{D})$, as shown in Theorem 5.

Theorem 5. *Given any $i \in M$ and a real constant $\epsilon > 0$, it follows that $\Pi^\epsilon[x_i, u; \mathbb{D}]$ is an x_i-to-u $(\mathbb{D}, \epsilon)$-ARCP, for any $x_i \in \mathcal{P}(e_i)$ and $u \in S \setminus \{x_i\}$.*

Theorem 6. *Given any $i \in M$, a real constant $\epsilon > 0$ and a point $x_i \in \mathcal{P}(e_i)$, it follows that for any $u \in S \setminus \{x_i\}$ and two points $\mathcal{L}_i, \mathcal{U}_i \in \mathcal{P}(e_i) \setminus \{u\}$ satisfying $c_i(\mathcal{L}_i) \le c_i(x_i) \le c_i(\mathcal{U}_i)$,*

$$\min_{\mathcal{L}_i, \mathcal{U}_i \neq u} \{C^\epsilon(\mathcal{L}_i, u; \mathbb{D}),\ C^\epsilon(\mathcal{U}_i, u; \mathbb{D})\} \le (1 + \epsilon)C(x_i, u; \mathbb{D}). \tag{19}$$

Given any $i \in M$, it is evident that t_i^1 and t_i^2 are both a point on e_i and satisfy $c_i(t_i^1) \le c_i(x_i) \le c_i(t_i^2)$ for any $x_i \in \mathcal{P}(e_i)$. Specifically, we substitute $\mathcal{L}_i = t_i^1$ and $\mathcal{U}_i = t_i^2$ into Theorem 6 to obtain the following corollary.

Corollary 3. *Given any $i \in M$ and a real constant $\epsilon > 0$, it follows that for any $x_i \in \mathcal{P}(e_i)$ and $u \in S \setminus \{x_i\}$*

$$\min_{t_i^1, t_i^2 \neq u} \{C^\epsilon(t_i^1, u; \mathbb{D}),\ C^\epsilon(t_i^2, u; \mathbb{D})\} \le (1 + \epsilon)C(x_i, u; \mathbb{D}). \tag{20}$$

Moreover, when we are given a node $u \in S$, we consider all x-to-u $\mathbb{D}$-RCPs with $x \in \mathcal{P} \setminus \{u\}$ and use $\mathcal{F}(u, \mathcal{P}; \mathbb{D})$ to denote the cost of the cheapest x-to-u $\mathbb{D}$-RCP. Let u^* be a node which maximizes $\mathcal{F}(u, \mathcal{P}; \mathbb{D})$. We have

$$\mathcal{F}(u, \mathcal{P}; \mathbb{D}) = \min_{x \in \mathcal{P} \setminus \{u\}} C(x, u; \mathbb{D}), \quad \forall u \in S, \tag{21}$$

and

$$\mathcal{F}(u^*, \mathcal{P}; \mathbb{D}) = \max_{u \in S} \mathcal{F}(u, \mathcal{P}; \mathbb{D}). \tag{22}$$

On the other hand, we consider all v-to-u $(\mathbb{D}, \epsilon)$-ARCP with $v \in T'$. Let $\mathcal{G}(u, T'; \mathbb{D}, \epsilon)$ denote the cost of the cheapest v-to-u $(\mathbb{D}, \epsilon)$-ARCP. Let u^ϵ be a node which maximizes $\mathcal{G}(u, T'; \mathbb{D}, \epsilon)$. We have

$$\mathcal{G}(u, T'; \mathbb{D}, \epsilon) = \min_{v \in T' \setminus \{u\}} C^\epsilon(v, u; \mathbb{D}), \quad \forall u \in S, \tag{23}$$

and

$$\mathcal{G}(u^\epsilon, T'; \mathbb{D}, \epsilon) = \max_{u \in S} \mathcal{G}(u, T'; \mathbb{D}, \epsilon). \tag{24}$$

Lemma 5. $\mathcal{F}(u^*, \mathcal{P}; \mathbb{D}) \le f(x^*, S; \mathbb{D})$.

Moreover, combining Eqs. (6), (7), (23) and (24), we use the similar way to the proof of Lemma 5 to obtain the following corollary.

Corollary 4. $\mathcal{G}(u^\epsilon, T'; \mathbb{D}, \epsilon) \le g(v^\epsilon, S; \mathbb{D}, \epsilon)$.

Obviously, $C^\epsilon(v, v; \mathbb{D}) = 0$ for any node $v \in T'$. All $C^\epsilon(v, u; \mathbb{D})$'s produced by applying AlgRV1CP to RV1CP over S and T' form a $|T'| \times |S|$ matrix. We call $(v^\star, u^\star)$ a *saddle point* of the matrix (also called a saddle point of $C^\epsilon(v, u; \mathbb{D})$), if $(v^\star, u^\star)$ satisfies that for any pair of nodes $v \in T'$ and $u \in S$

$$C^\epsilon(v^\star, u; \mathbb{D}) \le C^\epsilon(v^\star, u^\star; \mathbb{D}) \le C^\epsilon(v, u^\star; \mathbb{D}). \tag{25}$$

Theorem 7. *Given any* $\epsilon > 0$, $\mathcal{G}(u^\epsilon, T'; \mathbb{D}, \epsilon) = g(v^\epsilon, S; \mathbb{D}, \epsilon)$ *if and only if* $C^\epsilon(v, u; \mathbb{D})$ *has a saddle point.*

In the following, we present Theorem 8 to show the relationship of $\mathcal{F}(u, \mathcal{P}; \mathbb{D})$ and $\mathcal{G}(u, T'; \mathbb{D}, \epsilon)$ for any $u \in S$, which will play an important role in the subsequent approximation analysis.

Theorem 8. *Given any constant* $\epsilon > 0$, *for any* $u \in S$

$$\mathcal{G}(u, T'; \mathbb{D}, \epsilon) \le (1 + \epsilon)\mathcal{F}(u, \mathcal{P}; \mathbb{D}). \tag{26}$$

4.2 Approximation Algorithms

Let J represent an input instance of RA1CP, including an undirected graph $G = (V, E, c, d)$, a node subset $S \subseteq V$ and a point subset $\mathcal{P}$, and a constant $\mathbb{D} \ge 0$. Let I' be the instance of RV1CP derived from replacing $\mathcal{P}$ with T' into J. We use $\mathrm{OPT}(J)$ to denote an optimal solution to RA1CP, and use $\mathrm{A}(I', \epsilon)$ to denote the solution produced by applying AlgRV1CP to I' for a given $\epsilon > 0$.

Given any $u \in S$ and $x \in \mathcal{P}, x \neq u$, the combination of Theorems 4 and 5 implies that $\Pi^\epsilon[x, u; \mathbb{D}]$ is not only a simple path but also an x-to-u $(\mathbb{D}, \epsilon)$-ARCP. We have

$$\begin{aligned}&\min_{x\in\mathcal{P}} \max_{u\in S\setminus\{x\}} C^\epsilon(x, u; \mathbb{D})\\ &\leq (1+\epsilon) \min_{x\in\mathcal{P}} \max_{u\in S\setminus\{x\}} C(x, u; \mathbb{D})\\ &= (1+\epsilon) f(x^*, S; \mathbb{D}).\end{aligned}$$

Although we can compute $C^\epsilon(x, u; \mathbb{D})$, it is impossible for us to traverse all points $x \in \mathcal{P}$ to determine x^ϵ such that the cost of the most costly x^ϵ-to-u $(\mathbb{D}, \epsilon)$-ARCP is minimized. Fortunately, $\mathrm{A}(I', \epsilon)$ is also a $(1+\epsilon)$-approximation of RA1CP when the output $C^\epsilon(v, u; \mathbb{D})$ obtained by applying AlgRV1CP to I' and $\epsilon > 0$ satisfies some property, see Theorem 9. This essentially provides us with a shortcut to compute a $(1+\epsilon)$-approximation of RA1CP.

Theorem 9. *Provided that $C^\epsilon(v, u; \mathbb{D})$ produced by AlgRV1CP has a saddle point, AlgRV1CP can find a $(1+\epsilon)$-approximation of RA1CP on $G = (V, E, c, d)$.*

Theorem 9 means that AlgRV1CP is an FPTAS for RA1CP when $C^\epsilon(v, u; \mathbb{D})$ produced by AlgRV1CP has a saddle point. However, it is unfortunate that Theorem 7 implies that AlgRV1CP cannot find a $(1+\epsilon)$-approximation of RA1CP when $C^\epsilon(v, u; \mathbb{D})$ has no saddle point. In fact, the solution produced by AlgRV1CP may be *arbitrarily bad* in the worst case. A counterexample is shown below in Fig. 2, where $S = V$ and $\mathcal{P} = \mathcal{P}(G)$, and the delay bound is $\mathbb{D} = 5$ and γ is a sufficiently small positive number. Every edge has a pair of weights (c', d'), where c' is its cost and d' is its delay. Clearly, for any $\epsilon > 0$, there exists exactly one or no $(5, \epsilon)$-ARCP between any pair of nodes. Furthermore, v_3 cannot reach v_4 and v_5, v_4 cannot reach v_3 and v_5, and v_5 cannot reach v_3 and v_4, within delay bound 5. We use a 5×5 matrix below to store all values of $C^\epsilon(v_i, v_j; 5)$, whose row label is i and column label is j.

−	2γ	1	γ	3γ
2γ	−	2γ	2	3γ
1	2γ	−	−	−
γ	2	−	−	−
3γ	3γ	−	−	−

We know this matrix has no saddle point by using TEST given subsequently. Hence, we only need to consider two nodes v_1 and v_2. We verify that

$$f(v_1, S; 5) = C(v_1, v_3; 5) = 1, \quad \text{and} \quad f(v_2, S; 5) = C(v_2, v_4; 5) = 2.$$

Thus,

$$f(v^*, S; 5) = \min\{f(v_1, S; 5),\ f(v_2, S; 5)\} = 1.$$

Next, we consider the points on edges. Obviously, the points on edges (v_1, v_5) and (v_1, v_4) cannot reach v_3, and (v_2, v_5) and (v_2, v_3) cannot reach v_4, respectively within delay bound 5. So, we only need to consider the points on (v_1, v_2), (v_1, v_3) and (v_2, v_4). Let x_1 be a point on (v_1, v_3) with delay 1.5 to v_1, x_2 be a point on (v_2, v_4) with delay 1.5 to v_2, and x_3 be a point on (v_1, v_2) with delay 1.5 to v_1. Clearly, the cost of the section between x_1 and v_1 is 0.3, that between x_2 and v_2 is 0.6, and that between x_3 and v_1 is γ. We have

$$\begin{aligned} f(x_1, S; 5) &= C(x_1, v_3; 5) = 0.7, \\ f(x_2, S; 5) &= C(x_2, v_4; 5) = 1.4, \\ f(x_3, S; 5) &= C(x_3, v_5; 5) = \gamma + 3\gamma = 4\gamma. \end{aligned}$$

Thus,

$$f(x^*, S; 5) = \min\{f(x_1, S; 5),\ f(x_2, S; 5),\ f(x_3, S; 5)\} = 4\gamma.$$

Hence, $f(\mathrm{OPT}(I'), S; 5) = 1$ and $f(\mathrm{OPT}(J), S; 5) = 4\gamma$. Therefore,

$$\frac{f(\mathrm{A}(I', \epsilon), S; 5)}{f(\mathrm{OPT}(J), S; 5)} \geq \frac{f(\mathrm{OPT}(I'), S; 5)}{f(\mathrm{OPT}(J), S; 5)} = \frac{1}{4\gamma} \to \infty\ (\gamma \to 0).$$

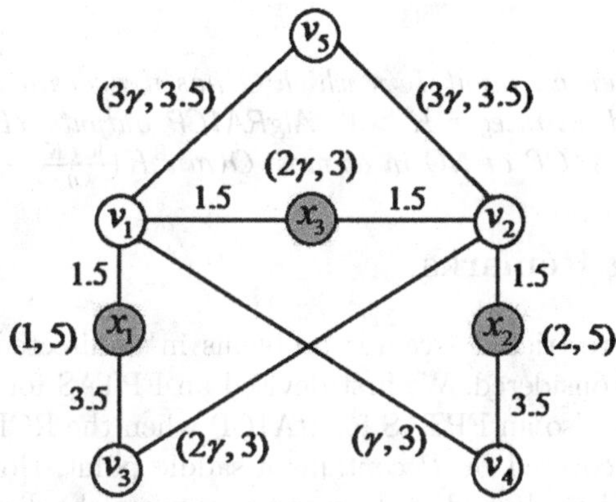

Fig. 2. A counterexample.

We note that the values of $C^\epsilon(v, u; \mathbb{D})$ relies on the value of ϵ. So, we can design an algorithm to test whether $C^\epsilon(v, u; \mathbb{D})$ has a saddle point for given $\epsilon > 0$. This testing algorithm is called TEST with parameter ϵ and described roughly as follows. TEST(ϵ) traverses all nodes $v \in T'$ one by one, finds $u^\star$ such that $C^\epsilon(v, u; \mathbb{D}) \leq C^\epsilon(v, u^\star; \mathbb{D}), \forall u \in S \setminus \{v\}$ for every node v, and decides whether $C^\epsilon(v', u^\star; \mathbb{D}) \geq C^\epsilon(v, u^\star; \mathbb{D}), \forall v' \in T' \setminus \{u^\star\}$. If $C^\epsilon(v, u; \mathbb{D})$ has a saddle point, then TEST(ϵ) outputs YES. Otherwise, TEST(ϵ) outputs NO. Clearly, TEST takes $O(n^2)$ time.

Given any $\eta > 0$ and an integer $K > 0$, we set a step-size parameter $\delta = \frac{\eta}{K}$. Based on above discussions, we can seek an approximation solution of RA1CP by using at most K times TEST with $\epsilon_k = k\delta, 1 \le k \le K$ as parameter. If there exist some values of k so that TEST(ϵ_k) outputs YES, then we can find the smallest one $\widetilde{k}$ and use AlgRV1CP$(I', \epsilon_{\widetilde{k}})$ to determine a $(1 + \epsilon_{\widetilde{k}})$-approximation of RA1CP. Otherwise, we say that we cannot find a $(1 + \eta)$-approximation and thus output NO. The idea can be described as algorithm AlgRA1CP.

Algorithm AlgRA1CP(J, η, K):

Input: an instance J, a real constant $\eta > 0$ and an integer $K > 0$;
Output: YES with ϵ_k and $A(I', \epsilon_k)$, or NO

$\delta := \frac{\eta}{K}$; $T' := \{t_i^1, t_i^2 | i \in M\}$;
Replace $\mathcal{P}$ with T' into J to obtain an instance I' of RV1CP;
for $k := 1$ to K **do**
  $\epsilon_k := k\delta$;
  Call AlgRV1CP(I', ϵ_k) to obtain $C^{\epsilon_k}(v, u; \mathbb{D})$ and $A(I', \epsilon_k)$;
  if TEST(ϵ_k)=YES **then**
    Stops; returns YES with ϵ_k and $A(I', \epsilon_k)$;
  endif
endfor
Returns NO

Theorem 10. *Given an input J in which G has n nodes and m edges, a real constant $\eta > 0$ and an integer $K > 0$, AlgRA1CP outputs YES and a $(1 + \eta)$-approximation of RA1CP or NO in at most $O(mn^3 K(\frac{\log K}{\eta} + \log\log n))$ time.*

5 Concluding Remarks

In this paper, two restricted 1-center problems in undirected graphs, RV1CP and RA1CP, were considered. We first devised an FPTAS for RV1CP, and further discovered it is also an FPTAS for RA1CP when the RCP distance matrix between S and T' covered by $\mathcal{P}$ contains a saddle point. However, its output solution may be arbitrarily bad in the worst case when the distance matrix contains no saddle point. Considering that the distance matrix varies with the value of ϵ, we developed a scaling algorithm to seek ϵ such that the FPTAS for RV1CP becomes one for RA1CP.

The bound on delay considered in this paper is uniform for all nodes in S. In fact, our approximation algorithms also can be applied to the scenarios where the delay bounds of nodes in S are nonuniform. Moreover, our algorithm AlgRV1CP can be easily extended to RV1CP on node-weighted undirected graphs.

Note that our scaling algorithm cannot guarantee to find such ϵ. Therefore, we suggest to further consider whether we can find more properties which can help to find such ϵ definitely. In addition, although we claim that AlgRV1CP can apply to the case of RV1CP with $S = T = V$ and AlgRA1CP can apply to the case of RA1CP with $S = V$ and $\mathcal{P} = \mathcal{P}(G)$, the question whether or not they are NP-hard remains open.

Acknowledgement. We thank the reviewers for their valuable comments and suggestions.

References

1. Bernstein, A.: Near linear time $(1+\epsilon)$-approximation for restricted shortest paths in undirected graphs. In: Proceedings of the 23th Annual ACM-SIAM Symposium on Discrete Algorithms, SODA 2012, pp. 189–201, Kyoto, Japan, January 2012
2. Eiselt, H.A., Marianov, V.: Foundations of Location Analysis. Springer, Heidelberg (2011)
3. Ergun, F., Sinha, R., Zhang, L.: An improved FPTAS for restricted shortest path. Inf. Proc. Lett. **83**(5), 287–291 (2002)
4. Fredman, M.L., Tarjan, R.E.: Fibonacci heaps and their uses in improved network optimization algorithms. J. ACM **34**(3), 596–615 (1987)
5. Garey, M.R., Johnson, D.S.: Computers and Intractability: A Guide to the Theory of NP-Completeness. Freeman, San Francisco (1979)
6. Hakimi, S.L.: Optimal locations of switching centers and medians of a graph. Oper. Res. **12**(3), 450–459 (1964)
7. Hakimi, S.L., Schmeichel, E.F., Pierce, J.G.: On p-centers in networks. Transport. Sci. **12**(1), 1–15 (1978)
8. Hassin, R.: Approximation schemes for the restricted shortest path problem. Math. Oper. Res. **17**, 36–42 (1992)
9. Karger, D.R., Koller, D., Phillips, S.J.: Finding the hidden path: time bounds for all-pairs shortest paths. SIAM J. Comput. **22**, 1199–1217 (1993)
10. Kariv, O., Hakimi, S.L.: An algorithmic approach to network location problems. I: the p-centers. SIAM J. Appl. Math. **22**, 1199–1217 (1993)
11. Lorenz, D., Raz, D.: A simple efficient approximation scheme for the restricted shortest path problem. Oper. Res. Lett. **28**, 213–219 (1999)
12. Pettie, S.: A new approach to all-pairs shortest paths on real-weighted graphs. Theor. Comp. Sci. **312**(1), 47–74 (2004)
13. Phillips, C.: The network inhibition problem. In: Proceedings of the 25th Annual ACM Symposium on the Theory of Computing (STOC 1993), pp. 776–785, San Diego, CA, May 1993
14. Tansel, B.C., Francis, R.L., Lowe, T.J.: Location on networks: a survey. part I: the p-center and p-median problems. Manag. Sci. **29**(4), 482–497 (1983)
15. Warburton, A.: Approximation of pareto optima in multiple-objective shortest path problems. Oper. Res. **35**, 70–79 (1987)

The Disjunctive Bondage Number and the Disjunctive Total Bondage Number of Graphs

Eunjeong Yi[✉]

Texas A&M University at Galveston, Galveston, TX 77553, USA
yie@tamug.edu

Abstract. Let G be a graph with vertex set $V(G)$ and edge set $E(G)$. A set $S \subseteq V(G)$ is a *disjunctive dominating set* of G if every vertex in $V(G) - S$ is adjacent to a vertex of S or has at least two vertices in S at distance two from it. For G with no isolated vertex, a set $S \subseteq V(G)$ is a *disjunctive total dominating set* of G if every vertex in G is adjacent to a vertex of S or has at least two vertices of S at distance two from it. The *disjunctive domination number* $\gamma^d(G)$ of G is the minimum cardinality over all disjunctive dominating sets of G, and the *disjunctive total domination number* $\gamma_t^d(G)$ of G is the minimum cardinality over all disjunctive total dominating sets of G. We define *disjunctive bondage number* of G to be the minimum cardinality among all subsets of edges $B \subseteq E(G)$ for which $\gamma^d(G - B) > \gamma^d(G)$. For G with no isolated vertex, we define *disjunctive total bondage number*, $b_t^d(G)$, of G to be the minimum cardinality among all subsets of edges $B' \subseteq E(G)$ satisfying $\gamma_t^d(G - B') > \gamma_t^d(G)$ and that $G - B'$ contains no isolated vertex; if no such subset B' exists, we define $b_t^d(G) = \infty$. In this paper, we initiate the study of the disjunctive (total) bondage number of graphs. We determine the disjunctive (total) bondage number of the Petersen graph, cycles, paths, and some complete multipartite graphs. We also obtain upper bounds of the disjunctive bondage number for trees and some Cartesian product graphs, and we show the existence of a tree T satisfying $b_t^d(T) = k$ for each positive integer k.

Keywords: Disjunctive domination · Disjunctive total domination · Disjunctive bondage number · Disjunctive total bondage number · The Petersen graph · Cycles · Paths · Complete multipartite graphs · Trees · Cartesian product graphs

1 Introduction

Let $G = (V(G), E(G))$ be a finite, simple, and undirected graph. An *empty graph* is a graph consisting of isolated vertices. The *Cartesian product* of two graphs G and H, denoted by $G \square H$, is the graph with the vertex set $V(G) \times V(H)$ such that (u, v) is adjacent to (u', v') if and only if (i) $u = u'$ and $vv' \in E(H)$ or (ii) $v = v'$ and $uu' \in E(G)$. The *distance* between two vertices $x, y \in V(G)$, denoted by $d_G(x, y)$, is the length of a shortest path between x and y in G. The *diameter* of

Z. Lu et al. (Eds.): COCOA 2015, LNCS 9486, pp. 660–675, 2015.
DOI: 10.1007/978-3-319-26626-8_48

a graph G, denoted by $diam(G)$, is $\max\{d_G(x,y) \mid x,y \in V(G)\}$. For $v \in V(G)$, let $N_G(v) = \{u \in V(G) \mid d_G(u,v) = 1\}$ and $N_G[v] = \{u \in V(G) \mid d_G(u,v) \leq 1\}$; similarly, for a positive integer k, let $N_G^k(v) = \{u \in V(G) \mid d_G(u,v) = k\}$ and $N_G^k[v] = \{u \in V(G) \mid d_G(u,v) \leq k\}$. For a set $S \subseteq V(G)$, its *open neighborhood* is the set $N_G(S) = \cup_{v \in S} N_G(v)$ and its *closed neighborhood* is the set $N_G[S] = N_G(S) \cup S$. The degree of a vertex $v \in V(G)$ is $\deg_G(v) = |N_G(v)|$. The *maximum degree* among the vertices of G is denoted by $\Delta(G)$, and the *minimum degree* among the vertices of G is denoted by $\delta(G)$. A *leaf* is a vertex of degree one, and a *support vertex* is a vertex that is adjacent to a leaf. We denote by C_n, P_n, K_n, respectively, the cycle, path, complete graph on n vertices.

Domination (total domination, respectively) is an extensively studied topic that models a network on resource allocation (resource allocation with redundancy being allowed in case of resource failure, respectively) at a particular node (see [8]). A set $S \subseteq V(G)$ is a *dominating set* of G if every vertex in $V(G) - S$ is adjacent to at least one vertex of S, and the *domination number*, $\gamma(G)$, of G is the minimum cardinality over all dominating sets of G. For G with no isolated vertex, a set $S \subseteq V(G)$ is a *total dominating set* of G if every vertex in G is adjacent to at least one vertex of S, and the *total domination number*, $\gamma_t(G)$, of G is the minimum cardinality over all total dominating sets of G. It is known that determining (total) domination number of a graph is NP-complete (see [3]). For surveys on the topic of domination, we refer to [5,6].

Recently, Goddard et al. [4] introduced *disjunctive domination*, and Henning and Naicker [8] introduced *disjunctive total domination*. The authors of [8] notes that disjunctive (total) domination can be viewed as a relaxation of the concept of (total) domination. Following [7], a set $S \subseteq V(G)$ is a *disjunctive dominating set* (DD-set) of G if every vertex in $V(G) - S$ is adjacent to a vertex of S or has at least two vertices of S at distance two from it; the *disjunctive domination number*, $\gamma^d(G)$, of G is the minimum cardinality over all DD-sets of G. Following [8], a set $S \subseteq V(G)$ is a *disjunctive total dominating set* (DTD-set) of G if every vertex in G is adjacent to a vertex of S or has at least two vertices of S at distance two from it; the *disjunctive total domination number*, $\gamma_t^d(G)$, of G is the minimum cardinality over all DTD-sets of G. We denote by $\gamma^d(G)$-set a minimum DD-set of G, and by $\gamma_t^d(G)$-set a minimum DTD-set of G. We say that a vertex $v \in V(G)$ is *disjunctively dominated* by a vertex set S if $v \in N_G[S]$ or v is at distance two from at least two vertices of S; similarly, we say that a vertex $v \in V(G)$ is *disjunctively totally dominated* by a vertex set S if $v \in N_G(S)$ or v is at distance two from at least two vertices of S. It is shown in [4] that there is a linear-time algorithm for computing $\gamma^d(T)$ for a tree T, and that determining $\gamma^d(G)$ of a general graph G is NP-hard.

Bauer et al. [1] introduced the concept of bondage number (called a 'domination-line-stability') as a measure of efficiency of domination in graphs. Fink et al. [2] officially introduced and studied bondage number as a graph parameter that measures the vulnerability of interconnection network. Kulli and Patwari [10] introduced the concept of total bondage number. The *bondage number*, $b(G)$, of a nonempty graph G is the minimum cardinality among all subsets of edges $B \subseteq E(G)$ for which $\gamma(G - B) > \gamma(G)$. If G contains no isolated vertex, the *total bondage number*,

$b_t(G)$, of G is the minimum cardinality among all subsets $B' \subseteq E(G)$ such that $G - B'$ has no isolated vertex and $\gamma_t(G - B') > \gamma_t(G)$. Hu and Xu [9] showed that determining the (total) bondage number for general graphs is an NP-hard problem. Bondage number is a graph parameter that has been studied extensively. For a survey article on the bondage number of graphs, we refer to [11].

In this paper, we initiate the study of disjunctive (total) bondage number of a graph. For a non-empty graph G, we define *disjunctive bondage number*, $b^d(G)$, of G to be the minimum cardinality among all subsets of edges $B \subseteq E(G)$ for which $\gamma^d(G - B) > \gamma^d(G)$. For G with no isolated vertex, we define *disjunctive total bondage number*, $b_t^d(G)$, of G to be the minimum cardinality among all subsets of edges $B' \subseteq E(G)$ satisfying the following: (1) $G - B'$ has no isolated vertex; (2) $\gamma_t^d(G - B') > \gamma_t^d(G)$. In the case that there is no such subset B', we define $b_t^d(G) = \infty$. We obtain exact values and bounds of the disjunctive (total) bondage number for some classes of graphs.

This paper is organized as follows. In Sect. 2, we recall some known results that are used in Sects. 3 and 4. In Sect. 3, we determine exact values of the disjunctive bondage number for the Petersen graph, cycles, paths, and complete k-partite graphs. We also obtain upper bounds of the disjunctive bondage number for trees and some Cartesian product of graphs. Further, we express the upper bound of the disjunctive bondage number of a graph in terms of degree sum of two vertices that are at distance at most two from each other. In Sect. 4, we determine disjunctive total bondage number for the Petersen graph, cycles, paths, and some complete k-partite graphs. We also show that, for a given positive integer k, there exists a tree T for which $b_t^d(T) = k$. In Sect. 5, we give an example showing that $\gamma(G) - \gamma^d(G)$ and $\gamma_t(G) - \gamma_t^d(G)$ can be arbitrarily large. We conclude with an open problem.

2 Preliminaries

In this section, we recall some known results which will be used in Sects. 3 and 4. First, we recall the disjunctive (total) domination number of some graphs. We begin with a result on the effect of an edge deletion on the disjunctive domination number of a graph.

Theorem 1. *[4] For any graph G, $\gamma^d(G-e) \leq \gamma^d(G)+1$ for any edge $e \in E(G)$.*

Theorem 2. *[4]*

(a) For the Petersen graph $\mathcal{P}$, $\gamma^d(\mathcal{P}) = 2$.
(b) For $n \geq 3$,

$$\gamma^d(C_n) = \begin{cases} 2 & \text{if } n = 4 \\ \lceil \frac{n}{4} \rceil & \text{otherwise.} \end{cases} \tag{1}$$

(c) For $n \geq 2$, $\gamma^d(P_n) = \lceil \frac{n+1}{4} \rceil$.
(d) For $n \geq 1$, $\gamma^d(P_n \Box P_2) = \lceil \frac{n+2}{3} \rceil$.

Theorem 3. *[8] Let $n \geq 3$.*

(a)

$$\gamma_t^d(C_n) = \begin{cases} \frac{2n}{5} & \text{if } n \equiv 0 \pmod 5 \\ \left\lceil \frac{2(n+1)}{5} \right\rceil & \text{otherwise.} \end{cases} \tag{2}$$

(b)

$$\gamma_t^d(P_n) = \begin{cases} \left\lceil \frac{2(n+1)}{5} \right\rceil + 1 & \text{if } n \equiv 1 \pmod 5) \\ \left\lceil \frac{2(n+1)}{5} \right\rceil & \text{otherwise.} \end{cases} \tag{3}$$

Next, we recall a couple of results on the bondage number.

Theorem 4. *[2]*

(a) For $n \geq 3$,

$$b(C_n) = \begin{cases} 3 & \text{if } n \equiv 1 \pmod 3), \\ 2 & \text{otherwise.} \end{cases} \tag{4}$$

(b) If $G = K_{a_1,a_1,\ldots,a_k}$ is a complete k-partite graph ($k \geq 2$) with $a_k = \max\{a_i : 1 \leq i \leq k\}$, where s is the number of partite sets consisting of one element, then

$$b(G) = \begin{cases} \lceil \frac{s}{2} \rceil & \text{if } s \neq 0, \\ 2k-1 & \text{if } a_i = 2 \text{ for each } i \in \{1,2,\ldots,k\}, \\ \sum_{i=1}^{k-1} a_i & \text{otherwise.} \end{cases} \tag{5}$$

3 The Disjunctive Bondage Number of Graphs

3.1 Some Exact Values

In this section, we determine disjunctive bondage number for the Petersen graph, cycles, paths, and complete k-partite graphs. We begin with some observations.

Observation 5. *Let G be a nonempty graph. Then*

(a) $\gamma^d(G) = 1$ if and only if $\Delta(G) = |V(G)| - 1$;
(b) if $\Delta(G) = |V(G)| - 1$, then $b^d(G) = b(G)$;
(c) if $\gamma(G) = 2$, then $\gamma^d(G) = 2$ and $b^d(G) \geq b(G)$.

Proposition 1. *For the Petersen graph $\mathcal{P}$, $b^d(\mathcal{P}) = 3$.*

Proof. Let the Petersen graph $\mathcal{P}$ be given by two 5-cycles $u_1u_2 \ldots u_5u_1$, $w_1w_2 \ldots w_5w_1$, and five additional edges u_1w_1, u_2w_3, u_3w_5, u_4w_2, u_5w_4 (see Fig. 1). By Theorem 2(a), $\gamma^d(\mathcal{P}) = 2$. For $B \subseteq E(\mathcal{P})$, let $H = \mathcal{P} - B$.

First, let $|B| = 1$. Since $\mathcal{P}$ is edge-transitive, take $B = \{u_1u_2\}$; then $\gamma^d(H) = 2 = \gamma^d(\mathcal{P})$ since $S = \{u_3, u_5\}$ is a $\gamma^d(H)$-set. So, $b^d(\mathcal{P}) \geq 2$.

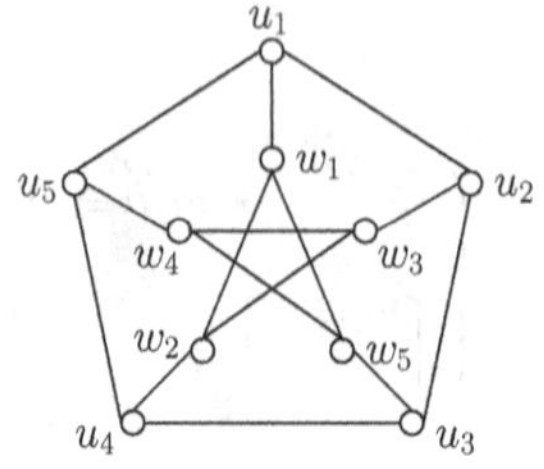

Fig. 1. The Petersen graph $\mathcal{P}$ and its labeling of vertices

Second, let $|B| = 2$. If two edges in B are adjacent, say $B = \{u_1u_2, u_2u_3\}$, then $S = \{w_2, w_4\}$ is a $\gamma^d(H)$-set; thus $\gamma^d(H) = 2 = \gamma^d(\mathcal{P})$. If two edges in B are not adjacent but have a common edge, say $B = \{u_1u_2, u_3u_4\}$, then $S = \{w_2, w_5\}$ is a $\gamma^d(H)$-set; thus $\gamma^d(H) = 2 = \gamma^d(\mathcal{P})$. If two edges in B are neither adjacent nor have a common edge, say $B = \{u_1u_2, w_4w_5\}$, then $S = \{u_3, u_5\}$ is a $\gamma^d(H)$-set; thus $\gamma^d(H) = 2 = \gamma^d(\mathcal{P})$. So, $b^d(\mathcal{P}) \geq 3$.

Next, let $|B| = 3$. If $B = \{u_1u_2, u_1u_5, u_1w_1\}$, then $H = \mathcal{P} - B$ consists of an isolated vertex and a connected graph H' with $\triangle(H') \neq |V(H')| - 1$; thus, $\gamma^d(H) \geq 3 = 1 + \gamma^d(\mathcal{P})$. So, $b^d(\mathcal{P}) \leq 3$. Therefore, $b^d(\mathcal{P}) = 3$. □

Proposition 2. *For $n \geq 3$,*

$$b^d(C_n) = \begin{cases} 3 & \text{if } n = 4, \\ 1 & \text{if } n \neq 4 \text{ and } n \equiv 0 \pmod 4, \\ 2 & \text{if } n \equiv 1, 2, 3 \pmod 4. \end{cases} \tag{6}$$

Proof. Let C_n be the n-cycle given by $v_1v_2 \ldots v_nv_1$, where $n \geq 3$.

Case 1: $n \equiv 0 \pmod 4$. We write $n = 4k$, where $k \geq 1$. First, let $k = 1$; then $\gamma(C_4) = 2 = \gamma^d(C_4)$. Since $b(C_4) = 3$ by Theorem 4(a), $b^d(C_4) \geq 3$ by Observation 5(c). Let $H = C_4 - \{v_1v_2, v_2v_3, v_3v_4\}$; then $\gamma^d(H) = 3 > \gamma^d(C_4)$, and hence $b^d(C_4) \leq 3$. Thus, $b^d(C_4) = 3$. Next, let $k \geq 2$. Then $\gamma^d(P_{4k}) = 1 + \gamma^d(C_{4k})$ by Theorem 2(b)(c); thus, $b^d(C_n) = 1$ for $n \neq 4$ and $n \equiv 0 \pmod 4$.

Case 2: $n \equiv 1, 2, 3 \pmod 4$. Since $\gamma^d(C_n) = \gamma^d(P_n)$ for $n \not\equiv 0 \pmod 4$ by Theorem 2(b)(c), $b^d(C_n) \geq 2$. Let $H = C_n - \{v_1v_2, v_2v_3\}$; then H consists of K_1 and P_{n-1}. If $n = 4k+1$ $(k \geq 1)$, then $\gamma^d(H) = \gamma^d(K_1) + \gamma^d(P_{4k}) = 1 + (k+1) = 1 + \gamma^d(C_{4k+1})$; thus $b^d(C_{4k+1}) \leq 2$. If $n = 4k+2$ $(k \geq 1)$, then $\gamma^d(H) = \gamma^d(K_1) + \gamma^d(P_{4k+1}) = 1 + (k+1) = 1 + \gamma^d(C_{4k+2})$; thus $b^d(C_{4k+2}) \leq 2$. If $n = 4k+3$ $(k \geq 0)$, then $\gamma^d(H) = \gamma^d(K_1) + \gamma^d(P_{4k+2}) = 1 + (k+1) = 1 + \gamma^d(C_{4k+3})$; thus $b^d(C_{4k+3}) \leq 2$. Therefore, $b^d(C_n) = 2$ for $n \equiv 1, 2, 3 \pmod 4$. □

Proposition 3. *For $n \geq 2$,*

$$b^d(P_n) = \begin{cases} 2 & \text{if } n = 4, \\ 1 & \text{otherwise.} \end{cases} \tag{7}$$

Proof. Let P_n be the n-path given by $v_1v_2\dots v_n$, where $n \geq 2$.

Case 1: $n \equiv 0 \pmod 4$. First, we consider for $n = 4$. For $B \subseteq E(P_4)$, if $|B| = 1$, then $P_4 - B$ is isomorphic to either $2K_2$ or $K_1 \cup P_3$; in each case, $\gamma^d(P_4 - B) = 2 = \gamma^d(P_4)$. So, $b^d(P_4) \geq 2$. Since $\gamma^d(P_4 - \{v_1v_2, v_2v_3\}) = 3 > \gamma^d(P_4)$, $b^d(P_4) \leq 2$. Thus, $b^d(P_4) = 2$. Next, let $n = 4k$ for $k \geq 2$. Let $H = P_{4k} - \{v_4v_5\}$; then H consists of P_4 and P_{4k-4}. Since $\gamma^d(H) = \gamma^d(P_4) + \gamma^d(P_{4k-4}) = 2 + k = 1 + (k+1) = 1 + \gamma^d(P_{4k})$ by Theorem 2(c), $b^d(P_n) = 1$ for $n \neq 4$ and $n \equiv 0 \pmod 4$.

Case 2: $n \equiv 1, 2, 3 \pmod 4$. Let $H = P_n - \{v_1v_2\}$; then H consists of K_1 and P_{n-1}. If $n = 4k+1$ $(k \geq 1)$, then $\gamma^d(H) = \gamma^d(K_1) + \gamma^d(P_{4k}) = 1 + (k+1) = 1 + \gamma^d(P_{4k+1})$ by Theorem 2(c). If $n = 4k+2$ $(k \geq 0)$, then $\gamma^d(H) = \gamma^d(K_1) + \gamma^d(P_{4k+1}) = 1 + (k+1) = 1 + \gamma^d(P_{4k+2})$ by Theorem 2(c). If $n = 4k+3$ $(k \geq 0)$, then $\gamma^d(H) = \gamma^d(K_1) + \gamma^d(P_{4k+2}) = 1 + (k+1) = 1 + \gamma^d(P_{4k+3})$ by Theorem 2(c). In each case, $b^d(P_n) = 1$ for $n \equiv 1, 2, 3 \pmod 4$. □

Proposition 4. *For $k \geq 2$, let $G = K_{a_1, a_1, \dots, a_k}$ be a complete k-partite graph with $a_k = \max\{a_i : 1 \leq i \leq k\}$, and let s be the number of partite sets consisting of one element. Then*

$$b^d(G) = \begin{cases} \lceil \frac{s}{2} \rceil & \text{if } s \neq 0 \\ 2k - 1 & \text{if } a_i = 2 \text{ for each } i \in \{1, 2, \dots, k\} \\ \sum_{i=1}^{k-1} a_i & \text{otherwise.} \end{cases} \tag{8}$$

Proof. For $k \geq 2$, let $G = K_{a_1, a_1, \dots, a_k}$ be a complete k-partite graph, where $a_k \geq a_{k-1} \geq \dots \geq a_1$. Let $V(G)$ be partitioned into k-partite sets $V_1, V_2, \dots, V_k$ such that $|V_i| = a_i$ for each $i \in \{1, 2, \dots, k\}$; further, let $V_i = \{u_{i1}, u_{i2}, \dots, u_{ia_i}\}$. Let s be the number of partite sets consisting of one element.

If $s \neq 0$, then $\Delta(G) = |V(G)| - 1$. By Observation 5(b) and Theorem 4(b), $b^d(G) = \lceil \frac{s}{2} \rceil$.

Next, let $s = 0$; then $a_i \geq 2$ for each $i \in \{1, 2, \dots, k\}$ and $\gamma(G) = 2 = \gamma^d(G)$. If $a_i = 2$ for each $i \in \{1, 2, \dots, k\}$, let $B = \{u_{11}u_{i1}, u_{11}u_{i2} : 2 \leq i \leq k\} \cup \{u_{12}u_{21}\}$ and let $H = G - B$; then $|B| = 2k - 1$, and H consists of K_1 and a connected graph H', where $\Delta(H') \neq |V(H')| - 1$. So, $\gamma^d(H) \geq 3 = 1 + \gamma^d(G)$; thus, $b^d(G) \leq 2k - 1$. By Observation 5(c) and Theorem 4(b), $b^d(G) = 2k - 1$. If $a_i \geq 3$ for some $i \in \{1, 2, \dots, k\}$, let $B' = \{u_{k1}u_{ij} : 1 \leq i \leq k-1, 1 \leq j \leq a_i\}$ and let $H = G - B'$; then $|B'| = \sum_{i=1}^{k-1} a_i$ and H is isomorphic to $K_1 \cup K_{a_1, a_2, \dots, a_k - 1}$. Since $a_k - 1 \geq 2$, $\gamma^d(H) = 3 = 1 + \gamma^d(G)$; thus $b^d(G) \leq \sum_{i=1}^{k-1} a_i$. By Observation 5(c) and Theorem 4(b), $b^d(G) = \sum_{i=1}^{k-1} a_i$. □

3.2 Some Upper Bounds

In this section, we obtain upper bounds of the disjunctive bondage number of some classes of graphs. We begin with the following useful lemma.

Lemma 1. *Let T be a tree. If u is a support vertex of degree two in T, then there exists a $\gamma^d(T)$-set containing u.*

Proof. Let u be a support vertex of a tree T with $\deg_T(u) = 2$. Let $N_T(u) = \{\ell, w\}$, where ℓ is a leaf neighbor of u. If w is a leaf, then $T = P_3$ and $\{u\}$ is a unique $\gamma^d(T)$-set. So, let w be a non-leaf neighbor of u, and let S be a $\gamma^d(T)$-set. Then $S \cap \{\ell, u\} \neq \emptyset$; otherwise, $\ell \notin N_T(S)$ and ℓ is at distance two from at most one vertex in S, and hence ℓ fails to be disjunctively dominated by S. If $\ell \in S$, then $(S - \{\ell\}) \cup \{u\}$ is also a $\gamma^d(T)$-set. Thus, there exists $\gamma^d(T)$-set containing u. □

Theorem 6. *If T is a nontrivial tree, then $b^d(T) \leq 2$.*

Proof. Let T be a tree of order $n \geq 2$. If $n \in \{2, 3\}$, then $b^d(T) = 1$. So, let $n \geq 4$, and let S be a $\gamma^d(T)$-set.

Case 1: T has a support vertex that is adjacent to at least two leaves. Let u be a support vertex in T and let $L_u = \{\ell_1, \ell_2, \ldots, \ell_k\}$, where $k \geq 2$, be the set of leaves that are adjacent to u in T. If $\ell_i \in S$ for some $i \in \{1, 2, \ldots, k\}$, then $(S - \{\ell_i\}) \cup \{u\}$ is also a $\gamma^d(T)$-set. So, we may assume that $S \cap L_u = \emptyset$. Let $H = T - \{u\ell_1\}$ and let S_H be a $\gamma^d(H)$-set. First, suppose that $u \in S$. If $k \geq 3$, then clearly $u \in S_H$; if $k = 2$, then u becomes a support vertex of degree two in H, and thus $u \in S_H$ by Lemma 1. So, $\gamma^d(H) = 1 + \gamma^d(T)$; thus, $b^d(T) = 1$. Second, suppose that $u \notin S$. Since $S \cap L_u = \emptyset$, u must have two or more non-leaf neighbors, say w_1 and w_2, such that $\{w_1, w_2\} \subseteq S$; then $\gamma^d(H) = 1 + \gamma^d(T)$, and hence $b^d(T) = 1$.

Case 2: Each support vertex of T is adjacent to exactly one leaf. Notice that T has a degree-two support vertex. Let u be a support vertex of degree two in T with $N_T(u) = \{\ell, w\}$, where ℓ is a leaf and w is a non-leaf neighbor of u. By Lemma 1, we may assume that $u \in S$. Let $H = T - \{u\ell, uw\}$ and let $T' = T - \{\ell, u\}$; further, let S' be a $\gamma^d(T')$-set. We will show that $b^d(T) \leq 2$. Assume, to the contrary, that $b^d(T) > 2$. Then $\gamma^d(H) = \gamma^d(T)$ and $\gamma^d(H) = \gamma^d(T') + 2$; thus $|S| = |S'| + 2$. Since $S' \cup \{u\}$ is a disjunctive dominating set for T, $|S| \leq |S'| + 1$, a contradiction. So, $b^d(T) \leq 2$. □

Based on the proof of Theorem 6, we have the following

Remark 1. Let T be a nontrivial tree.

(a) If T has a support vertex that is adjacent to at least two leaves, then $b^d(T) = 1$.
(b) There exists T satisfying $b^d(T) = 2$ when each support vertex of a tree T is adjacent to exactly one leaf. For example, let T be a tree obtained from a star $K_{1,k}$, where $k \geq 3$, by subdividing each edge exactly once (see Fig. 2). Then $S = \{v_i : 1 \leq i \leq k\}$ is a $\gamma^d(T)$-set with $|S| = k$, where $k \geq 3$. Let $T_1 = T - \{v_1 w_1\}$ and $T_2 = T - \{uv_1\}$. Then $S_i = \{v_j : 2 \leq j \leq k\} \cup \{w_1\}$ is a $\gamma^d(T_i)$-set with $|S_i| = |S|$, where $i \in \{1, 2\}$. So, $b^d(T) \geq 2$. If we let $T' = T - \{uv_1, v_1 w_1\}$, then $S' = \{v_j : 2 \leq j \leq k\} \cup \{v_1, w_1\}$ is a $\gamma^d(T')$-set with $|S'| = |S| + 1$; thus, $b^d(T) \leq 2$. Therefore, $b^d(T) = 2$.

Next, we examine the bondage number of $P_n \square P_2$, $C_n \square P_2$, and $K_n \square K_n$, respectively.

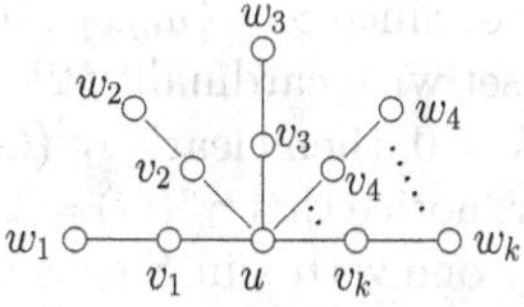

Fig. 2. A tree T with $b^d(T) = 2$

Proposition 5. *For $n \geq 3$, $b^d(P_n \Box P_2) \leq 2$. Moreover, if $n \equiv 1 \pmod 3$ and $n \geq 4$, then $b^d(P_n \Box P_2) = 1$.*

Proof. For $n \geq 3$, let $G = P_n \Box P_2$ be given by two n-paths $u_1 u_2 \dots u_n$ and $w_1 w_2 \dots w_n$, together with n edges joining u_i and w_i, where $i \in \{1, 2, \dots, n\}$. For $B \subseteq E(G)$, let $H = G - B$.

Case 1: $n = 3k$, where $k \geq 1$. By Theorem 2(d), $\gamma^d(G) = k + 1$. If $B = \{u_2u_3, w_2w_3\}$ with $|B| = 2$, then $H = (P_2 \Box P_2) \cup (P_{3k-2} \Box P_2)$. Since $\gamma^d(H) = 2 + k = 1 + (k+1) = 1 + \gamma^d(G)$ by Theorem 2(d), $b^d(G) \leq 2$.

Case 2: $n = 3k + 1$, where $k \geq 1$. By Theorem 2(d), $\gamma^d(G) = k + 1$. If $B = \{u_2u_3, w_2w_3\}$ with $|B| = 2$, then $H = (P_2 \Box P_2) \cup (P_{3k-1} \Box P_2)$ and $\gamma^d(H) = 2 + (k+1) = 2 + \gamma^d(G)$. By Theorem 1, $b^d(G) \leq 1$; thus $b^d(G) = 1$.

Case 3: $n = 3k + 2$, where $k \geq 1$. By Theorem 2(d), $\gamma^d(G) = k + 2$. If $B = \{u_2u_3, w_2w_3\}$, then $H = (P_2 \Box P_2) \cup (P_{3k} \Box P_2)$. Since $\gamma^d(H) = 2 + (k+1) = 1 + (k+2) = 1 + \gamma^d(G)$, $b^d(G) \leq 2$. □

Next, we determine $\gamma^d(C_n \Box P_2)$, which will be used in examining the bondage number of $C_n \Box P_2$.

Proposition 6. *For $n \geq 3$,*

$$\gamma^d(C_n \Box P_2) = \begin{cases} \frac{n}{3} + 1 & \text{if } n \equiv 3 \pmod 6, \\ \lceil \frac{n}{3} \rceil & \text{otherwise.} \end{cases} \tag{9}$$

Proof. For $n \geq 3$, let $G = C_n \Box P_2$ be given by two distinct n-cycles $u_1 u_2 \dots u_n u_1$ and $w_1 w_2 \dots w_n w_1$, together with n edges joining u_i and w_i, where $i \in \{1, 2, \dots, n\}$. Let S be a $\gamma^d(G)$-set. Since each vertex in S can dominate up to 4 vertices and has up to four vertices at distance two from it, $|S| \geq \frac{|V(G)|}{6}$. Without loss of generality, we may assume that $u_1 \in S$.

Case 1: $n = 6k$, where $k \geq 1$. In this case, $|S| \geq 2k$. Since $S = \{u_{6i+1} : 0 \leq i \leq k-1\} \cup \{w_{6j+4} : 0 \leq j \leq k-1\}$ forms a DD-set with cardinality $2k$, $\gamma^d(G) \leq 2k$. Thus, $\gamma^d(G) = 2k$.

Case 2: $n = 6k + 1$ or $n = 6k + 2$, where $k \geq 1$. In this case, $|S| \geq 2k + 1$. Since $S = \{u_{6i+1} : 0 \leq i \leq k\} \cup \{w_{6j+4} : 0 \leq j \leq k - 1\}$ forms a DD-set with cardinality $2k + 1$, $\gamma^d(G) \leq 2k + 1$. Thus, $\gamma^d(G) = 2k + 1$.

Case 3: $n = 6k+3$, where $k \geq 0$. Since $S = \{u_{6i+1} : 0 \leq i \leq k\} \cup \{w_{6j+4} : 0 \leq j \leq k-1\} \cup \{w_{6k+3}\}$ forms a DD-set with cardinality $2k+2$, $\gamma^d(G) \leq 2k+2$. We will show that $\gamma^d(G) \geq 2k+2$. If $k = 0$, then clearly $\gamma^d(C_3 \Box P_2) = 2$. So, let $k \geq 1$. In this case, $|S| \geq 2k+1$. Further, notice that $\gamma^d(G) = 2k+1$ only if each vertex of G is either dominated by exactly one vertex in S or is at distance two from exactly two vertices in S, but not both. Let $A_0 = C_{6k+3} \Box P_2$. Since $u_1 \in S$, by removing all vertices in $N_G[u_1] = \{u_1, u_2, u_{6k+3}, w_1\}$ and their incident edges from A_0, we obtain a derived graph A_1 with vertices in $N_G^2(u_2) = \{u_3, u_{6k+2}, w_2, w_{6k+3}\}$ being half circles. This forces w_4 and w_{6k+1} of A_1 to be in S. From A_1, by deleting vertices in $N_G[\{w_4, w_{6k+1}\}] = \{u_4, w_3, w_4, w_5, u_{6k+1}, w_{6k}, w_{6k+1}, w_{6k+2}\}$ and vertices in $\{u_3, w_2, u_{6k+2}, w_{6k+3}\}$ (the set of vertices at distance two from u_1 and, either w_4 or w_{6k+1}) together with their incident edges, we obtain a second derived graph A_2 with vertices in $\{u_5, w_6, u_{6k}, w_{6k-1}\}$ being half circles. After k iterations, we obtain A_k that is isomorphic to Fig. 3. Clearly, A_k requires three vertices to disjunctively dominate $V(A_k)$; thus $\gamma^d(G) \geq 2k+2$. Therefore, $\gamma^d(G) = 2k+2$.

Case 4: $n = 6k+4$ or $n = 6k+5$, where $k \geq 0$. In this case, $|S| \geq 2k+2$. Since $S = \{u_{6i+1} : 0 \leq i \leq k\} \cup \{w_{6j+4} : 0 \leq j \leq k\}$ forms a DD-set with cardinality $2k+2$, $\gamma^d(G) \leq 2k+2$. Thus, $\gamma^d(G) = 2k+2$. □

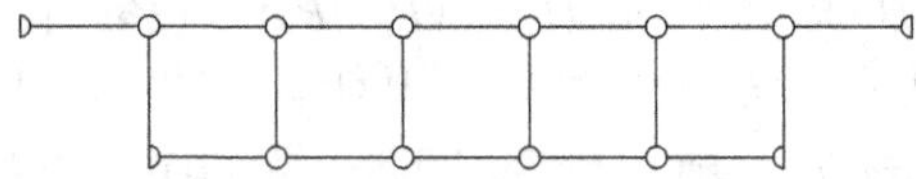

Fig. 3. A_k when $G = C_{6k+3} \Box P_2$

Proposition 7. *For $n \geq 3$, $b^d(C_n \Box P_2) \in \{2, 3, 4\}$. Furthermore, $b^d(C_n \Box P_2) = 2$ if $n \equiv 0, 2, 5 \pmod 6$, and $b^d(C_n \Box P_2) \leq 3$ if $n \equiv 1, 4 \pmod 6$.*

Proof. For $n \geq 3$, let $G = C_n \Box P_2$ be given by two distinct n-cycles $u_1 u_2 \ldots u_n u_1$ and $w_1 w_2 \ldots w_n w_1$, together with n edges joining u_i and w_i, where $i \in \{1, \ldots, n\}$. For $B \subseteq E(G)$, let $H = G - B$. If $|B| = 1$, then one can easily check that $\gamma^d(H) = \gamma^d(G)$; thus $b^d(G) \geq 2$.

Case 1: $n \equiv 0, 2, 5 \pmod 6$. If $B = \{u_1 u_2, w_1 w_2\}$ with $|B| = 2$, then $H = P_n \Box P_2$ and $\gamma^d(H) = 1 + \gamma^d(G)$ by Theorem 2(d) and Proposition 6; thus, $b^d(G) \leq 2$. Therefore, $b^d(G) = 2$.

Case 2: $n \equiv 1 \pmod 6$. We write $n = 6k+1$, where $k \geq 1$; then $\gamma^d(G) = 2k+1$ by Proposition 6. If $B = \{u_1 u_2, w_1 w_2, u_3 u_4, w_3 w_4\}$ with $|B| = 4$, then $H = (P_2 \Box P_2) \cup (P_{6k-1} \Box P_2)$ and $\gamma^d(H) = 2 + (2k+1) = 2 + \gamma^d(G)$ by Theorem 2(d) and Proposition 6; thus, $b^d(G) \leq 3$ by Theorem 1.

Case 3: $n \equiv 3 \pmod 6$. We write $n = 6k+3$, where $k \geq 0$; then $\gamma^d(G) = 2k+2$ by Proposition 6.

Subcase 3.1: $k = 0$. In this case, $G = C_3 \Box P_2$ and $\gamma^d(G) = 2$. If $B = \{u_1 u_2, u_2 u_3, u_2 w_2\}$ with $|B| = 3$, then H consists of an isolated vertex and a connected graph H' with $\Delta(H') \neq |V(H')| - 1$; thus $\gamma^d(H) \geq 3 = 1 + \gamma^d(G)$. So, $b^d(G) \leq 3$.

Subcase 3.2: $k \geq 1$. If $B = \{u_1u_2, w_1w_2, u_3u_4, w_3w_4\}$ with $|B| = 4$, then $H = (P_2 \square P_2) \cup (P_{6k+1} \square P_2)$ and $\gamma^d(H) = 2 + (2k+1) = 1 + \gamma^d(G)$ by Theorem 2(d) and Proposition 6; thus, $b^d(G) \leq 4$.

Case 4: $n \equiv 4$ *(mod 6).* We write $n = 6k + 4$, where $k \geq 0$; then $\gamma^d(G) = 2k + 2$ by Proposition 6.

Subcase 4.1: $k = 0$. In this case, $G = C_4 \square P_2$ and $\gamma^d(G) = 2$. If $B = \{u_1u_2, u_2u_3, u_2w_2\}$ with $|B| = 3$, then H consists of an isolated vertex and a connected graph H' with $\Delta(H') \neq |V(H')| - 1$; thus $\gamma^d(H) \geq 3 = 1 + \gamma^d(G)$. So, $b^d(G) \leq 3$.

Subcase 4.2: $k \geq 1$. If $B = \{u_1u_2, w_1w_2, u_6u_7, w_6w_7\}$ with $|B| = 4$, then $H = (P_5 \square P_2) \cup (P_{6k-1} \square P_2)$ and $\gamma^d(H) = 3 + (2k+1) = 2 + \gamma^d(G)$ by Theorem 2(d) and Proposition 6; thus, $b^d(G) \leq 3$ by Theorem 1. □

Proposition 8. *For* $n \geq 3$, $b^d(K_n \square K_n) \leq 2(n-1)$.

Proof. For $n \geq 3$, let $G = K_n \square K_n$; then $\gamma^d(G) = 2$ since $diam(G) = 2$. Let E_v be the set of edges that are incident to a vertex v in G. If we let $H = G - E_v$, then $\gamma^d(H) \geq 3 = 1 + \gamma^d(G)$; thus $b^d(G) \leq |E_v| = 2(n-1)$. □

Next, we express the upper bound of the disjunctive bondage number of a graph in terms of degree sum of two vertices that are at distance at most two from each other. We begin with the following observation.

Observation 7. *For a graph* G, *let* $B \subseteq E(G)$ *be a set of edges of* G *with* $|B| = k$. *If* $H = G - B$ *with* $b^d(H) = 1$, *then* $b^d(G) \leq k + 1$.

Theorem 8. *If* G *is a nonempty graph, then* $b^d(G) \leq \min\{\deg_G(u) + \deg_G(v) - 1 : d_G(u,v) \leq 2\}$.

Proof. Let $\ell = \min\{\deg_G(u) + \deg_G(v) - 1 : d_G(u,v) \leq 2\}$, and let E_w be the set of edges that are incident to a vertex w in G. Let x and y be two distinct vertices in G with $d_G(x,y) \leq 2$ satisfying $\deg_G(x) + \deg_G(y) - 1 = \ell$.

Case 1: $d_G(x,y) = 1$. Let $H_1 = G - (E_x \cup E_y - \{xy\})$; notice that $|E_x \cup E_y - \{xy\}| = \deg_G(x) + \deg_G(y) - 2$. Since $\gamma^d(H_1 - \{xy\}) = 1 + \gamma^d(H_1)$, $b^d(H_1) = 1$. By Observation 7, $b^d(G) \leq \ell$.

Case 2: $d_G(x,y) = 2$. For some $z \in N_G(x) \cap N_G(y)$, let $H_2 = G - (E_x \cup E_y - \{xz, yz\})$; notice that $|E_x \cup E_y - \{xz, yz\}| = \deg_G(x) + \deg_G(y) - 2$. Since $\gamma^d(H_2 - \{xz, yz\}) = 2 + \gamma^d(H_2)$, $b^d(H_2) = 1$ by Theorem 1. By Observation 7, $b^d(G) \leq \ell$. □

Theorem 8 implies $b^d(G) \leq \Delta(G) + \delta(G) - 1$, for a nonempty connected graph G.

4 The Disjunctive Total Bondage Number of Some Graphs

In this section, we determine disjunctive total bondage number for the Petersen graph, cycles, paths, and some complete k-partite graphs. We also show that, for a given positive integer k, there exists a tree T for which $b_t^d(T) = k$.

Observation 9. *Let G be a connected graph of order at least two. If $diam(G) \in \{1,2\}$, then $\gamma_t^d(G) = 2$.*

Proof. Let G be a connected graph of order at least two. If $diam(G) = 1$, then G is a complete graph; thus $\gamma_t^d(G) = 2$. Next, suppose that $diam(G) = 2$. Let $S \subseteq V(G)$ be a set consisting of any two adjacent vertices. Then each vertex in $N(S)$ is adjacent to a vertex in S, and each vertex in $V(G) - N(S)$ is at distance two from both vertices in S by the condition that $diam(G) = 2$; thus, $\gamma_t^d(G) \leq 2$. Since $\gamma_t^d(G) \geq 2$, we have $\gamma_t^d(G) = 2$. □

Proposition 9. *For the Petersen graph $\mathcal{P}$, $b_t^d(\mathcal{P}) = 2$.*

Proof. Let $\mathcal{P}$ be the Petersen graph as in Fig. 1. Since $diam(\mathcal{P}) = 2$, $\gamma_t^d(\mathcal{P}) = 2$ by Observation 9. For $B \subseteq E(\mathcal{P})$, let $H = \mathcal{P} - B$. If $|B| = 1$, say $B = \{u_1u_2\}$ without loss of generality (by the edge-transitivity of $\mathcal{P}$), then $\gamma_t^d(H) = 2 = \gamma_t^d(\mathcal{P})$ since $S = \{w_4, w_5\}$ is a $\gamma_t^d(H)$-set; thus $b_t^d(\mathcal{P}) \geq 2$.

Next, let $B = \{u_1u_2, u_2u_3\}$ with $|B| = 2$. Let S be a $\gamma_t^d(H)$-set of cardinality two. In order for u_2 to be disjunctively totally dominated by S, $N_H(u_2) = \{w_3\} \subseteq S$ or $N_H^2(u_2) = \{w_2, w_4\} \subseteq S$; here, $\{w_2, w_4\}$ cannot be a $\gamma_t^d(H)$-set, since $w_2w_4 \notin E(H)$. So, suppose that $w_3 \in S$. If $S = \{u_2, w_3\}$, then $u_1 \notin N_H(S)$ and u_1 is at distance at least three from each vertex in S. If $S = \{w_3, w_4\}$ or $S = \{w_3, w_2\}$, say the former without loss of generality, then $u_1 \notin N_H(S)$ and $d_H(w_3, u_1) = 3$. So, in each case, S fails to be a $\gamma_t^d(H)$-set; thus $\gamma_t^d(H) \geq 3$. Since $\{w_2, w_3, w_4\}$ forms a DTD-set for H, $\gamma_t^d(H) = 3 = 1 + \gamma_t^d(\mathcal{P})$. Therefore, $b_t^d(\mathcal{P}) = 2$. □

Proposition 10. *For $n \geq 4$,*

$$b_t^d(C_n) = \begin{cases} 1 & \text{if } n \equiv 0, 1 \pmod 5 \\ 2 & \text{if } n \equiv 2, 3, 4 \pmod 5, \end{cases} \tag{10}$$

and $b_t^d(C_3) = \infty$.

Proof. Let C_n be the n-cycle given by $v_1v_2 \ldots v_nv_1$. One can easily cheek that $b_t^d(C_3) = \infty$. So, let $n \geq 4$.

Case 1: $n \equiv 0, 1 \pmod 5$. Let $H = C_n - \{v_1v_2\}$; then $H = P_n$. If $n \equiv 0 \pmod 5$, then $\gamma_t^d(H) = \gamma_t^d(P_n) = \lceil \frac{2(n+1)}{5} \rceil = \frac{2n}{5} + 1 = \gamma_t^d(C_n) + 1$ by Theorem 3. If $n \equiv 1 \pmod 5$, then $\gamma_t^d(H) = \gamma_t^d(P_n) = \lceil \frac{2(n+1)}{5} \rceil + 1 = \gamma_t^d(C_n) + 1$ by Theorem 3. So, $b_t^d(C_n) = 1$ for $n \equiv 0, 1 \pmod 5$.

Case 2: $n \equiv 2, 3, 4 \pmod 5$. By Theorem 3, $\gamma_t^d(C_n) = \gamma_t^d(P_n)$ for $n \geq 4$; thus $b_t^d(C_n) \geq 2$. Let $H = C_n - \{v_1v_2, v_3v_4\}$; then H consists of two paths, P_2 and P_{n-2}. First, we consider $n \equiv 2 \pmod 5$. Let $n = 5k + 2$ and $n' = n - 2 = 5k$, where $k \geq 1$. Then $\gamma_t^d(H) = \gamma_t^d(P_2) + \gamma_t^d(P_{n'}) = 2 + \lceil \frac{2(n'+1)}{5} \rceil = 2 + (2k + 1) = 1 + (2k + 2) = 1 + \lceil \frac{2(n+1)}{5} \rceil = 1 + \gamma_t^d(C_n)$ by Theorem 3. Second, we consider $n \equiv 3 \pmod 5$. Let $n = 5k + 3$ and $n' = n - 2 = 5k + 1$, where $k \geq 1$. Then $\gamma_t^d(H) = \gamma_t^d(P_2) + \gamma_t^d(P_{n'}) = 3 + \lceil \frac{2(n'+1)}{5} \rceil = 3 + (2k + 1) = 2 + (2k + 2) =$

$2+\lceil\frac{2(n+1)}{5}\rceil = 2+\gamma_t^d(C_n)$ by Theorem 3. Third, we consider $n \equiv 4 \pmod 5$. Let $n = 5k+4$ and $n' = n-2 = 5k+2$, where $k \geq 0$. Then $\gamma_t^d(H) = \gamma_t^d(P_2) + \gamma_t^d(P_{n'}) = 2+\lceil\frac{2(n'+1)}{5}\rceil = 2+(2k+2) = 2+\lceil\frac{2(n+1)}{5}\rceil = 2+\gamma_t^d(C_n)$ by Theorem 3. So, $b_t^d(C_n) = 2$ for $n \equiv 2,3,4 \pmod 5$, where $n \geq 4$. □

Proposition 11. *For $n \geq 4$,*

$$b_t^d(P_n) = \begin{cases} 2 & \text{if } n = 6 \\ 1 & \text{otherwise,} \end{cases} \tag{11}$$

and $b_t^d(P_2) = b_t^d(P_3) = \infty$.

Proof. Let P_n be the n-path given by $v_1v_2\ldots v_n$. One can easily see that $b_t^d(P_2) = b_t^d(P_3) = \infty$. So, let $n \geq 4$.

Case 1: $n \equiv 1 \pmod 5$. Notice that $n \geq 6$ in this case. First, we consider $n = 6$. Then, for any $H = P_6 - e$ satisfying $e \in \{v_2v_3, v_3v_4, v_4v_5\}$, we have $\gamma_t^d(H) = 4 = \gamma_t^d(P_6)$ by Theorem 3(b). So, $b_t^d(P_6) \geq 2$. If $H' = P_6 - \{v_2v_3, v_4v_5\}$, then $\gamma_t^d(H') = 3\gamma_t^d(P_2) = 6 = 2+\gamma_t^d(P_6)$; thus, $b_t^d(P_6) = 2$. Next, let $n \neq 6$ and let $H = P_n - \{v_5v_6\}$; then H consists of two paths, P_5 and P_{n-5}. Let $n' = n-5 = 5k+1$ and $n = 5(k+1)+1$. Then $\gamma_t^d(H) = \gamma_t^d(P_5) + \gamma_t^d(P_{n'}) = 3+(\lceil\frac{2(n'+1)}{5}\rceil+1) = 3+(2k+2) = 1+(2k+4) = 1+(\lceil\frac{2(n+1)}{5}\rceil+1) = 1+\gamma_t^d(P_n)$ by Theorem 3(b). So, $b_t^d(P_n) = 1$ for $n \equiv 1 \pmod 5$ and $n \neq 6$.

Case 2: $n \equiv 0,2,3,4 \pmod 5$. Let $H = P_n - \{v_2v_3\}$; then H consists of two paths, P_2 and P_{n-2}. Let $n' = n-2$. If $n \equiv 0 \pmod 5$, i.e., $n = 5k$ and $n' = 5k-2$ for $k \geq 1$, then $\gamma_t^d(H) = \gamma_t^d(P_2)+\gamma_t^d(P_{n'}) = 2+\lceil\frac{2(n'+1)}{5}\rceil = 2+2k = 1+(2k+1) = 1+\lceil\frac{2(n+1)}{5}\rceil = 1+\gamma_t^d(P_n)$ by Theorem 3(b). If $n \equiv 2 \pmod 5$, i.e., $n = 5k+2$ and $n' = 5k$ for $k \geq 1$, then $\gamma_t^d(H) = \gamma_t^d(P_2) + \gamma_t^d(P_{n'}) = 2+\lceil\frac{2(n'+1)}{5}\rceil = 2+(2k+1) = 1+(2k+2) = 1+\lceil\frac{2(n+1)}{5}\rceil = 1+\gamma_t^d(P_n)$ by Theorem 3(b). If $n \equiv 3 \pmod 5$, i.e., $n = 5k+3$ and $n' = 5k+1$ for $k \geq 1$, then $\gamma_t^d(H) = \gamma_t^d(P_2)+\gamma_t^d(P_{n'}) = 2+(\lceil\frac{2(n'+1)}{5}\rceil+1) = 2+(2k+2) = 2+\lceil\frac{2(n+1)}{5}\rceil = 2+\gamma_t^d(P_n)$ by Theorem 3(b). If $n \equiv 4 \pmod 5$, i.e., $n = 5k+4$ and $n' = 5k+2$ for $k \geq 0$, then $\gamma_t^d(H) = \gamma_t^d(P_2) + \gamma_t^d(P_{n'}) = 2+\lceil\frac{2(n'+1)}{5}\rceil = 2+(2k+2) = 2+\lceil\frac{2(n+1)}{5}\rceil = 2+\gamma_t^d(P_n)$ by Theorem 3(b). So, $b_t^d(P_n) = 1$ for $n \equiv 0,2,3,4 \pmod 5$, where $n \geq 4$. □

Next, we determine the disjunctive total bondage number of complete bi-partite graphs.

Proposition 12. *Let $G = K_{a,b}$ be a complete bi-partite graph. Then*

$$b_t^d(G) = \begin{cases} \min\{a,b\} & \text{if } 1 \notin \{a,b\} \\ \infty & \text{if } 1 \in \{a,b\}. \end{cases} \tag{12}$$

Proof. Let $G = K_{a,b}$ be a complete bi-partite graph; notice that $\gamma_t^d(G) = 2$ and that every $\gamma_t^d(G)$-set consists of one vertex from each partite set. Let $V(G)$ be partitioned into $V_1 = \{u_{11}, u_{12}, \ldots, u_{1a}\}$ and $V_2 = \{u_{21}, u_{22}, \ldots, u_{2b}\}$; we may assume that $b \geq a$ by relabeling if necessary.

First, let $a = 1$; then $G = K_{1,b}$ is the star on $b+1$ vertices, where $b \geq 1$. For any $e \in E(G)$, $G - e$ contains an isolated vertex. Thus, $b_t^d(G) = \infty$.

Second, let $a \geq 2$. For $B \subseteq E(G)$, let $H = G - B$. If $|B| < a$, then there exist a vertex, say v, in V_1 with $N_H(v) = V_2$ and a vertex, say w, in V_2 with $N_H(w) = V_1$; thus $\{v, w\}$ forms a DTD-set for H and hence $b_t^d(G) \geq a$. If $B = \{u_{1i}u_{2i} : 1 \leq i \leq a\}$, then, for each $i \in \{1, 2 \ldots, a\}$, $N_H(u_{1i}) = V_2 - \{u_{2i}\}$ and $d_H(u_{1i}, u_{2i}) \geq 3$; thus $\gamma_t^d(H) \geq 3 = 1 + \gamma_t^d(G)$, and hence $b_t^d(G) = a$. □

We apply a method similar to the one used in proving Proposition 12 to obtain the following result.

Lemma 2. *For $k \geq 2$, let $G = K_{a_1, a_2, \ldots, a_k}$ be a complete k-partite graph of order $\sum_{i=1}^{k} a_i$, where $a_i \geq 2$ for each $i \in \{1, 2, \ldots, k\}$ and $a_k = \max\{a_i : 1 \leq i \leq k\}$. For $B \subseteq E(G)$, let $H = G - B$. If S is a $\gamma_t(H)$-set with $|S| \neq 2$, then $|B| \geq \sum_{i=1}^{k-1}(k-i)a_i$.*

Proof. Let $G = K_{a_1, a_2, \ldots, a_k}$ be a complete k-partite graph of order $\sum_{i=1}^{k} a_i$, where $k \geq 2$; then $\gamma_t(G) = 2$. Let $V(G)$ be partitioned into k-partite sets $V_1, V_2, \ldots, V_k$; let $V_i = \{u_{i,1}, u_{i,2}, \ldots, u_{i,a_i}\}$ for each $i \in \{1, 2, \ldots, k\}$, and let $a_k \geq a_{k-1} \geq \ldots \geq a_1 \geq 2$ by relabeling if necessary. For $B \subseteq E(G)$, let $H = G - B$. We note that if S is a total dominating set for G (or for H) with cardinality two, then the two vertices of S must belong to different partite sets in G. Let S be a $\gamma_t(H)$-set with $|S| \neq 2$, i.e., $|S| \geq 3$. Then, with a possible exception of exactly one partite set, say V_t for some $t \in \{1, 2, \ldots, k\}$, every vertex of V_i $(i \neq t)$ must fail to be adjacent to at least one vertex of V_j, for each $j \in \{1, 2, \ldots, k\} - \{i\}$, in H, i.e., one should delete from G at least one edge that is incident to both a vertex in V_i and a vertex in each V_j $(j \in \{1, 2, \ldots, k\} - \{i\})$; so, one should delete $(k-1)a_i$ edges, from G, that are incident to a vertex in V_i. So, the minimum number of edges one needs to delete from G, to satisfy $|S| \geq 3$, is when each $u \in V_i \subseteq V(G) - V_k$ satisfies $N_H(u) \cap V_j \neq V_j$ for each $j \in \{1, 2, \ldots, k-1\} - \{i\}$. Notice that (i) in order for $N_H(u_{1,j}) \cap V_p \neq V_p$ for each $p \in \{2, \ldots, k\}$, one should delete $(k-1)a_1$ edges from G that are incident to a vertex in V_1 and a vertex in V_p $(p \in \{2, \ldots, k\})$; (ii) in order for $N_H(u_{2,j}) \cap V_q \neq V_q$ for each $q \in \{1, 2, \ldots, k\} - \{2\}$, one should delete $(k-2)a_2$ edges from G that are incident to a vertex in V_2 and a vertex in V_q $(q \in \{3, \ldots, k\})$, since a_1 edges that are incident to both a vertex in V_1 and a vertex in V_2 are already deleted in the step described in (i); (iii) by continuing the process, in order for $N_H(u_{k-1,j}) \cap V_r \neq V_r$ for each $r \in \{1, 2, \ldots, k\} - \{k-1\}$, one should delete a_{k-1} edges from G that are incident to a vertex in V_{k-1} and a vertex in V_k, since the edges that are incident to both a vertex in V_{k-1} and a vertex in each V_i $(i \in \{1, 2 \ldots, k-2\})$ are already deleted in the previous steps. Thus, the minimum number of edges one needs to delete from G, to satisfy $|S| \geq 3$, is $(k-1)a_1 + (k-2)a_2 + \ldots + a_{k-1} = \sum_{i=1}^{k-1}(k-i)a_i$; thus $|B| \geq \sum_{i=1}^{k-1}(k-i)a_i$. □

Theorem 10. *For $k \geq 2$, let $G = K_{a_1,a_2,\ldots,a_k}$ be a complete k-partite graph of order $\sum_{i=1}^{k} a_i$, where $a_i \geq 2$ for each $i \in \{1,2,\ldots,k\}$ and $a_k = \max\{a_i : 1 \leq i \leq k\}$. Then $b_t(G) = \sum_{i=1}^{k-1}(k-i)a_i$.*

Proof. Let G be a complete k-partite graph as described in the current theorem, where $k \geq 2$; then $\gamma_t(G) = 2$. Let $V(G)$ be partitioned into k-partite sets $V_1, V_2, \ldots, V_k$; let $V_i = \{u_{i,1}, u_{i,2}, \ldots, u_{i,a_i}\}$ for each $i \in \{1,2,\ldots,k\}$, and let $a_k \geq a_{k-1} \geq \ldots \geq a_1 \geq 2$ by relabeling if necessary. For $B \subseteq E(G)$, let $H = G - B$. By Lemma 2, $b_t(G) \geq \sum_{i=1}^{k-1}(k-i)a_i$. Let $B = \cup_{i=1}^{k-1}(\cup_{j=i+1}^{k}\{u_{i,1}u_{j,1}, u_{i,2}u_{j,2}, \ldots, u_{i,a_i}u_{j,a_i}\})$; notice that $|B| = \sum_{i=1}^{k-1}(k-i)a_i$. We will show that $\gamma_t(H) > \gamma_t(G)$. Assume, to the contrary, that S is a $\gamma_t(H)$-set with $|S| = 2$; then the two vertices of S belong to different partite sets in G. We may assume that $u_{i,x} \in V_i \cap S$ for some $i \in \{1,2,\ldots,k-1\}$. Then $S \cap V_\alpha = \emptyset$ for each $\alpha < i$, since, for each vertex $u_{\alpha,y} \in V_\alpha$, $u_{\alpha,y}u_{i,y} \notin E(H)$ and $u_{i,x}u_{i,y} \notin E(H)$. Also, notice that $S \cap V_\beta = \emptyset$ for each $\beta > i$: (i) for each $j \in \{1,2,\ldots,a_i\}$, $u_{\beta,j}u_{i,j} \notin E(H)$ and $u_{i,x}u_{i,j} \notin E(H)$; (ii) for each $p > a_i$, $u_{\beta,p}u_{\beta,x} \notin E(H)$ and $u_{i,x}u_{\beta,x} \notin E(H)$. So, any vertex set consisting of two adjacent vertices in H fails to be a total dominating set of H; thus $\gamma_t(H) > 2$. So, $\gamma_t(H) \geq 3 = 1 + \gamma_t(G)$, and hence $b_t(G) \leq \sum_{i=1}^{k-1}(k-i)a_i$. Therefore, $b_t(G) = \sum_{i=1}^{k-1}(k-i)a_i$. □

Proposition 13. *For $k \geq 2$, let $G = K_{a_1,a_2,\ldots,a_k}$ be a complete k-partite graph of order $\sum_{i=1}^{k} a_i$, where $a_i \geq 2$ for each $i \in \{1,2,\ldots,k\}$ and $a_k = \max\{a_i : 1 \leq i \leq k\}$. Then $b_t^d(G) \geq \sum_{i=1}^{k-1}(k-i)a_i$. Moreover, if $a_i = 2$ for each $i \in \{1,2,\ldots,k\}$, then $b_t^d(G) = k(k-1)$.*

Proof. Let $G = K_{a_1,a_2,\ldots,a_k}$ be a complete k-partite graph, where $k \geq 2$; further, assume that $a_k \geq a_{k-1} \geq \ldots \geq a_1 \geq 2$ by relabeling if necessary. Then $\gamma_t^d(G) = 2$; here, we note that $b_t^d(G) \geq b_t(G)$ when $\gamma_t^d(G) = 2 = \gamma_t(G)$. By Lemma 2, $b_t^d(G) \geq \sum_{i=1}^{k-1}(k-i)a_i$. Next, let $a_i = 2$ for each $i \in \{1,2,\ldots,k\}$. Let $V(G)$ be partitioned into k-partite sets $V_1, V_2, \ldots, V_k$; let $V_i = \{u_{i,1}, u_{i,2}\}$ for each $i \in \{1,2,\ldots,k\}$. For $B \subseteq E(G)$, let $H = G - B$. Let $B = \cup_{i=1}^{k-1}(\cup_{j=i+1}^{k}\{u_{i,1}u_{j,1}, u_{i,2}u_{j,2}\})$; notice that $|B| = \sum_{i=1}^{k-1} 2(k-i) = \sum_{i=1}^{k-1} 2k - \sum_{i=1}^{k-1} 2i = 2k(k-1) - (k-1)k = k(k-1)$. By Theorem 10, $b_t^d(G) \geq b_t(G) = k(k-1)$. Let D be a $\gamma_t^d(H)$-set. Since H is regular and vertex-transitive, we may assume that $u_{1,1} \in D$. Then $N_H(u_{1,1}) = \{u_{j,2} : 2 \leq j \leq k\}$. If $k = 2$, then $H = 2K_2$, and hence $\gamma_t^d(H) = 4 = 2 + \gamma_t^d(G)$. If $k \geq 3$, then $d_H(u_{1,1}, u_{1,2}) = 3$ and $d_H(u_{j,2}, u_{1,2}) = 2$ for each $j \in \{2,\ldots,k\}$; thus, any vertex set consisting of $u_{1,1}$ and exactly one vertex in $N_H(u_{1,1})$ fails to be a DTD-set for H, i.e., $|D| \geq 3$. In each case, $\gamma_t^d(H) \geq 3 = 1 + \gamma_t^d(G)$, and hence $b_t^d(G) \leq k(k-1)$. Therefore, $b_t^d(G) = k(k-1)$ when $a_i = 2$ for each $i \in \{1,2,\ldots,k\}$. □

Next, we give an example showing that, for a given positive integer k, there exists a tree T for which $b_t^d(T) = k$.

Remark 2. For a given positive integer k, there exists a tree T for which $b_t^d(T) = k$. For $k \geq 2$, let T be a tree obtained from the star $S_{k+1} = K_{1,k+1}$ on $k+2$ vertices by subdividing each edge exactly twice; let v_0 be the central vertex of

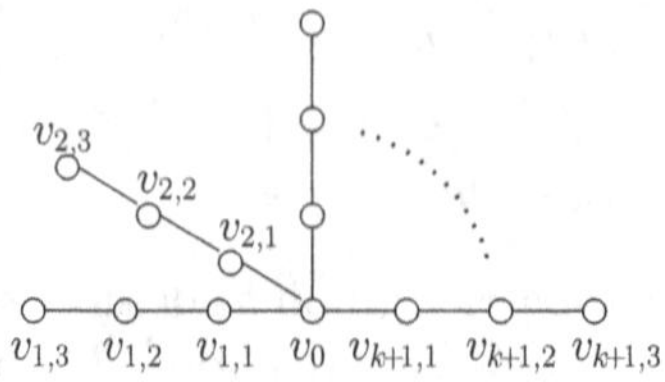

Fig. 4. A tree T with $b_t^d(T) = k$

degree $k+1$, and let each of the 4-path rooted at v_0 be given by $v_0, v_{i,1}, v_{i,2}, v_{i,3}$, where $i \in \{1, 2, \ldots, k+1\}$ (see Fig. 4).

First, we show that $\gamma_t^d(T) = 2k + 2$. Let S be a $\gamma_t^d(T)$-set. For each $i \in \{1, 2, \ldots, k+1\}$, (i) in order for $v_{i,3}$ to be disjunctively totally dominated by S, $v_{i,2} \in S$; (ii) in order for $v_{i,2}$ to be disjunctively totally dominated by S, $S \cap \{v_{i,1}, v_{i,3}\} \neq \emptyset$. So, $\gamma_t^d(T) \geq 2(k+1)$. Since $\{v_{i,1}, v_{i,2} : 1 \leq i \leq k+1\}$ forms a DTD-set for T, $\gamma_t^d(T) \leq 2(k+1)$. Thus, $\gamma_t^d(T) = 2k+2$.

Second, we show that $b_t^d(T) = k$. For $B \subseteq E(T)$, let $H = T - B$. If $B = \{v_{i,1}v_{i,2} : 1 \leq i \leq k\}$ with $|B| = k$, then $H = kP_2 \cup T'$, where T' is a tree obtained from S_{k+1} by subdividing exactly one edge exactly twice, and $\gamma_t^d(H) = k\gamma_t^d(P_2) + \gamma_t^d(T') = 2k+3 = 1 + \gamma_t^d(T)$; thus $b_t^d(T) \leq k$. So, suppose that $|B| = k-1$. Then, for each $i \in \{1, 2, \ldots, k+1\}$, (i) $v_{i,2}v_{i,3} \notin B$ since $b_t^d(T) < \infty$; (ii) $|B \cap \{v_0v_{i,1}, v_{i,1}v_{i,2}\}| \leq 1$. We may assume that $B = \{v_0v_{i,1} : 1 \leq i \leq s\} \cup \{v_{i,1}v_{i,2} : s+1 \leq i \leq k-1\}$ by relabeling if necessary, where $s = 0$ (i.e., $B = \{v_{i,1}v_{i,2} : 1 \leq i \leq k-1\}$) or $s = k-1$ (i.e., $B = \{v_0v_{i,1} : 1 \leq i \leq k-1\}$) are possibilities. Then $H = sP_3 \cup (k-s-1)P_2 \cup T^*$, where T^* can be viewed as a tree obtained from S_{k-s+1} by subdividing exactly two edges exactly twice. Notice that $\gamma_t^d(H) = s\gamma_t^d(P_3) + (k-s-1)\gamma_t^d(P_2) + \gamma_t^d(T^*) = 2s + 2(k-s-1) + 4 = 2k+2 = \gamma_t^d(T)$; here, $\{v_{k,1}, v_{k,2}, v_{k+1,1}, v_{k+1,2}\}$ forms a $\gamma_t^d(T^*)$-set. So, $b_t^d(T) \geq k$. Thus, $b_t^d(T) = k$ for $k \geq 2$.

5 Closing Remark

It is known that $\gamma^d(G) \leq \gamma(G)$ for any graph (see [4]) and that $\gamma_t^d(G) \leq \gamma_t(G)$ for any graph G containing no isolated vertex (see [8]). We note that both $\gamma(G) - \gamma^d(G)$ and $\gamma_t(G) - \gamma_t^d(G)$ can be arbitrarily large: for example, if $G = K_n \square K_n$ for $n \geq 3$, then one can easily check that $\gamma(G) = \gamma_t(G) = n$ and $\gamma^d(G) = \gamma_t^d(G) = 2$. It would be an interesting problem to investigate the relation between $b(G)$ and $b^d(G)$, as well as the relation between $b_t(G)$ and $b_t^d(G)$, if any. Moreover, it would be worth studying the computational complexity of determining the disjunctive (total) bondage number of general graphs.

Acknowledgement. The author wishes to thank the anonymous referees for some constructive and helpful comments and suggestions.

References

1. Bauer, D., Harary, F., Nieminen, J., Suffel, C.L.: Domination alteration sets in graphs. Discrete Math. **47**, 153–161 (1983)
2. Fink, J.F., Jacobson, M.S., Kinch, L.F., Roberts, J.: The bondage number of a graph. Discrete Math. **86**, 47–57 (1990)
3. Garey, M.R., Johnson, D.S.: Computers and Intractability: A Guide to the Theory of NP-Completeness. Freeman, New York (1979)
4. Goddard, W., Henning, M.A., McPillan, C.A.: The disjunctive domination number of a graph. Quaest. Math. **37**(4), 547–561 (2014)
5. Haynes, T.W., Hedetniemi, S.T., Slater, P.J.: Fundamentals of Domination in Graphs. Marcel Dekker, New York (1998)
6. Henning, M.A.: A survey of selected recent results on total domination in graphs. Discrete Math. **309**, 32–63 (2009)
7. Henning, M.A., Marcon, S.A.: Domination versus disjunctive domination in trees. Discrete Appl. Math. **184**, 171–177 (2015)
8. Henning, M.A., Naicker, V.: Disjunctive total domination in graphs. J. Comb. Optim. DOI: 10.1007/s10878-014-9811-4
9. Hu, F.T., Xu, J.M.: Complexity of bondage and reinforcement. J. Complex. **28**(2), 192–201 (2012)
10. Kulli, V.R., Patwari, D.K.: The total bondage number of a graph. In: Kulli, V.R. (ed.) Advances in Graph Theory, pp 227–235. Vishwa, Gulbarga (1991)
11. Xu, J.M.: On bondage numbers of graphs: a survey with some comments. Int. J. Comb. **2013**, 13 (2013). Article ID: 595210

Edge-Disjoint Packing of Stars and Cycles

Minghui Jiang[1], Ge Xia[2], and Yong Zhang[3](✉)

[1] Department of Computer Science, Utah State University, Logan, UT 84322, USA
mjiang@cc.usu.edu
[2] Department of Computer Science, Lafayette College, Easton, PA 18042, USA
xiag@lafayette.edu
[3] Department of Computer Science and Information Technology, Kutztown University, Kutztown, PA 19530, USA
zhang@kutztown.edu

Abstract. We study the parameterized complexity of two graph packing problems, EDGE-DISJOINT k-PACKING OF s-STARS and EDGE-DISJOINT k-PACKING OF s-CYCLES. With respect to the choice of parameters, we show that although the two problems are FPT with both k and s as parameters, they are unlikely to be fixed-parameter tractable when parameterized by only k or only s. In terms of kernelization complexity, we show that EDGE-DISJOINT k-PACKING OF s-STARS has a kernel with size polynomial in both k and s, but in contrast, unless NP $\subseteq$ coNP/poly, EDGE-DISJOINT k-PACKING OF s-CYCLES does not have a kernel with size polynomial in both k and s, and moreover does not have a kernel with size polynomial in s for any fixed k. Specifically, (1) from the negative direction, we show that EDGE-DISJOINT k-PACKING OF s-STARS is W[1]-hard with parameter k in general graphs, and that EDGE-DISJOINT k-PACKING OF s-CYCLES is W[1]-hard with parameter k and NP-hard for any even $s \geq 4$ in bipartite graphs; (2) from the positive direction, we show that EDGE-DISJOINT k-PACKING OF s-STARS admits a ks^2 kernel, and that EDGE-DISJOINT k-PACKING OF 4-CYCLES admits a $96k^2$ kernel in general graphs and a $96k$ kernel in planar graphs.

1 Introduction

Given two undirected input graphs G and H, the edge-disjoint graph packing problem is to find in G the maximum number of edge-disjoint subgraphs isomorphic to H. Many combinatorial problems such as the chromatic index problem and the independent set problem can be viewed as edge packing problems. The reliability of a computer network can be efficiently bounded using techniques based on edge packing [6,7].

We use *s-star* to denote the complete bipartite graph $K_{1,s}$, and *s-cycle* to denote the simple cycle graph C_s of length s. In this paper we study the edge-disjoint packing of s-stars and s-cycles in the context of *parameterized complexity*. A *parameterized problem* is a set of instances of the form (x, k), where x is the input instance and k is a nonnegative integer called the *parameter*. A parameterized problem is said to be *fixed parameter tractable* if there is an algorithm that

Z. Lu et al. (Eds.): COCOA 2015, LNCS 9486, pp. 676–687, 2015.
DOI: 10.1007/978-3-319-26626-8_49

solves the problem in time $f(k)|x|^{O(1)}$, where f is a computable function solely dependent on k, and $|x|$ is the size of the input instance. When dealing with NP-hard problems in practice, *kernelization* is a very useful preprocessing technique. The idea of kernelization is to design data reduction rules to reduce the input instance to an equivalent *kernel* of smaller size. Formally, the *kernelization* of a parameterized problem is a reduction to a *problem kernel*, that is, to apply a polynomial-time algorithm to transform any input instance (x, k) to an equivalent reduced instance (x', k') such that $k' \leq k$ and $|x'| \leq g(k)$ for some function g solely dependent on k. It is known that a parameterized problem is fixed parameter tractable if and only if the problem is kernelizable. We refer interested readers to [8, 14] for more details. We define the parameterized version of these two problems as follows. Given an undirected input graph G and two integers k and s, EDGE-DISJOINT k-PACKING OF s-STARS is to determine whether G has an edge-disjoint packing of s-stars of size at least k, EDGE-DISJOINT k-PACKING OF s-CYCLES is to determine whether G has an edge-disjoint packing of s-cycles of size at least k.

EDGE-DISJOINT k-PACKING OF s-STARS is solvable in linear time when $s = 2$ [19] and NP-complete when $s \geq 3$ [9]. Heath and Vergara [16] showed that EDGE-DISJOINT k-PACKING OF s-STARS is polynomial time solvable in trees and outerplanar graphs. On the other hand, they showed that when $s \geq 3$, EDGE-DISJOINT k-PACKING OF s-STARS is NP-complete in planar graphs. The vertex disjoint version of s-star packing was first studied by Prieto and Sloper [22]. They gave an $O(k^2)$ kernel for k-VERTEX DISJOINT s-STAR PACKING for the cases when $s \geq 3$ and a $15k$ kernel for the case when $s = 2$. The two kernelization results were further improved to $O(k)$ by Fellows et al. [11] and $7k$ by Wang et al. [23].

The main application of EDGE-DISJOINT k-PACKING OF s-CYCLES lies in computational biology [2]. Holyer [17] showed that EDGE-DISJOINT k-PACKING OF s-CYCLES is NP-complete in general graphs. Heath and Vergara [16] showed that it remains NP-complete in planar graphs. Mathieson, Prieto and Shaw [20] gave a $4k$ problem kernel for EDGE-DISJOINT k-PACKING OF s-CYCLES when $s = 3$. This kernel was recently improved to $3.5k$ by Yang [24]. Bodlaender, Thomassé and Yeo [4] studied a similar problem k-EDGE DISJOINT CYCLE PACKING, in which the cycle packing may contain cycles of different and arbitrary lengths. They give an $O(k \log k)$ kernel for this version of the cycle packing problem. As for the vertex-disjoint version of s-cycle packing, Fellows et al. [12] implies an $O(k^3)$ kernel for vertex-disjoint triangle packing, which was improved to $45k^2$ by Moser [21]. Guo and Niedermeier [15] gave a $732k$ kernel for vertex-disjoint triangle packing in planar graphs.

Prieto and Sloper [22] posed as an open problem whether problem kernels and fixed-parameter algorithms can be obtained for the edge-disjoint version of packing s-stars with $s \geq 3$. Mathieson, Prieto and Shaw [20] posed as an open problem whether problem kernels can be obtained for edge-disjoint s-cycle packing with $s \geq 4$.

In this paper we study the parameterized complexity as well as the classical computational complexity of Edge-Disjoint k-Packing of s-Stars and Edge-Disjoint k-Packing of s-Cycles. Our results are grouped into the next two sections, with the various hardness results in Sect. 2 and the polynomial kernelization results in Sect. 3.

The two problems are both fixed-parameter tractable when parameterized by both k and s, and admit $2^{O(ks)}$-time algorithms as implied by a k-FPT algorithm[1] for the general edge-disjoint graph packing problem [5]. On the other hand, the NP-hardness of the two problems for any constant $s \geq 3$ [9,17] implies that unless P = NP, they are not FPT with only s as parameter. We show that with only k as parameter, the two problems are both W[1]-hard and hence are not FPT either, unless FPT = W[1]. Moreover, Edge-Disjoint k-Packing of s-Cycles is W[1]-hard even in bipartite graphs when parameterized by k. We also study the classical computational complexity of Edge-Disjoint k-Packing of s-Cycles in specific graph classes, and show that it remains NP-complete when restricted to bipartite graphs and balanced 2-interval graphs.

The $2^{O(ks)}$-time algorithms for the two problems imply that they have kernels with size exponential in both k and s. It is natural to ask whether kernels with size polynomial in k and s are possible. Fellows et al. gave an $O(k^r)$ kernel for r-Set Packing [13, Lemma 6], which implies an $O(k^s)$ kernel[2] for the general graph packing problem of determining whether a graph G contains k edge-disjoint copies of another graph H with s vertices, for any fixed s. Note that the result of Fellows et al. [13] does not imply a size-$O(k^s)$ kernel for the general graph packing problem with both k and s as parameters, because the enumeration of all copies of H (s-stars or s-cycles in our setting) in graph packing into s-sets in set packing requires $n^{O(s)}$ preprocessing time. Also, the size of the implied kernel is only polynomial in k when s is fixed.

We show that Edge-Disjoint k-Packing of s-Stars in fact has a kernel with size polynomial in both k and s, but in contrast, unless NP $\subseteq$ coNP/poly, Edge-Disjoint k-Packing of s-Cycles does not have a kernel with size polynomial in both k and s, and moreover does not have a kernel with size polynomial in s for any fixed k. Our kernelization results include a ks^2 kernel for Edge-Disjoint k-Packing of s-Stars in general graphs, a $96k^2$ kernel for

[1] For the general problem of deciding whether a graph G contains k edge-disjoint copies of a graph H, the running time of this algorithm [5, Corollary 5.7] is $4^{rk+O(\log^3(rk))}n^{r+1}$, where n and r are the numbers of vertices in G and H, respectively. The n^{r+1} factor in the running time accounts for the time for solving the subproblem of determining whether G contains H as a subgraph, by brute force. The brute-force algorithm for this subproblem can be replaced by faster algorithms in our setting, with polynomial running time when H is an s-star, or $2^{O(s)}\text{poly}(n)$ time when H is an s-cycle [1].

[2] We remark that in our setting of Edge-Disjoint k-Packing of s-Cycles, since any two different s-cycles share at most $s-2$ edges, the base case of $f(0,k)=1$ in the analysis of [13, page 170] can be upgraded to $f(1,k)=1$, which saves 1 in the exponent and yields an $O(k^{s-1})$ kernel.

EDGE-DISJOINT k-PACKING OF 4-CYCLES in general graphs, and a $96k$ kernel for EDGE-DISJOINT k-PACKING OF 4-CYCLES in planar graphs.

2 Hardness Results

We first study the complexities of EDGE-DISJOINT k-PACKING OF s-STARS and EDGE-DISJOINT k-PACKING OF s-CYCLES with various choices of parameters. The NP-hardness of both problems for any constant $s \geq 3$ [9,17] implies that they are not s-FPT unless P = NP. In the following two theorems we show that the two problems are not k-FPT unless FPT = W[1].

Theorem 1. EDGE-DISJOINT k-PACKING OF s-STARS *is W[1] [1]-hard with parameter k.*

Proof. By an FPT reduction from k-INDEPENDENT SET. Given a graph G with $n \geq 2$ vertices, we construct a graph G' by connecting each vertex of degree d in G to $n - d$ distinct dummy vertices. Then G has an independent set of size k if and only if G' has k edge-disjoint n-stars. □

Theorem 2. EDGE-DISJOINT k-PACKING OF s-STARS[3] *is W[1]-hard with parameter k, even in bipartite graphs.*

Proof. We first prove the W[1]-hardness of EDGE-DISJOINT k-PACKING OF s-CYCLES in general graphs, by an FPT reduction from BIN PACKING. Given a set of $\hat{n}$ items of integer sizes a_i, $1 \leq i \leq \hat{n}$, and $\hat{\kappa}$ bins each of integer size b, BIN PACKING is the problem of deciding whether the set of $\hat{n}$ items can be partitioned into $\hat{\kappa}$ subsets, such that the total size of the items in each subset is at most b. Our reduction is from a restricted version of BIN PACKING where all integers a_i and b are encoded in unary and moreover $\hat{\kappa} \cdot b = \sum_{i=1}^{\hat{n}} a_i$. Note that under this latter condition, the problem simply asks whether the set of $\hat{n}$ items can be partitioned into $\hat{\kappa}$ subsets, such that the total size of the items in each subset is exactly b. Jansen et al. [18] proved that BIN PACKING is W[1]-hard with parameter $\hat{\kappa}$ even for this restricted version.

Put $\hat{m} = \hat{n}^3(2\hat{n} - 1)(\hat{n} + 1) + 1$. We construct a graph G with $\hat{\kappa} \cdot b \cdot \hat{m} + \hat{m} - 1$ edges in $\hat{n} + \hat{n}^3(2\hat{n} - 1)$ edge-disjoint paths:

- $\hat{n}$ vertex-disjoint *long* paths $p_i \to q_i$ of length $a_i \cdot \hat{m}$, $1 \leq i \leq \hat{n}$;
- $\hat{n}(2\hat{n} - 1)$ vertex-disjoint (except the two endpoints) *short* paths including $\hat{n}$ paths of each length $2, 3 \ldots, 2\hat{n}$ between q_i and p_j for each of the $\hat{n}^2$ pairs (i, j) with $1 \leq i, j \leq \hat{n}$.

Note that the total length of the $\hat{n}^3(2\hat{n} - 1)$ short paths is $\hat{n}^3(2\hat{n} - 1) \cdot (\hat{n} + 1) = \hat{m} - 1$. Set $k = \hat{\kappa}$ and $s = b \cdot \hat{m} + 2\hat{n}$. We claim that the $\hat{n}$ items of sizes a_i

[3] We remark that the construction in our proof of Theorem 2 can be easily adapted to prove the hardness of all four variants of related problems on vertex/edge-disjoint cycles/paths.

can be partitioned into $\hat{\kappa}$ subsets each of total size b if and only if G contains k edge-disjoint s-cycles.

We first prove the direct implication of the claim. Suppose that the $\hat{n}$ items of sizes a_i can be partitioned into $\hat{\kappa}$ subsets each of total size b. For each subset of t items, a cycle of total length $s = b \cdot \hat{m} + 2\hat{n}$ in G can be composed of t long paths and t short paths in an alternating pattern, where the t long paths correspond to the t items, and the t short paths include $t-1$ paths of the length 2 and one path of length $2\hat{n} - 2(t-1)$. Since there are $\hat{n} \geq t$ edge-disjoint short paths of each length $2, 3 \ldots, 2\hat{n}$ between each pair of endpoints of the long paths, these $k = \hat{\kappa}$ cycles can be easily made edge-disjoint.

We next prove the reverse implication of the claim. Suppose that G contains k edge-disjoint s-cycles. Since the total length of all short paths is less than $\hat{m}$ and the length of each long path is a multiple of $\hat{m}$, each cycle of length $s = b \cdot \hat{m} + 2\hat{n}$ where $2\hat{n} < \hat{m}$ must include a certain number of long paths with total length $b \cdot \hat{m}$, and the items corresponding to these long paths must have total size b. Thus the k edge-disjoint s-cycles correspond to $\hat{\kappa}$ disjoint subsets of items each of total size b.

This proves the W[1]-hardness of EDGE-DISJOINT k-PACKING OF s-CYCLES in general graphs. We show that the problem remains W[1]-hard in bipartite graphs by an FPT reduction from the same problem in general graphs. For any graph G, let G_2 be the bipartite graph obtained by subdividing each edge of G into two edges. Then G has k edge-disjoint s-cycles if and only if G_2 has k edge-disjoint $2s$-cycles. Note the parameter k remains the same. □

Thus it is necessary that we use both k and s as parameters. Recall that a result of Chen et al. [5, Corollary 5.7] implies a $2^{O(ks)}$-time algorithm for EDGE-DISJOINT k-PACKING OF s-CYCLES, which in turn implies a kernel exponential in both k and s for this problem. It is natural to ask whether a polynomial kernel is possible. In the next theorem we show that this is unlikely to be the case.

Theorem 3. EDGE-DISJOINT k-PACKING OF s-CYCLES *does not have a kernel with size polynomial in both k and s unless NP $\subseteq$ coNP/poly.*

Proof. Suppose there is a kernelization algorithm that computes a kernel with size polynomial in both k and s for EDGE-DISJOINT k-PACKING OF s-CYCLES. Then for $k = 1$, the same algorithm computes a kernel with size polynomial in s only for the problem of finding a single s-cycle. Since deciding whether a graph has a cycle of a certain length is well-known to have no polynomial kernel with the cycle length as parameter unless NP $\subseteq$ coNP/poly (see for example Corollary 6 in [4]), the theorem follows. □

Recall that for any fixed s, a result of Fellows et al. [13, Lemma 6] implies that EDGE-DISJOINT k-PACKING OF s-CYCLES has a kernel with size polynomial in k. It is natural to ask whether this is also true for the symmetric case, i.e., whether EDGE-DISJOINT k-PACKING OF s-CYCLES has a kernel with size polynomial in s for any fixed k. In the next theorem we show that this is again unlikely to be the case.

Theorem 4. *For any fixed $k \geq 1$,* Edge-Disjoint k-Packing of s-Cycles *does not have a kernel with size polynomial in s unless NP $\subseteq$ coNP/poly.*

Proof. We use the framework developed in Bodlaender, Jansen, and Kratsch [3]. Corollary 3.6 in [3] states that if an NP-hard problem L AND/OR-cross-composes into a parameterized problem Q, then Q does not admit a polynomial kernel unless NP $\subseteq$ coNP/poly and the polynomial hierarchy collapses. Let L be the NP-hard problem Hamiltonian Path. Let Q be Edge-Disjoint k-Packing of s-Cycles parameterized by s. Let $k \leq 1$ be a constant. We construct an OR-cross-composition from L into Q as follows. Given t input graphs $G_1, \ldots, G_t$ of L with each graph G_i containing n vertices. Construct the input graph H for Q as the disjoint union of G_is, $1 \leq i \leq t$, and $k-1$ copies of C_n. Let $s = n$. Clearly H has a k-edge disjoint s-cycle packing if and only if one of the G_is has a Hamiltonian path. □

Next we study the complexities of Edge-Disjoint k-Packing of s-Cycles in bipartite graphs and 2-interval graphs. Recall that a bipartite graph has no odd cycles.

Theorem 5. Edge-Disjoint k-Packing of s-Cycles *in bipartite graphs is NP-hard for any even $s \geq 4$.*

Proof. It is known that this problem is NP-hard for any $s \geq 3$ in general graphs [17]. For any graph G, let G_2 be the bipartite graph obtained by subdividing each edge of G into two edges. Then G has k edge-disjoint s-cycles if and only if G_2 has k edge-disjoint $2s$-cycles. It follows from [17] that this problem is NP-hard for any even $s \geq 6$ in bipartite graphs. The NP-hardness of this problem for $s = 4$ in bipartite graphs can be proved in the following by a similar reduction as in [10].

It is known that deciding whether a cubic graph G is edge-3-colorable is NP-complete [17]. Equivalently, deciding whether a cubic graph G admits an edge-partition into 3 perfect matchings is NP-complete. Given a cubic graph G with n vertices, let G_1 be the graph with three dummy vertices v_1, v_2, and v_3 that are connected to every vertex in G. We claim that G is edge-3-colorable if and only if G_1 admits an edge-partition into $3n/2$ edge-disjoint triangles.

For the direction implication of the claim, suppose that the cubic graph G is edge-3-colorable and hence admits an edge-partition into 3 perfect matchings. Then the $n/2$ edges in any one of the 3 perfect matchings and the n edges incident to any one of the three dummy vertices form $n/2$ edge-disjoint triangles. In total we have $3n/2$ edge-disjoint triangles.

For the reverse implication of the claim, suppose that G_1 admits an edge partition into $3n/2$ edge-disjoint triangles. Since the three dummy vertices are disjoint and each of them has degree n, the $3n/2$ edge-disjoint triangles must consist of exactly $n/2$ triangles incident to each dummy vertex. For each dummy vertex v, the $n/2$ triangles incident to v contain $n/2$ disjoint edges not incident to v, which form a perfect matching in G. Thus G is edge-3-colorable.

Let G_2 be the bipartite graph obtained from G_1 by subdividing each edge uv in G (which is in G_1 too) into a path of two edges. We claim that G is edge-3-colorable if and only if the edges of G_2 can be partitioned into $3n/2$ edge-disjoint 4-cycles.

The direct implication of this claim for G_2 can be proved in a similar way as the direct implication of the claim for G_1. For the reverse implication of the claim, suppose that G_2 admits an edge partition into $3n/2$ edge-disjoint 4-cycles. Observe that each 4-cycle in G_2 must contain exactly two vertices u and v from G. Moreover, it must be one of two types: either (i) consisting of the four edges from u and v to two dummy vertices, or (ii) consisting of the two-edge path corresponding to the edge uv in G and the two edges from u and v to one dummy vertex. Since there are exactly $3n/2$ two-edge paths between vertices from G, it follows that the $3n/2$ edge-disjoint 4-cycles are all of the second type. Then the same argument as before shows that G is edge-3-colorable. □

Given a family of sets $\mathcal{F} = \{S_1, \ldots, S_n\}$, the *intersection graph* $\Omega(\mathcal{F})$ of $\mathcal{F}$ is the graph with $\mathcal{F}$ as the vertex set and with two different vertices S_i and S_j adjacent if and only if $S_i \cap S_j \neq \emptyset$. We call $\mathcal{F}$ a *representation* of the graph $\Omega(\mathcal{F})$. Let $t \geq 2$ be an integer. A *t-interval* is the union of t disjoint intervals in the real line. A *t-track interval* is the union of t disjoint intervals on t disjoint parallel lines called tracks, one interval on each track. A *t-interval graph* (resp. *t-track interval graph*) is the intersection graph of a family of t-intervals (resp. t-track intervals). Note that the t disjoint tracks for a t-track interval graph can be viewed as t disjoint "host intervals" in the real line for a t-interval graph. If a t-interval graph has a representation in which the t disjoint intervals of each t-interval have the same length (although the intervals from different t-intervals may have different lengths), then the graph is a *balanced t-interval graph.* Similarly we define balanced t-track interval graphs.

Theorem 6. EDGE-DISJOINT k-PACKING OF s-CYCLES *is NP-hard in balanced 2-interval graphs for any $s \geq 3$, and is NP-hard in balanced 2-track interval graphs for any even $s \geq 4$.*

Proof. We adapt the NP-hardness proof of k-EDGE DISJOINT 3-CYCLE PACKING in split graphs [10]. The reduction is from EDGE-DISJOINT k-PACKING OF s-CYCLES in general graphs, which is known to be NP-complete [17]. We first show the proof for the case $s = 4$. Given an input graph $G = (V, E)$ with n vertices and m edges, we construct a graph $G' = (V', E')$ where

$$V' = V \cup \{w_{uv}^1, w_{uv}^2 \mid uv \notin E\}$$

$$E' = E \cup \{uv, uw_{uv}^1, w_{uv}^1 w_{uv}^2, w_{uv}^2 v \mid uv \notin E\}.$$

Basically we add to G' a 4-cycle for each nonadjacent pair of vertices in G. It is easy to show that G has a 4-cycle packing of size k if and only if G' has a 4-cycle packing of size $k + \binom{n}{2} - m$.

Next we show that G' is a balanced 2-track interval graph. Note that the n vertices in G form a clique in G' and the added vertices form chains of length 3 connecting pairs of vertices in the clique. We construct a family F of balanced 2-track intervals as shown in Fig. 1 to represent G'.

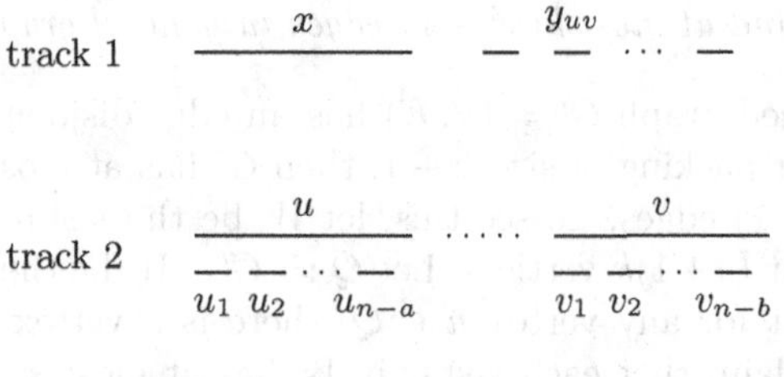

Fig. 1. An illustration of the construction of $G'_{\mathcal{F}}$ in Theorem 6.

On track 1 there are $\binom{n}{2} - m + 1$ disjoint intervals. The first interval, labeled by x, has length n; the other intervals, labeled by y_{uv}, have length 1. On track 2 there are two rows of intervals. The first row has n disjoint intervals of length n, one for each vertex in V. For a vertex $u \in V$, if the degree of u in G is a, then there are $n - a$ disjoint intervals $u_1, u_2, \ldots, u_{n-a}$ of length 1 on the second row, all intersecting with the interval for u in the first row.

For every vertex $u \in V$, add a 2-track interval (x, u) to $\mathcal{F}$. For a vertex $u \in V$ with degree a, fix a one-to-one correspondence between vertices nonadjacent to u in G and intervals u_i with $1 \leq i \leq n - a$. For each pair of nonadjacent vertices u and v in G, let u_i be the interval corresponding to v and v_j be the interval corresponding to u based on the one-to-one correspondence, we add two 2-track intervals (y_{uv}, u_i) and (y_{uv}, v_j) to $\mathcal{F}$. In total there are $n + 2(\binom{n}{2} - m)$ balanced 2-track intervals in $\mathcal{F}$.

The proof technique and the interval construction naturally extend to other values of s. More intervals can be added to track 1 and track 2 to represent larger cycles. For odd s, however, only a balanced 2-interval graph can be constructed since one of the 2-intervals has to be on the same track. In particular, for $s = 3$, each pair of 2-track intervals (y_{uv}, u_i) and (y_{uv}, v_j) is contracted into a single 2-interval (u_i, v_j). □

3 Kernelization Results

Fellows et al. [13] give an $O(k^r)$ kernel for general graph edge-disjoint packing problems based on ideas developed for r-Set Packing. This implies an $O(k^s)$ problem kernel for Edge-Disjoint k-Packing of s-Stars and Edge-Disjoint k-Packing of s-Cycles for any fixed s. In this section we present improved kernelization bounds.

3.1 Kernelization for Edge-Disjoint k-Packing of s-Stars

We show that Edge-Disjoint k-Packing of s-Stars has a ks^2 kernel in general graphs.

Reduction Rule 1. Remove any edge and vertex that do not belong to any s-star.

Theorem 7. EDGE-DISJOINT k-PACKING OF s-STARS *admits a kernel with at most* ks^2 *vertices and at most* $k(s^2-1)$ *edges in general graphs.*

Proof. If the reduced graph $G' = (V, E)$ has an edge-disjoint s-star packing of size k but no s-star packing of size $k+1$, then G' has at most ks^2 vertices and at most $k(s^2+s-1)$ edges. To see this, let W be the s-star packing of size k, then W has at most $(s+1)k$ vertices. Let $Q := G' - W$ be the rest of the graph. First we claim that for any vertex $u \in Q$, there is a vertex $v \in W$ such that $uv \in E$. Next we claim that each s-star in W has at most s^2-s-1 neighbors in Q. To see this, note that every vertex in W has at most $s-1$ neighbors in Q, since otherwise it leads to another edge-disjoint s-star not in the packing. The situation where an s-star has the maximum number of neighbors in Q is where the center of the s-star has $s-1$ neighbors in Q and each leaf node of the s-star has $s-2$ neighbors in Q. Therefore the total number of vertices is at most $k(s^2-s-1) + k(s+1) = ks^2$.

We use $V[W]$ and $E[W]$ to denote the vertex set and the edge set of the packing W. There are exactly ks edges in $E[W]$. For a vertex $u \in V[W]$, consider those edges adjacent to u that are not in $E[W]$. By the same argument as in the preceding paragraph, the total *unused* degree of the $s+1$ vertices of each s-star is at most $s-1+s(s-2) = s^2-s-1$. Thus there are at most $k(s^2-s-1)$ unused edges incident to the vertices in W. This also includes the edges between W and Q. Moreover, note that Q induces an independent set. To see this, suppose there is an edge between two vertices in Q, then by Reduction Rule 1 this edge must belong to some s-star. However, this s-star is not in the packing W, contradicting that W is maximum. Therefore the total number of edges is at most $ks + k(s^2-s-1) = k(s^2-1)$. □

3.2 Kernelization for Edge-Disjoint k-Packing of 4-Cycles

We next study the kernelization of EDGE-DISJOINT k-PACKING OF s-CYCLES when $s = 4$. We first show that EDGE-DISJOINT k-PACKING OF 4-CYCLES has a quadratic kernel in general graphs.

Reduction Rule 1. Delete all vertices and edges that do not belong to any 4-cycle.

We enumerate all 4-cycles in G and use a greedy algorithm to compute a maximal set of 4-cycles such that any pair of 4-cycles in the maximal set share at most one edge. During the computation of this maximal set we keep track of the number of 4-cycles that share the same edge and apply the following reduction rule whenever applicable.

Reduction Rule 2. If the number of 4-cycles in the maximal set that share an edge uv is greater than $4k-4$, then remove uv from G, and decrease k by 1.

Reduction Rule 1 is obviously correct. To see the correctness of Reduction Rule 2, let G and G' be the graphs before and after deleting uv. Then G has a k-packing if and only if G' has a $(k-1)$-packing. The forward direction of this claim is obvious. For the other direction, note that the 4-cycles in the maximal set that share the edge uv have all distinct edges except uv. If G' has a $(k-1)$-packing, then each 4-cycle in the packing uses edges in at most four different 4-cycles in G that share the edge uv. There is at least one 4-cycle left in the maximal set that can be added to the packing, yielding a k-packing in G.

Restart the computation of the maximal set after each application of Reduction Rule 2. Let S be the final maximal set when Reduction Rule 2 is no longer applicable and W be the subgraph of G' induced by vertices of S. Let $Q := G' - W$ be the rest of the graph. Then we have the following four claims.

Claim 1. *If the number of 4-cycles in S is greater than $16k^2$, then at least k of them are edge-disjoint.*

Proof. By Reduction Rule 2, the number of 4-cycles in S that share an edge is at most $4k-4$. So each 4-cycle in S shares edges with at most $4(4k-4-1) = 16k-20$ other 4-cycles in S. If the number of 4-cycles in S is greater than $16k^2 > (k-1)(16k-20+1)+1$, then even a greedy algorithm can find a subset of at least k 4-cycles in S that are edge-disjoint. □

Claim 2. *Q induces an independent set.*

Proof. If there is an edge in the graph induced by Q, then this edge must belong to a 4-cycle. This 4-cycle will share at most one edge with 4-cycles in S, and therefore should be included in W due to maximality of S. □

We call a vertex u in Q the *cycle-mate* of a 4-cycle C in S if u belongs to a 4-cycle C' that shares two consecutive edges with C.

Claim 3. *Every vertex in Q is the cycle-mate of at least one 4-cycle in S.*

Proof. By Reduction Rule 1, a vertex u in Q must belong to at least one 4-cycle C'. Since at least two edges of C' are between W and Q, the 4-cycle C' shares at most two edges with any 4-cycle in S. If C' shares less than two edges with each 4-cycle in S, then it should be included in S. Thus C' must share exactly two consecutive edges with a 4-cycle C in S. □

Claim 4. *Every 4-cycle in S has at most two cycle-mates in Q, one for each pair of diagonal vertices.*

Proof. If a 4-cycle C in S has two cycle-mates in Q that share consecutive edges with C between the same pair of diagonal vertices, then the two cycle-mates together with four edges between W and Q will form a 4-cycle which should be included in S. □

By Claim 1, we can assume without loss of generality that the number of 4-cycles in S is at most $16k^2$ and hence the number of vertices in W is at most $64k^2$. Then by Claims 3 and 4, the number of vertices in Q is at most $32k^2$. Combining the sizes of W and Q, we have the following theorem.

Theorem 8. Edge-Disjoint k-Packing of 4-Cycles *admits a kernel with at most* $96k^2$ *vertices in general graphs.*

When restricted to planar graphs, the kernel size for Edge-Disjoint k-Packing of 4-Cycles can be further improved to linear. Given a 4-cycle packing P, we call a 4-cycle C an *uncovered 4-cycle* if none of the edges in C is in the edge sets of P. If P is a maximum 4-cycle packing, then there should not be any uncovered 4-cycle. We call a 4-cycle *dangling* if only one edge in the 4-cycle is shared with other 4-cycles.

Reduction Rule 1. Delete all vertices and edges that do not belong to any 4-cycle.

Reduction Rule 2. If there is a dangling 4-cycle, then add this 4-cycle in the solution set, delete it from G, decrease k by 1.

Reduction Rule 1 is obviously correct. The correctness of Reduction Rule 2 follows by a simple replacement argument: Consider any edge-disjoint maximum 4-cycle packing and any dangling 4-cycle C. If this packing does not include C, then it must contains exactly one 4-cycle C' that intersects C. By removing C' and adding C, we obtain another edge-disjoint maximum 4-cycle packing.

Theorem 9. Edge-Disjoint k-Packing of 4-Cycles *admits a kernel with at most* $96k$ *vertices in planar graphs.*

References

1. Alon, N., Yuster, R., Zwick, U.: Color-coding. J. ACM **42**(4), 844–856 (1995)
2. Bafna, V., Pevzner, P.A.: Genome rearrangements and sorting by reversals. SIAM J. Comput. **25**(2), 272–289 (1996)
3. Bodlaender, H.L., Jansen, B.M.P., Kratsch, S.: Kernelization lower bounds by cross-composition. SIAM J. Discrete Math. **28**(1), 277–305 (2014)
4. Bodlaender, H.L., Thomassé, S., Yeo, A.: Kernel bounds for disjoint cycles and disjoint paths. Theor. Comput. Sci. **412**(35), 4570–4578 (2011)
5. Chen, J., Kneis, J., Lu, S., Molle, D., Richter, S., Rossmanith, P., Sze, S., Zhang, F.: Randomized divide-and-conquer: improved path, matching, and packing algorithms. SIAM J. Comput. **38**(6), 2526–2547 (2009)
6. Colbourn, C.J.: The Combinatorics of Network Reliability. Oxford University Press Inc., New York (1987)
7. Colbourn, C.J.: Edge-packings of graphs and network reliability. Discrete Math. **72**, 49–61 (1988)
8. Downey, R.G., Fellows, M.R.: Parameterized Complexity. Springer, New York (1999)
9. Dyer, M.E., Frieze, A.M.: On the complexity of partitioning graphs into connected subgraphs. Discrete Appl. Math. **10**(2), 139–153 (1985)
10. Feder, T., Subi, C.S.: Packing edge-disjoint triangles in given graphs. Electron. Colloquium Comput. Complex. (ECCC) **19**, 13 (2012)
11. Fellows, M.R., Guo, J., Moser, H., Niedermeier, R.: A generalization of Nemhauser and Trotter's local optimization theorem. J. Comput. Syst. Sci. **77**(6), 1141–1158 (2011)

12. Fellows, M.R., Heggernes, P., Rosamond, F.A., Sloper, C., Telle, J.A.: Finding k disjoint triangles in an arbitrary graph. In: Proceedings of the 30th International Workshop on Graph-Theoretic Concepts in Computer Science, pp. 235–244 (2004)
13. Fellows, M.R., Knauer, C., Nishimura, N., Ragde, P., Rosamond, F., Stege, U., Thilikos, D.M., Whitesides, S.: Faster fixed-parameter tractable algorithms for matching and packing problems. Algorithmica **52**(2), 167–176 (2008)
14. Guo, J., Niedermeier, R.: Invitation to data reduction and problem kernelization. SIGACT News **38**(1), 31–45 (2007)
15. Guo, J., Niedermeier, R.: Linear problem kernels for NP-hard problems on planar graphs. In: Arge, L., Cachin, C., Jurdziński, T., Tarlecki, A. (eds.) ICALP 2007. LNCS, vol. 4596, pp. 375–386. Springer, Heidelberg (2007)
16. Heath, L.S., Vergara, J.P.C.: Edge-packing in planar graphs. Theor. Comput. Syst. **31**, 629–662 (1998)
17. Holyer, I.: The NP-completeness of some edge-partition problems. SIAM J. Comput. **10**(4), 713–717 (1981)
18. Jansen, K., Kratsch, S., Marx, D., Schlotter, I.: Bin packing with fixed number of bins revisited. J. Comput. Syst. Sci. **79**, 39–49 (2013)
19. Masuyama, S., Ibaraki, T.: Chain packing in graphs. Algorithmica **6**, 826–839 (1991)
20. Mathieson, L., Prieto, E., Shaw, P.: Packing edge disjoint triangles: a parameterized view. In: Downey, R.G., Fellows, M.R., Dehne, F. (eds.) IWPEC 2004. LNCS, vol. 3162, pp. 127–137. Springer, Heidelberg (2004)
21. Moser, H.: A problem kernelization for graph packing. In 35th International Conference on Current Trends in Theory and Practice of Computer Science, pp. 401–412 (2009)
22. Prieto, E., Sloper, C.: Looking at the stars. Theor. Comput. Sci. **351**(3), 437–445 (2006)
23. Wang, J., Ning, D., Feng, Q., Chen, J.: An improved parameterized algorithm for a generalized matching problem. In: Agrawal, M., Du, D.-Z., Duan, Z., Li, A. (eds.) TAMC 2008. LNCS, vol. 4978, pp. 212–222. Springer, Heidelberg (2008)
24. Yang, Y.: Towards optimal kernel for edge-disjoint triangle packing. Inf. Process. Lett. **114**(7), 344–348 (2014)

Dynamic Minimum Bichromatic Separating Circle

Bogdan Armaselu(✉) and Ovidiu Daescu

Department of Computer Science, The University of Texas at Dallas, Richardson, TX, USA
{bxa120530,daescu}@utdallas.edu

Abstract. In the Minimum Bichromatic Separating Circle problem, we are given a set of n red points R and a set of m blue points B, and we want to find the smallest circle C that contains all red points and the smallest possible number of blue points. In this paper we study the Dynamic Minimum Bichromatic Separating Circle and present data structures that allow to perform efficient dynamic operations on the input points, specifically for adding and removing blue points. We first present a unified data structure that allows both additions and deletions. Then, we present data structures that allow to report an optimal circle in logarithmic time when only insertions are allowed, at the expense of an increase in update time. Our results are the first known for the Dynamic Minimum Bichromatic Separating Circle problem.

Keywords: Dynamic · Bichromatic · Separating circle

1 Introduction

In the Minimum Bichromatic Separating Circle (MBSC) problem, the input is a set of n red points R and a set of m blue points B, and the goal is to find the smallest circle C that contains all red points and the smallest possible number of blue points. In general, the optimal solution is not unique. In this paper we consider the Dynamic version of the MBSC problem, termed the Dynamic Minimum Bichromatic Separating Circle (OMBSC) problem. In the OMBSC problem the goal is to maintain the optimal solution under addition and deletion of red and blue points. In this work address only the addition and removal of blue points, while the set of red points is fixed at input.

According to Bitner, Daescu et al. [3], the motivation of studying the MBSC problem comes form military missions (e.g. one is given a set of enemy targets (red points) and wants to destroy all enemy targets while minimizing other damage, marked as blue points). In practice, it is often possible that the data is dynamically updated, which in turn rises the need for fast updates on the optimal solution.

This research was partially supported by NSF award IIP1439718 and CPRIT award RP150164.

Z. Lu et al. (Eds.): COCOA 2015, LNCS 9486, pp. 688–697, 2015.
DOI: 10.1007/978-3-319-26626-8_50

1.1 Related Work

In 2010, Bitner, Daescu et al. [3] studied the (static) MBSC problem and presented two solutions for finding the optimal circle, based on the static Farthest Point Voronoi Diagram of the red points ($FVD(R)$), one with $O((m+n)\ logn + m^{1.5} \log^{O(1)} m)$ time and $O(m^{1.5} \log^{O(1)} m)$ space based on circular range queries, and one with $O(mn \log m + n \log n)$ time and $O(m+n)$ space based on sweeping $FVD(R)$. In the same year, in a different paper, Daescu et al. [6] gave a linear programming appraoch for the problem, with an expected running time of $O(N + k^{11/4} N^{1/4} \log^{O(1)} N)$, where $N = m + n$ and k is the number of blue points in the MBSC. More recently, Cheung, Daescu et al. [7] solved the Kinetic MSC problem, in which points can move along linear trajectories, and the goal is to find the locus of the MBSC over time.

Barbay et al. [1] considered a related problem, where circles are replaced by boxes and points have associated weights. Given a set P of n points in d-dimensional space, where each point p in P is associated a real weight $w(p)$, the maximum-weight box problem is to find an axis-aligned box B maximizing $\sum_{p \in (P \cap B)} w(p)$. In the plane, they describe an algorithm with $O(n^2)$ worst case running time. They further show that the maximum-weight box problem can be solved in $O(n^d)$ time for any constant $d \geq 2$. If the points are partitioned in red and blue based on their weight (positive or negative), with n red points and m blue points, then they give an algorithm that runs in $(n \min\{n, m\})$ time. Very recently, Bereg and Daescu [2] considered a similar problem for circles in the plane. They study the problem of finding a circle of smallest radius such that the total weight of the points covered by the circle is maximized and present an algorithm with polynomial time depending on the number of points with positive and negative weight. They also consider a restricted version of the problem where the center of the circle should be on a given line and give an algorithm that runs in $O(n(m+n)\log(m+n))$ time using $O(m+n)$ space. Moreover, for this version, if all positively weighted points are required to be included within the circle, they provide an algorithm that runs in $O((n+m)\log(n+m))$ time.

Another related problem, the geometric version of the red-blue set cover for unit squares, was very recently considered by Chan and Hu [4]. In this version, given a red point set, a blue point set, and a set of objects (unit squares) the goal is to chose a subset of objects that cover all red points while minimizing the number of blue points covered. They prove that this problem is NP-hard even when the objects are unit squares in the plane and give a polynomial-time approximation scheme.

1.2 Our Results

In this paper we address only the insertion and deletion of blue points (that is, the set of red points is known at the input) and present novel data structures to handle these operations. Our first solution is a unified data structure that allows both insertion and removal of blue points and can handle updates and queries in $O(n \log m + m)$ time. Then we present data structures that allow to report an optimal circle in logarithmic time when only insertions are allowed, at the expense of higher times for updating the data structures. Specifically, the update time is $O(nm\log(mn))$. This makes sense since the update can be done as a "backstage" operation, assuming that queries arise at a reasonable rate in practice. Our algorithms are the first reported for the OMBSC problem.

We assume that, having initially available n red points and m blue points, the $O(mn\log m + n\log n)$ time and $O(m+n)$ space static algorithm of [3] has been executed and the set M of $O(n)$ minimum separating circles is available.

Let $FVD(R)$ be the FVD of the set of red points.

Definition 1. Let e be an edge of $FVD(R)$. For any $q \in e$, we denote by $C(q)$ the separating circle centered in q and passing through the two red points defining e.

Definition 2. [3] Let e be an edge of $FVD(R)$. A point $q \in e$ is an *enter* (resp. *exit*) event point if its corresponding blue point is included (resp. excluded) from the circle centered at q and passing through the two red points defining e, as we sweep along e in increasing order of circle radius.

Definition 3. Let $e_{i,j}$ be an edge of $FVD(R)$ defined by two red points p_i and p_j. For any $q, r \in e_{i,j}$, we say that q is **to the left of** r, and we write $q < r$, if the circle centered at q and defined by p_i and p_j has a smaller radius than the one centered in r and defined by p_i and p_j.

2 A Unified Approach for Insertion/Deletion of Blue Points

In this section we present an unified approach that allows for efficient insertion and deletion of blue points.

Let m_k be the number of blue points in an MBSC of M when m blue points are present. Let $m(C)$ be the number of blue points in a circle C. For every edge e of the $FVD(R)$, and for each exit event point $q \in e$, assume we have stored a number m_q, denoting the number of blue points included in $C(q)$. Let q_e be the leftmost exit event point such that m_{q_e} is minimum for all $q \in e$. If one of the current circles in the MBSC set M is centered on e then obviously $m_q = m_k$, otherwise $m_q \geq m_k$.

Insertion (Algorithm 1a). Suppose we want to add a new blue point p. For every edge e of $FVD(R)$, let M_e be the circle with minimum radius and number of blue points among all circles centered on e and passing through the two red points defining e. We compute a possible new candidate circle M'_e after the insertion of p. To do that, we first compute the event point $s \in e$ corresponding to p. Then, based on the location of s, there are only five possible cases to consider.

(1) $s \notin e$ (see Fig. 1 for an illustration). In this case, M_e remains the same. We set $M'_e = M_e$.

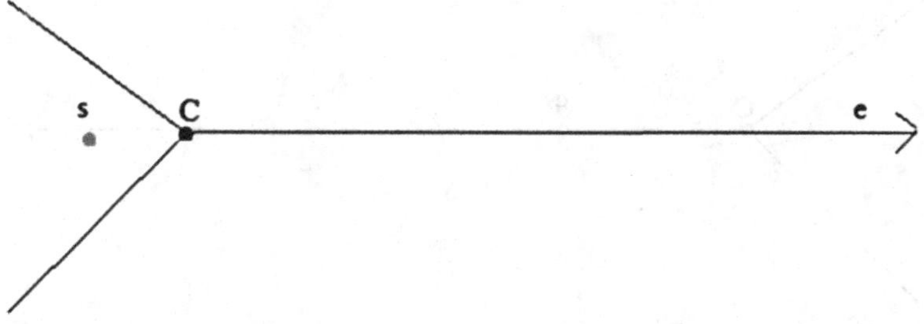

Fig. 1. Case 1: $s \notin e$

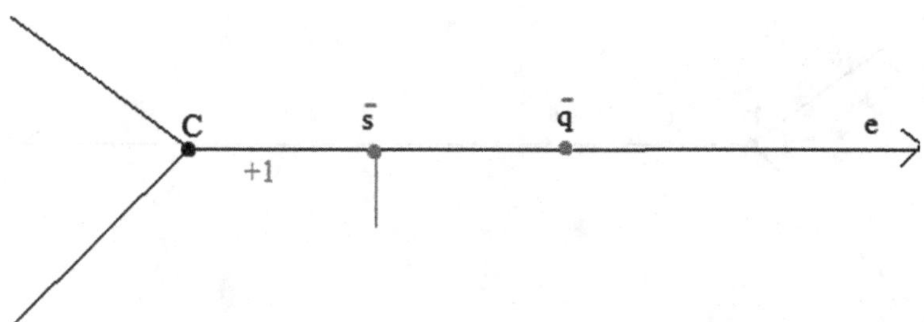

Fig. 2. Case 2: $s < q$ and s is an exit event point

(2) $s < q$ and s is an exit event point (see Fig. 2). In this case, only the event points $t < s$ will have their m_t increased. We set $M'_e = M_e$.
(3) $q < s$ and s is an enter event point (see Fig. 3). In this case, only the event points $s < t$ will have their m_t increased. We set $M'_e = M_e$.
(4) $s < q$ and s is an enter event point (see Fig. 4). In this case, we find the leftmost exit event point $t < s$ such that $m_t = m_q + 1$. If such a point exists, then we set $M'_e = C(t)$. Otherwise we set $M'_e = M_e$.
(5) $q < s$ and s is an exit event point (see Fig. 5). In this case, we need to find the leftmost exit event point $t \geq s$ such that $m_t = m_q$. If such point exists, then we set $M'_e = C(t)$, otherwise we set $M'_e = M_e$.

After all edges e are treated, we keep only the candidate circles M''s for which $m(M')$ is minimum.

Deletion (Algorithm 1b). Suppose we want to remove a blue point p. Similarly as for addition of a blue point, there are only 5 possible cases to consider.

(1) $s \notin e$ (see Fig. 1 for an illustration). In this case, M_e remains the same. We set $M'_e = M_e$.
(2) $q < s$ and s is an exit event point (see Fig. 6). In this case, the event points $t < s$ will have their m_t decreased, including q. We set $M'_e = M_e$.
(3) $s < q$ and s is an enter event point (see Fig. 7). In this case, the event points $s < t$ will have their m_t decreased, including q. We set $M'_e = M_e$.

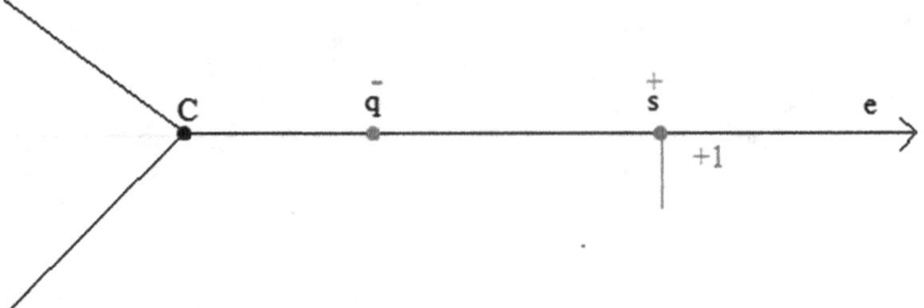

Fig. 3. Case 3: $q < s$ and s is an enter event point

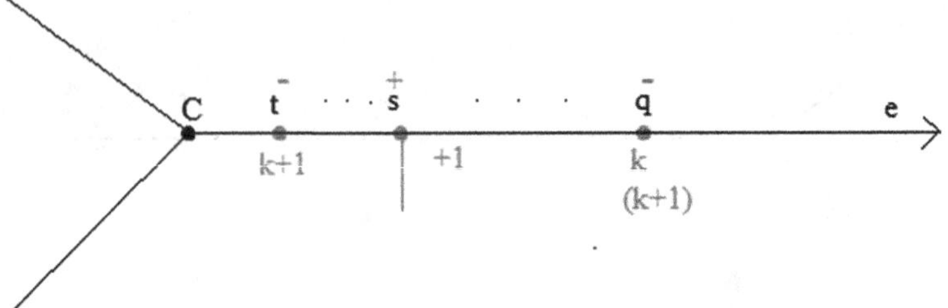

Fig. 4. Case 4: $s < q$ and s is an enter event point

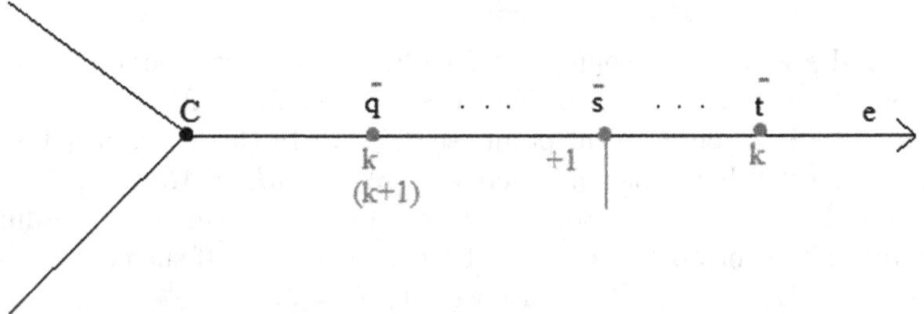

Fig. 5. Case 5: $q < s$ and s is an exit event point

(4) $q < s$ and s is an enter event point (see Fig. 8). In this case, we need to find the leftmost exit event point $t \geq s$ such that $m_t = m_q$. If such point exists, then we set $M'_e = C(t)$, otherwise we set $M'_e = M_e$.
(5) $s < q$ and s is an exit event point (see Fig. 9). In this case, we need to find the leftmost exit event point $t < s$ such that $m_t = m_q + 1$. If such point exists, then we set $M'_e = M(t)$, otherwise we set $M'_e = M_e$.

Theorem 1. Insertion (resp., deletion) of a blue point can be performed in $O(n \log m + m)$ time each, using **Algorithm 1a** (resp., **Algorithm 1b**).

Proof. For each edge e of the $FVD(R)$, we store the following:

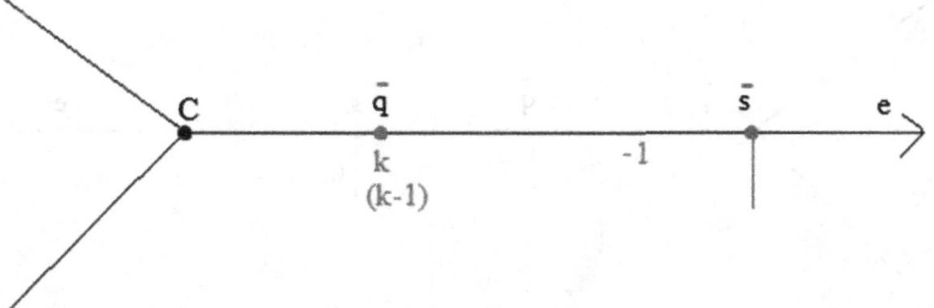

Fig. 6. Case 2: $q < s$ and s is an exit event point

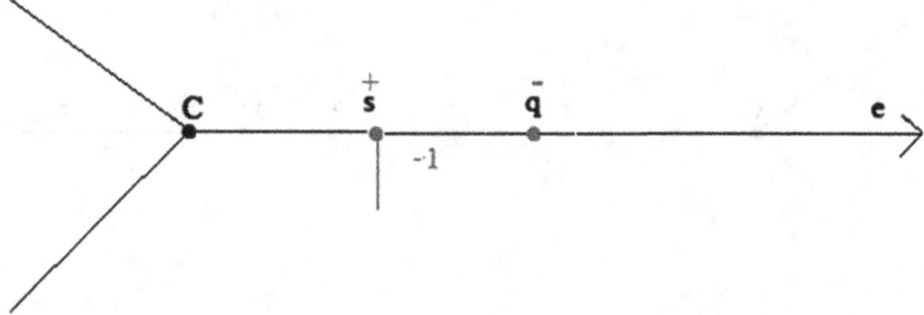

Fig. 7. Case 3: $s < q$ and s is an enter event point

1. the leftmost exit event point q such that $C(q)$ is minimum for all $q \in e$
2. for each exit event point q, store m_q
3. for each exit event point corresponding to a MBSC, store the following:
   3.1. the leftmost exit event point $l(q) < q$ such that $m_{l(q)} = m_q + 1$
   3.2. the leftmost exit event point $r(q) \geq q$ such that $m_{r(q)} = m_q$.

The event points on e are stored in a binary search tree, and hence, simply inserting an event point s can be done in $O(\log m)$ time. After inserting s, we treat cases (1) and (3) in $O(1)$ time. Cases (2) and (5) require $O(m)$ time each, as we need to find $l(s)$ and $r(s)$, and we do that by sweeping along the edge. We also need to increment (decrement) m_q for each exit event point q between $l(s)$ and s. Also, if $l(q)$ or $(r(q))$ is null, we may need to set $l(q)$ (or $r(q)$) $= s$. However, as stated in [3], every blue point has an exit event point on at most one edge of the $FVD(R)$. Hence, cases (2) and (5) occur only once. Case (4) occurs $O(n)$ times, but is treated in $O(1)$ time for each occurrence, as we already know $l(q)$ and $r(q)$ for each edge e. In order to update the data structure, we need $O(m + n)$ time to compute m_s for the edge where s is an exit event point. We require another $O(m)$ time to update all other m_q's, as there are $O(m)$ of them. We also spend $O(m + n)$ time updating $l(q)$ and $r(q)$ and computing $l(s)$ and $r(s)$, over all edges of $FVD(R)$. Hence, the insertion (deletion) of a blue point takes $O(n \log m + m)$ time. □

Notice that our update solution is polynomially better than recomputing the optimal solution from scratch, using the algorithms in [3].

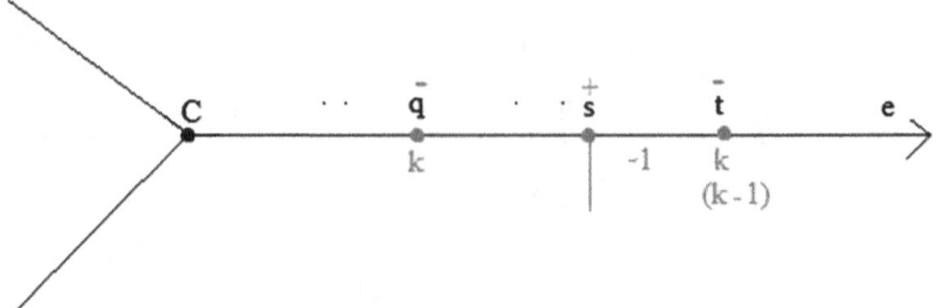

Fig. 8. Case 4: $q < s$ and s is an enter event point

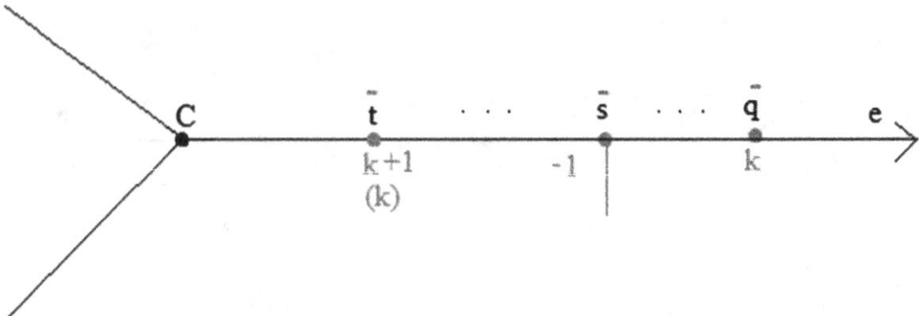

Fig. 9. Case 5: $s < q$ and s is an exit event point

3 Logarithmic Query for Insertions

In this section we present data structures that allow to report an optimal circle in logarithmic time when only insertions are allowed, at the expense $O(nm\log(mn))$ time for updating the data structures. This makes sense in practice since the update can be done as a "backstage" operation, assuming that queries arise at a reasonable rate.

Algorithm 2. We maintain all enclosing circles defined by event points on the edges of $FVD(R)$ in a balanced binary search tree like data structure, T, with $O(m)$ nodes. The keys of T are the number of blue points i, and the values are lists T^i of enclosing circles with i blue points. Each T^i is stored in a balanced binary search tree like data structure that is an extension of the Offline Ball Exclusion Search Data Structure (OLBES) introduced in [5]. Specifically, the circles are stored at the leaves of T^i, in sorted order by radius, along with the edge of $FVD(R)$ that they are centered on. Each T^i contains $O(n)$ circles. The inner nodes store intersections of circles, that are leaf descendants of that node. See Fig. 10 for an illustration. Given a blue point p to insert, we do the following.

1. Let k be the smallest key in T.
2. Perform an off-line ball exclusion search (OLBES) on T^k with query point p, to obtain the smallest circle in T^k not containing p.

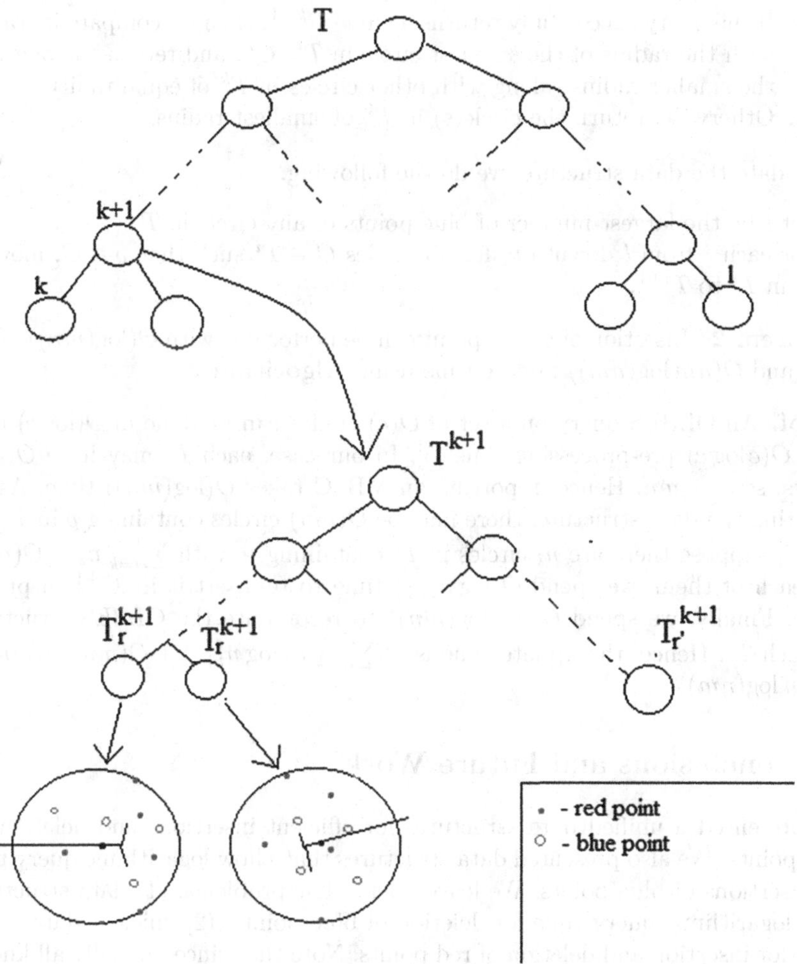

Fig. 10. Balanced binary search tree T stores the enclosing circles, sorted by number of blue points. Each node i points to another balanced binary search tree T^i containing circles with i blue points, sorted by radius. The circles are maintained along with the corresponding edge on $FVD(R)$ (Color figure online).

3. If the OLBES query successfully returns a circle C^{k*}, report it as MBSC. (Note: we can also report all circles in T^k of equal radii, in postorder following C^{k*}, if all optimal solutions are needed.)
4. Otherwise (i.e. all circles in T^k contain p, so they would have $k+1$ blue points after the insertion):
   4.1. We perform an OLBES on T^{k+1} with query point p

4.2. If the query successfully returns a circle C^{k+1}, then we compare its radius with the radius of the smallest circle in T^k, C^k, and return the one with the smaller radius (along with other circles in T^k of equal radii)
4.3. Otherwise, return the circle(s) in T^k of smallest radius.

To update the data structure, we do the following.

1. Let l be the largest number of blue points of any circle in T
2. For each i from l downto k, for all circles $C \in T^i$ such that $p \in C$, move C from T^i to T^{i+1}.

Theorem 2. Insertion of a blue point can be performed with $O(\log(mn))$ query time and $O(nm\log(nm))$ update time using **Algorithm 2**.

Proof. An OLBES query on a set of $O(c)$ circles can be done in $O(\log c)$ time with $O(c\log c)$ pre-processing time [5]. In our case, each T^i may have $O(mn)$ circles, so $c = mn$. Hence, reporting an MBSC takes $O(\log(mn))$ time. As for updating the data structure, there may be $O(nm)$ circles containing p in T. For each i, suppose there are n_i circles in T^i containing p, with $\sum_{i=k}^{l} n_i = O(nm)$. For each of them, we spend $O(\log n_{i+1})$ time to re-insert it in T^{i+1} in proper order. Finally, we spend $O(mn\log(mn))$ to recompute the OLBES structures for each T^i. Hence, the update time is $O(\sum_{i=k}^{l} n_i \log n_{i+1}) + O(mn\log(mn) = O(nm\log(nm))$. □

4 Conclusions and Future Work

We presented a unified data structures for efficient insertions and deletions of blue points. We also presented data structures that allow logarithmic query time for insertions of blue points. We leave open a few problems: (1) data structures with logarithmic query time for deletion of blue points. (2) efficient data structures for insertion and deletion of red points. Note that since virtually all known algorithms for finding the MBSC (static, kinetic, and dynamic versions) rely on the $FVD(R)$ it seems that solutions for insertion/deletion of red points would rely on the dynamic FVD, which is known to have $\Omega(n))$ worst case update time. (3) unified data structures for insertion and deletion of both blue and red points.

References

1. Barbay, J., Chan, T.M., Navarro, G., Pérez-Lantero, P.: Maximum-weight planar boxes in $O(n^2)$ time (and better). Inf. Process. Lett. **114**(8), 437–445 (2014)
2. Bereg, S., Daescu, O., Zivanic, M., Rozario, T.: Smallest maximum-weight circle for weighted points in the plane. In: Gervasi, O., Murgante, B., Misra, S., Gavrilova, M.L., Rocha, A.M.A.C., Torre, C., Taniar, D., Apduhan, B.O. (eds.) ICCSA 2015. LNCS, vol. 9156, pp. 244–253. Springer, Heidelberg (2015)
3. Bitner, S., Cheung, Y. K., Daescu, O.: Minimum separating circle for bichromatic points in the plane, International Symposium on Voronoi Diagrams in Science and Engineering (2010)

4. Chan, T.M., Hu, N.: Geometric red-blue set cover for unit squares and related problems. Comput. Geom. **48**(5), 380–385 (2015)
5. Chen, D.Z., Daescu, O., Hershberger, J., Kogge, P.M., Mi, N., Snoeyink, J.: Polygonal path simplification with angle constraints. In: Proceedings Computational Geometry: Theory and Applications (2004)
6. Cheung, Y.K., Daescu, O.: Minimum separating circle for bichromatic points by linear programming. In: 20th Annual Fall Workshop on Computational Geometry (2010)
7. Cheung, Y.K., Daescu, O., Zivanic, M.: Kinetic red-blue minimum separating circle. In: Wang, W., Zhu, X., Du, D.-Z. (eds.) COCOA 2011. LNCS, vol. 6831, pp. 448–463. Springer, Heidelberg (2011)

Miscellaneous

Searching Graph Communities by Modularity Maximization via Convex Optimization

Yuqing Zhu[1](✉), Chengyu Sun[1], Deying Li[2], Cong Chen[3], and Yinfeng Xu[3]

[1] Department of Computer Science, California State University, Los Angeles, CA 90032, USA
{yuqing.zhu,csun}@calstatela.edu

[2] School of Information, Renmin University of China, Beijing 100872, People's Republic of China
deyingli@ruc.edu.cn

[3] School of Management, Xi'an Jiaotong University, Xi'an 710049, Shaanxi, People's Republic of China
{chencong2779,yfxu}@stu.xjtu.edu.cn

Abstract. Communities in networks are the densely knit groups of individuals. Newman suggested *modularity* - a natural measure of the quality of community partition, and several community detection strategies aiming on maximizing the modularity have been proposed. In this paper, we give a new combinatorial model for modularity maximization problem, and introduce a convex optimization based rounding algorithm. Importantly, even given the maximum number of wanted communities, our solution is still capable of maximizing the modularity and obtaining the upper bound on the best possible solution.

1 Introduction

Many systems of mutual interacting entities can be expressed as networks, which consist of *nodes or vertices* and *edges* between them. For example, *wireless sensor networks* are spatially distributed sensors to monitor physical or environmental conditions, and to cooperatively pass their data through the network to a main location. *social networks* represent the individuals and their social connections, such as friendships, colleagueship, etc. *Metabolic networks* is the sets of metabolic and physical processes which determining cells' physiological and biochemical properties. They model enzymes and metabolites and their chemical reactions.

The study of networks has experienced a enormous surge of interest in the last decades. In recent, the boom of "big data" pushes the network study to a new era, since unlike the traditional data, the network data is unstructured and correlated. More and more people join, aiming to understand network data more deeply and design scalable algorithms for large scale networks.

In networks (graphs), *communities* are subsets of nodes within which the connections are dense, but between which the connections are sparser. Many complex networks have community or modular structure, and the detection and characterization of communities in networks has interested people immensely,

Z. Lu et al. (Eds.): COCOA 2015, LNCS 9486, pp. 701–708, 2015.
DOI: 10.1007/978-3-319-26626-8_51

for community identification is extreme useful for analyzing and understanding networks. For instance, communities within world wide web are possibly corresponding to sets of web pages focusing on related topic, communities within biological networks may refer to same functions, and communities within social networks may share common interests are locations. Far another instance, people can use community to compress the networks when they are of large-scale, since individuals in same communities share common properties. In addition, communities can be used to optimize system performance. Nowadays data handled by on-line social networking sites (Facebook, Twitter etc.) is extremely large, and multiple machines are employed for storing data and handling online communications. The communication cost between machines affects system performance a lot. Processing individuals in a same community on a same machine can reduce community cost among machines, and thus accelerate the system response.

Due to the huge importance of determining community structure in graphs, people have been witnessing the bloom of community research in computer science, biology, physics, sociology and economics, the bloom lasts for more than a decade and is still thriving. Agarwal and Kempe in [1] studied community partitioning as a modularity optimization problem, and proposed two algorithms basing on linear programing and vector programming techniques. Zhang et al. focused on how to partition a network into as many qualified communities as possible while maintaining simple community identification in [6], they formulated their model as a linear integer programming problem and designed a qualified min-cut based bisecting algorithm. In [2], Bi et al. studied a new problem: community expansion caused by influence propagation. Their model is inspired by the charged system in the physic, and two objective functions are proposed for choosing proper candidates to enlarge a community. Linear programming approach is designed to maximize those two objective functions. Lu et al. in [4] aimed at partitioning a social network into K disjoint communities such that the sum of influence propagation within these communities is maximized. They developed an optimal algorithm for the case where $K = 2$, and proved that the problem is NP-hard for general K. Zhu et al. in [7] defined the mutual closeness and strangeness between each node pair in a social network, and formulate community partition problem into a semidefinite program, considering both the tightness of the same community and the looseness across different communities. The mathematical models and the objective functions are given, as well as the performance bounds of the proposed algorithms.

In this paper, we focus the most common used community detection criterion: modularity. We devise a new combinatorial model for modularity maximization problem, and introduce a convex optimization based rounding algorithm. Given an integer K, our solution can give qualified community partitioning such that the number of communities obtained not exceeding K. What is more, ours is the first technique that can track the upper bound of best modularity when partitioning the graph into at most K communities.

2 Preliminaries

We represent the network as an undirected graph $G = (V, E)$ with $|V| = n, |E| = m$, and denote $A = [a_{u,v}]$ as the adjacency matrix of G. According to the definition of adjacency matrix, for undirected graph $a_{u,v} = a_{v,u} = 1$ if and only if there exists an edge between u and v, and $a_{u,v} = a_{v,u} = 1$ otherwise. The degree of a node v is represented by d_v.

The problem is to partition V into several disjoint communities, denote the aimed number of communities as K, and use $\mathcal{C} = \{C_1, ..., C_K\} = \{(V_1, E_1), ..., (V_K, E_K)\}$ to represent the partition. Modularity [5] of a partition $\mathcal{C} = \{C_1, ..., C_K\}$ is the total number of edges within the communities, minus the expected number of such edges in a randomized network with the same size and the same degree sequence. If we use $\overline{V}_k$ to denote $V \setminus V_k$, the complementary node set of V_k, and $\Theta(V_k, V_k) = \frac{1}{2}\sum_{u\in V_k}\sum_{v\in V_k} a_{uv}$, and $\Theta(V_i, \overline{V}_i) = \sum_{u\in V_i}\sum_{v\in \overline{V}_i} a_{uv}$, then the modularity Q is defined as

$$Q = \sum_{k=1}^{K}\left[\frac{\Theta(V_k, V_k)}{m} - \left(\frac{\Theta(V_k, V_k) + \Theta(V_k, \overline{V}_k)}{m}\right)^2\right] = \sum_{k=1}^{K} Q_i.$$

A large number of community detection methods have been proposed by optimizing Q. It is still so far the best indicator to measure the quality of obtained community partition. Q is a number from the interval $[-1, 1]$, and usually higher Q indicates better community partition.

In mathematical optimization, convex optimization is a subfield of maximizing convex functions based on convex constraint functions. A convex function f satisfies the condition that $f(\alpha x + \beta y) \leq \alpha f(x) + \beta f(y)$ for all vectors $x, y \in \mathbf{R}^n$ and all $\alpha, \beta \in \mathbf{R}$ with $\alpha + \beta = 1, \alpha \geq 0, \beta \geq 0$. A convex optimization problem is of the form

$$\begin{aligned} \min \quad & f_0(x) \\ \text{s. t.} \quad & f_i(x) \leq b_i, \quad i = 1, ..., m, \end{aligned} \tag{1}$$

where the functions $f_0, ..., f_m : \mathbf{R}^n \to \mathbf{R}$ are convex.

3 Problem Definition

Given a network and a integer K, our problem is to partition network into K non-overlapping communities such that the modularity Q is maximized. Note that for our problem K is not a necessary input, if K cannot be decided yet, our solution finds the best value of K such that Q is maximized.

3.1 The Convex Optimization Problem

As we introduced before, n and m are respectively the numbers of nodes and edges in the network. Note that if K is provided then it must be no greater than n. We use variable z_{lk} to denote whether edge e_l belongs to the k-th community, and $1 \le l \le m$, $1 \le k \le K$. Denote the two ends of edge e_l as v_i and v_j, let x_{ik} be a binary variable indicating whether node v_i is in community k. It is in common sense that if an edge is in community k then its two ends must be also in this community, therefore we have the following constraints:

$$z_{lk} \le x_{ik} \text{ and } z_{lk} \le x_{jk},$$

also if one edge is not in community k then at least one of its two edges is not in community k, and we use constraint

$$z_{lk} \ge x_{ik} + x_{jk} - 1$$

to consider this.

Let us then check $\Theta(V_k, V_k)$ and $\Theta(V_k, V_k) + \Theta(V_k, \overline{V}_k)$ to compute modularity Q. The former is the number of edges in community k, and it is easy to find out that $\Theta(V_k, V_k) = \sum_{l=1}^{m} z_{lk}$. The latter is the total number of edges that are adjacent to some nodes in community k, and $\sum_{j=1}^{n} \sum_{i=1}^{n} x_{ik} a_{ij}$ $(*)$ seems to be a good answer. However, since adjacency matrix is symmetric for an undirected graph, in expression $(*)$ the edges within community k are counted twice, therefore, $\Theta(V_k, V_k) + \Theta(V_k, \overline{V}_k) = \sum_{i=1}^{n} x_{ik} a_{ij} - \sum_{l=1}^{m} z_{lk}$.

Therefore, our community searching problem to maximize the modularity can be formulated as following programming:

$$
\begin{aligned}
\max \quad & \sum_{k=1}^{K} \left[\frac{\sum_{l=1}^{m} z_{lk}}{m} - \left(\frac{\sum_{j=1}^{n} \sum_{i=1}^{n} x_{ik} a_{ij} - \sum_{l=1}^{m} z_{lk}}{m} \right)^2 \right] \\
\text{s. t.} \quad & \sum_{k=1}^{K} x_{ik} = 1 \text{ for all } i, \\
& z_{lk} \le x_{ik} \\
& z_{lk} \le x_{jk} \\
& x_{ik} + y_{jk} - 1 \le z_{lk} \\
& \qquad\qquad\qquad\qquad \text{for every edge } e_l \text{ and its two ends } i, j, \\
& \sum_{i=1}^{n} x_{ik} \ge 1 \text{ for all } k \\
& x_{ik} \in \{0,1\}, z_{lk} \in \{0,1\} \text{ for all } i, l, k.
\end{aligned} \tag{2}
$$

In [3] it has been proved even if $K = 2$, to find out the best partition into 2 communities is strongly NP-hard, this means directly solving problem (2) is difficult. However, it doesn't mean we can only use simple heuristics. Note that in (2), all variables are a binary value being 0 or 1, and we can relax this condition to allow variables be any value in interval $[0, 1]$. Employing this relaxation and doing some equivalent transformation, following programming is obtained:

$$\min \sum_{k=1}^{K}\left[\left(\sum_{j=1}^{n}\sum_{i=1}^{n} x_{ik}a_{ij} - \sum_{l=1}^{m} z_{lk}\right)^2 - m\sum_{l=1}^{m} z_{lk}\right]$$

$$\begin{aligned}
\text{s. t.}\quad & \sum_{k=1}^{K} x_{ik} = 1 \text{ for all } i,\\
& z_{lk} - x_{ik} \le 0\\
& z_{lk} - x_{jk} \le 0\\
& x_{ik} + y_{jk} - z_{lk} \le 1\\
& \qquad\qquad\qquad \text{for every } l \text{ and its two ends } i, j,\\
& -\sum_{i=1}^{n} x_{ik} \le -1 \text{ for all } k\\
& x_{ik} \in [0,1], z_{lk} \in [0,1] \text{ for all } i, l, k.
\end{aligned} \tag{3}$$

For Problem (3), though it looks not exactly of the form of (1) at the first glance, it is a convex optimization.

Theorem 1. *Problem (3) is a convex optimization problem.*

Proof. Note that totally there are $(m+n)k$ variables ($m \cdot k$ variables of x and $n \cdot k$ variables of z) in (3), we first create a column vector $Y \in \mathbf{R}^{(m+n)k\times 1}$ that contains all variables.

$$Y = \begin{matrix}[\, x_{11}, x_{21}, ..., x_{n1}, z_{11}, z_{21}, ..., z_{m1},\\ x_{12}, x_{22}, ..., x_{n2}, z_{12}, z_{22}, ..., z_{m2},\\ ...\\ x_{1k}, x_{2k}, ..., x_{nk}, z_{1k}, z_{2k}, ..., z_{mk}\,]^{\mathrm{T}}\end{matrix}$$

First we prove that the objective function of problem (3) is a convex function w.r.t. Y. We create K row vectors A_k in $\mathbf{R}^{1\times(m+n)k}$, where

$$A_k = \begin{matrix}[\underbrace{0,\ 0,\ ...\ ,\ 0}_{(n+m)(k-1)},\\ \sum_{j=1}^{n} a_{1j},\ \sum_{j=1}^{n} a_{2j},\ ...\ ,\ \sum_{j=1}^{n} a_{nj},\ \underbrace{-1,\ -1,\ ...\ ,\ -1}_{m},\\ \underbrace{0,\ 0,\ ...\ ,\ 0}_{(n+m)(K-k)}],\end{matrix}$$

and K row vectors B_k in $\mathbf{R}^{1\times(m+n)k}$, where

$$B_k = [\underbrace{0,\ 0,\ ...\ ,\ 0}_{(n+m)(k-1)},\ \underbrace{0,\ 0,\ ...\ ,\ 0}_{n},\ \underbrace{-m,\ -m,\ ...\ ,\ -m}_{m},\ \underbrace{0,\ 0,\ ...\ ,\ 0}_{(n+m)(K-k)}],$$

then the objective function of (3) can be represented as

$$\sum_{k=1}^{K}[(A_kY)^2 - B_kY], \tag{4}$$

since obviously for any single $1 \leq k \leq K$, $(A_kY)^2 - B_kY$ is a convex function w.r.t. Y, and (4) which is a positive weighted sum of them is also a convex function w.r.t. Y.

For the constraints of (3), all the left sides of them are linear functions which are convex, therefore (3) is a convex optimization problem.

3.2 CAR - Convex Optimization Based Accurate Rounding Algorithm

Problem (3) can be solved in polynomial time with an error. However, the solution of (3) is not directly applicable of Problem (2), since the variables may be fractional. Therefore, some rounding techniques are needed for rounding fractional values into binary values.

The following is our solution which is called *Convex optimization Accurate Rounding (CAR)*, which is based on rounding up the edges. The algorithm heuristically solves Problem (3) and finds out the highest edge variable $z_{l_0k_0}$ and make it 1. This is to assign edge e_{l_0} to community k_0, as well as i_0, j_0, the two ends of e_{l_0}. This procedure means: first, $z_{l_0k_0}, x_{i_0k_0}, x_{j_0k_0}$ will all be changed to 1, second, we assume that the communities are not overlap, therefore e_{l_0} cannot be in any other communities, and for all $k \leq K, k \neq k_0$, z_{l_0k}, x_{i_0k}, and x_{j_0k} must be set to 0. Steps 5–9 in Algorithm 1 correspond to the treatment of assigning e_{l_0} to community k_0. Note that in step 1 when we put l_r to community 1, similar treatment has to be done, but for the sake of simplicity we omit it in the algorithm's pseudocode.

Like many rounding techniques, we put randomization into the algorithm, which is the first step of Algorithm 1 that randomly choose an edge and put it into community 1, this improves the algorithm's performance.

We assume that the input graph of CAR contains no isolated nodes, since each isolated node is just a community itself, and before applying CAR to any graph, the isolated nodes can be identified and removed easily and quickly. Note that CAR takes K, the maximum number of target communities as the input. People may argue how to decide K will be a problem, however, K is considered to ease the use of CAR. When people have their limitation of K, they can assign K some integer between 1 and n, if they don't, CAR can automatically decide the best value K.

Convex problem can be solved in polynomial time with an error ϵ. Let us use $f_\epsilon()$ to denote the function of time needed, then for Problem (3), whose input size is $K(n+m)$, the time needed is $f_\epsilon(K(n+m))$. Each round we pick the greatest value among z_{lk}, and at most $\min\{n, m\}$ rounds is needed. Therefore the overall time complexity of Algorithm 1 is $mf_\epsilon(K(n+m))$, usually $f_\epsilon(K(n+m))$ is more than $(K(n+m))^2$, therefore, Algorithm 1 takes $\Omega(\min\{n, m\}K^2(n+m)^2)$ polynomial time, which we admit, is not a very quick method. However, CAR can be further improved by designing quicker rounding techniques, and CAR provides better partition than past algorithms. An important benefit of CAR, is

Algorithm 1. CAR Algorithm

Input: A graph $G = (V, E)$ with no isolated vertices, and an integer $K \in [1, n]$ and $n = |V|$.
Output: A partition of G containing at most K non-overlapping communities.
1: Randomly choose an edge l_r and assign it into community 1;
2: Solve updated convex minimization problem (3);
3: However, in most cases, the variables x and z are not 1 or 0. Therefore, we choose the maximum $z_{l_0 k_0}$ satisfying that: Let v_{i_0} and v_{j_0} be the two ends of l_0, none of the two nodes has been assigned to any community;
4: **if** such $z_{l_0 k_0}$ exists **then**
5: $\quad z_{l_0 k_0} \leftarrow 1$, $x_{i_0 k_0} \leftarrow 1$, $x_{j_0 k_0} \leftarrow 1$; $\quad$ $//i_0$ and j_0 are the two ends of l_0
6: $\quad$ **for** $k := 1;\ k \neq k_0, k \leq K;\ k := k + 1$ **do**
7: $\quad\quad z_{l_0 k} \leftarrow 0$;
8: $\quad\quad x_{i_0 k} \leftarrow 0$, $x_{j_0 k} \leftarrow 0$;
9: $\quad$ **end for**
10: $\quad$ Start over from step 1 with the decided variables;
11: **else**
12: $\quad$ For all remaining nodes v_i that have not been assigned to any community, assign v_i to community b, where $b = \arg_{1 \leq k \leq K} \max\{x_{ik}\}$.
13: **end if**

that it provides an upper bound on the best solution of partitioning into K communities. This bound is stronger than that provided by LP rounding algorithm in [1], for our bound is specified for partitions given the maximum number of communities, as well as the global upper bound for all possible partitions, but the bound in [1] is only the global upper bound.

4 Conclusion

Though people argue that modularity is not accurate enough to judge if a community partitioning is good, without a doubt, it is still the most widely used and precise criterion for testing the community partitioning quality. In this paper, we study community partitioning under a constraint: number of communities does not exceed a given number K. Our aim is to maximize the modularity. We devise a new combinatorial model, and use a convex optimization based rounding algorithm to solve it. Our solution not only gives qualified community partitioning result, it is the first one that tracks the upper bound of best modularity for community partitioning into at most K communities.

References

1. Agarwal, G., Kempe, D.: Modularity-maximizing graph communities via mathematical programming. Eur. Phys. J. B **66**(3), 409–418 (2008)
2. Bi, Y., Weili, W., Zhu, Y., Fan, L., Wang, A.: A nature-inspired influence propagation model for the community expansion problem. J. Comb. Optim. **28**(3), 513–528 (2014)

3. Brandes, U., Delling, D., Gaertler, M., Gorke, R., Hoefer, M., Nikoloski, Z., Wagner, D.: On modularity clustering. IEEE Trans. Knowl. Data Eng. **20**(2), 172–188 (2008)
4. Lu, Z., Zhu, Y., Li, W., Wu, W., Cheng, X.: Influence-based community partition for social networks. Comput. Soc. Netw. **1**(1) (2014)
5. Newman, M.E.J., Girvan, M.: Finding and evaluating community structure in networks. Phys. Rev. E **69**, 026113 (2004)
6. Zhang, X.-S., Li, Z., Wang, R.-S., Wang, Y.: A combinatorial model and algorithm for globally searching community structure in complex networks. J. Comb. Optim. **23**(4), 425–442 (2012)
7. Zhu, Y., Li, D., Wen, X., Weili, W., Fan, L., Willson, J.: Mutual-relationship-based community partitioning for social networks. IEEE Trans. Emerg. Top. Comput. **2**(4), 436–447 (2014)

A New Tractable Case of the QAP with a Robinson Matrix

Eranda Çela[1], Vladimir G. Deineko[2](✉), and Gerhard J. Woeginger[3]

[1] TU Graz, Graz, Austria
[2] Warwick Business School, Coventry, UK
V.Deineko@warwick.ac.uk
[3] TU Eindhoven, Eindhoven, The Netherlands

Abstract. We consider a new polynomially solvable case of the well-known Quadratic Assignment Problem (QAP). One of the two matrices involved is a Robinsonian dissimilarity with an additional structural property: this matrix can be represented as a conic combination of cut matrices in a certain normal form. The other matrix is a monotone anti-Monge matrix.

Keywords: Combinatorial optimization · Quadratic assignment · Robinsonian · Cut matrix · Monge matrix · Kalmanson matrix

1 Introduction

In this paper we investigate the Quadratic Assignment Problem (QAP), which is a well-known problem in combinatorial optimization; we refer the reader to the book [7] by Çela and the book [4] by Burkard, Dell'Amico and Martello for comprehensive surveys on the QAP. The QAP in Koopmans-Beckmann form [20] takes as input two $n \times n$ square matrices $A = (a_{ij})$ and $B = (b_{ij})$ with real entries. The goal is to find a permutation π that minimizes the objective function

$$Z_\pi(A, B) := \sum_{i=1}^{n} \sum_{j=1}^{n} a_{\pi(i)\pi(j)}\, b_{ij}. \tag{1}$$

Here π ranges over the set S_n of all permutations of $\{1, 2, \ldots, n\}$. In general, the QAP is extremely difficult to solve and hard to approximate. One branch of research on the QAP concentrates on the algorithmic behavior of strongly structured special cases; see for instance Burkard et al. [3], Deineko and Woeginger [15], Çela et al. [10], or Çela, Deineko and Woeginger [8] for typical results in this direction. In our short paper we follow recent developments and represent several new results in this exciting area of research.

Recently, Laurent and Seminaroti [21] have introduced a new tractable special case of the QAP, where matrix A is a *Robinsonian* dissimilarity and B is a special *Toeplitz* matrix.

Z. Lu et al. (Eds.): COCOA 2015, LNCS 9486, pp. 709–720, 2015.
DOI: 10.1007/978-3-319-26626-8_52

- A symmetric matrix $A = (a_{ij})$ is a *Robinsonian* dissimilarity, if for all $i < j < k$ it satisfies the conditions $a_{ik} \geq \max\{a_{ij}, a_{jk}\}$; in words, the entries in the matrix are placed in non-decreasing order in each row and column when moving away from the main diagonal.
- A symmetric matrix $A = (a_{ij})$ is a *Robinsonian* similarity, if for all $i < j < k$ it satisfies the conditions $a_{ik} \leq \max\{a_{ij}, a_{jk}\}$.

In what follows we also assume that the diagonal elements are not important and set $a_{ii} = 0$, for all i. Since 1951, when these specially structured matrices were first introduced by Robinson [26] for an analysis of archaeological data, they have been widely used in combinatorial data analysis; see the books [16,17,22,23] and the surveys [2,6] for examples of various applications of Robinsonian structures in quantitative psychology, analysis of DNA sequences, cluster analysis, etc. In what follows, we will deal with Robinsonian dissimilarities and we will simply call them *Robinson* matrices.

- A matrix $B = (b_{ij})$ is called a Toeplitz matrix if it has constant entries along each of its diagonals; in other words, there exist $2n - 1$ real numbers $t_{1-n}, \ldots, t_0, \ldots, t_{n-1}$ such that $b_{ij} = t_{i-j}$. A symmetric Toeplitz matrix with $t_0 = 0$ and $t_1 \geq t_2 \geq \ldots \geq t_{n-1}$ will be called *simple* Toeplitz matrix.

Laurent and Seminaroti [21] considered the *QAP* with a simple Toeplitz matrix B and Robinson matrix A, and proved that the optimal value of *QAP*(A, B) is attained on the identity permutation.

Cela, Deineko and Woeginger [9] considered recently another well-solvable case of the QAP with a special Robinson matrix. We now need some further definitions to proceed.

- For a $q \times q$ matrix $P = (p_{ij})$, we say that an $n \times n$ matrix $B = (b_{ij})$ is a *block matrix with block pattern* P if the following holds: (i) there exists a partition of the row and column set $\{1, \ldots, n\}$ into q (possibly empty) sets $I_1, \ldots, I_q$ such that for $1 \leq k \leq q - 1$ all elements of set I_k are smaller than all elements of set I_{k+1}; (ii) for all indices i and j with $1 \leq i, j \leq n$ and $i \in I_k$ and $j \in I_\ell$, we have $b_{ij} = p_{k\ell}$. The sets $I_1, \ldots, I_q$ form the so-called row and column blocks of matrix B. If it is clear from context we will sometimes refer to the sets as the *blocks* of matrix B.
- A *cut matrix* B is a block matrix whose pattern matrix has 0's along the main diagonal and 1's everywhere else. A cut matrix is in *CDW normal form*, if its block sizes are in non-decreasing order with $|I_1| \leq |I_2| \leq \cdots \leq |I_q|$.

It is easy to see that any cut matrix is a Robinson matrix. So it follows from [21] that if A is a cut matrix, and B is a simple Toeplitz matrix (as defined above), then the QAP is solved by the identity permutation. Cela, Deineko and Woeginger [9] on the other hand have shown that in case matrix A is a cut matrix in CDW normal form and matrix B is a monotone anti-Monge matrix, then the QAP is again solved by the identity permutation.

- An $n \times n$ matrix $B = (b_{ij})$ is *monotone*, if $b_{ij} \leq b_{i,j+1}$ and $b_{ij} \leq b_{i+1,j}$ holds for all i, j, that is, if the entries in every row and every column are in non-decreasing order.

– Matrix B is an *anti-Monge matrix*, if its entries are non-negative and satisfy the anti-Monge inequalities

$$b_{ij} + b_{rs} \ \geq \ b_{is} + b_{rj} \qquad \text{for } 1 \leq i < r \leq n \text{ and } 1 \leq j < s \leq n. \qquad (2)$$

In other words, in every 2×2 submatrix the sum of the entries on the main diagonal dominates the sum of the entries on the other diagonal.

This Monge property essentially dates back to the work of Gaspard Monge [24] in the 18th century. Much research has been done on the effects of Monge structures in combinatorial optimization, and we refer the reader to the survey [5] by Burkard, Klinz and Rudolf for more information on Monge and anti-Monge structures.

A straightforward generalisation of the results of [9] is to consider the QAP with matrix A being a Robinson matrix which is obtained as a *conic* combination of cut matrices in CDW normal form, that is, as a linear combination of such matrices with non-negative weight coefficients. The QAP with a such obtained matrix A and a monotone anti-Monge matrix B is obviously solved by the identity. For this observation to be meaningful, one need to know how to recognize this special subclass of Robinson matrices, in other words to know how to resolve the following problem:

Given an $n \times n$ Robinson matrix, can it be represented as a conic combination of cut matrices in CDW normal form?

As an illustrative example we consider the following Robinson matrix:

$$C = \begin{pmatrix} 0 & 1 & 2 & 3 & 3 & 3 \\ 1 & 0 & 2 & 3 & 3 & 3 \\ 2 & 2 & 0 & 2 & 3 & 3 \\ 3 & 3 & 2 & 0 & 2 & 2 \\ 3 & 3 & 3 & 2 & 0 & 1 \\ 3 & 3 & 3 & 2 & 1 & 0 \end{pmatrix}$$

which is obtained as a sum of three cut-matrices $C = C_1 + C_2 + C_3$; here matrix C_1 has the three blocks $\{1,2,3\}$, $\{4\}$, $\{5,6\}$, matrix C_2 has three blocks $\{1,2\}$, $\{3\}$, $\{4,5,6\}$, and matrix C_3 has five blocks $\{1\}$, $\{2\}$, $\{3,4\}$, $\{5\}$, and $\{6\}$. As none of these matrices above is a cut matrix in CDW normal form, there are no reasons to assume that the QAP with C and a monotone anti-Monge matrix B is solved by the identity permutation. Later we will show that C can indeed be represented as a conic combination of cut matrices in CDW normal form, and hence the corresponding QAP is solved by the identity permutation.

To further investigate the structure behind the considered special subclass of Robinson matrices we refer to another well-known class of specially structured matrices.

– A symmetric matrix (c_{ij}) is called a Kalmanson matrix, if it satisfies the conditions

$$c_{ij} + c_{kl} \leq c_{ik} + c_{jl} \qquad (3)$$

$$c_{ik} + c_{jl} \geq c_{il} + c_{jk} \qquad (4)$$

for all i,j,k and l with $1 \leq i < j < k < l \leq n$.

In 1975 Kenneth Kalmanson [18] proved that if the distance matrix in the travelling salesman problem satisfies these conditions, then it is solved by the identity permutation. Later Kalmanson matrices appeared in polynomially solvable cases of such problems as the prize-collecting TSP [11], the master tour problem [14], the Steiner tree problem [19], the three-dimensional matching problem [25], and the quadratic assignment problem [7,15] (to be also discussed in the final section of this paper).

Lemma 1. *Any cut matrix C satisfies conditions* (3)–(4) *and hence is a Kalmanson matrix.*

Proof. As was mentioned above, any cut matrix is a Robinson matrix. Therefore inequality (3) rearranged as $(c_{ik} - c_{ij}) + (c_{jl} - c_{kl}) \geq 0$ follows immediately from the monotonicity of rows and columns when moving from the main diagonal.

Since C is a binary matrix, it is easy to enumerate all possible values of items when condition (4) is violated. All of these cases but one are eliminated by analogous monotonicity arguments. The only case that is different is the case with $c_{ik} = 0$, $c_{jl} = 0$, $c_{jk} = 0$, and $c_{il} = 1$. This case is also impossible because of the block structure of cut matrix C. This completes the proof of the lemma.□

Further Useful Definitions. An $n \times n$ matrix $A = (a_{ij})$ is a *sum matrix*, if there exist real numbers $\alpha_1, \ldots, \alpha_n$ and $\beta_1, \ldots, \beta_n$ such that

$$a_{ij} = \alpha_i + \beta_j \qquad \text{for } 1 \leq i, j \leq n. \tag{5}$$

Matrix A is a *constant matrix*, if all elements in the matrix are the same. It can easily be seen that a constant matrix is just a special case of a sum matrix. It was mentioned above that all items on the main diagonal in a Robinson matrix are zeros. In some cases though it would be useful to redefine the diagonal elements. Hence the following definitions. Matrix A is a *weak sum matrix*, if A can be turned into a sum matrix by appropriately changing the entries on its main diagonal. Matrix A is a *weak constant matrix*, if A can be turned into a constant matrix by appropriately changing the entries on its main diagonal.

2 Conic Representation of Specially Structured Matrices

2.1 Cut Weights and Specially Structured Matrices

In this section, we investigate the structure of matrices which are both Kalmanson matrices and Robinson dissimilarities. We show that any matrix in this class can be represented as a sum of a constant matrix and a conic combination of cut matrices.

Lemma 2. *([13, 14]) A symmetric $n \times n$ matrix C is a Kalmanson matrix if and only if*

$$c_{i,j+1} + c_{i+1,j} \leq c_{ij} + c_{i+1,j+1} \;\; \forall i,j : 1 \leq i \leq n-3,\; i+2 \leq j \leq n-1, \tag{6}$$

$$c_{i,1} + c_{i+1,n} \leq c_{in} + c_{i+1,1} \;\; \forall i,j :\; 2 \leq i \leq n-2. \tag{7}$$

The statement above will be used for further structural characterisations of Kalmason matrices. For the characterisation we will make use of special cut matrices. A cut matrix $A^{kl} = (a_{ij})$, $k < l$, contains one block of size $(k - l + 1)$ with $a_{ij} = 0$ for $k \le i, j \le l$, and all other $n - k + l - 1$ blocks of size 1.

Note that for an $n \times n$ cut matrix A^{kl}, $2 \le k < l < n$, there is only one strict inequality in (6):

$$a_{k-1,l} + a_{k,l+1} > a_{kl} + a_{k-1,l+1},$$

and there is only one strict inequality in (7) for $A^{1,k-1}$ and A^{kn}, $3 \le k < n$:

$$a_{k-1,1} + a_{kn} < a_{k1} + a_{k-1,n}.$$

Lemma 3. *A symmetric $n \times n$ matrix C is a Kalmanson matrix if and only if it can be represented as a linear combination of a weak sum matrix S and cut matrices A^{kl}:*

$$C = S + \sum_{i=1}^{n-3} \sum_{j=i+2}^{n-1} \delta_{i+1,j} A^{i+1,j} + \sum_{i=2}^{n-2} (\alpha_i A^{1,i} + \beta_i A^{i+1,n}) \qquad (8)$$

where $\delta_{i+1,j} = (-c_{i,j+1} - c_{i+1,j} + c_{ij} + c_{i+1,j+1})$, $\alpha_i = c_{i+1,1} - c_{i,1}$, $\beta_i = c_{in} - c_{i+1,n}$ and $\delta_{i+1,j} \ge 0$, $\alpha_i + \beta_i \ge 0$.

We will refer to coefficients α, β, and δ as cut-weights. Similar structural properties of Kalmanson matrices in terms of cuts and cut-weights have also been studied in [1, 12]. In both papers though the authors suggested algorithms for calculating the cut-weights while we provide simple analytical expressions for the weights.

Proof. It can easily be checked that any sum matrix, a cut matrix A^{kl}, and a linear combination $\alpha_i A^{1,i} + \beta_i A^{i+1,n}$ with $\alpha_i + \beta_i \ge 0$ are Kalmanson matrices, and therefore any matrix defined as (8) is a Kalmanson matrix.

Assume now that C is a Kalmanson matrix. Let i and j, $1 \le i < j < n$ be two indices from a strict inequality in (6): $c_{i,j+1} + c_{i+1,j} < c_{ij} + c_{i+1,j+1}$ (for equalities in (6)–(7) the corresponding coefficients in (8) are zeros and have no influence). In the illustration below the corresponding four items in the matrix are highlighted bold (note that all elements shown are above the main diagonal):

$$C = \begin{pmatrix} & & & \cdots & & & \\ \cdots & c_{i,p} & \cdots & c_{i,j-1} & \mathbf{c}_{i,j} & \mathbf{c}_{i,j+1} & \cdots \\ \cdots & c_{i+1,p} & \cdots & c_{i+1,j-1} & \mathbf{c}_{i+1,j} & \mathbf{c}_{i+1,j+1} & \cdots \\ & & & \cdots & & & \\ \cdots & c_{q,p} & \cdots & c_{q,j-1} & c_{q,j} & c_{q,j+1} & \cdots \\ & & & \cdots & & & \end{pmatrix}$$

Consider now matrix $C' = C - \delta_{i+1,j}A^{i+1,j}$. To simplify the illustration we use notation $\Delta := \delta_{i+1,j}$:

$$C' = \begin{pmatrix} & & \cdots & & \\ \ldots c_{i,p} - \Delta \ldots & c_{i,j-1} - \Delta & \mathbf{c}_{i,j} - \Delta & \mathbf{c}_{i,j+1} - \Delta & \ldots \\ \ldots \quad c_{i+1,p} \quad \ldots & c_{i+1,j-1} & \mathbf{c}_{i+1,j} & \mathbf{c}_{i+1,j+1} - \Delta & \ldots \\ & & \cdots & & \\ \ldots \quad c_{q,p} \quad \ldots & c_{q,j-1} & c_{q,j} & c_{q,j+1} - \Delta & \ldots \\ & & \cdots & & \end{pmatrix}$$

Simple algebraic transformations show that in matrix $C' = (c'_{jj})$ we have $c'_{i,j+1} + c'_{i+1,j} = c'_{ij} + c'_{i+1,j+1}$ and none of the other inequalities in (6) is influenced.

Assume now that system (7) contains some strict inequalities. We pick up an index i, $2 \le i \le n-2$, such that $c_{i1} + c_{i+1,n} < c_{i+1,1} + c_{in}$:

$$C = \begin{pmatrix} & & \cdots & & \\ \mathbf{c}_{i1} & c_{i2} \ \ldots \ 0 & \ldots & c_{i,n-1} & \mathbf{c}_{in} \\ \mathbf{c}_{i+1,1} & c_{i+1,2} \ \ldots \ 0 & \ldots & c_{i+1,n-1} & \mathbf{c}_{i+1,n} \\ & & \cdots & & \end{pmatrix}$$

Consider matrix $C' = C - \alpha \times A^{1i} - \beta \times A^{i+1,n}$ (for simplicity we omit index i for α and β here):

$$C' = \begin{pmatrix} & & \cdots & & \\ \mathbf{c}_{i1} - \beta & c_{i2} - \beta & \ldots c_{i,n-1} - \beta - \alpha & \mathbf{c}_{in} - \beta - \alpha \\ \mathbf{c}_{i+1,1} - \beta - \alpha & c_{i+1,2} - \beta - \alpha \ldots & c_{i+1,n-1} - \alpha & \mathbf{c}_{i+1,n} - \alpha \\ & & \cdots & & \end{pmatrix}$$

It can be easily checked that $c'_{i1} + c'_{i+1,n} = c'_{i+1,1} + c'_{in}$.

So eventually we get a transformed matrix without strict inequalities in the system (6)–(7). It can be shown that such a matrix is a weak sum matrix. The lemma is proved. □

Lemma 4. *A symmetric $n \times n$ Kalmanson matrix C is a Robinson dissimilarity if and only if it can be represented as a linear combination of a constant matrix Z and cut matrices A^{kl}:*

$$C = Z + \sum_{i=1}^{n-3} \sum_{j=i+2}^{n-1} \delta_{i+1,j} A^{i+1,j} + \sum_{i=2}^{n-1} \alpha_i A^{1,i} + \sum_{i=1}^{n-2} \beta_i A^{i+1,n} \qquad (9)$$

where $\delta_{i+1,j} = (-c_{i,j+1} - c_{i+1,j} + c_{ij} + c_{i+1,j+1})$, $\alpha_i = c_{i+1,1} - c_{i,1}$, $\beta_i = c_{in} - c_{i+1,n}$ and $\delta_{i+1,j} \ge 0$, $\alpha_i \ge 0$, $\beta_i \ge 0$.

Proof. As the proof of the (if)-part of the lemma is straightforward, we will concentrate now on the (only if)-part. In the proof we will use the previous lemma and the fact that C is a Robinson dissimilarity. In particular, by the definition of the dissimilarity we have $\alpha_i \ge 0$ and $\beta_i \ge 0$. We will show that each step of eliminating strict inequalities in (6)–(7) keeps the matrix in the set of Robinson dissimilarities.

Consider first the transformation $C' = C - \Delta A^{i+1,j}$, where $\Delta = \delta_{i+1,j}$. We claim that $c_{ip} - \Delta \geq c_{i+1,p}$ for all $p = i+2, \ldots, j$. Let $\Lambda_p = c_{ip} - c_{i+1,p}$, $p = i+2, \ldots, j+1$. Since C is a Robinson dissimilarity, we have $\Lambda_p \geq 0$. Since C is a Kalmanson matrix, we have $c_{ip} + c_{i+1,j+1} - c_{i+1,p} - c_{i,j+1} = \Lambda_p - \Lambda_j \geq 0$ and $\Lambda_p \geq \Lambda_j$. It is easy to see that $c_{ip} - \Delta - c_{i+1,p} = \Lambda_p - \Lambda_j + \Lambda_{j+1} \geq 0$. The claim is proved.

The claim that $c_{q,j+1} - \Delta \geq c_{qj}$ for all $q = i+1, \ldots, j-1$ is proved in a similar way. So the new matrix C' is a Robinson dissimilarity.

If we consider transformations $C' = C - \alpha_i A^{1,i}$ or $C' = C - \beta_i A^{i+1,n}$, the Kalmanson inequalities (6) ensure that C' is a Robinson dissimilarity.

What is left to prove is that $Z = S - \alpha_{n-1} A^{1,n-1} - \beta_1 A^{2,n}$ is a constant matrix, where S is a sum matrix from presentation (8) (note the difference in summations in (8) and (9)).

By construction, S is a weak sum matrix which is also a Robinson dissimilarity. It can be shown that a weak symmetric sum matrix is a Robinson dissimilarity if and only if $u_1 \geq u_2 = u_3 = \ldots = u_{n-1} \leq u_n$. It is easy to see that $S - (u_n - u_{n-1}) \times A^{1,n-1} - (u_1 - u_2) \times A^{2,n} = S - \alpha_{n-1} A^{1,n-1} - \beta_1 A^{2,n}$ is a constant matrix. This completes the proof of the lemma. □

Corollary 1. *Given a cut matrix C with m blocks; k of these blocks ($k \leq m$) contain more than one element: $I_1, I_2, \ldots, I_k$, $|I_j| > 1$, $\forall j$, and $I_1 = \{i_1 = 1, \ldots, j_1\}$, $I_2 = \{i_2, \ldots, j_2\}$, ..., $I_k = \{i_k, \ldots, j_k\}$, $i_k \geq j_{k-1} + 1$ and $i_k < j_k$, it can be represented as $C = Z + \sum_{l=1}^{l=k} A^{i_l j_l}$, where $Z = (z_{ij})$ with $z_{ij} = -(k-1)$ for $i \neq j$.*

2.2 Recognizing Conic Combinations of Cut Matrices in CDW Normal Form

Given a matrix C which is both Kalmanson matrix and Robinson matrix, it is a question of interest whether the matrix can be represented as a conic combination of cut matrices in CDW normal form. Note that a constant matrix is a special cut matrix in CDW normal form with all blocks of length one. To formulate a simple rule of recognising this special subclass of Kalmanson (and Robinson) matrices, we will define an $n \times n$ symmetric cut-weight matrix $D(C) = (d_{ij})$ with $d_{ij} = \delta_{ij} = c_{i-1,j} + c_{i,j+1} - c_{ij} - c_{i-1,j+1}$ for $2 \leq i < j \leq n-1$, and $d_{1,i} = \alpha_i = c_{i+1,1} - c_{i1}$, $i = 2, \ldots, n-1$, $d_{in} = \beta_{i-1} = c_{i-1,n} - c_{in}$, $i = 2, \ldots, n-1$. The elements which are not defined are irrelevant to further considerations, and set to be zeros.

Assume that matrix C is a conic combination of cut matrices in CDW normal form. For each of these cut matrices the representation as described in Corollary 1 has the following properties: $2 \leq |I_1| \leq |I_2| \leq \ldots \leq |I_k|$, $\cup_{l=1}^{l=k} I_l = \{i_1, i_1 + 1, \ldots, n\}$ and $I_l = \{j_{l-1}, \ldots, j_l\}$. Note that we pick up here only the blocks with more than one elements. The cut-weight matrix for each of such matrices contains only k non-zero elements with $d_{i_1, j_1} = 1$ for each block.

We illustrate a $n \times n$ cut matrix in CDW normal form on a graph with $n+1$ nodes on a line, with the numbers of nodes increasing from left to right; see Fig. 1 for an illustration. For non-zero cut weights with $d_{i_1,j_1} = 1$ we introduce an edge that connects nodes i_1 and $j_1 + 1$. For each node the length of an edge (defined as the number of nodes covered) entering the node from left can not be greater than the length of the edge going from this node to a node on the right.

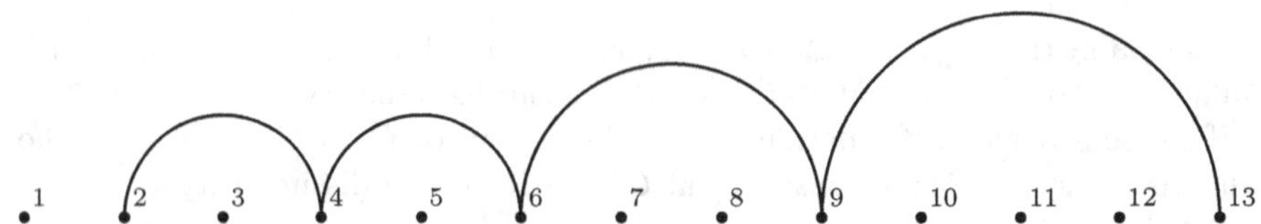

Fig. 1. Illustration for a block representation of a 12×12 cut matrix in CDW normal form with the blocks $\{2,3\}$, $\{4,5\}$, $\{6,7,8\}$, and $\{9,10,11,12\}$.

Theorem 1. *Given a symmetric $n \times n$ Kalmanson matrix C which is also a Robinson dissimilarity with the cut-weight matrix $D(C)$, it can be represented as a conic combination of cut matrices in CDW normal form if and only if the following inequalities hold*

$$\sum_{i=1}^{l} d_{ik} \leq \sum_{j=2k+1-l}^{n} d_{k+1,j} \tag{10}$$

for $k = 2, \ldots, n-1$ and $l = 1, \ldots, k-1$.

The right hand sum in (10) is zero if $2k + 1 - l > n$. This is particular means that $d_{ik} = 0$ for $k = \lceil n/2 \rceil, \ldots, n-1$ and $i = 1, \ldots, 2k - n$.

Proof. For simplicity (and without loss of generality) we assume that weight coefficients in the conic combination mentioned below are integers. It yields immediately that the entries in corresponding cut-weight matrix are also integers. Let C be a matrix which is a conic combination of cut matrices in CDW normal form, and $D(C) = (d_{ij})$ is the corresponding cut weight matrix. Inequality (10) can be rearranged as

$$d_{lk} \leq \sum_{j=2k+1-l}^{n} d_{k+1,j} - \sum_{i=1}^{l-1} d_{ik}. \tag{11}$$

The value d_{lk} can be viewed as the number of edges connecting nodes l and $k+1$ and this number can not be greater than the number of edges going out of node $k+1$ and having the length of at least $(k+1-l)$, which is calculated as the right hand side in (10). If there are non-zero weights d_{ik} with $i < l$, i.e. there are edges entering $k+1$ with the lengths greater than $(k+1-l)$, then d_{lk} has to be compared with the number of edges calculated as shown in right hand side of (11).

Assume now that the entries in a cut matrix D for a matrix C satisfies the inequalities (10). We build an auxiliary multi-graph which we refer as the *cut weight graph* with $n+1$ nodes and d_{ij} edges connecting nodes i and $j+1$ for each $d_{ij} > 0$.

We build a conic representation of C as follows. We start with node $n+1$ in the cut weight graph, and build the path from right to left choosing on each step an edge with the largest length. Let it be the path on nodes $i_1 < i_2 < \ldots < i_k = n+1$. It follows from (10) that cut weight d_{i_1,i_2-1} corresponding to edge (i_1, i_2) is the smallest among all cut weights for the edges in the path. The path found describes a cut matrix in CDW normal form with the blocks $\{i_1, i_1+1, \ldots, i_2-1\}$, ..., $\{i_{k-1}+1, \ldots, i_k = n+1\}$. We add this matrix multiplied by the weight coefficient d_{i_1,i_2-1} to the conic representation of C. We delete all d_{i1,i_2-1} copies of the found path from the cut weight graph and repeat the steps until the graph is empty. □

An Illustrative Example. Consider matrix C which is in the class of Kalmanson and Robinson matrices:

$$C = \begin{pmatrix} 0&1&2&3&3&3\\ 1&0&2&3&3&3\\ 2&2&0&2&3&3\\ 3&3&2&0&2&2\\ 3&3&3&2&0&1\\ 3&3&3&2&1&0 \end{pmatrix}$$

We first illustrate the proofs of Lemmas 3 and 4 and show how to represent C as a conic representation of cut matrices A^{kl}.

Note that for matrix C there is only one strict inequality in system (6) $c_{25} + c_{34} < c_{24} + c_{35}$, and three strict inequalities in system (7): $c_{i1} + c_{i+1,6} < c_{i6} + c_{i+1,1}$ with $i = 2, 3, 4$.

We first eliminate the strict inequality from (6) by subtracting from C a cut matrix from the conic representation of C. If in the initial matrix C we have $c_{24} + c_{35} - c_{25} - c_{34} = d_{24} = 1 > 0$, then in the new transformed matrix $C' = C - A^{34}$ we will have $c'_{24} + c'_{35} - c'_{25} - c'_{34} = 0$:

$$C' = C - A^{34} = \begin{pmatrix} 0&0&1&2&2&2\\ 0&0&1&2&2&2\\ 1&1&0&2&2&2\\ 2&2&2&0&1&1\\ 2&2&2&1&0&0\\ 2&2&2&1&0&0 \end{pmatrix}$$

We have now $\alpha_2 = c'_{32} - c'_{21} = 1$, $\beta_2 = c'_{26} - c'_{36} = 0$, therefore the next step of transformation gives a new matrix with only two strict inequalities in (7):

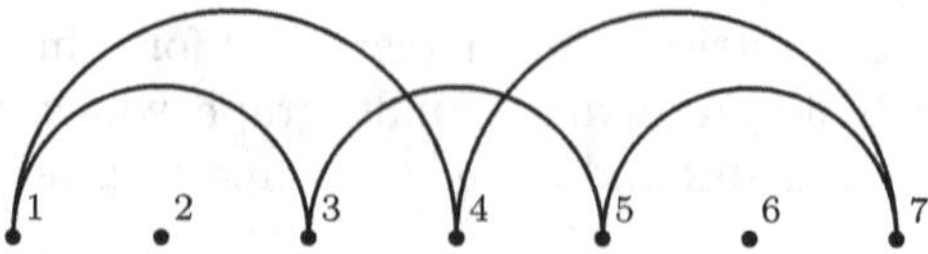

Fig. 2. Cut weight graph for the illustrative example

$$C - A^{34} - A^{12} = \begin{pmatrix} 0 & 0 & 0 & 1 & 1 & 1 \\ 0 & 0 & 0 & 1 & 1 & 1 \\ 0 & 0 & 0 & 1 & 1 & 1 \\ 1 & 1 & 1 & 0 & 0 & 0 \\ 1 & 1 & 1 & 0 & 0 & -1 \\ 1 & 1 & 1 & 0 & -1 & 0 \end{pmatrix}$$

In the next step $\alpha_3 = 1$, $\beta_3 = 1$,

$$C - A^{34} - A^{12} - A^{13} - A^{46} = \begin{pmatrix} 0 & -1 & -1 & -1 & -1 & -1 \\ -1 & 0 & -1 & -1 & -1 & -1 \\ -1 & -1 & 0 & -1 & -1 & -1 \\ -1 & -1 & -1 & 0 & -1 & -1 \\ -1 & -1 & -1 & -1 & 0 & -2 \\ -1 & -1 & -1 & -1 & -2 & 0 \end{pmatrix}$$

And eventually we get

$$C - A^{34} - A^{12} - A^{13} - A^{46} - A^{56} = \begin{pmatrix} 0 & -2 & -2 & -2 & -2 & -2 \\ -2 & 0 & -2 & -2 & -2 & -2 \\ -2 & -2 & 0 & -2 & -2 & -2 \\ -2 & -2 & -2 & 0 & -2 & -2 \\ -2 & -2 & -2 & -2 & 0 & -2 \\ -2 & -2 & -2 & -2 & -2 & 0 \end{pmatrix}$$

The cut weight matrix D contains five non-zero entries

$$D = \begin{pmatrix} 0 & 1 & 1 & 0 & 0 & 0 \\ 0 & 0 & 0 & 0 & 0 & 0 \\ 0 & 0 & 0 & 1 & 0 & 0 \\ 0 & 0 & 0 & 0 & 0 & 1 \\ 0 & 0 & 0 & 0 & 0 & 1 \end{pmatrix}$$

which correspond to the five edges in the cut weight graph shown on Fig. 2. It is clear now that matrix C can be represented as a sum of a weak constant matrix $Z = (z_{ij})$ and two cut matrices in CDW normal form, one matrix has two blocks $\{1, 2, 3\}$ and $\{4, 5, 6\}$, and the other matrix has tree blocks $\{1, 2\}$, $\{3, 4\}$, and $\{5, 6\}$.

3 Conclusions

In this paper we have introduced a new class of specially structured matrices. A matrix belongs to this special class if it can be represented as a conic combination of cut matrices in CDW normal form. We considered a new solvable case of the QAP, where one of the underlying matrices is in the introduced special class of matrices, and the other matrix is a monotone anti-Monge matrix. It follows from [9] that an optimal solution to this special case of the QAP is attained on the identity permutation.

Our algorithmic main contribution is a recognition algorithm for the new class of matrices, which is in fact a subclass of Robinsonian dissimilarities and a subclass of Kalmanson matrices. Laurent and Seminaroti have recently shown in [21] that the QAP with one matrix being a Robinson matrix, and the other matrix being a special class of Toeplitz matrices (see Sect. 1) is solved to optimality by the identity permutation.

Deineko and Woeginger [15] have shown that the QAP with one matrix being a Kalmanson matrix, and the other matrix being from a special sub-class of Toeplitz matrices is again solved to optimality by the identity permutation. An $n \times n$ symmetic Toepliz matrix B considered in [15] is defined as $b_{ji} = b_{ij} = t_{j-i}$, $i < j$, with $t_i = t_{n-i}$ for $i = 1, 2, \ldots, \lfloor n/2 \rfloor$ and $t_{i-1} \geq t_i$ for $i = 2, \ldots, \lfloor n/2 \rfloor$.

Our results allow us now to combine the three solvable QAP cases into a single new one: if one matrix in the QAP is a conic representation of cut matrices in CDW normal form and the other matrix is a conic combination of (i) an anti-Monge monotone matrix [9], (ii) a Toeplitz matrix of Laurent and Seminaroti [21], and (iii) a Toeplitz matrix of Deineko and Woeginger [15], then the QAP is still polynomially solvable and the identity is an optimal permutation. A simple linear programming technique can be used to recognize whether a symmetric matrix can be represented/approximated as a weighted sum of matrices from the three special classes mentioned above.

Acknowledgements. Vladimir Deineko acknowledges support by Warwick University's Centre for Discrete Mathematics and Its Applications (DIMAP). Gerhard Woeginger acknowledges support by the Zwaartekracht NETWORKS grant of NWO, and by the Alexander von Humboldt Foundation, Bonn, Germany.

References

1. Bandelt, H.J., Dress, A.W.M.: A canonical decomposition theory for metrics of a finite set. Adv. Math. **92**, 47–105 (1992)
2. Barthèlemy, J.P., Brucker, F., Osswald, C.: Combinatorial optimization and hierarchical classifications. 4OR **2**, 179–219 (2004)
3. Burkard, R.E., Çela, E., Rote, G., Woeginger, G.J.: The quadratic assignment problem with a monotone anti-Monge and a symmetric Toeplitz matrix: easy and hard cases. Math. Program. **B82**, 125–158 (1998)
4. Burkard, R.E., Dell'Amico, M., Martello, S.: Assignment Problems. SIAM, Philadelphia (2009)
5. Burkard, R.E., Klinz, B., Rudolf, R.: Perspectives of Monge properties in optimization. Discrete Appl. Math. **70**, 95–161 (1996)
6. Brucker, F.: Subdominant theory in numerical taxonomy. Discrete Appl. Math. **154**, 1085–1099 (2006)

7. Çela, E.: The Quadratic Assignment Problem: Theory and Algorithms. Kluwer Academic Publishers, Dordrecht (1998)
8. Çela, E., Deineko, V.G., Woeginger, G.J.: Another well-solvable case of the QAP: maximizing the job completion time variance. Oper. Res. Lett. **40**, 356–359 (2012)
9. Çela, E., Deineko, V.G., Woeginger, G.J.: Well-solvable cases of the QAP with block-structured matrices. Discrete Appl. Math. **186**, 56–65 (2015)
10. Çela, E., Schmuck, N., Wimer, S., Woeginger, G.J.: The Wiener maximum quadratic assignment problem. Discrete Optim. **8**, 411–416 (2011)
11. Coene, S., Filippi, C., Spieksma, F.C.R., Stevanato, E.: Balancing profits and costs on trees. Networks **61**, 200–211 (2013)
12. Christopher, G., Farach, M., Trick, M.: The structure of circular decomposable metrics. In: Díaz, J. (ed.) ESA 1996. LNCS, vol. 1136, pp. 406–418. Springer, Heidelberg (1996)
13. Deineko, V.G., Rudolf, R., Van der Veen, J.A.A., Woeginger, G.J.: Three easy special cases of the Euclidean travelling salesman problem. RAIRO Oper. Res. **31**, 343–362 (1997)
14. Deineko, V.G., Rudolf, R., Woeginger, G.J.: Sometimes travelling is easy: the master tour problem. SIAM J. Discrete Math. **11**, 81–93 (1998)
15. Deineko, V.G., Woeginger, G.J.: A solvable case of the quadratic assignment problem. Oper. Res. Lett. **22**, 13–17 (1998)
16. Hubert, L.J.: Assignment Methods in Combinatorial Data Analysis. Marcel Dekker, New York (1987)
17. Hubert, L., Arabie, P., Meulman, J.: Combinatorial Data Analysis: Optimization by Dynamic Programming. SIAM, Philadelphia (2001)
18. Kalmanson, K.: Edgeconvex circuits and the traveling salesman problem. Can. J. Math. **27**, 1000–1010 (1975)
19. Klinz, B., Woeginger, G.J.: The Steiner tree problem in Kalmanson matrices and in Circulant matrices. J. Comb. Optim. **3**, 51–58 (1999)
20. Koopmans, T.C., Beckmann, M.J.: Assignment problems and the location of economic activities. Econometrica **25**, 53–76 (1957)
21. Laurent, M., Seminaroti, M.: The quadratic assignment problem is easy for Robinsonian matrices with Touplitz structure. Oper. Res. Lett. **43**, 103–109 (2015)
22. Mirkin, B., Rodin, S.: Graphs and Genes. Springer, Berlin (1984)
23. Mirkin, B.: Mathematical Classification and Clustering. Kluwer, Dordrecht (1996)
24. Monge, G.: Mémoires sur la théorie des déblais et des remblais. In: Histoire de l'Academie Royale des Sciences, Année M. DCCLXXXI, avec les Mémoires de Mathématique et de Physique, pour la même Année, Tirés des Registres de cette Académie, Paris, pp. 666–704 (1781)
25. Polyakovskiy, S., Spieksma, F.C.R., Woeginger, G.J.: The three-dimensional matching problem in Kalmanson matrices. J. Comb. Optim. **26**, 1–9 (2013)
26. Robinson, W.S.: A method for chronological ordering archaeological deposits. Am. Antiq. **16**, 293–301 (1951)

An Online Model of Berth and Quay Crane Integrated Allocation in Container Terminals

Feifeng Zheng[1](✉), Longliang Qiao[1], and Ming Liu[2]

[1] Glorious Sun School of Business and Management, Donghua University, Shanghai 200051, People's Republic of China
ffzheng@dhu.edu.cn

[2] School of Economics and Management, Tongji University, Shanghai 200092, People's Republic of China

Abstract. Due to the frequent arrivals of real-time vessels beyond those being well planned in many container terminals, this paper studies an online over-list model of integrated allocation of berths and quay cranes in a container terminal. We consider a hybrid berth which consists of three adjacent small berths together with five quay cranes. The objective is to minimize the makespan or the maximum completion time of container vessels. Our focus is the case with two types of vessels in size which require different numbers of berths and QCs in service, and our main contribution is an optimal 4/3-competitive online algorithm.

Keywords: Online scheduling · Container terminal · Competitive algorithm

1 Introduction

In rapid development of Chinese container ports, one of the key problems in improving service efficiency of terminal resources is to make and fulfill a flexible and reliable service schedule for arrival vessels. To reduce the idle rate of terminal resources and increase service revenue, some ports such as Shanghai Yangshan deepwater port and Shenzhen Shekou Mawan terminal accept lots of daily real-time vessels beyond those preplanned vessels. Thus the assignment of resource as well as the schedule of loading and unloading containers are much affected by the dynamic arrivals of these vessels. It needs to make daily replanning or adjustment on assigning berths, quay cranes (abbr. QC) and other resources.

Recently, there is a growing interest in berth and quay crane allocation (see Vis and de Koster [1]; Stahlbock and Vob, [2], Bierwirth and Meise [3]). The issue of assigning berth and service time to container vessels for loading and unloading containers is referred to as the Berth Allocation Problem (abbr. BAP). The transhipment of containers between a vessel and the quay is generally performed by QCs which are mounted on rail tracks alongside the berths, and the service time of the vessel is inversely proportional to the number of assigned QCs. The assignment of QCs to some berths for serving the vessels is called

Z. Lu et al. (Eds.): COCOA 2015, LNCS 9486, pp. 721–730, 2015.
DOI: 10.1007/978-3-319-26626-8_53

the Quay Crane Assignment Problem (abbr. QCAP). Since BAP and QCAP are much interrelated in practice, it is observed a trend towards an integrated solution of berth and QCs, i.e., the BAP-QCAP problem (see Carlo et al. [4]).

There are *discrete*, *continuous* and *hybrid* berths as classified in Imai et al. [5]. In *hybrid* berth layout, the quay is partitioned into berths such that a small vessel occupies a single berth while a large vessel occupies two or more berths. Due to geographic condition limitations, the hybrid berth layout is common in China such as in Shenzhen Shekou Mawan Terminal. By our field survey at the terminal, the confirmed information of real-time vessels is generally received a few days or even half a day before their arrivals. A service schedule is quickly produced based on FCFS (first-come-first-serve) rule and executed soon. In this paper we model the above FCFS-based resource assignment for serving vessels as the online over-list BAP-QCAP problem and focus on the hybrid berth layout.

1.1 Related Work

Most literature investigated the offline version of the BAP-QCAP problem such that the full information of all the vessels is known at the beginning. Bierwirth and Meisel [3] made a classification on the release time of a vessel. In a *static* model all the vessels are available at the beginning, while in a *dynamic* model vessels are of different release times. Lokuge and Alahakoon [6] studied the dynamic BAP-QCAP problem in hybrid berth layout where the time to serve a vessel is related to the number of assigned QCs and berthing position. They proposed a multi-agent system which constitutes a feedback loop integration of BAP and QCAP for minimizing total waiting times and total tardiness. Park and Kim [7] and Theofanis et al. [8] considered the static BAP-QCAP problem in continuous berth layout, giving optimization models of berth and QCs assignment. Giallombardo et al. [9] presented a mixed integer programming and a tabu search algorithm for the dynamic BAP-QCAP problem with discrete berths. Blazewicz et al. [10], Zhen et al. [11] and Chen et al. [12] furthered the study by considering various objectives including the makespan.

Zhang et al. [13] introduced the online theory into the area of container vessel scheduling. They studied both over-list and over-time versions of online QC assignment. For the objective of the makespan, they presented an asymptotically optimal $m/\lceil \log_2(m+1) \rceil$-competitive algorithm for the over-list model and a 3-competitive algorithm for the over-time model where m is the number of QCs.

This paper studies the online over-list BAP-QCAP problem in hybrid berth layout with three berths and five QCs. The contributions of this work are twofold: (a) consider the impact of real-time vessels in seaside operations. (b) propose an implementable method for berth and QC assignment. The remainder of the paper is as follows. Section 2 describes the problem and gives some notations. In Sect. 3 we present an online algorithm together with its basic properties. Section 4 gives the proof on the competitiveness of the algorithm. In Sect. 5 we prove a matching lower bound, and finally Sect. 6 concludes this work.

2 Problem Description and Basic Notations

We are given three berths b_1, b_2, b_3 in a line and accordingly five identical QCs $q_1, q_2, \ldots, q_5$ from left to right. Assume that there are exactly two types of vessels in size each of which corresponds to a service request with uniform processing load. More precisely, a small request with a load of one requires a single berth, while a large request with a load of $\Delta(\geq 2)$ occupies two neighbouring berths. For QC assignment, it is assumed that a small (or large) request is served by up to two (or four) QCs simultaneously. Similar to the assumption on the processing time of a vessel in Lokuge and Alahakoon [6], we roughly assume that the actual processing time of a small (or large) request r_i is equal to $1/m_i$ (or Δ/m_i) units of time where m_i is the number of QCs assigned to the request.

A sequence of requests $\mathcal{I} = (r_1, r_2, \ldots, r_n)$ $(n \geq 1)$ are to be released over list, i.e., to be released one by one at time zero. On the release of each request r_i, a scheduler has to decide immediately its service resource combination, i.e., the specific berth(s) and QC(s), as well as the processing time interval for the request. The objective is to minimize the makespan or the completion time of the last completed request. Similar to the quadruple notation scheme in Bierwirth and Meisel [3], we denote the considered problem as $hybr, 3B, 5QC|online - over - list|BAP - QCAP|C_{max}$ where the term *hybr* means it is in the hybrid berth layout, terms 3B and 5QC denote the scenario with 3 small berths and 5 QCs, $online - over - list$ means the requests are released in over-list, $BAP - QCAP$ represents the integrated allocation of berths and QCs, and the last term C_{max} denotes the objective function, i.e., makespan.

Below we give some notations related to requests in a schedule produced by an online algorithm. On the release of any request r_i $(i \geq 1)$, let $t_{i,1}$ be the earliest time by which at least one berth has completed all of its currently assigned requests, $t_{i,2}$ (or $t_{i,3}$) the earliest time by which two (or three) consecutive berths have completed all of their currently assigned requests, and let $C_{i,j}$ $(1 \leq j \leq 3)$ be the earliest time by which berth b_j has completed all of its currently assigned requests. Let s_i, e_i be the start and end time of request r_i respectively. For notational convenience, we sometimes denote by $r_i = 1$ and $r_i = \Delta$ for the case where r_i is a small request respectively, a large one.

As to be revealed later on, the assignment of a large request r_i (≥ 2) may induce a forced idle time segment on berth b_1 or b_2, i.e., an idle time segment $[e_k, s_i)$ for some $1 \leq k < i$. If $0 < s_i - e_k < 1/2$, then any small request released later cannot be satisfied in $[e_k, s_i)$ and we say it is a *waste time segment*, denoted by T_w; otherwise if $s_i - e_k \geq 1/2$, then it is an *available idle time segment*, denoted by T_a. Set $T_a = [t_1, t_2)$ where $t_1 = e_k$ and $t_2 = s_i$. Initially, $T_w = T_a = [0, 0)$.

The performance of an online algorithm $\mathcal{A}$ is often evaluated by parameter *competitive ratio* as defined below (see Borodin [14]). For any request input instance $\mathcal{I}$, let $C_\sigma(\mathcal{I}), C^*(\mathcal{I})$ be the makespan of schedule σ produced by $\mathcal{A}$ and that of an optimal schedule respectively. Then we say $\mathcal{A}$ has a competitive ratio of ρ or $\mathcal{A}$ is ρ-competitive where

$$\rho = \inf_r \left\{ r | \frac{C_\sigma(\mathcal{I})}{C^*(\mathcal{I})} \leq r \right\}.$$

3 The Online Algorithm and Its Basic Properties

In this section we present an online algorithm named GLR (Greedy for Large Request) and some properties in any schedule it produces. GLR predetermines the locations of the QCs which provides service without moving between berths so that it only needs to decide berth assignment for each request. Its basic idea is to greedily assign each large request to two berths with four QCs except two cases where $r_1 = \Delta$ and where $r_1 = 1$ and $r_2 = \Delta$. In the algorithm, the leftmost two QCs q_1, q_2 are dedicated to the leftmost berth b_1, and q_3, q_4 are dedicated to the middle berth b_2, while the rightmost QC q_5 processes requests assigned to b_3. So, a large request assigned to the leftmost (or rightmost) two berths is processed by four (or three) QCs simultaneously and consumes $\Delta/4$ (or $\Delta/3$) units of time. Define id as the index of berth with a T_a segment.

3.1 Algorithm Description

Below is the detailed description of the online algorithm.

Algorithm GLR:

On the release of request r_i $(i \geq 1)$, GLR behaves in two cases.
Case 1. $r_i = \Delta$. Divide this case into three subcases in each of which r_i is assigned to the leftmost two berths b_1, b_2 unless it is specified.

- **Case 1.1.** $t_{i,1} = t_{i,2} \leq t_{i,3}$. If $t_{i,2} = 0$, set $s_i = 0$ and assign r_i to berths b_2, b_3. Otherwise set $s_i = t_{i,3}$ (or $s_i = t_{i,1}$) if $C_{i,1} = t_{i,3}$ (or $C_{i,3} = t_{i,3} > t_{i,1}$). Set $T_a = [t_{i,1}, t_{i,3})$ and $id = 2$ provided that $C_{i,1} = t_{i,3} = t_{i,1} + 1/2$.
- **Case 1.2.** $t_{i,1} < t_{i,2} = t_{i,3}$. Set $s_i = t_{i,3}$. If $t_{i,3} - C_{i,1} \geq 1/2$, set $T_a = [C_{i,1}, t_{i,3})$ and $id = 1$.
- **Case 1.3.** $t_{i,1} < t_{i,2} < t_{i,3}$. If $C_{i,3} = t_{i,3}$, set $s_i = t_{i,2}$; otherwise set $s_i = C_{i,1} = t_{i,3}$, and set $T_a = [t_{i,2}, t_{i,3})$ and $id = 2$ given that $t_{i,3} - t_{i,2} = 1/2$.

Case 2. $r_i = 1$. If $T_a = [t_1, t_2) \neq [0, 0)$, set $s_i = t_1$, assign r_i to berth b_{id}, and reset $T_a = [0, 0)$ (or $T_a = [t_1 + 1/2, t_2)$) provided that $t_2 - t_1 < 1$ (or $1 \leq t_2 - t_1$). Otherwise if $T_a = [0, 0)$, consider the following five subcases.

- **Case 2.1.** $t_{i,1} = t_{i,2} = t_{i,3}$. Set $s_i = t_{i,1}$ and assign r_i to berth b_1.
- **Case 2.2.** $t_{i,1} = t_{i,2} < t_{i,3}$. Set $s_i = t_{i,1}$, and assign r_i to berth b_2.
- **Case 2.3.** $C_{i,1} = t_{i,1} < t_{i,2} = t_{i,3}$. It is the same as in Case 2.1.
- **Case 2.4.** $C_{i,3} = t_{i,1} < t_{i,2} = t_{i,3}$. If $t_{i,2} - t_{i,1} \geq 1/2$, set $s_i = t_{i,1}$ and assign r_i to berth b_3; otherwise set $s_i = t_{i,3}$ and assign r_i to b_1.
- **Case 2.5.** $t_{i,1} < t_{i,2} < t_{i,3}$. Set $s_i = t_{i,2}$, and assign r_i to berth b_2.

Figure 1 is an illustration of the algorithm where each rectangle represents a large (or small) request if it occupies two berths (or a single berth) in the horizontal axis, and its height denotes the actual processing time of the request. We observe that in Cases 1.1, 1.2 and 1.3 of the algorithm, the assignment of a large request r_i may induce a T_a or T_w. By Case 2 of the algorithm, there is at

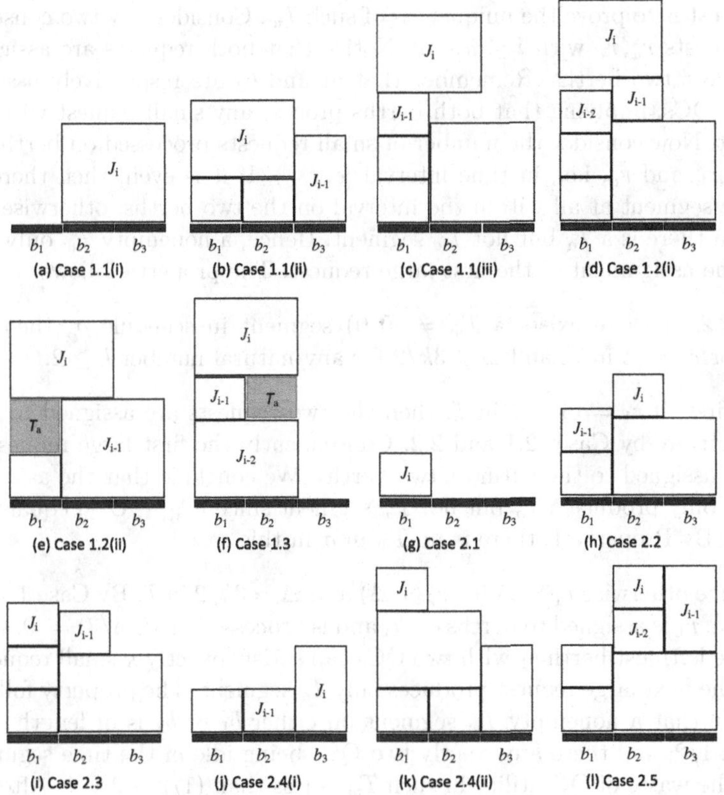

Fig. 1. An illustration of algorithm GLR

most one T_a on the release of any request. For T_w, Case 1.1 implies that if exactly three QCs are idle at time $t_{i,1}$ and $0 < t_{i,3} - t_{i,1} < 1/2$, then $T_w = [t_{i,1}, t_{i,3})$ occurs on berth b_2. Similarly, Case 1.2 (or Case 1.3) tells that if $0 < t_{i,3} - C_{i,1} < 1/2$ (or $C_{i,3} = t_{i,3}$), then $T_w = [C_{i,1}, t_{i,3})$ (or $[t_{i,1}, t_{i,2})$) occurs on b_1.

3.2 Basic Properties

For any request input sequence $\mathcal{I} = \{r_1, r_2, \ldots, r_n\}$, let σ be the schedule produced by GLR.

Property 1. There is at most one waste time segment $T_w \neq [0, 0)$ in schedule σ, and it is due to assigning the first and the second large requests in $\mathcal{I}$ to the rightmost and leftmost two berths, respectively.

Proof. By Case 1 of the algorithm and previous analysis on T_w, a nonempty T_w segment, if exists, is due to assigning the first large request r_j to the rightmost two berths and later on the second large request r_i to the leftmost two berths. Note that T_w may occur on either b_1 or b_2.

The rest is to prove the uniqueness of such T_w. Consider any two consecutive large requests r_u, r_v with $i \leq u < v$. Notice that both requests are assigned to the leftmost two berths. Remember that b_1 and b_2 are respectively associated with two QCs, implying that both berths process any small request with half a time unit. Now consider the number of small requests processed on berths b_1, b_2 between r_u and r_v, i.e., in time interval $[e_u, s_v)$. If it is even, then there is no idle time segment at all within the interval on the two berths; otherwise if it is odd, then there is a T_a but not T_w segment. Hence, a nonempty T_w only occurs during the assignment of the first large request. The property follows. □

Property 2. If there exists a $T_w \neq [0, 0)$ segment in schedule σ, then either $r_1 = \Delta$ or $r_2 = \Delta$ in $\mathcal{I}$, and $\Delta \neq 3k/2$ for any natural number $k \geq 2$.

Proof. First, if $r_1 = r_2 = 1$ in $\mathcal{I}$, then the two requests are assigned to b_1 and b_2 respectively by Cases 2.1 and 2.2. Consequently the first large request r_i, if exists, is assigned to the leftmost two berths. We conclude that the assignment of r_i can only produce a T_a but not T_w segment since $|C_{i,1} - C_{i,2}|$ equals either 1/2 or 0. By Property 1, there is no T_w in σ in this case.

Assume otherwise $r_1 = \Delta$ (or $r_2 = \Delta$) and $\Delta = 3k/2$ in $\mathcal{I}$. By Case 1.1 of the algorithm, r_1 is assigned to berths b_2, b_3 and is processed in $[0, \Delta/3) = [0, k/2)$, in which the leftmost berth b_1 with two QCs can satisfy exactly k small requests. It implies the next large request produces only T_a segment. The property follows. □

Notice that a nonempty T_w segment on either b_1 or b_2 is of length strictly less than 1/2, and there are exactly two QCs being idle in the time segment. It implies the waste of QC utility in such T_w is less than $(1/2) * 2 = 1$. After GLR assigns the last request in $\mathcal{I}$, we denote by $T_j \neq [0, 0)$ (or $T_j = [0, 0)$) the case with the existence (or nonexistence) of T_j $(j = w, a)$ in the schedule σ.

4 Competitive Analysis of Algorithm GLR

Denote by $C_\sigma(\mathcal{I})$ the makespan of σ produced by GLR in sequence $\mathcal{I}$, and $C^*(\mathcal{I})$ that of a schedule produced by an optimal offline algorithm OPT. In the following we first present two lemmas on the ratio of $C_\sigma(\mathcal{I})/C^*(\mathcal{I})$ for two specific cases of $\mathcal{I}$, and then prove GLR's competitiveness.

Lemma 1. *If $r_1 = r_2 = 1$ and $n \geq 4$ in sequence $\mathcal{I}$, then $C_\sigma(\mathcal{I})/C^*(\mathcal{I}) \leq 4/3$.*

Proof. By the lemma condition $r_1 = r_2 = 1$ and Property 2, we have $T_w = [0, 0)$ in σ. For the last request r_n in $\mathcal{I}$, we have $r_n = \Delta$ (or $r_n = 1$) if it is assigned by Case 1 (or Case 2) of algorithm GLR. There are five cases below.

Case 1. $r_n = \Delta$ and it is assigned to berths b_1, b_2 by Case 1 of the algorithm. Assume that the last k $(1 \leq k \leq n - 2)$ requests in $\mathcal{I}$ are large, while r_{n-k} is a small request. If $T_a = [0, 0)$ and thus the left four QCs on berths b_1, b_2 are kept busy during $[0, C_\sigma(\mathcal{I}))$, then the total workload in $\mathcal{I}$ is at least $4C_\sigma(\mathcal{I})$ and $C^*(\mathcal{I}) \geq 4C_\sigma(\mathcal{I})/5$. Otherwise if $T_a = [t_1, t_2) \neq [0, 0)$, then $t_2 - t_1 = 1/2$ by the proof of Property 2 and $C_{n,3} = C_{n-k+1,3} \geq 1$ by Case 2.4 of the algorithm. Moreover, we claim by the lemma condition $r_1 = r_2 = 1$ that Case 1.3 of the algorithm is not true for request r_{n-k+1}, implying $s_{n-k+1} = t_{n-k+1,3}$. $C_\sigma(\mathcal{I}) = t_{n-k+1,3} + k\Delta/4$ and $C^*(\mathcal{I}) \geq (4C_\sigma(\mathcal{I}) - 1 + C_{n,3})/5 \geq 4C_\sigma(\mathcal{I})/5$.

Case 2. $r_n = 1$ and it is assigned by Cases 2, 2.3 or 2.5 of the algorithm. We claim by the lemma condition $r_1 = r_2 = 1$ that the left four QCs on berths b_1, b_2 are kept busy during $[0, C_\sigma(\mathcal{I}))$, and again $C_\sigma(\mathcal{I})/C^*(\mathcal{I}) \leq 5/4$.

Case 3. $r_n(= 1)$ is assigned by Case 2.1 of the algorithm. Then $t_{n,1} = t_{n,3} \geq 1$, $C_\sigma(\mathcal{I}) = t_{n,1} + 1/2$ and $C^*(\mathcal{I}) \geq t_{n,1} + 1/5$. So, $C_\sigma(\mathcal{I})/C^*(\mathcal{I}) \leq 5/4$.

Case 4. $r_n(= 1)$ is assigned by Case 2.2 of the algorithm, implying $T_a = [0, 0)$. If there is no large request in $\mathcal{I}$, then $C^*(\mathcal{I}) = C_\sigma(\mathcal{I})$; otherwise if $\mathcal{I}$ contains at least one large request, then $t_{n,2} \geq 1/2 + \Delta/4 \geq 1$, $C_\sigma(\mathcal{I}) = t_{n,2} + 1/2$ and $C^*(\mathcal{I}) > t_{n,2} + 1/5$, implying a ratio less than 5/4.

Case 5. $r_n(= 1)$ is assigned by Case 2.4 of the algorithm. By the lemma condition $r_1 = r_2 = 1$, we have $C_{n,1} = C_{n,2} = t_{n,2}$ and $C_{n,3} = t_{n,1}$ in this case.

Case 5.1. $t_{n,2} - t_{n,1} \geq 1/2$ and then r_n is assigned to berth b_3. If $\mathcal{I}$ contains no large requests, then $C^*(\mathcal{I}) = C_\sigma(\mathcal{I})$. Otherwise if $\mathcal{I}$ contains at least one large request, then $t_{n,1} \geq 1$. If $C_\sigma(\mathcal{I}) = t_{n,2}$, then $C^*(\mathcal{I}) > 4C_\sigma(\mathcal{I})/5$; otherwise if $C_\sigma(\mathcal{I}) = t_{n,1} + 1$, then $C^*(\mathcal{I}) \geq t_{n,2} \geq t_{n,1} + 1/2$ and thus $C_\sigma(\mathcal{I})/C^*(\mathcal{I}) \leq 4/3$.

Case 5.2. $t_{n,2} - t_{n,1} < 1/2$ and then r_n is assigned to berth b_1. By the case condition $t_{n,2} - t_{n,1} < 1/2$, we claim that $\mathcal{I}$ contains at least one large request and $t_{n,1} \geq 1$. $C_\sigma(\mathcal{I}) = t_{n,2} + 1/2$, and $C^*(\mathcal{I}) \geq (4t_{n,2} + 1 + t_{n,1})/5 \geq 4C_\sigma(\mathcal{I})/5$.

The lemma follows. □

Lemma 2. *If $r_1 = \Delta$ or $r_2 = \Delta$ and $n \geq 4$ in sequence $\mathcal{I}$ where $\Delta \neq 3k/2$ for any natural number $k \geq 2$, then $C_\sigma(\mathcal{I})/C^*(\mathcal{I}) \leq 4/3$.*

Proof. By the lemma condition r_1 (or r_2)$= \Delta$ with $\Delta \neq 3k/2$, if there is a single large request in $\mathcal{I}$, then $T_w = T_a = [0, 0)$ and it is straightforward that $C_\sigma(\mathcal{I})/C^*(\mathcal{I}) \leq 4/3$ no matter $C_\sigma(\mathcal{I}) = \Delta/3$ or not. In the following we focus on the scenario where there are at least two large requests in $\mathcal{I}$, and we consider three cases by whether $T_w = [0, 0)$ or $T_a = [0, 0)$ in schedule σ. Remember that a T_w segment induces waste of QC utility strictly less than one.

Case 1. $T_w \neq [0, 0)$ and $T_a = [t_1, t_2) \neq [0, 0)$. The case condition implies that there are at least three large requests in $\mathcal{I}$ and $C_\sigma(\mathcal{I})$ is determined by the completion time of the last large request. Let r_i and r_j $(j < i \leq n)$ be the large requests whose assignments induce the T_a and T_w segment, respectively. $t_{n,1} = C_{n,3} \geq \Delta/3 + 1 \geq 5/3$ and $C_\sigma(\mathcal{I}) \geq \Delta/3 + \Delta/4 + 1/2 + \Delta/4 = 5\Delta/6 + 1/2 \geq 13/6$ due to the existence of T_w and T_a. Notice that $t_2 - t_1 = 1/2$, and the T_a and T_w segments waste less than two units of QC utility. $C^*(\mathcal{I}) > (4C_\sigma(\mathcal{I}) - 2 + t_{n,1})/5 \geq (4C_\sigma(\mathcal{I}) - 1/3)/5$, and $C_\sigma(\mathcal{I})/C^*(\mathcal{I}) < 5C_\sigma(\mathcal{I})/(4C_\sigma(\mathcal{I}) - 1/3) \leq 13/10$.

Case 2. $T_w = [0, 0)$ and $T_a = [t_1, t_2) \neq [0, 0)$. In this case we claim that T_a is on berth b_1 and within the time interval of processing the first large request on the rightmost two berths, i.e., $t_2 = \Delta/3$ and $t_1 \geq 0$. Moreover, all the small requests in $\mathcal{I}$ are assigned to berth b_1 and processed in $[0, t_1)$. Assume that there are $k \geq 2$ large requests in $\mathcal{I}$. Then $C_\sigma(\mathcal{I}) = (k-1)\Delta/4 + \Delta/3$ while $C^*(\mathcal{I}) \geq k\Delta/4$, and thus $C_\sigma(\mathcal{I})/C^*(\mathcal{I}) < 4/3$ due to $k \geq 2$.

Case 3. $T_w \neq [0,0)$ and $T_a = [0,0)$.

Case 3.1. $r_n = \Delta$. Note that the first large request in $\mathcal{I}$ is completed at time $\Delta/3$ on berths b_2, b_3. If r_n starts at time $s_n = \Delta/3$, then $C_\sigma(\mathcal{I}) = s_n + \Delta/4 = 7\Delta/12$ while $C^*(\mathcal{I}) \geq \Delta/4 + \Delta/4 = 6C_\sigma(\mathcal{I})/7$. Otherwise if $s_n > \Delta/3 \geq 2/3$, then before r_n the leftmost berth b_1 either processes another large request not earlier than $\Delta/3$ or serves at least two small requests from time 0. In both cases $s_n \geq 1$. $C_\sigma(\mathcal{I}) = e_n \geq 1 + \Delta/4 \geq 3/2$, and $C^*(\mathcal{I}) > (4C_\sigma(\mathcal{I}) - 1 + C_{n,3}))/5 \geq 4C_\sigma(\mathcal{I})/5 - 1/15$ due to $C_{n,3} \geq \Delta/3 \geq 2/3$. Thus $C_\sigma(\mathcal{I})/C^*(\mathcal{I}) < 45/34 < 4/3$.

Case 3.2. $r_n = 1$. If r_n is assigned to some T_a segment, inducing a T_w for the remaining idle time of the T_a, then we claim that $C_\sigma(\mathcal{I}) = \Delta/3 + (k-1)\Delta/4$ where $k \geq 2$ is the number of large requests in $\mathcal{I}$. $C^*(\mathcal{I}) \geq k\Delta/4$, and the ratio is at most 7/6. Below we consider the rest five subcases on the assignment of r_n.

Case 3.2.1. r_n is assigned by Case 2.1 of the algorithm, implying $t_{n,1} = t_{n,3}$. In this case $t_{n,3} \geq \Delta/3 + 1 \geq 5/3$ due to assigning the first large request to berths b_2, b_3. $C_\sigma(\mathcal{I}) = t_{n,3} + 1/2$, and $C^*(\mathcal{I}) > t_{n,3}$ since r_n cannot be satisfied in the T_w, implying a ratio less than 4/3.

Case 3.2.2. r_n is assigned by Case 2.2 of the algorithm, implying $C_\sigma(\mathcal{I}) = t_{n,2} + 1/2$. For the case where $t_{n,3} = C_{n,3}$, if $t_{n,2} = \Delta/4 + \Delta/3$, then $C^*(\mathcal{I}) \geq \max\{(\Delta+\Delta)/4+1/2, 2\Delta/3\}$; otherwise berth b_1 has served at least two small (or two large) requests before r_n with $C_{n,1} = t_{n,2} \geq 1+\Delta/4$ while $C^*(\mathcal{I}) > t_{n,2}$ with the same argument as in Case 3.2.1. With $\Delta \geq 2$ the ratio in the above both cases is less than 4/3. For the other case where $t_{n,3} = C_{n,1}$, we have $t_{n,3} = t_{n,2} + 1/2$, $t_{n,2} = C_{n,3} \geq \Delta/3 + 1$, and thus $C^*(\mathcal{I}) > (5t_{n,2} + 1)/5 > 3C_\sigma(\mathcal{I})/4$.

Case 3.2.3. r_n is assigned by Case 2.3 of the algorithm. In this case $C_\sigma(\mathcal{I}) = t_{n,2} = t_{n,1}+1/2 \geq \Delta/3+\Delta/4+1/2$. Notice that Case 2.3 of the algorithm implies $C_{n,1} = t_{n,1} < C_{n,3}$ and thus $t_{n,2} - C_{n,3} < 1/2$. $C^*(\mathcal{I}) > (4t_{n,2} - 1 + C_{n,3})/5 > (5t_{n,2} - 3/2)/5$, and $C_\sigma(\mathcal{I})/C^*(\mathcal{I}) < 5t_{n,2}/(5t_{n,2} - 3/2) < 50/41 < 4/3$.

Case 3.2.4. r_n is assigned by Case 2.4 of the algorithm. Since $T_w \neq [0,0)$, $C_{n,1} = C_{n,2} = t_{n,2}$ and there are at least two small and two large requests in $\mathcal{I}$. We first consider the case where $t_{n,2} - t_{n,1} < 1/2$ and then r_n is assigned to berth b_1. In this case $t_{n,2} > t_{n,1} \geq \Delta/3 + 1 \geq 5/3$ and $C_\sigma(\mathcal{I}) = t_{n,2} + 1/2$. $C^*(\mathcal{I}) > (4t_{n,2} + t_{n,1})/5 \geq 4t_{n,2}/5 + 1/3$, implying a ratio less than 13/10.

Consider the other case where $t_{n,2} - t_{n,1} \geq 1/2$ and r_n is assigned to berth b_3. If $C_\sigma(\mathcal{I}) = t_{n,2}$, then $C^*(\mathcal{I}) > 4C_\sigma(\mathcal{I})/5$ since the total workload in σ is more than $4C_\sigma(\mathcal{I})$. Otherwise if $C_\sigma(\mathcal{I}) = t_{n,1} + 1$, then $C^*(\mathcal{I}) > (4t_{n,2} + t_{n,1})/5$ (or $C^*(\mathcal{I}) \geq \min\{2\Delta/3, 2\}$) provided that $t_{n,1} \geq \Delta/3+1$ (or $t_{n,1} = \Delta/3$). Together with $t_{n,2} - t_{n,1} \geq 1/2$ and $\Delta \geq 2$, the ratio $C_\sigma(\mathcal{I})/C^*(\mathcal{I}) \leq 25/19 < 4/3$.

Case 3.2.5. r_n is assigned by Case 2.5 of the algorithm. Since $T_w \neq [0,0)$, $t_{n,1} = C_{n,3}$ and $C_\sigma(\mathcal{I}) = t_{n,3} = t_{n,2} + 1/2 < t_{n,1} + 1$ in this case. So, $C^*(\mathcal{I}) > (4t_{n,3} + t_{n,1} - 1)/5 > t_{n,3} - 2/5$, and $C_\sigma(\mathcal{I})/C^*(\mathcal{I}) < 4/3$ due to $t_{n,3} \geq \Delta/3 + \Delta/4 + 1/2 \geq 5/3$.

The lemma follows. □

Based on the above lemmas, we prove the competitiveness of algorithm GLR in the following.

Theorem 1. *For problem* $hybr, 3B, 5QC|online - over - list|BAP - QCAP|C_{max}$*, GLR is 4/3-competitive.*

Proof. Given any request input sequence $\mathcal{I} = \{r_1, r_2, \ldots, r_n\}$ and the schedule σ produced by GLR. If $1 \leq n \leq 3$, it is trivial to verify that $C_\sigma(\mathcal{I})/C^*(\mathcal{I}) \leq 4/3$ for any combination of small and large requests.

In the remaining we focus on the scenario where $n \geq 4$. For the cases where $r_1 = r_2 = 1$ and where r_1 (or r_2) $= \Delta$ with $\Delta \neq 3k/2$ for $k \geq 2$, $C_\sigma(\mathcal{I})/C^*(\mathcal{I}) \leq 4/3$ is straightforward by Lemmas 1 and 2. For the other case where r_1 (or r_2)$= \Delta$ with $\Delta = 3k/2$, we have $T_w = [0, 0)$ in σ by Property 2. no matter whether one of the first two requests is a large one. No matter $T_a = [0, 0)$ or not, the reasoning is similar to that in the proof of Lemma 2 and the same bound of 4/3 holds. The theorem follows. □

5 A Matching Lower Bound

Below we present a matching lowing bound for the considered problem.

Theorem 2. *For problem* $hybr, 3B, 5QC|online - over - list|BAP - QCAP|C_{max}$*, any deterministic online algorithm cannot be better than 4/3-competitive.*

Proof. To prove the theorem, it suffices to construct a request input sequence $\mathcal{I}$ with at most two requests to make any deterministic online algorithm $\mathcal{A}$ be at best 4/3-competitive. Let $C_\mathcal{A}(\mathcal{I})$ and $C^*(\mathcal{I})$ be the makespan of the schedule produced by $\mathcal{A}$ and by an optimal offline algorithm OPT in $\mathcal{I}$, respectively.

The first request $r_1 = 1$. It is then processed by either one or two QCs since it is a small request. If $\mathcal{A}$ processes the request with a single QC from some time $t \geq 0$, then no more requests arrive and $\mathcal{I}$=(r_1). In this case $C_\mathcal{A}(\mathcal{I}) \geq 1$, while OPT processes r_1 with two QCs from time 0, resulting in $C^*(\mathcal{I}) = 1/2$ and $C_\mathcal{A}(\mathcal{I})/C^*(\mathcal{I}) \geq 2$. Otherwise if $\mathcal{A}$ processes r_1 with two QCs, then there releases the next and last request $r_2 = \Delta$ and $\mathcal{I}$=(r_1, r_2). $\mathcal{A}$ processes r_2 with either three or four QCs, $C_\mathcal{A}(\mathcal{I}) \geq \min\{\Delta/3, 1/2 + \Delta/4\}$. OPT processes respectively r_1 and r_2 by one QC and four QCs simultaneously from time 0, and thus $C^*(\mathcal{I}) = \max\{1, \Delta/4\}$. Setting $\Delta = 6$, we have $C_\mathcal{A}(\mathcal{I}) \geq 2$ and $C^*(\mathcal{I}) = 3/2$. Hence, $C_\mathcal{A}(\mathcal{I})/C^*(\mathcal{I}) \geq 4/3$. The theorem is established. □

6 Conclusion

This paper studies an online integrated allocation of berths and quay cranes in a container terminal. We consider the hybrid layout of quay and focus on the scenario with a small number of berths and cranes. To meet the requirement of quick response to real-time vessels, we present an online over-list model, and mainly present an online deterministic algorithm, which is proved optimally 4/3-competitive. One of the interesting problems in further research is to consider the problem in the continuous berth layout scenario or to consider the case with nonuniform load for the same type of request.

Acknowledgements. This work was partially supported by the National Natural Science Foundation of China under Grants 71172189, 71071123 and 61221063, Program for New Century Excellent Talents in University (NCET-12-0824), DHU Distinguished Young Professor Program (A201305), and the Fundamental Research Funds for the Central Universities (2232013D3-46).

References

1. Vis, I.F.A., de Koster, R.: Transshipment of containers at a container terminal: an overview. Eur. J. Oper. Res. **147**(1), 1–16 (2003)
2. Stahlbock, R., VoB, S.: Operations research at container terminals: a literature update. OR Spectrum **30**(1), 1–52 (2008)
3. Bierwirth, C., Meisel, F.: A survey of berth allocation and quay crane scheduling problems in container terminals. Eur. J. Oper. Res. **202**(3), 615–627 (2010)
4. Carlo, H.J., Vis, I.F.A., Roodbergen, K.J.: Seaside operations in container terminals: literature overview, trends, and research directions. Flex. Serv. Manuf. J. **27**(2–3), 224–262 (2015)
5. Imai, A., Sun, X., Nishimura, E., Papadimitriou, S.: Berth allocation in a container port: using a continuous location space approach. Transp. Res. Part B **39**(3), 199–221 (2005)
6. Lokuge, P., Alahakoon, D.: Improving the adaptability in automated vessel scheduling in container ports using intelligent software agents. Eur. J. Oper. Res. **177**(3), 1985–2015 (2007)
7. Park, Y.M., Kim, K.H.: A scheduling method for berth and quay cranes. OR Spectrum **25**(1), 1–23 (2003)
8. Theofanis, S., Golias, M., Boile, M.: Berth and quay crane scheduling: a formulation reflecting service deadlines and productivity agreements. In: TRANSTEC 2007, pp. 124–140. Czech Technical University, Prague (2007)
9. Giallombardo, G., Moccia, L., Salani, M., Vacca, I.: Modeling and solving the tactical berth allocation problem. Transp. Res. Part B **44**(2), 232–245 (2010)
10. Blazewicz, J., Cheng, T.C.E., Machowiak, M., Oguz, C.: Berth and quay crane allocation: a moldable task scheduling model. J. Oper. Res. Soc. **62**, 1189–1197 (2011)
11. Zhen, L., Chew, E.P., Lee, L.H.: An integrated model for berth template and yard template planning in transshipment hubs. Transp. Sci. **45**(4), 483–504 (2011)
12. Chen, J.H., Lee, D.H., Cao, J.X.: A combinatorial benders' cuts algorithm for the quayside operation problem at container terminals. Transp. Res. Part E **48**(1), 266–275 (2012)
13. Zhang, L.L., Khammuang, K., Wirth, A.: On-line scheduling with non-crossing constraints. Oper. Res. Lett. **36**(5), 579–583 (2008)
14. Borodin, A., El-Yaniv, R.: Online Computation and Competitive Analysis. Cambridge University Press, New York (1998)

On the Minimal Constraint Satisfaction Problem: Complexity and Generation

Guillaume Escamocher(✉) and Barry O'Sullivan

Department of Computer Science, Insight Centre for Data Analytics, University College Cork, Cork, Ireland
{guillaume.escamocher,barry.osullivan}@insight-centre.org

Abstract. The Minimal Constraint Satisfaction Problem, or Minimal CSP for short, arises in a number of real-world applications, most notably in constraint-based product configuration. Despite its very permissive structure, it is NP-hard, even when bounding the size of the domains by $d \geq 9$. Yet very little is known about the Minimal CSP beyond that. Our contribution through this paper is twofold. Firstly, we generalize the complexity result to any value of d. We prove that the Minimal CSP remains NP-hard for $d \geq 3$, as well as for $d = 2$ if the arity k of the instances is strictly greater than 2. Our complexity result can be seen as providing a *dichotomy theorem* for the Minimal CSP. Secondly, we build a generator that can create Minimal CSP instances of any size, using the constrainedness as a parameter. Our generator can be used to study behaviors that are typical of NP-hard problems, such as the presence of a phase transition, in the case of the Minimal CSP.

1 Introduction

An instance of the Minimal Constraint Satisfaction Problem, or Minimal CSP for short, is a CSP instance where all allowed k-tuples are part of at least one solution, with k being the arity of the instance, that is the size of the scope of the constraints. Since all Minimal CSP instances are satisfiable, solving such an instance does not refer to the decision problem of determining whether it has a solution, but to the exemplification of a solution.

Minimal CSP is often found 'naturally' in configuration problems [8]. A seller might want to offer its customers a large degree of customization. If, for example, the product sold is a car, some possible options might be the color of the vehicle and whether it is automatic or manual. If after choosing "automatic", "red" remains a valid option for the color parameter, then it is preferable that at least one red automatic car can be configured. The Minimal CSP can answer a number of queries relevant to product configuration in polynomial time [6], such as whether a solution exists that satisfies a given unary constraint, or whether an assignment to k variables is consistent in a Minimal CSP where all constraints are defined over k-tuples of the variables. These queries can be answered simply by inspecting the constraints of the problem instance. However, answering queries over arbitrary assignments to the variables remains hard, which has given rise

Z. Lu et al. (Eds.): COCOA 2015, LNCS 9486, pp. 731–745, 2015.
DOI: 10.1007/978-3-319-26626-8_54

to many studies of the use of automata and decision diagrams to reason about the solution sets of complex configuration problems [2].

Gottlob has shown that computing a solution to a binary Minimal CSP is NP-Hard [6]. While considerable work has been done on the study of problem hardness and instance generation for many classes of CSPs [1,10,11], nothing has been done in the case of Minimal CSPs, which are an interesting class of problem in practice. In this paper, we show that when bounding the size of the domains by a constant d and the arity of the constraints by a constant k, Minimal CSP and the general CSP are NP-hard for the exact same values of d and k. We also present an algorithm that can be used to generate Minimal CSP instances of any size, while retaining control over some parameters like the tightness of the constraints. Since tightness often reveals many statistical properties of CSPs, like for example phase transition behaviors, many empirical analysis of such problems use it as a parameter [3,5]. We tested our generator to determine if it was refined enough to detect the diverse properties held by the Minimal CSP, and found that it could indeed expose how the Minimal CSP behaves: like most other NP-hard problems [7], Minimal CSP exhibits an easy-hard-easy pattern as constraint tightness is varied.

The remainder of the paper is organized as follows. In the next Section, we formally define the Minimal CSP and present our complexity results, and in particular a dichotomy theorem, that generalizes Gottlob's complexity result [6] to instances with very small domain sizes. In Sect. 3.1 we provide some preliminary definitions needed for the description of our generator. Then in Sect. 3.2, we present the algorithm itself, which is able to generate in an efficient way Minimal CSP instances of a given size; tightness, that is the average number of incompatibilities in a constraint, is used as a parameter. In Sect. 3.3, we show some empirical results that we obtained when solving Minimal CSP instances generated with this method, revealing the presence of a peak of difficulty. Finally, we conclude by summarizing our contributions and outlining some future work in this area.

2 The Minimal CSP: Definitions and Complexity

2.1 Definitions

We recall the definition of the Constraint Satisfaction Problem, or CSP.

Definition 1 (CSP). *A* CSP instance I *comprises:*

1. *A set* $V = \{v_1, \ldots, v_n\}$ *of* n *variables.*
2. *A set* $A = \{A_{v_1}, \ldots, A_{v_n}\}$ *of* n *domains. For all* $i \in [1, n]$, $A_{v_i} = \{a_1, \ldots, a_{d_i}\}$ *contains the* d_i *possible values for the variable* v_i.
3. *A set* $C = \{C_1, \ldots, C_m\}$ *of* m *constraints. To each constraint* C_i *is associated a scope* $W_i = \{w_1, \ldots, w_{k_i}\} \subseteq V$ *and a set* U_i *of* k_i*-tuples from* $[A_{w_1}] \times [A_{w_2}] \times \cdots \times [A_{w_{k_i}}]$. *We say that these tuples are* allowed, *or* compatible, *for the scope* W_i, *that the tuples from* $[A_{w_1}] \times [A_{w_2}] \times \cdots \times [A_{w_{k_i}}]$ *that are not in* U_i *are*

forbidden, or incompatible, *for the scope* W_i *and that* k_i *is the* arity *of the constraint* C_i. *Let* $C_j \in C$ *be the constraint with the highest arity. We say that the arity of* C_j *is the arity of the instance.*

A partial solution *to* I *on* $W = \{w_1, \ldots, w_r\} \subseteq V$ *is a r-tuple* $b = \{b_1, \ldots, b_r\}$ *such that* $\forall i \in [1, r]$, $b_i \in A_{w_i}$ *and* $\forall j \in [1, m]$, *such that* k_j *is the arity of* C_j, *there is no* k_j*-tuple* $u \subseteq b$ *that is incompatible on* S_j. *A* solution *to* I *is a partial solution on* V.

We now formally define the Minimal Constraint Satisfaction Problem, or Minimal CSP.

Definition 2 (Minimal CSP). *A CSP instance* $I = \{V, A, C\}$ *is a* Minimal CSP instance *if and only if:* $\forall C_i \in C$, $\forall u$ *such that* u *is a compatible tuple on the scope of* C_i, *there is at least one solution* S *to* I *such that* S *contains* u.

In this paper, we only consider Minimal CSP instances with non-empty domains and such that each value in each domain belongs to at least one compatible tuple.

2.2 Complexity

Not only are Minimal CSP instances always satisfiable, but they typically contain many solutions. A Minimal CSP instance will contain as least as many solutions as there are allowed k-tuples in its least constrained constraint. For this reason, Minimal CSP instances will often be very trivial to solve. Yet, computing a solution to a Minimal CSP instance is NP-hard [6].[1] The proof given in [6] is a reduction from 3-SAT to a set of CSP instances M_9 such that for each instance $I \in M_9$:

- I is either a Minimal CSP instance or unsatisfiable.
- I contains at most 9 values in each domain.

This is stronger than just NP-hardness, the actual result in [6] is that computing a solution to a Minimal CSP instance is NP-hard, even when bounding the size of the domains by a fixed integer $d \geq 9$. We now further generalize this result to any $d \geq 3$:

Proposition 1. *Computing a solution to a Minimal CSP instance is NP-hard, even when bounding the size of the domains by a fixed integer* $d \geq 3$.

Proof. Suppose that we have a k-ary Minimal CSP instance I. Let v be a variable in I such that the domain of v is of size $d_v > 3$. We replace v by two new variables v_1 and v_2 to obtain a new instance I' defined as follow:

[1] We are aware that finding a solution to a Minimal CSP is both a *search* problem (find a solution) and a *promise* problem (the input CSP is satisfiable because it is minimal), rather that a *decision* problem. However, we use this terminology in the same manner as Gottlob [6], where a thorough discussion of the matter can be found.

1. With $\{a_1, a_2, \ldots, a_{d_v}\}$ being the domain of v, we set the domain of v_1 to $\{a_1, a_2, x\}$ and the domain of v_2 to $\{\overline{x}, a_3, \ldots, a_{d_v}\}$.
2. Let $B = \{b_1, b_2, b_3, \ldots, b_k\}$ be a k-tuple such that b_1 is in the domain of v_1 and b_2 is in the domain of v_2. Then B is compatible if and only if ($b_1 = x$, $b_2 \neq \overline{x}$ and there exists some value b' such that the k-tuple $B' = \{b', b_2, b_3, \ldots, b_k\}$ is compatible in I) or ($b_1 \neq x$, $b_2 = \overline{x}$ and there exists some value b' such that the k-tuple $B' = \{b_1, b', b_3, \ldots, b_k\}$ is compatible in I).
3. Let b be a value not in the domain of v_1 or v_2. Let $B = \{b, b', b_3, \ldots, b_k\}$ be a k-tuple such that b' is in the domain of v_1 and B does not contain any value from the domain of v_2. If $b' = a_i$ for $1 \leq i \leq 2$, then B is compatible in I' if and only if B was compatible in I. If $b' = x$, let $B_i = \{b, a_i, b_3, \ldots, b_k\}$ for $3 \leq i \leq d_v$. B is compatible in I' if and only if one of the B_i was compatible in I.
4. Let b be a value not in the domain of v_1 or v_2. Let $B = \{b, b', b_3, \ldots, b_k\}$ be a k-tuple such that b' is in the domain of v_2 and B does not contain any value from the domain of v_1. If $b' = a_i$ for $3 \leq i \leq d_v$, then B is compatible in I' if and only if B was compatible in I. If $b' = x$, let $B_i = \{b, a_i, b_3, \ldots, b_k\}$ for $1 \leq i \leq 2$. B is compatible in I' if and only if one of the B_i was compatible in I.
5. All other k-tuples in I' remain as they were in I.

Figure 1 illustrates an example of the transformation for $k = 2$ and $d_v = 4$. A_v, A_{v_1} and A_{v_2} denote the domains of v, v_1 and v_2 respectively. The continuous lines represent compatibility edges, while the dashed lines represent incompatibility edges.

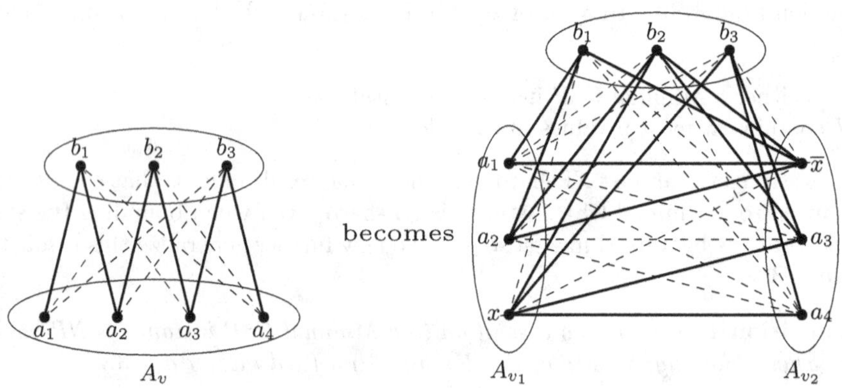

Fig. 1. Transforming a variable v into two variables v_1 and v_2 with smaller domains.

In order to obtain the desired result, we have to prove that I' is a Minimal CSP instance and that finding a solution for I' allows us to find in polynomial time a solution for I.

1. I' is a Minimal CSP instance:
   We have to prove that each compatible k-tuple in I' is in a solution for I'. Let $B = \{b_1, b_2, b_3, \ldots, b_k\}$ be a compatible k-tuple in I' such that:
   - Either b_1 is in the domain A_{v_1} of v_1, or B does not contain any value from A_{v_1}.
   - Either b_2 is in the domain A_{v_2} of v_2, or B does not contain any value from A_{v_2}.

   We necessarily have one of the following seven cases:
   (a) $b_1 = x$ and $b_2 = a_i$ for some $i \in [3, \ldots, d_v]$: from the second point in the definition of I', we know that there is some compatible k-tuple $B' = \{b', a_i, b_3, \ldots, b_k\}$ in I. Since I is a Minimal CSP instance, there is some solution S for I such that B' belongs to S. Let $S' = S \cup \{x\}$. By construction, S' is a solution for I' containing B.
   (b) $b_1 = a_i$ for some $i \in [1, 2]$ and $b_2 = \overline{x}$: same argument as for (a).
   (c) $b_1 = x$ and $b_2 \notin A_{v_2}$: from the third point in the definition of I', we know that there is some compatible k-tuple $B_i = \{a_i, b_2, b_3, \ldots, b_k\}$ in I, with $i \in [3, \ldots, d_v]$. Since I is a Minimal CSP instance, there is some solution S for I such that B_i belongs to S. Let $S' = S \cup \{x\}$. By construction, S' is a solution for I' containing B.
   (d) $b_1 \notin A_{v_1}$ and $b_2 = \overline{x}$: same argument as for (c), using the fourth point in the definition of I' instead of the third.
   (e) $b_1 \in A_{v_1}$, $b_1 \neq x$ and $b_2 \notin A_{v_2}$: from the third point in the definition of I', we know that B is compatible in I too. Since I is a Minimal CSP instance, there is some solution S for I such that B belongs to S. Let $S' = S \cup \{x\}$. By construction, S' is a solution for I' containing B.
   (f) $b_1 \notin A_{v_1}$, $b_2 \in A_{v_2}$ and $b_2 \neq \overline{x}$: same argument as for (e), using the fourth point in the definition of I' instead of the third.
   (g) $b_1 \notin A_{v_1}$ and $b_2 \notin A_{v_2}$: from the fifth point in the definition of I', we know that B is compatible in I too. Since I is a Minimal CSP instance, there is some solution S for I such that B belongs to S. Let a be the point of S in A_v. If $a = a_1$ or $a = a_2$, then let $S' = S \cup \{\overline{x}\}$. Otherwise, let $S' = S \cup \{x\}$. By construction, S' is a solution for I' containing B.

   So if B is a compatible k-tuple of I', then B is in a solution for I'.
2. Finding a solution for I from a solution for I':
   Suppose that we have a solution S' for I'. From the second point in the definition of I', we know that there is no compatible k-tuple containing both x and $\overline{x}$. Therefore, S' must contain one of the a_i. Let $S = S' \backslash \{x\}$ (or $S = S' \backslash \{\overline{x}\}$ if $\overline{x} \in S'$). By construction, S is a solution for I.

The domain size of v_1 is $3 < d_v$, and the domain size of v_2 is $d_v - 1 < d_v$. Therefore, if we have a variable with a domain of size strictly greater than 3, then we can replace it by two variables with strictly smaller domain size. By iteratively applying this operation until all domains have a size of 3 or less, we can reduce I to a Minimal CSP instance where the size of all domains is bounded by 3. Since computing a solution to a Minimal CSP instance is NP-hard [6], we have the result. □

In the case of Boolean domains, we also have NP-hardness if the arity of the instances is $k \geq 3$.

Proposition 2. *For all $k > 2$, k-ary Minimal CSP is NP-hard, even when bounding the size of the domains by 2.*

The proof is based on the transformation from a domain $A = \{a_1, a_2, a_3\}$ of size 3 to three domains $A_1 = \{a_1, \overline{a_1}\}$, $A_2 = \{a_2, \overline{a_2}\}$, $A_3 = \{a_3, \overline{a_3}\}$, each of size 2.

Proof. Let $k > 2$. From Proposition 1, we know that computing a solution to a k-ary Minimal CSP instance is NP-hard, even when bounding the size of the domains by 3. Therefore, we just need to reduce the k-ary Minimal CSP with domain size bounded by 3 to the k-ary Minimal CSP with domain size bounded by 2. Let I be a k-ary Minimal CSP instance such that the size of each domain in I is bounded by 3. Let v be a variable in I such that the domain of v is of size 3. We replace v by three new variables v_1, v_2 and v_3 to obtain a new instance I' defined as follows:

1. With $\{a_1, a_2, a_3\}$ being the domain of v, we set the domain of v_i to be $A_{v_i} = \{a_i, \overline{a_i}\}$ for $1 \leq i \leq 3$.
2. All k-tuples in I' containing both a_i and a_j for some $1 \leq i \neq j \leq 3$ are incompatible.
3. Let $B = \{\overline{a_i}, \overline{a_j}, b_1, b_2, \ldots, b_{k-2}\}$ be a k-tuple for some $1 \leq i \neq j \leq 3$. Let h be the integer between 1 and 3 such that $h \neq i$ and $h \neq j$. B is compatible if and only if there is some compatible k-tuple in I containing a_h and all the b_g for $1 \leq g \leq k-2$. Note that this covers the particular cases when a_h is one of the b_g (in which case B is compatible if and only if there is a compatible k-tuple in I containing all the b_g) and when $\overline{a_h}$ is one of the b_g (in which case B is incompatible because $\overline{a_h}$ does not appear in I).
4. Let $B = \{a_i, b_1, b_2, \ldots, b_{k-1}\}$ be a k-tuple such that $1 \leq i \leq 3$ and no b_j is in the domain of v_1, v_2 or v_3. B is compatible in I' if and only if B is compatible in I.
5. Let $B = \{\overline{a_i}, b_1, b_2, \ldots, b_{k-1}\}$ be a k-tuple such that $1 \leq i \leq 3$ and no b_j is in the domain of v_1, v_2 or v_3. B is compatible if and only if there is some $j \neq i$, with $1 \leq j \leq 3$, such that $\{a_j, b_1, b_2, \ldots, b_{k-1}\}$ is compatible in I.
6. Let $B = \{a_i, \overline{a_j}, b_1, b_2, \ldots, b_{k-2}\}$ be a k-tuple such that $1 \leq i \neq j \leq 3$ and no b_h is in the domain of v_1, v_2, or v_3. B is compatible if and only if there is some b_{k-1} such that $\{a_i, b_{k-1}, b_1, b_2, \ldots, b_{k-2}\}$ is compatible in I.
7. All other k-tuples in I' remain as they were in I.

In order to obtain the desired result, we have to prove that I' is a Minimal CSP instance and that finding a solution for I' allows us to find in polynomial time a solution for I.

1. I' is a Minimal CSP instance:
   We have to prove that each compatible k-tuple in I' is in a solution for I'. Let B be a compatible k-tuple in I'. We necesarily have one of the six following cases:

(a) $B = \{a_i, b_1, b_2, \ldots, b_{k-1}\}$ with $1 \leq i \leq 3$ and neither b_j in the domain of v_1, v_2 or v_3. Without loss of generality, we assume that $i = 1$. From the fourth point in the definition of I', we know that B is also compatible in I. Since I is a Minimal CSP instance, there is some solution S for I such that B belongs to S. Let $S' = S \cup \{\overline{a_2}, \overline{a_3}\}$. By construction, S' is a solution for I' containing B.

(b) $B = \{a_i, \overline{a_j}, b_1, b_2, \ldots, b_{k-2}\}$ with $1 \leq i \neq j \leq 3$ and no b_h in the domain of v_1, v_2 or v_3. Without loss of generality, we assume that $i = 1$. From the sixth point in the definition of I', we know that there is a compatible triple in I containing a_1 and all b_h for $1 \leq h \leq k-1$. Since I is a Minimal CSP instance, there is some solution S for I such that both a_1 and all the b_h belong to S. Let $S' = S \cup \{\overline{a_2}, \overline{a_3}\}$. By construction, S' is a solution for I' containing B.

(c) $B = \{a_i, \overline{a_j}, \overline{a_h}, b_1, b_2, \ldots, b_{k-3}\}$ with $1 \leq i, j, h \leq 3$ and i, j, h all distinct. Without loss of generality, we assume that $i = 1$. From the third point in the definition of I', we know that there is a compatible k-tuple in I containing a_1 and all the b_g for $1 \leq g \leq k - 3$. Since I is a Minimal CSP instance, there is a solution S for I such that a_i and all the b_g belong to S. Let $S' = S \cup \{\overline{a_2}, \overline{a_3}\}$. By construction, S' is a solution for I' containing B.

(d) $B = \{\overline{a_i}, b_1, b_2, \ldots, b_{k-1}\}$ with $1 \leq i \leq 3$ and no b_j in the domain of v_1, v_2 or v_3. Without loss of generality, we assume that $i = 1$. From the fifth point in the definition of I', either $\{a_2, b_1, b_2, \ldots, b_{k-1}\}$ or $\{a_3, b_1, b_2, \ldots, b_{k-1}\}$ is compatible in I. Without loss of generality, we assume the former. Since I is a Minimal CSP instance, there is some solution S for I such that $\{a_2, b_1, b_2, dots, b_{k-1}\}$ belongs to S. Let $S' = S \cup \{\overline{a_1}, \overline{a_3}\}$. By construction, S' is a solution for I' containing B.

(e) $B = \{\overline{a_i}, \overline{a_j}, b_1, b_2, \ldots, b_{k-2}\}$ with $1 \leq i \neq j \leq 3$ and no b_h not in the domain of v_1, v_2 or v_3. Without loss of generality, we assume that $i = 1$ and $j = 2$. From the third point in the definition of I', there is a compatible triple in I containing a_3 and all the b_h for $1 \leq h \leq k - 2$. Since I is a Minimal CSP instance, there is some solution S for I such that a_3 and all the b_h belong to S. Let $S' = S \cup \{\overline{a_1}, \overline{a_2}\}$. By construction, S' is a solution for I' containing B.

(f) $B = \{b_1, b_2, \ldots, b_k\}$ with no b_i being in the domain of v_j, for any $1 \leq i, j \leq 3$. From the seventh point in the definition of I', we know that B is compatible in I too. Since I is a Minimal CSP instance, there is some solution S for I such that B belongs to S. Let a_i be the point of S in the domain of v. Without loss of generality, we assume that $i = 1$. Let $S' = S \cup \{\overline{a_2}, \overline{a_3}\}$. By construction, S' is a solution for I' containing B.

So if B is a compatible k-tuple of I', then B is in a solution for I'.

2. Finding a solution for I from a solution for I':
Suppose that we have a solution S' for I'. From the second and third points in the definition of I', we know that S' must contain a_i, $\overline{a_j}$ and $\overline{a_h}$, for some distinct i, j and h between 1 and 3. Without loss of generality, we assume that $i = 1$. Let $S = S' \backslash \{\overline{a_2}, \overline{a_3}\}$. By construction, S is a solution for I.

Therefore, if we have a variable with a domain of size 3, then we can replace it by three variables with domains of size 2. By iteratively applying this operation until all domains have a size of 2 or less, we can reduce I to a Minimal CSP instance where the size of all domains is bounded by 2. So we have the result. □

The binary Boolean Minimal CSP is polynomial, since the more general binary Boolean CSP can be trivially reduced to 2-SAT, which is polynomial [9]. Combined with Propositions 1 and 2, and with the triviality of CSP instances consisting entirely of single-valued variables, we have the main Theorem:

Theorem 1 (The Minimal CSP Dichotomy Theorem). *k-ary Minimal CSP when the size of the domains is bounded by d is NP-hard if and only if ($d \geq 3$ or ($d = 2$ and $k \geq 3$)).*

The result is summarized in Table 1.

Table 1. Complexity of the k-ary Minimal CSP with domains of size bounded by d.

d \ k	2	≥ 3
1	tractable	tractable
2	tractable	NP-hard
≥ 3	NP-hard	NP-hard

Corollary 1. *When bounding the size of the domains by a constant d and the arity of the constraints by a constant k, Minimal CSP and the general CSP are NP-hard for the exact same values of d and k.*

3 Generating Minimal CSP Instances

3.1 Preliminary Notions

The unique properties defining Minimal CSP instances are strongly global. Random general CSP instances can be generated constraint after constraint. Even random satisfiable CSP instances, which are also defined by a global property, still have considerable leeway for local modifications [1]. In contrast, a slight change of one given constraint in a Minimal CSP instance can jeopardize its minimality. Therefore, it is not as straightforward to generate Minimal CSP instances, compared to many other kinds of CSP instances.

Since the NP-hardness proofs for the Minimal CSP are all constructive, what would appear to be a method of creating Minimal CSP instances is to reduce general CSP instances to Minimal CSP instances. Unfortunately, this results in instances that are far too large. For example, the reduction used in [6] transforms satisfiable 3-SAT instances with n clauses into Minimal CSP instances with

$1000n$ variables of domain size 9. Therefore, it is not practical to look for hard Minimal CSP instances this way.

Another intuitive idea would be to minimalize random satisfiable CSP instances, that is to only keep the compatibilities of a given satisfiable CSP that are in a solution. Indeed, every satisfiable CSP instance contains a unique minimalized version of itself. However, this approach is neither practical nor efficient. While it avoids too large instances, it does not allow for an effective control of the size of the instances. In particular, if the original satisfiable CSP instance has only very few solutions, then the resulting Minimal CSP instance will be very small, and its resolution trivial. Furthermore, minimalizing an arbitrary CSP instance is NP-hard [6]. Since the output of minimalization is not certain and the effort required to minimalize an instance is too high, another generation method is needed.

Since in a Minimal CSP instance every single compatibility is part of a solution, any Minimal CSP instance I's compatibilities can be defined by a set S of solutions, such that each solution from S is a solution to I, and any compatibility of I belongs to one of the solutions in S. The solutions from S can be intersecting, and for a given Minimal CSP instance I, the set S is in general not unique. The general intuition behind our algorithm is to start with a Minimal CSP instance with the fewest possible number of solutions in the set S, then add solutions to the set until we reach the desired constrainedness. We now define some notions that we will be using when describing our generator.

Definition 3 (CSP size when domains are equal). *Let I be a k-ary CSP instance with n variables, such that the domain of each variable contains exactly d values. Then we say that I is* of size (d, n, k).

When the domains of the variables do not all have the same size, we have the following definition:

Definition 4 (CSP size when domains are different). *Let I be a k-ary CSP instance with n variables, such that the domain of each variable contains at most d values. Then we say that I is* of size (d^-, n, k).

Consider Minimal CSP instances of size (d, n, k) for some given d, n and k. Without loss of generality, we can assume that the domain of each variable contains the integers 1 through d. The Minimal CSP instance I of this size with the fewest compatibilities will be the one defined by a set $S = \{s_1, \ldots, s_d\}$ of d solutions, such that for each i, s_i is the solution where all variables take the value i. In this particular case, the set S defining I is unique, and the total number of solutions to I is the same as the number of solutions in S, d. The tightness of I is $\frac{d^k-d}{d^k} = 1 - \frac{1}{d^{k-1}}$, because there are $d^k - d$ incompatibilities in each constraint.

Definition 5 (Bare Minimal CSP when domains are equal). *Let I be a CSP instance of size (d, n, k). We say that I is a* bare *Minimal CSP instance if for all k-tuple of variables $C = \{v_{i_1}, \ldots, v_{i_k}\}$, for all k-tuple of assignments $A = \{(v_{i_1}, a_{i_1}), \ldots, (v_{i_k}, a_{i_k})\}$, A is compatible if and only if $a_{i_j} = a_{i_{j'}}$ for all $1 \le j < j' \le n$.*

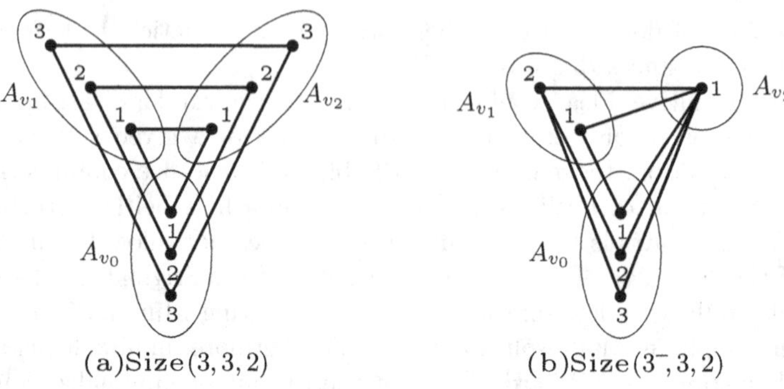

Fig. 2. Two examples of bare Minimal CSP instances.

An example of a bare binary Minimal CSP instance with three variables v_0, v_1 and v_2 and with all domains A_{v_0}, A_{v_1} and A_{v_2} of same size $d = 3$ is given in Fig. 2(a). In this figure, continuous lines represent compatibilities, and if there is no straight line between two values then the values are incompatible.

The notion of bare Minimal CSP instance can be generalized to CSP instances where the domains of the variables have different sizes.

Definition 6 (Bare Minimal CSP when domains are different). *Let I be a CSP instance of size (d^-, n, k). We say that I is a* bare *Minimal CSP instance if for all k-tuple of variables $C = \{v_{i_1}, \ldots, v_{i_k}\}$, for all k-tuple of assignments $A = \{(v_{i_1}, a_{i_1}), \ldots, (v_{i_k}, a_{i_k})\}$, A is compatible if and only if for all $1 \leq j < j' \leq n$ at least one of the following is true:*

- $a_{i_j} = a_{i_{j'}}$
- $a_{i_j} \leq a_{i_{j'}}$ *and the domain of v_{i_j} is of size a_{i_j}*
- $a_{i_j} \geq a_{i_{j'}}$ *and the domain of $v_{i_{j'}}$ is of size $a_{i_{j'}}$*

By noticing that in such a Minimal CSP instance the number of compatibilities in each constraint is equal to the size of the largest domain, it is easy to verify that a bare Minimal CSP instance of size (d^-, n, k) has the fewest possible compatibilities for a Minimal CSP instance of this size.

An example of bare binary Minimal CSP instance with three variables v_0, v_1 and v_2 and with the associated domains A_{v_0}, A_{v_1} and A_{v_2} of different sizes is given in Fig. 2(b). Like in Fig. 2(a), continuous lines represent compatibilities, and if there is no straight line between two values then the values are incompatible.

3.2 The Generator

Suppose that we want a Minimal CSP instance of size (d, n, k), such that the average tightness of each constraint is t, for some given t. What we do is generate a bare Minimal CSP instance I of this size, then we add solutions to I until we have the desired tightness. This is illustrated by Algorithm 1.

```
Data: d, n, k integers, t ∈ [0,1]
Result: A Minimal CSP instance of size (d,n,k), and with a tightness equal to
        t' ∈ [t − 1/d^k, t].
Generate a bare Minimal CSP instance I;
t_i ← tightness of I;
while t_i > t do
    Add solution to I;
    t_i ← tightness of I;
end
return I
```

Algorithm 1. Minimal CSP Instance Generator

```
Data: d, n, k integers, t ∈ [0,1]
Result: A Minimal CSP instance of size (d,n,k), and with a tightness equal to
        t' ∈ [t − 1/d^k, t].
Generate a bare Minimal CSP instance I;
for i ← 1 to n do
    randomize the ordering of the values in the domain of v_i;
end
t_I ← tightness of I;
while t_I > t do
    C = {v_i1, ..., v_ik} ← most constrained constraint in I;
    A = {v_i1 : a_i1, ..., v_ik : a_ik} ← random incompatibility k-tuple in C;
    Add solution containing A to I;
    t_I ← tightness of I;
end
return I
```

Algorithm 2. Improved Minimal CSP Instance Generator

A few adjustments must be made to the algorithm in order to obtain the optimal result. Specifically, we must randomize the order of the values in each domain after generating the bare Minimal CSP instance, so that looking at the first instantiation in the lexicographical order does not yield a trivial solution. Also, the solution that is added at each iteration of the algorithm must be built around one of the incompatibilities in the most constrained constraint. This serves two purposes. Firstly, building a solution around an incompatibility ensures that the number of compatibilities strictly increases at each iteration, so that no iteration is wasted and the algorithm always terminates. Secondly, choosing the most constrained constraint ensures that the tightness of each constraint is as balanced as possible. Algorithm 2 illustrates the improvements made to the original algorithm.

Adding one solution to an instance can, and will in the first iterations, add several compatibilities at once. Therefore, the tightness of the resulting instance will not always be exactly equal to the desired value t. However, adding one solution only adds at most one compatibility per constraint. Therefore, in the worst case, the tightness of the resulting instance will deviate from t by $\frac{1}{d^k}$. The number of compatibilities added at each iteration will decrease along with the tightness. During the first iterations, adding one solution will add one compatibility in most constraints. If the desired tightness t is low, then during the last iterations adding one solution will only add the one compatibility obtained from picking the most constrained constraint. In the general case, adding one solution to a CSP instance I will add in average t_I compatibilities in each constraint, with t_I being the tightness of I. Therefore, the tightness of the resulting instance will, on average, deviate from the desired tightness t by $\frac{t}{d^k}$.

3.3 Behavior of the Minimal CSP

An important question is whether our generator is refined enough to observe any interesting property held by the Minimal CSP. To this end, we used our algorithm to generate Minimal CSP instances with varying numbers of compatibilities, then solved them with two state of the art solvers. We found that we could observe the same behavior found in most other NP-hard problems, that is an easy-hard-easy pattern in the difficulty of the instances when varying the constrainedness.

We, therefore, looked for the phase transition behind this phenomenon. The classical notion of phase transition relies on satisfiability: loosely constrained CSP instances are usually satisfiable and easy to solve, while highly constrained CSP instances are rarely satisfiable, albeit also easy to solve. The transition from the former set of instances to the latter is very sharp, and includes what appears to be the hardest to solve CSP instances.

However, because of the nature of the Minimal CSP, we cannot rely directly on satisfiability to find an adequate phase transition. Other approaches, like for example counting the number of backbone variables [1], also fail to adapt to the Minimal CSP. In order to avoid this problem, we introduced the notion of p-step instance, specifically tailored to be able to be used in the case of the Minimal CSP.

What we call a p-step instance of a minimal instance I of size (d, n, k), formally defined in Definition 7, is the sub-instance of size $(d^-, n - p, k)$ obtained from I after making p random assignments while propagating $(1, k-1)$-consistency [4] after each assignment.

Definition 7 (p-Step Instance). *Let I and I' be two CSP instances. Let I_{cons} be the instance obtained from I by propagating $(1, k-1)$-consistency [4]. For all p, we say that I' is a* p-step *instance of I if any of the following is true:*

- *$p = 0$ and $I' = I_{cons}$.*
- *$p = 1$, all the domains of I_{cons} contain at most one value, and $I' = I_{cons}$.*
- *$p = 1$ and I' is the instance obtained from I_{cons} by assigning some value to some variable v in I_{cons} such that the domain of v contains at least two values, then propagating $(1, k-1)$-consistency.*
- *$p > 1$ and there is a CSP instance I'' such that I'' is a $(p-1)$-step instance of I and I' is a 1-step instance of I''.*

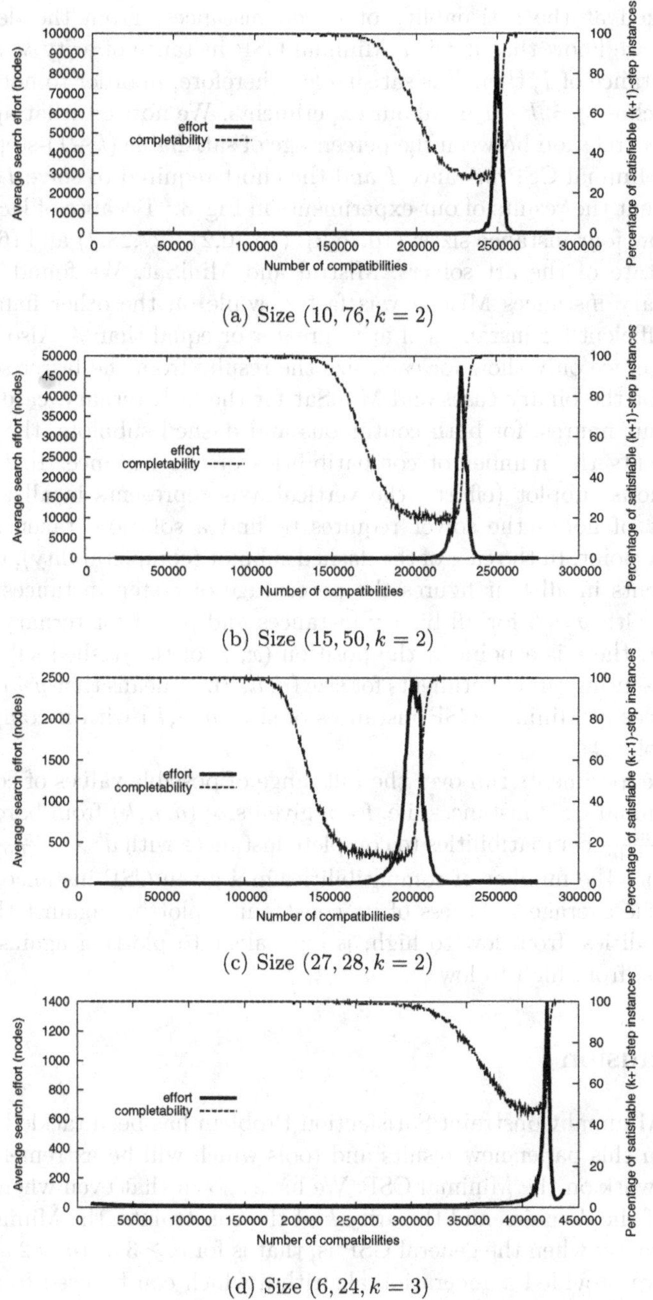

(a) Size $(10, 76, k = 2)$

(b) Size $(15, 50, k = 2)$

(c) Size $(27, 28, k = 2)$

(d) Size $(6, 24, k = 3)$

Fig. 3. Comparison between the difficulty of a given Minimal CSP instance and the percentage of its $(k + 1)$-step instances that are satisfiable, for various problem sizes.

We looked at the satisfiablity of p-step instances. From the definition of minimality, we know that if I is a Minimal CSP instance of arity k, and if I' is a k-step instance of I, then I' is satisfiable. Therefore, in order to not get trivial results, we chose $p = k + 1$ in all our experiments. We noticed what appeared to be a stong correlation between the percentage of satisfiable $(k+1)$-step instances of a given Minimal CSP instance I and the effort required to solve I.

We present the results of our experiments in Fig. 3.[2] Because of lack of space, we only show four instance sizes: (10,76,2), (15,50,2), (27,28,2) and (6,24,3). We used two state of the art solvers: Mistral and MiniSat. We found that when solving binary instances Mistral was faster, while on the other hand MiniSat was more efficient for instances of arity greater or equal than 3. Also because of lack of space, we only show for each size the results from the faster solver, that is Mistral for the binary cases and MiniSat for the last, ternary, case.

In all four figures, for both continuous and dashed subplots, the horizontal axis represents the number of compatibilities of the instance. In the case of the continuous subplot (effort), the vertical axis represents in all four figures the number of nodes the solver requires to find a solution, taken across 100 instances by point. In the case of the dashed subplot (completability), the vertical axis represents in all four figures the percentage of p-step instances that have a solution, with $p = 3$ for all binary instances and $p = 4$ for ternary instances. Therefore, if there is a point at the position (x, y) of the dashed subplot in the figure representing our experiments for size (d, n, k), it means that $y\%$ of the $k+1$-step instances of Minimal CSP instances of size (d, n, k) with x compatibilities are satisfiable.

All our experiments run over the full range of possible values of compatibilities for Minimal CSP instances, i.e. for a given size (d, n, k) from bare instances with $d \times \frac{n!}{(n-k)!k!}$ compatibilities to complete instances with $d^k \times \frac{n!}{(n-k)!k!}$ compatibilities. Since the number of compatibilities in a given CSP instance is directly related to the average tightness of its constraints, plotting against the number of compatibilities, from low to high, is equivalent to plotting against the constrainedness, from high to low.

4 Conclusion

While the Minimal Constraint Satisfaction Problem has been tackled before, we presented in this paper new results and tools which will be extremely useful in any future work on the Minimal CSP. We have shown that even when bounding the size d of the domains and the arity k of the constraints, the Minimal CSP is NP-hard exactly when the general CSP is, that is for $d \geq 3$ or ($d = 2$ and $k \geq 3$). We also have provided a generation algorithm which can be used to create and parameterize Minimal CSP instances of any size. We ran our generator in order to verify that it is refined enough to be able to expose any interesting property

[2] Our experiments are run under CentOS 6.6, on two Intel processors (1.33 Ghz each), and with 12 GB of DDR2 FB-DIMM RAM.

held by the Minimal CSP. This led us to the empirical discovery of an apparent correlation between the effort required to solve a given Minimal CSP instance and the percentage of its $(k+1)$-step instances that are satisfiable, with p-step instances being a new notion that we introduced.

The work we have done in this paper opens up a number of future avenues of research. In particular, one could look at the relation between the completability of $(k+1)$-step instances and the effort required to find a solution to Minimal CSP instances, and find out how the former influences the latter. Alternatively, one could further study this new notion of p-step instance, by using for example higher values for p or by applying the concept to other kinds of CSP instances. Indeed, one major asset of p-step instances is that their definition is not restricted to the Minimal CSP, nor any specific subclass of the CSP.

Acknowledgments. This publication has emanated from research conducted with the financial support of Science Foundation Ireland (SFI) under Grant Number SFI/12/RC/2289.

References

1. Achlioptas, D., Gomes, C., Kautz, H., Selman, B.: Generating satisfiable problem instances. In: Kautz, H.A., Porter, B.W. (eds.) Proceedings of AAAI, pp. 256–261. AAAI Press/The MIT Press (2000)
2. Amilhastre, J., Fargier, H., Marquis, P.: Consistency restoration and explanations in dynamic csps application to configuration. Artif. Intell. **135**(1–2), 199–234 (2002)
3. Clark, D.A., Frank, J., Gent, I.P., MacIntyre, E., Tomov, N., Walsh, T.: Local search and the number of solutions. In: Freuder, Eugene C. (ed.) CP 1996. LNCS, vol. 1118. Springer, Heidelberg (1996)
4. Freuder, E.C.: A sufficient condition for backtrack-bounded search. J. ACM **32**(4), 755–761 (1985)
5. Gent, I.P., MacIntyre, E., Prosser, P., Walsh, T.: The constrainedness of search. In: Clancey, W.J., Weld, D.S. (eds.) Proceedings of AAAI, pp. 246–252. AAAI Press/The MIT Press (1996)
6. Gottlob, G.: On minimal constraint networks. Artif. Intell. **191–192**, 42–60 (2012)
7. Hogg, T., Huberman, B.A., Williams, C.P.: Phase transitions and the search problem. Artif. Intell. **81**(1–2), 1–15 (1996)
8. Junker, U.: Configuration. In: Handbook of Constraint Programming. Foundations of Artificial Intelligence, pp. 837–873. Elsevier (2006)
9. Schaefer, T.J.: The complexity of satisfiability problems. In: Lipton, R.J., Burkhard, W.A., Savitch, W.J., Friedman, E.P., Aho, A.V. (eds.) Proceedings of the 10th Annual ACM Symposium on Theory of Computing, 1–3 May 1978, San Diego, pp. 216–226. ACM (1978)
10. Xu, K., Li, W.: Exact phase transitions in random constraint satisfaction problems. J. Artif. Intell. Res. (JAIR) **12**, 93–103 (2000)
11. Xu, K., Li, W.: Many hard examples in exact phase transitions. Theor. Comput. Sci. **355**(3), 291–302 (2006)

Algebraic Theory on Shortest Paths for All Flows

Tadao Takaoka(✉)

Department of Computer Science, University of Canterbury, Christchurch, New Zealand
tad@cosc.canterbury.ac.nz

Abstract. As a mathematical model for the passenger routing problem for ticketing in a railway network, we consider a shortest path problem for a directed graph with edges labeled with a cost and a capacity. The problem is to push flow f from a specified source to all other vertices with the minimum cost for all f values. If there are t different capacity values, we can solve the single source shortest path problem for all f t times in $O(tm + tn \log n) = O(m^2)$ time when $t = m$. We improve this time to $O(cmn)$ if edge costs are non-negative integers bounded by c.

Keywords: Information sharing · Shortest path problem for all flows · Priority queue · Limited edge cost

1 Introduction

As the routing problem for passengers buying tickets in a railway network, we consider a network optimization problem such that each edge has two quantities associated; cost and capacity. We want to maximize a flow from a specified source vertex s to a destination vertex v with the minimum cost. Here we have two objectives; flow amount and path cost. Both cannot be optimized at the same time. Let us call the minimum cost path the shortest path. We need to compute the shortest path for the given flow value f for all possible f. For the routing problem in a train network, suppose f passengers want to travel together in a group from a station specified as the source vertex s to the destination station expressed by vertex v. On the way they may need to change trains at several stations. The capacity of an edge corresponds to the remaining number of seats on the train and the cost corresponds to the fare. Let d be the cost of a path from s to v and f be the flow (unsplittable) from s to v. The pair (d, f) is called a df-pair.

Another example is a computer network. Here vertices correspond to hub computers and edges correspond to the links. Capacities are band-widths and flows are packet sizes to be sent. It is regarded as better if packets are transmitted together to prevent packet loss and recovery.

If $d \leq d'$ and $f \geq f'$, (d, f) is better than (d', f'), the latter being represented by the former. Otherwise they are incomparable. We only need to compute incomparable df-pairs.

Z. Lu et al. (Eds.): COCOA 2015, LNCS 9486, pp. 746–757, 2015.
DOI: 10.1007/978-3-319-26626-8_55

Similar problems in the literature are the multi (bi)-objective shortest path problem [11] and the minimum cost flow problem [1]. In the former, the two objectives are similar; additive costs over paths. In our problem, they are cost and capacity. In the latter, the flow can be split over several paths to minimize the cost. In our model, a flow cannot be divided. Unsplittable flow is studied in a few papers such as [4,9], in which flow amounts are considered and costs are not.

The network optimization model in this paper is simple enough to be used by network practitioners, but to the author's knowledge there is no algorithmic or theoretical analysis on this model in the literature apart from the recent [16–18]. There are many papers on train routing such as [3,20]. They are more practical with many constraints such as shunting, coupling/decoupling, safety margin and connection with other trains in addition to cost and capacity. Their problems are mostly NP-hard, requiring some sort of meta-heuristic or human intervention. Our problem is much simpler; cost and capacity. Our efficient algorithm should be used as a subroutine in a complicated system.

The algorithmic technique is viewed as information sharing described in [14], which solves the all pairs shortest path problem efficiently. More specifically, for a graph with n vertices, the single source shortest path problem is solved n times by changing the source n times, where they share common resources obtained in advance as preprocessing or during the course of computation. In our problem, we solve the single source shortest path problem for all flow amounts at simultaneously utilizing some common data structures.

To prepare for the later development, we describe the single source shortest path problem for all flows (SSSP-AF) in the following. Let $G = (V, E)$ be a directed graph where $V = \{v_1, \cdots, v_n\}$ and $E \subseteq V \times V$. Let $|E| = m$. The cost and capacity of edge (u, v) is a non-negative real number denoted by $cost(u, v)$ and a positive real number $cap(u, v)$ respectively. We specify a vertex, s, as the source. A shortest path from s to vertex v is a path such that the sum of edge costs of this path is the minimum among all paths from s to v. The minimum cost is also called the shortest distance. The single source shortest path problem (SSSP) is to compute shortest paths from s to all other vertices. The bottleneck (value) of a path is the minimum capacity of all edges on the path. The bottleneck of the pair of vertices (u, v) is the maximum bottleneck of all paths from u to v. Such a path is called the bottleneck path from u to v. The single source bottleneck path (SSBP) problem is to compute the bottleneck path from s to all vertices v. The bottleneck from s to v is the maximum flow value of a simple path from s to v. Those two problems are well studied. For the bottleneck path problem the readers are referred to [13] for single source and [10] for all pairs.

If we flow a smaller amount from s to v, there may be a shorter path from s to v. Thus it makes sense to compute the shortest path from s to v for all possible flows for all vertices v. We compute a tuple of pairs (d, f) for each v where d is the shortest distance of a path that can push f to v.

The problem can be solved by removing edges one by one. Suppose there are t capacity values cap_1, ..., cap_t in this order. The simplest algorithm looks like:

for $i = 1$ to t do begin
  Remove edges whose capacity is less than cap_i,
  for the flow f such that $cap_i \leq f < cap_{i+1}$.
  Solve the single source shortest path problem for the resultant graph.
end

If we use a Fibonacci heap [12] for the priority queue for the frontier set, the complexity of this algorithm becomes $O(tm + tn \log n)$, including the time for sorting capacities. This trivial upper bound is our starting point. The above algorithm works for edge costs of non-negative real numbers. In this paper, we improve the complexity when edge costs are non-negative integers bounded by a small positive constant c, achieving $O(\min\{t, cn\}m+cn^2) \leq O(cmn)$. The trivial complexity above is strongly polynomial, while our complexity looks pseudo polynomial. The point here is that we have a speed-up when c is small. When $t = O(m)$ and $m = O(n^2)$, the trivial complexity hits $O(n^4)$, called quartic, while our complexity $O(cn^3)$ can stay sub-quartic when $c = o(n)$. Similar studies are done on the all pairs shortest path problem (APSP) with integer edge costs such as [2,15,21], who investigated up to what value of c we can stay in sub-cubic for the APSP complexity. The best bound for such c is $O(n^{0.624})$ if we use the Coppersmith-Winograd matrix multiplication algorithm. Recent studies improve this bound slightly.

The rest of the paper is as follows: In Sects. 2 and 3, SSSP and SSBP are described in a pedagogical way so that we can see how they can be combined to solve the SSSP-AF problem. In Sect. 4, SSSP-AF is solved with the data structure of one dimensional bucket system. The complexity in this section is already known. In Sect. 5, we improve the complexity in Sect. 4 by introducing another data structure and enhancing the one-dimensional bucket system. This section is the major contribution of the paper. In Sect. 6, we define the single source bottleneck path for all costs (SSBP-AC) problem. Although we can design an algorithm for this problem on its own, we show the problem can be solved as a by-product of the algorithm in Sect. 5. Section 7 concludes the paper. We use up-right fonts for some long names of variable and functions for readability.

2 Single Source Shortest Path Problem

We describe Dijkstra's algorithm [8] below in our style. The set S, called the solution set, is the set of vertices to which the shortest distances have been finalized by the algorithm. The set F, called the frontier set, is the set of vertices which is outside S and can be reached from S by a single edge. We note that the distances to vertices in F can be limited in a small band when edge costs are bounded by a small integer.

Let $OUT(v) = \{w|(v, w) \in E\}$. The solution (the shortest distances from s) is in the array d at the end of the computation. To simplify presentation, only the shortest path distances are calculated, not the shortest paths. We assume all vertices are reachable from the source. Paths are given by a sequence of vertices such that for two successive vertices u and v, there is an edge (u, v). We list two invariants maintained by Algorithm 1 below.

(1) S is the set of vertices v to which shortest distances are worked out in $d[v]$.
(2) If v is in F, $d]v]$ is the distance of the shortest path that lies in S except for the end point v itself.

Lemma 1. *The invariants (1) and (2) are kept through Algorithm 1.*

Proof. Lemma is true before while. Suppose (1) is true immediately after line 4. If there is a shorter path to v after line 5 via another vertex, say u, in F, which must exist to reach v, then $dist[u]$ is shorter than $dist[v]$, which is a contradiction. After v is included in S, all w in F or in $V - S - F$ are updated with the smallest possible $dist[w]$. Thus (2) is preserved.

Throughout the paper, comments are given in the pseudo codes of the algorithms by the double slash for readability.

Algorithm 1
1. $S = \emptyset$
2. $dist[s] = 0$; $dist[v] = \infty$ for all $v \neq s$
3. $F = \{s\}$
4. while F is not empty do begin
5. v =delete-min(F) // with key $dist[v]$
6. $S = S \cup \{v\}$
7. for $w \in OUT(v)$ do
8. if $w \notin S$ then
9. if $w \in F$ then $dist[w] = \min\{dist[w], dist[v] + cost(v, w)\}$ //decrease-key
10. else begin $dist[w] = dist[v] + cost(v, w)$; $F = F \cup \{w\}$ end // insert
11. end

At the end of computation F becomes empty and S becomes V, giving the solution in *dist*. We use a simple data structure of one-dimensional bucket system with array Q. $Q[i]$ is a list of items whose key value is i. Items in our case are vertices. We observe delete-min or delete-max operations can be done in $O(cn)$ time in total where cn is the size of Q, and decrease-key or increase-key, and insert can be done in $O(1)$ time per operation.

Theorem 1. *Algorithm 1 solves the SSSP in $O(m + cn)$ time [7].*

Proof. We use a one-dimensional bucket system for the priority queue following Dial's idea [7], where total delete-min takes $O(cn)$ time and each insertion and decrease-key takes $O(1)$ time. Suppose there are m_i edges from vertex v_i. Summation of $m_iO(1)$ gives the result, where $m = m_1 + ... + m_n$.

3 Single Source Bottleneck Path Problem

We modify Algorithm 1 slightly for the single source bottleneck path problem. Note that we can push flow f from s to v through a path whose bottleneck value is f. The bottleneck path is sometimes called the widest path, where the capacity of an edge is viewed as the width. The solution set S and frontier set F are similarly defined.

(1) S is the set of vertices v to which maximum flows are worked out in $f[v]$.
(2) If v is in F, $flow]v]$ is the flow of the path with the maximum flow to v that lies in S except for the end point v itself.

We use array "*flow*" instead of "*dist*" in the following. Capacities, which are non-negative real numbers, are sorted and normalized to integers $1, .., t, t+1$, where there are t different capacity values in the graph. Note that $t \leq m$.

Algorithm 2
1. $S = \emptyset$
2. $flow[s] = t + 1$; $flow[v] = 0$ for all $v \neq s$
3. $F = \{s\}$
4. while F is not empty do begin
5. $\quad v$ =delete-max(F) // with key $flow[v]$
6. $\quad S = S \cup \{v\}$
7. $\quad$ for w in $OUT(v)$ do
8. $\quad\quad$ if $w \notin S$ then
9. $\quad\quad\quad$ if $w \in F$ then $flow[w] = \max\{flow[w], \min\{flow[v], cap(v,w)\}\}$ // increase-key
10. $\quad\quad\quad$ else begin $flow[w] = \min\{flow[v], cap(v,w)\}$; $F = F \cup \{w\}$ end // insert
11. end

Lemma 2. *Invariants (1) and (2) are kept through the iteration in the while loop. Proof omitted.*

Theorem 2. *After normalization of capacities, Algorithm 2 solves the SSBP in $O(m + t) = O(m)$ time. Proof omitted.*

4 Single Source Shortest Paths for All Flows

We parameterize Dijkstra's algorithm with the flow value f. Array $dist[v]$ is extended to $dist[v, f]$ whose intuitive meaning is the distance of the shortest path that can push f to v. The solution set S is extended to $S(f)$, meaning the solution set for the SSSP for the flow value f. Data structure Q is used for F such that items (v, f) are kept in the list at $Q[dist[v, f]]$, that is, $dist[v, f]$ is the key. The idea is to solve $t + 1$ SSSP's in parallel with the shared data structure Q.

Algorithm 3
Main data structures
$dist[v, f]$: currently shortest distance of path from source s to vertex v that can push flow f.
Q : a one-dimensional array of lists of items (v, f). If $Q[d]$ includes (v, f), $dist[v, f]$ is d. In each list the same vertex may appear more than once with different f. This part is improved in Algorithm 4.
Pointer array indexed by (v, f) : $pointer[v, f]$ points to item (v, f) in Q. Capacities are normalized to $1, ..., t, t+1$. 1. $S[f] = \emptyset$ for $f = 1, ..., t, t+1$ // t+1 is for infinity

```
2. dist[s, f] = 0 for f = 1, ..., t, t+1; Other dist are initialized to ∞
3. Q[0] = {(s, t+1)}
4. while Q is not empty do begin
5.    (v, f) =delete-min(Q) // with key dist[v, f]
6.    S[f] = S[f] ∪ {v}
7.    for w in OUT(v) do begin
8.        d* = dist[v, f] + cost(v, w) // candidate distance for w
9.        f* = min{f, cap(v, w)} // candidate flow for w
10.       if w is not in S[f*] then
11.         if (w, f*) ∈ Q then dist[w, f*] = min{dist[w, f*], d*} //decrease-key
12.           else begin dist[w, f*] = d*; Q = Q ∪ {(w, f*)} end // insert
13.   end
14. end.
```

The following lemma is obvious.

Lemma 3. *In Algorithm 3, we have* $f \leq f' \Rightarrow S(f) \supseteq S(f')$

We establish two assertions similar to those in the previous sections.

(1) For all v in $S[f]$, $dist[v, f]$ is the distance of the shortest path from s to v that can push flow f.
(2) For all (v, f) in Q, $dist[v, f]$ is the distance of the shortest path from s to v that can push flow f whose vertices are in $S[f]$ except for the end point v.

Lemma 4. *The above invariants (1) and (2) are kept through iterations by the while-loop.*

Proof. Proof is based on induction through the while-loop. Before the while-loop (1) and (2) are obviously true. Suppose there is a shorter path to v via u in Q that can push flow f at line 5. This means $dist[u, f] < dist[v, f]$, which is a contradiction to line 5 that chose $dist[v, f]$ as minimum. To push flow f^* via v after v is included in $S[f]$, we update all (w, f^*) in Q with possible shorter distances via v, or include (w, f^*) if it was outside Q with the new distance via v. Note that $S(f^*) \supseteq S(f)$ from Lemma 3, meaning the path in $S(f)$ for (w, f^*) is included in $S(f^*)$ except for w. Thus at the end of one iteration (2) is preserved.

Theorem 3. *Algorithm 3 solves SSSP-AF in* $O(tm + cn)$ *time [16].*

Proof. The correctness is seen from the fact that at the end of the algorithm the set *S(f)* includes all v to which flow f can be pushed. The time is analysed from delete-min and decrease-key/insert. The former takes *O(cn)*. The latter takes *O(tm)*, because each vertex v_i joins *S(f)*'s at most t times and decrease-key/insert takes $O(tm_i)$ for each v_i, where $|OUT(v_i)| = m_i$, resulting in *O(tm)* over summation on i. Note that all *(v, f)* in Q are distinct so that we have at most t such *(v, f)*'s in Q for each v.

The following monotone property is obvious and can be used for printing.

Lemma 5. *It holds for finalized distances that* $f \leq f' \Rightarrow dist[v, f] \leq dist[v, f']$.

From this lemma, we can list up incomparable *df*-pairs in non-decreasing order for each v

```
for each v do
  for f = 1 to t do
    if dist[v, f] < ∞ and dist[v, f] ≠ dist[v, f − 1] then Print (v, dist[v, f], f)
```

In [5,6], the simple one-dimensional bucket system is generalized to the k-level cascading bucket system for SSSP. The following is a very brief sketch of the data structure. Now the initial key value $d[v] = cost(s, v)$ is given like a radix-p number, where only x_{k-1} may exceed p

$$d[v] = x_{k-1}p^{k-1} + ... + x_1 p + x_0 \quad (0 \leq x_0, x_1, ..., x_{k-2} \leq p - 1, \quad 0 \leq x_{k-1} \leq \lceil c/p^{k-1} \rceil - 1) \text{ for some k.} \tag{1}$$

The data structure has k levels of buckets. At the i-th level for each i, there are p buckets. Let i be the largest index of non-zero x_i. Item v is inserted into the x_i-th bucket at level i for all v in the frontier. During the computation, we maintain the items in the appropriate buckets based on the current value of $d[v]$. Let a_i, called the active pointer, be the smallest index of a non-empty bucket in level i. The role of a_i is to skip many empty buckets at level i. The base for level i, B_i, and the range for the j-th bucket at level i, R_j, are defined by

$$B_i = a_{k-1}p^{k-1} + ... + a_{i+1}p^{i+1}, R_j = [B_i + jp^i, B_i + (j + 1)p^i - 1]$$

If item v is in level i for $d[v] = x_{k-1}p^{k-1} + ... + x_1 p + x_0$, i is the largest index such that $a_{k-1} = x_{k-1}$, ..., $a_{i+1} = x_{i+1}$ and $a_i \neq x_i$. Items move from a higher level to a lower level and from a higher bucket to a lower bucket in the same level. The minimum can be found by scanning for a non-empty level and then the first non-empty bucket. Decrease-key can be done by moving the item in the data structure. Insert can be done by putting the item at the largest level and follow decrease-key. Suppose we solve t SSSP's. In [14], it is shown that t SSSP's can be solved in $O(tm + tn \log(c/t))$ time with this data structure. In [14], the data structure is used for the all pairs shortest path problem, where $t = n$, achieving the complexity of $O(mn + n^2 \log(c/n))$. We can use the data structure for the SSSP-AF problem where t SSSP's are solved in $O(tm + tn\log(c/t))$ time. This complexity is good when c is large with a better second term, but when t is large, the first term of $O(tm)$ is outstanding. The same thing can be said of Thorup's data structure [19] that spends $O(tm + tn\log\log c)$ time when applied to our problem. We try to improve the first term in the next section.

We note at this stage that in the list Q[d] for some d, there might be items (v, f) and (v, f') such that $f \neq f'$ for some v. The following section is to prevent this duplication of items for the same v with more formalism.

5 A Faster Algorithm for SSSP-AF

Definition 1. *Natural order $\leq_n$ is defined on df-pairs, (d, f) and (d', f'), by*

$$(d, f) \leq_n (d', f') \Rightarrow d \leq d' \wedge f \leq f'$$

Merit order $\leq_m$ is defined on (d, f) and (d', f') by

$$(d, f) \leq_m (d', f') \Rightarrow d' \leq d \wedge f \leq f'$$

The natural order represents a numerical order while the merit order specifies which is better for our objective.

Definition 2. *For v in $S[f]$, pair $(dist[v, f], f)$ is said to be Pareto optimal at v if there is no pair $(dist[v, f'], f')$ such that v is in $S(f')$ and $(dist[v, f'], f') >_m (dist[v, f], f)$. In other words, $(dist[v, f], f)$ is Pareto optimal if there is no better df-pair so far at v.*

The priority queue Q is augmented by array *flow*, which is initialized to all 0. $flow[v, d]$ is the maximum flow so far from s to v with cost d. We maintain each list $Q[d]$ such that each v appears at most once in the list. If (v, f) is to be inserted to list $Q[d]$, where $d = dist[v, f]$, $flow[v, d]$ is consulted. If $f \leq flow[v, d]$, this insertion is ignored. If not, $(v, flow[v, d])$ is deleted from Q, (v, f) is inserted and $flow[v, d]$ is updated with f. Decrease-key(v, f) is to perform $delete(v, f)$ and $insert(v, f)$ with the new distance. We maintain pointers for each pair (v, f) to locate (v, f) in Q in $O(1)$ time. Delete-min takes $O(cn)$ time in total.

Algorithm 4
Main data structures
$dist[v, f]$: same as Algorithm 3
Q : a one-dimensional array of lists (buckets) of items (v, f). If $Q[d]$ includes (v, f), $dist[v, f]$ is d. In each list every vertex appears at most once.
$flow$: $flow[v, d]$ gives the maximum flow that can be pushed from s to v through a path in $S(f)$ except v with cost d. The size of $flow$ is $O(cn^2)$.
Pointer array indexed by (v, f) : same as Algorithm 3
0. $dist[v, f]$ are initialized to ∞ for all $v \neq s$ and f
1. $S[f] = \emptyset$ for $f = 1, ..., t, t + 1$; // $t + 1$ is for infinity

```
2. dist[s, f] = 0 for f = 1, ..., t, t + 1; flow[s, d] = 0 for all d
3. Q[0] = {(s, t + 1)}
4. while Q is not empty do begin
5.   (v, f) =delete-min(Q) // with key dist[v, f]
6.   S[f] = S[f] ∪ {v}
7.   for w in OUT(v) do begin
8.       d* = dist[v, f] + cost(v, w) // candidate distance for w
9.       f* = min{f, cap(v, w)} // candidate flow for w
10.      if w is not in S[f*] then
11.         if (w, f*) is in Q then begin
12.            dist[w, f*] = min{dist[w, f*], d*}
13.            decrease-key(w, f*)
14.            flow[w, d*] = max{flow[w, d*], f*}
15.         end
16.         else begin
17.            dist[w, f*] = d*
18.            insert(w, f*)
19.            flow[w, d*] = f*
20.         end // if-else
21.   end // for
22. end // while
23. procedure insert(w, f*)
24. begin
25.   if f* > flow[w, d*] then begin
26.      delete(w, flow[w, d*])
27.      Q = Q ∪ (w, f*) // insert with key dist[w, f*]
28.   end
29. end
30. procedure decrease-key(w, f*)
31. begin delete(w, f*); insert(w, f*) end
```

The loop invariants (1) and (2) in the previous section holds for Algorithm 4 as well. In addition we have the following lemma, which is similar to (2).

Lemma 6. *For all v and d, let $f = flow[v, d]$. If (v, f) is in Q, f is the maximum flow of the path with cost d that can push f from s to v whose vertices are in $S[f]$ except for the end point v.*

Proof. Suppose this invariant holds at the beginning of the while loop. After v is included in $S[f]$, $flow[w, d^*]$ is updated at lines 14 and 19. Note that we have $f^* \leq f$ and thus $S[f^*] \supseteq S[f]$ from Lemma 3. Thus the path is in $S(f^*)$ except for the end point and the lemma holds for $f^* = flow[w, d^*]$ as well.

Lemma 7. *At each iteration of while loop, pair $(dist[v, f], f)$ is Pareto optimal for any $(v, f) \in S[f]$ at line 5.*

Proof. Suppose the statement is true at the beginning of each iteration. We perform one more iteration. Suppose $(dist[v, f], f)$ is not Pareto optimal for some v and f at the end of the iteration. Then for some $(dist[v, f'], f')$ we have $(dist[v, f'], f') >_m (dist[v, f], f)$, which means

$$(dist[v, f'] < dist[v, f] \wedge f' \geq f) \vee (dist[v, f'] \leq dist[v, f] \wedge f' > f).$$

By a simple calculation, this is equivalent to

$$(dist[v, f'] < dist[v, f] \wedge f' \geq f) \vee (dist[v, f'] = dist[v, f] \wedge f' > f).$$

This contradicts the fact that $dist[v, f]$ is the distance of the shortest path that can push f to v, or the fact that (v, f) is updated (in the form of (w, f^*)) with the maximum possible f by consulting $flow[v, dist[v, f]]$. Lemma 6 guarantees f is the maximum flow to v with cost $dist[v, f]$.

In the following lemma we abbreviate $(dist[v, f], f)$ as (d, f).

Lemma 8. *All Pareto optimal df-pairs at any v can be sorted in increasing natural order. Furthermore if $(d, f) \leq_n (d', f')$, we have $d < d'$ and $f < f'$.*

Proof. If not sorted in natural order, there must be (d, f) and (d', f') at v such that $d > d'$ and $f \leq f'$ or $d \leq d'$ and $f > f'$. Then $(d, f) <_m (d', f')$ or $(d, f) >_m (d', f')$, a contradiction to Pareto optimal. The latter half can be seen as follows: Suppose there are (d, f) and (d', f') at v such that $d = d'$ or $f = f'$, which is a contradiction to Pareto optimal.

Theorem 4. *Algorithm 4 solves SSSP-AF in $O(min\{t, cn\}m + cn^2)$ time.*

Proof. Correctness is similar to that of Theorem 3. We measure the complexity by the number of accesses to major data structures. If a one-dimensional bucket system is used for Q, the total time for scanning the array for delete-min is $O(cn)$. For edge inspection at line 7, we observe pair (v, f) at line 5 brings Pareto optimal $(dist[v, f], f)$ at v. The size of the Pareto optimal solution at each v is bounded by $\min\{t, cn\}$ from the previous lemma. Thus the number of edge inspections for decrease-key and insert at line 7 is bounded by $\min\{t, cn\}m_i$ for vertex v_i. Summation over i can give us the time for decrease-key and insert being $O(\min\{t, cn\}m)$. The initialization for array $dist$, array $flow$ and Boolean arrays for membership of $S[f^*]$ and Q used at lines 10 and 11 takes $O(cn^2 + tn)$. Thus the total time is given by $O(\min\{t, cn\}m + cn + (cn + t)n) = O(\min\{t, cn\}m + (cn + t)n) = O(\min\{t, cn\}m + cn^2)$.

Corollary. The SSSP-AF problem with edge costs bounded by c can be solved in $O(cmn)$ time. If the cost is a unit, it can be solved in $O(mn)$ time.

Note. We could use the cascading bucket system to improve delete-min operations with $O(tn \log(c/t))$, but cannot improve the complexity of $O(cn^2)$ for the initialization of $flow$. It is open whether we can improve this time for initialization. In a way we improved the complexity of the first term at the higher cost of the second term, resulting in a better overall complexity.

6 Single Source Bottleneck Paths for All Costs Problem (SSBP-AC)

For every given cost d, we work out maximum flow $flow[v, d]$ that can be pushed from s to v for all v and d. Although we can design an algorithm for this problem by swapping the roles of distance and flow in Algorithm 4, Algorithm 4 already solves this problem in array $flow$. Following the monotone property of $flow$ similar to Lemma 5, the solution can be printed by

for each v do
  for $d = 1$ to cn do if $flow[v, d] > 0$ then
    if $flow[v, d] \neq flow[v, d-1]$ then Print $(v, d, flow[v, d])$

7 Concluding Remarks

We improved time complexities of the SSSP-AF problem for special types of graphs using the idea of information sharing. When $c = O(n)$ or costs are real numbers, the complexity is standing at $O(n^4)$ with only n as variable. It is open whether sub-quartic is possible.

There are some possibilities to extend our idea to improve time complexities for the all pairs shortest paths for all flows (APSP-AF) problem. Another direction may be to seek some possibility of using our algorithm as a subroutine for the flow augmenting path for the minimum cost flow problem, where flow splitting is allowed.

Our network model is simple enough to be incorporated into a larger networks such as any transportation network and a local area communication network as our algorithm is quite efficient. Our algorithm will be useful when used as a subroutine of a larger system.

References

1. Ahuja, R.K., Magnanti, T.L., Orlin, J.B.: Network Flows: Theory, Algorithms, and Applications. Prentice-Hall, Englewood Cliffs (1993)
2. Alon, N., Galil, Z., Margalit, O.: On the exponent of the all pairs shortest path problem. In: Proceedings of 32th IEEE FOCS, pp. 569–575 (1991). Also JCSS 54, 255–262 (1997)
3. Anderegg, L., Eidenbenz, S., Gentenbein, M., Stamm, C., Taylor, D.S., Weber, B., Widmeyer, P.: Routing algorithms: concepts. design choices, and practical considerations. In: ALENEX 2003, pp. 106–118 (2003)
4. Chakrabarti, A., Chekuri, C., Gupta, A., Kumar, A.: Approximation algorithms for the unsplittable flow problem. Algorithmica **47**(1), 53–78 (2007)
5. Cherkassky, B.V., Goldberg, A.V., Radzik, T.: Shortest paths algorithms: theory and experimental evaluation. Math. Prog. **73**, 129–174 (1996)
6. Denardo, E.V., Fox, B.L.: Shortest-route methods: I. Reaching, pruning, and buckets. Oper. Res. **27**, 161–186 (1979)

7. Dial, R.B.: Algorithm 360: shortest path forest with topological ordering. Commun. ACM **12**, 632–633 (1969)
8. Dijkstra, E.W.: A note on two problems in connexion with graphs. Numer. Math. **1**(269–271), 1343–1345 (1959)
9. Dinits, Y., Garg, N., Goemans, N.: On the single-source unsplittable flow problem. Combinatorica **19**(1), 17–41 (1999). Springer
10. Duan, R., Pettie, S.: Fast algorithms for (max, min)-matrix multiplication and bottleneck shortest paths. In: Proceedings of the 20th Annual ACM-SIAM Symposium on Discrete Algorithms (SODA 2009), pp. 384–391 (2009)
11. Ehrgott, M.: Multicriteria Optimization. Springer, Berlin (2005)
12. Fredman, M.L., Tarjan, R.E.: Fibonacci heaps and their uses in improved network optimization algorithms. J. ACM **34**, 596–615 (1987)
13. Gabow, H.N., Tarjan, R.E.: Algorithms for two bottleneck optimization problems. J. Algorithms **9**(3), 411–417 (1988)
14. Takaoka, T.: Sharing information for the all pairs shortest path problem. Theor. Comput. Sci. **520**, 43–50 (2014)
15. Takaoka, T.: Subcubic cost algorithms for the all pairs shortest path problem. Algorithmica **20**(3), 309–318 (1998)
16. Shinn, T.-W., Takaoka, T.: Combining the shortest paths and the bottleneck paths problems. In: ACSC, pp. 13–18 (2014)
17. Shinn, T.-W., Takaoka, T.: Some extensions of the bottleneck paths problem. In: Pal, S.P., Sadakane, K. (eds.) WALCOM 2014. LNCS, vol. 8344, pp. 176–187. Springer, Heidelberg (2014)
18. Shinn, T.-W., Takaoka, T.: Combining all pairs shortest paths and all pairs bottleneck paths problems. In: Pardo, A., Viola, A. (eds.) LATIN 2014. LNCS, vol. 8392, pp. 226–237. Springer, Heidelberg (2014)
19. Thorup, M.: Integer priority queues with decrease key in constant time and the single source shortest paths problem. In: STOC 2003, pp. 149–158 (2003)
20. Zwanevelt, P.J., Kroon, L.G., Romeijn, H.E., Salomon, M., Dauzere-Peres, S., Van Hoesel, S.P.M., Ambergen, H.W.: Routing trains through railway stations: model formulation and algorithms. Transp. Sci. **30**(3), 181–194 (1996)
21. Zwick, U.: All pairs shortest paths using bridging sets and rectangular matrix multiplication. J. ACM **49**(3), 289–317 (2002)

The Minimum Acceptable Violation Ranking of Alternatives from Voters' Ordinal Rankings

Kelin Luo[1,2](✉) and Yinfeng Xu[1,2]

[1] School of Management, Xi'an Jiaotong University, Xi'an, China
luokelin@stu.xjtu.edu.cn
[2] State Key Lab for Manufacturing System Engineering, Xi'an 710049, China

Abstract. Motived by applications of ordinal ranking and adjustment consensus in group decision making, we study the problem of aggregating all voters' ordinal ranking on a set of alternatives into an adjusted "consensus" ranking. In this problem, every voter ranks a set of alternatives respectively, and we know the adjustment acceptability. The problem is to find an optimal ordinal ranking which minimizes the sum of voter's acceptable violation. We analyse this problem by utilizing both pairwise preference and order-based ranking, and develop a branch-and-bound algorithm to solve this problem. The effectiveness and efficiency of this algorithm are verified with a small example and numerical experiments.

Keywords: Group decisions · Ordinal ranking · Acceptability index · Branch-and-bound algorithm

1 Introduction

Group decisions are widely used in many areas: the president election, the Oscar film review, rank projects, and so on. In the group decision setting, every voter ranks a subset of alternatives based on his or her own preference, and the chairman has to combine all the individual rankings together in order to get a "consensus" ranking.

The traditional group decision problem based on ordinal ranking has been studied for over 200 years. Borda [1] firstly proposed this concept by using the average ranks assigned by voters. We know that there are many paradoxes. Kendall [2] represented a pairwise preference model to describe group decision problem, and Wei [3] tried to solve it by using algebraic method. Then Blin [4] and other scholars [5–8] extended the initial violation distance among the rankings of all candidates or alternatives through variety approaches.

One of the most famous and widely used models for representing preference own to Kemeny and Snell (KS) [5]. Their model aggregated the individual preference difference by pairwise comparisons of the form "alternative a is preferred to alternative b, alternative a is preferred to alternative c, etc". The objective function is to minimize the total violations (or differences) between every voters preference and "consensus" ranking preference. They named this problem

Z. Lu et al. (Eds.): COCOA 2015, LNCS 9486, pp. 758–770, 2015.
DOI: 10.1007/978-3-319-26626-8_56

as Minimum Violation Ranking (MVR) problem. Cook and Saipe [9] used a median ranking method to solve this MVR problem. Jonathan *et al.* [10] used an equivalent positive KS (PKS) form and expressed this problem as a network for solving it more efficiently. Cook *et al.* [11] used heuristics to find the order ranking with a minimum violation. Furthermore, Cook and Kress [12] gave us some transformed models based on the Date Envelopment Model (DEA) framework. Rademaker and De Baets [13] formulated a new ranking procedure for the initial MVR problem and gave some great properties of this ranking procedure. In addition, there are many transformed MVR problem, such as Cook *et al.* [14] came up with a partial MVR and solved it with a branch and bound algorithm.

We will study this problem from a new aspect. We know that people probably have violation (or difference) acceptability when they are involved in a group decision environment. It means that people will probably ignore the violation (or difference) if the violation is not significant. This concept has been used in group adjustment consensus decisions for ten years [15–17]. Considering voters' acceptability could make people accept the final order easily. However, few researchers apply it to traditional discrete group decision problem. We attempt to model a new MVR problem with an acceptability index and give a procedure to solve it.

This problem can be titled as Minimum Acceptable Violation Ranking (MAVR) problem. We present the acceptable violation based on pairwise preference decision matrix and ranking decision matrix. We also permit every voter to adjust their preferences less than or equal to a constant step before we calculate the violation. The unique objective is to find a ranking C for which the total violation between C and the set of all voters adjusted rankings is minimized.

While the MVR problem is NP-hard for it is equivalent either to a minimum feedback edge set problem [18], or a problem of which the group preference matrix has the property of being an upper and lower triangular [19]. Both of them are NP-complete or NP-hard. The MAVR problem is more difficult because we have to identify whether voters violate the acceptability index or not. Thus, branch-and-bound algorithm is a reasonable method to solve this problem.

The rest of the paper is organized as follows. In Sect. 2, we define the MAVR model and give some equivalent formulations based on different distance forms. In Sect. 3, we present a branch-and-bound ranking procedure and give an example to illustrate the algorithm. In Sect. 4, we report a set of numerical experiments to evaluate the proposed procedure. We conclude the main results in Sect. 5.

2 Formulation

In this section, we propose a formulation of MAVR problem which is based on presenting a set of definitions and two transformations.

2.1 Preliminaries

We start by presenting some basic concepts and network formations of MAVR problem. Let $X = \{x_1, x_2, ..., x_m\}$ be a discrete set of alternatives. Let $E =$

$\{e_1, e_2, ..., e_m\}$ be the set of voters that participate in the group decision making. Let $R^l = (r^l_{ij})_{m*m}(l = 1, 2, ..., n)$ be the ranking decision matrix given by voter $e_l \in E$, where (r^l_{ij}) presents that the ranking order of alternative $x_i \in X$ is jth $(j \in \{1, 2, ..., m\})$, and

$$r^l_{ij} = \begin{cases} 1 & \text{if voter } e_l \text{ deems that alternative } x_i \text{ is } j\text{th}, \\ 0 & \text{otherwise}, \end{cases}$$

$$\text{s.t.} \quad \begin{array}{ll} \sum_{i=1}^{m} r^l_{ij} = 1 & l = 1, ..., n, j = 1, ..., m, \\ \sum_{j=1}^{m} r^l_{ij} = 1 & l = 1, ..., n, i = 1, ..., m. \end{array}$$

Let $A^l = (a^l_{ij})_{m*m}(l = 1, 2, ..., n)$ be the preference decision matrix which we assume that the pairwise preference are given by the voter e_l, where a^l_{ij} represents the preference between $x_i \in X$ and $x_j \in X$ [5,10], and

$$a^l_{ij} = \begin{cases} 1 & \text{if voter } e_l \text{ deems that alternative } x_i \text{ is preferred to alternative } x_j, \\ 0 & \text{otherwise}. \end{cases}$$

A^l can be derived from R_l. For example, suppose voter e_l' ranking decision matrix is R^l. It means that alternative x_1 is first, x_2 is second, and x_3 is third, thus we have

$$R^l = \begin{pmatrix} 1 & 0 & 0 \\ 0 & 1 & 0 \\ 0 & 0 & 1 \end{pmatrix} \Longrightarrow A^l = \begin{pmatrix} 0 & 1 & 1 \\ 0 & 0 & 1 \\ 0 & 0 & 0 \end{pmatrix}.$$

Generally, Positive Kemeny and Snell (PKS) distance [10] is used for representing the difference (or violation) of two preference decision matrixes $A^l = (a^l_{ij})$ and $C = (c_{ij})$ [11–17]:

$$d_{PKS}(A^l, C) = \sum_i \sum_j \left| a^l_{ij} - c_{ij} \right|.$$

Definition 2.1 (*Ranking Violation*). Given two ranking decision matrix $R^l = (r^l_{ij})$ and $\tilde{C} = (\tilde{c}_{ij})$, the ranking violation (or difference) of this two ranking decision matrix is given by:

$$d_{RD}(R^l, \tilde{C}) = \sum_i \sum_j v_{ij}$$

$$\text{s.t. } v_{ij} = \begin{cases} 1 & (\sum_{i_1} i_1 * r^l_{i,i_1} - \sum_{j_1} j_1 * r^l_{j,j_1}) * (\sum_{i_1} i_1 * \tilde{c}_{i,i_1} - \sum_{j_1} j_1 * \tilde{c}_{j,j_1}) < 0, \\ 0 & \text{otherwise}. \end{cases}$$

If $A^l = (a^l_{ij})$ is derived from $R^l = (r^l_{ij})$, and $C = (c_{ij})$ is derived from $\tilde{C} = (\tilde{c}_{ij})$, we have $d_{RD}(R^l, \tilde{C}) = d_{PKS}(A^l, C)$.

2.2 Minimum Acceptable Violation Ranking Model

Definition 2.2 (*Acceptable Violation*). Given a set of preference decision rankings $\{A^1, A^2, ..., A^n\}$, the acceptable violation of a ranking C(in matrix representation) is given by

$$V(C) = \sum_{l=1}^{n} \sum_{i=1}^{m} \sum_{j=1}^{m} |b_{ij}^l - c_{ij}|$$

$$\text{s.t.} \qquad \textstyle\sum_i^m \sum_j^m |a_{ij}^l - b_{ij}^l| \le \alpha \quad l = 1, ..., n,$$
$$c_{ij} \text{ satisfies transitivity property} \quad i = 1, ..., m,\ j = 1, ..., m.$$

The above definition can also be presented as below. It's similar to the linear adjustment consensus group decision linear model [16].

$$V(C) = \sum_{l=1}^{n} \sum_{i=1}^{m} \sum_{j=1}^{m} |b_{ij}^l - a_{ij}^l|$$

$$\text{s.t.} \qquad \textstyle\sum_i^m \sum_j^m |c_{ij} - b_{ij}^l| \le \alpha \quad l = 1, ..., n,$$
$$c_{ij} \text{ satisfies transitivity property} \quad i = 1, ..., m, j = 1, ..., m.$$

where $A^l = (a_{ij}^l)$ is the initial preference matrix of voter e_l, $B^l = (b_{ij}^l)$ is the adjusted preference matrix of voter e_l after k^l adjustments ($k^l \leqslant \alpha$), $C = (c_{ij})$ is a *consensus preference decision matrix*, and α is the *violation acceptability index* which denotes the established adjustment threshold. α must be even because for each pair of violation, we have $|0 - 1| + |1 - 0| = 2$.

Definition 2.3 (*Minimum Acceptable Violation Ranking-PKS (MAVR-PKS)*). A preference decision matrix C^* is an optimal matrix if it minimizes the acceptable violation $V(C)$ over all possible C. The problem is given by

$$\min V(C) \tag{2.1}$$

$$\text{s.t.} \qquad \textstyle\sum_i^m \sum_j^m |a_{ij}^l - b_{ij}^l| \le \alpha \quad l = 1, ..., n,$$
$$c_{ij} \text{ satisfies transitivity property} \quad i = 1, ..., m,\ j = 1, ..., m.$$

Definition 2.4 (*Violating Voters and No-violating Voters*). Given a set of preference decision rankings $\{A^1, A^2, ..., A^n\}$, and a consensus ranking C. If voter e_l's violation (PKS distance between A^l and C) is not less than α, e_l is a *violating voter*; otherwise, e_l is a *no-violating voter*.

In Lemma 2.1, we will simplify objective notation by removing the adjusted preference decision matrix $B^l = (b_{ij}^l)$.

Lemma 2.1. Problem (2.1) is equivalent to

$$\min_{c_{ij}} \sum_{l=1}^{n} \ max\{\sum_{i=1}^{m} \sum_{j=1}^{m} |a_{ij}^l - c_{ij}|, \alpha\}. \tag{2.2}$$

Proof. We firstly prove that Problem (2.1) is equivalent to

$$\min_{c_{ij}} \sum_{l=1}^{n} [max\{\sum_{i=1}^{m}\sum_{j=1}^{m} \left|a_{ij}^l - c_{ij}\right|, \alpha\} - \alpha]. \tag{2.3}$$

The MAVR model allows all voters to adjust their preferences before calculating the violation distance. It means that if voter e_l is a violating voter, minimizing objective requires the adjusted violation do equals α; otherwise, the adjusted violation equals to $d_{PKS}(A^l, C)$.

(i) For a violating voter, we have $\sum_{i=1}^{m}\sum_{j=1}^{m} \left|a_{ij}^l - c_{ij}\right| \geq \alpha$. To minimize the violation between B^l and C, we know that

$$\sum_{i=1}^{m}\sum_{j=1}^{m} \left|a_{ij}^l - b_{ij}^l\right| = \alpha.$$

So we have

$$\sum_{i=1}^{m}\sum_{j=1}^{m} \left|b_{ij}^l - c_{ij}\right| = \sum_{i=1}^{m}\sum_{j=1}^{m} \left|a_{ij}^l - c_{ij}\right| - \alpha = max\{\sum_{i=1}^{m}\sum_{j=1}^{m} \left|a_{ij}^l - c_{ij}\right|, \alpha\} - \alpha.$$

(ii) For a no-violating voter, we have $\sum_{i=1}^{m}\sum_{j=1}^{m} \left|a_{ij}^l - c_{ij}\right| < \alpha$. To minimize the violation between B^l and C, we know that

$$\sum_{i=1}^{m}\sum_{j=1}^{m} \left|a_{ij}^l - b_{ij}^l\right| = \sum_{i=1}^{m}\sum_{j=1}^{m} \left|a_{ij}^l - c_{ij}\right|.$$

So we have

$$\sum_{i=1}^{m}\sum_{j=1}^{m} \left|b_{ij}^l - c_{ij}\right| = 0 = max\{\sum_{i=1}^{m}\sum_{j=1}^{m} \left|a_{ij}^l - c_{ij}\right|, \alpha\} - \alpha.$$

Obviously, adding a constant to or removing a constant from the objective function will not affect the result. Hence, we add $n * \alpha$ to transfer problem (2.3) to problem (2.2).

This proves that Problem (2.1) and Problem (2.2) are equivalent. □

The greatest difficulty in solving this problem is to deal with the requirement that the elements c_{ij} in matrix C must satisfy transitivity property [18,19]. The optimal *MAVR* also need to satisfy this property.

Definition 2.5 *(Acceptable Ranking Distance(ARD)).* The acceptable ranking distance between ranking decision matrix R^l and $\tilde{C}$ is

$$d_{ARD}(R^l, \tilde{\mathrm{C}}) = \sum_{i}\sum_{j} w_{ij}^l$$

$$\text{s.t.} \qquad w_{ij}^l = \begin{cases} 1 & v_{ij}^l = 1, d_{RD}(R^l, \tilde{C}) \geq \alpha, \\ \frac{\alpha}{d_{RD}(R^l, \tilde{C})} & v_{ij}^l = 1, 0 < d_{RD}(R^l, \tilde{C}) < \alpha, \\ 0 & \text{otherwise}, \end{cases}$$

which is presented by ranking decision matrix, where $d_{RD}(R^l, \tilde{C}) = \sum_i \sum_j v_{ij}^l$, and $v_{ij}^l = \begin{cases} 1 & (\sum_{i_1} i_1 * r_{i,i_1}^l - \sum_{j_1} j_1 * r_{j,j_1}^l) * (\sum_{i_1} i_1 * \tilde{c}_{i,i_1} - \sum_{j_1} j_1 * \tilde{c}_{j,j_1}) < 0, \\ 0 & \text{otherwise.} \end{cases}$

Definition 2.6 *(Minimum Acceptable Violation Ranking-Acceptable Ranking Distance (MAVR-ARD)).* Given a set of rankings $\{R^1, R^2, ..., R^n\}$, the minimum acceptable violation ranking $\tilde{C}^*$ is given by $\min_{\tilde{c}_{ij}} \sum_l d_{ARD}(R^l, \tilde{C})$.

The elements $\tilde{c}_{ij}$ in the ranking matrix $\tilde{C}$ do not need to satisfy transitivity property.

Lemma 2.2. Problem (2.2) is equivalent to

$$\min_{\tilde{c}_{ij}} \sum_l d_{ARD}(R^l, \tilde{C}) \tag{2.4}$$

$$\text{s.t.} \qquad \begin{array}{ll} \sum_{i=1}^m \tilde{c}_{ij}^l = 1, & l = 1, ..., n, j = 1, ..., m, \\ \sum_{j=1}^m \tilde{c}_{ij}^l = 1, & l = 1, ..., n, i = 1, ..., m. \end{array}$$

Proof. According to Definition (2.5), we have the following.

(i) For an arbitrary voter e_l, if $d_{ARD}(R^l, \tilde{C}) \geq \alpha$ holds,

For any pair of alternatives (x_i, x_j), x_i is preferred to x_j both in R^l *and* $\tilde{C}$, we have $w_{ij}^l = \left|a_{ij}^l - c_{ij}\right| = 0$.

For any pair of alternatives (x_i, x_j), x_i is preferred to x_j in R^l but not in $\tilde{C}$; or x_i is preferred to x_j in $\tilde{C}$ but not in R^l, we have $w_{ij}^l = \left|a_{ij}^l - c_{ij}\right| = 1$.

Therefore, $d_{ARD}(R^l, \tilde{C}) = \sum_i \sum_j w_{ij}^l = max\{\sum_i \sum_j \left|a_{ij}^l - c_{ij}\right|, \alpha\}$ holds.

(ii) For an arbitrary voter e_l, if $d_{ARD}(R^l, \tilde{C}) < \alpha$ holds,

For any pair of alternatives (x_i, x_j), x_i is preferred to x_j both in R^l *and* $\tilde{C}$, we have $w_{ij} = \left|a_{ij}^l - c_{ij}\right| = 0$.

For any pair of alternatives (x_i, x_j), x_i is preferred to x_j in R^l but not in $\tilde{C}$; otherwise x_i is preferred to x_j in $\tilde{C}$ but not in R^l, we have $w_{ij}^l = \frac{\alpha}{d_{RD}(R^l, \tilde{C})}$.

Therefore, $d_{ARD}(R^l, \tilde{C}) = \alpha = max\{\sum_i \sum_j \left|a_{ij}^l - c_{ij}\right|, \alpha\}$ holds.

For all of voters $e_l \in E$, $\sum_l d_{ARD}(R^l, \tilde{C}) = \sum_l \ max\{\sum_{i=1}^m \sum_{j=1}^m \left|a_{ij}^l - c_{ij}\right|, \alpha\}$ holds.

Thus, Problem (2.2) and Problem (2.4) are equivalent. □

3 Ranking Procedure

In this section, we will examine some properties of MAVR and apply *branch-and-bound ranking algorithm* to derive the best solution of the MAVR problem. Note that the optimal consensus ranking may be found out before the algorithm "walks through" all possible rankings. Additionally, we'll show a simple example.

3.1 Branch-and-Bound Ranking Procedure

We start by presenting some definitions and propositions that are derived from Sect. 2.2.

Definition 3.1. Given a set of rankings $\{R^1, R^2, ..., R^n\}$, if $\exists x_i \in X = \{x_1, x_2, ..., x_m\}$ and $\forall e_l \in E = \{e_1, e_2, ..., e_n\}$, $\tilde{a}_{i1}^l = 1$ or $\tilde{a}_{im}^l = 1$, define x_i as the absolutely first or last alternative.

Proposition 3.1 *(Separate property). For any given set of rankings, removing the absolutely first or last alternatives initially and add them back finally will not change the optimal ranking.*

For absolutely first or last alternative is easy to identify, we do not consider such a condition.

Before introducing the branch-and-bound algorithm, we give the following definitions.

Definition 3.2. Given a set of ranking $\{A^1, A^2, ..., A^n\}$ which comes from ranking decision matrix set $\{R^1, R^2, .., R^n\}$, denote $s_{ij} = \sum_l a_{ij}^l$ as the aggregated preference between x_i and x_j.

Definition 3.3. For a ranking $\tilde{C} = \{x_1 \succ x_2 ... \succ x_m\}$, if $\tilde{C}' = \{x_1 \succ x_2 ... \succ x_i\} (i < m)$, $\tilde{C}'$ is a prefix ranking of $\tilde{C}$.

Given a set of ranking $\{R^1, R^2, ..., R^n\}$, acceptability index α, and $\tilde{C}'$, we divide alternative set X into two parts: X_1 is the set of alternatives ranked by $\tilde{C}'$, and X_2 is the set of left alternatives. According to Definition 2.4, the voter set E can be separated into two sets: set E_1 includes violating voters, while set E_2 includes no-violating voters.

Definition 3.4. A *lower bound* with the prefix ranking $\tilde{C}'$ is given by

$$\underline{M} = M_0 + M_1 \tag{3.1}$$

and an *upper bound* can be represented by

$$\overline{M} = M_0 + M_2, \tag{3.2}$$

where

$$M_0 = \sum_{e_l \in E_1} \sum_i \sum_j w_{ij}^l + \|E_2\| * \alpha \qquad x_i, x_j \in X;\ \text{one of } x_i, x_j \in X_1,$$

$$M_1 = \sum_i \sum_j min\{s_{ij}, s_{ji}\} \qquad x_i, x_j \in X_2, e_l \in E_1,$$

$$M_2 = \sum_i \sum_j max\{s_{ij}, s_{ji}\} - 2 * (max^1\{s_{ij}\} + max^2\{s_{ij}\}) + 2 * (min^1\{s_{ij}\} + min^2\{s_{ij}\}) \qquad x_i, x_j \in X_2, e_l \in E,$$

where $min^1\{s_{ij}\}$ and $min^2\{s_{ij}\}$ represent the first and second lowest collected preference $s_{ij}(x_i, x_j \in X_2)$; $max^1\{s_{ij}\}$ and $max^2\{s_{ij}\}$ represent the first and second greatest collected preference $s_{ij}(x_i, x_j \in X_2)$.

We give a brief illustration about $\underline{M}$ as follows.

If $e_l \in E_1$, when at least one of $x_i, x_j \in X_1$, $\sum_l \sum_i \sum_j w_{ij}^l = \sum_l \sum_i \sum_j w_{ij}^l$; when $x_i, x_j \in X_2$, $x_i, x_j \in X_1$, $\sum_i \sum_j min\{s_{ij}, s_{ji}\} \leq \sum_l \sum_i \sum_j w_{ij}^l$.

If $e_l \in E_2$, we assume that all voters in E_2 are no-violating voters. So we have $\|E_2\| * \alpha \leq \sum_l max\{\sum_{i=1}^m \sum_{j=1}^m w_{ij}^l, \alpha\}(x_i, x_j \in X)$;

Hence, $M_0 + M_1$ is a lower bound.

Similarly, We give a brief illustration about $\overline{M}$ as follows.

If at least one of $x_i, x_j \in X_1$ and $e_l \in E$, we have $\sum_{e_l \in E_1} \sum_i \sum_j w_{ij}^l + \|E_2\| * \alpha > \sum_{e_l \in E} \sum_i \sum_j w_{ij}^l$.

If $x_i, x_j \in X_2$ and $e_l \in E$, we have $\sum_i \sum_j max\{s_{ij}, s_{ji}\} - 2 * (max^1\{s_{ij}\} + max^2\{s_{ij}\}) + 2 * (min^1\{s_{ij}\} + min^2\{s_{ij}\}) \geq \sum_l \sum_i \sum_j w_{ij}^l$.

Hence, $M_0 + M_1$ is a upper bound.

Proposition 3.2 *(Uninterchangeability Property). For an adjacent pair of the prefix order I (say, $x_i \succ x_j$) and I' (say, $x_j \succ x_i$), if both $s_{ij} < s_{ji}$ and $M_0(I) > M_0(I')$ hold, I must not be a prefix order of $\tilde{C}^*$.*

Proof. Assume that I is a prefix order of the minimum acceptance violation ranking $\tilde{C}^I$, we have $V(\tilde{C}^I) \leqslant V(\tilde{C}^{I'})$, $I' \subseteq \tilde{C}^{I'}$ and the other rankings of $\tilde{C}^{I'}$ are the same with $\tilde{C}^I$.

Denote $W^l(I) = d_{RD}(R^l, I)$ and $W(I) = \sum_l W^l(I)$, at least one of $x_i, x_j \in X_1$, where $d_{RD}(R^l, I)$ has been denoted by Definition 2.1.

We know that $M_0(I) > M_0(I')$ and $W^l(I') = W^l(I) - 2$ (or $W^l(I') = W^l(I) + 2$) because the only difference is the prefix order itself. At the same time, the number of $W^l(I') = W^l(I) - 2$ is not less than the number of $W^l(I') = W^l(I) + 2$ because of $s_{ji} > s_{ij}$.

For giving α, there are three cases because violation $W^l(I)$ is even.

Case 1: If $W^l(I) \geqslant \alpha + 2$, we know that $V(\tilde{C}^I/I) = V(\tilde{C}^{I'}/I')$ because these voters have been listed in E_1. Due to $M_0(I) > M_0(I')$, we have $V(\tilde{C}^I) > V(\tilde{C}^{I'})$.

Case 2: If $W^l(I) = \alpha$, we know that the violations of I and I' will not change when there is no additional violation. If there are some additional violations, when $W^l(I') = W^l(I) - 2$, we have $V(\tilde{C}^I/I) = V(\tilde{C}^{I'}/I') + 2$; when $W^l(I') = W^l(I) + 2$, we have $V(\tilde{C}^I/I) = V(\tilde{C}^{I'}/I')$. As we know that the number of $W^l(I') = W^l(I) - 2$ is not less than the number of $W^l(I') = W^l(I) + 2$, so $V(\tilde{C}^I/I) > V(\tilde{C}^{I'}/I')$. Due to $M_0(I) > M_0(I')$, we have $V(\tilde{C}^I) > V(\tilde{C}^{I'})$.

Case 3: If $W^l(I) \leq \alpha - 2$, we know that the violations of I and I' will not change when there is no additional violation. If there are some additional violation, when $W^l(I') = W^l(I) - 2$, we have $V(\tilde{C}^I/I) = V(\tilde{C}^{I'}/I') + 2$; when $W^l(I') = W^l(I) + 2$, we have $V(\tilde{C}^I/I) = V(\tilde{C}^{I'}/I') + 2$. As we know that the number of $W^l(I') = W^l(I) - 2$ is not less than the number of $W^l(I') = W^l(I) + 2$, so $V(\tilde{C}^I/I) > V(\tilde{C}^{I'}/I')$. Due to $M_0(I) > M_0(I')$, we have $V(\tilde{C}^I) > V(\tilde{C}^{I'})$.

Summing up all voters' violation, we have $V(\tilde{C}^I) > V(\tilde{C}^{I'})$. It contradicts with the assumption.

Thus order I will not be a prefix order of $\tilde{C}^*$, this proves the proposition. □

In the algorithm below, we start with an empty ranking. Then adding each alternative to the first node one by one, and calculate the lower bound and upper bound for later, and then cut off some branches based on branching principle or Proposition 3.2. Note that we start from the branches with the lowest bound and check all other branches' upper bounds when we add a new alternative.

We present an algorithm to generate MAVR as follows.

Algorithm—Branch-and-bound ranking algorithm

Input: Set of alternatives $X = \{x_1, x_2, ..., x_m\}$ and voters' ranking decision matrix $\{R^1, R^2, ..., R^n\}$.

Step 0 (Initialization): $\tilde{C}' = \phi(j = 0)$. Calculate its lower bound $\underline{M}(\tilde{C}')$ and upper bound $\overline{M}(\tilde{C}')$, and store them as the branch-and-bound tree root node. If the lower bound equals to the upper bound, go to Step 3; Otherwise, define this node as an active node and go to Step 1.

Step 1 (Selecting the node to branch form): Select the one with the lowest lower bound from the set of nodes of the branch and bound tree. If its upper bound is less than the current best known lower bound, go to Step 3.

Step 2: For all eligible branches one by one.

Step 2a (Branching): Construct node $\tilde{c}_{ij} = 1(i \in X_2, j = j + 1)$ to selected node $\tilde{C}'$. $\tilde{C}'$ includes the alternatives which have been ranked already.

Step 2b (Bounding 1): Calculate these new nodes' lower bounds and upper bounds. If one' lower bound is greater than currently best known minimum upper bound or solution, cut off this branch.

Step 2c (Create a new node): Add these c_{ij} to $\tilde{C}'$ as active nodes to the tree and store $\underline{M}(\tilde{C}')$ with nodes.

Step 2d (Bounding 2) (Uninterchangeability property): Use the Proposition 3.2 to cut off some branches which have both smaller s_{ij} and larger M_0 than its pair branch.

Step 3 (Termination): If there are still some active nodes or the number of the left alternatives is greater than 2, returen to Step 1; otherwise, select the smaller s_{ij} of the left two alternatives, and add this order ranking behind the termination ranking.

3.2 A Numerical Example

We consider an example with *acceptability index* $\alpha = 4$, $n = 10$ voters, and $m = 4$ alternatives:

Voters' frequency	Alternatives	Ranking
4	$\{1,2,3,4\}$	$3 \succ 4 \succ 2 \succ 1$
3	$\{1,2,3,4\}$	$4 \succ 3 \succ 2 \succ 1$
2	$\{1,2,3,4\}$	$2 \succ 1 \succ 3 \succ 4$
1	$\{1,2,3,4\}$	$2 \succ 3 \succ 1 \succ 4$

The table below represented the summarized pairwise preference:

$$s_{ij} = \begin{pmatrix} 0 & 0 & 2 & 3 \\ 10 & 0 & 3 & 3 \\ 8 & 7 & 0 & 7 \\ 7 & 7 & 3 & 0 \end{pmatrix}$$

Initialization: $\tilde{C}' = \phi$. Calculate $M_0 = 4 * 10 = 40$, $M_1 = 0$, and $M_2 = 2 * (0 + 2 + 7 + 7 + 7 + 7) = 60$. We obtain a lower bound $\underline{M} = M_0 + M_1 = 40$, and an upper bound $\overline{M} = M_0 + M_2 = 40 + 60 = 100$.

Branching: We process the root node with the empty ranking one by one as shown in Fig. 1, we check them all:

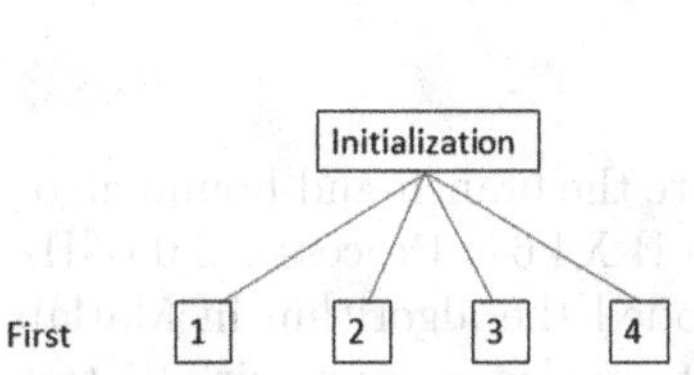

Fig. 1. First iteration of minimum acceptable violation ranking

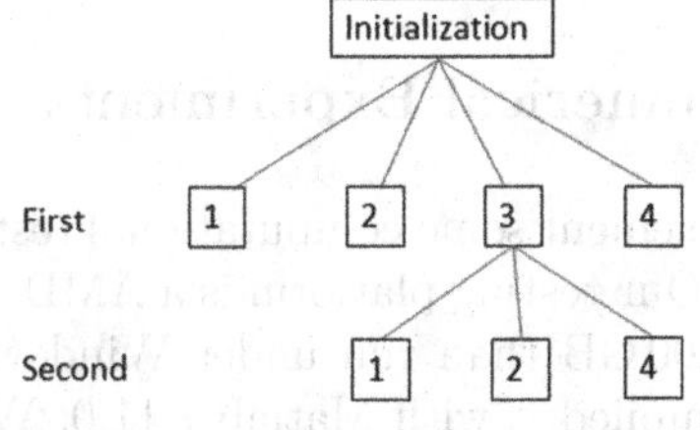

Fig. 2. Second iteration of minimum acceptable violation ranking

- *Adding alternative 1:*
  Let $\tilde{c}_{11} = 1$, we have $\tilde{C}' = \{1\}$. Calculate $M_0 = 4 * 6 + 3 * 6 + 2 * \alpha = 52$, $M_1 = 2 * (1 + 1 + 3) = 10$, and $M_2 = 2 * (3 + 3 + 7) = 26$. We obtain a lower bound $\underline{M} = 62$, and an upper bound $\overline{M} = 78$.
- *Adding alternative 2:*
  Let $\tilde{c}_{21} = 1$, we have $\tilde{C}' = \{2\}$, Calculate $M_0 = 40$, $M_1 = 6$, and $M_2 = 24$. We obtain a lower bound $\underline{M} = 46$, and an upper bound $\overline{M} = 64$.
- *Adding alternative 3:*
  Let $\tilde{c}_{31} = 1$, we have $\tilde{C}' = \{3\}$, Calculate $M_0 = 40$, $M_1 = 0$, and $M_2 = 20$. We obtain a lower bound $\underline{M} = 40$, and an upper bound $\overline{M} = 60$.
- *Adding alternative 4:*
  Let $\tilde{c}_{41} = 1$, we have $\tilde{C}' = \{4\}$, Calculate $M_0 = 46$, $M_1 = 2$, and $M_2 = 16$. We obtain a lower bound $\underline{M} = 48$, and an upper bound $\overline{M} = 62$.
  We could not bound any point via the rule $max\{\underline{M}\} < min\{\overline{M}\}$.

Branching: Ranking all the nodes by their $\underline{M}$, we process the root node one by one again as shown in Fig. 2. We check them all:

- *Adding alternative 2 to* $\tilde{C}' = \{3\}$, *for now we have* $\tilde{C}' = \{3, 2\}$:
  Calculate $M_0 = 4 * \alpha + 3 * \alpha + 2 * 4 + 1 * \alpha = 40$, $M_1 = 0$, and $M_2 = 2 * 3 = 6$. We obtain a lower bound $\underline{M} = 40$, and an upper bound $\overline{M} = 46$.
- *Adding alternative 4 to* $\tilde{C}' = \{3\}$,*for now we have* $\tilde{C}' = \{3, 4\}$:
  We obtain a lower bound $\underline{M} = 50$. We do not need to calculate upper bound and add it to the branch-and-bound tree because the lower bound is greater than present $min\{\overline{M}\} = 46$.
- *Adding alternative 1 to* $\tilde{C}' = \{3\}$, *for now we have* $\tilde{C}' = \{3, 1\}$:
  We obtain a lower bound $\underline{M} = 52$. We do not need to calculate upper bound and add it to the branch-and-bound tree because the lower bound is greater than present $min\{\overline{M}\} = 46$.

Bounding 1: For $\underline{M}(\tilde{C}' = \{1\}) = 62$, $\underline{M}(\tilde{C}' = \{2\}) = 46$, and $\underline{M}(\tilde{C}' = \{4\}) = 48$ is not less than $\overline{M}(\tilde{C}' = \{3, 2\}) = 46$, we cut off these branches from the branch-and-bound tree.

Bounding 2 (Uninterchangeability property): For $s_{41} = 7 > s_{14} = 3$ and $M_0(\tilde{C}' = \{3, 2, 4, 1\}) = 44 < M_0(\tilde{C}' = \{3, 2, 1, 4\}) = 46$, we cut off $\tilde{C}' = \{3, 2, 1, 4\}$.

Hence, we know that the MAVR is $3 \succ 2 \succ 4 \succ 1$.

4 Numerical Experiments

We implement some computational tests to evaluate the branch-and-bound algorithm. Our testing platform is a AMD Athlon(tm) II X4 640 Processor 3.00 GHz with 4.00 GB than run under Windows 7. We coded the algorithm in Matlab and compiled it with Matlab 7.11.0. We construct six classes, each with 25 test instances, and different number of alternatives or voters as shown in Table 1. For each of the 150 instances in Table 1, we created 3 different sets of voters decision matrixes which have different normally distributed noise $N(0, \sigma^2)$, with $\sigma^2 = 1, 4$ or 9. In Table 2 we present the average and worst case running time (in seconds) for each of the six classes and three levels of noises. It's apparent in Table 2 that we are able to obtain the optimal ranking for every instance within 5 seconds. Although the number of alternatives is not high($m \leq 5$), it matches with most of practical scenarios.

Additionally, we have three conclusions. Firstly, the fact that every class with different level of noise take a similar amount of time to process implies that our algorithm is insensitive to the level of noise. Secondly, as the size of alternatives increases (Class C, D and E), the processing time increases quickly because the pairwise preference comparisons increase with factorial growth. Thirdly, Classes A, B, and C take a similar amount of processing time, which implies that our algorithm is insensitive to the amount of number of voters. Our method do have a good performance.

Table 1. Six classes of the test problems

Class	No. of alternatives	No. of pairwise comparisons	No. of voters
A	3	6	60
B	3	6	120
C	3	6	240
D	4	24	240
E	5	120	240
F	5	120	2400

Table 2. Running time (s) of the algorithm and percentage of the problems solved

class	$\sigma^2 = 1$		$\sigma^2 = 4$		$\sigma^2 = 9$	
	Avg.(max) running time	% solved to optimality	Avg.(max) running time	% solved to optimality	Avg. (max) running time	% solved to optimality
A	< 0.01(< 0.01)	100	< 0.01(< 0.01)	100	< 0.01(< 0.01)	100
B	< 0.01(< 0.01)	100	< 0.01(< 0.01)	100	< 0.01(< 0.01)	100
C	< 0.01(< 0.01)	100	< 0.01(< 0.01)	100	< 0.01(< 0.01)	100
D	< 0.02(< 0.04)	100	< 0.02(< 0.04)	100	< 0.02(< 0.04)	100
E	< 0.2(< 0.3)	100	< 0.2(< 0.3)	100	< 0.2(< 0.3)	100
F	< 3(< 6)	100	< 3(< 6)	100	< 3(< 6)	100

5 Conclusions

In this paper, we considered that voters are more willing to accept the final ranking because, on one hand, most of voters' violation will not exceed the acceptability index, and on the other, the total extra violation is minimized. Additionally, the numerical results shown that the processing time will be shorter than MVR problem because the acceptability index is helpful to bound branches.

We proposed the Minimum Acceptable Violation "Consensus" Ranking problem in group decision making. To simplify the MAVR problem, we came up with two transformations. On one hand, we removed the adjusted preference matrixes $\{B^1, B^1, ..., B^m\}$ from the model, and on the other hand, we using acceptable ranking violation instead of PKS to represent the model so that the matrix of the final ranking do not need to be transitive. Based on Problem 2.4 and the uninterchangeability property, we presented a branch-and-bound algorithm for obtaining an optimal ranking and demonstrated the algorithm is effective and efficiency. For all experiments, the processing time is less than 5 s. This paper helps users attain an "acceptable consensus ranking" in practical group decision problems with a clear procedure.

Acknowledgments. This paper was supported by the National Natural Science Foundation of China (No. 61221063), the Program for Changjiang Scholars and Innovative Research Team in University (IRT1173).

References

1. De Borda J.C.: Mmoire sur les lections au scrutin (1781)
2. Kendall, M.G.: Rank correlation methods, Biometrika (1990)
3. Wei, T.-H.: The algebraic foundations of ranking theory, University of Cambridge (1952)
4. Blin, J.M.: A linear assignment formulation of the multiattribute decision problem. RAIRO - Oper. Res. - Recherche Opérationnelle **10**(V2), 21–32 (1976)
5. Kemeny, J.G., Snell, L.J.: Preference ranking: an axiomatic approach. In: Mathematical Models in the Social Sciences, pp. 9–23 (1962)
6. Cook, W.D., Seiford, L.M.: Priority ranking and consensus formation. Manage. Sci. **24**(16), 1721–1732 (1978)
7. Cook, W.D., Seiford, L.M.: On the Borda-Kendall consensus method for priority ranking problems. Manage. Sci. **28**(6), 621–637 (1982)
8. Ali, I., Cook, W.D., Kress, M.: On the minimum violations ranking of a tournament. Manage. Sci. **32**(6), 660–672 (1986)
9. Cook, W.D., Saipe, A.L.: Committee approach to priority planning: the median ranking. Cahiers du Centre dtudes de Recherche Oprationnelle **18**(3), 337–351 (1976)
10. Barzilai, J., Cook, W.D., Kress, M.: A generalized network formulation of the pairwise comparison consensus ranking model. Manage. Sci. **32**(8), 1007–1014 (1986)
11. Cook, W.D., Golan, I., Kress, M.: Heuristics for ranking players in a round robin tournament. Comput. Oper. Res. **15**(2), 135–144 (1988)
12. Cook, W.D., Kress, M.: A data envelopment model for aggregating preference rankings. Manage. Sci. **36**(11), 1302–1310 (1990)
13. Rademaker, M., De Baets, B.: A ranking procedure based on a natural monotonicity constraint. Inf. Fusion **17**, 74–82 (2014)
14. Cook, W.D., Golany, B., Penn, M., et al.: Creating a consensus ranking of proposals from reviewers partial ordinal rankings. Comput. Oper. Res. **34**(4), 954–965 (2007)
15. Ben-Arieh, D., Chen, Z.: Linguistic labels aggregation and consensus measure for autocratic decision-making using group recommendations. IEEE Trans. Syst. Man Cybern. Part A Syst. Hum. **36**(3), 558–568 (2006)
16. Ben-Arieh, D., Easton, T.: Multi-criteria group consensus under linear cost opinion elasticity. Decis. Supp. Syst. **43**(3), 713–721 (2007)
17. Ben-Arieh, D., Easton, T., Evans, B.: Minimum cost consensus with quadratic cost functions. IEEE Trans. Syst. Man Cybern. Part A Syst. Hum. **39**(1), 210–217 (2009)
18. Isaak, G., Narayan, D.A.: Complete classification of all tournaments having a disjoint union of directed paths as a minimum feedback arc set. J. Graph Theor. **45**(1), 2847 (2004)
19. Bader, D., Cong, G.: Graph algorithms. Encycl. Parallel Comput. **11**(5), 467–492 (2011)

Listing Center Strings Under the Edit Distance Metric

Hiromitsu Maji(✉) and Taisuke Izumi

Graduate School of Engineering, Nagoya Institute of Technology,
Nagoya 466-8555, Japan
cke17607@stn.nitech.ac.jp, t-izumi@nitech.ac.jp

Abstract. Given a set W of k strings of length n over an alphabet Σ, the center string of W is defined as the string w such that the maximum distance to w of all strings in W is minimized under some specified metric. We present a new algorithm for the decision version of this problem under the edit distance metric. Given a threshold parameter d, the algorithm lists all the strings such that the distance from any input string is bounded by d in $O((3^d(d+2))^k dk|\Sigma|n + Mn)$ time, where M is the number of the output strings. To the best of our knowledge, this is the first FPT algorithm for the center string under the edit distance metric (even as a finding algorithm). By a slight modification, we also obtain an algorithm listing length-l common subsequences of W, which runs in $O((n-l)^{k+1}k|\Sigma|l + Ml)$ time.

1 Introduction

Finding a common structure from a given set of strings is recognized as one of the important problems in computational biology. A *center string* (or equivalently, *closest string*) is the one that minimizes the maximum distance of all strings in a input set W under some specific metric. The decision version of the center string problem under metric δ, which is the primary problem considered in this paper, is formalized as follows:

Input: A set W of k strings of length n over an alphabet Σ, and a threshold value $d \in \mathbb{N}$.
Output: A string w such that $\delta(w, w') \leq d$ holds for any $w' \in W$ if it exists. Otherwise the value of "FALSE".

In the definition above, the distance metric is not concretely defined. Usually it is chosen according to applications. Popular metrics useful in many applications are *Hamming distance* and *edit distance*. Unfortunately, for both metrics, the center string problem is NP-complete [4], and thus we need some sort of relaxation for attacking this problem. In this paper, we consider fixed-parameter algorithms for the center string problem under the edit distance metric. However, unfortunately again, that problem is W[1]-hard with respect to the number of

This work is supported in part by KAKENHI No. 15H00852 and 25289227.

Z. Lu et al. (Eds.): COCOA 2015, LNCS 9486, pp. 771–782, 2015.
DOI: 10.1007/978-3-319-26626-8_57

input strings k [16], which implies that the problem is unlikely to have an algorithm with a running time such as $O(f(k) \cdot poly(n))$. The currently best algorithm is the one by Nicolas et al. [16], which achieves $O(|\Sigma|n^k)$ time bound.

To circumvent the hardness results above, we focus on the fixed-parameter tractability for parameters both d and k. That is, we explore the algorithms having a running time with the form of $O(f(d,k) \cdot poly(n))$. The primary contribution of this paper is that such an algorithm actually exists. Our new algorithm finds a center string in $O((3^d(d+2))^k dk|\Sigma|n)$ time . To the best of our knowledge, this is the first FPT algorithm for finding center string problem under the edit distance metric. By a simple extension of this finding algorithm , we propose an algorithm to list all the solutions in the output-sensitive manner: The algorithm lists all the center strings for the edit distance metric in $O((3^d(d+2))^k dk|\Sigma|n + Mn)$ time, where M is the number of the output strings. Since in typical scenarios the center string problem is considered with a small d, our algorithms are practically more useful than the previous one.

The algorithms are constructed with a new dynamic-programming strategy, where the DP table records the distance information around diagonal vertices in alignment graphs. Interestingly, we can utilize the same strategy to solve another problem. Our second result is an algorithm listing length-l common subsequences of all input strings. The time complexity of this algorithm is $O((n-l)^{k+1}|\Sigma|l + Ml)$. Note that the longest common subsequence (LCS) problem, which is the optimization version of the length-l common subsequence problem, is known to be NP-complete [14], and W[1]-hard for parameter k [17]. On the other hand, finding length-l common subsequences trivially allows an $O(|\Sigma|^l poly(n,k))$-time algorithm by checking all strings of length l. That is, it is fixed-parameter tractable for parameter l in the case of constant-size alphabets. Furthermore, Irving and Fraser shows two algorithms of finding a LCS of length at least l in $O((n-l)^{k-1}kn)$ and $O((n-l)^{k-1}kl + k|\Sigma|n)$ times respectively [9]. That is, finding a common subsequence with length near to n is also fixed-parameter tractable (for $n-l$). Our algorithm can be seen as a listing version of the two algorithms by Irving and Fraser.

The paper is organized as follows: In Sect. 2, we present the prior work related to the topics of this paper. Section 4 provides our algorithm for the center string problem. Its extension to the LCS problem is considered in Sect. 5 Finally, we conclude this paper with a future direction in Sect. 6.

2 Related Work

The center string problem for the Hamming distance metric (often called *closest string problem*) is extensively studied. In general, that problem is NP-complete [6,11], but allows a fixed-parameter algorithm with respect to d. Following the first FPT-algorithm by Gramm et al. [7], a number of papers improved the time complexity [3,13,18]. The closest substring problem, which is a generalized version of the closest string problem, is also well studied. Interestingly the closest substring problem is W[1]-hard with respect to both d and k even if the alphabet is binary [15]. Marx shows an efficient algorithm for computing the closest substring for small d and/or k (but it is not an FPT algorithm) [15].

Compared to the Hamming distance metric, the center string problem under the edit distance metric is less studied. As we stated in the introduction, the paper by Nicolas and Rivals is only the one explicitly considering that setting [16]. The case of other metrics is considered in [5].

The longest common subsequence (LCS) problem for multiple strings is regarded as a special case of the center string for the edit distance metrics. It is equivalent to the center string problem for the edit distance metric with substitution cost two. About exact solutions for the LCS problem, a few papers propose several algorithms with different characteristics [8,9]. The LCS problem for some restricted instances is considered in [1,2].

Another variant of the center string problem is the median string problem, which requires to find the string minimizing the sum of the distance to each input string. While the median string under the Hamming distance metric is easily solvable in polynomial time, the case for edit distance is known to be NP-complete [4], and W[1]-hard for parameter k [16].

Approximated solutions for the problems introduced above are also investigated [11,12]. PTASs are allowed for the closest (sub)string problem [12], but the longest common subsequence problem has no polynomial-time algorithm with any approximation ratio better than n^c for some constant $c > 0$ unless P = NP [10]. No polynomial-time approximated solution for the center string problem under the edit distance metric is known so far.

3 Preliminaries

3.1 Edit Distance

We denote the alphabet by Σ. An element in Σ^* is called *string*. The length of a string w is denoted by $|w|$, and the i-th character of w is denoted by $w[i]$ ($1 \leq i \leq |w|$). The operator $\circ$ means the concatenation of two strings (or characters). For $w \in \Sigma^*$, let $ta(w)$ be the string obtained by removing the first character of w. That is, $w = w[1] \circ ta(w)$. Letting W be a set of strings, we define $W \circ x = \{w \circ x | w \in W\}$.

The *edit distance* $ED(w_1, w_2)$ between two strings w_1 and w_2 is defined as follows:

$$ED(w_1, w_2) = \begin{cases} \max\{|w_1|, |w_2|\} \\ \qquad\qquad\qquad\qquad\qquad (\text{if } |w_1| = 0 \vee |w_2| = 0) \\ \min\{ED(ta(w_2), ta(w_1)) + c(w_1[1], w_2[1]), \\ \quad ED(ta(w_1), w_2) + 1, ED(w_1, w_2) + 1\} \\ \qquad\qquad\qquad\qquad\qquad (\text{otherwise}), \end{cases}$$

where $c(a, b)$ is the function returning zero if $a = b$ or one otherwise. Note that while we assume that $c(a, b)$ is uniform (i.e.,the substitution cost does not depend on target characters), our algorithm can be applied to the case of non-uniform cost functions (as long as it returns an integer value).

It is well-known that the computation of the edit distance between two strings w_1 and w_2 can be reduced to the shortest path problem for some directed acyclic graph $G(w_1, w_2) = (V, E, f)$, called *alignment graph*, which defined as follows (Fig. 1):

- $V = \{v_{i,j} | i, j \in [0, n]\}$.
- For any $i, j \in [0, n]$, $v_{i,j}$ has a directed edge to each vertex $v_{i+1,j}$, $v_{i,j+1}$, and $v_{i+1,j+1}$ if it exists.
- $f(e) = \begin{cases} 0 \text{ if } e = (v_{i,j}, v_{i+1,j+1}) \text{ for some } i, j \in [0, n] \text{ and } w_1[i] = w_2[j] \\ 1 \text{ otherwise.} \end{cases}$

An edge $e = (v_{i,j}, v_{i',j'}) \in E$ is called a *horizontal*, *vertical*, or *diagonal* edge if $i = i'$, $j = j'$, or $(i \neq i' \wedge j \neq j')$ holds respectively. The set of vertices $\{v_{i,j} | j \in [0, n]\}$ and $\{v_{i,j} | i \in [0, n]\}$ are called *i-th row* and *j-th column* respectively. The distance between two vertices u and v is denoted by $dist(u, v)$. In particular, if $u = v_{0,0}$, we omit the first argument and describe $dist(u)$ for short. The following theorem is a classical fact.

Theorem 1. $dist(v_{n,n}) = ED(w_1, w_2)$.

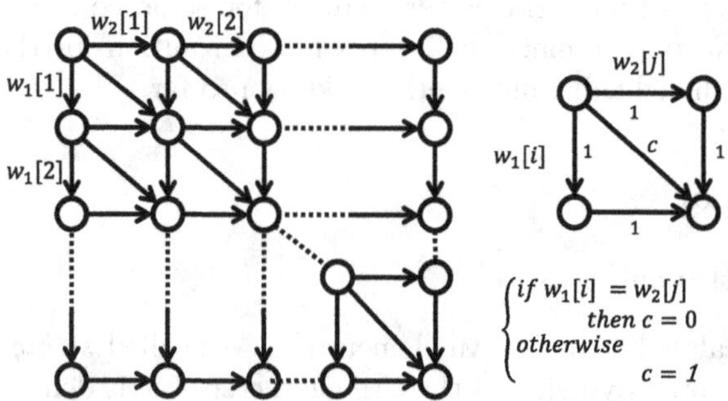

Fig. 1. Alignment graph $G(w_1, w_2)$

The *band* of alignment graphs is defined as the set of vertices $\{v_{i,j} || i - j| \leq d/2\}$[1]. We also define $h(j)$ as the intersection size of the j-th column and the band. That is, let $h(j) = \min\{j + d/2, d, (n - j) + d/2\}$. For ease of explanation, we give aliases to each vertex in the band: The vertices of the j-th column in the band are called $u_{0,j}, u_{1,j}, \ldots, u_{h(j),j}$ from the upper side. The notations above are illustrated in Fig. 2.

[1] The definition of the band depends on the value of d. Hence it may be more precise to include that dependency in the notation (e.g., calling d-band). However, to avoid the complication of notations, we treat the value of d as a certain kind of "global constant." Actually, in the following argument, we introduce several definitions dependent on d with no explicit description of the dependency.

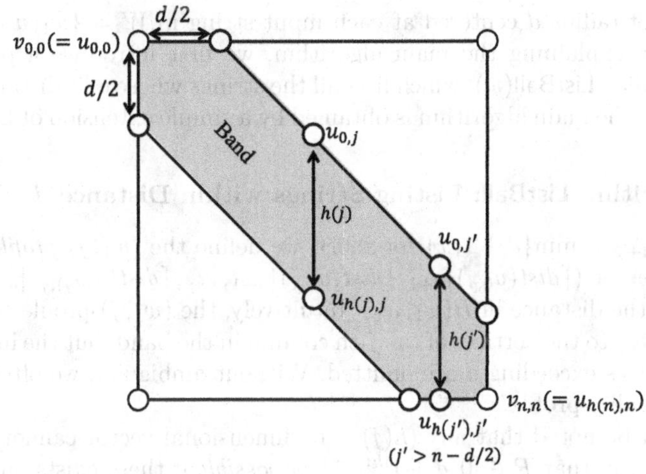

Fig. 2. Band and vertex aliasing

We have the following lemma:

Lemma 1. *Given $w_1, w_2 \in \Sigma^n$ satisfying $ED(w_1, w_2) \leq d$, all the vertices constituting the shortest path from $v_{0,0}$ to $v_{n,n}$ in $G(w_1, w_2)$ are contained in the band.*

Proof. The shortest path from $v_{i,j}$ to $v_{n,n}$ must contain at least $|i - j|$ non-diagonal edges, all of which have weight one. Thus its length is more than or equal to $|i-j|$. Similarly, the length of the shortest path from $v_{0,0}$ to $v_{i,j}$ is also more than or equal to $|i - j|$. If $|i - j| > d/2$ (i.e., $v_{i,j}$ is out of the band), the length of any path from $v_{0,0}$ to $v_{n,n}$ via $v_{i,j}$ is more than d. It follows that $v_{i,j}$ cannot be contained in the shortest path because $dist(v_{n,n}) = ED(w_1, w_2) \leq d$. □

The lemma above implies that it suffices to consider the subgraph of $G(w_1, w_2)$ induced by the band because we only care about the paths from $v_{0,0}$ to $v_{n,n}$ of length at most d. We denote that induced subgraph by $B(w_1, w_2)$. The terminologies and notations introduced for $G(w_1, w_2)$ are also used for $B(w_1, w_2)$. In the following argument, we sometimes treat $B(w_1, w_2)$ for some string w_2 of length less than n. So we extend the definition of $B(w_1, w_2)$: For strings $w_1 \in \Sigma^n$ and $w_2 \in \Sigma^m$ such that $m < n$, we define $B(w_1, w_2)$ as the one in which edge $e = (v_{i,j}, v_{i',j'})$ for $j < m$ has the weight according to the original function f, and all other edges have weight one.

4 Listing Center Strings

In this section, we propose an algorithm for the center string problem, called ListCenter(W). The core idea of ListCenter(W) is to compute the intersection of the k balls of radius d centered at each input string in $W = \{w_1, w_2, \cdots, w_k\}$. Thus, before explaining the main algorithm, we first introduce a preliminary algorithm called ListBall(w), which lists all the strings whose edit distance from w is at most d. The main algorithm is obtained by a simple extension of ListBall(w).

4.1 Algorithm ListBall: Listing Strings within Distance d

Letting $\lceil x \rceil_{d+1} = \min\{d+1, x\}$ for short, we define the (w_1, j)-*profile* of string w_2 as the vector $(\lceil dist(u_{0,j}) \rceil_{d+1}, \lceil dist(u_{1,j}) \rceil_{d+1}, \ldots, \lceil dist(u_{h(j),j}) \rceil_{d+1})$, where $dist(u_{i,j})$ is the distance in $B(w_1, w_2)$. Intuitively, the (w_1, j)-profile of w_2 is the distance vector to the vertices of the j-th column in the band, but the information about distances exceeding d are omitted. Without ambiguity, we often omit w_1 and simply call j-profile.

It should be noted that any $(h(j)+1)$-dimensional vector cannot become a j-profile. we say that $P \in [0, d+1]^{h(j)+1}$ is *possible* if there exists $w_1, w_2 \in \Sigma^n$ such that P becomes the (w_1, j)-profile of w_2. We can show a necessary condition for the possibility of P:

Lemma 2. *If* $P = (p_0, p_1, \ldots, p_{h(j)}) \in [0, d+1]^{h(j)+1}$ *is a possible* j*-profile,* $|p_i - p_{i+1}| \leq 1$ *holds for any* $i \in [0, h(j)-1]$.

Proof. Since $dist(u_{i+1,j}) - dist(u_{i,j}) \leq 1$ obviously holds because edge $(u_{i,j}, u_{i+1,j})$ has weight one, it suffices to show $dist(u_{i,j}) - dist(u_{i+1,j}) \leq 1$. Let $u_{0,0} = x_0, \cdots, x_r = u_{i+1,j}$ be the shortest path from $u_{0,0}$ to $u_{i+1,j}$ in $B(w_1, w_2)$, x_c be the last vertex contained in the i-th row, and $dist(x_c) = l$ (see Fig. 3). Since x_c is reachable to $u_{i,j}$ only traversing horizontal edges, $dist(u_{i,j}) \leq l + (r-c)$ holds. The shortest path from x_c to $u_{i+1,j}$ can contain at most one diagonal edge, its length is at least $r - c - 1$, and thus $dist(u_{i+1,j}) \geq l + (r-c-1)$ holds. Consequently, we have $dist(u_{i,j}) - dist(u_{i+1,j}) \leq 1$. □

Let $\mathcal{P}_j$ be the set of the sequences in $[0, d+1]^{h(j)+1}$ satisfying the condition of Lemma 2. Clearly $\mathcal{P}_j$ contains all the possible profiles. From Lemma 2, we have the following corollary:

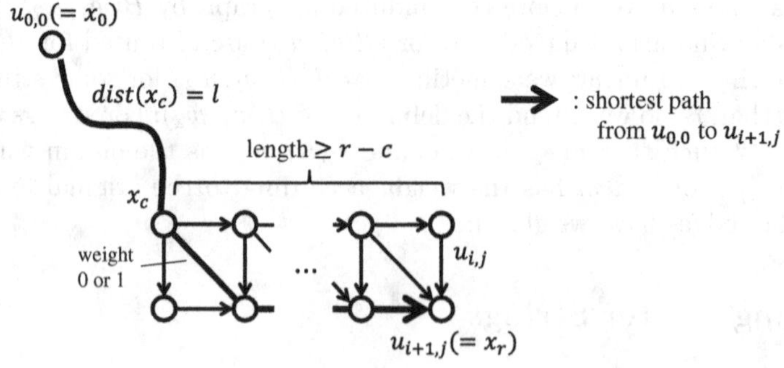

Fig. 3. Proof of Lemma 2

Corollary 1. *For any $j \in [0,n]$, $|\mathcal{P}_j| \leq 3^d(d+2)$.*

We also have the following corollary from the definition of profiles.

Corollary 2. *Any string w has 0-profile $P_{init} = (0,1,2,\ldots,h(0))$.*

For $P = (p_0, p_1, \ldots, p_{h(j)}) \in \mathcal{P}_j$ and $w \in \Sigma^n$, we define $S^w_{P,j}$ as the set of length-j strings having P as its (w,j)-profile. Let $\mathcal{P}_{term} = \{(p_0, p_1, \ldots, p_{h(n)}) \in [0, d+1]^{h(n)+1} | p_{h(n)} \leq d\}$. Then the union $\cup_{P\in\mathcal{P}_{term}} S^w_{P,n}$ is the set of the strings to be listed as the computation result. The core idea of $\mathsf{ListBall}(w)$ is to compute $S^w_{P,j}$ for any P and j via dynamic programming. To lead the DP recurrence formula, we introduce one more notation defined as follows: Let $j \in [0, n-1]$ and $w \in \Sigma^n$. For two vectors $P = (p_0, p_1, \ldots, p_{h(j)}) \in [0, d+1]^{h(j)+1}$ and $Q = (q_0, q_1, \ldots, q_{h(j+1)}) \in [0, d+1]^{h(j+1)+1}$, we say that *$P$ is connected to Q with $(x, w) \in \Sigma \times \Sigma^*$, denoted by* $P \overset{w,x,j}{\longmapsto} Q$, if there exists a string w' such that $w'[j+1] = x$ and P and Q are respectively the (w,j)-profile and $(w, j+1)$-profile of w'. The key fact to obtain the recurrence formula is the next lemma:

Lemma 3. *Fixing $w \in \Sigma^n$, the $(j+1)$-profile Q satisfying $P \overset{w,x,j}{\longmapsto} Q$ is uniquely determined from P and x in $O(d)$ time.*

Proof. The uniqueness of Q is obvious because any shortest path to a vertex in the $(j+1)$-th column must pass a vertex in the j-th column in $B(w,*)$. Thus we prove that Q can be computed in $O(d)$ time. In graph $B(w,*)$, we can determine the weights of all the edges between jth and $(j+1)$-th columns by x. Thus we can compute Q by calculating the distances up to d from $v_{0,0}$ to the vertices in the $(j+1)$-th column, provided distances up to d to the vertices in the j-th column. For any $i' \in [0,n]$, the predecessor of $v_{i',j+1}$ in the shortest path from $v_{0,0}$ to $v_{i',j+1}$ is either $v_{i'-1,j}$, $v_{i',j}$, or $v_{i'-1,j+1}$. So if the distances (up to d) to those vertices are already known, the shortest path to $v_{i',j+1}$ (with length up to d) can be computed in a constant time. This implies that the values of the $(j+1)$-profile can be fixed from the upper side sequentially (i.e., in the order of $u_{j+1,0}, u_{j+1,1}, \ldots, u_{j+1,h(j+1)}$). Since $h(j+1) = O(d)$ holds, we can compute Q in $O(d)$ time. The lemma is proved. □

This lemma implies that if we know the j-profile of a string w, we can know the $(j+1)$-profile of $w \circ x$ for any $x \in \Sigma$. Conversely, if we want to know some (unknown) string $w = w' \circ x$ having some $(j+1)$-profile, it suffices to idenfity w' and its j-profile. This fact induces the following recurrence formula.

$$S^w_{Q,j+1} = \bigcup_{\substack{x\in\Sigma \\ P:P\overset{w,x,j}{\longmapsto}Q}} S^w_{P,j} \circ x. \tag{1}$$

Now we are ready to explain the algorithm, which consists of the following two steps:

- The first step of the algorithm is to construct the edge-labeled DAG $\Gamma = (V_\Gamma, E_\Gamma, f_\Gamma)$ defined as follows:

- $V_\Gamma = (\cup_{j=0}^{n}\{(P,j) | P \in \mathcal{P}_j\} \cup \{t\}$, where t is the special sink vertex. We also give alias s to the vertex $(P_{init}, 0)$.
- A vertex (P, j) is connected to $(Q, j+1)$ by an edge with label x if $P \overset{w,x,j}{\longmapsto} Q$. Note that if two or more characters x satisfy $P \overset{w,x,j}{\longmapsto} Q$, (P, j) and $(Q, j+1)$ are connected by multiedges. Finally, we add edges from all the vertices (P, n) satisfying $P \in \mathcal{P}_{term}$ to t with the null-character label.

– For any s-t path $X = e_0, e_1, \ldots e_n$ in Γ, we define $\gamma(X)$ as the string formed by traversing X (i.e., $\gamma(X) = f_\Gamma(e_0) \circ f_\Gamma(e_1) \circ \cdots \circ f_\Gamma(e_n)$). The second step of the algorithm is to output $\gamma(X)$ for each s-t path X in Γ.

The correctness of this algorithm relies on the following lemma:

Lemma 4. *For any $(P, j) \in V_\Gamma$ and a string $w_2 \in \Sigma^j$, $w_2 \in S^{w_1}_{P,j}$ holds if and only if there exists a path X from s to (P, j) such that $w_2 = \gamma(X)$ holds.*

Proof. The proof is done by induction on j. (**Basis**): It is obvious from Corollary 2. (**Induction step**): Suppose as the induction hypothesis that the lemma holds for $j = j' - 1$ and consider the case of $j = j'$. We prove only the direction of $\Rightarrow$ because the opposite direction ($\Leftarrow$) is easily proved by following backward the argument. Let $w_2 = w_2' \circ x$. The recurrence formula (Eq. 1) implies that if $w_2 \in S^{w_1}_{P,j'}$, there exists a $(j'-1)$-profile P' such that $w_2' \in S^{w_1}_{P',j'-1}$ and $P' \overset{w,x,j'-1}{\longmapsto} P$ hold. Then, by the induction hypothesis, there exists a path X' from s to (P', j') and $w_2' = \gamma(X')$. So there exists a path X to (P, j') from s and $\gamma(X) = w_2' \circ x$. The lemma is proved. □

We consider the running time of the algorithm. In the first step, for each vertex (P, i), the algorithm needs to compute all the pairs (x, Q) such that $P \overset{w,x,i}{\longmapsto} Q$ holds. From Lemma 3, it can be computed in $O(|\Sigma|d)$ time. From Corollary 1, we also have $|V_\Gamma| = O(3^d dn)$. Thus the total running time of the first step is $O(3^d(d^2|\Sigma|n)$. For the second step, all s-t paths can be enumerated by the naive recursion after pruning the vertices from which t is unreachable. Its running time is $O(Mn + |E_\Gamma|)$, where M is the number of paths enumerated. Finally, we have the following theorem.

Theorem 2. *Given $w \in \Sigma^n$, algorithm* ListBall(w) *lists all the strings whose distance from w is less than or equal to d in $O(3^d d^2|\Sigma|n + Mn)$ time.*

4.2 Listing Center Strings

By extending algorithm ListBall(w), we construct the main algorithm ListCenter(W). The primary idea is that given $W = \{w_1, w_2, \ldots, w^k\}$, ListCenter$(W)$ concurrently runs ListBall(w_i) for each input $w_i \in W$. We give the detailed explanation below:

A k-tuple of profiles $\mathbf{P} = (P_1, P_2, \ldots, P_k) \in \mathcal{P}_j^k$ is called the (W, j)-*profile* of a string w if P_i is w's (w_i, j)-profile for any $i \in [1, k]$. The notation $\mathbf{P} \overset{W,x,j}{\longmapsto} \mathbf{Q}$

for $\mathbf{P} = (P_1, P_2, \ldots, P_k) \in \mathcal{P}_j^k$ and $\mathbf{Q} = (Q_1, Q_2, \ldots, Q_k) \in \mathcal{P}_j^k$ means that there exists $w \in \Sigma^n$ and $j \in [0, n]$ such that $\mathbf{P}$ and $\mathbf{Q}$ are w's (W, j)-profile and $(W, j+1)$-profile respectively and $w[j] = x$ holds.

The remaining structure of $\mathsf{ListCenter}(W)$ is almost the same as $\mathsf{ListBall}(w)$, but the definition of profiles and its connectivity relationship are replaced by the ones above. More precisely, the algorithm utilizes the graph Γ^k defined as follows, instead of Γ:

- $V_\Gamma = (\cup_{i=0}^n \{(\mathbf{P}, i) | \mathbf{P} \in \mathcal{P}_i^k\} \cup \{t\}$, where t is the special sink vertex. We also give alias s to $((P_{init}, P_{init}, \ldots, P_{init}), 0)$.
- A vertex $(\mathbf{P}, i)$ is connected to $(\mathbf{Q}, i+1)$ by an edge with label x if $\mathbf{P} \overset{W,x,i}{\longmapsto} \mathbf{Q}$. Note that if two or more characters x satisfy $\mathbf{P} \overset{W,x,i}{\longmapsto} \mathbf{Q}$, $(\mathbf{P}, i)$ and $(\mathbf{Q}, i+1)$ are connected by multiedges. Finally, we add the edges from all the vertices $(\mathbf{P}, n)$ satisfying $\mathbf{P} \in \mathcal{P}_{term}^k$ to t with the null-character label.

It is not difficult to prove that the string $\gamma(X)$ corresponding to a s-t path X in Γ^k has a distance at most d to each string $w_i \in W$. Thus by enumerating all s-t paths we can list all center strings. We bound the running time of this algorithm. The analysis of the second step completely follows that for $\mathsf{ListBall}(w)$. For the first step, the size of Γ^k is larger than Γ. The number of vertices in Γ^k is $O((3^d(d+2))^k n)$. In addition, the computation of outgoing edges for each vertex takes obviously k times of the case for $\mathsf{ListBall}(w)$, i.e., $O(k|\Sigma|d)$ time. Hence the total running time of the first step is $O((3^d(d+2))^k dk|\Sigma|n)$. Consequently we have the following main theorem.

Theorem 3. *Algorithm ListCenter(W) lists all center strings for W under the edit distance metric in $O((3^d(d+2))^k dk|\Sigma|n + Mn)$ time, where M is the number of output strings.*

5 Listing Common Subsequences

A subsequence of a string $w \in \Sigma^n$ is any string obtained from w by deleting several characters. We denote by $Sub(w)$ the set of all subsequences of w. The decision version of the *longest common subsequence problem* (LCS) is defined as follows:

> **Input**: A set $W = \{w_1, w_2, \cdots, w_k\}$ of k strings over Σ of length n, and a threshold value $l \in \mathbb{N}$.
> **Output**: A string $w \in \cap_{i=1}^k Sub(w_i)$ such that $|w| \geq l$ if it exists. Otherwise the value of "FALSE".

Let $\bar{l} = n - l$ for short. In this section we show an algorithm called $\mathsf{ListLCS}_l(W)$, which is an algorithm listing common subsequences of length l. This algorithm is obtained by a refinement of $\mathsf{ListCenter}(W)$. For $w_1 \in \Sigma^n$ and $w_2 \in \Sigma^l$, we construct the *LCS alignment graph* $G_{LCS}(w_1, w_2) = (V_{LCS}, E_{LCS})$ as follows:

- $V_{LCS} = \{v_{i,j} | i \in [0,n], j \in [0,l]\}$.
- For any $i \in [0, n-1]$ and $j \in [0,l]$, add $e = (v_{i,j}, v_{i,j+1})$. In addition, add $e = (v_{i,j}, v_{i+1,j+1})$ if $w_1[i] = w_2[j]$.

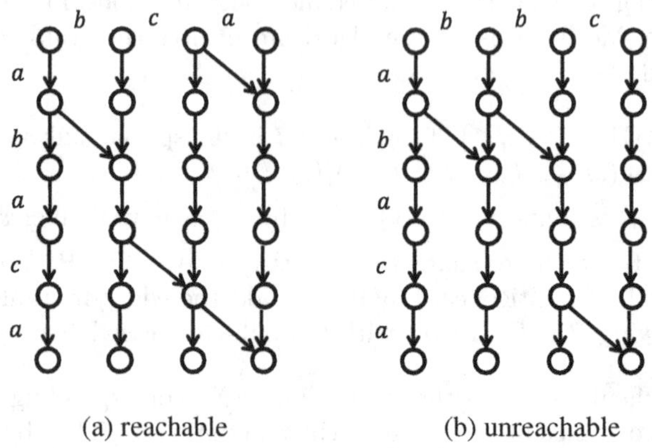

(a) reachable (b) unreachable

Fig. 4. Two examples of LCS alignment graphs.

Note that LCS alignment graphs are unweighted. It is not difficult to prove the following lemma:

Lemma 5. *A string w_2 is a subsequence of a string w_1 if and only if $v_{0,0}$ is reachable to $v_{n,l}$ in $G_{LCS}(w_1, w_2)$.*

Two examples of LCS alignment graphs, which correspond to reachable and unreachable cases respectively, are shown in Fig. 4. We define the band of LCS alignment graphs as the set of vertices $\{v_{i,j} | j \leq i \leq j + \bar{l}, j \in [0,n]\}$ (see Fig. 5). The following lemma is analogous to Lemma 1 in the center-string case.

Lemma 6. *Any vertex $v_{i,j}$ out of the band is either unreachable to $v_{0,0}$ or $v_{n,l}$.*

Similarly as the center string case, let $B_{LCS}(w_1, w_2)$ be the subgraph of $G_{LCS}(w_1, w_2)$ induced by the band. Now we introduce the refined definition of profiles: The (w_1, j)-profile of w_2 is a binary $(\bar{l}+1)$-dimensional vector representing the reachability from $v_{0,0}$ to each vertex in the j-th column That is, a vertex $v_{i+j,j}$ is reachable from $v_{0,0}$ in $B_{LCS}(w_1, w_2)$ if and only if the (w_1, j)-profile $P \in [0,1]^{\bar{l}+1}$ of w_2 satisfies $P[i] = 1$. Since $v_{i,j'}$ for $j' > j$ is reachable from $v_{0,0}$ when $v_{i,j}$ is reachable from $v_{0,0}$, any possible j-profile can be represented as the concatenation of an all-zero sequence followed by an all-one sequence. Furthermore, we do not have to consider the all-zero vector as a profile. Therefore, the total number of possible j-profiles is at most $\bar{l}$. We set $\mathcal{P}_j$ to all possible j-profiles.

The remaining part of algorithm $\mathsf{ListLCS}_l(W)$ is almost the same as $\mathsf{ListCenter}(W)$. Following the definition of profiles above, we construct the graph Γ^k

and enumerate all s-t paths in Γ^k. Only the difference is the design of the source and the edges incoming to the sink in Γ^k. In the context of listing common subsequences, the $(\bar{l}+1)$-dimensional all-one vector is the unique possible 0-profile, and thus s is set to it. The vertices in $\{(P,l)|P[\bar{l}]=1\}$ is adjacent to t.

By the analysis similar with Sect. 4, we can bound the running time of this algorithm as follows:

Theorem 4. *For any set of k strings $W = \{w^1, w^2, \dots w^k\}$, algorithm ListLCS$_l(W)$ enumerates all length-l common sequences in $O(\bar{l}^{k+1}k|\Sigma|l+Ml)$ time, where M is the number of output strings.*

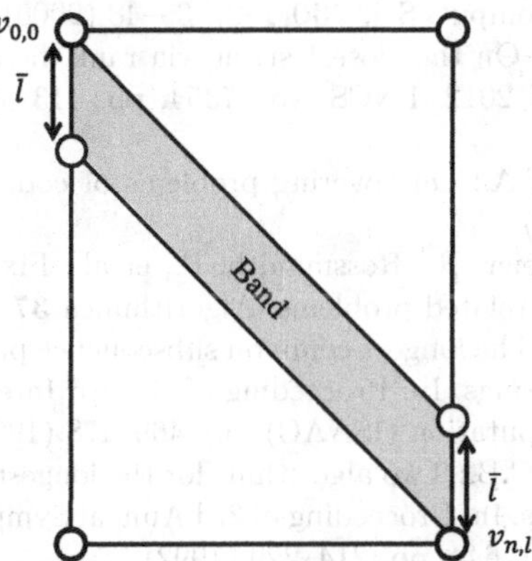

Fig. 5. Band of LCS alignment graphs.

6 Concluding Remarks

In this paper, we presented two algorithms called ListCenter(W) and ListLCS$_l(W)$. Algorithm ListCenter enumerates all the center strings for given k strings and a threshold distance d in $O((3^d(d+2))^k dk|\Sigma|n+Mn)$ time. In addition, this algorithm finds one solution in $O((3^d(d+2))^k dk|\Sigma|n)$ time, which is the first FPT algorithm for the center string problem under the edit distance metric. Algorithm ListLCS$_l$ is designed with the same framework as ListCenter, which enumerates length-l common subsequences for k strings in $O(\bar{l}^{k+1}k|\Sigma|l+Ml)$ time.

On the parameterized complexity of the center string problem under the edit distance metric is surprisingly less studied. An important open problem is to show the fixed-parameter (in)tractability with respect to d only. While the authors conjecture W[1]-hardness of that setting, the proof is still missing. Even if it is actually W[1]-hard, the exploration of faster algorithms (for example, running in $O(d^k \cdot poly(n)$ time) is also an interesting open problem).

References

1. Blin, G., Bonizzoni, P., Dondi, R., Sikora, F.: On the parameterized complexity of the repetition free longest common subsequence problem. Inf. Process. Lett. **112**(7), 272–276 (2012)
2. Blin, G., Bulteau, L., Jiang, M., Tejada, P.J., Vialette, S.: Hardness of longest common subsequence for sequences with bounded run-lengths. In: Kärkkäinen, J., Stoye, J. (eds.) CPM 2012. LNCS, vol. 7354, pp. 138–148. Springer, Heidelberg (2012)
3. Chen, Z.-Z., Wang, L.: Fast exact algorithms for the closest string and substring problems with application to the planted (l, d)-motif model. IEEE/ACM Trans. Comput. Biol. Bioinf. **8**(5), 1400–1410 (2011)
4. de la Higuera, C., Casacuberta, F.: Topology of strings: median string is NP-complete. Theoret. Comput. Sci. **230**(1–2), 39–48 (2000)
5. Dinu, L.P., Popa, A.: On the closest string via rank distance. In: Kärkkäinen, J., Stoye, J. (eds.) CPM 2012. LNCS, vol. 7354, pp. 413–426. Springer, Heidelberg (2012)
6. Frances, M., Litman, A.: On covering problems of codes. Theor. Comput. Syst. **30**(2), 113–119 (1997)
7. Gramm, J., Niedermeier, R., Rossmanith, P., et al.: Fixed-parameter algorithms for closest string and related problems. Algorithmica **37**(1), 25–42 (2003)
8. Hakata, K., Imai, H.: The longest common subsequence problem for small alphabet size between many strings. In: Proceeding of the 3rd International Symposium on Algorithms and Computation (ISAAC), pp. 469–478 (1992)
9. Irving, R.W., Fraser, C.B.: Two algorithms for the longest common subsequence of three (or more) strings. In: Proceeding of 3rd Annual Symposium on Combinatorial Pattern Matching, vol. 644, pp. 214–229 (1992)
10. Jiang, T., Li, M.: On the approximation of shortest common supersequencesand longest common subsequences. SIAM J. Comput. **24**(5), 1122–1139 (1995)
11. Lanctot, J.K., Li, M., Ma, B., Wang, S., Zhang, L.: Distinguishing string selection problems. Inf. Comput. **185**(1), 41–55 (2003)
12. Li, M., Ma, B., Wang, L.: On the closest string and substring problems. J. ACM **49**(2), 157–171 (2002)
13. Ma, B., Sun, X.: More efficient algorithms for closest string and substring problems. SIAM J. Comput. **39**(4), 1432–1443 (2009)
14. Maier, D.: The complexity of some problems on subsequences and supersequences. J. ACM **25**(2), 322–336 (1978)
15. Marx, D.: Closest substring problems with small distances. SIAM J. Comput. **38**(4), 1382–1410 (2008)
16. Nicolas, F., Rivals, E.: Hardness results for the center and median string problems under the weighted and unweighted edit distances. J. Discrete Algorithms **3**(2–4), 390–415 (2005)
17. Pietrzak, K.: On the parameterized complexity of the fixed alphabet shortest common supersequence and longest common subsequence problems. J. Comput. Syst. Sci. **67**(4), 757–771 (2003)
18. Wang, L., Zhu, B.: Efficient algorithms for the closest string and distinguishing string selection problems. In: Deng, X., Hopcroft, J.E., Xue, J. (eds.) FAW 2009. LNCS, vol. 5598, pp. 261–270. Springer, Heidelberg (2009)

Online Scheduling for Electricity Cost in Smart Grid

Xin Feng[1,2,3](✉), Yinfeng Xu[1,2,3], and Feifeng Zheng[4]

[1] School of Management, Xi'an Jiaotong University, Xi'an 710049, Shaanxi, China
fengxin.xjtu@stu.xjtu.edu.cn

[2] State Key Lab for Manufacturing Systems Engineering, Xi'an 710049, Shaanxi, China

[3] Ministry of Education Key Lab for Process Control and Efficiency Engineering, Xi'an 710049, Shaanxi, China

[4] Glorious Sun School of Business and Management, Donghua University, Shanghai 200051, People's Republic of China

Abstract. This paper studies an online scheduling problem in the smart grid, which is arised in demand response management under the scenario with real-time communication between the grid operator and consumers. Consumers send the power requests online over-list. The request is released with a limited set of timeslots. Only one of the timeslots in the set can this request be served by the operator. In a timeslot, the electricity cost consumed to serve the requests is a quadratic function of the load in it. Our aim is to find a best possible online schedule which generates the minimal total electricity cost. In this paper, we propose a greedy algorithm of this problem which is 2-competitive. Besides, we prove our algorithm is optimal.

Keywords: Online scheduling · Smart grid · Demand response · Greedy algorithm

1 Introduction

The smart grid is one of the major challenges in the 21st century [4], which concentrates on harnessing information and communication technologies to improve the electric grid flexibility and reliability. The information of power requests is communicated between the consumers and the grid operator via IP addressable components over the internet [11]. In the smart grid, the sudden augment of voltage fluctuations increases the possibility of power outage, and thus reduce the grid reliability. Besides, the demand in peak hours needs to be satisfied by supplementary generated power, which are often more expensive than the regular generated power. In this way, smoothing the power demand profile, which is called the *demand response management*, can improves its reliability and reduces the cost of operating the grid. Thus the basic aim in demand response management is to alleviate peak load by transferring non-emergency power demands at

Z. Lu et al. (Eds.): COCOA 2015, LNCS 9486, pp. 783–793, 2015.
DOI: 10.1007/978-3-319-26626-8_58

off-peak-load time intervals [8]. We refer the reader to Hamilton and Gulhar [6], Ipakchi [7] and Lui et al. [11] for a comprehensive review on smart grid.

The scheduling in the smart grid is to assign the power requests in a schedule where the power demand profile is as smooth as possible. Koutsopoulos and Tassiulas [8] proved this problem is NP-hard when the requested is scheduled non-preemptively. Logenthiran et al. [10] proposed heuristics to solve this problem for demand side management. Sousa et al. [14] considered a multi-objective methodology with both minimization of the operation cost and the minimization of the voltage magnitude difference.

In practice, demand response management is executed in real-time. The information of power requests from consumers are sent to the operator one by one in a sequence. Before the operator receives a power request, he has no information about this request. The operator needs to execute the demand response management without any knowledge of the future. As a traditional online load balancing problem, Azar [1] proposed a parallel machine scheduling model to minimize the maximum load of the machines. This problem is similar to this paper if we treat the machines as timeslots. The difference is that the aim of online load balancing problem is to minimize the maximum load of machines. Most of the researches studied the stochastic model for online scheduling in smart grid, in which the power requests are released to the grid operator according to a Poisson process [8,9]. Narayanaswamy et al. [13] applied online convex optimization framework in smart grid. Lu et al. [12] studied the competitive online algorithms for energy generation in microgrids with intermittent energy sources and co-generation. Georgiadis and Papatriantafilou [5] studied the deterministic online scheduling model within a given time interval to minimize the maximum load on the timeslots. They proposed a greedy algorithm which is $(\lceil \log n \rceil + 1)$-competitive, where $n = 48$.

In this paper, we assume the time is divided into integral timeslots. The request is released to the operator online-over-list. Each request can be served in serval determined timeslots. The operator needs to control the system in the smart grid with the minimum electricity cost. Similar to Burcea et al. [3], we assume that the power requirement and the duration of service requested are both unit-size.

The rest of this paper is organized as follows. In Sect. 2, we describe the problem. In Sect. 3, we present an online algorithm and propose the competitive analysis. Finally, Sect. 4 concludes this paper.

2 Preliminaries

Consider an online scheduling problem in the smart grid. The time is divided into integral timeslots $T = \{1, 2, \cdots\}$. The input is a set of power requests $\mathcal{R} = \{r_1, r_2, \cdots\}$, which are released online over-list. All the power requests have the unit-size energy requirement and unit-size service duration. The information of a given request r_i is a set of timeslots $I_i \subseteq T$, which cannot be learned until r_i is released. r_i can only be assigned in one of the timeslots in I_i. The load of a timeslot t is denoted as $l(t)$, which is the total number of requests assigned in it. The electricity cost consumed in a timeslot is a convex function of its load. Inspired by Shahidehpour et al. [15], the electricity cost consumed in t is expressed as

$$C(t) = a[l(t)]^2 + bl(t) + c \tag{1}$$

As for the total electricity cost, $\sum C(t) = a\sum [l(t)]^2 + b\sum l(t) + c \cdot m$, where m is the number of occupied timeslots. In this paper, the evaluation of algorithms is to compare the value of the corresponding total electricity cost. We ignore the monomial item $b\sum l(t)$ since the total loads $\sum l(t)$ counts for the number of requests which is a determined value when we compare the value of the total electricity cost of two schedules. For the constant item $c \cdot m$, it restraints operator to assign requests to more timeslots. Since there are huge amounts of power requests sent to the operator rapidly in the smart grid in practice, it is not common to achieve an idle timeslot which has no power load. We ignore the constant item in this hypothesis. Thus the electricity cost consumed in t is modified as

$$C(t) = [l(t)]^2 \tag{2}$$

The aim in this paper is to find an online schedule in the smart grid with the minimal possible total electricity cost. The performance of an online schedule is generally evaluated by the competitive ratio (see [2]). Translated into our problem terminology, for any power requests input instance I, let $C(I)$ be the objective value of schedule produced by an online algorithm, and $OPT(I)$ that of schedule by an optimal off-line algorithm which has the information of all requests at the beginning. Then the algorithm is ρ-competitive if $C(I) \leq \rho OPT(I) + \varepsilon$ holds for any I where $\rho \geq 1$ is some constant and ε is an arbitrary real number. ρ is also called the competitive ratio of the algorithm.

3 Algorithm and Competitive Analysis

In this section, we propose an online greedy algorithm to obtain a preferable smart grid schedule. The principle of this algorithm is to reduce the peak load by diffusing the power requests into as many timeslots as possible.

Algorithm 1. Greedy algorithm

When a power request r_i is released with the set of available timeslots I_i, assign it into the timeslot which has the minimal load

---

We use σ_g and C_g to denote the online schedule of power requests and the corresponding electricity cost obtained by Greedy algorithm. Assume in σ_g, there are n occupied timeslots, each of which is assigned at least one power request. Besides these, the timeslot which has no request assigned is called an idle timeslot. We present the competitive ratio of Greedy algorithm by deriving the worst scenario of σ_g.

Lemma 1. *For each occupied timeslot in σ_g, there is at most one power request which can be served in an idle timeslot.*

Proof. Suppose in an arbitrary timeslot t_1, there are two power requests r_i and r_j, which can be served in another two idle timeslots t_2 and t_3 respectively, i.e., $t_2 \in I_i$ and $t_3 \in I_j$. Without loss of generality, r_i is assumed to be released earlier than r_j. Requests are always assigned to the available timeslot which has the minimal load in σ_g. Since $t_3 \in I_j$ and $l(t_1) > l(t_3) = 0$, r_j must be assigned into t_3 other than t_1, which contradicts our assumption. The lemma follows. □

Since there are totally n occupied timeslots in σ_g, there are at most n power requests which can be moved into the idle timeslots. We call these requests as *dummy requests.* For these dummy requests, the following lemma is helpful to derive the worst scenario of σ_g.

Lemma 2. *Under the worst scenario of σ_g, there must be n dummy requests, which are assigned in n idle timeslots in the optimal schedule.*

Proof. Figure 1(a) is an illustration of σ_g. The electricity cost of σ_g is

$$C_g = \sum_{i=1}^{n} [l(i)]^2 \tag{3}$$

The optimal schedule is shown in Fig. 1(b). In the optimal schedule, the load in the timeslot t is denote as $l^*(t)$. Suppose there is a dummy request r_i which is not assigned into an idle timeslot in the optimal schedule. The optimal electricity cost is

$$C_{opt} = \sum_{i=1}^{2n-1} [l^*(i)]^2 \tag{4}$$

In contrast, if I_i is assigned in the idle timeslot $2n$ as shown in Fig. 1(c), the electricity cost then changes to be

$$C' = \sum_{i=1}^{t-1} [l^*(i)]^2 + [l^*(t) - 1]^2 + \sum_{i=t+1}^{2n-1} [l^*(i)]^2 + 1 \tag{5}$$

$C_{opt} - C' = [l^*(t)]^2 - [l^*(t) - 1]^2 - 1 \geq 0$, which contradicts our assumption that C_{opt} is the minimum electricity cost. The lemma follows. □

According to Lemma 2, we can derive the worst scenario of σ_g by generate n dummy requests which are assigned in n idle timeslots in the optimal schedule. However, there are still some other requests left, i.e., $l(t) - 1$ requests in timeslot t, which can only be served in the n non-idle timeslots. We call these requests as *regular requests.* To derive the worst scenario of the regular requests' assignment, we regrade the loads $l(1), l(2), \cdots, l(n)$ in σ_g in descending orders as $l_1, l_2, \cdots, l_n$. The corresponding timeslot of l_i is denoted as τ_i. Thus

$$C_g = l_1^2 + l_2^2 + \cdots + l_n^2 \tag{6}$$

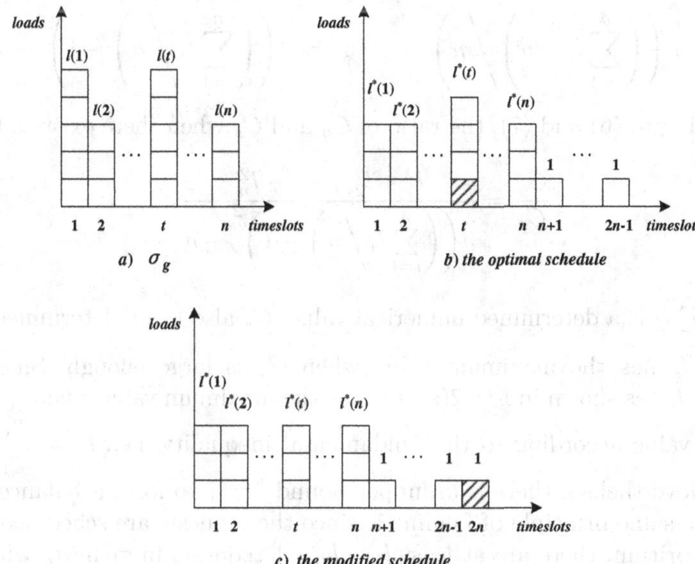

Fig. 1. Illustration of the proof of Lemma 2

If the regular requests can be assigned in a schedule where the load in each timeslot from τ_1 to τ_n are all the same, this schedule is called a *balance schedule.* It can be easily concluded that the electricity cost has a minimum value when the left $\sum_{i=1}^{n}(l_i-1)$ regular requests can form a balance schedule from τ_1 to τ_n. In other words, if there exists a balance schedule for the regular requests, this balance schedule must be the optimal schedule.

Besides, we consider a special schedule σ_s where the requests can be split arbitrarily to its available timeslots. For example, a unit-size requirement request can be split into two timeslots, the one with 1/3 power requirement and the other with 2/3. In this way, σ_s can get a balance schedule more easily than the optimal schedule of our problem. Denote the electricity of this special schedule as C_s. It can be concluded that $C_s \leq C_{opt}$. Thus, if we derive the maximum ratio ρ of C_g and C_s, ρ is also the competitive ratio of the Greedy algorithm since $\frac{C_g}{C_{opt}} \leq \frac{C_g}{C_s} \leq \rho$.

To derive the worst scenario of σ_g, we divided it into two cases by whether the left $\sum_{i=1}^{n}(l_i-1)$ regular requests can form a balance schedule in σ_s. We first analyze the case where the regular requests can form a balance schedule, and obtain the competitive ratio of the Greedy algorithm in this case.

If the left $\sum_{i=1}^{n}(l_i-1)$ requests can form a balance schedule in σ_s as shown in Fig. 2(b),

$$C_s = n\left(\left(\left(\sum_{i=1}^{n} l_i - n\right)\Big/ n\right)\right)^2 + n \cdot 1^2 = n\left(\left(\sum_{i=1}^{n} l_i \Big/ n\right) - 1\right)^2 + n \quad (7)$$

According to (6) and (7), the ratio of C_g and C_s when there exists a balance schedule is

$$\frac{C_g}{C_s} = \frac{l_1^2 + l_2^2 + \cdots + l_n^2}{n\left(\left(\sum_{i=1}^{n} l_i \Big/ n\right) - 1\right)^2 + n} \quad (8)$$

Since $\sum_{i=1}^{n} l_i$ is a determined numerical value, C_s also has a determined value. Thus C_g/C_s has the maximum value when C_g is large enough. Since $l_1 \geq l_2 \geq \cdots \geq l_n$, as shown in Fig. 2(a), C_g has the maximum value when l_1 has the maximum value according to the fundamental inequality, i.e., $l_1 = \sum_{i=1}^{n} l_i, l_2 = \cdots = l_n$. Nevertheless, there is an upper bound for l_1 to form a balance schedule. For the same principle of Lemma 1, since the requests are scheduled by the Greedy algorithm, there are at least $l_1 - l_2 - 1$ requests in τ_1 in σ_g which can only be assigned in τ_1. Figure 2(a) is the illustration of the worst scenario of σ_g. Let $a_1 = l_1 - l_2 - 1, a_2 = l_2 = l_3 = \cdots = l_n$. According Fig. 2(a) and (b),

$$a_1 = \frac{na_2}{n-1} + \frac{1}{n-1} \quad (9)$$

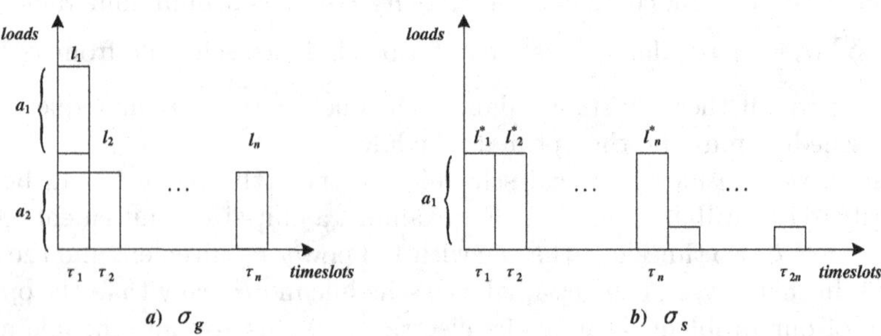

Fig. 2. Illustration of the worst scenario of σ_g

Lemma 3. *If there exists a balance schedule in σ_s, there are exactly two timeslots occupied in σ_g under the worst scenario.*

Proof. Assume there are n timeslots occupied in $\sigma_g, n \geq 2$. We prove this lemma by showing that adding a timeslot in σ_g reduces the ratio of C_g and C_s.

As shown in Fig. 3(a), a timeslot τ_{n+1} is added with the load $a_2 + 1$. a_1 then changes to be $a_1' = \frac{(n+1)a_2}{n} + \frac{1}{n}$. Thus the electricity cost of σ_g changes to be

$$C_g' = C_g + (a_2 + 1)^2 + ({a'}_1 + a_2 + 1)^2 - (a_1 + a_2 + 1)^2 \quad (10)$$

For the electricity cost in σ_s,

$$C_s' = C_s + 1 + (a'_1)^2 + n(a'_1)^2 - na_1^2 \tag{11}$$

Let $\Delta_1 = (a_2+1)^2 + (a'_1 + a_2 + 1)^2 - (a_1 + a_2 + 1)^2$ and $\Delta_2 = 1 + (a'_1)^2 + n(a'_1)^2 - na_1^2$. Since $\Delta_2 - \Delta_1 = 2a_2 + \frac{(a_2+1)^2}{n(n-1)} > 0$, $\frac{C'_g}{C'_s} = \frac{C_g+\Delta_1}{C_s+\Delta_2} < \frac{C_g}{C_s}$. It can be concluded that adding a timeslot in σ_g reduces the ratio of C_g and C_s. Since $n \geq 2$, to obtain the maximum ratio of C_g and C_s, there is only two occupied timeslots in σ_g under the worst scenario. □

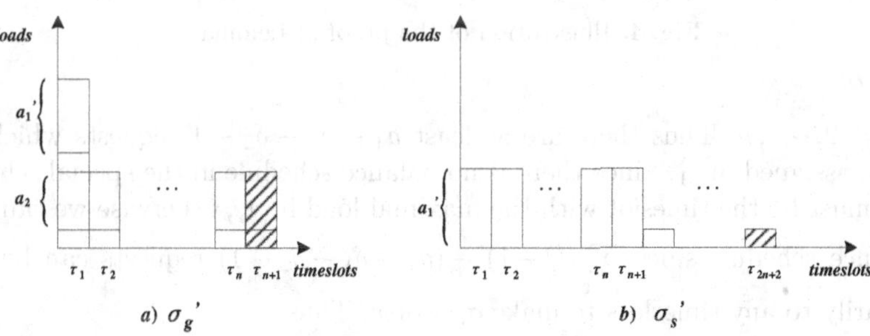

Fig. 3. Illustration of the proof of Lemma 3

Lemma 4. *If there exists a balance schedule, the competitive ratio of the Greedy algorithm is 2.*

According to Lemma 3, there are two occupied time slots in σ_g under the worst scenario. When $n = 2$, $a_1 = \frac{na_2}{n-1} + \frac{1}{n-1} = 2a_2 + 1$. We ignore the item 1 so that $a_1 = 2a_2$ to form the balance schedule.

$$\rho_1 = \frac{C_g}{C_s} = \frac{(3a_2+1)^2 + (a_2+1)^2}{2(2a_2)^2 + 2} \leq \frac{5a_2^2 + 4a_2 + 1}{4a_2^2 + 1}$$

When $a_2 = \frac{1+\sqrt{65}}{16}$, ρ_1 has the maximum value. However, since we have assumed that the requirement of the power requests are all unit-size, a_2 must be an integer. Thus $\rho_1 \leq 2$. □

We have proved that the competitive ratio of the Greedy algorithm is 2 when there exists a balance schedule. In the case where there are some redundant requests so that the regular requests cannot form a balance schedule. For these redundant requests, we have the following lemma.

Lemma 5. *If there are some redundant requests which cannot be assigned into a balance schedule, only the requests which are assigned in τ_1 can make σ_g worse.*

Proof. Suppose there is a schedule in which the requests can form a balance schedule. Thus the profile of worst scenario of σ_g is shown in Fig. 2(a). In this case, we add some redundant requests, i.e. δ_i requests in τ_i as shown in Fig. 4(a). For convenient calculation without disturbing our analysis, let $\delta_i = \delta_2$

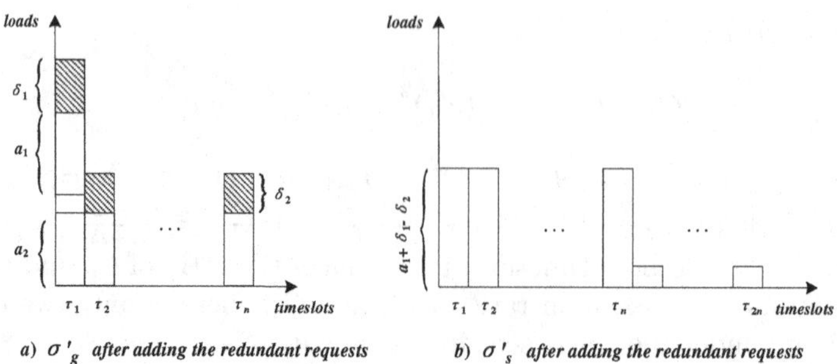

Fig. 4. Illustration of the proof of Lemma 5

for $i = 2, \cdots, n$. Thus there are at least $a_1 + \delta_1 - \delta_2 - 1$ requests which can only be assigned in τ_1. Since there is no balance schedule in the special schedule σ_s, τ_1 must be the timeslot with the maximal load in σ_s, otherwise we can form a balance schedule since $\sum_{i=1}^{n}(l_i - 1) - (a_1 + \delta_1 - \delta_2 - 1)$ requests can be split arbitrarily to any timeslots to make σ_g worse. Thus

$$a_1 + \delta_1 + \delta_2 \geq \frac{na_2 + n\delta_2 + 1}{n-1} \tag{12}$$

Since $a_1 = \frac{na_2+1}{n-1}$,

$$\delta_1 \geq \frac{2n-1}{n-1}\delta_2 \tag{13}$$

For the electricity cost in σ_g,

$$C'_g = C_g + \left[(l_1 + \delta_1)^2 - l_1^2\right] + (n-1)\left[(l_2 + \delta_2)^2 - l_2^2\right] \tag{14}$$

For the electricity cost in σ_s,

$$C'_s = C_s + n\left[(a_1 + \delta_1 - \delta_2)^2 - a_1^2\right] \tag{15}$$

Let $\alpha_1 = \frac{(l_1+\delta_1)^2 - l_1^2}{(a_1+\delta_1-\delta_2)^2 - a_1^2} = \frac{2l_1\delta_1+\delta_1^2}{(a_1+\delta_1-\delta_2)^2-a_1^2}$, and $\alpha_2 = \frac{(l_2+\delta_2)^2-l_2^2}{(a_1+\delta_1-\delta_2)^2-a_1^2} = \frac{2l_2\delta_2+\delta_2^2}{(a_1+\delta_1-\delta_2)^2-a_1^2}$. Thus $\frac{C'_g}{C'_s} \leq \max\left\{\frac{C_g}{C_s}, \alpha_1, \alpha_2\right\}$, in which α_1 counts for the influence of the added requests in τ_1 while α_2 counts for that in τ_i, $i = 2, \cdots, n$.

According to inequality (10), $\alpha_1 \geq \alpha_2$. Thus only the requests which are assigned in τ_1 can make σ_g worse. □

Lemma 6. *If the regular requests cannot form a balance schedule, the competitive ratio of the Greedy algorithm is 2.*

Proof. Before the redundant requests are added, there exists a balance schedule. The corresponding electricity cost obtained by the Greedy algorithm and the special schedule are denoted as $\widetilde{C}_g$ and $\widetilde{C}_s$ respectively. According to Lemma 5, the redundant requests are assigned in τ_1 in σ_g under the worst scenario. Thus,

$$\begin{aligned}\rho_2 &= \frac{\widetilde{C}_g + (a_1 + a_2 + 1 + \delta)^2 - (a_1 + a_2 + 1)^2}{\widetilde{C}_{opt} + (a_1 + \delta)^2 - a_1^2} \\ &= \frac{\widetilde{C}_g + \delta^2 + 2\delta a_1 + 2(a_2 + 1)\delta}{\widetilde{C}_{opt} + \delta^2 + 2\delta a_1}\end{aligned} \tag{16}$$

Since $\frac{\widetilde{C}_g+\delta^2}{\widetilde{C}_{opt}+\delta^2} < \frac{\widetilde{C}_g}{\widetilde{C}_{opt}} = 2$ and $\frac{2\delta a_1+2(a_2+1)\delta}{2\delta a_1} = 1 + \frac{n-a_2-2}{na_2+1} \leq 2$, $\rho_2 \leq 2$. □

According to Lemmas 4 and 6, we obtain Theorem 1 as follows.

Theorem 1. *The Greedy algorithm for online scheduling problem for electricity cost in smart grid is 2-competitive.*

Theorem 2. *Greedy algorithm is the optimal algorithm for online scheduling problem for electricity cost in smart grid.*

Proof. To prove the theorem, it suffices to construct a power requests instance to make any online algorithm performs poorly which has no competitive ratio less than 2.

The first two requests r_1 and r_2 are released with available timeslots $I_1 = \{1, 3\}$ and $I_2 = \{2, 4\}$. Without loss of generality, we assign r_1 and r_2 in timeslots 1 and 2 respectively. After that, r_3 and r_4 are released consecutively with $I_3 = I_4 = \{1, 2\}$.

Case 1. r_3 and r_4 are assigned at the same timeslot, i.e. timeslot 1. In this case, no more power requests are released. The electricity cost in this case is $C = 3^2 + 1 = 10$. In the optimal schedule, r_1, r_2, r_3 and r_4 are assigned in timeslots 3, 4, 1 and 2 respectively, $C_{opt} = 4$. In this case,

$$C/C_{opt} = 2.5 \tag{17}$$

Case 2. r_3 and r_4 are assigned in different timeslots, i.e. r_3 in timeslot 1 and r_4 in timeslot 2. In this case, r_5 and r_6 with $I_5 = I_6 = \{1\}$ are released. The electricity cost in this case is $C = 4^2 + 2^2 = 20$. In the optimal schedule, the operator serves r_5 and r_6 in timeslot 1, r_3 and r_4 in timeslot 2. Besides, r_1 and r_2 are assigned in timeslots 3 and 4. $C_{opt} = 2^2 + 2^2 + 2 = 10$. In this case,

$$C/C_{opt} = 2 \tag{18}$$

From the two cases above, it can be concluded that there is no online algorithm which can achieve a competitive ratio less than 2. Thus the Greedy algorithm is the optimal algorithm for online scheduling problem for electricity cost in smart grid. □

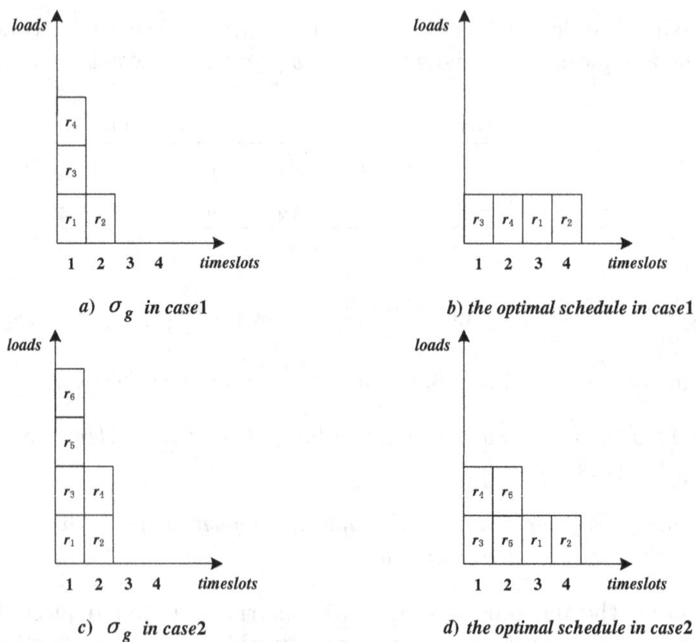

Fig. 5. Illustration of the proof of Theorem 2

4 Conclusions and Remarks

In this paper we investigate an online scheduling problem in the smart grid to minimize the total electricity cost, in which the power requirement and duration of requests are both unit-size. The requests are released online-over-list which can be assigned into serval integral timeslots. We propose a 2-competitive greedy algorithm which the best possible algorithm for this problem. The competitive analysis is performed by deriving the worst scenario of the schedule obtained by our algorithm. As we assumed that the power requirement and service duration of requests are both unit-size, an obvious research direction is to extend this research when requests have arbitrary power requirement and service duration. Besides, it is considerable to analyze how the scenario where the information is partially predictable can help the operation in smart grid.

Acknowledgements. This work was partially supported by the National Natural Science Foundation of China under Grants 71172189, 71071123 and 61221063, Program for Changjiang Scholars and Innovative Research Team in University (No. IRT1173), New Century Excellent Talents in University (NCET-12-0824), and the Fundamental Research Funds for the Central Universities.

References

1. Azar, Y.: On-line load balancing. In: Fiat, A. (ed.) Online Algorithms 1996. LNCS, vol. 1442, pp. 178–195. Springer, Heidelberg (1998)
2. Borodin, A., El-Yaniv, R.: Online Computation and Competitive Analysis, ch. 1. Cambridge University Press, Cambridge (1998)
3. Burcea, M., Hon, W.-K., Liu, H.-H., Wong, P.W.H., Yau, D.K.Y.: Scheduling for electricity cost in smart grid. In: Widmayer, P., Xu, Y., Zhu, B. (eds.) COCOA 2013. LNCS, vol. 8287, pp. 306–317. Springer, Heidelberg (2013)
4. Europen Commission: Europen smart grids technology platform (2006)
5. Georgiadis, G., Papatriantafilou, M.: A greedy algorithm for the unforecasted energy dispatch problemwith storage in smart grids. In: Proceedings of the Fourth International Conference on Future Energy Systems, pp. 273–274. ACM (2013)
6. Hamilton, K., Gulhar, N.: Taking demand response to the next level. IEEE Power Energ. Mag. **8**(3), 60–65 (2010)
7. Ipakchi, A., Albuyeh, F.: Grid of the future. IEEE Power Energ. Mag. **7**(2), 52–62 (2009)
8. Koutsopoulos, I., Tassiulas, L.: Control and optimization meet the smart power grid: scheduling of power demands for optimal energy management. In: Proc. eEnergy, pp. 41–50 (2011)
9. Koutsopoulos, I., Tassiulas, L.: Optimal control policies for power demand scheduling in the smart grid. IEEE J. Sel. Areas Commun. **30**(6), 1049–1060 (2012)
10. Logenthiran, T., Srinivasan, D., Shun, T.Z.: Demand side management in smart grid using heuristic optimization. IEEE Trans. Smart Grid **3**(3), 1244–1252 (2012)
11. Lui, T., Stirling, W., Marcy, H.: Get smart. IEEE Power Energ. Mag. **8**(3), 66–78 (2010)
12. Lu, L., Tu, J., Chau, C.K., Chen, M., Lin, X.: Online energy generation scheduling for microgrids with intermittent energy sources and co-generation, vol. 41, pp. 53–66. ACM (2013)
13. Narayanaswamy, B., Garg, V.K., Jayram, T.S.: Online optimization for the smart (micro) grid. In: Proceedings of the 3rd International Conference on Future Energy Systems: Where Energy, Computing and Communication Meet. ACM (2012)
14. Sousa, T., Morais, H., Vale, Z., Castro, R.: A multi-objective optimization of the active and reactive resource scheduling at a distribution level in a smart grid context. Energy **85**, 236–250 (2015)
15. Shahidehpour, M., Yamin, H., Li, Z.: Market Operations in Electric Power Systems: Forecasting, Scheduling, and Risk Management, 1st edn. Wiley-IEEE Press, New York (2002)

Proportional Cost Buyback Problem with Weight Bounds

Yasushi Kawase[1(✉)], Xin Han[2], and Kazuhisa Makino[3]

[1] Department of Social Engineering, Tokyo Institute of Technology, Tokyo, Japan
kawase.y.ab@m.titech.ac.jp

[2] Software School, Dalian University of Technology, Dalian, China
hanxin@dlut.edu.cn

[3] Research Institute for Mathematical Science, Kyoto University, Kyoto, Japan
makino@kurims.kyoto-u.ac.jp

Abstract. In this paper, we study the *proportional cost buyback problem*. The input is a sequence of elements $e_1, e_2, \ldots, e_n$, each of which has a weight $w(e_i)$. We assume that weights have an upper and a lower bound, i.e., $l \leq w(e_i) \leq u$ for any i. Given the ith element e_i, we either accept e_i or reject it with no cost, subject to some constraint on the set of accepted elements. During the iterations, we could cancel some previously accepted elements at a cost that is proportional to the total weight of them. Our goal is to maximize the profit, i.e., the sum of the weights of elements kept until the end minus the total cancellation cost occurred. We consider the matroid and unweighted knapsack constraints. For either case, we construct optimal online algorithms and prove that they are the best possible.

1 Introduction

In this paper, we study *proportional cost buyback problem*. The buyback problem was introduced in [3,7] as a model of selling advertisement online with a buyback option. The input for the problem is a sequence of elements (or bidders) $e_1, e_2, \ldots, e_n$, each of which has a weight (or bid) $w(e_i)$. Given the ith element e_i, we either accept e_i or reject it with no cost, subject to some constraint on the set of accepted elements. During the iterations, we could cancel some of the previously accepted elements by paying cancellation fees. Our goal is to maximize the profit, i.e., the sum of the weights of elements kept until the end minus the total cancellation cost occurred.

Examples of cancellation costs are compensatory payment, paperwork cost, and shipping charge. Some are proportional to the total weight of canceled elements, and some depend only on the number of canceled elements. In this paper, we consider the former case.

Related Work. Before studying the buyback problem, the removal online problem, i.e., the buyback problem without cancellation cost, was investigated

Z. Lu et al. (Eds.): COCOA 2015, LNCS 9486, pp. 794–808, 2015.
DOI: 10.1007/978-3-319-26626-8_59

[6,8,10,11,15]. For example, Iwama and Taketomi [12] showed that the unweighted knapsack problem is $\frac{1+\sqrt{5}}{2} \approx 1.618$-competitive, where knapsack problem is called unweighted if the value of each item is equal to its size. We remark that the problem has unbounded competitive ratio, for weighted or unremovable cases [12,13].

The buyback problem with proportional cost was studied in [1–5,7,9]. In this model, the cancellation cost of each element e_i is proportional to its weight, i.e., it is $f \cdot w(e_i)$, where $f > 0$ is a fixed constant called *buyback factor*. Babaioff *et al.* [3] and Constantin *et al.* [7] showed that the problem is $(1+2f+2\sqrt{f(1+f)})$-competitive with the single-element constraint. Babaioff *et al.* [3] also showed that the problem has the competitive ratio $(1+2f+2\sqrt{f(1+f)})$ with the matroid constraint. Ashwinkumar and Kleinberg [2] showed that the buyback problem for the matroid constraint is randomized $-W(\frac{-1}{e(1+f)})$ competitive against an oblivious adversary. Here W denote Lambert's W function, defined as the inverse of the function $g(z) = ze^z$ for $z \leq -1$.

Ashwinkumar [1] extended their results to the intersection of k matroids and showed that it is $k(1+f)\left(1+\sqrt{1-\frac{1}{k(1+f)}}\right)^2$-competitive. Babaioff *et al.* [3,4] also studied the buyback problem with the weighted knapsack constraint. They showed that if the largest element is of size at most γ, where $0 < \gamma < 1$, then the competitive ratio is $(1+2f+2\sqrt{f(1+f)})$ with respect to the optimum solution for the knapsack problem with capacity $(1-2\gamma)$. Han *et al.* [9] studied the buyback problem with the unweighted knapsack constraint. They proved that the problem is $\max\left\{2, \frac{1+f+\sqrt{f^2+2f+5}}{2}\right\}$-competitive.

Unit cost buyback problem, i.e., the buyback problem with fixed cancellation cost for each canceled element, was introduced by Han *et al.* [9]. In their model, the cancellation cost of each element is a fixed constant $c > 0$. They showed a tight competitive ratio for the unweighted knapsack constraint case with the assumption that every element has a weight at least c, since in many applications, the cancellation cost is not higher than its bid. Kawase *et al.* [14] studied the unit cost buyback problem with weight bounds. Let $u > l > 0$ be an upper and a lower bound of element weights, i.e., $l \leq w(e_i) \leq u$ for any element e_i. They provide optimal competitive ratio for the single-element, the matroid, or the unweighted knapsack constraint with parameters l and u.

Our Results. In this paper, we study the proportional cost buyback problem for the single-element constraint, the matroid constraint, or the unweighted knapsack constraint with *weight bounds.* In this model, the cancellation cost of each element e_i is proportional to its weight, i.e., it is $f \cdot w(e_i)$, where $w(e_i)$ denotes the weight of e_i and $f > 0$ is the buyback factor.

Let $u > l > 0$ be an upper and a lower bound of element weights, i.e., $l \leq w(e_i) \leq u$ for any element e_i. We show that the proportional cost buyback problem for the single-element or matroid constraint has the competitive ratio $\nu(l, u, f)$ which is defined below.

Let $\phi_\rho(n)$ satisfy the following recurrence relation

$$\phi_\rho(n) = \begin{cases} l, & (n=1), \\ \rho(\phi_\rho(n-1) - f \cdot \sum_{i=1}^{n-2} \phi_\rho(i)) & (n = 2, 3, \dots). \end{cases}$$

Thus, we have

$$\begin{cases} \phi_\rho(1) = l, \ \phi_\rho(2) = l\rho, \\ \phi_\rho(n+1) - (\rho+1)\phi_\rho(n) + \rho(1+f)\phi_\rho(n-1) = 0 & (n = 2, 3, \dots). \end{cases}$$

Define $\nu(l, u, f)$ the smallest value ρ (> 1) and satisfies that

$$\phi_\rho(n_\rho) = u \ \text{ where } \ n_\rho = \min\{n \in \mathbb{Z}_{++} \mid \phi_\rho(n) \geq \phi_\rho(n+1)\}.$$

Due to the space limitation, the proof of the existence of n_ρ and $\nu(l, u, f)$ is omitted. For example, the competitive ratio $\nu(l, u, f)$ for $(l, u) = (0.5, 1.0)$ and $(u, f) = (1.0, 0.2)$ are given in Fig. 1.

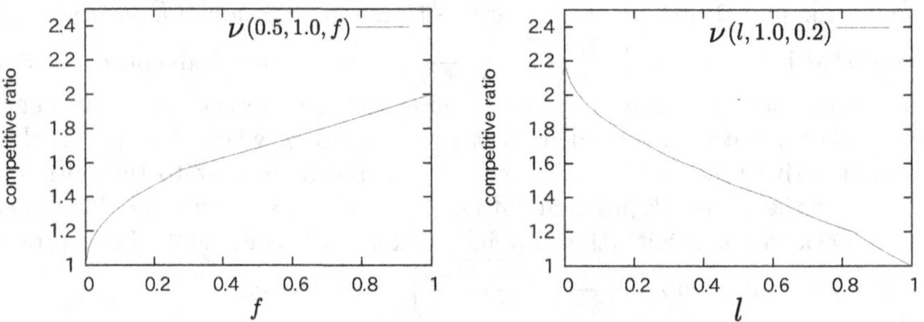

Fig. 1. The competitive ratio $\nu(l, u, f)$ of the single-element or matroid constraint, where the left and right figures represent the ratios for $(l, u) = (0.5, 1.0)$ and $(u, f) = (1.0, 0.2)$, respectively. We can see that the ratio $\nu(l, u, f)$ is monotone increasing for f and monotone decreasing for l.

We mention that the competitive ratio is independent of positive scaling for l and u, i.e., we have $\nu(l, u, f) = \nu(l/\alpha, u/\alpha, f)$ for any $u > l > 0$ and $\alpha > 0$. Moreover, we have $\lim_{u/l \to \infty} \nu(l, u, f) = 1 + 2f + 2\sqrt{f(1+f)}$, which corresponds to the competitive ratio of the single-element case without upper and lower bounds of weights provided by Babaioff *et al.* [3] and Constantin *et al.* [7].

For the unweighted knapsack constraint, let $u = 1$ and let ζ be a function defined by

$$\zeta(l, f) = \begin{cases} \nu(l, 1, f) & (l > 1/2, \ f < (1-l)/l), \\ 2 & \left({0 \leq l < 1/2, \ 0 < f \leq \max\{1/2, \frac{4l-1}{2l}\} \atop \text{or } \ l = 1/2, \ 1/4 \leq f \leq 1}\right), \\ \frac{\sqrt{9f^2+8f+8}+3f}{2} & (l = 1/2, \ 0 < f \leq 1/4), \\ \frac{1+f+\sqrt{f^2+2f+5}}{2} & (0 \leq l \leq 1/3, \ 1/2 \leq f \leq \frac{l^2-3l+1}{l(1-l)}), \\ \frac{fl+\sqrt{f^2l^2+4l}}{2l} & (\max\{\frac{l^2-3l+1}{l(1-l)}, \frac{4l-1}{2l}\} \leq f \leq \frac{1-l}{l}), \\ 1/l & (f \geq (1-l)/l) \end{cases}$$

when $l \in [0, 1]$ and $f > 0$. Then, we show that the problem is $\zeta(l, f)$-competitive when the size of the knapsack is one. For example, the competitive ratios $\zeta(l, f)$ for $l = 1/2$ and $f = 1/2$ are given in Fig. 2. We note that the competitive ratio $\zeta(l, f)$ consists of 6 parts, which depends on l and f (see Fig. 3). On the other hand, the competitive ratio for the unit cost case consists of infinitely many parts (Kawase *et al.* [14]).

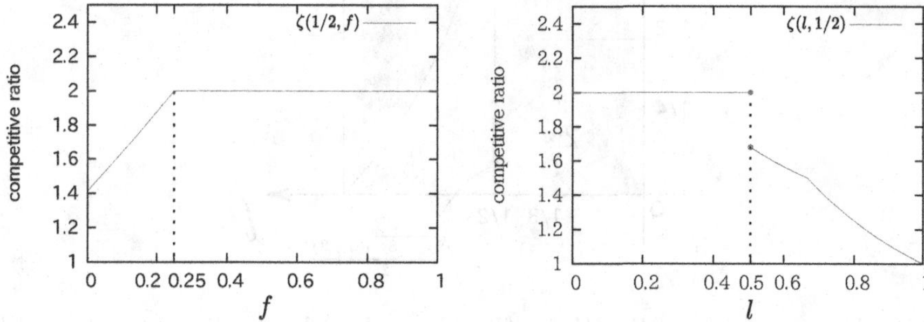

Fig. 2. The competitive ratio $\zeta(l, f)$ of the unweighted knapsack constraint, where the left and right figures represent the ratios for $l = 1/2$ and $f = 1/2$, respectively.

The rest of the paper is organized as follows. In Sect. 2 we formally define the proportional cost buyback problem. In Sect. 3, we handle the matroid constraint, and in Sect. 4, we discuss the knapsack constraint.

2 Preliminaries

In this section, we formally define the proportional cost buyback problem. Let $(E, \mathcal{I})$ be an independence system, i.e., E is a finite set and $\mathcal{I}$ is a family of subsets of E, and $J \subseteq I \in \mathcal{I} \Rightarrow J \in \mathcal{I}$. Elements from $E = \{e_1, \ldots, e_n\}$ are presented to an algorithm in a sequential manner, and when an element e_k is presented, it must be accepted or rejected immediately. Each element e_i is associated with a weight $w(e_i)$. During the iterations, the algorithm could cancel some of the previously accepted elements.

Let B_k be the set of selected elements at the end of kth round. Then $B_k \subseteq B_{k-1} \cup \{e_k\}$ and $B_k \in \mathcal{I}$. The algorithm must run based only on the weights $w(e_i)$ $(1 \leq i \leq k)$ and the feasibility of subsets $T \subseteq \{e_1, \ldots, e_k\}$. The utility of the algorithm is the total weight of the accepted elements minus the penalty paid to the canceled elements. The cancellation cost for each element e_i is proportional to its weight, i.e., $f \cdot w(e_i)$, where f is the buyback factor [3].

Let $B = B_n$ be the final set held by the algorithm and $R = (\bigcup_i B_i) \setminus B$ be the set of elements canceled. Then the utility of the algorithm is defined as $\sum_{e \in B} w(e) - f \cdot \sum_{e \in R} w(e)$.

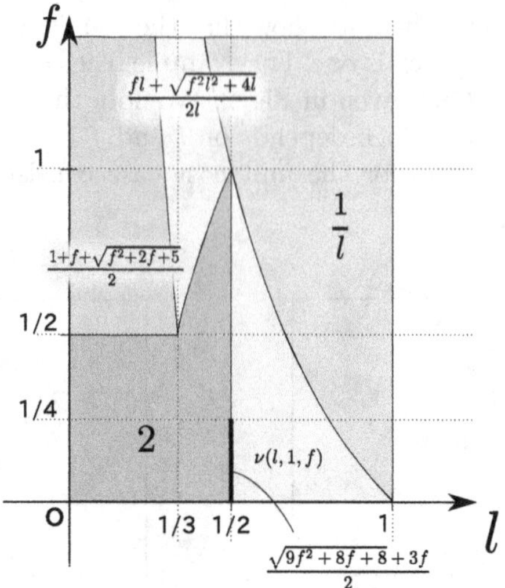

Fig. 3. The competitive ratio $\zeta(l, f)$ of the unweighted knapsack constraints, which consists of 6 parts depending on l and f.

3 The Matroid Constraint

In this section, we consider the matroid case with upper and lower bound of weights, where the constraint $\mathcal{I}$ is an arbitrary independence family of matroid $M = (E, \mathcal{I})$, and each element e_i has weight $w(e_i)$ such that $l \le w(e_i) \le 1$.

3.1 An Optimal Online Algorithm

We show that the competitive ratio for this problem is at most $\nu(l, u, f)$ by proving that Algorithm 1 is $\nu(l, u, f)$-competitive. We define $\phi(n)$ as

$$\phi(n) := \phi_{\nu(l,u,f)}(n)$$

and n_* as

$$n_* := n_{\nu(l,u,f)} \quad (= \min\{n \in \mathbb{Z}_{++} \mid \phi(n) \ge \phi(n+1)\})$$

where $\mathbb{Z}_{++}$ denotes the set of nonnegative integers. Here we note that $l = \phi(1) < \phi(2) < \cdots < \phi(n_*) = u$. In the algorithm, let e_i be the element given in the ith round, and let B_i be the set of selected elements at the end of the ith round. We denote by $w(B_i)$ the total value of elements in B_i. If e_i is accepted, $(k_i - 1)$ means how many elements the algorithm canceled to accept it, and if rejected $k_i = 0$. $p(e_i)$ denote the element canceled when the algorithm accepts e_i, if exists. For example, if the input sequence is $\{e_1, e_2, e_3, e_4, e_5\}$ and the algorithm runs as $B_1 = \{e_1\}$, $B_2 = \{e_1, e_2\}$, $B_3 = \{e_3, e_2\}$, $B_4 = \{e_3, e_2\}$, $B_5 = \{e_5, e_2\}$ then $k_1 = 1$, $k_2 = 1$, $k_3 = 2$, $k_4 = 0$, $k_5 = 3$ and $p(e_3) = e_1$, $p(e_5) = e_3$.

Algorithm 1. Matroid Case

```
B_0 ← ∅
for all elements e_i, in order of arrival, do
    if B_{i-1} ∪ {e_i} ∈ I then B_i ← B_{i-1} ∪ {e_i}, k_i ← 1
    else let e_j be the element with smallest value (φ(k_j+1)/φ(k_j)) · w(e_j) such that B_{i-1} ∪ {e_i} \ {e_j} ∈ I
        if w(e_i) > (φ(k_j+1)/φ(k_j)) · w(e_j) then B_i ← B_{i-1} ∪ {e_i} \ {e_j}, k_i ← k_j + 1, p(e_i) ← e_j
        else B_i ← B_{i-1} and k_i ← 0
end for
```

Theorem 1. *The online Algorithm 1 is $\nu(l, u, f)$-competitive.*

Proof. We first remark that we have $k_i < n_*$ for each i since $w(e_i) > \phi(k_i)$ if $k_i \geq 2$ and $\phi(n_*) = u \geq w(e_i)$. Let OPT denote an optimal solution for the offline problem whose input sequence is $e_1, \ldots, e_n$, and let $B_n = \{e_{b_1}, e_{b_2}, \ldots, e_{b_h}\}$. If each element e_i has a weight

$$w'(e_i) = \begin{cases} \frac{\phi(k_i+1)}{\phi(k_i)} \cdot w(e_i) & (k_i \geq 1), \\ w(e_i) & (k_i = 0) \end{cases}$$

then B_n is a maximum-weight base of the matroid M because Algorithm 1 can be seen as a matroid greedy algorithm for the weight w'. Thus, $w(OPT) \leq \sum_{i=1}^{h} w'(e_{b_i}) = \sum_{i=1}^{h} \frac{\phi(k_{b_i}+1)}{\phi(k_{b_i})} w(e_{b_i})$ since $w'(e_i) \geq w(e_i)$ for each element e_i. For each $e_i \in B_n$, let $R_i = \{r_1^i, r_2^i, \ldots, r_{k_i-1}^i\}$ with $r_{k_i-1}^i = p(e_i)$ and $r_j^i = p(r_{j+1}^i)$ for $j = 1, 2, \ldots, k_i - 2$, and let $R = \bigcup_{e_i \in B_n} R_i$ be the elements canceled by the algorithm. Then $w(r_{j+1}^i) > \frac{\phi(j+1)}{\phi(j)} w(r_j^i)$ for $j = 1, 2, \ldots, k_i - 2$, and $w(e_i) > \frac{\phi(k_i)}{\phi(k_i-1)} w(r_{k_i-1}^i)$. Thus, for $j = 1, 2, \ldots, k_i - 2$, we obtain

$$w(r_j^i) < \frac{\phi(j)}{\phi(j+1)} \cdot \frac{\phi(j+1)}{\phi(j+2)} \cdot \cdots \cdot \frac{\phi(k_i-1)}{\phi(k_i)} w(e_i) = \frac{\phi(j)}{\phi(k_i)} w(e_i).$$

Therefore, the competitive ratio is at most

$$\begin{aligned} \frac{w(OPT)}{w(B_n) - f \cdot w(R)} &\leq \frac{\sum_{i=1}^{h} \frac{\phi(k_{b_i}+1)}{\phi(k_{b_i})} \cdot w(e_{b_i})}{\sum_{i=1}^{h} \left(w(e_{b_i}) - f \cdot \sum_{j=1}^{k_{b_i}-1} \frac{\phi(j)}{\phi(k_{b_i})} \cdot w(e_{b_i}) \right)} \\ &\leq \max_{i=1}^{h} \frac{\frac{\phi(k_{b_i}+1)}{\phi(k_{b_i})} \cdot w(e_{b_i})}{w(e_{b_i}) - f \cdot \sum_{j=1}^{k_{b_i}-1} \frac{\phi(j)}{\phi(k_{b_i})} \cdot w(e_{b_i})} \\ &= \max_{i=1}^{h} \frac{\phi(k_{b_i}+1)}{\phi(k_{b_i}) - f \cdot \sum_{j=1}^{k_{b_i}-1} \phi(j)} = \nu(l, u, f). \end{aligned}$$

□

3.2 Lower Bound

In this subsection, we show that $\nu(l, u, f)$ is also a lower bound for the competitive ratio of the problem even if the constraint is the single-element one. We note that the single-element constraint is a special case of matroid constraint, i.e., the uniform matroid of rank 1.

Theorem 2. *There exists no online algorithm with competitive ratio less than $\nu(l, u, f)$ for the proportional cost buyback problem with the single-element constraint.*

Proof. Let A denote an online algorithm chosen arbitrarily. Our adversary requests the sequence of elements whose weights are

$$\phi(1), \phi(2), \dots, \phi(n_*), \tag{1}$$

until A rejects some element in (1).

If A rejects the element with weight $\phi(1)$, then the competitive ratio of A becomes infinite. On the other hand, if A rejects the element with weight $\phi(k+1)$ for some $k > 1$, A cancels $k-1$ elements with weights $\phi(1), \phi(2), \dots, \phi(k-1)$ and the competitive ratio is at least

$$\frac{\phi(k+1)}{\phi(k) - f \cdot \sum_{i=1}^{k-1} \phi(i)} = \nu(l, u, f). \tag{2}$$

Finally, if A accepts all the elements in (1), then the competitive ratio is at least

$$\frac{\phi(n_*)}{\phi(n_*) - f \cdot \sum_{i=1}^{n_*-1} \phi(i)} = \frac{\phi(n_*)}{\phi(n_*+1)} \cdot \nu(l, u, f) \geq \nu(l, u, f).$$ □

4 The Unweighted Knapsack Constraint

In this section, we consider the unweighted knapsack case with upper and lower bound of weight, where the constraint $\mathcal{I} = \{T \subseteq E \mid \sum_{e \in T} w(e) \leq 1\}$ is the set of elements whose total weight is at most one, and each element e_i has weight $w(e_i)$ such that $l \leq w(e_i) \leq 1$.

4.1 Optimal Online Algorithms

In this subsection, we show the upper bound on the competitive ratio for the problem.

Theorem 3. *There exists a $\zeta(l, f)$-competitive algorithm for the proportional cost buyback problem with the unweighted knapsack constraint.*

We consider four cases: the case $f \geq \frac{1-l}{l}$ or $l = 1/2,\ f \geq 1/4$ in Theorem 4, the case $l > 1/2$ in Theorem 5, the case $l = 1/2$ and $0 < f < 1$ in Theorem 6, and the remaining case $l < 1/2$ and $f < \frac{1-l}{l}$ in Theorem 7.

Theorem 4. *There exists a $1/l$-competitive algorithm for the proportional cost buyback problem with the unweighted knapsack constraint.*

Proof. Consider an online algorithm that takes the first element e_1 and rejects the remaining elements. Since $w(e_1) \geq l$ and the optimal offline value is at most 1, the competitive ratio is at most $1/l$. □

Theorem 5. *There exists a $\nu(l, 1, f)$-competitive algorithm for the proportional cost buyback problem with the unweighted knapsack constraint if $l > 1/2$.*

Proof. This follows from Theorem 1 since we can hold only one element in the knapsack at the same time. □

For $l = 1/2$ and $0 < f < 1/4$, we use Algorithm 2. Let e_i be the element given in the ith round. Define by B_i the set of selected elements at the end of ith round, and by $w(B_i)$ the total weight in B_i. Note that, we can keep two elements only when their sizes are 1/2.

Algorithm 2. Removal at most Twice

1: $B_0 \leftarrow \emptyset$
2: **for all** elements e_i, in order of arrival, **do**
3:   **if** $w(B_{i-1}) + w(e_i) \leq 1$ **then**
4:     $B_i \leftarrow B_{i-1} \cup \{e_i\}$
5:     **if** $|B_i| = 2$ **then** STOP
6:     **else if** $w(B_i) \geq \frac{\sqrt{9f^2+8f+8}-3f}{4(1+f)}$ **then** STOP
7:   **else if** $w(e_i) \geq \frac{\sqrt{9f^2+8f+8}-f}{4}$ **then** $B_i \leftarrow \{e_i\}$ and STOP
8:   **else if** $w(e_i) = 1/2$ **then** $B_i \leftarrow \{e_i\}$
9:   **else** $B_i \leftarrow B_{i-1}$
10: **end for**

Here STOP denotes that the algorithm rejects the elements after this round.

Theorem 6. *The online Algorithm 2 is $\frac{\sqrt{9f^2+8f+8}+3f}{2}$-competitive for the proportional cost buyback problem with the unweighted knapsack constraint if $l = 1/2$ and $0 < f < 1/4$.*

Proof. If the algorithm stops at the fifth line, then the algorithm canceled nothing or one element with size at most $\frac{\sqrt{9f^2+8f+8}-3f}{4(1+f)}$. Thus, the competitive ratio is at most

$$\frac{1}{1 - f \cdot \frac{\sqrt{9f^2+8f+8}-3f}{4(1+f)}} \leq \frac{1}{\frac{\sqrt{9f^2+8f+8}-f}{4} - f \cdot \left(\frac{\sqrt{9f^2+8f+8}-3f}{4(1+f)} + \frac{1}{2}\right)}$$
$$= \frac{\sqrt{9f^2+8f+8}+3f}{2}.$$

where the first inequality holds by $\sqrt{9f^2+8f+8}-f \leq 4$ for $0 < f < 1/4$.

If the algorithm stops at the sixth line, then the competitive ratio is at most

$$\frac{1}{\frac{\sqrt{9f^2+8f+8}-3f}{4(1+f)}} = \frac{\sqrt{9f^2+8f+8}+3f}{2}$$

since the algorithm has never canceled elements.

If the algorithm stops at the seventh line, then the competitive ratio is at most

$$\frac{1}{\frac{\sqrt{9f^2+8f+8}-f}{4} - f \cdot \left(\frac{\sqrt{9f^2+8f+8}-3f}{4(1+f)} + \frac{1}{2}\right)} = \frac{\sqrt{9f^2+8f+8}+3f}{2}$$

since it cancels at most two elements with size $1/2$ and smaller than $\frac{\sqrt{f^2-2f+2}-f}{2}$.

If the algorithm has never stopped (at the fifth, sixth, or seventh line), then every element has size smaller than $\frac{\sqrt{9f^2+8f+8}-f}{4}$ and there exists at most one element with size $1/2$. Thus, the competitive ratio is at most

$$\frac{\frac{\sqrt{9f^2+8f+8}-f}{4}}{\frac{1}{2} - f \cdot \left(\frac{\sqrt{9f^2+8f+8}-3f}{4(1+f)}\right)} = \frac{\sqrt{9f^2+8f+8}+3f}{2}$$

since it cancels at most one element with size at most $(\sqrt{9f^2+8f+8}-3f)/(4(1+f))$. □

In the rest of this subsection, we would like to show that Algorithm 3 is $\zeta(l,f)$-competitive for $f < \frac{1-l}{l}$ and $l < 1/2$. The main ideas of the algorithm are: (i) it rejects elements (with no cost) many times, but in at most one round, it cancels some elements from the knapsack. (ii) some elements are canceled from the knapsack, only when the total value in the resulting knapsack gets high enough to guarantee the optimal competitive ratio. We mention that the algorithm is the same as Algorithm 1 in [9] when $f \leq \frac{l^2-3l+1}{l(1-l)}$ and $l < 1/3$, that is $\zeta(l,f) = \max\{2, \frac{1+f+\sqrt{f^2+2f+5}}{2l}\}$.

Let e_i be the element given in the ith round. Define by B_{i-1} the set of elements in the knapsack at the beginning of ith round, and by $w(B_{i-1})$ the total weight in B_{i-1}.

Lemma 1. *If $w(B_{i-1}) + w(e_i) > 1$ and some $B'_{i-1} \subseteq B_{i-1}$ satisfies $\zeta(l,f) \cdot w(B_{i-1}) < w(B'_{i-1}) + w(e_i) \leq 1$, then the sixth line of Algorithm 3 is executed in the ith round.*

Proof. We consider the following three cases.

Case 1: $f \leq \frac{l^2-3l+1}{l(1-l)}$ and $l \leq 1/3$. It holds by Han *et al.* [9], Lemma 2.

Algorithm 3.

$B_0 = \emptyset$
for all elements e_i, in order of arrival, **do**
  if $w(B_{i-1}) + w(e_i) \leq 1$ **then**
    $B_i \leftarrow B_{i-1} \cup \{e_i\}$
    if $w(B_i) \geq 1/\zeta(l, f)$ **then** STOP
  else if $\exists B'_{i-1} \subseteq B_{i-1}$ s.t. $\frac{1}{\zeta(l,f)} + f \cdot (w(B_{i-1}) - w(B'_{i-1})) < w(B'_{i-1}) + w(e_i) \leq 1$
    then $B_i \leftarrow B'_{i-1} \cup \{e_i\}$ and STOP
  else $B_i \leftarrow B_{i-1}$
end for

Here STOP denotes that the algorithm rejects the elements after this round.

Case 2: $f \leq \frac{4l-1}{2l}$ and $1/3 \leq l < 1/2$. Since $\zeta(l, f) \cdot w(B_{i-1}) < w(B'_{i-1}) + w(e_i)$, $w(B_{i-1}) \geq l$ and $\zeta(l, f) = 2$ we obtain

$$
\begin{aligned}
&\frac{1}{\zeta(l,f)} + f \cdot (w(B_{i-1}) - w(B'_{i-1})) \\
&\quad = \frac{1}{2} + f \cdot (w(B_{i-1}) - w(B'_{i-1})) \\
&\quad \leq \frac{w(B'_{i-1}) + w(e_i)}{4l} + f \cdot \left(\frac{w(B'_{i-1}) + w(e_i)}{2} - w(B'_{i-1}) \right) \\
&\quad = \left(\frac{1}{4l} - \frac{f}{2} \right) w(B_{i-1'}) + \left(\frac{1}{4l} + \frac{f}{2} \right) w(e_i).
\end{aligned}
$$

As $f \leq \frac{4l-1}{2l}$, we have $\frac{1}{4l} - \frac{f}{2} \leq \frac{1}{4l} + \frac{f}{2} \leq 1$.

Case 3: $\max\{\frac{l^2-3l+1}{l(1-l)}, \frac{4l-1}{2l}\} \leq f \leq (1-l)/l$. Since $\zeta(l, f) \cdot w(B_{i-1}) < w(B'_{i-1}) + w(e_i)$ and $w(B_{i-1}) \geq l$, we obtain

$$
\begin{aligned}
&\frac{1}{\zeta(l,f)} + f \cdot (w(B_{i-1}) - w(B'_{i-1})) \\
&\quad \leq \frac{w(B'_{i-1}) + w(e_i)}{l \cdot \zeta^2(l,f)} + f \cdot \left(\frac{w(B'_{i-1}) + w(e_i)}{\zeta(l,f)} - w(B'_{i-1}) \right) \\
&\quad = \left(\frac{1 + l \cdot f \cdot \zeta(l,f)}{l \cdot \zeta^2(l,f)} - f \right) w(B'_{i-1}) + \left(\frac{1 + l \cdot f \cdot \zeta(l,f)}{l \cdot \zeta^2(l,f)} \right) w(e_i).
\end{aligned}
$$

As $l \cdot \zeta^2(l, f) = 1 + l \cdot f \cdot \zeta(l, f)$ by the definition of $\zeta(l, f)$, we have

$$
\frac{1 + l \cdot f \cdot \zeta(l,f)}{l \cdot \zeta^2(l,f)} - f \leq \frac{1 + l \cdot f \cdot \zeta(l,f)}{l \cdot \zeta^2(l,f)} = 1
$$

□

Let OPT denote an optimal solution for the offline problem whose input sequence is $e_1, \ldots, e_i$.

Lemma 2. *If* $w(B_i) < 1/\zeta(l, f)$ *then we have* $|OPT \setminus B_i| \leq 1$.

Proof. B_i contains all the elements at most 1/2 seen so far, since $w(B_i) < 1/\zeta(l,f) \leq 1/2$. Thus, any element $u \in OPT \setminus B_i$ has size greater than $1 - 1/\zeta(l,f) \geq 1/2$. Therefore, $|OPT \setminus B_i| \leq 1$ holds since the capacity of the knapsack is 1. □

Theorem 7. *The online Algorithm 3 is $\zeta(l,f)$-competitive.*

Proof. Suppose that the condition in the sixth line is satisfied in round k. Then it holds that $\frac{1}{\zeta(l,f)} + f \cdot (w(B_{k-1}) - w(B'_{k-1})) < w(B'_{k-1}) + w(e_k) = w(B_k)$. Since $w(B_j) = w(B_k)$ holds for all $j \geq k$, we have

$$\frac{w(OPT)}{w(B_i) - f \cdot (w(B_{k-1}) - w(B'_{k-1}))} \leq \frac{1}{w(B_k) - f \cdot (w(B_{k-1}) - w(B'_{k-1}))} < \zeta(l,f).$$

We next assume that the condition in sixth line has never been satisfied. If $w(B_i) \geq 1/\zeta(l,f)$, we have the competitive ratio $w(OPT)/w(B_i) \leq 1/w(B_i) \leq \zeta(l,f)$. On the other hand, if $w(B_i) < 1/\zeta(l,f)$, $|OPT \setminus B_i| = 0$ or 1 holds by Lemma 2. If $|OPT \setminus B_i| = 0$, we obtain the competitive ratio 1. Otherwise (i.e., $OPT \setminus B_i = \{e_l\}$ for some l), Lemma 1 implies that $\zeta(l,f) \cdot w(B_{l-1}) \geq w(B'_{l-1}) + w(e_l)$ for $B'_{l-1} = OPT \cap B_{l-1}$. Therefore, we obtain

$$\begin{aligned}\frac{w(OPT)}{w(B_i)} &\leq \frac{w(B'_{l-1}) + w(e_l) + w(B_i \setminus B_{l-1})}{w(B_{l-1}) + w(B_i \setminus B_{l-1})} \\ &\leq \max\left\{\frac{w(B'_{l-1}) + w(e_l)}{w(B_{l-1})}, \frac{w(B_i \setminus B_{l-1})}{w(B_i \setminus B_{l-1})}\right\} \leq \zeta(l,f).\end{aligned}$$

□

Before concluding this subsection, we remark that the condition in the sixth line can be checked efficiently.

Proposition 1. *We can check the condition in the sixth line of Algorithm 3 in $O(|B_{i-1}| + 2^{\zeta^2(l,f)})$ time.*

Proof. This proposition is almost the same as Proposition 5 in Han *et al.* [9].

Let $x = \frac{1}{1+f}\left(\frac{1}{\zeta(l,f)} + fw(B_{i-1}) - w(e_i)\right)$ and $y = 1 - w(e_i)$. Our goal is to decide whether there exists $B'_{i-1} \subseteq B_{i-1}$ such that $x < w(B'_{i-1}) \leq y$ in $O(|B_{i-1}| + 2^{\zeta^2(l,f)})$ time. As $w(B_{i-1}) < 1/\zeta(l,f)$, $w(e_i) \leq 1$, and $\zeta^2(l,f) \geq (1+f)\zeta(l,f) + 1$ by the definition of $\zeta(l,f)$, we get

$$\begin{aligned} y - x &= 1 - \frac{1}{\zeta(l,f)(1+f)} - \frac{f}{1+f}(w(e_i) + w(B_{i-1})) \\ &> 1 - \frac{1}{\zeta(l,f)(1+f)} - \frac{f}{1+f}(1 + \frac{1}{\zeta(l,f)}) \\ &= \frac{\zeta(l,f) - 1 - f}{\zeta(l,f)(1+f)} \geq \frac{\zeta(l,f)}{\zeta^2(l,f) - 1} - \frac{1}{\zeta(l,f)} = \frac{1}{\zeta^3(l,f) - \zeta(l,f)} \geq \frac{1}{\zeta^3(l,f)}. \end{aligned} \quad (3)$$

Let $B_{i-1} = \{b_1, b_2, \ldots, b_m\}$ satisfy $w(b_1) \geq \cdots \geq w(b_k) \geq y - x > w(b_{k+1}) \geq \cdots \geq w(b_m)$. Then we claim the existence of B'_{i-1} is equivalent to the existence of $A \subseteq \{b_1, b_2, \ldots, b_k\}$ such that $x - \sum_{i=k+1}^{m} w(b_i) < w(A) \leq y$. If such an A exists, then $B'_{i-1} = A \cup \{b_{k+1}, \ldots, b_l\}$ satisfies the conditions, where $l = \min\{l \geq k \mid w(A) + \sum_{i=k+1}^{l} w(b_i) > x\}$. On the other hand, if there exists B'_{i-1} such that $x < w(B'_{i-1}) \leq y$, then $A = B'_{i-1} \setminus \{b_{k+1}, \ldots, b_m\}$ satisfies $x - \sum_{i=k+1}^{m} w(b_i) < w(A) \leq y$.

Therefore we need to check the condition $x - \sum_{i=k+1}^{m} w(b_i) < w(A) \leq y$ for at most $2^k < 2^{\zeta^2(l,f)}$ subsets, since $k \leq w(B_{i-1})/(y-x) < \zeta^2(l,f)$ holds by $w(B_{i-1}) < 1/\zeta(l,f)$ and $y - x > 1/\zeta^3(l,f)$. Thus, we can check the condition in the sixth line in $O(|B_{i-1}| + 2^{\zeta^2(l,f)})$. □

4.2 Lower Bound for the Knapsack Constraint Case

In this subsection, we show a lower bound of the competitive ratio $\zeta(f,l)$ for the proportional cost buyback problem with knapsack constraint.

Theorem 8. *There exists no online algorithm with a competitive ratio less than $\zeta(l,f)$ for the proportional cost buyback problem with the unweighted knapsack constraint.*

Due to the space limitation, we prove only the case $l < 1/2$ and $f < \frac{1-l}{l}$. The proof of the other case is omitted due to space restriction. For the case $l < 1/2$ and $f < \frac{1-l}{l}$, we consider three subcases: the case $f \leq \max\{\frac{1}{2}, \frac{4l-1}{2l}\}$ in Theorem 9, the case $\max\{\frac{l^2-3l+1}{l(1-l)}, \frac{4l-1}{2l}\} \leq f \leq \frac{1-l}{l}$ in Theorem 10, and the case $\frac{1}{2} \leq f \leq \frac{l^2-3l+1}{l(1-l)}$ in Theorem 11.

Theorem 9. *If $l < 1/2$, there exists no online algorithm with a competitive ratio less than 2 for the proportional cost buyback problem with the unweighted knapsack constraint.*

Proof. We can obtain this theorem with the adversary in Han *et al.* [9] Theorem 1, Case 1.

Let A denote an online algorithm chosen arbitrarily. For a sufficiently small $\varepsilon\,(> 0)$, our adversary requests the sequence of elements whose weights are

$$\frac{1}{2} + \varepsilon,\ \frac{1}{2} + \frac{\varepsilon}{2},\ \ldots,\ \frac{1}{2} + \frac{\varepsilon}{\lceil 1/f \rceil + 1}, \tag{4}$$

until A rejects some element in (4). If A rejects the element with weight $\frac{1}{2} + \varepsilon$, then the adversary stops the input sequence. On the other hand, if it rejects the element with weight $\frac{1}{2} + \frac{\varepsilon}{k}$ for some $k > 1$, then the adversary requests an element with weight $\frac{1}{2} - \frac{\varepsilon}{k}$ and stops the input sequence.

We first note that algorithm A must take the first element, since otherwise the competitive ratio of A becomes infinite. After the first round, A always keeps exactly one element in the knapsack, since all the elements in (4) have weight larger than $\frac{1}{2}$ (i.e., a half of the knapsack capacity) and for any $j < k$ we have $(\frac{1}{2} + \frac{\varepsilon}{j}) + (\frac{1}{2} - \frac{\varepsilon}{k}) > 1$. This implies that A cancels the old element from the knapsack to accept a new element. If A rejects $\frac{1}{2} + \frac{\varepsilon}{k}$ for some $k > 1$, the competitive ratio is at least $1/\left(\frac{1}{2} + \varepsilon\right)$, which approaches 2 as $\varepsilon \to 0$. Finally, if A rejects no element in (4), then its profit is

$$\frac{1}{2} + \frac{\varepsilon}{\lceil 1/f \rceil + 1} - f \cdot \sum_{i=1}^{\lceil 1/f \rceil} \left(\frac{1}{2} + \frac{\varepsilon}{i}\right) \leq \frac{1}{2} + \varepsilon - f \cdot \frac{1}{f} \cdot \frac{1}{2} = \varepsilon$$

while the optimal profit for the offline problem is $\frac{1}{2} + \varepsilon$, which completes the proof. □

Theorem 10. *If* $\max\{\frac{l^2-3l+1}{l(1-l)}, \frac{4l-1}{2l}\} < f \leq \frac{1-l}{l}$ $(0 \leq l \leq 1/2)$ *there exists no online algorithm with a competitive ratio less than*

$$\frac{fl + \sqrt{f^2 l^2 + 4l}}{2l}$$

for the proportional cost buyback problem with the unweighted knapsack constraint.

Proof. Let A denote an online algorithm chosen arbitrarily, and let

$$y = \frac{fl + \sqrt{f^2 l^2 + 4l}}{2}.$$

As $\frac{l^2-3l+1}{l(l-1)} < f$ and $0 \leq l \leq 1/2$, we have $y + l > 1$ and $y \geq \sqrt{l} > l$. Our adversary requests the following sequence of elements

$$l,\, y,\, 1 \tag{5}$$

until A rejects some element in (5), and if A rejects the element then the adversary immediately stops the input sequence.

Note that A must accept the first element l, since otherwise, the competitive ratio becomes infinite. If A rejects the second element, then the competitive ratio is at least

$$\frac{y}{l} = \frac{fl + \sqrt{f^2 l^2 + 4l}}{2l}$$

If A takes the second element y (and cancels the first element), the competitive ratio is at least

$$\frac{1}{y - f \cdot l} = \frac{fl + \sqrt{f^2 l^2 + 4l}}{2l}$$

since $y - f \cdot l \geq 1 - f \cdot (l + y)$ by $f > \max\{\frac{l^2-3l+1}{l(1-l)}, \frac{4l-1}{2l}\} \geq \frac{-3l+\sqrt{l^2+8l}}{4l}$. □

Theorem 11. *If* $1/2 \le f \le \frac{l^2-3l+1}{l(1-l)}$ *and* $0 \le l \le 1/3$ *there exists no online algorithm with a competitive ratio less than*

$$\frac{1+f+\sqrt{f^2+2f+5}}{2}$$

for the proportional cost buyback problem with the unweighted knapsack constraint.

Proof. Let A denote an online algorithm chosen arbitrarily, and let

$$x = \frac{3+f-\sqrt{f^2+2f+5}}{2(1+f)} = \frac{2}{3+f+\sqrt{f^2+2f+5}}.$$

As $1/2 \le f \le \frac{l^2-3l+1}{l(1-l)}$, it holds that $l \le x \le 1/3$. For a sufficiently small $\varepsilon\ (> 0)$, our adversary requests the following sequence of elements

$$x,\ 1-x+\varepsilon,\ 1-x, \tag{6}$$

until A rejects some element in (6), and if A rejects the element then the adversary immediately stops the input sequence.

Note that A must accept the first element x, since otherwise, the competitive ratio becomes infinite. If A rejects the second element, then the competitive ratio is at least

$$\frac{1-x+\varepsilon}{x} \ge \frac{1-x}{x} = \frac{1+f+\sqrt{f^2+2f+5}}{2}.$$

If A takes the second element $1-x+\varepsilon$ (and cancels the first element), the competitive ratio is at least

$$\frac{1}{1-x+\varepsilon-f\cdot x} \to \frac{1}{1-x-f\cdot x} = \frac{1+f+\sqrt{f^2+2f+5}}{2}$$

as $\varepsilon \to 0$. □

Acknowledgments. The first author is supported by JSPS KAKENHI Grant Number 26887014 and JST, ERATO, Kawarabayashi Large Graph Project. The second author is supported by NFSC(11571060), RGC (HKU716412E) and the Fundamental Research Funds for the Central Universities (DUT15LK10). The last author is supported by JSPS KAKENHI Grant.

References

1. Ashwinkumar, B.V.: Buyback problem - approximate matroid intersection with cancellation costs. In: Proceedings of the 38th International Colloquium on Automata, Language and Programming (2011)
2. Ashwinkumar, B.V., Kleinberg, R.: Randomized online algorithms for the buyback problem. In: Leonardi, S. (ed.) WINE 2009. LNCS, vol. 5929, pp. 529–536. Springer, Heidelberg (2009)

3. Babaioff, M., Hartline, J.D., Kleinberg, R.D.: Selling banner ads: online algorithms with buyback. In: Proceedings of the 4th Workshop on Ad Auctions (2008)
4. Babaioff, M., Hartline, J.D., Kleinberg, R.D.: Selling ad campaigns: online algorithms with cancellations. In: Proceedings of the 10th ACM Conference on Electronic Commerce, pp. 61–70 (2009)
5. Biyalogorsky, E., Carmon, Z., Fruchter, G.E., Gerstner, E.: Research note: overselling with opportunistic cancellations. Marketing Science **18**(4), 605–610 (1999)
6. Buchbinder, N., Feldman, M., Schwartz, R.: Online submodular maximization with preemption. In: Symposium on Discrete Algorithms, pp. 1202–1216 (2015)
7. Constantin, F., Feldman, J., Muthukrishnan, S., Pál, M.: An online mechanism for ad slot reservations with cancellations. In: Proceedings of the twentieth Annual ACM-SIAM Symposium on Discrete Algorithms, pp. 1265–1274 (2009)
8. Han, X., Kawase, Y., Makino, K.: Randomized algorithms for removable online knapsack problems. In: Frontiers in Algorithmics and Algorithmic Aspects in Information and Management (2013)
9. Han, X., Kawase, Y., Makino, K.: Online unweighted knapsack problem with removal cost. Algorithmica **70**, 76–91 (2014)
10. Han, X., Kawase, Y., Makino, K., Guo, H.: Online knapsack problem under convex function. Theor. Comput. Sci. **540–5419**, 62–69 (2014)
11. Han, X., Makino, K.: Online minimization knapsack problem. In: Bampis, E., Jansen, K. (eds.) WAOA 2009. LNCS, vol. 5893, pp. 182–193. Springer, Heidelberg (2010)
12. Iwama, K., Taketomi, S.: Removable online knapsack problems. In: Widmayer, P., Triguero, F., Morales, R., Hennessy, M., Eidenbenz, S., Conejo, R. (eds.) ICALP 2002. LNCS, vol. 2380, p. 293. Springer, Heidelberg (2002)
13. Iwama, K., Zhang, G.: Optimal resource augmentations for online knapsack. In: Charikar, M., Jansen, K., Reingold, O., Rolim, J.D.P. (eds.) RANDOM 2007 and APPROX 2007. LNCS, vol. 4627, pp. 180–188. Springer, Heidelberg (2007)
14. Kawase, Y., Han, X., Makino, K.: Unit cost buyback problem. In: Cai, L., Cheng, S.-W., Lam, T.-W. (eds.) Algorithms and Computation. LNCS, vol. 8283, pp. 435–445. Springer, Heidelberg (2013)
15. Zhou, Y., Chakrabarty, D., Lukose, R.: Budget constrained bidding in keyword auctions and online knapsack problems. In: Papadimitriou, C., Zhang, S. (eds.) WINE 2008. LNCS, vol. 5385, pp. 566–576. Springer, Heidelberg (2008)

Author Index

Printed in the United States
By Bookmasters